AFE63683

Geological studies in the Klamath Mountains province, California and Oregon: A volume in honor of William P. Irwin

Edited by

Arthur W. Snoke
Department of Geology and Geophysics
University of Wyoming
Laramie, Wyoming 82071-3006
USA

Calvin G. Barnes
Department of Geosciences
Texas Tech University
Lubbock, Texas 79409-1053
USA

THE GEOLOGICAL SOCIETY OF AMERICA

Special Paper 410

3300 Penrose Place, P.O. Box 9140 ▪ Boulder, Colorado 80301-9140, USA

2006

Published by The Geological Society of America, Inc.
3300 Penrose Place, P.O. Box 9140, Boulder, Colorado 80301-9140, USA
www.geosociety.org

Printed in USA.

GSA Books Science Editors: Abhijit Basu and Marion E. Bickford

Library of Congress Cataloging-in-Publication Data

Geological studies in the Klamath Mountains province, California, and Oregon :
a volume in honor of William P. Irwin / edited by Arthur W. Snoke, Calvin G. Barnes.
p. cm. (Special paper ; 410)
Includes bibliographic references and index.
ISBN-13 978-0-8137-2410-2 (pbk.)
ISBN-10 0-8137-2410-4 (pbk.)
1. Geology—Klamath Mountains Region (Calif. and Or.). 2. Geology, Structural—Klamath Mountains Region (Calif. and Or.). 3. Geophysics—Klamath Mountains Region (Calif. and Or.). I. Snoke, Arthur W. II. Barnes, Calvin G., 1951–. Special papers (Geological Society of America) ; 410.

QE89.G455 2006
557.95′2—dc22

2006041328

Cover: A view to the southeast across the Klamath Mountains province from the Pacific Crest Trail, with Lily Pad Lake and Goodbye Lake Ridge in the foreground (chiefly amphibolite of the Rattlesnake Creek terrane), Kangaroo Mountain area, California, and snow-covered Mt. Shasta (14,162 ft [4319 m]) of the southern Cascade Range in the background.

10 9 8 7 6 5 4 3 2 1

Contents

CD-ROM Contents

Associated Chapter	Item
Dedication	"Geologic map of northwestern California" (scale 1:500,000) by William P. Irwin (1960). This geologic map was Plate 1 in *Geologic reconnaissance of the northern Coast Ranges and Klamath Mountains, California:* California Division of Mines Bulletin 179. (Also GSA Data Repository item 2006195.)
Snoke and Barnes (Chapter 1); Allen and Barnes (Chapter 11)	Fossil localities of the Rattlesnake Creek, western and eastern Hayfork, and North Fork terranes of the Klamath Mountains by William P. Irwin and Charles D. Blome, 2004, 1 sheet, scale 1:500,000 + 45 p. (Also GSA Data Repository item 2006196.)
Snoke and Barnes (Chapter 1)	Correlation of the Klamath Mountains and Sierra Nevada compiled by William P. Irwin, 2003, 2 sheets. (Also GSA Data Repository item 2006197.)
Yule et al. (Chapter 3)	Figure 2 as a continuous image.
Allen and Barnes (Chapter 11)	Appendix tables 1–3. (Also GSA Data Repository item 2006198.)
Bushey et al. (Chapter 14)	Plate 1. Geologic map of the northern part of the Bear Mountain intrusive complex, Klamath Mountains, California. (Also GSA Data Repository item 2006199.)
McFadden et al. (Chapter 16)	Plate 1. Geologic map of the Bear Peak intrusive complex, Siskiyou County California. (Also GSA Data Repository item 2006200.)
Barnes et al. (Chapter 17)	Appendix 1. Petrography. Appendix 2. Whole-rock geochemical data, location and rock type, oxygen isotope data, notes. Appendix 3. Analytical methods. (Also GSA Data Repository item 2006201.)
Aalto (Chapter 22)	Preliminary map of selected post-Nevadan geologic features of the Klamath Mountains and adjacent areas, California and Oregon compiled by William P. Irwin, 1997, 1 sheet, scale 1:500,000 + 29 p. (Also GSA Data Repository item 2006202.)

Additional Material

A bibliography of Klamath Mountains geology, California and Oregon, listing authors from Aalto to Zucca for the years 1849 to mid-2003 compiled by William P. Irwin: U.S. Geological Survey Open-File Report 03-306 (also available in the GSA Data Repository as item 2006203).

Dedication

William P. Irwin: Career and Contributions

This volume is dedicated to William P. "Porter" Irwin, Scientist Emeritus, U.S. Geological Survey, for his broad-ranging and fundamental contributions to the geology of western North America, such as his groundbreaking syntheses of the tectonic evolution of the northern California Coast Ranges and Klamath Mountains of California and Oregon.

William P. Irwin was born on September 22, 1919, in Springfield, Illinois. He graduated from Springfield High School in 1937. Porter entered the New Mexico School of Mines, Socorro, in 1939 to study geological engineering. His schooling was interrupted when he enlisted in the U.S. Army on April 28, 1942. While waiting to be called for active duty, Porter worked for the U.S. Geological Survey for several months, assisting Richard H. Jahns in mapping the beryllium- and tungsten-rich skarn deposits of the Iron Mountain district of southwestern New Mexico. The Army then sent him to an air corps training center in Jackson, Tennessee, and subsequently to Jefferson Barracks, Missouri, where he was discharged on June 16, 1943. He returned to New Mexico to work with Jahns and others of the U.S. Geological Survey, mapping mica-bearing pegmatite deposits of the Petaca District. In January 1944, Porter returned to the School of Mines at Socorro to finish his academic studies and received his B.S. degree in geological engineering on April 28, 1944. Following graduation, he was employed by the U.S. Geological Survey as a Geologist P-1, and as part of the strategic minerals investigations, he mapped mica-bearing pegmatites in the Petaca District of New Mexico, Spruce Pine District of North Carolina, and Keene District of New Hampshire. His employment with the U.S. Geological Survey was terminated November 4, 1945, by a survey-wide reduction in force.

In 1946, he worked as a field geologist for Transcontinental Resources, Ltd., Vancouver, British Columbia, Canada, and for over a year, he examined both placer and lode gold and silver mines in the Yukon Territory. In late 1948, he decided to continue his education at the graduate-school level and entered the California Institute of Technology, Pasadena, where he again worked with Richard H. Jahns. Interestingly, his M.S. thesis was not on pegmatite deposits but focused on the stratigraphy of the Upper Tick Canyon area, Los Angeles County, California. Porter graduated from Caltech with an M.S. degree in 1950.

Returning to the Mineral Deposits Branch of the U.S. Geological Survey in January 1951, he was based at the Old Mint Building in San Francisco. From there he carried out field studies throughout California and Nevada on various mineral commodities, including Pb, Zn, Cu, Mn, W, Fe, and Be. In 1953, he was designated the project chief of the Trinity River Basin project, with E.H. Bailey as his immediate supervisor. This project inspired Porter's interest in the geological evolution of the northern California Coast Ranges and adjacent Klamath Mountains. He and Ed Bailey collaborated on several fundamental studies on the Franciscan Complex, an enormous Mesozoic–Tertiary accretionary complex that includes broad tracts of mélange. Irwin's studies in the northern Coast Ranges led to the recognition that the "Franciscan group" includes age-equivalent rocks to the well-stratified and fossiliferous Great Valley sequence, and eventually, to the discovery that a fundamental regional fault, the "Coast Range fault," separates these two structurally dissimilar but age-equivalent units. An important by-product of Irwin's regional work in northern California was the publication in 1960 of the landmark study titled *Geological reconnaissance of the northern Coast Ranges and Klamath Mountains, California* as California Division of Mines Bulletin 179. Accompanying the text of this important bulletin was Plate 1: "Geologic map of northwestern California" (see the CD-ROM accompanying this volume or the GSA Data Repository[1]). This geologic map was part compilation and part original geologic mapping, and the total area covered by the study was ~19,000 square miles (~49,200 km^2), including most of the northern Coast Ranges and Klamath Mountains province of California as well as the western part of the Sacramento Valley province. During much of the mapping, Porter was assisted in the field by Donald B. Tatlock. As the first synthesis of the complex geology of the Klamath Mountains, Bulletin 179 was built on the pioneering work of J.S. Diller, O.H. Hershey, and others.

[1]GSA Data Repository item 2006195, Geologic map of northwestern California, is available online at www.geosociety.org/pubs/ft2006.htm, or on request from editing@geosociety.org or Documents Secretary, GSA, P.O. Box 9140, Boulder, CO 80301, USA.

Figure 1. Porter Irwin in his office at the U.S. Geological Survey, Menlo Park, California (1981).

In late December 1953, Porter was in the first group of geologists to move to the survey's new campus in Menlo Park, California. From the mid-1950s until 1973, he was a geologist in the Mineral Deposits Branch and subsequently in the Pacific Coast Branch in Menlo Park. During this important interval of his career, he carried out basic geologic mapping in the southern Klamath Mountains and environs. The geologic maps he worked on during that interval are summarized below in the selected bibliography. An important tectonic concept that resulted from Irwin's extensive mapping of "alpine-type" ultramafic rocks in the northern Coast Ranges and Klamath Mountains was his recognition that these ultramafic rocks were commonly tectonically emplaced during orogeny and typically associated with regional thrust faults. His paper with Peter W. Lipman on the Trinity ultramafic sheet in 1962 clearly challenged the then-existing dogma in the U.S. Geological Survey and elsewhere that such ultramafic rocks were parts of large intrusive bodies that had evolved from the fractionation of basaltic magmas. Subsequently, an even more fundamental contribution to the tectonic evolution of the western North American Cordillera arose from his regional geologic studies. In 1972, Porter recognized that the "western Paleozoic and Triassic belt" that he had delineated on his 1960 geologic map was composed of "terranes," which he referred to as "an association of geologic features, such as stratigraphic formations, intrusive rocks, mineral deposits, and tectonic history, some or all of which lend a distinguishing character to a particular tract of rocks and which differ from those of an adjacent terrane," which he considered to be juxtaposed along faults (Irwin, 1972, p. C103). Although it would take a half-dozen or more years for the full impact of the terrane concept to permeate the psyche of many tectonicists studying the geologic evolution of the western North America Cordillera, the recognition of terranes would eventually become a central theme in the interpretation of ancient orogens dominated by accretionary tectonics.

Once Porter recognized that terranes were the fundamental building blocks of accreted orogens, he turned his attention to various detailed studies involving paleontology, paleomagnetism, ophiolites, and plutonic belts to decipher terrane accretion and the distribution and correlation of terranes. Throughout these studies, Porter always emphasized and used fundamental, field-based data: geologic maps, regional cross-sections, large-scale structural features, fossil assemblages and distributions, geochronology of plutonic belts, and distribution of mineral deposits. His great talent is to carefully and logically compile these

Figure 2. Porter Irwin in the field near Wintoon Flat along the South Fork of the Trinity River, Hyampom 15′ quadrangle (1974).

Figure 3. Preston E. Hotz and Porter Irwin on the high-level erosion surface near Hoopa Valley (1969?).

basic geologic data into a regional tectonic framework. An important culmination of his many years of study of the geology of the Klamath Mountains province was the publication of his *Geologic Map of the Klamath Mountains, California and Oregon* as U.S. Geological Survey Miscellaneous Investigations Series Map I-2148 in 1994. This important geologic map is a classic synthesis of the evolution of an accreted orogenic belt. As Porter has pointed out, the overall areal extent of the Klamath Mountains province is comparable to the Western Alps of Europe, although the quantity of detailed geological studies and their overall coverage is considerably less in the province than in the Alps. During the past 50 years, Porter Irwin has provided the geologic community with a rich description of a fascinating geologic province that records a history from the Neoproterozoic to Present. This volume is a tribute to his enormous contributions to the geology of the Klamath Mountains, and it is also a recognition of his fundamental contributions to the understanding of the development of mountain belts in general.

Porter formally retired from the U.S. Geological Survey at the end of December 1988. However, he has continued his studies of the tectonics of central and northern California as Scientist Emeritus, Earthquake Hazards Team, Menlo Park, California. Since his official retirement, his interest in scientific research has not faded, and he continues to investigate the details of accretionary orogens, especially the distribution of fossiliferous rocks in terranes and the potential correlation of various terranes throughout the western North American Cordillera. Porter, we thank you for your insight, encouragement, and friendship over the many years that we have known you.

Arthur W. Snoke
Calvin G. Barnes

SELECTED BIBLIOGRAPHY OF WILLIAM P. IRWIN

Albers, J.P., Kinkel, A.R., Jr., Drake, A.A., and Irwin, W.P., 1964, Geologic map of the French Gulch quadrangle, California: Washington, D.C., U.S. Geological Survey Geologic Map GQ-336, scale 1:62,500, 1 sheet.

Ando, C.J., Irwin, W.P., Jones, D.L., and Saleeby, J.B. 1983, The ophiolitic North Fork terrane in the Salmon River region, central Klamath Mountains, California: Geological Society of America Bulletin, v. 94, p. 236–252, doi: 10.1130/0016-7606(1983)94<236:TONFTI>2.0.CO;2.

Bailey, E.H., and Irwin, W.P., 1959, K-feldspar content of Jurassic and Cretaceous graywacke of the northern Coast Ranges and Sacramento Valley, California: American Association of Petroleum Geologists Bulletin, v. 43, p. 2797–2809.

Bailey, E.H., Irwin, W.P., and Jones, D.L., 1964, Franciscan and related rocks and their significance in the geology of western California: San Francisco, California Division of Mines and Geology Bulletin 183, 177 p.

Blake, M.C., Jr., Irwin, W.P., and Coleman, R.G., 1967, Upside-down metamorphic zonation, blueschist facies, along a regional thrust in California and Oregon: Washington, D.C., U.S. Geological Survey Professional Paper 575-C, p. C1–C9.

Blake, M.C., Jr., Harwood, D.S., Helley, E.J., Irwin, W.P., Jayko, A.S., and Jones, D.L., 1999, Geologic map of the Red Bluff 30′ x 60′ quadrangle, California: Reston, Virginia, U.S. Geological Survey Map I-2542, scale 1:250,000, 1 sheet with 15p. pamphlet.

Blome, C.D., and Irwin, W.P., 1983, Tectonic significance of late Paleozoic to Jurassic radiolarians from the North Fork terrane, Klamath Mountains, California, *in* Stevens, C.H., ed., Pre-Jurassic rocks in western North American suspect terranes: Los Angeles, Pacific Section Society of Economic Paleontologists and Mineralogists, p. 77–89.

Irwin, W.P., 1957, Franciscan group in Coast Ranges and its equivalents in Sacramento Valley, California: American Association of Petroleum Geologists Bulletin, v. 41, p. 2284–2297.

Irwin, W.P., 1960, Geologic reconnaissance of the northern Coast Ranges and Klamath Mountains, California, with a summary of the mineral resources: San Francisco, California Division of Mines Bulletin 179, 80 p.

Irwin, W.P., 1960, Relations between Abrams Mica Schist and Salmon Hornblende Schist in Weaverville quadrangle, California, *in* Geological survey research 1960: Short papers in the Geological Sciences: Washington, D.C., U.S. Geological Survey Professional Paper 400-B, B315–B316.

Irwin, W.P., 1963, Preliminary geologic map of the Weaverville quadrangle, California: Washington, D.C., U.S. Geological Survey Mineral Investigations Field Studies Map MF-275, scale 1:62,500, 1 sheet.

Irwin, W.P., 1964, Late Mesozoic orogenies in the ultramafic belts of northwestern California and southwestern Oregon, *in* Geological survey research 1964: Washington, D.C., U.S. Geological Survey Professional Paper 501-C, p. C1–C9.

Irwin, W.P., 1966, Geology of the Klamath Mountains province, *in* Bailey, E.H., ed., Geology of northern California: San Francisco, California Division of Mines and Geology, p. 19–38.

Irwin, W.P., 1972, Terranes of the western Paleozoic and Triassic belt in the southern Klamath Mountains, California, *in* Geological survey research 1972: Washington, D.C., U.S. Geological Survey Professional Paper 800-C, p. C103–C111.

Irwin, W.P., 1974, Reconnaissance geologic map of the Hayfork quadrangle, Trinity County, California: Reston, Virginia, U.S. Geological Survey Miscellaneous Field Studies Map MF-576, scale 1:62,500, 1 sheet.

Irwin, W.P., 1977, Review of Paleozoic rocks of the Klamath Mountains, *in* Stewart, J.H., et al., eds., Paleozoic paleogeography of the western United States I: Los Angeles, Pacific Section, Society of Economic Paleontologists and Mineralogists, p. 441–454.

Irwin, W.P., 1977, Ophiolitic terranes of California, Oregon, and Nevada, *in* Coleman, R.G., and Irwin, W.P., eds., North American ophiolites: Portland, Oregon Department of Geology and Mineral Industries Bulletin 95, p. 75–92.

Irwin, W.P., 1981, Tectonic accretion of the Klamath Mountains, *in* Ernst, W.G., ed., The geotectonic development of California—Rubey volume I: Englewood Cliffs, New Jersey, Prentice-Hall, p. 29–70.

Irwin, W.P., 1985, Age and tectonics of plutonic belts in accreted terranes of the Klamath Mountains, California and Oregon, *in* Howell, D.G., ed., Tectonostratigraphic terranes of the circum-Pacific region: Houston, Circum-Pacific Council for Energy and Mineral Resources, Earth Sciences Series 1, p. 187–199.

Irwin, W.P., 1985, Reconnaissance geologic map of the Hyampon quadrangle, Trinity County, California: Reston, Virginia, U.S. Geological Survey Miscellaneous Field Studies Map MF-1809, scale 1:62,500, 1 sheet.

Irwin, W.P., 1989, Cryptic tectonic domains of the Klamath Mountains, California and Oregon: Engineering Geology, v. 27, p. 433–448, doi: 10.1016/0013-7952(89)90040-9.

Irwin, W.P., 1990, Geology and plate-tectonic development, *in* Wallace, R.E., ed., The San Andreas fault system, California: Reston, Virginia, U.S. Geological Survey Professional Paper 1515, p. 60–80.

Irwin, W.P., 1994, Geologic map of the Klamath Mountains, California: Reston, Virginia, U.S. Geological Survey Miscellaneous Investigations Series Map I-2148, scale 1:500,000, 2 sheets.

Irwin, W.P., 1997, Preliminary map of selected post-Nevadan geologic features of the Klamath Mountains and adjacent areas, California and Oregon: Reston, Virginia, U.S. Geological Survey, Open-File Report 97-465, scale 1:500,000, with 29p. pamphlet.

Irwin, W.P., 2003, Correlation of the Klamath Mountains and Sierra Nevada: Sheet 1; Map showing accreted terranes and plutons of the Klamath Mountains and Sierra Nevada, scale 1:1,000,000, and Sheet 2; Successive accretionary episodes of the Klamath Mountains and northern part of the Sierra Nevada: Reston, Virginia, U.S. Geological Survey Open-File Report 02-490, 2 sheets.

Irwin, W.P., 2003, A bibliography of Klamath Mountains geology, California and Oregon, listing authors from Aalto to Zucca for the years 1849 to mid-2003: Reston, Virginia, U.S. Geological Survey Open-File Report 03-306, 123 p.

Irwin, W.P., and Barnes, I., 1975, Effects of geologic structure and metamorphic fluids on seismic behavior of the San Andreas fault system in central and northern California: Geology, v. 3, p. 713–716, doi: 10.1130/0091-7613(1975)3<713:EOGSAM>2.0.CO;2.

Irwin, W.P., and Barnes, I., 1980, Tectonic relations of carbon dioxide discharges and earthquakes: Journal of Geophysical Research, v. 85, p. 3115–3121.

Irwin, W.P., and Blome, C.D., 2004, Map showing fossil localities of the Rattlesnake Creek, western and eastern Hayfork, and North Fork terranes of the Klamath Mountains, California and Oregon: Reston, Virginia, U.S. Geological Survey Open-File Report 2004-1094, scale 1:500,000, with 45 p. pamphlet.

Irwin, W.P., and Galanis, S.P., Jr., 1976, Map showing limestones and selected fossil localities in the Klamath Mountains province, California and Oregon: Reston, Virginia, U.S. Geological Survey Miscellaneous Field Studies Map MF-749, scale 1:500,000, 1 sheet.

Irwin, W.P., and Lipman, P.W., 1962, A regional ultramafic sheet in eastern Klamath Mountains, California: Washington, D.C., U.S. Geological Survey Professional Paper 450-C, p. 18–21.

Irwin, W.P., and Tatlock, D.B., 1955, Reconnaissance of northwestern California, a progress report: Geological Society of America Bulletin, v. 66, part 2, p. 1653–1654.

Irwin, W.P., and Wooden, J.L., 1999, Plutons and accretionary episodes of the Klamath Mountains, California and Oregon: Reston, Virginia, U.S. Geological Survey Open-File Report 99-374, scale 1:500,000, 1 sheet.

Irwin, W.P., and Wooden, J.L., 2001, Map showing plutons and accreted terranes of the Sierra Nevada, California, with a tabulation of U/Pb isotopic ages: Reston, Virginia, U.S. Geological Survey Open-File Report 01-229, scale 1:1,000,000, 1 sheet.

Irwin, W.P., Wolfe, E.W., Blake, M.C., Jr., and Cunningham, C.G., Jr., 1974, Geologic map of the Pickett Peak quadrangle, Trinity County, California: Reston, Virginia, U.S. Geological Survey Map GQ-1111, scale 1:62,500, 1 sheet.

Irwin, W.P., Jones, D.L., and Pessagno, E.A., Jr., 1977, Significance of Mesozoic radiolarians from the pre-Nevadan rocks of the southern Klamath Mountains, California: Geology, v. 5, p. 557–562, doi: 10.1130/0091-7613(1977)5<557:SOMRFT>2.0.CO;2.

Irwin, W.P., Jones, D.L., and Kaplan, T.A., 1978, Radiolarians from pre-Nevadan rocks of the Klamath Mountains, California and Oregon, *in* Howell, D.G., and McDougall, K.A., eds., Mesozoic paleogeography of the western United States II: Los Angeles, Pacific Section, Society of Economic Paleontologists and Mineralogists, p. 303–310.

Irwin, W.P., Jones, D.L., and Blome, C.D., 1982, Map showing samples radiolarian localities in the western Paleozoic and Triassic belt, Klamath Mountains, California: Reston, Virginia, U.S. Geological Survey Miscellaneous Field Studies Map MF-1399, scale 1:250,000, 1 sheet.

Irwin, W.P., Wardlaw, B.R., and Kaplan, T.A., 1983, Conodonts of the western Paleozoic and Triassic belt, Klamath Mountains, California and Oregon: Journal of Paleontology, v. 57, p. 1030–1039.

Irwin, W.P., Yule, J.D., Court, B.L., Snoke, A.W., Stern, L.A., and Copeland, W.B., 1985, Reconnaissance geologic map of the Dubakella Mountain quadrangle, Trinity, Shasta, and Tehama counties, California: Reston, Virginia, U.S. Geological Survey Miscellaneous Field Studies Map MF-1808, scale 1:62,500, 1 sheet.

Jones, D.L., and Irwin, W.P., 1971, Structural implications of an offset Early Cretaceous shoreline in northern California: Geological Society of America Bulletin, v. 82, p. 815–822.

Jones, D.L., Irwin, W.P., and Ovenshine, A.T., 1972, Southeast Alaska—A displaced continental fragment? *in* Geological Survey Research 1972: Washington, D.C., U.S. Geological Survey Professional Paper 800-B, p. 211–217.

Lanphere, M.A., and Irwin, W.P., 1987, In search of the Abrams Post Office, Trinity County: California Geology, May 1987, p. 99–103.

Lanphere, M.A., Irwin, W.P., and Hotz, P.E., 1968, Isotopic age of the Nevadan orogeny and older plutonic and metamorphic events in the Klamath Mountains, California: Geological Society of America Bulletin, v. 79, p. 1027–1052.

Lanphere, M.A., Blake, M.C., Jr., and Irwin, W.P., 1978, Early Cretaceous metamorphic age of the South Fork Mountain Schist in the northern Coast Ranges of California: American Journal of Science, v. 278, p. 798–815.

Mankinen, E.A., and Irwin, W.P., 1982, Paleomagnetic study of some Cretaceous and Tertiary sedimentary rocks of the Klamath Mountains province, California: Geology, v. 10, p. 82–87, doi: 10.1130/0091-7613(1982)10<82:PSOSCA>2.0.CO;2.

Mankinen, E.A., and Irwin, W.P., 1990, Review of paleomagnetic data from the Klamath Mountains, Blue Mountains, and Sierra Nevada: Implications for paleogeographic reconstructions, *in* Harwood, D.S., and Miller, M.M., eds., Paleozoic and early Mesozoic paleogeographic relations; Sierra Nevada, Klamath Mountains, and related terranes: Boulder, Colorado, Geological Society of America Special Paper 255, p. 397–409.

Mankinen, E.A., Irwin, W.P., and Grommé, C.S., 1984, Implications of paleomagnetism for the tectonic history of the eastern Klamath and related terranes in California and Oregon, *in* Nilsen, T.H., ed., Geology of the Upper Cretaceous Hornbrook Formation, Oregon and California: Los Angeles, Pacific Section, Society of Economic Paleontologists and Mineralogists, v. 42, p. 221–229.

Mankinen, E.A., Irwin, W.P., and Grommé, C.S., 1988, Paleomagnetic results from the Shasta Bally plutonic belt in the Klamath Mountains province, northern California: Geophysical Research Letters, v. 15, p. 56–59.

Mankinen, E.A., Irwin, W.P., and Grommé, C.S., 1989, Paleomagnetic study of the eastern Klamath terrane, California, and implications for the tectonic history of the Klamath Mountains province: Journal of Geophysical Research, v. 94, p. 10,444–10,472.

Mankinen, E.A., Irwin, W.P., and Blome, C.D., 1996, Far-traveled Permian chert of the North Fork terrane, Klamath Mountains, California: Tectonics, v. 15, p. 314–328, doi: 10.1029/95TC03054.

Schweickert, R.A., and Irwin, W.P., 1989, Extensional faulting in southern Klamath Mountains, California: Tectonics, v. 8, p. 135–149.

Acknowledgments

We thank the following reviewers for their timely and thorough reviews of the manuscripts that comprise this volume. Their contribution of effort and time to improve the quality of the volume is gratefully acknowledged.

James S. Beard, Virginia Museum of Natural History, Martinsville

Charles D. Blome, U.S. Geological Survey, Denver

Drew S. Coleman, University of North Carolina, Chapel Hill

Fabrice Cordey, C.N.R.S., Université Claude Bernard Lyon 1, France

Darrel S. Cowan, University of Washington, Seattle

Howard W. Day, University of California, Davis

Susan M. DeBari, Western Washington University, Bellingham

John F. Dewey, University of California, Davis

William R. Dickinson, University of Arizona, Tucson

Michael J. Dorais, Brigham Young University, Salt Lake City, Utah

James P. Evans, Utah State University, Logan

Todd C. Feeley, Montana State University, Bozeman

Mark L. Ferns, Oregon Department of Geology and Mineral Industries, Baker City

Kenneth Fowler, ExxonMobil Development Company, Houston, Texas

Carol D. Frost, University of Wyoming, Laramie

Sarah R. Garlick, University of Wyoming, Laramie

Gary H. Girty, San Diego State University, California

John W. Goodge, University of Minnesota, Duluth

Laurel B. Goodwin, University of Wisconsin, Madison

Gary G. Gray, ExxonMobil Upstream Research Company, Houston, Texas

Timothy W. Grover, Castleton State College, Vermont

Bradley R. Hacker, University of California, Santa Barbara

James W. Hawkins, Scripps Institution of Oceanography, La Jolla, California

Mark A. Helper, University of Texas, Austin

John P. Hogan, University of Missouri, Rolla

Jose M. Hurtado, University of Texas, El Paso

Raymond V. Ingersoll, University of California, Los Angeles

Helen M. Lang, West Virginia University, Morgantown

Daniel R. Lux, University of Maine, Orono

Glenn J. MacPherson, Smithsonian Institution, National Museum of Natural History, Washington, D.C.

James M. Mattinson, University of California, Santa Barbara

Nancy J. McMillan, New Mexico State University, Las Cruces

Rodney V. Metcalf, University of Nevada, Las Vegas

Jonathan S. Miller, San Jose State University, California

Robert B. Miller, San Jose State University, California

Clyde J. Northrup, Boise State University, Idaho

P. Jonathan Patchett, University of Arizona, Tucson

Scott R. Paterson, University of Southern California, Los Angeles

Tore Prestvik, Norwegian Institute of Science and Technology, Trondheim

Scott D. Samson, Syracuse University, New York

Mark D. Schmitz, Boise State University, Idaho

John W. Shervais, Utah State University, Logan

John Wakabayashi, Hayward, California

Craig M. White, Boise State University, Idaho

James E. Wright, University of Georgia, Athens

Sandra J. Wyld, University of Georgia, Athens

Aaron S. Yoshinobu, Texas Tech University, Lubbock

Geological Society of America
Special Paper 410
2006

The development of tectonic concepts for the Klamath Mountains province, California and Oregon

Arthur W. Snoke*
Department of Geology and Geophysics, Dept. 3006, University of Wyoming, 1000 East University Avenue, Laramie, Wyoming 82071, USA
Calvin G. Barnes
Department of Geosciences, Texas Tech University, Lubbock, Texas 79409-1053, USA

ABSTRACT

The Klamath Mountains province of northwestern California and southwestern Oregon is a classic example of a mountain belt that developed by the tectonic accretion of rock assemblages of oceanic affinity during progressive crustal growth along an active continental margin. Consequently, the Klamath Mountains province has served as an important model for the definition and application of the terrane concept as applied to the evolution of Phanerozoic orogenic belts. Early regional studies divided the Klamath Mountains province into four arcuate lithic belts of contrasting age (from east to west): the eastern Klamath, central metamorphic, western Paleozoic and Triassic, and western Jurassic belts. The lithic belts are bounded by regional thrust faults that commonly include ophiolitic assemblages in the hanging-wall block. The age of thrusting is a complex problem because of structural overprinting, but generally the age of regional thrust faulting is older in eastern parts of the province and younger to the west. The lithic belts were subsequently subdivided into many tectonostratigraphic terranes, and these lithotectonic units are always fault-bounded. Few of the regional faults are fossil subduction zones, but multiple episodes of high pressure–low temperature (blueschist-facies) metamorphism are recognized in the Klamath Mountains province. The tectonostratigraphic terranes of the Klamath Mountains province are intruded by many composite, mafic to felsic, arc-related plutons, some of which reach batholithic dimensions. Many of these plutonic bodies were emplaced during the Jurassic; however, radiometric dates ranging from Neoproterozoic through Early Cretaceous have been determined from (meta)plutonic rocks of the Klamath Mountains province. The orogenic evolution of the province apparently involved the alternation of contraction and extension, as exemplified by the Jurassic history of the province. Widespread Middle Jurassic plutonism and metamorphism is associated with a poorly understood contractional history followed by the development of the Preston Peak–Josephine ophiolite and Upper Jurassic Galice Formation in a probable transtensional inter-arc basin. During the Late Jurassic Nevadan orogeny, this basin collapsed, and rocks of the Galice Formation were thrust beneath the Rattlesnake Creek terrane along the Orleans fault. During this regional deformation, the

*E-mail: snoke@uwyo.edu.

Snoke, A.W., and Barnes, C.G., 2006, The development of tectonic concepts for the Klamath Mountains province, California and Oregon *in* Snoke, A.W., and Barnes, C.G., eds., Geological studies in the Klamath Mountains province, California and Oregon: A volume in honor of William P. Irwin: Geological Society of America Special Paper 410, p. 1–29, doi: 10.1130/2006.2410(01). For permission to copy, contact editing@geosociety.org.

Galice Formation experienced polyphase deformation and was metamorphosed under lower greenschist-facies conditions. Immediately following thrusting, the hanging-wall and footwall blocks of the Orleans fault were intruded by a suite of composite, mafic to felsic plutons (i.e., western Klamath plutonic suite) that have oceanic-arc geochemical and isotopic characteristics, indicating a subduction-zone petrogenesis for the magmas. The western boundary of the Klamath Mountains province is a regional thrust fault that emplaced the rocks of the province above Early Cretaceous blueschist-facies rocks (South Fork Mountain Schist) of the Franciscan Complex. Neogene structural doming is manifested in the north-central Klamath Mountains by the Condrey Mountain window, which exposes the high pressure–low temperature Condrey Mountain Schist framed by chiefly amphibolite-facies metamorphic rocks of the Rattlesnake Creek terrane.

Keywords: Klamath Mountains, northwestern California, southwestern Oregon, history of geology, thrust faults, ophiolite, igneous and metamorphic rocks, Jurassic orogenies, tectonostratigraphic terranes, Tethyan fauna, blueschist-facies metamorphism, intraoceanic arcs, accretionary subduction complexes, continental growth, accretionary orogen, Neogene structural dome

INTRODUCTION

The Klamath Mountains province of northwestern California and southwestern Oregon is the largest exposure of pre-Tertiary rocks between the northern Sierra Nevada and North Cascade core of Washington State (Figs. 1 and 2). Furthermore, the Klamath Mountains province is an archetypal example of a mountain belt developed by the progressive tectonic accretion of oceanic rocks (Figs. 3 and 4). Perhaps because the Klamath Mountains province is part of a megabelt of accreted terranes that stretches from Alaska through Mexico and beyond (Dickinson, 2004; Monger et al., 2005; Snoke, 2005) the size of the Klamath Mountains province sensu stricto is commonly overlooked, but its area is comparable to that of the Western Alps of Europe (Fig. 5).

The Klamath Mountains province is bounded on the west by the eastward-dipping Coast Range thrust or equivalent fault, beneath which lies the Franciscan Complex (Fig. 3). The geologic character of the eastern margin of the Klamath Mountains province is variable, and rocks of the province may extend to the east in the subsurface (Fuis et al., 1987). Immediately north and south of the California-Oregon border, rocks of the Klamath Mountains province are unconformably overlain by sedimentary rocks of the Cretaceous Hornbrook Formation (Nilsen, 1984) that in turn are overlain by Tertiary volcanic and volcanogenic rocks of the Western Cascade Group. As the eastern margin is traced south, Tertiary rocks overlie the Klamath Mountains province, with no exposed Cretaceous rocks. Far to the south, fossiliferous Cretaceous sedimentary rocks of the Great Valley sequence unconformably overlie the rocks of the Klamath Mountains province (including the Early Cretaceous Shasta Bally batholith); this locality at the south end of the Klamath Mountains province (Blake et al., 1999) is historically important in the development of the Lower Cretaceous timescale (e.g., Curtis et al., 1958). In Oregon, the northeastern margin of the Klamath Mountains province is generally faulted, with Tertiary rocks juxtaposed against various rock units of the Klamath Mountains province.

The fundamental structural character of the Klamath Mountains province is a system of fault-bounded, imbricated plates of oceanic-affinity rocks that dip eastward in a regional sense but are locally folded and cut by younger, high-angle faults. This imbricated pattern of lithotectonic units is interpreted as a manifestation of progressive accretion of oceanic rocks along an active continental margin during early Paleozoic through Late Jurassic time. If the tectonic accretion history of the adjacent northern California Coast Ranges is considered, the exposed accretionary history spans into the middle Cenozoic, and active tectonic accretion is ongoing along the Pacific margin in the form of the Cascadia subduction zone. The ages of the lithotectonic units range from Neoproterozoic remnants in the eastern Klamath Mountains (Mankinen et al., 2002; Lindsley-Griffin et al., 2003, this volume) to Upper Jurassic in the western Klamath Mountains (Diller, 1903; Irwin, 1960, 1994). The age of the lithotectonic units generally decreases to the west as well as structurally downward. Consequently, many workers have viewed the Klamath Mountains province as an excellent example of thin-skinned accretionary tectonics in which successive arc-related lithotectonic units were accreted by thrust-fault imbrication with older, inboard, amalgamated terranes (Davis, 1968; Snoke, 1977; Saleeby et al., 1982; Wright, 1982; Harper and Wright, 1984; Wright and Fahan, 1988).

The fault-bounded, lithotectonic units are intruded by many plutons that range in age from early Paleozoic to Early Cretaceous, in size from stocks to batholiths, and in composition from ultramafic to silicic. Besides petrogenetic signifi-

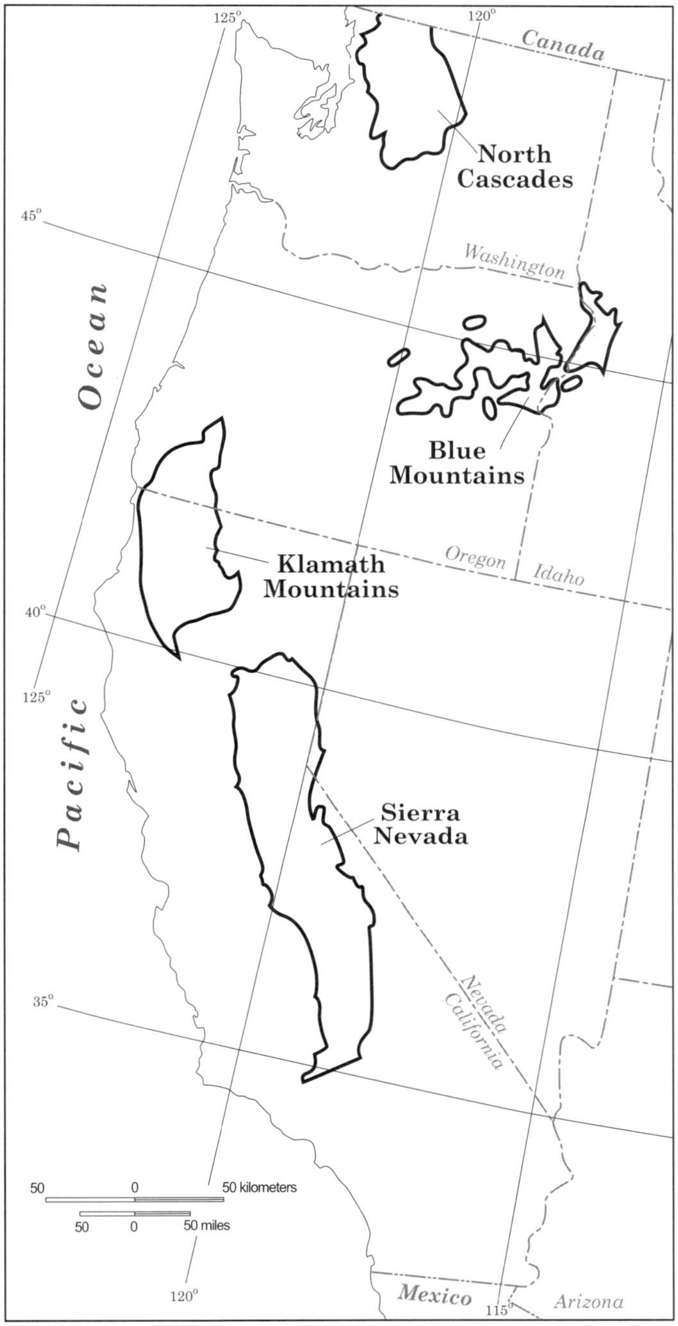

Figure 1. Geologic sketch map of a part of the western North American Cordillera showing the position of the Klamath Mountains province with respect to the Sierra Nevada (central California), Blue Mountains (northeast Oregon), and North Cascades (north-central Washington State). Adapted from King (1967, 1970).

cance, the plutons are important in regard to bracketing the age of deformation in the Klamath Mountains province (Lanphere et al., 1968; Davis et al., 1978; Wright and Fahan, 1988). Some plutons, such as the ca. 139-Ma Grants Pass pluton (Fig. 6), are classic examples of post-thrusting (postkinematic) emplacement because they intruded and metamorphosed both the hanging-wall and footwall blocks on either side of a regional thrust fault (Orleans fault in this example). Other plutonic complexes (ca. 162-Ma Wooley Creek batholith and comagmatic Slinkard pluton) are apparently "rootless" in that they occur in the hanging wall of a regional thrust plate with a displacement history younger than the emplacement/crystallization ages of the intrusive plutonic bodies (Barnes et al., 1986). The relationship among deformation, magmatism, and metamorphism through time during the evolution of the Klamath Mountains province is fundamental to understanding basic orogenic processes along an active continental margin. Furthermore, the Klamath Mountains province provides an excellent opportunity to better understand the accretion and growth of continental lithosphere as well as its modification during progressive orogenesis.

SOME EARLY GEOLOGIC OBSERVATIONS AND CONTRIBUTIONS

Among the first geologists to explore the Klamath Mountains province were those associated with J.D. Whitney's Geological Survey of California. Although their specific comments on the geology of the Klamath Mountains province were minor compared to their more detailed exploration of the adjacent Cascade Range (e.g., environs surrounding Mount Shasta and Lassen Peak), in 1863 William H. Brewer and Clarence King made a rapid reconnaissance geologic traverse from Yreka to Crescent City at the Pacific coast (Whitney, 1865, p. 356–363). The extensive geologic mapping of J.S. Diller (Fig. 7) of the U.S. Geological Survey (USGS) represented the first systematic exploration of the Klamath Mountains province. Diller's many publications on the Klamath Mountains province as well as the adjacent Oregon Coast Ranges had a profound influence on all subsequent studies, and some of his basic conclusions are still accepted today (e.g., correlation of the Upper Jurassic Galice Formation with the Mariposa Formation of the western Sierra Nevada [see MacDonald et al., this volume; Gray, this volume], recognition of important high-level erosion surfaces in the Klamath Mountains and adjacent Coast Ranges [Diller, 1902; see Aalto, this volume]). Furthermore, Diller established a stratigraphic framework for the eastern Klamath Mountains (Diller, 1906), where a coherent stratigraphy ranges from mid-Paleozoic to mid-Jurassic time and is now recognized as one of the longest-lived oceanic-arc sequences in the western North American Cordillera (Lapierre et al., 1985, 1987; Miller, 1989; Charvet et al., 1990; Miller and Harwood, 1990; Dickinson, 2000). When J.S. Diller retired from the USGS on December 31, 1923, at the age of 73, he had studied large portions of the Klamath Mountains province and Coast Ranges of northwestern California and southwestern Oregon as well as completed important studies on Crater Lake, Mount Shasta, and Lassen Peak. His contributions to the geologic mapping of the region include the following USGS folios: Roseberg (no. 49, 1898), Coos Bay

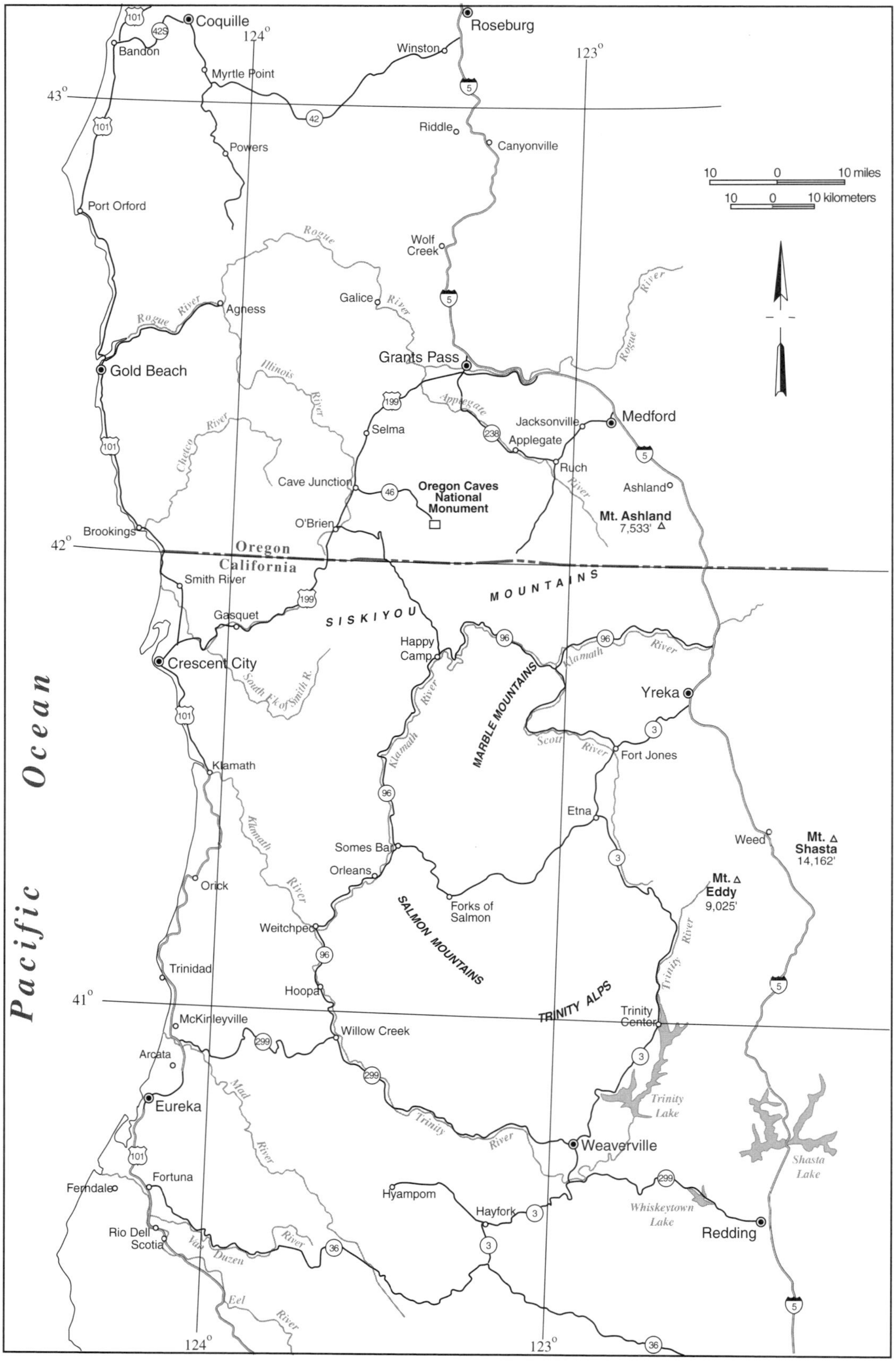

Figure 2. Map of the Klamath Mountains province showing geographic localities cited in the text.

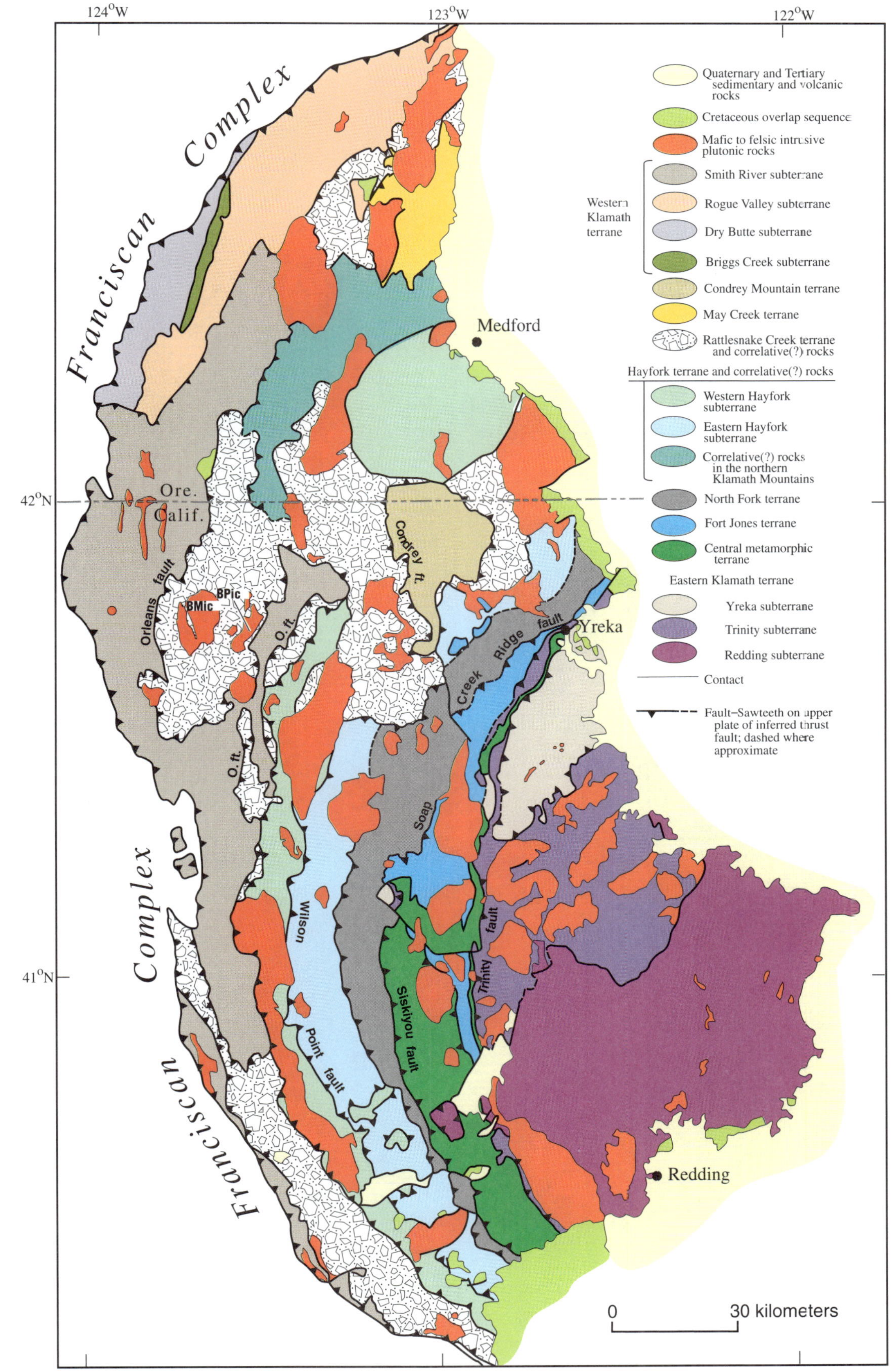

Figure 3. Geologic sketch map of the tectonostratigraphic terranes of the Klamath Mountains province, California and Oregon. BMic—Bear Mountain intrusive complex; BPic—Bear Peak intrusive complex; O. ft.—Orleans fault. Modified from Irwin (1994).

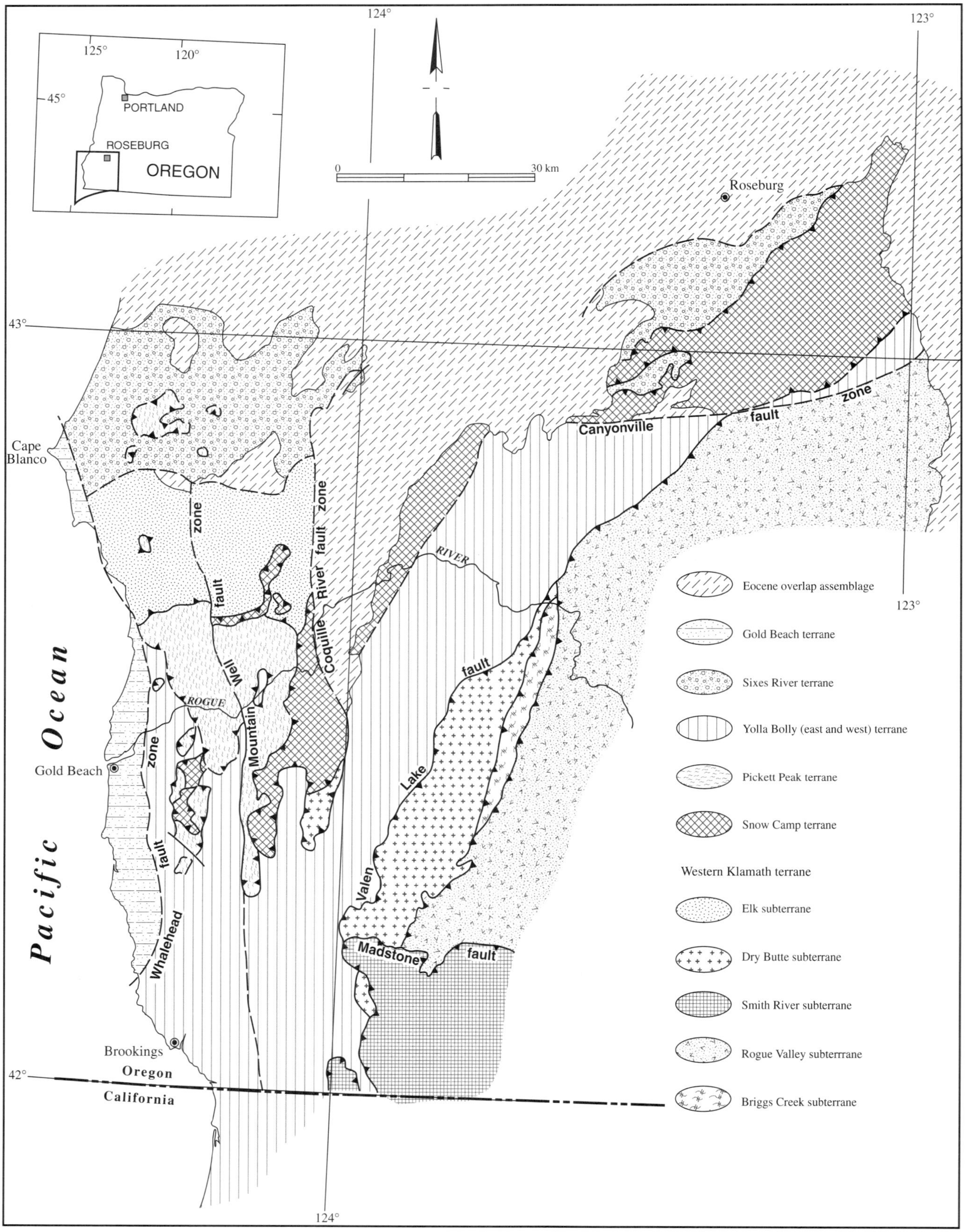

Figure 4. Geologic sketch map of the tectonostratigraphic terranes in southwestern Oregon. After Blake et al. (1985).

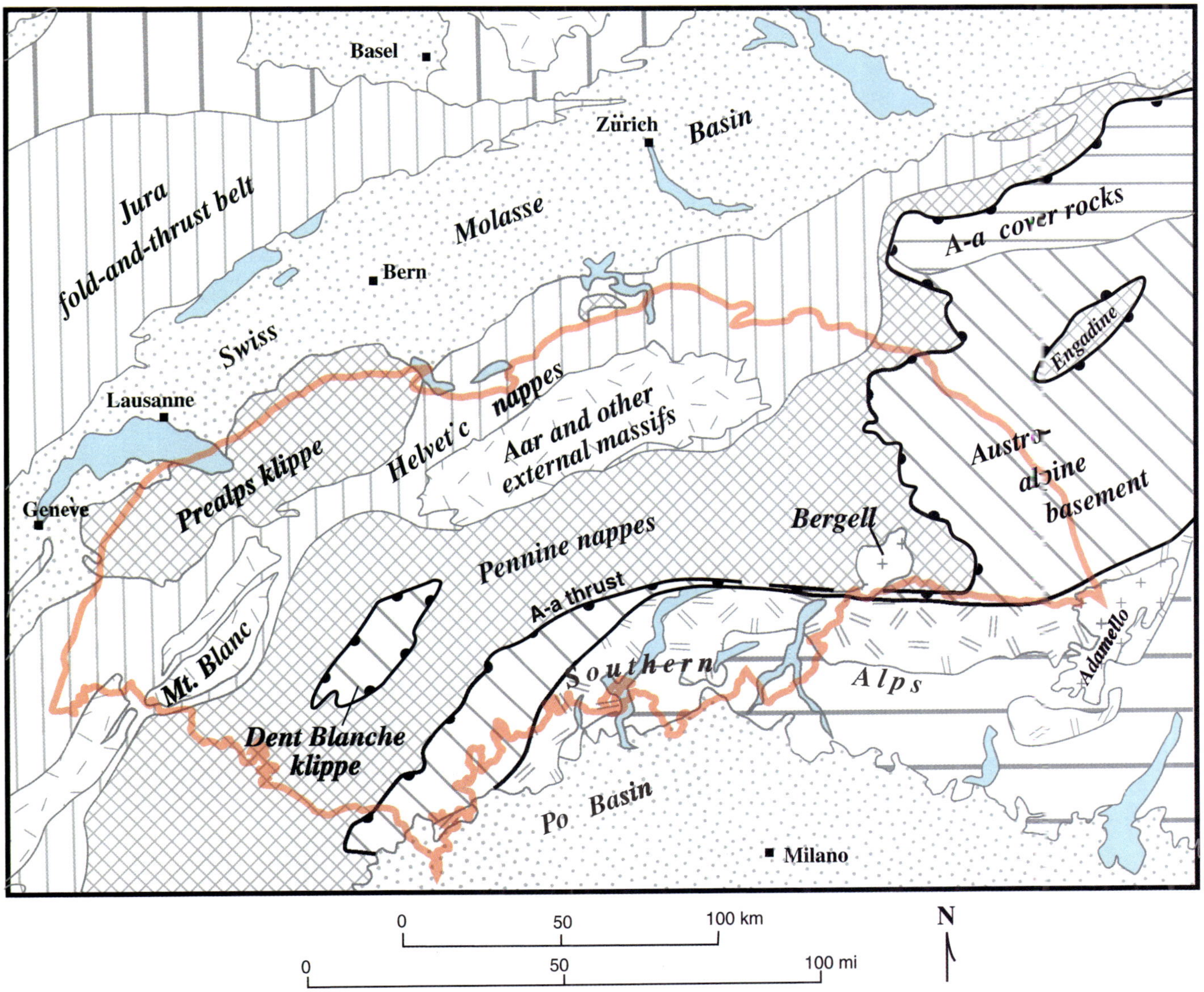

Figure 5. Comparison of the Klamath Mountains province (outlined in red and turned sideways such that north is to the right) with the Western Alps of Europe (inspired by a comment made by W.P. Irwin to AWS). A-a—Austro-alpine. Geology of the Western Alps modified from Ramsay (1991); outline of the Klamath Mountains province from Irwin (1994).

(no. 73, 1901), Port Orford (no. 89, 1903), Redding (no. 138, 1906), and Riddle (no. 218, 1924).

An important contemporary of Diller's was Oscar H. Hershey (Fig. 8), a consulting geologist, who was an associate of Andrew C. Lawson, professor of geology at the University of California–Berkeley. After Hershey's death in 1939, Lawson (1941) wrote the "Memorial to Oscar H. Hershey" that was published in the *Proceedings of the Geological Society of America for 1940*. Hershey, born in 1874 in Pennsylvania, was not formally trained in geology or any of the sciences but was a careful observer as well as a diligent reader. He was also fond of walking and made numerous excursions throughout Illinois and neighboring states after his family moved to Freeport, Illinois, in 1891. In his early twenties, Hershey began to write papers on the physiography and geology of Illinois. Eventually, after geologic excursions in the Ozark Mountains and Panama, he settled in California and began extensive field studies of the Klamath Mountains. During 1901–1906, fifteen of his papers on aspects of the geology of the Klamath Mountains were published in various periodicals on geology. His last paper on the Klamath Mountains, a one-page note with a geologic sketch map of Del Norte County, was published in 1911, but this short contribution showed the approximate position of the Orleans fault, now known to be a fundamental structural feature in the western Klam-

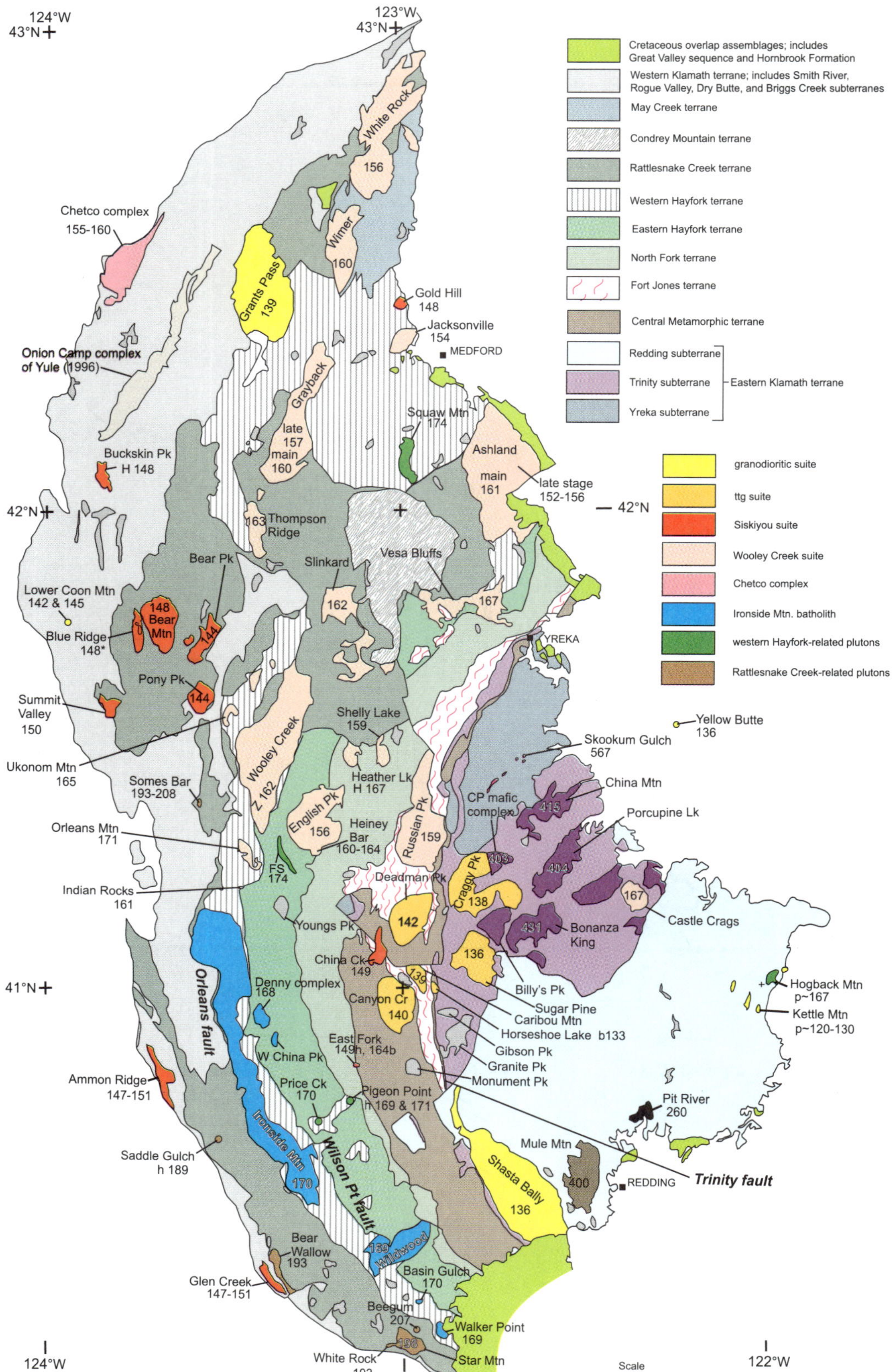

Figure 6. Geologic map of the Klamath Mountains province after Irwin and Wooden (1999) showing principal tectonostratigraphic terranes and plutons color coded according to age group. In the key, "ttg suite" refers to Early Cretaceous tonalite-trondhjemite-granodiorite plutons. Numbers associated with each pluton are ages in Ma. Ages were determined by U-Pb (zircon) unless noted: t indicates U-Pb on titanite, h and b indicate K-Ar ages on hornblende and biotite, respectively, and H and p indicate $^{40}Ar/^{39}Ar$ ages or hornblende and plagioclase, respectively. Data are from Allen and Barnes (this volume), Chamberlain et al. (this volume), and sources summarized in Irwin and Wooden (1999). The age of the Blue Ridge pluton is a minimum age determined on the basis of a cross-cutting tonalitic pluton (Chamberlain et al., this volume). Ck—Creek; CP—Craggy Peak; FS—forks of Salmon pluton; Lk—Lake; Pk—Peak.

Figure 7. J.S. Diller (1850–1928). See Collier (1929).

ath Mountains (see below for a more extensive discussion of the Orleans fault). An early synthesis of the Klamath Mountains province was Hershey's (1901) article titled "Metamorphic formations of northwestern California," published in *The American Geologist.* In this classic paper, Hershey classified the pre-Cretaceous rocks of the Klamath Mountains into "seven great formations" (exclusive of the intrusive granitic rocks) and provided the first detailed descriptions of the "Abrams Mica Schist" and "Salmon Hornblende Schist" (see additional discussions later in this chapter).

RECOGNITION OF THE LITHIC BELTS

As outlined above, early studies in the Klamath Mountains province by Diller and Hershey as well as the work of N.E.A. Hinds in the southern Klamath Mountains (1932, 1933, 1935) and F.G. Wells and associates in the Oregon Klamath Mountains (F.G. Wells et al., 1940, 1949; Wells and Walker, 1953; Wells, 1956) provided much stratigraphic and structural data regarding the geologic evolution of this complex, remote, and rugged terrane. Some of these geologic studies, especially those carried out by members of the USGS, were related to the assessment of various mineral commodities, for example, gold, chromite, copper, and manganese. Nevertheless, no regional synthesis of the Klamath Mountains province existed, and Irwin's (1960) *Geologic reconnaissance of the northern California Coast Ranges and Klamath Mountains, California,* published as California Division of Mines Bulletin 179, represented an enormous advancement in understanding the fundamental relationships between the Klamath Mountains and Coast Ranges as well as among the diverse group of rocks of contrasting age, deformational history, and metamorphism that constitute the Klamath Mountains province (Fig. 9). This survey was based on a compilation of all previous geologic studies but also involved numerous reconnaissance traverses in key areas throughout the California Klamath Mountains.

An important concept that Irwin (1960) introduced in his synthesis of the California Klamath Mountains and adjacent northern Coast Ranges was the recognition of distinct lithic belts within these geologic provinces (Fig. 10). Within the Klamath Mountains province, Irwin (1960) designated four lithic belts from east to west: (1) eastern Paleozoic belt, (2) central metamorphic belt, (3) western Paleozoic and Triassic belt, and (4) western Jurassic belt. The delineation of these belts was based on a combination of the available age data (especially fossil material) and lithology. The eastern Paleozoic belt is the most fossiliferous, with numerous fossil localities that span a considerable part of the Paleozoic and Mesozoic. Fossil localities in the western Jurassic belt are sparse and the collections small; but

Figure 8. Oscar H. Hershey (1874–1939). See Lawson (1941).

Figure 9. W.P. Irwin and D.B. Tatlock with a map compilation of the geology of the northern Coast Ranges and Klamath Mountains, California, prepared for a presentation (Irwin and Tatlock, 1955) at the Geological Society of America Cordilleran Section Meeting in Berkeley, California (April 28–30, 1955). This early report formed much of the basis for *California Division of Mines Bulletin 179* published in 1960. Photograph taken at the USGS offices in Menlo Park, California, in 1955.

the age of these rocks, including the Galice Formation of Diller (1907), is well known, based on the scattered occurrences of Late Oxfordian to middle Kimmeridgian *Buchia concentrica* (Sowerby) (Imlay, 1959). The original definition of the western Paleozoic and Triassic belt (Irwin, 1960) was chiefly based on locally common fossil occurrences in the southern Klamath Mountains. The fossils indicated a great range in ages, from Devonian through Triassic. More recently, Early and Middle Jurassic radiolarians were also reported from these rocks (see Irwin et al., 1977, 1978, 1982; Blome and Irwin, 1983; Irwin and Blome, on the CD-ROM accompanying this volume and in the GSA Data Repository[1]), and it is clear that the Paleozoic fossils are predominantly found in tectonic or olistostromal blocks in mélange (e.g., Goodge and Renne, 1993).

No fossil localities occur in the central metamorphic belt; in fact, some early workers considered these chiefly amphibolite-facies rocks as possibly Precambrian in age (e.g., Hershey, 1901, p. 243–245; Hinds, 1933, p. 84), whereas Irwin (1960, his plate 1) labeled them as "pre-Silurian." Irwin (1960) also included various schists exposed in the environs of Condrey Mountain (north-central Klamath Mountains) as part of the central metamorphic belt (Fig. 10). Subsequently Hotz (1967, 1979), and more recently Helper (1986), Helper et al. (1989), and Saleeby and Harper (1993) have demonstrated that the Condrey Mountain Schist is unrelated to—and significantly younger than—the rocks of the central metamorphic belt.

Davis and Lipman (1962) significantly revised the structural sequence and consequently the stratigraphic nomenclature of Irwin's (1960) central metamorphic belt by abandoning the name "Abrams Mica Schist" (Hershey, 1901) and redefining these rocks as two distinct metasedimentary units: Stuart Fork and Grouse Ridge Formations, separated by the mafic metavolcanic Salmon Formation ("Salmon Hornblende Schist" of Hershey, 1901). According to the structural interpretation of Davis and Lipman (1962), the Stuart Fork Formation is structurally below the Salmon Formation, and these contrasting metamorphic units are separated by a folded, regional thrust fault (i.e., Siskiyou thrust zone of Davis, 1968). The formal stratigraphic nomenclatural status of the Grouse Ridge Formation remains controversial, and Lanphere and Irwin (1987) argued that the name "Abrams Schist" should be retained for the amphibolite-facies, chiefly metasedimentary unit (i.e., the Grouse Ridge Formation of Davis and Lipman, 1962) that overlies the "Salmon Schist."

Goodge (1989a,b, 1995) studied the Stuart Fork Formation in detail, employing modern structural analysis and thermobarometry; he recognized that this unit was deformed and metamorphosed under high pressure–low temperature metamorphic conditions ($P > 6$–11 kb and $T = 250$–400 °C; Goodge, 1995), most likely related to subduction in an accretionary complex. K-Ar radiometric dating reported by Hotz et al. (1977) from the Stuart Fork Formation near Yreka, California, indicated a Late Triassic age (ca. 220 Ma) for the blueschist-facies metamorphism, significantly younger than the Devonian age (ca. 380 Ma) based on Rb-Sr and K-Ar radiometric dating reported by Lanphere et al. (1968) on rocks and minerals from the "Abrams Mica Schist" and "Salmon Hornblende Schist."

In the pamphlet that accommodated the *Preliminary geologic map of the Kerby quadrangle, Oregon* (1948, scale 1:96,000), Wells et al. (1949, p. 3) defined the Triassic(?) Applegate Group as: "A thick assemblage of metamorphosed volcanic rocks with lens-shaped interbeds of argillite, chert, quartzite, conglomerate, and marble." The Applegate Group has since been shown to consist primarily of rocks that may be correlated to terranes recognized in the southern Klamath Mountains province and potentially to the May Creek Schist and related rocks (see below). Many of the rocks of the classic

[1]GSA Data Repository item 2006196, Fossil localities of Rattlesnake Creek, western and eastern Hayfork, and North Fork terranes of the Klamath Mountains, is available online at www.geosociety.org/pubs/ft2006.htm, or on request from editing@geosociety.org or Documents Secretary, GSA, P.O. Box 9140, Boulder, CO 80301, USA.

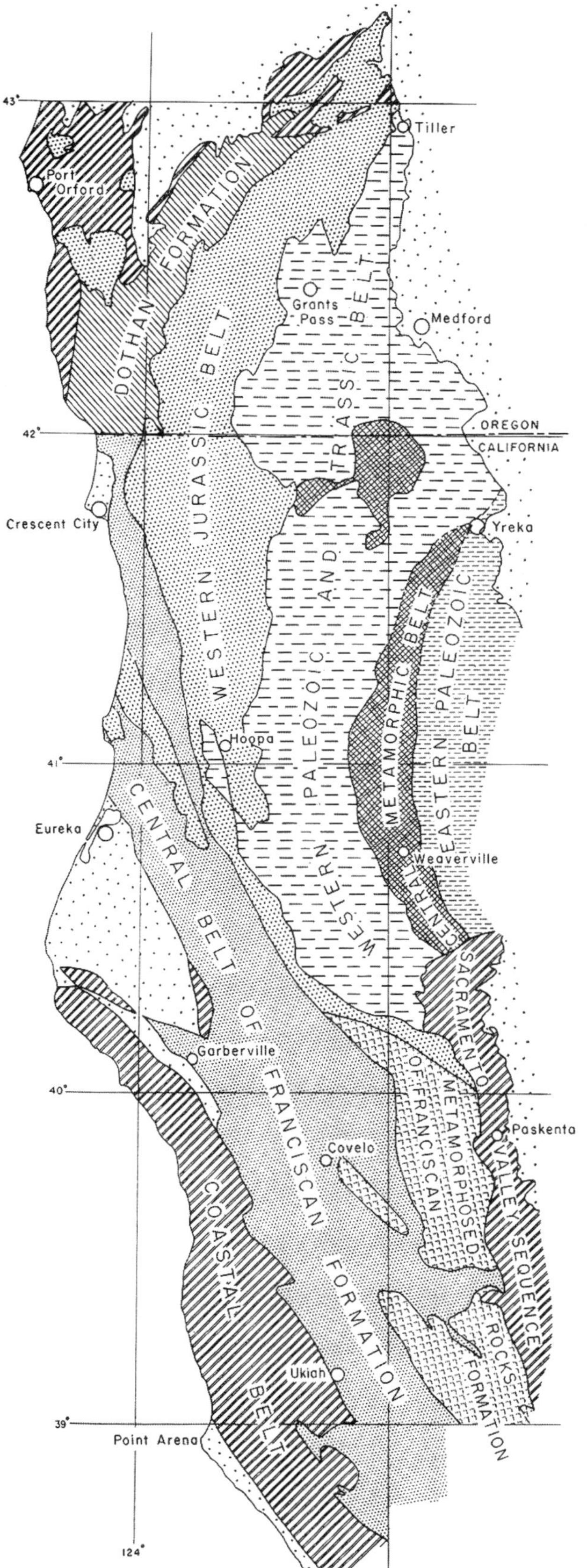

Applegate Group of Wells et al. (1949) are lithologically identical to the western Hayfork terrane of Wright (1982) and contain igneous and detrital (igneous) hornblende grains with identical $^{40}Ar/^{39}Ar$ ages (Donato et al., 1996).

Southeast of the Wimer pluton (Fig. 6) the "Applegate Group" is thrusted above medium- to high-grade semipelitic schist and amphibolitic metabasite (Donato, 1991a,b), collectively referred to as the "May Creek terrane" (Silberling et al., 1987; Irwin, 1994). The amphibolitic unit lacks a formal name; the metasedimentary rocks are called the "May Creek Schist" (Donato, 1991a,b; also see Kays, 1968, 1970, 1995). Donato (1992) described the boundary between the May Creek Schist and structurally underlying amphibolitic unit as an 800- to 1500-m-thick ductile shear zone. Surprisingly, this shear zone is one of the few tectonic boundaries in the Klamath Mountains province that has been studied in detail in regard to its kinematic evolution based on mesoscopic and microscopic structural features. Within the metasedimentary rocks of the hanging-wall plate (i.e., May Creek Schist), Donato (1992) recognized S-C fabrics and asymmetric porphyroclasts and determined quartz [c]-axes patterns in twelve oriented samples. All these data indicate northwest-directed (N38–45°W) thrusting of the May Creek Schist above the amphibolitic unit. However, the age of this thrusting cannot be Nevadan (i.e., Late Jurassic; see later section on Jurassic orogenies for additional discussion), as initially speculated by Donato (1992), because the mapped thrust-fault zone is intruded by the ca. 160-Ma Wimer pluton (Donato, 1991b; Yule, 1996). This ductile shear zone is thus a contractional structural feature that must predate the Late Jurassic. Furthermore, the protolith age of the May Creek terrane is unknown, and its relationship with adjacent tectonostratigraphic terranes of the Klamath Mountains province is everywhere tectonic (fault-bounded) (Silberling et al., 1987; Irwin, 1994; Fig. 3).

SIGNIFICANCE OF ALPINE-TYPE ULTRAMAFIC ROCKS AND THE OPHIOLITE CONCEPT

The recognition that bodies of peridotite-dunite, serpentinite and/or metaserpentinite occur in belts in many orogens on Earth was well known in the early part of the twentieth century (Hess, 1955). The Australian geologist W.N. Benson (1926) referred to such ultramafic rocks in orogenic belts as "alpine-type" ultramafic rocks and inferred an intrusive mode of emplacement. Many subsequent writers delineated the differences between alpine-type ultramafic rocks and stratiform ultramafic complexes, such as the Stillwater Complex, Montana (e.g., see Thayer, 1960, for a comparison). Nevertheless, Robert S. Dietz's (1963) speculative suggestion that alpine-type serpentinites were a slice of oceanic crust tectonically emplaced into

Figure 10. Distribution of lithic belts of the Klamath Mountains province. After Irwin (1960, his fig. 3).

an orogenic belt initiated significant rethinking of the tectonic significance of the so-called "alpine-type" ultramafic rocks. The Dutch geologist W.P. de Roever (1957) had made a similar suggestion about the significance of alpine-type ultramafic rocks in orogens, but his essay was written in German, published in *Geologische Rundschau,* and apparently largely unnoticed by English-speaking geologists. The seminal paper by Harry H. Hess (1962) titled "History of ocean basins" had previously argued that oceanic crust consisted chiefly of serpentinized peridotite (see fig. 7 *in* Hess 1962) and thus also set the stage for Dietz's interpretation of alpine-type ultramafic rocks as tectonic slices of oceanic crust in an orogen.

In this light, the Klamath Mountains played a key role in understanding the tectonic significance of ultramafic belts in orogens. Beginning before World War II and continuing throughout the war years and into the early 1950s, economic geologists, especially members of the USGS, mapped and studied the many chromite deposits associated with the alpine-type ultramafic rocks of the Klamath Mountains (e.g., Rynearson and Smith, 1940; Wells et al., 1946; Wells and Cater, 1950). These early studies were important in determining the broad distribution of the ultramafic rocks in this mountain belt; in many cases, these geologists recognized that mafic rocks (e.g., gabbro, diabase) were associated with these ultramafic rocks. The concept that the ultramafic rocks were part of an intrusive body pervaded the thinking of virtually all geologists who studied such orogenic ultramafic rocks throughout the 1950s and into the early 1960s. The recognition that regional sheets of ultramafic and related rocks (Irwin and Lipman, 1962; Lipman, 1964) occurred along the boundaries of the lithic belts defined by Irwin (1960) eventually stimulated a significant shift in thinking regarding the emplacement of alpine-type ultramafic rocks in the Klamath Mountains province and elsewhere. In contrast to the ruling hypothesis that alpine-type ultramafic rocks were intrusive bodies, Irwin (1964) argued that they were part of the hanging wall of regional-scale allochthons that occurred as imbricated thrust sheets throughout the Klamath Mountains province.

The first publication that specifically used the term "ophiolite" to describe an ultramafic-mafic complex in California was an abstract written by Stephen P. Bezore (1969) in regard to an ophiolitic sequence exposed in the northern California Coast Ranges near Saint Helena. In this abstract, Bezore described a pseudostratiform sequence of ultramafic-mafic rocks that ranged from (1) a basal peridotite unit through (2) an intermixed zone of ultramafic and gabbroic rocks to (3) massive gabbro and in turn overlain by (4) diabase breccia. The basal contact with the Franciscan Complex was interpreted as a tectonic boundary, manifested by a zone of mélange, whereas the contact with the overlying Knoxville Formation was interpreted as a depositional contact. This early description of an ophiolitic sequence quickly led to the recognition of similar sequences at many localities within the California Coast Ranges (Bailey et al., 1970), and these ophiolitic sequences were in general referred to as the "Coast Range ophiolite." In the adjacent Klamath Mountains, Snoke (1971) recognized an ophiolitic sequence in the Preston Peak area that had originally been included in Irwin's (1960) "western Paleozoic and Triassic belt." During the late 1970s and early 1980s, numerous detailed studies of ophiolitic suites from the Klamath Mountains province were published (e.g., Lindsley-Griffin, 1977; Snoke, 1977; Snoke et al., 1977; Harper, 1980, 1984; Quick, 1981, Ando et al., 1983). Many of these ophiolitic sequences were subsequently radiometrically dated (summarized in Saleeby, 1990) and/or their age was established by radiolarians from sedimentary rocks that form interpillow sediment or rest in depositional contact immediately above the lavas (e.g., Pessagno and Blome, 1990; Pessagno et al., 2000).

Of all the ophiolites exposed in the Klamath Mountains province, the Josephine ophiolite is the best-documented sequence because of the many studies by Gregory D. Harper and coworkers (e.g., Harper, 1980, 1984, 2003; Harper and Wright, 1984; Harper et al., 1985, 1988, 1990, 1994; Saleeby and Harper, 1993). Other workers have also made significant contributions on the petrogenesis, deformation features and mechanisms, and microfabric of the ultramafic rocks forming the basal unit of the Josephine ophiolite (Loney and Himmelberg, 1976; Dick, 1977; Evans, 1987; Kelemen and Dick, 1995). A probable continuation of the Josephine ophiolite (i.e., Devils Elbow ophiolite remnant) has been recognized in the southernmost Klamath Mountains (Wright and Wyld, 1986; Wyld and Wright, 1988). Plagiogranite from the Josephine ophiolite yielded a U-Pb zircon age of 163 ± 5 Ma (table 2, sample 6 *in* Harper et al., 1994), whereas Wright and Wyld (1986, their table 1, fig. 3) reported a U-Pb zircon age of 164 ± 1 Ma from the Devils Elbow ophiolitic remnant in the southern Klamath Mountains.

The Josephine ophiolite (Harper, 1984) is an "ideal ophiolite" in the sense that it corresponds closely to the classic definition of ophiolite as published in *Geotimes* (Anonymous, 1972) after the first Penrose Field Conference on ophiolites and has been commonly cited in ophiolite literature for the past thirty or more years. As pointed out by some recent ophiolite workers (e.g., Dilek, 2003), this classic definition, established in the early days of the application of the plate tectonic paradigm to orogenic belts, is too confining in its emphasis on a pseudostratigraphic character, and many ophiolitic assemblages are more complex than the definition admits (e.g., ophiolitic mélange —see Gansser, 1974) or are polygenetic (Saleeby, 1982). Nevertheless, the Josephine ophiolite fits the classic definition as well as any ophiolite in the Klamath Mountains. An important aspect of the Josephine ophiolite is that its early history is exceptionally well preserved along its margins, even though the Josephine ophiolite and the marginal "rift facies" were involved in regional contraction manifested by the Late Jurassic Nevadan orogeny (see detailed discussion later in this chapter). The rift facies of the Josephine ophiolite was originally identified along the eastern margin of the Late Jurassic basin that is floored by the Josephine ophiolite. In the Preston Peak area, Snoke (1977) described a mafic complex associated with the Preston Peak ophiolite. This mafic complex consists of many small intrusive bodies of metadiabase and/or metagabbro as well as numerous metadiabase dikes (i.e., a mafic dike swarm). Saleeby and Harper (1993, their table 1, sample 3) reported a U-Pb zircon age of 164 ± 4 Ma on a quartz diorite dike, a scarce member of the mafic complex of the Preston Peak ophiolite, indicating that the dike swarm was similar in age or slightly older than the Josephine ophiolite (ca. 164–162 Ma). Yule et al. (1992) and Yule and Saleeby (1993) argued that mafic dikes intrusive into the Rattlesnake Creek terrane on the west side of the western Jurassic belt (i.e., part of the Onion Camp complex; Yule, 1996) are a remnant of the rifted *western* margin of the basin (also see Yule et al., this volume) and consequently equivalent to the mafic complex of the Preston Peak ophiolite of the eastern margin of the basin. If this interpretation is correct, the Josephine ophiolite and its rifted margins represent the most complete ophiolitic sequence in North America. A somewhat similar scenario has been proposed for the development of the Rocas Verdes ophiolites of southernmost South America (Stern and De Wit, 2003, and references therein) but in this setting, the rifting occurred within continental crust rather than in an older ensimatic terrane (i.e., Rattlesnake Creek terrane).

The tectonic setting of the rifting associated with the development of the Josephine ophiolite is still uncertain. Recently, Harper (2003) interpreted the development of the Josephine ophiolite as a propagating rift analogous to the rifting that formed the modern Lau Basin of the western Pacific Ocean (see his fig. 9). However, previously, Harper et al. (1985) argued that the rifting associated with the development of the Josephine ophiolite developed during regional, oblique extension associated with a broad zone of concurrent strike-slip faulting and extension (intra-arc transform fault), as presently manifested in the Andaman Sea north of Sumatra.

Another ophiolitic assemblage that has a long history of study by various workers is the Trinity ultramafic-mafic complex (Trinity subterrane of Irwin, 1994) exposed in the eastern Klamath Mountains (Lipman, 1964; Lindsley-Griffin, 1977, 1983; Quick, 1981; Boudier et al., 1989; Lindsley-Griffin and Griffin, 1991; Wallin and Metcalf, 1998; Metcalf et al., 2000; Mankinen et al., 2002). This suite of ultramafic and mafic rocks is the oldest ophiolitic assemblage in the Klamath Mountains province but is also a polygenetic suite of rocks that has yielded a broad range of ages—from Neoproterozoic through Early Devonian (Mattinson and Hopson, 1972; Jacobsen et al., 1984; Wallin and Metcalf, 1998; Metcalf et al., 2000; Mankinen et al., 2002; Lindsley-Griffin et al., 2003). Multiple hypotheses have been advanced for this assemblage; the most recent interpretation is that it is a supra-subduction ophiolite formed during forearc rifting in an incipient intraoceanic arc (Wallin and Metcalf, 1998; Metcalf et al., 2000). A Sm-Nd mineral isochron of 472 ± 32 Ma on plagioclase peridotite (Jacobsen et al., 1984) is interpreted to date the lithospheric emplacement of the Trinity peridotite. This peridotite massif was subsequently intruded by a suite of gabbroic plutons that range in age from Early Silurian to Early Devonian (431–404 Ma) (Wallin and Metcalf, 1998). Wallin and Metcalf (1998) and Metcalf et al. (2000) argued that the Early Devonian Copley Greenstone and Balaklala Rhyolite of the Redding section may be a downfaulted or erosional remnant of the volcanic carapace related to these Siluro-Devonian plutonic rocks (see their figs. 7 and 12, respectively). Wallin and Metcalf (1998) and Metcalf et al. (2000) also argued that the Trinity ultramafic-mafic complex was part of a west-facing intraoceanic arc system, with the Central Metamorphic terrane interpreted as a remnant of underthrusted oceanic lithosphere. Dickinson (2000) proposed an opposite subduction-zone geometry for these terranes, in which the Trinity ultramafic-mafic complex was part of an east-facing subduction system that migrated southeastward during slab rollback. The Late Devonian volcanic assemblages of the northern Sierra Nevada are interpreted as younger manifestations of this southeast-facing island-arc system.

Perhaps the most enigmatic ophiolitic assemblage in the Klamath Mountains province is the Rattlesnake Creek terrane (Irwin, 1972; Petersen, 1982; Rawson and Petersen, 1982; Hill, 1984; Gorman, 1985; Gray, 1985; Donato, 1987, 1989; Wright and Wyld, 1994). The Rattlesnake Creek terrane is best known in the southern Klamath Mountains, where Irwin (1972) originally delineated it as a mappable subdivision of his "western Paleozoic and Triassic belt." Irwin (1972) recognized the ophiolitic nature of this heterogeneous assemblage of rocks as well as its overall mélange structural character. Wright and Wyld (1994) demonstrated the polygenetic evolution of this terrane. They recognized that the Rattlesnake Creek terrane consists of a widespread basement of ophiolitic mélange overlain locally by various "cover sequences," which exhibit characteristics of an oceanic-arc regime. Furthermore, Wright and Wyld (1994) argued that the ophiolitic mélange developed in a fracture zone (i.e., oceanic transform fault zone) that predates the Late Triassic. The first cover sequence is interpreted to be an intraoceanic Late Triassic–Early Jurassic arc tied to eastward-dipping subduction zone(s) developed over the fracture-zone mélange (see their fig. 10). During the Middle Jurassic, the Rattlesnake Creek terrane apparently served as ensimatic basement for a second oceanic-arc cover sequence referred to as the "western Hayfork terrane" (Wright, 1982; Wright and Fahan, 1988; Donato et al., 1996). During regional Middle Jurassic contraction (prior to ca. 170 Ma), this oceanic-arc system collapsed, and the eastern Hayfork terrane (part of a long-lived accretionary subduction complex; see Wright, 1982) was thrust on top of the western Hayfork terrane along the Wilson Point fault (Fig. 3). In turn, the western Hayfork terrane was apparently thrust (i.e., Salt Creek thrust) on top of the Rattlesnake Creek terrane (fig. 10 *in* Wright and Wyld, 1994). The Wilson Point thrust is intruded by the ca. 170-Ma Ironside Mountain batholith (Barnes et al., this volume, Chapter 10) and thus is a manifestation of regional contraction in the Klamath Mountains province associated with a cryptic Middle Jurassic orogenic history (see later section on Jurassic orogenies).

In the north-central Klamath Mountains, rocks correlated

with the Rattlesnake Creek terrane (Marble Mountain terrane of Blake et al., 1982) become progressively metamorphosed from greenschist facies to locally as high-grade as granulite facies (Barrows, 1969; Medaris, 1975; Rawson and Petersen, 1982; Donato, 1987, 1989; Coleman et al., 1988). The highest metamorphic grade equivalents of the Rattlesnake Creek terrane frame the Condrey Mountain window, a Neogene structural dome (Mortimer and Coleman, 1985) that exposes the high pressure–low temperature Condrey Mountain Schist (Hotz, 1967, 1979; Helper, 1986). The high-grade rocks of the Rattlesnake Creek terrane are penetrated by numerous late Middle Jurassic plutonic complexes, but the Condrey Mountain Schist is *not* intruded by plutonic rocks of this age. The correlation of the high-grade rocks of the western Paleozoic and Triassic belt with the Rattlesnake Creek terrane and the abundance of large, penecontemporaneous plutons (e.g., Hacker et al., 1995) suggest that the anomalous high-grade metamorphism of the Rattlesnake Creek terrane may be a manifestation of late Middle Jurassic regional contact metamorphism. In this interpretation, intrusive arc magmas play a fundamental role in the advection of heat into the Rattlesnake Creek terrane in the north-central part of the Klamath Mountains province. Such a simple model to explain the increase in metamorphic grade of the Rattlesnake Creek terrane is complicated by the recognition that regional thrusting predated emplacement of the Middle Jurassic plutons (Donato et al., 1982), and the regional metamorphism apparently involved an increase in pressure as well as temperature (Medaris, 1975; Grover, 1984; Lieberman and Rice, 1986). The evolution of the Rattlesnake Creek ophiolitic mélange and its progressive metamorphism to upper amphibolite-facies conditions and locally even higher metamorphic grade is still a major unsolved problem in the Klamath Mountains province (see section on important problems for future study).

RECOGNITION OF REGIONAL THRUST FAULTS

An analysis of each individual thrust-fault system in the Klamath Mountains province is beyond the scope of this summary, and in some cases, the data are insufficient to attempt such an analysis. However, the Orleans fault, originally recognized and named by Hershey (1906, 1911) has been studied in many places along its trace throughout the western Klamath Mountains (Klein, 1977; Snoke, 1977; Petersen, 1982; Hill, 1984; Gorman, 1985; Gray, 1985; Jachens et al., 1986; Cashman, 1988; Saleeby and Harper, 1993; Gray, this volume). A synthesis of these data thus serves as a useful introduction to the complexity of the thrust-fault systems of the Klamath Mountains province. Along the Klamath River from near Happy Camp to Orleans, California (Fig. 2), the Orleans fault is exposed in the canyon walls. It forms an erosional re-entrant through the hanging-wall upper plate near Happy Camp (Klein, 1977; Petersen, 1982; Fig. 3) and underlies a large klippe of hanging-wall rocks north of Orleans (Gray, 1985). The rocks of the hanging wall consist of the Rattlesnake Creek terrane, western Hayfork terrane, and various intrusive plutonic rocks, whereas the footwall rocks are the Upper Jurassic Galice Formation of the Smith River subterrane (i.e., a subdivision of the western Klamath terrane). Near Happy Camp, the rocks of the Rattlesnake Creek terrane (Lower Triassic to Early Jurassic and older[?]) reach amphibolite-facies metamorphic conditions (Petersen, 1982; Hill, 1984; Donato, 1987, 1989; Coleman et al., 1988), whereas the footwall rocks are lower greenschist facies but penetratively strained to Textural Zone 2 or 2+ (Gray, 1985, this volume; Cashman, 1988). Thus the Orleans fault emplaced older rocks onto younger rocks and, at least locally, high-grade metamorphic rocks onto lower-grade metamorphic rocks. These hanging-wall–footwall relationships indicate a contractional fault system.

A gravity survey (Jachens et al., 1986) was carried out to determine the subsurface extent of the Orleans fault in the west-central Klamath Mountains. These data support the interpretation that the Orleans fault is an original low-angle regional thrust fault that has been subsequently folded by broad, kilometer-scale upright folds. Furthermore, a suite of Late Jurassic plutons (western Klamath suite; see Barnes et al., this volume, Chapter 17) locally intrude rocks of both the hanging wall and footwall of the Orleans fault. These field relationships therefore provide tight age constraints on the developmental history of this regional thrust system (Harper et al., 1994). The plutons range in age from ca. 151 to 144 Ma, whereas detritial zircon grains as young as ca. 153 Ma have been identified in metagraywackes from the Upper Jurassic Galice Formation (Miller et al., 2003). The data therefore constrain the movement history of the Orleans fault as between 151 and 153 Ma and support long-standing interpretations that this structural feature is part of the classic Late Jurassic Nevadan orogeny (see section on Jurassic orogenies for a complete discussion). The regional thrust faulting manifested by the Orleans fault and development of penetrative cleavage in the Upper Jurassic Galice Formation are the quintessential structural elements of the Nevadan orogeny in the Klamath Mountains province.

The significance of the regional Orleans thrust system in regard to the overall Late Jurassic plate tectonic framework of western North America remains a fundamental tectonic question. The lack of high pressure–low temperature metamorphism associated with the Upper Jurassic Galice Formation argues against a fossil subduction-zone interpretation for this fault system, which can be traced for the length of the Klamath Mountains province (Fig. 3). It could be an intraplate fault that is a manifestation of outboard Late Jurassic subduction (Davis et al., 1978). Late Jurassic blueschist- and amphibolite-facies rocks occur as tectonic blocks in some mélanges of the Franciscan Complex (Coleman and Lanphere, 1971), and the Late Jurassic plutons of the western Klamath suite have arc geochemical characteristics (Barnes et al., this volume, Chapter 17). These geologic features suggest that subduction was ongoing along the margin of western North America during the Late Jurassic (ca. 150 Ma). The presence of rift facies of the Josephine ophiolite on both the east and west margins of the Late Jurassic Josephine ophiolite basin (i.e., depocenter for the Upper Jurassic Galice Formation) is perhaps the most definitive geologic relationship that eliminates the Orleans fault as a direct manifestation of subduction-zone underthrusting. The Orleans fault, therefore, is a contractional fault system that has been localized along the eastern flank of the Josephine marginal rift basin. This fault system has led to significant underthrusting of the Smith River subterrane beneath the Rattlesnake Creek terrane and other terranes situated to the east. The estimated 100 km of contraction accommodated by this structure (Jachens et al., 1986) is quite considerable but still is far shy of the thousands of kilometers associated with fossil subduction-zone systems.

Older thrust faults occur in the hanging-wall plate of the Orleans fault, but their ages are not well constrained. For example, Barnes et al. (1986) argued that the thrust-fault system that frames the Condrey Mountain window (Fig. 3) is younger than ca. 162 Ma. This interpretation is based on the U-Pb radiometric age of the Slinkard pluton that occurs in the hanging-wall plate of this low-angle fault system but also contains a high-temperature foliation subparallel to the fault. This fault system, which emplaced high-grade metamorphic rocks of the Rattlesnake Creek terrane (Marble Mountain terrane of Blake et al., 1982; also see Donato et al., 1982; Donato, 1987, 1989; Coleman et al., 1988) above the greenschist-blueschist Condrey Mountain terrane, is chiefly exposed because of Neogene doming in this part of the Klamath Mountains province (Mortimer and Coleman, 1985). Neither this fault system nor the Condrey Mountain Schist has been identified farther to the west, suggesting that the Orleans fault truncated the older thrust-fault system (fig. 2 *in* Saleeby and Harper, 1993) during the Nevadan orogeny.

The ages of thrust faults exposed farther to the east of the Orleans fault are even more broadly constrained and are commonly only bracketed by the youngest rocks involved in thrusting and the oldest pluton that intruded both the hanging-wall and footwall plates of these thrust faults (Davis et al., 1978; fig. 9 *in* Wright and Fahan, 1988). An important exception is the Trinity thrust (Fig. 3), which has been interpreted as a high-temperature shear zone that formed during the emplacement of the Trinity peridotite (Peacock and Norris, 1989). Devonian (ca. 380 Ma) Rb-Sr radiometric dates from minerals and rocks of the underlying Grouse Ridge Formation of the Central Metamorphic terrane are commonly interpreted as dating the emplacement of the Trinity peridotite.

One final topic related to the character of the thrust faults of the Klamath Mountains province concerns the various faults that bound several terranes, which Irwin (1972) included as part of his western Paleozoic and Triassic belt. These tectonostratigraphic terranes are the Fort Jones, North Fork, and eastern Hayfork terranes (Fig. 3). Each terrane is bounded by an eastward-dipping regional thrust fault. The Fort Jones terrane includes ca. 220-Ma blueschist-facies rocks (Hotz et al., 1977; Goodge, 1989a,b), the North Fork terrane is ophiolitic (Ando et al., 1983), and the eastern Hayfork is a mélange containing Tethyan fauna (Irwin and Galanis, 1976; Nestell et al., 1981; Miller, 1987; Miller and Wright, 1987). One possible interpretation of these three terranes is that they make up an accretionary subduction complex that began to evolve in Late Triassic time and continued into Middle Jurassic time (Wright, 1982; Dickinson, 2000). Jurassic radiolarians have been recovered from both the North Fork and eastern Hayfork terranes (Irwin et al., 1977, 1978, 1982; Blome and Irwin, 1983; Irwin and Blome, on the CD-ROM accompanying this volume and in the GSA Data Repository [see footnote 1]).

TECTONOSTRATIGRAPHIC TERRANE CONCEPT

An early application of plate tectonic theory to the North American Cordilleran orogen—and in particular, the Klamath Mountains and northern California Coast Ranges—was Warren Hamilton's (1969) classic paper titled "Mesozoic California and the underflow of Pacific mantle." In this important synthesis of California geology, Hamilton established the framework for a complete reevaluation of the geology of the Klamath Mountains and environs. In particular, he recognized the importance of distinct tectonic assemblages of rocks, such as island arcs and accretionary prisms and associated mélanges.

Although Hamilton (1969) did not employ the term "terrane" to describe his various tectonic assemblages in the California Coast Ranges, Klamath Mountains, and western Sierra Nevada, he clearly analyzed the geologic evolution of these provinces in terms of distinct lithotectonic units. W.P. Irwin (1972) introduced the term "terrane" into modern geologic literature and first delineated a group of distinct "terranes" in the southern Klamath Mountains to better understand the heterogeneous and complex "western Paleozoic and Triassic belt" of his earlier synthesis. Irwin (1972) recognized three distinct terranes (from east to west): North Fork, Hayfork, and Rattlesnake Creek terranes (Fig. 11). Irwin (1972, p. C103) stated the following regarding his use of the term "terrane" "refers to an association of geologic features, such as stratigraphic formations, intrusive rocks, mineral deposits, and tectonic history, some or all of which lend a distinguishing character to a particular tract of rocks and which differ from those of an adjacent terrane."

Subsequently, Wright (1982) recognized that the Hayfork terrane of Irwin (1972) was composite in lithology, age, and inferred tectonic setting. He subdivided the original Hayfork terrane into the eastern and western Hayfork terranes. The former consists chiefly of tectonic mélange and broken formation; Wright (1982) interpreted it as part of a Permo–Triassic accretionary subduction complex. The latter consists chiefly of volcaniclastic deposits, which were dated as Middle Jurassic (177–168 Ma), based on conventional K-Ar hornblende dates (table 5 *in* Wright and Fahan, 1988). Wright (1982) interpreted this terrane as deposited in an intraoceanic-arc setting.

Although the published record clearly indicates that Irwin's (1972) recognition of terranes in the southern Klamath Moun-

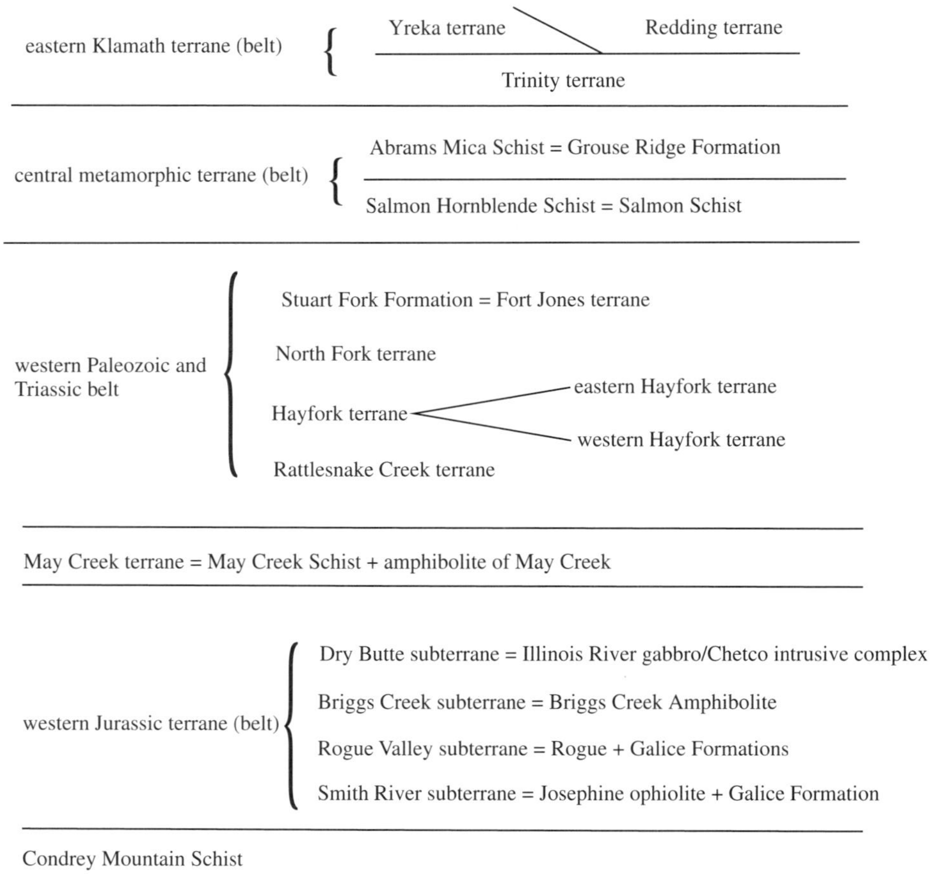

Figure 11. A comparison of lithic belts as defined by Irwin (1960) and subsequent terrane nomenclature (see text for pertinent references) for the Klamath Mountains province.

tains was fundamental in analyzing the tectonic development of that mountain belt, the concept of terrane analysis as applied to orogens in general was not widely used by the geologic community until the 1980s. In this light, the paper by Peter J. Coney, David L. Jones, and James W.H. Monger, published in 1980 in *Nature* is commonly recognized as the most significant contribution to interpretation of the North American Cordillera as a collage of allochthonous terranes. Gradually the term "terrane" was modified into the more descriptive term "tectonostratigraphic terrane." In a book on the terranes of the circum-Pacific region, Howell et al. (1985, p. 4) defined "tectonostratigraphic terrane" as "a fault-bounded package of rocks of regional extent characterized by a geologic history which differs from that of neighboring terranes."

Gray (1986) introduced the term "native terrane" for tectonostratigraphic terranes that have specific geologic and paleontological ties to western North America, and he used several tectonostratigraphic terranes of the Klamath Mountains as examples, specifically, the terranes of Irwin's (1960) "western Paleozoic and Triassic belt" (Fig. 10). This term has been partially supplanted by the term "pericratonic terrane," defined as a terrane that formed near a craton (i.e., North America), based on facies or clastic component provenance (e.g., Monger and Nokleberg, 1996). Furthermore, a growing body of geochemical and isotopic studies indicate a Precambrian (cratonic) input into some accreted tectonostratigraphic terranes along the length of the Cordillera (Centeno-Garcia et al., 1993; Patchett and Gehrels, 1998; Unterschutz et al., 2002; Frost et al., this volume). In the Klamath Mountains province, Nd isotope data for argillaceous rocks from the Rattlesnake Creek terrane and Upper Jurassic Galice Formation of the western Klamath terrane indicate a significant input of Precambrian debris during deposition of portions of these accreted terranes (Frost et al., this volume). Detritial zircon studies from upper Paleozoic strata of the eastern Klamath terrane (Miller and Saleeby, 1991; Gehrels and Miller, 2000) and Upper Jurassic Galice Formation (Miller and Saleeby, 1995) also indicate the input of Precambrian zircon grains during deposition. Likewise, xenocrystic Paleoproterozoic (ca. 1.7 Ga) zircon grains have been recognized in the Upper Jurassic Devils Elbow ophiolite remnant of the southern Klamath Mountains (Wright and Wyld, 1986).

The North Fork terrane provides an informative example of another possible native or pericratonic terrane. Wright (1982) showed that (meta)sedimentary rocks of the North Fork and eastern Hayfork terranes were similar in lithologic character and age and showed evidence of craton-derived (hemipelagic) sedimentary provenance. He suggested that the two units were part of what is now referred to as a supra-subduction-zone complex in which the coeval arc lay in what is now the eastern Klamath belt (Redding section). On the basis of detailed mapping, geochemical, and geochronological studies in the central Klamath Mountains province, Ernst (1999) supported the supra-subduction- zone interpretation. However, Ernst and his co-workers found that North Fork metabasaltic rocks in the Sawyers Bar area consisted of island-arc basalt (IAB) and ocean island basalt (OIB), which in turn suggest that the North Fork terrane was at least in part an arc terrane. On the basis of OIB geochemistry, Mortimer (1985) correlated similar metabasaltic rocks in the Yreka area with the North Fork terrane. In this volume, Scherer et al. present new geochemical data for North Fork mafic metavolcanic rocks from the southern Klamath Mountains province. They found OIB-type basalts, as seen elsewhere in the terrane, but also mid-ocean ridge basalts (MORB). The presence of MORB in the southern North Fork terrane may not be consistent with a pericratonic origin and was interpreted by Scherer et al. (this volume) to indicate a distal tectonic setting. It is presently unclear whether the North Fork terrane represents oceanic crust whose petrologic character changed dramatically along strike, or whether the North Fork terrane in the southern Klamath Mountains province is a distinct tectonic fragment compared to the more northern parts of the terrane.

Since the advent of the terrane concept in the interpretation of orogenic belts, paleontological data have played an important role in recognizing the potential exotic nature of a specific terrane. In the Klamath Mountains province, the recognition of Tethyan fauna in several terranes was considered especially significant (Irwin, 1972). Permian limestones containing verbeekinid fusulinids have played a key role in understanding the distribution of Tethyan fauna in the western North American Cordillera (Monger and Ross, 1971). Two fragmentary, subparallel belts of tectonostratigraphic terranes containing contrasting Permian fauna have been recognized along the length of the western North American Cordillera and designated the McCloud and Cache Creek belts (Miller, 1987). The Cache Creek belt lies generally west of the McCloud belt, and blueschist-facies metamorphic rocks are commonly an element in the former (Hotz et al., 1977). The presence of blueschist-facies rocks and the commonly chaotic nature of the terranes that are part of the Cache Creek belt have led to an interpretation of these terranes as components of an accretionary complex developed during subduction. The presence of Early and Middle Jurassic radiolarians (e.g., Irwin et al., 1977, 1978, 1982; Blome and Irwin, 1983) in some of the terranes of the Cache Creek belt indicate that the development of the accretionary complex continued nto the early Mesozoic. In the Klamath Mountains province, the tectonostratigraphic terranes that have been classified as part of the Cache Creek belt are the Fort Jones and eastern Hayfork terranes. To the northeast, in the Blue Mountains province, the Baker terrane (Silberling et al., 1987) is part of the Cache Creek belt; to the south, the Calaveras Complex (Schweickert et al., 1977) is also interpreted as belonging to the Cache Creek belt (Miller, 1987), although no fossils of Tethyan affinity have ever been recovered from this rock assemblage.

IMPORTANCE OF PLUTONIC STUDIES

The geologic map that Irwin (1960) compiled to accompany his geologic reconnaissance of the California Klamath Mountains and adjacent northern Coast Ranges clearly demonstrated that distinct plutons, some of batholithic dimensions, intruded all the lithic belts of the Klamath Mountains. In contrast, except for the so-called "alpine-type ultramafic rocks," such plutonic rocks do not occur in the northern California Coast Ranges. When Irwin's map was published (1960), none of the intrusive plutonic complexes in the Klamath Mountains province had been studied in detail, although Curtis et al. (1958) had dated hornblende and biotite from the Shasta Bally pluton using the K-Ar method. This early interest in the Shasta Bally pluton arose because it is depositionally overlain by fossiliferous, Lower Cretaceous sedimentary rocks of the Great Valley sequence, which places a maximum absolute age on these rocks.

Early, detailed petrologic studies of Klamath Mountains province plutons focused on well-exposed intrusions in the Trinity Alps region (south-central Klamath Mountains province). These studies are exemplified by the work of Davis (1963) and Lipman (1963), who showed that individual plutons consist of distinct, mappable intrusive units and that magma emplacement styles ranged from forceful emplacement to stoping. These observations were emphasized in the compilation study of the Trinity Alps region (Davis et al., 1965) that showed pluton distribution, structural character, and contact relations. Davis et al. (1965) also showed that many plutons truncated regional tectonic boundaries, which emphasized the need for dating individual intrusive events. Another important observation from these early studies was the abundance of tonalitic to trondhjemitic rocks in the Trinity Alps region and the similarity of Klamath tonalitic plutons with those of the western Sierra Nevada Metamorphic belt (e.g., Hietanen, 1951, 1976; Compton, 1955; Clark, 1964, 1976; also see Irwin, 2003, on the CD-ROM accompanying this volume and the GSA Data Repository,[2] and Day and Bickford, 2004, for more recent summaries). In the Klamath Mountains province, this sequence of primarily tonalitic to trondhjemitic plutons spans a narrow range of geologic time (ca. 142–136 Ma; Allen and Barnes, this vol-

[2]GSA Data Repository item 2006197, Correlation of the Klamath Mountains and Sierra Nevada, is available online at www.geosociety.org/pubs/ft2006.htm, or on request from editing@geosociety.org or Documents Secretary, GSA, P.O. Box 9140, Boulder, CO 80301, USA.

ume) and is related to partial melting of deep-seated metabasic rocks (Barnes et al., 1996). As discussed below and in Allen and Barnes (this volume), other plutonic suites in the Klamath Mountains province display distinct compositional and petrogenetic features.

Few geochronological data were available prior to the study by Lanphere et al. (1968). They reported K-Ar and Rb-Sr ages of plutonic and metamorphic rocks from the California Klamath Mountains province and showed that pluton ages ranged from at least as old as Silurian to as young as Cretaceous. On the basis of ages and petrographic and chemical compositions, they recognized three distinct plutonic belts, which represent the foundation of modern views of plutonic suites in the province. These results were expanded to include the entire Klamath Mountains province by Hotz (1971), who enlarged the coverage of K-Ar ages and presented a large petrographic and geochemical dataset. Hotz (1971) recognized the predominantly calcic to calc-alkalic nature of most of the plutons, which was consistent with their position west of Moore's (1959) "quartz diorite line." However, Lanphere et al. (1968) and Hotz (1971) also recognized the potassic nature of a distinct group of plutons associated with the Ironside Mountain batholith—compositions that are otherwise absent west of the quartz diorite line (see Barnes et al., this volume, Chapter 10). At this time, four plutonic belts were recognized: southwestern Ironside Mountain belt (pyroxene-rich mildly to strongly potassic plutons), "northern plutonic area" (diverse calcic to calc-alkalic plutons), Trinity Mountains belt (tonalitic and trondhjemitic plutons), and the eastern belt (diverse plutons, some of which had Triassic and Silurian K-Ar ages; Fig. 6).

Published accounts of Klamath Mountains province plutons languished until 1980. With the appearance of field and geochemical results from the Castle Crags pluton (Vennum, 1980), Bear Mountain intrusive complex (Snoke et al., 1981), and Wooley Creek batholith (Barnes, 1983) and the isotopic data of Masi et al. (1981), studies of Klamath plutons entered the modern era. These petrologic and geochemical studies were accompanied by the rapid growth and refinement of the geochronological database, in which U-Pb (zircon) ages supplanted K-Ar and Rb-Sr data, and the renewed recognition of the importance of "pinning plutons" with regard to tectonic history. Irwin (1985) published a synthesis of the geochronological and tectonic data, in which he recognized eleven plutonic belts and distinguished between pre- and post-amalgamation plutons. This synthesis was necessarily preliminary because many plutons lacked reliable U-Pb (zircon) ages. It was revisited in 1999 (Irwin and Wooden, 1999) with a greatly expanded geochronological dataset. At this writing, only a handful of plutons in the Klamath Mountains province lack radiometric age information.

The ages of plutonic rocks in the Klamath Mountains province were nowhere more important than for those that intrude the terranes of the eastern Klamath Mountains. The K-Ar dating (summarized in Hotz, 1971) indicated the possible presence of Silurian and Cretaceous plutons and Mattinson and Hopson (1972) reported U-Pb (zircon) ages of ca. 455 Ma from trondhjemitic boulders in a conglomerate and ca. 480 Ma from a gabbroic body along the northern margin of the Trinity peridotite. The Paleozoic ages, bimodal nature (gabbro and trondhjemite), and common mafic dike swarms (Petersen et al., 1991; Wallin and Metcalf, 1998) led some workers (e.g., Brouxel and Lapierre, 1988; Brouxel et al., 1988) to propose a genetic relationship between the Trinity peridotite and Paleozoic plutons. However, Wallin and coworkers (Wallin et al., 1995; Wallin and Metcalf, 1998) showed that Paleozoic plutons in the eastern Klamath Mountains are primarily Silurian and early Devonian and therefore younger than the Trinity peridotite. Detailed geochemical study of these plutonic systems led these workers (Wallin and Metcalf, 1998; Metcalf et al., 2000) to suggest that the Silurian–Devonian plutons formed in an oceanic, protoarc setting. Wallin et al. (1995) also dated a Neoproterozoic metagabbro in tectonic contact with the Trinity peridotite, one indication of the complexity and polygenetic nature of the Trinity terrane.

As isotopic and trace-element data for Jurassic and Early Cretaceous plutons increased (e.g., Masi et al., 1981; Barnes et al., 1990, 1992; Gribble et al., 1990), it was possible to refine the compositional distinctions recognized by Lanphere et al. (1968) and Hotz (1971). Barnes et al. (1992) showed that isotopic distinctions between plutonic suites may be related to their age and tectonic history. This concept is further refined in this volume (Allen and Barnes; Barnes et al., this volume, Chapter 10, Chapter 17). In these contributions, plutonic subdivisions are recognized on the basis of age; Nd, Sr, and oxygen isotope ratios; rare-earth element patterns; inferred source regions; and petrogenetic processes. Specifically, these data show that (1) most plutons show evidence for intense and complex interaction between mantle-derived magmas and crustal rocks (or magmas), (2) crustal melting was widespread during at least some Mesozoic igneous episodes, (3) changes in magmatic compositions were related to discrete tectonic events (thrust faulting) $\pm$ thermal maturation history, and (4) many plutons contain zircons inherited from their source region(s) or incorporated as xenocrysts during ascent (see section on problems for future research).

Many plutons in the Klamath Mountains province still lack detailed petrogenetic study (e.g., Chetco Complex, Grants Pass, Gold Hill, Vesa Bluffs, Sugarpine, Canyon Creek, Shasta Bally, and Wildwood plutons). This omission is unfortunate, for reasons cited above and because these plutons are excellent examples of precursor activity to the giant pulses of Cretaceous magmatism in the Sierra Nevada and Idaho batholiths. Moreover, because Klamath plutons typically lack the deformation and thermal alteration commonly associated with Sierran Cretaceous activity, they preserve mineral assemblages conducive to detailed analysis of P, T, and compositional variations (e.g., Barnes, 1987; Cotkin and Medaris, 1993).

IMPORTANCE OF METAMORPHIC STUDIES

The central metamorphic belt of Irwin (1960), now referred to as the "Central Metamorphic terrane" (see Irwin, 1994), occurs as a fault-bounded slice of upper greenschist- to amphibolite-facies metavolcanic and metasedimentary rocks in the eastern part of the Klamath Mountains (Fig. 3). This terrane is bound by the Trinity thrust system on the east and Siskiyou thrust of Davis (1968) on the west. The hanging wall of the Trinity thrust is the ophiolitic Trinity ultramafic-mafic complex, discussed earlier in this chapter. The footwall of the Siskiyou thrust is referred to as the North Fork terrane (Irwin, 1972) (Fig. 3). The Stuart Fork Formation of Davis and Lipman (1962) (discussed below as an example of high pressure–low temperature metamorphism in the Klamath Mountains province) is locally sandwiched between the Siskiyou and Soap Creek Ridge faults (Fig. 3). The Stuart Fork Formation is now considered part of the Fort Jones terrane (Irwin, 1994). The Stuart Fork Formation is commonly interpreted as part of an early Mesozoic accretionary subduction complex (Wright, 1982; Goodge, 1989a; Dickinson, 2000).

As previously discussed in this chapter, the protolith age of the Central Metamorphic terrane is uncertain, and unit nomenclature (i.e., Abrams Mica Schist versus Grouse Ridge Formation) and the structural (stratigraphic) sequence in this belt (terrane) have been controversial. Clearly, a significant aspect of this terrane is its high-grade metamorphic character compared to the surrounding terranes (Hershey, 1901; Hinds, 1932, 1933). Moreover, the grade of metamorphism across the terrane decreases from amphibolite facies adjacent to the Trinity thrust (and peridotite) through albite-epidote amphibolite facies to upper greenschist facies near the structural base of the terrane (Davis et al., 1965). This inverted metamorphic gradient was interpreted to result from emplacement of the Trinity peridotite (Peacock and Norris, 1989), perhaps during intraoceanic thrusting (Boudier et al., 1982). Such gradients are commonly associated with metamorphic soles that form during emplacement of ophiolitic allochthons, such as the Bay of Islands ophiolite of Newfoundland (Malpas, 1979) or the Semail ophiolite of Oman (Searle and Malpas, 1980; Searle and Cox, 1999). The Devonian Rb-Sr isochron age reported by Lanphere et al. (1968) and Hotz (1977) is consistent with this interpretation, such that the Central Metamorphic terrane has been thought to represent oceanic rocks accreted to the eastern Klamath terrane during mid-Paleozoic subduction. Barrow and Metcalf (this volume) illustrate the MORB-like character of Salmon Schist metabasites, which supports an oceanic origin for the terrane.

However, $^{40}Ar/^{39}Ar$ cooling ages of hornblende from the Salmon Schist (Barrow and Metcalf, this volume) are Early Permian (ca. 274 Ma). These cooling ages and the observation that the temperature of the Trinity peridotite was below amphibolite-facies conditions at the time of faulting complicate the simple picture of the Central Metamorphic terrane as oceanic crust subducted during development of the Silurian–Devonian proto-arc plutons. Barrow and Metcalf (this volume) suggest that the cold-over-hot geometry of the Trinity fault (cold Trinity peridotite over hot Salmon Schist) may indicate that the final motion on the fault was normal sense. This suggestion does not negate the idea that the Central Metamorphic terrane represents oceanic crust subducted (underthrust) during Silurian–Devonian time, but it does require later normal reactivation of a cold Trinity fault system.

The Central Metamorphic terrane is not the only fault-bounded slice of high-grade metamorphic rocks associated with an ophiolitic assemblage in the Klamath Mountains province. A distinct dynamothermal metamorphic sole also has been documented along part of the western margin of the Josephine ophiolite, near Vulcan Peak, Oregon (Harper et al., 1990). The development of this metamorphic sole is atypical of the emplacement history described for the Bay of Islands (Newfoundland) or Semail (Oman) ophiolitic allochthons in that the Josephine ophiolite was apparently emplaced by the underthrusting of an active magmatic arc (Harper et al., 1996).

High-grade tectonic inclusions with K-Ar and $^{40}Ar/^{39}Ar$ ages of ca. 190 Ma are well known from the Rattlesnake Creek terrane (Snoke, 1977; Gorman, 1985; Wright and Wyld, 1994). Other allochthonous fragments of high-grade metamorphic rocks include the Briggs Creek amphibolite of Garcia (1979) (or Briggs Creek subterrane of Blake, 1984; Blake et al., 1985) and Big Craggies amphibolite exposed in southwestern Oregon (Coleman, 1972).

As previously noted, parts of the Rattlesnake Creek terrane are metamorphosed to amphibolite facies or locally even to granulite facies (e.g., Barrows, 1969; Medaris, 1975). Such rocks are widespread in the Marble Mountains area and along the Siskiyou Divide (north of Seiad Valley, California), and some workers once referred to these high-grade rocks as the "Marble Mountain terrane" (Blake et al., 1982; see discussion by Donato, 1987, 1989) in California and, in Oregon, the mélange of Dutchman Peak and rocks of Seiad Valley (Smith et al., 1982). For simplicity, most recent compilations of the Klamath Mountains province (e.g., Irwin, 1994) consider these rocks as part of the Rattlesnake Creek terrane (also see Donato et al., 1996). Radiometric ages of these high-grade rocks are in the 153–146 Ma range (see summaries in Hacker and Ernst, 1993, and Hacker et al., 1995). These ages are enigmatic because the high-grade Rattlesnake Creek terrane rocks are intruded by plutons as old as the Slinkard and Grayback plutons, with U-Pb (zircon) ages of ca. 162–161 Ma and $^{40}Ar/^{39}Ar$ cooling ages of ca. 167–148 Ma (Hacker and Ernst, 1993; Hacker et al., 1995).

Another area of anomalously high metamorphic grade occurs in the northernmost Klamath Mountains of Oregon and was originally called the "May Creek formation" in Diller and Kay (1924). These rocks are now referred to as the "May Creek terrane" (Silberling et al., 1987; Irwin, 1994), although they may simply be a high-grade area within the Rattlesnake Creek terrane. Detailed studies of the May Creek terrane are few, but Kays (1968, 1970, 1995) and Donato (1991a,b) have documented metamorphic grade variations and structural style in these rocks. Donato (1991b, 1992) recognized an important high-temperature ductile shear zone (i.e., low-angle thrust-fault zone) in the May Creek terrane where the May Creek Schist is thrusted on top of an unnamed amphibolite unit. This ductile thrust zone is intruded by the ca. 160-Ma Wimer pluton (Donato, 1991b, 1992; Yule, 1996), and consequently, the age of metamorphism and ductile faulting within the May Creek terrane is apparently late Middle Jurassic or older. Thus the May Creek terrane is an example of high-grade metamorphism and contractional deformation that predate the Late Jurassic Nevadan orogeny and is another manifestation of the enigmatic Middle Jurassic orogenic history of the Klamath Mountains province.

One of the most interesting and tectonically significant aspects of the metamorphic history of the Klamath Mountains province is that at least three distinct high pressure–low temperature (HP–LT) metamorphic events have been recognized in the province. Furthermore, a fourth episode of Early Cretaceous HP–LT metamorphism (Lanphere et al., 1978) is well known in the Pickett Peak terrane, part of the Franciscan Complex, immediately west of the Klamath Mountains province. The oldest episode of HP–LT metamorphism is in the eastern Klamath Mountains, and it is manifested as the schist of Skookum Gulch metamorphosed under blueschist-facies conditions ($T = 275$ °C and $P = 7.0$ kb; Cotkin, 1987) in the Late Ordovician (Cotkin et al., 1992). Lawsonite + glaucophane schist is interlayered with other schists and dolomitic marble. The glaucophane schist is well foliated and consists of lawsonite porphyroblasts set in a fine-grained matrix of Na-amphibole, chlorite, quartz, albite, and calcite (Cotkin, 1987). The schist of Skookum Gulch is the only early Paleozoic blueschist locality in western North America (Patrick and Day, 1995) and apparently formed in an accretionary prism during Late Ordovician subduction (Cotkin, 1992). However, the tectonic relationships of these early Paleozoic blueschist-facies rocks with other early Paleozoic elements of the western North American Cordillera are vague and potentially suggest an initial origin outside the Cordilleran realm for the schist of Skookum Gulch and perhaps other rock units of the Yreka terrane (Harms et al., 2003; Wright and Wyld, 2003). The Stuart Fork terrane (Goodge, 1989a,b) or Fort Jones terrane (Irwin, 1994) includes Late Triassic (ca. 220 Ma) blueschist-facies rocks, whereas the metamorphic age of the HP–LT metamorphic rocks exposed in the Condrey Mountain window apparently spans from Late Jurassic into mid-Cretaceous time (Helper, 1986, personal commun. 2005; Helper et al., 1989; Hacker et al., 1995). The Condrey Mountain Schist thus may be a composite unit, reflecting accretion (i.e., tectonic underplating) of progressively younger oceanic rocks.

The last topic that we discuss in regard to metamorphic studies in the Klamath Mountains province is the metamorphic evolution of schists that occur immediately west of the province and are part of the Pickett Peak terrane (Blake, 1984; Blake et al., 1985), the easternmost terrane in the Coast Ranges. In California, the Pickett Peak terrane consists of the South Fork Mountain Schist (Blake et al., 1967) and Valentine Spring Formation (Worrall, 1981), whereas in southwestern Oregon, this terrane consists of the Colebrooke Schist (Coleman, 1972) and an underlying serpentinite-matrix mélange that locally contains tectonic blocks of high-grade blueschist- and amphibolite-facies rocks. We include a brief discussion of these low-grade schists in this synthesis of the Klamath Mountains province, because their HP–LT metamorphism has been commonly interpreted as forming when the Klamath Mountains province allochthon was juxtaposed above the rocks of the Coast Ranges (Blake et al., 1967; Lanphere et al., 1978). Furthermore, these rocks represent the fourth episode of HP–LT metamorphism developed in association with the Klamath Mountains province, and their overall character is reminiscent of other, pelitic to wacke, schist units in the Klamath Mountains province (e.g., Condrey Mountain Schist). The South Fork Mountain and Colebrooke Schists are Textural Zone 3 (Blake et al., 1967) quartz–albite–white mica–chlorite ± lawsonite schists (metagraywacke protolith), whereas associated metabasalts (Chinquapin Metabasalt Member of the South Fork Mountain Schist) consist chiefly of the assemblage albite–chlorite–actinolite–epidote but also include blue-amphibole-bearing varieties. Although available geochronological data are not conclusive, the South Fork Mountain Schist was probably metamorphosed in the Early Cretaceous (ca. 125 Ma) (Lanphere et al., 1978). A similar age of metamorphism is also probable for the Colebrooke Schist in southwestern Oregon (Coleman, 1972). An interesting complication concerning metamorphism in the Franciscan Complex (which includes the Pickett Peak terrane) is the ca. 150-Ma age for the high-grade amphibolite-, eclogite-, and blueschist-facies blocks in mélange of the Franciscan Complex (Coleman and Lanphere, 1971). This apparent age of subduction-zone metamorphism is synchronous with the emplacement of the Orleans (thrust) fault in the adjacent Klamath Mountains province as well as the classic Nevadan orogeny (see below for a full discussion of this Late Jurassic orogeny).

JURASSIC OROGENIES

The Nevadian (later modified to "Nevadan," see Hinds, 1932, p. 378, 1935, p. 331–333) orogeny was defined by Blackwelder (1914, p. 643–645) as a Middle to Late Jurassic orogenic event that affected much of the western margin of the North American Cordillera. His type locality was the Sierra Nevada, but Blackwelder envisioned that Nevadan orogenic effects could be traced from the Alaska Range to Baja California and western Mexico. In California, Taliaferro (1942) recognized that two events were commonly included in the Nevadan orogeny: severe folding of bedding and widespread plutonic intrusion. However, he argued that these orogenic effects were separated in time and a distinction should be made between them. Based on field relations between folded Jurassic strata of the western

Sierra Nevada and intrusive plutonic rocks, Taliaferro (1942) concluded that the folding had occurred prior to the last, widespread intrusion of granitic rocks of the Sierra Nevada batholith. Furthermore, he proposed that the early phase of the Nevadan orogeny (folding and thrusting) could be tightly bracketed as post-late Oxfordian–early Kimmeridgian (age of the Mariposa Formation as well as the Galice Formation of the Klamath Mountains) but prior to the Tithonian (age of the Knoxville Formation; i.e., basal unit of the Great Valley sequence). These geologic field relationships indicated that the "restricted Nevadan orogeny" was Late Jurassic in age. Subsequently, radiometric dating demonstrated that the widespread plutonism of the Sierra Nevada batholith was Late Cretaceous in age (Evernden et al., 1957; Curtis et al., 1958; Evernden and Kistler, 1970), and this magmatism was not considered part of the Nevadan orogeny. Curtis et al. (1958) suggested that the widespread granitic plutonism of the Sierra Nevada and Salinian block (i.e., Santa Lucia Range, Gabilan Range, etc.) be referred to as the "Santa Lucian orogeny," but this name never took hold in the literature of California geology.

The restricted definition of the Nevadan orogeny, as proposed by Taliaferro (1942), has been widely used in the Klamath Mountains and western Sierra Nevada to interpret important regional orogenic features in these mountain belts (e.g., Lanphere et al., 1968; Schweickert and Cowan, 1975; Snoke, 1977; Davis et al., 1978; Saleeby et al., 1982; Harper and Wright, 1984; Schweickert et al., 1984). However, as detailed geologic mapping was coordinated with radiometric dating (especially U-Pb zircon dating), it also became clear that important orogenic events had occurred prior to the Late Jurassic Nevadan orogeny (e.g., Lanphere et al., 1968). In a paper that greatly expanded the U-Pb zircon geochronological database for the southern Klamath Mountains, Wright and Fahan (1988) argued that Middle Jurassic orogenic events, including regional thrusting, folding, and metamorphism, were widely developed during ca. 169–161 Ma. Geologic mapping and metamorphic studies by Coleman et al. (1988) further documented the importance of Middle Jurassic orogenic events in the north-central Klamath Mountains, and these workers referred to this orogenic episode as the "Siskiyou event," subsequently modified to "Siskiyou orogeny" (e.g., Hacker et al., 1995). However, as cited above, Hacker et al. (1995) showed that rocks considered the high-grade manifestations of the Siskiyou orogeny (i.e., high-grade rocks of the Rattlesnake Creek terrane) had $^{40}Ar/^{39}Ar$ cooling temperatures younger than cooling temperatures of plutons intruded during and/or after the regional high-grade metamorphic event. Furthermore, in many cases, these late Middle Jurassic plutons also imposed distinct mineralogical and/or textural effects of contact metamorphism on the adjacent Rattlesnake Creek terrane host rocks. Hacker et al. (1995) also noted the similarity in these cooling ages (152–148 Ma) with the approximate ages of Nevadan thrusting. This similarity suggests either that: (1) high-grade Rattlesnake Creek terrane did not cool below the ^{40}Ar hornblende closure temperature until it was refrigerated by Nevadan thrusting, or (2) the high-grade metamorphic rocks of the Rattlesnake Creek terrane were reheated after peak metamorphic conditions and after emplacement of large, late Middle Jurassic plutons. The fact that the plutons have older cooling ages, and that ca. 150-Ma resetting of magmatic hornblendes in the western Hayfork terrane has not been observed, makes possibility (1) problematic, and possibility (2) highly unlikely.

In summary, it is apparent that some orogenic events in the Klamath Mountains province, such as the "restricted Nevadan orogeny" encompass a specific set of structural, metamorphic, and magmatic events. However, it is also apparent that such orogenic events as the Siskiyou orogeny may be difficult to characterize, particularly in the absence of a well-defined and/or readily characterized thermal or tectonic signature. It is entirely possible that other high-grade metamorphic events in the Klamath Mountains province—specifically, metamorphism of the May Creek Schist and related amphibolites (Kays, 1970; Donato, 1991a,b, 1992) and the Central Metamorphic terrane (Davis et al., 1965; Lanphere et al., 1968; Peacock and Norris, 1989; Barrow and Metcalf, this volume)—have similarly complex thermal and tectonic histories.

IMPORTANT PROBLEMS FOR FUTURE STUDY

The following are important questions and possibly new directions that need to be addressed in future geologic studies in the Klamath Mountains province:

1. What were the facing directions of the oceanic arcs that form various terranes in the Klamath Mountains province? How can the polarity of these arcs be determined and tested through geologic and geochemical data?
2. What was the provenance of clastic metasedimentary rocks in various Klamath Mountains province terranes? New, albeit scant, isotopic data (e.g., Frost et al., this volume) suggest that even "oceanic" terranes contain cratonic sedimentary components. What is the importance and/or the volume of these cratonic materials, and what do these data tell us about the paleotectonic setting of various terranes and their subsequent translation (e.g., Wyld and Wright, 2001)? What can be learned from detailed study of detrital zircon in metasandstones as well as hemipelagic and pelagic units? Nearly every terrane in the Klamath Mountains province contains such rocks; they may reveal significant changes in sediment sources that can be correlated to tectonic history.
3. How did the crust of the Klamath Mountains province respond to contractional deformation? Evidence for ~8-kb differentiation of late Middle Jurassic magmas suggests thick crust, at least some time during amalgamation and accretion of the terranes of the Klamath Mountains province. Was thick crust ephemeral, followed quickly by exhumation and/or tectonic erosion?

4. Does evidence for thickened crust (i.e., thrust faulting: Donato et al., 1982; Donato, 1992) during late(?) Middle Jurassic time relate to the high-grade parts of the Rattlesnake Creek terrane? Is it possible that these high-grade zones were widespread in the deep crust but only locally preserved by syn- or post-Nevadan exhumation in the region around the Condrey Mountain dome? An associated conundrum with the high-grade rocks of the Rattlesnake Creek terrane is the range of $^{40}Ar/^{39}Ar$ ages available from these rocks. Although these high-grade rocks are widely intruded by late to post-synkinematic late Middle Jurassic plutons, the metamorphic rocks invariably yield significantly younger $^{40}Ar/^{39}Ar$ ages that must reflect a complex cooling history subsequent to both regional metamorphism and pluton emplacement.
5. Recent recognition of the importance of inherited/assimilated zircons in many Klamath plutons (Harper et al., 1994; Allen and Barnes, this volume; Chamberlain et al., this volume) and the correlation of zircon inheritance with isotopic systematics suggest that these plutons may be sensitive probes of the crust through which they passed. Will application of detailed, modern isotopic and dating methods to plutons from distinct tectonic panels provide a record of changes in crustal and mantle architecture through time?
6. What can modern geomorphologic and geochronological methods tell us about the Neogene and Quaternary history of the Klamath Mountains province, and particularly about landform development and the high-standing, deeply incised nature of the Klamath Mountains province?

ACKNOWLEDGMENTS

The senior author (AWS) of this chapter was introduced to the geology of the Klamath Mountains by W. Porter Irwin during the summer of 1968, when he served as a field assistant for Irwin during geologic mapping in the southern Klamath Mountains. It was during that memorable summer that Porter Irwin began to formalize his ideas about useful mapping techniques to delineate lithotectonic units within the western Paleozoic and Triassic belt that he had originally delineated and described in Irwin (1960). In addition to Porter's interest and support, we have benefited from collaboration and interactions with numerous Klamath workers, too many to mention here. In particular, we especially note C.M. Allen, R.G. Coleman, M.M. Donato, W.G. Ernst, M.L. Ferns, G.G. Gray, B.R. Hacker, G.D. Harper, M.A. Kays, R.W. Kistler, J.B. Saleeby, and J.E. Wright. Our studies in the Klamath Mountains have been most recently supported by National Science Foundation grants EAR-9902807 to AWS and EAR-9902912 to CGB. In addition, CGB acknowledges the support of Texas Tech University during his long-term geological studies in the Klamath Mountains province. We are grateful to D.S. Cowan, J.F. Dewey, W.R. Dickinson, S.R. Garlick, and G.G. Gray for helpful reviews of the manuscript; however, errors of omission and interpretation are our own.

REFERENCES CITED

Aalto, K.R., 2006, this volume, The Klamath peneplain: A review of J.S. Diller's classic erosion surface, *in* Snoke, A.W., and Barnes, C.G., eds., Geological studies in the Klamath Mountains province, California and Oregon: A volume in honor of William P. Irwin: Boulder, Colorado, Geological Society of America Special Paper 410, doi: 10.1130/2006.2410(22).

Allen, C.M., and Barnes, C.G., 2006, this volume, Ages and some cryptic sources of Mesozoic plutonic rocks in the Klamath Mountains, California and Oregon, *in* Snoke, A.W., and Barnes, C.G., eds., Geological studies in the Klamath Mountains province, California and Oregon: A volume in honor of William P. Irwin: Boulder, Colorado, Geological Society of America Special Paper 410, doi:10.1130/2006.2410(11).

Ando, C.J., Irwin, W.P., Jones, D.L., and Saleeby, J.B., 1983, The ophiolitic North Fork terrane in the Salmon River region, central Klamath Mountains, California: Geological Society of America Bulletin, v. 94, p. 236–252, doi: 10.1130/0016-7606(1983)94<236:TONFTI>2.0.CO;2.

Anonymous, 1972, Penrose Field Conference on ophiolites: Geotimes, v. 17, p. 24–25.

Bailey, E.H., Blake, M.C., Jr., and Jones, D.L., 1970, Onland Mesozoic oceanic crust in California Coast Ranges: Reston, Virginia, U.S. Geological Survey Professional Paper 700-C, p. C70–C81.

Barnes, C.G., 1983, Petrology and upward zonation of the Wooley Creek batholith, Klamath Mountains, California: Journal of Petrology, v. 24, p. 495–537.

Barnes, C.G., 1987, Mineralogy of the Wooley Creek batholith, Slinkard pluton, and related dikes, Klamath Mountains, northern California: American Mineralogist, v. 72, p. 879–901.

Barnes, C.G., Allen, C.M., and Saleeby, J.B., 1986, Open- and closed-system characteristics of a tilted plutonic system, Klamath Mountains, California: Journal of Geophysical Research, v. 91, p. 6073–6090.

Barnes, C.G., Allen, C.M., Hoover, J.D., and Brigham, R.H., 1990, Magmatic components of a tilted plutonic system, Klamath Mountains, California, *in* Anderson, J.L., ed., The nature and origin of Cordilleran magmatism: Boulder, Colorado, Geological Society of America Memoir 174, p. 331–346.

Barnes, C.G., Petersen, S.W., Kistler, R.W., Prestvik, T., and Sundvoll, B., 1992, Tectonic implications of isotopic variation among Jurassic and early Cretaceous plutons, Klamath Mountains: Geological Society of America Bulletin, v. 104, p. 117–126, doi: 10.1130/0016-7606(1992)104<0117:TIOIVA>2.3.CO;2.

Barnes, C.G., Petersen, S.W., Kistler, R.W., Murray, R., and Kays, M.A., 1996, Source and tectonic implications of tonalite-trondhjemite magmatism in the Klamath Mountains: Contributions to Mineralogy and Petrology, v. 123, p. 40–60, doi: 10.1007/s004100050142.

Barnes, C.G., Mars, E.V., Swapp, S., and Frost, C.D., 2006, this volume (Chapter 10), Petrology and geochemistry of the Middle Jurassic Ironside Mountain batholith: Evolution of potassic magmas in a primitive arc setting, *in* Snoke, A.W., and Barnes, C.G., eds., Geological studies in the Klamath Mountains province, California and Oregon: A volume in honor of William P. Irwin: Boulder, Colorado, Geological Society of America Special Paper 410, doi: 10.1130/2006.2410(10).

Barnes, C.G., Snoke, A.W., Harper, G.D., Frost, C.D., McFadden, R.R., Bushey, J.C., and Barnes, M.A.W., 2006, this volume (Chapter 17), Arc plutonism following regional thrusting: Petrology and geochemistry of syn- and post-Nevadan plutons in the Siskiyou Mountains, Klamath Mountains province, California, *in* Snoke, A.W., and Barnes, C.G., eds., Geological studies in the Klamath Mountains province, California and Oregon: A volume in honor of William P. Irwin: Boulder, Colorado, Geological Society of America Special Paper 410, doi: 10.1130/2006.2410(17).

Barrow, W.M., and Metcalf, R.V., 2006, this volume, A reevaluation of the paleotectonic significance of the Paleozoic Central Metamorphic terrane, eastern Klamath Mountains, California: New constraints from trace element geochemistry and $^{40}Ar/^{39}Ar$ thermochronology, *in* Snoke, A.W., and Barnes, C.G., eds., Geological studies in the Klamath Mountains province,

California and Oregon: A volume in honor of William P. Irwin: Boulder, Colorado, Geological Society of America Special Paper 410, doi: 10.1130/2006.2410(19).

Barrows, A.G., 1969, The petrogenesis of an alpine ultramafic complex in the Klamath Mountains, Siskiyou County, California: Boulder, Colorado, Geological Society of America Abstracts with Programs for 1969, pt. 3 (Cordilleran Section), p. 4.

Benson, W.N., 1926, The tectonic conditions accompanying the intrusion of basic and ultrabasic igneous rocks: U.S. National Academy of Sciences, Memoir, v. 1, p. 1–90.

Bezore, S.P., 1969, The Mount Saint Helena ultramafic-mafic complex of northern California Coast Ranges: Boulder, Colorado, Geological Society of America Abstracts with Programs for 1969, part 3 (Cordilleran Section), p. 5–6.

Blackwelder, E., 1914, A summary of the orogenic epochs in the geologic history of North America: Journal of Geology, v. 22, p. 633–654.

Blake, M.C., Jr., 1984, Tectonostratigraphic terranes in southwestern Oregon, *in* Nilsen, T.H., ed., Geology of the Upper Cretaceous Hornbrook Formation, Oregon and California: Los Angeles, California, Society of Economic Paleontologists and Mineralogists, Pacific Section, v. 42, p. 159–165.

Blake, M.C., Jr., Irwin, W.P., and Coleman, R.G., 1967, Upside-down metamorphic zonation, blueschist facies, along a regional thrust in California and Oregon: Reston, Virginia, U.S. Geological Survey Professional Paper 575–C, p. C1–C9.

Blake, M.C., Jr., Howell, D.G., and Jones, D.L., 1982, Preliminary tectonostratigraphic terrane map of California: Reston, Virginia, U.S. Geological Survey Open-File Report 82–593.

Blake, M.C., Jr., Engebretson, D.C., Jayko, A.S., and Jones, D.L., 1985, Tectonostratigraphic terranes in southwest Oregon, *in* Howell, D.G., ed., Tectonostratigraphic terranes of the Circum-Pacific region: Houston, Circum-Pacific Council for Energy and Mineral Resources, Earth Sciences Series 1, p. 147–157.

Blake, M.C., Jr., Harwood, D.S., Helley, E.J., Irwin, W.P., Jayko, A.S., and Jones, D.L., 1999, Geologic map of the Red Bluff 30′ × 60′ quadrangle, California: Reston, Virginia, U.S. Geological Survey Geologic Map I-2542, scale 1:100,000, 1 sheet with 15 p. pamphlet.

Blome, C.D., and Irwin, W.P., 1983, Tectonic significance of late Paleozoic to Jurassic radiolarians from the North Fork terrane, Klamath Mountains, California, *in* Stevens, C.H., ed., Pre-Jurassic rocks in western North American suspect terranes: Los Angeles, Pacific Section, Society of Economic Paleontologists and Mineralogists, p. 77–89.

Boudier, F., Nicolas, A., and Bouchez, J.L., 1982, Kinematics of oceanic thrusting and subduction from basal sections of ophiolites: Nature, v. 296, p. 825–828, doi: 10.1038/296825a0.

Boudier, F., Le Sueur, E., and Nicolas, A., 1989, Structure of an atypical ophiolite; The Trinity complex, eastern Klamath Mountains, California: Geological Society of America Bulletin, v. 101, p. 820–833, doi: 10.1130/0016-7606(1989)101<0820:SOAAOT>2.3.CO;2.

Brouxel, M., and Lapierre, H., 1988, Geochemical study of an early Paleozoic island-arc–back-arc basin system. Part 1: The Trinity ophiolite (northern California): Geological Society of America Bulletin, v. 100, p. 1111–1119, doi: 10.1130/0016-7606(1988)100<1111:GSOAEP>2.3.CO;2.

Brouxel, M., Lapierre, H., Michard, A., and Albarede, F., 1988, Geochemical study of an early Paleozoic island-arc–back-arc basin system. Part 2: Eastern Klamath, early to middle Paleozoic island-arc volcanic rocks (northern California): Geological Society of America Bulletin, v. 100, p. 1120–1130, doi: 10.1130/0016-7606(1988)100<1120:GSOAEP>2.3.CO;2.

Cashman, S.M., 1988, Finite-strain patterns of Nevadan deformation, western Klamath Mountains, California: Geology, v. 16, p. 839–843.

Centeno-Garcia, E., Ruiz, J., Coney, P.J., Patchett, P.J., and Ortega-Gutiérrez, F., 1993, Guerrero terrane of Mexico: Its role in the Southern Cordillera from new geochemical data: Geology, v. 21, p. 419–422, doi: 10.1130/0091-7613(1993)021<0419:GTOMIR>2.3.CO;2.

Chamberlain, K.R., Snoke, A.W., Barnes, C.G., and Bushey, J.C., 2006, this volume, New U-Pb radiometric dates of the Bear Mountain intrusive complex, Klamath Mountains, California, *in* Snoke, A.W., and Barnes, C.G., eds., Geological studies in the Klamath Mountains province, California and Oregon: A volume in honor of William P. Irwin: Boulder, Colorado, Geological Society of America Special Paper 410, doi: 10.1130/2006.2410(15).

Charvet, J., Lapierre, H., Rouer, O., Coulon, C., Campos, C., Marin, P., and Lecuyer, C., 1990, Tectono-magmatic evolution of Paleozoic and early Mesozoic rocks in the eastern Klamath Mountains, California, and Blue Mountains, eastern Oregon–western Idaho, *in* Harwood, D.S., and Miller, M.M., eds., Paleozoic and early Mesozoic paleogeographic relations; Sierra Nevada, Klamath Mountains, and related terranes: Boulder, Colorado, Geological Society of America Special Paper 255, p. 255–276.

Clark, L.D., 1964, Stratigraphy and structure of part of the western Sierra Nevada metamorphic belt: Washington, D.C., U.S. Geological Survey Professional Paper 410, 70 p.

Clark, L.D., 1976, Stratigraphy of the north half of the western Sierra Nevada metamorphic belt, California: Reston, Virginia, U.S. Geological Survey Professional Paper 923, 26 p.

Coleman, R.G., 1972, The Colebrooke Schist of southwestern Oregon and its relation to the tectonic evolution of the region: Washington, D.C., U.S. Geological Survey Bulletin 1339, 61 p.

Coleman, R.G., and Lanphere, M.A., 1971, Distribution and age of high-grade blueschists, associated eclogites, and amphibolites from Oregon and California: Geological Society of America Bulletin, v. 82, p. 2397–2412.

Coleman, R.G., Manning, C.E., Mortimer, N., Donato, M.M., and Hill, L.B., 1988, Tectonic and regional metamorphic framework of the Klamath Mountains and adjacent Coast Ranges, California and Oregon, *in* Ernst, W.G., ed., Metamorphism and crustal evolution of the western United States—Rubey volume VII: Englewood Cliffs, New Jersey, Prentice Hall, p. 1059–1097.

Collier, A.J., 1929, Memorial to Joseph Silas Diller: Geological Society of America Bulletin, v. 40, p. 61–79.

Compton, R.R., 1955, Trondhjemite batholith near Bidwell Bar, California: Geological Society of America Bulletin, v. 66, p. 9–44.

Coney, P.J., Jones, D.L., and Monger, J.W.H., 1980, Cordilleran suspect terranes: Nature, v. 288, p. 329–333, doi: 10.1038/288329a0.

Cotkin, S.J., 1987, Conditions of metamorphism in an Early Paleozoic blueschist, schist of Skookum Gulch, northern California: Contributions to Mineralogy and Petrology, v. 96, p. 192–200, doi: 10.1007/BF00375233.

Cotkin, S.J., 1992, Ordovician-Silurian tectonism in northern California: The Callahan event: Geology, v. 20, p. 821–824, doi: 10.1130/0091-7613(1992)020<0821:OSTINC>2.3.CO;2.

Cotkin, S.J., and Medaris, L.G., Jr., 1993, Evaluation of the crystallization conditions for the calcalkaline Russian Peak intrusive complex, Klamath Mountains, northern California: Journal of Petrology, v. 34, p. 543–571.

Cotkin, S.J., Cotkin, M.L., and Armstrong, R.L., 1992, Early Paleozoic blueschist from the schist of Skookum Gulch, eastern Klamath Mountains, northern California: Journal of Geology, v. 100, p. 323–338.

Curtis, G.H., Evernden, J.F., and Lipson, J.I., 1958, Age determinations of some granitic rocks in California by the potassium-argon method: San Francisco, California Division of Mines Special Report 54, 16 p.

Davis, G.A., 1963, Structure and mode of emplacement of Caribou Mountain pluton, Klamath Mountains, California: Geological Society of America Bulletin, v. 74, p. 331–348.

Davis, G.A., 1968, Westward thrust faulting in the south-central Klamath Mountains, California: Geological Society of America Bulletin, v. 79, p. 911–934.

Davis, G.A., and Lipman, P.W., 1962, Revised structural sequence of pre-Cretaceous metamorphic rocks in the southern Klamath Mountains, California: Geological Society of America Bulletin, v. 73, p. 1547–1552.

Davis, G.A., Holdaway, M.J., Lipman, P.W., and Romey, W.D., 1965, Structure, metamorphism, and plutonism in the south-central Klamath Mountains, California: Geological Society of America Bulletin, v. 76, p. 933–966.

Davis, G.A., Monger, J.W.H., and Burchfiel, B.C., 1978, Mesozoic construction of the Cordilleran "collage," central British Columbia to central California, *in* Howell, D.G., and McDougall, K.A., eds., Mesozoic paleogeography of the western United States, Pacific Coast Paleogeography Symposium 2: Los Angeles, Pacific Section, Society of Economic Paleontologists and Mineralogists, p. 1–32.

Day, H.W., and Bickford, M.E., 2004, Tectonic setting of the Jurassic Smartville and Slate Creek complexes, northern Sierra Nevada, California: Geological Society of America Bulletin, v. 116, p. 1515–1528, doi: 10.1130/B25416.1.

de Roever, W.P., 1957, Sind die alpinotypen Peridotitmassen vielleicht tektonisch verfrachtete Bruchstücke der Peridotitschale? Geologische Rundschau, v. 46, p. 137–146, doi: 10.1007/BF01802890.

Dick, H.J.B., 1977, Partial melting in the Josephine peridotite, I. The effect on mineral composition and its consequence for geobarometry and geothermometry: American Journal of Science, v. 277, p. 801–832.

Dickinson, W.R., 2000, Geodynamic interpretation of Paleozoic tectonic trends oriented oblique to the Mesozoic Klamath-Sierran continental margin in California, *in* Soreghan, M.J., and Gehrels, G.E., eds., Paleozoic and Triassic paleogeography and tectonics of western Nevada and northern California: Boulder, Colorado, Geological Society of America Special Paper 347, p. 209–245.

Dickinson, W.R., 2004, Evolution of the North American Cordillera: Annual Review of Earth and Planetary Sciences, v. 32, p. 13–45, doi: 10.1146/annurev.earth.32.101802.120257.

Dietz, R.S., 1963, Alpine serpentinites as oceanic rind fragments: Geological Society of America Bulletin, v. 74, p. 947–952.

Dilek, Y., 2003, Ophiolite concept and its evolution, *in* Dilek, Y., and Newcomb, S., eds., Ophiolite concept and the evolution of geological thought: Boulder, Colorado, Geological Society of America Special Paper 373, p. 1–16.

Diller, J.S., 1898, Description of the Roseberg quadrangle [Oregon]: Washington, D.C., U.S. Geological Survey, Geological Atlas, folio 49, 4 p. + maps.

Diller, J.S., 1901, Description of the Coos Bay quadrangle [Oregon]: Washington, D.C., U.S. Geological Survey, Geological Atlas, folio 73, 5 p. + maps.

Diller, J.S., 1902, Topographic development of the Klamath Mountains: Washington, D.C., U.S. Geological Survey Bulletin 196, 69 p.

Diller, J.S., 1903, Description of the Port Orford quadrangle [Oregon]: Washington, D.C., U.S. Geological Survey, Geological Atlas, folio 89, 6 p. + maps.

Diller, J.S., 1906, Description of the Redding quadrangle [California]: Washington, D.C., U.S. Geological Survey, Geological Atlas, folio 138, 14 p. + maps.

Diller, J.S., 1907, The Mesozoic sediments of southwestern Oregon: American Journal of Science, 4th series, v. 23, p. 401–421.

Diller, J.S., and Kay, G.F., 1924, Riddle quadrangle [Oregon]: Washington, D.C., U.S. Geological Survey, Geological Atlas, folio 218, 8 p. + maps.

Donato, M.M., 1987, Evolution of an ophiolitic tectonic mélange, northern California Klamath Mountains, USA: Geological Society of America Bulletin, v. 98, p. 448–464, doi: 10.1130/0016-7606(1987)98<448:EOAOTM>2.0.CO;2.

Donato, M.M., 1989, Metamorphism of an ophiolitic tectonic mélange, northern California Klamath Mountains, USA: Journal of Metamorphic Geology, v. 7, p. 515–528.

Donato, M.M., 1991a, Geochemical recognition of a captured back-arc basin metabasaltic complex, southwestern Oregon: Journal of Geology, v. 99, p. 711–728.

Donato, M.M., 1991b, Geologic map showing part of the May Creek Schist and related rocks, Jackson County, Oregon: Reston, Virginia, U.S. Geological Survey Map, MF-2171, scale 1:24,000, 1 sheet with 10 p. pamphlet.

Donato, M.M., 1992, A newly recognized ductile shear zone in the northern Klamath Mountains, Oregon—Implications for Nevadan accretion: Reston, Virginia, U.S. Geological Survey Bulletin 2028, 10 p.

Donato, M.M., Barnes, C.G., Coleman, R.G., Ernst, W.G., and Kays, M.A., 1982, Geological map of the Marble Mountain wilderness, Siskiyou County, California: Reston, Virginia, U.S. Geological Survey Miscellaneous Field Studies Map MF 1452-A, scale 1:48,000, 1 sheet.

Donato, M.M., Barnes, C.G., and Tomlinson, S.L., 1996, The enigmatic Applegate Group of southwestern Oregon: Age, correlation, and tectonic affinity: Oregon Geology, v. 58, p. 79–91.

Ernst, W.G., 1999, Mesozoic petrotectonic development of the Sawyers Bar suprasubduction-zone arc, central Klamath Mountains, northern California: Geological Society of America Bulletin, v. 111, p. 1217–1232, doi: 10.1130/0016-7606(1999)111<1217:MPDOTS>2.3.CO;2.

Evans, J.G., 1987, Deformation of the Josephine Peridotite, California and Oregon: Reston, Virginia, U.S. Geological Survey Professional Paper 1378, 45 p.

Evernden, J.F., and Kistler, R.W., 1970, Chronology of emplacement of Mesozoic batholithic complexes in California and western Nevada: Washington, D.C., U.S. Geological Survey Professional Paper 623, 42 p.

Evernden, J.F., Curtis, G.H., and Lipson, J.I., 1957, Potassium-argon dating of igneous rocks: American Association of Petroleum Geologists Bulletin, v. 43, p. 2120–2127.

Frost, C.D., Barnes, C.G., and Snokes, A.W., 2006, this volume, Nd and Sr isotopic data from argillaceous rocks of the Galice Formation and Rattlesnake Creek terrane, Klamath Mountains: Evidence for the input of Precambrian sources, *in* Snoke, A.W., and Barnes, C.G., eds., Geological studies in the Klamath Mountains province, California and Oregon: A volume in honor of William P. Irwin: Boulder, Colorado, Geological Society of America Special Paper 410, doi: 10.1130/2006.2410(5).

Fuis, G.S., Zucca, J.J., Mooney, W.D., and Milkereit, B., 1987, A geologic interpretation of seismic-refraction results in northeastern California: Geological Society of America Bulletin, v. 98, p. 53–65, doi: 10.1130/0016-7606(1987)98<53:AGIOSR>2.0.CO;2.

Gansser, A., 1974, The ophiolitic mélange, a world-wide problem on Tethyan examples: Eclogae Geologicae Helvetiae, v. 67, p. 479–507.

Garcia, M.O., 1979, Petrology of the Rogue and Galice Formations, Klamath Mountains, Oregon: Identification of a Jurassic island arc sequence: Journal of Geology, v. 86, p. 29–41.

Gehrels, G.E., and Miller, M.M., 2000, Detrital zircon geochronologic study of upper Paleozoic strata in the eastern Klamath terrane, northern California, *in* Soreghan, M.J., and Gehrels, G.E., eds., Paleozoic and Triassic paleogeography and tectonics of western Nevada and northern California: Boulder, Colorado, Geological Society of America Special Paper 347, p. 99–107.

Goodge, J.W., 1989a, Evolving early Mesozoic convergent margin deformation, central Klamath Mountains, northern California: Tectonics, v. 8, p. 845–864.

Goodge, J.W., 1989b, Polyphase metamorphic evolution of a Late Triassic subduction complex, Klamath Mountains, northern California: American Journal of Science, v. 289, p. 874–943.

Goodge, J.W., 1995, Pre-Middle Jurassic accretionary metamorphism in the southern Klamath Mountains of northern California, USA: Journal of Metamorphic Geology, v. 13, p. 93–110.

Goodge, J.W., and Renne, P.R., 1993, Mid-Paleozoic olistoliths in eastern Hayfork terrane mélange, Klamath Mountains: Implications for late Paleozoic–early Mesozoic Cordilleran forearc development: Tectonics, v. 12, p. 279–289.

Gorman, C.M., III, 1985, Geology, geochemistry and geochronology of the Rattlesnake Creek terrane, west-central Klamath Mountains, California [M.S. thesis]: Salt Lake City, University of Utah, 111 p.

Gray, G.G., 1985, Structural, geochronologic, and depositional history of the western Klamath Mountains, California and Oregon: Implications for the early to middle Mesozoic tectonic evolution of the western North American Cordillera [Ph.D. dissertation]: Austin, University of Texas, 161 p.

Gray, G.G., 1986, Native terranes of the central Klamath Mountains, California: Tectonics, v. 5, p. 1043–1054.

Gray, G.G., 2006, this volume, Structural and tectonic evolution of the western Jurassic belt along the Klamath River corridor, Klamath Mountains, California, *in* Snoke, A.W., and Barnes, C.G., eds., Geological studies in the Klamath Mountains province, California and Oregon: A volume in honor of William P. Irwin: Boulder, Colorado, Geological Society of America Special Paper 410, doi: 10.1130/2006.2410(07).

Gribble, R.F., Barnes, C.G., Donato, M.M., Hoover, J.D., and Kistler, R.W., 1990, Geochemistry and intrusive history of the Ashland pluton, Klamath Mountains, California and Oregon: Journal of Petrology, v. 31, p. 883–923.

Grover, T.W., 1984, Progressive metamorphism west of the Condrey Mountain dome, north-central Klamath Mountains, northern California [M.S. thesis]: Eugene, University of Oregon, 129 p.

Hacker, B.R., and Ernst, W.G., 1993, Jurassic orogeny in the Klamath Mountains: A geochronological analysis, *in* Dunn, G., and McDougall, K., eds., Mesozoic paleogeography of the western United States—II: Los Angeles, Pacific Section, Society of Economic Paleontologists and Mineralogists, p. 37–59.

Hacker, B.R., Donato, M.M., Barnes, C.G., McWilliams, M.O., and Ernst, W.G., 1995, Timescales of orogeny: Jurassic construction of the Klamath Mountains: Tectonics, v. 14, p. 677–703, doi: 10.1029/94TC02454.

Hamilton, W., 1969, Mesozoic California and the underflow of Pacific mantle: Geological Society of America Bulletin, v. 80, p. 2409–2430.

Harms, T.A., Dewey, J.F., and Mange, M.A., 2003, A model for the Appalachian origin of Paleozoic terranes in the North American Cordillera: Boulder, Colorado, Geological Society of America Abstracts with Programs, v. 35, no. 6, p. 113.

Harper, G.D., 1980, The Josephine ophiolite—Remains of a Late Jurassic marginal basin in northwestern California: Geology, v. 8, p. 333–337, doi: 10.1130/0091-7613(1980)8<333:TJOOAL>2.0.CO;2.

Harper, G.D., 1984, The Josephine ophiolite: Geological Society of America Bulletin, v. 95, p. 1009–1026, doi: 10.1130/0016-7606(1984)95<1009:TJONC>2.0.CO;2.

Harper, G.D., 2003, Fe-Ti basalts and propagating-rift tectonics in the Josephine ophiolite: Geological Society of America Bulletin, v. 115, p. 771–787, doi: 10.1130/0016-7606(2003)115<0771:FBAPTI>2.0.CO;2.

Harper, G.D., and Wright, J.E., 1984, Middle to Late Jurassic tectonic evolution of the Klamath Mountains, California-Oregon: Tectonics, v. 3, p. 759–772.

Harper, G.D., Saleeby, J.B., and Norman, E.A.S., 1985, Geometry and tectonic setting of sea-floor spreading for the Josephine ophiolite, and implications for Jurassic accretionary events along the California margin, *in* Howell, D.G., ed., Tectonostratigraphic terranes of the Circum-Pacific region: Houston, Circum-Pacific Council for Energy and Mineral Resources, Earth Science Series 1, p. 239–257.

Harper, G.D., Bowman, J.R., and Kuhns, R., 1988, A field, chemical, and stable isotope study of subseafloor metamorphism of the Josephine ophiolite, California-Oregon: Journal of Geophysical Research, v. 85, p. 4625–4656.

Harper, G.D., Grady, K., and Wakabayashi, J., 1990, A structural study of a metamorphic sole beneath the Josephine ophiolite, western Klamath terrane, California-Oregon, *in* Harwood, D.S., and Miller, M.M., eds., Paleozoic and early Mesozoic paleogeographic relations; Sierra Nevada, Klamath Mountains, and related terranes: Boulder, Colorado, Geological Society of America Special Paper 255, p. 379–396.

Harper, G.D., Saleeby, J.B., and Heizler, M., 1994, Formation and emplacement of the Josephine ophiolite and the Nevadan orogeny in the Klamath Mountains, California-Oregon: U/Pb zircon and $^{40}Ar/^{39}Ar$ geochronology: Journal of Geophysical Research, v. 99, p. 4293–4321, doi: 10.1029/93JB02061.

Harper, G.D., Grady, K., and Coulton, A.J., 1996, Origin of the amphibolite "sole" of the Josephine ophiolite: Emplacement of a cold ophiolite over a hot arc: Tectonics, v. 16, p. 296–313.

Helper, M.A., 1986, Deformation and high P/T metamorphism in the central part of the Condrey Mountain window, north-central Klamath Mountains, California and Oregon, *in* Evans, B.W., and Brown, E.H., eds., Blueschists and eclogites: Boulder, Colorado, Geological Society of America Memoir 164, p. 125–141.

Helper, M.A., Walker, N.W., and McDowell, F.W., 1989, Early Cretaceous metamorphic ages and Middle Jurassic U-Pb zircon protolith ages for the Condrey Mountain Schist, Klamath Mtns., NW Calif. and SW Oregon: Boulder, Colorado, Geological Society of America Abstracts with Programs, v. 21, no. 5, p. 92.

Hershey, O.H., 1901, Metamorphic formations of northwestern California: American Geologist, v. 27, p. 225–245.

Hershey, O.H., 1906, Some western Klamath stratigraphy: American Journal of Science, 4th series, v. 21, p. 58–66.

Hershey, O.H., 1911, Del Norte County [California] geology: Mining and Scientific Press, v. 102, p. 468.

Hess, H.H., 1955, Serpentines, orogeny, and epeirogeny, *in* Poldervaart, A., ed., Crust of the Earth (A symposium): New York, Geological Society of America Special Paper 62, p. 391–407.

Hess, H.H., 1962, History of ocean basins, *in* Engel, A.E.J., James, H.L., and Leonard, B.F., eds., Petrologic studies: A volume in honor of A.F. Buddington: New York, Geological Society of America, p. 599–620.

Hietanen, A., 1951, Metamorphic and igneous rocks of the Merrimac area, Plumas National Forest, California: Geological Society of America Bulletin, v. 62, p. 565–608.

Hietanen, A., 1976, Metamorphism and plutonism around the Middle and South Forks of the Feather River, California: Reston, Virginia, U.S. Geological Survey Professional Paper 920, 30 p.

Hill, L.B., 1984, A tectonic and metamorphic history of the north-central Klamath Mountains, California [Ph.D. dissertation]: Stanford, California, Stanford University, 248 p.

Hinds, N.E.A., 1932, Paleozoic eruptive rocks of the southern Klamath Mountains: University of California Publications in Geological Sciences, v. 20, no. 11, p. 375–410.

Hinds, N.E.A., 1933, Geologic formations of the Redding-Weaverville districts, northern California: California Journal of Mines and Geology, v. 29, no. 1–2, p. 76–122.

Hinds, N.E.A., 1935, Mesozoic and Cenozoic eruptive rocks of the southern Klamath Mountains, California: University of California Publications in Geological Sciences, v. 23, no. 11, p. 313–380.

Hotz, P.E., 1967, Geologic map of the Condrey Mountain quadrangle and parts of the Seiad Valley and Hornbrook quadrangles: Washington, D.C., U.S. Geological Survey Geologic Quadrangle Map GQ-618, scale 1:62,500, 1 sheet.

Hotz, P.E., 1971, Plutonic rocks of the Klamath Mountains, California and Oregon: Washington, D.C., U.S. Geological Survey Professional Paper 684-B, 20 p.

Hotz, P.E., 1977, Geology of the Yreka quadrangle, Siskiyou County, California: Reston, Virginia, U.S. Geological Survey Bulletin 1436, 72 p.

Hotz, P.E., 1979, Regional metamorphism in the Condrey Mountain quadrangle, north-central Klamath Mountains, California: Reston, Virginia, U.S. Geological Survey Professional Paper 1086, 25 p.

Hotz, P.E., Lanphere, M.A., and Swanson, D.A., 1977, Triassic blueschist from northern California and north-central Oregon: Geology, v. 5, p. 659–663, doi: 10.1130/0091-7613(1977)5<659:TBFNCA>2.0.CO;2.

Howell, D.G., Jones, D.L., and Schermer, E.L., 1985, Tectonostratigraphic terranes of the Circum-Pacific region, *in* Howell, D.G., ed., Tectonostratigraphic terranes of the Circum-Pacific region: Houston, Circum-Pacific Council for Energy and Mineral Resources, Earth Sciences Series 1, p. 3–30.

Imlay, R.W., 1959, Succession and speciation of the pelecypod *Aucella:* Washington, D.C., U.S. Geological Survey Professional Paper 314-G, p. 155–169, pls. 16–19.

Irwin, W.P., 1960, Geologic reconnaissance of the northern Coast Ranges and Klamath Mountains, California, with a summary of the mineral resources: San Francisco, California Division of Mines Bulletin 179, 80 p.

Irwin, W.P., 1964, Late Mesozoic orogenies in the ultramafic belts of northwestern California and southwestern Oregon: Washington, D.C., U.S. Geological Survey Professional Paper 501-C, p. C1–C9.

Irwin, W.P., 1972, Terranes of the western Paleozoic and Triassic belt in the southern Klamath Mountains, California: Washington, D.C., U.S. Geological Survey Professional Paper 800-C, p. C103–C111.

Irwin, W.P., 1985, Age and tectonics of plutonic belts in accreted terranes of the Klamath Mountains, California and Oregon, *in* Howell, D.G., ed., Tectonostratigraphic terranes of the circum-Pacific region: Houston, Circum-Pacific Council for Energy and Mineral Resources, Earth Sciences Series 1, p. 187–199.

Irwin, W.P., 1994, Geologic map of the Klamath Mountains, California: Reston, Virginia, U.S. Geological Survey Miscellaneous Investigations Series Map I-2148, scale 1:500,000, 2 sheets.

Irwin, W.P., and Galanis, S.P., Jr., 1976, Map showing limestones and selected fossil localities in the Klamath Mountains province, California and Oregon: Reston, Virginia, U.S. Geological Survey Miscellaneous Field Studies Map MF-749, scale 1:500,000, 1 sheet.

Irwin, W.P., and Lipman, P.W., 1962, A regional ultramafic sheet in eastern Klamath Mountains, California: Washington, D.C., U.S. Geological Survey Professional Paper 450-C, p. 18–21.

Irwin, W.P., and Tatlock, D.B., 1955, Reconnaissance of northwestern California, a progress report: Geological Society of America Bulletin, v. 66, no. 12, pt. 2, p. 1653–1654.

Irwin, W.P., and Wooden, J.L., 1999, Plutons and accretionary episodes of the Klamath Mountains, California and Oregon: Reston, Virginia, U.S. Geological Survey Open-file Report 99–374, 1 sheet.

Irwin, W.P., Jones, D.L., and Pessagno, E.A., Jr., 1977, Significance of Mesozoic radiolarians from the pre-Nevadan rocks of the southern Klamath Mountains, California: Geology, v. 5, p. 557–562, doi: 10.1130/0091-7613(1977)5<557:SOMRFT>2.0.CO;2.

Irwin, W.P., Jones, D.L., and Kaplan, T.A., 1978, Radiolarians from pre-Nevadan rocks of the Klamath Mountains, California and Oregon, *in* Howell, D.G., and McDougall, K.A., eds., Mesozoic paleogeography of the western United States, Pacific Coast Paleogeography Symposium 2: Los Angeles, Pacific Section, Society of Economic Paleontologists and Mineralogists, p. 303–310.

Irwin, W.P., Jones, D.L., and Blome, C.D., 1982, Map showing samples radiolarian localities in the western Paleozoic and Triassic belt, Klamath Mountains, California: Reston, Virginia, U.S. Geological Survey Miscellaneous Field Studies, Map MF-1399, scale 1:250,000, 1 sheet.

Jachens, R.C., Barnes, C.G., and Donato, M.M., 1986, Subsurface configuration of the Orleans fault: Implications for deformation in the western Klamath Mountains, California: Geological Society of America Bulletin, v. 97, p. 388–395, doi: 10.1130/0016-7606(1986)97<388:SCOTOF> 2.0.CO;2.

Jacobsen, S.B., Quick, J.E., and Wasserburg, G.J., 1984, A Nd and Sr isotopic study of the Trinity peridotite: Implications for mantle evolution: Earth and Planetary Science Letters, v. 68, p. 361–378, doi: 10.1016/0012-821X(84)90122-5.

Kays, M.A., 1968, Zones of alpine tectonism and metamorphism, Klamath Mountains, southwestern Oregon: Journal of Geology, v. 76, p. 17–36.

Kays, M.A., 1970, Mesozoic metamorphism, May Creek Schist belt, Klamath Mountains, Oregon: Geological Society of America Bulletin, v. 81, p. 2743–2758.

Kays, M.A., 1995, Metamorphism in the northern Klamath Mountains, Oregon, *in* Miller, D.M., and Busby, C., eds., Jurassic magmatism and tectonics of the North American Cordillera: Boulder, Colorado, Geological Society of America Special Paper 299, p. 173–190.

Kelemen, P.B., and Dick, H.J.B., 1995, Focused melt flow and localized deformation in the upper mantle: Juxtaposition of replacive dunite and ductile shear zones in the Josephine peridotite, SW Oregon: Journal of Geophysical Research, v. 100, p. 423–438, doi: 10.1029/94JB02063.

King, P.B., [1967] 1970, Tectonic features, *in* The national atlas of the United States: Washington, D.C., U.S. Department of the Interior, Geological Survey, p. 70–71, scale 1:7,500,000.

Klein, C.W., 1977, Thrust plates of the north-central Klamath Mountains near Happy Camp, California, *in* Short contributions to California geology: Sacramento, California Division of Mines and Geology Special Report 129, p. 23–26.

Lanphere, M.A., and Irwin, W.P., 1987, In search of the Abrams post office, Trinity County: California Geology, May, p. 99–103.

Lanphere, M.A., Irwin, W.P., and Hotz, P.E., 1968, Isotopic age of the Nevadan orogeny and older plutonic and metamorphic events in the Klamath Mountains, California: Geological Society of America Bulletin, v. 79, p. 1027–1052.

Lanphere, M.A., Blake, M.C., Jr., and Irwin, W.P., 1978, Early Cretaceous metamorphic age of the South Fork Mountain Schist in the northern Coast Ranges of California: American Journal of Science, v. 278, p. 798–815.

Lapierre, H., Albarède, F., Albers, J., Cabanis, B., and Coulon, C., 1985, Early Devonian volcanism in the eastern Klamath Mountains, California: Evidence for an immature island arc: Canadian Journal of Earth Sciences, v. 22, p. 214–226.

Lapierre, H., Brouxel, M., Albarede, F., Coulon, C., Lecuyer, C., Martin, P., Mascle, G., and Rouer, O., 1987, Paleozoic and lower Mesozoic magmas from the eastern Klamath Mountains (north California) and the geodynamic evolution of northwestern America: Tectonophysics, v. 140, p. 155–177, doi: 10.1016/0040-1951(87)90227-7.

Lawson, A.C., 1941, Memorial to Oscar H. Hershey: Proceedings of the Geological Society of America, v. 1940, p. 195–200.

Lieberman, J.E., and Rice, J.M., 1986, Petrology of marble and peridotite in the Seiad ultramafic complex, northern California, USA: Journal of Metamorphic Geology, v. 4, p. 179–199.

Lindsley-Griffin, N., 1977, The Trinity ophiolite, Klamath Mountains, California, *in* Coleman, R.G., and Irwin, W.P., eds., North American ophiolites: Portland, State of Oregon Department of Geology and Minerals Industry Bulletin 95, p. 107–120.

Lindsley-Griffin, N., 1991, The Trinity Complex: A polygenetic ophiolitic assemblage, *in* Cooper, J.D., and Stevens, C.H., eds., Paleozoic paleogeography of the western United States—II: Los Angeles, Pacific Section, Society of Economic Paleontologists and Mineralogists, p. 589–608.

Lindsley-Griffin, N., and Griffin, J.R., 1983, The Trinity terrane: An early Paleozoic microplate assemblage, *in* Stevens, C.H., ed., Pre-Jurassic rocks in western North American suspect terranes: Los Angeles, Pacific Section, Society of Economic Paleontologists and Mineralogists, p. 63–75.

Lindsley-Griffin, N., Griffin, J.R., and Farmer, J.D., 2003, Significance of Ediacaran cyclomedusids and other Pacific rim biota in the Yreka terrane, eastern Klamath Mountains, California: Geological Society of America Abstracts with Programs, v. 35, p. 14.

Lindsley-Griffin, N., Griffin, J.R., Farmer, J.D., Sivers, E.A., Bruckno, B., and Tozer, M.K., 2006, this volume, Ediacaran cyclomedusoids and the paleogeographic setting of the Neoproterozoic–early Paleozoic Yreka and Trinity terranes, eastern Klamath Mountains, California, *in* Snoke, A.W., and Barnes, C.G., eds., Geological studies in the Klamath Mountains province, California and Oregon: A volume in honor of William P. Irwin: Boulder, Colorado, Geological Society of America Special Paper 410, doi: 10.1130/2006.2410(20).

Lipman, P.W., 1963, Gibson Peak pluton—A discordant composite intrusion in the southeastern Trinity Alps, northern California: Geological Society of America Bulletin, v. 74, p. 1259–1280.

Lipman, P.W., 1964, Structure and origin of an ultramafic pluton in the Klamath Mountains, California: American Journal of Science, v. 262, p. 199–222.

Loney, R.A., and Himmelberg, G.R., 1976, Structure of the Vulcan Peak alpine-type peridotite, southwestern Oregon: Geological Society of America Bulletin, v. 87, p. 259–274, doi: 10.1130/0016-7606(1976)87<259:SOTVPA> 2.0.CO;2.

MacDonald, J.H., Jr., Harper, G.D., and Zhu, B., 2006, this volume, Petrology, geochemistry, and provenance of the Galice Formation, Klamath Mountains, Oregon and California, *in* Snoke, A.W., and Barnes, C.G., eds., Geological studies in the Klamath Mountains province, California and Oregon: A volume in honor of William P. Irwin: Boulder, Colorado, Geological Society of America Special Paper 410, doi: 10.1130/2006.2410(04).

Malpas, J., 1979, The dynamothermal aureole of the Bay of Islands ophiolite suite: Canadian Journal of Earth Sciences, v. 16, p. 2081–2101.

Mankinen, E.A., Lindsley-Griffin, N., and Griffin, J.R., 2002, Concordant paleolatitudes for Neoproterozoic ophiolitic rocks of the Trinity Complex,

Klamath Mountains, California: Journal of Geophysical Research, v. 107, B10, p. 2254, doi: 10.1029/2001JB001623.

Masi, U., O'Neil, J.R., and Kistler, R.W., 1981, Stable isotope systematics in Mesozoic granites of central and northern California and southwestern Oregon: Contributions to Mineralogy and Petrology, v. 76, p. 116–126, doi: 10.1007/BF00373691.

Mattinson, J.M., and Hopson, C.A., 1972, Paleozoic ophiolitic complexes in Washington and northern California: Carnegie Institute Year Book, v. 71, p. 578–583.

Medaris, L.G., Jr., 1975, Coexisting spinel and silicates in alpine peridotites of the granulite facies: Geochimica et Cosmochimica Acta, v. 39, p. 947–958, doi: 10.1016/0016-7037(75)90040-X.

Metcalf, R.V., Wallin, E.T., Willse, K.R., and Muller, E.R., 2000, Geology and geochemistry of the ophiolitic Trinity terrane, California: Evidence of middle Paleozoic depleted supra-subduction zone magmatism in a proto-arc setting, *in* Dilek, Y., et al., eds., Ophiolities and ocean crust: New insights from field studies and the ocean drilling program: Boulder, Colorado, Geological Society of America Special Paper 349, p. 403–418.

Miller, J., Miller, R., Wooden, J., and Harper, G., 2003, Geochronologic links between the Ingalls ophiolite, North Cascades, Washington and the Josephine ophiolite, Klamath Mts., Oregon and California: Geological Society of America Abstracts with Programs, v. 35, p. 113.

Miller, M.M., 1987, Dispersed remnants of a northeast Pacific fringing arc: Upper Paleozoic terranes of Permian McCloud faunal affinity, western U.S.: Tectonics, v. 6, p. 807–830.

Miller, M.M., 1989, Intra-arc sedimentation and tectonism: Late Paleozoic evolution of the eastern Klamath terrane, California: Geological Society of America Bulletin, v. 101, p. 170–187, doi: 10.1130/0016-7606(1989)101 <0170:IASATL>2.3.CO;2.

Miller, M.M., and Harwood, D.S., 1990, Paleogeographic setting of upper Paleozoic rocks in the northern Sierra and eastern Klamath terranes, northern California, *in* Harwood, D.S., and Miller, M.M., eds., Paleozoic and early Mesozoic paleogeographic relations; Sierra Nevada, Klamath Mountains, and related terranes: Boulder, Colorado, Geological Society of America Special Paper 255, p. 175–192.

Miller, M.M., and Saleeby, J.B., 1991, Continental detrital zircon in Carboniferous ensimatic arc rocks, Bragdon Formation, eastern Klamath terrane northern California: Geological Society of America Bulletin, v. 103, p. 268–276.

Miller, M.M., and Saleeby, J.B., 1995, U-Pb geochronology of detrital zircon from Upper Jurassic synorogenic turbidites, Galice Formation, and related rocks, western Klamath Mountains: Correlation and Klamath Mountains provenance: Journal of Geophysical Research, v. 100, p. 18,045–18,058.

Miller, M.M., and Wright, J.E., 1987, Paleogeographic implications of Permian Tethyan corals from the Klamath Mountains, California: Geology, v. 15, p. 266–269, doi: 10.1130/0091-7613(1987)15<266:PIOPTC>2.0.CO;2.

Monger, J.W.H., and Nokleberg, W.J., 1996, Evolution of the northern North American Cordillera: Generation, fragmentation, displacement and accretion of successive North American plate-margins arcs, *in* Coyner, A.R., and Fahey, P.L., eds., Geology and ore deposits of the American Cordillera: Geological Society of Nevada Symposium Proceedings, Reno/Sparks, Nevada, April 1995, p. 1133–1152.

Monger, J.W.H., and Ross, C.A., 1971, Distribution of fusulinaceans in the western Canadian Cordillera: Canadian Journal of Earth Sciences, v. 8, p. 259–278.

Monger, J.W.H., Price, R.A., and Nokleberg, W.J., 2005, Northern Cordillera, *in* Selley, R.C., et al., eds., Encyclopedia of Geology, v. 4: Oxford, Academic Press, p. 36–47.

Moore, J.G., 1959, The quartz diorite boundary line in the western United States: Journal of Geology, v. 67, p. 198–210.

Mortimer, N., 1985, Structural and metamorphic aspects of Middle Jurassic terrane juxtapostion, northeastern Klamath Mountains, California, *in* Howell, D.G., ed., Tectonostratigraphic terranes of the Circum-Pacific region: Houston, Circum-Pacific Council for Energy and Mineral Resources, Earth Sciences Series 1, p. 201–214.

Mortimer, N., and Coleman, R.G., 1985, A Neogene structural dome in the Klamath Mountains, California: Geology, v. 13, p. 253–256, doi: 10.1130/0091-7613(1985)13<253:ANSDIT>2.0.CO;2.

Nestell, M.K., Irwin, W.P., and Albers, J.P., 1981, Later Permian (Early Djulfian) Tethyan foraminifera from the southern Klamath Mountains, California: Geological Society of America Abstracts with Programs, v. 13, no. 7, p. 519.

Nilsen, T.H., 1984, Stratigraphy, sedimentology, and tectonic framework of the Upper Cretaceous Hornbrook Formation, Oregon and California, *in* Nilsen, T.H., ed., Geology of the Upper Cretaceous Hornbrook Formation, Oregon and California: Los Angeles, Pacific Section, Society of Economic Paleontologists and Mineralogists, v. 42, p. 51–88.

Patchett, P.J., and Gehrels, G.E., 1998, Continental influence on Canadian Cordilleran terranes from Nd isotopic study, and significance for crustal growth processes: Journal of Geology, v. 106, p. 269–280.

Patrick, B.E., and Day, H.W., 1995, Cordilleran high-pressure metamorphic terranes: Progress and problems: Journal of Metamorphic Geology, v. 13, p. 1–8.

Peacock, S.M., and Norris, P.J., 1989, Metamorphic evolution of the central metamorphic belt, Klamath province, California: An inverted metamorphic gradient beneath the Trinity peridotite: Journal of Metamorphic Geology, v. 7, p. 191–209.

Pessagno, E.A., Jr., and Blome, C.D., 1990, Implications of new Jurassic stratigraphic, geochronometric, and paleolatitudinal data from the western Klamath terrane (Smith River and Rogue Valley subterranes): Geology, v. 18, p. 665–668, doi: 10.1130/0091-7613(1990)018<0665:IONJSG>2.3.CO;2.

Pessagno, E.A., Jr., Hull, D.M., and Hopson, C., 2000, Tectonostratigraphic significance of sedimentary strata occurring within and above the Coast Range ophiolite (California Coast Ranges) and the Josephine ophiolite (Klamath Mountains), northwestern California, *in* Dilek, Y., Moores, E.M., Ethon, D., and Nicolas, A., eds., Ophiolites and oceanic crust: New insights from field studies and the ocean drilling program: Boulder, Colorado, Geological Society of America Special Paper 349, p. 383–394.

Petersen, S.W., 1982, Geology and petrology around Titus Ridge, north-central Klamath Mountains, California [M.S. thesis]: Eugene, University of Oregon, 73 p.

Petersen, S.W., Barnes, C.G., and Hoover, J.D., 1991, The Billy's Peak mafic complex of the Trinity sheet, California: Roots of a Paleozoic island arc, *in* Cooper, J.D., and Stevens, C.H., eds., Paleozoic paleogeography of the western United States—II: Los Angeles, Pacific Section, Society of Economic Paleontologist and Mineralogists, v. 67, p. 625–633.

Quick, J.E., 1981, Petrology and petrogenesis of the Trinity peridotite, an upper mantle diapir in the eastern Klamath Mountains, northern California: Journal of Geophysical Research, v. 86, p. 11,837–11,863.

Ramsay, J.G., 1991, Main tectonic units of the European Alps (Fig. 1), *in* Field Guide to excursions: A cross section of the Swiss Alps: Zürich, Switzerland, Conference on the geometry of naturally deformed rocks, p. 3.

Rawson, S.A., and Petersen, S.W., 1982, Structural and lithologic equivalence of the Rattlesnake Creek terrane and high-grade rocks in the western Paleozoic and Triassic belt, north central Klamath Mountains, California: Geological Society of America Abstracts with Programs, v. 14, no. 4, p. 226.

Rynearson, G.A., and Smith, C.T., 1940, Chromite deposits of the Seiad quadrangle, Siskiyou County, California: Washington, D.C., U.S. Geological Survey Bulletin 922-J, p. 281–306.

Saleeby, J.B., 1982, Polygenetic ophiolite belt of the California Sierra Nevada: Geochronological and tectonostratigraphic development: Journal of Geophysical Research, v. 87, p. 1803–1824.

Saleeby, J.B., 1990, Geochronological and tectonostratigraphic framework of Sierran-Klamath ophiolitic assemblages, *in* Harwood, D.S., and Miller, M.M., eds., Paleozoic and early Mesozoic paleogeographic relations; Sierra Nevada, Klamath Mountains, and related terranes: Boulder, Colorado, Geological Society of America Special Paper 255, p. 93–114.

Saleeby, J.B., and Harper, G.D., 1993, Tectonic relations between the Galice Formation and the Condrey Mountain Schist, Klamath Mountains, northern California, *in* Dunn, G.C., and McDougall, K.A., eds., Mesozoic Paleogeography of the western United States—II: Los Angeles, Pacific Section, Society of Economic Paleontologists and Mineralogists, v. 71, p. 61–80.

Saleeby, J.B., Harper, G.D., Snoke, A.W., and Sharp, W.D., 1982, Time relations and structural-stratigraphic patterns in ophiolite accretion, west-central Klamath Mountains, California: Journal of Geophysical Research, v. 87, p. 3831–3848.

Scherer, H.H., Snow, C.A., and Ernst, W.G., 2006, this volume, Geologic-petrochemical comparison of early Mesozoic mafic arc terranes: Western Paleozoic and Triassic belt, Klamath Mountains, and Jura–Triassic arc belt, Sierran Foothills, *in* Snoke, A.W., and Barnes, C.G., eds., Geological studies in the Klamath Mountains province, California and Oregon: A volume in honor of William P. Irwin: Boulder, Colorado, Geological Society of America Special Paper 410, doi: 10.1130/2006.2410(18).

Schweickert, R.A., and Cowan, D.S., 1975, Early Mesozoic tectonic evolution of the western Sierra Nevada, California: Geological Society of America Bulletin, v. 86, p. 1329–1336, doi: 10.1130/0016-7606(1975)86<1329:EMTEOT>2.0.CO;2.

Schweickert, R.A., Saleeby, J.B., Tobisch, O.T., and Wright, W.H., III, 1977, Paleotectonic and paleogeographic significance of the Calaveras Complex, western Sierra Nevada, California, *in* Stewart, J.H., et al., eds., Paleozoic paleogeography of the western United States, Pacific Coast Paleogeography Symposium 1: Los Angeles, Pacific Section, Society of Economic Paleontologists and Mineralogists, p. 381–394.

Schweickert, R.A., Bogen, N.L., Girty, G.H., Hanson, R.E., and Meguerian, C., 1984, Timing and structural expressions of the Nevadan orogeny, Sierra Nevada, California: Geological Society of America Bulletin, v. 95, p. 967–979, doi: 10.1130/0016-7606(1984)95<967:TASEOT>2.0.CO;2.

Searle, M.P., and Cox, J., 1999, Tectonic setting, origin, and obduction of the Oman ophiolite: Geological Society of America Bulletin, v. 111, p. 104–122, doi: 10.1130/0016-7606(1999)111<0104:TSOAOO>2.3.CO;2.

Searle, M.P., and Malpas, J., 1980, The structure and metamorphism of rocks beneath the Semail ophiolite of Oman and their significance in ophiolite obduction: Transactions of the Royal Society of Edinburgh: Earth Sciences, v. 71, p. 247–262.

Silberling, N.J., Jones, D.L., Blake, M.C., Jr., and Howell, D.G., 1987, Lithotectonic terrane map of the western conterminous United States: Reston, Virginia, U.S. Geological Survey Miscellaneous Field Studies Map MF-1874-C, scale 1:2,500,000, 1 sheet.

Smith, J.G., Page, N.J., Johnson, M.G., Moring, B.C., and Gray, F., 1982, Preliminary geologic map of the Medford 1° × 2° sheet, Oregon: Reston, Virginia, U.S. Geological Survey Open-File Report 82-955, scale 1:250,000, 1 sheet.

Snoke, A.W., 1971, A thrust plate of ophiolitic rocks in the Preston Peak area, Klamath Mountains, northwestern California: Geological Society of America Abstracts with Programs, v. 3, p. 198.

Snoke, A.W., 1977, A thrust plate of ophiolitic rocks in the Preston Peak area, Klamath Mountains, California: Geological Society of America Bulletin, v. 88, p. 1641–1659, doi: 10.1130/0016-7606(1977)88<1641:ATPOOR>2.0.CO;2.

Snoke, A.W., 2005, Southern Cordillera, *in* Selley, R.C., Cocks, L.R.M., and Plimer, I.R., eds., Encyclopedia of Geology, v. 4: Oxford, Academic Press, p. 48–61.

Snoke, A.W., Bowman, H.R., and Hebert, A.J., 1977, The Preston Peak ophiolite, Klamath Mountains, California, an ancient immature island arc: Petrochemical evidence, *in* Short Contributions to California Geology: Sacramento, California Division of Mines and Geology Special Report 129, p. 67–79.

Snoke, A.W., Quick, J.E., and Bowman, H.R., 1981, Bear Mountain igneous complex, Klamath Mountains, California: An ultrabasic to silicic calc-alkaline suite: Journal of Petrology, v. 22, p. 501–552.

Stern, C.R., and De Wit, M.J., 2003, Rocas Verdes ophiolites, southernmost South America: Remnants of progressive stages of development of oceanic-type crust in a continental margin back-arc basin, *in* Dilek, Y., and Robinson, P.T., eds., Ophiolites in Earth history: London, Geological Society Special Publication 218, p. 665–683.

Taliaferro, N.L., 1942, Geologic history and correlation of the Jurassic of southwestern Oregon and California: Geological Society of America Bulletin, v. 53, p. 71–112.

Thayer, T.P., 1960, Some critical differences between alpine-type and stratiform peridotite–gabbro complexes: International Geological Congress, Report of the 21st Session, Norden, Copenhagen, pt. 13, p. 247–259.

Unterschutz, J.L.E., Creaser, R.A., Erdmer, P., Thompson, R.I., and Daughtry, K.L., 2002, North America margin origin of Quesnel terrane strata in the southern Canadian Cordillera: Inferences from geochemical and Nd isotopic characteristics of Triassic metasedimentary rocks: Geological Society of America Bulletin, v. 114, p. 462–475, doi: 10.1130/0016-7606(2002)114<0462:NAMOOQ>2.0.CO;2.

Vennum, W.R., 1980, Petrology of the Castle Crags pluton, Klamath Mountains, California: Geological Society of America Bulletin, v. 91, pt. 2, p. 1332–1393, doi: 10.1130/0016-7606(1980)91<255:POTCCP>2.0.CO;2.

Wallin, E.T., and Metcalf, R.V., 1998, Supra-subduction zone ophiolite formed in an extensional forearc: Trinity terrane, Klamath Mountains, California: Journal of Geology, v. 106, p. 591–608.

Wallin, E.T., Coleman, D.S., Lindsley-Griffin, N., and Potter, A.W., 1995, Silurian plutonism in the Trinity terrane (Neoproterozoic and Ordovician), Klamath Mountains, California, USA: Tectonics, v. 14, p. 1007–1013.

Wells, F.G., 1956, Geology of the Medford quadrangle, Oregon-California: Washington, D.C., U.S. Geological Survey Geologic Quadrangle Map GQ-89, scale 1:96,000, 1 sheet.

Wells, F.G., and Cater, F.W., Jr., 1950, Chromite deposits of Siskiyou County, California: California Division of Mines Bulletin 134, pt. 1, p. 77–127.

Wells, F.G., and Walker, G.W., 1953, Geology of the Galice quadrangle, Oregon: Washington, D.C., U.S. Geological Survey Geologic Quadrangle Map GQ–25, scale 1:62,500, 1 sheet.

Wells, F.G., et al., 1940, Preliminary geologic map of the Grants Pass quadrangle, Oregon: Portland, Oregon Department of Geology and Mineral Industries, scale 1:96,000, 1 sheet.

Wells, F.G., Cater, F.W., Jr., and Rynearson, G.A., 1946, Chromite deposits of Del Norte County, California: San Francisco, California Division of Mines Bulletin 134, pt. 1, p. 1–76.

Wells, F.G., Hotz, P.E., and Cater, F.W., Jr., 1949, Preliminary description of the geology of the Kerby quadrangle, Oregon: Portland, Oregon Department of Geology and Mineral Industries Bulletin, v. 40, 23 p., map scale 1:96,000.

Whitney, J.D., 1865, Section XII.—The northern counties:—Shasta, Tehama, Siskiyou, Del Norte, and Trinity, *in* Report of progress and synopsis of the field work from 1860 to 1864: Sacramento, Geological Survey of California, p. 316–363.

Worrall, D.M., 1981, Imbricate low-angle faulting in uppermost Franciscan rocks, Yolla Bolly area, northern California: Geological Society of America Bulletin, v. 92, p. 703–729, doi: 10.1130/0016-7606(1981)92<703:ILFIUF>2.0.CO;2.

Wright, J.E., 1982, Permo-Triassic accretionary subduction complex, southwestern Klamath Mountains, northern California: Journal of Geophysical Research, v. 87, p. 3805–3818.

Wright, J.E., and Fahan, M.R., 1988, An expanded view of Jurassic orogenesis in the western United States Cordillera: Middle Jurassic (pre-Nevadan) regional metamorphism and thrust faulting within an active arc environment, Klamath Mountains, California: Geological Society of America Bulletin, v. 100, p. 859–876, doi: 10.1130/0016-7606(1988)100<0859:AEVOJO>2.3.CO;2.

Wright, J.E., and Wyld, S.J., 1986, Significance of xenocrystic Precambrian zircon contained within the southern continuation of the Josephine ophiolite: Devils Elbow ophiolite remnant, Klamath Mountains, northern California: Geology, v. 14, p. 671–674, doi: 10.1130/0091-7613(1986)14<671:SOXPZC>2.0.CO;2.

Wright, J.E., and Wyld, S.J., 1994, The Rattlesnake Creek terrane, Klamath Mountains, California: An early Mesozoic volcanic arc and its basement of tectonically disrupted oceanic crust: Geological Society of America Bulletin, v. 106, p. 1033–1056, doi: 10.1130/0016-7606(1994)106<1033:TRCTKM>2.3.CO;2.

Wright, J.E., and Wyld, S.J., 2003, Appalachian, Gondwanan, Cordilleran interactions: A new geodynamic model for the Paleozoic tectonic evolution of the North American Cordillera: Geological Society of America Abstracts with Programs, v. 35, no. 6, p. 557.

Wyld, S.J., and Wright, J.E., 1988, The Devils Elbow ophiolite remnant and overlying Galice Formation: New constraints on the Middle to Late Jurassic evolution of the Klamath Mountains, California: Geological Society of America Bulletin, v. 100, p. 29–44, doi: 10.1130/0016-7606(1988)100<0029:TDEORA>2.3.CO;2.

Wyld, S.J., and Wright, J.E., 2001, New evidence for Cretaceous strike-slip faulting in the United States Cordillera and implications for terrane-displacement, deformation patterns, and plutonism: American Journal of Science, v. 301, p. 150–181.

Yule, J.D., 1996, Geologic and tectonic evolution of Jurassic marginal ocean basin lithosphere, Klamath Mountains, Oregon [Ph.D. dissertation]: Pasadena, California Institute of Technology, 308 p.

Yule, J.D., and Saleeby, J.B., 1993, Highly extended oceanic lithosphere: The basement and wallrocks for the Late Jurassic Rogue-Chetco oceanic arc, Oregon Klamath Mountains: Geological Society of America Programs with Abstracts, v. 25, no. 5, p. 169.

Yule, J.D., Saleeby, J.B., Jones, D.L., and Silk, M., 1992, Correlation of basement terranes across the Late Jurassic Josephine inter-arc basin, southwestern Oregon and northern California: Geological Society of America Abstracts with Programs, v. 24, no. 5, p. 93.

Yule, J.D., Saleeby, J.B., and Barnes, C.G., 2006, this volume, A rift-edge facies of the Late Jurassic Rogue-Chetco arc and Josephine ophiolite, Klamath Mountains, Oregon, *in* Snoke, A.W., and Barnes, C.G., eds., Geological studies in the Klamath Mountains province, California and Oregon: A volume in honor of William P. Irwin: Boulder, Colorado, Geological Society of America Special Paper 410, doi: 10.1130/2006.2410(03).

Manuscript Accepted by the Society 10 January 2006

Printed in the USA

Geological Society of America
Special Paper 410
2006

Faunal evidence for the tectonic transport of Jurassic terranes in Oregon, California, and Mexico

Emile A. Pessagno Jr.*
Department of Geosciences, University of Texas at Dallas, P.O. Box 830688, Richardson, Texas 75083-0688, USA

ABSTRACT

Irwin's (1972) terrane concept fathered tectonostratigraphy and has done much to decipher complex geology throughout the Circum-Pacific region, Caribbean, and elsewhere in the world. The present chapter analyzes the tectonostratigraphy of seemingly unrelated Jurassic terranes utilizing an accurate chronostratigraphic framework. Faunal paleolatitudinal data, paleomagnetic data (where available), and microfacies analysis have been utilized in the tectonostratigraphic interpretations.

Four terranes/subterranes are analyzed herein: (1) the Izee terrane (Blue Mountains, northeastern Oregon), (2) the Rogue Valley subterrane (southwestern Oregon), (3) the Smith River subterrane (northwestern California), and (4) the San Pedro del Gallo terrane (east-central and central Mexico, western Cuba). Paleolatitudinal data indicate that all of these terranes/subterranes have undergone tectonic transport.

The Izee terrane originated at low paleolatitudes (Central Tethyan province) during the Late Triassic and migrated to higher paleolatitudes (Southern Boreal province: 30° N) by the late Bathonian (Middle Jurassic). Northward movement is postulated to have occurred along a megashear analogous to the present-day San Andreas fault system.

The Rogue Valley and Smith River subterranes both originated at low paleolatitudes (Central Tethyan province) during the Callovian and were transported to higher paleolatitudes (Southern Boreal province) by the the middle Oxfordian (early Late Jurassic).

The Huayacocotla remnant of the San Pedro del Gallo terrane originated at high Southern Boreal paleolatitudes (30–40° N) during the late Bathonian or early Callovian and, subsequently was transported from northwest to southeast along the west side of the Walper megashear. By the latest Tithonian (Late Jurassic), the Huayacocotla remnant had been transported to low paleolatitudes (Central Tethyan province).

Unconformities in the Bathonian to Callovian interval in the Izee terrane(?) and in the Huayacocotla remnant of the San Pedro del Gallo terrane are believed to reflect the breakup of Pangea. As demonstrated by the author in previous reports, each remnant of the San Pedro del Gallo terrane shows the same paleobathymetric fingerprint: (1) marine deposition at inner neritic depths during the Callovian to early Oxfordian (Middle to early Late Jurassic), (2) marine deposition at outer neritic depths during the late Oxfordian (Late Jurassic), and (3) sudden deepening to upper abyssal depths

*E-mail: pessagno@utdallas.edu.

Pessagno, E.A., Jr., 2006, Faunal evidence for the tectonic transport of Jurassic terranes in Oregon, California, and Mexico, *in* Snoke, A.W., and Barnes, C.G., eds., Geological studies in the Klamath Mountains province, California and Oregon: A volume in honor of William P. Irwin: Geological Society of America Special Paper 410, p. 31–52, doi: 10.1130/2006.2410(02). For permission to copy, contact editing@geosociety.org.

from the early Kimmeridgian (Late Jurassic) until the end of the Cretaceous. The sudden deepening event (outer neritic to upper abyssal) during the early Kimmeridgian is associated with a disconformity and hiatus in all San Pedro del Gallo remnants (early Kimmeridgian strata overlie middle Oxfordian strata). This event reflects the opening of the Gulf of Mexico.

Combined faunal and floral data indicate that the San Pedro del Gallo terrane was in a back-arc position at approximately the same latitude as the Foothills terrane of the Sierra Nevada during the middle Oxfordian. Given the faunal data as well as some paleomagnetic data, it is probable that these San Pedro del Gallo remnants might represent some of the missing Upper Jurassic back-arc deposits from farther north (i.e., western Nevada?).

In the Smith River subterrane, the disconformable contact between the volcanopelagic facies and Galice Formation sensu lato reflects a sudden influx of siliciclastic turbidite from a Jurassic volcanoplutonic arc source area and from older rocks of the accreted continental margin that lay inboard of the Nevadan island-arc complex. The same event is represented by the deposition of the siliciclastic turbidite of the Monte del Oro and Mariposa Formations (Foothills terrane, western Sierra Nevada) and by that of the Galice Formation sensu stricto (Rogue Valley subterrane).

Keywords: Jurassic, Nevadan orogeny, radiolarians, paleolatitudes, paleobiogeography, western Klamath terrane, Izee terrane, San Pedro del Gallo terrane, Klamath Mountains, Blue Mountains, Mexico

INTRODUCTION

The terrane concept was first advanced by W. Porter Irwin (U.S. Geological Survey, Menlo Park, California) in 1972. In a report titled "Terranes of the western Paleozoic and Triassic belt in the southern Klamath Mountains, California," Irwin (1972, C103) wrote: "The western Paleozoic and Triassic belt in the southern Klamath Mountains is herein subdivided into three concentric subbelts of rocks that are referred to, from west to east, as the Rattlesnake Creek, Hayfork, and North Fork terranes. The term "terrane" as used herein refers to an association of geologic features, such as stratigraphic formations, intrusive rocks, mineral deposits, and tectonic history, some or all of which lend a distinguishing character to a particular tract of rocks and which differ from those of an adjacent terrane."

Irwin's (1972) terrane concept provided the key to unraveling the complex geology of the Circum-Pacific margin. The collage of terranes comprising the Circum-Pacific margin includes both autochthonous terranes showing little evidence of tectonic transport as well as allochthonous terranes, such as the Wrangellia terrane (Jones et al., 1977) and San Pedro del Gallo terrane (Pessagno et al., 1999), which have been displaced thousands of kilometers from their point of origin.

The purpose of this chapter is to examine the paleontological record of selected terranes within the Nevadan[1] complex of California and Oregon (Blue Mountains: northeastern Oregon; Klamath Mountains: northwestern California and southwestern Oregon), and Mexico (east-central and central) for evidence of tectonic transport and to determine whether these terranes share a common geologic history.

[1]*Editors' note:* In this chapter, Pessagno uses the term "Nevadan" more broadly (i.e., Middle Jurassic–early Late Jurassic time) than some other authors have used in this volume (cf. Snoke and Barnes, this volume).

BIOPALEOGEOGRAPHY AND TECTONOSTRATIGRAPHY

Studies by Taylor et al. (1984), Pessagno and Blome (1986), Pessagno et al. (1986, 1993a,b, 1999, 2001), Montgomery et al. (1992, 1994a,b), and Pessagno and Hull (2003) established the importance of faunal and floral data in paleogeographic reconstructions in North America and the Caribbean. It is now well established that the recognition of displaced tectonostratigraphic terranes depends on paleolatitudinal data derived from paleontology and paleomagnetism. Existing paleomagnetic data can commonly be constrained by faunal and floral data. In some cases, such paleontological data can be used to determine whether a tectonostratigraphic terrane originated in the northern or southern hemisphere or in the eastern or western Pacific. The Wrangellia terrane of Jones et al. (1977) is a classic example of constraints placed on paleomagnetic data by paleontological data. Jones et al. (1977) indicated that the Wrangellia terrane originated 15° north or south of the Triassic paleoequator. Investigations by Tipper (1981), Taylor et al. (1984), and Pessagno and Blome (1986) indicate that both the molluscan and radiolarian assemblages were predominantly Tethyan in origin during the Late Triassic and Early Jurassic. The discovery of very rare Boreal ammonites (amaltheids) in the upper Pliensbachian portion of the Maude Formation (Queen Charlotte Islands, British Columbia) by Taylor et al. (1984) indicates that Wrangellia was situated in the northern hemisphere. In addition, the presence of the Tethyan pectenacid *Weyla* in Lower Jurassic strata of the Queen Charlotte remnant and other remnants of the Wrangellia terrane demonstrates that this terrane originated in the eastern Pacific (Smith, 1980; Taylor et al., 1984; Pessagno and Blome, 1986; Pessagno et al., 1986). *Weyla* is unknown from the western Pacific.

Radiolarian Paleolatitudinal Model of Pessagno and Blome (1986)

As noted by Pessagno et al. (1999) and more recently by Pessagno and Hull (2003), it is now apparent from our analyses of radiolarian faunal data from North America and elsewhere in the world that Radiolaria can be utilized in paleobiogeographic investigations and to monitor the tectonic transport of terranes both in the northern and southern hemispheres (Figs. 1 and 2).

Pessagno et al. (1999, p. 126) stated that "In the Northern Hemisphere Pessagno and Blome (1986), Pessagno et al. (1986, 1987, 1993a,b), and Hull (1991, 1995) divided the Tethyan Realm into a Central Tethyan Province characterized by a radiolarian assemblage with high pantanelliid abundance and diversity and the absence of *Parvicingula/Praeparvicingula* and into a Northern Tethyan Province with high pantanelliid abundance and diversity and common *Parvicingula/Praeparvicingula* (Text-Figure 7 [Fig. 1 herein]). The Boreal Realm was subdivided into a Southern Boreal Province and a Northern Boreal Province. The Southern Boreal Province is characterized by a sharp decline in pantanelliid abundance and diversity and by the abundance and diversity of species of *Parvicingula/Praeparvicingula;* the Northern Boreal radiolarian assemblage is distinguished by abundant *Parvicingula/Praeparvicingula* and by its total lack of pantanelliids." The boundary between the Tethyan and Boreal realms is placed at ~30° N. The boundary between the Central Tethyan and Northern Tethyan provinces is established by associated paleomagnetic data at ~22° N. The province boundary was placed at ~22° N because of the presence of Central Tethyan Berriasian fauna occurring at 20° N (Blake Bahama Basin: Ogg, 1983; Baumgartner, 1984; Pessagno et al., 1987). Moreover, this placement of the boundary is also supported by data from the Philippines (Yeh and Cheng, 1996; Pessagno et al., 1999). It should be pointed out that quantitative paleomagnetic data for the Phanerozoic is totally dependent on biostratigraphically derived chronostratigraphic data: the Phanerozoic chronostratigraphic scale is defined by biozones.

Pessagno and Blome (1986) originally proposed that the model for the southern hemisphere is the mirror image of that in the northern hemisphere. New data from the southern hemisphere has substantiated this model from Argentina (Pujana, 1989, 1991, 1993, 1996), New Zealand (Aita and Grant-Mackie, 1992), Antarctica (Kiessling, 1995, 1999; Kiessling and Scasso, 1996), and the Sula Islands (Pessagno and Hull, 2003).

As noted by Pessagno et al. (1999), Pessagno and Hull (2003), and Pessagno and Martin (2003), Hagstrum and Murchey (1996) in a report titled "Paleomagnetism of Jurassic radiolarian chert above the Coast Range ophiolite at Stanley Mountain, California and implications for its paleogeographic origins" challenged the validity of the methodology of Pessagno and Blome (1986) and Pessagno et al. (1987, 1993a,b). This criti-

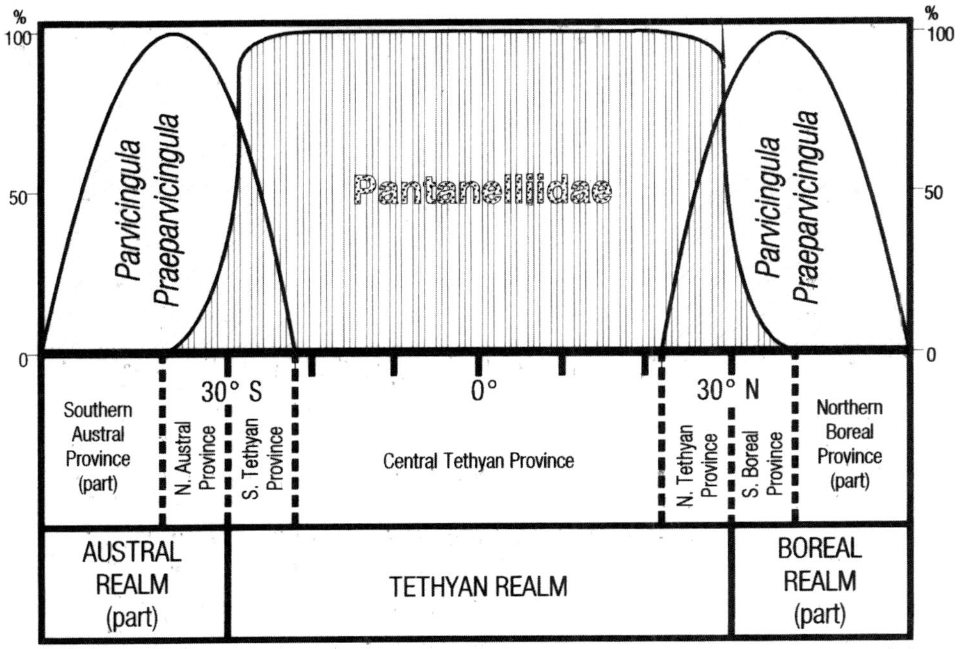

Figure 1. Paleolatitudinal model based on distribution of selected Radiolaria from Jurassic and Lower Cretaceous. From Pessagno et al. (1993a).

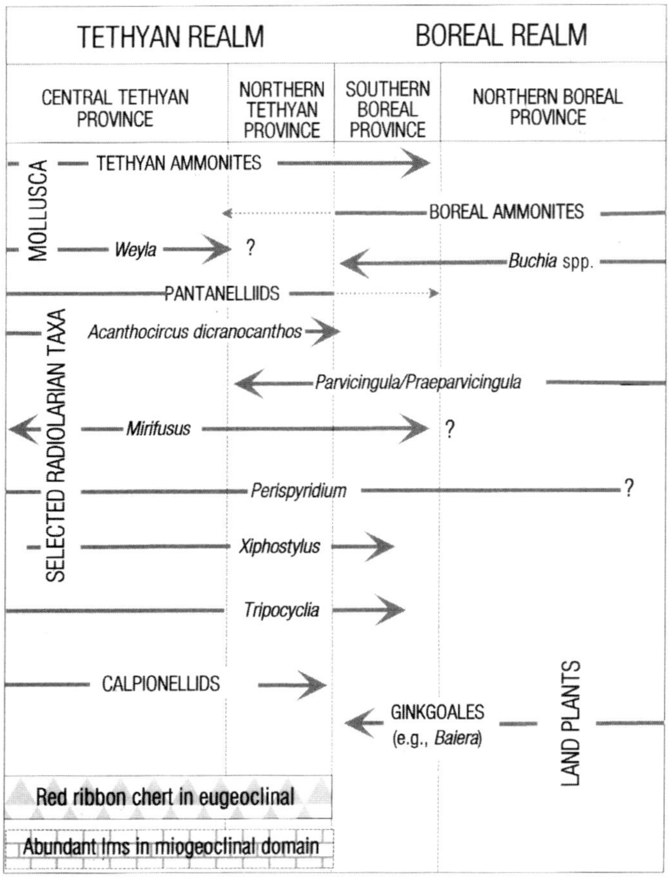

Figure 2. Multiple criteria for use in paleogeographic reconstructions in the northern hemisphere. All available tools should be utilized in addition to floral and faunal analysis. Some of the more obvious of these include paleomagnetism and physical stratigraphy (e.g., red ribbon cherts = Tethyan realm; abundance of coal = Boreal or Austral realms). From Pessagno and Hull (2003).

cism has been fully addressed by Pessagno et al. (1999) and Pessagno and Hull (2003). The discussion is detailed and lengthy and will not be repeated here.

Paleolatitudinal Reconstructions Using Multiple Criteria

Paleobiogeographic reconstructions based on the radiolarian assemblage can commonly stand alone. The tenet stressed in this chapter as well as in previous reports by the author (e.g., Pessagno et al., 1999) is that paleogeographic reconstructions should use all criteria available and should not focus on any single facet (e.g., analysis of only the ammonite assemblage). However, in many accreted oceanic terranes, macrofossils are absent. Yet it is commonly possible to determine the relative paleolatitudinal position of a given terrane from the analysis of the radiolarian assemblage alone. Conversely, in successions where megafossil assemblage is well known and the radiolarian assemblage has not been studied, it is sometimes possible to predict the paleolatitudinal fingerprint of the radiolarian assemblage from that of the macrofossil assemblage.

This was, in fact, the case in the author's initial studies of the San Pedro del Gallo area. Because there is a mixture of such Boreal ammonites as *Amoeboceras* sp. cf. *alterans* (von Busch) (Burckhardt, 1930–1931; Imlay, 1980) and *Buchia* associated with Tethyan ammonites, it was possible to determine that the radiolarian assemblage, if present in these strata, would be assignable to the Southern Boreal province. This prediction proved to be valid.

EVIDENCE OF TECTONIC TRANSPORT OF NEVADAN AND ASSOCIATED TERRANES

Introduction

Pessagno and Blome (1986) considered all of the terranes of the Blue Mountains province of northeastern Oregon and western Idaho to be dismembered components of a single island-arc complex. These workers also presented tectonostratigraphic data to demonstrate that calc-alkaline successions in the Nevadan complex of California and western Nevada reflect the same paleolatitudinal and paleolongitudinal fingerprint as that displayed by the terranes of the Blue Mountains province during the Late Triassic, Early Jurassic, and Middle Jurassic. The presence of the Tethyan pectenacid *Weyla* in the Early Jurassic strata of the Blue Mountains province, Sierra Nevada, and Klamath Mountains, for example, demonstrates an eastern Pacific origin (Taylor et al., 1984; Pessagno and Blome, 1986). Furthermore, Pessagno and Blome (1986) suggested that the calc-alkaline Jurassic terranes of the Blue Mountains province, Sierra Nevada, and western Nevada were part of the same island-arc complex, which originated at near equatorial latitudes in the eastern Pacific and was transported northward along a mega-shear system analogous to the present-day San Andreas fault system. This island-arc complex has been referred to as the "Jurassic volcanoplutonic arc" or "Nevadan island arc" by Pessagno et al. (2001).

Izee Terrane, East-Central Oregon

The Izee terrane (= Mesozoic Clastic terrane of Dickinson, 1979; Pessagno and Blome, 1986) is the best studied of all of the terranes in the Blue Mountains province and is discussed in detail below. The Izee terrane includes rocks of Late Triassic (late Karnian) to Middle Jurassic (middle Callovian) age. It lies between the central mélange terrane to the north and Seven Devils terrane and Huntington arc terrane to the south (Fig. 3). Pessagno and Blome suggested that the early Callovian to early Oxfordian Coon Hollow Formation is genetically related to strata in the Izee terrane (e.g., the Lonesome Formation; Fig. 4). The Coon Hollow Formation overlies the Triassic Seven Devils Group with angular unconformity in the Seven Devils terrane.

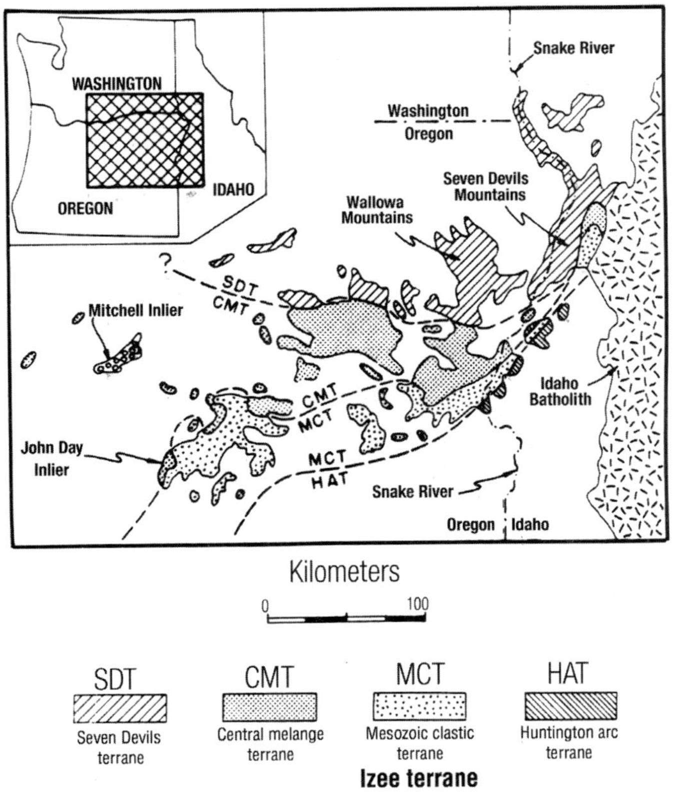

Figure 3. Map showing distribution of geologic terranes in the Blue Mountains province of northeastern Oregon and western Idaho. From Dickinson (1979) and Pessagno and Blome (1986).

The Izee terrane consists mainly of a siliciclastic/volcaniclastic succession characterized by turbidite. More detailed lithostratigraphic descriptions of these units were presented by Pessagno and Blome (1986) and Pessagno et al. (1986). Volcanic rocks include spilitic basalt and andesite flows of Late Triassic age (Karnian and late middle Norian; e.g., Begg Member of Vester and Field Creek Formations), scarce andesite flows of Early Jurassic age (e.g., late Pliensbachian Nicely Formation), and andesitic tuff breccia and tuff of Middle Jurassic age (Bajocian Silvies Member of Snowshoe Formation and lower Callovian Trowbridge Formation). Pessagno and Blome (1986) suggested that the Late Triassic phase of volcanism (spilitic basalt flows and andesites of Vester and Fields Creek Formations) represented near-axis island-arc volcanism, whereas the Early Jurassic to Middle Jurassic phase of volcanism represented off-axis island-arc volcanism. In the Greater Antilles (e.g., Puerto Rico: see Pessagno, 1960) island-arc volcanism occurred from Early Cretaceous (Aptian(?) and Albian) to early middle Eocene along a northwest-to-southeast axis, which gradually shifted northward (see Christman, 1953, 1973 for the Lesser Antilles).

No paleomagnetic data are yet available for the Izee terrane. However, it should be noted that Hillhouse et al. (1982) demonstrated that the calc-alkaline volcanic rocks of the Upper Triassic Huntington Formation (Brooks, 1979) formed 18° (±4°) north or south of the Late Triassic paleoequator in the neighboring Huntington arc terrane.

Abundant paleontological data from the Izee terrane are available from studies by Imlay (1964, 1968, 1973, 1980, 1981), Dickinson and Vigrass (1965), Pessagno and Blome (1980, 1982, 1986), Pessagno and Whalen (1982), Blome (1983, 1984a), and Pessagno et al. (1986, 1987, 1989). A detailed analysis of the megafossil and radiolarian assemblages was presented by Pessagno and Blome (1986). A summary of the faunal data presented by Pessagno and Blome is as follows:

1. Central Tethyan faunas predominated from the Late Triassic (late Karnian–late middle Norian) to the Early Jurassic (middle Toarcian): Brisbois Formation, Rail Cabin Argillite, Graylock Formation, Suplee Formation, Nicely Formation, Hyde Formation, and Warm Springs Member of the Snowshoe Formation (part).
2. Northern Tethyan faunas predominated from the Early Jurassic (middle Toarcian) to the Middle Jurassic (late Bajocian): Warms Springs (part), Silvies, and Schoolhouse members of Snowshoe Formation.
3. Southern Boreal faunas predominated during the Middle Jurassic (late Bathonian to middle Callovian): South Fork Member of Snowshoe Formation, Trowbridge Formation, and Lonesome Formation.

As noted by Pessagno and Blome (1986), the presence of the pectenacid *Weyla* in the Early Jurassic Suplee Formation is not only a good indicator for the Tethyan paleolatitudes, but it is also indicative of an eastern Pacific origin for the Izee terrane. All Late Triassic and Early Jurassic (Karnian to middle Toarcian) Central Tethyan radiolarian faunas are characterized by the abundance and diversity of pantanelliid taxa and by the absence of *Praeparvicingula* (Fig. 1). The change in faunal composition from the Central Tethyan province to Northern Tethyan province occurred during the middle Toarcian (Early Jurassic), with the first occurrence of *Praeparvicingula* in the Warms Spring Member of the Snowshoe Formation. The Northern Tethyan radiolarian assemblage is still characterized by abundant and diversified pantanelliids. During the late Bajocian (Middle Jurassic), pantanelliid diversity began to suddenly drop. The shift from Northern Tethyan province to Southern Boreal province occurs in the late Bathonian (Middle Jurassic) and is well documented by both megafossils and Radiolaria occurring at the top of the South Fork Member of the Snowshoe Formation at Izee and in the overlying Trowbridge and Lonesome Formations. As noted by Pessagno and Blome (1986), the late Bathonian to middle Callovian ammonite assemblage is less diverse and contains such genera as *Torricelliceras* Buckman, *Kepplerites* Neumayr and Uhlig, and *Pseudocadoceras* Buckman, which are characteristic of the Boreal realm. Pessagno and Blome noted that, together with these forms, there are such

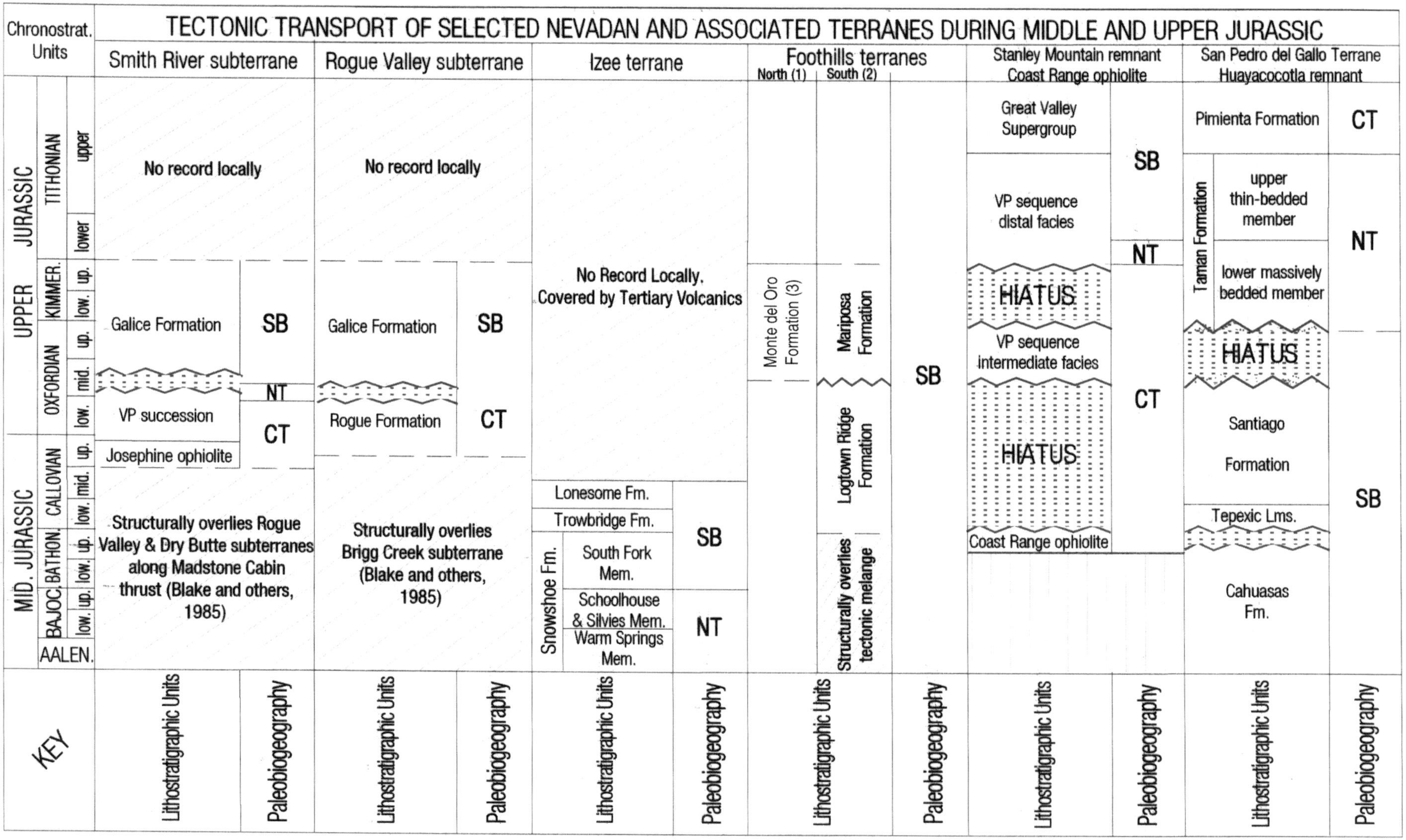

Figure 4. Tectonic transport of associated North American Jurassic terranes. In regard to the northern Sierra Nevada, it should be noted that the Lower Jurassic (upper Sinemurain) Lilac Argillite and Hardgrave Sanstone at Mount Jura near Taylorsville, California, contain the Tethyan eastern Pacific pectenacid *Weyla alata* (Imlay, 1980, p. 57). This taxon occurs abundantly in the upper Pliensbachian Suplee Formation of the Izee terrane (Imlay, 1980; Pessagno and Blome, 1986; Pessagno et al., 1986). The Lower Jurassic (Pliensbachian) Thompson Limestone contains the Tethyan bivalve *Plicatostylus,* which is also known from the Robertson Formation in the Izee terrane. Middle Jurassic (Bajocian) strata at Mount Jura contain essentially the same Tethyan ammonite assemblage as that occurring in the Snowshoe Formation of the Izee terrane.

genera as *Lilloetia* Crickmay that are known only from the boreal eastern Pacific and others, such as *Paratoceras* Spath, that are predominantly Tethyan. The association of boreal ammonites with those that also occur in the Tethyan realm is characteristic of the Southern Boreal province of Pessagno and Blome (1986) and Pessagno et al. (1999; Fig. 2). The radiolarian assemblage during the late Bathonian (Southfork Member of Snowshoe Formation) is characterized by very low pantanelliid diversity and abundance and by the abundance of *Praeparvicingula*. This trend continues into the Callovian (Lonesome Formation). It should be noted that many of the radiolarian species that occur in the Lonesome Formation also occur in the Callovian Sheilikof Formation of southern Alaska (Blome, 1984b).

The paleolatitudinal and paleolongitudinal faunal data indicate that the Izee terrane originated in the eastern Pacific during the Late Triassic at Central Tethyan latitudes and was gradually transported to Northern Tethyan latitudes by the Early Jurassic (middle Toarcian). The change in paleolatitudinal position from Northern Tethyan province to Southern Boreal province between the late Bajocian and late Bathonian was abrupt and represented accelerated tectonic transport of the Izee terrane and the breakup of Pangea (Fig. 4). Pessagno and Blome (1986) postulated that the Izee terrane and other Nevadan arc remnants were carried northward along a megashear analogous to the present-day San Andreas fault system. Such a megashear may be related to those well documented by Longoria (1984) and Pessagno et al. (1999) from east-central and north-central Mexico, and therefore to the rifting and subsequent seafloor spreading that formed the Gulf of Mexico and moved North America away from South America (Buffler et al., 1980; Dickinson and Coney, 1980; Pessagno and Martin, 2003).

Western Klamath Terrane

The western Klamath terrane of Blake et al. (1985; equivalent to the western Jurassic terrane of Irwin, 1985) includes Middle and Upper Jurassic rocks that crop out along the western margin of the Klamath Mountains and in a large outlier along the coast (southwestern Oregon). Blake et al. (1985) subdivided the western Klamath terrane into five distinct subterranes, which were each characterized by a different substrate. Two of these subterranes—the Rogue Valley subterrane and Smith River subterrane (Fig. 5)—are discussed herein (see Pessagno et al., 1993a).

Rogue Valley Subterrane. The Rogue Valley subterrane (Figs. 5 and 6) structurally overlies the Briggs Creek subterrane (Blake et al., 1985). It includes the type Rogue Formation (Wells and Walker, 1953) and overlying type Galice Formation (Diller, 1907). The lithostratigraphic characteristics of these two formations are summarized in Figure 6.

The **Rogue Formation** (Wells and Walker, 1953) consists of a thick succession (4500 m+) of calc-alkaline rocks, which include dacitic tuff and tuff breccia, andesitic to basaltic flows, tuffaceous chert, and volcanic wacke (Fig. 6). As noted by Pessagno et al. (1993a), the thickness given here represents only an apparent thickness, in that the Rogue is tightly folded. Blake et al. (1985) indicated that the Rogue was deposited on ophiolitic basement. Pessagno et al. (1993a) suggested that this ophiolitic basement, although now badly dismembered, might represent an extension of the Josephine ophiolite (see Smith River subterrane below and Fig. 7).

The Rogue Formation in its type area along the Rogue River Valley is conformably overlain by the Galice Formation sensu stricto. No megafossils are known from the Rogue Formation. However, we observed poorly preserved Central Tethyan Radiolaria in fine-grained tuff and tuffaceous chert samples just below the contact with the Galice Formation (loc. RR-89-8 of Pessagno et al., 1993a: entrance to the shaft of Almeda mine, Fig. 8). The sudden change from the volcaniclastic strata to the siliciclastic rocks of the overlying Galice Formation sensu stricto, together with the sudden change in faunal province from Central Tethyan to Southern Boreal, indicates a probable disconformity and slight hiatus between these two units.

As noted by Pessagno et al. (1993a), the geochronological assignment of the Rogue Formation can only be determined by superposition; that is, the top of the Rogue Formation is middle middle Oxfordian or older (see the next paragraphs). They stated that if unequivocal field evidence can be offered that proves that the Rogue Formation rests in sedimentary contact above the Josephine ophiolite (JO), the age of the base of the Rogue Formation must be no older than 164 Ma (Callovian, i.e., oldest reliable age on the JO; cf. Wright and Wyld, 1986).

The Galice Formation was originally named and described by Diller (1907, p. 403–407) for exposures along Galice Creek, a tributary of the Rogue River (Fig. 8). In its type area, the **Galice Formation sensu stricto** includes a thick siliciclastic turbidite succession consisting predominantly of black, slaty, siliceous mudstone ("argillite") and medium-gray, commonly calcareous metagraywacke together with minor amounts of black chert, conglomerate, and limestone (very scarce). The thickness of the Galice Formation is difficult to estimate because of the structural complexity of the area and because the top of the unit is generally obscured by faulting. However, estimates of thickness generally range from 3000 to 4500 m. As noted by Pessagno et al. (1993a), the Galice Formation is structurally thickened by large tight folds.

At the Almeda mine along the Rogue River (Fig. 8), *Buchia concentrica* was found together with the ammonite *Perisphinctes* (*Dichotomosphinctes*) in the basal Galice Formation adjacent to its conformable contact with the underlying Rogue Formation (Diller, 1907; Imlay, 1961, 1980). As noted by Pessagno et al. (1993a), the concurrence of these two taxa is indicative of the middle part of the middle Oxfordian (see also Pessagno et al., 1986; Pessagno and Blome 1988, 1990). *Buchia concentrica* occurs sporadically throughout the Galice in its type area (Diller, 1907; Imlay, 1961). However, because of the structural complexity of the Rogue Valley subterrane, it is not clear whether this taxon ranges to the top of the Galice. Because *Buchia con-*

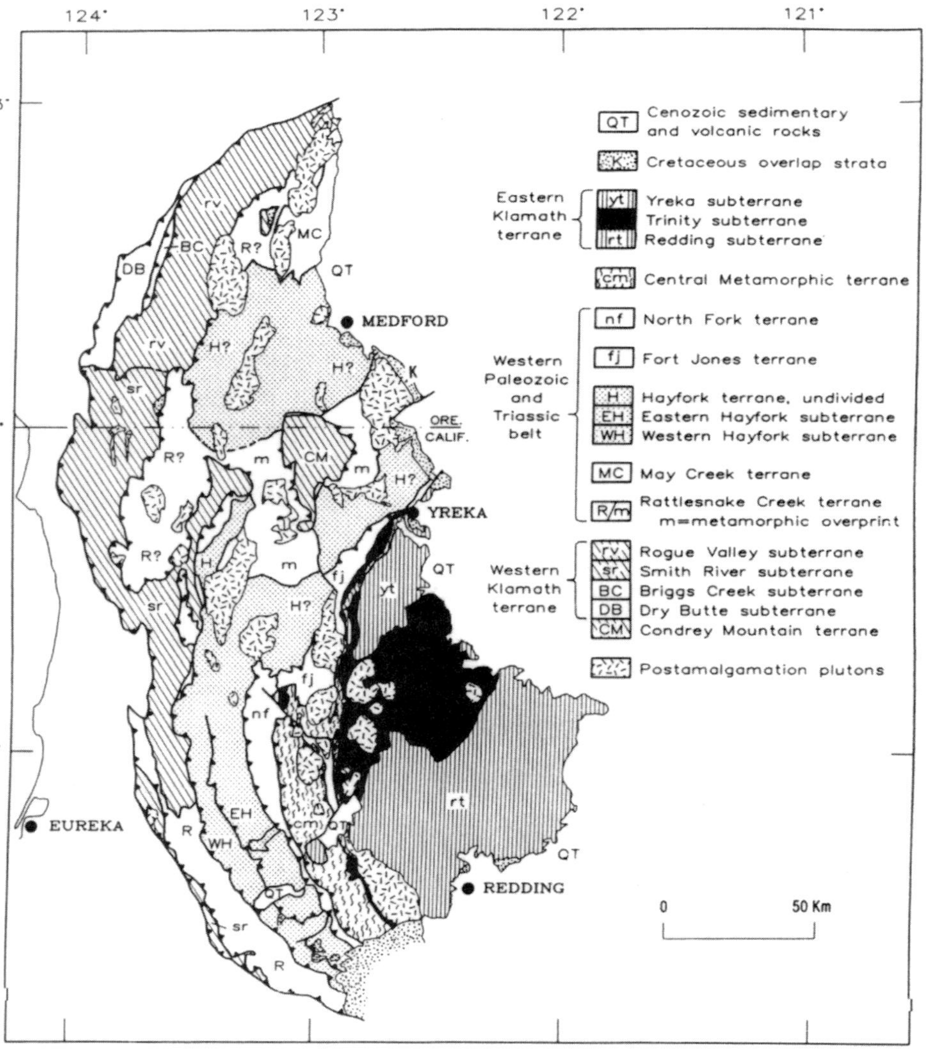

Figure 5. Map showing terranes, post-amalgamation plutons, and overlap assemblages of the Klamath Mountains. Note that "**K**" (Cretaceous overlap strata) in Irwin's key = Great Valley Supergroup. From Pessagno et al. (1993a); modified from Irwin (1989).

centrica ranges as high as the late Kimmeridgian elsewhere in North America (Imlay, 1980; Pessagno et al., 1999), it is possible that the top of the Galice Formation is that young.

Investigations by Pessagno et al. (1993a) established that Radiolaria occur in finer-grained clastic strata, thin-bedded chert, and sparse limestone. Well-preserved calcified Radiolaria (including *Mirifusus* spp. and *Praeparvicingula* spp.) occur in small lenticular masses of dark-gray, micritic pelagic limestone interbedded with dark-gray siliceous mudstone near the base of the unit at the Almeda mine (Fig. 8).

The association of the boreal bivalve *Buchia concentrica* with *Praeparvicingula* and the "Tethyan" ammonite *Perisphinctes* (*Dichotomosphinctes*) in the lower Galice Formation sensu stricto is indicative of Southern Boreal paleolatitudes (Figs. 1 and 2).

Smith River Subterrane. The Smith River subterrane (southwestern Oregon and northwestern California) structurally overlies the Rogue Valley and Dry Butte subterranes along the Madstone Cabin thrust (Blake et al., 1985). Previous studies by Pessagno et al. (1993a) of the Smith River subterrane have focused primarily on the area mapped by Harper (1983, 1984) in northwestern California and in an adjacent area bridging the Oregon-California border (Turner-Albright mine; Figs. 9 and 10). According to Harper (1983, 1984) the rocks of the Smith River subterrane experienced prehnite-pumpellyite to lower greenschist facies metamorphism.

The **volcanic member of the JO** was examined by Pessagno et al. (1993a) at the Turner-Albright mine along the Oregon-California border (Figs. 7 and 10). At this locality, these workers observed common masses of light-green to gray tuffaceous radiolarian chert interbedded with the pillow lavas of the volcanic member of the Josephine ophiolite. Along the Middle Fork of the Smith River (along strike and 10–20 km to west-southwest of Harper, 1983, his locality 1; JO 6-79 in Fig. 9), Saleeby (1987) obtained a concordant U-Pb age of 162 ± 1 Ma (zircon in plagiogranite) from the ophiolite.

The depositional contact between the volcanic member of

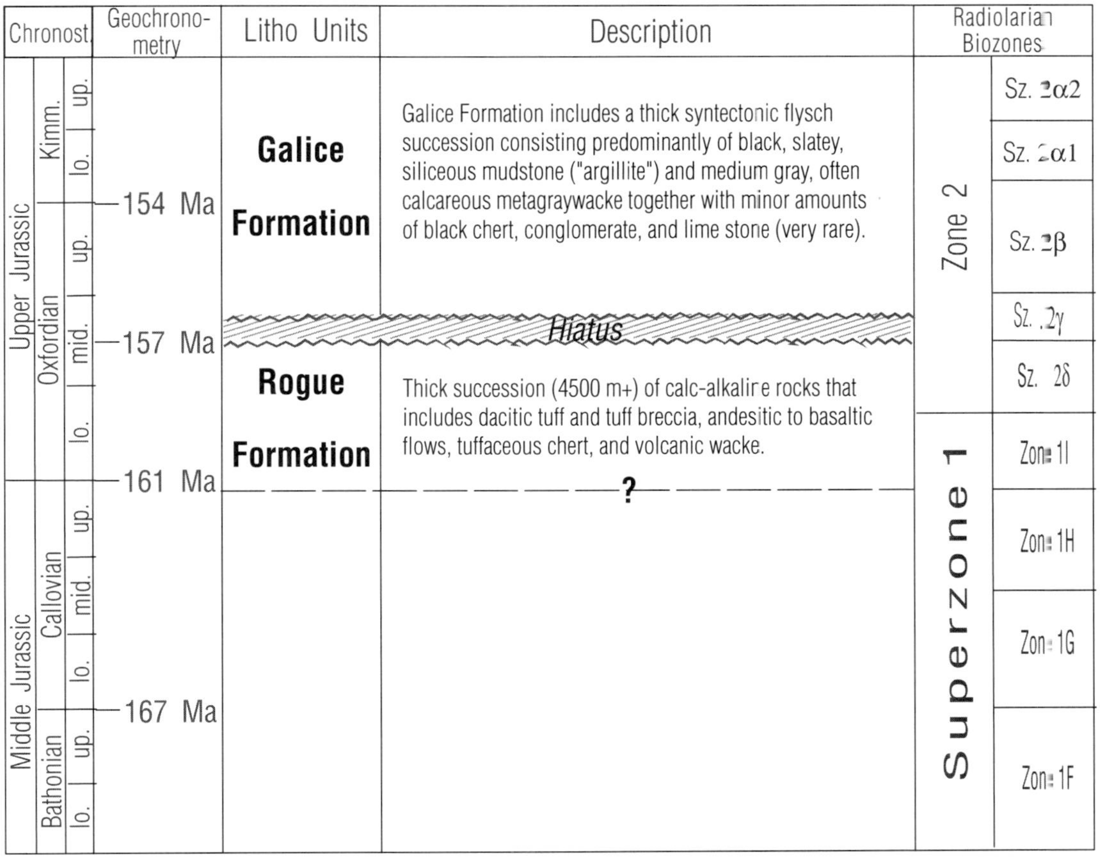

Figure 6. Stratigraphic summary for Rogue Valley subterrane. Kimm.—Kimmeridgian. Time scale after Gradstein et al. (1994).

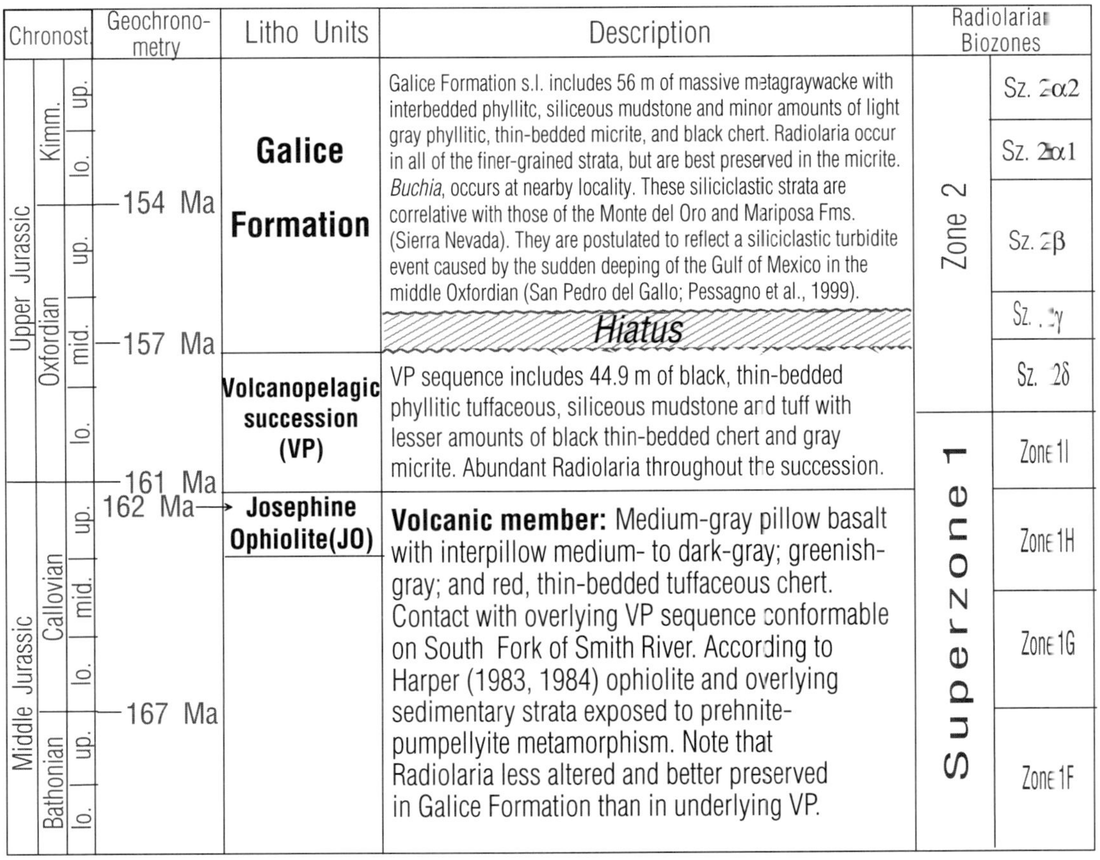

Figure 7. Stratigraphic summary for Smith River subterrane. Kimm.—Kimmeridgian; s.l.—sensu lato. Time scale after Gradstein et al. (1994).

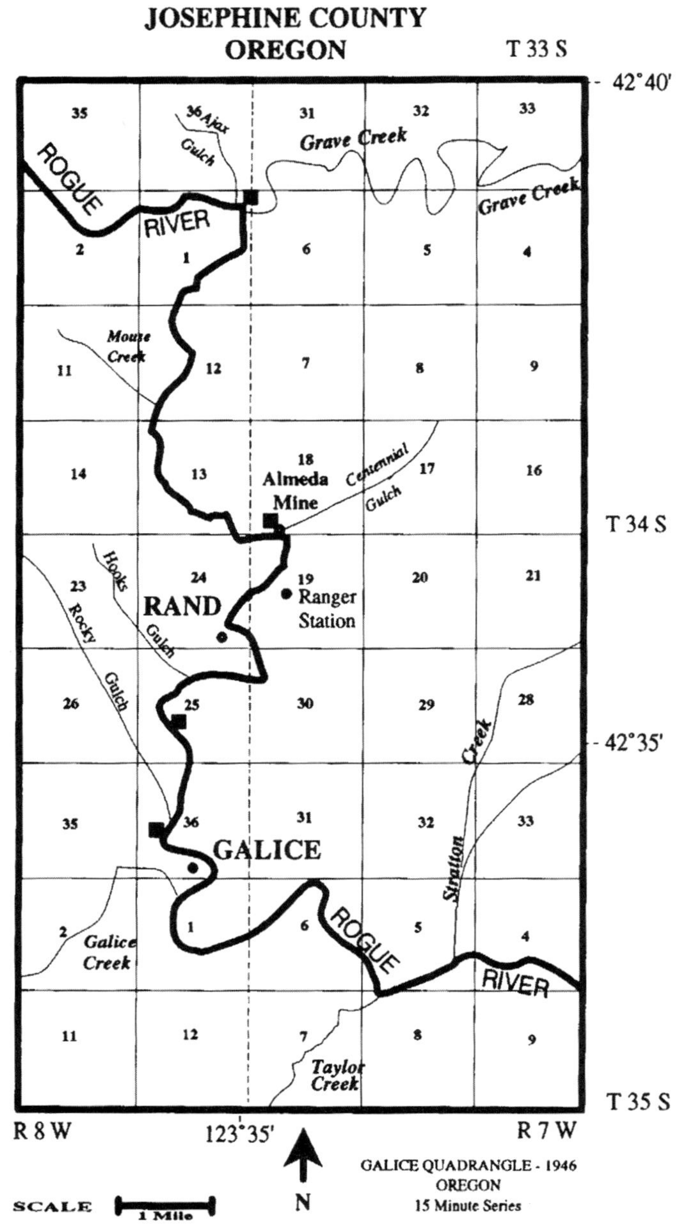

Figure 8. Index map for Rogue River Valley showing the location of Almeda mine. From Pessagno et al. (1993a).

the JO and overlying sedimentary strata is well exposed along the Middle Fork of the Smith River just upstream from the mouth of Little Jones Creek (Fig. 9). At this locality, the sedimentary succession (100.9 m) consists of two units (Fig. 10): a lower volcanopelagic succession and an upper siliciclastic unit succession characterized by turbidites (Galice Formation sensu lato).

According to Pessagno et al. (1993a) the **volcanopelagic (VP) succession** includes 44.9 m of black, thin-bedded, phyllitic, tuffaceous, siliceous mudstone (tuffaceous "argillite") and tuffs, which are interbedded with lesser black, thin-bedded chert and thin-bedded, lenticular masses of phyllitic, medium-gray pelagic limestone. The VP strata conformably overlie the volcanic member of the JO. Radiolaria occur throughout the VP succession, but are well preserved only in the pelagic limestone. No megafossils are known from the VP succession at this locality or, for that matter, at any other locality in the Smith River subterrane. The presence of Radiolaria and the lack of calcareous microfossils and megafossils within the VP succession suggest deposition below the calcite compensation depth (CCD) at abyssal depths. Radiolarian biostratigraphic data from the VP succession indicate that these strata are assignable to Superzone 1: Zone 1I and Zone 2, Subzone 2 delta (upper Callovian to middle middle Oxfordian; Pessagno et al., 1993a, p. 112).

According to Pessagno et al. (1993a), the **Galice Formation sensu lato** consists of 56 m of massive metagraywacke with interbedded, thick intervals of black, phyllitic, siliceous mudstone, and minor amounts of light-gray, phyllitic, thin-bedded pelagic limestone and thin-bedded black chert (Figs. 7 and 9). The siliceous mudstone beds in the upper 15.2 m of the siliciclastic turbidite succession contain common, small rip-up clasts of light-gray pelagic limestone. Radiolaria occur in all of the finer-grained strata. However, they are well preserved only in the pelagic limestone. *Buchia concentrica* (middle Oxfordian to upper Kimmeridgian) was noted from Galice sensu lato strata by Diller (1907) and Harper (1983) at nearby localities along Shelly and Monkey Creeks to the north and west of the exposure on the Middle Fork of the Smith River (Fig. 9). The presence of calcareous megafossils and microfossils (e.g., foraminifera in pelagic limestone) suggests deposition above the CCD at upper abyssal depths. Radiolaria assignable to Zone 2, Subzone 2 gamma (upper part: middle middle Oxfordian) were recovered from the Galice Formation sensu lato in the Rogue Valley subterrane (Pessagno et al., 1993a).

Conclusions

The following conclusions can be drawn from stratigraphic data from the Rogue Valley and Smith River subterranes:

1. In the Rogue Valley subterrane, Pessagno et al. (1993a) observed Central Tethyan Radiolaria at the top of the Rogue Formation. The presence of *Buchia* in association with *Persisphinctes* (*Dichotomosphinctes*) in the basal Galice Formation, together with abundant specimens of *Praeparvicingula,* indicates that these strata were deposited in the Southern Boreal province (Fig. 2). The remainder of the type Galice contains *Buchia* but lacks Radiolaria and other megafossils. The data at hand suggest that the Rogue Valley subterrane was subjected to northward tectonic transport similar to that displayed by the Smith River subterrane (see below). The sudden change from Central Tethyan to Southern Boreal faunas (radiolaria and megafossils) at the contact between the Rogue Formation and the overlying Galice Formation at the Almeda mine suggests the

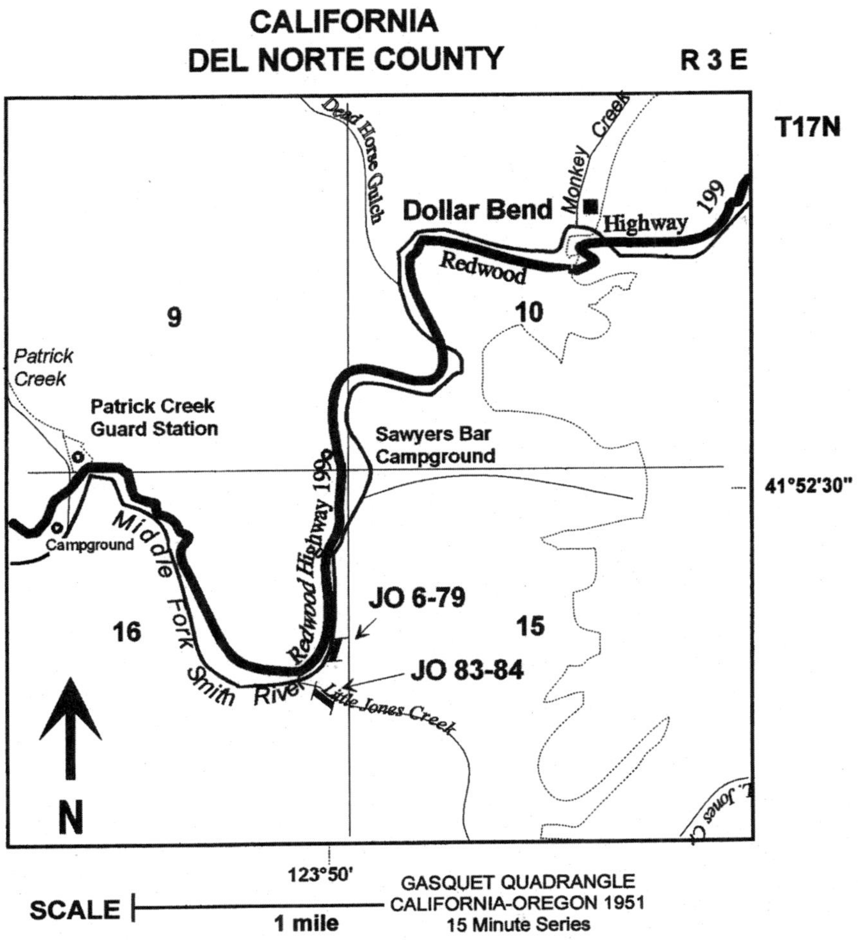

Figure 9. Index map for sample localities along the Middle Fork of the Smith River, northwestern California. From Pessagno et al. (1993a).

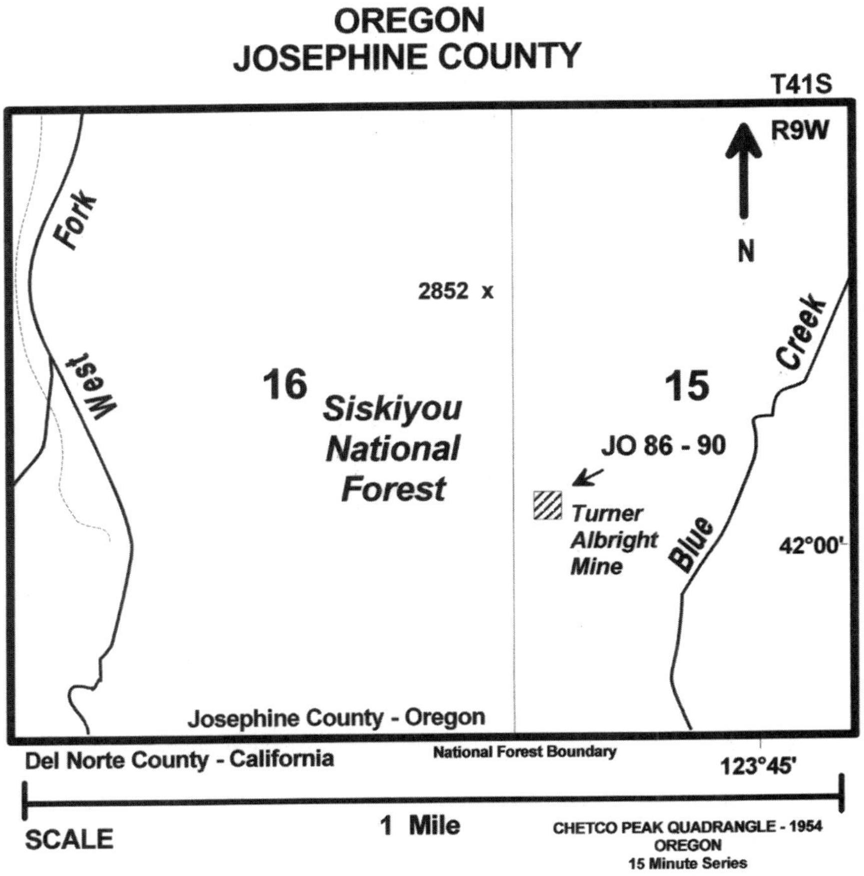

Figure 10. Index map for sample localities in the volcanic member of the Josephine ophiolite (JO) at the Turner-Albright mine, northwestern California and southwestern Oregon. From Pessagno et al. (1993a).

presence of an unconformity and accompanying hiatus (Rogue Northern Tethyan strata missing).

2. In the Smith River subterrane, the volcanic member of the JO (Turner-Albright mine) and the lower 39 m of the VP succession (Middle Fork of the Smith River near Little Jones Creek) contain a radiolarian assemblage characteristic of the Tethyan realm, Central Tethyan province. The radiolarian assemblage occurring within the volcanic member of the JO can be directly related to a concordant U-Pb age of 162 ± 1 Ma (zircon in plagiogranite) obtained by Saleeby (1987) on the ophiolite.
3. The upper 5 m of the VP succession contain a Northern Tethyan radiolarian assemblage (Fig. 1).
4. All of the overlying strata of the Galice Formation sensu lato contain a Southern Boreal radiolarian assemblage with abundant *Praeparvicingula* and a common-to-rare, poorly diversified pantanelliid assemblage characterized by the presence of *Gorgansium* Pessagno and Blome (Fig. 1). In addition, the boreal bivalve *Buchia concentrica* was reported in Galice sensu lato strata by Diller (1907), Harper (1983), and Pessagno et al. (1993a) at localities along Shelly and Monkey Creeks to the north of the site along the Middle Fork of the Smith River (Figs. 2 and 9). No paleomagnetic data are presently available from this succession.
5. Faunal data from strata within the volcanic member of the JO and above the JO indicate that the ophiolite originated at Central Tethyan paleolatitudes during the latest Callovian (ca. 162 Ma) and was carried northward to Northern Tethyan and Southern Boreal paleolatitudes during the early and middle Oxfordian (Figs. 1 and 2). If the JO formed at a back-arc spreading center, as suggested by Saleeby et al. (1982), it is likely that the rate of movement northward was not much greater than that of the continent. It is conceivable that the ophiolite originated near the boundary between the Central and Northern Tethyan provinces. The presence of similar tuffaceous cherts both within the volcanic member of the ophiolite and within the volcanopelagic succession tends to support a back-arc or forearc origin for the JO.
6. Pessagno et al. (1993a) observed that a dramatic difference in preservation occurs between radiolarian specimens extracted from pelagic limestone within the VP succession and from the overlying Galice sensu lato Radiolaria from the VP succession tend to be considerably more recrystallized and deformed than those from the overlying Galice. These easily observable differences in the preservation of the Radiolaria suggest that the JO and VP succession had been exposed to a period of diagenetic change (metamorphism?) prior to the deposition of the Galice. The dramatic difference in radiolarian preservation above and below the VP–Galice sensu lato contact taken together with the sudden change from hemipelagic (including calc-alkaline tuff and tuffaceous chert) sedimentation (VP) to siliciclastic turbidite sedimentation (Galice sensu lato) suggest that this contact is unconformable (Fig. 4). Galice sensu lato strata contain a well-preserved, poorly diversified Southern Boreal radiolarian assemblage with rare pantanelliids (Fig. 1). The abrupt shift from VP strata containing Northern Tethyan radiolarians to Galice sensu lato strata containing higher latitude Southern Boreal faunas may suggest that the unconformity noted above is accompanied by a slight hiatus.

San Pedro del Gallo Terrane

Introduction. The downfall of many tectonostratigraphic studies in Mexico in recent years has been the failure of workers to familiarize themselves with nomenclatural problems surrounding Mexican lithostratigraphic units and to examine these units in the field at their type localities (Pessagno and Martin, 2003). Investigations by Pessagno et al. (1984, 1987, 1993b, 1999) revealed major discrepancies between the description of rock units in the literature and actual field observations.

New Jurassic and Early Cretaceous stratigraphic data (Martin, 1996; Meng, 1997; Pessagno et al., 1999) from terranes (e.g., San Pedro del Gallo terrane: Pessagno et al., 1993a, 1999, 2001; Parral terrane: Campa, 1983) in central Mexico situated southwest of the Walper megashear demonstrate similar records of paleobathymetry and tectonic transport (Figs. 11 and 12). As demonstrated by Pessagno et al. (1999), each remnant of the San Pedro del Gallo terrane shows the same paleobathymetric fingerprint (Fig. 11) that includes: (1) marine deposition at inner neritic depths during the Callovian to early Oxfordian (Middle to Late Jurassic); (2) marine deposition at outer neritic depths during the late Oxfordian (Late Jurassic); (3) sudden deepening to bathyal or upper abyssal depths (i.e., below aragonite compensation depth [ACD]) from the early Kimmeridgian (Late Jurassic) until the end of the Cretaceous. Although varying in detail, each remnant of the San Pedro del Gallo terrane that has been examined in the region to the southwest of the Walper megashear shows evidence of tectonic transport from higher to lower latitudes during the late Middle Jurassic, Late Jurassic, and Early Cretaceous (Pessagno et al., 1999, Pessagno and Martin, 2003; Figures 4, 11, and 13). For example, the paleolatitudinal signature of the San Pedro del Gallo remnant of the San Pedro del Gallo terrane (Durango) supplied by faunal data (Radiolaria and megafossils) and preliminary paleomagnetic data indicates that this terrane was transported tectonically from higher paleolatitudes (Southern Boreal province: ~40° N) during the Late Jurassic (Oxfordian) to lower paleolatitudes (Tethyan realm: Northern Tethyan province) by the Early Cretaceous (Berriasian). The Jurassic and Lower Cretaceous successions at Mazapil (Zacatecas), Sierra de la Caja (Zacatecas), Sierra de Zuloaga and adjacent Sierra de Sombretillo (Zacatecas), Symon (Durango), and Sierra de Catorce (San Luis Potosi) are all genetically related to that at San Pedro del Gallo (Figs. 12

and 13). These successions represent dismembered remnants of the San Pedro del Gallo terrane. Farther to the southeast (San Luis Potosi, Hidalgo, Veracruz, and Puebla) in the Huayacocotla segment of the Sierra Madre Oriental, previous investigations indicated tectonic transport from Southern Boreal paleolatitudes (>30° N) during the Callovian to Northern Tethyan paleolatitudes (22–29° N) during the Kimmeridgian and Tithonian and to Central Tethyan paleolatitudes (<22° N) during the latest Tithonian (Late Jurassic) and the Berriasian (Early Cretaceous; Pessagno et al., 1984, 1987, 1993b, 1999; Pessagno and Martin, 2003).

Huayacocotla Remnant. In this area, the Mesozoic succession begins with the deposition of the Lower Jurassic (Sinemurian to lower Pliensbachian) Huayacocotla Formation, which consists of black shale, mudstone, and graywacke. A rich ammonite assemblage occurs in the lower and middle parts of the unit (Burckhardt, 1930–1931; Erben, 1956; Imlay, 1980). Bivalves and land plants have been recorded from the upper part.

As noted by Pessagno et al. (1999) and Pessagno and Martin (2003), the Huayacocotla Formation overlies the Mississippian to Lower Permian strata of the Guacamaya Formation with a marked hiatus associated with an angular unconformity. It is significant that Nestell (1979) recovered fusulinids with South American affinities from the Guacamaya Formation. Fusulinids with South American affinities also occur in the Guacamaya Formation at Peregrina Canyon near Ciudad Victoria (Tamaulipas). The thickness of the unit varies from 560 to 1200 m.

As indicated by Pessagno et al. (1999), all references to the presence of the Upper Triassic Huizachal Formation (continental red beds) in this area are erroneous. The Huizachal has largely been confused with the Middle Jurassic Cahuasas Formation, which also consists of continental red beds (cf., Imlay, 1980).

The Cahuasas Formation (Figs. 4, 11, and 14) consists of 40–1200 m of red arkosic sandstone, conglomerate, and shale that rest with angular unconformity on the Huayacocotla Formation (Imlay, 1980, p. 49). Imlay indicated that the Cahuasas must be older than Callovian because where it crops out on the surface, it lies disconformably below marine beds of early to middle Callovian age. In the subsurface, the Cahuasas underlies latest Bathonian to early Callovian marine shale of the Palo Blanco Formation (see Palo Blanco below). Imlay indicated that the Cahuasas must be younger than Toarcian (Early Jurassic), in that it passes downward into plant-bearing beds that are early Middle Jurassic.

The Cahuasas Formation is overlain disconformably in surface outcrops by the inner neritic early Callovian Tepexic Limestone. The Tepexic is a calcarenite containing common-to-abundant *Liogryphaea nebrascaensis* and such ammonites as *Neuqueniceras neogaeum* and *Reineckeia* (Cantú-Chapa, 1969; Imlay, 1980, p. 50). In the subsurface and at some surface localities, an inner neritic black shale unit, the Palo Blanco Formation (Cantú-Chapa, 1969, p. 5; Imlay, 1980, p. 49), underlies the Tepexic Limestone and rests disconformably on the Cahuasas. The Palo Blanco Formation contains the late Bathonian to early Callovian boreal ammonite *Kepplerites* (Cantú-Chapa, 1969, p. 5; Imlay, 1980). In the United States, this ammonite has been observed no farther south than east-central Oregon, Montana, and Wyoming (Imlay, 1980).

As indicated in Figure 14, the Tepexic Limestone is overlain conformably by silty black shale, siltstone, and silty micritic limestone constituting the Santiago Formation (middle Callovian to upper Oxfordian). The lower and middle parts of the Santiago contain bivalves (e.g., small *Ostrea;* personal observations) and ammonites. Microfacies analysis by Longoria (1984) indicated that most of the strata of the Santiago Formation were deposited at inner neritic depths. The uppermost (upper Oxfordian) part of the Santiago Formation (e.g., at Taman, San Luis Potosi) contains common Radiolaria as well as ammonites (Pessagno et al., 1987). These Santiago strata reflect the same sudden change in water depths from inner neritic to outermost neritic during the late Oxfordian that was recorded by Pessagno et al. (1999) in the San Pedro del Gallo and Mazapil remnants. The Santiago Formation is disconformably overlain by the Taman Formation (sensu Pessagno et al., 1984, 1987).

Pessagno et al. (1987) divided the Taman Formation (thickness ~30–60 m) into two informal units (Figs. 4, 11, and 14): (1) a massively bedded to medium-bedded micritic limestone member (lower Kimmeridgian to upper Tithonian) and (2) a thin-bedded micritic limestone member (upper Tithonian). Pessagno et al. (1984, 1987) noted that both members of the Taman Formation contain profusely abundant and highly diversified Radiolaria, rare foraminifera (chiefly Textulariina), common siliceous sponge spicules, common pectenacids (*Aulcomyella*), occasional ammonite aptychi, and rare ammonites. Pessagno et al. (1999) concluded that the abundance of Radiolaria, together with the sparse benthonic foraminiferal assemblage and the rarity of ammonites (aragonite), suggests that Taman strata were deposited at upper abyssal depths at or somewhat below the ACD (see the microfacies analysis in Longoria, 1984).

As shown in Figures 4, 11, and 14, the Taman is overlain conformably by the latest Tithonian (Late Jurassic) to Berriasian (Early Cretaceous) Pimienta Formation (sensu Pessagno et al., 1984, 1987) and overlain conformably by the Chapulhuacán Limestone (Berriasian to Valanginian). Pessagno et al. (1999) indicated that the Pimienta Formation consists of 200–400 m of light-gray, thin-bedded micritic limestone with thick shale intervals, thin-bedded black radiolarian chert, and light-green vitric tuff. In addition, it contains abundant Radiolaria, calpionellids, siliceous sponge spicules, and common ammonites (Fig. 12). Pessagno et al. (1999) concluded that Pimienta deposition likewise took place at upper abyssal depths somewhat above the ACD (Fig. 12).

Pessagno et al. (1999) indicated that the Chapulhuacán Limestones consists of ~30 m of medium to massively bedded, very fine-grained, cream to light-gray micrite with frequent nodules of black chert. As noted in Figure 12, this unit contains abundant Radiolaria, calpionellids, nannoconids, and planktonic

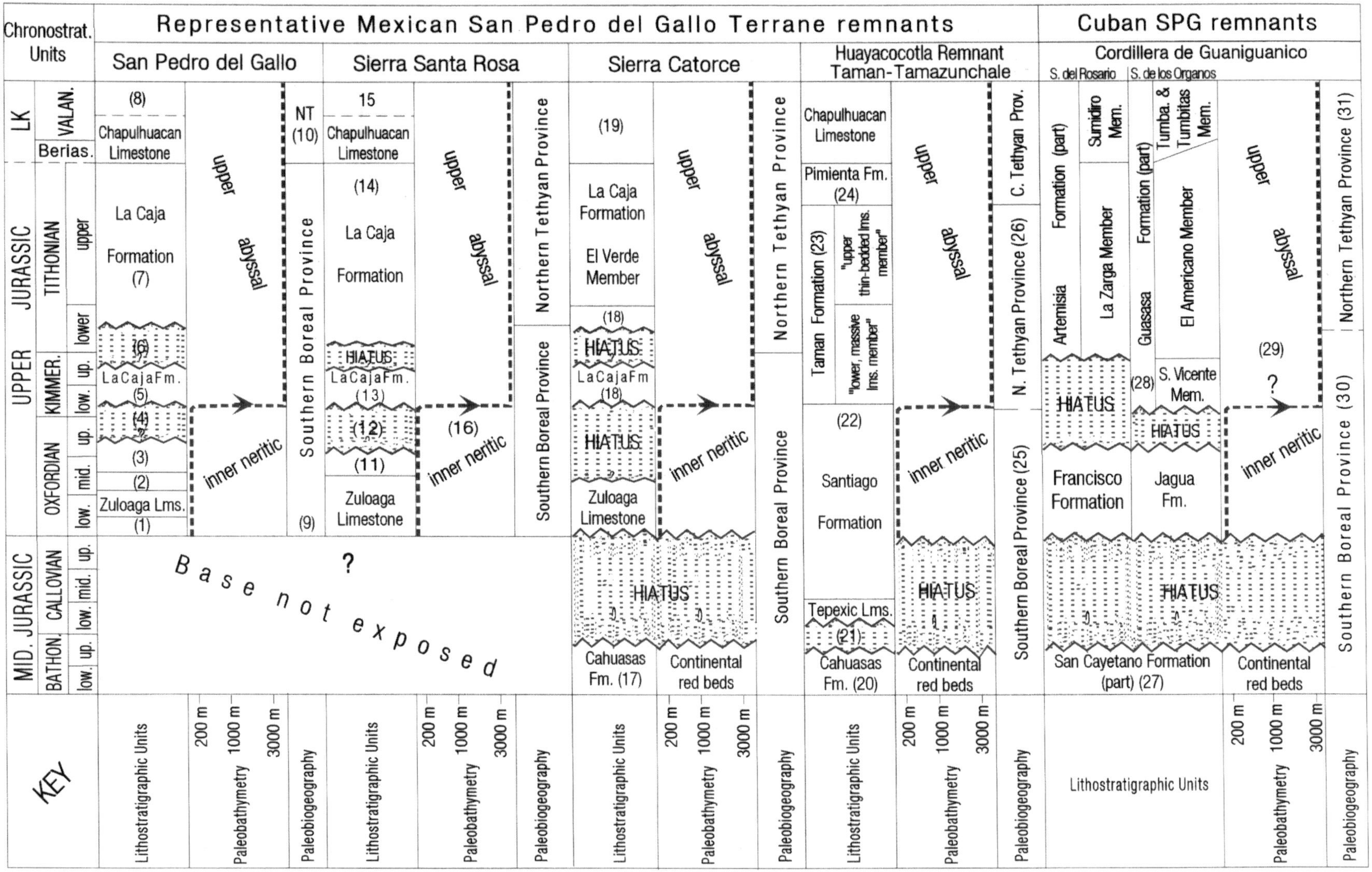

Figure 11. Correlation of lithostratigraphic and chronostratigraphic units with paleobathymetry and paleobiogeographic position in selected San Pedro del Gallo terrane remnants in Mexico and Cuba. All mentions of figures in this caption refer to those in Pessagno et al. (1999). Key: (1) "lower quartzite unit" of Burckhardt (1910). Unfossiliferous massively bedded, white to pink sandstone. (2) "upper quartzite unit" of Burckhardt (1910). Unfossiliferous massively bedded sandstone. Overlies Burckhardt's "nerineid limestone" = Zuloaga Limestone of Imlay (1938). (3) Unnamed pink silty limestone, mudstone, and siltstone. Contains *Buchia,* common ammonites, and Radiolaria (upper part only). Chronostratigraphically significant megafossils include the ammonites *Dichotomosphinctes* and *Discosphinctes* and the *Buchia concentrica* (middle to upper Oxfordian). (4) Lower Kimmeridgian *Ataxioceras* Zone and probably part of *Idoceras* Zone (ammonites) missing. Upper part of upper Oxfordian probably missing. Regional unconformity in much of western North America. Corresponds approximately to onset of deposition of flysch during middle Oxfordian times in Klamath Mountains of northwestern California and southwestern Oregon (Galice Formation) and in Sierra Nevada (Mariposa Formation, Monte del Oro Formation). See Pessagno et al. (1993a). Possibly reflecting pre-Nevadian orogenic pulse in backarc domain. (5) "Lower shale member" of La Caja Formation = lower part of "Capas de San Pedro" of Burckhardt (1910). Dark gray calcareous to siliceous mudstone with micrite nodules containing abundant Radiolaria, common ammonites, and *Buchia.* Basal strata assignable to *Idoceras* Zone (upper half of

lower Kimmeridgian) and radiolarian Subzone 2 alpha-1 (Meng, 1997). (6) Regional unconformity in western North America, recognizable in Nevadian back-arc terranes (e.g., all San Pedro del Gallo remnants in Mexico and in Cuba) and in Nevadian forearc terranes (e.g., volcanopelagic [VP] strata overlying Stanley Mountain remnant of Coast Range ophiolite, San Luis Obispo County, California, and Point Sal remnant of Coast Range ophiolite, Santa Barbara County, California). See Hull (1991, 1995); Hull et al. (1993); Hopson et al., (1996). (7) Includes upper part of "Capas de San Pedro" of Burckhardt (1910) and "chert," "upper shale," and "Cerro Panteon quarry unit 2" members of La Caja Formation herein (See Figs. 8 and 10 for more detailed litho description, biostratigraphic data, and chronostratigraphic data). Note that the La Caja Formation at San Pedro del Gallo was miscorrelated lithostratigraphically by Imlay (1939) with his La Casita Formation. All members of La Caja Formation at San Pedro del Gallo show abundant Radiolaria. Sudden influx of silty wacke at Cerro Panteon and at La Peña (10 km. north of San Pedro del Gallo) reflects onset of Nevadian orogeny. Contact of Great Valley Supergroup (flysch) and underlying VP sequence at Stanley Mountain above Stanley Mountain remnant of Coast Range Ophiolite occurs in lower part of Subzone 4 alpha (uppermost upper Tithonian). At San Pedro del Gallo, equivalent strata contain the ammonite *Durangites* and *Buchia piochii.* (8) Unnamed thin-bedded, tan to pink mudstone and micrite with common ammonites and calcified Radiolaria. These strata overlie the massive to medium bedded tan micrites of the Chapulhuacan Formation (type area in Taman-Tamazunchale area) to southeast. Imlay (1937) miscorrelated these strata with shallow neritic Taraises Formation (type area, Sierra de Parras). (9) Paleomagnetic data ("upper quartzite unit") from Ogg indicates ~40° N or S of Jurassic paleoequator (Pessagno, 1995). Presence of the Boreal megafossils *Buchia concentrica* and *Amoeboceras* sp. in overlying Oxfordian strata associated with such Tethyan ammonites as *Dichotomospinctes* indicates Southern Boreal Realm. Overlying La Caja Formation with *Buchia concentrica, B. rugosa, B. mosquensis,* and *B. piochii* associated with Southern Boreal radiolarian assemblage characterized by high diversity and abundance of *Parvicingula* and *Praeparvicingula.* Upper part of La Caja Formation with *Buchia* associated with *Parvicingula/Praeparvicingula* and abundant calpionellids (Tethyan: See Pessagno et al., 1996). Late Tithonian portion of La Caja Formation formed approximately at boundary between Boreal Realm and Tethyan Realm. (10) Abundant calpionellids, together with lack of *Buchia* and presence of only Tethyan ammonites, suggests Northern Tethyan Province (Meng, 1997). (11) Units G and F of Figure 13. (12) Regional unconformity noted in (4) above. (13) Unit E (Fig. 11). The discovery of the lower Kimmeridgian ammonite *Idoceras* in a limestone nodule 1.5 m below contact between units E and F indicates that the uppermost part of unit F is lower Kimmeridgian (identification by Dr. Cantú-Chapa, Instituto Politecnico Nacional, Mexico); the middle Oxfordian ammonite *Dichotomosphinctes* was recovered 7 m below this horizon (identification by Dr. Cantú-Chapa, Instituto Politecnico Nacional, Mexico). (14) Units E (part)–B of Figure 13. Note that silty wacke occurs in upper part of Unit B in lower Subzone 4 alpha. This horizon occurs below final occurrence of *Durangites* and *Substeueroceras;* identification by Dr. Cantú-Chapa, Instituto Politecnico Nacional, Mexico). (15) Unnamed limonitic mudstone and limestone of Burckhardt (1930–1931 = Burckhardt's unit B). (16) Change from inner neritic to outer neritic occurs in upper part of unit F (Fig. 13). Martin (1996) noted the first occurrence of common Radiolaria in upper unit F. This horizon may also correspond to regional unconformity noted in 12 above. (17) Called "La Joya Formation" by Imlay (1980). Lower Jurassic strata below this unit contain ammonites and probably are equivalent to the Huayacocotla Group. See Imlay (1980). (18) El Pastor Member of La Caja Formation (Verma and Westermann, 1973). Massively bedded medium-gray micrite with thin-bedded black radiolarian chert and wacke. Wacke often with displaced shallow neritic megafossils. Overlying El Verde member consisting of thin-bedded dark-gray micrite and black radiolarian chert together with wacke. Graded-bedding and displaced shallow water fossils noted in wacke. (19) Incorrectly correlated with shallow neritic La Taraises Formation by Verma and Westermann (1973). (20) Overlies Huayacocotla Group in Huayacocotla remnant. (21) Cantú-Chapa (1969) recovered the Boreal ammonite *Kepplerites* in the subsurface of the Huayacocotla remnant from the shallow neritic Palo Blanco Formation. This ammonite is common in terranes in the Sierra Nevada, in the Izee Terrane of east-central Oregon, and in western terranes north to Alaska. It is indicative of the uppermost Bathonian or lower Callovian (Imlay, 1980). On the surface, the Cahuasas Formation is overlain by the Tepexic calcarenite (See Fig. 13 for more detailed description). (22) All but uppermost part of Santiago Formation contain an inner neritic molluscan assemblage. Common Radiolaria (including *Praeparvicingula*) first occur near top of unit (Pessagno et al., 1987). (23) Taman Formation characterized by abundant Radiolaria, common pectenacids, and rare ammonites. The rarity of ammonites suggests deposition in the upper abyssal depth zone, just below the depth of composition of aragonite. See Pessagno et al. (1987). Co-occurring throughout the Taman Formation in the area south of Taman (e.g., near Huauchinango Puebla) are discontinuous masses of inner neritic calcarenite (San Andres Member; Cantú-Chapa, 1969, 1971; Imlay, 1980) that either represent shallow neritic carbonate sedimentation on seamounts or turbidites. Definition of Taman Formation following that of Pessagno et al. (1987). (24) Pimienta Formation differs from Taman Formation by consisting of light to medium-gray, thin-bedded micrite interbedded with black radiolarian chert and occasional layers of green vitric tuff. (25) See annotation 21 above. Taman Formation with rich Northern Tethyan radiolarian assemblage, including *Parvicingula, Praeparvicingula,* and abundant diversified pantanelliids associated with Tethyan ammonites and calpionellids (Tithonian). See Pessagno et al. (1987). (26) Pimienta Formation and overlying Chapulhuacan Limestone with abundant calpionellids, Tethyan ammonites, and lacking *Parvicingula/Praeparvicingula.* Central Tethyan Province. (27) Data from Haczewski (1976), Imlay (1980), Lewis and Draper (1990), and Myczynski and Pszczólkowski (1994). Imlay indicates that marine bivalves of probable Middle Jurassic age occur in upper part of this unit. (28) San Vincente Member (Myczynski, 1994) is anomalous and appears to be an analogue of San Andres Member of Taman Formation in Huayacocotla remnant. See annotation 23 above. (29) Radiolarian-rich strata comprising all of the Artemisa Formation and most of Guasasa Formation (except for San Vincente; see annotation 28 above) were deposited at upper abyssal depths either above or slightly above the calcite compensation depth of aragonite. (30) Presence of *Parvicingula/Praeparvicingula* in the radiolarian assemblage associated with Tethyan ammonites and *Buchia* (Myczynski, 1994) indicates that the Cuban San Pedro del Gallo terrane remnants were at Southern Boreal paleolatitudes. (31) Northern Tethyan paleolatitudes indicated by same association as in annotation 30, but with common to abundant calpionellids (Myczynski and Pszczólkowski, 1994). From Pessagno et al. (1999).

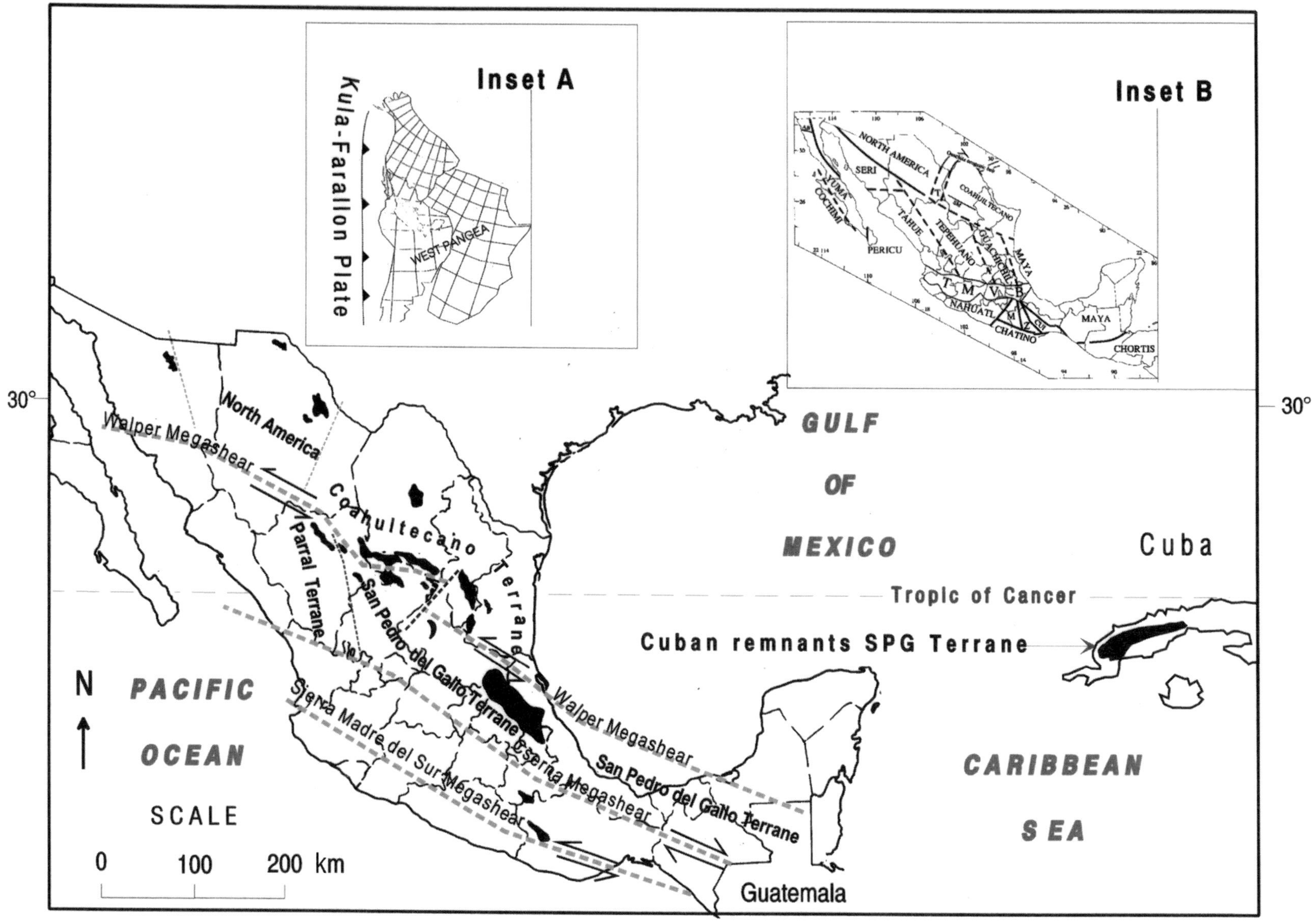

Figure 12. Map showing distribution of terranes referred to in text. Distribution of megashears largely follows that of Longoria (1984). Offset of Walper megashear differs, based on re-analysis of stratigraphic data during current study. Inset A, view of western Pangea showing geometric fit of Atlantic continents. The position of Mexico is outlined to illustrate overlap with South America. Approximately 70–80% of Mexico is overlapped by South America. Most workers utilize large strike-slip faults to bring Mexico to its present overlap position in their plate-tectonic reconstructions. Inset B, terrane map of Mexico of Sedlock et al. (1993). From Pessagno et al. (1999).

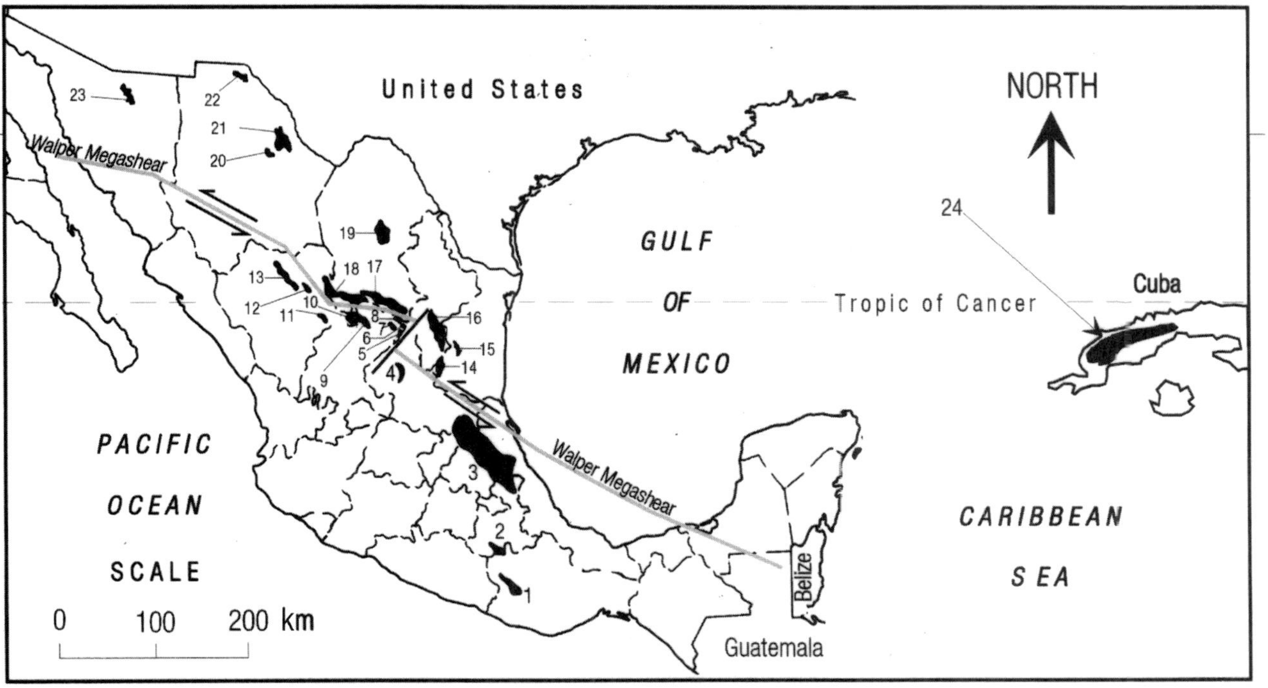

Figure 13. Index map showing important Jurassic localities in Mexico and Cuba. The most important localities for this report are 3–18, 24. Key to localities: (1) Tlaxiaco: Sierra Madre del Sur, Oaxaca. (2) Pletalcingo: Sierra Madre del Sur, Puebla. (3) Huayacocotla anticlinorium: Taman-Tamazunchale, San Luis Potosi; Huayacocotla, Veracruz; Huachinango, Puebla. (4) Sierra Catorce, San Luis Potosi. (5) Sierra Santa Rosa, Zacatecas. (6) Sierra de la Caja, Zacatecas. (7) Sierra Cadnelaria, Zacatecas. (8) Sierra Sombretillo and Sierra Zuloaga, Zacatecas. (9) Sierra de Ramirez, Zacatecas-Durango. (10) Sierra de Chivo, Durango. (11) Sierra de Palotes, Durango. (12) San Pedro del Gallo, Durango. (13) Santa Maria del Oro, Sierra de La Zarca, Durango. (14) Sierra Vieja-Arroyo Doctor, Tamaulipas. (15) Huizachal Anticlinorium, Tamaulipas. (16) Sierra Galeana-Iturbide, Nuevo Leon. (17) Sierra de Parras, Coahuila. (18) Sierra de Jimulco, Coahuila. (19) Sierra Menchaca, Cohuila. (20) Sierra Plomosas-Place de Guadalupe, Chihuahua. (21) Sierra El Cuchillo Parado, Chihuahua. (22) Sierra de Samalayuca, Chihuahua. (23) Sierra de Cucurpe, Sonora. (24) Cordillera de Guaniguanico, Cuba. Base map partly derived from that in Salvador et al. (1992). Modified from Pessagno et al. (1999).

foraminifera, and rare ammonites at most localities (Pessagno's observations and those of Longoria, 1984). Chapulhuacán strata were also deposited at upper abyssal depths somewhat above the ACD (Fig. 12). Deposition continued at these depths during the remainder of the Cretaceous in the Huayacocotla remnant as well as all other remnants of the San Pedro del Gallo terrane.

The following conclusions can be drawn from abundant faunal data derived from the Jurassic and Cretaceous strata of the Huayacocotla remnant of the San Pedro del Gall terrane:

1. The upper Bathonian (Middle Jurassic) to upper Oxfordian (Upper Jurassic) part of the succession contains such Boreal megafossils as the ammonite *Kepplerites* in the Palo Blanco Formation (Cantú-Chapa, 1969; Imlay, 1980; Pessagno et al., 1999). In western North America, this ammonite is known from Middle Jurassic strata from Southern Boreal strata in the upper Bathonian part of the Snowshoe Formation, Izee terrane, east-central Oregon, and from Boreal Middle Jurassic strata as far north as Alaska (see Imlay, 1980; Pessagno and Blome, 1986; Pessagno et al., 1986, 1987, 1999).
2. The lower Kimmeridgian to upper Tithonian (Upper Jurassic) part of the succession (Taman Formation sensu Pessagno et al., 1984, 1987) contains a rich Northern Tethyan radiolarian assemblage characterized by the abundance and diversity of pantanelliids and by the presence of common-to-abundant *Parvicingula* and *Praeparvicingula.* Calpionellids (Tethyan) occur in the upper Tithonian part of the Taman Formation (see Pessagno et al., 1987; Figs. 1 and 2). Moreover, the megafossil assemblage is Tethyan in aspect (see Imlay, 1980; Cantú-Chapa, 1989; Figs. 4 and 14).
3. The Pimienta Formation (as well as the overlying Chapulhuacán Limestone) is characterized by a Tethyan ammonite assemblage and by a microfossil assemblage including abundant calpionellids and nannoconids lacking the radiolarian genera *Parvicingula/Praeparvicingula.* This association of faunal elements is indicative of the Central Tethyan province (Figs. 1 and 2).
4. These data indicate that the Huayacocotla remnant of the San Pedro del Gallo terrane underwent tectonic transport from Southern Boreal paleolatitudes (>30° N) during the late Bathonian (Middle Jurassic) to Northern Tethyan paleolatitudes by the early Kimmeridgian (Late Jurassic) to Central

Litho Unit	Description	Age	Diagnostic Faunal Elements	Paleobathymetry	Faunal Realm/Province
Chapulhuacan Limestone	Medium to massively bedded very fine-grained cream to light gray micrite with abundant Radiolaria, calpionellids, and rare ammonites. Micrite with black chert nodules and lenses. Thickness = ~ 30 m.	Berriasian to Valanginian	*Subthurmannia* sp., *Neolissoceras* sp., *Spiticeras* sp., *Thurmanniceras* sp.	Abyssal	Central Tethyan Province
Pimienta Formation	Thin-bedded cream colored to light gray micrite interbedded with dark gray shale, common black radiolarian chert, and light green vitric tuff. Micrite with abundant Radiolaria and calpionellids together with rare ammonites and common sponge spicules. Thickness = ~ 200 m.	late Tithonian	*Paradontoceras* aff. *callistoides* *Durangites*, *Substeueroceras* + Subzone 4α Radiolaria at some localities to south		
Taman Formation T = 200-500 m: "upper thin-bedded limestone member"	Thin-bedded dark gray to medium gray micrite with thick interbeds of dark to medium gray shale. Shale layers with abundant micrite nodules. Micrite nodules and beds with abundant Subzone 4β Radiolaria siliceous sponge spicules and common ammonites.	late Tithonian	*Durangites*, *Kossmatia*, *Salinites grossicostatum* + very abundant Subzone 4β Radiolaria.	Upper	Northern Tethyan Province
Taman Formation T = 200-500 m: "lower massively bedded limestone member"	Massively bedded dark gray to medium gray micrite interbedded with thin-bedded to medium bedded shale. Upper part of unit with numerous dark gray micrite nodules. Massive micrites and micrite nodules with abundant Radiolaria rare to common ammonites and common pectenacids (*Aulacomyella*). Hyaline calpionellids occur in basal Zone 4, Subzone 4β strata at same horizon as lower Tithonian ammonite *Mazapilites*.	early Kimmeridgian to early Tithonian	*Ataxioceras*, *Idoceras*, *"Glochiceras fialar"* *Mazapilites*, *Virgatosphinctes* + Radiolaria assignable to Subzone 2α1, Zone 3, & Subzone 4β		
Santiago Formation	Silty black shale, mudstone, micrte. Containing ammonites. Radiolaria occurring in upper-most part. Thickness = ~169 m.	early Callovian to late Oxfordian	*Reineckeia*, *Dichotomosphinctes*, *Discosphinctes*, *Ochetoceras*	Outer Neritic / Inner Neritic	Northern Tethyan Province / Southern Boreal Province
Tepexic Limestone	Calcarenite containing ammonites and bivalves. Thicknes = ~ 39 m.	late early Callovian	*Reineckeia* *Neuqueniceras*	Inner Neritic	Southern Boreal Province
Cahuasas Formation	Continental red beds. Dominantly red shale, siltstone, sandstone, and conglomerate. Commonly cross-bedded. Overlies Lower Jurassic (Sinemurian) strata of Huayacocotla Group. The latter unit contains Sinemurian ammonites. Thickness = 40-1200 m. (1)	Bathonian to Bajocian	Fossil plants	NON MARINE	

Figure 14. Stratigraphic summary for Taman-Tamazunchale area, Huayacocotla remnant (part), San Luis Potosi, Mexico. In the subsurface, the Cahuasas Formation is overlain by the marine (inner neritic) black shale of the Palo Blanco Formation (Cantú-Chapa, 1969). The Palo Blanco Formation is important because it is the first marine strata in the Gulf of Mexico province overlying Middle Jurassic continental red beds. This unit also contains the Boreal upper Bathonian to lower Callovian ammonite *Kepplerites*. In most surface sections, the Palo Blanco is missing and the lower Callovian Tepexic Limestone overlies the Cahuasas. From Pessagno et al. (1999).

Tethyan paleolatitudes (<22° N) during the latest Tithonian (Late Jurassic; Figs. 1, 2, 4, and 11). It should be noted that the San Pedro del Gallo remnant of the San Pedro del Gallo terrane remained at Southern Boreal paleolatitudes until the end of the Jurassic (late Tithonian). Pessagno et al. (1999) showed that the macrofossil and radiolarian assemblage coupled with preliminary paleomagnetic data indicate that this terrane remnant originated at Southern Boreal paleolatitudes (~40° N; Pessagno et al., 1999) during the Oxfordian. The appearance of abundant calpionellid, coupled with the presence of *Buchia* and the presence of common *Parvicingula* and *Praeparvicingula* demonstrate that the San Pedro del Gallo remnant had been transported close to the boundary (~30° N) between the Northern Tethyan and Southern Boreal provinces by the latest Tithonian (Late Jurassic). The lack of such Boreal elements as *Buchia* in overlying Early Cretaceous strata indicates transport of the San Pedro del Gallo remnant to the Northern Tethyan province (between 22° and 30° N) by the Berriasian.

5. As suggested by Pessagno et al. (1999) and Pessagno and Martin (2003), the abrupt change in paleobathymetry from neritic to upper abyssal depths between the middle Oxfordian and early Kimmeridgian in the Huayacocotla succession and, indeed, all San Pedro del Gallo remnants, is a direct reflection of the opening up of the Gulf of Mexico (Fig. 11).

SUMMARY AND CONCLUSIONS

1. Tectonic transport of Jurassic terranes during the Jurassic occurred along megashears analogous to the present-day San Andreas fault system. Data presented in this chapter indicate both south-to-north and north-to-south displacement.
2. Faunal data and paleomagnetic data (where available) from terranes in the Blue Mountains province of east-central Oregon and western Idaho (e.g., Izee terrane) indicate an origin in the Central Tethyan province during the Late Triassic (Carnian). The Izee terrane, for example, was gradually transported northward to the Northern Tethyan province (~22–29° N; see Figs. 1, 2, and 4) by the middle Toarcian (Early Jurassic; Pessagno and Blome, 1986). Faunal paleolatitudinal data from the Izee terrane indicate that it resided in the Northern Tethyan province until the late Bajocian (Middle Jurassic). The rapid decline in pantanelliid diversity (Radiolaria) during the late Bajocian suggests an acceleration of tectonic transport along a major megashear (presently unnamed) and reflects the beginning of the breakup of Pangea. Both radiolarian and ammonite faunal data clearly indicate that the Izee terrane had been transported to the Southern Boreal province (>30° N; Pessagno and Blome, 1986) by the late Bathonian (Middle Jurassic).
3. That no marine Early Bathonian strata are known from the Izee terrane suggests the presence of an unconformity and associated hiatus (e.g., South Fork Member of Snowshoe Formation; Fig. 4). A definite disconformity and hiatus of similar magnitude occurs in between Bajocian and upper Bathonian to lower Callovian strata in the Huayacocotla remnant of the San Pedro del Gallo terrane. Elsewhere in North America, early Bathonian strata are known from: (1) northeastern Guerrero and northwestern Oaxaca, Mexico (Imlay, 1980) in an area situated between the Sierra Madre del Sur megashear and the Cserna megashear of Longoria (1984; Fig. 12); (2) the Western Interior of the United States (Imlay, 1980); (3) the Canadian Rocky Mountains (Imlay, 1980); and (4) southern Alaska (Imlay, 1980). As noted by Pessagno et al. (1999), the early Bathonian ammonite assemblage occurring in northeastern Guerrero and northwestern Oaxaca have a strong South American (Andean) affinity (Burckhardt, 1927, 1930–1931; Arkell, 1956; Imlay, 1980; Sandoval and Westermann, 1986; Sandoval et al., 1990; von Hillebrandt et al., 1992). Pessagno et al. (1999) concluded that the failure to find Middle Jurassic ammonite fauna with strong Andean affiliation elsewhere in Mexico and North America suggests that the Middle Jurassic (Bajocian to Callovian) succession in these states has undergone southeast-to-northwest tectonic transport along a more outboard megashear paralleling the Walper megashear (i.e., Cserna megashear of Longoria, 1984; see Fig. 12).
4. Pessagno and Blome (1986) suggested that the Late Triassic phase of volcanism (spilitic basalt flows and andesites of Vester and Fields Creek Formations) represented near-axis island-arc volcanism, whereas the Early Jurassic to Middle Jurassic phase of volcanism represented off-axis island-arc volcanism. Data from the Caribbean demonstrate that such a shift of volcanism indicates a deepening of the subduction zone resulting in the shift of the subduction toward the concave side of the island arc (see Christman, 1953, 1973; Pessagno, 1960).
5. The San Pedro del Gallo terrane was interpreted by Pessagno et al. (1999) to be a back-arc portion of the Nevadan island-arc complex. Paleolatitudinal data derived from abundant microfossils and megafossils indicate that the San Pedro del Gallo terrane (e.g., Huyayacocotla remnant) was transported from northwest to southeast during the Jurassic along the Walper megashear (Fig. 12). In the Huayacocotla remnant of the San Pedro del Gallo terrane, Southern Boreal ammonites (e.g., *Kepplerites*) occur in the upper Bathonian to lower Callovian strata of the Palo Blanco Formation (Cantú-Chapa, 1969; Imlay, 1980). Elsewhere in North America, *Kepplerites* is known from the Izee terrane (east-central Oregon), Western Interior (Montana and Saskatchewan), the Canadian Rockies, and northward to southern Alaska (see Imlay, 1980). *Kepplerites* is also known from Greenland (Imlay, 1980).
6. During the late Bathonian and early Callovian, the Huayacocotla remnant of the San Pedro del Gallo terrane and Izee terrane were both at Southern Boreal latitudes (>30° N).
7. Faunal data indicate that by the late Oxfordian or early Kimmeridgian (Late Jurassic), the Huayacocotla remnant had been transported to the Northern Tethyan paleolatitudes (~22–29° N; see Figs. 1, 2, and 4) and remained at these latitudes until the late Tithonian (Late Jurassic; see Figs. 4 and

12). A shift to Central Tethyan latitudes occurred during the latest Tithonian.

8. The unconformity and hiatus between the Tepexic Limestone or Palo Blanco Formation and the underlying Cahuasas Formation (Fig. 4) are interpreted to reflect the breakup of Pangea.
9. Faunal data indicate that both the Rogue Valley and Smith River subterranes (western Klamath terrane) originated in the Central Tethyan province during the late Callovian and were transported northward to Southern Boreal paleolatitudes by the middle Oxfordian (Late Jurassic; see Figs. 1, 2, and 4).
10. According to Pessagno and Martin (2003), the combined faunal and floral data demonstrate that the San Pedro del Gallo terrane was in a back-arc position at approximately the same latitude as the Foothills terrane of the Sierra Nevada during the middle Oxfordian (Fig. 4). Given the faunal data (radiolarians and macrofossils, e.g., *Buchia,* ammonites) as well as some paleomagnetic data from Ogg (in Pessagno et al., 1999), it is probable that these San Pedro del Gallo remnants might represent some of the missing Upper Jurassic back-arc from farther north (western Nevada?). Even though the San Pedro del Gallo terrane was situated at higher latitudes, Pessagno and Martin (2003) suggest that its stratigraphy reflects the opening of the Gulf of Mexico orogenic event. This thesis is supported by the abrupt change in paleobathymetry from neritic to upper abyssal depths between the middle Oxfordian and early Kimmeridgian in the successions of all San Pedro del Gallo remnants (a direct reflection of this event; Figs. 4 and 12) and by the presence of a prominent disconformity between the Caja Formation and unnamed siltstones in the San Pedro del Gallo and Mazapil remnants (Fig. 12) and between the Taman and Santiago Formations in the Huayacocotla anticlinorium (Figs. 12 and 13). The sudden influx of siliciclastic rocks in the Sierra Nevada (Mariposa and Monte del Oro Formations) and Klamath Mountains (Galice Formation) during the middle Oxfordian may also be a reflection of this event (Imlay, 1980; Pessagno et al., 1993a, 2001; Pessagno and Martin, 2003).

REFERENCES CITED

Aita, Y., and Grant-Mackie, J.A., 1992, Late Jurassic Radiolaria from Kowhai Point Siltstone, Murihku terrane, North Island, New Zealand, *in* Ishizaki, K., and Saito, T., eds., In centenary of Japanese micropaleontology: Tokyo, Terra Scientific Publishing, p. 375–382.

Arkell, W.J., 1956, Jurassic geology of the world: London, Oliver and Boyd, 806 p.

Baumgartner, P.O., 1984, Middle Jurassic–Early Cretaceous low latitude radiolarian zonation based on unitary association and age of Tethyan radiolarites: Eclogae Geologicae Helvetiae, v. 80, p. 831–879.

Blake, M.C., Jr., Engebretson, D.C., Jayko, A.S., and Jones, D.L., 1985, Tectonostratigraphic terranes in southwest Oregon, *in* Howell, D.G., ed., Tectonostratigraphic terranes of the Circum-Pacific region: Houston, Circum-Pacific Council for Energy and Mineral Resources, Earth Sciences Series 1, p. 147–157.

Blome, C.D., 1983, Upper Triassic Capnuchosphaeridae and Capnodocinae (Radiolaria) from east-central Oregon: Micropaleontology, v. 29, p. 11–49.

Blome, C.D., 1984a, Upper Triassic Radiolaria and radiolarian zonation for North America: Bulletins of American Paleontology, v. 84, no. 318, p. 1–98, 11 pls.

Blome, C.D., 1984b, Middle Jurassic (Callovian) Radiolaria from southern Alaska and eastern Oregon: Micropaleontology, v. 30, p. 343–389, pls. 1–16.

Brooks, H.C., 1979, Plate tectonics and geologic history of the Blue Mountains: Oregon Geology, v. 41, p. 71–80.

Buffler, R.T., Watkins, J.S., Sahub, F.J., and Worzel, J.L., 1980, Structure and early geologic history of the deep central Gulf of Mexico Basin, *in* Pilger, R.H., Jr., ed., The origin of the Gulf of Mexico and early opening of the central North Atlantic Ocean: Proceedings of a symposium at Louisiana State University, School of Geoscience, Louisiana State University, Baton Rouge, p. 1–16.

Burckhardt, C., 1910, Estudio geologico de la región de San Pedro del Gallo (Durango): Instituto Geologia de México, Parergones tomo III, v. 6, p. 307–357, 1 carta, 1 lámina, 1 mapa.

Burckhardt, C., 1927, Cefalópodos del Jurásico medio de Oaxaca y Guerrero: Instituto Geologia México Boletín, v. 47, 108 p., 34 pls.

Burckhardt, C., 1930–1931, Étude synthétique sur le Mésozoïque Méxicain: Soc. paléont. Suisse Mémoire, v. 49, p. 1–123, 32 figs. (1930); v. 50, p. 123–280, 22 figs. (1931).

Campa, M.F., 1983, The Mexican thrust belt, *in* Howell, D.G., ed., Tectonostratigraphic terranes of the Circum-Pacific region: Houston, Circum-Pacific Council for Energy and Mineral Resources, Earth Sciences Series 1, p. 299–313.

Cantú-Chapa, A., 1969, Estratigrafía del Jurásico medio-superior del subsuelo de Poza Rica, Ver.: Instituto Mexicana Petróleo Revista v. 1, no. 1, p. 3–9.

Cantú-Chapa, A., 1971, La serie Huasteca (Jurásico medio-superior) del centro este de Mexico: Instituto Mexicana Petróleo Revista v. 3, no. 2, p. 17–40.

Cantú-Chapa, A., 1989, Precisiones sobre el limite Jurasico–Cretacico en el subsuelo del este de México: Revista Mexicana Paleontologia v. 2, no. 1, p. 26–69.

Christman, R.A., 1953, Geology of St. Bartholomew, St. Martin, and Anguilla, Lesser Antilles: Geological Society of America Bulletin, v. 64, p. 65–96.

Christman, R.A., 1973, Volcanic geology of southwestern Antigua, B.W.I., *in* Shagam, R., et al., eds., Studies in Earth and space sciences: Geological Society of America Memoir 132, p. 439–448.

Dickinson, W.R., 1979, Mesozoic forearc basin in central Oregon: Geology, v. 7, p. 166–170, doi: 10.1130/0091-7613(1979)7<166:MFBICO>2.0.CO;2.

Dickinson, W.R., and Coney, P.J., 1980, Plate tectonic constraints on the origin of the Gulf of Mexico, *in* Pilger, R.H., Jr., ed., The origin of the Gulf of Mexico and early opening of the central North Atlantic Ocean: Proceedings of a symposium at Louisiana State University, School of Geoscience, Louisiana State University, Baton Rouge, p. 27–36.

Dickinson, W.R., and Vigrass, L.W., 1965, Geology of the Suplee-Izee area, Crook, Grant, and Harney Counties, Oregon: Oregon Department of Geology and Mineral Industries Bulletin, v. 58, p. 1–109.

Diller, J.S., 1907, The Mesozoic sediments of southwestern Oregon: American Journal of Science, v. 3, p. 401–421.

Erben, H.K., 1956, El Jurásico Inferior de México y sus amonitas: México D. F., XX Congreso Geológico Internacional, 393 p., 41 láminas, 5 mapas.

Gradstein, F.M., Agterberg, F.P., Ogg, J.G., Hardenbol, J., van Veen, P., Thierry, T., and Huang, Z., 1994, A Mesozoic time scale: Journal of Geophysical Research, v. 99, p. 24,051–24,074, doi: 10.1029/94JB01889.

Haczewski, G., 1976, Sedimentological reconnaissance of the San Cayetano Formation; An accumulative continental margin in the Jurassic of western Cuba: Acta Geologica Polonica, v. 26, p. 331–353.

Hagstrum, J.T., and Murchey, B.L., 1996, Paleomagnetism of Jurassic radiolarian chert above the Coast Range ophiolite at Stanley Mountain, California, and its implications for its paleogeographic origins: Geological Society of America Bulletin, v. 108, p. 643–652, doi: 10.1130/0016-7606(1996)108<0643:POJRCA>2.3.CO;2.

Harper, G.D., 1983, A depositional contact between the Galice Formation and a Late Jurassic ophiolite in northwestern California and southwestern Oregon: Oregon Geology, v. 45, p. 3–9.

Harper, G.D., 1984, Josephine ophiolite, northwestern California: Geological Society of America Bulletin, v. 95, p. 1009–1026, doi: 10.1130/0016-7606(1984)95<1009:TJONC>2.0.CO;2.

von Hillebrandt, A., Smith, P., Westermann, G.E.G., and Callomon, J.H., 1992, Ammonite zones of the Circum-Pacific region, *in* Westermann, G.E.G., ed., The Jurassic of the Circum-Pacific: Cambridge, Cambridge University Press, p. 247–272.

Hillhouse, J.W., Gromme, C.S., and Vallier, T.L., 1982, Geomagnetism and Mesozoic tectonics of the Seven Devils volcanic arc in northeastern Oregon: Journal of Geophysical Research, v. 87, p. 3777–3794.

Hopson, C.A., Pessagno, E.A., Jr., Mattinson, J.M., Luyendyk, B.P., Beebe, W., Hull, D.M., Muñoz, I.M., and Blome, C.D., 1996, Coast Range ophiolite as paleoequatorial mid-ocean lithosphere, *in* Dickinson, W.R., Hopson, C.A., and Saleeby, J.B., Alternate origins of the Coast Range ophiolite (California): Introduction and implications: GSA Today, v. 6, no. 2, p. 5–7.

Hull, D.M., 1991, Upper Jurassic radiolarian biostratigraphy of the lower member of the Taman Formation, east-central Mexico and of volcanopelagic strata overlying the Coast Range ophiolite, Stanley Mountain, southern California Coast Ranges [Ph.D. dissertation]: University of Texas at Dallas, 696 p.

Hull, D.M., 1995, Morphologic diversity and paleogeographic significance of the Family Parvicingulidae (radiolaria): Micropaleontology, v. 41, p. 1–48.

Hull, D.M., Pessagno, E.A., Jr., Hopson, C.A., Blome, C.D., and Muñoz, I.M., 1993, Chronostratigraphic assignment of volcanopelagic strata above the Coast Range ophiolite, *in* Dunn, G.C., and McDougall, K.A., eds., Mesozoic Paleogeography of the western United States—II: Los Angeles, Pacific Section, Society of Economic Paleontologists and Mineralogists, v. 71, p. 151–170.

Imlay, R.W., 1937, Geology of the middle part of the Sierra de Parras, Coahuila, Mexico: Geological Society of America Bulletin, v. 48, p. 587–630.

Imlay, R.W., 1938, Studies in the Mexican geosyncline: Geological Society of America Bulletin, v. 49, p. 1651–1694.

Imlay, R.W., 1939, Upper Jurassic ammonites from Mexico: Geological Society of America Bulletin, v. 50, p. 1–77.

Imlay, R.W., 1961, Late Jurassic ammonites from the western Sierra Nevada, California: Washington, D.C., U.S. Geological Survey Professional Paper 374-D, p. D1–D30, 6 pls.

Imlay, R.W., 1964, Upper Jurassic mollusks from eastern Oregon and western Idaho: Washington, D.C., U.S. Geological Survey Professional Paper 483-D, p. 1–21.

Imlay, R.W., 1968, Lower Jurassic (Pliensbachian and Toarcian) ammonites from eastern Oregon and western Idaho: Washington, D.C., U.S. Geological Survey Professional Paper 593-C, p. C1–C51.

Imlay, R.W., 1973. Middle Jurassic ammonites from eastern Oregon: Washington, D.C., U.S. Geological Survey Professional Paper 756, p. 1–100.

Imlay, R.W., 1980, Jurassic paleobiogeography of conterminous United States in its continental setting: Reston, Virginia, U.S. Geological Survey Professional Paper 1062, 134 p.

Imlay, R.W., 1981, Jurassic (Bathonian and Callovian) ammonites in eastern Oregon and western Idaho: Reston, Virginia, U.S. Geological Survey Professional Paper 1142, 24 p.

Irwin, W.P., 1972, Terranes of the western Paleozoic and Triassic belt in the southern Klamath Mountains, California: Washington, D.C., U.S. Geological Survey Professional Paper 800-C, p. C103–C111.

Irwin, W.P., 1985, Age and tectonics of plutonic belts in accreted terranes of the Klamath Mountains, California and Oregon, *in* Howell, D.G., ed., Tectonostratigraphic terranes of the Circum-Pacific region: Houston, Circum-Pacific Council for Energy and Mineral Resources, Earth Sciences Series 1, p. 187–199.

Irwin, W.P., 1989, Terranes of the Klamath Mountains, California and Oregon, *in* Tectonic evolution of northern California: Washington, D.C., American Geophysical Union, 28th International Geological Congress, Field Trip Guidebook T108, p. 19–32.

Jones, D.L., Silberling, N.J., and Hillhouse, J., 1977, Wrangellia—A displaced terrane in northwestern North America: Canadian Journal of Earth Sciences, v. 14, p. 2565–2577.

Kiessling, W., 1995, Palökologische Verwertbarkeit Ober–Jurassich Unterkretazischer Radiolarienfaunen mit Beispielen aus Antarktis, Oman, und Süalpen [Ph.D. dissertation]: Erlangen, Germany, Erlangen University, 465 p.

Kiessling, W., 1999, Late Jurassic radiolarians from Antarctic Peninsula: Micropaleontology, v. 45, suppl. 1, p. 1–96, 15 pls., appendices 1 and 2.

Kiessling, W., and Scasso, R., 1996, Ecological perspectives of Late Jurassic radiolarian faunas from the Antarctic Peninsula, *in* Riccardi, A.C., ed., Advances in Jurassic research: Zürich, Transtech Publications, p. 317–326.

Lewis, J.F., and Draper, G., 1990, Geology and tectonic evolution of the northern Caribbean margin, *in* Dengo, G., and Case, J.E., eds., The Caribbean region: Boulder, Colorado, Geological Society of America, Geology of North America, v. H, p. 77–140.

Longoria, J.F., 1984, Mesozoic tectostratigraphic domains in east-central Mexico, *in* Westermann, G.E.G., ed., Jurassic–Cretaceous biochronology and paleogeography of North America: St. Johns, Newfoundland, Geological Association of Canada Special Paper 27, p. 65–76.

Martin, C.B., 1996, Tectonostratigraphic analysis of the Upper Jurassic through Lower Cretaceous in the Mazapil, Sierra de Jimulco, and San Pedro del Gallo regions of north-central Mexico [M.S. thesis]: Dallas, University of Texas, 148 p.

Meng, X., 1997, Radiolarian biostratigraphy of the Upper Jurassic of San Pedro del Gallo terrane, north-central Mexico, and the Lower Cretaceous of Nooksack Group, Nooksack terrane, northwestern Washington [Ph.D. dissertation]: Dallas, University of Texas, 352 p.

Montgomery, H.A., Pessagno, E.A., Jr., and Muñoz, I., 1992, Jurassic (Tithonian) Radiolaria from La Désirade (Lesser Antilles): Preliminary paleontological and tectonic implications: Tectonics, v. 11, p. 1426–1432.

Montgomery, H.A., Pessagno, E.A., Jr., and Pindell, J.L., 1994a, A 195 Ma terrane in a 165 Ma sea: Pacific origin of Caribbean Plate: GSA Today, v. 4, no. 1, p. 3–6.

Montgomery, H.A., Pessagno, E.A., Jr., Lewis, J.A., and Schellekens, J.H., 1994b, Paleogeography of the Jurassic fragments in the Caribbean: Tectonics, v. 13, p. 725–732, doi: 10.1029/94TC00455.

Myczynski, R., 1994, Caribbean ammonite assemblages from Upper Jurassic–Lower Cretaceous sequences of Cuba: Studia Geologica Polonica, v. 105, p. 91–108.

Myczynski, R., and Pszczólkowski, A., 1994, Tithonian stratigraphy and microfacies in the Sierra del Rosario, western Cuba: Studia Geologica Polonica, v. 105, p. 7–38.

Nestell, M.K., 1979, Lower Permian fusulinids from the vicinity of Tianguistengo, Hidalgo, Mexico: Texas Academy of Sciences Meeting, Arlington, Texas, 1978, Program of Abstracts, p. 40.

Ogg, J.G., 1983, Magnetostratigraphy of the Upper Jurassic and lowest Cretaceous sediments, Deep Sea Drilling Site 534, western North Atlantic, *in* Orlofsky, S., ed., Initial reports of the Deep Sea Drilling Project: Washington, D.C., U.S. Government Printing Office, v. 76, p. 685–697.

Pessagno, E.A., Jr., 1960, Geology of the Ponce-Coamo area, Puerto Rico [Ph.D. dissertation]: Princeton, New Jersey, Princeton University, 147 p.

Pessagno, E.A., Jr., 1995, Stratigraphic evidence for southwest to northeast movement along the west side of the Walper Lineament: Geological Society of America Abstracts with Programs, v. 27, p. A-75.

Pessagno, E.A., Jr., and Blome, C.D., 1980, Upper Triassic and Jurassic Pantanelliinae from California, Oregon and British Columbia: Micropaleontology, v. 26, p. 225–273, 11 pls.

Pessagno, E.A., Jr., and Blome, C.D., 1982, Bizarre Nassellariina (Radiolaria) from the Middle and Upper Jurassic of North America: Micropaleontology, v. 28, p. 289–318.

Pessagno, E.A., Jr., and Blome, C.D., 1986, Faunal affinities and tectonogenesis of Mesozoic rocks in the Blue Mountains province of eastern Oregon and western Idaho, *in* Vallier, T.L., and Brooks, H.C., eds., Geology of the Blue Mountains region of Oregon, Idaho, and Washington. Geologic implications of Paleozoic and Mesozoic paleontology and biostratigraphy, Blue Mountains province, Oregon and Idaho: Reston, Virginia, U.S. Geological Survey Professional Paper 1435, p. 65–78.

Pessagno, E.A., Jr., and Blome, C.D., 1988. Biostratigraphic, chronostratigraphic, and U/Pb geochronometric data from the Rogue and Galice Formations, western Klamath terrane, (Oregon and California)—Their bearing on the age of the Oxfordian–Kimmeridgian boundary and the *Mirifusus* first occurrence event: 2nd International Symposium on Jurassic Stratigraphy, September 1987, Lisbon, Portugal, v. 1, p. 477–490.

Pessagno, E.A., Jr., and Blome, C.D., 1990, Implications of new Jurassic stratigraphic, geochronometric, and paleolatitudinal data from the western Klamath terrane (Smith River and Rogue Valley subterranes): Geology, v. 18, p. 665–668, doi: 10.1130/0091-7613(1990)018<0665:IONJSG>2.3.CO;2.

Pessagno, E.A., Jr., and Hull, D.M., 2003, Upper Jurassic (middle Oxfordian) Radiolaria from the Sula Islands (East Indies): Their taxonomic, biostratigraphic, chronostratigraphic, and paleobiogeographic significance: Micropaleontology, v. 48, p. 229–256, 14 pls.

Pessagno, E.A., Jr., and Martin, C.B., 2003, Tectonostratigraphic evidence for the origin of the Gulf of Mexico, *in* Bartolini, C., et al., eds., The Circum-Gulf of Mexico and the Caribbean: Hydrocarbon habitats, basin formation, and plate tectonics: Tulsa, Oklahoma, American Association of Petroleum Geologists Memoir 79, p. 46–74.

Pessagno, E.A., Jr., and Whalen, P.A., 1982, Lower and Middle Jurassic Radiolaria (multicyrtid Nassellariina) from California, east-central Oregon, and the Queen Charlotte Islands, B.C.: Micropaleontology, v. 28, p. 111–169, 13 pls.

Pessagno, E.A., Jr., Blome, C.D., and Longoria, J.F., 1984, A revised radiolarian zonation for the Upper Jurassic of western North America: Bulletins of American Paleontology, v. 87, p. 1–51, 5 pls.

Pessagno, E.A., Jr., Whalen, P.A., and Yeh, K.-Y., 1986, Jurassic Nassellariina (Radiolaria) from North American geological terranes: Bulletins of American Paleontology, v. 91, p. 1–75.

Pessagno, E.A., Jr., Longoria, J.F., MacLeod, N., and Six, W.M., Jr., 1987, Upper Jurassic (Kimmeridgian–upper Tithonian) Pantanelliidae from the Taman Formation, east-central Mexico: Tectonostratigraphic, chronostratigraphic, and phylogenetic implications, *in* Culver, S.J., ed., Studies of North American Jurassic Radiolaria, Part I: Washington, D.C., Cushman Foundation for Foraminiferal Research Special Publication 23, p. 1–51, 7 pls.

Pessagno, E.A., Jr., Six, W.M., Jr., and Yang, Q., 1989, The Xiphostylidae Haeckel and Parvivaccidae, n. fam., (Radiolaria) from the North American Jurassic: Micropaleontology, v. 35, p. 193–255, 10 pls.

Pessagno, E.A., Jr., Blome, C.D., Hull, D.M., and Six, W.M., Jr., 1993a, Middle and Upper Jurassic radiolaria from the western Klamath terrane, Smith River subterrane, northwestern California—Their biostratigraphic, chronostratigraphic, geochronologic, and paleolatitudinal significance: Micropaleontology, v. 39, p. 93–166.

Pessagno, E.A., Jr., Hull, D.M., Longoria, J.F., and Kelldorf, M.E., 1993b, Tectonostratigraphic significance of the San Pedro del Gallo area, Durango, western Mexico, *in* Dunn, G.C., and McDougall, K.A., eds., Mesozoic paleogeography of the western United States—II: Los Angeles, Pacific Section, Society of Economic Paleontologists and Mineralogists, v. 71, p. 141–156.

Pessagno, E.A., Jr., Cantú-Chapa, A., Hull, D.M., Kelldorf, M., Longoria, J.F., Martin, C.B., Meng, X., Montgomery, H., Ogg, J.G., and Urrutia-Fucugauchi, J., 1999, Stratigraphic evidence for northwest to southeast tectonic transport of Jurassic terranes in central Mexico and the Caribbean (western Cuba), *in* Mann, P., ed. (series ed., Hsü, K.J.), Caribbean basins, Sedimentary basins of the World—4: Amsterdam, Elsevier Science, p. 123–149.

Pessagno, E.A., Jr., Hopson, C.A., and Hull, D.M., 2001, Tectonostratigraphic significance of sedimentary strata occurring within and above the Coast Range ophiolite (California Coast Ranges) and the Josephine ophiolite (Klamath Mountains), northwestern California, *in* Dilik, Y., Moores, E.M., Elthon, D., and Nicolas, A., eds., Ophiolites and Oceanic Crust: New Insights from Field Studies and Ocean Drilling Program: Boulder, Colorado, Geological Society of America Special Paper 349, p. 383–394.

Pujana, I., 1989, Stratigraphical distribution of the multicyrtid Nassellariina (Radiolaria) at the Jurassic-Cretaceous boundary in the Neuquén Basin, Argentina: Zbl. Geol. Paläontographica. I, no. 5–6, p. 1043–1052.

Pujana, I., 1991, Pantanelliidae (Radiolaria) from the Tithonian of the Vaca Muerta Formation, Neuquén, Argentina: Neues Jahrbuch für Geologie und Paläontologie Abhandlungen, v. 180, no. 3, p. 391–408.

Pujana, I., 1993, Middle Jurassic (Bathonian–Callovian) Radiolaria from Chacay Melehue, Cordillera del Viento, Province of Neuquén, Argentina [M.S. thesis]: Dallas, University of Texas, 87 p.

Pujana, I., 1996, Occurrence of Vallupinae (Radiolaria) in the Neuquén Basin: Biostratigraphic implications, *in* Riccardi, A.C., ed., Advances in Jurassic Research: GeoResearch Forum, v. 1, no. 2, p. 459–456.

Saleeby, J.B., 1987, Discordance patterns in Pb/U zircon ages of the Sierra Nevada and Klamath Mountains: EOS (Transactions, American Geophysical Union), v. 68, no. 44, p. 1514–1515.

Saleeby, J.B., Harper, G.D., Snoke, A.W., and Sharp, W.D., 1982, Time relations and structural-stratigraphic patterns in ophiolite accretion, west-central Klamath Mountains, California: Journal of Geophysical Research, v. 87, p. 3831–3848.

Salvador, A., Westermann, G.E.G., Olóritz, F., Gordon, M.B., and Gursky, H.J., 1992, Meso-America, *in* Westermann, G.E.G., ed., The Jurassic of the Circum-Pacific: Cambridge, Cambridge University Press, p. 93–121.

Sandoval, J., and Westermann, G.E.G., 1988, The Bajocian (Jurassic) ammonite fauna of Oaxaca, Mexico: Journal of Paleontology, v. 60, p. 1220–1271.

Sandoval, J., Westermann, G.E.G., and Marshall, M.C., 1990, Ammonite fauna, stratigraphy, and ecology of the Bathonian–Callovian (Jurassic) Tecocoynca Group, south Mexico: Paleontographica, v. A210, p. 93–149.

Sedlock, R.L., Ortega-Gutiérrez, L., and Speed, R.C., 1993, Tectonostratigraphic terranes and tectonic evolution of Mexico: Boulder, Colorado, Geological Society of America Special Paper 278, 153 p.

Smith, P.L., 1980, Correlation of the members of the Jurassic Snowshoe Formation in the Izee basin of east-central Oregon: Canadian Journal of Science, v. 17, p. 1603–1608.

Snoke, A.W., and Barnes, C.G., 2006, this volume, The development of tectonic concepts for the Klamath Mountains province, California and Oregon, *in* Snoke, A.W., and Barnes, C.G., eds., Geological studies in the Klamath Mountains province, California and Oregon: A volume in honor of William P. Irwin: Geological Society of America Special Paper 410, doi: 10.1130/2006.2410(01).

Taylor, D.G., Callomon, J.H., Hall, R., Smith, P.L., Tipper, H.W., and Westermann, G.E.G., 1984, Jurassic ammonite biogeography of western North America: The tectonic implications, *in* Westermann, G.E.G., ed., Jurassic–Cretaceous biochronology and paleogeography of North America: Geological Association of Canada Special Paper 27, p. 121–141.

Tipper, H.W., 1981, Offset of an upper Pliensbachian geographic zonation in the North American Cordillera by transcurrent fault movement: Canadian Journal of Earth Sciences, v. 18, p. 1788–1792.

Verma, H.M., and Westermann, G.E.G., 1973, The Tithonian (Jurassic) ammonite fauna and stratigraphy of Sierra de Catorce, San Luis Potosi, Mexico: Bulletins of American Paleontology, v. 63, p. 107–278, 34 pls.

Wells, F.G., and Walker, G.W., 1953, Geology of the Galice quadrangle, Oregon: Washington, D.C., U.S. Geological Survey Geologic Quadrangle Map GQ-25, scale 1:62,500, 1 sheet.

Wright, J.E., and Wyld, S.J., 1986, Significance of xenocrystic Pre-Cambrian zircon contained within the southern continuation of the Josephine ophiolite: Devils Elbow ophiolite remnant, Klamath Mountains, northern California: Geology, v. 14, p. 671–674, doi: 10.1130/0091-7613(1986)14<671:SOXPZC>2.0.CO;2.

Yeh, K.-Y., and Cheng, Y., 1996, Jurassic radiolarians from the northwest coast of Busuanga Island, North Palawan block, Philippines: Micropaleontology, v. 42, p. 93–124, 12 pls.

Manuscript Accepted by the Society 10 January 2006

Printed in the USA

Geological Society of America
Special Paper 410
2006

A rift-edge facies of the Late Jurassic Rogue–Chetco arc and Josephine ophiolite, Klamath Mountains, Oregon

J. Douglas Yule*
Jason B. Saleeby
Division of Geological and Planetary Sciences, California Institute of Technology, Pasadena, California 91125, USA
Calvin G. Barnes
Department of Geosciences, Texas Tech University, Lubbock, Texas 79409, USA

ABSTRACT

The western Jurassic belt of the Klamath Mountains represents one of the Earth's best-preserved exposures of ancient marginal ocean basin lithosphere and chiefly consists of the coeval Rogue–Chetco volcanic-plutonic oceanic arc and Josephine ophiolite. This Late Jurassic ocean basin is hypothesized to have formed in response to rifting that initiated at ca. 165 Ma along the western margin of North America, disrupting a Middle Jurassic arc that had been constructed on older Klamath terranes and forming a marginal ocean basin with an active arc, inter-arc basin, and remnant arc. Previous workers characterized a "rift-edge" facies in the remnant-arc region. This chapter describes field, age, and geochemical data that suggest that a similar rift-edge facies exists in the vicinity of the active arc, on the opposite side of the marginal basin.

The rift-edge facies in the active arc setting consists of two main lithotectonic units, herein named informally as the Onion Camp complex and Fiddler Mountain olistostrome. The Onion Camp complex is partly composed of a characteristic metabasalt and red chert association. Red chert yielded scarce radiolarians of Triassic(?) and Early Jurassic age. A distinct chert-pebble conglomerate occurs at scarce localities within metasedimentary rocks. Concordant, composite bodies of amphibolite and serpentinized peridotite represent another distinctive feature of the Onion Camp complex. The metamorphic and lithologic features of the Onion Camp complex are similar to the lower mélange unit of the Rattlesnake Creek terrane, and the units are interpreted to be correlative. The Fiddler Mountain olistostrome is composed of Late Jurassic (Kimmeridgian?) pelagic and hemipelagic rocks interlayered with ophiolite-clast breccia and megabreccia, similar in character to olistostromal deposits associated with the rift-edge facies of the remnant arc. The occurrence of the Rattlesnake Creek terrane and an associated olistostromal deposit within the western Jurassic belt of southwestern Oregon may therefore represent the rift-edge facies in the active arc setting, at the transition between the Rogue–Chetco arc and Josephine ophiolite,

*Present address: California State University, Department of Geological Sciences, Northridge, California 91330-8266, USA; j.d.yule@csun.edu.

Yule, J.D., Saleeby, J.B., and Barnes, C.G., 2006, A rift-edge facies of the Late Jurassic Rogue–Chetco arc and Josephine ophiolite, Klamath Mountains, Oregon, *in* Snoke, A.W., and Barnes, C.G., eds., Geological studies in the Klamath Mountains province, California and Oregon: A volume in honor of William P. Irwin: Geological Society of America Special Paper 410, p. 53–76, doi: 10.1130/2006.2410(03). For permission to copy, contact editing@geosociety.org.

further corroborating previous models for the Late Jurassic tectonic evolution of the Klamath Mountains.

Keywords: ancient marginal ocean basin, rifting, Rogue–Chetco volcanic-plutonic arc, Josephine ophiolite, olistostromal deposits, radiolarians, Rattlesnake Creek terrane, Late Jurassic, Klamath Mountains, Oregon

INTRODUCTION

Marginal ocean basins of the western Pacific commonly form in extension-dominated tectonic settings above subduction zones where subduction of relatively old oceanic lithosphere occurs (e.g., Curray et al., 1979; Stern and Bloomer, 1992; Lagabrielle et al., 1997). They consist of oceanic lithosphere that is composite in nature and may record multiple magmatic and/or deformational episodes (Pearce et al., 1984; Parson and Wright, 1996; Lagabrielle et al., 1997; Hawkins, 2003). Primary petrotectonic elements of these ocean basins include active arcs, remnant arcs, and spreading centers that can occur in fore-arc, inter-arc, and/or back-arc settings (Karig, 1971; Uyeda, 1977). Lithosphere generated at marginal ocean basin spreading centers therefore can separate "matching" rifted fragments of older oceanic lithosphere that typically occurs beneath the remnant and active magmatic arcs (Fig. 1).

Ancient marginal ocean basin lithosphere exposed in the western Jurassic belt of the Klamath Mountains (Fig. 1) chiefly consists of the accreted Rogue–Chetco arc and Josephine ophiolite (Garcia, 1979, 1982; Harper, 1980, 1984) that apparently formed in the Late Jurassic when a Middle Jurassic arc system rifted apart, leaving a remnant arc along the western edge of North America (e.g., Snoke, 1977; Saleeby et al., 1982; Harper and Wright, 1984). Snoke (1977) described a Late Jurassic mafic complex near Preston Peak that intruded the Rattlesnake Creek terrane, a basement unit of the Middle Jurassic arc, and interpreted the Preston Peak region to represent a rift-edge facies of the Josephine ophiolite. The identification of a rift-edge facies in the remnant arc setting implies that a similar facies may exist on the opposite side of the marginal ocean basin—perhaps in the active arc setting.

A clue that a rift-edge facies might exist in the active arc setting was provided by Roure and DeWever (1983), who reported a Late Triassic age from red chert collected to the west of Kerby, Oregon (Fig. 1). Paradoxically, Late Triassic to Early Jurassic chert occurs commonly in the Rattlesnake Creek terrane and related terranes exposed in the remnant arc setting to the east of, and structurally above, the western Jurassic belt (e.g., Irwin et al., 1982). The primary objective of this chapter, therefore, is to describe the rocks associated with the Triassic chert reported by Roure and DeWever (1983) and to explore the stratigraphic and structural relationships between these uncharacteristic rocks of the western Jurassic belt and rocks of the Rogue–Chetco arc and Josephine ophiolite. The occurrence of rocks similar to the Rattlesnake Creek terrane within the western Jurassic belt supports previous tectonic models that interpret the Rogue–Chetco arc and Josephine ophiolite as integral parts of a Late Jurassic marginal ocean basin that formed in response to rifting of older Klamath Mountains terranes (Snoke, 1977; Saleeby et al., 1982; Harper and Wright, 1984; Harper, 2003).

GEOLOGIC AND TECTONIC FRAMEWORK

Regional, east-dipping thrust faults bound the western Jurassic belt of southwestern Oregon. To the east, the Orleans/Preston Peak thrust forms an upper tectonic boundary where rocks of the western Paleozoic and Triassic belt form a hanging-wall sheet above the western Jurassic belt (e.g., Irwin, 1960, 1966, 1972; Davis, 1969; Snoke, 1977; Hamilton, 1978; Burchfiel and Davis, 1981). To the west, rocks of the Franciscan Complex are underthrusted beneath the Valen Lake thrust, the western tectonic boundary of the western Jurassic belt (Ramp, 1977; Page et al., 1981; Blake et al., 1985; Fig. 1). Numerous northeast–southwest-striking faults and folds deform rocks of the western Jurassic belt, probably as a result of imbrication and emplacement of the large thrust sheets emplaced between ca. 155 and 150 Ma, the interpreted age of the Nevadan orogeny at this latitude (Saleeby et al., 1982; Harper and Wright, 1984; Harper et al., 1994).

The western Jurassic belt can be subdivided into three Late Jurassic petrotectonic units: the Josephine ophiolite (Harper, 1984, 2003); Rogue–Chetco volcano-plutonic arc complex and related metamorphic wall rocks (Jorgenson, 1970; Hotz, 1971; Dick, 1976; Loney and Himmelberg, 1977; Garcia, 1979, 1982; Saleeby, 1984; Coleman and Lanphere, 1991; Yule, 1996); and the Upper Jurassic Galice Formation, a flysch sequence, which overlaps both the ophiolite and arc-volcanic rocks (Harper, 1980; Miller and Saleeby, 1995). These units are part of a Late

Figure 1. Regional geologic map of the northern Klamath Mountains province and Oregon Coast Ranges. Outline shows area of Figure 2. Pluton names are: BK—Buckskin Peak, BM—Bear Mountain, BP—Bear Peak, CC—Chetco complex, GB—Grayback, GP—Grants Pass, PP—Pony Peak, RP—Russian Peak, S—Slinkard, SM—Squaw Mountain, SV—Summit Valley, TR—Thompson Ridge, VB—Vesa Bluffs, W—Wimer, WC—Wooley Creek, and WR—White Rock. Jurassic stages are: Kim.—Kimmeridgian, Oxf.—Oxfordian. Modified from Ramp (1977), Ramp and Peterson (1979), Coleman and Lanphere (1991), Irwin (1994), and Hacker et al. (1995).

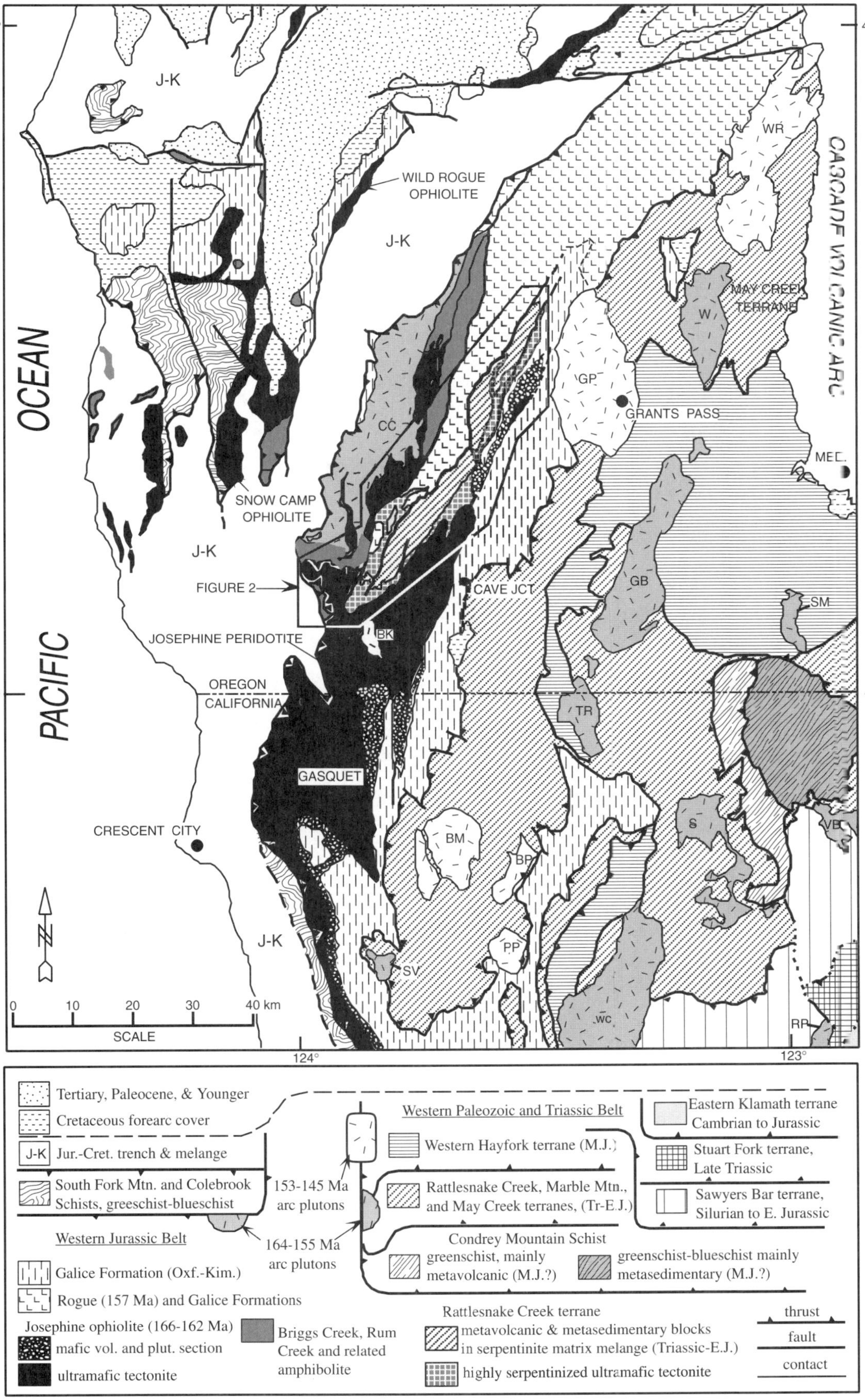

43°
J-K
WILD ROGUE OPHIOLITE
J-K
WR
CASCADE VOLCANIC ARC
MAY CREEK TERRANE
W
OCEAN
GP
GRANTS PASS
CC
MED.
SNOW CAMP OPHIOLITE
J-K
FIGURE 2
CAVE JCT
GB
SM
JOSEPHINE PERIDOTITE
BK
PACIFIC
OREGON
CALIFORNIA
42°
TR
GASQUET
CRESCENT CITY
BM
BP
S
VB
J-K
PP
SV
wc
RP
0
10
20
30
40 km
SCALE
124°
123°
Tertiary, Paleocene, & Younger
Cretaceous forearc cover
J-K Jur.-Cret. trench & melange
South Fork Mtn. and Colebrook Schists, greeschist-blueschist
Western Jurassic Belt
Galice Formation (Oxf.-Kim.)
Rogue (157 Ma) and Galice Formations
Josephine ophiolite (166-162 Ma)
mafic vol. and plut. section
ultramafic tectonite
153-145 Ma arc plutons
164-155 Ma arc plutons
Briggs Creek, Rum Creek and related amphibolite
Western Paleozoic and Triassic Belt
Western Hayfork terrane (M.J.)
Rattlesnake Creek, Marble Mtn., and May Creek terranes, (Tr-E.J.)
Condrey Mountain Schist
greenschist, mainly metavolcanic (M.J.?)
greenschist-blueschist mainly metasedimentary (M.J.?)
Rattlesnake Creek terrane
metavolcanic & metasedimentary blocks in serpentinite matrix melange (Triassic-E.J.)
highly serpentinized ultramafic tectonite
Eastern Klamath terrane Cambrian to Jurassic
Stuart Fork terrane, Late Triassic
Sawyers Bar terrane, Silurian to E. Jurassic
thrust
fault
contact

Jurassic marginal ocean basin assemblage that is constructed of oceanic-arc plutonic and intra-arc ophiolitic crust and mantle rocks, and subsequently was overlapped by laterally equivalent arc-volcanic rocks and flysch deposits (Saleeby et al., 1982; Harper and Wright, 1984; Harper et al., 1994; Harper, 2003).

This study identifies two additional petrotectonic units, the Onion Camp complex and Fiddler Mountain olistostrome, which were previously mapped as part of the Rogue–Chetco arc complex. These new units share a striking resemblance with rocks of the Rattlesnake Creek terrane (e.g., Irwin, 1972; Snoke, 1977; Wright and Wyld, 1994) and Lems Ridge olistostrome (Ohr, 1987), respectively. Both the Rattlesnake Creek terrane and olistostromal deposits appear to be correlative with components of the western Paleozoic and Triassic belt in the hanging wall of the Orleans/Preston Peak thrust to the east of the study area (Snoke, 1977; Wright and Fahan, 1988; Hacker et al., 1995; Fig. 1).

ONION CAMP COMPLEX

Overview

The Onion Camp complex is named for an association of rocks at Onion Camp, northwest of Fiddler Mountain, that occur along a 3- to 5-km-wide, ~40-km-long, northeast–southwest-striking belt, extending from near Canyon Peak on the southwest to the Rogue River on the northeast (Fig. 2). In addition, isolated patches of the Onion Camp complex occur ~4 km to the southeast of Oak Flats, along the western margin of the Chetco plutonic complex (Fig. 2). Hill slopes underlain by the complex are very steep and densely vegetated, so exposure is limited to weathered roadcuts and ridgelines. The best exposures occur along the Illinois and Rogue Rivers, where up to 500-m-long outcrops can be studied. Other excellent exposures of Onion Camp complex rocks occur in road gravel quarries near Onion Camp and Onion Mountain and in cirque exposures surrounding Babyfoot Lake (Fig. 2).

Rocks of the Onion Camp complex were previously included as part of the Rogue Formation by Wells and Walker (1953). Subsequent workers noted differences between the type Rogue Formation and rocks near Onion Camp but did not subdivide the unit (Wise, 1969; Dick, 1976; Ramp, 1977, 1984; Ramp and Peterson, 1979; Page et al., 1981). Roure and DeWever (1983) collected red chert from near Onion Camp (F. Roure, personal commun., 1993) that yielded a Triassic age, an unusually old age for rocks of the western Jurassic belt, suggesting that rocks here may have experienced a more complex geologic history than previously recognized.

The Onion Camp complex consists primarily of mafic metavolcanic rocks, metachert, argillaceous metasedimentary rocks, a heterogeneous mafic intrusive complex, amphibolite and amphibolite gneiss, and serpentinized harzburgite and dunite. Less common rock types include hornblende schist, garnet-mica quartzite, and lenses of variably sheared serpentinite.

Rock Units

Metavolcanic Rocks. Metavolcanic rocks compose most of the Onion Camp complex. Fresh samples exhibit a light green to dark gray-green hue, a manifestation of the abundant ferromagnesian, greenschist-facies minerals that characterize these rocks. Lithotypes can be classified into four main groups: (1) metamorphosed mafic lavas, locally amygdaloidal and pillowed (Fig. 3); (2) lava breccias (plagioclase > clinopyroxene-phyric clasts); (3) metamorphosed mafic volcaniclastic rocks (mainly pyroclastic deposits) consisting of fine-grained mafic tuffs and tuff breccias with highly flattened and altered lithic clasts and lapilli; and (4) metamorphosed, shallow-level mafic intrusive bodies and dikes (chilled dike and sill margins are present at a few localities). Massive lavas are the most common rocks, with subordinate amounts of fragmental volcanic rocks; shallow mafic intrusions and dikes are least common, comprising <10% of the metavolcanic rocks of the Onion Camp complex. The average chemical composition of the Onion Camp complex metavolcanic rocks is basaltic andesite (Yule, 1996).

The metavolcanic rocks of the Onion Camp complex consist chiefly of lower greenschist-facies minerals. Original igneous phases are commonly altered, partly or wholly, to various hydrous phases, but relict plagioclase ($>An_{50}$) and/or augite are present in many samples. The typical greenschist-facies assemblage is albite, quartz, actinolite, epidote, chlorite, white mica, and calcite. Pyrite and sphene are common accessory phases. Pumpellyite is present in some mafic lava samples. Abundant prehnite, quartz-prehnite, and quartz-calcite veins cut all exposures.

Layering in fine-grained fragmental volcanic rocks is discontinuous, probably due to shearing that transposed layering into parallelism with the regional northeast-striking, southeast-dipping foliation. Elongate pillow lobes (Fig. 3) may record paleohorizontal; however, like the fragmental rocks, the elongation orientation is parallel to the regional foliation. Nonclastic rocks are typically massive, but break along a cleavage that is parallel to the regional foliation. Tuffaceous rocks vary texturally and are ubiquitously metamorphosed; protoliths include crystal tuff, lithic lapilli tuff, and crystal-lithic tuff. Some of these rocks have been converted to chlorite schist.

Red Chert. Red chert blocks and lenses up to 10-m across crop out near metamorphosed pillowed and massive lavas, and are most common near Onion Camp, ~2 km to the west of Fiddler Mountain and Onion Mountain (Fig. 2). Red and whitish-red chert blocks commonly show transposed compositional layering and a sugary texture indicative of complete recrystallization. Interlayers of maroon chert 2- to 5-cm thick show original compositional layering that is oblique to the regional foliation. The maroon layers commonly contain clear, sub-mm-scale ovoids, which probably represent slightly recrystallized radiolarians.

Metasedimentary Rocks. Northeast–southwest-striking belts of metasedimentary rocks are most common along the eastern margin of the Onion Camp complex. The most common rock type is siliceous argillite with lesser amounts of impure red, white, and gray chert; tuffaceous siltstone; and rare polymictic sandstone, conglomerate, and conglomeratic sandstone. Colors range from white to greenish gray, with local red to reddish-brown interlayers. All metasedimentary rocks break along an imperfect slaty cleavage.

Within the larger metasedimentary belts, distinctive but scarce polymictic conglomeratic sandstone layers occur as m to several-m-thick interbeds in predominantly wacke, siltstone, and siliceous argillite. Poorly to moderately sorted sand- and pebble- to cobble-sized rock clasts are subangular to subrounded. The coarse-grained conglomerates are generally matrix supported. Clasts in the sandstones and conglomerates are diverse and include wacke, siltstone, siliceous argillite, recrystallized chert (with relict radiolarians), sugary quartzite, monocrystalline quartz, mafic volcanic rocks (with porphyritic, intergranular, and spherulitic textures), and recrystallized limestone. Siliceous argillite and chert are by far the most abundant clast types. Volcanic clasts, though present, are subordinate in volume to sedimentary clasts. The conglomeratic sandstone is therefore distinct from other polymict deposits in the region that primarily consist of mafic volcanic and/or shallow-level intrusive clasts; for example, see the description of Fiddler Mountain olistostrome later in this chapter.

The metasedimentary rocks of the Onion Camp complex consist primarily of quartz, phyllosilicates, and opaque oxides. Secondary minerals are calcite, albite, and pyrite; accessory minerals include zircon, apatite, and sphene. Tuffaceous layers contain chlorite, albite porphyroclasts, and minor epidote.

Original stratification is rarely preserved, but when present, it is generally parallel or nearly parallel to the regional foliation and rock cleavage. Lithologic layering tends to pinch out along strike, probably reflecting a high degree of bedding transposition. Boudinage of chert and silicic argillite layers is common. The rocks are well indurated, intensely strained, and structurally interlayered.

Amphibolite and Amphibolite Gneiss. Discontinuous belts of foliated-to-massive amphibolite and amphibolite gneiss with subordinate impure quartzite occur throughout the Onion Camp complex (Fig. 2). Quartzose layers are most abundant in the amphibolite masses near Squaw Mountain, where they occur as infolded layers within outcrops consisting mostly of hornblende schist and amphibolite gneiss. In hand samples and in thin sections, the quartzites consist of sugary-textured, fine- to medium-grained, banded, well-foliated and lineated rocks. Previous researchers have described these rocks as gneissic migmatite (Wells et al., 1949), metagabbro (Ramp, 1977), and gabbroic rocks and dike complexes (Page et al., 1981) and have interpreted their origin as small intrusions into the Rogue Formation or its metamorphosed equivalent. However, the findings of this study suggest that the protolith materials were greenschist-facies, mafic metavolcanic, and subordinate metasedimentary rocks of the Onion Camp complex.

All occurrences of amphibolite within the Onion Camp complex are at least partially bounded by variably sheared and serpentinized peridotite (Fig. 2). Local slivers of sheared serpentinite are contained entirely within the amphibolite. Small, isolated bodies of amphibolite have highly sheared and discordant contacts with serpentinized peridotite and show ubiquitous, retrograde greenschist-facies and lower-grade alterations. In contrast, contacts between the larger amphibolitic belts and serpentinized peridotite are weakly sheared, and rocks on either side of the contact have subparallel foliations. This finding suggests that these amphibolite and peridotite units comprise coherent, composite blocks.

The amphibolite consists of hornblende and plagioclase with subordinate epidote, sphene, and Fe-Ti opaque oxides (mostly ilmenite). Common alteration minerals include chlorite, epidote, clinozoisite, prehnite, green amphibole, white mica, and quartz. Hornblende usually comprises 60–80% of the rock. Plagioclase is generally untwinned and ranges in composition from albite to intermediate oligoclase in epidote-absent amphibolites. Impure quartzite consists of 60–90% quartz with metamorphic cummingtonite, almandine, biotite, white mica, and opaque oxides.

At Squaw Mountain, a distinct change from greenschist-facies to epidote-amphibolite-facies assemblages occurs across a narrow (<2-m-wide) zone (Fig. 2). In this zone, actinolite-chlorite schist, metabasalt, and impure chert abruptly change to hornblende schist, gneissic amphibolite, and impure meta-quartzite. This is coincident with a change in metamorphic assemblage from actinolite + chlorite + epidote + albite to hornblende + epidote + albite. In addition, the rocks become more coarsely crystalline across this boundary. A few exposures show plagiogranite dikelets and segregations cutting the highest-grade amphibolites.

Structural features include a faint to well-developed foliation that shows a faint to well-developed intersection lineation parallel to the hinge lines of tight to isoclinal folds. These folds are probably synmetamorphic and developed during amphibolite-facies metamorphism. Postmetamorphic, close to isoclinal folds deform the foliation and earlier folds.

Serpentinized Peridotite. Ultrabasic rocks of the Onion Camp complex occur as blocks, small lenses, and smears of sheared serpentinite (slickenite) and serpentinized peridotite that show a sharp to gradational contact with, but are distinct from, the Josephine peridotite (Fig. 2). This gradation is typically marked in the field by a distinct change in hill slope vegetation that is conspicuous on air photos and is easily visible in the field (Fig. 4A). The Onion Camp complex serpentinite forms smooth, subdued topography with sparse to grassy vegetation that is subject to landslides and other downslope movements, whereas the Josephine peridotite forms blocky, resistant outcrops and rugged topography. Distinctions between the Onion Camp complex and Josephine peridotite can also be made on the basis of different crosscutting relations, metamorphic mineral assemblages, and structural features. Contacts between

Explanation of Geologic Map and Cross Section

LATE JURASSIC

CROSS-STITCHING PLUTONIC ROCKS

Jh mafic to intermediate hornblende-phyric dikes and small intrusions; U-Pb, K-Ar, and Ar-Ar ages cluster at ca. 150 Ma (Yule, 1996; Dick, 1976; and Harper et al., 1994).

QUATERNARY

SURFICIAL DEPOSITS

Q Unconsolidated silt, sand, and gravel deposits occupying modern stream and river channels, and associated low terraces and flood plains.

LATE MIDDLE AND LATE JURASSIC

INTER-ARC BASIN VOLCANIC AND SEDIMENTARY ROCKS

Jr Rogue Formation; massive to well-bedded, basaltic to andesitic volcanogenic turbidites, volcanic breccias, and flows; 157 Ma (U-Pb zircon, Saleeby, 1984), 153 Ma (Ar/Ar hornblende, Yule, 1996).

Jg Galice Formation; basal hemipelagic section of Callovian age overlain by flysch sequence of Oxfordian-Kimmeridgian age (Pessango and Blome, 1990).

Jf Fiddler Mountain olistostrome; polymictic conglomerate, breccia, and megabreccia - ophiolite-clast types dominate, with subordinate chert and serpentine-clast types - interlayered with well-bedded sequences of chert and argillite (thin dashed lines near Fiddler Mtn. (FM)). Chert yields poorly preserved Late Jurassic (Kimmeridgian?) radiolaria (Yule et al., 1992).

CHETCO PLUTONIC COMPLEX

Jt weakly foliated tonalite-trondhjemite sill-like complex occupying high structural levels of plutonic complex (157 Ma, U-Pb zircon, Yule, 1996).

Jgd main phase of complex; foliated to massive biotite-hornblende quartz diorite, hornblende gabbro-diorite, and norite (160 Ma, U-Pb zircon, Yule, 1996).

Jog layered olivine gabbro and two-pyroxene gabbro forming the core of the plutonic complex.

Jfl marginal facies consisting of flaser quartz diorite, diorite, and gabbro with shallowly plunging lineations and steeply dipping foliations.

JOSEPHINE OPHIOLITE

Jj cumulate gabbro, massive gabbro, sheeted mafic dikes, and pillow lavas and lava breccias; interpreted as age equivalent with the Josephine ophiolite of Harper (1984); 162-166 Ma, U-Pb, zircon, Saleeby (1984).

Jp variably serpentinized tectonized harzburgite and dunite, interpreted as mantle lithosphere equivalent of Josephine ophiolite (Dick, 1976; Harper, 1984).

METAMORPHIC WALLROCKS

Jb Briggs Creek amphibolite (e.g., Coleman and Lanphere, 1991); amphibolite gneiss, schist, and impure quartzite (156-158 Ma, Ar/Ar amphibole, Yule, 1996).

Ja amphibolite and amphibolite gneiss, biotite-hornblende schist, impure quartzite and metagabbro found in metamorphic sole of Madstone Cabin thrust (148-153 Ma, Ar/Ar muscovite and hornblende, and U-Pb zircon, Harper et al., 1994).

PRE-MIDDLE JURASSIC

ONION CAMP COMPLEX

Jm heterogeneous mafic intrusive complex, consisting of diabase, gabbro, and sheeted mafic dikes; most extensive outcrop belts occur in areas near Whetstone Butte (WB), with local, smaller belts (not shown) occurring throughout the metavolcanic and metasedimentary sequence (JTrvs); age unknown, but may be broadly correlative with Josephine ophiolite gabbro, mafic dikes.

complexly folded outcrop belts of amphibolite and amphibolite gneiss, biotite-hornblende schist, and impure quartzite transitional to greenschist facies metavolcanic and metasedimentary rocks of complex, most common near Squaw Mtn (SM; 170-173 Ma, Ar/Ar hornblende, Yule 1996).

highly serpentinized and sheared dunite, harzburgite, and sparse websterite; cut by numerous intermediate to mafic dikes, and occasional rodingite dikes; plagiogranite dikes yield U-Pb zircon ages of 173 and 175 Ma (Table 2, this chapter; Yule, 1996).

TRIASSIC (?) AND JURASSIC

JTrvs greenschist facies metavolcanic and metasedimentary rocks, including massive greenstone, pillow lava, tuff and tuff-breccia, argillite and red chert; chert samples yield mostly amorphous radiolaria with scarce Triassic(?) and Jurassic forms (Yule et al., 1992; C. Blome, personal communication. 1992).

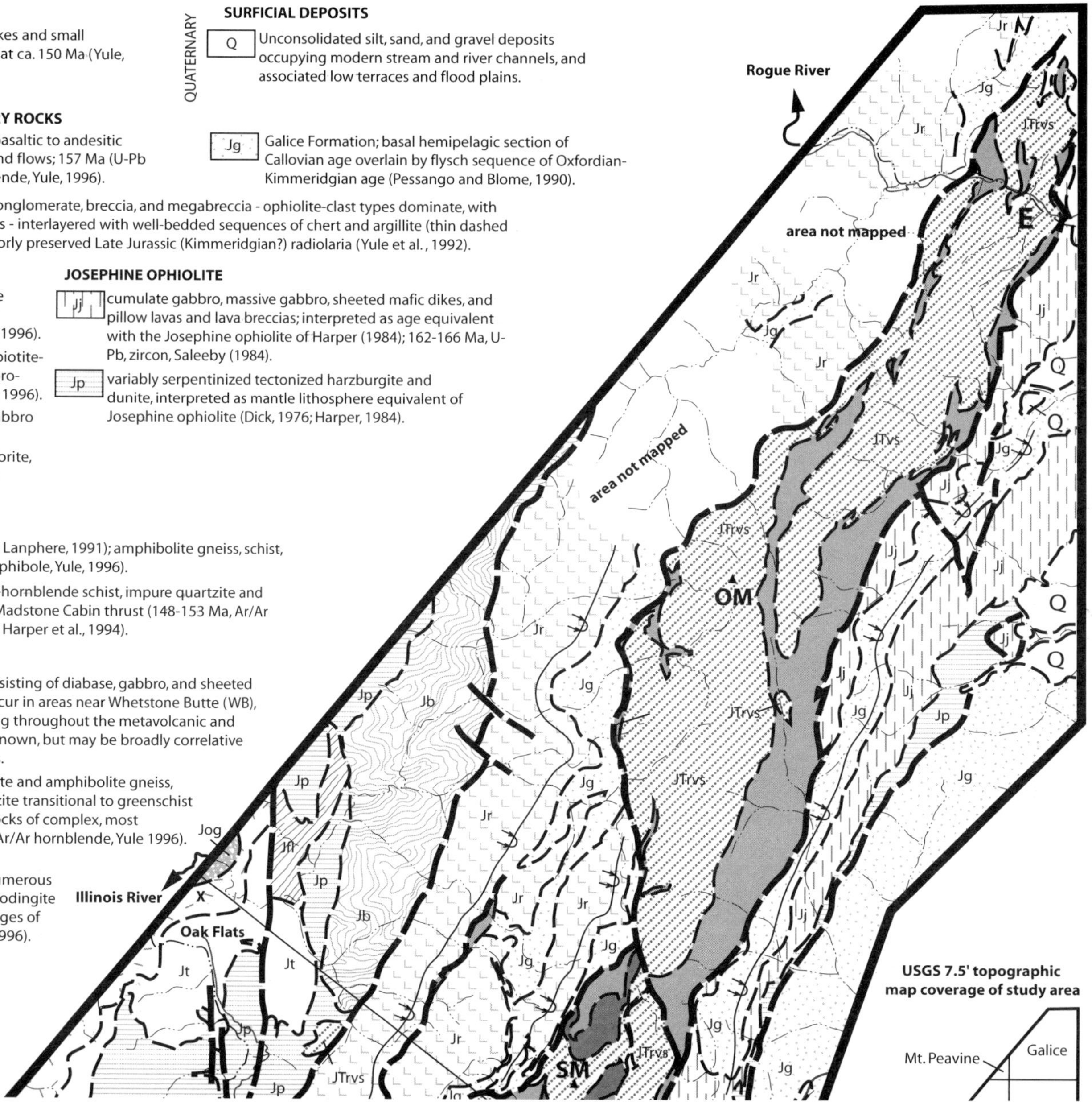

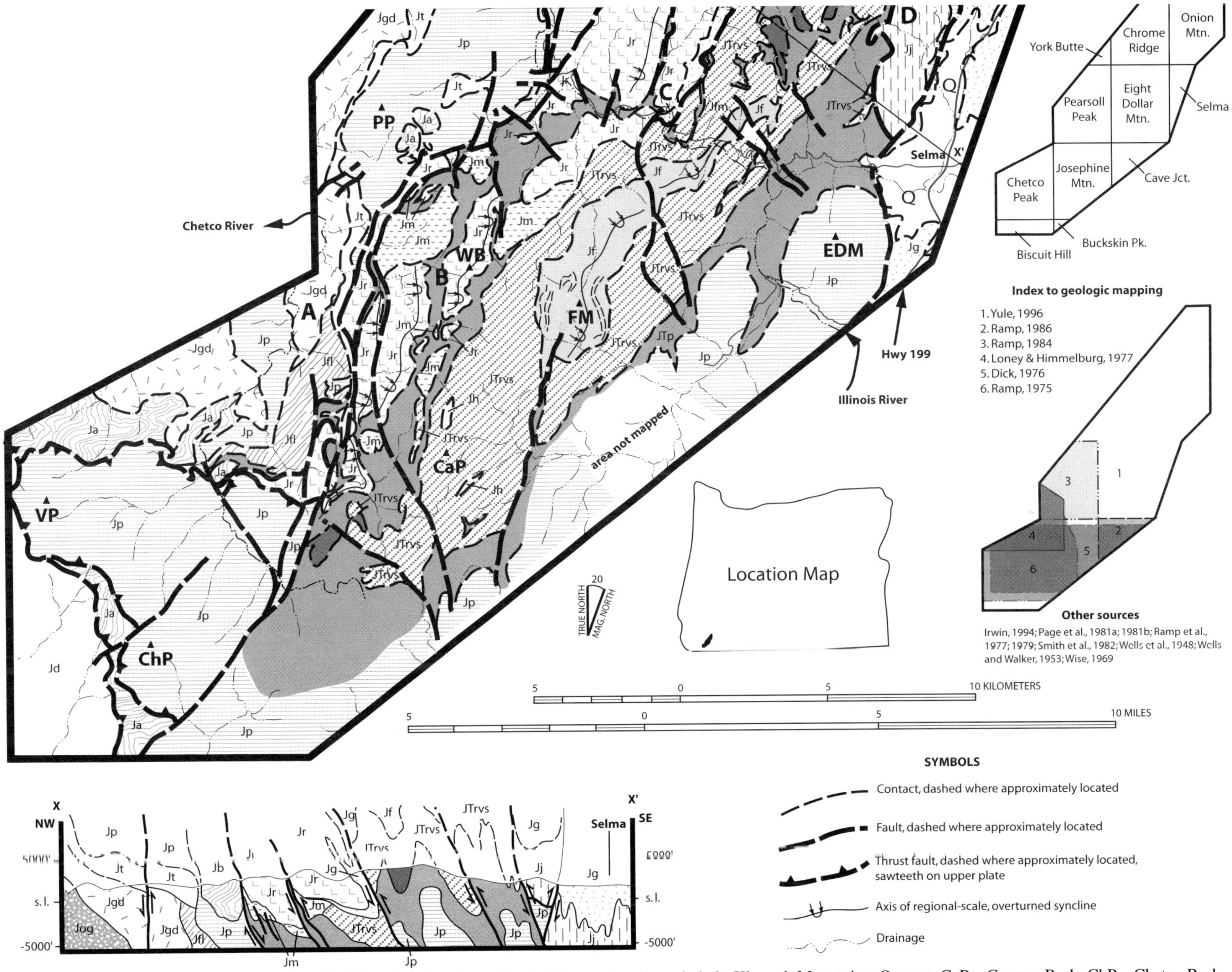

Figure 2. Geologic map, cross-section along X–X′, and explanation of part of the western Jurassic belt, Klamath Mountains, Oregon. CaP—Canyon Peak, ChP—Chetco Peak, EDM—Eight Dollar Mountain, FM—Fiddler Mountain, OM—Onion Mountain, PP—Pearsoll Peak, SM—Squaw Mountain, WB—Whetstone Butte, VP—Vulcan Peak. Bold-faced letters A–E refer to locations of columnar sections shown in Figure 15. This figure is also on the CD-ROM accompanying this volume as a continuous image.

Figure 3. Photograph of river polished pillow lavas of the Onion Camp complex, locality EDM-83 (see Table 3A for latitude and longitude), north bank of the Illinois River.

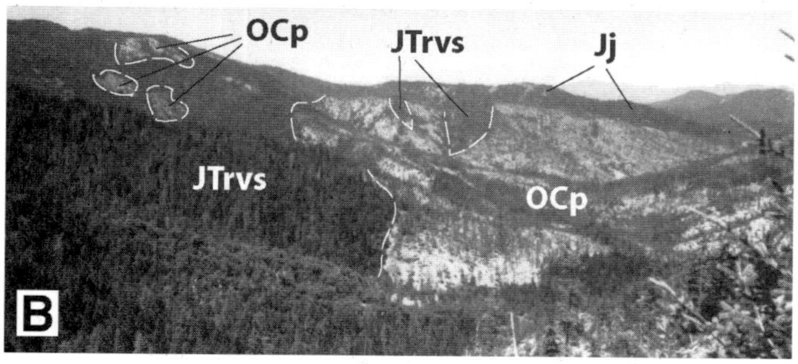

Figure 4. (A) Photograph of Eight Dollar Mountain, view toward the northeast. Vegetation change, indicated by dashed line on the west flank of Eight Dollar Mountain, marks the boundary between highly serpentinized peridotite of the Onion Camp complex (OCp) and relatively unserpentinized Josephine peridotite (Jp). Valley to the northeast is underlain by Galice Formation (Jg). (B) Photograph looking upstream along the Illinois River, view to the northeast. Heavily vegetated slopes are underlain by metavolcanic and metasedimentary rocks of the Onion Camp complex (JTrvs) and Josephine ophiolite (Jj), and the relatively sparsely vegetated hillslopes are underlain by OCp. Small patches of barren slopes in the upper left are underlain by complexly folded serpentinite (OCp) near Squaw Mountain (Fig. 2). Small densely vegetated patches in the center are underlain by JTrvs and surrounded by OCp.

serpentinite and other units vary from a nonconformity beneath deposits of the Rogue Formation and Fiddler Mountain olistostrome to a variably sheared boundary with greenschist- to amphibolite-grade metamorphosed units of the Onion Camp complex (Figs. 2 and 4B).

Outcrops of Onion Camp complex serpentinite and serpentinized peridotite range from massive and blocky, unaffected by shearing, to incoherent, highly sheared, platy serpentinite. The massive outcrops generally weather brown to reddish orange, with weathered surfaces studded by aggregates of primary minerals, mostly to wholly pseudomorphosed by serpentine-group minerals. Some massive weathered exposures are blue-gray in color. Sheared serpentinite (slickenite) varies in appearance, depending on the degree of shearing, ranging from moderately sheared, dark green or blue-green slickenite to highly sheared, polished black or metallic blue slickenite. Shearing is most intense along faults, in axial-surface regions of folds, and near contacts with other units. The central portions of large blocks of serpentinized peridotite tend to be the least altered, but they still contain small lenses and smears of slickenite.

Harzburgite, dunite, and scarce pyroxenite were the probable compositions of the original ultramafic protoliths of the Onion Camp complex. In many of these rocks, magnesian olivine probably formed ~75–95% of the pre-serpentinized ultramafic rocks. Kernels of orthopyroxene and clinopyroxene are occasionally seen in thin section, but are scarce because of the degree of alteration and deformation. Chromian spinel octahedra are visible on weathered surfaces of massive outcrops, but are largely pseudomorphosed by magnetite.

Alteration minerals comprise >80% of all Onion Camp complex ultramafic rocks. Relict olivine grains are scarce, being most common in the more massive serpentinized ultramafic rocks. Orthopyroxene is generally bastitic or uralitic. Low-temperature serpentine-group minerals (lizardite and chrysotile) generally comprise >75% of the rocks and occur as massive replacement minerals as well as in intricate vein networks indicative of multiple episodes of serpentinization and alteration. Higher temperature minerals (>300 °C), such as antigorite and tremolite, occur as fragmented, irregular patches in sheared serpentinite and occasionally border fragmented mafic dikes. Magnetite is ubiquitous and provides an indication of the degree of serpentinization, with the more highly serpentinized rocks being most strongly magnetic.

Onion Camp complex serpentinite and serpentinized peridotite record four distinct mineral and structural associations that formed as metamorphic temperature decreased. The earliest association is rare and is characterized by a mesoscopic mylonitic foliation defined by flattened and elongated aggregates of primary (now pseudomorphosed) minerals. Weathered rock faces that display the mylonitic fabric show multi-grain porphyroclastic augens of olivine and pyroxene in a fine-grained, xenoblastic matrix of mostly olivine. In thin section, the porphyroclasts invariably show pronounced strain effects, such as kink bands, deformed exsolution lamellae, and undulatory extinction.

Figure 5. (A) Photograph of a cut slab of serpentinized peridotite of the Onion Camp complex showing the metamorphic foliation that parallels foliation in gneissic amphibolite. The pale bands consist of light-green serpentine-group minerals (probably lizardite). The medium-gray bastite pseudomorphs of pyroxene grains in a dark-gray matrix of thoroughly serpentinized olivine. (B) Photograph of a cut slab of brecciated serpentinized peridotite collected near Squaw Mountain. The sample was collected from a 1- to 2-m-wide outcrop of cemented breccia in otherwise intact serpentinized peridotite.

The second association is defined by a serpentine foliation and lineation that is best developed in massive outcrops near the amphibolite-peridotite contact at Squaw Mountain (Fig. 2), but also occurs in scattered exposures throughout the unit. The foliation is defined by semi-continuous bands of light green serpentine spaced at ~1–3 mm intervals in an otherwise dark green, massively serpentinized peridotite. The alternating light green and dark green bands give the serpentinite a gneissose appearance (Fig. 5A). A poor to moderate cleavage spaced ~1 cm apart intersects the light green serpentine foliation at an angle of ~25°. The resulting intersection lineation and foliation are parallel to similar features in the adjoining amphibolite bodies.

The third association is characterized by ubiquitous, low-temperature static serpentinization, intricate vein networks, and cemented breccia zones. These features cut the mylonitic and serpentinite gneissic fabrics. Locally, some serpentinites exhibit an intricate patchwork of veins and fragmental textures (Fig. 5B). These textures, alterations, fragmentation, and veining relations are similar to those that occur in the mafic intrusive complex. Replacement minerals in massive serpentinite include lizardite and chrysotile, with secondary uralite, and magnetite. Secondary alteration products and vein minerals include lizardite; chrysotile; and local calcite, malachite, limonite, hematite, barite, and various sulfide-group minerals, including pyrrhotite, chalcopyrite, chalcocite, and sphalerite. Zones with high concentrations of sulfides ~4 km to the southeast of Pearsoll Peak (Fig. 2) have been prospected for Au and Cu (Ramp, 1984).

The latest structural and textural features of the serpentinite are sets of anastomosing shear bands that parallel the northeast-striking, southeast-dipping regional foliation, most commonly observed along faults and in axial surface regions of folds. Shear surfaces either display a black to dark green lustrous polish or contain linear, stepped serpentine fibers that indicate the shear direction. At outcrop scale, low-strain shear zones have a widely spaced, anastomosing schistocity (0.5–1.0 m wide) and equidimensional augens, whereas high-strain zones generally have a centimeter-scale, subparallel schistosity.

Mafic Intrusive Suite. A heterogeneous assemblage of hydrothermally altered, mafic intrusive rocks occurs in close association with the metavolcanic and metasedimentary rocks of the Onion Camp complex. Some of the larger bodies are mapped in Figure 2; numerous smaller outcrops are not shown but occur throughout the complex. The rocks are massive, nonschistose, and extremely tough, and the penetrative foliation that characterizes the metavolcanic and metasedimentary rocks is absent. The mafic intrusive suite is best observed along the roadcuts and cliff exposures in the Hellgate Canyon portion of the Rogue River; in the roadcut and ridgeline exposures to the west of Fiddler Mountain (Fig. 2); and along the trails, ridgelines, and drainages south of Canyon Peak.

The mafic intrusive suite consists predominantly of metadiabase and aphanitic greenstone with subordinate altered gabbro, diorite, and rare plagiogranite. Chilled dike-on-dike margins are common in some outcrops. Metamorphism of these rocks is characterized by a static, ubiquitous greenschist-facies and sub-greenschist-facies hydrothermal alteration. The primary mineral constituents in mafic rocks are actinolitic amphibole, altered plagioclase, epidote, chlorite, sulfide minerals, and partially to wholly altered ilmenite (leucoxene). Relict cores of clinopyroxene and hornblende are locally preserved in gabbro and diorite. Plagiogranite contains extensively altered plagioclase (saussurite), quartz, chlorite (replacing biotite), and minor Fe-Ti opaque oxides.

Distinctive features of the mafic intrusive suite include intricate networks of vein systems and patchworks of healed, fine- to coarse-grained breccia hosted in otherwise nondeformed, hydrothermally altered rock. Vein minerals are diverse and include epidote, epidote (and clinozoisite) + quartz, prehnite, prehnite + quartz, quartz + epidote + chlorite, and quartz + calcite. The breccia zones are locally sheared but typically lack penetrative deformation. Breccia clasts match the adjacent host rock type and are generally encased in vein materials and subordinate amounts of cataclastic matrix, probably the result of hydrofracturing. This distinctive hydrofracture texture is most commonly observed in the mafic intrusive suite but also occurs locally in other units of the Onion Camp complex and in the Josephine ophiolite.

Field relations support an intrusive relationship between the mafic intrusive suite and metavolcanic and metasedimentary sequences of the Onion Camp complex. In rare instances, the mafic intrusive suite shows chilled margins against metavolcanic and metasedimentary rocks. In addition, individual mafic dikes commonly intrude other units of the complex.

Dikes. A wide range of dikes cuts rocks of the Onion Camp complex, including pegmatitic websterite and gabbro, plagiogranite, diabase, aphanitic basalt, and acicular hornblende-phyric diorite. Rodingite dikes are scarce, suggesting that widespread serpentinization occurred prior to dike emplacement.

Dike orientations are random and probably reflect multiple episodes of intrusion and subsequent deformation. The dikes are rarely continuous for more than 10 m and essentially occur as megaboudins encased in serpentinite. Websterite and gabbro dikes contain clinopyroxene crystals as much as 50 cm in length but are now largely replaced by fibrous amphibole (uralite). Relatively unaltered mafic and intermediate dikes intrude the amphibolites. The mafic dikes display textures and mineralogy similar to rocks of the mafic intrusive suite, and the younger intermediate dikes are hornblende-phyric, typically with acicular crystals.

Crosscutting relations between dikes, though equivocal, suggest at least four episodes of dike intrusion: (1) highly altered and disrupted plagiogranite dikes; (2) diabasic and aphanitic basaltic dikes that resemble sheeted mafic dike complexes characteristic of some ophiolitic suites; (3) websterite and gabbro dikes that are similar in appearance to plutonic rocks of the Chetco plutonic complex; and (4) relatively unaltered, acicular hornblende-phyric dikes.

Metamorphic Conditions

The metavolcanic rocks of the Onion Camp complex contain mineral assemblages chiefly typical of greenschist-facies conditions (Liou et al., 1974; Apted and Liou, 1983). The absence of sodic amphibole or lawsonite precludes blueschist-facies metamorphic conditions. Pumpellyite occurs in a few samples and may indicate local pumpellyite-actinolite-facies conditions (Brown, 1977; Liou et al., 1987; Frey et al., 1991).

The metasedimentary rocks of the Onion Camp complex also contain mineral assemblages indicative of greenschist-facies conditions. The absence of biotite but the common occurrence of white mica + stilpnomelane ± actinolite suggests temperatures below 400 °C (Brown, 1975; Yardley, 1989).

Amphibolite and amphibolite gneiss show the following parageneses: (1) brown hornblende + oligoclase + clinopyroxene + titanite, (2) green hornblende + oligoclase + titanite, and (3) green hornblende + epidote + albite + titanite. Quartzite shows the following parageneses: (4) quartz + cummingtonite + biotite + almandine, and (5) quartz + biotite + white mica + almandine. These assemblages indicate epidote-amphibolite– to upper-amphibolite-facies metamorphic conditions of >4 kb and ~550–650 °C (Thompson, 1976; Apted and Liou, 1983). Greenschist and lower-metamorphic-grade alteration and vein mineral assemblages are very similar to those noted for the mafic intrusive suite.

Lizardite and chrysotile dominate the metamorphic mineral assemblages associated with the serpentine gneiss and suggest lower greenschist-facies conditions. The close proximity of the serpentinite gneiss to amphibolitic rocks and their apparent shared structural features suggest that the serpentinite gneiss may have experienced higher-metamorphic-grade conditions that were completely retrograded to the current greenschist-facies mineral assemblages. Alternatively, the greenschist-facies conditions may be primary, with a tectonic contact between the amphibolites and serpentinite gneiss. Similar amphibolite–serpentinized peridotite relations are described from parts of the western Paleozoic and Triassic belt (Snoke, 1977; Donato, 1992; Kays, 1992).

Structural Fabric Elements

Structural and textural features of rocks in the Onion Camp complex distinguish four deformational episodes. The first episode, recorded only locally in the serpentinized peridotite, is defined by a pseudomorphed blastomylonitic texture interpreted as recording deformation during subsolidous plastic flow of mantle lithosphere beneath a spreading center. Presumably, this deformation accompanied the construction of an ophiolite or primitive-arc sequence represented by the metasedimentary and metavolcanic rocks.

The second episode is characterized by greenschist- to amphibolite-facies regional metamorphism. The primary structural feature is a well- to moderately developed gneissose to schistose foliation (S_1) defined by the alignment of metamorphic mineral grains; for example, mica, platy amphibole, and quartz ribbons (Figs. 6A and 7A). Tight to isoclinal intrafolial folds are locally preserved in outcrops comprised of mm-scale laminated, siliceous, and tuffaceous meta-argillite (Fig. 6B). Relatively inconspicuous linear elements were only locally observed in some amphibolites. However, a lineation ($L_{1\times2}$) is common in schistose rocks, produced by an intersection of S_1 and a second foliation/rock cleavage (S_2) formed during episode four (Figs. 7B and 8). At least two and possibly three generations of open to isoclinal folds are evident in some chert out-

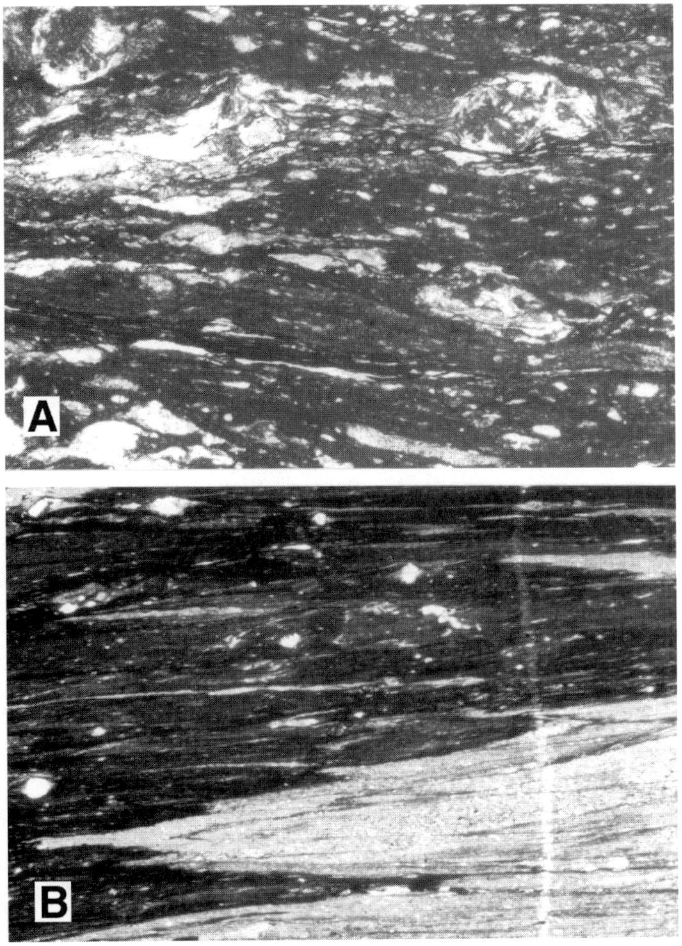

Figure 6. Photomicrographs of Onion Camp complex rocks, polarized light, horizontal field of view ~3 mm. (A) Strongly foliated, greenschist-facies metatuff with primary metamorphic minerals chlorite-actinolite-quartz-albite-oxide. (B) Highly strained silicic meta-argillite showing a tightly folded compositional boundary with axial surface foliation between quartz-rich and argillaceous layers.

crops, but it is unclear to which deformational episode these relate. Stereographic plots of S_1 and $L_{1\times2}$ (Fig. 7A and B) illustrate the apparent dispersal of amphibolite foliations and intersection lineations, probably the result of later deformational events.

The third episode is characterized by brittle fragmentation, cataclasis, and pervasive greenschist-facies and lower-grade alterations that affected all the rocks of the complex but are most common in the mafic intrusive suite. Zones of brecciation are commonly cemented by calcite and prehnite.

The fourth episode is characterized by regional to outcrop-scale, greenschist-facies folds, foliation, and faults. Structures include a northeast-striking, southeast-dipping foliation (S_2; Fig. 7C) defined by the alignment of metamorphic minerals, such as chlorite, and by flattening of crystal and lithic fragments; a coplanar, locally developed syn-metamorphic foliation (S_3) defined by a weakly developed spaced cleavage; syn-metamorphic, shallowly northeast-southwest plunging, tight to near-isoclinal F_1 and F_2 folds (Fig. 7D, E, and F); and postmetamorphic, moderately east plunging, open F_3 folds, and kink bands (Fig. 7F). Foliations S_2 and S_3 are well developed to absent, depending on the rock type. The S_3 foliation is scarce and is restricted to the slaty rocks, where it occurs in the axial-surface regions of tight F_2 folds, which fold S_2. The syn-metamorphic folds and linear structural features of the Onion Camp complex area are generally northeast–southwest-trending. F_1 folds commonly have attenuated and truncated limbs.

Fossil Ages

A sample of red chert (sample CP-5; Table 1) collected for radiolarians near Onion Camp yielded mostly amorphous forms, with the exception of a possible Late Triassic radiolarian form (*Canoptum* sp., M. Silk, written commun., 1991; Yule et al., 1992). Two additional samples collected from the same locality yielded scarce Jurassic, but no Triassic, forms (C. Blome, personal commun., 1993). These data support a Late Triassic to Jurassic age for chert near Onion Camp, consistent with chert ages from the Rattlesnake Creek terrane in the hanging wall of the Orleans/Preston Peak thrust (Irwin et al., 1982).

Radiometric Ages

Radiometric ages from Onion Camp complex rocks obtained for this study include two $^{40}Ar/^{39}Ar$ ages and two U-Pb zircon ages (Table 2). Brown pleochroic hornblende separated from two amphibolite gneiss samples collected near Squaw Mountain yield $^{40}Ar/^{39}Ar$ plateau cooling ages of 173 ± 0.6 Ma (EDM-10) and 169.6 ± 0.7 Ma (EDM-81A) (Hacker et al., 1995). These data suggest the amphibolites cooled below a temperature of ~500 °C at ca. 170 Ma.

U-Pb zircon ages were obtained from two plagiogranite dikes that intrude serpentinized peridotite near Whetsone Butte (JM-10 and JM-11; Fig. 2). Sample JM-11 yields a concordant age of 173 ± 1 Ma and sample JM-10 shows slight Pb loss with an intercept age of 175 ± 2 Ma (Fig. 9). Peridotite of the Onion Camp complex therefore predates ca. 173 Ma; in addition, the apparent lack of rodingitization of the plagiogranite dikes suggests that at least one serpentinization event occurred prior to ca. 173 Ma.

Geochemical Data

A wealth of geochemical data exists for metavolcanic rocks from the various lithotectonic units of the Klamath Mountains province (Barnes et al., 1995). The data set is complete enough

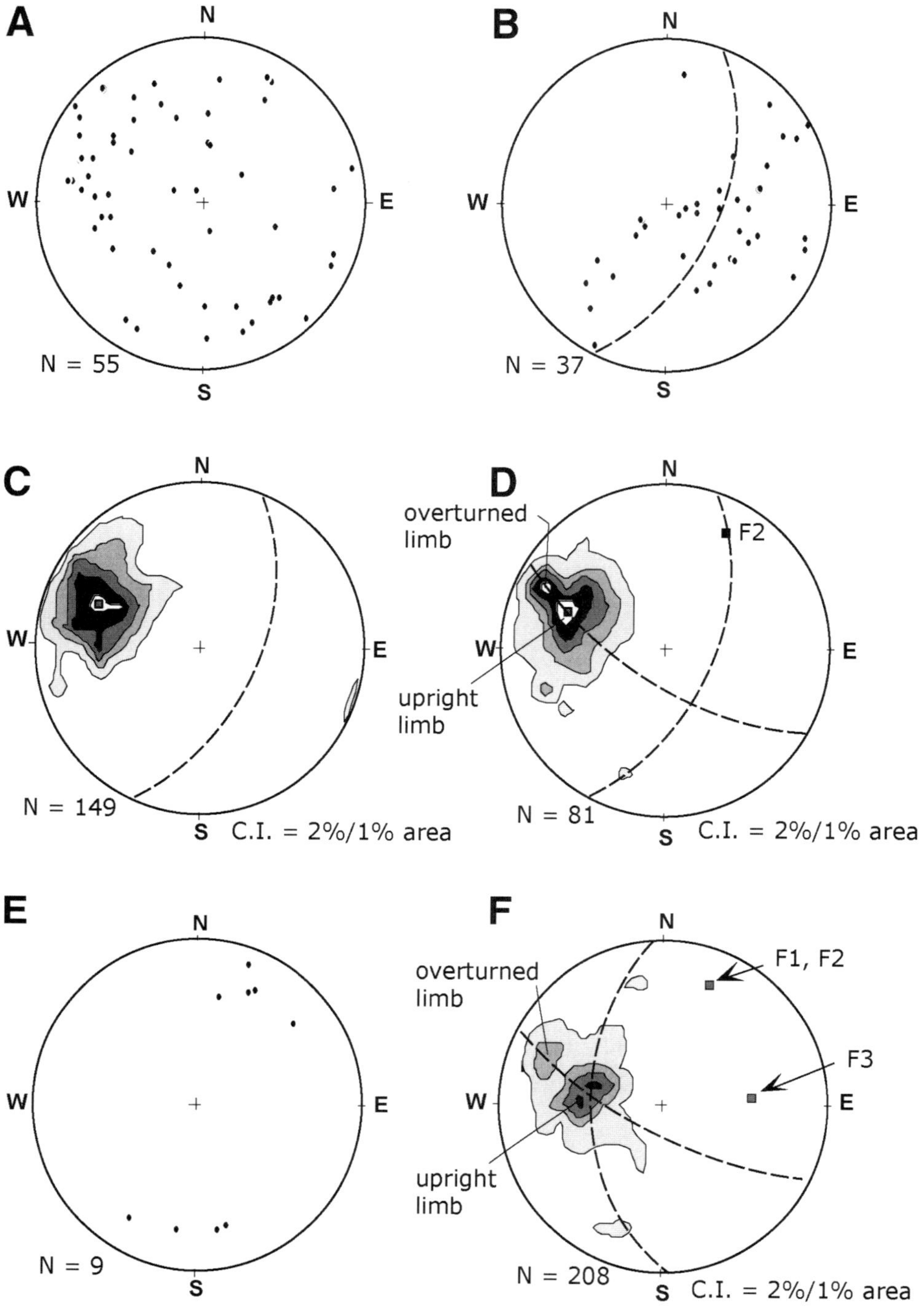

Figure 7. Lower hemisphere, equal-area stereographic plots of structural data. Onion Camp complex. (A) Poles to foliation in amphibolite (S_1); (B) intersection lineation ($L_{1\times2}$) measured on S_2 foliation planes (great circle = average S_2 orientation); (C) poles to regional foliation (S_2) and average S_2 orientation (great circle); (D) Rogue and Galice Formations and Fiddle Mountain olistostrome: poles to S_2, average S_2 (northeast-striking great circle), and northwest-striking great circle whose pole defines average F_2 fold; (E) Galice Formation: gently plunging north-northeast and south-southwest F_1 and F_2 hinge lines; (F) Rogue Formation: poles to bedding with two primary maxima define limbs of northeast-plunging, northwest-vergent, tight F_1 and F_2 folds (pole to northwest-striking great circle); secondary maxima define limbs of east-plunging, upright, open F_3 folds (pole to north-striking great circle). C.I.—contour interval.

that the characteristic geochemical signatures of numerous lithotectonic units are well defined. Three rock associations—the crustal sequence of the Josephine ophiolite, the Rogue Formation, and the Rattlesnake Creek terrane—exhibit important differences in their diagnostic geochemical signatures.

Ten greenstone and pillow lava samples from the metavolcanic sequence were analyzed for major and trace element abundances (Table 3), with rare-earth element (REE) abundances determined for two of these samples (Table 4). Though the major elements may be mobile during metamorphism, most trace and rare-earth elements are considered immobile at greenschist and lower metamorphic grades (e.g., Shervais, 1982).

The data in Barnes et al. (1995) provide criteria to evaluate the possible correlatives to the Onion Camp complex. In support of the age and lithologic similarities, the geochemical data in Tables 3 and 4 and Figure 10 exhibit trends that are quite similar

Figure 8. Photograph of steeply dipping metachert in the Onion Camp complex (rock hammer for scale). Note down-dip lineation, produced by an intersection of S_1 and a rock cleavage parallel to the later, regional foliation (S_2), typical of metavolcanic and metasedimentary rocks of the complex.

to those of the lower mélange unit of the Rattlesnake Creek terrane (Wright and Wyld, 1994). The mafic rocks of the Rattlesnake Creek terrane and Onion Camp complex fall in the compositional range of 46–55% SiO_2 and display a wide range of FeO/MgO values and generally high TiO_2 concentrations (Wright and Wyld, 1994; Barnes et al., 1995). Also, the Onion Camp complex metavolcanic rocks plot within the same fields as the lower mélange unit of the Rattlesnake Creek terrane on trace element discrimination diagram plots (Fig. 10). REE data (Fig. 11) from metavolcanic rocks of the Onion Camp complex show a strong enrichment in light REE typical of an alkalic, within-plate basalt (WPB) (pillow basalt sample EDM-83B), and a flat REE pattern typical of normal tholeiitic mid-ocean-ridge basalt (N-MORB) (greenstone sample OM-11); very similar to the red chert-basalt breccia and metabasalt association, respectively, of the lower mélange unit, Rattlesnake Creek terrane (Wright and Wyld, 1994).

Possible Broad Correlations

The rocks of the Onion Camp complex comprise a lithologically distinct petrotectonic unit that previously was unrecognized in the western Jurassic belt. Age relations show that the red chert-basalt association is Triassic(?) to Early Jurassic in age, and postmetamorphic dikes indicate that the serpentinized peridotite predates ca. 173 Ma. Distinctive lithotypes of the complex include the N-MORB and WPB metabasalts (usually associated with red chert), a scarce chert-clast conglomerate, and an amphibolite and serpentinized peridotite association. These features of the Onion Camp complex are also common to the Rattlesnake Creek terrane of the western Paleozoic and Triassic belt (Irwin, 1972), and are strikingly similar to the lower mélange unit of the Rattlesnake Creek terrane (Wright and Wyld, 1994). Heterogeneous mafic complexes that intrude the Onion Camp complex (Yule, 1996; this study) and the Rattlesnake Creek terrane near Preston Peak (Snoke, 1977) further support a correlation between the Onion Camp complex and Rattlesnake Creek terrane.

FIDDLER MOUNTAIN OLISTOSTROME

Overview

The Fiddler Mountain olistostrome is named for natural exposures on the ridgelines and flanks of Fiddler Mountain. Additional exposures occur along the Babyfoot Lake–Onion Camp U.S. Forest Service access road (Fig. 2). The widest part of the outcrop belt coincides with Fiddler Mountain. Isolated patches of the olistostrome are not shown on Figure 2, but their locations are shown on maps in Yule (1996). The olistostrome occupies a position at the base of the interfingering and overlapping Rogue and Galice Formations, and is deposited on a variable substratum comprised of both the Onion Camp complex and northern outliers of the Josephine ophiolite.

TABLE 1. RADIOLARIAN FOSSIL SAMPLES AND LOCALITIES

Station	Description	Latitude (N) and longitude (W) (°/′/″)	Interpreted age	Species list
1. CP-37	Tightly folded white-gray chert; N46E, 31SE.	42/14/28.5 123/47/37.5	Late Jurassic (Upper Kimmeridgian–Lower Tithonion?)	*Lupherium* sp. *Praeconocaryomma magnimamma* Large spheres (unidentifiable)
2. CP-5	Locality of Roure and DeWever (1983). Isolated exposure of well-bedded red chert; N20E, 39SE.	42/14/19 123/46/52.5	Upper Triassic(?)–Bajocian	Probable *Conoptums*

TABLE 2A. URANIUM-LEAD ISOTOPIC DATA

Size (µm)†	Zircon fraction properties††	Wt. (mg)@	U (ppm)	$^{206}Pb^*$ (ppm)	Atomic ratios $^{206}Pb/^{204}Pb$	$^{206}Pb^*/^{238}U$	$^{207}Pb^*/^{235}U$	$^{207}Pb^*/^{206}Pb^*$	Apparent ages (m.y.)§ $^{206}Pb^*/^{238}U$	$^{207}Pb^*/^{235}U$	$^{207}Pb^*/^{206}Pb^*$
93DY-JM-10: Plagiogranite dike cutting serpentinized peridotite of the Onion Camp complex (123°48′35.5″W, 42°14′35.5″N).											
<45[2]	Clear, Eu > Fr, 2:1 to 5:1	2.3	379.8	8.37	13,361	0.02547(17)	0.17444	0.04970(12)	162.1	163.3	181 (6)
45-64[2]	Clear, Eu > Fr, 2:1 to 5:1	1.4	726.6	16.69	12,111	0.02654(18)	0.18107	0.04950(55)	168.9	169.0	171 (26)
64-80[2]	Clear, Eu > Fr, 2:1 to 5:1	0.8	562.3	13.21	8013	0.02760(17)	0.18817	0.04947(18)	175.5	175.1	170 (8)
91DY-JM-11: Plagiogranite dike cutting serpentinized peridotite of the Onion Camp complex (123°48′35″W, 42°14′45″N).											
45-64[3]	1:1 to 4:1 w/clear incl., Eu > Fr	2.3	198.3	4.67	9,855	0.02723(15)	0.18604	0.04957(13)	173.2	173.3	175 (6)
64-100[3]	1:1 to 4:1 w/clear incl., Eu > Fr	1.3	215.2	5.08	8,002	0.02725(15)	0.18650	0.04965(15)	173.4	173.6	178 (7)

† Fractions separated by grain size and magnetic properties. Magnetic properties are given as nonmagnetic split on Frantz Isodynamic Separator for 1.7 amps with side/front slopes of [1]0.5/20, [2]1/20, [3]2/20, or [4]5/20. Dissolution and chemical extraction technique modified from Krogh (1973).

†† All samples are hand-picked to 99.9% purity prior to dissolution. Abbreviations: Eu = euhedral grains; Fr = fragmented grains; ratios given are the respective length:width ratios of euhedral zircon grains; incl. = inclusion; Lt = light.

@ Due to uncertainties in weight of sample, U-Pb concentration may be in error, use of mixed ^{205}Pb-^{230}Th-^{235}U spike ensures that age of sample is correct.

* Radiogenic/nonradiogenic correction based upon 40 picogram blank Pb (1:18.78;15.61;38.50) and initial Pb approximations: 206/204 = 18.6; 207/204 = 15.6; 208/204 = 38.2.

§ Decay constants used in age calculation: $\lambda^{238}U = 1.55125 \times 10^{-10}$, $^{235}U = 9.98465 \times 10^{-10}$ (Jaffey and others, 1971); $^{238}U/^{235}U$ atom = 137.88. Uncertainties (±) in radiogenic ratios calculated by quadratic sum of total derivatives of ^{238}U and $^{206}Pb^*$ concentration and $^{207}Pb^*/^{206}Pb^*$ equations with error differentials defined as follows: (1) Isotope ratio determinations from standard errors (β/n) of mass spectrometer runs plus uncertainties in fractionation corrections based on multiple runs of NBS 981, 983, and U500 standards; (2) Spike concentrations from range of deviations in multiple calibrations with normal solutions; (3) Spike compositions from external precisions of multiple isotope ratio determinations; (4) Uncertainty in natural $^{238}U/^{235}U$ from Chen and Wasserburg (1981); and (5) Nonradiogenic Pb isotopic compositions from uncertainties in isotope ratio determinations of blank Pb and uncertainties in composition of initial Pb from estimates of regional variations based on references given above and consideration of rock type.

TABLE 2B. $^{40}Ar/^{39}Ar$ APPARENT AGES

Lithologic unit	Station	Latitude (N) and longitude (W) (°/′/″)	Description	Mineral	Plateau age (Ma)
Onion Camp complex	91-EDM-81A	42/20/40N 123/39/40W	Squaw Mountain; amphibolite gneiss	Hornblende	169.6 ± 0.5
	90-EDM-10	42/18/58N 123/40/46W	Squaw Mountain; hornblende schist	Hornblende	173.1 ± 0.6

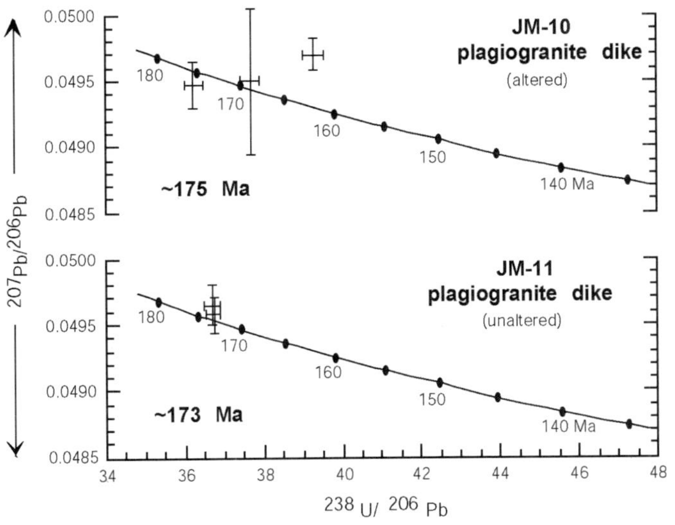

Figure 9. Tera-Wasserburg plots of U-Pb and Pb-Pb radiogenic age data (Tera and Wasserburg, 1972) for plagiogranite dikes cutting serpentinized peridotite of the Onion Camp complex (JM-10 and JM-11).

Rock Units

Metasedimentary Polymictic Breccia and Conglomerate. Poorly exposed breccia, conglomerate, and sandstone occur throughout the olistostromal unit (Fig. 12). The dominant clast types are aphyric to porphyritic green to gray metabasalt, fine- to medium-grained metadiabase, and medium- to coarse-grained gabbro. Secondary clast types include chert (red and gray), slaty argillite, serpentinite, and serpentinized peridotite and scarce occurrences of quartzite, chlorite schist, phyllonite, and amphibolite. In cases where the olistostromal rocks are deposited upon serpentinized peridotite, the breccias consist entirely of serpentinite and serpentinized peridotite clasts (Fig. 13). A rare but distinctive ophicalcite breccia, exposed in a roadcut near locality "C" on Figure 2, consists of matrix-supported serpentine clasts in a hematitic and calcareous serpentine mud matrix. The matrix materials generally consist of clay- to angular sand-size and coarser grains derived from the same material as the clasts.

Grain size ranges from silt- to boulder-sized, with coarse pebble- and fine cobble-sized the most common. Matrix in the coarse pebble and fine cobble breccias consists of silt- to fine pebble-sized material, similar in composition to the larger clasts. Matrix:clast ratios are highly variable but are typically >1:10. Fine cobble and coarser breccias are typically massive, but layering is common in pebble breccias, conglomerates, and sands (Fig. 12).

Megabreccia. Decameter- and larger-sized (≤100-m) blocks of gabbro, diabase, and basalt are exposed along the south flank of Fiddler Mountain and comprise ~50% of the complex. Altered and fragmented gabbro and diabase are the most common megabreccia clast lithology. The largest blocks often consist of sheeted mafic dikes, commonly with gabbro and/or basaltic screens. The size of these blocks gives the impression that the unit is a heterolithologic intrusive complex or a tectonic mélange. However, zones of relatively nonresistant, polymictic-clast breccias occur between the large-scale blocks, and indicate that the large blocks are entirely part of a sedimentary breccia/megabreccia deposit.

Metachert and Slaty Argillite. Horizons of white to gray bedded chert and slaty argillite comprise <10% of the olistostrome and occur as 1- to 20-m-thick, chert-rich intervals that can be traced laterally for up to 100 m. The chert is mostly recrystallized, but oval to round vugs, mm-sized and finer, are evident in broken faces of gray chert. Samples processed for radiolarian yield a poorly constrained Late Jurassic (Kimmeridgian?) age for these deposits (Yule et al., 1992; Table 1).

Provenance and Depositional Environment

Clasts correlate with rock types found in the Onion Camp complex and the northern outliers of the Josephine ophiolite. The primary outcrop belt overlies the Onion Camp complex (Fig. 2), but isolated outcrops too small to be shown on Figure 2 overlie northern outliers of Josephine ophiolite. Both could be the source terrane for the olistostromal deposits. The Fiddler Mountain olistostrome, therefore, is interpreted to have originated as a talus breccia along a marine escarpment that exposed the Onion Camp complex and/or Josephine ophiolite, perhaps along a transform fault in the Josephine marginal basin. Pelagic sedimentary rocks would have accumulated between episodes of catastrophic deposition of the megabreccia and breccia deposits.

Metamorphic Conditions

The metasedimentary rocks of the Fiddler Mountain olistostrome consist chiefly of prehnite-pumpellyite to lowermost greenschist-facies mineral phases (Liou et al., 1974; Apted and Liou, 1983). Original igneous mineral phases in clasts are commonly altered, partly or wholly, to various hydrous phases, but relict calcic plagioclase and pyroxene (augite?) are present in some samples. The typical mineral assemblage for mafic clasts is albite + quartz + epidote + chlorite + white mica + calcite + prehnite + opaque minerals, and for argillaceous rocks is quartz + albite + white mica + clinozoisite + chlorite + opaque minerals. Pumpellyite is scarce, but was identified in some metabasaltic clasts. Pyrite is a common accessory mineral phase in gray, slaty argillite.

Ubiquitous lower greenschist-facies static alteration is present in all clast types and probably records hydrothermal alteration of the bedrock prior to incorporation of the clasts into the sedimentary breccia deposit. This interpretation is supported by the observation that regional foliation and shear planes cut veins and fractures associated with the static alteration.

TABLE 3A. GEOCHEMICAL SAMPLES AND LOCALITIES

Station	Lithology	Latitude (N) (°/′/″)	Longitude (W) (°/′/″)	Remarks
CP-3	Pillow lava	42/14/11	123/47/46	Collected from quarry near Onion Camp.
OM-8	Pillow lava	42/26/26	123/35/44	
EDM-83A	Pillow lava	42/17/20	123/37/33	River-polished exposure of pillow lavas adjacent the Illinois River.
EDM-83B	Pillow lava	42/17/20	123/37/33	River-polished exposure of pillow lavas adjacent the Illinois River.
EDM-87	Pillow lava	42/17/48	123/43/42	
G-14C	Pillow lava	42/32/37	123/30/40	Roadcut opposite highway overlook at Hellgate Canyon, Rogue River gorge.
OM-11	Massive greenstone	42/27/55	123/35/44	Collected from quarry near Onion Mountain.
OM-14	Massive greenstone	42/29/52	123/32/36	
G-14A	Massive greenstone	42/32/37	123/30/40	Roadcut opposite highway overlook at Hellgate Canyon, Rogue River gorge.
CP-45	Massive greenstone	42/13/12	123/48/22	Collected from cirque wall above Babyfoot Lake.
CR-22	Massive greenstone	42/25/38	123/37/33	
JM-11	Dike	42/14/45	123/48/35	Zircon sample (ca. 173–174 Ma). Plagiogranite dike cutting Onion Camp complex peridotite near Whetstone Butte.

TABLE 3B. MAJOR AND TRACE ELEMENT ANALYSES FROM THE ONION CAMP COMPLEX

	Pillow lava						Massive Greenstone					Dike
Sample	CP-3	OM-8	EDM-83A	EDM-83B#	EDM-87	G-14C	OM-11#	OM-14	G-14A	CP-45	CR-22	JM-11†
SiO_2	48.51	48.54	46.31	45.68	52.12	52.49	54.06	53.98	49.02	50.14	49.52	76.63
TiO_2	2.66	1.38	2.20	2.06	0.70	0.77	1.18	0.51	1.27	0.93	1.38	0.22
Al_2O_3	14.45	14.57	15.36	15.39	16.33	15.45	13.47	14.79	15.89	16.19	12.12	13.33
$Fe_2O_3^*$	13.10	12.24	10.75	9.65	10.26	8.38	7.82	7.99	9.25	8.89	13.70	1.12
MnO	0.20	0.21	0.16	0.15	0.15	0.13	0.17	0.14	0.15	0.15	0.22	0.02
MgO	4.47	5.93	6.70	5.00	6.61	9.12	7.70	8.23	6.46	6.63	7.17	0.53
CaO	8.86	10.05	9.73	11.42	5.44	7.42	8.57	8.45	9.69	13.14	8.86	2.27
Na_2O	4.58	3.68	4.03	4.44	4.77	4.14	4.09	3.23	3.68	2.21	2.83	4.68
K_2O	0.27	0.42	0.30	0.48	0.43	0.12	0.75	0.34	0.14	0.03	0.22	1.06
P_2O_5	0.39	0.20	0.39	0.38	0.09	0.09	0.16	0.08	0.16	0.10	0.11	0.03
LOI	2.84	2.43	4.44	6.33	3.41	2.74	2.20	2.99	2.66	3.69	3.09	0.68
Total	100.33	99.66	100.37	100.97	100.31	100.85	100.16	100.75	98.37	102.11	99.22	100.57
Rb	7	9	6	8	7	3	13	8	4	2	7	22
Sr	225	245	361	419	145	137	90	156	256	56	100	121
Zr	197	81	156	146	44	57	99	36	104	65	68	105
Y	43.3	39.7	26.1	24.7	23.8	22.1	29.6	16.5	32.8	24.5	32.8	10.9
Nb	28	4	30	32	1	1	4	b.d.	2	b.d.	4	5
Ba	70	46	132	201	123	27	62	155	33	11	26	355
Sc	33.4	43.8	24.4	21.9	35.6	38.7	28.4	36.2	35.5	37.6	49.3	2.3
V	285	279	209	204	296	223	212	228	245	195	325	16
Cr	64	212	200	179	111	547	247	341	233	209	109	11
Ni	44	53	138	121	91	252	97	101	90	75	50	20
Cu	55	23	51	41	60	75	53	74	52	104	85	n.d.
Zn	115	83	78	74	87	57	55	51	63	46	103	9
Ti	15,947	8271	12,350	13,178	4180	4598	7070	3081	7641	5551	8273	1338

Notes: Major element data given in wt%; minor and trace element data in ppm. Analyses done at Texas Tech University, Lubbock, by inductively coupled plasma mass spectrometry, inductively coupled atomic emission mass spectrometry, and, for Rb, by flame emission spectroscopy. Some of the trace elements (Sr, Ba, Zr, and Y) were analyzed twice; once with dilute solutions used for major elements and again with concentrated solution used for the other trace elements. Values from the two runs were averaged with the following exceptions: (1) for high Sr values, the trace element run values were used; (2) for high Ba values, the major element run values were used; and (3) for Y, the trace element runs were used because of a better fit to standards. b.d.—below detection limits; Greenst.—greenstone; LOI—loss on ignition; n.d.—no data available.

*Total Fe content is given as Fe_2O_3.

†U/Pb geochronologic samples.

#Samples with rare-earth element analyses (see Tables 2–5).

TABLE 4. RARE-EARTH ELEMENT COMPOSITIONS OF SELECTED BASALTS

Sample	La	Ce	Nd	Sm	Eu	Tb	Yb	Lu
92-DY-EDM-83B	21.5	41.4	23	5.3	1.50	0.81	2.07	0.31
92DY-OM-11	4.2	11.1	8.9	2.9	1.00	0.67	2.51	0.40

Notes: Data in ppm. Samples are from Onion Camp complex.

The overall metamorphism of the Fiddler Mountain olistostrome may be best described as a low-pressure, low-temperature type regional metamorphism associated with strong penetrative deformation.

Structural Elements

Structures in the Fiddler Mountain olistostrome are similar in orientation and style to those observed in the Rogue and Galice Formations, units that interfinger and overlap the olistostromal units and suggests that the Fiddler Mountain, Rogue, and Galice units share the same deformational history. For example, S_2 foliations in the olistostrome strike northeast and dip southeast and are coplanar with those observed in the Rogue and Galice Formations (Fig. 7D). In addition, bedded chert exposed immediately to the north of Fiddler Mountain in the hinge region of the synclinoria (Fig. 2) contains F_1 and F_2 folds that trend north-northeast to south-southwest and plunge <15°, similar to those observed in the Galice Formation (Fig. 7E). Chert and argillite layers near Fiddler Mountain define a synclinorium with an overturned, steeply southeast-dipping western limb and an upright, moderately southeast-dipping eastern limb similar to the F_1 and F_2 folds observed in the Rogue Formation (Fig. 7F). The chert and argillite are also deformed by east-west open (F_3) folds (Fig. 7F). Large-offset faults are difficult to recognize in the field, but numerous small-offset faults can be observed cutting sand, conglomerate, and pebble breccia layers (Fig. 12B).

Possible Regional Correlations

Deposits similar in description to the Fiddler Mountain olistostrome occur at various localities throughout the Klamath Mountains and California Coast Ranges; for example, the Lems Ridge olistostrome (Ohr, 1987); the Devils Elbow ophiolite (Wyld and Wright, 1988); and the Preston Peak ophiolite (Snoke, 1977). In these locations, and in this study area, coarse, polymict breccia/megabreccia deposits generally overlie Late Jurassic ophiolitic crust and underlie and interfinger with the basal section of the Galice flysch; in addition, the olistostromal deposits are spatially associated with the outcrop belts of the Rattlesnake Creek terrane and/or the Josephine ophiolite. These relations suggest that localized areas of considerable topographic relief occurred within the Late Jurassic Josephine basin, probably as a result of extensional and/or transform faulting.

DISCUSSION

The northeast-striking belt of Rattlesnake Creek terrane contained within the western Jurassic belt of Oregon corroborates previous models for the Late Jurassic tectonic evolution of the Rogue–Chetco/Josephine paired arc/ophiolite basin (Snoke, 1977; Saleeby et al., 1982; Harper and Wright, 1984). This model invokes rifting of preexisting Klamath terranes—primarily the western Paleozoic and Triassic belt—to produce an active arc, inter-arc basin, and remnant arc triad along the margin of western North America (Fig. 14A). This model predicts that fragment(s) of older oceanic crust may occur as rift-screens within ophiolitic crust and/or as constructs for the active arc. It is, therefore, not surprising to discover a fragment of the Rattlesnake Creek terrane within the western Jurassic belt, beneath and adjacent to the Rogue–Chetco arc forming the "active arc" rift-edge facies, perhaps a matching piece to the "remnant arc" rift-edge facies of Snoke (1977) exposed in the Preston Peak region. This finding illustrates the complexity of the stratigraphic and structural relations in the Rogue–Chetco/Josephine basin prior to their accretion along the western margin of North America in Late Jurassic time.

Basement-Level Exposures in the Rogue–Chetco/Josephine Basin

The structural features and framework of marginal ocean basins is known from studies in modern settings (e.g., Karig, 1971; Uyeda, 1977; Curray et al., 1979; Pearce et al., 1984; Stern and Bloomer, 1992; Parson and Wright, 1996; Lagabrielle et al., 1997). However, modern marginal ocean basin lithosphere is concealed by water or buried beneath thick surficial deposits. Without direct observation, several questions regarding the tectonic framework and evolution of marginal ocean basin lithosphere remain unanswered. For example, relatively little is known regarding (1) the age of the mantle lithosphere relative to the overlying crustal sequences and the nature of the boundary between lithospheric and asthenospheric mantle; (2) the nature of the transition between arc and oceanic crust; (3) the character of the relatively thick crust and mantle lithosphere in oceanic arcs and remnant arcs; and (4) the relative similarities and/or differences between active and remnant arc basement lithotypes. The relatively deep-level exposures of the Rogue–Chetco/Josephine marginal ocean basin provide an opportunity to explore these relatively poorly understood relations through direct observation.

The lithotectonic elements of the Rogue–Chetco/Josephine ocean basin can be grouped into three levels (Figs. 14B and 15): (1) a basal, upper-mantle lithosphere level composed of two distinct types of variably serpentinized peridotite; (2) an intermediate lower- to mid-crustal level composed of a heterogeneous complex of arc-plutonic rocks and variably metamorphosed mafic volcanic and metasedimentary rocks; and (3) an upper-crustal level composed of arc-derived flysch and olis-

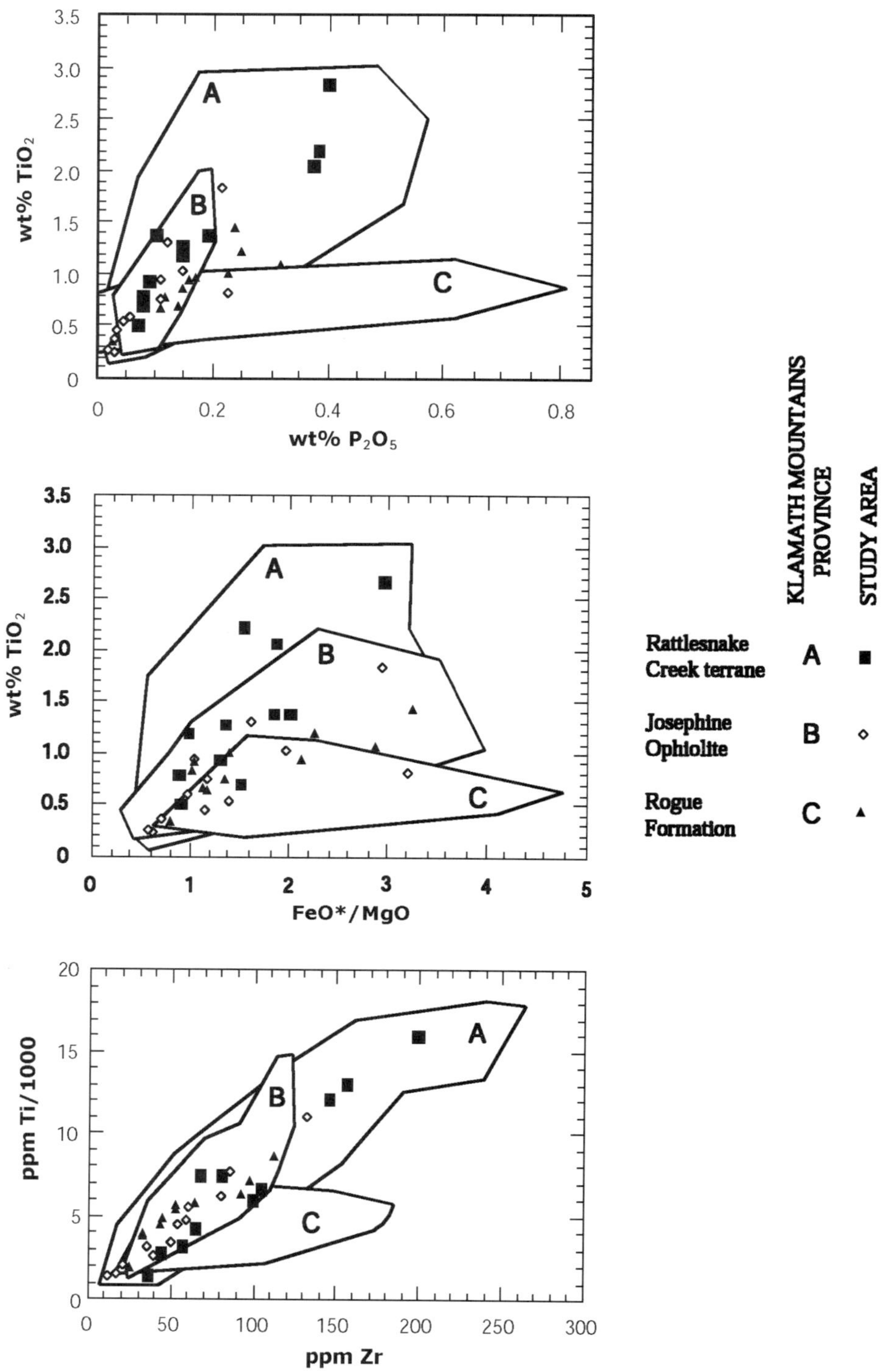

Figure 10. Plots of P_2O_5, FeO*/MgO, and Zr vs. Ti abundance/concentrations for basaltic rocks of the study area. Fields A, B, and C are fields for the entire Klamath Mountains province (Barnes et al., 1995) from rocks of the Rattlesnake Creek terrane, the Josephine ophiolite, and the Rogue Formation, respectively.

tostromal deposits. The basal level contains peridotite that is spatially and temporally linked to either the Josephine ophiolite (Harper, 1984) or Onion Camp complex (this study). The intermediate level consists of the Chetco plutonic complex and related metamorphic rocks (Jorgenson, 1970; Hotz, 1971; Dick, 1976; Loney and Himmelberg, 1977; Garcia, 1979, 1982; Saleeby, 1984; Coleman and Lanphere, 1991; Yule, 1996), Onion Camp complex (this study), and mafic sequence of the Josephine ophiolite (Harper, 1984, 2003). The upper level consists of the Rogue Formation (Garcia, 1979, 1982), Galice Formation (Harper, 1980; Miller and Saleeby, 1995), and Fiddler Mountain olistostrome (this study). Therefore, at least eight lithotectonic elements comprise the Late Jurassic lithosphere of the western Jurassic terrane.

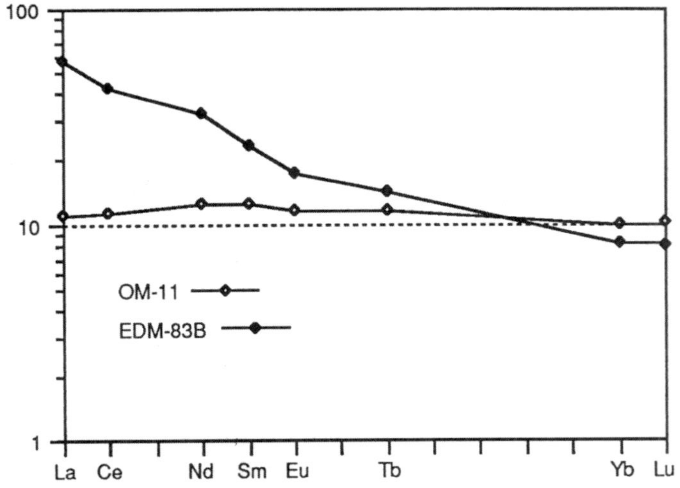

Figure 11. REE abundances of two Onion Camp complex basalts (in ppm), normalized to chondrite values, showing two distinctly different patterns. Sample OM-11 exhibits a flat, MORB-like pattern and sample EDM-83B exhibits a light REE-enriched pattern, typical of alkalic within-plate basalts.

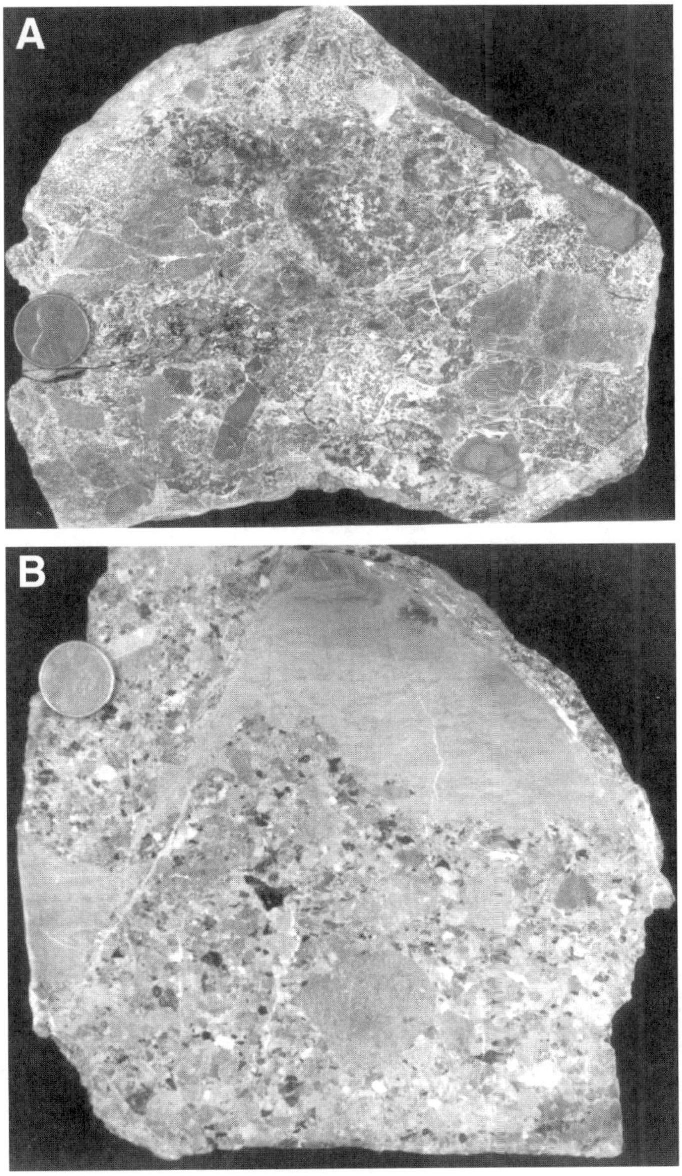

Figure 12. Photographs of slabs of Fiddler Mountain olistostrome rocks. (A) Ophiolite-clast conglomerate, common constituent of the Fiddler Mountain olistostrome, clasts consist entirely of gabbro, diabase, and greenstone. (B) Ophiolite-clast conglomerate with a medium-grained silty-sand interlayer, clast types are diabase, greenstone, chert, and secondary chert (small white clasts) and serpentinite (e.g., black clast near the center of the photo). Note the small fault offset of the silty sand layer.

Two Types of Peridotite

The Josephine peridotite is exposed over ~700 km^2 of the California-Oregon border region. The peridotite is thought to represent the residue of partial melting (Dick, 1976) and presumably formed during the genesis of the Late Jurassic Josephine ophiolite sequence (Harper, 1984). However, the northern and northeastern part of the Josephine peridotite sheet may consist of two distinct types of peridotite (Table 5; Figs. 1 and 2). Field relations and crosscutting ca. 173-Ma dikes (Yule, 1996; this chapter) in peridotite associated with the Onion Camp complex indicate that at least part of the peridotite massif is pre-Middle Jurassic in age. The fact that the ca. 173-Ma plagiogranite dikes are not rodingitized indicates that the Onion Camp complex peridotite experienced at least one episode of serpentinization prior to dike emplacement. The age of serpentinization is not constrained but may be as old as Late Triassic to Early Jurassic based on its association with metavolcanic rocks and chert of this age and on its tentative correlation with the lower mélange unit of the Rattlesnake Creek terrane (Wright and Wyld, 1994).

The northern region of the Josephine peridotite sheet is bounded by, and, near Onion Mountain, appears to be infolded with, greenschist- and amphibolite-facies rocks of the Onion Camp complex (Figs. 2 and 4). Field relations show a relatively sharp "serpentinization front" ~0.5–2 km from the crustal rocks (Fig. 2) that in places appears to mark a transition between highly serpentinized peridotite, characteristic of the Onion Camp complex, and little to moderately serpentinized peridotite, characteristic of the Josephine peridotite. No noticeable shearing or structural discontinuity was recognized along this serpentinization boundary. South and east of this boundary, relatively fresh peridotite can be traced ~30 km to the Smith River region, where the type locality of the crustal sequence of the Josephine ophiolite is exposed (Harper, 1984). In the type area, radiometric ages constrain the age of the ophiolite to be 162–164 Ma (Saleeby et al., 1982; Harper, 2003). The boundary between Onion Camp complex and Josephine peridotite is therefore interpreted to be the serpentinization front, which separates

Figure 13. Photograph of monolithogic, serpentinite-clast, and debris flow breccia located ~3 km to the northwest of Whetstone Butte (Fig. 2). Deposit occurs at the base of the overlapping Rogue Formation, which nonconformably overlie serpentinized peridotite of the Onion Camp complex.

>173-Ma, highly serpentinized peridotite from ca. 164-Ma, weakly serpentinized peridotite.

There are at least two explanations for the difference in apparent ages of the Onion Camp complex and Josephine peridotites. One considers the two peridotites to represent a >9 m.y. continuum of formation, from >173 Ma to ca. 164 Ma, perhaps spanning the full history of ophiolite genesis at Josephine basin spreading centers. Interestingly, dikes of ca. 172 Ma in northern California have been attributed to the onset of Josephine ophiolite genesis (Saleeby and Harper, 1993). However, this hypothesis cannot explain the structural and apparent stratigraphic association of >173-Ma peridotite with other rocks of the Onion Camp complex as old as Early Jurassic and perhaps Triassic. In the alternative explanation, the two peridotites are interpreted as forming at two distinct times. The Onion Camp complex peridotite formed in Triassic and Early Jurassic times and is associated with the crust rocks of the Onion Camp complex. It represents a rifted fragment or screen of preexisting oceanic lithosphere within the Late Jurassic Rogue–Chetco/Josephine basin (Fig. 14). The close association of the upper mantle and crustal rocks of the Onion Camp complex is therefore interpreted to be a composite fragment of preexisting Klamath lithosphere (Rat-

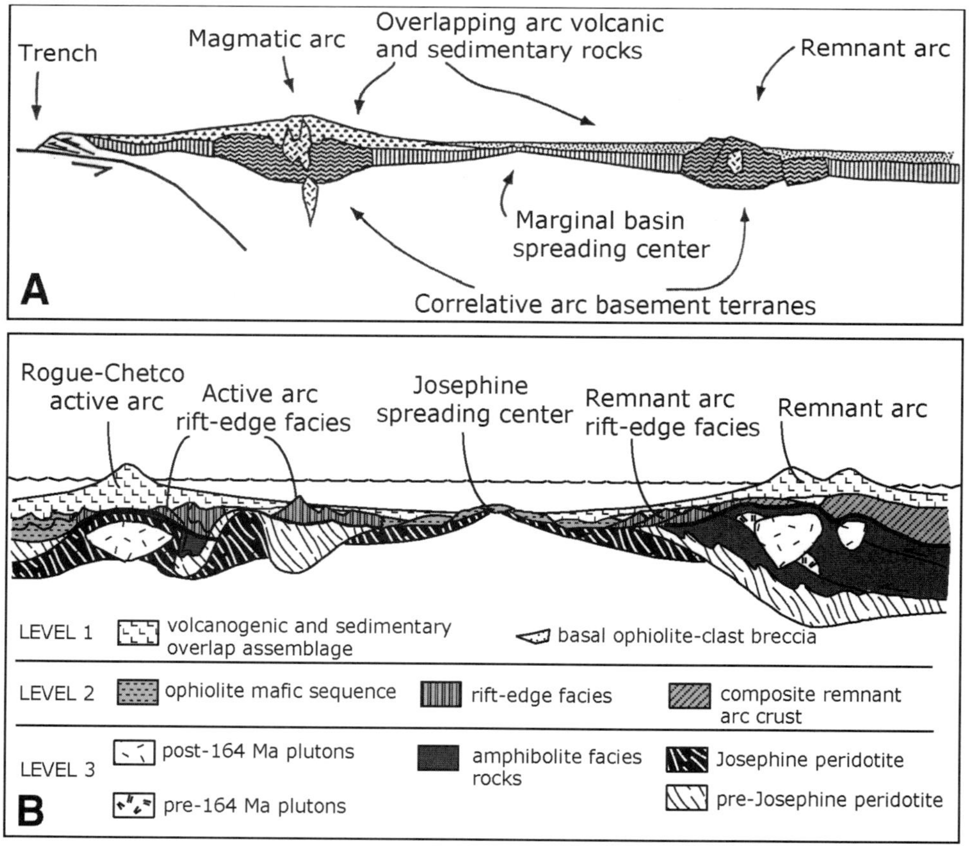

Figure 14. (A) Generalized cross-section through a marginal ocean basin (modified from Karig, 1971). (B) Simplified cross-section across the reconstructed (pre-accretion) Rogue–Chetco/Josephine marginal ocean basin system at ca. 155–160 Ma; vertical exaggeration ~3:1. The distance across the inter-arc basin is minimized to give equal emphasis to the three lithotectonic elements, the Rogue–Chetco active arc, Josephine inter-arc basin ophiolite, and remnant arc.

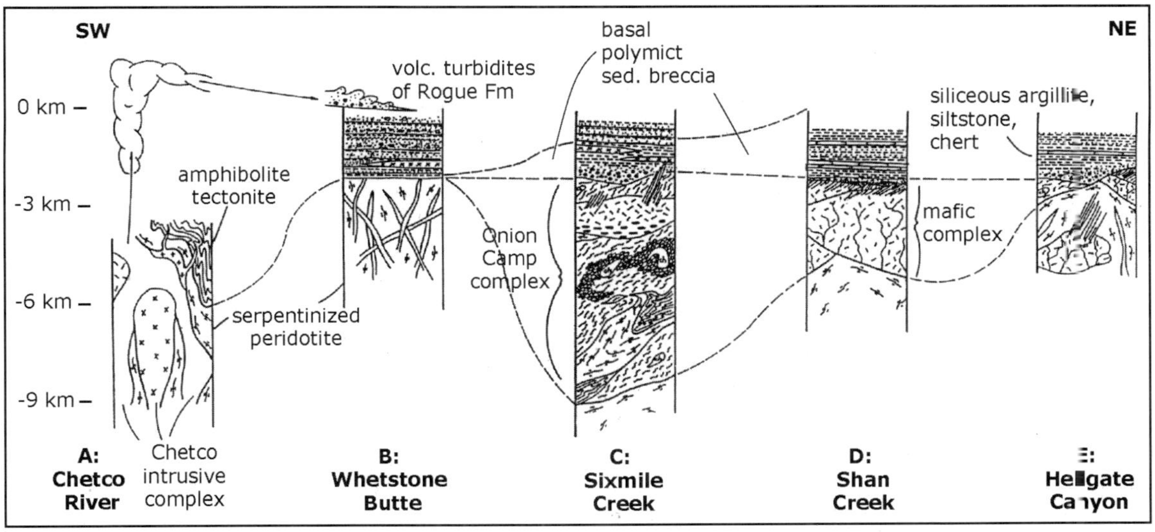

Figure 15. Diagrammatic columnar sections from five locations within the study area showing the three-tiered stratigraphy of the region (see Fig. 2 for locations of columns A–E). Upper layer is comprised of arc volcanogenic flysch (Rogue and Galice Formations) and Fiddler Mountain olistostrome. Intermediate layer of variable thickness is comprised of the Onion Camp complex and Josephine ophiolite. Deep layer is comprised of variably serpentinized peridotite, locally intruded by dikes. The Chetco plutonic complex intrudes and metamorphoses the middle and deep layers. Fm—formation.

tlesnake Creek terrane) that rifted from the margin of the Klamath Mountain province as the Josephine basin opened. In this scenario, it is conceivable that the contact between the Josephine and Onion Camp complex peridotites is the lithospheric/asthenospheric boundary prior to Josephine-age spreading.

Rift-Edge Facies

Most general cross-section views of modern marginal ocean basins (Fig. 14) show sharp boundaries that separate inter-arc basin crust from active arc or remnant arc crust. However, Snoke (1977) described a more gradual transition or "rift-edge" facies for rocks in the remnant arc setting, specifically the "Preston Peak ophiolite" (see Ohr, 1987; Snoke and Barnes, this volume). The rift-edge concept applies in very similar ways to the Onion Camp complex and Fiddler Mountain olistostrome of the western Jurassic belt described in this chapter. These units therefore are thought to represent the rift-edge facies for rocks in the active arc setting (Yule and Saleeby, 1993; Yule, 1996; Fig. 14).

TABLE 5. DISTINGUISHING FEATURES OF THE JOSEPHINE AND RATTLESNAKE CREEK TERRANE PERIDOTITE MASSES

Rattlesnake Creek terrane (RCT)	Josephine peridotite
1. Completely to moderately serpentinized (>50%).	1. Moderately to nonserpentinized, with pristine harzburgite and dunite locally common.
2. Amphibolite facies penetrative fabric, now mostly serpentinized; associated with Middle Jurassic (ca. 173 Ma) amphibolite tectonites.	2. Absence of amphibolite facies penetrative fabric.
3. Exposed on the seafloor, as evidenced by the nonconformable overlap of Rogue Formation strata.	3. No evidence of exposure on seafloor.
4. Anastomosing regional Nevadan (155–150 Ma) shear foliations common.	4. Anastomosing regional Nevadan (155–150 Ma) shear foliations rare.
5. At least Middle Jurassic in age, but may be Triassic in age by association with crustal rocks of the RCT. Cut by numerous generations of dikes ranging in age from ca. 175 to 150 Ma (Yule, 1996). Podiform chromite rarely present.	5. Probably Late Jurassic in age, the age of the crustal sequence of the Josephine ophiolite. Less commonly intruded by dikes whose ages range from ca. 160 to 150 Ma (Yule, 1996). Podiform chromite locally abundant.
6. Rodingite dikes rare.	6. Rodingite dikes common.

Distinctive features of the rift-edge facies in the western Jurassic belt, in the "active arc" setting include: (1) a heterogeneous mafic complex, chiefly consisting of gabbro and sheeted diabase and basalt dikes that intrude other units of the Onion Camp complex; (2) extensive vein and vein-cemented-breccia networks, and lower- and sub-greenschist-facies static alteration that affect rocks of the mafic complex and locally affect other units of the Onion Camp complex; and (3) ophiolite-clast megabreccia and breccia deposits interlayered with argillite and Late Jurassic (Kimmeridgian?) chert that overlap a basement mélange comprised of the Onion Camp complex and Josephine ophiolite. These features are consistent with a rift-dominated tectonic setting. For example, the abundance of sheeted dikes in the mafic complex suggests that extension played a key role during intrusion of the complex. Heat from shallow levels of the mafic complex probably drove hydrothermal systems that produced the vein networks and static alteration that occur throughout the Onion Camp complex. In addition, rifting could have produced ophiolite-cored escarpments that shed coarse olistostromal deposits.

Identifying a rift-edge facies beneath and/or adjacent to the active and remnant arcs, on either side of the Josephine basin (Fig. 14), should help constrain paleogeographic models for the Josephine ophiolite. The current position of the Onion Camp complex and Fiddler Mountain olistostrome, the active arc rift-edge facies, is ~50 km to the north of the Preston Peak ophiolite and Lems Ridge olistostrome, the remnant arc rift-edge facies (Fig. 13). Restoring a minimum of 40 km of east–west apparent shortening across the Preston Peak thrust (Harper et al., 1994) would place the two rift-edge facies at least 60 km to the northwest and southeast of each other, respectively (Fig. 13). Lateral displacement across the Preston Peak fault is not constrained; left- and right-oblique thrust motion across the fault would increase and decrease the apparent offset of these two belts, respectively. This model suggests that the Josephine basin may have opened during northwest–southeast-directed extension, a slight modification to the previous paleogeographic models (Pessagno and Blome, 1990; Alexander and Harper, 1992) that invoke north–south-directed spreading.

CONCLUSIONS

Two previously unrecognized lithotectonic units are present in the western Jurassic belt in the Oregon Klamath Mountains. These units, named the Onion Camp complex and Fiddler Mountain olistostrome, are elongate, lithosphere-scale blocks along the transition zone between the Late Jurassic Rogue–Chetco oceanic island arc and Josephine ophiolite. The Onion Camp complex is characterized by (1) N-MORB and WPB mafic metavolcanic rocks and associated Triassic(?) and Early Jurassic red chert; (2) scarce, but distinctive chert-clast conglomerate lenses in metasedimentary sequences; and (3) complexly folded regions of greenschist- to amphibolite-facies metavolcanic, metasedimentary, and serpentinized peridotite. The $^{40}Ar/^{39}Ar$ cooling ages of hornblende in amphibolite suggest that the amphibolite-grade metamorphism ended by ca. 169 Ma. Plagiogranite dikes with ca. 173-Ma U-Pb zircon ages cut the serpentinized peridotite. The Fiddler Mountain olistostrome consists of pelagic to hemipelagic rocks interlayered with ophiolite-clast conglomerate, breccia, and megabreccia. Radiolarians from chert interbeds indicate that the olistostrome was deposited in Late Jurassic time.

The Onion Camp complex bears a striking resemblance to rocks of the Rattlesnake Creek terrane of the western Paleozoic and Triassic belt. The Fiddler Mountain olistostrome shares strong similarities with olistostrome deposits at Lems Ridge and Devils Elbow in the western Paleozoic and Triassic belt, and with other olistostromal units of the western Jurassic belt (Ohr, 1987; Wyld and Wright, 1988). Together the Onion Camp complex and Fiddler Mountain olistostrome comprise a rift-edge facies, formed when preexisting Klamath terranes rifted to form the Late Jurassic Rogue–Chetco/Josephine marginal ocean basin. This conclusion is consistent with previous models for the tectonic evolution of the western Jurassic belt put forth by Snoke (1977), Saleeby et al. (1982), and Harper and Wright (1984).

ACKNOWLEDGMENTS

We thank Robert B. Miller and Arthur W. Snoke for their helpful reviews. Many field visits and conversations with Greg Harper were invaluable. Conversations with W.P. Irwin were of great assistance for this study. A 1983 National Association of Geology Teachers summer internship working for W.P. Irwin served as JDY's introduction to the geology of the Klamath Mountains in California; an introduction that sparked interest in the Oregon Klamath Mountains and eventually led to this study. A field visit by L. Ramp in the earliest stages of this project was greatly appreciated. Much of the field mapping in the Kalmiopsis Wilderness could not have been completed without the assistance of B. Yule. We thank B.R. Hacker for obtaining $^{40}Ar/^{39}Ar$ age determinations and M.O. McWilliams for the use of his lab at Stanford University. We also thank C. Blome and M. Silk, who kindly made their unpublished radiolarian fossil age determinations available to us. This work was supported by a grant from the Oregon Department of Geology and Mineral Industries, Geological Society of America Research grants nos. 4596–90, 4842–91, and 5089–92 to JDY, by National Science Foundation grants EAR-9117103 and EAR-8720141 to CGB. Contribution 9132 of the Division of Geological and Planetary Sciences, California Institute of Technology, Pasadena, California.

REFERENCES CITED

Alexander, R.J., and Harper, G.D., 1992, The Josephine ophiolite, an ancient analog for oceanic lithosphere formed at intermediate-spreading ridges, *in* Parsons, B., and Browning, P., eds., Ophiolites and their modern oceanic analogues: Oxford, Blackwell Scientific Publications, p. 3–38.

Apted, M.J., and Liou, J.G., 1983, Phase relations among greenschist, epidote-amphibolite, and amphibolite in a basaltic system: American Journal of Science, v. 283-A, p. 328–354.

Barnes, C.G., Donato, M.M., Barnes, M.A., Yule, J.D., Hacker, B.R., and Helper, M.A., 1995, Geochemical compositions of metavolcanic and metasedimentary rocks, western Jurassic and western Paleozoic and Triassic belts, Klamath Mountains, Oregon and California: Reston, Virginia, U.S. Geological Survey Open-File Report 95–227-A, 63 p.

Blake, M.C., Jr., Engebretson, D.C., Jayko, A.S., and Jones, D.L., 1985, Tectonostratigraphic terranes in southwest Oregon, *in* Howell, D.G., ed., Tectonostratigraphic terranes of the Circum-Pacific region: Houston, Circum-Pacific Council for Energy and Mineral Resources, Earth Science Series 1, p. 147–157.

Brown, E.H., 1975, A petrogenetic grid for reactions producing biotite and other Al-Fe-Mg silicates in the greenschist facies: Journal of Petrology, v. 16, p. 258–271.

Brown, E.H., 1977, Phase equilibria among pumpellyite, lawsonite, epidote and associated minerals in low grade metamorphic rocks: Contributions to Mineralogy and Petrology, v. 64, p. 123–136, doi: 10.1007/BF00371507.

Burchfiel, B.C., and Davis, G.A., 1981, Triassic and Jurassic tectonic evolution of the Klamath Mountains–Sierra Nevada geologic terrane, *in* Ernst, W.G, ed., The geotectonic development of California—Rubey Volume 1: Englewood Cliffs, New Jersey, Prentice-Hall, p. 50–70.

Chen, J.H., and Wasserburg, G.J., 1981, Isotopic determination of uranium in picomole and subpicomole quantities: Analytical Chemistry, v. 53, p. 2060–2067, doi: 10.1021/ac00236a027.

Coleman, R.G., and Lanphere, M.A., 1991, The Briggs Creek amphibolite, Klamath Mountains, Oregon: Its origin and dispersal: New Zealand Journal of Geology and Geophysics, v. 34, p. 271–284.

Curray, J.R., Moore, D.G., Lawver, L.A., Raitt, R.W., and Henry, M., 1979, Tectonics of the Andaman Sea and Burma, *in* Watkins, J.S., et al., eds., Geological and geophysical investigations of continental margins: American Association of Petroleum Geologists Memoir 29, p. 189–198.

Davis, G.A., 1969, Tectonic correlations, Klamath Mountains and western Sierra Nevada, California: Geological Society of America Bulletin, v. 80, p. 1095–1108.

Dick, H.J.B., 1976, The origin and emplacement of the Josephine Peridotite of southwestern Oregon [Ph.D. dissertation]: New Haven, Connecticut, Yale University, 409 p.

Donato, M.M., 1992, A newly recognized ductile shear zone in the northern Klamath Mountains, Oregon—Implications for Nevadan accretion: Reston, Virginia, U.S. Geological Survey Bulletin 2028, 10 p.

Frey, M., de Capitani, C., and Liou, J., 1991, A new petrogenetic grid for low-grade metabasites: Journal of Metamorphic Geology, v. 9, p. 497–509.

Garcia, M.O., 1979, Petrology of the Rogue and Galice Formations, Klamath Mountains, Oregon: Identification of a Jurassic island-arc sequence: Journal of Geology, v. 87, p. 29–41.

Garcia, M.O., 1982, Petrology of the Rogue River island-arc complex, southwest Oregon: American Journal of Science, v. 282, p. 783–807.

Hacker, B.R., Donato, M.M., Barnes, C.G., McWilliams, M.O., and Ernst, W.G., 1995, Timescales of orogeny: Jurassic construction of the Klamath Mountains: Tectonics, v. 14, p. 677–703, doi: 10.1029/94TC02454.

Hamilton, W., 1978, Mesozoic tectonics of the western United States, *in* Howell, D.G., and McDougall, K.A., eds., Mesozoic paleogeography of the western United States, Pacific Coast Paleogeography Symposium 2: Los Angeles, Pacific Section, Society of Economic Paleontologists and Mineralogists, p. 33–70.

Harper, G.D., 1980, The Josephine ophiolite-remains of a late Jurassic marginal basin in northwestern California: Geology, v. 8, p. 333–337, doi: 10.1130/0091-7613(1980)8<333:TJOOAL>2.0.CO;2.

Harper, G.D., 1984, The Josephine ophiolite: Geological Society of America Bulletin, v. 95, p. 1009–1026, doi: 10.1130/0016-7606(1984)95<1009:TJONC>2.0.CO;2.

Harper, G.D., 2003, Fe-Ti basalts and propagating rift tectonics in the Josephine ophiolite: Geological Society of America Bulletin, v. 115, p. 771–787, doi: 10.1130/0016-7606(2003)115<0771:FBAPTI>2.0.CO;2.

Harper, G.D., and Wright, J.E., 1984, Middle to Late Jurassic tectonic evolution of the Klamath Mountains, California-Oregon: Tectonics, v. 3, p. 759–772.

Harper, G.D., Saleeby, J.B., and Heizler, M., 1994, Formation and emplacement of the Josephine ophiolite, and the age of the Nevadan orogeny in the Klamath Mountains, California-Oregon: U/Pb zircon and $^{40}Ar/^{39}Ar$ geochronology: Journal of Geophysical Research, v. 99, p. 4293–4321, doi: 10.1029/93JB02061.

Hawkins, J.W., 2003, Geology of suprasubduction zones; Implications for the origin of ophiolites, *in* Dilek, Y., and Newcomb, S., eds., Ophiolite concept and the evolution of geologic thought: Boulder, Colorado, Geological Society of American Special Paper 373, p. 227–268.

Hotz, P.E., 1971, Geology of lode gold districts in the Klamath Mountains, California and Oregon: Washington, D.C., U.S. Geological Survey Bulletin 1290, 91 p.

Irwin, W.P., 1960, Geologic reconnaissance of the northern Coast Ranges and Klamath Mountains, California, with a summary of the mineral resources: California Division of Mines Bulletin 179, 80 p.

Irwin, W.P., 1966, Geology of the Klamath Mountains province, *in* Bailey, E.H., ed., Geology of northern California: California Division of Mines Bulletin 190, p. 19–38.

Irwin, W.P., 1972, Terranes of the western Paleozoic and Triassic belt in the southern Klamath Mountains, California: Washington, D.C., U.S. Geological Survey Professional Paper 800-C, p. C103–C11 .

Irwin, W.P., 1994, Geologic map of the Klamath Mountains, California: Reston, Virginia, U.S. Geological Survey Miscellaneous Investigations Series Map I–2148, scale 1:500,000, 2 sheets.

Irwin, W.P., Jones, D.L., and Blome, C.D., 1982, Map showing radiolarian localities in the western Paleozoic and Triassic belt, Klamath Mountains, California: Reston, Virginia, U.S. Geological Survey Miscellaneous Field Investigations MF-1399, 1 sheet.

Jaffey, A.H., Flynn, K.F., Glendenin, L.E., Bentley, W.C., and Essling, A.M., 1971, Precision measurement of half-lives and specific activities of ^{235}U and ^{238}U: Physical Review, v. 4, p. 1889–1906.

Jorgenson, D.B., 1970, Petrology and origin of the Illinois River gabbro, a part of the Josephine peridotite-gabbro complex, Klamath Mountains, southwestern Oregon [Ph.D. dissertation]: University of California, Santa Barbara, 195 p.

Karig, D.E., 1971, Origin and development of marginal basins in the western Pacific: Journal of Geophysical Research, v. 76, p. 2542–2560.

Kays, M.A., 1992, Geologic guide for the northern Klamath Mountains, 1, Cow Creek to Red Mountain: Oregon Geology, v. 54, no. 2, p. 27–33.

Krogh, T.E., 1973, A low contamination method for the hydrothermal decomposition of zircon and extraction of U and Pb for isotopic age determinations: Geochimica et Cosmochimica Acta, v. 37, p. 485–494, doi: 10.1016/0016-7037(73)90213-5.

Lagabrielle, Y., Goslin, J., Martin, H., Thirot, J.-L., and Auzende, J.-M., 1997, Multiple active spreading centres in the hot North Fiji Basin (Southwest Pacific); A possible model for Archaean seafloor dynamics? Earth and Planetary Science Letters, v. 149, p. 1–13, doi: 10.1016/S0012-821X(97)00060-5.

Liou, J.G., Kuniyoshi, S., and Ito, K., 1974, Experimental studies of the phase relationships between greenschist and amphibolite in a basaltic system: American Journal of Science, v. 274, p. 613–632.

Liou, J.G., Maruyamam, S., and Cho, M., 1987, Very low-grade metamorphism of volcanic and volcaniclastic rocks: Mineral assemblages and mineral facies, *in* Frey, M., ed., Low temperature metamorphism: Glasgow, Blackie and Son Limited, p. 57–113.

Loney, R.A., and Himmelberg, G.R., 1977, The geology of the gabbroic complex along the northern border of the Josephine peridotite, Vulcan Peak area, southwestern Oregon: U.S. Geological Survey Journal of Research, v. 5, p. 761–781.

Miller, M.M., and Saleeby, J.B., 1995, U-Pb geochronology of detrital zircon from Upper Jurassic synorogenic turbidites, Galice Formation, and related rocks, western Klamath Mountains: Correlation and Klamath Mountains provenance: Journal of Geophysical Research, v. 100, p. 18,045–18,058, doi: 10.1029/95JB00761.

Ohr, M., 1987, Geology, geochemistry and geochronology of the Lems Ridge olistostrome, Klamath Mountains, California [M.S. thesis]: Albany, State University of New York, 278 p.

Page, N.J., Gray, F., Cannon, J.K., Foose, M.P., Lipin, B., Moring, B.C., Nicholson, S.W., Sawline, M.G., Till, A., and Ziemianski, W.P., 1981, Geologic map of the Kalmiopsis Wilderness Area, Oregon: Reston, Virginia, U.S. Geological Survey Miscellaneous Field Studies Map MF-1240-A, 1 sheet.

Parson, L.M., and Wright, I.C., 1996, The Lau-Hauvre-Taupo back-arc basin: A southward propagating, multi-stage evolution from rifting to spreading: Tectonophysics, v. 263, p. 1–22, doi: 10.1016/S0040-1951(96)00029-7.

Pearce, J.A., Lippard, S.J., and Roberts, S., 1984, Characteristics and tectonic significance of suprasubduction zone ophiolites, *in* Kokelaar, B.P., and Howells, M.F., eds., Marginal basin geology: London, Blackwell Scientific Publications, Geological Society Special Publication 16, p. 77–94.

Pessagno, E.A., Jr., and Blome, C.D., 1990, Implications of new Jurassic stratigraphic, geochronometric, and paleolatitudinal data from the western Klamath terrane (Smith River and Rogue Valley subterranes): Geology, v. 18, p. 665–668, doi: 10.1130/0091-7613(1990)018<0665:IONJSG>2.3.CO;2.

Ramp, L., 1977, Geology, mineral resources, and rock material of Curry County, Oregon: Oregon Department of Geology and Mineral Industries Bulletin 88, 47 p.

Ramp, L., 1984, Geologic map of the southeast quarter of the Pearsoll Peak quadrangle, Curry and Josephine counties, Oregon: Oregon Department of Geology and Mineral Industries Geologic Map Series GMS-30, scale 1:24,000.

Ramp, L., 1986, Geologic map of the northwest quarter of the Cave Junction quadrangle, Josephine County, Oregon: Oregon Department of Geology and Mineral Industries Geologic Map Series GMS-38, scale 1:24,000, 1 sheet.

Ramp, L., and Peterson, N.V., 1979, Geology and mineral resources of Josephine County, Oregon: Oregon Department of Geology and Mineral Industries Bulletin 100, 45 p.

Roure, F., and DeWever, P., 1983, Découverte de radiolarites du Trias dans l'unité occidentale des Klamath, sud-ouest de l'Oregon, U.S.A.: Conséquences sur l'âge des péridotite de Josphene: Comptes Rendus de l'Académie des Sciences (Paris), v. 297, p. 161–164.

Saleeby, J.B., 1984, Pb/U zircon ages from the Rogue River area, western Jurassic belt, Klamath Mountains, Oregon: Geological Society of America Abstracts with Programs, v. 16, no. 5, p. 331.

Saleeby, J.B., and Harper, G.D., 1993, Tectonic relations between the Galice Formation and the Condrey Mountain Schist, Klamath Mountains, northern California, *in* Dunn, G.C., and McDougall, K.A., eds., Mesozoic paleogeography of the western United States—II: Los Angeles, Pacific Section, Society of Economic Paleontologists and Mineralogists, v. 71, p. 61–80.

Saleeby, J.B., Harper, G.D., Snoke, A.W., and Sharp, W.D., 1982, Time relations and structural-stratigraphic patterns in ophiolite accretion, west central Klamath Mountains, California: Journal of Geophysical Research, v. 87, p. 3831–3848.

Shervais, J.W., 1982, Ti-V plots and the petrogenesis of modern and ophiolitic lavas: Earth and Planetary Science Letters, v. 59, p. 101–118, doi: 10.1016/0012-821X(82)90120-0.

Snoke, A.W., 1977, A thrust plate of ophiolitic rocks in the Preston Peak area, Klamath Mountains, California: Geological Society of America Bulletin, v. 88, p. 1641–1659, doi: 10.1130/0016-7606(1977)88<1641:ATPOOR>2.0.CO;2.

Snoke, A.W., and Barnes, C.G., 2006, this volume, The development of tectonic concepts for the Klamath Mountains province, California and Oregon, *in* Snoke, A.W., and Barnes, C.G., eds., Geological studies in the Klamath Mountains Province, California and Oregon: A volume in honor of William P. Irwin: Geological Society of America Special Paper 410, doi: 10.1130/2006.2410(01).

Stern, R.J., and Bloomer, S.H., 1992, Subduction zone infancy; Examples from the Eocene Izu-Bonin-Mariana and Jurassic California arc: Geological Society of America Bulletin, v. 104, p. 1621–1636, doi: 10.1130/0016-7606(1992)104<1621:SZIEFT>2.3.CO;2.

Tera, F., and Wasserburg, G.J., 1972, U-Th-Pb systematics in three Apollo 14 basalts and the problems of initial Pb in lunar rocks: Earth and Planetary Science Letters, v. 14, p. 281–304, doi: 10.1016/0012-821X(72)90128-8.

Thompson, A.B., 1976, Mineral reactions in pelitic rocks: I. Prediction of P–T–X (Fe–Mg) phase relations: American Journal of Science, v. 276, p. 401–424.

Uyeda, S., 1977, Some basic problems in the trench-arc-back arc system, *in* Talwani, M., and Pitman, W.C., III, eds., Island arcs, deep trenches, and back-arc basins: Washington, D.C., American Geophysical Union, Maurice Ewing Series v. 1, p. 1–14.

Wells, F.G., Hotz, P.E., and Cater, F.W., Jr., 1949, Preliminary description of the geology of the Kerby quadrangle, Oregon: Oregon Department of Geology and Mineral Industries Bulletin 40, 23 p.

Wells, F.G., and Walker, G.W., 1953, Geology of the Galice quadrangle, Oregon: Washington, D.C., U.S. Geological Survey Quadrangle Map GQ-25, scale 1:62,500, 1 sheet.

Wise, J.P., 1969, Geology and petrography of a portion of the Jurassic Galice Formation, Babyfoot Lake area, southwestern Oregon [M.S. thesis]: Pocatello, Idaho State University, 108 p.

Wright, J.E., and Fahan, M.R., 1988, An expanded view of Jurassic orogenesis in the western United States Cordillera: Middle Jurassic (pre-Nevadan) regional metamorphism and thrust faulting within an active arc environment, Klamath Mountains, California: Geological Society of America Bulletin, v. 100, p. 859–876, doi: 10.1130/0016-7606(1988)100<0859:AEVOJO>2.3.CO;2.

Wright, J.E., and Wyld, S.J., 1994, The Rattlesnake Creek terrane, Klamath Mountains, California: An early Mesozoic volcanic arc and its basement of tectonically disrupted oceanic crust: Geological Society of America Bulletin, v. 106, p. 1033–1056, doi: 10.1130/0016-7606(1994)106<1033:TRCTKM>2.3.CO;2.

Wyld, S.J., and Wright, J.E., 1988, The Devils Elbow ophiolite remnant and overlying Galice Formation: New constraints on the Middle to Late Jurassic evolution of the Klamath Mountains, California: Geological Society of America Bulletin, v. 100, p. 29–44, doi: 10.1130/0016-7606(1988)100<0029:TDEORA>2.3.CO;2.

Yardley, B.W.D., 1989, An introduction to metamorphic petrology: Essex, England, Longman Scientific and Technical, 248 p.

Yule, J.D., 1996, Geologic and tectonic evolution of Jurassic marginal ocean basin lithosphere, Klamath Mountains, Oregon [Ph.D. dissertation]: Pasadena, California Institute of Technology, 308 p.

Yule, J.D., and Saleeby, J.B., 1993, Highly extended oceanic lithosphere: The basement and wallrocks for the Late Jurassic Rogue–Chetco oceanic arc, Oregon Klamath Mountains: Geological Society of America Abstracts with Programs, v. 25, p. 169.

Yule, J.D., Saleeby, J.B., Jones, D.L., and Silk, M., 1992, Correlation of basement terranes across the Late Jurassic Josephine inter-arc basin, southwestern Oregon and northern California: Geological Society of America Abstracts with Programs, v. 24, p. 93.

Manuscript Accepted by the Society 10 January 2006

Printed in the USA

Geological Society of America
Special Paper 410
2006

Petrology, geochemistry, and provenance of the Galice Formation, Klamath Mountains, Oregon and California

James H. MacDonald Jr.*
Gregory D. Harper
Bin Zhu
Department of Earth and Atmospheric Sciences,
State University of New York, Albany, New York 12222, USA

ABSTRACT

The Upper Jurassic Galice Formation of the Klamath Mountains, Oregon-California, overlies the ca. 162-Ma Josephine ophiolite and the slightly younger Rogue–Chetco volcano-plutonic arc complex. The Galice Formation that overlies the Josephine ophiolite consists of a siliceous hemipelagic sequence, which grades upward into a thick turbidite sequence. Bedded hemipelagic rocks and scarce sandstone, however, also occur at several localities within the Josephine ophiolite pillow basalts. Corrected paleoflow current data suggest that the Galice Formation was derived predominantly from the east and north. Detrital modes of sandstones from the Galice Formation indicate an arc source as well as a predominantly chert-argillite source with minor metamorphic rocks. A sandstone located ~20 m below the top of the Josephine ophiolite has detrital modes and heavy mineral suites similar to the turbidite sandstones. Detrital Cr-spinel compositions from the turbidite and intra-pillow lava sandstones are also similar, indicating supra-subduction zone mantle peridotite and volcanic sources. Published detrital zircon data from a turbidite sandstone chiefly give a bimodal age distribution of 153 Ma and ca. 227 Ma but with a minor Proterozoic component. Whole-rock geochemistry from intra-pillow lava sedimentary rocks, the hemipelagic sequence, and the turbidites suggest a mixture between mafic and cratonic sources. It is suggested that the source area for the intra-pillow lava sedimentary rocks, hemipelagic sequence, and turbidites resulted from the mixing of arc and accreted terranes. These data indicate that the source areas for the Galice Formation were already established by ca. 162 Ma, probably during a Middle Jurassic orogeny that predated formation of the Josephine basin.

Keywords: Upper Jurassic, Galice Formation, hemipelagic sequence, turbidites, provenance, detrital zircons, Josephine ophiolite, California, Oregon

*Email: macdonal@atmos.albany.edu.

MacDonald, J.H., Jr., Harper, G.D., and Zhu, B., 2006, Petrology, geochemistry, and provenance of the Galice Formation, Klamath Mountains, Oregon and California, *in* Snoke, A.W., and Barnes, C.G., eds., Geological studies in the Klamath Mountains province, California and Oregon: A volume in honor of William P. Irwin: Geological Society of America Special Paper 410, p. 77–101, doi: 10.1130/2006.2410(04). For permission to copy, contact editing@geosociety.org.

INTRODUCTION

The Upper Jurassic Galice Formation occurs within the western Klamath terrane (western Jurassic belt) of the Klamath Mountains, Oregon-California (Fig. 1). It consists predominantly of slate and lesser metasandstone, but volcanic (predominantly volcaniclastic) members are locally present. The type section of the Galice Formation conformably overlies both the ca. 160–157-Ma Rogue–Chetco volcano-plutonic arc complex and ca. 164–162-Ma Josephine ophiolite (Harper et al., 1994; Fig. 1). A lower hemipelagic sequence is present where the Galice Formation overlies the Josephine ophiolite (Fig. 2). The Galice Formation also occurs within the Elk outlier (Fig. 1), where it is in fault contact with sheeted dikes that are correlated with the Josephine ophiolite (Giaramita and Harper, this volume).

The western Klamath terrane is the youngest of a series of eastward-dipping, imbricated thrust sheets within the Klamath Mountains (Irwin, 1964, 1966, 1994; Burchfiel and Davis, 1981). The roof thrust of the western Klamath terrane, the Orleans thrust (Fig. 1), is a major crustal boundary with geological and geophysical data suggesting 40 to >100 km of displacement, respectively (Jachens et al., 1986). The basal thrust, the Madstone Cabin thrust (Fig. 1), juxtaposes the Josephine ophiolite over the Chetco intrusive complex of similar age (Dick, 1976; Harper et al., 1994; Yule, 1996). The Galice Formation is regionally metamorphosed from prehnite-pumpellyite to lower greenschist facies (Harper et al., 1988) and is unconformably overlain by Lower Cretaceous nonmarine and marine conglomerate and sandstone (Harper et al., 1994).

The Galice Formation and its basement formed in a suprasubduction setting, as indicated by the presence of the underlying Rogue–Chetco arc complex and members of similar lithology in the Galice Formation (Garcia, 1982) and a high proportion of arc detritus in metasandstones (Snoke, 1977; Harper, 1984; Wyld and Wright, 1988; Harper et al., 1994). The Galice basin apparently formed as a result of extension within a west-facing arc, ultimately resulting in seafloor spreading that produced the Josephine ophiolite (Snoke, 1977; Saleeby et al., 1982; Harper and Wright, 1984; Wyld and Wright, 1988; Harper et al., 1994; Yule et al., this volume). Seafloor spreading was postulated to have occurred in a back-arc basin behind the Rogue–Chetco arc, but the presence of boninites in the Josephine ophiolite and the apparently slightly younger age of the Rogue–Chetco arc led Harper (2004) to suggest that arc rifting and initial seafloor spreading took place in the fore-arc. Similarly, the modern Lau back-arc basin apparently first formed in the fore-arc, but became situated in the back-arc after a trenchward jump in the arc axis (Parson and Wright, 1996). Collapse of the basin occurred during the Late Jurassic Nevadan orogeny by underthrusting of the Galice Formation and its basement beneath older terranes of the Klamath Mountains, resulting in low-grade regional metamorphism (Snoke, 1977, Saleeby et al., 1982; Harper and Wright, 1984). Geochronologic and biostratigraphic data suggest that the Galice Formation was deposited

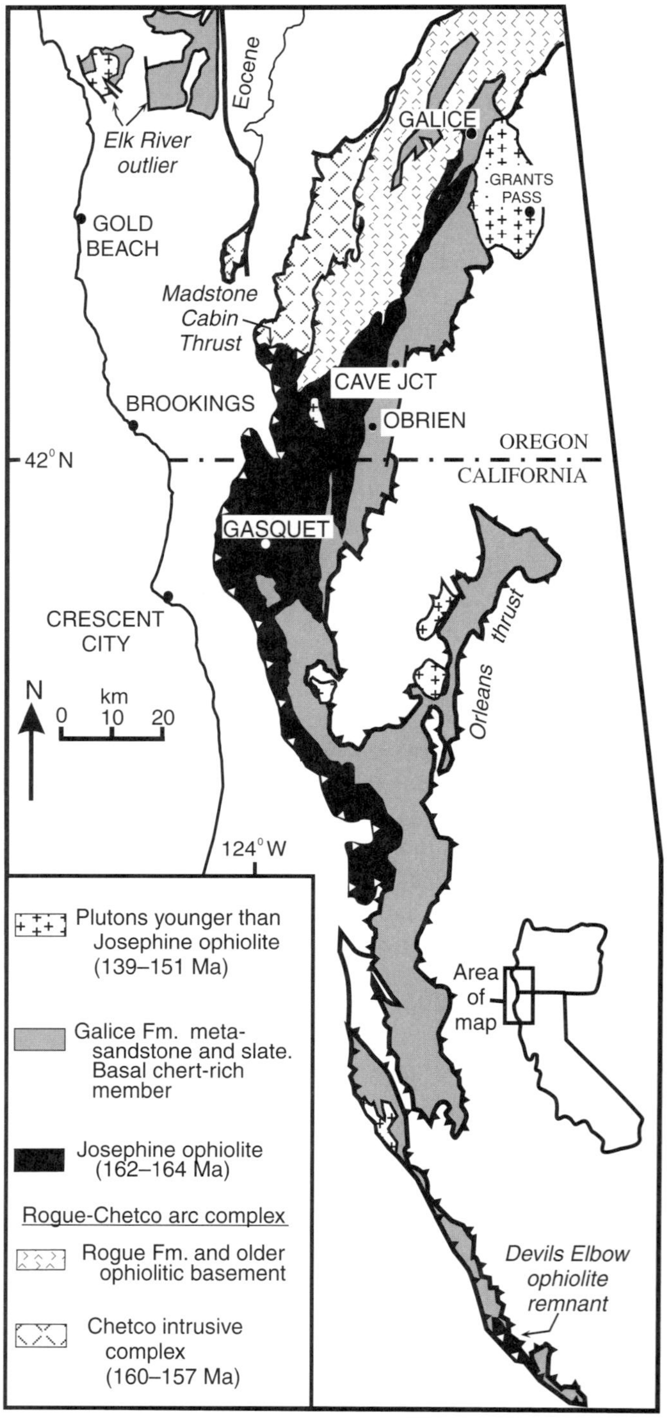

Figure 1. Generalized geologic map of western Klamath terrane, including Galice Formation. Modified from Wyld and Wright (1988), Miller and Saleeby (1995), Harper (2003), and Giaramita and Harper (this volume).

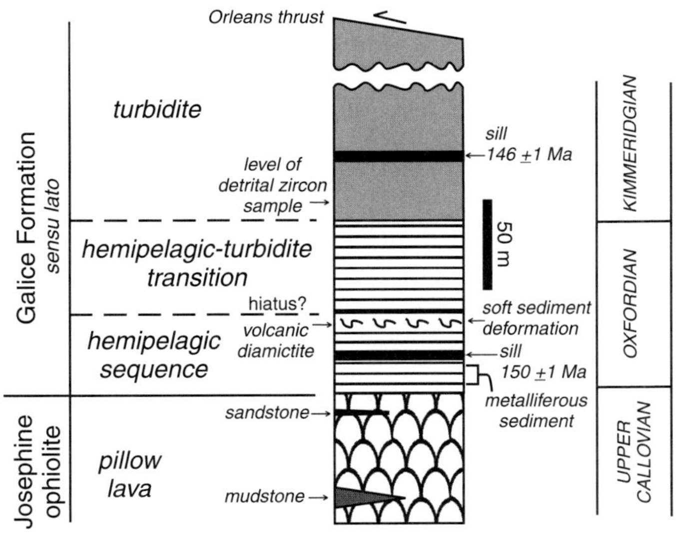

Figure 2. Sedimentary sequence overlying ca. 162-Ma Josephine ophiolite. Biostratigraphic ages are from Pessagno and Blome (1990) and Pessagno et al. (2000), and radiometric ages are from Harper et al. (1994). Location of metalliferous sediments is from Pinto-Auso and Harper (1985). Detrital-zircon sample (ca. 153- and ca. 227-Ma peaks; Miller et al., 2003) is from a locality of *Buchia concentrica* of middle Oxfordian to late Kimmeridgian age (Imlay et al., 1959). Modified from Harper (1994).

during the Nevadan orogeny, perhaps in a trenchlike basin along the western edge of the Klamath Mountains (Harper et al., 1994). The large displacement along the Orleans thrust suggests that the Josephine ophiolite was partially subducted during the Nevadan orogeny.

This chapter focuses on the provenance of the Galice Formation, based on new and existing data for detrital modes, chemistry of detrital Cr-spinel, whole-rock sedimentary geochemistry, and ages of detrital zircon. In addition, a comparison is made between the upper Galice Formation and clastic sedimentary rocks in the basal hemipelagic sequence overlying and interbedded with the Josephine ophiolite, including a newly discovered sandstone bed within the pillow lava unit.

GALICE FORMATION

Diller (1907) originally named the section of slate, sandstone, and minor conglomerate exposed on Galice Creek in southwest Oregon, west of Grants Pass, the "Galice Formation" (Fig. 1). Wells and Walker (1953) mapped the Galice 7.5-min quadrangle and recognized volcanic members lithologically similar to the underlying Rogue Formation. Cater and Wells (1953) correlated sedimentary and volcanic rocks in northern California (Gasquet 7.5-min quadrangle) with the Galice Formation, and Irwin (1960, 1966) showed the Galice Formation extending the entire length of the Klamath Mountains province. Although volcanic members are present in the type Galice Formation, Vail (1977) and Harper (1980, 1984) showed that a volcanic member mapped by Wells et al. (1949) and Cater and Wells (1953) is actually the upper part of a complete ophiolite complex (Josephine ophiolite). Volcanic members are generally absent in the Galice Formation that overlies the Josephine ophiolite (Harper, 1984), but one is present within the Galice Formation overlying the Devils Elbow remnant of the Josephine ophiolite in the southern Klamath Mountains (Wyld and Wright, 1988).

Harper (1980, 1994) suggested that the Galice Formation can be subdivided into three units (Fig. 2). The lower unit represents a hemipelagic sequence, whereas the upper unit is comprised of a turbidite sequence consisting of interbedded sandstone, siltstone, and radiolarian argillite. Lying between these two units is an ~55-m-thick transitional unit that represents the transition from hemipelagic to turbidite deposition (Fig. 2).

The ages of the Josephine ophiolite and Rogue–Chetco arc, which conformably underlie the Galice Formation, were initially thought to be the same (ca. 157 Ma; Saleeby et al., 1982; Saleeby, 1984). Based on this, Harper (1984) and Pinto-Auso and Harper (1985) included the hemipelagic sequence directly overlying the Josephine ophiolite within the Galice Formation. Subsequent biostratigraphic (Pessagno and Blome, 1990) and radiometric data (Harper et al. 1994) indicate, however, that the hemipelagic sequence is older than the type Galice Formation. This finding led Pessagno et al. (1993, 2000) to exclude the hemipelagic sequence (their "volcanopelagic succession") from the Galice Formation. Pessagno et al. (1993, 2000) suggested a hiatus occurs ~50 m stratigraphically above the Josephine ophiolite that separates the turbidite part of the Galice Formation from the underlying hemipelagic sequence (Fig. 2). They cited the much poorer preservation of radiolarians below this contact in support of their interpretation. They further suggested that the hemipelagic sequence is lithologically and genetically unrelated to the overlying turbidite; this suggestion, however, is inconsistent with petrographic and geochemical data presented below. Our observations show that bedding in rocks beneath the apparent hiatus is disrupted over a thickness of ~10 m below the hemipelagic-turbidite transition (Fig. 2). Beds are discontinuous, but not folded, and are cut by numerous small low-angle faults. This disruption appears to have occurred before complete lithification (Harper, this volume), because deformed beds and some small faults are cut by ca. 150-Ma syn-Nevadan dikes, some of which have amoeboid margins suggestive of intrusion into wet sediment (Harper, this volume).

For the purposes of this chapter, we include the hemipelagic sequence along with the turbidite in the Galice Formation (sensu lato), because the transitional unit (Fig. 2) is clearly lithologically gradational in character between the hemipelagic sequence and the Galice turbidite. Furthermore, as discussed below, sandstones in the transitional unit and in the lower part of the basal turbidite are very similar to the scarce sandstones that occur within both the hemipelagic sequence and pillow lavas of the Josephine ophiolite.

Structures in the Galice Formation that formed during the Late Jurassic Nevadan orogeny include slaty cleavage and associated overturned folds, stretching lineations, fibrous extension veins, and small thrust faults (Kays, 1968; Snoke, 1977; Harper, 1980, 1984; Norman, 1984; Gray, 1985, this volume; Wyld, 1985; Harper, this volume). The slaty cleavage and associated flattening of sand and pebble grains varies from very weak, as in the area of Cave Junction, Oregon (Jones, 1988; Fig. 1), to very strong. Strain data (i.e., Harper, 1980; Cashman, 1988; Jones, 1988) show a southward increase in strain from the Cave Junction area into northern California, coincident with an increase in metamorphic grade from prehnite-pumpellyite to lower greenschist facies (Harper, 1980; Harper et al., 1988). Paleomagnetic data suggest that the Galice Formation in southwestern Oregon may have undergone clockwise rotation of as much as 100° (Schultz and Levi, 1983; Bogen, 1986; Harper and Park, 1986).

Sedimentary Rocks within the Josephine Ophiolite

Pure, light-green chert is common between pillows in pillow lavas throughout the Josephine ophiolite. Although uncommon, bedded sedimentary rocks up to 5 m in thickness locally occur within the pillow lava unit of the Josephine ophiolite (Pinto-Auso and Harper, 1985; Kuhns and Baitis, 1987; Harper et al., 1988; Zierenberg et al., 1988; Figs. 2 and A-1). Radiolarians from these rocks indicate a late Callovian age, consistent with the ca. 162-Ma age of the Josephine ophiolite (Pessagno et al., 2000; Fig. 1). These sedimentary rocks are interbedded with Josephine lavas as well as separate mineralized lavas and massive-sulfide deposits from overlying lava flows (Pinto-Auso and Harper, 1985; Kuhns and Baitis, 1987; Zierenberg et al., 1988). Although most of the intra-pillow lava sedimentary rocks are bedded, some in the Turner-Albright mine area on the Oregon-California border southwest of Cave Junction, Oregon, are diamictites that probably formed as debris flows (Fig. A-1). These deposits and the presence of abundant talus breccias suggest the presence of fault scarps during formation of the Josephine ophiolite (Kuhns and Baitis, 1987; Zierenberg et al., 1988). The diamictites are very similar in hand sample, and in terms of geochemistry and color (black and green; Fig. A-1) to samples analyzed from the hemipelagic sequence. They are dominantly argillites, with green and black varieties interbedded on millimeter to centimeter scales. Where the regional metamorphic grade is lower-greenschist facies, they are slaty. Petrographic observations and geochemistry indicate that they are a mixture of mud, radiolarians, and hydrothermal sediment (Pinto-Auso and Harper, 1985; Kuhns and Baitis, 1987; Zierenberg et al., 1988; Figs. 2 and A-1). Less common rock types in the bedded sequences in the ophiolite include red radiolarian argillite (or slate) and chert. Detrital muscovite observed in black argillite lamella indicates a terrigenous component. The green argillite lamellae are probably tuffaceous, as shown for similar, better-studied green argillites in the hemipelagic sequence (Pinto-Auso and Harper, 1985).

An uncommon 2- to 10-cm-thick sandstone bed was found beneath the uppermost lava flow of the Josephine ophiolite, ~20 m below the depositional contact with the hemipelagic sequence in the type section of the pillow lava unit (Harper et al., 2002; MacDonald et al., 2004; Fig. 2). Harper (1994) originally identified this sandstone as a tuff, based on its characteristics in hand sample, including green color, but subsequent petrographic observations indicate the presence of terrigenous detritus, as discussed below. The green color is apparently due to ocean-ridge hydrothermal alteration, which has affected the entire pillow lava unit up to the depositional contact at this locality (Harper, 1995). Several meters along strike of where it was originally discovered, this bed has a medium-gray color similar to sandstones in the overlying Galice Formation (Harper et al., 1988).

Hemipelagic Sequence

The hemipelagic sequence is ~45-m thick and consists predominantly of green to black slaty radiolarian argillite with lesser green to black radiolarian chert (Pinto-Auso and Harper, 1985; Pessagno et al., 2000). Argillite in the hemipelagic sequence is a mixture of radiolarians, terrigenous and tuffaceous detritus, and hydrothermal sediment (Pinto-Auso and Harper, 1985; Kuhns and Baitis, 1987; Harper et al., 1988). The tuffaceous component is commonly evident from the presence of angular plagioclase, quartz (some bipyramidal; Fig. 3A) and, uncommonly, altered glass, silt, and sand (Fig. 3B). A terrigenous component from the presence of detrital muscovite is evident in thin section. The Fe/(Fe + Mn + Al) ratios indicate a relatively small metalliferous component derived from hydrothermal springs (Fig. 4), even for those overlying massive-sulfide deposits of hydrothermal origin (Kuhns and Baitis, 1987; Zierenberg et al., 1988).

Elevated metal contents are common in samples from the hemipelagic sequence, especially 8–23 m above the depositional contact with the underlying Josephine ophiolite, and some samples have sufficiently high Fe/(Fe + Mn + Al) to be classified as metalliferous (Pinto-Auso and Harper, 1985; Fig. 4). In the field, metalliferous sedimentary rocks are indicated by a shiny manganiferous coating or, less commonly, by a red color. The geochemistry and petrography of the metalliferous samples indicate that they all have substantial clastic detritus and variable amounts of radiolarians (Pinto-Auso and Harper, 1985). In addition, most of the rocks in the hemipelagic sequence, as well as those interbedded with pillow lavas of the Josephine ophiolite, have Fe/(Fe + Mn + Al) that is elevated relative to terrigenous sediments (Fig. 4), implying a small metalliferous component. Metalliferous sediments typically directly overlie ophiolites and modern ocean crust (e.g., Barrett et al., 1987). Pinto-Auso and Harper (1985) suggested that the stratigraphic

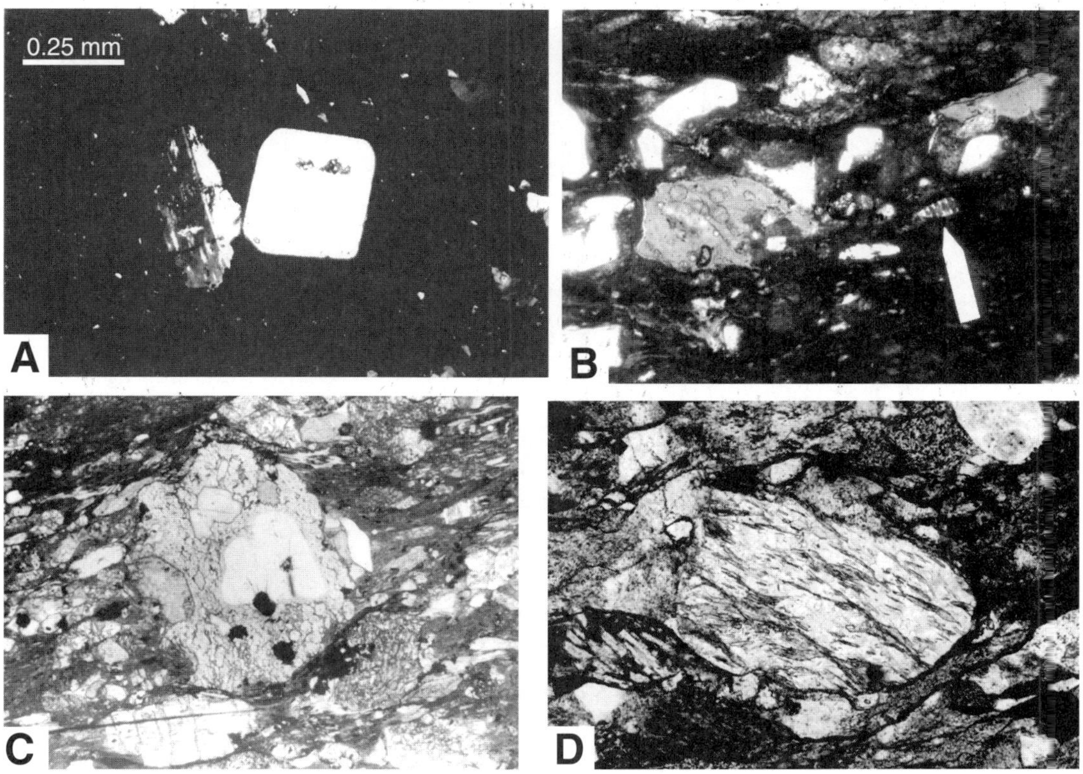

Figure 3. (A) Photomicrograph of a tuffaceous chert showing bipyramidal (volcanic) quartz and plagioclase. A 0.25-mm scale is located on figure. Crossed polarizers. (B) Photomicrograph of tuffaceous radiolarian argillite from the hemipelagic sequence showing well-preserved, cone-shaped Radiolaria (arrow), altered vesicular glass (left center), and angular quartz grains (clear). Plane light. Horizontal scale is 2 mm. (C) Photomicrograph of metasandstone from the Galice Formation showing felsic volcanic clast with resorbed quartz phenocryst. Colorless high-relief grain below this clast is augite. Note foliation in the matrix. Horizontal scale is ~1.5 mm. (D) Photomicrograph of metasandstone from the Galice Formation showing a clast of quartz-mica schist. At the lower left of this grain is a mafic volcanic clast. Dark seams are parallel to foliation. Horizontal scale is ~0.8 mm.

position of the metalliferous sediments, 8–23 m above the Josephine ophiolite, was the result of low-temperature, off-axis springs analogous to those along the flanks of the Galapagos Rise. Harper (2003) proposed that the metalliferous horizon was the result of the passage of a propagating spreading center, with the metalliferous component representing distal fallout from high-temperature, on-axis hot springs. The massive-sulfide deposits show that hydrothermal springs that could produce Fe- and Mn-rich metalliferous sediments were present during eruption of the Josephine pillow lavas. Thus, the low metal content of intra-pillow lava sedimentary rocks, including those directly overlying the massive sulfide deposits, and of sedimentary rocks comprising the lower 8 m of the hemipelagic sequence is apparently the result of dilution by abundant terrigenous mud.

Sandstone and scarce volcanic pebbly mudstone are uncommon rock types in the hemipelagic sequence (Pinto-Auso and Harper, 1985; Harper, 1994; Pessagno et al., 2000). The pebbly mudstone occurs ~40 m above the depositional contact with the Josephine ophiolite. It contains volcanic clasts up to 20 cm in diameter and has a radiolarian-tuffaceous matrix containing biotite (Harper, 1994). Sandstone was observed at two localities within the hemipelagic sequence; one within an isolated outcrop but the other ~7–8 m above the contact with the Josephine ophiolite (Pinto-Auso and Harper, 1985; Fig. 5). The sandstone beds are graded (Fig. 5) and compositionally very similar to those in the basal part of the turbidite sequence, with ~95% of the grains consisting of lithic volcanic rock fragments, plagioclase, and clinopyroxene. Other clasts, especially in the very coarse fraction, are predominantly chert and siliceous argillite. Scouring of underlying chert is evident along the base of one of these sandstone beds; this bed also contains a 3-cm angular rip-up clast of underlying chert (Fig. 5). Pessagno and Blome (1990) also reported the presence of "pelagic limestone," but these occur as nodules and probably formed by replacement of radiolarian argillite. Formation of the nodules occurred before formation of slaty cleavage (i.e., diagenetic) because radio-

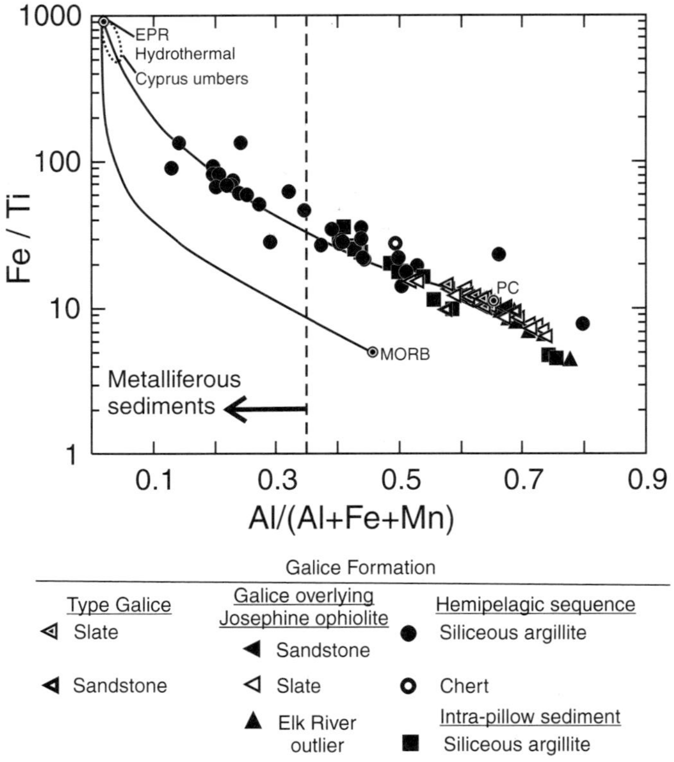

Figure 4. Plot illustrating apparent mixing relationships for sedimentary rocks within Josephine ophiolite, hemipelagic sequence, and Galice Formation. Mixing curves are for East Pacific Rise (EPR), hydrothermal sediment, and Pacific pelagic clay (PC), and EPR hydrothermal sediment and mid-ocean ridge basalt (MORB; values from Barrett, 1981). Also shown is field for Margi umbers, Cyprus, which are pure metalliferous sediment (Ravizza et al., 1999). Figure modified from Pinto-Auso and Harper (1985) and Harper et al. (1988).

larians and other spherical features (pellets?) in the nodules are undeformed.

The age range of the hemipelagic sequence is late Callovian to middle Oxfordian, according to the radiolarian biostratigraphy of Pessagno et al. (1993, 2000; Fig. 2). The absolute age range is from ca. 162 (age of the underlying ophiolite) to ca. 157 Ma. The younger age limit is based on correlation of the top of the hemipelagic sequence to radiolarian tuffs at the top of the Rogue Formation; Saleeby (1984) reports a Pb/U zircon age of 157 ± 2 Ma for volcanic rocks beneath the radiolarian tuffs, so the age of the top of the hemipelagic sequence may be somewhat less than 157 Ma. The apparently missing section at the top of the hemipelagic sequence (Fig. 2), suggested by radiolarian biostratigraphy, may be a disconformity (Pessagno and Blome, 1990). Although it may indeed be a disconformity, rocks immediately beneath the contact show considerable disruption of bedding, probably before complete lithification (Harper, this volume; Fig. 2). Thus, the apparent hiatus may be due to normal faulting or to submarine landsliding prior to deposition of the hemipelagic-turbidite transition unit.

Transition Zone

An ~55-m-thick transition zone exists between the lower hemipelagic sequence and the upper turbidite of that part of the Galice Formation that overlies the Josephine ophiolite (Harper, 1994; Fig. 2). This zone consists predominantly of radiolarian argillite with minor sandstone. The proportion of radiolarians and the proportion and size of clasts (mostly silt) are generally higher than those of radiolarian argillites in the hemipelagic sequence, and there is a negligible hydrothermal component (Pinto-Auso and Harper, 1985). Radiolarian-bearing limestone nodules occur throughout the transition zone, being abundant at the top of the zone (Harper, 1994). The lower contact for the transition zone is defined only for one section located along the Middle Fork of the Smith River, a sequence that includes the type section of the pillow lava unit of the Josephine ophiolite (Harper, 1994, 2003). In this section, which was studied by Pessagno and Blome (1990), as well as by Pinto-Auso and Harper (1985), the base of the transition zone is marked by the lowest occurrence of sandstone beds above the Josephine ophiolite.

Pessagno and Blome (1990) included rocks of the transition zone in the Galice turbidite, but we reserve use of the term "turbidite" for rocks above what we call the transition zone (Fig. 2). The above authors used the lowest occurrence of sandstones, as

Figure 5. Graded bed from hemipelagic sequence of the Galice Formation. Bed beneath sandstone is a radiolarian chert. On the back side of this sample there is a 3-cm subangular rip-up clast of underlying sediment located within sandstone.

well as the presence of the hiatus discussed in the preceding paragraph, to define the contact between the hemipelagic sequence (their "volcanopelagic unit") and the transition zone (their "Galice Formation sensu lato"). As noted above, a sandstone bed occurs within the pillow lava unit of the ophiolite (at this locality); at other localities, sandstone beds occur within the hemipelagic sequence (Fig. 5). Thus, the use of the first sandstone beds to define the base of the transition zone in the section (Fig. 2) is only valid for this particular location. We interpret the contact to be gradational, with an upward increase in proportion of sandstone beds and decreasing proportion of radiolarians in argillites.

According to the biostratigraphy of Pessagno et al. (1993, 2000), the transition zone ranges in age from middle Oxfordian to middle Kimmeridgian. Recently, Miller et al. (2003) estimated the age of deposition of a massive sandstone from the base of the overlying turbidite sequence to be ca. 153 Ma, based on the age of the youngest detrital zircon age spectrum (Miller et al., 2003). Thus, deposition of the transition zone appears to have ranged from ca. 157 to ca. 153 Ma.

Turbidite of the Galice Formation

The base of the turbidite sequence, as defined in this chapter, occurs ~100 m above the top of the Josephine ophiolite at the locality represented in Figure 2. The Galice Formation in its type area, where it overlies the Rogue Formation, consists entirely of turbidite (i.e., no hemipelagic sequence) that interfingers with various volcanic members. The Galice turbidite consists predominantly of slate and thin-bedded to massive metasandstone, along with scarce pebble conglomerate. Sandstones are predominantly feldspathic litharenites, are commonly graded, show scouring along their bases, and display partial to complete Bouma sequences. Load and flame structures, sole marks, and mud rip-up clasts are locally common. Trace fossils from slates include *Chondrites, Cosmophorae,* and *Spirophycus,* which indicate abyssal, or perhaps bathyal, water depths (A.A. Ekdale, written commun., 1980). In the section represented in Figure 2, and extending at least 10 km farther north and 1 km farther south, a 200- to 300-m-thick massive sandstone unit overlies the transition zone between the hemipelagic and turbidite sequences (Cater and Wells, 1953; Harper, 1980). The thickness of the turbidite cannot be accurately determined, due to repetition by folding and faulting, but the original stratigraphic thickness was probably at least several kilometers, based on its large outcrop area (Fig. 1).

The Galice turbidite, in both its type area and where it overlies the Josephine ophiolite, contains the bivalve *Buchia concentrica,* which has a known age range of middle Oxfordian to late Kimmeridgian (Imlay et al., 1959; Imlay, 1980). The age of the base of the turbidite based on the biostratigraphy of Pessagno et al. (1993, 2000) is middle Kimmeridgian. The absolute age of the turbidite is constrained by the ca. 153-Ma age of the youngest detrital zircons (Miller et al., 2003) and the age of crosscutting dikes, sills, and small plutons, which are as old as ca. 150 Ma (Harper et al., 1994).

PALEOFLOW DATA

Paleoflow data were obtained by Park-Jones (1988) for the type Galice Formation overlying the Rogue Formation west of Grants Pass, Oregon, and by us for the Galice Formation overlying the Josephine ophiolite where it is exposed in Rough and Ready Creek, southwest of Cave Junction, Oregon. Folds in the Galice Formation usually are plunging, so paleoflow data were restored by first rotating the fold axis (usually bedding-cleavage intersection) to horizontal, followed by rotation of bedding to horizontal. The resulting bidirectional and unidirectional flow directions are plotted in Figure 6. Predominantly west paleocurrent directions (in present-day coordinates) were found for the type Galice Formation (Park-Jones, 1988; Fig. 6), compared to predominantly north directions for the Galice Formation overlying the Josephine ophiolite (Fig. 6). The paleocurrent data probably need to be corrected for clockwise rotation of the Klamath Mountains, which has been suggested to explain paleomagnetic declination anomalies (Mankinen and Irwin, 1982; Schultz and Levi, 1983; Bogen, 1986; Renne and Scott, 1988; Smith and Harper, 1993). Some workers (e.g., Mankinen and Irwin, 1982; Bogen, 1986) argued that the entire Klamath Mountains rotated as a single block, whereas others have argued that the arcuate shape of the Klamath Mountains is the result of oroclinal bending, possibly related to formation of the Columbia embayment (Renne and Scott, 1988; Saleeby and Harper, 1993). The strike of slaty cleavage at both localities for which there are paleocurrent data is ~020°–030° (Park-Jones, 1988; this study). Figure 6B and D show the paleocurrent data after correcting for an assumed clockwise rotation of 65°, which assumes the western Klamath terrane had an original trend of ~340°, parallel to the overall structure of correlative rocks in the Sierra Nevada foothills, which have not been rotated (Bogen et al., 1985; Frei, 1986). The inferred ~65° rotation for the Galice Formation in southwest Oregon is less than the ~100° and ~78° clockwise rotations inferred for a volcanic member in the type Galice section and for the Grants Pass pluton that intrudes the Galice Formation (Schultz and Levi, 1983; Bogen, 1986), respectively, and the ~100° for the Josephine ophiolite (Smith and Harper, 1993). The actual amount of clockwise rotation may be less than suggested by the paleomagnetic data because of unknown amounts of post-Nevadan tilting, observed in Cretaceous sedimentary rocks elsewhere in the west-central and northwest Klamath Mountains, and possibly because of insufficient magnetic cleaning (Harper and Park, 1986; Renne and Scott, 1986).

DETRITAL MODES

Sandstones within the Galice Formation are generally lithic wackes and feldspatholithic wackes (Snoke, 1977; Harper, 1980; Wyld, 1985). Sandstones are generally foliated (e.g., Fig.

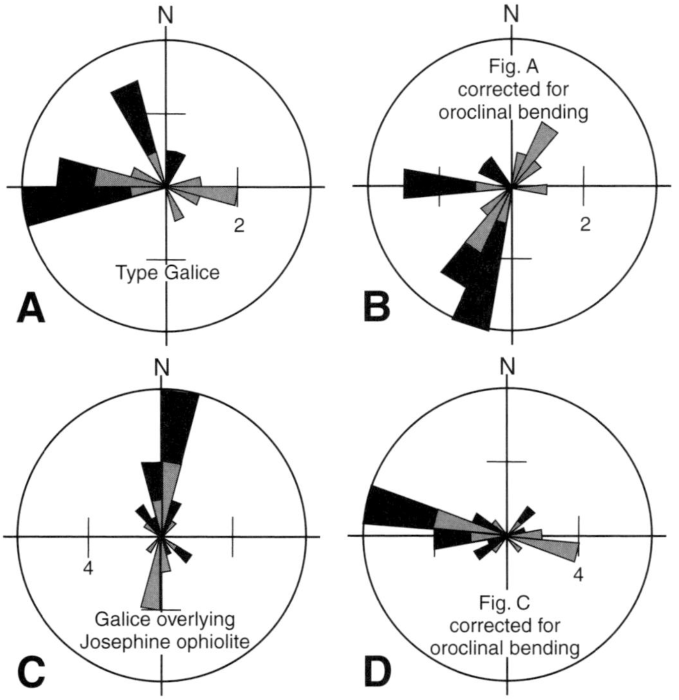

Figure 6. Rose diagrams displaying paleoflow current directions. (A) Type area of the Galice Formation, corrected for folding (from Park-Jones, 1988). (B) Type area of Galice, also corrected for inferred ~65° oroclinal bending of the Klamath Mountains. (C) Galice Formation that overlies Josephine ophiolite, corrected for folding. (D) Galice Formation that overlies Josephine ophiolite, also corrected for inferred ~65° oroclinal bending of Klamath Mountains. Unidirectional data displayed in black, bimodal data are gray. Intervals of 15° are used. Note the different scale in A and B, and C and D.

3C and D) but range from strongly foliated to unfoliated (Fig. 5), as in the Cave Junction to Obrien, Oregon, area. Clastic textures, however, are still evident in all samples. The most abundant detrital components are siliceous argillite and chert (both commonly with vestiges of radiolaria), plagioclase, and volcanic rock fragments. Less common clast types include monocrystalline and polycrystalline quartz; whereas phyllite, quartz-mica schist (Figs. 3C and D), foliated polycrystalline quartz, and siltstone or sandstone were observed with even less frequency (Snoke, 1972; Harper, 1980; Wyld, 1985). Harper (1980) and Norman (1984) determined that no K-feldspar is present in Galice sandstones that they studied, using both standard thin-section observations as well as staining. We found no K-feldspar by staining an additional thirty samples from both the type Galice Formation and the Galice Formation overlying the Josephine ophiolite near the Oregon-California border. Wyld (1985), however, reported very minor amounts (<1%) of K-feldspar in sandstones from the southernmost Klamath Mountains (Fig. 1).

Detrital modes for Galice sandstones were determined by Harper (1980), Norman (1984), and Wyld (1985). Medium- to coarse-grained sandstones were point-counted following the procedures of Dickinson (1970) and the terminology of Ingersoll and Suczek (1979) as well as Ingersoll et al. (1984). The detrital modes should be considered semiquantitative, due to partial recrystallization and, in many samples, deformation during low-grade metamorphism. Matrix generally constitutes >20%, and in some samples >30%, of the counted points. Such matrix contents are much higher than those for modern turbidites and are probably the result of alteration of lithic fragments (e.g., Dickinson, 1970, 1985). Quantitative modal analysis is also compromised by the textural similarity of many felsic volcanic rock fragments to chert; only Harper (1980) used feldspar staining to remove this uncertainty.

Structural complexity of the Galice Formation does not allow for accurate determination of stratigraphic height above basement. Harper (1980) and Wyld (1985), however, had sufficient stratigraphic control to broadly group their samples into basal, lower, and middle-upper.

Harper (1980) found that 62–92% of the volcanic rock fragments in samples of Galice sandstone have microlitic textures that are typical of andesites. Altered glass (replaced by chlorite) ranges from ~1–23% of volcanic clasts, although most samples have <10%. The proportion of felsitic clasts is highly variable, ranging from 3 to 33% of volcanic clasts. Clasts with lathwork texture (mafic) are much less common, never comprising >4% of the volcanic clasts.

The lithic- and feldspathic-rich nature of the sandstones is evident on various triangular diagrams (Fig. 7). Some sandstones plot as moderately quartz-rich on the quartz-feldspar-lithic (Q–F–L) diagram (Fig. 7A), but chert dominates the Q fraction, and monocrystalline quartz (Q_m) contents are low, as evident on the monocrystalline quartz-feldspar-lithic diagram (Q_m–F–L_t; Fig. 7B). Polycrystalline quartz (Q_p), consisting of quartz sandstone, and quartz siltstone, is generally a minor component (Fig. 7C), but abundant chert and argillite results in high values of lithic sedimentary clasts (L_{sm} in Fig. 7D; L_{sm} = sedimentary lithic + metasedimentary lithics, L_v = volcanic lithic). Detrital modes of sandstones for the Galice Formation from its type area (Harper, 1980), near the Oregon-California border (Harper, 1980; Norman, 1984), and from where it overlies the Devils Elbow remnant of the Josephine ophiolite in the southern Klamath Mountains, are similar (Fig. 7). Those samples overlying the Devils Elbow remnant, however, tend to be less rich in lithic fragments (Fig. 7A and B) and have more monocrystalline quartz (Fig. 7B).

A trend of increasing proportion of total quartz (Q; mostly chert), Q_m, Q_p, and L_{sm} (mostly siliceous argillite and chert) with increasing stratigraphic height is evident in Figure 7 for samples overlying the Devils Elbow remnant of the Josephine ophiolite in the southern Klamath Mountains (Wyld, 1985). A similar trend is evident for the Galice Formation overlying the Josephine ophiolite in its type area in northwestern California

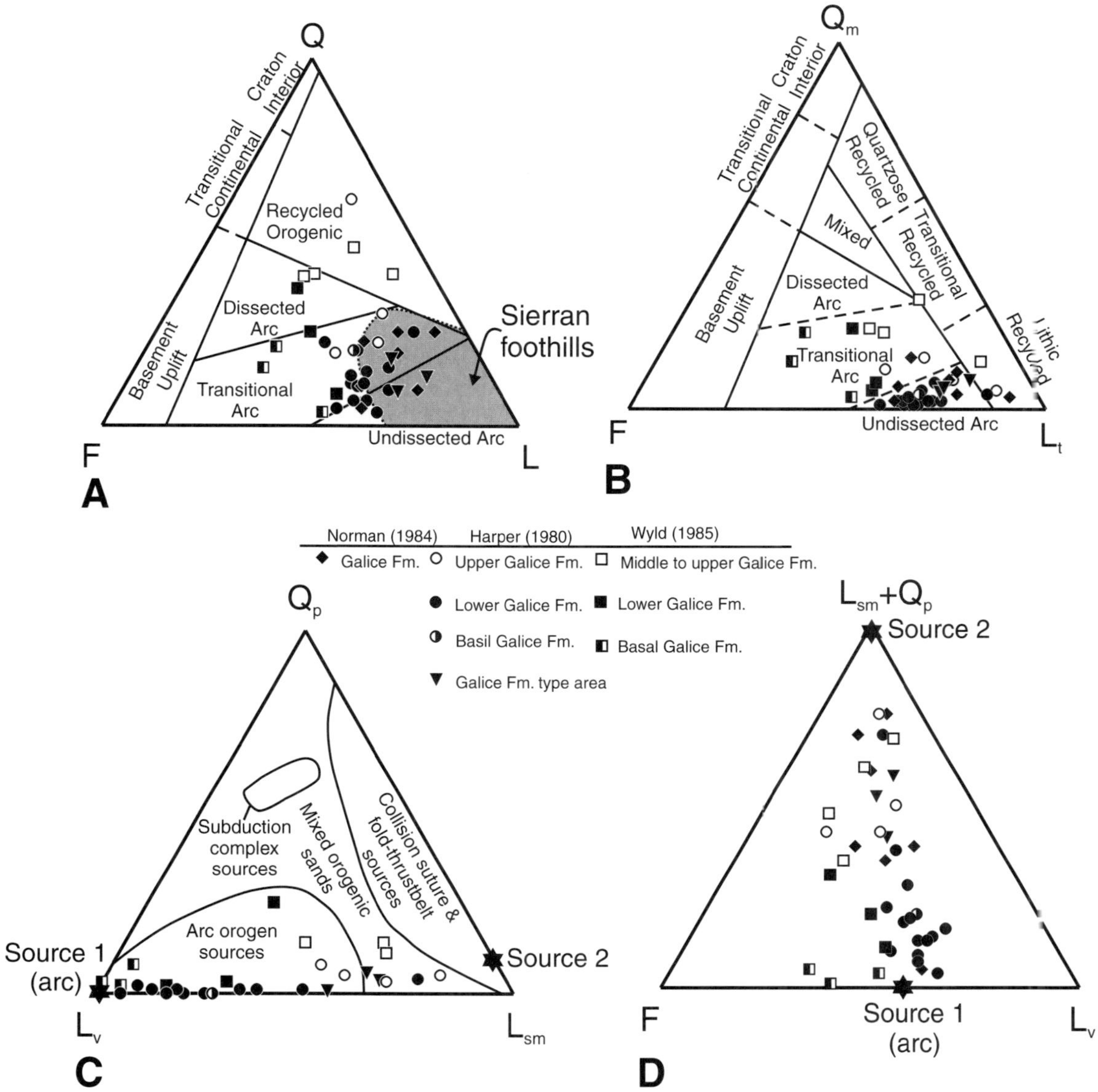

Figure 7. Triangular plots of point-count data for sandstones from Galice Formation in its type area (Harper, 1980), where it overlies the Josephine ophiolite near the Oregon-California border (Harper, 1980; Norman, 1984), and overlying Devils Elbow remnant of the Josephine ophiolite in the southern Klamath Mountains (Wyld, 1985). Q—monocrystalline quartz + polycrystalline quartz + chert; F—feldspar (all plagioclase in these samples); L—unstable lithics; Q_m—monocrystalline quartz; L_t—total polycrystalline lithic fragments, including stable quartzose; Q_p—chert + polycrystalline quartz; L_v—lithic volcanics; L_{sm}—sedimentary + metasedimentary lithics. (A) Q–F–L diagram after Dickinson et al. (1983). Shaded field is for correlative rocks in the Sierra Nevada foothills, including the Mariposa Formation (Behrman and Parkinson; 1978). (B) Q_m–F–L_t diagram after Dickinson et al. (1983). (C) Q_p–L_v–L_{sm} diagram after Dickinson (1985). Inferred end-member sources for Galice sandstones are indicated. (D) L_{sm}+Q_p–F–L_v diagram.

(Harper, 1980; Norman, 1984), except that some lower Galice sandstones have modes similar to those typical of the upper Galice (Fig. 7). This similarity is not the result of uncertainties in stratigraphic position of samples, because sandstones having high L_{sm} + Q_p ratios in the lower Galice Formation were found interbedded with those having low L_{sm} + Q_p (Harper, 1980). Future work might determine whether these two detrital modes are correlated with different paleocurrent directions. The linear trends on the Q_p–L_v–L_{sm} and L_{sm}+Q_p–F–L_v diagrams (Fig. 7C and D) suggest mixing of two sources, one of which has a relatively constant F/L_v ratio (Fig. 7). The relatively constant F/L_v, the very common occurrence of plagioclase as phenocrysts in volcanic rock clasts, and the near absence of plutonic rocks in pebble conglomerates (Wyld, 1985; Seiders, 1991, written commun., 1992) indicate that the plagioclase in the Galice sandstones is virtually all volcanic.

A sandstone that occurs within the upper pillow lavas of the Josephine ophiolite is too altered, including abundant patchy prehnite formed during ocean-ridge hydrothermal metamorphism (Harper, 1995), for accurate determination of a mode. Nevertheless, it is clearly similar to those within the lower turbidite of the Galice Formation, consisting of abundant volcanic (mostly microlitic) rock fragments, less common chert, and siliceous argillite clasts. The only major difference between this sandstone and that of the basal turbidite is a substantially lower plagioclase content than Galice sandstones, but this may be due to the replacement of plagioclase by prehnite.

Wyld (1985) and Seiders (1991, written commun., 1992) obtained detrital modes for conglomerates from the Galice Formation by point counting (Fig. 8). In their modal analysis, Wyld (1985) counted clasts larger than 1 mm, whereas Seiders (1991) counted clasts in the 0.5–6 cm diameter range. Galice conglomerates are rich in chert (C) and generally have very few quartzose sandstone clasts (Q; Fig. 8). The igneous component, which consists almost essentially of volcanic clasts, varies from near 0 to ~20%. This result implies a much lower proportion of volcanic clasts than in associated sandstones. Such a grain-size bias is evident in individual sandstone beds that have pebbly bases. Two of Wyld's (1985) conglomerate samples from the lower Galice Formation in the southern Klamath Mountains have much higher Q and are transitional between typical Galice Formation chert-rich conglomerates and correlative conglomerates of the Sierra Nevadan foothills (Fig. 8).

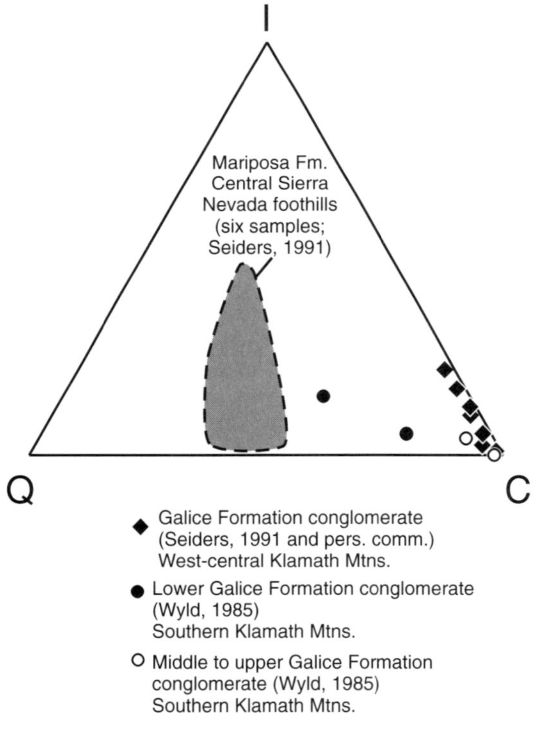

Figure 8. Conglomerate point count data for the Galice Formation and the Mariposa Formation in the central Sierran foothills illustrating provenance differences. I—igneous rocks; Q—quartz sandstone and quartzite; C—chert; Fm—formation. Data from Wyld (1985) and Seiders (1991, written commun., 1992).

HEAVY MINERALS

Occurrence

Snoke (1972, 1977), Harper (1980, 1984), and Wyld (1985) reported heavy-mineral assemblages in the Galice Formation. The most common heavy minerals are zircon (euhedral to very well rounded), tourmaline, apatite, biotite, muscovite, Cr-spinel, and, in volcanic-rich basal turbidite sandstones, clinopyroxene and hornblende. Less common to rare are garnet, epidote, staurolite, and glaucophane. The 2- to 5-cm-thick sandstone bed within the Josephine pillow lava unit has abundant clinopyroxene (~5 modal %), with lesser hornblende, Cr-spinel, and rare glaucophane (violet to clear pleochroism); zircon has not yet been found in this sample; these heavy minerals, as well as the detrital modes, are similar to those of volcanic-rich sandstones within the hemipelagic-turbidite transition zone and basal turbidite.

Cr-Spinel

McLennan et al. (1993) suggested that Cr-spinel most likely indicates an ophiolitic source area. Spinels can occur in any primitive mafic volcanic rocks, however, as well as in any type of ultramafic rock. Cookenboo et al. (1997) and Lee (1999) stressed the importance of using the chemical composition of detrital Cr-spinels to infer rock types in the source area. Lee (1999) suggested that Cr-spinels are significant in paleogeographic reconstructions because their deposition occurs within proximity of their source area.

Spinels from four sandstones were analyzed for this study. One sample is from the bed within the Josephine ophiolite pillow lava unit (sample PC-2b; Table 1), another sample from the basal Galice turbidite overlying the ophiolite (sample LJC-23; Table 1 and Fig. 2) and the other two samples from the turbidite sequence (samples DB-1 and CJ-7; Table 1 and Fig. 2). The Cr-spinels were analyzed with a JEOL 733 Superprobe at the Department of Earth and Environmental Sciences, Rensselaer Polytechnic Institute, Troy, New York, and with a JEOL 8900 electron microprobe at Binghamton University, Binghamton, New York. The elements Ti, Al, Cr, Mn, Mg, Ni, and Fe were analyzed using a 15-kV accelerating voltage, 15-nÅ beam current, and a 1-μm beam diameter. Elements Al, Cr, Mg, and Fe were counted for 40 seconds whereas Ti, Mn, and Ni were counted for 100 seconds each. Sample USNM 117075 from Tiebaghi Mine, New Caledonia, was used as an Al, Cr, Mg, and Fe standard at Rensselaer Polytechnic Institute. Other elemental standards used at Rensselaer Polytechnic Institute were rutile for Ti, tephroite for Mn, and diopside glass for Ni. Analy-

TABLE 1. REPRESENTATIVE ANALYSES OF Cr-RICH SPINEL

Sample	Locality*	TiO_2	Al_2O_3	Cr_2O_3	MnO	MgO	NiO	FeO	Fe_2O_3	Total	Cr^{3+}	Fe^{3+}	Al^{3+}	Mg#	Cr#
LJC-23†	Turbidite	0.63	21.09	39.66	0.35	9.67	0.15	18.01	10.35	99.91	0.48	0.13	0.38	0.44	0.56
LJC-23†	Turbidite	0.09	22.16	44.75	0.27	11.81	0.05	15.62	4.39	99.14	0.55	0.05	0.40	0.54	0.58
LJC-23†	Turbidite	b.d.#	33.28	31.93	0.21	13.77	0.10	14.03	5.26	98.64	0.37	0.06	0.57	0.60	0.39
LJC-23†	Turbidite	0.18	31.04	33.36	0.28	12.73	0.06	15.30	6.55	99.50	0.39	0.08	0.54	0.56	0.42
LJC-23†	Turbidite	b.d.#	16.11	50.77	0.40	9.15	0.08	18.69	4.48	99.76	0.64	0.05	0.30	0.44	0.68
LJC-23†	Turbidite	0.46	27.38	32.36	0.29	12.13	0.14	15.14	11.89	99.78	0.37	0.15	0.47	0.52	0.44
LJC-23†	Turbidite	0.21	12.80	51.33	0.30	9.99	0.06	16.31	9.01	100.01	0.64	0.12	0.24	0.47	0.73
LJC-23†	Turbidite	0.85	28.47	33.28	0.29	11.75	0.09	16.64	8.36	99.73	0.39	0.10	0.50	0.51	0.44
LJC-23†	Turbidite	b.d.#	31.39	35.26	0.21	14.06	0.15	13.53	3.84	98.49	0.41	0.04	0.55	0.62	0.43
LJC-23†	Turbidite	0.12	32.20	35.53	0.21	14.12	0.12	14.21	3.59	100.09	0.41	0.04	0.55	0.61	0.43
LJC-23†	Turbidite	0.21	16.26	45.95	0.39	8.96	0.05	17.99	9.71	99.52	0.57	0.13	0.30	0.42	0.65
LJC-23†	Turbidite	0.20	16.40	49.25	0.34	9.01	0.05	18.87	5.89	100.01	0.62	0.07	0.31	0.43	0.67
LJC-23†	Turbidite	b.d.#	13.87	47.99	0.38	9.33	0.09	16.81	11.30	99.82	0.59	0.15	0.26	0.44	0.70
LJC-23†	Turbidite	0.11	28.49	34.49	0.23	12.16	0.16	15.31	8.53	99.48	0.40	0.10	0.49	0.54	0.45
LJC-23†	Turbidite	0.14	19.36	46.28	0.29	11.44	0.13	15.27	6.22	99.13	0.57	0.08	0.35	0.53	0.62
LJC-23†	Turbidite	b.d.#	31.68	36.11	0.24	14.39	0.15	13.39	3.01	98.99	0.42	0.03	0.55	0.63	0.43
LJC-23†	Turbidite	1.40	19.93	37.83	0.26	12.01	0.24	14.69	12.80	99.16	0.46	0.18	0.36	0.52	0.56
LJC-23†	Turbidite	0.11	10.87	54.23	0.36	7.80	0.06	19.44	7.15	100.02	0.70	0.09	0.21	0.38	0.77
LJC-23†	Turbidite	1.31	19.86	38.52	0.22	11.83	0.14	15.13	12.59	99.61	0.47	0.17	0.36	0.52	0.57
CJ-7†	Turbidite	b.d.#	8.51	58.22	0.34	8.41	0.01	18.53	6.61	100.71	0.75	0.08	0.16	0.41	0.82
DB-1§	Turbidite	0.04	35.73	29.77	0.15	14.87	0.15	5.99	11.99	98.70	20.37	8.38	18.30	0.66	0.52
DB-1§	Turbidite	0.11	27.22	35.25	0.27	12.47	0.19	8.21	16.43	100.15	24.12	11.49	14.40	0.54	0.63
DB-1§	Turbidite	0.19	16.82	41.40	0.42	8.88	0.03	10.15	20.29	98.17	28.33	14.19	8.30	0.40	0.76
DB-1§	Turbidite	0.03	29.13	38.60	0.29	14.39	0.09	5.78	11.56	99.87	26.41	8.09	15.41	0.66	0.63
DB-1§	Turbidite	0.07	8.01	56.85	0.46	7.64	0.03	8.84	17.68	99.58	38.90	12.37	4.24	0.40	0.90
DB-1§	Turbidite	0.24	15.64	41.66	0.36	7.99	0.07	11.29	22.59	99.85	28.50	15.80	8.28	0.35	0.78
DB-1§	Turbidite	0.02	30.84	38.16	0.30	13.68	0.16	5.92	11.84	100.91	26.11	8.28	16.32	0.64	0.62
PC-2b§	Intra-pillow	0.14	12.96	43.93	0.53	5.53	0.10	11.04	22.09	96.32	30.06	15.45	6.36	0.28	0.81
PC-2b§	Intra-pillow	0.24	14.44	49.40	0.33	14.67	0.13	6.75	13.50	99.47	33.80	9.44	7.64	0.63	0.82
PC-2b†	Intra-pillow	b.d.#	29.97	36.50	0.26	12.42	0.09	15.92	4.29	99.45	0.43	0.05	0.52	0.55	0.45
PC-2b†	Intra-pillow	b.d.#	13.96	54.91	0.37	9.85	0.09	17.86	0.48	97.58	0.72	0.01	0.27	0.49	0.73
PC-2b†	Intra-pillow	0.25	14.89	50.36	0.37	9.20	0.04	18.16	6.37	99.64	0.64	0.08	0.28	0.44	0.69
PC-2b†	Intra-pillow	0.13	27.22	35.01	0.33	11.33	0.15	15.96	8.25	98.37	0.42	0.10	0.48	0.51	0.46
PC-2b†	Intra-pillow	b.d.#	29.25	37.70	0.28	12.25	0.09	16.28	3.79	99.67	0.44	0.04	0.51	0.55	0.46
PC-2b†	Intra-pillow	b.d.#	16.56	51.52	0.36	10.26	0.05	17.37	2.95	99.13	0.65	0.03	0.31	0.49	0.68
PC-2b†	Intra-pillow	1.31	17.76	23.55	0.30	7.66	0.22	20.10	29.57	100.48	0.27	0.42	0.30	0.32	0.47
PC-2b†	Intra-pillow	b.d.#	30.27	36.97	0.26	12.32	0.13	16.48	3.80	100.23	0.43	0.04	0.53	0.55	0.45
PC-2b†	Intra-pillow	b.d.#	28.07	39.18	0.16	12.61	0.06	15.71	3.22	99.10	0.47	0.04	0.50	0.57	0.48

Note: The ferric iron content of each analysis was determined by assuming stoichiometry, following the methods of Barnes and Roeder (2001).
*From Figure 2.
†Sample analyzed at Rensselaer Polytechnic Institute.
§Sample analyzed at Binghamton University.
#b.d.—below detection limits.

sis of the standard USNM 117075 was done throughout the session at Rensselaer Polytechnic Institute to assess proper calibration. Standards used at Binghamton University were TiO_2 for Ti, Al_2O_3 for Al, chromite for Cr, spessartine garnet for Mn, MgO for Mg, Ni-metal for Ni, and hematite for Fe.

The compositions of detrital spinels from the sandstone within the Josephine ophiolite pillow lava unit and the sandstone from the basal turbidite are given in Table 1 and are plotted in Figures 9 and 10. The detrital spinels show a wide range of compositions. Some plot entirely within the field for mantle peridotites, whereas others plot entirely in the field for volcanic spinels, including the mid-ocean ridge basalt (MORB) field and between the MORB and ocean-island—or within-plate—basalt (OIB) fields (Fig. 9). Other detrital spinels plot in the area of overlap for volcanic and mantle peridotite spinels (Fig. 9).

Those detrital spinels identified as originating from mantle peridotites in Figure 9 have Cr/(Cr + Al) ratios (Cr#) that overlap those of abyssal (mid-ocean ridge origin) and supra-subduction zone (SSZ) peridotites (Fig. 10A). All, however, have Mg/(Mg + Fe^{2+}) ratios that are lower than those for abyssal peridotite spinels; the Mg/(Mg + Fe^{2+}) ratios for all the detrital mantle spinels fall within the field for SSZ peridotites,

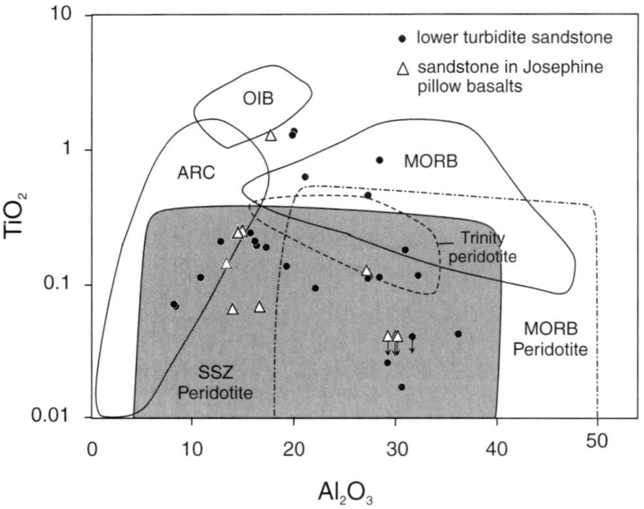

Figure 9. Al_2O_3 vs. TiO_2 diagram showing detrital Cr-spinels from an intra-pillow sandstone in the Josephine ophiolite and a turbidite sandstone of the Galice Formation. Fields are from Kamenetsky et al. (2001). This diagram is a good discriminator for tectonic setting as well as for distinguishing volcanic from peridotite spinels. OIB—ocean-island (within-plate) basalt; MORB—mid-ocean ridge basalt; SSZ—supra-subduction zone. Also shown is the field for spinels from the Trinity peridotite (Quick, 1981).

a result that implies that they are all derived from SSZ ophiolites (Fig. 10A).

Those detrital spinels identified as volcanic (MORB to MORB-OIB) in Figure 9 have Cr/(Cr + Al) ratios similar to the upper range of most modern MORB, but are shifted to lower Mg/(Mg + Fe^{2+}) ratios (Fig. 10B). This shift to lower Mg/(Mg + Fe^{2+}) ratios is probably due, in part, to the OIB component that is indicated in Figure 9, as OIB basalts have Mg/(Mg + Fe^{2+}) ratios that extend to much lower values. However, it may also be due to crystal fractionation of olivine, which causes the Mg/(Mg + Fe^{2+}) of spinel to increase (Dick and Bullen, 1984; Kamenetsky et al., 2001; Fig. 10B).

Figure 10C shows those detrital Cr-spinels that could have been derived from either mantle peridotite or volcanic rocks based on the classification shown in Figure 9. Most of these spinels, as well as some of the peridotite spinels (Fig. 10A), have high Cr/(Cr + Al) ratios, indicative of high degrees of partial melting, which is typical of arcs (Dick and Bullen, 1984). Four of the detrital spinels from the turbidite sample and one from the intra-pillow sample have Cr/(Cr + Al) ratios of ~0.8, which is characteristic of boninites (Fig. 10C) and mantle that has undergone very high degrees of melting or mantle that has interacted with a boninitic magma (e.g., Dick and Bullen, 1984; Harper, 2004).

AGE OF SOURCE AREAS

The age of detrital zircons within Galice sandstones was studied by Miller and Saleeby (1995) and Miller et al. (2003)

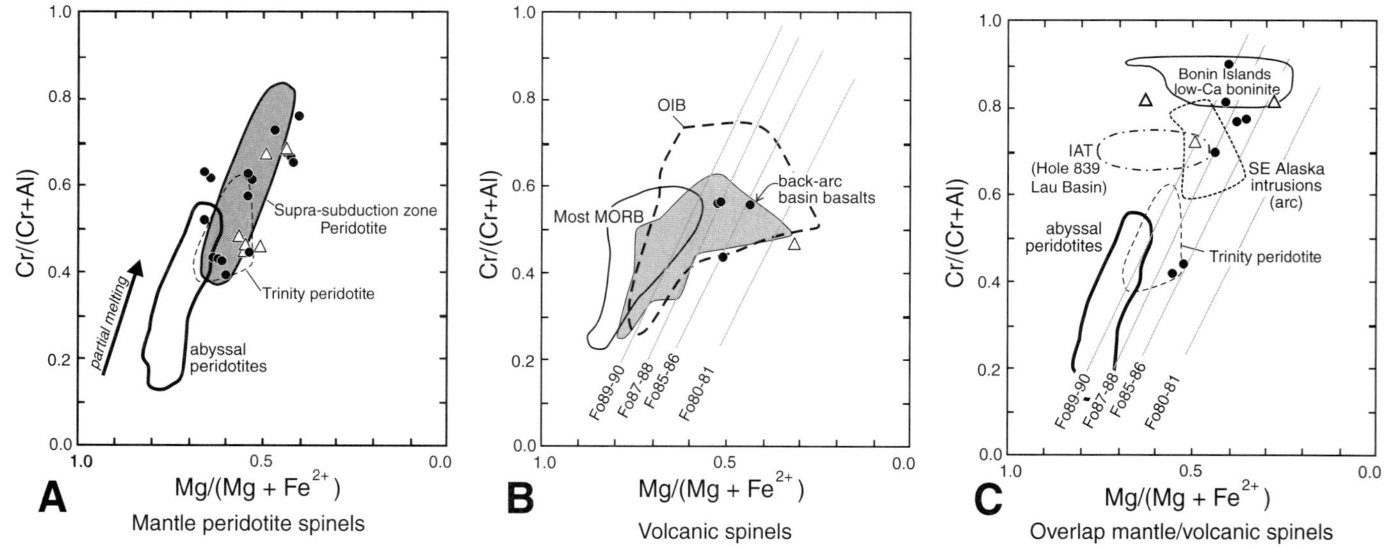

Figure 10. Cr/(Cr + Al) vs. Mg/(Mg + Fe^{2+}) diagram for detrital Cr-spinels from a sandstone within the pillow lavas of the Josephine ophiolite and a sandstone from the basal turbidite sequence overlying the ophiolite. Symbols are the same as in Figure 9. Field for Trinity peridotite from Quick (1981). Fields for MORB, southeastern Alaskan intrusions, and abyssal peridotites (MORB residue) from Dick and Bullen (1984); field for back-arc basin basalts from Dick and Bullen (1984), Hawkins and Melchoir (1985), Saunders and Tarney (1979; one sample); field for Hole 839 in the Lau Basin is from Allan (1994); field for low-Ca boninites is from Umino (1986); field for OIB from Kamenetsky et al. (2001); and field for supra-subduction zone peridotite is modified from Parkinson and Pearce (1998). Olivine isopleths are from Kamenetsky et al. (2001). A: Spinels derived from peridotite (as determined from Fig. 9). B: Spinels of volcanic origin (as determined from Fig. 9). C: Spinels that could be either volcanic or peridotitic origin (as determined from Fig. 9).

through U/Pb age dating. Miller and Saleeby (1995), using multiple grain fractions from various localities throughout the Galice Formation, obtained U-Pb data that plot as a chord on a concordia diagram, with an average upper intercept of ca. 1600 Ma and a lower intercept of ca. 215 Ma, although the youngest zircons approach ca. 170 Ma. Surprisingly, Miller and Saleeby (1995) found no difference in the age distributions of euhedral and well-rounded (recycled) zircon populations. All four of their Galice samples show the same spread on a concordia diagram, which implies no major variations in the age of detrital zircons in these samples. A fifth sample gives an older upper-intercept age; however, this sample is not from the Galice Formation but from the Lems Ridge olistostrome that underlies the Galice, which probably represents a fragment of older Klamath basement in the western Klamath terrane that is correlative with the lower Mesozoic Rattlesnake Creek terrane (Ohr, 1987).

Recently, Miller et al. (2003) reported ion-microprobe single-crystal U/Pb ages for detrital zircons from a Galice Formation sandstone near the base of the turbidite sequence at a *Buchia concentrica* locality. Most zircons from this sample are Late Jurassic, averaging ca. 153 Ma. Zircons of ca. 227 Ma are common, and a few Paleozoic and Proterozoic zircons were found. The high proportion of volcanic clasts in the sandstone, including altered glass, indicates erosion of an active arc; thus, the ca. 153-Ma age most likely approximates the time of turbidite deposition, consistent with the Late Jurassic age of the Galice turbidite.

The Nd and Sr isotopic values for slates from the Galice Formation turbidite sequence are rather uniform and indicate derivation from both Precambrian continental rocks and young mantle-derived rocks (Frost et al., this volume). These workers also found that the continental isotopic signal is greater in slates than in sandstones of the turbidite sequence, a relationship that may be consistent with differences in the whole-rock geochemistry discussed below.

Two chert pebbles from the Galice Formation were dated as Late Triassic using radiolaria (D.L. Jones, personal commun., 1979; Harper, 1980). One of the pebbles is from a volcanic-rich sandstone near the base of the turbidite sequence overlying the Josephine ophiolite, and the other pebble is from a conglomerate near the base of the type Galice Formation overlying the Rogue Formation.

GEOCHEMISTRY

Major- and trace-element geochemistry of sedimentary rocks can provide invaluable information about the original tectonic setting and the provenance of a sedimentary unit (Roser and Korsch, 1986; McLennan et al., 1993). Unlike petrographic methods for tectonic discrimination that are limited to using coarser grain sizes, chemical analyses of sedimentary rocks have no grain-size restrictions (Roser and Korsch, 1986). McLennan (1989) notes, however, that concentrations of heavy minerals in sandstones can affect rare-earth-element (REE) abundances and

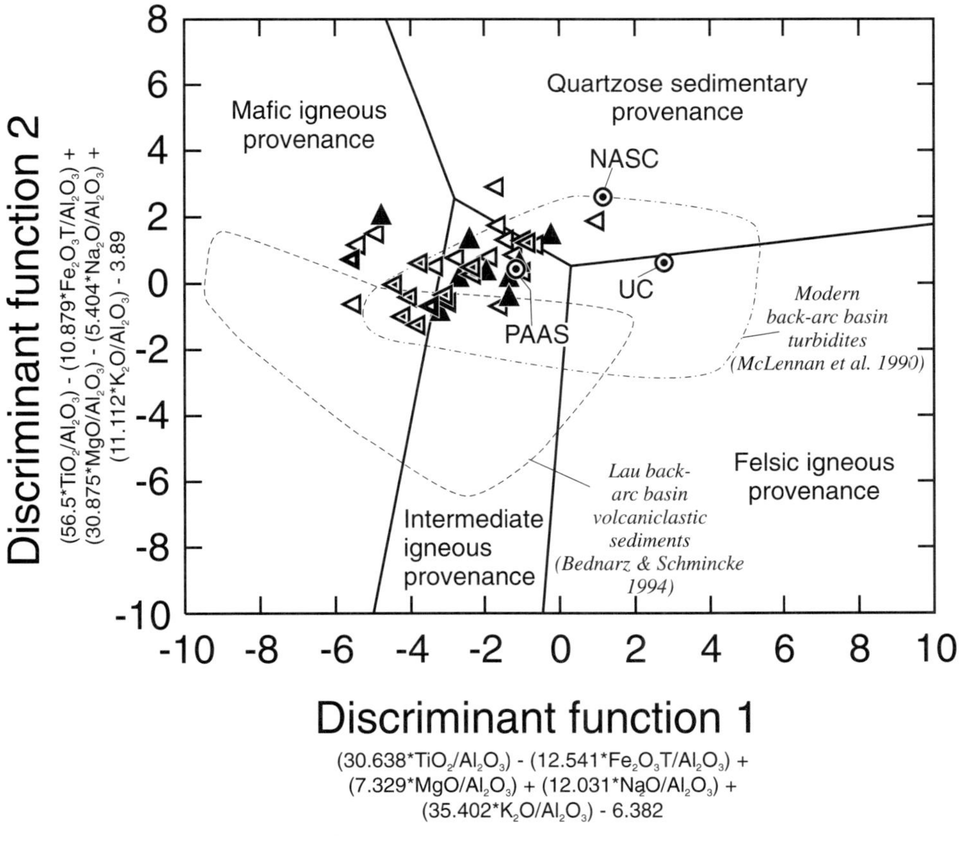

Figure 11. Provenance discriminant-function diagram of Roser and Korsch (1988). Only samples from the Galice turbidite are plotted. Siliceous argillite and chert from the hemipelagic sequence and Josephine ophiolite are not plotted because of their Fe-rich hemipelagic component. Also plotted are the average values for North American shale composite (NASC), post-Archean Australian shale composite (PAAS), upper continental crust (UC; Gromet et al., 1984; McLennan, 1989; McLennan, 2001), and fields for modern back-arc basin turbidites (McLennan et al., 1990) and volcaniclastic sediments from the Lau back arc (Bednarz and Schmincke, 1994). Symbols are the same as in Figure 5.

ratios, and that abundances of elements such as rare earths are diluted by the greater concentration of quartz in sandstones. Biogenic $CaCO_3$ or SiO_2 can also dilute concentrations of other elements within a sample (Roser and Korsch, 1986, 1988). Diagenetic reactions related to burial can affect silicate and calcite abundances of sandstones (Galloway, 1974). Thus, caution should be used when making interpretations based on geochemical analyses of sedimentary rocks, and ratios should be used when possible. Because we do not have CO_2 values for our analyses, we cannot correct for dilution resulting from $CaCO_3$. We have, thus, followed the advice of Roser and Korsch (1986, 1988) and recalculated major-element values on a CaO-free basis before plotting on some diagrams (e.g., Fig. 11).

Published and unpublished geochemical data for the Galice Formation are available from other studies (Coleman, 1972; Pinto-Auso and Harper, 1985; Kuhns and Baitis, 1987; Park-Jones, 1988; Zierenberg et al., 1988; Barnes et al., 1995; Frost et al., this volume; Table 2). In addition, we report a chemical analysis for a sandstone sample (D25C) that was analyzed along with samples of Park-Jones (1988) by X-ray fluorescence at McGill University. Only major-element data of Coleman (1972) are used, as his trace-element data are semiquantitative; sample

TABLE 2. SELECTED ANALYSES OF GALICE FORMATION SAMPLES

Sample*	22	24	30	33	65	89	102	139	147	37G	38G
Lithology	Slate	Slate	Slate	Slate	Slate	Slate	Slate	Slate	Slate	Graywacke	Graywacke
Locality	Type area	Type area	Type area	Type area	Type area	Type area	Type area	Type area	Type area	Type area	Type area
SiO_2	60.57	62.41	63.95	60.23	63.72	62.95	61.95	66.21	63.49	65.77	73.57
TiO_2	0.80	0.72	0.71	0.80	0.72	0.79	0.82	0.60	0.70	0.71	0.76
Al_2O_3	17.49	17.12	15.86	18.29	16.29	17.15	16.93	13.35	15.52	16.75	11.26
Fe_2O_3T†	8.24	7.27	6.60	7.91	7.53	6.99	7.33	7.37	7.49	5.80	6.32
MnO	0.16	0.06	0.07	0.07	0.03	0.08	0.02	0.09	0.04	0.02	0.02
MgO	2.49	2.70	3.03	2.60	2.41	2.82	2.87	2.71	3.29	1.92	2.13
CaO	0.76	0.70	0.64	0.14	0.29	0.33	0.33	1.66	0.54	0.02	0.02
Na_2O	2.18	2.00	2.21	0.84	1.59	1.78	2.24	1.26	1.99	1.08	0.37
K2O	2.23	2.32	2.80	2.49	2.01	2.56	2.47	1.84	1.85	2.06	1.29
P_2O_5	0.32	0.32	0.12	0.14	0.20	0.19	0.21	0.61	0.38	0.08	0.12
LOI	5.32	5.07	4.26	6.75	5.47	4.72	5.26	4.13	5.05	6.13	4.47
Total	100.56	100.69	100.25	100.26	100.26	100.36	100.43	99.83	100.34	100.34	100.33
Ba§	638	584	856	756	439	726	674	494	597	635	383
Rb	70	71	86	83	61	84	71	54	53	52	39
Sr	147	135	172	86	117	171	105	116	114	60	42
Pb	12	15	20	16	13	19	17	12	13	12	14
Y	147	27	24	34	27	36	28	30	25	29	12
Zr	159	152	145	178	143	173	176	131	144	144	174
Ni	54	60	47	85	50	23	75	48	39	50	81
Cr	112	101	110	104	100	125	148	85	120	198	281
V	190	180	197	196	200	180	200	174	213	142	156
Cs	4.38	5.07	n.d.	n.d.	n.d.	n.d.	n.d.	n.d.	2.34	n.d.	1.32
Sc	24	24	n.d.	n.d.	n.d.	n.d.	n.d.	n.d.	27.2	n.d.	20.28
Th	9.83	9.46	n.d.	n.d.	n.d.	n.d.	n.d.	n.d.	6.91	n.d.	6.59
Hf	4.69	4.41	n.d.	n.d.	n.d.	n.d.	n.d.	n.d.	4.68	n.d.	5.43
Ta	0.90	0.95	n.d.	n.d.	n.d.	n.d.	n.d.	n.d.	0.61	n.d.	0.67
La	31.02	29.08	n.d.	n.d.	n.d.	n.d.	n.d.	n.d.	25.48	n.d.	25.19
Ce	62.64	60.26	n.d.	n.d.	n.d.	n.d.	n.d.	n.d.	50.86	n.d.	45.36
Nd	33.60	31.85	n.d.	n.d.	n.d.	n.d.	n.d.	n.d.	24.67	n.d.	18.66
Sm	7.71	7.02	n.d.	n.d.	n.d.	n.d.	n.d.	n.d.	6.28	n.d.	4.16
Eu	1.53	1.42	n.d.	n.d.	n.d.	n.d.	n.d.	n.d.	1.68	n.d.	0.89
Tb	1.15	1.09	n.d.	n.d.	n.d.	n.d.	n.d.	n.d.	0.93	n.d.	0.47
Yb	4.32	3.76	n.d.	n.d.	n.d.	n.d.	n.d.	n.d.	3.39	n.d.	2.15
Lu	0.54	0.47	n.d.	n.d.	n.d.	n.d.	n.d.	n.d.	0.47	n.d.	0.28

Notes: JO—overlies Josephine ophiolite; LOI—loss on ignition; n.d.—no data.

*Data from Park-Jones (1988) except sample D25c, which is previously unpublished.

†Fe_2O_3T—total iron as Fe_2O_3.

§If all trace elements reported, analysis of Ba, Rb, Sr, Pb,Y, Zr, Ni, Cr, and V by X-ray fluorescence; others by instrumental neutron activation analysis.

B6C from Pinto-Auso and Harper (1985) was not used because of its extremely high CaO concentration.

A metalliferous component in sediments is best detected using the ratio Al/(Al + Fe + Mn), which is <0.35 for modern metalliferous sediments and >0.5 for nonmetalliferous pelagic clay (Bostrom and Peterson, 1969; Bostrom, 1973). This component can be derived either from hydrothermal springs or precipitation from seawater (as with manganese nodules). Negative, rather than positive, Ce anomalies on REE patterns for samples overlying and within the Josephine ophiolite indicate a hydrothermal origin. Figure 4, from Pinto-Auso and Harper (1985), shows that siliceous argillite and chert samples from the hemipelagic sequence and pillow lava unit of the Josephine ophiolite fall along a mixing line between modern terrigenous and low-temperature hydrothermal sediments. As discussed above, samples from within the pillow lava unit of the Josephine ophiolite and from the lower 8 m of the hemipelagic sequence are not metalliferous, although most have a metalliferous component (i.e., Al/[Al + Fe + Mn] <0.5); in fact, the least metalliferous hemipelagic sediments actually occur within the pillow lavas. This is somewhat surprising, because stratiform massive-sulfide deposits of hydrothermal origin are present in the

G1	G2	G3	G4	G5	G6	G7	G8	G9	G10	D25c
Slate	Slate	Slate	Slate	Slate	Slate	Slate	Slate	Slate	Slate	Graywacke
JO	JO	JO	JO	JO	JO	JO	JO	JO	JO	JO
62.21	61.20	63.24	62.22	66.01	62.12	61.58	68.76	67.42	61.06	64.12
0.77	0.77	0.84	0.94	0.76	0.77	0.87	0.62	0.59	0.87	0.63
17.17	15.90	18.52	19.23	15.29	16.06	17.49	14.05	14.16	18.55	15.31
5.02	8.01	5.11	5.86	7.99	7.99	6.49	6.17	6.95	5.75	5.62
0.02	0.04	0.01	0.04	0.06	0.04	0.03	0.03	0.05	0.02	0.05
1.91	2.70	2.07	2.86	2.81	2.79	2.68	2.38	2.21	2.00	2.17
0.33	0.61	0.30	0.01	0.16	1.03	0.36	0.20	0.42	0.06	5.34
1.30	1.61	0.72	n.d.	0.64	1.05	0.03	1.65	1.84	0.27	1.98
3.68	2.86	4.00	3.38	2.49	2.78	3.47	2.11	1.84	3.74	1.78
0.20	0.22	0.02	0.09	0.21	0.22	0.18	0.16	0.20	0.18	0.08
7.16	5.75	5.43	5.42	4.26	5.70	6.08	3.86	4.46	7.76	3.25
99.77	99.67	100.26	100.05	100.68	100.55	99.26	99.99	100.14	100.26	100.33
949	781	1232	1487	887	758	923	748	591	1227	612
125	88	139	124	77	92	112	68	57	123	n.d.
89	95	90	72	83	122	82	92	121	51	n.d.
n.d.	11	31	27	34	35	25	34	32	34	n.d.
26	26	29	18	26	28	22	14	18	40	n.d.
157	169	180	181	173	170	166	141	143	178	n.d.
67	60	57	60	81	77	63	40	66	53	65
99	110	129	172	124	112	122	111	110	152	164
177	159	210	252	157	170	206	154	106	220	162
12.24	8.42	n.d.	n.d.	n.d.	n.d.	n.d.	n.d.	n.d.	n.d.	n.d.
26.19	21.44	n.d.	n.d.	n.d.	n.d.	n.d.	n.d.	n.d.	n.d.	n.d.
13.29	10.41	n.d.	n.d.	n.d.	n.d.	n.d.	n.d.	n.d.	n.d.	n.d.
5.06	5.78	n.d.	n.d.	n.d.	n.d.	n.d.	n.d.	n.d.	n.d.	n.d.
1.16	0.87	n.d.	n.d.	n.d.	n.d.	n.d.	n.d.	n.d.	n.d.	n.d.
37.51	30.76	n.d.	n.d.	n.d.	n.d.	n.d.	n.d.	n.d.	n.d.	n.d.
70.23	63.52	n.d.	n.d.	n.d.	n.d.	n.d.	n.d.	n.d.	n.d.	n.d.
34.72	31.49	n.d.	n.d.	n.d.	n.d.	n.d.	n.d.	n.d.	n.d.	n.d.
7.41	7.01	n.d.	n.d.	n.d.	n.d.	n.d.	n.d.	n.d.	n.d.	n.d.
0.99	1.42	n.d.	n.d.	n.d.	n.d.	n.d.	n.d.	n.d.	n.d.	n.d.
0.99	1.05	n.d.	n.d.	n.d.	n.d.	n.d.	n.d.	n.d.	n.d.	n.d.
4.19	3.66	n.d.	n.d.	n.d.	n.d.	n.d.	n.d.	n.d.	n.d.	n.d.
0.55	0.49	n.d.	n.d.	n.d.	n.d.	n.d.	n.d.	n.d.	n.d.	n.d.

Josephine ophiolite (Kuhns and Baitis, 1987; Zierenberg et al., 1988), but are apparently the result of dilution by abundant terrigenous detritus, as indicated by Figure 4. "Off-axis" metalliferous sediments are present 8–23 m above the base of the hemipelagic sequence (Fig. 2), either due to precipitation from low-temperature, off-axis springs (Pinto-Auso and Harper, 1985) or from distal fallout of high-temperature springs at a younger, propagating spreading center (Harper, 2003). All slate and sandstone samples from the Galice turbidite, including those from the Elk outlier, have Al/(Al + Fe + Mn) ratios >0.5 (Fig. 4), indicating a negligible metalliferous component.

Interpretations about source area and tectonic setting based on sediment geochemistry can be made from several plots (Figs. 11, 12, and 13). The majority of the turbidite samples, including those from the Elk outlier, plot in the intermediate-igneous and mafic-igneous provenance fields on Figure 11 (Roser and Korsch, 1988), which uses linear equations of major-element ratios except for CaO and SiO_2 (because they can be elevated by biogenic contributions). The remaining turbidite samples plot in and near the quartzose sedimentary provenance field (Fig. 11). Samples from the hemipelagic sequence and pillow lava unit are not plotted on Figure 11 because their high-Fe hydrothermal components cause these samples to plot far to the left. Figure 11 also shows that Galice turbidite samples fall within the field for back-arc basin turbidites of McLennan et al. (1990) and are displaced away from average continental crust (UC) and sediments largely derived from continental crust (NASC and PAAS in Fig. 11).

Figure 12 uses the trace elements Th, Zr, and Sc, which are generally immobile during metamorphism. This diagram is useful in that sediments typical of passive continental margins have elevated Zr/Sc ratios due to zircon enrichment through sediment recycling (McLennan et al., 1993; Fig. 12). Active-margin samples, however, plot along a trend between mafic and continental sources. Figure 12 shows that the Galice Formation samples plot along this trend, indicating insignificant sediment recycling, although the presence of well-rounded detrital zircons in sandstones indicates some recycling has occurred (Harper, 1980; Miller and Saleeby, 1995). Figure 11 suggests that the Galice samples were derived, on average, from sources that are more mafic than typical continental crust; perhaps from the mixing of continental and intermediate arc material. The chert sample from the hemipelagic sequence above the Josephine pillow lavas plots at significantly higher Zr/Sc values compared to other samples (Fig. 12); although the nonbiogenic component in the chert should consist of terrigenous or tuffaceous material similar to that in the hemipelagic rocks and thus plot along the same trend as other Galice samples. Hence, it may not be valid to plot cherts on this diagram (McLennan et al., 1993). Moreover, the analytical uncertainties for Zr and Sc are relatively high for this particular sample since they are so diluted by SiO_2.

A more complete comparison of Galice samples with post-Archean Australian shale (PAAS), derived largely from continental sources, can be made by normalizing immobile trace and rare-earth elements to PAAS (Fig. 13). McLennan (1989) suggested that samples should be normalized to PAAS rather than to North American shale composite (NASC), due to the possible inclusion of abnormal samples in NASC. Reference samples are shown in Figure 13A and B for comparison with those from the Galice Formation. The NASC and average upper continental crust generally have flat PAAS-normalized patterns (Fig. 13A), reflecting the dominantly continental sources for the shale

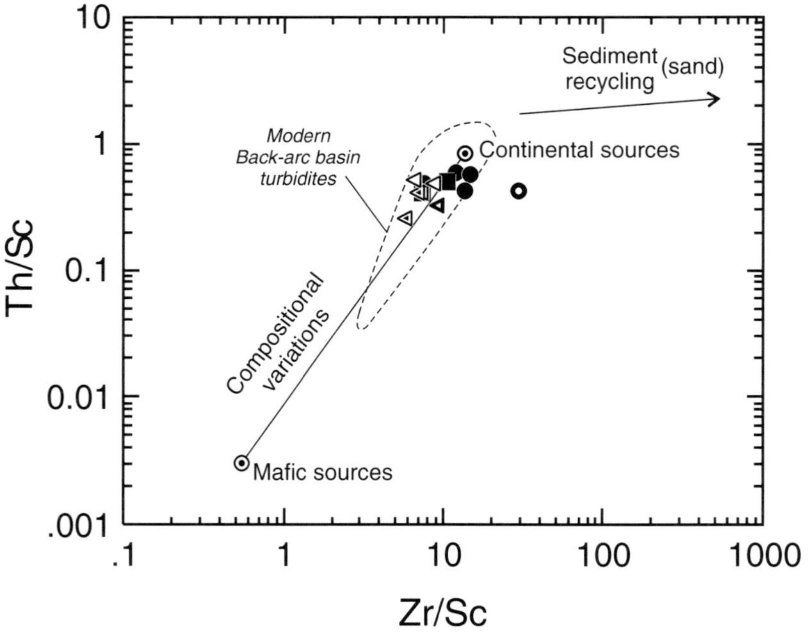

Figure 12. Th/Sc vs. Zr/Sc plot modified from McLennan et al. (1993) by the addition of average upper continental crust (representing cratonic sources; McLennan, 2001), a modern island-arc tholeiite (representing mafic sources; Pearce et al., 1995), and the field for modern back-arc basin turbidities (McLennan et al., 1990). Symbols are the same as in Figure 4.

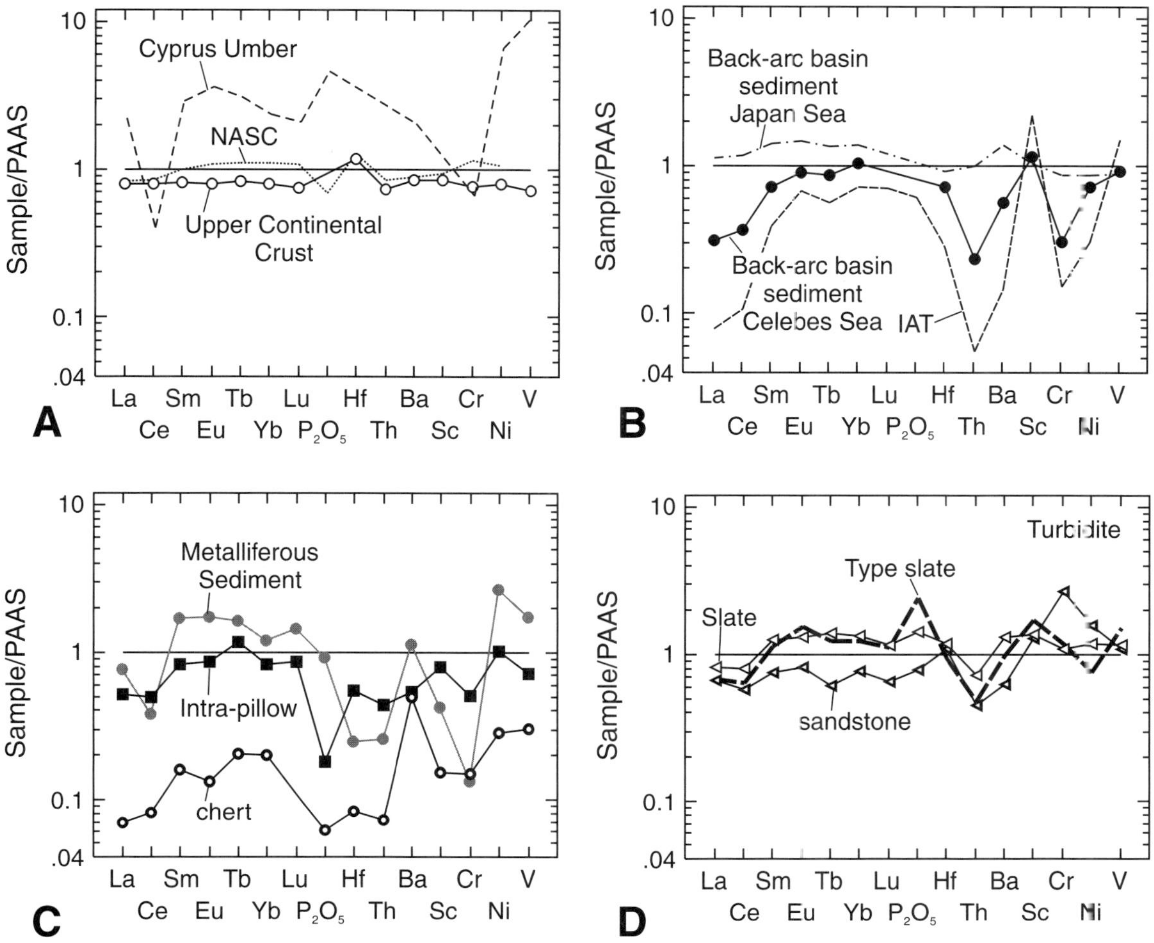

Figure 13. Post-Archean Australian shale composite (PAAS) normalized diagrams. (A) North American shale composite (NASC) from Gromet et al. (1984), Cyprus umber from Ravizza et al. (1999), and upper continental crust values from McLennan (2001). (B) Back-arc basin sediments (Japan and Celebes Seas) from McLennan et al. (1990) and modern island-arc tholeiite (IAT) from Pearce et al. (1995). (C) Chert from the hemipelagic sequence and typical metalliferous and intra-pillow argillites. (D) Typical slate and sandstone from the turbidite overlying the Josephine ophiolite, and a typical slate from the type area of the Galice Formation. Normalization values are from McLennan (1989).

composites. A Cyprus umber, which is essentially pure metalliferous sediment, shows enrichment with respect to PAAS, especially in Ni and V, except for negative Ce and Cr anomalies (Fig. 13A). The back-arc basin sediment from the Celebes Sea and an average island-arc tholeiite (IAT) have remarkably similar PAAS-normalized patterns (Fig. 13B); both are depleted in the light REEs compared to PAAS, have negative Hf, Th, and Cr anomalies, and have positive Sc anomalies (Fig. 13B). Back-arc basin sediment from the Japan Sea has a flatter PAAS-normalized pattern and a positive Ba anomaly (Fig. 13B), which is probably due to a large continental, or calc-alkaline volcanic, source component; this sample displays a slight negative slope from La through Eu, a result that is similar to the IAT sample and the Celebes Sea back-arc basin sediment (Fig. 13B).

Galice Formation slate from the type area and from the turbidite sequence overlying the Josephine ophiolite has similar patterns when normalized to PAAS (Fig. 13C). They show depletion in light REEs, a negative Th anomaly and a positive P_2O_5 anomaly. A sandstone from the basal part of the turbidite overlying the Josephine ophiolite has a similar pattern to the slates, except for a negative Ba anomaly and positive Hf, Sc, and Cr anomalies (Fig. 13D). The Cr anomaly is probably from detrital Cr-spinel, as it is common in thin section, and the Sc anomaly is probably from the presence of detrital augite, which is abundant in this and most sandstone from the basal turbidite overlying the Josephine ophiolite (Harper, 1980).

The intra-pillow siliceous argillite, which is interbedded with pillow lavas of the Josephine ophiolite, has a similar pattern to slate from the Galice Formation turbidite (Fig. 13C), suggesting a similar terrigenous source. This sample, however, has a negative rather than a positive P_2O_5 anomaly; however, because only one of the slates has a significant P_2O_5 anomaly and the sandstone has none, P may have been mobile during diagenesis and metamorphism. The pattern for the sandstone sample (Fig. 13D) is shifted downward relative to the slates, a relationship that is readily explained by dilution from biogenic silica, which is evident in thin section from the presence of radiolarians. As expected, this dilution effect is even greater for the chert from the hemipelagic sequence overlying the Josephine ophiolite (Fig. 13C). The pattern for the chert is similar to that for the intrapillow siliceous argillite (Fig. 13C), suggesting a similar terrigenous component, although the chert has a prominent positive Ba anomaly of unknown origin (seawater?), similar to the metalliferous sediment. The metalliferous sediment sample has a pattern that has similarities to both the slates of the Galice turbidite and to the Cyprus umber (e.g., negative Ce and Cr anomalies); this observation is consistent with the conclusion of Pinto-Auso and Harper (1985) that the metalliferous sediments in the hemipelagic sequence all have substantial terrigenous components, as evident from elevated Al/(Al + Fe + Mn) ratios (Fig. 4).

DISCUSSION

Provenance

Plots of detrital modes (Fig. 7), sediment geochemistry (Figs. 11, 12, and 13), and Nd and Sr isotopes (Frost et al., this volume) show trends for Galice turbidite samples that can be explained by mixing between two sources. The abundance of volcanic rock fragments in sandstones and conglomerates (Figs. 7 and 8) indicates that one of these sources is volcanic, and this source is almost certainly a magmatic arc. The non-arc source could represent basement of the arc, or a geographically distinct continental-like source.

The microlitic texture of most volcanic clasts suggest that the arc source was composed primarily of andesite, although the abundance of detrital clinopyroxene (up to 5 modal %) is perhaps more consistent with a basaltic-andesite or basalt source. A negative slope for light REE, negative Th anomaly, and positive Eu and Sc anomalies for Galice turbidite samples on a PAAS-normalized diagram (Fig. 13) are also suggestive of a mafic source. In addition, the trend in Figure 11 extends from the field for mafic, rather than intermediate, igneous provenances. The Nd and Sr isotope data for slates from the Galice Formation suggest that the arc was ensimatic (i.e., built on non-continental basement). The high lithic content and generally low modal monocrystalline quartz in sandstones (Fig. 7B), along with the paucity of plutonic clasts in conglomerates (<3%, nearly all <1%; Wyld, 1985; Seiders, 1991, written commun., 1992) indicate that the arc was essentially undissected. Many of the sandstones plot in the transitional (partially dissected) arc field of the Q–F–L diagram (Fig. 7A), but this is a result of the inclusion of the chert clasts within Q. Nevertheless, several samples, especially those from the southern Klamath Mountains (Wyld, 1985), have sufficiently low total lithic clasts, relative to quartz and feldspar that they plot in the transitional arc field on a Q_m–F–L_t diagram (Fig. 7B). This may also be artificial, however, for the following reasons: Wyld's (1985) samples generally have >30% matrix, which is far higher than modern turbidites suggesting a post-depositional reduction of the L_t/F and L_t/Q_m ratios through the breakdown of lithic clasts; Wyld's (1985) reported "matrix" in her point-count tables includes both orthomatrix and pseudomatrix (i.e., Dickinson, 1970), and thus lithic clasts are underrepresented in the Q_m–F–L_t diagram; and Wyld's (1985) conglomerate samples have <3% plutonic clasts. Ingersoll (1978, 1983) suggested that pseudomatrix >20% should be a cautionary sign when analyzing point-count data.

Detrital modes of sandstone and conglomerate indicate that the non-arc component (source 2 on Fig. 7C and D) consisted almost entirely of chert and siliceous argillite. The inferred end member corresponding to this source component is essentially a lithic recycled component, in the terms of Figure 7B. The scarce metamorphic rock fragments (phyllite, quartz-mica tectonite, and quartzite) are attributed to this source, as are heavy minerals that are typically metamorphic in origin (garnet, glaucophane, epidote, and muscovite). The staurolite is originating from an unknown source area (Snoke, 1977; Harper, 1980). Ophiolitic rocks were also part of the non-arc source, as evident from detrital Cr-spinel compositions that indicate the presence of mantle peridotites, at least some of which were part of supra-subduction zone ophiolites (Figs. 9 and 10). Cr-spinels with compositions that indicate derivation from MORB and transitional MORB-OIB basalts (Fig. 9) were probably derived from non-SSZ ophiolites. A positive Cr anomaly on a PAAS-normalized diagram for a Galice sandstone (Fig. 13) is readily interpreted as due to the presence of detrital Cr-spinel. In contrast, an ophiolitic signature (high Cr or Ni) is not evident in the slates.

Extrapolation of the trend on Figure 11 suggests that the non-arc sources lie in the field of quartzose sedimentary province. The Th/Sc versus Zr/Sc diagram is compatible with this interpretation in that many of the samples plot close to continental sources (Fig. 12), but this diagram also shows that highly mature sands resulting from recycling, such as those formed along passive rifted margins, are not a significant component of Galice sandstones. Mature sandstones were present in the source area, however, based on the presence of very well rounded zircon in sandstone. Precambrian zircon is also present in the euhedral zircon population from Galice sandstones (Miller and Saleeby, 1995), indicating that at least some Precambrian sediment is first cycle. Overall, Miller et al. (2003) found that only a small fraction of the detrital zircons are Precambrian. Frost et al. (this volume) conclude from Nd and Sr isotope data that Galice slates have a significantly larger Precambrian continental component than do sandstones.

Snoke (1977) concluded that much of the detritus in Galice sandstones could be readily derived from older, more easterly terranes in the Klamath Mountains. This hypothesis is consistent with the west-directed flow indicated by paleocurrent data from the Galice Formation where it overlies the Josephine ophiolite (Harper, 1980; Fig. 6D). The Klamath Mountains contain abundant chert and siliceous argillite, much of it Triassic (like two pebbles from the Galice Formation), as well as metamorphic and ophiolitic rocks (e.g., Irwin, 1966, 1972; Davis et al., 1978; Saleeby and Busby-Spera, 1992). The basement to all but the eastern Klamath arc rocks, including the Rogue–Chetco arc (Yule, 1996; Yule et al., this volume), is the early Mesozoic Rattlesnake Creek terrane (Wright and Wyld, 1994; Hacker et al., 1995). The Rattlesnake Creek terrane contains Triassic chert as well as mafic volcanic rocks that could have supplied MORB to MORB-OIB affinity detrital Cr-spinel (Fig. 9), although OIB affinity volcanic rocks are also present in the Sawyers Bar terrane of the central Klamath Mountains (Hacker et al., 1993). Units in the Klamath Mountains that could have supplied metamorphic detritus include the greenschist- to amphibolite-facies central metamorphic belt (e.g., Irwin, 1966; Davis et al., 1978). As for detrital glaucophane, the blueschist-facies Stuart Fork Terrane (Hotz et al., 1977) and Skookum Gulch Schist (Cotkin et al., 1992), as well as blueschist blocks in mélange of the Eastern Hayfork unit in the Sawyers Bar terrane (Wright, 1982; Hacker and Ernst, 1993), could have been the source(s). The large supra-subduction zone Trinity ophiolite could have supplied detrital peridotitic and boninitic Cr-spinel (Quick, 1981; Metcalf et al., 2000; Fig. 10).

Possible contributors of arc detritus from east of the Galice Formation include Middle Jurassic rocks in the eastern Klamath Mountains (e.g., Renne and Scott, 1988), the ca. 177- to 168-Ma Hayfork arc (western Hayfork terrane and related plutonic rocks; Wright and Fahan, 1988), and the ca. 159- to 165-Ma Wooley Creek plutonic belt (e.g., Hacker et al., 1995). These are not the appropriate age to have provided the ca. 153-Ma detrital zircons in the Galice Formation, nor are they likely to have yielded volcanic glass that is abundant in the volcanic-rich sandstones of the basal turbidite. Three plutons along the northeastern margin of the Klamath Mountains, however, may be the subvolcanic remains of a ca. 153-Ma zircon source. Irwin and Wooden (1999) reported 154- and 156-Ma U/Pb zircon ages for the Jacksonville and White Rock plutons in the northern Klamath Mountains, respectively, and Yule et al. (this volume) interpret an age of ca. 152–156 Ma from complex U/Pb zircon age data for the younger of two phases of the Ashland pluton. Frost et al. (this volume) suggest that the Rogue–Chetco arc (Fig. 1) was the source of the ca. 153-Ma volcanic detritus in Galice sandstones and slates, but this seems unlikely because: (1) the arc axis, probably represented by the plutonic Chetco complex lies west of the Galice Formation, inconsistent with paleocurrent data; (2) U/Pb zircon ages for the Chetco Complex are too old, ranging from 160 to 157 Ma (Yule, 1996; Yule et al., this volume); (3) the Rogue Formation underlies the Galice Formation and is thus too old, although one $^{40}Ar/^{39}Ar$ hornblende age is 153.4 Ma (Hacker et al., 1995); and (4) some of the components in the sandstones (e.g., glaucophane) are not known from the basement rocks of the Rogue–Chetco arc complex, which are dominantly ophiolitic (Yule, 1996; Yule et al., this volume). Volcanic members in the Galice Formation, which include coarse breccias, were obviously derived from nearby volcanic sources, but volcanic detritus in the Galice sandstones and slates had to have been mixed with non-arc sources before deposition.

Derivation of the ca. 227-Ma detrital zircons within the Galice Formation from the Klamath Mountain terranes to the east does appear unlikely, because there is no known source in the Klamath Mountains. The only known igneous rocks of the appropriate age in the Klamath Mountains are dacites in the Pit Formation of the eastern Klamath Mountains (Albers and Robertson, 1961), but these apparently are barren of zircon (J.E. Wright, written commun., 2004). Hacker et al. (1993) infer that arc rocks of this age are present in the Sawyers Bar terrane in the central Klamath Mountains, but even if they are, these rocks are predominantly mafic and unlikely to have yielded much zircon. The ca. 227-Ma detrital zircon, however, might be present in Upper Triassic to Middle Jurassic clastic rocks of the Klamath Mountains. A pre-Cretaceous reconstruction by Wyld and Wright (2001) places the Pine Nut terrane and the basinal Luning assemblage just east of the Klamath Mountains. Both the Pine Nut terrane and the Luning assemblage contain strata with peak detrital zircon ages of ca. 225 Ma (Manuszak et al., 2000), in addition to rocks of ca. 230 Ma in the former (Dilles and Wright, 1988); thus, these terranes are potential sources of the ca. 227-Ma detrital zircon in Galice sandstones.

South-directed paleocurrent directions in the type Galice Formation (Fig. 6B) suggest that the main source area for the Galice Formation may have been north of the present Klamath Mountains. Klamath rocks are covered by Cenozoic volcanic rocks to the north, but chert, including Triassic chert, (e.g., Pessagno and Whalen, 1982), and supra-subduction zone ophiolite peridotite (Bishop, 1995) are present in northeastern Oregon. Significant volumes of ca. 153-Ma volcanic rocks do not appear to be present (e.g., Saleeby and Busby-Spera, 1992), although some andesitic dikes and the Sunrise Butte pluton have K-Ar ages in the range of 150–158 Ma (Johnson et al., 1995). Only a few probable Late Jurassic arc rocks are exposed farther north in Washington (Miller et al., 1993). Possible source rocks for the ca. 227-Ma detrital zircons in the Galice may be present in the Blue Mountains, Oregon, Washington, and Idaho (e.g., 262- to 219-Ma Sparta complex and Carnian to Norian volcanic rocks of the Olds Ferry terrane; Avé Lallemant, 1995). The monocrystalline quartz-poor Galice sandstones are unlikely to have had a source to the south, because Paleozoic quartz sandstones in the central and southern Sierra Nevada contain zircons much older (>2.2 Ga) than detrital zircons in the Galice Formation (Harding et al., 2000).

Regional Variations

Miller et al. (2003) found that detrital-zircon age populations in a sandstone overlying the Ingalls ophiolitic complex in the Cascade Mountains of Washington State were similar to ages from a sandstone of the Galice Formation, with peaks at ca. 153 and ca. 227 Ma. The Ingalls sandstone, however, contains no Precambrian zircon. No modal petrographic data are available for the Ingalls sandstones, although the types of clasts are similar to those in the Galice Formation, including Cr-spinel (Southwick, 1974; Miller et al., 1993).

Terranes similar to those of the western Klamath terrane are present in the Sierra Nevada region (Davis, 1969; Irwin, 2003). Conglomerates from the Mariposa Formation of the central Sierra Nevada foothills, which has long been correlated to the Galice Formation (e.g., Diller, 1907; Davis, 1969), are distinct from Galice conglomerates in that they have a much higher proportion of quartzose clasts (Seiders and Blome, 1988; Seiders, 1991, written commun., 1992; Fig. 8). This observation suggests a Sierran rather than Klamath Mountains source for the Mariposa Formation (Seiders and Blome, 1988), even though paleocurrent data reported by Bogen (1984) indicate south-southeast transport. Most of the conglomerates in the southernmost Galice Formation (Wyld, 1985; Fig. 1) are similar to those from the Galice Formation farther north, in which pebbles consist almost entirely of chert, argillite, or volcanic rocks, but some have elevated contents of quartzose clasts and are thus compositionally transitional to conglomerates of the Mariposa Formation (Fig. 8).

Temporal Variations and Tectonic Implications

The ca. 162- to 153-Ma sedimentary rocks of the hemipelagic sequence, the hemipelagic-turbidite transition, and those within the Josephine ophiolite have similarities to the turbidite (Figs. 11, 12, and 13), except for the presence of a metalliferous (hydrothermal) component in some samples (Fig. 4). There is a striking similarity in petrography, and especially detrital Cr-spinel compositions (Figs. 9 and 10), between the thin sandstone bed within pillow lavas of the Josephine ophiolite (ca. 162 Ma) and sandstones of the lower Galice turbidite (ca. 153 Ma). This observation and similarities in the geochemistry of siliceous argillites and Galice slates suggest that the source area for the Galice turbidite was already established by ca. 162 Ma. This implies that the Galice source area was probably produced during ca. 165-Ma thrusting ("Siskiyou orogeny") that preceded formation of the Josephine ophiolite (Wright and Fahan, 1988; Hacker et al., 1995; Fig. 14). The inferred basin and source area relationships just after this episode of shortening are shown in Figure 14A.

Tectonic Setting of Deposition

The tectonic setting at ca. 160 Ma for the western Klamath Mountains is well constrained (e.g., Saleeby et al., 1982; Harper and Wright, 1984; Wyld and Wright, 1988; Hacker et al., 1995; Yule, 1996; Yule et al., this volume), and a model in which the Josephine ophiolite is located within a back-arc basin behind the

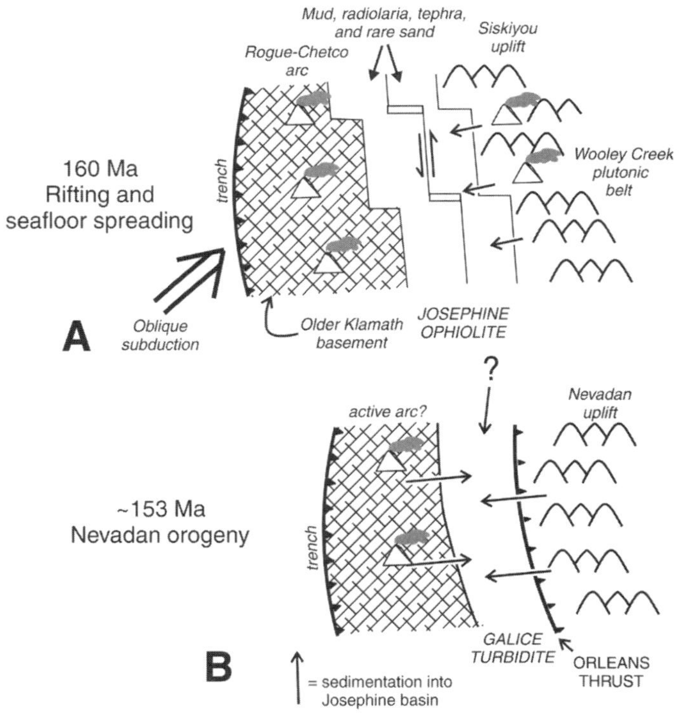

Figure 14. Tectonic model for the generation and emplacement of the Galice Formation and underlying Josephine ophiolite. The inferred oblique-subduction model is similar to the modern Andaman Sea. Arrows indicate probable paleocurrent flow directions. Modified from Harper and Wright (1984), Harper et al. (1985), Wyld and Wright (1988), and Harper et al. (1994). (A) Paleogeography prior to deposition of Galice turbidite. (B) Paleogeography during deposition of Galice turbidite.

west-facing Rogue–Chetco arc at ca. 160 Ma (Fig. 14A) is generally accepted (e.g., Dickinson et al., 1996). The ca. 177- to 168-Ma western Hayfork arc would represent a remnant arc at this time, and a tectonic highland is inferred because of a ca. 165-Ma shortening event (Wright and Fahan, 1988), sometimes called the "Siskiyou orogeny." Seafloor spreading in the back-arc basin is inferred to have been transform-dominated (Fig. 14A), based on the east-west orientation (in modern coordinates) of spreading centers inferred from sheeted dike orientations and other structural data (Harper et al., 1985; Alexander and Harper, 1992). Parts of both sides of this basin were underlain by older rifted Klamath crust, in places cut by Josephine-age mafic complexes; the Rogue–Chetco arc was built on older Klamath crust rifted away during seafloor spreading in the back-arc basin (Saleeby et al., 1982; Harper and Wright, 1984; Wyld and Wright, 1988; Yule, 1996; Yule et al., 1996).

Deposition of the Galice turbidite was interpreted by Harper and Wright (1984) and Wyld and Wright (1988) to have occurred in this basin before the Nevadan orogeny, with volcanic members derived from the Rogue–Chetco arc from the west (Fig. 14A). This is probably the explanation for the high terrigenous content of sediments within the Josephine pillow lava unit and hemipelagic sequence, with the terrigenous sediment derived from the nearby "Siskiyou" highlands.

Based on the closeness in age of the Galice turbidite with the oldest plutons that cut the roof (Orleans) thrust, which may have >100-km displacement, Harper et al. (1994) reinterpreted the Galice turbidite as a synorogenic deposit (Fig. 14B). In this scenario, the turbidite was derived from Late Jurassic uplift related to underthrusting of the Josephine ophiolite. The area of uplift is inferred to have been similar to that affected by the pre-Josephine ("Siskiyou") deformation, which would account for the very similar clasts compositions and detrital modes of the sandstone within the Josephine pillow lavas (ca. 162 Ma) and the basal Galice turbidite (ca. 153 Ma). The great increase in amount of sandstone at the top of the hemipelagic-turbidite transition is based on the peak ca. 153-Ma age of detrital zircon in a sandstone from the basal turbidite (Miller et al., 2003) as well as other geochronologic constraints (Harper et al., 1994). Widespread ca. 150-Ma $^{40}Ar/^{39}Ar$ hornblende and mica cooling ages in the central Klamath Mountains (Hacker et al., 1995) may record Nevadan underthrusting and uplift.

The presence of the apparent hiatus at the top of the hemipelagic sequence (Pessagno and Blome, 1990; Pessagno et al., 2000), which is underlain by a zone of pre-cleavage deformation (Harper, this volume; Fig. 2), is enigmatic. As discussed above, this contact could be a minor unconformity, a normal fault, or a submarine landslide surface. A possible explanation for this surface is that it is a disconformity resulting from nonsedimentation and/or submarine erosion on a flexural bulge in front of the trench as the ophiolite and overlying Galice were thrust beneath older terranes of the Klamath Mountains during the Nevadan orogeny (Fig. 14). Passage of crust over a flexural bulge in front of a trench can result in an unconformity and normal faulting (e.g., Rowley and Kidd, 1981). Underthrusting of the Josephine ophiolite, which involved at least 40 km of displacement, is broadly analogous to subduction. In addition, relief created by this faulting could have resulted in a submarine landslide that removed strata prior to deposition of the hemipelagic-turbidite unit.

The quartzite-rich conglomerates in the Mariposa Formation in the Sierra Nevada foothills (Fig. 8) may indicate that the Mariposa and associated Late Jurassic arc rocks have not been translated appreciably relative to older rocks in the Sierra Nevada. If this suggestion is correct, then the Mariposa should contain >2.2-Ga detrital zircon known to be present in older rocks of the Sierra Nevada (Harding et al., 2000). Similarly, the Galice sandstones and conglomerates appear to be derived largely from the Klamath Mountains, except that there is no known Klamath source for ca. 227-Ma detrital zircons, and the location that provided ca. 153-Ma zircons is uncertain. Additional dating of zircon in the Klamath Mountains and elsewhere in the Cordillera might better constrain source areas for Galice sedimentary rocks and, in turn, better constrain models for Late Jurassic tectonics and paleogeography of western North America.

ACKNOWLEDGMENTS

Much of the data are from GDH's Ph.D. thesis at the University of California, Berkeley. Preparation of this manuscript was funded by National Science Foundation Grant EAR-0003444 to GDH. The Geology Department at San Jose State University kindly allowed JHM office space during preparation of this chapter. The Cr-spinel probe work was funded by the Department of Earth and Atmospheric Sciences, State University of New York (SUNY) Albany and a SUNY Albany Faculty Research Award Program grant to GDH. Reviews by Gary H. Girty and Raymond V. Ingersoll greatly improved this manuscript. Art Snoke proved to be a first-class editor. Helpful discussions were provided by Calvin G. Barnes, James E. Wright, and Bill Kidd. John Delano kindly allowed use of his lab for sample preparation, and Bill Blackburn was a great help to JHM on the microprobe.

APPENDIX

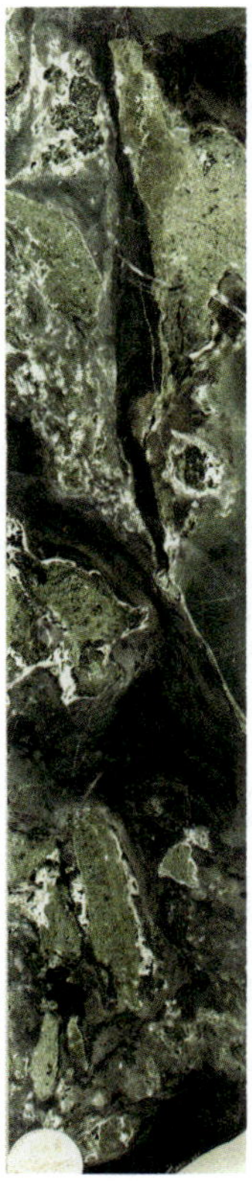

Figure A-1. Drill core sample from the Turner-Albright mine showing siliceous argillite that is interbedded with lavas of the Josephine ophiolite. The black and green laminations are interpreted as terrigenous and tuffaceous in origin, respectively. The alternating black and green colors are also typical of the hemipelagic sequence overlying the Josephine ophiolite. Note coin for scale.

REFERENCES CITED

Albers, J.P., and Robertson, J.F., 1961, Geology and ore deposits of East Shasta copper-zinc district, Shasta County, California: Washington, D.C., U.S. Geological Survey Professional Paper 338, 107 p.

Alexander, R., and Harper, G.D., 1992, The Josephine ophiolite, an ancient analog for oceanic lithosphere formed at slow/intermediate spreading centers, *in* Parson, B., et al., eds., Ophiolites and their modern oceanic analogues: London, Blackwell Scientific Publications, Geological Society Special Publication 60, p. 3–38.

Allan, J.F., 1994, Cr-spinel in depleted basalts from the Lau Basin backarc; Petrogenetic history from Mg-Fe crystal-liquid exchange, *in* Hawkins, J., et al., eds., Proceedings of the Ocean Drilling Program, Scientific results, volume 135: College Station, Texas, Ocean Drilling Program, p. 565–583.

Avé Lallemant, H.G., 1995, Pre-Cretaceous tectonic evolution of the Blue Mountains province, northeastern Oregon, *in* Vallier, T. L., and Brooks, H.C., eds., Geology of the Blue Mountains region of Oregon, Idaho, and Washington: Petrology and tectonic evolution of pre-Tertiary rocks of the Blue Mountains region: Reston, Virginia, U.S. Geological Survey Professional Paper 1438, p. 271–304.

Barnes, C.G., Donato, M.M., Barnes, M.A., Yule, J.D., Hacker, B.R., and Helper, M.A., 1995, Geochemical compositions of metavolcanic and metasedimentary rocks, western Jurassic and western Paleozoic and Triassic belts, Klamath Mountains, Oregon and California: Reston, Virginia, U.S. Geological Survey Open-File Report OF 95–0227-A, 63 p.

Barnes, S.J., and Roeder, P.L., 2001, The range of spinel compositions in terrestrial mafic and ultramafic rocks: Journal of Petrology, v. 42, p. 2279–2302, doi: 10.1093/petrology/42.12.2279.

Barrett, T.J., 1981, Chemistry and mineralogy of Jurassic bedded chert overlying ophiolites in the North Apennines, Italy: Chemical Geology, v. 34, p. 289–317, doi: 10.1016/0009-2541(81)90118-2.

Barrett, T.J., Taylor, P.N., and Lugowski, J., 1987, Metalliferous sediments from DSDP Leg 92; the East Pacific Rise transect: Geochimica et Cosmochimica Acta, v. 51, p. 2241–2253, doi: 10.1016/0016-7037(87)90278-X.

Bednarz, U., and Schmincke, H.-U., 1994, Composition and origin of volcaniclastic sediments in the Lau Basin (Southwest Pacific), Leg 135 (sites 834–839), *in* Hawkins, J., et al., eds., Proceedings of the Ocean Drilling Program, scientific results, volume 135: College Station, Texas, Ocean Drilling Program, p. 51–74.

Behrman, P.G., and Parkinson, G.A., 1978, Paleogeographic significance of the Callovian to Kimmeridgian strata, central Sierra Nevada Foothills, California, *in* Howell, D.G., and McDougall, K., eds., Mesozoic paleogeography of the western United States, Pacific Coast Paleogeography Symposium 2: Los Angeles, Pacific Section, Society of Economic Paleontologists and Mineralogists, p. 349–360.

Bishop, E.M., 1995, Mafic and ultramafic rocks of the Baker terrane, eastern Oregon, and their implications for terrane origin, *in* Vallier, T.L., and Brooks, H.C., eds., Geology of the Blue Mountains region of Oregon, Idaho, and Washington: Petrology and tectonic evolution of pre-Tertiary rocks of the Blue Mountains region: Reston, Virginia, U.S. Geological Survey Professional Paper 1438, p. 221–245.

Bogen, N.L., 1984, Stratigraphy and sedimentary petrology of the Upper Jurassic Mariposa Formation, western Sierra Nevada, California, *in* Crouch, J.K., and Bachman, S.B., eds., Tectonics and sedimentation along the California margin: Los Angeles, Pacific Section, Society of Economic Paleontologists and Mineralogists Field Trip Guidebook 38, p. 119–134.

Bogen, N.L., 1986, Paleomagnetism of the Upper Jurassic Galice Formation, southwestern Oregon; Evidence for differential rotation of the eastern and western Klamath Mountains: Geology, v. 14, p. 335–338, doi: 10.1130/0091-7613(1986)14<335:POTUJG>2.0.CO;2.

Bogen, N.L., Kent, D.V., and Schweickert, R.A., 1985, Paleomagnetism of Jurassic rocks in the western Sierra Nevada metamorphic belt and its bearing on the structural evolution of the Sierra Nevada Block: Journal of Geophysical Research, v. 90, p. 4627–4638.

Bostrom, K., 1973, The origin and fate of ferromanganoan active ridge sediments: Stockholm Contributions in Geology, v. 27, p. 149–243.

Bostrom, K., and Peterson, M.N.A., 1969, The origin of aluminum-poor ferromanganoan sediments in areas of high heat flow on the East Pacific Rise: Marine Geology, v. 7, p. 427–447, doi: 10.1016/0025-3227(69)90016-4.

Burchfiel, B.C., and Davis, G.A., 1981, Triassic and Jurassic tectonic evolution of the Klamath Mountains–Sierra Nevada geologic terrane, *in* Ernst, W.G, ed., The geotectonic development of California—Rubey Volume 1: Englewood Cliffs, New Jersey, Prentice-Hall, p. 50–70.

Cashman, S.M., 1988, Finite-strain patterns of Nevadan deformation, western Klamath Mountains, California: Geology, v. 16, p. 839–843, doi: 10.1130/0091-7613(1988)016<0839:FSPOND>2.3.CO;2.

Cater, F.W., Jr., and Wells, F.G., 1953, Geology and mineral resources of the Gasquet Quadrangle, California-Oregon: Washington, D.C., U.S. Geological Survey Bulletin 995-C, p. 79–133.

Coleman, R.G., 1972, The Colebrooke Schist of southwestern Oregon and its relation to the tectonic evolution of the region: Washington, D.C., U.S. Geological Survey Bulletin 1339, 61 p.

Cookenboo, H.O., Bustin, R.M., and Wilks, K.R., 1997, Detrital chromian spinel compositions used to reconstruct the tectonic setting of provenance; implications for orogeny in the Canadian Cordillera: Journal of Sedimentary Research, v. 67, p. 116–123.

Cotkin, S.J., Cotkin, M.L., and Armstrong, R.L., 1992, Early Paleozoic blueschist from the schist of Skookum Gulch, eastern Klamath Mountains, Northern California: Journal of Geology, v. 100, p. 323–338.

Davis, G.A., 1969, Tectonic correlations, Klamath mountains and western Sierra Nevada, California: Geological Society of America Bulletin, v. 80, p. 1095–1108.

Davis, G.A., Monger, J.W.H., and Burchfiel, B.C., 1978, Mesozoic construction of the Cordilleran "collage," central British Columbia to central California, *in* Howell, D.G., and McDougall, K., eds., Mesozoic paleogeography of the western United States, Pacific Coast Paleogeography Symposium 2: Los Angeles, Pacific Section, Society of Economic Paleontologists and Mineralogists, p. 1–32.

Dick, H.J.B., 1976, The origin and emplacement of the Josephine Peridotite of southwestern Oregon [Ph.D. dissertation]: New Haven, Connecticut, Yale University, 409 p.

Dick, H.J.B., and Bullen, T., 1984, Chromian spinel as a petrogenetic indicator in abyssal and alpine-type peridotites and spatially associated lavas: Contributions to Mineralogy and Petrology, v. 86, p. 54–76, doi: 10.1007/BF00373711.

Dickinson, W.R., 1970, Interpreting detrital modes of graywacke and arkose: Journal of Sedimentary Petrology, v. 40, p. 695–707.

Dickinson, W.R., 1985, Interpreting provenance relations from detrital modes of sandstones, *in* Zuffa, G.G., ed., Provenance of arenites, NATO ASI Series C: Mathematical and Physical Sciences v. 148, p. 333–361.

Dickinson, W.R., Beard, L.S., Brakenridge, G.R., Erjavec, J.L., Ferguson, R.C., Inman, K.F., Knepp, R.A., Lindberg, F.A., and Ryberg, P.T., 1983, Provenance of North American Phanerozoic sandstones in relation to tectonic setting: Geological Society of America Bulletin, v. 94, p. 222–235.

Dickinson, W.R., Hopson, C.A., Saleeby, J.B., Schweickert, R.A., Ingersoll, R.V., Pessagno, E.A., Jr., Mattinson, J.M., Luyendyk, B.P., Beebe, W., Hull, D.M., Munoz, I.M., and Blome, C.D., 1996, Alternate origins of the Coast Range Ophiolite (California); Introduction and implications: GSA Today, v. 6, p. 1–10.

Diller, J.S., 1907, The Mesozoic sediments of southwestern Oregon: American Journal of Science, v. 23, p. 401–421.

Dilles, J.H., and Wright, J.E., 1988, The chronology of early Mesozoic arc magmatism in the Yerington District of western Nevada and its regional implications: Geological Society of America Bulletin, v. 100, p. 644–652, doi: 10.1130/0016-7606(1988)100<0644:TCOEMA>2.3.CO;2.

Frei, L.S., 1986, Additional paleomagnetic results from the Sierra Nevada; Further constraints on Basin and Range extension and northward displacement in the western United States: Geological Society of America

Bulletin, v. 97, p. 840–849, doi: 10.1130/0016-7606(1986)97<840:APRFTS>2.0.CO;2.

Frost, C.D., Barnes, C.G., and Snoke, A.W., 2006, this volume, Nd and Sr isotopic data from argillaceous rocks of the Galice Formation and Rattlesnake Creek terrane, Klamath Mountains: Evidence for the input of Precambrian sources, *in* Snoke, A.W., and Barnes, C.G., eds., Geological studies in the Klamath Mountains province, California and Oregon: A volume in honor of William P. Irwin: Boulder, Colorado, Geological Society of America Special Paper 410, doi: 10.1130/2006.2410(05).

Galloway, W.E., 1974, Deposition and diagenetic alteration of sandstone in northeast Pacific arc-related basins; implications for graywacke genesis: Geological Society of America Bulletin, v. 85, p. 379–390, doi: 10.1130/0016-7606(1974)85<379:DADAOS>2.0.CO;2.

Garcia, M.O., 1982, Petrology of the Rogue River island-arc complex, southwest Oregon: American Journal of Science, v. 282, p. 783–807.

Giaramita, M.J., and Harper, G.D., 2006, this volume, Geochemistry of ophiolitic rocks associated with the western part of the Elk outlier of the western Klamath terrane, southwestern Oregon, *in* Snoke, A.W., and Barnes, C.G., eds., Geological studies in the Klamath Mountains province, California and Oregon: A volume in honor of William P. Irwin: Boulder, Colorado, Geological Society of America Special Paper 410, doi: 10.1130/2006.2410(08).

Gray, G.G., 1985, Structural, geochronologic, and depositional history of the western Klamath Mountains, California and Oregon; Implications for the early to middle Mesozoic tectonic evolution of the western North American Cordillera [Ph.D. dissertation]: Austin, University of Texas, 224 p.

Gray, G.G., 2006, this volume, Structural and tectonic evolution of the western Jurassic belt along the Klamath River corridor, Klamath Mountains, California, *in* Snoke, A.W., and Barnes, C.G., eds., Geological studies in the Klamath Mountains province, California and Oregon: A volume in honor of William P. Irwin: Boulder, Colorado, Geological Society of America Special Paper 410, doi: 10.1130/2006.2410(07).

Gromet, L.P., Dymek, R.F., Haskin, L.A., and Korotev, R.L., 1984, The "North American shale composite"; Its compilation, major and trace element characteristics: Geochimica et Cosmochimica Acta, v. 48, p. 2469–2482, doi: 10.1016/0016-7037(84)90298-9.

Hacker, B.R., and Ernst, W.G., 1993, Jurassic orogeny in the Klamath Mountains: A geochronological analysis, *in* Dunn, G., and McDougall, K., eds., Mesozoic paleogeography of the western United States—II: Los Angeles, Pacific Section, Society of Economic Paleontologists and Mineralogists, p. 37–59.

Hacker, B.R., Ernst, W.G., and McWilliams, M.O., 1993, Genesis and evolution of a Permian-Jurassic magmatic arc/accretionary wedge, and reevaluation of terranes in the central Klamath Mountains: Tectonics, v. 12, p. 387–409.

Hacker, B.R., Donato, M.M., Barnes, C.G., McWilliams, M.O., and Ernst, W.G., 1995, Timescales of orogeny: Jurassic construction of the Klamath Mountains: Tectonics, v. 14, p. 677–703, doi: 10.1029/94TC02454.

Harding, J.P., Gehrels, G.E., Harwood, D.S., and Girty, G.H., 2000, Detrital zircon geochronology of the Shoo Fly Complex, northern Sierra Terrane, northeastern California, *in* Soreghan, M.J., and Gehrels, G.E., eds., Paleozoic and Triassic paleogeography and tectonics of western Nevada and northern California: Boulder, Colorado, Geological Society of America Special Paper 347, p. 43–55.

Harper, G.D., 1980, Structure and petrology of the Josephine ophiolite and overlying metasedimentary rocks, northwestern California [Ph.D. dissertation]: Berkeley, University of California, 260 p.

Harper, G.D., 1984, The Josephine ophiolite, northwestern California: Geological Society of America Bulletin, v. 95, p. 1,009–1,026.

Harper, G.D., 1994, A review of hemipelagic and turbidite sedimentation associated with the Josephine Ophiolite, California: Ofioliti, v. 19, p. 397–411.

Harper, G.D., 1995, Pumpellyosite and prehnitite associated with epidosite in the Josephine ophiolite—Ca metasomatism during upwelling of hydrothermal fluids at a spreading axis, *in* Schiffman, P., and Day, H., eds., Low-grade metamorphism of mafic rocks: Boulder, Colorado, Geological Society of America Special Paper 296, p. 101–122.

Harper, G.D., 2003, Fe-Ti basalts and propagating rift tectonics in the Josephine Ophiolite: Geological Society of America Bulletin, v. 115, p. 771–787, doi: 10.1130/0016-7606(2003)115<0771:FBAPTI>2.0.CO;2.

Harper, G.D., 2004, Tectonic implications of boninite, arc tholeiite, and MORB magma types in the Josephine Ophiolite, California-Oregon, *in* Dilek, Y. and Robinson, P.T., eds., Ophiolites in Earth history: London, Blackwell Scientific Publications, Geological Society Special Publication 218, p. 207–229.

Harper, G.D., 2006, this volume, Structure of syn-Nevadan dikes and their relationship to deformation of the Galice Formation, western Klamath terrane, northwestern California, *in* Snoke, A.W., and Barnes, C.G., eds., Geological studies in the Klamath Mountains province, California and Oregon: A volume in honor of William P. Irwin: Boulder, Colorado, Geological Society of America Special Paper 410, doi: 10.1130/2006.2410(06).

Harper, G.D., and Park, R., 1986, Paleomagnetism of the Upper Jurassic Galice Formation, southwestern Oregon; Evidence for differential rotation of the eastern and western Klamath Mountains: Comment: Geology, v. 14, p. 1049–1050, doi: 10.1130/0091-7613(1986)14<1049:CAROPO>2.0.CO;2.

Harper, G.D., and Wright, J.E., 1984, Middle to Late Jurassic tectonic evolution of the Klamath Mountains, California-Oregon: Tectonics, v. 3, p. 759–772.

Harper, G.D., Saleeby, J.B., and Norman, E.A.S., 1985, Geometry and tectonic setting of sea-floor spreading for the Josephine ophiolite, and implications for Jurassic accretionary events along the California margin, *in* Howell, D.G., ed., Tectonostratigraphic terranes of the Circum-Pacific region: Houston, Circum-Pacific Council for Energy and Mineral Resources, Earth Science Series 1, p. 239–257.

Harper, G.D., Bowman, J.R., and Kuhns, R., 1988, Field, chemical, and isotopic aspects of submarine hydrothermal metamorphism of the Josephine ophiolite, Klamath Mountains, California-Oregon: Journal of Geophysical Research, v. 93, p. 4,625–4,657.

Harper, G.D., Saleeby, J.B., and Heizler, M., 1994, Formation and emplacement of the Josephine ophiolite and the age of the Nevadan Orogeny in the Klamath Mountains, California-Oregon: U/Pb zircon and $^{40}Ar/^{39}Ar$ geochronology: Journal of Geophysical Research, v. 99, p. 4293–4321, doi: 10.1029/93JB02061.

Harper, G.D., Giaramita, M.J., and Kosanke, S.B., 2002, Josephine and Coast Range Ophiolites, Oregon and California, *in* Moore, G.W., ed., Field guide to the geologic processes in Cascadia: Portland, Oregon, 98th Annual Meeting of the Cordilleran Section of the Geological Society of America Guidebook, p. 1–22.

Hawkins, J.W., and Melchior, J.T., 1985, Petrology of Mariana Trough and Lau Basin basalts: Journal of Geophysical Research, v. 90, p. 11,431–11,468.

Hotz, P.E., Lanphere, M.A., and Swanson, D.A., 1977, Triassic blueschist from northern California and north-central Oregon: Geology, v. 5, p. 659–663, doi: 10.1130/0091-7613(1977)5<659:TBFNCA>2.0.CO;2.

Imlay, R.W., Dole, H.M., Peck, D.L., and Wells, F.G., 1959, Relations of certain Upper Jurassic and Lower Cretaceous formations in southwestern Oregon: American Association of Petroleum Geologists Bulletin, v. 43, p. 2770–2785.

Ingersoll, R.V., 1978, Petrofacies and petrologic evolution of the Late Cretaceous fore-arc basin, northern and central California: Journal of Geology, v. 86, p. 335–352.

Ingersoll, R.V., 1983, Petrofacies and provenance of late Mesozoic forearc basin, northern and central California: American Association of Petroleum Geologists Bulletin, v. 67, p. 1125–1142.

Ingersoll, R.V., and Suczek, C.A., 1979, Petrology and provenance of Neogene sand from Nicobar and Bengal fans, DSDP sites 211 and 218: Journal of Sedimentary Petrology, v. 49, p. 1217–1228.

Ingersoll, R.V., Fullard, T.F., Ford, R.L., Grimm, J.P., Pickle, J.D., and Sares, S.W., 1984, The effect of grain size on detrital modes: A test of the Gazzi-Dickinson point-counting method: Journal of Sedimentary Petrology, v. 54, p. 103–116.

Irwin, W.P., 1960, Geologic reconnaissance of the northern Coast Ranges and Klamath Mountains, California, with a summary of the mineral resources: San Francisco, California Division of Mines and Geology Bulletin 179, 80 p.

Irwin, W.P., 1964, Late Mesozoic orogenies in the ultramafic belts of northwestern California and southwestern Oregon: Washington, D.C., U.S. Geological Survey Professional Paper 501-C, p. C1–C9.

Irwin, W.P., 1966, Geology of the Klamath Mountains province, *in* Bailey, E.H., ed., Geology of northern California: San Francisco, California Division of Mines and Geology Bulletin 190, p. 19–38.

Irwin, W.P., 1972, Terranes of the Western Paleozoic and Triassic belt in the southern Klamath Mountains, California: Washington, D.C., U.S. Geological Survey Professional Paper 800-C, p. C103–C111.

Irwin, W.P., 1994, Geologic map of the Klamath Mountains, California and Oregon: Reston, Virginia, U.S. Geological Survey Miscellaneous Field Investigations, I-2148, scale 1:500,000, 1 sheet.

Irwin, W.P., 2003, Correlation of the Klamath Mountains and Sierra Nevada: Reston, Virginia, U.S. Geological Survey Open-File Report 02–0490, 2 sheets.

Irwin, W.P., and Wooden, J.L., 1999, Plutons and accretionary episodes of the Klamath Mountains, California and Oregon: Reston, Virginia, U.S. Geological Survey Open-File Report 99–0374, 1 sheet.

Jachens, R.C., Barnes, C.G., and Donato, M.M., 1986, Subsurface configuration of the Orleans Fault; Implications for deformation in the western Klamath Mountains, California: Geological Society of America Bulletin, v. 97, p. 388–395, doi: 10.1130/0016-7606(1986)97<388:SCOTOF>2.0.CO;2.

Johnson, K., Walton, C., Barnes, C.G., and Kistler, R.W., 1995, Time-dependent geochemical variations of Jurassic and Cretaceous plutons in the Blue Mountains, northeastern Oregon: Geological Society of America Abstracts with Programs, v. 27, p. 435.

Jones, R.F., 1988, Structural geology of the northern Galice Formation, western Klamath Mountains, Oregon and California [M.S. thesis]: Albany, State University of New York, 211 p.

Kamenetsky, V.S., Crawford, A.J., and Meffre, S., 2001, Factors controlling chemistry of magmatic spinel; An empirical study of associated olivine, Cr-spinel and melt inclusions from primitive rocks: Journal of Petrology, v. 42, p. 655–671, doi: 10.1093/petrology/42.4.655.

Kays, M.A., 1968, Zones of alpine tectonism and metamorphism, Klamath Mountains, southwestern Oregon: Journal of Geology, v. 76, p. 17–36.

Kuhns, R.J., and Baitis, H.W., 1987, Preliminary study of the Turner Albright Zn-Cu-Ag-Au-Co massive sulfide deposit, Josephine County, Oregon: Economic Geology and the Bulletin of the Society of Economic Geologists, v. 82, p. 1362–1376.

Lee, Y.I., 1999, Geotectonic significance of detrital chromian spinel; A review: Geoscience Journal, v. 3, p. 23–29.

MacDonald, J.H., Jr., Harper, G.D., Miller, J.S., and Zhu, B., 2004, Petrology, provenance, and further age control of the Galice Formation, Klamath Mountains, Oregon-California: Geological Society of America Abstracts with Programs, v. 36, p. 36–37.

Mankinen, E.A., and Irwin, W.P., 1982, Paleomagnetic study of some Cretaceous and Tertiary sedimentary rocks of the Klamath Mountains province, California: Geology, v. 10, p. 82–87, doi: 10.1130/0091-7613(1982)10<82:PSOSCA>2.0.CO;2.

Manuszak, J.D., Satterfield, J.I., and Gehrels, G.E., 2000, Detrital zircon geochronology of Upper Triassic strata in western Nevada, *in* Soreghan, M.J., and Gehrels, G.E., eds., Paleozoic and Triassic paleogeography and tectonics of western Nevada and northern California: Boulder, Colorado, Geological Society of America Special Paper 347, p. 109–118.

McLennan, S.M., 1989, Rare earth elements in sedimentary rocks; Influence of provenance and sedimentary processes: Reviews in Mineralogy, v. 21, p. 169–200.

McLennan, S.M., 2001, Relationships between the trace element composition of sedimentary rocks and upper continental crust: Geochemistry, Geophysics, Geosystems—G^3, v. 2, doi: 2000GC000109.

McLennan, S.M., Taylor, S.R., McCulloch, M.T., and Maynard, J.B., 1990, Geochemical and Nd-Sr isotopic composition of deep-sea turbidites; Crustal evolution and plate tectonic associations: Geochimica et Cosmochimica Acta, v. 54, p. 2015–2050, doi: 10.1016/0016-7037(90)90269-Q.

McLennan, S.M., Hemming, S., McDaniel, D.K., and Hanson, G.N., 1993, Geochemical approaches to sedimentation, provenance, and tectonics, *in* Johnson, M.J., and Basu, A., eds., Processes controlling the composition of clastic sediments: Boulder, Colorado, Geological Society of America Special Paper 284, p. 21–40.

Metcalf, R.V., Wallin, E.T., Willse, K.R., and Muller, E.R., 2000, Geology and geochemistry of the ophiolitic Trinity terrane, California; Evidence of middle Paleozoic depleted supra-subduction zone magmatism in a proto-arc setting, *in* Dilek, Y., et al., Ophiolites and oceanic crust; New insights from field studies and the Ocean Drilling Program: Boulder, Colorado, Geological Society of America Special Paper 349, p. 403–418.

Miller, M.M., and Saleeby, J.B., 1995, U-Pb geochronology of detrital zircon from Upper Jurassic synorogenic turbidites, Galice Formation, and related rocks, western Klamath Mountains; Correlation and Klamath Mountains provenance: Journal of Geophysical Research, v. 100, p. 18,045–18,058, doi: 10.1029/95JB00761.

Miller, J.S., Miller, R.B., Wooden, J.L., and Harper, G.D., 2003, Geochronologic links between the Ingalls Ophiolite, North Cascades, Washington and the Josephine Ophiolite, Klamath Mts., Oregon and California: Geological Society of America Abstracts with Programs, v. 35, no. 6, p. 113.

Miller, R.B., Mattinson, J.M., Funk, S.G., Hopson, C.A., and Treat, C.L., 1993, Tectonic evolution of Mesozoic rocks in the southern and central Washington Cascades, *in* Dunn, G., and McDougall, K., eds., Mesozoic Paleogeography of the western United States—II: Los Angeles, Pacific Section, Society of Economic Paleontologists and Mineralogists, v. 71, p. 81–98.

Norman, E.A.S., 1984, The structure and petrology of the Summit Valley area, Klamath Mountains, California [M.S. thesis]: Salt Lake City, University of Utah, 148 p.

Ohr, M., 1987, Geology, geochemistry, and geochronology of the Lems Ridge olistostrome, Klamath Mountains, California [M.S. thesis]: Albany, State University of New York, 247 p.

Park-Jones, R., 1988, Sedimentology, structure, and geochemistry of the Galice Formation: Sediment fill of a back-arc basin and island arc in the western Klamath Mountains [M.S. thesis]: Albany, State University of New York, 165 p.

Parkinson, I.J., and Pearce, J.A., 1998, Peridotites from the Izu-Bonin-Mariana forearc (ODP Leg 125); Evidence for mantle melting and melt-mantle interaction in a supra-subduction zone setting: Journal of Petrology, v. 39, p. 1577–1618, doi: 10.1093/petrology/39.9.1577.

Parson, L.M., and Wright, I.C., 1996, The Lau-Hauvre-Taupo back-arc basin: A southward-propagating, multi-stage evolution from rifting to spreading: Tectonophysics, v. 263, p. 1–22, doi: 10.1016/S0040-1951(96)00029-7.

Pearce, J.A., Baker, P.E., Harvey, P.K., and Luff, I.W., 1995, Geochemical evidence for subduction fluxes, mantle melting and fractional crystallization beneath the South Sandwich island arc: Journal of Petrology, v. 36, p. 1073–1109.

Pessagno, E.A., Jr., and Blome, C.D., 1990, Implications of new Jurassic stratigraphic, geochronometric, and paleolatitudinal data from the Western Klamath terrane (Smith River and Rogue Valley subterranes): Geology, v. 18, p. 665–668.

Pessagno, E.A., Jr., and Whalen, P.A., 1982, Lower and Middle Jurassic Radiolaria (multicyrtid Nassellariina) from California, east-central Oregon and the Queen Charlotte Islands, B.C.: Micropaleontology, v. 28, p. 111–169.

Pessagno, E.A., Jr., Blome, C.D., Hull, D., and Six, W.M., Jr., 1993, Middle and Upper Jurassic radiolaria from the western Klamath terrane, Smith River subterrane, northwestern California: Their biostratigraphic, chronostratigraphic, geochronologic, and paleolatitudinal significance: Micropaleontology, v. 39, p. 93–166.

Pessagno, E.A., Jr., Hull, D.M., and Hopson, C.A., 2000, Tectonostratigraphic significance of sedimentary strata occurring within and above the Coast Range Ophiolite (California Coast Ranges) and the Josephine Ophiolite (Klamath Mountains), northwestern California, *in* Dilek, Y., et al., eds., Ophiolites and oceanic crust; New insights from field studies and the Ocean Drilling Program: Boulder, Colorado, Geological Society of America Special Paper 349, p. 383–394.

Pinto-Auso, M., and Harper, G.D., 1985, Sedimentation, metallogenesis, and tectonic origin of the basal Galice Formation overlying the Josephine ophiolite, northwestern California: Journal of Geology, v. 93, p. 713–725.

Quick, J.E., 1981, Petrology and petrogenesis of the Trinity Peridotite, an upper mantle diapir in the eastern Klamath Mountains, northern California: Journal of Geophysical Research, v. 86, p. 11,837–11,863.

Ravizza, G., Sherrell, R.M., Field, M.P., and Pickett, E.A., 1999, Geochemistry of the Margi umbers, Cyprus, and the Os isotope composition of Cretaceous seawater: Geology, v. 27, p. 971–974, doi: 10.1130/0091-7613(1999)027<0971:GOTMUC>2.3.CO;2.

Renne, P.R., and Scott, G.R., 1986, Paleomagnetism of the Upper Jurassic Galice Formation, southwestern Oregon: Evidence for differential rotation of the eastern and western Klamath Mountains: Comment: Geology, v. 14, p. 1048–1049, doi: 10.1130/0091-7613(1986)14<1048:CAROPO>2.0.CO;2.

Renne, P.R., and Scott, G.R., 1988, Structural chronology, oroclinal deformation, and tectonic evolution of the southeastern Klamath Mountains, California: Tectonics, v. 7, p. 1223–1242.

Roser, B.P., and Korsch, R.J., 1986, Determination of tectonic setting of sandstone-mudstone suites using SiO_2 content and K_2O/Na_2O ratios: Journal of Geology, v. 94, p. 635–650.

Roser, B.P., and Korsch, R.J., 1988, Provenance signatures of sandstone-mudstone suites determined using discriminant function analysis of major-element data: Chemical Geology, v. 67, p. 119–139, doi: 10.1016/0009-2541(88)90010-1.

Rowley, D.B., and Kidd, W.S.F., 1981, Stratigraphic relationships and detrital composition of the medial Ordovician flysch of western New England; Implications for the tectonic evolution of the Taconic Orogeny: Journal of Geology, v. 89, p. 199–218.

Saleeby, J.B., 1984, Pb/U zircon ages from the Rogue River area, western Jurassic belt Klamath Mountains, Oregon: Geological Society of America Abstracts with Programs, v. 16, p. 331.

Saleeby, J.B., and Busby-Spera, C., 1992, Early Mesozoic tectonic evolution of the western U.S. Cordillera, *in* Burchfiel, B.C., Lipman, P.W., and Zoback, M.L., eds., The Cordilleran Orogen; Conterminous U.S.: Boulder, Colorado, Geological Society of America, The Geology of North America, v. G-3, p. 107–168.

Saleeby, J.B., and Harper, G.D., 1993, Tectonic relations between the Galice Formation and the Condrey Mountain Schist, Klamath Mountains, northern California, *in* Dunn, G.C., and McDougall, K.A., eds., Mesozoic Paleogeography of the western United States—II: Los Angeles, Pacific Section, Society of Economic Paleontologists and Mineralogists, v. 71, p. 61–80.

Saleeby, J.B., Harper, G.D., Snoke, A.W., and Sharp, W., 1982, Time relations and structural-stratigraphic patterns in ophiolite accretion, west-central Klamath Mountains, California: Journal of Geophysical Research, v. 87, p. 3831–3848.

Saunders, A.D., and Tarney, J., 1979, The geochemistry of basalts from a back-arc spreading centre in the East Scotia Sea: Geochimica et Cosmochimica Acta, v. 43, p. 555–572, doi: 10.1016/0016-7037(79)90165-0.

Schultz, K.L., and Levi, S., 1983, Paleomagnetism of middle Jurassic plutons of the north-central Klamath Mountains: Geological Society of America Abstracts with Programs, v. 15, p. 427.

Seiders, V.M., 1991, Conglomerate stratigraphy and tectonics in the Franciscan assemblage of northern California and implications for Cordilleran tectonics: Reston, Virginia, U.S. Geological Survey Open-File Report PFR 91–50, 10 p.

Seiders, V.M., and Blome, C.D., 1988, Implications of upper Mesozoic conglomerate for suspect terrane in western California and adjacent areas: Geological Society of America Bulletin, v. 100, p. 374–391, doi: 10.1130/0016-7606(1988)100<0374:IOUMCF>2.3.CO;2.

Smith, G.M., and Harper, G.D., 1993, Paleomagnetism of the Josephine ophiolite: EOS, Transactions, American Geophysical Union, v. 74, p. 214.

Snoke, A.W., 1972, The petrology and structure of the Preston Peak area, Del Norte and Siskiyou counties, California [Ph.D. dissertation]: Stanford, California, Stanford University, 274 p.

Snoke, A.W., 1977, A thrust plate of ophiolitic rocks in the Preston Peak area, Klamath Mountains, California: Geological Society of America Bulletin, v. 88, p. 1641–1659, doi: 10.1130/0016-7606(1977)88<1641:ATPOOR>2.0.CO;2.

Southwick, D.L., 1974, Geology of the alpine-type ultramafic complex near Mount Stuart, Washington: Geological Society of America Bulletin, v. 85, p. 391–402, doi: 10.1130/0016-7606(1974)85<391:GOTAUC>2.0.CO;2.

Umino, S., 1986, Magma mixing in boninite sequence of Chichijima, Bonin Islands: Journal of Volcanology and Geothermal Research, v. 29, p. 125–157, doi: 10.1016/0377-0273(86)90042-9.

Vail, S.G., 1977, Geology and geochemistry of the Oregon Mountain area, southwest Oregon and northern California [Ph.D. dissertation]: Corvallis, Oregon State University, 159 p.

Wells, F.G., and Walker, G.W., 1953, Geology of the Galice quadrangle, Oregon: Washington, D.C., U.S. Geological Survey Geological Quadrangle Map GQ-25, scale 1:62,500, 1 sheet.

Wells, F.G., Hotz, P.E., and Cater, F.W., 1949, Preliminary description of the geology of the Kerby quadrangle: Portland, Oregon Department of Geology and Mineral Industries Bulletin, v. 40, 23 p.

Wright, J.E., 1982, Permo-Triassic accretionary subduction complex, southwestern Klamath Mountains, northern California: Journal of Geophysical Research, v. 87, p. 3805–3818.

Wright, J.E., and Fahan, M.R., 1988, An expanded view of Jurassic orogenesis in the western United States Cordillera: Middle Jurassic (pre-Nevadan) regional metamorphism and thrust faulting within an active arc environment, Klamath Mountains, California: Geological Society of America Bulletin, v. 100, p. 859–876, doi: 10.1130/0016-7606(1988)100<0859:AEVOJO>2.3.CO;2.

Wright, J.E., and Wyld, S.J., 1994, The Rattlesnake Creek terrane, Klamath Mountains, California: An early Mesozoic volcanic arc and its basement of tectonically disrupted oceanic crust: Geological Society of America Bulletin, v. 106, p. 1033–1056, doi: 10.1130/0016-7606(1994)106<1033:TRCTKM>2.3.CO;2.

Wyld, S.J., 1985, Geology of the western Jurassic Belt, South Fork Trinity River Area, Klamath Mountains, California [M.S. thesis]: Berkeley, University of California, 168 p.

Wyld, S.J., and Wright, J.E., 1988, The Devils Elbow ophiolite remnant and overlying Galice Formation: New constraints on the Middle to Late Jurassic evolution of the Klamath Mountains, California: Geological Society of America Bulletin, v. 100, p. 29–44, doi: 10.1130/0016-7606(1988)100<0029:TDEORA>2.3.CO;2.

Wyld, S.J., and Wright, J.E., 2001, New evidence for Cretaceous strike-slip faulting in the United States Cordillera and implications for terrane-displacement, deformation patterns, and plutonism: American Journal of Science, v. 301, p. 150–181.

Yule, J.D., 1996, Geologic and tectonic evolution of Jurassic marginal ocean basin lithosphere, Klamath Mountains, Oregon [Ph.D. dissertation]: Pasadena, California Institute of Technology, 308 p.

Yule, J.D., Saleeby, J.B., and Barnes, C.G., 1996, The 175–135 Ma plutons of the Klamath Mountains province reinterpreted as a single oceanic arc batholithic system: Geological Society of America Abstracts with Programs, v. 28, no. 5, p. 128.

Yule, J.D., Saleeby, J.B., and Barnes, C.G., 2006, this volume, A rift-edge facies of the Late Jurassic Rogue–Chetco arc and Josephine ophiolite, Klamath Mountains, Oregon, *in* Snoke, A.W., and Barnes, C.G., eds., Geological studies in the Klamath Mountains province, California and Oregon: A volume in honor of William P. Irwin: Boulder, Colorado, Geological Society of America Special Paper 410, doi: 10.1130/2006.2410(03).

Zierenberg, R.A., Shanks, W.C., III, Seyfried, W.E., Jr., Koski, R.A., and Strickler, M.D., 1988, Mineralization, alteration, and hydrothermal metamorphism of the ophiolite-hosted Turner-Albright sulfide deposit, southwestern Oregon: Journal of Geophysical Research, v. 93, p. 4657–4674.

Manuscript Accepted by the Society 10 January 2006

Geological Society of America
Special Paper 410
2006

Nd and Sr isotopic data from argillaceous rocks of the Galice Formation and Rattlesnake Creek terrane, Klamath Mountains: Evidence for the input of Precambrian sources

Carol D. Frost*
Department of Geology and Geophysics, Dept. 3006, 1000 East University Avenue, University of Wyoming, Laramie, Wyoming 82071, USA
Calvin G. Barnes
Department of Geosciences, Texas Tech University, Lubbock, Texas 79409-1053, USA
Arthur W. Snoke
Department of Geology and Geophysics, Dept. 3006, 1000 East University Avenue, University of Wyoming, Laramie, Wyoming 82071, USA

ABSTRACT

Nd and Sr isotopic data are presented for argillaceous rocks from two terranes in the Klamath Mountains: the western Klamath terrane and Rattlesnake Creek terrane. In the collage of terranes exposed in the Klamath Mountains, these terranes are located farthest outboard from the North American craton and include (meta)igneous rocks of clear oceanic affinity. Nevertheless, the argillaceous rocks from these terranes preserve a strong isotopic signal of terrigenous sedimentary input. The lowermost portion of the Rattlesnake Creek terrane, the serpentinite-matrix mélange, is interpreted to have formed within an oceanic fracture zone. The argillaceous sediment incorporated into this tectonic assemblage was probably derived partly from eolian dust eroded from the continents and partly from local juvenile detritus shed from topographic highs along the fracture zone. The Upper Triassic–Lower Jurassic "cover sequence" of the Rattlesnake Creek terrane has been interpreted as an oceanic-arc assemblage, but the argillaceous rocks of the cover sequence have the most negative ε_{Nd} (–8.3) and radiogenic $^{87}Sr/^{86}Sr$ (0.7114) of any samples analyzed in this study. We infer that cratonic sediment was delivered to the depocenter of the Rattlesnake Creek terrane arc, probably transported by river systems. This situation suggests proximity to a continental landmass during arc magmatism. The Galice Formation, a thick turbiditic sequence above the Late Jurassic Josephine ophiolite, appears to be composed of detritus shed from both the Rogue–Chetco oceanic arc on the west (in present geographic coordinates) and previously accreted Klamath Mountains terranes and/or North American craton to the east. The continental isotopic signal is stronger in the argillaceous rocks than in the (meta)graywackes, suggesting that the finer-grained rocks contain a greater proportion of cratonic debris, material that may have been

*E-mail: frost@uwyo.edu.

Frost, C.D., Barnes, C.G., and Snoke, A.W., 2006, Nd and Sr isotopic data from argillaceous rocks of the Galice Formation and Rattlesnake Creek terrane, Klamath Mountains: Evidence for the input of Precambrian sources, *in* Snoke, A.W., and Barnes, C.G., eds., Geological studies in the Klamath Mountains province, California and Oregon: A volume in honor of William P. Irwin: Geological Society of America Special Paper 410, p. 103–120, doi: 10.1130/2006.2410(05). For permission to copy, contact editing@geosociety.org.

reduced to mud-sized particles during sediment recycling. The presence of continental-derived sediment in these otherwise ensimatic terranes indicates that although continental crustal growth by accretion of oceanic terranes is an important process, such accreted terranes commonly are not composed entirely of juvenile crust.

Keywords: Nd isotopes, continental crustal growth, Rattlesnake Creek terrane, western Klamath terrane, Galice Formation, Jurassic

INTRODUCTION

An important mechanism of continental crustal growth during at least the past 2.5 billion years is accretion of juvenile terranes of oceanic affinity to the margins of continents. Large amounts of juvenile crust have been documented within the 1.95- to 1.65-Ga terranes of North America, Greenland, and northern Europe, 1.2- to 1.0-Ga Grenville orogenic belt, Arabian-Nubian shield, and the Proterozoic crust of Australia (see Samson and Patchett, 1991, and references therein). Phanerozoic examples of crustal addition by terrane accretion include the western North American Cordillera (Coney et al., 1980; Silberling et al., 1987; Armstrong, 1988; Samson and Patchett, 1991; Creaser et al., 1997; Patchett and Gehrels, 1998; Unterschutz et al., 2002), Appalachians (e.g., Secor et al., 1983), New Zealand (Frost and Coombs, 1989), and other less well-documented examples. Nd and Sr isotopic studies in the Canadian Cordillera and New Zealand have shown that parts of these collages include terranes that are composed almost entirely of oceanic rocks that formed from mantle sources, such as Stikine, Wrangellia, and Alexander terranes of the Canadian Cordillera and Brook Street terrane of New Zealand. As such, these terranes provide unequivocal evidence for Phanerozoic continental crustal addition. Other terranes within these collages are composed of rocks that reflect a mixture of juvenile and continental sources, and modeling the amount of continental crustal growth represented by these terranes is complex.

The Klamath Mountains of northwestern California and southeastern Oregon preserve a collage of tectonostratigraphic terranes of oceanic affinity and thus are a significant site of Phanerozoic crustal growth in the U.S. Cordillera. Isotopic studies of intrusive igneous rocks of the Klamath Mountains show that these magmatic rocks are juvenile (e.g., Paleozoic plutons in the eastern Klamath terrane: Petersen et al., 1991; Wallin and Metcalf, 1998; Middle Jurassic Ironside Mountain batholith: Barnes et al., this volume, Chapter 10) or have significant juvenile components (e.g., late Middle Jurassic "Wooley Creek belt" plutons; Masi et al., 1981; Barnes et al., 1990, 1992a, 1995; Gribble et al., 1990). However, isotopic data from (meta)sedimentary rocks of the terranes in the Klamath Mountains are also critical in terrane affinity and accretion history. Even in terranes that chiefly include isotopically juvenile igneous rocks, the (meta)sedimentary rocks may have more complex provenance that includes substantial recycled crustal components (e.g., Patchett and Gehrels, 1998).

The present study provides the first extensive Sr and Nd isotopic data set for metasedimentary rocks from the Klamath Mountains. We have chosen to focus on two of the terranes that lie farthest outboard from the North American craton, which are composed almost entirely of rocks of clear oceanic affinity. The first, the western Klamath terrane, includes the Josephine ophiolite and overlying turbiditic sequence of the Upper Jurassic Galice Formation. The second, the Rattlesnake Creek terrane, has been interpreted as initially forming in an oceanic fracture zone setting upon which a younger (Late Triassic–Early Jurassic) oceanic arc was built. As such, both settings are candidate sites of new crust formation. Nevertheless, our results show that sediment of continental origin is present in argillaceous rocks of both terranes. The presence of continental detritus in these oceanic terranes is significant in terms of the paleogeographic and tectonic positions of the terranes, the origins of their detrital sediments, and their contribution to Phanerozoic continental crustal growth.

GEOLOGIC SETTING

The Klamath Mountains of northwestern California and southwestern Oregon are a tectonic collage of tectonostratigraphic terranes of oceanic affinity imbricated along regional thrust faults (Irwin, 1972, 1994). Roughly, the age of thrust faulting is younger from east to west (Fig. 1), from the mid-Paleozoic through the Late Jurassic (Davis et al., 1978; Irwin and Wooden, 1999). During the Early Cretaceous, the westernmost thrust, the South Fork fault, emplaced the Klamath Mountain province above Jurassic and Cretaceous rocks of the Coast Ranges, including the South Fork Mountain Schist of the Pickett Peak terrane. Argillaceous rocks are abundant in many of the terranes of the Klamath Mountains and are especially widespread in the Upper Jurassic Galice Formation of the western Klamath terrane as well as within the Rattlesnake Creek terrane.

The argillaceous rocks of both the Galice Formation and Rattlesnake Creek terrane apparently formed in close spatial and temporal association with oceanic arcs, but in contrasting

→

Figure 1. Geologic sketch map of the tectonostratigraphic terranes of the Klamath Mountains province, California and Oregon. The area of Figure 2 is outlined. BMic—Bear Mountain intrusive complex, BPic—Bear Peak intrusive complex. Modified from Irwin (1994).

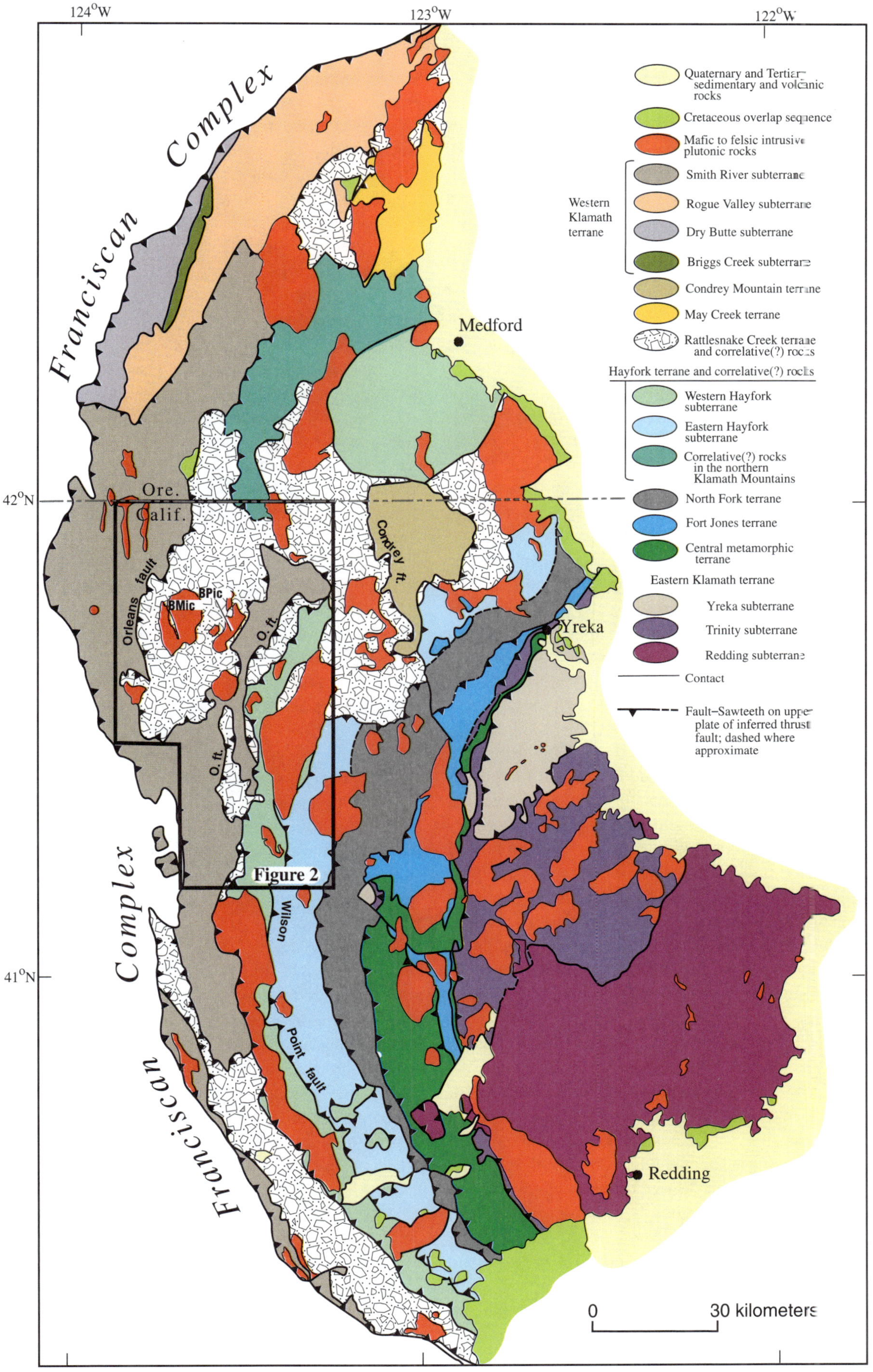
124°W
123°W
122°W
42°N
41°N
Franciscan Complex
Complex
Franciscan
Medford
Yreka
Redding
Ore.
Calif.
Orleans fault
BPic
BMic
O. ft.
O. ft.
Condrey ft.
Wilson Point fault
Figure 2
Quaternary and Tertiary sedimentary and volcanic rocks
Cretaceous overlap sequence
Mafic to felsic intrusive plutonic rocks
Western Klamath terrane
Smith River subterrane
Rogue Valley subterrane
Dry Butte subterrane
Briggs Creek subterrane
Condrey Mountain terrane
May Creek terrane
Rattlesnake Creek terrane and correlative(?) rocks
Hayfork terrane and correlative(?) rocks
Western Hayfork subterrane
Eastern Hayfork subterrane
Correlative(?) rocks in the northern Klamath Mountains
North Fork terrane
Fort Jones terrane
Central metamorphic terrane
Eastern Klamath terrane
Yreka subterrane
Trinity subterrane
Redding subterrane
Contact
Fault–Sawteeth on upper plate of inferred thrust fault; dashed where approximate
0
30 kilometers

settings with regard to coeval arc magmatism. The Galice Formation is a thick Upper Jurassic turbidite sequence above the Late Jurassic Josephine ophiolite and equivalent sequences (e.g., Devils Elbow ophiolite remnant; Wyld and Wright, 1988). In the Oregon Klamath Mountains, the Galice Formation interfingers with volcanogenic rocks of the Rogue–Chetco oceanic arc (Garcia, 1979; Harper, 1983), which is also stratigraphically above the Josephine ophiolite. The age of the Galice Formation is constrained to be younger than the youngest population of detrital zircons it contains (ca. 153 Ma; Miller et al., 2003), yet older than the 151 ± 2- and 150 ± 2-Ma dikes that intrude it (Saleeby et al., 1982). Paleontological data indicate that the Galice Formation is Middle Oxfordian–Lower Kimmeridgian, based on *Buchia concentrica* (Sowerby) (Diller, 1907; Imlay, 1959, 1980). The depositional setting of the Galice Formation has been commonly interpreted to be in an inter-arc or back-arc basin (Snoke, 1977; Harper, 1980; Harper and Wright, 1984), and the structural development of this basin may be related to transtensional deformation within an older, west-facing, oceanic-arc terrane (Harper et al., 1985). This rifted, older arc terrane includes Rattlesnake Creek terrane basement (Donato et al., 1996; Yule, 1996), the Middle Jurassic western Hayfork arc sequence, and late Middle Jurassic calc-alkalic plutons (e.g., Barnes et al., 1992b). In this light, the source of Galice detritus could be from the rifted older arc and/or late Middle Jurassic igneous rocks, the coeval arc (Rogue–Chetco arc), or from still older previously accreted terranes of the Klamath Mountains.

The provenance of Galice detritus was first addressed by Snoke (1977), who studied the petrography of the metagraywackes of the Galice Formation, including the detrital, heavy-mineral assemblage. Snoke (1977) concluded that a substantial amount of the detritus in the metagraywackes of the Galice Formation could be derived from older, more easterly terranes of the Klamath Mountains. This conclusion was supported by Harper (1983, 1984) and Park-Jones (1988), who also noted a diverse heavy mineral assemblage as well as paleocurrent indicators of eastward derivation of the bulk of the detritus. These authors also recognized juvenile volcaniclastic debris presumably derived from the coeval Rogue–Chetco oceanic arc, west of the Galice basin.

In the southwestern Klamath Mountains, Wright and Wyld (1986) reported xenocrystic zircon in plagiogranite of the Devils Elbow ophiolite remnant (the southern continuation of the Josephine ophiolite). This finding was the first evidence of recycled cratonic detritus in a Klamath Mountain ensimatic environment, in this case, a supra-subduction ophiolite. Miller and Saleeby (1987, 1995) then showed that metagraywackes of the Galice Formation contain detrital Precambrian zircons with an average upper intercept of 1583 Ma, with the oldest $^{207}Pb/^{206}Pb$ age at 2079 Ma. Recently, Miller et al. (2003) also found sparse Precambrian detrital zircons in the metagraywackes of the Galice Formation using ion probe U-Pb dating. In summary, a substantial body of diverse data links the tectonic setting of the Galice Formation to previously accreted terranes of the Klamath Mountains and to the coeval fringing Rogue arc. Detrital zircon geochronologic studies establish these same links, but also indicate the presence of Precambrian-age detritus whose origins are uncertain.

In contrast to the Galice Formation, little is known about the provenance of Rattlesnake Creek terrane metasedimentary rocks. The Rattlesnake Creek terrane was defined by Irwin (1972) as an ophiolitic mélange. However, Wright and Wyld (1994) recognized a pseudostratigraphy within the terrane, consisting of a disrupted oceanic assemblage (serpentinite-matrix mélange) overlain by coherent Upper Triassic and Lower Jurassic metavolcanic and metasedimentary rocks, which they referred to as a "cover sequence." The contact between the underlying serpentinite-matrix mélange and overlying cover sequence was interpreted as an unconformity, although this contact is commonly overprinted by brittle faulting. Wright and Wyld (1994) also documented a diverse suite of intrusive rocks (gabbroic to quartz dioritic) that intrude the mélange and cover sequence; these plutonic rocks range in age from 207 to 193 Ma (U-Pb zircon).

Wright and Wyld (1994) argued that the serpentinite-matrix mélange unit formed in an oceanic fracture zone setting, whereas the overlying coherent cover sequence is an oceanic-arc sequence related to eastward-dipping subduction. The paleogeographic setting of this oceanic arc was inferred to be near western North America on the basis of terrigenous detritus in Rattlesnake Creek terrane wackes, arenites, and coarser-grained detrital rocks.

In summary, the Galice Formation and Rattlesnake Creek terrane are assemblages of rocks that clearly developed in oceanic or ensimatic settings. However, both units also appear to reflect proximity to a continental margin that was a source of terrigenous sedimentary input, including sediments with ancient cratonic origins.

SAMPLES

Geochemical and isotopic data were obtained on nine argillaceous rock samples from the Galice Formation and ten meta-argillite samples from the Rattlesnake Creek terrane. Sample locations were chosen to provide a broad geographic distribution and are summarized in Figure 2 with geographic coordinates listed in the Appendix. We also attempted to obtain samples from different stratigraphic levels within these units, but complex folding and faulting preclude a detailed determination of stratigraphic position.

Figure 2. Generalized geologic map of the north-central Klamath Mountains, California, showing the approximate locations of samples analyzed in this study. The base for this sample-location map is Wagner and Saucedo (1987).

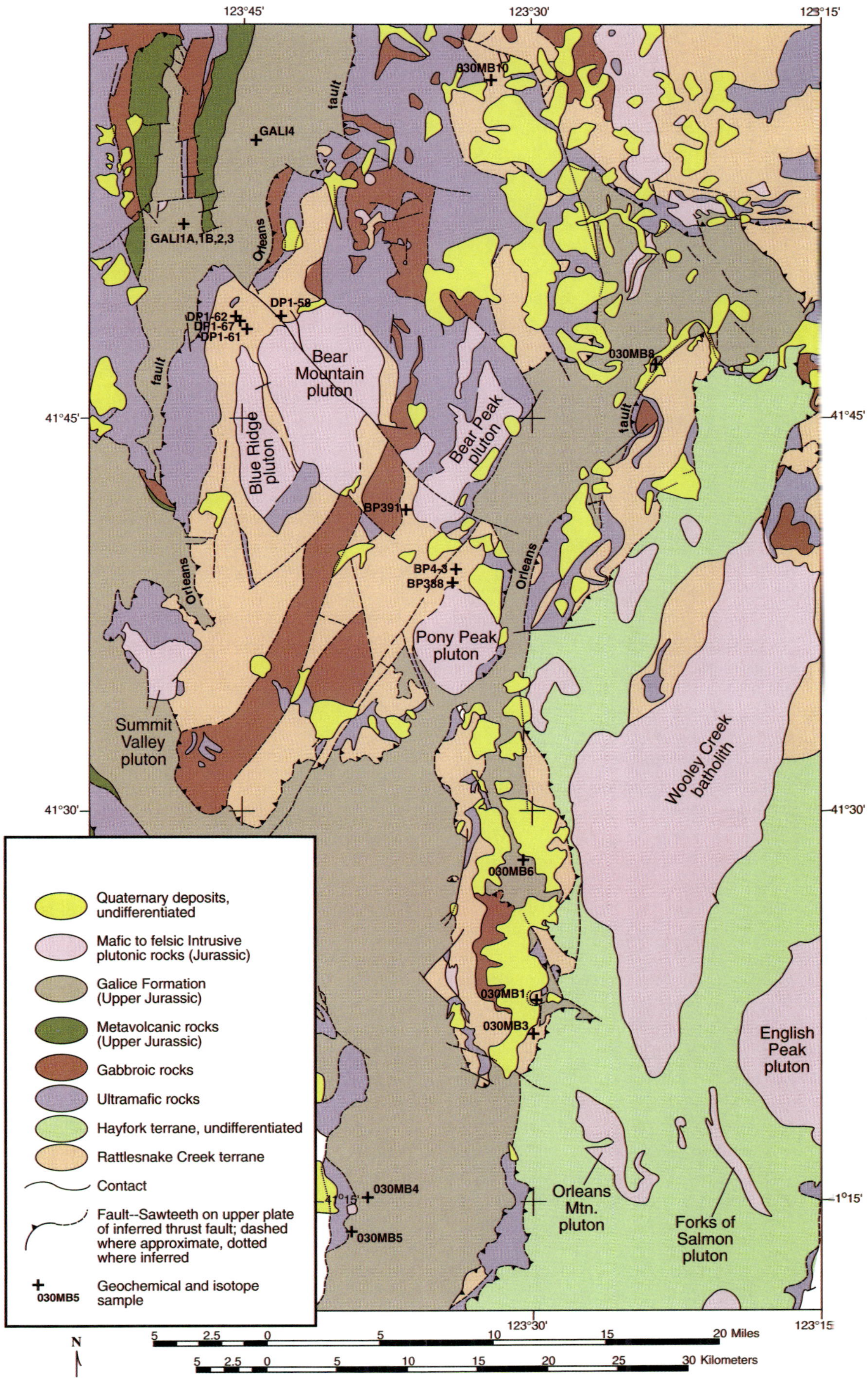

123°45'
123°30'
123°15'
41°45'
41°30'
1°15'
41°15'
830MB10
GALI4
GALI1A,1B,2,3
DP1-58
DP1-62
DP1-67
DP1-61
030MB8
BP391
BP4-3
BP388
030MB6
030MB1
030MB3
030MB4
030MB5
fault
Orleans
Bear Mountain pluton
Blue Ridge pluton
Bear Peak pluton
Pony Peak pluton
Summit Valley pluton
Wooley Creek batholith
English Peak pluton
Orleans Mtn. pluton
Forks of Salmon pluton
Quaternary deposits, undifferentiated
Mafic to felsic Intrusive plutonic rocks (Jurassic)
Galice Formation (Upper Jurassic)
Metavolcanic rocks (Upper Jurassic)
Gabbroic rocks
Ultramafic rocks
Hayfork terrane, undifferentiated
Rattlesnake Creek terrane
Contact
Fault--Sawteeth on upper plate of inferred thrust fault; dashed where approximate, dotted where inferred
Geochemical and isotope sample
N
5 2.5 0 5 10 15 20 Miles
5 2.5 0 5 10 15 20 25 30 Kilometers

Galice Formation

Argillaceous rocks from the Galice Formation are typically dark gray to black slates to phyllites, locally pyrite bearing. The samples are listed in Tables 1 and 2 from north to south, geographically. Sample GALI4 was collected in the drainage of the Middle Fork of the Smith River, whereas GALI1A–3 were collected in the Siskiyou Fork of the Smith River. These samples appear to represent structurally, and presumably, stratigraphically, the upper part of the Galice Formation. Sample 03OMB6 was collected along California State Highway 96 near the entrance to the Marble Mountain Ranch. Sample 03OMB4 was also collected along California State Highway 96, south of Orleans near the confluence of Slate Creek with the Klamath River; and 03OMB5 was collected along the Klamath River near the Aikens Creek campground. Both 03OMB4 and 03OMB5 are interpreted as samples near the base of the Galice Formation, and metabasaltic rocks correlated with part of the Late Jurassic Josephine ophiolite underlie the Galice Formation south of Aikens Creek campground along the Klamath River.

Sample 03OMB8 is a crenulated, graphitic phyllite from the confluence of Little Grider Creek with the Klamath River, south of the town of Happy Camp, California. The outcrop was mapped as Galice Formation by Klein (1975, 1977) and Petersen (1982) but is surrounded by Quaternary deposits. The Orleans fault or an equivalent structure lies immediately to the east of the locality; the hanging-wall block consists of a diverse suite of rocks of the Rattlesnake Creek terrane, including ultramafic and mafic rocks metamorphosed at upper greenschist- to lower amphibolite-facies conditions (Petersen, 1982). Interestingly, Klein (1977), Hill (1985), and Saleeby and Harper (1993) show complex faulting in this area with thrust slices of different lithologic character. In Figure 2 of Hill (1985), a thrust slice of rocks correlated with the Condrey Mountain terrane (Hotz, 1979; Helper, 1986) is shown sandwiched between structurally overlying rocks of the Rattlesnake Creek terrane and structurally underlying rocks of the Galice Formation (western Klamath terrane). Likewise, a regional cross-section through the area (Saleeby and Harper, 1993, their Fig. 2) shows multiple thrust slices of rocks perhaps equivalent to the Condrey Mountain terrane in the same structural position. Thus, sample 03OMB8 was collected in a structurally complex zone near a major, regional thrust fault, and its terrane affinity is not certain. This sample may be part of the Galice Formation as previously mapped (Klein, 1975, 1977; Petersen, 1982); however, it could conceivably be a rock unit related to the older Condrey Mountain terrane. Another possibility, based strictly on the phyllitic nature of the sample, is a correlation with the South Fork Mountain Schist of the Pickett Peak terrane. However, such a correlation would require the imbrication of the Pickett Peak terrane *above* the western Klamath terrane, a relationship not reported at any other locality in the Klamath Mountains.

Rattlesnake Creek Terrane

Samples from the Rattlesnake Creek terrane were chosen to include meta-argillite from blocks within the serpentinite-matrix mélange as well as argillaceous units that form part of the stratigraphy of coherent metavolcanic and metasedimentary sequences (i.e., "cover sequences" after the usage of Wright and Wyld, 1994). In the west-central Klamath Mountains, Snoke (1977) mapped a distinctive conglomerate-grit unit, characterized by a diverse suite of clasts (see his Figs. 9 and 10, p. 1652). This unit is part of a coherent metavolcanic and metasedimentary sequence (his Bear Basin Road sequence), which is now considered to be a Rattlesnake Creek terrane cover sequence (Bushey et al., this volume). Several of our meta-argillite samples (DP1-58, 61, 62, and 67) from the Rattlesnake Creek terrane were collected from this unit. Other cover sequence samples include BP4-3 and BP388, collected north of the Pony Peak pluton (Fig. 2). Samples from interpreted mélange blocks include 03OMB10, collected ~1 km south of Little Grayback Mountain; BP39, a contact metamorphosed black argillite collected 0.5 km west of the Bear Peak pluton; 03OMB1 from the southeast side of Sugarloaf Mountain near the confluence of the Salmon and Klamath Rivers; and 03OMB3, collected along California State Highway 96 ~7.5 km north of Orleans, California.

RESULTS

Methods

Bulk geochemical analysis was performed at Texas Tech University by inductively coupled plasma-atomic emission spectroscopy for major elements, plus Sr, Zr, Y, Nb, Ba, Sc, Cu, Cr, and Zn. Rb was analyzed by flame emission. Nd and Sr isotopic compositions were determined at the University of Wyoming by thermal ionization mass spectrometry. Concentrations of Sm, Nd, Rb, and Sr were determined by isotope dilution on aliquots of the same sample dissolved for isotope ratio measurements. Further analytical details are given in the footnotes to Tables 1 and 2.

Geochemical Data

Galice Formation and Rattlesnake Creek terrane samples have geochemical compositions typical of argillaceous rocks (Table 1), characterized by high alumina contents (all but two samples have Al_2O_3 between 12 and 17 wt. %). As expected, Galice argillaceous rocks have higher Al_2O_3 and lower SiO_2 contents than found in Galice Formation metagraywackes (see MacDonald et al., this volume). Although their geochemical compositions overlap considerably, it is possible to distinguish argillaceous rocks of the Galice Formation from the meta-argillites of the Rattlesnake Creek terrane. As shown in Figure 3, argillaceous rocks of the Galice Formation have generally

TABLE 1. GEOCHEMICAL DATA FOR ARGILLACEOUS ROCKS OF THE KLAMATH MOUNTAINS, CALIFORNIA

										Rattlesnake Creek terrane argillites, deposited ca. 210 Ma									
	Galice argillites, deposited 153 Ma									*Cover sequence*						*Melange blocks*			
Sample	GALI4	GALI1A	GALI1B	GALI2	GALI3	030MB6	030MB4	030MB5	030MB8	DP1-58	DP1-61	DP1-62	DP1-67	BP4-3	BP388	030MB10	BP391	030MB1	030MB3
SiO_2	60.46	60.60	60.10	59.87	67.30	61.95	62.00	63.77	63.82	68.87	65.38	75.85	81.71	68.45	69.81	66.06	61.23	63.12	65.70
TiO_2	0.82	0.75	0.72	0.78	0.62	0.78	0.81	0.84	0.66	0.78	0.68	0.60	0.28	0.60	0.82	0.67	1.03	0.93	0.67
Al_2O_3	16.98	15.82	14.74	15.82	12.13	16.01	16.62	17.12	13.90	12.49	13.45	10.00	6.18	12.99	12.55	13.48	16.19	12.84	13.72
$Fe_2O_3^t$	7.27	8.68	5.34	7.90	8.18	7.31	7.03	4.62	7.10	6.46	6.11	4.68	4.59	5.77	5.80	5.92	6.94	7.65	6.77
MnO	0.04	0.07	0.03	0.04	0.04	0.06	0.10	0.04	0.08	0.05	0.06	0.04	0.05	0.13	0.11	0.07	0.23	0.37	0.07
MgO	2.92	3.44	2.36	3.20	3.42	3.46	3.83	2.56	3.26	2.69	3.39	2.42	2.31	2.66	2.68	3.28	2.90	5.52	3.41
CaO	0.21	0.95	0.13	0.54	0.59	1.40	0.94	0.86	2.02	0.88	1.76	0.82	0.62	1.62	0.86	0.94	2.52	1.29	1.28
Na_2O	1.35	1.99	1.76	1.88	1.30	2.01	1.13	2.30	1.93	1.13	1.95	0.56	0.75	1.97	1.08	2.19	3.83	1.75	0.98
K_2O	3.39	2.39	2.56	2.43	1.54	2.64	2.79	3.15	1.97	2.86	3.20	2.39	1.49	2.54	3.19	2.64	2.67	1.65	3.28
P_2O_5	0.21	0.70	0.14	0.27	0.60	0.18	0.19	0.19	0.19	0.08	0.19	0.07	0.04	0.20	0.06	0.20	0.22	0.14	0.25
LOI	6.31	5.20	7.23	5.83	4.18	5.09	4.71	4.57	5.39	2.96	1.17	1.93	1.31	2.41	3.08	4.15	2.25	4.44	3.73
TOTAL	99.95	100.60	95.11	98.56	99.89	100.89	100.15	100.02	100.32	99.25	97.33	99.36	99.34	99.34	100.04	99.60	100.01	99.70	99.85
Rb	107	75	80	78	47					113	98	80	51						
Sr	78	161	85	111	98	151	179	155	89	125	52	70	100	327	89	188	424	76	40
Zr	172	147	221	154	145	179	162	192	149	145	139	135	99	116	158	128	185	135	121
Y	27	33	23	26	29	26	21	27	22	25	15	20	18	23	22	22	24	26	22
Nb	10	6	6	7	7	9	9	10	9	8	11	6	4	7	11	6	9	9	7
Ba	1341	885	907	965	586	1085	815	1017	764	846	1203	1322	491	2038	975	1738	1120	747	645
Sc	21	22	15	22	17	18	21	21	17	16	17	14	11	19	17	21	23	23	21
V	203	177	158	217	155	141	172	190	131	98	147	85	71	173	129	330	170	176	200
Cr	99	97	105	110	95	95	89	93	69	66	108	65	36	50	60	104	44	94	112
Ni						31	55	4	31					30	22	4	32	67	47
Cu	40	85	23	65	24	37	24	11	50	88	72	76	167	51	96	24	138	2	56
Zn	100	189	83	140	161	148	175	125	143	122	161	123	153	131	138	89	279	189	225
Be	5	4	4	4	3					5	5	5	0						

Notes: Samples were fused in lithium metaborate and dissolved in 50 ml of 5% HCl solution for inductively coupled plasma-atomic emission spectroscopy and flame emission analysis. Relative uncertainties at the 95% confidence level are <1% for Si and Al; <2% for V; <3% for Ti, Fe, Mn, Ca, Na, Sr, Ba, and Zr; <5% for K, Mg, Y, and Rb; <10% for P, Sc, Cu and Zn; and <15% for Nb, Cr, and Ni. LOI—loss on ignition.

TABLE 2. Nd AND Sr ISOTOPIC DATA FOR ARGILLACEOUS ROCKS OF THE KLAMATH MOUNTAINS, CALIFORNIA

Sample	Rb (ppm)	Sr (ppm)	$^{87}Rb/^{86}Sr$	$^{87}Sr/^{86}Sr$	$^{87}Sr/^{86}Sr$ (153 Ma)	$^{87}Sr/^{86}Sr$ (210 Ma)	Sm (ppm)	Nd (ppm)	$^{147}Sm/^{144}Nd$	$^{143}Nd/^{144}Nd$	$^{143}Nd/^{144}Nd$ (153 Ma)	Initial ε_{Nd} (153 Ma)	$^{143}Nd/^{144}Nd$ (210 Ma)	Initial ε_{Nd} (210 Ma)	Nd model age (Ga)
Galice argillites, deposited 153 Ma															
Gali 4	98.45	70.30	4.0554	0.71575	0.706934		3.925	20.67	0.11484	0.512427	0.512312	–2.5			1.12
Gali 1A	42.93	114.95	1.0809	0.70912	0.706765		5.298	23.09	0.13872	0.512497	0.512359	–1.6			1.33
Gali 1B	48.57	64.79	2.1703	0.71130	0.706579		3.354	18.33	0.11068	0.512505	0.512394	–0.9			0.96
Gali 2	48.92	84.82	1.6697	0.71051	0.706875		3.021	14.06	0.13000	0.512467	0.512337	–2.0			1.25
Gali 3*	29.64	97	0.8895	0.70908	0.707144		5.436	23.89	0.13760	0.512511	0.512373	–1.3			1.28
03OMB6	87.66	150.93	1.6811	0.70958	0.705925		5.296	27.58	0.11612	0.512504	0.512387	–1.1			1.01
03OMB4	94.55	181.75	1.5057	0.70920	0.705930		4.351	23.90	0.11008	0.512437	0.512327	–2.2			1.05
03OMB5	112.46	159.34	2.0427	0.70950	0.705059		5.329	24.55	0.13126	0.512509	0.512378	–1.2			1.19
03OMB8§	59.54	82.21	2.0967	0.71260	0.708044	0.706343	5.189	26.10	0.12023	0.512201	0.512080	–7.0	0.512036	–6.48	1.55
Rattlesnake Creek terrane argillites, deposited ca. 210 My															
Cover sequence															
DP1-58	91.09	126.13	2.0913	0.71404	0.709491	0.707794	5.181	30.78	0.10178	0.512166	0.512064	–7.4	0.512026	–6.67	1.34
DP1-61	77.78	178.34	1.2627	0.71270	0.709955	0.708931	5.230	29.21	0.10825	0.512093	0.511985	–8.9	0.511944	–8.26	1.53
DP1-62	48.90	42.66	3.3217	0.72131	0.714083	0.711388	2.335	11.48	0.12300	0.512182	0.512059	–7.5	0.512013	–6.92	1.63
DP1-67	32.09	71.64	1.2969	0.71238	0.709560	0.708510	2.383	10.48	0.13744	0.512325	0.512188	–4.9	0.512136	–4.52	1.67
BP4-3	99.53	485.65	0.5931	0.70733	0.706040	0.705559	5.790	26.95	0.12991	0.512563	0.512433	–0.2	0.512385	0.34	1.07
BP388	114.74	88.91	3.7381	0.71747	0.709337	0.706303	4.013	21.15	0.11474	0.512223	0.512108	–6.5	0.512065	–5.91	1.43
Melange blocks															
03OMB10	77.13	200.80	1.1116	0.70869	0.706273	0.705371	4.926	22.59	0.13188	0.512590	0.512458	0.3	0.512409	0.80	1.05
BP391	85.32	419.55	0.5885	0.70786	0.706578	0.706101	3.897	17.46	0.13495	0.512793	0.512658	4.2	0.512607	4.68	0.69
03OMB1	54.01	84.48	1.8508	0.71078	0.706754	0.705252	5.419	25.42	0.12892	0.512453	0.512324	–2.3	0.512275	–1.80	1.26
03OMB3	76.87	37.27	5.9749	0.71796	0.704965	0.700116	5.576	26.53	0.12712	0.512729	0.512602	3.1	0.512555	3.65	0.74

Notes: Approximately 80 to 100 mg of each sample was dissolved in $HF\text{-}HNO_3$. After conversion to chlorides, one-third of the sample was spiked with ^{87}Rb, ^{84}Sr, ^{149}Sm, and ^{146}Nd. Rb, Sr, and REEs were separated by conventional cation-exchange procedures. Sm and Nd were further separated in di-ethyl-hexyl orthophosphoric acid columns.

All isotopic measurements were made on a VG Sector multi-collector mass spectrometer at the University of Wyoming. An average $^{87}Sr/^{86}Sr$ isotopic ratio of 0.710251 ± 20 (2 standard deviations) was measured for NBS 987 Sr, and an average $^{143}Nd/^{144}Nd$ ratio of 0.511846 ± 11 (2 standard deviations) was measured for the La Jolla Nd standard. Uncertainties in Sr isotopic ratio measurements are ±0.00002 and uncertainties in Nd isotopic ratio measurements are ± 0.00001 (2 standard deviations). Blanks are 50 pg for Rb, Sr, Nd, and Sm, and no blank correction was made. Uncertainties in Rb, Sr, Nd, and Sm concentrations are ±2% of the measured value; uncertainties on initial ε_{Nd} are ± 0.3 epsilon units. The Nd model ages are calculated based upon the depleted mantle model of Goldstein et al. (1984).

*Sr concentration for Gali 3 by inductively coupled plasma-atomic emission spectroscopy (Table 1).

§Initial Nd and Sr isotopic ratios for 03OMB8 calculated for 153 Ma and for 210 Ma, due to the uncertain assignment of this sample to the Galice Formation.

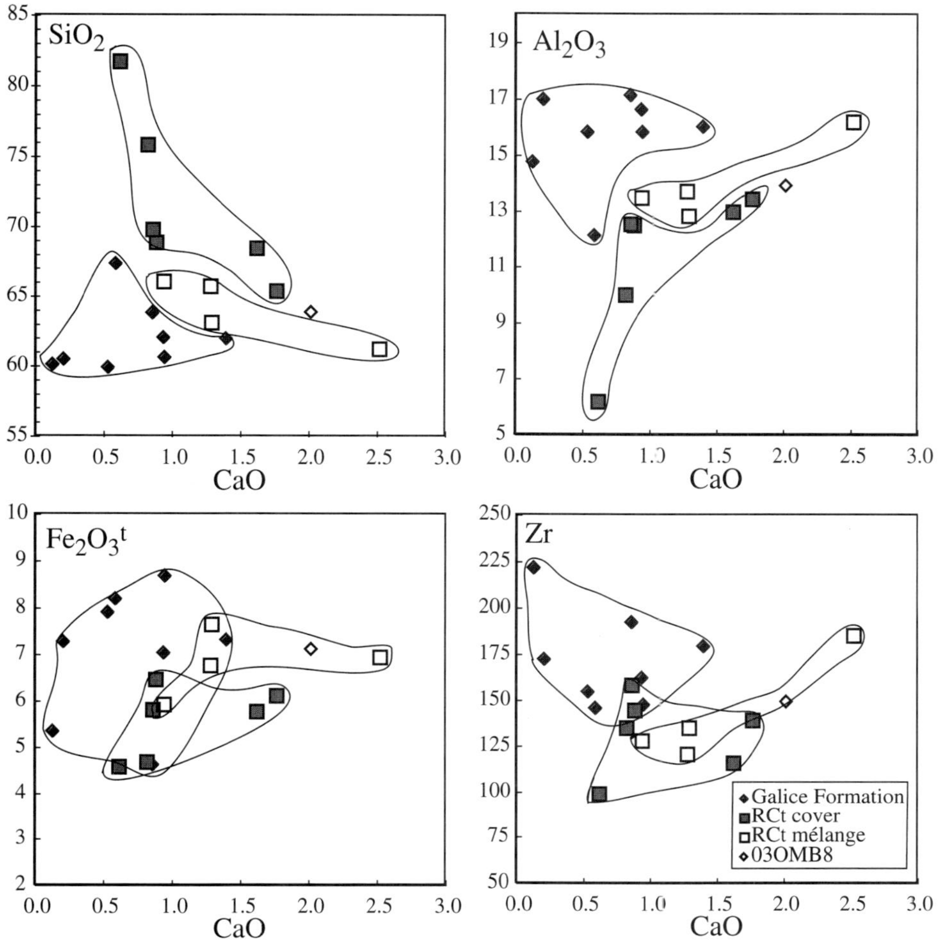

Figure 3. Geochemical compositions of Galice Formation argillites and Rattlesnake Creek terrane meta-argillites. Galice argillites are distinguished by lower SiO_2 and CaO, and higher $Fe_2O_3{}^t$, Al_2O_3, and Zr compared to Rattlesnake Creek terrane samples. The meta-argillite from the Rattlesnake Creek terrane cover sequences is more siliceous and has lower Al_2O_3 and $Fe_2O_3{}^t$ than does meta-argillite from the serpentinite-matrix mélange.

lower SiO_2 and CaO, and higher $Fe_2O_3{}^t$, Al_2O_3, and Zr abundances. Meta-argillite from the inferred cover sequences of the Rattlesnake Creek terrane is more siliceous and has correspondingly lower Al_2O_3 and $Fe_2O_3{}^t$ than does meta-argillite from the serpentinite-matrix mélange.

Two samples from the cover sequences of the Rattlesnake Creek terrane, DP1-62 and DP1-67, have higher SiO_2 and lower Al_2O_3 than found in all other samples from this study. These rocks are best described as siliceous or cherty meta-argillite, and there is a complete spectrum from meta-argillite to siliceous argillite to impure chert in the Rattlesnake Creek terrane. This range is true of meta-argillite from the serpentinite-matrix mélange as well as metamorphosed argillaceous rocks from the inferred cover sequences.

Sample 03OMB8, collected from an outcrop that has been mapped as Galice Formation (Klein, 1975, 1977; Petersen, 1982), is geochemically distinct from the other analyzed Galice samples. Although 03OMB8 falls within fields defined by the analyzed Rattlesnake Creek terrane samples, it has experienced lower-grade metamorphism than Rattlesnake Creek terrane metasedimentary rocks in the adjacent hanging-wall block of the Orleans fault. We thus conclude that either this sample has an unusual composition for argillaceous rocks of the Galice Formation, or as discussed previously, this sample may be a fault-bounded slice of Condrey Mountain terrane, a third argillaceous unit that is widely exposed in a structural window (Hotz, 1979; Helper, 1986) east-northeast of Happy Camp, California.

Isotopic Data

Sm-Nd and Rb-Sr isotopic data are presented in Table 2; the ages used to calculate initial isotopic ratios are 153 Ma for samples from the Galice Formation and 210 Ma for samples from the Rattlesnake Creek terrane, respectively. Isotopic data for metamorphosed argillaceous rocks from the Galice Formation are more uniform than are the data for meta-argillite samples from the Rattlesnake Creek terrane. Galice samples (apart from 03OMB8) have initial $^{87}Sr/^{86}Sr$ of 0.7051–0.7071 and initial ε_{Nd} of –0.9 to –2.5 (Fig. 4). Meta-argillites of the Rattlesnake Creek terrane exhibit greater ranges both in initial $^{87}Sr/^{86}Sr$ (0.7001–0.7114) and initial ε_{Nd} (+4.7 to –8.3). The calculated initial $^{87}Sr/^{86}Sr$ ratio of 0.7001 is unrealistically low, suggesting that Rb and/or Sr mobility has affected this and possibly other samples. Sample 03OMB8 is distinct from the other Galice samples, falling within the range defined by Rattlesnake Creek terrane samples with an initial Sr isotopic ratio of 0.7080 and initial ε_{Nd} of –7.0. The range in $^{147}Sm/^{144}Nd$ exhibited both by Galice Formation and Rattlesnake Creek terrane samples ($^{147}Sm/^{144}Nd$ = 0.1018–0.1387) extends to higher ratios than are typical of upper continental crust ($^{147}Sm/^{144}Nd$ = 0.090–0.125).

Meta-argillites from the Rattlesnake Creek terrane cover sequence have in general higher initial $^{87}Sr/^{86}Sr$ and more negative initial ε_{Nd} than do meta-argillites from the serpentinite-matrix mélange. Meta-argillites from the cover sequences also tend to have lower P_2O_5 than found in meta-argillite samples from the Rattlesnake Creek terrane serpentinite-matrix mélange (Fig. 5A). Cover sequence meta-argillite sample BP4–3 is the exception; it falls within the range of Nd and Sr isotopic compositions defined by the meta-argillite samples collected from inferred mélange blocks. Possibly we have misidentified this sample as part of the cover sequence, and it might instead have been collected from a very large mélange block. However, this sample has much higher Sr and Ba than other cover sequence argillites (Table 2 and Fig. 5B) and one of the highest CaO and Na_2O contents. These geochemical traits, which are consistent with a greater volcanogenic component, appear to correlate with its more radiogenic initial Nd isotopic ratio. It is possible, then, that this meta-argillite from the cover sequence incorporates a larger proportion of young, oceanic detritus than other meta-argillites from the cover sequence. A similar correlation of high Sr with radiogenic initial ε_{Nd} is observed for the meta-argillite sample from the serpentinite-matrix mélange that exhibits the most radiogenic Nd isotopic ratio, sample BP391 (Fig. 5B).

DISCUSSION

Provenance of Galice Formation Argillaceous Rocks

The metamorphosed argillaceous rocks of the Galice Formation analyzed in this study are distinguished by their geochemical and isotopic homogeneity, despite their varying geographic locations and stratigraphic positions. The Sm-Nd isotopic data preserve evidence both for ancient continental crust and younger crust and/or mantle-derived components in these argillaceous rocks. Evidence for incorporation of the latter is preserved in the $^{147}Sm/^{144}Nd$ ratios of the samples, which range from 0.1101 to 0.1387. Half of the samples have ratios

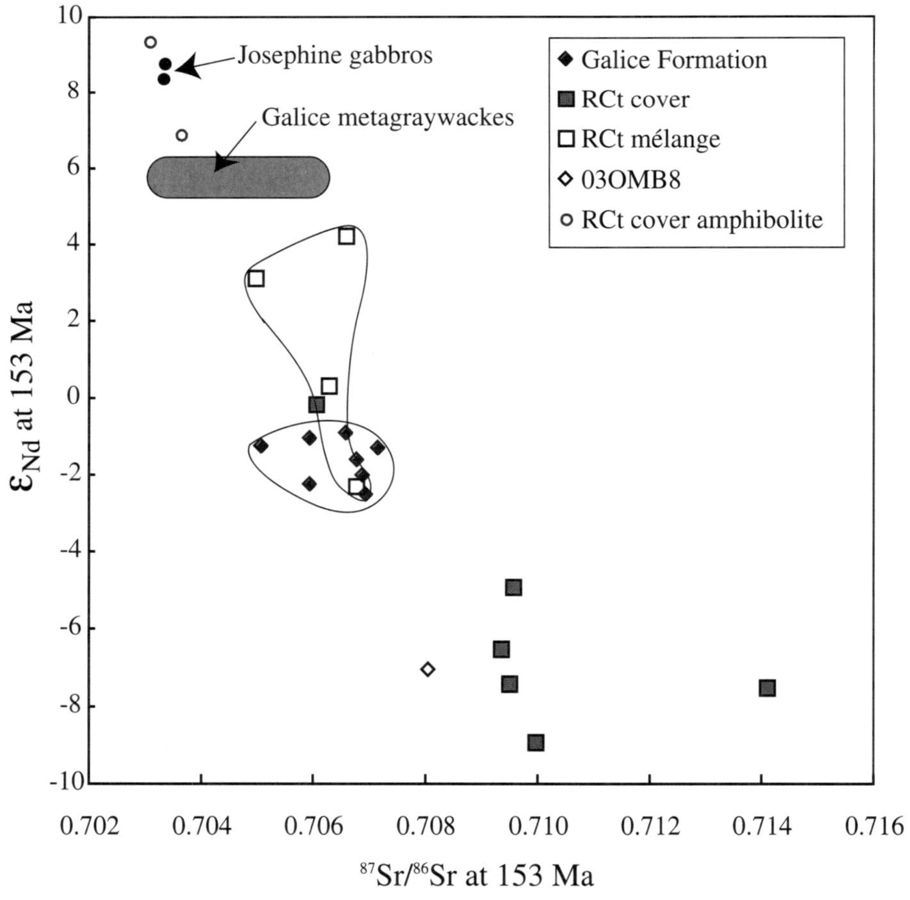

Figure 4. Nd and Sr isotopic data for Galice Formation and Rattlesnake Creek terrane metasedimentary rocks calculated at 153 Ma, the estimated time of Galice deposition. Galice Formation argillites have relatively uniform initial ε_{Nd} of –0.9 to –2.5, and initial $^{87}Sr/^{86}Sr$ of 0.7051–0.7071, distinct from the initial ε_{Nd} and $^{87}Sr/^{86}Sr$ of Galice (meta)graywackes. Rattlesnake Creek terrane meta-argillites have less uniform initial ε_{Nd} and $^{87}Sr/^{86}Sr$, and in general, cover sequence samples have higher initial $^{87}Sr/^{86}Sr$ and lower initial ε_{Nd} than do samples from the serpentinite-matrix mélange. No metasedimentary sample has a depleted mantle signature like those exhibited by gabbros from the Josephine ophiolite (Shaw and Wasserburg, 1984) or Rattlesnake Creek terrane amphibolites from the cover sequence (C.D. Frost, unpublished data, 2003).

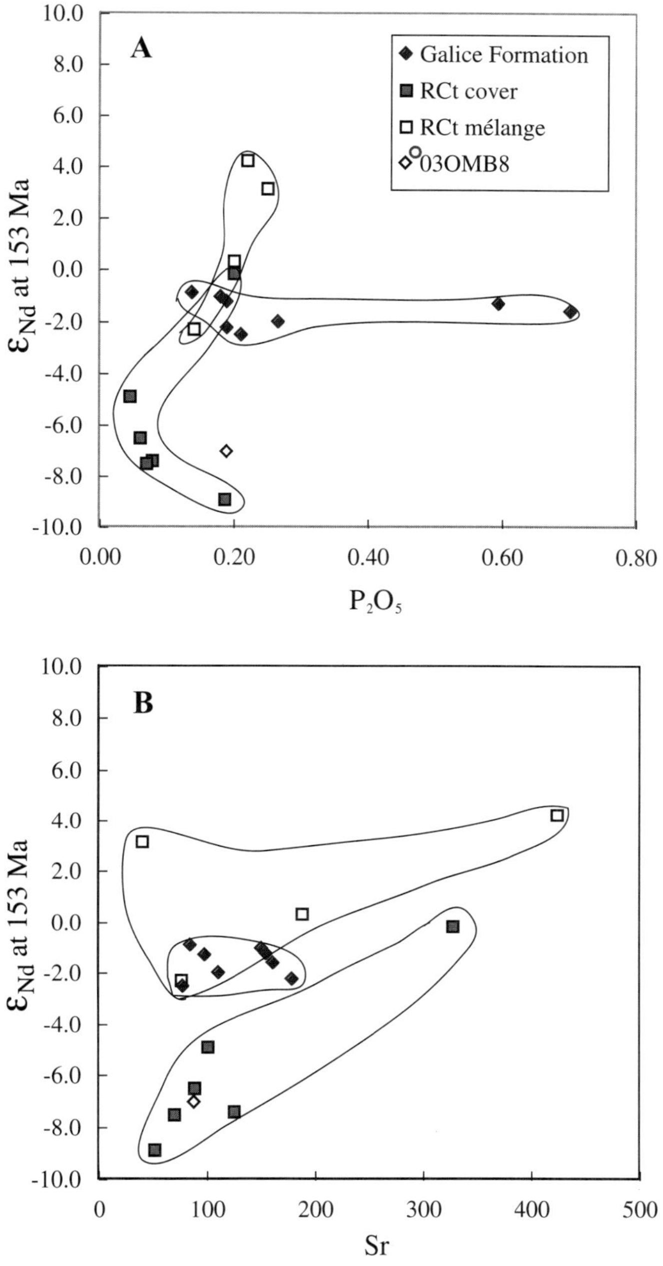

Figure 5. (A) Plot of P_2O_5 content (in wt. %) vs. ε_{Nd} at 153 Ma, showing generally lower P_2O_5 content of Rattlesnake Creek terrane cover sequence meta-argillites compared to Rattlesnake Creek terrane mélange and Galice argillaceous rocks. (B) Plot of Sr content (in ppm) vs. ε_{Nd} at 153 Ma, showing that the most radiogenic samples within the Rattlesnake Creek terrane cover and the Rattlesnake Creek terrane mélange also have high Sr and suggesting that these samples may have incorporated a greater volcanogenic component than the other samples within each group.

higher than 0.125 and thus exceed $^{147}Sm/^{144}Nd$ ratios typical of continental crust (Fig. 6). However, the samples have slightly negative initial ε_{Nd} and Mesoproterozoic depleted mantle Nd model ages of 0.96–1.25 Ga. These model ages represent the weighted average crust-formation age of the sources that eroded and contributed detritus to the Galice Formation. They strongly imply that although the argillaceous rocks of the Galice Formation were deposited upon the Late Jurassic Josephine ophiolite, their sediment sources were not restricted to proximal arc materials but instead included detritus ultimately derived from Precambrian continental crust. The initial Sr isotopic compositions of the samples also suggest that they are a mixture of continental and mantle-derived sources: initial $^{87}Sr/^{86}Sr$ of between 0.7051 and 0.7071 are more radiogenic than depleted mantle and island-arc magmas, but less radiogenic than a North American cratonic source (Fig. 6).

Limited Nd and Sr isotopic data from metagraywackes of the Galice Formation provide an instructive comparison with our meta-argillite data. Initial Sr isotopic ratios of metagraywackes reported by Coleman (1972: $^{87}Sr/^{86}Sr = 0.7030–0.7066$) are on average less radiogenic than those of the meta-argillites. An analysis of Galice metagraywacke reported by Miller et al. (2003) yielded an initial ε_{Nd} of +5.6, considerably more radiogenic than the initial ε_{Nd} of –0.9 to –2.5 obtained from our argillaceous samples (Fig. 4). The Miller et al. (2003) sample also contained an abundant population of detrital zircons with an age of 153 Ma. Both the Sr and Nd isotopic compositions and the detrital zircon ages suggest that Galice Formation metagraywackes incorporate detritus from a Late Jurassic oceanic arc source, probably the Rogue–Chetco arc. Miller et al. (2003) documented a less well-defined peak in detrital zircon ages at 227 Ma and sparse older zircon, indicating that the metagraywacke also incorporated detritus from older sources, including sediment that ultimately was eroded from the North American Precambrian basement (Miller et al., 2003). Interpreting U-Pb detrital zircon ages from a larger number of Galice metagraywacke samples, Miller and Saleeby (1995) also concluded that the Galice Formation was deposited within the range of terrigenous input from a recycled continental source, and identified the older Klamath terranes as suitable sediment sources. The data presented in our study are consistent with the hypothesis of these workers that the turbidites of the Galice Formation contain both a young arc-derived component and an older cratonic source. However, the more negative ε_{Nd} of the argillaceous rocks compared to the metagraywacke sample analyzed by Miller et al. (2003) and more radiogenic initial Sr isotopic compositions of the argillaceous rocks compared to the metagraywackes analyzed by Coleman (1972) suggest that the proportions of detritus from various sources are different in the argillaceous rocks compared to the metagraywackes in this turbidite sequence.

Differences in initial ε_{Nd} for coarse-grained and fine-grained portions of turbidite flows have been documented in both modern (McLennan et al., 1989) and ancient (Frost and Coombs, 1989) environments. These differences may arise if

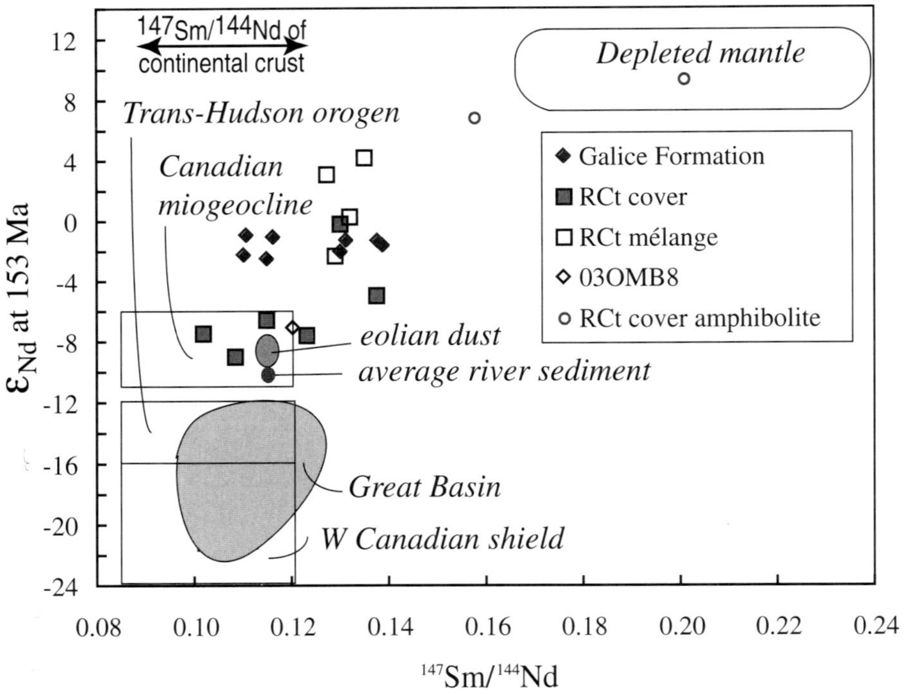

Figure 6. Plots of $^{147}Sm/^{144}Nd$ data vs. ε_{Nd} at 153 Ma for Galice and Rattlesnake Creek terrane meta-argillite samples and possible sediment sources. The Klamath metasedimentary samples have Sm-Nd isotopic characteristics that are intermediate between depleted mantle and North American crust. Fields for the latter include Neoproterozoic metasedimentary rocks of the Great Basin (Farmer and Ball, 1997), western Canadian shield and Trans-Hudson orogen (Chauvel et al., 1987; Boghossian et al., 1996), and Devonian to Jurassic Canadian miogeocline (Boghossian et al., 1996; Patchett and Gehrels, 1998). Rattlesnake Creek terrane cover sequence meta-argillites have Sm-Nd isotopic characteristics comparable to eolian dust and particulates from major river systems (Goldstein et al., 1984). Galice Formation argillite samples and meta-argillite from Rattlesnake Creek terrane serpentinite-matrix mélange have incorporated a greater proportion of juvenile detritus.

different source areas are shedding debris of contrasting grain size; for example, if coarser debris is eroding from proximal arcs and fine-grained material is carried from more distant source areas (i.e., as suspended particulate in rivers; Frost and Coombs, 1989). A second factor is the concentration or preferential breakdown of certain components such as volcanic glass or unstable minerals during weathering and transport, and their incorporation in the finer-grained sediment (McLennan et al., 1989). A third mechanism is the preferential incorporation of heavy minerals in the coarser-grained sediment during sedimentary transport and sorting. A modest enrichment of minerals such as zircon or monazite in the sand-sized portion of a turbidite flow can significantly affect its Sm and Nd budget and hence its Nd isotopic mass balance (Frost and Winston, 1987). Because the argillaceous rocks have the more unradiogenic initial ε_{Nd} compared to the metagraywacke sample, we conclude that the finer-grained rocks contain a greater proportion of recycled cratonic debris, material that may have been reduced to mud-sized particles during repeated episodes of erosion, transport, and deposition.

It is noteworthy, however, that in some cases, no correlation is observed between grain size and ε_{Nd} value. For example, in a study of Quesnel terrane strata in the Canadian Cordillera Unterschutz et al. (2002) documented variable ε_{Nd} for metasedimentary rocks of all grain sizes.

SEDIMENT SOURCES FOR META-ARGILLITES FROM THE RATTLESNAKE CREEK TERRANE MÉLANGE BASEMENT AND COVER SEQUENCE

With one exception, meta-argillites from the Rattlesnake Creek terrane cover sequence have initial Sr and Nd isotopic compositions that are distinct from meta-argillite samples from the serpentinite-matrix mélange. Meta-argillites from Rattlesnake Creek terrane cover sequences are characterized by radiogenic initial $^{87}Sr/^{86}Sr$ of 0.7063–0.7114, negative initial ε_{Nd} from –4.5 to –8.3, and depleted mantle Nd model ages of 1.34–1.67 Ga. Meta-argillite samples from serpentinite-matrix mélange, however, have less radiogenic initial Sr isotope ratios (≤0.7061), less negative initial ε_{Nd} (+4.7 to –1.8), and younger model ages (0.69–1.26 Ga). As discussed below, these isotopic differences reflect a shift in the provenance of the argillaceous rocks from the basement to the cover portions of the Rattlesnake Creek terrane.

Wright and Wyld (1994) hypothesized that the mélange basement was formed by disruption of oceanic crust far removed from arc or continental landmasses, perhaps in an oceanic fracture zone setting. The blocks within the mélange are composed of rock types that make up the oceanic crust and uppermost mantle, including pillow basalt and diabase, gabbro, pyroxenite/peridotite, and amphibolite. The sedimentary rocks are restricted to chert, argillite, and limestone; no blocks contain clastic material of terrigenous or volcanic-arc provenance. The positive to only slightly negative initial ε_{Nd} and unradiogenic Sr isotopic compositions of the argillites analyzed in this study are consistent with an oceanic tectonic setting. Moreover, the samples all have elevated $^{147}Sm/^{144}Nd$ compared to typical continental crust (Fig. 6). However, no argillite sample has radiogenic Nd and unradiogenic Sr isotopic compositions of contemporary depleted mantle, so even in this oceanic setting, the sediments must include continental components. The absence of siliceous clastic rocks among the sedimentary blocks of the Rattlesnake Creek terrane mélange basement argues against its formation within the range of terrigenous sedimentation. Although this distance is highly variable, depending upon the amount of sediment being discharged into the oceans by rivers, the width of the continental shelf and slope, and the potential for transport of sediment long distances in submarine canyons and along trenches, Patchett and Gehrels (1998) suggest that 1500 km is a reasonable limit. For sedimentation occurring beyond this distance, far-traveled hemipelagic sediment or eolian dust are potential sources of continental material.

The Nd isotopic compositions and Sm/Nd ratios of modern river particulates and eolian dusts are very uniform and yield depleted mantle Nd model ages of 1.70 ± 0.35 Ga (Goldstein et al., 1984; Jones et al., 1994). Correcting for radiogenic growth since Late Triassic time, this continental source of detritus would have had an initial ε_{Nd} of approximately –8 to –10 at the time the Rattlesnake Creek terrane mélange basement was deposited (Fig. 6). Assuming that continental material with this Nd isotopic composition is an appropriate estimate of far-traveled detritus that could have been incorporated into Rattlesnake Creek terrane basement argillite blocks, and taking into account the higher Nd concentration in atmospheric dusts and river particulates compared to oceanic crust, it is possible to estimate the proportion of continental material present in Rattlesnake Creek terrane basement argillite samples. Accordingly, the range of initial ε_{Nd} of the blocks analyzed (ε_{Nd} = +4.7 to –1.8) can be modeled as mixtures of ~15–45% continental material and the remainder of juvenile oceanic sources. This result contrasts with the less radiogenic ε_{Nd} values of modern pelagic sediment in the central North Pacific Ocean. Jones et al. (1994) concluded that this pelagic sediment is composed almost entirely of eolian sediment and includes little to no juvenile detritus. In contrast, the high proportion of mantle-derived sources in Rattlesnake Creek terrane basement argillite blocks suggests that the inferred oceanic fracture zone setting of these argillites must have produced abundant argillaceous detritus of juvenile origin that was mixed in varying proportions with far-traveled continental sediment.

The Rattlesnake Creek terrane cover sequence that unconformably overlies the mélange basement was interpreted by Wright and Wyld (1994) as a volcanic-arc suite. The igneous rocks of the cover sequence have geochemical signatures suggesting an island-arc environment (Wright and Wyld, 1994). The Nd and Sr isotopic data from amphibolitic rocks (contact-metamorphosed metabasalt) of the cover sequence (C.D. Frost, unpublished data; Figs. 4 and 6) confirm the arc signature of these rocks. The quartzose and argillaceous nature of the sedimentary rocks of the cover sequence suggested to Wright and Wyld (1994) that the metasedimentary rocks of the cover sequence were likely derived from a terrigeneous source, possibly Paleozoic rocks of western North America. The isotopic data for Rattlesnake Creek terrane meta-argillite samples presented here—namely, more radiogenic initial $^{87}Sr/^{86}Sr$, more negative initial ε_{Nd}, and generally lower $^{147}Sm/^{144}Nd$ for meta-argillites of the cover sequence compared to those of the mélange basement (Figs. 4 and 6)—are consistent with this general hypothesis that continental input is more pronounced in the cover sequence than in the mélange basement.

We note a similarity both in absolute ε_{Nd} values and variability between the argillaceous rocks of the Rattlesnake Creek terrane cover sequence and Late Quaternary pelagic sediments from the Circum-Pacific rim analyzed by Jones et al. (1994). The latter were suggested to be composed of volcanic ash, hemipelagic sediments derived from volcanic arcs, and eolian loess. These fine-grained sediments are less radiogenic than modern turbidites from the same locations (McLennan et al., 1990). Jones et al. (1994) explained this discrepancy by noting that the turbidites are probably dominated by debris from island-arc sources, but that the finer pelagic sediment would contain both material from these sources of sediment and more distally derived continental loess.

Our sample set from the Rattlesnake Creek terrane cover sequence includes samples with a range of SiO_2 and Al_2O_3 contents that vary from true argillite to cherty or siliceous argillite. The sample with the isotopic characteristics suggesting the greatest proportion of juvenile arc component, BP4-3, has geochemical characteristics expected of clastic rocks with a significant volcanogenic component, including relatively high Sr, CaO, Na_2O, and Al_2O_3. Sample DP1-67, which has the second highest initial ε_{Nd} after sample BP4-3, is the most siliceous of the Rattlesnake Creek terrane cover sequence samples. Thus it appears that the range in isotopic compositions of meta-argillite samples from the Rattlesnake Creek terrane cover sequence reflects varying proportions of local arc and more distal continental components, the latter being more pronounced in the finer-grained, less siliceous samples.

Implications for Tectonic Evolution of Klamath Terranes

The lithologic and isotopic characteristics of the blocks of metamorphosed argillaceous rocks in the Rattlesnake Creek terrane are consistent with development of a serpentinite-matrix mélange distal from a continent. In addition, the presence of (meta)limestone blocks in the mélange suggests significant seafloor topography and the possibility that oceanic basement was exposed and available for weathering, and consequently was a source of juvenile detritus in the argillaceous rocks. These conclusions are consistent with the higher ε_{Nd} and lower initial $^{87}Sr/^{86}Sr$ ratios among these samples than would be expected if their source was entirely derived from a continental source. Thus, our data are consistent with the oceanic fracture-zone setting postulated by Wright and Wyld (1994) for the serpentinite-matrix mélange unit of the Rattlesnake Creek terrane (Fig. 7A).

In contrast, the cover sequence of the Rattlesnake Creek terrane contains abundant evidence for deposition in an oceanic-arc setting (Wright and Wyld, 1994). However, argillaceous samples from the cover sequence exhibit Nd isotopic compositions typical of continental detritus, with little evidence for admixture of primitive-arc detritus (Fig. 6). The distinctly "continental" isotopic compositions of the argillites of the cover sequence may be explained by an absence of fine-grained arc

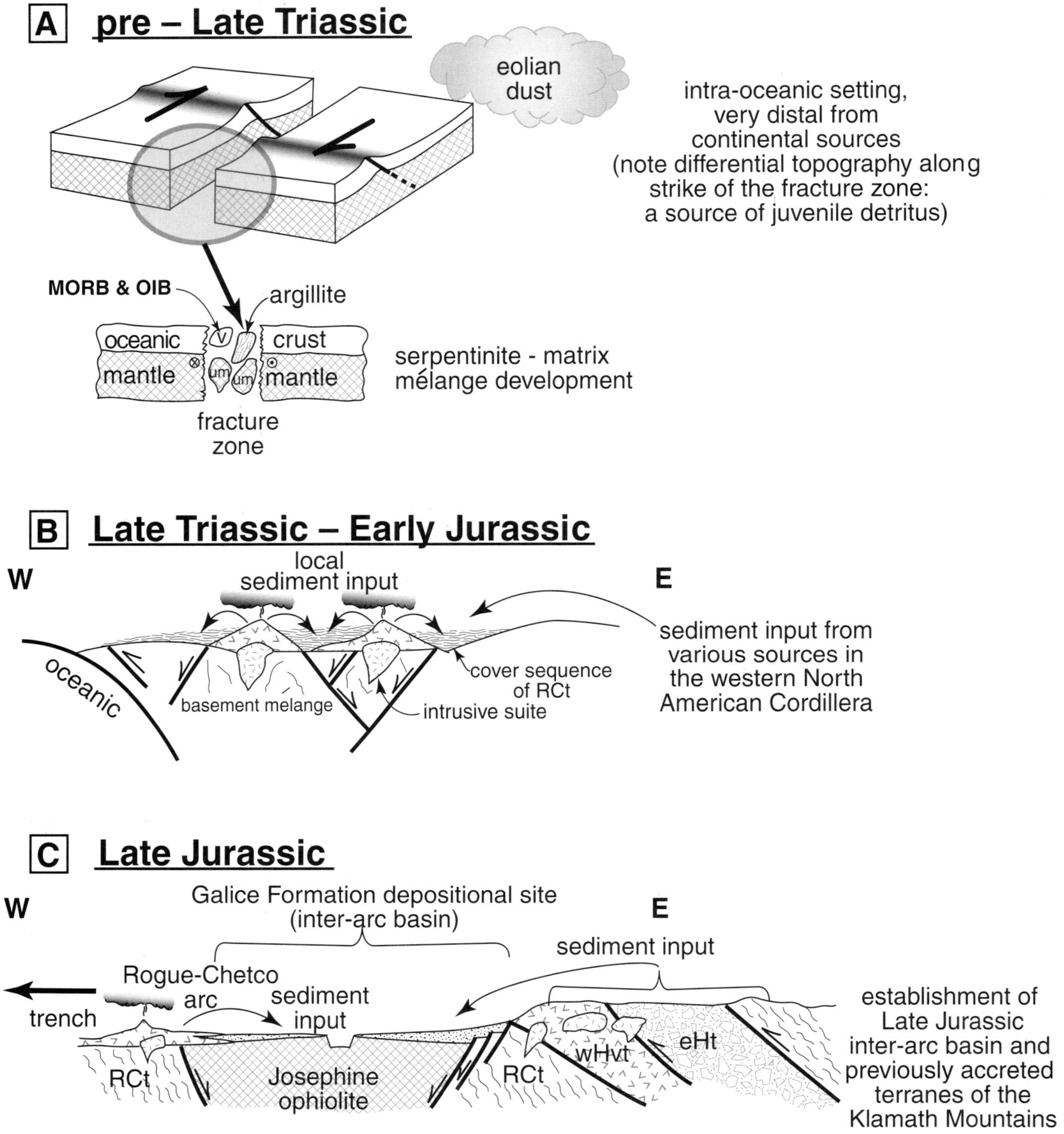

Figure 7. Inferred tectonic settings (pre-Late Triassic through Late Jurassic) for the argillaceous rocks analyzed in this study. (A) Development of serpentinite-matrix mélange (basement of the Rattlesnake Creek terrane) in a pre-Late Triassic oceanic fracture zone; (B) Late Triassic–Early Jurassic oceanic-arc setting for deposition of the cover sequence of the Rattlesnake Creek terrane. (C) Development of the Late Jurassic, inter-arc Galice basin through transtensional rifting of older, previously accreted terranes of the Klamath Mountains.

detritus. For example, if the arc was composed primarily of lava erupted subaqueously rather than of abundant submarine debris-flow or epiclastic deposits, then little of this juvenile arc component may be incorporated into the cover sequence meta-argillites. An alternative explanation may lie in our sampling technique: we specifically tried to sample the finest grained rocks available at each locality. If the arc detritus was incorporated preferentially into sand-sized sediment, then such detrital rocks were preferentially excluded from the sample set.

By Middle Jurassic time, sedimentary rocks of the western Hayfork arc terrane were deposited directly on Rattlesnake Creek terrane rocks (Wright and Fahan, 1988; Donato et al., 1996). The western Hayfork terrane contains quartz-rich sedimentary rocks that suggest a position proximal to a continental landmass (Barnes et al., this volume, Chapter 10), presumably North America. Paleomagnetic data (Mankinen et al., 1989) show that by Middle Jurassic time, the Klamath province was close to its present latitudinal position relative to North America, although minor shifts are permissible. Therefore, the isotopic data support the concept that the Rattlesnake Creek terrane mélange formed in an oceanic setting. By the time the cover sequence was deposited, the Rattlesnake Creek terrane was presumably near a continental-affinity landmass (Fig. 7B). It subsequently became the basement for a near-margin, albeit oceanic-affinity, arc (western Hayfork terrane) in Middle Jurassic time. This inferred scenario thus suggests that the Middle Jurassic western Hayfork arc was not "exotic" to western North America and favors that this Middle Jurassic arc was west-facing (relative to present geographic coordinates).

The isotopic composition of metamorphosed argillaceous rocks from the Upper Jurassic Galice Formation stand in marked contrast to the meta-argillites of the Rattlesnake Creek terrane cover sequence. The relative homogeneity and intermediate compositions of the argillaceous rocks of the Galice Formation suggest significant and uniform mixing of arc and continental detritus. Whether the continental detritus was primary (i.e., directly from a continental drainage) or recycled (e.g., from terranes in the eastern Klamath Mountains) is more difficult to determine and will probably require thorough sampling of a variety of stratigraphic units (Fig. 7C). At least two explanations for the well-mixed nature of the argillaceous rocks of the Galice Formation are possible: (1) the continental component was part of the suspended load from riverine input direct from North America and was mixed with fine-grained, relatively juvenile input from proximal, outboard Rogue–Chetco arc as well as the Middle Jurassic inboard arc components (e.g., western Hayfork terrane); and (2) the source was entirely from the Klamath province, in which case mixing occurred in the source regions, which already contained arc and continental detritus (Miller and Saleeby, 1991; Gehrels and Miller, 2000; Wallin et al., 2000). In the latter case, no new input from continental drainages is necessary. There is no evidence for drainage systems that crossed the Great Basin during Jurassic time. Therefore, the first explanation implies either that the Klamath province was in a more southerly paleolatitude during Galice deposition, as suggested by Wyld and Wright (2001), or that wind-borne detritus associated with Early Jurassic Navajo/Aztec–type sands reached the drainages that fed the Galice basin. In contrast to the first alternative, no implications about paleolatitude may be made from the second explanation because no external source of sedimentary detritus is necessary.

CONCLUSIONS

This study represents the first extensive Nd and Sr isotopic examination of metasedimentary rocks from the Klamath Mountains. Our results from the two terranes farthest outboard of the North American craton show that all of the sedimentary rocks of these oceanic terranes contain a significant continental component. The pre-Late Triassic serpentinite-matrix mélange of the Rattlesnake Creek terrane is interpreted as forming in an oceanic fracture zone. The continental component in its argillaceous rocks was apparently supplied by eolian processes. In contrast, the cover sequence of the Rattlesnake Creek terrane was deposited when the terrane was more proximal to a continental landmass, presumably North America, and its metasedimentary rocks are mixtures of varying proportions of volcanic-arc debris and river-borne terrigenous sediment. The Nd and Sr isotopic compositions of the argillaceous rocks of the Upper Jurassic Galice Formation are remarkably uniform, a particularly striking feature in view of the likelihood that sediment was shed into the Galice basin from both the previously accreted terranes of the Klamath Mountains as well as the outboard, west-facing Rogue–Chetco oceanic arc.

The isotopic data presented here are consistent with the general observation that sedimentary rocks from accreted terranes preserve a stronger signal of continental input than do the contemporaneous igneous rocks within the terrane. Although this observation has been made previously (i.e. Samson et al., 1991; Patchett and Gehrels, 1998), the results of this study suggest that variations in the homogeneity of the isotopic compositions also preserve important information. In particular, the striking homogeneity of Nd and Sr isotopic compositions of Galice argillaceous metasedimentary samples contrasts with the variability in Nd isotopic composition observed in metasedimentary rocks of other oceanic terranes. For example, the variability of initial ε_{Nd} composition of strata from the Quesnel terrane was explained by differing proportions of primitive and evolved sources present along the North American continental margin (Unterschutz et al., 2002). This terrane was interpreted to lie on or near the North American margin, such that there was no opportunity for homogenization of these sources by sedimentary processes. In contrast, the remarkable uniformity of isotopic compositions of Galice Formation argillaceous rocks requires that arc and continental detritus were well homogenized prior to deposition upon the Josephine ophiolite. Although the ability of major rivers to homogenize the isotopic composition of sediments they transport is well known (Goldstein et al., 1984), the processes resulting in isotopic uniformity of sediment in an arc terrane setting clearly require further investigation.

Although the present study represents an important first step, a more detailed understanding of sediment sources, the importance of sediment recycling, and paleogeographic and tectonic positions of Klamath Mountains terranes requires additional detailed investigations. In particular, isotopic characteristics of a variety of lithologic units from the other Klamath terranes will help to determine whether the continental material contributed to the western Klamath and Rattlesnake Creek terranes was derived from adjacent, previously accreted terranes, or whether it was delivered directly from the North American craton. A corollary study of the collage of time-correlative tectonostratigraphic terranes in the western Sierra Nevada would also be interesting and potentially informative in the evaluation of contrasting tectonic models for the Klamath Mountains and western Sierra Nevada. For example, comparison of lithologic, geochemical, and isotopic characteristics of the Upper Jurassic Galice and Mariposa Formations may help distinguish between postulated tectonic accretion scenarios (e.g., Schweickert and Cowan, 1975; Harper and Wright, 1984; Ingersoll and Schweickert, 1986). For nearly a hundred years since Diller (1907) originally suggested that the Galice and Mariposa Formations are correlative units, it has been recognized that a thorough understanding of the tectonic setting and depositional history of these important Upper Jurassic stratigraphic units is fundamental to the mid-Mesozoic history of the continental margin of western U.S. Cordillera.

ACKNOWLEDGMENTS

Melanie Barnes and Jim Lehane produced the bulk-rock chemical analyses. Partial support was provided by National Science Foundation grants EAR-9902807, EAR-9902912, EAR-9117103, and EAR-8720141. We thank Scott Samson and Jonathan Patchett for very helpful reviews.

APPENDIX. SAMPLE LOCALITIES WITH UTM COORDINATES

Sample	UTM coordinates (easting, northing)
Argillaceous rocks of the Rattlesnake Creek terrane: inferred cover sequence	
DP1-58	0440761, 4629350
DP1-61	0438312, 4628546
DP1-62	0437732, 4692112
DP1-67	0437865, 4629149
BP4-3	0453336, 4611678
BP388	0453163, 4610102
Argillaceous rocks of the Rattlesnake Creek terrane: inferred blocks in serpentinite-matrix mélange	
03OMB1	0458847, 4580800
03OMB3	0458350, 4578617
03OMB10	0455807, 4646294
BP391	0449840, 4615794
Argillaceous rocks of the Upper Jurassic Galice Formation	
GALI1A	434795, 4635549
GALI1B	434795, 4635549
GALI2	434848, 4635482
GALI3	434610, 4635529
GALI4	438774, 4642048
03OMB4	0446277, 4566604
03OMB5	0445357, 4564287
03OMB6	0457702, 4591098
03OMB8	0467288, 4625712

REFERENCES CITED

Armstrong, R.L., 1988, Mesozoic and early Cenozoic magmatic evolution of the Canadian Cordillera, *in* Clark, S.P., et al., eds., Processes in continental lithosphere deformation: Boulder, Colorado, Geological Society of America Special Paper 218, p. 55–91.

Barnes, C.G., Allen, C.M., Hoover, J.D., and Brigham, R.H., 1990, Magmatic components of a tilted plutonic system, Klamath Mountains, California, *in* Anderson, J.L., ed., The nature and origin of Cordilleran magmatism: Boulder, Colorado, Geological Society of America Memoir 174, p. 331–346.

Barnes, C.G., Petersen, S.W., Kistler, R.W., Prestvik, T., and Sundvoll, B., 1992a, Tectonic implications of isotopic variation among Jurassic and early Cretaceous plutons, Klamath Mountains: Geological Society of America Bulletin, v. 104, p. 117–126, doi: 10.1130/0016-7606(1992)104<0117:TIOIVA>2.3.CO;2.

Barnes, C.G., Barnes, M.A., and Kistler, R.W., 1992b, Petrology of the Caribou Mountain pluton, Klamath Mountains, California: Journal of Petrology, v. 33, p. 95–124.

Barnes, C.G., Johnson, K., Barnes, M.A., Prestvik, T., Kistler, R.W., and Sundvoll, B., 1995, The Grayback pluton: Magmatism in a Jurassic back-arc environment, Klamath Mountains, Oregon: Journal of Petrology, v. 36, p. 397–415.

Barnes, C.G., Mars, E.V., Swapp, S., and Frost, C.D., 2006, this volume (Chapter 10), Petrology and geochemistry of the Middle Jurassic Ironside Mountain batholith: Evolution of potassic magmas in a primitive arc setting, *in* Snoke, A.W., and Barnes, C.G., eds., Geological studies in the Klamath Mountains province, California and Oregon: A volume in honor of William P. Irwin: Boulder, Colorado, Geological Society of America Special Paper 410, doi: 10.1130/2006.2410(10).

Boghossian, N.D., Patchett, P.J., Ross, G.M., and Gehrels, G.E., 1996, Nd isotopes and the source of sediments in the miogeocline of the Canadian Cordillera: Journal of Geology, v. 104, p. 259–277.

Bushey, J.C., Snoke, A.W., Barnes, C.G., and Frost, C.D., 2006, this volume, Geology of the Bear Mountain intrusive complex, Klamath Mountains, California, *in* Snoke, A.W., and Barnes, C.G., eds., Geological studies in the Klamath Mountains province, California and Oregon: A volume in honor of William P. Irwin: Boulder, Colorado, Geological Society of America Special Paper 410, doi: 10.1130/2006.2410(14).

Chauvel, C., Arndt, N.T., Kielinzcuk, S., and Thom, A., 1987, Formation of Canadian 1.9 Ga old continental crust: Nd isotopic data: Canadian Journal of Earth Sciences, v. 24, p. 396–406.

Coleman, R.G., 1972, The Colebrook Schist of southwestern Oregon and its relation to the tectonic evolution of the region: Washington, D.C., U.S. Geological Survey Bulletin 1339, 61 p.

Coney, P.J., Jones, D.L., and Monger, J.W.H., 1980, Cordilleran suspect terranes: Nature, v. 288, p. 329–333, doi: 10.1038/288329a0.

Creaser, R.A., Erdmer, P., Stevens, R.A., and Grant, S.L., 1997, Tectonic affinity of Nisutlin and Anvil assemblage strata from the Teslin tectonic zone, northern Canadian Cordillera: Constraints from neodymium isotope and geochemical evidence: Tectonics, v. 16, p. 107–121, doi: 10.1029/96TC03317.

Davis, G.A., Monger, J.W.H., and Burchfiel, B.C., 1978, Mesozoic construction of the Cordilleran "collage," central British Columbia to central California, *in* Howell, D.G., and McDougall, K.A., eds., Mesozoic paleogeography of the western United States, Pacific Coast Paleogeography Symposium 2: Los Angeles, Pacific Section, Society of Economic Paleontologists and Mineralogists, p. 1–32.

Diller, J.S., 1907, The Mesozoic sediments of southwestern Oregon: American Journal of Science, fourth series, v. 23, p. 401–421.

Donato, M.M., Barnes, C.G., and Tomlinson, S.L., 1996, The enigmatic Applegate Group of southwestern Oregon: Age, correlation, and tectonic affinity: Oregon Geology, v. 58, p. 79–91.

Farmer, G.L., and Ball, T.T., 1997, Sources of Middle Proterozoic to Early Cambrian siliciclastic sedimentary rocks in the Great Basin: A Nd isotope study: Geological Society of America Bulletin, v. 109, p. 1193–1205, doi: 10.1130/0016-7606(1997)109<1193:SOMPTE>2.3.CO;2.

Frost, C.D., and Coombs, D.S., 1989, Nd isotope character of New Zealand sediments: Implications for terrane concepts and crustal evolution: American Journal of Science, v. 289, p. 744–770.

Frost, C.D., and Winston, D., 1987, Nd isotope systematics of coarse- and fine-grained sediments: Examples from the Middle Proterozoic belt—Purcell supergroup: Journal of Geology, v. 95, p. 309–327.

Garcia, M.O., 1979, Petrology of the Rogue and Galice Formations, Klamath Mountains, Oregon: Identification of a Jurassic island arc sequence: Journal of Geology, v. 86, p. 29–41.

Gehrels, G.E., and Miller, M.M., 2000, Detrital zircon geochronologic study of upper Paleozoic strata in the eastern Klamath terrane, northern California, *in* Soreghan, M.J., and Gehrels, G.E., eds., Paleozoic and Triassic paleogeography and tectonics of western Nevada and northern California: Boulder, Colorado, Geological Society of America Special Paper 347, p. 99–107.

Goldstein, S.L., O'Nions, R.K., and Hamilton, P.J., 1984, A Sm-Nd isotopic study of atmospheric dusts and particulates from major river systems: Earth and Planetary Science Letters, v. 70, p. 221–236, doi: 10.1016/0012-821X(84)90007-4.

Gribble, R.F., Barnes, C.G., Donato, M.M., Hoover, J.D., and Kistler, R.W., 1990, Geochemistry and intrusive history of the Ashland pluton, Klamath Mountains, California and Oregon: Journal of Petrology, v. 31, p. 883–923.

Harper, G.D., 1980, The Josephine ophiolite—Remains of a Late Jurassic marginal basin in northwestern California: Geology, v. 8, p. 333–337, doi: 10.1130/0091-7613(1980)<333:TJOOAL>2.0.CO;2.

Harper, G.D., 1983, A depositional contact between the Galice Formation and a Late Jurassic ophiolite in northwestern California and southwestern Oregon: Oregon Geology, v. 45, p. 3–7.

Harper, G.D., 1984, The Josephine ophiolite: Geological Society of America Bulletin, v. 95, p. 1009–1026, doi: 10.1130/0016-7606(1984)95<1009:TJONC>2.0.CO;2.

Harper, G.D., and Wright, J.E., 1984, Middle to Late Jurassic tectonic evolution of the Klamath Mountains, California-Oregon: Tectonics, v. 3, p. 759–772.

Harper, G.D., Saleeby, J.B., and Norman, E.A.S., 1985, Geometry and tectonic setting of sea-floor spreading for the Josephine ophiolite, and implications for Jurassic accretionary events along the California margin, *in* Howell, D.G., ed., Tectonostratigraphic terranes of the Circum-Pacific region: Houston, Circum-Pacific Council for Energy and Mineral Resources, Earth Science Series 1, p. 239–257.

Helper, M.A., 1986, Deformation and high P/T metamorphism in the central part of the Condrey Mountain window, north-central Klamath Mountains, California and Oregon, *in* Evans, B.W., and Brown, E.H., eds., Blueschists and eclogites: Boulder, Colorado, Geological Society of America Memoir 164, p. 125–141.

Hill, L.B., 1985, Metamorphic, deformational, and temporal constraints on terrane assembly, northern Klamath Mountains, California, *in* Howell, D.G., ed., Tectonostratigraphic terranes of the Circum-Pacific region: Houston, Circum-Pacific Council for Energy and Mineral Resources, Earth Science Series 1, p. 173–186.

Hotz, P.E., 1979, Regional metamorphism in the Condrey Mountain quadrangle, north-central Klamath Mountains, California: Reston, Virginia, U.S. Geological Survey Professional Paper 1086, 25 p.

Imlay, R.W., 1959, Succession and speciation of the pelecypod *Aucella:* Washington, D.C., U.S. Geological Survey Professional Paper 314-G, p. 155–169, pls. 16–19.

Imlay, R.W., 1980, Jurassic paleobiogeography of the conterminous United States in its continental setting: Reston, Virginia, U.S. Geological Survey Professional Paper 1062, 134 p.

Ingersoll, R.V., and Schweickert, R.A., 1986, A plate-tectonic model for Late Jurassic ophiolite genesis, Nevadan orogeny and forearc initiation, northern California: Tectonics, v. 5, p. 901–912.

Irwin, W.P., 1972, Terranes of the western Paleozoic and Triassic belt in the southern Klamath Mountains, California: Washington, D.C., U.S. Geological Survey Professional Paper 800–C, p. C103–C111.

Irwin, W.P., 1994, Geologic map of the Klamath Mountains, California and Oregon: Reston, Virginia, U.S. Geological Survey, Miscellaneous Investigations Map I-2148, scale 1:500,000, 2 sheets.

Irwin, W.P., and Wooden, J.L., 1999, Plutons and accretionary episodes of the Klamath Mountains, California and Oregon: Reston, Virginia, U.S. Geological Survey Open-File Report 99–374, 1 sheet.

Jones, C.E., Halliday, A.N., Rea, D.K., and Owen, R.M., 1994, Neodymium isotopic variations in North Pacific modern silicate sediment and the insignificance of detrital REE contributions to seawater: Earth and Planetary Science Letters, v. 127, p. 55–66, doi: 10.1016/0012-821X(94)90197-X.

Klein, C.W., 1975, Structure and petrology of a southeastern portion of the Happy Camp quadrangle, Siskiyou County, northwest California [Ph.D. dissertation]: Cambridge, Massachusetts, Harvard University, 288 p.

Klein, C.W., 1977, Thrust plates of the north-central Klamath Mountains near Happy Camp, California: California Division of Mines and Geology Special Report, v. 129, p. 23–26.

MacDonald, J.H., Jr., Harper, G.D., and Zhu, B., 2006, this volume, Petrology, geochemistry, and provenance of the Galice Formation, Klamath Mountains, Oregon and California, *in* Snoke, A.W., and Barnes, C.G., eds., Geological studies in the Klamath Mountains province, California and Oregon: A volume in honor of William P. Irwin: Boulder, Colorado, Geological Society of America Special Paper 410, doi: 10.1130/2006.2410(04).

Mankinen, E.A., Irwin, W.P., and Gromme, C.S., 1989, Paleomagnetic study of the eastern Klamath terrane, California, and implications for the tectonic history of the Klamath Mountains province: Journal of Geophysical Research, v. 94, p. 10,444–10,472.

Masi, U., O'Neil, J.R., and Kistler, R.W., 1981, Stable isotope systematics in Mesozoic granites of central and northern California and southwestern Oregon: Contributions to Mineralogy and Petrology, v. 76, p. 116–126, doi: 10.1007/BF00373691.

McLennan, S.M., McCulloch, M.T., Taylor, S.R., and Maynard, J.B., 1989, Effects of sedimentary sorting on neodymium isotopes in deep-sea turbidites: Nature, v. 337, p. 547–549, doi: 10.1038/337547a0.

McLennan, S.M., McCulloch, M.T., Taylor, S.R., and Maynard, J.B., 1990, Geochemical and Nd-Sr isotopic composition of deep-sea turbidites: Crustal evolution and plate tectonic associations: Geochimica et Cosmochimica Acta, v. 54, p. 2015–2050, doi: 10.1016/0016-7037(90 90269-Q.

Miller, J., Miller, R., Wooden, J., and Harper, G., 2003, Geochronologic links between the Ingalls ophiolite, North Cascades, Washington and the Josephine ophiolite, Klamath Mts., Oregon and California: Geological Society of America Abstracts with Programs, v. 35, p. 13.

Miller, M.M., and Saleeby, J.B., 1987, Detrital zircon studies of the Galice Formation: Common provenance of strata overlying the Josephine ophiolite and Rogue island arc, W. Klamath Mtns. terrane: Geological Society of America Abstracts with Programs, v. 19, p. 772–773.

Miller, M.M., and Saleeby, J.B., 1991, Continental detrital zircon in Carboniferous ensimatic arc rocks, Bragdon Formation, eastern Klamath terrane, northern California: Geological Society of America Bulletin, v. 103, p. 268–276, doi: 10.1130/0016-7606(1991)103<0268:CDZICE>2.3.CO;2.

Miller, M.M., and Saleeby, J.B., 1995, U-Pb geochronology of detrital zircon from Upper Jurassic synorogenic turbidites, Galice Formation, and related rocks, western Klamath Mountains: Correlation and Klamath Mountains provenance: Journal of Geophysical Research, v. 100, p. 18,045–18,058, doi: 10.1029/95JB00761.

Park-Jones, R., 1988, Sedimentology, structure and geochemistry of the Galice Formation: Sediment fill of a back-arc basin and island arc in the western Klamath Mountains [M.S. thesis]: Albany, State University of New York, 165 p.

Patchett, P.J., and Gehrels, G.E., 1998, Continental influence on Canadian Cordilleran terranes from Nd isotopic study, and significance for crustal growth processes: Journal of Geology, v. 106, p. 269–280.

Petersen, S.W., 1982, Geology and petrology around Titus Ridge, north-central Klamath Mountains, California [M.S. thesis]: Eugene, University of Oregon, 73 p.

Petersen, S.W., Barnes, C.G., and Hoover, J.D., 1991, The Billy's Peak mafic complex of the Trinity sheet, California: Roots of a Paleozoic island arc, *in* Cooper, J.D., and Stevens, C.H., eds., Paleozoic paleogeography of the western United States II: Los Angeles, Pacific Section, Society of Economic Paleontologists and Mineralogists, v. 67, p. 625–633.

Saleeby, J.B., and Harper, G.D., 1993, Tectonic relations between the Galice Formation and the Condrey Mountain Schist, Klamath Mountains, northern California, *in* Dunne, G.C., and McDougall, K.A., eds., Mesozoic paleogeography of the western United States II: Los Angeles, Pacific Section, Society of Economic Paleontologists and Mineralogists, v. 71, p. 61–80.

Saleeby, J.B., Harper, G.D., Snoke, A.W., and Sharp, W.D., 1982, Time relations and structural-stratigraphic patterns in ophiolite accretion, west-central Klamath Mountains, California: Journal of Geophysical Research, v. 87, p. 3831–3848.

Samson, S.D., and Patchett, P.J., 1991, The Canadian Cordillera as a modern analogue of Proterozoic crustal growth: Australian Journal of Earth Sciences, v. 38, p. 595–611.

Samson, S.D., Patchett, P.J., McClelland, W.C., and Gehrels, G.E., 1991, Nd isotopic characterization of metamorphic rocks in the Coast Mountains, Alaskan and Canadian Cordillera: Ancient crust bounded by juvenile terranes: Tectonics, v. 10, p. 770–780.

Schweickert, R.A., and Cowan, D.S., 1975, Early Mesozoic tectonic evolution of the western Sierra Nevada, California: Geological Society of America Bulletin, v. 86, p. 1329–1336, doi: 10.1130/0016-7606(1975)86<1329:EMTEOT>2.0.CO;2.

Secor, D.T., Jr., Samson, S.L., Snoke, A.W., and Palmer, A.R., 1983, Confirmation of the Carolina slate belt as an exotic terrane: Science, v. 221, p. 649–651.

Shaw, H.F., and Wasserburg, G.J., 1984, Isotopic constraints on the origin of Appalachian mafic complexes: American Journal of Science, v. 284, p. 319–329.

Silberling, N.J., Jones, D.L., Blake, M.C., Jr., and Howell, D.G., 1987, Lithotectonic terrane map of the western conterminous United States: Reston, Virginia, U.S. Geological Survey MF-1874-C, scale 1:2,500,000, with 20-p. pamphlet.

Snoke, A.W., 1977, A thrust plate of ophiolitic rocks in the Preston Peak area, Klamath Mountains, California: Geological Society of America Bulletin, v. 88, p. 1641–1659, doi: 10.1130/0016-7606(1977)88<1641:ATPOOR>2.0.CO;2.

Unterschutz, J.L.E., Creaser, R.A., Erdmer, P., Thompson, R.I., and Daughtry, K.L., 2002, North American margin origin of Quesnel terrane strata in the southern Canadian Cordillera: Inferences from geochemical and Nd isotopic characteristics of Triassic metasedimentary rocks: Geological Society of America Bulletin, v. 114, p. 462–475, doi: 10.1130/0016-7606(2002)114<0462:NAMOOQ>2.0.CO;2.

Wagner, D.L., and Saucedo, G.J., compilers, 1987, Geologic map of the Weed quadrangle, California: California Division of Mines and Geology Regional Geologic Map Series, Map 4A, scale 1:250,000.

Wallin, E.T., and Metcalf, R.V., 1998, Supra-subduction zone ophiolite formed in an extensional forearc: Trinity terrane, Klamath Mountains, California: Journal of Geology, v. 106, p. 591–608.

Wallin, E.T., Noto, R.C., and Gehrels, G.E., 2000, Provenance of the Antelope Mountain Quartzite, Yreka terrane, California: Evidence for large-scale late Paleozoic sinistral displacement along the North American Cordilleran margin and implications for the mid-Paleozoic fringing-arc model, *in* Soreghan, M.J., and Gehrels, G.E., eds., Paleozoic and Triassic paleogeography and tectonics of western Nevada and northern California: Boulder, Colorado, Geological Society of America Special Paper 347, p. 119–131.

Wright, J.E., and Fahan, M.R., 1988, An expanded view of Jurassic orogenesis in the western United States Cordillera: Middle Jurassic (pre-Nevadan) regional metamorphism and thrust faulting within an active arc environment, Klamath Mountains, California: Geological Society of America Bulletin, v. 100, p. 859–876, doi: 10.1130/0016-7606(1988)100<0859:AEVOJO>2.3.CO;2.

Wright, J.E., and Wyld, S.J., 1986, Significance of xenocrystic Precambrian zircon contained within the southern continuation of the Josephine ophiolite: Devils Elbow ophiolite remnant, Klamath Mountains, northern California: Geology, v. 14, p. 671–674, doi: 10.1130/0091-7613(1986)14<671:SOXPZC>2.0.CO;2.

Wright, J.E., and Wyld, S.J., 1994, The Rattlesnake Creek terrane, Klamath Mountains, California: An early Mesozoic volcanic arc and its basement of tectonically disrupted oceanic crust: Geological Society of America Bulletin, v. 106, p. 1033–1056, doi: 10.1130/0016-7606(1994)106<1033:TRCTKM>2.3.CO;2.

Wyld, S.J., and Wright, J.E., 1988, The Devils Elbow ophiolite remnant and overlying Galice Formation: New constraints on the Middle to Late Jurassic evolution of the Klamath Mountains, California: Geological Society of America Bulletin, v. 100, p. 29–44, doi: 10.1130/0016-7606(1988)100<0029:TDEORA>2.3.CO;2.

Wyld, S.J., and Wright, J.E., 2001, New evidence for Cretaceous strike-slip faulting in the United States Cordillera and implications for terrane-displacement, deformation patterns, and plutonism: American Journal of Science, v. 301, p. 150–181.

Yule, J.D., 1996, Geologic and tectonic evolution of Jurassic marginal ocean basin lithosphere, Klamath Mountains, Oregon [Ph.D. dissertation]: Pasadena, California Institute of Technology, 308 p.

Manuscript Accepted by the Society 10 January 2006

Printed in the USA

Geological Society of America
Special Paper 410
2006

Structure of syn-Nevadan dikes and their relationship to deformation of the Galice Formation, western Klamath terrane, northwestern California

Gregory D. Harper*
Department of Geological Sciences, State University of New York Albany, 1400 Washington Avenue, Albany, New York 12222, USA

ABSTRACT

The Galice Formation is characterized by slaty cleavage, overturned tight-to-isoclinal folds having variable hingeline orientations, and a south–southeast-trending stretching lineation formed during the Nevadan orogeny. Calc-alkaline dikes and sills (151–146 Ma) that intruded the Galice Formation and its basement (Josephine ophiolite) are regionally metamorphosed, and some are deformed; however, some plutons of this age also overprint slaty cleavage, suggesting syntectonic intrusion. Amoeboid margins on some sills suggest intrusion began prior to lithification of the Galice Formation. Some dikes are intruded into pre-existing small thrust faults that predate the slaty cleavage. Dikes show a wide range of orientations, and poles to dikes are consistently oriented at a high angle to poles to extension veins and to the stretching lineation in the Galice Formation. Poles to dikes define two quadrants on an equal-area, lower-hemisphere projection separated by planes oriented at right angles. These planes are analogous to nodal planes of a fault-plane solution, and thus allow determination of *P*- and *T*-axes. Restoration of structures to their original (Nevadan) orientation results in the *P*- and *T*-axes, stretching lineations, poles to extension veins, poles to small syn-cleavage faults, and poles to cleavage all being essentially coplanar with the "movement plane" that strikes to the northwest and dips steeply. The "fault-plane solution" derived from dike orientations indicates northwest-southeast contraction, consistent with slip directions for most small faults having slickenfibers. A wide range of fold hingeline orientations and slip directions on small pre-cleavage faults, however, may record early west-directed shortening.

Keywords: Upper Jurassic, Galice Formation, Nevadan orogeny, syn-tectonic dikes, western Klamath terrane, meta-igneous rocks

*E-mail: gdh@albany.edu.

Harper, G.D., 2006, Structure of syn-Nevadan dikes and their relationship to deformation of the Galice Formation, western Klamath terrane, northwestern California, *in* Snoke, A.W., and Barnes, C.G., eds., Geological studies in the Klamath Mountains province, California and Oregon: A volume in honor of William P. Irwin: Geological Society of America Special Paper 410, p. 121–140, doi: 10.1130/2006.2410(06). For permission to copy, contact editing@geosociety.org.

INTRODUCTION

The Nevadan orogeny involved large-scale thrust faulting and low-grade regional metamorphism in the western Klamath terrane (Fig. 1), affecting sedimentary rocks as young as Kimmeridgian (ca. 150 Ma). The end of the Nevadan orogeny is bracketed by clastic strata of the Great Valley Group (Myrtle Group in Oregon) as old as Valanginian (Fig. 2) that overlie the western Klamath terrane with great unconformity along the southern Klamath Mountains, in southwestern Oregon near Cave Junction (Fig. 1), and near the coast north of Gold Beach, Oregon (Irwin, 1966; Dott, 1971; Harper et al., 1994). The greatest amount of contraction was probably taken up along thrust faults, especially the Orleans thrust, which forms the roof of the western Klamath terrane (Fig. 1). Thrusting appears to have been essentially complete by ca. 150 Ma, but penetrative deformation continued to at least ca. 145 Ma and possibly until the end of regional metamorphism at ca. 135 Ma (Harper et al., 1994; Figs. 1 and 2). The Josephine ophiolite was emplaced during the Nevadan orogeny by being thrust, along with its

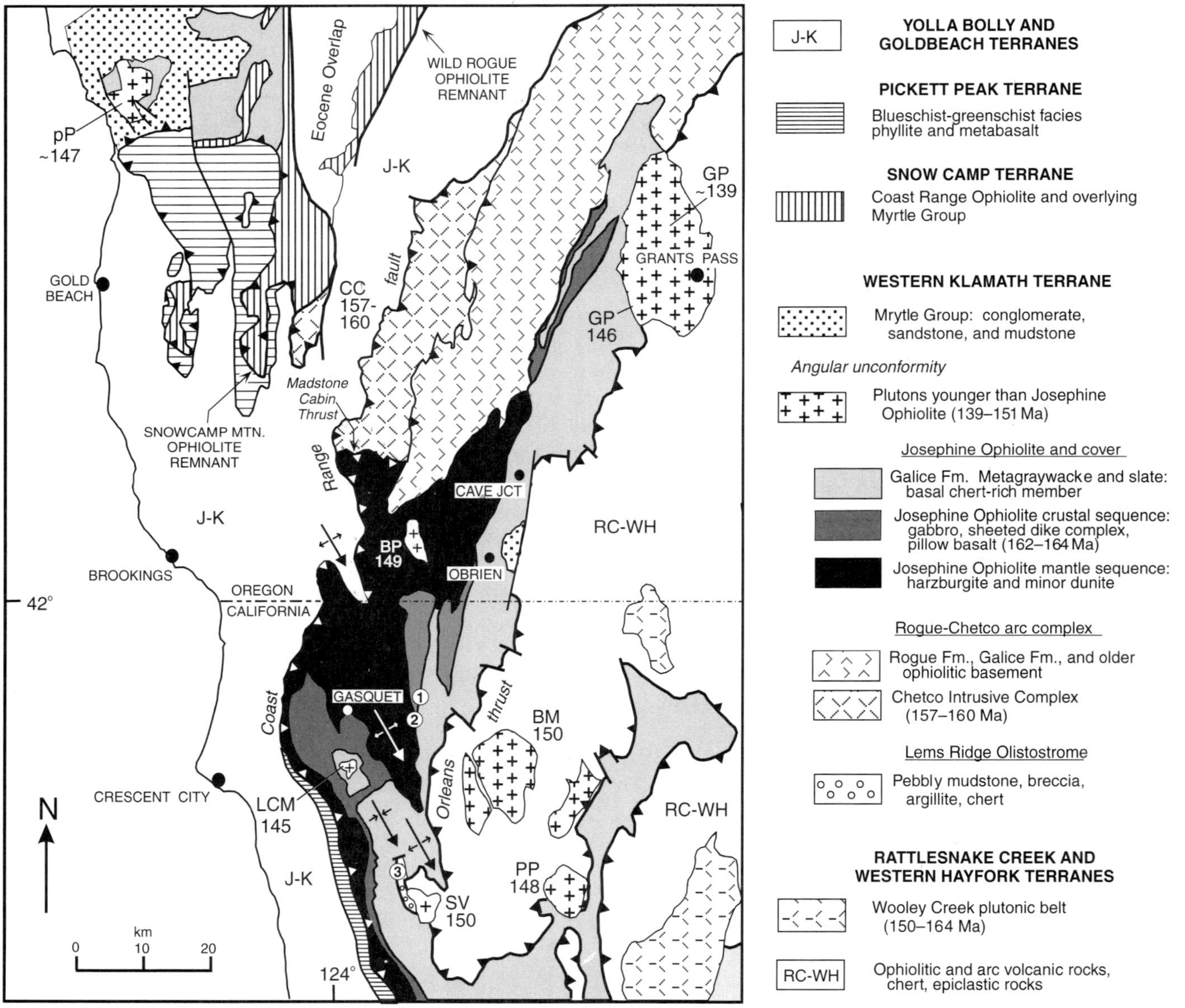

Figure 1. Generalized geologic map of the west-central Klamath Mountains. Circled numbers refer to localities discussed in the text. Abbreviations for plutons: BM—Bear Mountain; BP—Buckskin Peak; LCM—Lower Coon Mountain; GP—Grants Pass; PP—Pony Peak; pP—Pearse Peak; SV—Summit Valley. Ages for plutons are from Harper et al. (1994), Allen and Barnes (this volume), Barnes et al. (this volume, Chapter 17), Giaramita and Harper (this volume), Yule (1996), and Yule et al. (this volume). Map compiled from Dott (1971), Snoke (1977), Harper (1980, 1984), Saleeby et al. (1982), Norman (1984), Blake et al. (1985), and Yule (1996).

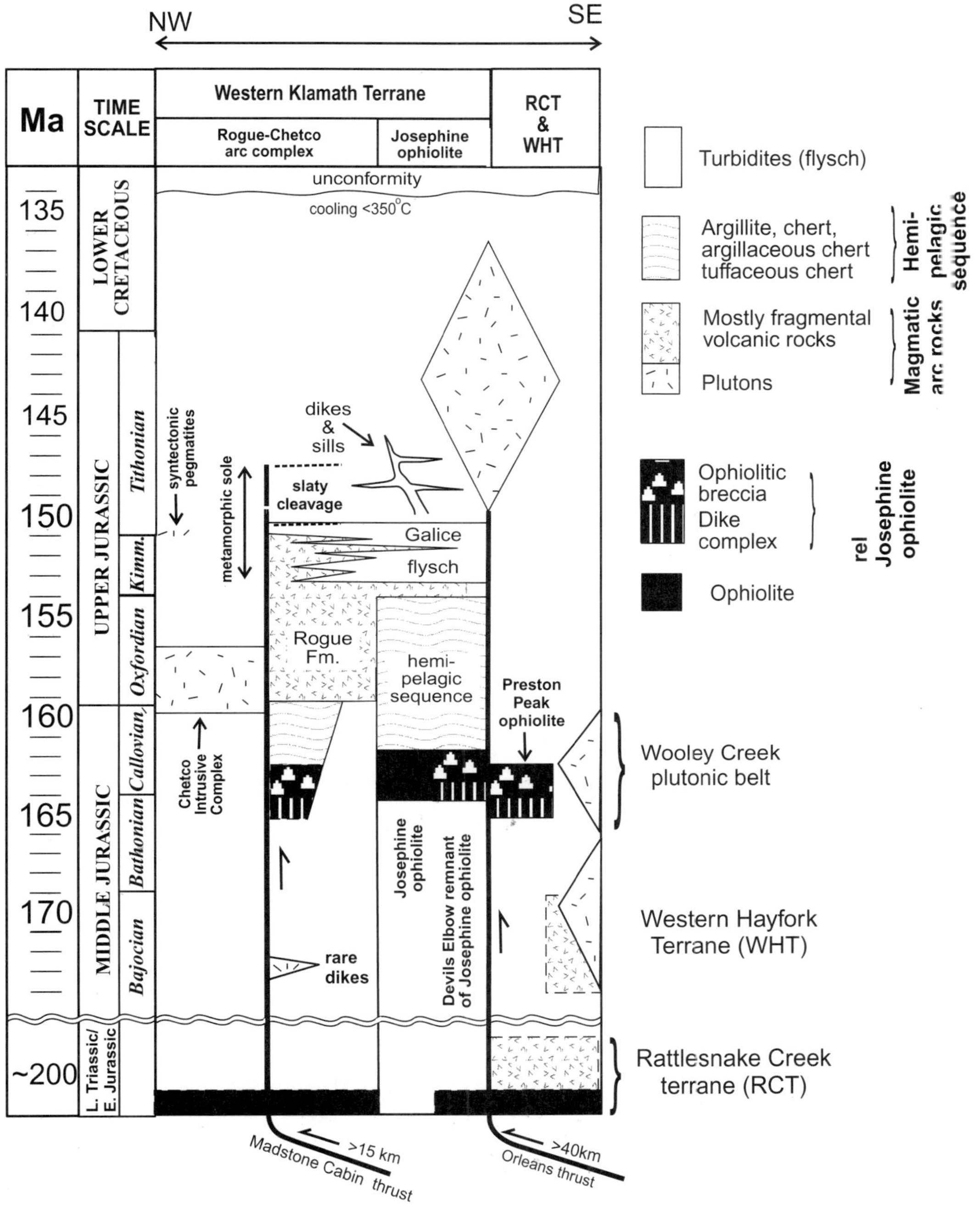

Figure 2. Summary of age data for the western Klamath terrane. Based on data and compilations in Harper et al. (1994), Hacker et al. (1995), Yule (1996), and Yule et al. (this volume).

sedimentary cover (Galice Formation), beneath older terranes of the Klamath Mountains. Underthrusting was accompanied by widespread cooling of structurally higher terranes at ca. 150 Ma (Hacker et al., 1995).

In a classic sense, orogenies were thought to include a culmination in pluton emplacement. Magmatism during the Nevadan orogeny in the Klamath Mountains, however, is volumetrically minor, although widespread, in the western Klamath Mountains. These include calc-alkaline dikes, sills, and small plutons ranging in age from ca. 151 to 144 Ma (Dick, 1976; Saleeby et al., 1982; Harper and Wright, 1984; Saleeby, 1984; Wyld, 1985; Ohr, 1987; Wright and Fahan, 1988; Harper et al., 1994; Hacker et al., 1995; Allen and Barnes, this volume; Chamberlain et al., this volume). Their ages and structural relations with the Galice Formation and thrust faults indicate they are syntectonic with respect to the Nevadan orogeny (Harper et al., 1994). Approximately 150-Ma syntectonic dikes also intrude the metamorphic "sole" beneath the basal (Madstone Cabin) thrust of the Josephine ophiolite (Harper et al., 1991, 1994, 1996; Figs. 1 and 2). The ca. 151- to 144-Ma intrusions are known only from the western Klamath Mountains, intruding either the western Klamath terrane or the structurally overlying Rattlesnake Creek terrane (Saleeby et al., 1982; Ohr, 1987; Hacker and Ernst, 1993; Harper et al., 1994; Hacker et al., 1995; Allen and Barnes, this volume; Barnes et al., this volume, Chapter 17; Fig. 1).

The syn-Nevadan intrusions have provided tight constraints on the age of deposition of the Galice Formation, timing of displacement along the Madstone Cabin and Orleans thrust faults, and regional metamorphism and deformation of the Galice Formation and its basement (Saleeby et al., 1982; Harper and Wright, 1984; Harper et al., 1994; Fig. 2). The purpose of this chapter is to describe structural relations between these intrusions, especially dikes and sills, and the Galice Formation, with a goal of better understanding the kinematics of Nevadan deformation in the Klamath Mountains.

REGIONAL GEOLOGY

The Klamath Mountains of northwestern California and southwestern Oregon consist of ophiolitic and island-arc terranes that young to the west and that are separated by thrust faults (e.g., Irwin, 1964, 1966, 1972, 1981, 1994; Burchfiel and Davis, 1981). One of the westernmost of these terranes is the western Klamath terrane (Fig. 1), originally named the "western Jurassic belt" by Irwin (1966), but renamed by Blake et al. (1985), who divided it into several subterranes. The western Klamath terrane consists of two tectonic elements, a Late Jurassic island-arc complex (Garcia, 1982; Harper and Wright, 1984; Riley, 1987; Yule, 1996), referred to here as the Rogue–Chetco arc complex, and the ca. 162- to 164-Ma Josephine ophiolite (Harper, 1984; Wyld and Wright, 1988; Harper et al., 1994). The Chetco complex (Hotz, 1971), also called the "Illinois River plutonic complex" (Yule, 1996) or "Illinois batholith" (Dick, 1976), forms the plutonic root of the Rogue–Chetco arc complex and has been dated by U-Pb zircon at 160 ± 1 to 157 ± 1 Ma (Yule, 1996; Yule et al., this volume; Fig. 2). The Rogue Formation, which has radiometric ages ranging from 157 to 153 Ma (Fig. 2), forms the volcanic part of the arc, although because it is predominantly volcaniclastic and conformably overlain by flysch, it probably represents a basinal fringe of the arc (Yule, 1996). The Rogue Formation unconformably overlies early Mesozoic ophiolitic rocks (Figs. 1 and 2), which have been correlated with the Rattlesnake Creek terrane that structurally overlies the western Klamath terrane. Similar rocks occur as country rock within the Chetco complex (Yule, 1996; Yule et al., this volume).

Both the Rogue Formation and the Josephine ophiolite (and its hemipelagic sedimentary cover) are conformably overlain by Late Jurassic (late Oxfordian to early Kimmeridgian) flysch of the Galice Formation (Wells and Walker, 1953; Harper, 1984; Figs. 1 and 2). Clast compositions suggest the flysch was derived from a volcanic arc and basement rocks of the Klamath Mountains (Snoke, 1977; Harper, 1980, 1983, 1984; MacDonald et al., this volume). Harper et al. (1994) suggested deposition of the Galice flysch was synorogenic, having been deposited in a trenchlike basin in front of the advancing Orleans thrust sheet. The Galice Formation contains volcaniclastic members in both the northern (Wells and Walker, 1953; Garcia, 1982; Park-Jones, 1988) and southern (Wyld and Wright, 1988) parts of the western Klamath terrane. Detrital zircon age populations have two distinct age peaks at ca. 153 and ca. 223 Ma (Miller et al., 2003), with a minor Paleoproterozoic component (Miller and Saleeby, 1995; Miller et al., 2003). The hemipelagic sequence overlying the Josephine ophiolite, as well as sediments locally intercalated with pillow lavas of the ophiolite, are hemipelagic rocks consisting of radiolarian argillite and argillaceous and tuffaceous chert (Pinto-Auso and Harper, 1985; MacDonald et al., this volume). In addition, a graywacke similar in composition to those of the Galice flysch is locally intercalated with pillow lavas (MacDonald et al., this volume).

The southernmost exposure of the Josephine ophiolite, the Devils Elbow remnant, appears to be built on older Klamath basement correlative with the Rattlesnake Creek terrane and has been interpreted as a rift-edge facies (Wyld and Wright, 1988; Figs. 1 and 2). Other rift-edge facies occur within the basement of Rogue Formation (Yule, 1996; Yule et al., this volume) and within the Rattlesnake Creek terrane structurally overlying the Galice Formation in northwestern California (Preston Peak ophiolite of Snoke, 1977; Saleeby et al., 1982; Fig. 2).

The Rogue Formation, Josephine ophiolite, and Galice Formation were regionally metamorphosed under conditions of the prehnite-pumpellyite to lower greenschist facies (Snoke, 1977; Harper, 1980; Garcia, 1982; Harper et al., 1988; Wyld and Wright, 1988). Some of the basement rocks of the Rogue–Chetco arc (e.g., Briggs Creek amphibolite) have been metamorphosed to amphibolite facies, and many of the intrusive rocks of the Chetco complex have deformation fabrics formed under amphibolite-facies conditions overprinted by mineral assemblages characteristic of the prehnite-pumpellyite to lower greenschist facies (Wells and Walker, 1953; Dick, 1976; Coleman and Lanphere, 1991; Harper et al., 1991; Yule, 1996).

The Orleans thrust forms the roof of the western Klamath terrane (Fig. 1). Its outcrop pattern indicates a minimum of 40-km displacement (Fig. 1), and gravity data suggest possibly as much as 110 km of offset (Jachens et al., 1986). The thrust is cut by the ca. 150- to 144-Ma Summit Valley and ca. 148-Ma Pony Peak plutons (Fig. 1), and the former contact metamorphosed both the upper and lower plates (Harper et al., 1994). Displacement on the thrust has been interpreted to be top-to-the-west (Irwin, 1966, 1981; Snoke, 1977; Gray, 1985, this volume; Wyld, 1985; Jones, 1988), and tectonic burial by underthrusting was apparently the cause of low-grade regional metamorphism in the western Klamath terrane (Harper et al., 1994). The basal (Madstone Cabin) thrust of the Josephine ophiolite is also

Nevadan in age (Dick, 1976; Harper et al., 1994) and apparently represents underthrusting of the hot intrusive rocks of the Chetco complex beneath the relatively cold Josephine ophiolite (Harper et al., 1996). Displacement on the basal thrust was synchronous with retrograde metamorphism of a metamorphic sole from amphibolite to lower greenschist facies (~500° and ~300 °C cooling ages of ca. 152–146 Ma, respectively), and the amphibolite sole beneath the thrust is intruded by sparse ca. 150-Ma late syntectonic pegmatite dikes (Harper et al., 1991, 1994, 1996; Fig. 2). Displacement on the Madstone Cabin thrust was top-to-the-northeast (Cannat and Boudier, 1985; Harper et al., 1991). This direction of displacement is at a high angle to that for the Orleans thrust, even after allowing for different amounts of oroclinal bending between northernmost California and southwestern Oregon as discussed below.

The upper plate of the Orleans thrust is the Rattlesnake Creek terrane, consisting of disrupted ophiolitic rocks overlain by a sequence of Late Triassic to Lower Jurassic volcanic, pelagic, and epiclastic rocks (Wright and Wyld, 1994). The basal ophiolitic rocks and overlying stratified sequence are interpreted to have formed in an intraoceanic fracture zone and fringing island arc, respectively (Wright and Wyld, 1994). The Rattlesnake Creek terrane is overlain by a mid-Jurassic arc complex, consisting of the western Hayfork terrane (Figs. 1 and 2) and related plutons, which was built on the Rattlesnake Creek terrane and affected by a ca. 165-Ma orogenic event (Wright and Fahan, 1988; Hacker et al., 1995). Subsequently, the 170- to 156-Ma Wooley Creek plutonic belt (Allen and Barnes, this volume) was intruded, overlapping in time with formation of the Josephine ophiolite (Fig. 2).

The position of the ca. 162- to 164-Ma Josephine ophiolite relative to the ca. 160- to 153-Ma Rogue–Chetco arc complex and the mid-Jurassic western Hayfork arc complex suggest that the Josephine ophiolite formed by rifting and back-arc spreading of an arc built on the western margin of North America, leaving the Hayfork arc behind as a remnant arc (Harper and Wright, 1984). The recognition of rocks correlative with the Rattlesnake Creek terrane as basement for the Rogue–Chetco arc (Fig. 2) strongly supports such a model (Yule et al., 1992, this volume; Yule, 1996). A variation on this model that is consistent with an inferred east-west orientation of the spreading axis for the Josephine ophiolite (Alexander and Harper, 1992) involves rifting and ophiolite generation along an intra-arc transform-ridge system, similar to the modern Andaman Sea (Harper et al., 1985; Saleeby, 1992). According to the current age data summarized in Figure 2, the Rogue–Chetco arc magmatism is slightly younger than the Josephine ophiolite, and Wooley Creek belt magmatism overlapped in time with Josephine spreading. This timing and a near absence of mid-Jurassic arc rocks in the Rogue–Chetco arc complex (Yule, 1996; Fig. 2) suggest that arc rifting and initial seafloor spreading occurred in the fore-arc. These events were apparently followed by a trenchward jump in the arc axis at ca. 160 Ma to form the Rogue–Chetco arc. This model is similar to that for the evolution of the modern Lau back-arc basin and Tonga arc in the southwest Pacific (e.g., Hawkins, 1995; Parson and Wright, 1996). Highly fractionated Fe- and Ti-rich basalts are present in the Josephine ophiolite and most likely formed by propagating spreading centers, also analogous to the modern Lau Basin and other back-arc basins (Pearce et al., 1994; Parson and Wright, 1996; Harper, 2003). The influx of turbidites (flysch) onto the Josephine ophiolite and Rogue–Chetco arc complex probably signals the onset of the Nevadan orogeny at ca. 155 Ma (Harper et al., 1994).

No intrusive rocks are known in the Josephine ophiolite or overlying Galice Formation between the time of ophiolite formation at ca. 162 Ma and intrusion of the ca. 151- to 145-Ma calc-alkaline dikes, sills, and small plutons. The beginning of intrusion of the latter was interpreted by Harper (1984) and Harper and Wright (1984) to be the result of thrusting of the Rogue–Chetco arc beneath the Josephine ophiolite, resulting in the ophiolite being situated over the axis of arc magmatism.

PETROGRAPHY

Dikes and Sills

Dikes and sills mostly range in thickness from ~12 cm to 2 m. Most are mafic to intermediate in composition, and they range in color from dark gray to medium grayish green. They are distinguished in the field by the presence of hornblende phenocrysts and/or acicular hornblende in the groundmass, in contrast to dikes of the Josephine ophiolite, which are nearly all aphyric and do not contain igneous hornblende. Grain size, shape and size of phenocrysts, and percentage of phenocrysts are highly variable. Most of the mafic to intermediate dikes and sills are porphyritic, typically with 10–25% phenocrysts, although a few are aphyric. The groundmass of most samples is highly altered, but relict groundmass textures are typically very fine grained and hypidiomorphic. Mafic samples have augite ± plagioclase (altered) ± hornblende ± orthopyroxene(?) pseudomorphs as phenocrysts, and brown hornblende, augite, and rarely relict calcic plagioclase (An_{65-70}) in the groundmass. In intermediate dikes and sills, the dominant phenocryst phase is brown hornblende, but clinopyroxene or plagioclase phenocrysts are present in some samples, and the groundmass contains relict brown hornblende, altered plagioclase, and up to 10% quartz. Felsic dikes are much less common and are readily recognized in the field by their light-green or light-gray color. They are porphyritic, with 1–30% phenocrysts consisting primarily of altered feldspar. Neither K-feldspar nor quartz was observed as phenocrysts. Groundmasses consist dominantly of anhedral quartz and anhedral to subhedral albite. All of the sampled felsic dikes are extensively altered.

Metamorphic mineral assemblages in the dikes and sills include chlorite, epidote, albite ± actinolite ± pumpellyite ± prehnite ± quartz, except in the southern part of the area of Figure 1, where pumpellyite and prehnite are absent. Many of the dikes that cut serpentinite in the Josephine ophiolite also contain sodic metamorphic minerals that, along with unusually high Na contents (Barnes et al., this volume, Chapter 17), suggest sodic metasomatism. Minerals identified in these unusual Na-rich rocks by optical microscope, electron microprobe, and X-ray diffraction include needles of Na-amphibole in the groundmass, the zeolites natrolite and cancrinite in the groundmass and in veins, and, in one sample, pectolite (Na-Ca pyroxenoid) fibers filling fractures of hornblende porphyroclasts. Fe-rich ophiolitic dikes in serpentinite also contain Na-Ca amphibole (Harper, 2003).

One light-gray sill in the basal Galice Formation has plagioclase phenocrysts (An_{58-60}) in a groundmass that is almost entirely replaced by calcite. This type of metasomatism of igneous dikes, probably caused by introduction of CO_2-rich fluids from the enclosing pelitic rocks during regional metamorphism, may be the origin of deformed calcite-quartz-chlorite dikes that are common in the Galice Formation (Snoke, 1972, 1977; Harper, 1980).

Veins consisting of fibrous quartz and calcite ± chlorite, and rarely of quartz + epidote + actinolite, are common in dikes and sills, and locally in the pillow lavas of the Josephine ophiolite. Similar style veins also occur in serpentinites of the Josephine ophiolite, consisting of serpentine ± dolomite (Coulton et al., 1995). These veins are considered to have formed as extension fractures during regional metamorphism and associated penetrative deformation for the following reasons: their orientation is normal to the stretching lineation in the Galice Formation, fibers are oriented normal to vein walls and have "crack-seal" morphology (Ramsay, 1980) in thin section, actinolite and epidote are present in some veins, and fluid inclusions indicate formation at ~300 °C (Milby, 1990). These veins are referred to below as "extension veins."

Plutonic Rocks

Plutons related to the Nevadan dikes and sills are highly variable in composition. A complete petrographic description is beyond the scope of this chapter, but the more salient features are discussed for some of the plutons. The subhorizontal, ≥300-m-thick, sill-like Lower Coon Mountain pluton intrudes the Galice Formation (Harper, 1980; Fig. 1). It consists predominantly (>90%) of ultramafic rocks, mostly olivine ± plagioclase clinopyroxenite, along with less abundant olivine two-pyroxene gabbro, and hornblende diorite. Rare undeformed granitic dikes intrude the Galice Formation along the margins of the pluton and have yielded U-Pb zircon ages of ca. 142 and 145 Ma (Saleeby et al., 1982; Harper et al., 1994). Ultramafic rocks are exposed from the base to the roof of the pluton, suggesting that it is a cumulate formed by filter pressing. No igneous layering or solid-state deformation fabric was observed in the pluton, and it has a narrow contact aureole consisting of pelitic hornfels with pseudomorphs after andalusite (replaced by muscovite) that overprints slaty cleavage in the Galice Formation. Partial recrystallization of the plutonic rocks to low-grade assemblages and retrograde metamorphism of the hornfels (Harper, 1980), along with a 135-Ma K-feldspar cooling (uplift) age (Harper et al., 1994) suggest the pluton was intruded during Nevadan regional metamorphism (ca. 150–135 Ma).

The Summit Valley plutonic complex, which intrudes the Orleans thrust (Fig. 1), is very heterogeneous, and shows complex crosscutting intrusive relationships (Griesau, 1992; Fig. 3A). A U-Pb zircon age of ca. 150 Ma and $^{40}Ar/^{39}Ar$ hornblende cooling age of ca. 144 Ma suggest the Summit Valley plutonic complex was formed over an interval of perhaps as much as 6 m.y. (Harper et al., 1994). The plutonic complex consists dominantly of hornblende gabbro, hornblendite, and olivine clinopyroxenite (Norman, 1984). Both the hanging wall and footwall of the Orleans thrust, consisting of serpentinite of the Rattlesnake Creek terrane and pebbly mudstone of the Lems Ridge olistostrome, respectively, show contact metamorphism. Although this plutonic complex pins the Orleans thrust, hypersolidus and subsolidus deformation fabrics are common, some of which are cut by dikes that are themselves deformed (Griesau, 1992; Fig. 3B). High-temperature shear zones up to 300-m wide consisting of banded mylonite (e.g., Fig. 3B) are present in the plutonic complex and consist of polygonal clinopyroxene, brown hornblende, and plagioclase (Griesau, 1992). Epidote and prehnite veins having slickenfibers are common and are interpreted by Griesau (1992) and Griesau and Harper (1992) to have formed during cooling of the complex to the temperature of the wall rocks, which were undergoing lower greenschist-facies regional metamorphism (Griesau, 1992). These veins, as well as high-temperature shear zones, are highly variable in orientation and sense of shear, with no simple geometric relationship to the Orleans thrust. Paleostress analysis, however, yielded consistent paleostress directions except that σ_2 and σ_3 paleostress directions are interchanged for earlier (high-T) and later (low-T) sets of shear zone data (Griesau, 1992). The 150-Ma zircon age of the Summit Valley plutonic complex was taken as a younger age bracket for displacement on the Orleans thrust. Because the many intrusive phases of the pluton overlapped high-temperature deformation, however, it is possible that the pluton was intruded during thrusting, and that contact metamorphism of the upper and lower plate occurred sometime after 150 Ma but before cooling below ~500 °C at 144 Ma.

The ca. 148- to 145-Ma Pony Peak pluton (Harper et al., 1994; Allen and Barnes, this volume), which also cuts the Orleans thrust (Fig. 1), is predominantly tonalite but ranges from gabbro to granodiorite. Syntectonic intrusion is suggested by deformation of an early phase of the pluton and a strongly foliated contact aureole in the Galice Formation (Harper et al., 1994). The ca. 149-Ma Buckskin Peak pluton, which intrudes peridotite of the Josephine ophiolite (Fig. 1), is a biotite-hornblende quartz diorite (Harper et al., 1994) but has received little study since being mapped by Wells et al. (1949). The ca. 146 to 139-Ma Grants Pass pluton (Fig. 1) has a dynamic contact aureole that overprints slaty cleavage in the Galice Formation (Harper et al., 1994; Yule, 1996).

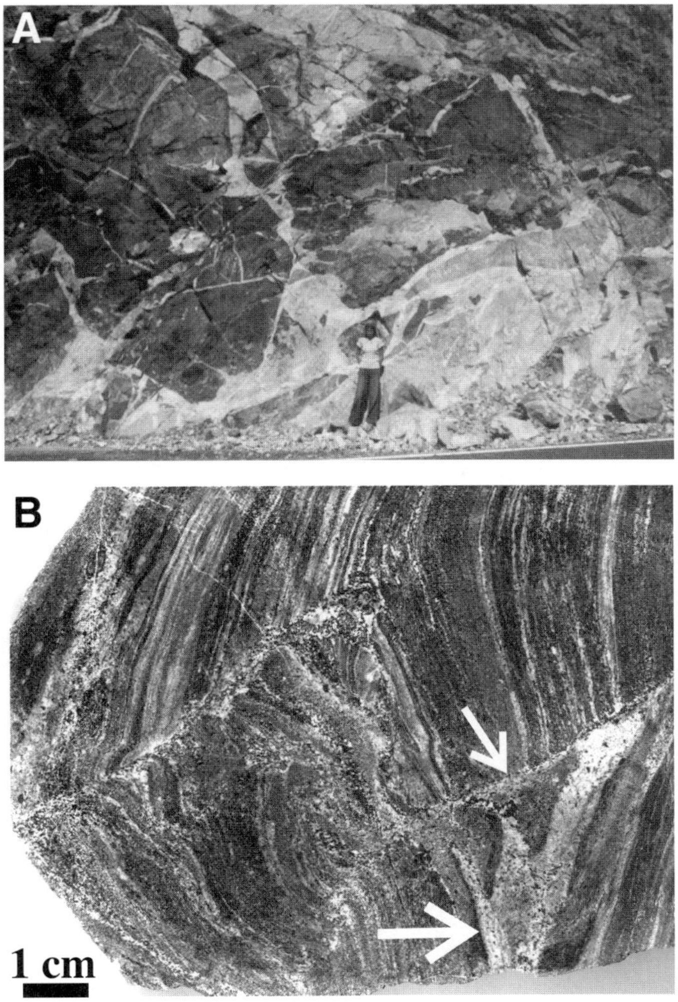

Figure 3. (A) Road-cut exposure of the Summit Valley plutonic complex showing complex intrusive relationships. Four phases are evident, each progressively lighter colored. (B) Mylonitic metadiorite from a 300-m-wide ductile shear zone in the Summit Valley plutonic complex showing evidence that intrusion overlapped in time with deformation. Undeformed diorite in pocket (large arrow) is bound by a small fault intruded by similar diorite (small arrow), suggesting that intrusion overlapped in time with faulting. Also note thin diorite dikelets and pockets oriented at low angles to mylonitic foliation.

STRUCTURE

Galice Formation

Nevadan deformation is most extensive in the Galice Formation. This unit is characterized by a generally steeply east-dipping slaty cleavage. A faint lineation is commonly evident on the cleavage plane, which in thin section is defined by elongated grains and fibers around pyrite framboids. This observation, along with the parallelism of the lineation with the long axis of grains in intercalated metagraywackes and pebble conglomerates, indicates that the lineation is a stretching lineation. The lineation plunges gently to moderately to the southeast throughout the study area.

Veins are common in dikes and sills that intrude the Galice Formation. They consist of fibers oriented normal to vein walls and showing internal banding parallel to the walls. These are characteristics of crack-seal veins, which are generally believed to record incremental strain through repeated hydraulic fracturing and mineral precipitation (Ramsay, 1980). Fibers recording incremental strain are also developed around spherical pyrite framboids in Galice slates. Within the plane of the cleavage, the macroscopic lineation is parallel to fibers around pyrite framboids, indicating that the direction of stretching remained constant during cleavage development (Jones, 1988).

Folds associated with the slaty cleavage (F_1) are tight to isoclinal and overturned. Orientations of lineations formed by cleavage-bedding intersection indicate that fold hinges are variable in orientation, although gentle to moderate plunges predominate (Kays, 1968; Snoke, 1977; Harper 1980; Norman, 1984; Gray, 1985, this volume; Jones, 1988). Slaty cleavage is generally axial planar, but it was observed to be slightly oblique to axial surfaces of F_1 folds (i.e., transected folds) along Rough and Ready Creek, between Cave Junction and O'Brien, Oregon (Fig. 1), where slaty cleavage is weakly developed. In the Cave Junction area, where strain is relatively low, many outcrops of the Galice Formation consist of broken formation along with a small amount of mélange (graywacke blocks in sheared argillite; Jones, 1988).

Strain in the Galice Formation increases with metamorphic grade southward from the area of Cave Junction, Oregon, to the southernmost part of the area shown in Figure 1 (Harper, 1980; Harper et al., 1988; Jones, 1988). Strain in the upper part of the pillow lava unit of the ophiolite also increases progressively southward from unstrained near the Oregon-California border, to flattened pillow breccias and weakly flattened pillow rims at locality 1 (Fig. 1), to highly flattened and locally phyllitic pillow lavas west and southwest of locality 3 (Fig. 1). In the Cave Junction area, slaty cleavage is generally weakly developed, and strains measured using fibers around pyrite framboids in sections cut both perpendicular and parallel to slaty cleavage indicate ≤33% flattening in slates; strains recorded by fibers around framboids are true strains unaffected by volume loss (Jones, 1988). Farther south, in the area east of Gasquet, California, slaty cleavage is well developed, and flattening in slates averages ~68%. Clasts in metagraywackes, pebble conglomerates, and volcanic breccias have shapes characteristic of S–L tectonites (Harper, 1980; Cashman, 1988) and thus apparently do not record plane strain as in the slates. Apparent S–L tectonites can form, however, by a combination of plane strain and volume loss (e.g., Ramsay and Wood, 1973); this possibility is considered likely in Galice metasedimentary rocks because textures in metagraywackes and pebble conglomerates suggest volume loss by pressure solution (Harper, 1980). Strain in the Galice Formation has also been reported to increase with proximity to the Orleans thrust, where slaty cleavage has, in places, been observed to intensify and progressively rotate toward parallelism with the thrust (Snoke, 1972, 1977; Gray, 1985, this volume; Wyld, 1985; Cashman, 1988).

Strain measured using fibers around pyrite framboids indicate that strain recorded by slates is coaxial (Jones, 1988). With one exception, incremental strain histories recorded by the fibers are similar for both upright and overturned limbs of folds studied by Jones (1988), indicating that most folding predated cleavage formation. This is also a likely explanation for the origin of transected folds in which slaty cleavage is at an angle to the axial surface of folds.

The Galice Formation, where it overlies the Josephine ophiolite (i.e., not in its type area west of Grants Pass, Oregon), is characterized by bedding-parallel slaty cleavage and infrequent outcrop- and map-scale overturned folds (Snoke, 1972; 1977; Harper, 1980; Norman, 1984; Gray, 1985, this volume; Wyld, 1985; Jones, 1988). Bedding and cleavage are commonly subparallel, and the location of map-scale folds is commonly only evident by the change from upright to overturned bedding in an otherwise homoclinal section. Where graded bedding is well preserved in areas of low strain, the hinges of folds are commonly faulted (Jones, 1988). Furthermore, outcrop-scale, hook-shaped folds with short faulted overturned limbs are common in the otherwise upright homoclinal metagraywacke unit in the lower part of the Galice Formation northeast of Gasquet (locality 1, Fig. 1) that is the focus of this chapter. Overall, the Galice Formation overlying the Josephine ophiolite is upright, with only ~15% of beds overturned (Jones, 1988).

The structure of the type Galice Formation in the Rogue River area, west of Grants Pass, Oregon (Fig. 1), is different in that map-scale overturned folds are common and systematic changes in cleavage-bedding relationships occur across the folds (Park-Jones, 1988). In addition, no outcrops of broken formation or mélange have been observed. The type Galice Formation is much farther from the Orleans thrust and was thus perhaps not as strongly affected by deformation associated with regional thrust faulting.

Folding of the slaty cleavage (F_2) in the Galice Formation is widespread and is locally associated with a crenulation cleavage (Harper, 1980; Norman, 1984; Gray, 1985, this volume; Wyld, 1985; Cashman, 1988; Jones, 1988). The south-southeast trend and gentle plunge of F_2 fold hinges, predominantly east-dipping axial surfaces, and reverse faults associated with the F_2 folds indicate a westward vergence (Harper, 1980; Norman, 1984; Gray, 1985, this volume; Jones, 1988). Some of the F_2 folds are reportedly cut by dikes similar to those discussed in this chapter (Gray, 1985, this volume). This cutting and neocrystallization of mica along the crenulation cleavage (Gray, 1985, this volume; Cashman, 1988; Jones, 1988) suggest that deformation associated with F_2 folding occurred during the later stages of the Nevadan orogeny (Gray, 1985, this volume; Harper et al., 1994).

Large-scale folds of the Josephine-Galice thrust sheet (Fig. 1), and locally of the Orleans thrust (Norman, 1984), have axes and axial surfaces similar in orientation to macroscopic F_2 folds and thus probably formed during the same deformation. The large-scale folds, however, appear to affect the fault contact between the Josephine ophiolite and underlying uppermost Jurassic to Lower Cretaceous Franciscan Complex (Dothan Formation of the Yolla Bolly terrane; Blake et al., 1985) near the Oregon-California border (Wells et al., 1949; Fig. 1). This suggests that at least some of the large-scale folds may be post-Nevadan, perhaps Cretaceous or early Tertiary in age (Harper, 1980).

Post-Nevadan open folding about a subhorizontal east–northeast-trending axis is recorded in Lower Cretaceous sedimentary rocks that unconformably overlie the Galice Formation near the Oregon-California border; dips are generally <15° but range up to ~30° (Jones, 1988). Post-Nevadan folding (F_3) is also evident from variations in the strike of the slaty cleavage and trend of F_1 and F_2 fold hinges (Harper, 1980; Norman, 1984; Jones, 1988). The large reentrant of the Orleans thrust (Fig. 1) may also be related to this folding. This late folding, as well as strike-slip displacement on cross faults, may be related to oroclinal bending of the Klamath Mountains that has been inferred from paleomagnetic data (e.g., Renne and Scott, 1988).

Most of the observations discussed in this chapter are based on a detailed study of the basal ~300 m of the Galice Formation exposed on the Middle Fork of the Smith River above the type section of the pillow lavas of the Josephine ophiolite near the mouth of Little Jones Creek, ~11 km east-northeast of Gasquet, California (Harper, 1983, 1984, 2003; Pinto-Auso and Harper, 1985; locality 1, Fig. 1). An advantage of studying the structure of this section is that it is homoclinal, well exposed, and nearly free of F_2 folds. Slaty cleavage in this section is oriented at ≤20° to bedding, dipping more steeply than bedding except in hinges of rare small folds in the basal hemipelagic sequence (Fig. 4) and in the hinges of meter-scale overturned hook-shaped folds above small faults in the overlying massive metagraywacke unit. Fluid inclusion data from quartz in an extension vein in pillow lavas at this locality, formed during regional metamorphism, indicate a temperature of ~330 °C, assuming a pressure of 2 kb (Milby, 1990). This temperature is consistent with prehnite-pumpellyite–facies mineral assemblages in syn-Nevadan dikes from this locality (Harper et al., 1988; Harper, 1995).

Structure of Syn-Nevadan Dikes and Sills

Syn-Nevadan dikes and sills are widespread and intrude the Galice Formation along the length of the western Klamath terrane as well as the Lems Ridge olistostrome, Josephine ophiolite, and Rogue Formation (Dick, 1976; Harper, 1984; Harper and Wright, 1984; Saleeby, 1984; Harper et al., 1985, 1994; Gray, 1985, this volume; Wyld, 1985; Ohr, 1987; Figs. 4–10). Dikes and sills commonly form swarms in the peridotite unit of the ophiolite, the basal Galice Formation, and the Lems Ridge olistostrome (Fig. 5). Syn-Nevadan dikes cut many sheared ser-

Figure 4. Two sills (light color) within basal hemipelagic sequence at locality 1 (Fig. 1). The lower sill shows boudinage with considerable thinning in a neck; the upper sill has also been boudinaged, but to a lesser extent (not evident in this photo). Note asymmetric fold in thin bed below hammer (upper left) showing vergence to the left. Bedding dips steeply to the upper left (east). Lower sill was dated by Harper et al. (1994) at 150 ± 1 Ma, and an analysis of a sample from this sill (D24) is given by Barnes et al. (this volume, Chapter 17).

pentinite fault zones in the Josephine peridotite, indicating that much of the serpentinization is relatively old (Coulton et al., 1995). Dikes and sills are far more abundant in the eastern part of the Josephine ophiolite and overlying Galice Formation than in the western half (Dick, 1976; Harper, 1980).

The amount of Nevadan deformation of the sills and dikes varies with the amount of strain in the Galice Formation, which increases southward from the Cave Junction, Oregon, area. This variation is apparently the result of increasing temperature, because regional metamorphic grade also increases southward (Harper, 1980; Harper et al., 1988). Sills and dikes in the vicinity of the Oregon border are undeformed; however, south of the state border, where slaty cleavage is well developed, these intrusive bodies are cut by fibrous extension veins (e.g., Fig. 6), and ductile deformation is evident in boudin necks (Fig. 4) and along some faults (Fig. 7). Some dikes south of Gasquet, California, where the metamorphic grade is higher (lower greenschist facies), are well foliated, including a felsic dike that yielded a U-Pb (zircon) age of 151 ± 3 Ma (sample C23; Saleeby et al., 1982; Barnes et al., this volume, Chapter 17). A >10 m-thick dike intruding foliated serpentinite in the Josephine Peridotite ~10 km north-northeast of Gasquet is, in places, mylonitic (Fig. 10).

Dikes, in general, are commonly used to infer the direction of regional extension, with the assumption that walls of the dikes are oriented normal to the direction of extension. Dikes commonly open obliquely, however, such as when they are intruded along pre-existing surfaces. Oblique extension must also have occurred along steps in dikes, which commonly form by propagation of en echelon cracks that are subsequently joined as the dike inflates (e.g., Pollard, 1988); steps are common in the dikes and sills of this study (e.g., Fig. 7). The true extension direction across a dike is best determined by the use of piercing points and descriptive geometry (e.g., Bussell, 1989).

Dikes were measured at three localities where they are abundant (localities 1, 2, and 3, Fig. 1). As is apparent in the swarm on Lems Ridge (locality 3, Fig. 1) shown in Figure 5, dikes have a wide range in orientation (Fig. 11A). The wall rocks of this swarm consist of massive, weakly foliated, pebbly mudstone of the Lems Ridge olistostrome (Harper et al., 1985, 1994; Ohr et al., 1986; Ohr, 1987), and thus no sills are present. The dikes are regionally metamorphosed but undeformed. Poles to dikes define a broad girdle on an equal-area net (Fig. 11A), corresponding to predominantly west-northwest strikes and dips ranging from subhorizontal to vertical, as evident in Figure 5. Steeply and gently dipping dikes are mutually crosscutting. Individual dikes are commonly segmented (Fig. 11A), indicating that the extension direction is, for at least some dikes, oblique to dike walls (extension in at least one segment of a segmented dike must be oblique). True extension directions, determined using the descriptive geometry technique of Bussell (1989), are indicated with an "E" in Figure 11A. These show that many extension directions are similar but are highly variable overall, ranging in plunge from shallow to subvertical. A weak slaty cleavage and rare fibrous extension veins in the wall rocks are broadly similar in orientation to regional trends in the Galice Formation and pillow lavas of the Josephine ophiolite (e.g., Fig. 11C and D). Some of the poles to dikes and some of the true extension directions are similar in orientation to the regional south-southeast trending, gently plunging extension direction associated with regional metamorphism, but most are substantially different in orientation (cf. Fig. 11A and C).

Figure 5. Dike swarm in massive pebbly mudstone of the Lems Ridge olistostrome (locality 3, Fig. 1). Both gently and steeply dipping dikes are evident. The light band near the right side of the photo consists of five subparallel dikes dipping steeply to the south (right) that cut the gently dipping dikes. The $^{40}Ar/^{39}Ar$ hornblende ages indicate these dikes are similar in age to dikes and sills cutting the Galice Formation (Ohr, 1987).

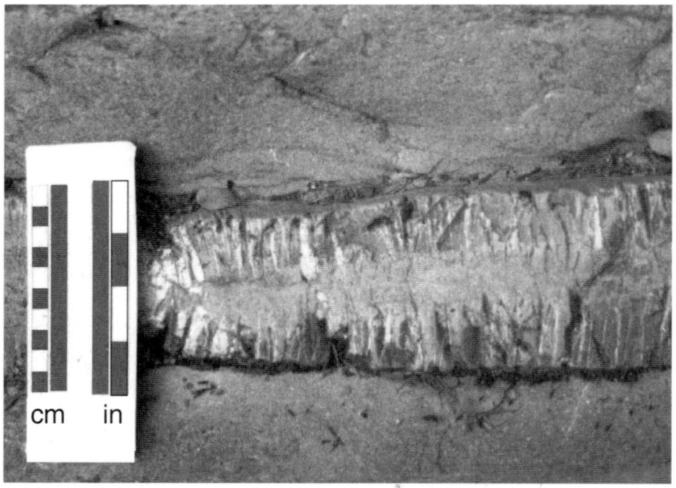

Figure 6. Dike intruding volcanic breccia of the Josephine ophiolite at locality 1 (Fig. 1), showing extension veins formed during regional metamorphism. Fibers in veins consist of quartz + epidote + actinolite. They are oriented normal to the vein walls and plunge gently to the right (southeast), parallel to the stretching direction in the overlying Galice Formation. Note that this extension direction is at a high angle to the dike margins, consistent with a change in the direction of extension with time. Subhorizontal surface; dike dips steeply into the outcrop.

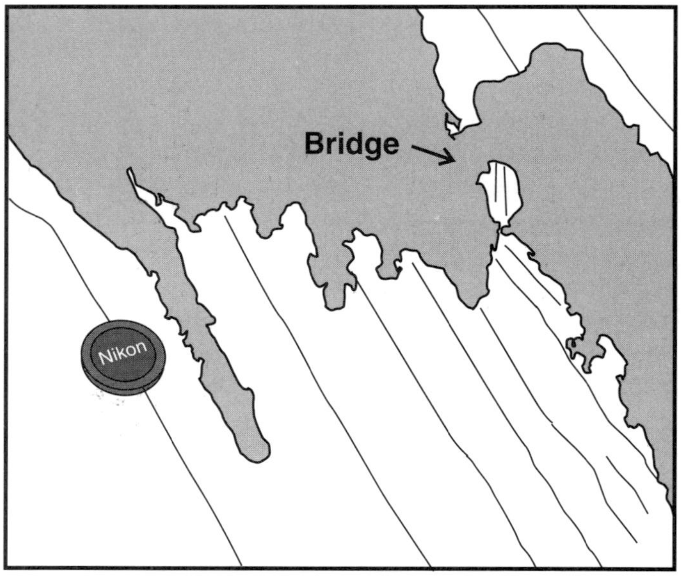

Figure 8. Sketch of a hornblende-phyric mafic sill in slate of the lower Galice Formation (locality 1, Fig. 1). Solid lines are traces of very thin beds. The amoeboid dike margins suggest intrusion before the wall rock was completely lithified. The dike consists of two main segments joined by a bridge within which bedding is rotated. This feature and the apophysis on the left (another segment) are typical of dikes that form by propagation of en echelon segments that are subsequently joined with bridges as the dike inflates (e.g., Pollard, 1988; Bussell, 1989). Lens cap is ~4.5 cm in diameter.

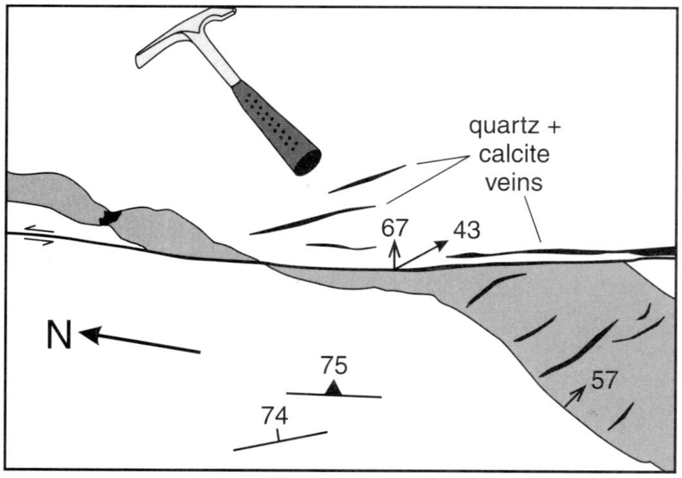

Figure 7. Sketch of subhorizontal surface showing a hornblende andesite dike intruding a steeply dipping metagraywacke bed in the Galice Formation (locality 1, Fig. 1). The dike is cut by a fault that is oriented subparallel to foliation in the metagraywacke. Open arrow on fault indicates dip and closed arrow shows trend and plunge of calcite + quartz slickenfibers; open arrow on dike indicates dike dip. Fibrous extension veins in dike (black) are also filled with quartz + calcite. Near the fault, the dike is boudinaged (upper left) and ductily deformed. Localization of ductile deformation of the dike and extension veins near the fault indicates the fault was active during penetrative deformation of the Galice Formation.

The poles to dikes in the lower part of the Galice Formation and underlying pillow lavas of the Josephine ophiolite at locality 1, along with those from a dike swarm in the underlying peridotite ~2 km to the south (locality 2, Fig. 1), are shown in Figure 11B–D. Bedding and slaty cleavage in this section dip steeply to the east (Fig. 11D). The stretching direction associated with the slaty cleavage plunges to the south-southeast, as indicated by fibrous extension veins, stretched pebbles, and lineation in slates (Fig. 11D). As with the Lems Ridge dike swarm, dike orientations are highly variable, and poles to dikes form a broad girdle (Fig. 11C and D). Many dikes have segments that vary widely in orientation (connected by arcs in Fig. 11), especially those in the pillow lavas, where some of the dikes intrude variably oriented small faults. Poles to dikes are generally oriented at a high angle to poles to fibrous extension veins, as evident in Figure 6, and to the stretching lineation in the Galice Formation (Fig. 11C and D).

Many of the dikes in the basal Galice Formation at locality 1 are intruded along small, steeply dipping faults (Fig. 9A and B) that strike north-northwest (Fig. 11F). Many other dikes in the Galice Formation, including clastic dikes and some dikes in the underlying pillow lavas, have similar orientations (Fig. 11C–E) and were probably also intruded along this set of faults. Some of these small faults are refracted as they pass through beds of different composition, indicating that they predate slaty

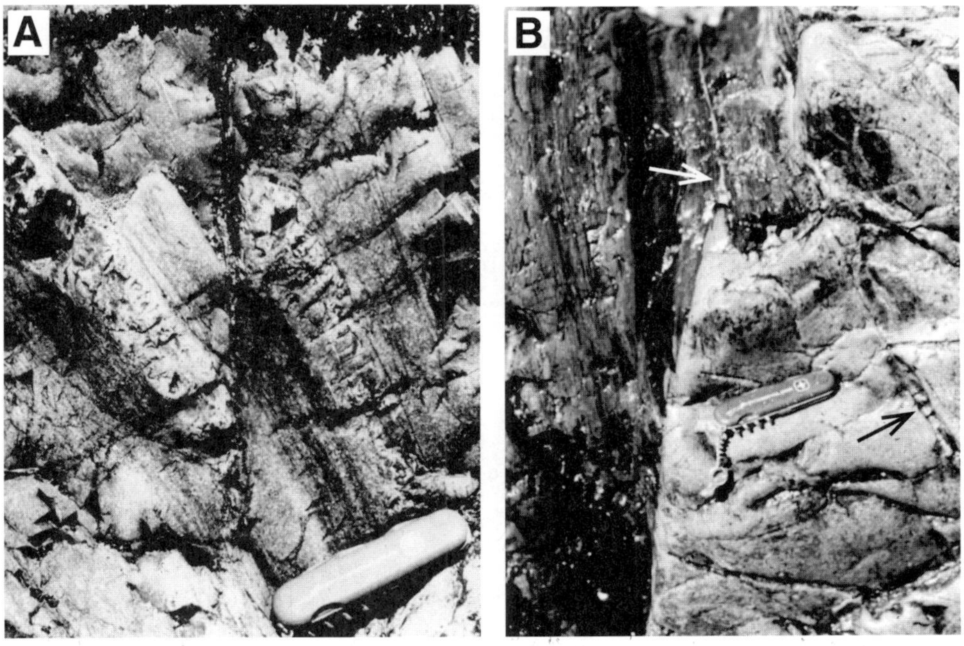

Figure 9. (A) Small, steeply dipping fault in chert (lighter bed) and siliceous slate in the basal Galice Formation, ~5 m above a depositional contact with pillow lavas of the Josephine ophiolite. Vertical face, view is toward the north. Slaty cleavage dips ~5° degrees more steeply than bedding. Drag of bedding indicates left (south) side moved down. Orientation of this and similar faults are shown in Figures 11F and 14A (restored). (B) Dike intruding basal Galice Formation ~4 m to the right (east) of outcrop shown in A. Lower two-thirds of photo is a vertical face, whereas the upper third is a horizontal surface. The dike margin has the same orientation as the fault in A. The white arrow points to an "overshoot" that can be traced into a small fault similar to that shown in A. Black arrow points to fibrous extension vein indicating south-southeast extension during regional metamorphism. These early faults predate slaty cleavage and have thrust displacements after correcting for post-cleavage folding (see text for discussion). Knife is ~7.5 cm in length.

Figure 10. Photomicrograph of mylonitic hornblende basalt dike that intrudes coherent foliated serpentinite ~10 km north-northeast of Gasquet, California. Porphyroclasts consist of relict brown hornblende phenocrysts with fractures filled by fibrous pectolite (Na-Ca pyroxenoid). The matrix consists of relict igneous hornblende, albite, sodic amphibole, and sphene. The presence of pectolite and sodic amphibole, in addition to a high Na_2O content (Barnes et al., this volume, Chapter 17), suggests that this dike has undergone substantial Na metasomatism.

cleavage. A pre-cleavage origin is also suggested by the presence of clastic dikes observed along two of the faults—one consisting of mud and the other of sand—as well as by the similar orientation of most clastic dikes and this set of early faults (cf. Fig. 11E and F). A pre-cleavage age for some igneous dikes and sills is suggested by the amoeboid-shaped margins of a sill in the Galice Formation at locality 1 (Fig. 8), a feature that is most likely the result of intrusion prior to complete lithification of the wall rock. However, slaty cleavage is overprinted by contact aureoles of plutons as old as ca. 148 Ma (Harper and Wright, 1984; Wright and Fahan, 1988), and the youngest dikes (ca. 146–142 Ma; Saleeby et al., 1982; Harper et al., 1994) are affected by regional metamorphism. Thus, dike intrusion predated and overlapped in time with regional metamorphism, which lasted until ca. 135 Ma (Harper et al., 1994).

Dikes are also commonly cut by faults that are inferred to have formed during regional metamorphism, as indicated by ductile deformation along the faults (Fig. 7) and/or slickenfibers on the faults consisting of the same minerals (quartz + calcite ± chlorite) that make up fibers in extension veins. Some dikes and sills also have similar slickenfibers on their margins, indicating faulting after intrusion. Because the dikes and sills are regionally

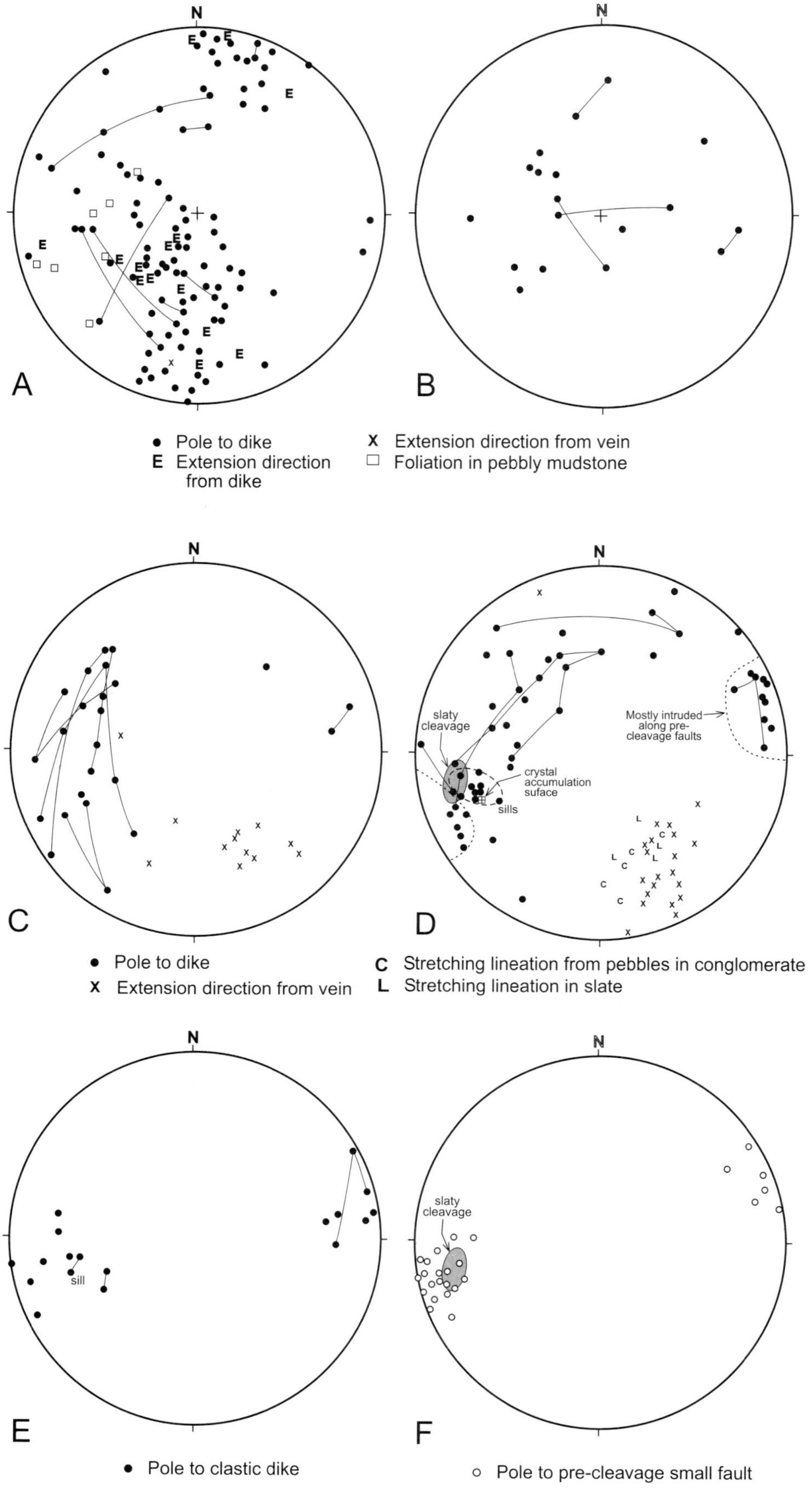

Figure 11. Equal-area, lower-hemisphere projections for dikes and structures. Each great-circle arc connects segments of an individual dike. Veins used to determine extension direction, formed during regional metamorphism, have fibers oriented normal to vein walls (e.g., Fig. 6) and show banding typical of crack-seal veins (Ramsay, 1980). (A) Equal-area, lower-hemisphere projection of poles to dikes and true extension directions in the Lems Ridge dike swarm shown in Figure 5. (B) Equal area, lower-hemisphere projection of poles to dikes for a dike swarm on Kelly Ridge (locality 2, Fig. 1) that cuts the upper part of the peridotite unit of the Josephine ophiolite. (C) Structures in the pillow lava unit of the Josephine ophiolite, locality 1 (Fig. 1). (D) Structures in the lower Galice Formation, locality 1. (E) Poles to clastic dikes in lower Galice Formation, locality 1. All are sand dikes except for one mud dike. These dikes are foliated and thus predate the end of Nevadan deformation. (F) Poles to small early faults (e.g., Fig. 9A). Some of these faults are refracted through beds of different composition, indicating they predate slaty cleavage.

metamorphosed, and in some cases deformed (Figs. 4, 6, 7, and 9), they were originally considered to be entirely pre-Nevadan (Harper, 1980; Saleeby et al., 1982; Harper and Wright, 1984). As discussed above, a syn-Nevadan age is based on radiometric ages overlapping those of plutons that intrude the Orleans thrust and Galice Formation with contact aureoles that overprint slaty cleavage in the Galice Formation. In addition, a weakly penetratively deformed (i.e., late syntectonic) dike that intrudes the amphibolite sole of the Josephine ophiolite yielded a U-Pb zircon age of 151 ± 1 Ma (Harper et al., 1994).

The distribution of poles to dikes in both the Lems Ridge swarm (Fig. 11A) and in the Galice Formation and underlying ophiolite at locality 1 (Fig. 11C and D) essentially falls into quadrants separated by two planes oriented at 90° (Fig. 12A and B). This pattern could be the result of intrusion of dikes along planes of many orientations within a constant stress field. For triaxial stress and infinitesimal strain, all planes having a component of extension will fall within two quadrants (e.g., Hobbs et al., 1976). The orthogonal planes separating the quadrants are planes of maximum shear stress, and the greatest and least principle stresses (σ_1 and σ_3) will bisect these planes. Interpretation of the distribution of dike orientations in Figure 12 in this manner seems reasonable in that many of the dikes and sills are clearly intruded along pre-existing surfaces. Two of the quadrants are shaded, analogous to a fault plane solution for an earthquake, and *P*- and *T*-axes (approximating σ_1 and σ_3) are shown (Fig. 12). The two localities yield similar solutions (Fig. 12A and B). An insufficient number of dikes is available for the dike swarm in peridotite at locality 2 (Fig. 11B) to derive a "fault-plane solution," but nearly all of the poles to these dikes fall within the shaded quadrants for locality 1. This result is not surprising, because this swarm occurs on the same east-dipping limb of a large fold. Validity of this paleostress analysis is supported by the near coincidence of the *T*-axis for the Lems Ridge swarm with the largest cluster of true extensions directions determined for dikes (Fig. 12A). True extension directions were

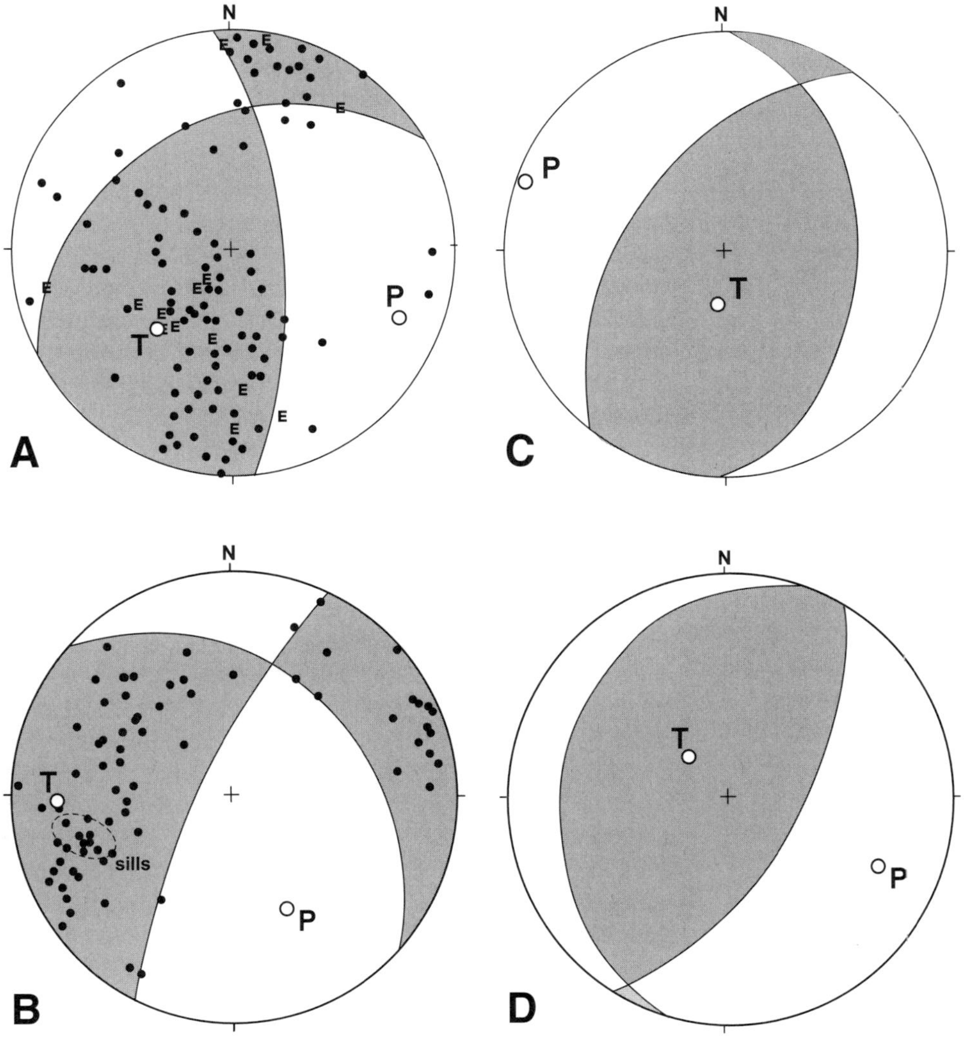

Figure 12. Visually best-fit orthogonal planes to distribution of poles to dikes (solid circles; equal-area, lower-hemisphere projection). Shading and *P*- and *T*-axes (open circles) are analogous to a fault plane solution for an earthquake. This solution assumes that all dikes formed under a similar strain field. (A) Dike swarm in Lems Ridge olistostrome (data from Fig. 11A). E—extension direction (from Fig. 11A). (B) Dikes and sills in the Galice Formation and underlying pillow lava unit of the Josephine ophiolite, locality 1 (data from Fig. 11C and D). (C) Same solution as in A, but corrected for post-Nevadan folding. The axis for tilt correction and the dip direction is well constrained because it is on the east dipping limb of a large gently plunging fold (Norman, 1984), but the amount of dip correction is not well constrained because the pebbly mudstone wall rock to the dike swarm is not bedded and foliation is variable in dip (Fig. 11A). (D) Same solution as in B, but corrected for folding assuming the crystal accumulation surface in the sill shown in Figure 13 was initially horizontal.

not determined for dikes at locality 1, however, so this comparison cannot be made for Figure 12B. Note that the *T*-axis is roughly normal to the stretching lineations and poles to extension veins (Fig. 12B) and is approximately coincident with the poles to slaty cleavage in the Galice Formation (Fig. 11D).

Uncertainties in the fault-plane solutions include not only uncertainties in the visual best fit of orthogonal planes, but also changes in the orientation of pre-cleavage dikes and sills in the Galice Formation (Fig. 11C) resulting from penetrative deformation of the metasedimentary wall rocks. The deformation should have caused dikes to rotate toward the plane of flattening (slaty cleavage), but dikes in the essentially undeformed pillow lavas (Fig. 11C) and peridotite (Fig. 11B) would not have been affected. Even allowing for these uncertainties, the overall form of the fault plane solution will be the same.

Restoration of Dike Orientations and Nevadan Structures

The original orientation of Nevadan dikes and structures has been changed by post-Nevadan folds, such as the large anticline whose hinge is near Gasquet, California (Fig. 1), across which slaty cleavage changes dip from steeply east-northeast to moderately west-southwest (Harper, 1980). The axis of rotation to restore the Lems Ridge dike swarm (locality 3) is well constrained because it is on the east-dipping limb of a large gently plunging syncline, but the dip of this limb is poorly constrained because of the massive nature of the pebbly mudstone and variable dip of a weakly developed slaty cleavage (Norman, 1984; Fig. 11A). The fault-plane solution in Figure 12A was restored assuming an east dip of ~40°. Although this is a rough estimate, variation of the tilt correction by ±20° will still result in a reverse-fault solution (Fig. 12C) with a steeply plunging *T*-axis. The subhorizontal, west–northwest-trending *P*-axis (and west-northwest strike of the plane containing the *P*- and *T*-axes) is consistent with west-northwest thrusting suggested by other indicators (discussed below).

A more precise reconstruction of the initial orientation of Nevadan structures at locality 1 may be possible using an accumulation surface defined by phenocrysts in a sill that intrudes metagraywacke (Fig. 13). The top of the accumulation zone is parallel, within the uncertainty of measurement (~5°), to bedding and oriented at ~20° to foliation in the metagraywacke wall rock (Fig. 11D). The sill is not foliated, but it is metamorphosed and cut by extension veins that have fibers oriented parallel to stretched grains in the host metagraywacke. Similar accumulation of phenocrysts parallel to bedding occurs in a sill in the Galice Formation far to the north, west of Grants Pass, Oregon (J.D. Yule, personal commun., 1992). The magma most likely had an initially low viscosity because (1) the sill is mafic (sample D26 of Barnes et al., this volume, Chapter 17); (2) the chilled margin is nearly free of phenocrysts; and (3) the magma was hydrous (which decreases viscosity), as is evident from the abundance of hornblende. This inferred low viscosity and the presence of abundant crystals in only the lower half of the sill are consistent with accumulation by gravitational settling. Thus, the crystal accumulation surface is interpreted as an approximate paleohorizontal surface, although an initial dip of even several tens of degrees cannot be ruled out. If this interpretation is correct, the essentially homoclinal section at locality 1, into which this sill is intruded, along with the conformably underlying Josephine ophiolite, were subhorizontal when the sill was intruded. Furthermore, the bedding-cleavage relationships in this outcrop suggest that slaty cleavage had an east dip of ~15–20° at this time, assuming the sill intruded after the slaty cleavage formed (or was not tilted prior to cleavage formation). This assumption is probably true, because its ~146 ± 1-Ma age (Harper et al., 1994) is similar to that for plutons having static contact aureoles overprinting slaty cleavage (Harper and Wright, 1984; Wright and Fahan, 1988; Harper et al., 1994; Allen and Barnes, this volume).

Structures in the lower Galice Formation at locality 1 were restored in two steps, assuming the crystal accumulation surface is a paleohorizontal reference. The first step restores the F_2-axis, which has an average trend and plunge of 176° and 12°, respectively, in the area of locality 1 (Harper, 1980), to horizontal. This operation was followed by rotation of the crystal accumulation surface, which is parallel to bedding, to horizontal. This proce-

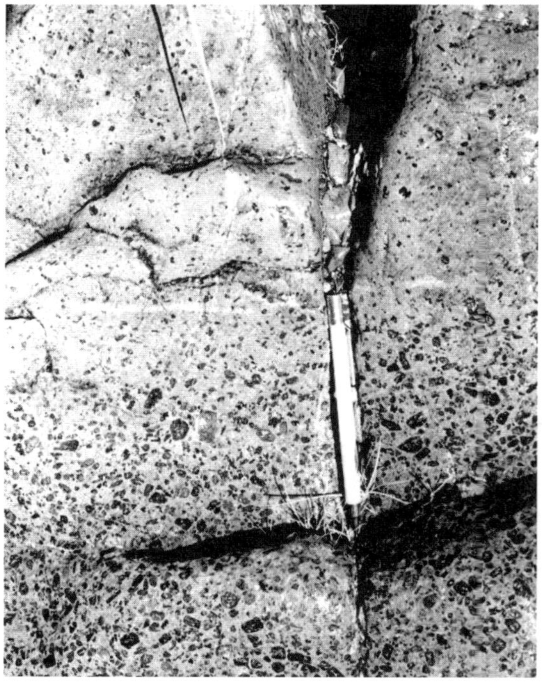

Figure 13. Accumulation of hornblende and pyroxene phenocrysts in a mafic sill in metagraywacke of the lower Galice Formation at locality 1, ~250 m above the top of the Josephine ophiolite. The top of the accumulation zone dips steeply into the surface of the outcrop and is parallel to bedding in the wall rocks. This sill is inferred to be syntectonic because it is regionally metamorphosed but has a 146 ± 1-Ma $^{40}Ar/^{39}Ar$ hornblende age similar to plutons that overprint the slaty cleavage in the Galice Formation (Harper et al., 1994).

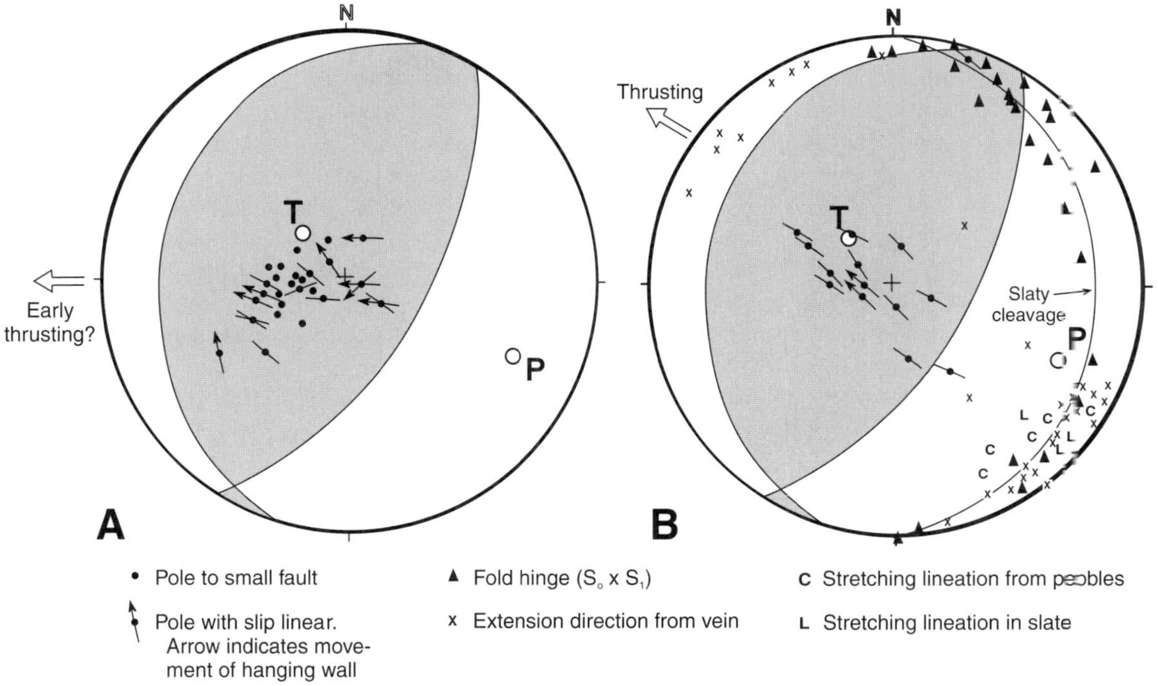

Figure 14. Selected structural data of Figure 11 (all from homoclinal section at locality 1) restored to original orientation, assuming that the crystal accumulation surface shown in Figure 12 was initially horizontal. Also shown on many of the faults is a slip linear (Marshak and Mitra, 1988), which is an arc segment of the great circle (movement plane) containing the pole of the fault and the slip direction (slickenfibers, for these faults); arrowheads indicate displacement of the hanging wall as determined from displaced markers (e.g., Fig. 11A) or steps on slickenfibers. Poles to faults that have two slip linears indicate the presence of two sets of slickenfibers. The fault-plane solution of Figure 12D, including *P*- and *T*-axes, is also shown. (A) Pre-cleavage faults. (B) Structures formed during penetrative deformation of the Galice Formation.

dure assumes that F_2-axes were initially horizontal, but the results are similar if bedding is restored to horizontal using a single rotation.

The restored fault-plane solution for locality 1 (Fig. 12D) is similar to that for the Lems Ridge dike swarm (Fig. 12C), with a steeply plunging *T*-axis and southeast-trending *P*-axis (Fig. 12B). The restored orientations of other structures at locality 1 are shown in Figure 14. For restored small faults, the slip direction based on slickenfibers is shown in Figure 14 by a slip linear, which is an arc segment parallel to the great circle ("movement plane") containing the pole of the fault and the slip direction (Marshak and Mitra, 1988). Arrowheads on the slip linears indicate displacement of the hanging wall. Some of the faults have more than one set of slickenfibers, and thus poles to some faults are shown with two slip linears (Fig. 14A).

The restored pre-cleavage faults have gentle to moderate eastward dips (Fig. 14A), although their orientation must have been modified to some degree by penetrative deformation of the Galice Formation, which would cause their poles to have rotated toward the pole to cleavage. Those pre-cleavage faults having known displacement based on offsets and stepped slickenfibers (e.g., Fig. 9A) are nearly all thrust faults in their restored orientations. Slip directions vary from west-southwest to west-northwest, and several have more than one set of slickenfibers (Fig. 9A). These observations suggest that the earliest direction of thrusting may have been top to the west-southwest (Fig. 14A). In contrast, faults showing ductile deformation near their margins (e.g., Fig. 7)—and thus inferred to have been active during penetrative deformation (i.e., syn-cleavage)—show top-to-the-northwest movement (Fig. 14B). Lack of curvature of fibers in pressure shadows around pyrite framboids in slates (Jones, 1988) and of fibers in extension veins indicates that the direction of extension was uniform over the time of penetrative deformation when this later set of faults is inferred to have been active.

Structures in the study area have probably been reoriented by vertical-axis clockwise rotation, as implied by paleomagnetic declination anomalies in southwestern Oregon (Schultz and Levi, 1983; Bogen, 1986; Smith and Harper, 1993). Some workers (e.g., Mankinen and Irwin, 1982; Bogen, 1986) argue that the entire Klamath Mountains rotated as a single block, but Renne and Scott (1988) found that the amount of rotation correlates with structural grain, suggesting that the arcuate shape of the Klamath Mountains is the result of oroclinal bending. The structural data in Figure 14 have not been corrected for vertical-axis rotations, but assuming structural trends in the Western Klamath terrane were originally ~340°, parallel to the overall structure of correlative rocks in the Sierra Nevada foothills, which apparently have not been rotated (Bogen et al., 1985; Frei, 1986), structures in the study area would need to be restored by a ~30° counterclockwise rotation. Much larger (90°) vertical-axis rotations are suggested by declination anomalies in the Josephine ophiolite (Smith and Harper, 1993), but the amount of rotation may be overestimated because of incomplete removal of secondary magnetization (J.D. Yule, written commun., 1994) and uncertainties in the amount of post-Nevadan tilt correction. Some post-Nevadan tilting has occurred, at least locally, because Early Cretaceous sediments unconformably overlying the Galice Formation near the Oregon-California border dip from 10° to 30° to the north (Jones, 1988). Unfortunately, preliminary data from these sedimentary rocks, which include red beds, indicate these rocks do not record a primary magnetism (R. Wells, personal commun., 1994).

KINEMATIC INTERPRETATION

Direction of Contraction

Both fault-plane solutions derived from the distribution of dike and sill orientations (Fig. 12C and D) are consistent in the interpretation that dike and sill (and related pluton) intrusion occurred during contraction (Gray, 1985, this volume; Harper et al., 1994). The *P*- and *T*-axis orientations of the fault-plane solutions (Fig. 12B) are consistent with southeast-northwest contraction, as are slip directions for small syn-cleavage faults at locality 1 and gently plunging north-northeast trend of most F_1 fold hinges (Fig. 14B). Northwest–southeast-directed contraction is distinct from east-west contraction inferred by previous workers from fold geometry (Gray, 1985, this volume; Jones, 1988), offsets on small faults beneath the Orleans thrust (Snoke, 1977), and angular relationships between slaty cleavage and the Orleans thrust (Gray, 1985, this volume). The latter was also used by Wyld (1985) to infer west-directed thrusting in the southern Klamath Mountains, but this inference is equivalent to northwest-directed thrusting in the study area if the Klamath Mountains underwent oroclinal bending as discussed above. A south component of dip of the Josephine–Galice thrust sheet (before post-Nevadan folding), which in turn implies a north–south component of slip if thrusting was purely dip-slip, is suggested by the southward increase in metamorphic grade in the study area. The strike of slaty cleavage after restoration, however, would be expected to be toward the northeast, rather than north (Fig. 14B). This apparent discrepancy may be due to uncertainties and assumptions in restoration of structures. Alternatively, it may reflect sinistral oblique-slip, rather than dip-slip motion during thrusting. Further work is needed on the Galice Formation directly beneath the Orleans thrust; sense-of-shear can probably be derived from composite foliations, which are evident in unoriented samples collected in the early 1970s by A.W. Snoke (personal commun., 2001) (Harper, 1980; Harper et al., 1988). Harper (1980) approximately located the transition from the prehnite-pumpellyite to lower greenschist facies ~18 km south-southeast of Gasquet, California. Examination of thin sections used in Harper's (1980) and Coulton's (1995) studies, along with a distinct color change evident on LANDSAT images (Raines and Wynn, 1982), indicates an approximately northeast-striking antigorite-in isograd in the Josephine peridotite at about the latitude of Gasquet, California. Assuming metamorphic grade correlates with depth of underthrusting, the strike of this isograd implies an originally southwest-dipping thrust sheet, which in turn implies top-to-the-northwest thrusting, assuming pure dip-slip displacement.

The dip directions of small, pre-cleavage thrust faults at locality 1 suggest the possibility of early west-directed, rather than northwest-directed, thrusting (Fig. 14A). These dip directions and the multiple sets of slickenfibers on several of the pre-cleavage faults, some of which indicate top-to-the-west and west-southwest, could be the result of a change in direction of contraction. Although these faults have very small displacements (e.g., Fig. 9A) and have only been observed by detailed study at one locality, pyrite framboids in overturned limbs of folds having thrust-out hinges indicate that nearly all folding, and presumably thrusting out of fold hinges, preceded formation of slaty cleavage (Jones, 1988). A change from west- to northwest-directed thrusting could also account for the wide range of hingeline orientations in the study area (Fig. 14B) and elsewhere in the Galice Formation (Harper, 1980; Norman, 1984; Jones, 1988), including its type area west of Grants Pass, Oregon (Park-Jones, 1988). Early folds having north–south-oriented hinges would have been reoriented about southwest–northeast-trending folds during penetrative deformation, resulting in a spread in fold hinge orientations. Transected folds (i.e., slaty cleavage not parallel to the axial surface) observed in an area of low strain south of Cave Junction, Oregon, may also be the result of a change in contraction direction. No direct observations of refolded folds or downward facing structures, however, have been observed, and the observed variation in fold hinge orientations evident in Figure 14B could also be due to rotation of fold hinges toward the stretching direction during noncoaxial strain. Strain in slates was found to be coaxial, however, and the same spread in F_1 fold hinges is observed in areas where strain is very low (Jones, 1988).

Contrasting Extension Directions Recorded by Dikes and Structures in the Galice Formation

Virtually all poles to dikes and sills at locality 1 are oriented at a high angle to the extension direction in the Galice Formation (stretching lineations and poles to fibrous extension veins; Fig. 11C–E). This discrepancy is particularly evident in Figure 14B, where the *T*-axis, derived from the distribution of dike orientations, is approximately normal to the poles to fibrous extension veins and stretching lineations. If the dikes and sills are indeed syntectonic, as suggested by the geochronologic data (Harper et al., 1994), this is a surprising result, because the extension veins are probably analogous to dikes in that they form as cracks that fill with fluid. One possibility is that nearly all dikes and sills at locality 1 predate slaty cleavage, consistent with the observation that all the dikes and sills have been regionally metamorphosed, many dikes and sills are cut by extension veins and/or are penetratively deformed (Figs. 4, 6, 7, and 10), and no dikes were observed that cut extension veins. If the dikes and sills at locality 1 are indeed mostly older than slaty cleavage, it implies that a major change in the stress regime occurred between diking and penetrative deformation.

In the Lems Ridge dike swarm, some of the subvertical dikes have true extension directions that are similar to the regional south-southeast extension direction in the Galice Formation (cf. Fig. 11A and D). Crosscutting relationships indicate that the subvertical dikes tend to be younger in this dike swarm, but some are cut by more shallowly dipping dikes. Perhaps the Lems Ridge swarm was intruded over a period of time in which the intermediate (σ_2) and minimum (σ_3) principle stresses flipped; indeed, such a flip of σ_2 and σ_3 is suggested by paleostress analysis of high- and low-temperature faults in the Summit Valley plutonic complex to the south (Griesau, 1992). The low-temperature (epidote coated) faults, which should be younger than the 144 ± 1-Ma $^{40}Ar/^{39}Ar$ hornblende cooling age (~500 °C), yielded a shallowly south–southeast-plunging σ_3 (both in the present and restored reference frames), similar in orientation to the stretching direction in the Galice Formation.

Flipping of stress directions during deformation could have resulted from increasing tectonic burial as the western Klamath terrane was thrust deeper and deeper beneath older terranes of the Klamath Mountains. The increased vertical load could have resulted in the vertical stress becoming the intermediate, and eventually the maximum, principle stress. Alternatively, penetrative deformation in the Galice Formation might actually record extensional, rather than contractional, deformation, which would be consistent with the gently dipping slaty cleavage and gently plunging stretching direction evident in Figure 14B. This type of deformation would require that the folds in the Galice Formation pre-date slaty cleavage, which is the conclusion of Jones (1988), based on studies of fibers around pyrite framboids. However, the following suggests deformation was contractional during cleavage formation: Jones (1988) found evidence that one fold developed during cleavage formation, most small faults at locality 1 (e.g., Fig. 7) interpreted to have been active during slaty cleavage development have thrust displacement (Fig. 14B), and there is an angular relationship between slaty cleavage and the Orleans thrust (i.e., steeper than the thrust and rotating toward the thrust; Gray, 1985, this volume; Wyld, 1985). Furthermore, a 135 ± 1-Ma cooling age (~350 °C; Harper et al., 1994) is not consistent with extensional unroofing during formation of slaty cleavage, in that the cleavage is overprinted by plutons as old as ca. 145 Ma (Wright and Fahan, 1988; Harper et al., 1994; Allen and Barnes, this volume). Finally, high-*T* (upper amphibolite-facies) shear zones, similar in appearance to the sample shown in Figure 2B also yielded a steeply plunging σ_1 (after correction for post-Nevadan folding) but with σ_3 in the position of σ_2 derived for the epidote-coated faults. The high-*T* shear zones should have formed between 150 ± Ma (U-Pb zircon), whereas the epidote-coated faults should have formed after ca. 144 Ma.

SUMMARY AND CONCLUSIONS

1. The structure of the Galice Formation where it overlies the Josephine ophiolite is characterized by upright bedding (~85% of beds), a generally east-dipping bedding-parallel slaty cleavage, a south–southeast-trending stretching lineation, uncommon tight-to-isoclinal folds with variably oriented hinges, plane strain in slates, and a southward increase in strain and regional metamorphic grade.
2. Calc-alkaline dikes and sills of age ca. 151–146 Ma are all regionally metamorphosed; some are deformed (Figs. 4, 7, and 9) and cut by synmetamorphic fibrous extension veins (Fig. 6). One sill appears to have been intruded prior to complete lithification of the Galice Formation (Fig. 8). These features suggest pre-tectonic intrusion, but syntectonic intrusion is suggested by previous studies showing that related plutons intrude the Orleans thrust, have contact aureoles that overprint slaty cleavage, or have dynamic contact aureoles (e.g., Harper et al., 1994).
3. Many dikes, as well as some clastic dikes, intruded along small, pre-cleavage thrust faults (Fig. 8). Many dikes also show evidence for oblique opening.
4. Poles to dikes are mostly oriented at a high angle to the stretching lineation and to poles to extension veins in the Galice Formation (Fig. 6). This relationship is puzzling if the dikes are, in fact, syntectonic. One possibility is that most dikes studied were intruded prior to penetrative deformation, followed by flipping of principle stress axes as the overburden increased as the western Klamath terrane was progressively thrust deeper beneath older terranes of the Klamath Mountains.
5. Crystal accumulation within a sill dated at 146 ± 1 Ma (Fig. 13) was used to estimate the original orientation of dikes and Nevadan structures. After restoration, slaty cleavage dips ~15–20° to the east, and the stretching lineation plunges shallowly to the south-southeast.
6. Distributions of poles to dikes at two localities were used to construct fault-plane solutions. The solutions are similar and, after correction for post-Nevadan folding, indicate dike intrusion during northwest-southeast contraction.
7. After correction for post-Nevadan folding, small pre-cleavage faults mostly show top-to-the-west displacements. Small syn-cleavage faults have northwest- and southeast-trending slip vectors, two of which have shear-sense indicators showing top-to-the-northwest displacement, parallel

to the trend of the stretching lineation and similar to the trend of *T*- and *P*-axes derived from dike orientations (Fig. 13B).

ACKNOWLEDGMENTS

Part of this research is from GDH's Ph.D. dissertation completed at the University of California at Berkeley (1980). Additional work was supported by National Science Foundation grant EAR-0003444. James MacDonald provided useful comments. Reviews by Kenneth Fowler and Mark Helper are appreciated and resulted in significant improvement of this chapter.

REFERENCES CITED

Alexander, R., and Harper, G.D., 1992, The Josephine ophiolite, an ancient analog for oceanic lithosphere formed at slow/intermediate spreading centers, *in* Parson, B., et al., eds., Ophiolites and their modern oceanic analogues: Oxford, Blackwell Scientific Publications, Journal of the Geological Society [London] Special Publication no. 60, p. 3–38.

Allen, C.M., and Barnes, C.G., 2006, this volume, Ages and some cryptic sources of Mesozoic plutonic rocks in the Klamath Mountains, California and Oregon, *in* Snoke, A.W., and Barnes, C.G., eds., Geological studies in the Klamath Mountains province, California and Oregon: A volume in honor of William P. Irwin: Boulder, Colorado, Geological Society of America Special Paper 410, doi:10.1130/2006.2410(11).

Barnes, C.G., Snoke, A.W., Harper, G.D., Frost, C.D., McFadden, R.R., Bushey, J.C., and Barnes, M.A.W., 2006, this volume (Chapter 17), Arc plutonism following regional thrusting: Petrology and geochemistry of syn- and post-Nevadan plutons in the Siskiyou Mountains, Klamath Mountains province, California, *in* Snoke, A.W., and Barnes, C.G., eds., Geological studies in the Klamath Mountains province, California and Oregon: A volume in honor of William P. Irwin: Boulder, Colorado, Geological Society of America Special Paper 410, doi: 10.1130/2006.2410(17).

Blake, M.C., Jr., Engebretson, D.C., Jayko, A.S., and Jones, D.L., 1985, Tectonostratigraphic terranes in southwest Oregon, *in* Howell, D.G., ed., Tectonostratigraphic terranes of the Circum-Pacific region: Houston, Circum-Pacific Council for Energy and Mineral Resources, Earth Science Series 1, p. 147–157.

Bogen, N.L., 1986, Paleomagnetism of the Upper Jurassic Galice Formation, southwestern Oregon; Evidence for differential rotation of the eastern and western Klamath Mountains: Geology, v. 14, p. 335–338, doi: 10.1130/0091-7613(1986)14<335:POTUJG>2.0.CO;2.

Bogen, N.L., Kent, D.V., and Schweickert, R.A., 1985, Paleomagnetism of Jurassic rocks in the western Sierra Nevada metamorphic belt and its bearing on the structural evolution of the Sierra Nevada block: Journal of Geophysical Research, v. 90, p. 4627–4638.

Burchfiel, B.C., and Davis, G.A., 1981, Triassic and Jurassic tectonic evolution of the Klamath Mountains–Sierra Nevada geologic terrane, *in* Ernst, W.G, ed., The geotectonic development of California—Rubey volume 1: Englewood Cliffs, New Jersey, Prentice-Hall, p. 50–70.

Bussell, M.A., 1989, A simple method for the determination of the dilation direction of intrusive sheets: Journal of Structural Geology, v. 11, p. 679–687, doi: 10.1016/0191-8141(89)90004-7.

Cannat, M., and Boudier, F., 1985, Structural study of intra-oceanic thrusting in the Klamath Mountains, Northern California; Implications on accretion geometry: Tectonics, v. 4, p. 435–452.

Cashman, S.M., 1988, Finite-strain patterns of Nevadan deformation, western Klamath Mountains, California: Geology, v. 16, p. 839–843, doi: 10.1130/0091-7613(1988)016<0839:FSPOND>2.3.CO;2.

Chamberlain, K.R., Snoke, A.W., Barnes, C.G., and Bushey, J.C., 2006, this volume, New U-Pb radiometric dates of the Bear Mountain intrusive complex, Klamath Mountains, California, *in* Snoke, A.W., and Barnes, C.G., eds., Geological studies in the Klamath Mountains province, California and Oregon: A volume in honor of William P. Irwin: Boulder, Colorado, Geological Society of America Special Paper 410, doi: 10.1130/2006.2410(15).

Coleman, R.G., and Lanphere, M.A., 1991, The Briggs Creek amphibolite, Klamath Mountains, Oregon: Its origin and dispersal: New Zealand Journal of Geology and Geophysics, v. 34, p. 271–284.

Coulton, A.J., 1995, Fault-related oceanic and emplacement-age serpentinization in the Josephine ophiolite of NW California and SW Oregon [Ph.D. dissertation]: Albany, State University of New York, 240 p.

Coulton, A.J., Harper, G.D., and O'Hanley, D.S., 1995, Oceanic versus emplacement-age serpentinization in the Josephine ophiolite: Implications for the nature of the Moho at intermediate and slow spreading ridges: Journal of Geophysical Research, v. 100, p. 22,245–22,260, doi: 10.1029/95JB02157.

Dick, H.J.B., 1976, The origin and emplacement of the Josephine Peridotite of southwestern Oregon [Ph.D. dissertation]: New Haven, Connecticut, Yale University, 409 p.

Dott, R.H., Jr., 1971, Geology of the southwestern Oregon coast west of the 124th meridan: Portland, Oregon Department of Geology and Mineral Industries Bulletin 69, 63 p.

Frei, L.S., 1986, Additional paleomagnetic results from the Sierra Nevada; Further constraints on Basin and Range extension and northward displacement in the western United States: Geological Society of America Bulletin, v. 97, p. 840–849, doi: 10.1130/0016-7606(1986)97<840:APRFTS>2.0.CO;2.

Garcia, M.O., 1982, Petrology of the Rogue River island-arc complex, southwest Oregon: American Journal of Science, v. 282, p. 783–807.

Giaramita, M.J., and Harper, G.D., 2006, this volume, Geochemistry of ophiolitic rocks associated with the western part of the Elk outlier of the western Klamath terrane, southwestern Oregon, *in* Snoke, A.W., and Barnes, C.G., eds., Geological studies in the Klamath Mountains province, California and Oregon: A volume in honor of William P. Irwin: Boulder, Colorado, Geological Society of America Special Paper 410, doi: 10.1130/2006.2410(08).

Gray, G.G., 1985, Structural, geochronologic, and depositional history of the western Klamath Mountains, California and Oregon; Implications for the early to middle Mesozoic tectonic evolution of the western North American Cordillera [Ph.D. dissertation]: Austin, University of Texas, 224 p.

Gray, G.G., 2006, this volume, Structural and tectonic evolution of the western Jurassic belt along the Klamath River corridor, Klamath Mountains, California, *in* Snoke, A.W., and Barnes, C.G., eds., Geological studies in the Klamath Mountains province, California and Oregon: A volume in honor of William P. Irwin: Boulder, Colorado, Geological Society of America Special Paper 410, doi: 10.1130/2006.2410(07).

Griesau, N., 1992, A kinematic study of the Summit Valley plutonic complex [M.S. thesis]: Albany, State University of New York, 133 p.

Griesau, N., and Harper, G.D., 1992, Deformation of a Late Jurassic plutonic complex intruding the Orleans thrust, west-central Klamath Mountains, California: Geological Society of America Abstracts with Programs, v. 24, p. 28.

Hacker, B.R., and Ernst, W.G., 1993, Jurassic orogeny in the Klamath Mountains: A geochronological analysis, *in* Dunn, G., and McDougall, K., eds., Mesozoic paleogeography of the western United States II: Los Angeles, Pacific Section, Society of Economic Paleontologists and Mineralogists, p. 37–59.

Hacker, B.R., Donato, M.M., Barnes, C.G., McWilliams, M.O., and Ernst, W.G., 1995, Timescales of orogeny: Jurassic construction of the Klamath Mountains: Tectonics, v. 14, p. 677–703, doi: 10.1029/94TC02454.

Harper, G.D., 1980, Structure and petrology of the Josephine ophiolite and overlying metasedimentary rocks, northwestern California [Ph.D. dissertation]: Berkeley, University of California, 260 p.

Harper, G.D., 1983, A depositional contact between the Galice Formation and a Late Jurassic ophiolite in northwestern California and southwestern Oregon: Oregon Geology, v. 45, p. 3–7.

Harper, G.D., 1984, The Josephine ophiolite, northwestern California: Geological Society of America Bulletin, v. 95, p. 1,009–1,026.

Harper, G.D., 1995, Pumpellyosite and prehnitite associated with epidosite in the Josephine ophiolite—Ca metasomatism during upwelling of hydrothermal fluids at a spreading axis, *in* Schiffman, P., and Day, H., eds., Low-grade metamorphism of mafic rocks: Boulder, Colorado, Geological Society of America Special Paper 296, p. 101–122.

Harper, G.D., 2003, Fe-Ti basalts and propagating rift tectonics in the Josephine ophiolite: Geological Society of America Bulletin, v. 115, p. 771–787, doi: 10.1130/0016-7606(2003)115<0771:FBAPTI>2.0.CO;2.

Harper, G.D., and Wright, J.E., 1984, Middle to Late Jurassic tectonic evolution of the Klamath Mountains, California-Oregon: Tectonics, v. 3, p. 759–772.

Harper, G.D., Saleeby, J.B., and Norman, E., 1985, Geometry and tectonic setting of sea-floor spreading for the Josephine ophiolite, and implications for Jurassic accretionary events along the California margin, *in* Howell, D.G., ed., Tectonostratigraphic terranes of the Circum-Pacific region: Houston, Circum-Pacific Council for Energy and Mineral Resources, Earth Science Series 1, p. 239–257.

Harper, G.D., Bowman, J.R., and Kuhns, R., 1988, Field, chemical, and isotopic aspects of submarine hydrothermal metamorphism of the Josephine ophiolite, Klamath Mountains, California-Oregon: Journal of Geophysical Research, v. 93, p. 4625–4657.

Harper, G.D., Grady, K., and Wakabayashi, J., 1991, A structural study of a metamorphic sole beneath the Josephine ophiolite, western Klamath terrane, California-Oregon, *in* Harwood, D.S., and Miller, M.M., eds., Paleozoic and early Mesozoic paleogeographic relations in the Klamath Mountains, Sierra Nevada, and related terranes: Boulder, Colorado, Geological Society of America Special Paper 255, p. 379–396.

Harper, G.D., Saleeby, J.B., and Heizler, M., 1994, Formation and emplacement of the Josephine ophiolite and the age of the Nevadan orogeny in the Klamath Mountains, California-Oregon: Pb/U zircon and $^{40}Ar/^{39}Ar$ geochronology: Journal of Geophysical Research, v. 99, p. 4293–4321, doi: 10.1029/93JB02061.

Harper, G.D., Grady, K., and Coulton, A.J., 1996, Origin of the amphibolite "sole" of the Josephine ophiolite; Emplacement of a cold ophiolite over a hot arc: Tectonics, v. 15, p. 296–313, doi: 10.1029/95TC02525.

Hawkins, J.W., 1995, The geology of the Lau Basin, *in* Taylor, B., ed., Back-arc basins: Tectonics and magmatism: New York, Plenum Press, p. 63–138.

Hobbs, B.E., Means, W.D., and Williams, P.F., 1976, An outline of structural geology: New York, John Wiley and Sons, 571 p.

Hotz, P.E., 1971, Plutonic rocks of the Klamath Mountains, California and Oregon: Washington, D.C., U.S. Geological Survey Bulletin 1290, 91 p.

Irwin, W.P., 1964, Late Mesozoic orogenies in the ultramafic belts of northwestern California and southwestern Oregon: Washington, D.C., U.S. Geological Survey Professional Paper 501-C, p. C1–C9.

Irwin, W.P., 1966, Geology of the Klamath Mountains province, *in* Bailey, E.H., ed., Geology of northern California: Sacramento, California Division of Mines and Geology Bulletin 190, p. 19–38.

Irwin, W.P., 1972, Terranes of the western Paleozoic and Triassic belt in the southern Klamath Mountains, California: Washington, D.C., U.S. Geological Survey Professional Paper 800-C, p. 103–111.

Irwin, W.P., 1981, Tectonic accretion of the Klamath Mountains, *in* Ernst, W.G., ed., The geotectonic development of California—Rubey volume 1: Englewood Cliffs, New Jersey, Prentice-Hall, p. 29–70.

Irwin, W.P., 1994, Geologic map of the Klamath Mountains, California and Oregon: Reston, Virginia, U.S. Geological Survey Miscellaneous Field Investigations, I-2148, scale 1:500,000, 2 sheets.

Jachens, R.C., Barnes, C.G., and Donato, M.M., 1986, Subsurface configuration of the Orleans Fault; Implications for deformation in the western Klamath Mountains, California: Geological Society of America Bulletin, v. 97, p. 388–395, doi: 10.1130/0016-7606(1986)97<388:SCOTOF>2.0.CO;2.

Jones, R.F., 1988, Structural geology of the northern Galice Formation, western Klamath Mountains, Oregon and California [M.S. thesis]: Albany, State University of New York, 211 p.

Kays, M.A., 1968, Zones of alpine tectonism and metamorphism, Klamath Mountains, southwestern Oregon: Journal of Geology, v. 76, p. 17–36.

MacDonald, J.H., Jr., Harper, G.D., and Zhu, B., 2006, this volume, Petrology, geochemistry, and provenance of the Galice Formation, Klamath Mountains, Oregon and California, *in* Snoke, A.W., and Barnes, C.G., eds., Geological studies in the Klamath Mountains province, California and Oregon: A volume in honor of William P. Irwin: Boulder, Colorado, Geological Society of America Special Paper 410, doi: 10.1130/2006.2410(04).

Mankinen, E.A., and Irwin, W.P., 1982, Paleomagnetic study of some Cretaceous and Tertiary sedimentary rocks of the Klamath Mountains province, California: Geology, v. 10, p. 82–87, doi: 10.1130/0091-7613(1982)10<82:PSOSCA>2.0.CO;2.

Marshak, S., and Mitra, G., 1988, Basic methods of structural geology: Englewood Cliffs, New Jersey, Prentice-Hall, 446 p.

Milby, B.J., 1990, A reconnaissance fluid inclusion study of oceanic crustal rocks from the Josephine ophiolite, California-Oregon [M.S. thesis]: Atlanta, Georgia State University, 97 p.

Miller, J.S., Miller, R.B., Wooden, J.L., and Harper, G.D., 2003, Geochronologic links between the Ingalls ophiolite, North Cascades, Washington and the Josephine ophiolite, Klamath Mts., Oregon and California: Geological Society of America Abstracts with Programs, v. 35, p. 113.

Miller, M.M., and Saleeby, J.B., 1995, U-Pb geochronology of detrital zircon from Upper Jurassic synorogenic turbidites, Galice Formation, and related rocks, western Klamath Mountains; Correlation and Klamath Mountains provenance: Journal of Geophysical Research, v. 100, p. 18,045–18,058, doi: 10.1029/95JB00761.

Norman, E.A.S., 1984, The structure and petrology of the Summit Valley area, Klamath Mountains, California [M.S. thesis]: Salt Lake City, University of Utah, 148 p.

Ohr, M., 1987, Geology, geochemistry, and geochronology of the Lems Ridge olistostrome, Klamath Mountains, California [M.S. thesis]: Albany, State University of New York, 247 p.

Ohr, M., Harper, G.D., and Wright, J.E., 1986, Fracture zone origin of the Lems Ridge olistostrome, NW California, Klamath Mountains: EOS, Transactions, American Geophysical Union, v. 67, p. 1210.

Park-Jones, R., 1988, Sedimentology, structure, and geochemistry of the Galice Formation: Sediment fill of a back-arc basin and island arc in the western Klamath Mountains [M.S. thesis]: Albany, State University of New York, 165 p.

Parson, L.M., and Wright, I.C., 1996, The Lau-Hauvre-Taupo back-arc basin: A southward-propagating, multi-stage evolution from rifting to spreading: Tectonophysics, v. 263, p. 1–22, doi: 10.1016/S0040-1951(96)00029-7.

Pearce, J.A., Ernewein, M., Bloomer, S.H., Parson, L.M., Murton, B.J., and Johnson, L.E., 1994, Geochemistry of Lau Basin volcanic rocks: Influence of ridge segmentation and arc proximity, *in* Mellie, J.L., ed., Volcanism associated with extension at consuming plate margins: London, Geological Society Special Publication 81, p. 53–75.

Pinto-Auso, M., and Harper, G.D., 1985, Sedimentation, metallogenesis, and tectonic origin of the basal Galice Formation overlying the Josephine ophiolite, northwestern California: Journal of Geology, v. 93, p. 713–725.

Pollard, D.D., 1988, Elementary fracture mechanics applied to the structural interpretation of dikes: Geological Survey of Canada Special Paper, v. 34, p. 5–24.

Raines, G.L., and Wynn, J.C., 1982, Mapping of ultramafic rocks in a heavily vegetated terrain using LANDSAT data: Economic Geology and the Bulletin of the Society of Economic Geologists, v. 77, p. 1755–1761.

Ramsay, J.G., 1980, The crack-seal mechanism of rock deformation: Nature, v. 284, p. 135–139, doi: 10.1038/284135a0.

Ramsay, J.G., and Wood, D.S., 1973, The geometric effects of volume change during deformation processes: Tectonophysics, v. 16, p. 263–277, doi: 10.1016/0040-1951(73)90015-2.

Renne, P.R., and Scott, G.R., 1988, Structural chronology, oroclinal deformation, and tectonic evolution of the southeastern Klamath Mountains, California: Tectonics, v. 7, p. 1223–1242.

Riley, T.A., 1987, The petrogenetic evolution of a Late Jurassic island arc: The Rogue Formation, Klamath Mountains, Oregon [M.S. thesis]: Stanford, California, Stanford University, 40 p.

Saleeby, J.B., 1984, Pb/U zircon ages from the Rogue River area, western Jurassic belt Klamath Mountains, Oregon: Geological Society of America Abstracts with Programs, v. 16, p. 331.

Saleeby, J.B., 1992, Petrotectonic and paleogeographic settings of the U.S. Cordilleran ophiolites, *in* Burchfiel, B.C., Lipman, P.W., and Zoback, M.L., eds., The Cordilleran Orogen: Conterminous U.S.: Boulder, Colorado, Geological Society of America, The Geology of North America, v. G-3, p. 653–682.

Saleeby, J.B., Harper, G.D., Snoke, A.W., and Sharp, W.D., 1982, Time relations and structural-stratigraphic patterns in ophiolite accretion, west-central Klamath Mountains, California: Journal of Geophysical Research, v. 87, p. 3831–3848.

Schultz, K.L., and Levi, S., 1983, Paleomagnetism of Middle Jurassic plutons of the north-central Klamath Mountains: Geological Society of America Abstracts with Programs, v. 15, p. 427.

Smith, G.M., and Harper, G.D., 1993, Paleomagnetism of the Josephine ophiolite: EOS, Transactions, American Geophysical Union, v. 74, p. 214.

Snoke, A.W., 1972, The petrology and structure of the Preston Peak area, Del Norte and Siskiyou counties, California [Ph.D. dissertation]: Stanford, California, Stanford University, 274 p.

Snoke, A.W., 1977, A thrust plate of ophiolitic rocks in the Preston Peak area, Klamath Mountains, California: Geological Society of America Bulletin, v. 88, p. 1641–1659, doi: 10.1130/0016-7606(1977)88<1641:ATPOOR>2.0.CO;2.

Wells, F.G., Hotz, P.E., and Cater, F.W., 1949, Preliminary description of the geology of the Kerby quadrangle: Portland, Oregon Department of Geology and Mineral Industries Bulletin 40, 23 p.

Wells, F.G., and Walker, G.W., 1953, Geology of the Galice quadrangle, Oregon: Washington, D.C., U.S. Geological Survey Geological Quadrangle Map GQ-25, scale 1:62,500, 1 sheet.

Wright, J.E., and Fahan, M.R., 1988, An expanded view of Jurassic orogenesis in the western United States Cordillera; Middle Jurassic (pre-Nevadan) regional metamorphism and thrust faulting within an active arc environment, Klamath Mountains, California: Geological Society of America Bulletin, v. 100, p. 859–876, doi: 10.1130/0016-7606(1988)100<0859:AEVOJO>2.3.CO;2.

Wright, J.E., and Wyld, S.J., 1994, The Rattlesnake Creek terrane, Klamath Mountains, California: An early Mesozoic volcanic arc and its basement of tectonically disrupted oceanic crust: Geological Society of America Bulletin, v. 106, p. 1033–1056, doi: 10.1130/0016-7606(1994)106<1033:TRCTKM>2.3.CO;2.

Wyld, S.J., 1985, Geology of the western Jurassic belt, South Fork Trinity River Area, Klamath Mountains, California [M.S. thesis]: Berkeley, University of California, 168 p.

Wyld, S.J., and Wright, J.E., 1988, The Devils Elbow ophiolite remnant and overlying Galice Formation: New constraints on the Middle to Late Jurassic evolution of the Klamath Mountains, California: Geological Society of America Bulletin, v. 100, p. 29–44, doi: 10.1130/0016-7606(1988)100<0029:TDEORA>2.3.CO;2.

Yule, J.D., 1996, Geologic and tectonic evolution of Jurassic marginal ocean basin lithosphere, Klamath Mountains, Oregon [Ph.D. dissertation]: Pasadena, California Institute of Technology, 308 p.

Yule, J.D., Saleeby, J.B., Jones, D.L., and Silk, M., 1992, Correlation of basement terranes across the Late Jurassic Josephine inter-arc basin, southwestern Oregon and northern California: Geological Society of America Abstracts with Programs, v. 24, p. 93.

Yule, J.D., Saleeby, J.B., and Barnes, C.G., 2006, this volume, A rift-edge facies of the Late Jurassic Rogue–Chetco arc and Josephine ophiolite, Klamath Mountains, Oregon, *in* Snoke, A.W., and Barnes, C.G., eds., Geological studies in the Klamath Mountains province, California and Oregon: A volume in honor of William P. Irwin: Boulder, Colorado, Geological Society of America Special Paper 410, doi: 10.1130/2006.2410(03).

Manuscript Accepted by the Society 10 January 2006

Printed in the USA

Geological Society of America
Special Paper 410
2006

Structural and tectonic evolution of the western Jurassic belt along the Klamath River corridor, Klamath Mountains, California

Gary G. Gray*
ExxonMobil Upstream Research Company, P.O. Box 2189, Houston, Texas 77252-2189, USA

ABSTRACT

Marginal basin flysch deposits of the western Jurassic belt of the Klamath Mountains were thrust eastward beneath the western Paleozoic and Triassic belt during the Late Jurassic Nevadan orogeny. Nevadan underthrusting created two generations of nearly coaxial north-trending folds within the western Jurassic belt rocks. These structures formed at chlorite-grade, greenschist-facies conditions and have accompanying pressure solution and mineral recrystallization. The geometry of the Nevadan structures suggests that the thrusting direction was roughly west–east in present coordinates. This direction is perpendicular to the regional strike of the bounding thrust faults and is consistent across 35 km of dip exposure. Post-Nevadan structures include locally developed strike-slip faults and related third- and fourth-generation folds. These features have associated quartz and calcite veins but lack the metamorphic mineral growth associated with the Nevadan structures. Relatively young, high-angle normal faults are very common and appear to be contemporaneous with Neogene uplift of the entire range.

Nevadan-age structures within the western Sierra Nevada Foothills terrane also formed in response to west–east thrusting. Global plate-circuit models suggest that the Nevadan Farallon–Pacific relative motion may have been orthogonal to the continental margin at the latitude of the Klamath Mountains. This convergence direction and the kinematic analyses suggest that the Klamath Mountains and Sierra Nevada Foothills were in their same relative orientation during the Nevadan orogeny.

Keywords: Galice Formation, Upper Jurassic, Nevadan orogeny, structural geology, kinematic analysis, geometric analysis, western Klamath Mountains

INTRODUCTION

The Nevadan orogeny has long been recognized as a Late Jurassic deformational event that affected Middle Oxfordian to mid-Kimmeridgian and older rocks within the Klamath Mountains and Sierra Nevada (Diller, 1907; Blackwelder, 1914; Hinds, 1934; Taliaferro, 1942; Wells et al., 1949; Schweickert et al., 1984; Fig. 1). The Nevadan is also part of a larger orogenic system that affected the entire western margin of North America and ultimately ties into the tectonic history of the Gulf of Mexico (e.g., Dickinson and Lawton, 2001).

Nevadan deformation in the Klamath Mountains is manifest

*E-mail: gary.g.gray@exxonmobil.com.

Gray, G.G., 2006, Structural and tectonic evolution of the western Jurassic belt along the Klamath River corridor, Klamath Mountains, California, *in* Snoke, A.W., and Barnes, C.G., eds., Geological studies in the Klamath Mountains province, California and Oregon: A volume in honor of William P. Irwin: Geological Society of America Special Paper 410, p. 141–151, doi: 10.1130/2006.2410(07). For permission to copy, contact editing@geosociety.org.

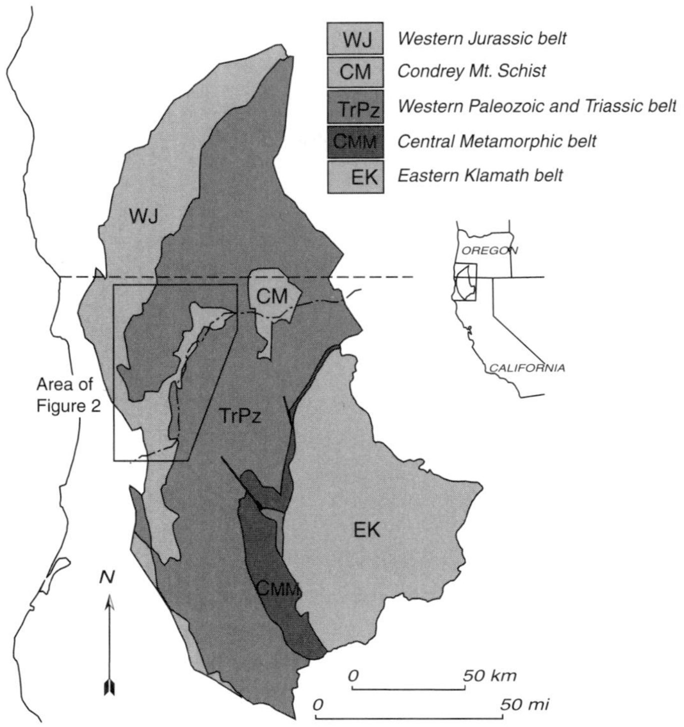

Figure 1. Lithotectonic belts of the Klamath Mountains and location map for the areas discussed in this chapter. Shading is used only to differentiate adjacent belts. Dash-dot line is the course of the Klamath River through the Klamath Mountains. Modified after Irwin (1960), Smith et al. (1982), and Wagner and Saucedo (1987).

principally in the rocks of the western Jurassic belt. The Middle Oxfordian to mid-Kimmeridgian flysch deposits within this belt were folded and metamorphosed to chlorite-grade, greenschist facies during this event. Irwin (1966) and Snoke (1977) recognized that older western Paleozoic and Triassic belt units are thrust over younger western Jurassic belt rocks in a klippenlike geometry near Preston Peak (Fig. 2). This interpretation is supported by gravity modeling that indicates low-density, western Jurassic beltlike rocks also extend farther eastward beneath the higher-density units of the western Paleozoic and Triassic belt (Jachens et al., 1986).

Very little of the Nevadan structural geology has been documented in the Klamath Mountains. The only published work dealing specifically with Nevadan structural geometries in the Klamath Mountains is that of Cornelius (1986), which covers a limited area of western Jurassic belt rocks south of the Preston Peak area (Fig. 2). Cashman (1988) documented thin-section–scale finite strains at various localities within the western Jurassic belt, but did not link these microscopic strains to the outcrop-scale structures and fold generations. Cornelius (1986) interpreted the Nevadan structures as having formed during west–east contraction, whereas Cashman (1988) interpreted the Nevadan deformation as having a significant component of right-lateral strike-slip.

Paleomagnetic inclinations from the Sierra Nevada Foothills terranes and Klamath Mountains rocks suggest that neither area has experienced a significant latitudinal shift relative to stable North America (Mankinen and Irwin, 1982; Mankinen et al., 1984; Bogen, 1986; Irwin and Mankinen, 1998; Mankinen et al., 2002). However, these data also suggest that the Klamath Mountains lay almost perpendicular (60–85° counterclockwise rotation) to the Sierra Nevada Foothills before and during the Nevadan orogeny. Post-Nevadan overlapping sedimentary rocks of Valanginian age, however, show no rotation (Mankinen et al., 1984; Bogen, 1986; Irwin and Mankinen, 1998). Although the significance of paleomagnetic data from metamorphic rocks remains unclear, these data do provide a hypothesis that is testable against the preserved Nevadan structural geometries.

Some of the geometric configurations and the kinematic regimes that follow from these data are shown schematically in Figure 3. In particular, if the Klamath Mountains and Sierra Nevada Foothills areas were oriented similar to their present geometry during the Nevadan, their structures should indicate similar kinematics, whether that was normal convergence or strike-slip (Fig. 3A). If the regions were at high angles to one another, structures within each area would have formed under different kinematic regimes, such that if one area experienced normal convergence, the other would experience transform-dominated movements (e.g., Fig. 3B and C). Alternatively, both areas may have experienced transform-dominated movements with opposite senses of shear (Fig. 3D).

This study documents the geometry of Nevadan structures within part of the western Jurassic belt of the Klamath Mountains. This area is compared to the Sierra Nevada Foothills as a test of both the relative orientations of the two blocks and the kinematics of Late Jurassic deformation along this portion of the western North American margin.

WESTERN JURASSIC BELT OF THE KLAMATH MOUNTAINS

The western Jurassic belt (Irwin, 1960) in the study area is an arcuate unit composed primarily of interbedded sandstone and shale flysch deposits (Galice Formation of Diller, 1907) deposited on the oceanic crust of the Josephine ophiolite (Harper, 1980a; Fig. 1). The lower and northwestern portions of this depositional sequence locally contain appreciable interbedded volcanic and volcaniclastic rocks. These volcanic rocks are called the Rogue Formation in Oregon (Diller, 1907). Metavolcanic rocks are also locally very abundant in the western Jurassic belt in California but have not been mapped as a separate unit (Fig. 2). The main western Jurassic belt lies west of the western Paleozoic and Triassic belt (Irwin, 1960), but it also crops out within a northeast-trending reentrant in the western Paleozoic and Triassic belt near Happy Camp, California (Figs. 1 and 2).

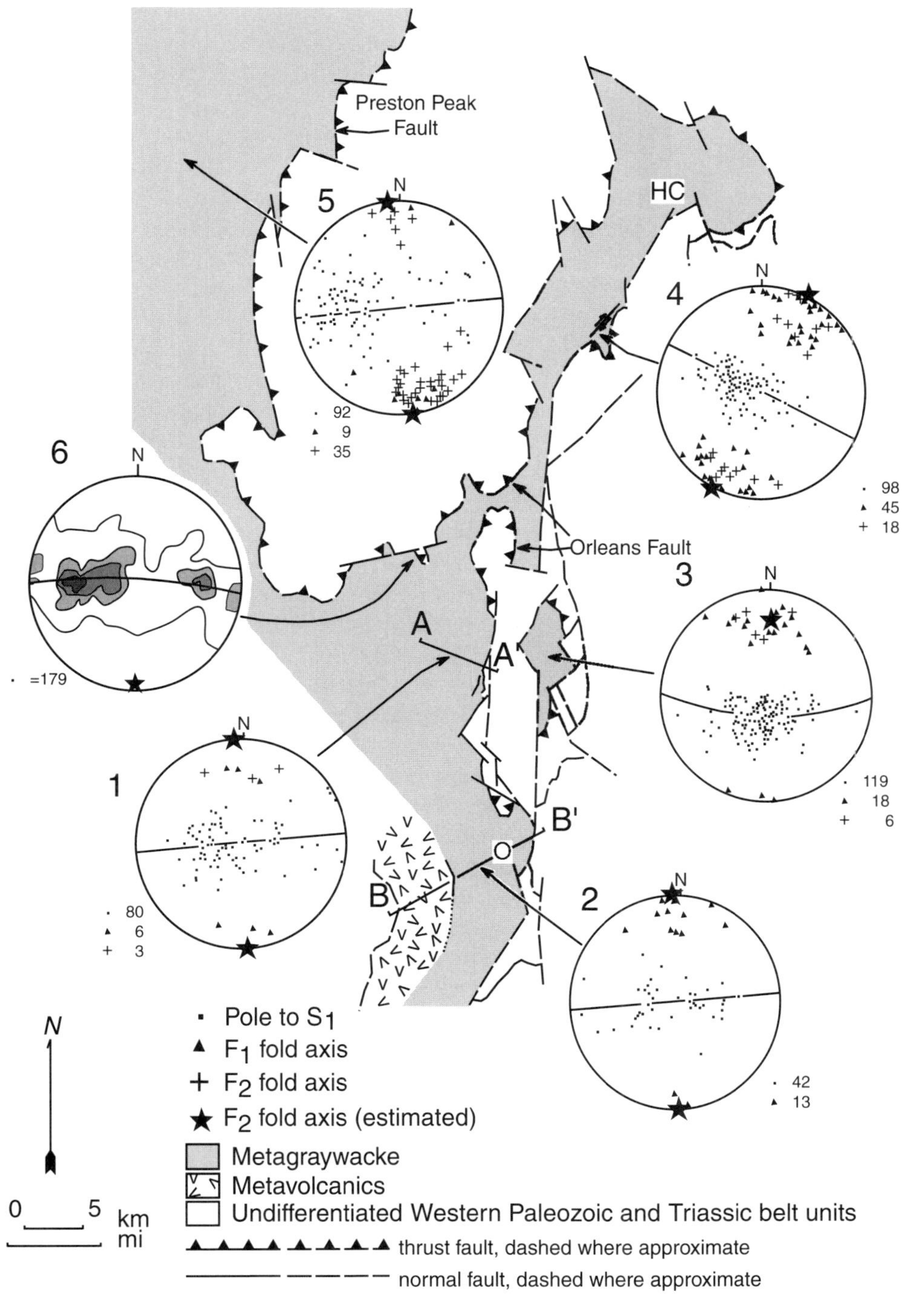

Figure 2. Generalized geology and structural data from the western Jurassic belt in the Orleans–Happy Camp and adjacent areas. Gray shaded areas are folded metasedimentary rocks that are correlative with the Galice Formation and are the primary focus of this chapter. The area with the random V pattern is predominantly metavolcanic rocks thought to be correlative to the Rogue Formation in Oregon. Unshaded areas are occupied by the western Paleozoic and Triassic belt rocks. Letters O and HC refer to the towns of Orleans and Happy Camp, respectively. Structural data at localities 1–4 from Gray (1985). Locality 5 is from Harper (1980b, Domain 8). The numbers and symbols at the lower left and right of these stereonets are the number of observations for each element plotted. Locality 6 is after Cornelius (1986), and is a smoothed contour plot of 179 poles to S cleavage. Contour values are 0, 3, 5, and 10%. Great circles on stereonets are estimated best fits to S_1 pole distributions and are perpendicular to the average S_2 fold axes plotted. Cross-sections A–A′ and B–B′ are presented in Figure 6. Modified from Wagner and Saucedo (1987).

The Galice Formation depositionally overlies the Josephine ophiolite at the latitude of the Oregon-California border. It has been dated as Middle Oxfordian to mid-Kimmeridgian on the basis of sporadic occurrences of the bivalve *Buchia concentrica* (Sowerby) (Harper, 1980a,b, 1983) and detrital zircon analysis (Miller et al., 2003). Albite granites from the ophiolite have been dated at ca. 157 Ma (U-Pb zircon: Saleeby et al., 1982; Harper et al., 1994).

All of the western Jurassic belt rocks have experienced lower greenschist-facies metamorphism, resulting in the widespread development of chlorite + epidote + white mica ± actinolite mineral assemblages. This metamorphism was dated at 146–151 Ma by Lanphere et al. (1968, 1978) and is synchronous with Nevadan deformation as determined by Saleeby et al. (1982), Harper and Wright (1984), and Harper et al. (1994).

The western Jurassic belt rocks have been interpreted as the

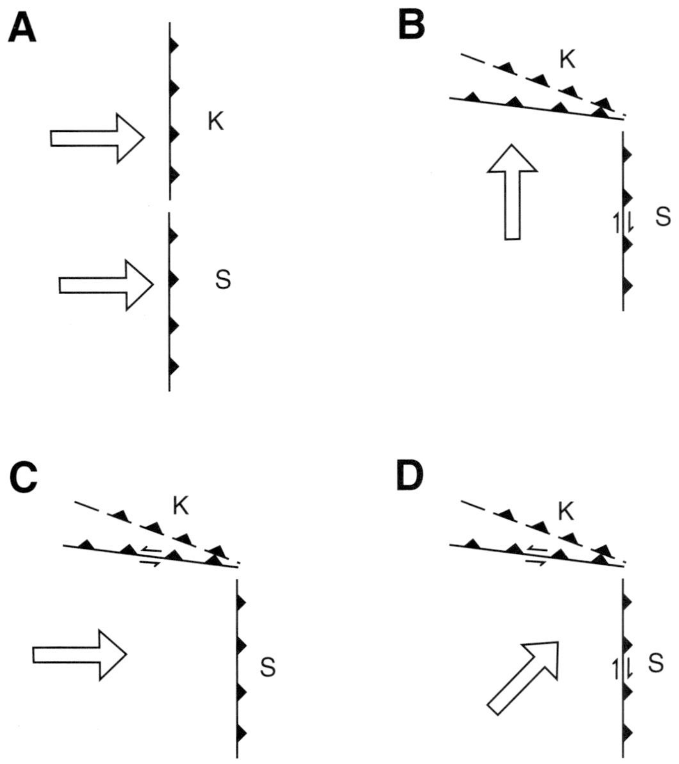

Figure 3. Possible plate interactions suggested by paleomagnetic and plate-motion studies. In B–D, the Klamath Mountains have been rotated 60° (dashed line) to 85° (solid line) counterclockwise relative to the Sierra Nevada to a pre-Nevadan position, as interpreted by Bogen (1986) and Irwin and Mankinen (1998). (A) Predicted kinematics using the Engebretson et al. (1985) relative motions and the present orientation of the Klamath Mountains and Sierra Nevada Foothills. In this case, both areas would show dominantly west-to-east contraction. If the plate convergence were not orthogonal, both areas would still have structural geometries reflecting a similar kinematic regime, be it right- or left-lateral. (B) Predicted kinematics using the Engebretson et al. (1985) relative plate motion and the paleomagnetically-oriented Klamath Mountains position. In this scenario, the Klamath Mountains would show a predominantly left-lateral sense of shear, whereas the Sierra Nevada Foothills would show a predominantly west-to-east thrust sense of shear. (C) Kinematic interactions using the paleomagnetically oriented Klamath Mountains and a hypothetical right-lateral relative motion between Farallon and North America. The Klamath Mountains would have a thrust sense of shear, whereas the Sierra Nevada Foothills would be predominantly right-lateral. (D) Hybrid model combining a reoriented Klamath block and hypothetical right-oblique Farallon–North America plate convergence. This scenario suggests both thrust and transcurrent structures should be present in both areas, but the Klamath Mountains should be left-lateral and the Sierra Nevada Foothills should be right-lateral. Note that a paleomagnetically reoriented Klamath block is incompatible with hypothetical left-lateral or left-oblique Pacific–North America plate motion, as it would predict extension in the Klamath Mountains (i.e., releasing-bend geometry), not convergence. K—Klamath Mountains, S—Sierra Nevada; large arrow—motion of Farallon–Pacific plate relative to North America. Barbs indicate thrust faulting, opposite arrows represent strike-slip faulting.

remains of a Late Jurassic marginal ocean basin and arc assemblage that was accreted to the continental margin near its site of origin (Snoke, 1977; Harper, 1980a,b, 1983, 1984; Harper and Wright, 1984; Miller and Saleeby, 1995). The overlying western Paleozoic and Triassic belt is a complex assemblage of ophiolitic, volcanic, and sedimentary rocks that was accreted to the North American continent in the Late Triassic (Gray, 1986; Hacker et al., 1995). These two belts of rocks are separated in the Orleans area, California, by the Orleans fault (Hershey, 1906). In the northern part of the study area, the Orleans fault was referred to as the Preston Peak fault (Snoke, 1977; Fig. 3). The Orleans fault truncates older terrane boundaries, slices the roots off many large Middle Jurassic plutons (Jachens et al., 1986), and has caused considerable remixing in the overlying western Paleozoic and Triassic belt units (Snoke, 1977; Saleeby et al., 1982; Barnes, 1983; Gray, 1986; Wright and Wyld, 1994; Hacker, et al., 1995).

There is a minimum of 35 km of overthrusting along the Orleans/Preston Peak fault system. This is the shortest distance from the trace of the Preston Peak fault, where the western Jurassic belt disappears, to the eastern edge of the western Jurassic belt within the Happy Camp reentrant (Gray, 1983; see also Harper et al., 1994). The Orleans/Preston Peak fault has a similar elevation at these two endpoints, indicating the fault is almost horizontal. Jachens et al. (1986) have used gravity models to estimate an additional 60-km horizontal displacement along this fault. Their models are consistent with the relationships at the surface trace of the Orleans fault, but it is not clear that all of the low-density rocks in the lower plate of this thrust are western Jurassic belt. Immediately east of the Happy Camp reentrant, older but similar low-density schists crop out in the Condrey Mountain window beneath high-density rocks of the western Paleozoic and Triassic belt in a relationship analogous to the Orleans fault (Klein, 1977; Helper, 1992; Saleeby and Harper, 1993). It is not known to what extent the Condrey Mountain rocks extend southward into the area modeled by Jachens et al. (1986). Therefore, these data bracket the displacement of the western Jurassic belt beneath the Orleans fault to somewhere between 35 and 95 km along a shallow, east-dipping thrust.

This study presents the results of a detailed examination of the structure of the western Jurassic belt at four sites (Fig. 2). In the majority of the study area, the western Jurassic belt is composed of laminated to thickly bedded turbidites and shales, which are ideal for preserving successive generations of structures. Unfortunately, the lack of distinct marker beds or an identifiable stratigraphic succession within this unit precludes a palinspastic restoration of the western Jurassic belt to its original thickness and width.

The thick flysch deposits are faulted against thick-bedded volcanic flows, breccias, and layered volcaniclastic rocks along the Klamath River (Fig. 2). This fault appears to have normal displacement, suggesting that the volcanic section is a deeper structural and/or stratigraphic unit than the flysch deposits. Both rock units in this belt are intruded by felsic to mafic dikes plugs and sills (Dick, 1973; Harper, 1980b; Gray, 1985, Harper et al., 1994; Hacker et al., 1995).

STRUCTURE OF THE WESTERN JURASSIC BELT

Three deformational events affected the western Jurassic belt in the study area during and after its juxtaposition with the western Paleozoic and Triassic belt. Two generations of nearly coaxial, west-vergent folds (F_1 and F_2) formed during the first deformation. The F_1 folds are asymmetric and tight to isoclinal (Fig. 4). They trend northward and have shallow average plunge. These folds are best seen in thinly interbedded siltstone/shale sequences, where they typically range in amplitude from ~0.3 to 3 m. In thick sandstone beds, the amplitude of F_1 folds may exceed 100 m. A well-developed, continuous, closely spaced pressure-solution cleavage, S_1, is axial planar to the F_1 folds and is present in both the siliciclastic and volcanic rock units throughout the western Jurassic belt. Muscovite and chlorite have grown parallel to this cleavage (Fig. 4). The amplitude/wavelength ratio of F_1 folds is high, typically greater than 10; therefore, the S_1 cleavage commonly is nearly parallel to bedding surfaces. Many F_1 folds have been dismembered along the S_1 axial surfaces. The sense of offset along these F_1 axial surfaces is predominantly top-to-the-west. The asymmetry of the early F_1 folds and the westward displacement of these folds along their axial surfaces are shown schematically in Figure 5.

The second-generation F_2 folds are generally open and upright in comparison to the F_1 folds. They also become asymmetric with east-dipping axial planes near the Orleans fault (Figs. 2 and 6). The F_2 folds range in wavelength and amplitude from tens to hundreds of meters and have an associated axial planar crenulation cleavage, S_2, which is best developed in the hinge areas of these folds but may be absent in the limbs. Pressure solution and growth of white mica in cleavage planes also accompanied formation of the S_2 crenulation cleavage (Fig. 7). The growth of micas parallel to S_1 and S_2 cleavages indicates that both formed during the lower greenschist-facies Nevadan metamorphism.

The set of folds that is synchronous with the Nevadan metamorphism is cut in many places by plugs, dikes, and sills, which give ca. 150-Ma ages (K-Ar and U-Pb; Dick, 1973; Saleeby et al., 1982; Snoke et al., 1982; Harper and Wright, 1984; Gray, 1985). For the rest of this chapter, structures referred to as Nevadan have evidence of metamorphic recrystallization and are cut by the 145- to 150-Ma dike suite. All younger structures are considered post-Nevadan.

The orientations of fold axes and cleavages from the western Jurassic belt are shown in Figure 2. The data on the stereonets 1–4 are from this study. The data on stereonet 5 is from Harper (1980b), and stereonet 6 is from Cornelius (1986). The F_1 and F_2 folds trend generally north–south, with shallow plunges in each area studied. At locality 4 (Fig. 8), the same folds trend northeast, and have been refolded around at least two additional fold axes (F_3 and F_4; see discussion of post-Nevadan structures below).

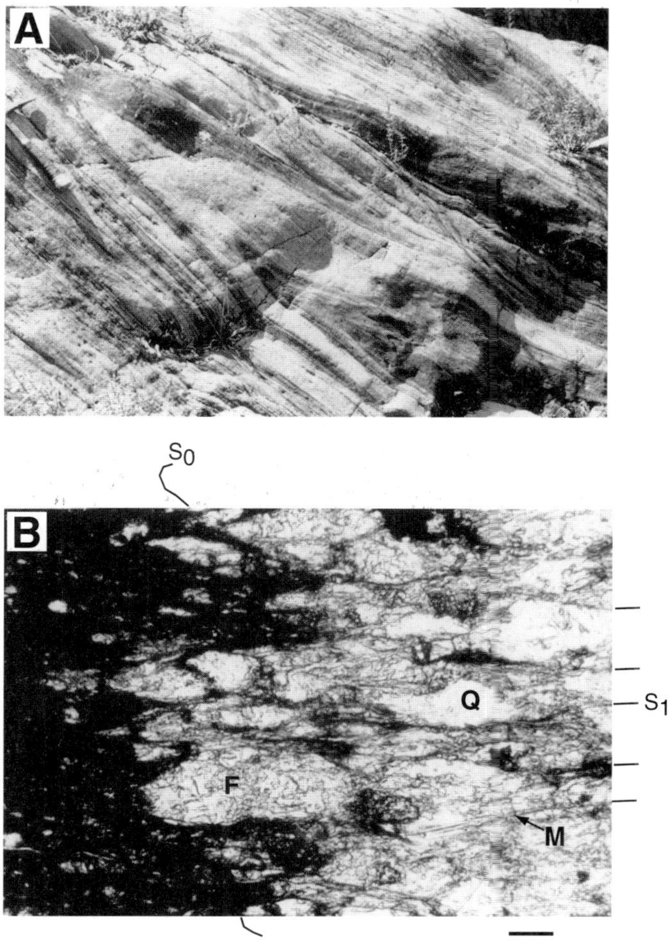

Figure 4. (A) Photograph of an isoclinal F_1 fold from locality 4 in Figure 2, looking north. This fold has an amplitude of 20 m and a wavelength <1 m. A ubiquitous axial planar cleavage (S_1) is developed in the western Jurassic belt rocks during this folding that is predominantly parallel to bedding. Hammer for scale is 33 cm long. (B) Photomicrograph of S_1 cleavage from the hinge zone of an F_1 fold. This view demonstrates the high degree of pressure solution experienced by the sand grains and the prominent recrystallization of white mica grains parallel to pressure solution surfaces. F—albitized feldspar partly replaced by white mica and minor calcite; M—muscovite; Q—quartz. Dark-light contrast is bedding (S_o). Scale bar is 0.01 mm.

A high degree of variability (up to 60°) in the trend of both fold generations is evident in the stereonets and in the field (Fig. 2). The scatter in F_1 axes is attributed in large part to the dismemberment of these folds along their axial surfaces (Fig. 5). Where F_1 and F_2 axes are observed together, they are not strictly coaxial, varying up to 30° in trend and plunge. A systematic variation between the trends of the two generations was not evident in this study. Despite these complicating factors, the S_1 data lie on fairly well-defined girdles. The estimated F_2 fold axis trends, approximately north at each locality, are close to the average trend and plunge for the F_1 folds, and this parallelism is also true

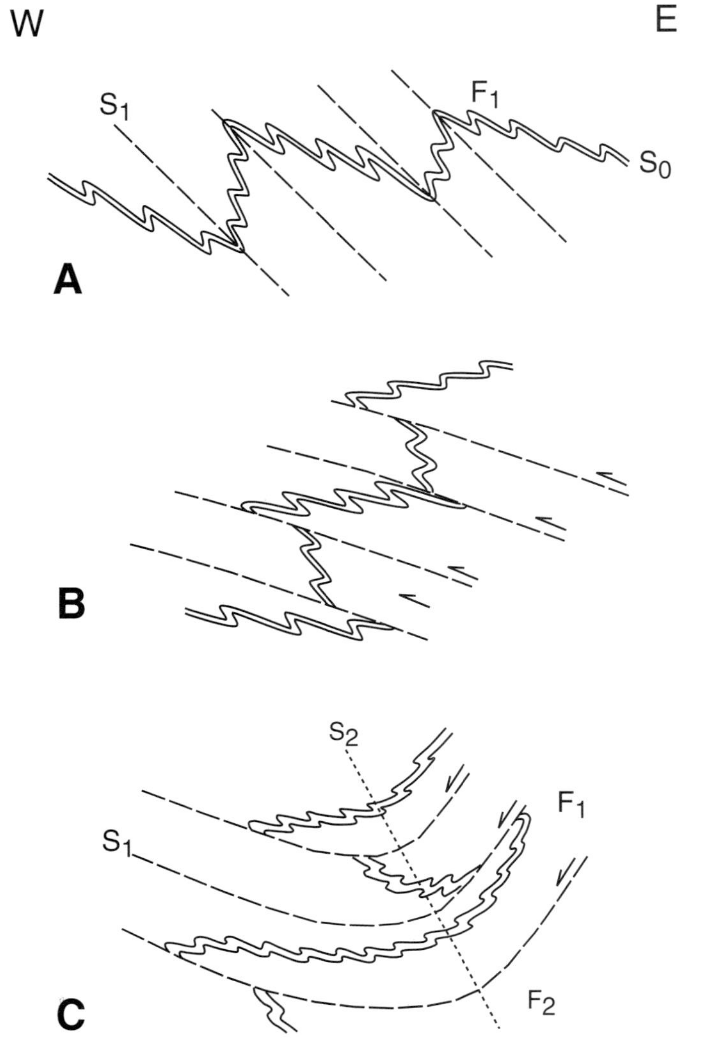

Figure 5. Schematic drawing depicting the development of Nevadan F_1 and F_2 structures in western Jurassic belt rocks. Solid lines are bedding, dashed lines are the S_1 cleavage, and dotted line is the S_2 cleavage. (A) Asymmetric first generation Nevadan folds, F_1, are westward-verging with penetrative axial-planar cleavage. (B) Folds rotated and began shearing along many S_1 surfaces during the development of the second fold set. In some areas, nearly all F_1 folds are dismembered in this way. Where it can be determined, transport of the displaced portion is top-to-the-west. (C) Simplified present relationships showing the superposition of F_2 folds on the disrupted F_1 structures. To the east of the S_2 surface, west-vergent F_1 structures are completely overturned and can appear to be east-vergent in the absence of tops indicators. In the field, S_2 folds can be found at many scales. Small-scale parasitic F_2 folds have been omitted from this diagram. Part C of this figure corresponds to the box shown on cross-section A–A′ in Figure 6.

for the northeast trend at locality 4. The similarity of the F_1 and F_2 trends throughout the study area and the similar metamorphic fabric associated with each of these sets of structures suggest that both formed under the same contractional regime. The north trend of the F_1 and F_2 folds indicates that contraction was predominantly east–west in the present coordinates.

Two cross-sections showing the variation in orientation of S_1 cleavage in the study area are presented in Figure 6. Section B–B′ shows the relatively large amplitude of F_2 structures. In both sections, F_2 folds become asymmetric with westward vergence near the Orleans fault, and S_2 dips to the east. Section A–A′ also shows thrust faults within the western Jurassic belt that cut the S_1 cleavage. Many younger normal faults are also evident on both sections.

In summary, the geometry of the Nevadan structures in the study area indicates that they formed in response to approxi-

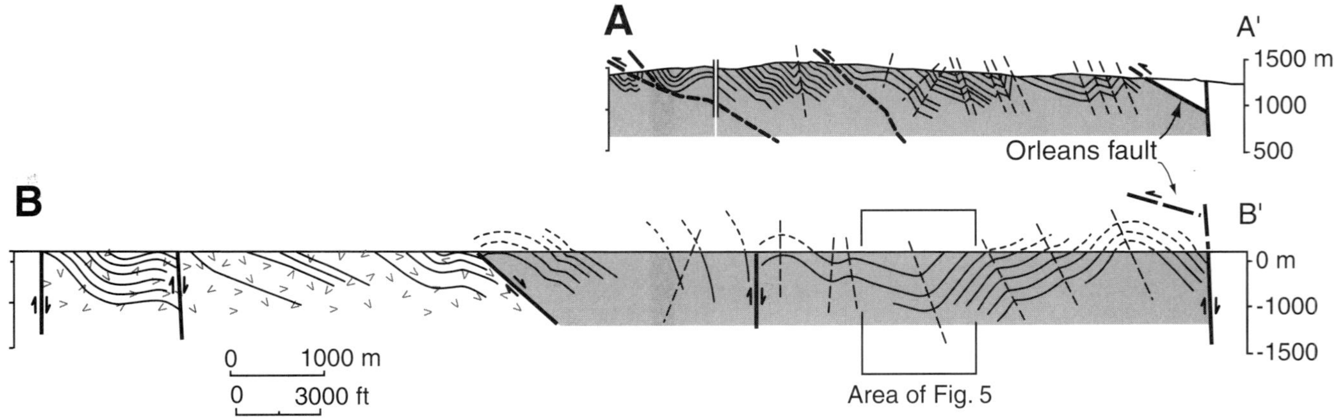

Figure 6. Interpreted cross-sections through the western Jurassic belt in the study area. The locations of the sections are shown in Figure 2. Solid lines represent the attitude of S_1 cleavages, dashed lines are S_2 cleavages, or S_1 cleavage projected upward. Gray shading represents the metagraywacke rocks, and the random V pattern represents the metavolcanic rocks. Section A–A′ is drawn along the Gasquet-Orleans Road and demonstrates the fold-and-thrust style deformation typical in the metagraywacke immediately below the Orleans fault. Note that some thrust faults apparently cut the S_1 cleavages. Section B–B′ traverses the entire Western Jurassic belt from Orleans to Weitchpec, California. The volcanic rocks present on the west end of this section also contain the S_1 cleavage. The vertical to slightly west-dipping S_2 cleavage in the center of this cross-section becomes east dipping toward the Orleans fault. The western Jurassic belt and Orleans fault in this area have been cut by several younger (Tertiary?), steeply dipping normal faults. The field data used to construct these cross-sections are from Gray (1985).

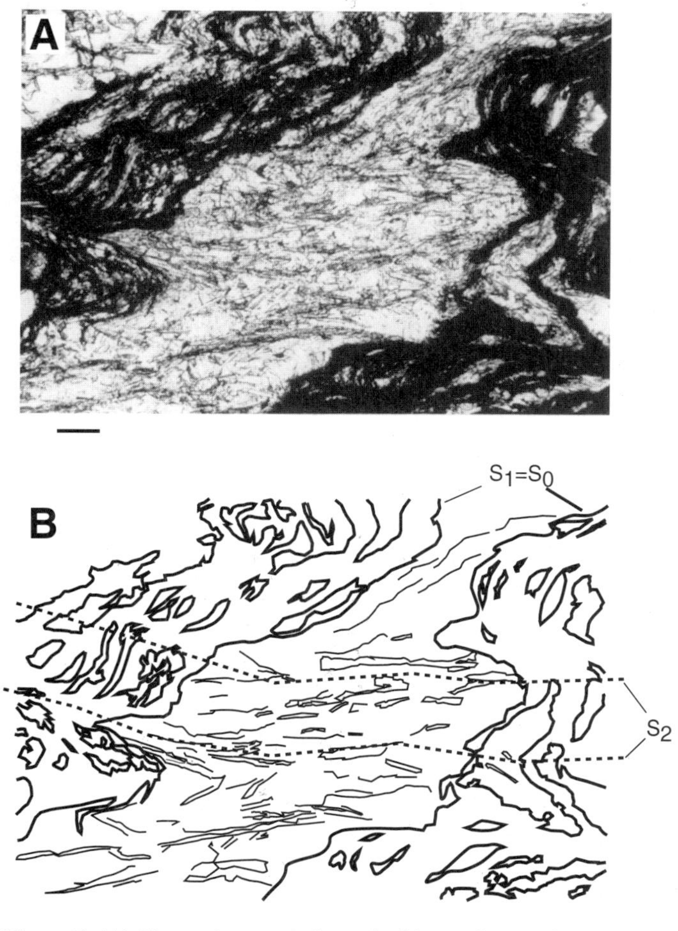

Figure 7. (A) Photomicrograph from the hinge of an F_2 fold, demonstrating that the S_2 axial-planar cleavage also has overprinting white mica growth. This growth indicates that it, too, developed synchronous with Nevadan metamorphism. Scale bar is 0.01 mm. (B) Interpreted line drawing of the photomicrograph. Heavy lines mark the S_0/S_1 surface between quartz and clay-rich beds. Thin lines outline mica crystals. Heavy dashed line is the S_2 cleavage surface. This sample has abundant, relatively large, white micas that have grown within the quartz selvages parallel to the S_2 cleavage. The sample has also experienced some later deformation that has slightly folded the S_2 cleavage.

mately east–west contraction in the present-day reference frame. There is nothing in these data or in the field observations to suggest that significant strike-slip displacement accompanied the Nevadan thrusting and metamorphism.

POST-NEVADAN STRUCTURES

Two types of post-Nevadan structures occur within the study area. The oldest of these are northeast-striking, right-lateral strike-slip faults and related folds restricted to a small area south of Happy Camp (Figs. 2 and 9). A portion of the area affected by these faults is shown in Figure 9. Northeast-directed right-lateral displacement has rotated the first- and second-generation folds into a northeast trend parallel to the faults and refolded them about horizontal and vertical F_3 and F_4 hinge lines (Fig. 9). This younger rotation is responsible for reorienting the north-trending F_1 and F_2 folds into their northeast trend at this locality. Most of the faults are shallowly east-dipping and occur as reactivations of the preexisting Nevadan cleavages. A small subset of these faults have left-lateral displacements and strike at high angle to the right-lateral faults in a classic Reidel shear geometry (Fig. 9).

These faults cut F_1 and F_2 folds and 150-Ma intrusive rocks and postdate the regional greenschist-facies metamorphism. These relationships, combined with the change in kinematic regime that they represent, suggest that these strike-slip faults postdate the Nevadan event. The parallelism of these strike-slip faults to the trend of the structural reentrant within the western Paleozoic and Triassic belt (Fig. 2) suggests that this system of faults may partly control the expression of this reentrant.

Regional uplift and normal faulting occurred during the last deformational event in this area, and have modified all the ear-

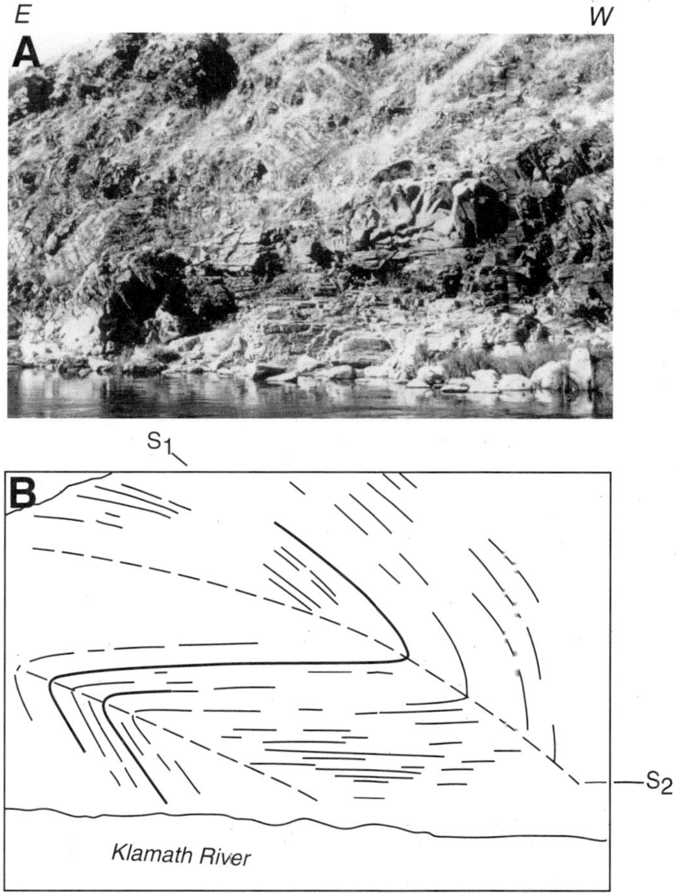

Figure 8. (A) Photograph of a small F_2 fold from locality 4 in Figure 2, looking south. Cliff is ~12 m high. The S_2 axial plane dips shallowly to the west, suggesting that this fold might be east-vergent. However, this F_2 fold has been overturned by later F_3 folding associated with local strike-slip faults. (B) Line drawing of photograph in A, showing the asymmetry of this fold.

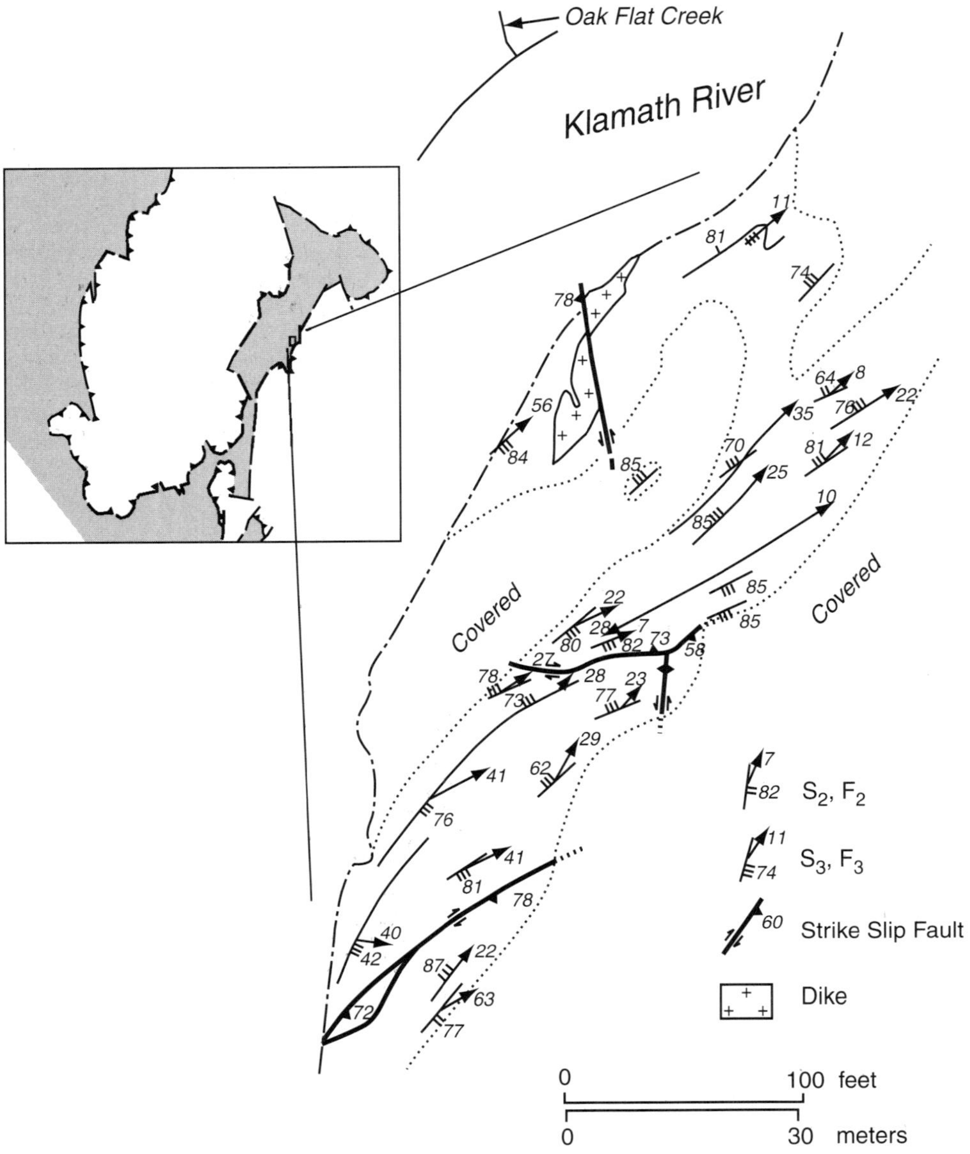

Figure 9. Detailed structure map from a part of locality 4, Figure 2. This area along the Klamath River contains locally abundant right- and left-lateral strike-slip faults. These faults cut and refold earlier thrust-related structures and post-date greenschist-facies metamorphism. The dike (plus symbols) is an undated plagioclase porphyry. From Gray (1985).

lier structures in many places (Fahan and Wright, 1984; Mortimer, 1985; Helper, 1986; Jachens et al., 1986; Calderwood and Armstrong, 1990). The uplift is centered on the Condrey Mountain dome, northeast of the study area, and the normal faults (and even some of the strike-slip faults) in the area may be radial to this feature (Fig. 2) (Barnes, 1983; Coleman and Helper, 1983; Mortimer and Coleman, 1985; Saleeby and Harper, 1993).

COMPARISON OF NEVADAN STRUCTURES IN THE KLAMATH MOUNTAINS AND SIERRA NEVADA FOOTHILLS

Nevadan-age structures in the Sierra Nevada Foothills terrane are generally much better studied than coeval structures in the Klamath Mountains. In gross aspect, the Nevadan-age faults of the Sierra Nevada Foothills dip much more steeply than do coeval faults in the Klamath Mountains (Clark, 1960; Davis, 1969; Day et al., 1985; Paterson and Vernon, 2001, and references therein). The steep orientation of the Sierra Nevada Foothill faults relative to the western Klamath belt faults suggests the Sierra Nevada Foothill structures may be exposed at a deeper structural level than the Klamath structures (e.g., Saleeby, 1981). Both of these areas may have undergone significant rotation from their original Nevadan attitude (e.g., Paterson and Vernon, 2001). Finally, the two areas may have formed under entirely different local strain patterns. For instance, the Sierran Foothills may have been predominantly a strike-slip regime and the Klamath Mountains may have been a thrust regime, as depicted in Figure 3B.

There is still disagreement regarding the kinematics of Nevadan deformation in the Sierra Nevada Foothills terrane (cf.

Patterson and Vernon, 2001, and Bell and Hickey, 1999). Historically, both strike-slip and dip-slip displacement have been interpreted as the predominant slip vector for the Nevadan structures in the Sierra Nevada Foothills (Best, 1963; Clark, 1964; Cebull, 1972; Avé Lallemant et al., 1977; Saleeby, 1977; Tobisch et al., 1977; Cebull and Russel, 1979; Nokleberg and Kistler, 1980; Paterson and Wainger, 1991; Paterson and Vernon, 2001).

One observation that might reconcile these two opposing points of view is the degree to which lithology controls the geometry and/or the amounts of strain accommodated by Nevadan-age structures (e.g., Paterson and Vernon, 2001). Within the central belt rocks of the Sierra Nevada, steeply plunging fold axes are primarily found within the less competent formations, including the greywacke-slate and mélange units, whereas adjacent metavolcanic units have horizontal axes (Best, 1963; Duffield and Sharp, 1975; Nokleberg and Kistler, 1980; Paterson and Redfield, 1984; Paterson and Vernon, 2001). The Nevadan fold axes in the less competent units form great-circle distributions that are parallel with the regional cleavage (Best, 1963; Clark, 1964; Duffield and Sharp, 1975; Nokleberg and Kistler, 1980; Paterson and Vernon, 2001). This pattern suggests that the fold axes have been rotated toward the direction of transport (e.g., Sanderson, 1973). This interpretation is consistent with strain studies in the area that also show maximum extension directions oriented near vertical (Tobisch et al., 1977; Paterson and Redfield, 1984; Paterson and Vernon, 2001). If this mechanism was operative during Nevadan deformation in the area, then both horizontal and "vertical" (i.e., great-circle distribution) fold axes may have formed during a single thrust event, as proposed by Paterson and Vernon (2001). Hacker (1993) also noted very similar geometrical relationships in slightly older Sierra Nevada Foothills structures.

DISCUSSION

The Nevadan-age structures in the western Jurassic belt of the Klamath Mountains and the Foothills terrane of the Sierra Nevada Mountains appear to reflect a relatively similar history. Late Jurassic flysch deposits in both areas have undergone a regional folding event, with minor heterogeneities developed locally. In both areas, this event produced at least one regionally consistent penetrative cleavage synchronous with greenschist-facies metamorphism. The trends of the regional folds in both areas are parallel to the bounding structures, and the major fold axes are near horizontal. The regional metamorphic grade is the same in the two areas, suggesting that the same structural (or crustal) level is exposed, and therefore the difference in attitude of the cleavages and bounding faults between the two areas is significant. Paterson and Vernon (2001) interpret the steep dips of the Sierra Nevada Foothills structures to reflect eastward tilting of these structures during late Nevadan or subsequent deformational events. This rotation may have reactivated the fault zones somewhat and added to the overall structural complexity of the system, but does not appear to have significantly altered the structural geometries of the units between the major bounding faults. The simplest kinematic interpretation of the Nevadan structural geometries for both areas is that they reflect relatively orthogonal contraction in the present reference frame. If this interpretation is true, then it follows that both areas were probably oriented similar to their current positions because significant deviation from these trends should induce different kinematic regimes in each area.

Global plate circuit models suggest that Pacific–North America relative motion during the Nevadan was nearly orthogonal convergence (Engebretson et al., 1985; Kelley and Engebretson, 1994). These models have a high degree of uncertainty, as very little Pacific seafloor of this age remains. It is also problematic to directly relate plate-motion vectors to the movement on individual structures. Within these limits, however, the models of Engebretson et al. (1985) are compatible with the interpretation of predominantly orthogonal contraction for the Nevadan systems in both the Klamath and Sierra Nevada Foothills. For the purposes of this general interpretation, it appears that the general plate convergence/kinematic model presented in Figure 3A is the most likely configuration of both the Klamath Mountains and Sierra Nevada during Late Jurassic Nevadan deformation.

ACKNOWLEDGMENTS

This chapter is the result of the author's dissertation research at the University of Texas at Austin and was supported by grants from the Geological Society of America, Sigma Xi, California Department of Mines and Geology, Chevron, and the University of Texas at Austin. Bill Carlson, John Maxwell, Eric Rosencrantz, Art Snoke, and especially Sharon Mosher are thanked for their help and encouragement in this effort. The U.S. Geological Survey, the U.S. Forest Service, and in particular, Dennis Inman, provided logistical support for the fieldwork portion. Cal Barnes, Mary Donato, Mark Helper, Greg Harper, Reinold Cornelius, and Jim Wright freely shared their ideas about Klamath tectonics and spent time discussing these relationships in the field. Scott Paterson, Tim Grover, and Cal Barnes are thanked for their patient and thorough reviews of this manuscript, although the author accepts all responsibility for any errors.

REFERENCES CITED

Avé Lallemant, H.G., Weisenberg, C.W., and Standlee, L.A., 1977, Structural development of the Melones fault zone, northeastern California: Geological Society of America Abstracts with Programs, v. 9, p. 383.

Barnes, C.G., 1983, Petrology and upward zonation of the Wooley Creek batholith, Klamath Mountains, California: Journal of Petrology, v. 24, p. 495–537.

Bell, T.H., and Hickey, K.A., 1999, Complex microstructures preserved in rocks with a simple matrix: Significance for deformation and metamorphic processes: Journal of Metamorphic Geology, v. 17, p. 521–535, doi: 10.1046/j.1525-1314.1999.00214.x.

Best, M.G., 1963, Petrology and structural analysis of metamorphic rocks in the southwestern Sierra Nevada Foothills, California: California University Publications in Geological Sciences, v. 42, p. 111–158.

Blackwelder, E., 1914, A summary of the orogenic epochs in the geologic history of North America: Journal of Geology, v. 22, p. 633–654.

Bogen, N.L., 1986, Paleomagnetism of metavolcanic rocks of the upper Jurassic Galice Formation, western Oregon: Evidence for differential rotation of the eastern and western Klamath Mountains: Geology, v. 14, p. 335–338, doi: 10.1130/0091-7613(1986)14<335:POTUJG>2.0.CO;2.

Calderwood, A.R., and Armstrong, R.L., 1990, Geochronologic and tectonic constraints on the onset and mechanism of Neogene uplift of the Klamath Mountains, northern California: EOS, American Geophysical Union Transactions, v. 71, no. 49, p. 1633.

Cashman, S.M., 1988, Finite-strain patterns of Nevadan deformation, western Klamath Mountains, California: Geology, v. 16, p. 839–843, doi: 10.1130/0091-7613(1988)016<0839:FSPOND>2.3.CO;2.

Cebull, S.E., 1972, Sense of displacement on the Foothills fault system: New evidence from the Melones fault zone, western Sierra Nevada, California: Geological Society of America Bulletin, v. 83, p. 1185–1190.

Cebull, S.E., and Russel, L.R., 1979, Role of Melones fault zone in the structural chronology of the north Yuba River area, western Sierra Nevada, California: Geological Society of America Bulletin, v. 90, part 1, p. 225–227, doi: 10.1130/0016-7606(1979)90<225:ROTMFZ>2.0.CO;2.

Clark, L.D., 1960, The Foothills fault system, western Sierra Nevada, California: Geological Society of America Bulletin, v. 71, p. 483–496.

Clark, L.D., 1964, Stratigraphy and structure of a part of the western Sierra Nevada metamorphic belt, California: Washington, D.C., U.S. Geological Survey Professional Paper 410, 70 p.

Coleman, R.G., and Helper, M.A., 1983, The significance of the Condrey Mountain dome in the evolution of the Klamath Mountains, California and Oregon: Geological Society of America Abstracts with Programs, v. 15, p. 294.

Cornelius, R.R., 1986, Geology and structure of the Rock Creek Butte area, Del Norte County, Klamath Mountains, California: California Geology, v. 39, p. 195–201.

Davis, G.A., 1969, Tectonic correlations, Klamath Mountains and western Sierra Nevada, California: Geological Society of America Bulletin, v. 80, p. 1095–1108.

Day, H.W., Moores, E.M., and Tuminas, A.C., 1985, Structure and tectonics of the northern Sierra Nevada: Geological Society of America Bulletin, v. 96, p. 436–450, doi: 10.1130/0016-7606(1985)96<436:SATOTN>2.0.CO;2.

Dick, H.J.B., 1973, K-Ar dating of intrusive rocks in the Josephine peridotite and Rogue Formation west of Cave Junction, southwestern Oregon: Geological Society of America Abstracts with Programs, v. 5, p. 33–34.

Dickinson, W.R., and Lawton, T.F., 2001, Carboniferous to Cretaceous assembly and fragmentation of Mexico: Geological Society of America Bulletin, v. 113, p. 1142–1160, doi: 10.1130/0016-7606(2001)113<1142:CTCAAF>2.0.CO;2.

Diller, J.S., 1907, Mesozoic sediments of southwestern Oregon: American Journal of Science, fourth series, v. 22, p. 401–422.

Duffield, W.A., and Sharp, R.V., 1975, Geology of the Sierra Foothills mélange and adjacent areas, Amador County, California: Reston, Virginia, U.S. Geological Survey Professional Paper 827, 30 p.

Engebretson, D.C., Cox, A., and Gordon, R.G., 1985, Relative motions between oceanic and continental plates in the Pacific basin: Boulder, Colorado, Geological Society of America Special Paper 206, 59 p.

Fahan, M.R., and Wright, J.E., 1984, The eastern Hayfork terrane, a Permo-Triassic subduction complex, southwestern Klamath Mountains: Geological Society of America Abstracts with Programs, v. 15, p. 272.

Gray, G.G., 1983, A kinematic analysis of thrust faulting, west-central Klamath Mountains, California: Geological Society of America Abstracts with Programs, v. 15, p. 585.

Gray, G.G., 1985, Structural, geochronologic, and depositional history of the western Klamath Mountains, California and Oregon: Implications for the early to middle Mesozoic tectonic evolution of the western North American Cordillera [Ph.D. dissertation]: Austin, University of Texas at Austin, 162 p.

Gray, G.G., 1986, Native terranes of the central Klamath Mountains, California: Tectonics, v. 5, p. 1043–1054.

Hacker, B.R., 1993, Evolution of the Sierra Nevada metamorphic belt: Petrological, structural and Ar/Ar constraints: Geological Society of America Bulletin, v. 105, p. 637–656, doi: 10.1130/0016-7606(1993)105<0637:EOTNSN>2.3.CO;2.

Hacker, B.R., Donato, M.M., Barnes, C.G., McWilliams, M.O., and Ernst, W.G., 1995, Timescales of orogeny, Jurassic construction of the Klamath Mountains: Tectonics, v. 14, p. 677–703, doi: 10.1029/94TC02454.

Harper, G.D., 1980a, The Josephine ophiolite—Remains of a Late Jurassic marginal basin in northwestern California: Geology, v. 8, p. 333–337, doi: 10.1130/0091-7613(1980)8<333:TJOOAL>2 0.CO;2.

Harper, G.D., 1980b, Structure and petrology of the Josephine ophiolite and overlying sedimentary rocks, northwestern California [Ph.D. dissertation]: Berkeley, University of California, 259 p.

Harper, G.D., 1983, A depositional contact between the Galice Formation and a Late Jurassic ophiolite in northwestern California and southwestern Oregon: Oregon Geology, v. 45, p. 3–9.

Harper, G.D., 1984, The Josephine ophiolite, northwestern California: Geological Society of America Bulletin, v. 95, p. 1009–1026, doi: 10.1130/0016-7606(1984)95<1009:TJONC>2.0.CO;2.

Harper, G.D., and Wright, J.E., 1984, Middle to Late Jurassic tectonic evolution of the Klamath Mountains, California–Oregon: Tectonics, v. 3, p. 759–772.

Harper, G.D., Saleeby, J.B., and Heizler, M., 1994, Formation and emplacement of the Josephine ophiolite and the Nevadan orogeny in the Klamath Mountains, California–Oregon; U/Pb zircon and $^{40}Ar/^{39}Ar$ geochronology: Journal of Geophysical Research, v. 99, p. 4293–4321, doi: 10.1029/93JB02061.

Helper, M.A., 1986, Deformation and high P/T metamorphism in the central part of the Condrey Mountain window, north-central Klamath Mountains, California and Oregon, *in* Evans, B.W., and Brown, E.H., eds., Blueschists and eclogites: Boulder, Colorado, Geological Society of America Memoir 164, p. 125–141.

Helper, M.A., 1992, Evidence for successive Late Jurassic–Early Cretaceous underplating during high P/T metamorphism of the Condrey Mountain Schist, central Klamath Mountains, California and Oregon: Geological Society of America Abstracts with Programs, v. 24, p. 33.

Hershey, O.H., 1906, Some western Klamath Stratigraphy: American Journal of Science, fourth series, v. 21, p. 58–66.

Hinds, N.E.A., 1934, The Jurassic age of the last granitoid intrusives in the Klamath Mountains and Sierra Nevada, California: American Journal of Science, v. 27, p. 182–193.

Irwin, W.P., 1960, Geologic reconnaissance of the northern Coast Ranges and Klamath Mountains, California, with a summary of the mineral resources: San Francisco, California Division of Mines and Geology Bulletin 179, 80 p.

Irwin, W.P., 1966, Geology of the Klamath Mountains province *in* Bailey, E.H., ed., Geology of northern California: San Francisco, California Division of Mines and Geology Bulletin 190, p.17–38.

Irwin, W.P., and Mankinen, E.A., 1998, Rotational and accretionary evolution of the Klamath Mountains, California and Oregon, from Devonian to present time: Reston, Virginia, U.S. Geological Survey Open File Report 98-114, 7 p.

Jachens, R.C., Barnes, C.G., and Donato, M.M., 1986, Subsurface configuration of the Orleans fault: Implications for deformation in the western Klamath Mountains, California: Geological Society of America Bulletin, v. 97, p. 388–395, doi: 10.1130/0016-7606(1986)97<388:SCOTOF>2.0.CO;2.

Kelley, K.P., and Engebretson, D.C., 1994, Updated relative plate motions and terrane trajectories for North America and oceanic plates: Cretaceous to Present: Geological Society of America Abstracts with Programs, v. 26, p. 459.

Klein, C.W., 1977, Thrust plates of the north-central Klamath Mountains near Happy Camp, California, *in* Short contributions to California Geology: Sacramento, California Division of Mines and Geology Special Report 129, p. 23–26.

Lanphere, M.A., Irwin, W.P., and Hotz, P.E., 1968, Isotopic age of the Nevadan orogeny and older plutonic and metamorphic events in the Klamath Mountains, California: Geological Society of America Bulletin, v. 79, p. 1027–1052.

Lanphere, M.A., Blake, M.C., Jr., and Irwin, W.P., 1978, Early Cretaceous metamorphic age of the South Fork Mountain schist in the northern Coast Ranges of California: American Journal of Science, v. 278, p. 798–815.

Mankinen, E.A., and Irwin, W.P., 1982, Paleomagnetic study of some Cretaceous and Tertiary sedimentary rocks of the Klamath Mountains province, California: Geology, v. 10, p. 82–87, doi: 10.1130/0091-7613(1982)10<82:PSOSCA>2.0.CO;2.

Mankinen, E.A., Irwin, W.P., and Grommé, C.S., 1984, Implications of paleomagnetism for the tectonic history of the eastern Klamath and related terranes in California and Oregon, *in* Nilsen, T.H., ed., Geology of the Upper Cretaceous Hornbrook Formation, Oregon and California: Los Angeles, Pacific Section, Society of Economic Paleontologists and Mineralogists, v. 42, p. 221–229.

Mankinen, E.A., Lindsley-Griffin, N., and Griffin, J.R., 2002, Concordant paleolatitudes for Neoproterozoic ophiolitic rocks of the Trinity complex, Klamath Mountains, California: Journal of Geophysical Research, v. 107, no. B10, p. 2254, doi: 10.1029/2001JB001623.

Miller, M.M., and Saleeby, J.B., 1995, U-Pb geochronology of detrital zircon from Upper Jurassic synorogenic turbidites, Galice Formation, and related rocks, western Klamath Mountains; Correlation and Klamath Mountains provenance: Journal of Geophysical Research, v. 100, p. 18,045–18,058, doi: 10.1029/95JB00761.

Miller, J., Miller, R., Wooden, J., and Harper, G., 2003, Geochronological links between the Ingalls ophiolite, North Cascades, Washington and the Josephine ophiolite, Klamath Mountains, Oregon and California: Geological Society of America Abstracts with Programs, v. 35, p. 113.

Mortimer, N., 1985, Constraints on the Permian to Jurassic evolution of the Klamath Mountains: Geological Society of America Abstracts with Programs, v. 17, p. 371.

Mortimer, N., and Coleman, R.G., 1985, A Neogene structural dome in the Klamath Mountains, California: Geology, v. 13, p. 253–256, doi: 10.1130/0091-7613(1985)13<253:ANSDIT>2.0.CO;2.

Nokleberg, W.J., and Kistler, R.J., 1980, Paleozoic and Mesozoic deformations in the central Sierra Nevada, California: Reston, Virginia, U.S. Geological Survey Professional Paper 1145, 24 p.

Paterson, S.R., and Redfield, T.F., 1984, Finite strains and structures across the Melones fault zone near Moccasin, California: Geological Society of America Abstracts with Programs, v. 16, p. 619.

Paterson, S.R., and Vernon, R.H., 2001, Inclusion trail patterns in porphyroblasts from Foothills terrane, California: A record of orogenesis or local strain heterogeneity? Journal of Metamorphic Geology, v. 19, p. 351–372, doi: 10.1046/j.0263-4929.2000.00315.x.

Paterson, S.R., and Wainger, L., 1991, The Melones Fault zone: A terrane bounding stretching fault: Tectonophysics, v. 194, p. 69–90, doi: 10.1016/0040-1951(91)90273-U.

Saleeby, J.B., 1977, Fracture zone tectonics, continental margin fragmentation and emplacement of the Kings-Kaweah ophiolite belt, southwest Sierra Nevada, California, *in* Coleman, R.G., and Irwin, W.P., eds., North American ophiolites: Portland, Oregon Department of Geology and Mineral Industries Bulletin 95, p. 141–160.

Saleeby, J.B., 1981, Ocean floor accretion and volcanoplutonic arc evolution of the Mesozoic Sierra Nevada, *in* Ernst, W.G., ed., The Geotectonic evolution of California—Rubey Volume 1: Englewood Cliffs, New Jersey, Prentice-Hall, p. 132–181.

Saleeby, J.B., and Harper, G.D., 1993, Tectonic relations between the Galice Formation and the Condrey Mountain Schist, Klamath Mountains, northern California, *in* Dunne, G.C., and McDougall, K.A., eds., Mesozoic paleogeography of the western United States II, Field Trip Guidebook: Los Angeles, Pacific Section, Society of Economic Paleontologists and Mineralogists, v. 71, p. 61–80.

Saleeby, J.B., Snoke, A.W., and Sharp, W.D., 1982, Time relations and structural-stratigraphic pattern in ophiolite accretion, west central Klamath Mountains, California: Journal of Geophysical Research, v. 87, p. 3831–3848.

Sanderson, P.J., 1973, The development of folds oblique to the regional trend: Tectonophysics, v. 16, p. 55–70.

Schweickert, R.A., Bogen, N.L., Girty, G.G., Hanson, R.E., and Meguerian, C., 1984, Timing and structural expression of the Nevadan orogeny, Sierra Nevada, California: Geological Society of America Bulletin, v. 95, p. 967–979, doi: 10.1130/0016-7606(1984)95<967:TASEOT>2.0.CO;2.

Smith, J.G., Page, N.J., Johnson, M.G., Moring, B.C., and Gray, F., 1982, Preliminary geologic map of the Medford 1°×2° quadrangle, Oregon and California: Reston, Virginia, U.S. Geological Survey Open-File Report OFR 82–955.

Snoke, A.W., 1977, A thrust plate of ophiolitic rocks in the Preston Peak area, Klamath Mountains, California: Geological Society of America Bulletin, v. 88, p. 1641–1659, doi: 10.1130/0016-7606(1977)88<1641:ATPOOR>2.0.CO;2.

Snoke, A.W., Sharp, W.D., Wright, J.E., and Saleeby, J.B., 1982, Significance of mid-Mesozoic peridotitic to dioritic intrusive complexes, Klamath Mountains–western Sierra Nevada, California: Geology, v. 10, p. 160–166, doi: 10.1130/0091-7613(1982)10<160:SOMPTD>2.0.CO;2.

Taliaferro, N.L., 1942, Geologic history and correlation of the Jurassic of southwestern Oregon and California: Geological Society of America Bulletin, v. 53, p. 71–112.

Tobisch, O.T., Fiske, R.S., Sacks, S., and Taniguchi, D., 1977, Strain in metamorphosed volcaniclastic rocks and its bearing on the evolution of orogenic belts: Geological Society of America Bulletin, v. 88, p. 23–40, doi: 10.1130/0016-7606(1977)88<23:SIMVRA>2.0.CO;2

Wagner, D.L., and Saucedo, G.J., compilers, 1987, Geologic map of the Weed quadrangle, California: Sacramento, California Division of Mines and Geology Regional Geologic Map Series, Map 4A, scale 1:250,000.

Wells, F.G., Hotz, P.E., and Cater, F.W., Jr., 1949, Preliminary description of the geology of the Kirby quadrangle, Oregon: Portland, Oregon Department of Geology and Mineral Industries Bulletin 40, 23 p.

Wright, J.E., and Wyld, S.J., 1994, The Rattlesnake Creek terrane, Klamath Mountains, California: An early Mesozoic volcanic arc and its basement of tectonically disrupted oceanic crust: Geological Society of America Bulletin, v. 106, p. 1033–1056, doi: 10.1130/0016-7606(1994)106<1033:TRCTKM>2.3.CO;2.

Manuscript Accepted by the Society 10 January 2005

Printed in the USA

Geological Society of America
Special Paper 410
2006

Geochemistry of ophiolitic rocks associated with the western part of the Elk outlier of the western Klamath terrane, southwestern Oregon

Mario J. Giaramita
Department of Physics and Geology, California State University, Stanislaus, 801 W. Monte Vista Avenue, Turlock, California 95382, USA
Gregory D. Harper
Department of Earth and Space Sciences, State University of New York Albany, 1400 Washington Avenue, Albany, New York 12222, USA

ABSTRACT

A newly discovered dike complex and serpentinite-matrix mélange may represent basement for the Upper Jurassic Galice Formation in the Elk outlier of the western Klamath terrane, which lies far to the west of the Klamath Mountains province. The fault-bounded dike complex consists of virtually 100% parallel dikes with scarce diorite screens and amphibole-rich hydrothermal metamorphic assemblages, all consistent with it being a sheeted dike complex of an ophiolite. Magmatic affinities of the dikes are transitional island-arc tholeiite (IAT) to mid-ocean ridge basalt (MORB) and boninitic, similar to the Josephine ophiolite that underlies the Galice Formation elsewhere. The mélange includes blocks of mafic volcanic (many pillowed) and hypabyssal rocks, igneous breccia, amphibolite, pyroxenite, peridotite, and metasandstone in a sheared serpentinite matrix. Volcanic rocks and amphibolites have MORB magmatic affinities, and some of the pillow lavas are unusual highly fractionated Fe-Ti basalts. Hypabyssal blocks have MORB, within-plate, and calc-alkaline affinities. The magmatic affinities of the mélange blocks are distinct from the dike complex, and thus the mélange and dike complex are unrelated petrogenetically. The mélange may belong to the underlying Pickett Peak terrane of the Franciscan Complex, or it may be part of the western Klamath terrane and correlative to Lower Mesozoic ophiolitic rocks found elsewhere in "rift-edge facies" of the Josephine ophiolite. The Elk outlier is far to the north of the part of the western Klamath terrane that contains the Josephine ophiolite. Thus, the presence of a remnant of the Josephine ophiolite in the Elk outlier would be consistent with Cretaceous dextral strike-slip displacement proposed by previous workers or with the Josephine extending farther north than previously recognized.

Keywords: Elk outlier, dike complex, serpentinite-matrix mélange, Pearse Peak pluton, Josephine ophiolite, geochemistry, boninites, IAT, MORB, western Klamath terrane, Oregon

*E-mail: marty@geology.csustan.edu.

Giaramita, M.J., and Harper, G.D., 2006, Geochemistry of ophiolitic rocks associated with the western part of the Elk outlier of the western Klamath terrane, southwestern Oregon, *in* Snoke, A.W., and Barnes, C.G., eds., Geological studies in the Klamath Mountains province, California and Oregon: A volume in honor of William P. Irwin: Geological Society of America Special Paper 410, p. 153–176, doi: 10.1130/2006.2410(08). For permission to copy, contact editing@geosociety.org.

INTRODUCTION

The presence of an outlier of the western Klamath terrane situated far to the west of the Klamath Mountains geologic province (Fig. 1) has long been known (Wells and Peck, 1961; Dott, 1965, 1971; Coleman, 1972; Blake et al., 1985a). This part of the western Klamath terrane, called the "Elk subterrane" by Blake et al. (1985a) and the "Elk outlier" of the western Klamath terrane by Jayko and Blake (1993), has received virtually no study since initial work by Dott (1965, 1966, 1971), Dott's students (Kaiser, 1962; Koch, 1966; Aalto, 1968), and Coleman (1972). The oldest previously mapped rock unit in the outlier is the Upper Jurassic Galice Formation. In this chapter, we describe a newly recognized sheeted(?) dike complex that may represent ophiolitic basement for the Galice Formation. In addition, we describe a serpentinite-matrix mélange that may either be part of the Galice basement or belong to the underlying Pickett Peak terrane.

The dike complex and serpentinite-matrix mélange are fault-bounded units. We characterize these units petrographically and geochemically to test whether they are related. In addition, the geochemistry is used to constrain the tectonic setting of magmatism for each unit and to compare with potentially correlative rocks elsewhere. The model for strike-slip displacement of the outlier during the Cretaceous (Jayko and Blake, 1993) is examined in light of the new data.

REGIONAL GEOLOGY

The Klamath Mountains consist of an east-dipping stack of ophiolitic and arc-affinity thrust sheets that generally young to the west (e.g., Irwin, 1966, 1981, 1994; Davis, 1969; Snoke, 1977; Burchfiel and Davis, 1981; Hacker et al., 1995; Fig. 1). West of the Klamath Mountains province, as defined by Irwin (1966), are the Pickett Peak, Yolla Bolly, and Gold Beach terranes of the Franciscan accretionary complex and the ophiolitic Snowcamp Mountain terrane (Fig. 1). All terranes, except for those of the Franciscan Complex, are intruded by Mesozoic plutons (e.g., Harper et al., 1994, 2000; Irwin, 1994; Hacker et al., 1995; Kosanke, 2000; Fig. 1).

Rattlesnake Creek and Western Hayfork Terranes

The Lower Mesozoic Rattlesnake Creek terrane structurally overlies the western Klamath terrane along the Orleans thrust (Snoke, 1977; Jachens et al., 1986; Wyld and Wright, 1988; Fig. 1). The Rattlesnake Creek terrane consists of a serpentinite-matrix mélange and peridotite massifs overlain by an intact sedimentary-volcanic sequence (Wright and Wyld, 1994). The mélange is cut by ca. 200- to 190-Ma intrusions that apparently fed volcanic rocks overlying the mélange. A volcano-plutonic arc complex built on Rattlesnake Creek terrane includes the 174- to 166-Ma western Hayfork terrane and 164- to 159-Ma Wooley Creek plutonic belt (Wright and Fahan, 1988; Hacker et al., 1995; Fig. 1).

The Preston Peak ophiolite (Snoke, 1977) occurs within the Rattlesnake Creek terrane near the Oregon-California border. It is a composite ophiolite consisting of Rattlesnake Creek terrane serpentinite-matrix mélange and a mafic complex of diabase dikes and diabase breccia. The mafic complex has been dated at 164 ± 1 Ma (Saleeby and Harper, 1993) and is thus substantially younger than the mélange and may be a "rift-edge facies" formed during rifting prior to seafloor spreading that formed the Josephine ophiolite (Saleeby et al., 1982; Harper et al., 1994; Yule et al., this volume).

Western Klamath Terrane

Klamath Mountains Province. The Upper Jurassic Galice Formation makes up much of the western Klamath terrane (Irwin, 1966; MacDonald et al., this volume; Fig. 1). It consists of dark slate, metagraywacke (lithic feldspathic wacke), and scarce conglomerate, and has sedimentary structures indicating deposition by turbidity currents. Scarce fossils of the bivalve *Buchia concentrica* indicate an age of upper Oxfordian to early Kimmeridgian.

In the southern part of the western Klamath terrane, the Galice Formation conformably overlies the ca. 162-Ma Josephine ophiolite and its hemipelagic cover (Harper, 1984; Harper et al., 1994; MacDonald et al., this volume). In contrast, in the northern part of the western Klamath terrane, the Galice Formation conformably overlies the Upper Jurassic (ca. 157–153 Ma) Rogue Formation, a thick sequence of submarine, predominantly volcaniclastic rocks considered to be part of an arc complex (Dick, 1976; Garcia, 1982; Harper et al., 1994; Yule, 1996; Yule et al., this volume).

The basement of the Rogue Formation consists of disrupted ophiolitic rocks that are lithologically and geochemically virtually identical to the basal ophiolitic unit of the Rattlesnake Creek terrane (Yule, 1996; Yule et al., this volume). In addition, mafic complexes similar in age and lithology to those of the Preston Peak ophiolite occur locally beneath the Rogue Formation. Along the north-central and northeastern parts of the Josephine ophiolite, the ophiolite appears to terminate against this older Rattlesnake Creek terrane–like basement, which includes local Josephine-age mafic complexes (rift-edge facies; Yule, 1996; Yule et al., this volume).

The 160- to 157-Ma, predominantly mafic Chetco intrusive complex (Illinois River plutonic complex of Yule, 1996) structurally underlies the basement of the Rogue Formation and the Josephine ophiolite (Fig. 1) and is generally considered to represent the plutonic roots of the Rogue–Chetco arc complex (Dick, 1976; Garcia, 1982; Harper et al., 1994; Yule, 1996; Yule et al., this volume). A retrograded amphibolite "metamorphic sole" is present along the thrust contact between the Josephine ophiolite and Chetco complex (Madstone Cabin thrust, Fig. 1)

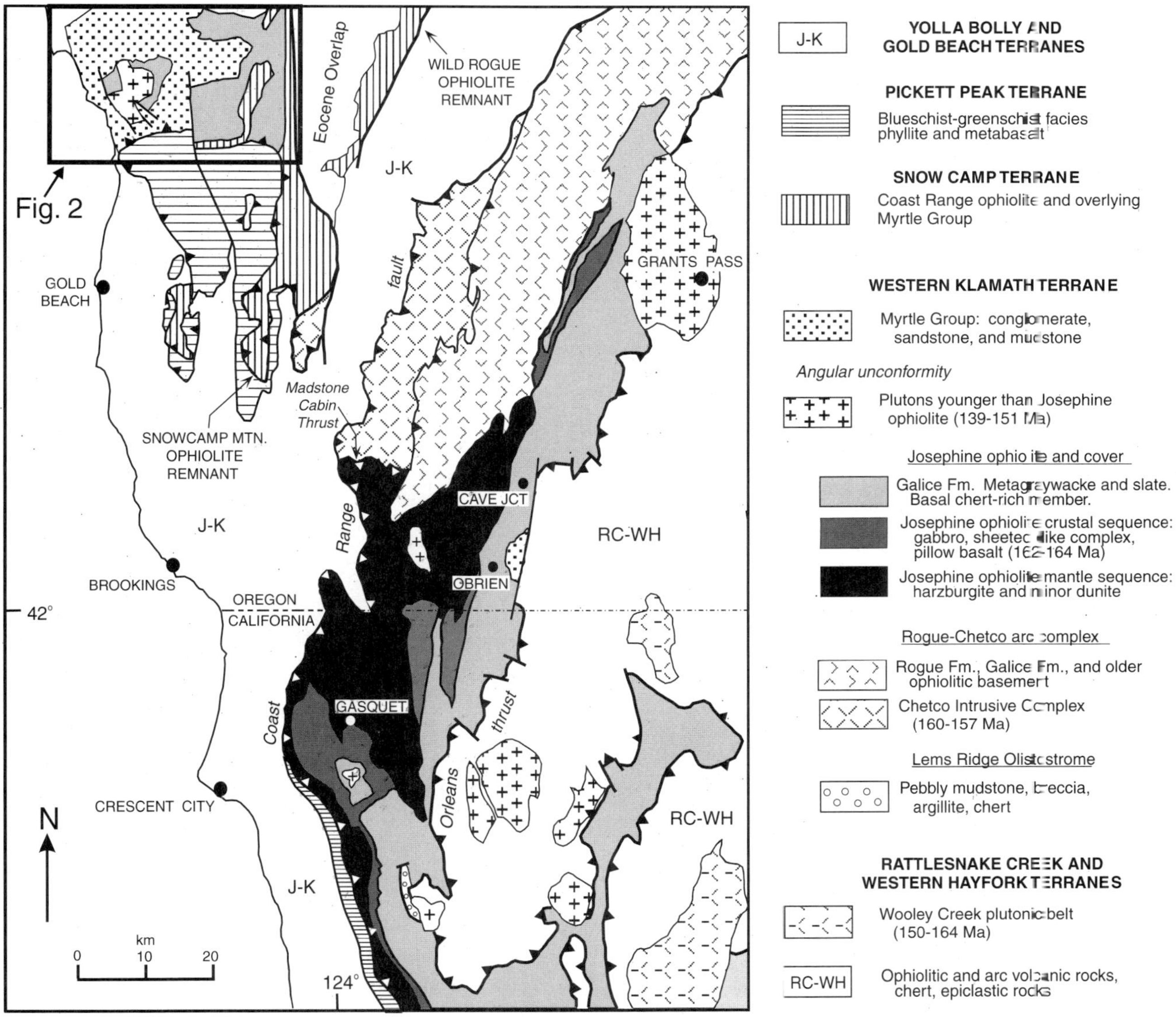

Figure 1. Generalized geologic map of the west-central Klamath Mountains. Compiled from Dott (1971), Snoke (1977), Harper (1980, 1984), Saleeby et al. (1982), Blake et al. (1985a), and Yule (1996).

and appears to have formed during thrusting of the ophiolite over the Chetco complex (Harper et al., 1990, 1994, 1996).

The Galice Formation, Josephine ophiolite, and Rogue Formation are intruded by numerous ca. 150- to 145-Ma calc-alkaline dikes, sills, and small plutons (Harper, 1980, 1984, this volume; Harper et al., 1994; Barnes et al., this volume, Chapter 17; Fig. 1). They are apparently syntectonic, in that dikes and sills in the Galice Formation are regionally metamorphosed and locally deformed, yet some of the related plutons cut the Orleans thrust (Fig. 1) or have contact aureoles that overprint slaty cleavage in the Galice (Harper et al., 1994). Displacement on the Orleans and Madstone Cabin thrusts is constrained to have occurred primarily between ca. 155 and 150 Ma (Harper et al., 1994). The thrusting and dynamothermal low-grade regional metamorphism of the western Klamath terrane are manifestations of the Late Jurassic Nevadan orogeny. A Nevadan angular unconformity is present east of Obrien, Oregon (Fig. 1), where Lower Cretaceous strata overlie the Galice formation (Harper et al., 1994).

Elk Outlier. The area of this study is within the Elk outlier, a western outlier of the western Klamath terrane near the Oregon Coast (Figs. 1 and 2). The outlier consists dominantly of slate, metagraywacke (lithic feldspathic wacke), and scarce pebble conglomerate correlated with the Galice Formation, based

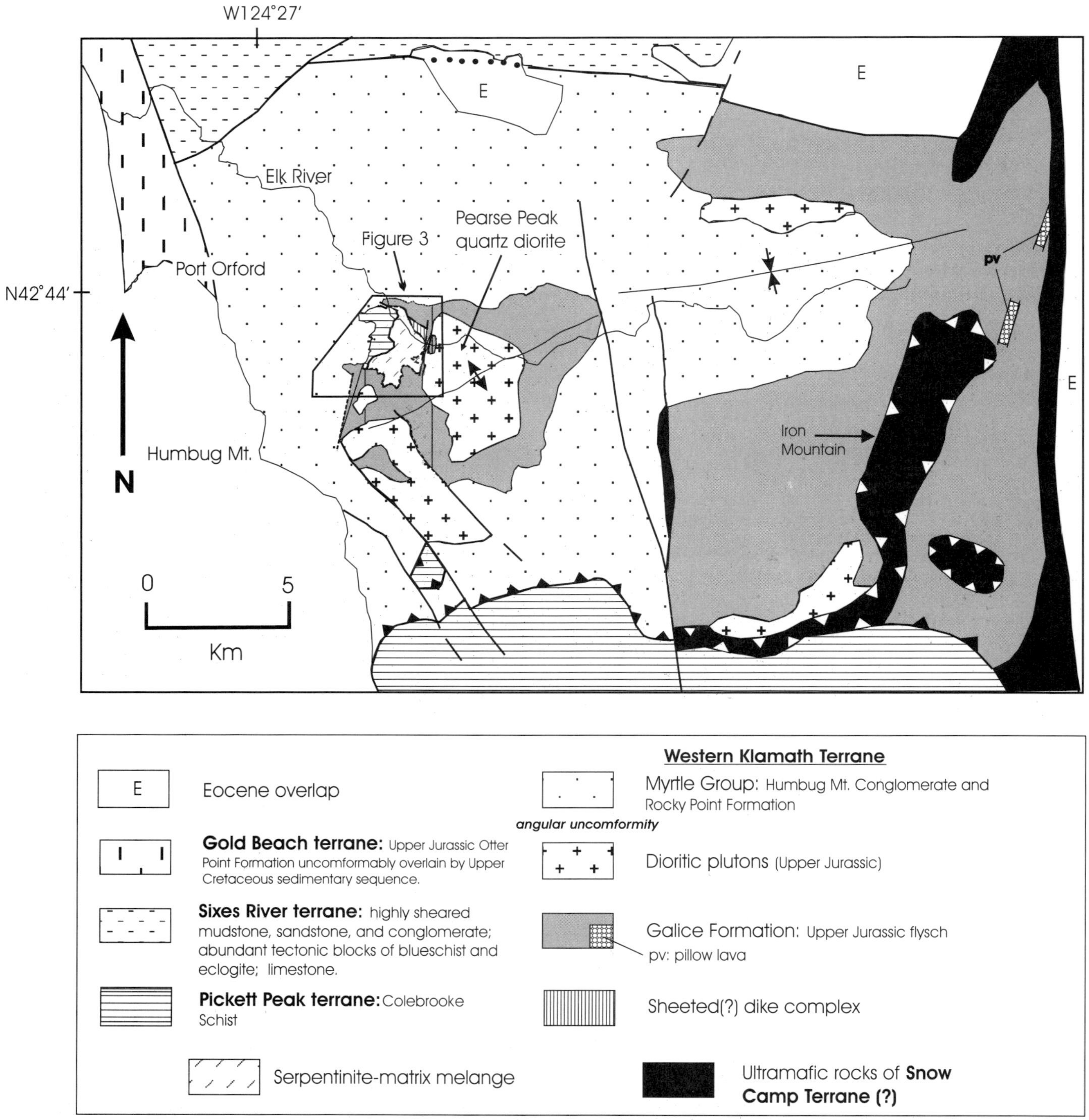

Figure 2. Geology and terranes of the Elk River outlier. Boxed region is the lower Elk River area (Fig. 3), which was remapped in this study. Compiled from Kaiser (1962), Dott (1971), Blake et al. (1985a), Walker and MacLeod (1991), and this study.

on lithologic similarities and the occurrence of the bivalve *Buchia concentrica* (Wells and Peck, 1961; Dott, 1966, 1971).

More than 2500 m of Lower Cretaceous strata correlated with the Myrtle Group (Humbug Mountain Conglomerate and Rocky Point Formation) unconformably overlie the Galice Formation in the Elk outlier (Dott, 1971). This is the same relationship observed in the Klamath Mountains near Obrien, Oregon, although the basal strata in the Elk outlier are slightly older (Valanginian versus Hauterivian; Harper et al., 1994).

Small dioritic plutons and dikes intrude and contact metamorphose the Galice Formation in the Elk outlier (Kaiser, 1962; Dott, 1965, 1966, 1971; Koch, 1966). Their upper age is bracketed by the age of the Galice Formation wall rocks (ca. 153–150 Ma; MacDonald et al., this volume), and their lower age is bracketed by unconformably overlying Valanginian strata (ca. 135 Ma) and 144 ± 7- and 148 ± 4-Ma K/Ar biotite cooling ages (Dott, 1965; ages recalculated using decay constants of Steiger and Jager, 1977). Thus, these intrusions are similar in age to the ca. 150- to 145-Ma dikes, sills, and plutons that cut the Galice Formation elsewhere (Harper et al., 1994; Harper, this volume).

Basement rocks for the Galice Formation in the Elk outlier have not been previously recognized. Pillow basalt and associated chert reported by Dott (1966, 1971) in the eastern part of the outlier (Fig. 2) perhaps suggest an ophiolitic basement. The Iron Mountain ultramafic massif (Fig. 2) may further represent ophiolite basement, although Blake et al. (1985a) interpreted it to be a klippe of the Snowcamp terrane.

Snowcamp Terrane

The Snowcamp terrane lies between the Klamath Mountains province and Elk outlier of the western Klamath terrane (Fig. 1). The terrane consists of an ophiolite overlain by uppermost Jurassic to Lower Cretaceous clastic strata of the Myrtle Group (Blake et al., 1985a). The Snowcamp Mountain remnant has all the components of a complete ophiolite, including massive peridotite, although it is highly faulted (Harper et al., 2000; Schoonmaker et al., this volume).

Blake et al. (1985a) correlated the ophiolite and overlying clastic strata of the Myrtle Group with the Coast Range ophiolite and overlying Great Valley Group, respectively, which are exposed in the California Coast Ranges. The Coast Range ophiolite is similar in age and geochemistry to the Josephine ophiolite (Kosanke, 2000; Metzger et al., 2002), but it has not been affected by regional dynamothermal metamorphism of the Nevadan orogeny and did not receive turbidite deposition until the latest Jurassic (Tithonian; Hopson et al., 1981). The Great Valley Group conformably overlies the Coast Range ophiolite and its pelagic cover, whereas the Great Valley Group and its correlative units in the Klamath Mountains overlie the Josephine ophiolite and Galice Formation with great unconformity (e.g., Harper et al., 1994; Dickinson et al., 1996).

Stratigraphically, the Snowcamp terrane appears to resemble the Coast Range ophiolite. However, the Wild Rogue remnant and its overlying volcaniclastic unit appear to have been metamorphosed to prehnite-pumpellyite facies, and deformed pegmatite dikes in the metagabbro yielded 148-Ma muscovite cooling ages (Kosanke, 2000), which are in the range of Nevadan regional metamorphism in the western Klamath terrane (Harper et al., 1994). Thus, an unconformity between unmetamorphosed Myrtle Group and the underlying volcaniclastic sequence and Wild Rogue ophiolite remnant may be present. Furthermore, the Snowcamp Mountain remnant shows several remarkable similarities to the Josephine ophiolite (Harper et al., 2000; Schoonmaker et al., this volume): (1) magmatic affinities of pillow lavas and sheeted dikes; (2) distinctive chemistry of Cr-spinel in peridotite and pillow lavas; (3) intrusion by ca. 150- to 145-Ma calc-alkaline plutons and dikes; and (4) a ca. 150-Ma retrograded amphibolite "sole" thrust between peridotite and the underlying Chetco complex. A plausible explanation for similarities of the Snowcamp terrane to both the Coast Range and Josephine ophiolites is that the Josephine and Coast Range are part of the same ophiolite belt, as argued by some workers (Saleeby, 1992; Dickinson et al., 1996).

Pickett Peak Terrane

The Pickett Peak terrane structurally underlies both the western Klamath terrane and Snowcamp terrane along the Coast Range fault (e.g., Blake et al., 1985a,b; Fig. 1). In southwestern Oregon, the Pickett Peak terrane includes the Colebrooke Schist (Coleman, 1972), consisting primarily of pelitic phyllite and minor semischist (metagraywacke). Pillow lavas are a minor but consistent component as blocks ranging in scale from tens of meters to kilometers (Coleman, 1972; Plake, 1989; Katrib, 2005), and green phyllites (metavolcanic) occur locally. Ultramafic rocks are rare, consisting of sheared serpentinite and talc schist.

The Colebrooke Schist is typically crenulated; it is also commonly sheared and contains abundant quartz veins. Pillow lavas are mostly undeformed, although some are foliated. Metamorphism is transitional greenschist-blueschist facies, and the metamorphic age is most likely Early Cretaceous (Coleman, 1972; Brown and Blake, 1987).

Roure and Blanchet (1983) mapped a serpentinite-matrix mélange adjacent to the Colebrooke Schist that Blake et al. (1985a) included as part of the Pickett Peak terrane. A serpentinite-matrix mélange was also recently mapped in the Snowcamp Mountain area, situated structurally between the Snowcamp terrane and underlying Colebrooke Schist (Katrib, 2005; Schoonmaker et al., this volume). Blocks in the mélange include pillow basalt, phyllite identical in appearance to that in the Colebrooke Schist, peridotite, hornblende diorite, and rare ultramafic cumulate. The phyllite is almost certainly derived from the Colebrooke Schist, but peridotite, diorite, peridotite, and ultramafic cumulate are rare or absent in the Colebrooke and may have been derived from the overlying Snowcamp terrane; in addition, pillow lavas are geochemically more similar to those of the Snowcamp terrane (Katrib, 2005). The mélange likely formed during underthrusting of the Colebrooke beneath the Snowcamp terrane, resulting in mixing of rocks from the headwall and footwall.

GEOLOGY OF THE STUDY AREA

Our mapping (Fig. 3) of the northwest lobe of the Pearse Peak Diorite is a substantial revision of previous geologic maps. Kaiser (1962) subdivided the Pearse Peak Diorite into two

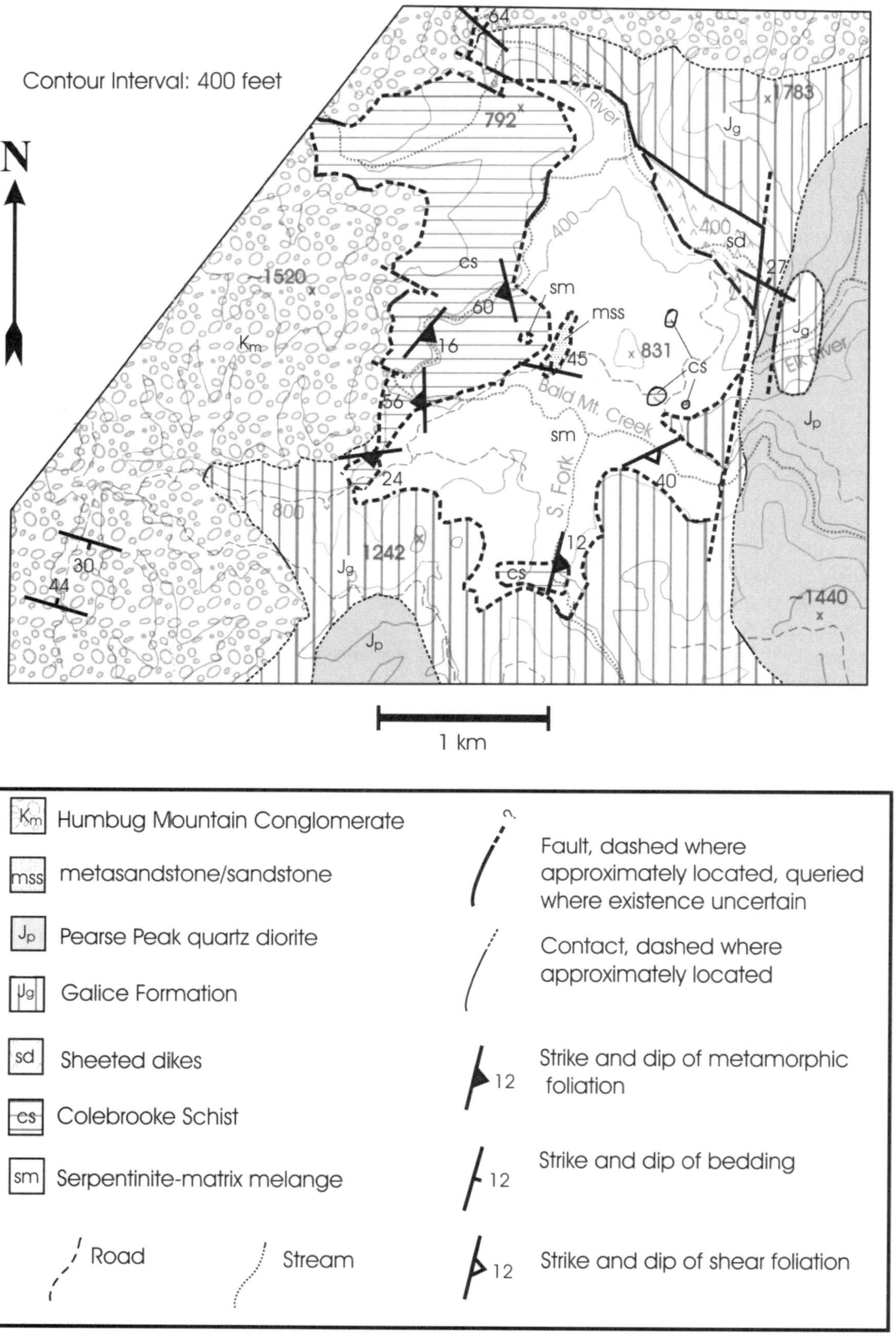

Figure 3. Geologic map of the study area along the lower Elk River, portions of the Port Orford and Father Mountain 7.5″ quadrangles, southwest Oregon. Boldfaced numbers (e.g., 831, 1520) are elevations above sea level.

phases: a biotite-hornblende quartz diorite phase, which constitutes the main mass of the pluton, and a diorite phase that comprises a northwestern lobe. Our mapping reveals that what Kaiser (1962) mapped as the northwestern lobe of the pluton actually consists of two lithotectonic units, neither of which are part of the Pearse Peak pluton: (1) a mafic dike complex and (2) serpentinite-matrix mélange (Fig. 3). The Pearse Peak pluton on our map and cross-sections includes only light-colored, hypidiomorphic granular, hornblende-biotite quartz diorite.

Dike Complex

The newly recognized dike complex consists of virtually 100% parallel dikes (Fig. 4), ~15 cm to 1 m thick, having distinct chilled margins. The rocks are gray-green in color and are commonly difficult to recognize as dikes except where they have been polished by stream flow. Abundant dike margins are well exposed in landslide blocks, up to 10 m in diameter, on the Elk River (Universal Trans Mercator coordinates [UTM] in meters, zone 10: 4729390N, 387512E; Fig. 3). One boulder contains a diorite screen intruded by two dikes with chilled margins (Fig. 4B). Recognizable dikes are also evident in some outcrops on the Elk River (e.g., UTM: 4729777N, 387448E) and along the Elk River road (4728533N, 386899E).

Altered primary minerals in dikes include augite (partially to completely altered to fibrous actinolite), plagioclase (altered to albite), and the Fe-Ti oxides (altered to sphene). Additional secondary minerals include calcite, chlorite, epidote, pumpellyite, prehnite, and quartz. The diorite screen has relict augite rimmed in places by brown, igneous(?) hornblende. Both are partially altered to actinolitic hornblende (pale brown to light-green pleochroism) and fibrous actinolite. Plagioclase is albitized and secondary chlorite is present. The dikes and screen are cut by prehnite + albite veins (Fig. 4B). The presence of actinolite and actinolitic hornblende suggests high-temperature hydrothermal metamorphism followed by retrogressive (chlorite, albite, prehnite, pumpellyite) metamorphism.

The dike complex has characteristics of the lower parts of ophiolitic sheeted dike complexes, including: (1) virtually 100% dikes, (2) parallel dike margins, (3) plutonic screens, and (4) retrograded amphibole-rich hydrothermal metamorphic mineral assemblages (e.g., Harper et al., 1988).

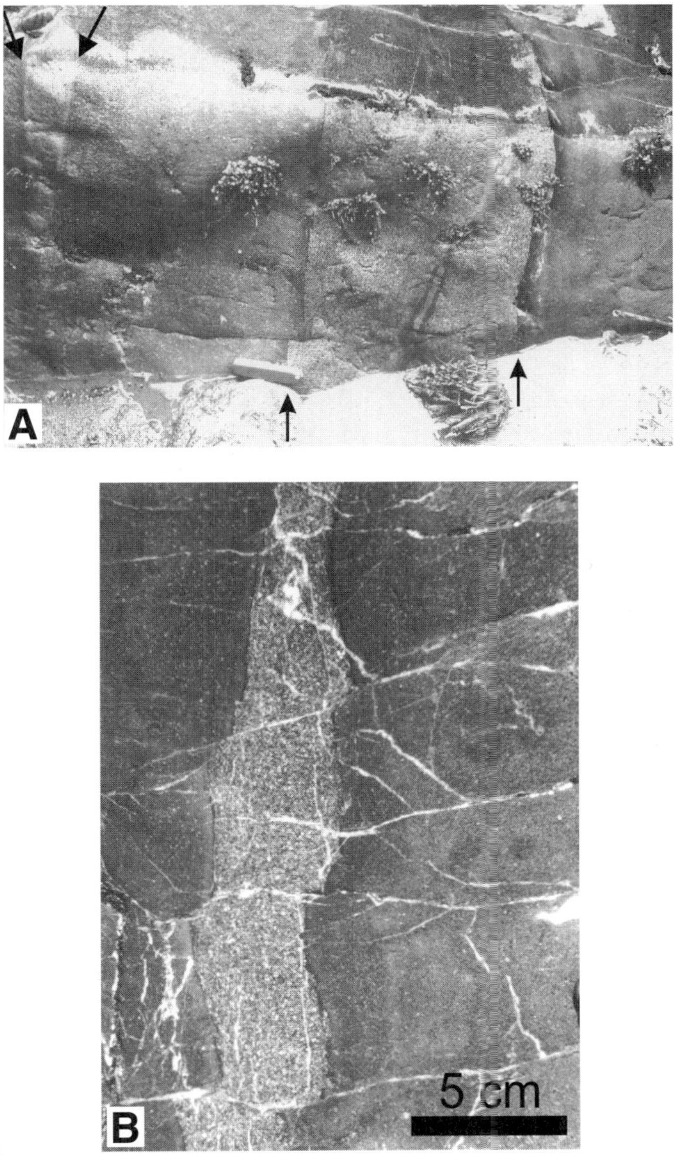

Figure 4. (A) Multiple parallel dikes in the sheeted(?) dike complex associated with the Elk River outlier. Swiss army knife in front center of photograph for scale. (B) Gabbro screen and several dike margins in the sheeted(?) dike complex associated with the Elk River outlier. Screen depicted in photograph was not analyzed geochemically.

Serpentinite-Matrix Mélange

Introduction. The mélange consists of a wide variety of blocks ranging from less than 1 m to hundreds of meters in diameter within a sheared serpentinite matrix. Petrography of selected mélange blocks is given in Appendix Table A-1. The relative proportion of blocks to matrix varies from nearly 100% blocks to 100% sheared serpentinite. Block-free areas are covered by grassy meadows that reveal sheared serpentinite in gulleys, whereas block-rich regions are densely forested. The shear fabric in the serpentinite matrix defines a foliation that generally varies considerably in orientation, even within a single outcrop.

Exposures showing contact relationships between block and matrix are scarce. Evidence that greenstone and other rock types occur as blocks in sheared serpentinite, rather than as separate rock units, includes: (1) rocks observed in rare outcrops as blocks wholly or partially encased in sheared serpentinite; (2) blocks in landslides having remnants of sheared serpentinite casings still attached; (3) the variety of rock types, including plutonic, volcanic, metamorphic, and sedimentary rocks, occurring within an outcrop belt of sheared serpentinite; and (4) isolated wooded areas underlain by greenstone within meadows underlain by sheared serpentinite.

Colebrooke Schist exposures occur within the area mapped as mélange (Fig. 3), but contacts are not exposed and blocks similar to the Colebrooke Schist were never observed as blocks in serpentinite. Thus, it is uncertain whether these exposures are blocks in the mélange or part of the adjacent Colebrooke Schist unit in fault contact with the mélange.

Igneous Blocks. Pillow lavas were recognized in several meter-scale exposures and in a 300-m-long canyon exposure on Bald Mountain Creek, a few 100 m upstream from its confluence with the Elk River (UTM: 4730212N, 386395E; Fig. 3). Pillow diameters vary from 2 to 30 cm. Dark green, originally glassy rinds are commonly evident on the pillows. No interpillow sediment was observed. Massive metabasalts, which have volcanic textures and mineral phases identical to the pillow lavas, but without observable pillow structure, are also present.

Phaneritic intrusives having average grain sizes of ~2 mm are present in the mélange. Some have cumulate textures, whereas others are hypidiomorphic granular. Relict augite is present in some samples; all contain amphibole (actinolite or actinolitic hornblende, some fibrous) and highly altered, albitized plagioclase. Secondary sphene and epidote are common. Greenstone blocks having fine-grained phaneritic texture are presumably hypabyssal intrusives. Some blocks exhibit partial to complete alteration to rodingite.

Igneous breccia contains clasts of mafic volcanic, hypabyssal, and plutonic rock fragments, which vary from millimeters to tens of centimeters in diameter. A dacite-breccia block exposed in Bald Mountain Creek (UTM: 4729940N, 386177E) consists almost exclusively of fragments of dacite and sand-sized grains derived from dacite.

What appear to be cumulate clinopyroxenite and olivine(?) clinopyroxenite are exposed in a grassy meadow underlain by sheared serpentinite (UTM: 4728978N, 386607E). Noncumulate peridotite also occurs as blocks within the sheared serpentinite matrix. Aligned bastite grains and stringers of Cr-spinel indicate the rocks are foliated and suggest that the peridotites are mantle tectonites from an ophiolite, although Cr-spinel in one of the samples shows both euhedral and embayed grains, suggesting that it may be a cumulate.

High-Grade Metamorphic Blocks. Upper amphibolite- to granulite-facies(?) two-pyroxene amphibolite is exposed in a grassy meadow underlain by sheared serpentinite on hill 831 (UTM: 4729279N, 386738; Fig. 3). Together with abundant brown hornblende, this rock contains augite and strongly pleochroic hypersthene. In addition, garnet amphibolite, some of which occurs as monolithologic breccia, crops out along the South Fork of Bald Mountain Creek (UTM: 4728495N, 386474E).

Sedimentary and Metasedimentary Rocks. A block of tuffaceous radiolarian chert was found in Bald Mountain Creek (UTM: 4728726N, 386667E). Abundant radiolarians, ~0.1–0.2 mm in diameter, are evident in thin section, as well as sparse detrital grains of quartz, plagioclase, and Fe-Ti oxides.

An essentially volcaniclastic, coarse-grained weakly foliated metagraywacke from Bald Mountain Creek (UTM: 4728877N, 386672E) contains felsic and andesitic volcanic rock fragments, plagioclase, altered glass including pumice, mudstone, and <2% each of chert, monocrystalline quartz, and augite. This sample is too weakly foliated to have been derived from the Colebrooke Schist and contains too few chert clasts compared to typical Galice graywackes; it is similar, however, to volcanic-rich graywackes of the basal Galice Formation elsewhere in the western Klamath terrane (Harper, 1984; MacDonald et al., this volume).

The three other sandstones samples come from an 800 × 400-m area mapped by Diller (1903) and Kaiser (1962) as belonging to the Cretaceous Myrtle Group ("mss" in Fig. 3). One sandstone has abundant subangular grains of chert-argillite, mudstone, volcanic rocks, plagioclase and ~15% modal monocrystalline quartz, which is similar to sandstones of the Myrtle Group (Koch, 1966). A second sandstone, however, is essentially volcaniclastic, with abundant plagioclase and scarce clasts of chert, quartzite, and mudstone; an abundance of chlorite, suturing of grain boundaries in quartz veins, and very fine needles of pumpellyite(?) suggest this sandstone may be a metagraywacke. The third sample is very fine grained and has a well-defined foliation and is thus clearly a metagraywacke; it is less well foliated than the Colebrooke metagraywackes, but appears to have a lower lithic content and higher modal monocrystalline quartz than do Galice metagraywackes (MacDonald et al., this volume). The contact of this "mss" unit (Fig. 3) with the serpentinite mélange matrix is not exposed. The mapped unit contains metagraywacke and graywacke, but does not contain conglomerate. The contact is not an uncomformity, because basal Humbug Mountain Conglomerate is characterized by massive boulder conglomerates (Koch, 1966). Consequently, the unit is mapped as fault-bounded; it could also possibly be a block in the mélange.

Dikes Cutting Sheared Serpentinite. Dacitic dikes cross-cut sheared serpentinite at two localities (e.g., 4728586N, 386000E). The contacts between the dikes and serpentinite are straight and dike width is <1 m. No rodingite is evident at the contacts, indicating the dikes postdate serpentinization.

Field Relationships

The study area is very heavily vegetated, and the best exposures are along the Elk River and its tributaries. Exposed contacts between map units are scarce, and many of the exposures are deeply weathered. Moreover, in the absence of observable contacts between sheared serpentinite and phaneritic igneous rocks within the mélange, it is difficult to determine whether these rocks are blocks or intrusive into the mélange.

The dike complex is bounded by high-angle faults. A faulted, rather than intrusive, contact with the mélange is indicated by brecciation of the dike complex along the contact and the apparent absence of mafic dikes cutting the mélange. The dike complex could be a block in the mélange, but its aerial extent is much greater than other blocks, and it seems unlikely that this single block would mark the fault contact on both its east and northeast borders with the Galice Formation (Fig. 3).

The study area contains two rock units consisting of low-grade metamorphic rocks: the Galice Formation of the western Klamath terrane and the Colebrooke Schist of the Pickett Peak terrane. The Galice Formation is, for the most part, contact metamorphosed to a dark-gray cordierite(?)-biotite hornfels in the study area. The hornfels is intruded by numerous intermediate-to-mafic dikes varying from ~0.3 to 10 m in thickness, which are not shown on the map because of difficulty in tracing them beyond stream and road-cut exposures. By contrast, the Colebrooke Schist is a highly crenulated, phyllitic, carbonaceous phyllite/schist that is generally sheared and cut by abundant quartz veins. The Colebrooke Schist shows neither the effects of contact metamorphism nor is it intruded by dikes or plutons.

The intrusive nature of the contact between the Galice Formation and Pearse Peak quartz diorite is evident from the contact metamorphic overprint on the Galice Formation and from blocks of the Galice "floating" in the pluton. In places, however, the two units are in fault contact; an obvious fault contact between the two units is exposed on the Elk River near the western margin of the pluton (Fig. 3) and has an attitude of N2°W, 56°E. The apparent absence of contact metamorphism of the serpentinite-matrix mélange suggests that the Pearse Peak pluton is in fault contact with the mélange.

Topographically, high exposures of Galice Formation bound the serpentinite-matrix mélange on three sides, suggesting that the Galice lies above and in low-angle contact with the mélange. The absence of contact metamorphism of the mélange suggests that the contact is faulted rather than depositional. The dip of the contact between the serpentinite-matrix mélange and Colebrooke Schist, however, could not be determined. The map pattern is consistent with the mélange coring a domical upwarp in the center of the study area (Fig. 3), in which case, the mélange dips west beneath the Colebrooke Schist. However, paucity of outcrops allow for the possibility that the contact dips to the east, which would place the mélange structurally on top of the Colebrooke Schist and beneath the Galice Formation.

The Humbug Mountain Conglomerate unconformably overlies the Galice Formation, and the angular unconformity is

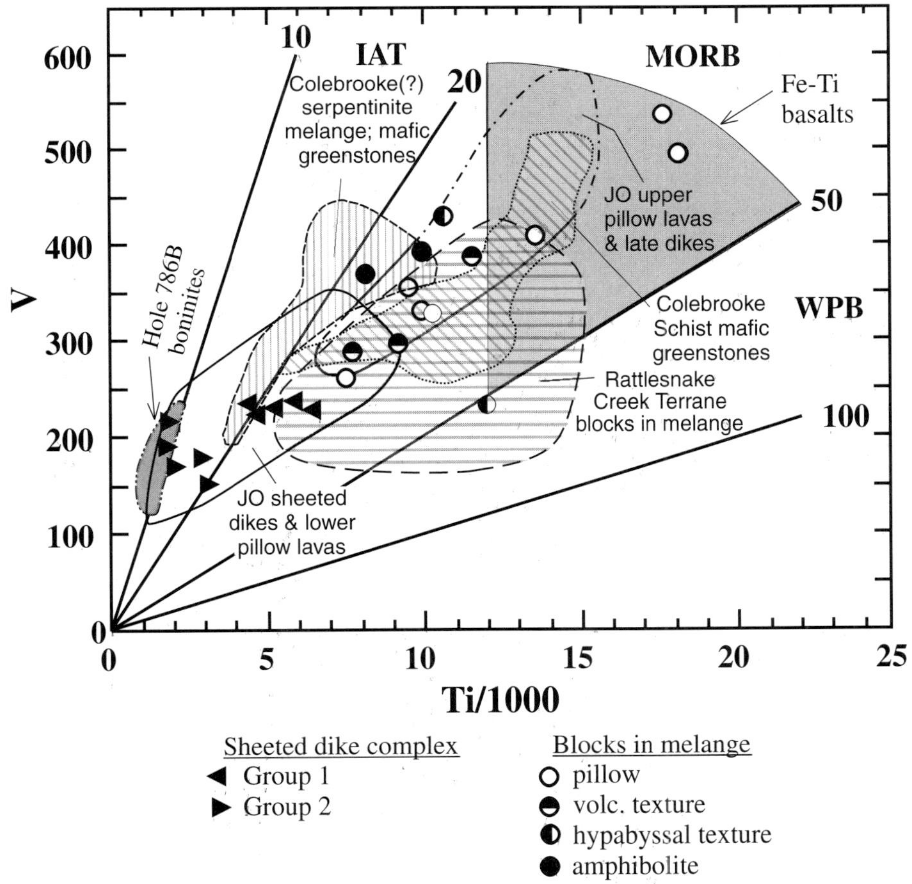

Figure 5. Ti vs. V diagram. The two gabbros in the data set samples are not plotted because their MORB-normalized patterns (Figs. 9 and 10) suggest they are cumulates. Two of the hypabyssal samples are not plotted because of their calc-alkaline affinities (Fig. 7). The absence of observed contacts between matrix and hypabyssal rocks permits the interpretation that they are intrusive into rather than blocks in the mélange. Calc-alkaline samples, based on Figure 7, not plotted. IAT—island-arc tholeiite, JO—Josephine ophiolite, MORB—mid-ocean ridge basalt, WPB—within-plate basalt. After Shervais (1982). Field for boninites from the Izu-Bonin forearc is from Murton et al. (1992), and field for highly fractionated MORB-affinity Fe-Ti basalts is based on definition by Sinton et al. (1983; >2 wt% TiO_2 and >12 wt% FeO^T). Fields for JO are from Harper (2003, 2004); field for Rattlesnake Creek terrane mafic mélange blocks is based on data in Gorman (1985) and Wright and Wyld (1994); field for Colebrooke greenstones is from Plake (1989) and Katrib (2005); field for Colebrooke(?) mélange blocks is from Katrib (2005).

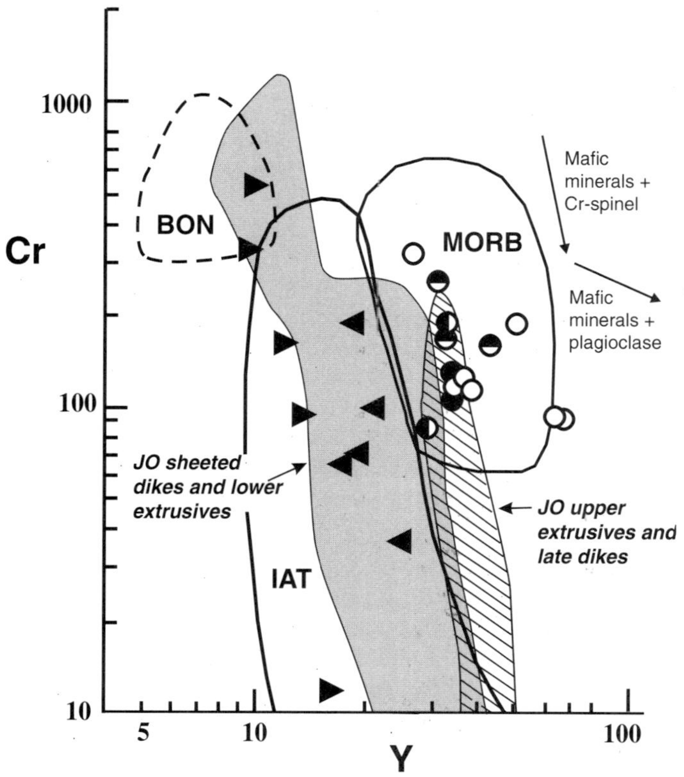

Figure 6. Cr vs. Y diagram. The two gabbros in the data set are not plotted because their MORB-normalized patterns (Figs. 9 and 10) suggest they are cumulates. Calc-alkaline samples, based on Figure 7, are not plotted. Hypabyssal rocks in mélange are possibly intrusive rather than blocks. See Figure 5 for key to symbols. BON—boninite field, based on data from Murton et al. (1992) for Izu-Bonin forearc; IAT—island-arc tholeiite, JO—Josephine ophiolite; MORB—mid-ocean ridge basalt. Modified from Pearce et al. (1984). Fields for JO are from Harper (2003, 2004).

exposed on the Elk River in the northwestern part of the study area (Fig. 3). However, the contact between the Humbug Mountain Conglomerate and Colebrooke Schist is a fault, based on the local presence of serpentinite along the contact. The outcrop pattern indicates that the fault has a shallow dip and places Humbug Mountain Conglomerate on top of the Colebrooke Schist. The relationship between the Humbug Mountain Conglomerate and serpentinite-matrix mélange is unknown.

GEOCHEMISTRY

Introduction

Samples from the dike complex and serpentinite-matrix mélange were analyzed and plotted on various discriminant diagrams (Figs. 5–10) to classify magma types, which give some constraints on tectonic environment of formation, determine whether the dike complex and mélange blocks are petrogenetically related, and allow evaluation of correlations with units outside the Elk outlier. X-ray fluorescence, inductively coupled plasma mass spectrometry, and loss on ignition analyses were performed at the Geoanalytical Lab, Washington State University (Johnson et al., 1999). Data are presented in Table 1 for samples from the dike complex, Table 2 for samples from blocks in the serpentinite-matrix mélange, and Table 3 for a dacite dike in the mélange and for a dike cutting the Galice Formation. UTM coordinates are provided in Table A-2.

Dike Complex

The ten dike samples from the dike complex are basaltic andesites and andesites, based on SiO_2 contents (Table 1). The dike samples are classified as basalts and basaltic andesites, however, using the immobile element ratios Nb/Y and Ti/Zr (Winchester and Floyd, 1976; Pearce, 1996), suggesting that the samples gained SiO_2 during ocean-ridge hydrothermal metamorphism. The samples have high Na_2O (three with >6 wt%), and five of the ten samples plot in the alkaline field on a SiO_2 versus $Na_2O + K_2O$ plot (not shown), even though ratios of immobile elements (e.g., Ta/Yb, Ti/V) indicate they are subalkaline. Thus, some of the samples apparently gained substantial Na_2O and, likely, SiO_2, probably during albitization of Ca-rich plagioclase. Negative Eu anomalies, which are usually taken as an indicator of plagioclase fractionation, are in some of the samples (Fig. 9B and C) almost certainly related to Ca loss during plagioclase alteration (e.g., Harper, 1995).

The dike samples fall into two groups of five samples each on many major and trace-element chemical variation diagrams. Group 1 samples (Table 1) fall in the field for island-arc tholeiites (IAT) on both the Y versus Cr (Fig. 6) and Th/Yb versus Ta/Yb (Fig. 7A) discriminant diagrams. Although Group 1 samples plot in the IAT field on most discriminant diagrams and have negative Ta anomalies characteristic of arc magmas (cf. Fig. 9A and B), four of the five samples have Ti/V ratios >20, characteristic of mid-ocean ridge basalt (MORB; Fig. 5). These data indicate that Group 1 samples have magmatic affinities that are transitional between IAT and MORB.

The linear trend on the Y versus Cr diagram for Group 1 samples is parallel to the fractionation trend resulting from removal of Cr-spinel and mafic minerals (Fig. 6), suggesting derivation from a similar parent magma. This hypothesis is also consistent with these samples plotting close together on the Th/Yb versus Ta/Yb diagram. With crystal fractionation of a magma, the level of incompatible elements increases, but the pattern on chondrite- and MORB-normalized diagrams change little, except for development of negative Eu and Ti anomalies related to plagioclase and Fe-Ti oxide fractionation, respectively (e.g., Pearce, 1983; Pearce and Parkinson, 1993). Thus the similar patterns for the five Group 1 samples (Fig. 9B) is also consistent with fractionation from a similar parent magma.

Group 2 samples are generally more depleted in high-field strength elements, such as Ti, Zr, Y, P, and heavy rare-earth

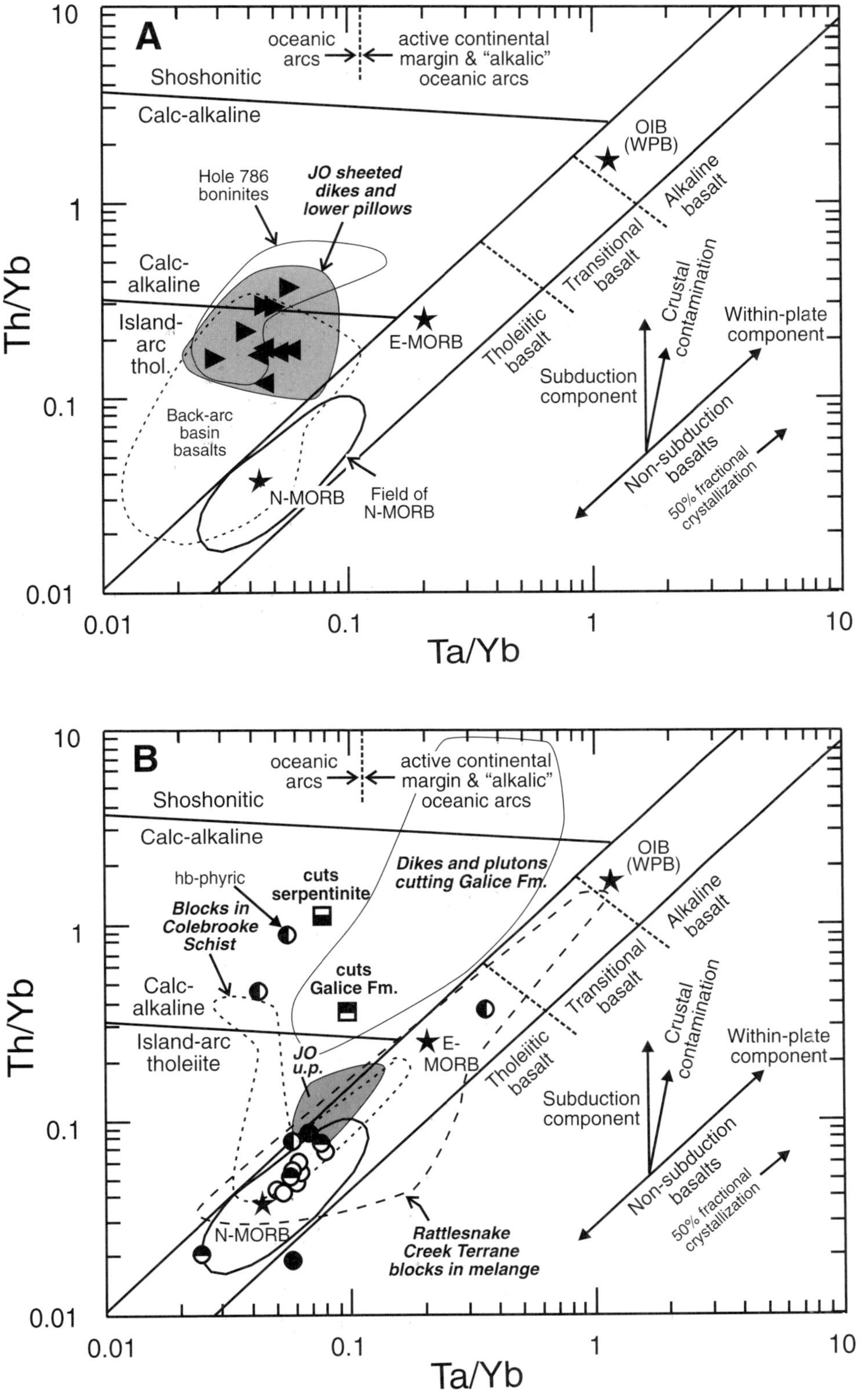

Figure 7. Th/Yb vs Ta/Yb discriminant diagram (Pearce, 1982). See Figure 5 for key to symbols. E-MORB—enriched mid-ocean ridge basalt, JO u.p.—upper pillow lava unit and late ophiolitic dikes in the Josephine ophiolite, N-MORB—normal MORB, OIB—ocean island basalt (within-plate basalt), WPB—within-plate basalt. (A) Samples from the sheeted dike complex of the Elk outlier (Table 1). A gabbro screen plots far off this diagram to the lower left, probably because it is a plagioclase-rich cumulate as suggested by MORB-normalized pattern (Fig. 9). Note similarity to JO sheeted dikes and lower pillow lavas; boninitic rocks in the JO also occur within the sheeted dike complex and lower pillow lavas (Harper, 2004). (B) Data for blocks from the serpentinite mélange associated with the Elk outlier (circles; see Fig. 5 for key to symbols); the gabbro plots in the IAT field but is not plotted here, as its MORB-normalized pattern (Fig. 10) suggests that it is a cumulate. In addition, two late dikes are plotted (squares), one cutting serpentinite and the other cutting the Galice Formation; these dikes are probably related to the ca. 145-Ma Pearse Peak Diorite (Fig. 2). Hypabyssal rocks in mélange are possibly intrusive rather than blocks. Fields for ca. 162-Ma Josephine ophiolite are from Harper (2003, 2004); Field for back-arc basins (Mariana Trough and Lau Basin) compiled by Harper (2003), based largely on data of Pearce et al. (1994). Field for Rattlesnake Creek terrane is from Wright and Wyld (1994), and field for blocks in the high P/T Colebrooke Schist structurally below the western Klamath terrane is from Katrib (2005). Field for dikes and plutons cutting the Galice Formation elsewhere in the western Klamath terrane, which are ca. 150–140 Ma, is based on data in Barnes et al. (this volume, Chapter 17).

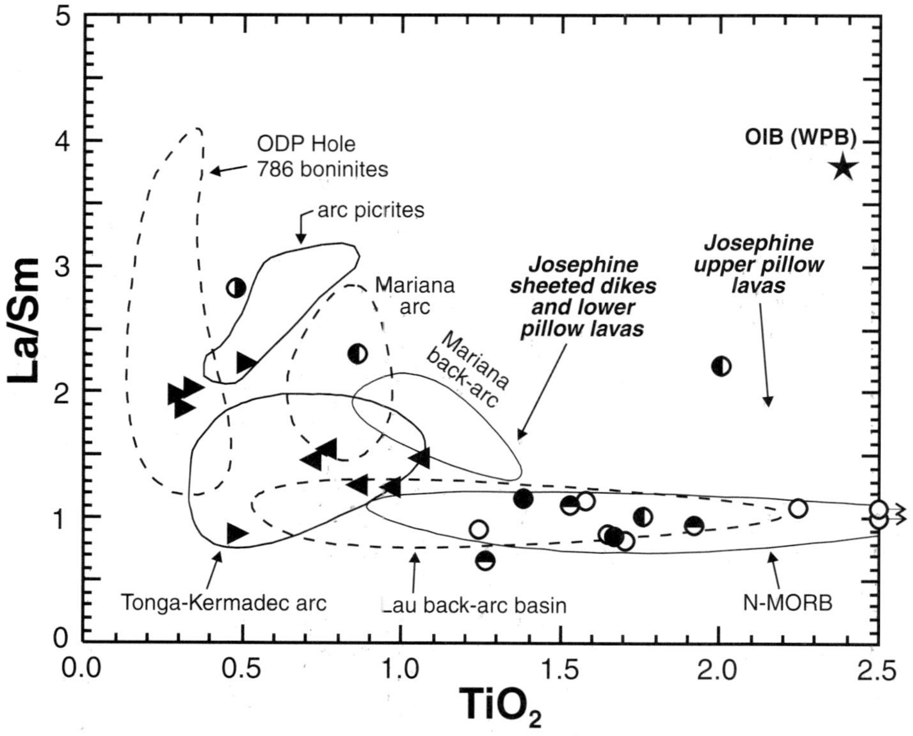

Figure 8. La/Sm vs. TiO_2 diagram with samples plotted from the Elk River outlier (symbols same as in Fig. 5). Fields for Josephine ophiolite are from Harper (2003, 2004). Boninites from ODP Hole 786 (Murton et al., 1992) include both low-Ca and high-Ca types. Arc picrites (compilation by Harper, 2004) are essentially primitive calc-alkaline basalts and less depleted than boninites as indicated by higher TiO_2. Large range in TiO_2 for N-MORB is due to tholeiitic fractionation; MORB with TiO_2 >2% are Fe-Ti basalts. Representative OIB is from Sun and McDonough (1989). This figure shows that the range of compositions of the Elk River samples is similar to the sheeted dike complex and lower pillow lavas of the Josephine ophiolite and that some samples are boninites. Note also that nearly all mélange blocks fall in the N-MORB field. Hypabyssal rocks in mélange are possibly intrusive rather than blocks. N-MORB—normal mid-oceanic ridge basalt, OIB—ocean-island basalt. Modified from Meffre et al. (1996) and Harper (2004).

elements (REEs; Table 1, Fig. 6), compared to Group 1 samples. Most of the REE patterns have a distinctive and unusual "U" shape (Fig. 9C); one sample, however, is very light REE (LREE) depleted. Positive and negative Eu anomalies are likely the result of Eu mobility related to Ca gain and loss, respectively, during hydrothermal metamorphism (e.g., Harper, 1995). Group 2 samples have greater Th enrichment and deeper negative Ta anomalies (cf. Fig. 9B and C), as well as generally lower Ti/V ratios (Fig. 5), compared to Group 1 samples, indicating clear affinities to arc magmas. Group 2 samples define a linear trend on a Y versus Cr diagram parallel to the Cr-spinel + mafic mineral fractionation trend (Fig. 6); this trend is parallel to, but at lower Y, than that for Group 1 samples, illustrating their more depleted nature. The trend for Group 2 samples on the Y versus Cr diagram is consistent with fractionation from a single parent magma. Crossing REE and MORB-normalized patterns (Fig. 9C), however, cannot result from fractionation from a single magma. The Th/Yb versus Ta/Yb diagram also indicates variation in magmas that is too great to be explained by fractionation alone (Fig. 7A). Based on the vectors on Figure 7, the trend defined by Group 2 lavas is the result of a similar amount of an arc component (Th/Yb) added to a variably depleted or enriched mantle (Ta/Yb).

Group 2 samples fall in the IAT and CA (calc-alkaline) fields on the Ta/Yb versus Th/Yb diagram (Fig. 7a), but also overlap the field for Tertiary boninites from the Izu-Bonin forearc, which includes both high-Ca and low-Ca boninites. The recent International Union of Geological Sciences definition of boninite is $SiO_2 > 52\%$, $MgO > 8\%$, and $TiO_2 < 0.5\%$ (Le Maitre et al., 2002). By this definition, two of the Group 2 samples are boninites, but MgO and SiO_2 may have been mobile during hydrothermal metamorphism. These two samples, however, also plot in the field for boninites on the Y versus Cr diagram (Fig. 6), have a Ti/V ratio <10 typical of boninites (Fig. 5), and have the U-shaped REE patterns common for boninites (cf. Fig. 9A and C). A La/Sm versus TiO_2 (Fig. 8) diagram is also useful for the recognition of boninites: this diagram illustrates the high degree of depletion of high-field strength elements, such as Ti, believed to result from remelting of residual mantle (i.e., secondary melting), with enrichment of large-ion lithophile elements, such as LREEs, believed to be contributed from slab-derived fluids (e.g., Crawford et al., 1989; Pearce and Peate, 1995). On this diagram, three of the Group 2 samples plot in the field defined by the Izu-Bonin boninites (Fig. 8). Although the Th/Yb versus Ta/Yb diagram (Fig. 7A) shows that the five samples cannot be derived from a single parent, they may have all formed by fractionation from a broadly similar magma. This possibility is suggested by the linear trend on the Y versus Cr diagram (Fig. 6), and (not shown) linear trends on MgO versus TiO_2, MgO versus SiO_2, and MgO versus Al_2O_3 diagrams. Thus, all the Group 2 samples can be considered "boninitic" in the broad sense.

The analyzed screen in the dike complex has ~56% SiO_2 and thus is a diorite, assuming SiO_2 has not been added during

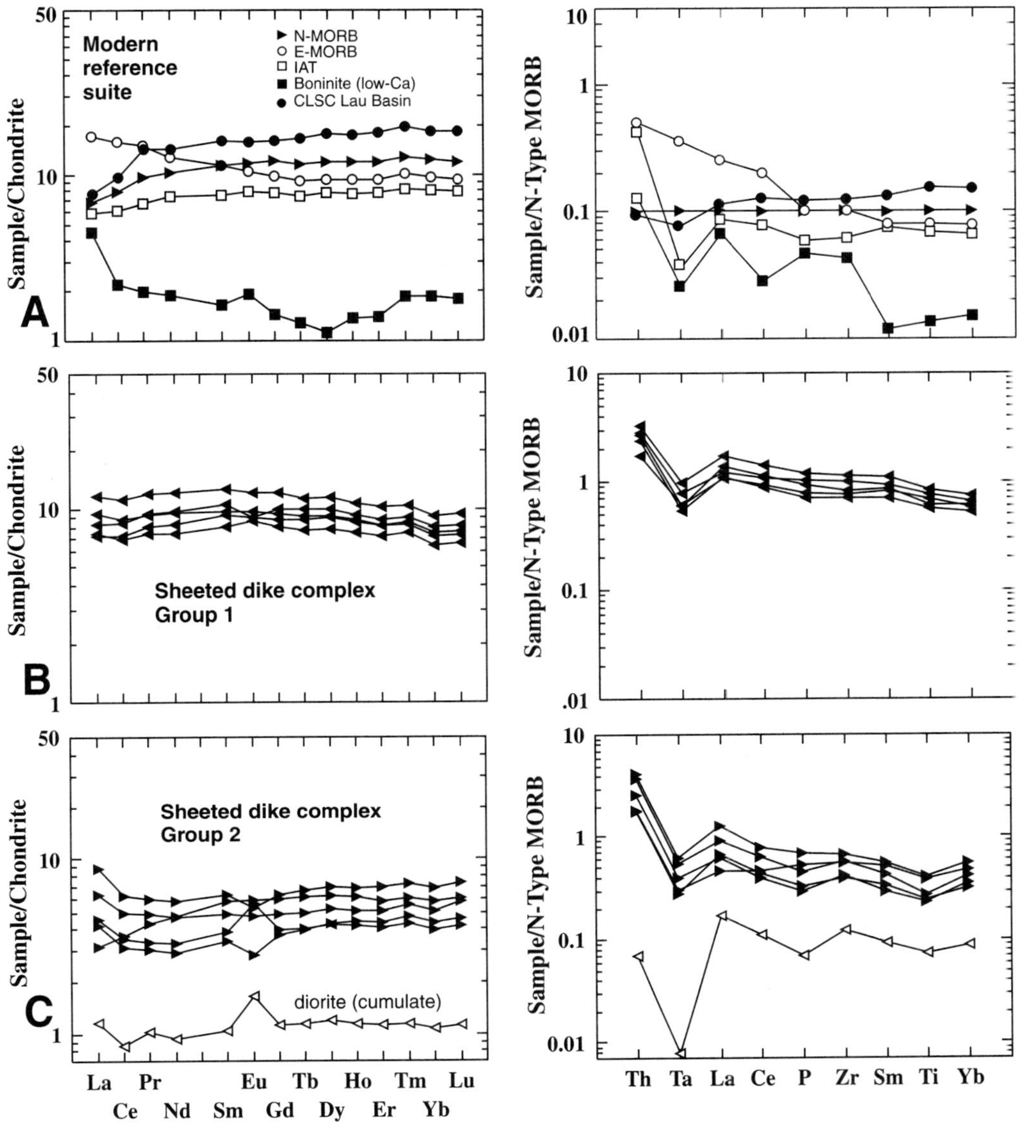

Figure 9. Chondrite-normalized and MORB-normalized diagrams. (A) Reference suite. (B) Samples of Group 1 of the sheeted dike complex. (C) Samples of Group 2 of the sheeted dike complex. Symbols are the same as in Figure 5. CLSC—central Lau spreading center, IAT—island-arc tholeiite, E-MORB—enriched mid-ocean ridge basalt, N-MORB—normal MORB.

hydrothermal metamorphism. The extremely low concentrations of high-field strength elements, such as Zr, Y, and heavy REEs, and positive Eu anomaly (Fig. 9) suggest that this sample is a plagioclase cumulate, because calcic plagioclase has very low high-field strength element concentrations but high Eu. This sample is not particularly enriched in Al_2O_3 (18.3 wt%) nor is it depleted in MgO (7.6 wt%), however, as would be expected for extensive plagioclase accumulation. Thus, the diorite's very low high-field strength elements, as well as its low Ti/V ratio (Fig. 5), most likely indicates derivation from a boninite magma.

Blocks in Serpentinite-Matrix Mélange

Chemical analyses of selected mélange blocks appear in Table 2. Pillow lavas, other mafic volcanic (based on textures) samples, and one hypabyssal block from the mélange fall in the field for MORB on all discriminant diagrams (Figs. 5–8). A second hypabyssal sample plots on the boundary between MORB and within-plate basalt (WPB; Figures 5 and 7B), and is probably best classified as transitional between enriched MORB (E-MORB) and WPB. The amphibolites also have MORB affinities (Figs. 5, 6, and 8).

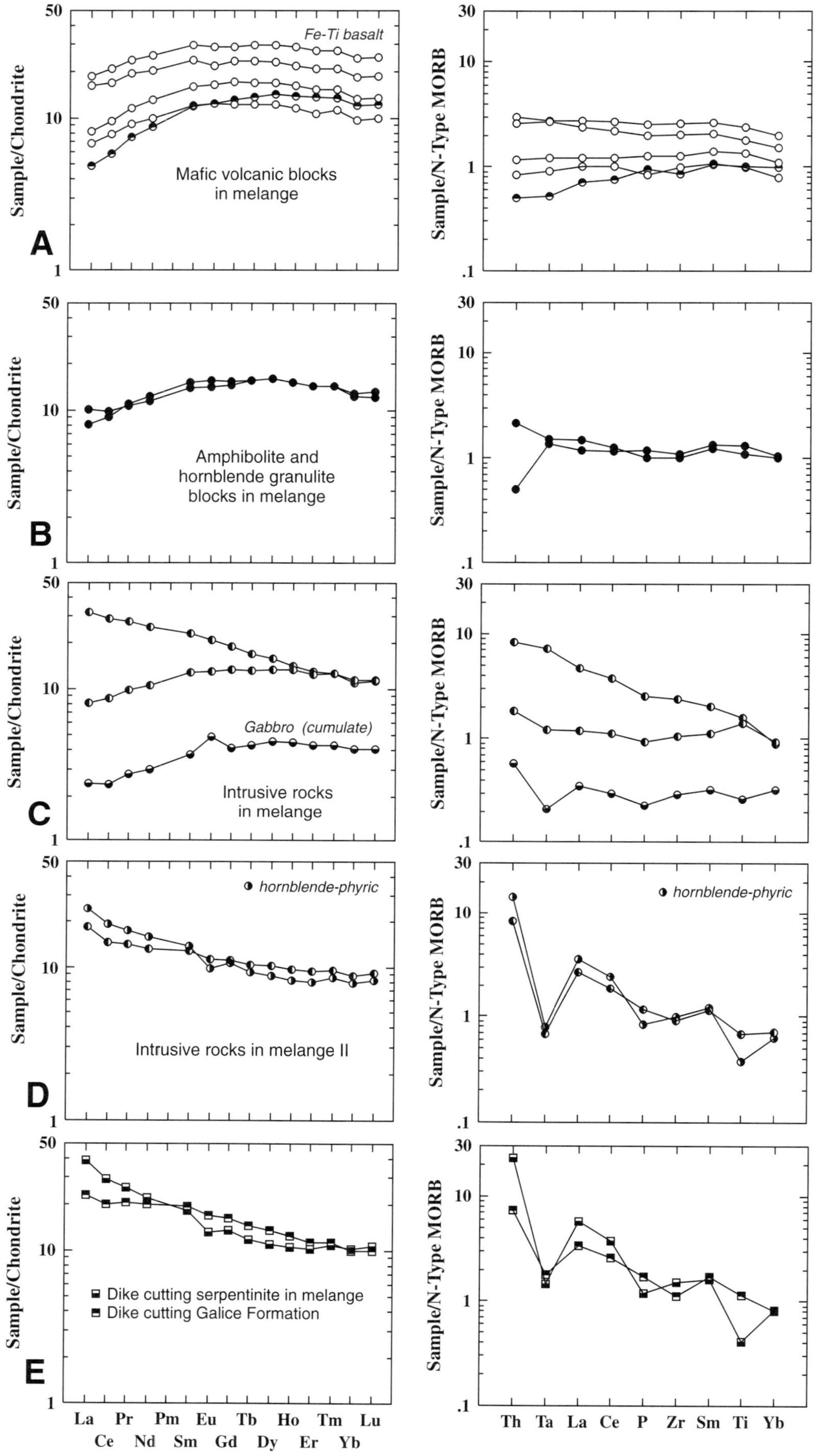

Fe-Ti basalt
Mafic volcanic blocks in melange
A
Amphibolite and hornblende granulite blocks in melange
B
Gabbro (cumulate)
Intrusive rocks in melange
C
hornblende-phyric
Intrusive rocks in melange II
D
Dike cutting serpentinite in melange
Dike cutting Galice Formation
E
Sample/Chondrite
Sample/N-Type MORB
La Ce Pr Nd Pm Sm Eu Gd Tb Dy Ho Er Tm Yb Lu
Th Ta La Ce P Zr Sm Ti Yb

TABLE 1. ANALYSES OF SHEETED DIKES

	Group 1					Group 2 and screen					
	Dike	Dike	Dike	Dike	Dike	Dike	Dike	Dike	Dike	Dike	Screen
Sample	00ER21A	00ER41A	00ER41B	00ER41C	01ER46A	00ER682A	00ER682B	00ER682C	02ER09B	02ER11A	00ER68A
SiO_2	56.17	54.47	54.94	54.07	54.50	58.36	56.32	60.94	59.83	60.38	55.85
TiO_2	0.97	0.86	0.77	0.72	1.06	0.32	0.30	0.35	0.49	0.52	0.09
Al_2O_3	16.13	15.67	15.92	16.14	16.52	11.87	13.56	14.03	14.74	16.89	18.27
$FeO^{T‡}$	7.69	8.77	8.31	7.53	8.40	7.80	6.98	5.64	6.79	5.07	3.91
MnO	0.14	0.21	0.19	0.15	0.16	0.21	0.15	0.16	0.20	0.13	0.08
MgO	6.21	8.24	7.62	6.77	6.41	9.18	10.83	6.14	6.87	3.08	7.61
CaO	6.69	6.54	6.80	9.85	6.29	7.05	7.10	5.70	4.39	6.29	8.79
Na_2O	5.55	3.80	4.46	4.64	5.97	4.98	4.51	6.80	6.49	7.48	4.69
K_2O	0.34	1.34	0.87	0.05	0.55	0.19	0.21	0.18	0.14	0.08	0.71
P_2O_5	0.12	0.09	0.11	0.08	0.14	0.04	0.04	0.05	0.06	0.08	0.01
Total§	100.00	100.00	100.00	100.00	100.00	100.00	100.00	100.00	100.00	100.00	100.00
LOI	2.48	2.49	2.63	3.19	4.02	1.15	2.34	0.93	2.71	0.99	2.89
Ba#	82	226	107	27	102	61	57	38	52	15	127
Rb	4	15	9	0	4	2	3	1	1	1	13
Sr	257	279	325	48	148	144	135	113	111	33	139
Y	20.6	18.4	19.0	17.2	24.3	9.6	10.1	12.0	13.4	16.2	2.8
Zr	74	56	60	52	84	29	31	42	41	50	9
Nb	1.4	1.0	1.1	0.9	1.7	0.5	0.6	0.8	0.5	1.0	0.0
Th	0.34	0.21	0.32	0.28	0.39	0.21	0.31	0.46	0.22	0.50	0.01
Pb	1.6	1.5	1.5	2.0	0.4	1.5	1.4	1.4	0.0	0.4	1.7
Ga	12	12	14	16	18	7	11	11	13	13	11
Zn	45	53	64	52	68	58	67	44	51	39	14
Cu	14	0	80	67	86	121	3	2	14	211	0
Ni	50	58	43	49	38	121	199	74	40	9	143
V	239	230	223	236	228	216	192	170	179	152	63
Cr	99	189	71	65	36	330	543	162	93	12	400
Hf	1.79	1.37	1.43	1.27	2.15	0.77	0.84	1.16	1.11	1.40	0.06
Cs	0.30	1.72	1.12	0.09	0.98	0.75	0.80	0.67	0.11	0.04	0.34
Sc	33	40	36	36	33	41	41	36	38	25	14
Ta	0.10	0.08	0.08	0.07	0.13	0.04	0.05	0.07	0.04	0.08	0.00
U	0.12	0.08	0.12	0.10	0.14	0.09	0.13	0.16	0.07	0.14	0.01
La	3.04	2.67	3.48	2.74	4.27	1.67	1.53	2.31	1.16	3.22	0.43
Ce	8.08	6.94	8.39	6.65	10.70	3.34	3.00	4.72	3.50	5.89	0.82
Pr	1.28	1.10	1.27	1.02	1.63	0.46	0.42	0.67	0.59	0.81	0.14
Nd	6.86	5.88	6.75	5.30	8.59	2.36	2.08	3.31	3.38	4.11	0.67
Sm	2.45	2.14	2.25	1.88	2.91	0.89	0.78	1.13	1.34	1.45	0.24
Eu	0.78	0.79	0.85	0.75	1.06	0.49	0.25	0.41	0.51	0.46	0.14
Gd	3.04	2.71	2.87	2.46	3.70	1.21	1.13	1.49	1.82	1.93	0.34
Tb	0.58	0.51	0.53	0.45	0.66	0.23	0.23	0.29	0.35	0.39	0.07
Dy	3.77	3.43	3.49	2.99	4.35	1.61	1.64	1.98	2.35	2.64	0.45
Ho	0.79	0.73	0.75	0.65	0.91	0.36	0.37	0.44	0.52	0.58	0.10
Er	2.19	2.03	2.04	1.81	2.54	1.01	1.09	1.28	1.44	1.72	0.28
Tm	0.32	0.30	0.31	0.27	0.37	0.15	0.17	0.20	0.21	0.26	0.04
Yb	2.01	1.78	1.88	1.61	2.26	0.98	1.09	1.27	1.43	1.70	0.27
Lu	0.31	0.28	0.29	0.25	0.36	0.16	0.18	0.22	0.23	0.28	0.04

Notes: Analyzed at Geoanalytic Lab, Washington State University. Major elements analyzed by X-ray fluorescence (XRF). LOI—loss on ignition.

§FeO^T = total iron as FeO.

‡Analyses normalized to 100% by Geoanalytic Lab, Washington State University.

#Trace elements in ppm. The elements Ba, Zr, Zn, Cu, Ni, V, Cr, and Sc were analyzed by XRF; all other trace and rare-earth elements by inductively coupled plasma mass spectrometry.

Figure 10. Chondrite-normalized and MORB-normalized diagrams for samples from the serpentinite-matrix mélange of the Elk River outlier. (A) Mafic volcanic blocks (open circles are pillowed). (B) Amphibolites. (C, D) Intrusive rocks in mélange; hypabyssal rocks and gabbro in mélange are possibly intrusive rather than blocks. (E) Dike cutting serpentinite of the mélange and dike cutting Galice Formation. Symbols are the same as in Figure 5 except where noted in diagram. N-MORB—normal mid-ocean ridge basalt.

TABLE 2. ANALYSES OF BLOCKS IN SERPENTINITE MELANGE

Sample	Pillow 00ER74A	Pillow 00ER74B	Pillow 00ER74C	Pillow 00ER93	Pillow 00ER94B	Pillow 01ER11B	Pillow 01BMC1A	Volcanic 01ER19A	Volcanic 01ER41	Volcanic 02ER49B	Hypabyssal 01ERKK2	Hypabyssal 01ER27A	Hypabyssal 01ER30C	Hypabyssal 02ER16A	Gabbro 00ER76B	Amphi-bolite 02ER51	Garnet phibolite 01ER16A
SiO_2	51.44	52.33	50.75	51.92	50.33	52.66	48.50	51.61	53.15	53.64	46.29	52.24	52.12	59.21	55.76	44.88	49.79
TiO_2	1.58	1.65	2.26	1.71	1.25	3.02	2.94	1.93	1.27	1.53	1.77	2.01	0.87	0.48	0.34	1.68	1.39
Al_2O_3	14.03	14.16	13.82	14.42	14.61	13.21	11.40	13.04	14.40	14.49	14.43	16.11	16.28	14.61	16.03	14.83	13.97
FeO^T	10.51	10.30	12.13	11.48	9.90	11.47	15.71	11.63	10.03	9.14	12.20	8.97	8.89	6.07	5.69	12.41	12.47
MnO	0.17	0.19	0.18	0.20	0.18	0.17	0.31	0.19	0.15	0.15	0.21	0.14	0.15	0.21	0.10	0.24	0.23
MgO	6.08	6.68	4.13	8.00	8.80	5.10	7.08	7.85	6.35	7.07	5.94	7.80	8.14	5.48	8.71	8.19	8.55
CaO	11.40	9.93	11.73	7.14	10.44	9.04	10.32	9.04	9.20	8.14	17.57	6.77	8.68	7.53	8.34	16.51	9.94
Na_2O	4.23	4.43	4.70	4.96	2.97	4.53	2.68	3.86	4.47	4.44	1.30	4.98	4.19	6.16	4.70	0.81	3.27
K_2O	0.42	0.18	0.08	0.04	1.42	0.51	0.78	0.67	0.88	1.26	0.18	0.68	0.56	0.15	0.30	0.31	0.27
P_2O_5	0.14	0.15	0.23	0.15	0.10	0.30	0.29	0.18	0.11	0.13	0.11	0.30	0.14	0.10	0.03	0.14	0.12
Total	100.00	100.00	100.00	100.00	100.00	100.00	100.00	100.00	100.00	100.00	100.00	100.00	100.00	100.00	100.00	100.00	100.00
LOI	3.65	3.31	2.72	3.26	3.07	1.52	1.85	2.70	2.11	2.10	2.11	4.11	3.99	1.08	2.26	3.29	1.62
Ba	45	35	33	44	655	113	271	830	137	2277	131	693	162	32	47	494	72
Rb	12	5	2	1	38	13	13	9	27	26	2	6	5	2	5	6	4
Sr	126	96	116	246	90	160	102	269	149	534	256	368	112	51	163	134	116
Y	35.0	36.4	49.0	37.4	26.4	63.0	66.1	42.4	32.5	30.5	29.1	32.8	22.3	19.7	9.7	33.6	34.7
Zr	95	98	149	93	73	190	185	115	64	96	78	176	68	73	22	81	75
Nb	2.6	2.7	4.8	2.1	1.6	4.8	4.6	2.9	0.9	2.9	2.2	13.9	1.3	1.7	0.4	2.5	2.9
Th	0.17	0.16	0.31	0.14	0.10	0.36	0.34	0.20	0.06	0.21	0.22	1.00	0.99	1.73	0.07	0.06	0.26
Pb	1.8	1.6	1.9	0.2	1.7	1.3	0.5	0.4	0.2	0.7	0.4	0.6	2.1	0.3	1.5	0.2	0.5
Ga	17	18	15	18	16	15	17	14	17	13	18	16	17	14	9	19	18
Zn	66	73	97	84	65	144	111	97	84	78	88	66	71	51	29	88	296
Cu	52	54	50	75	87	58	49	71	42	98	75	44	120	5	0	76	108
Ni	30	32	71	54	67	48	37	57	42	59	33	104	44	24	67	49	68
V	357	331	411	329	261	496	538	388	288	298	431	234	312	152	168	394	371
Cr	119	120	187	113	320	93	90	156	167	257	84	186	89	75	172	105	125
Hf	2.57	2.59	3.99	2.47	1.91	5.17	5.03	3.14	1.80	2.42	2.11	4.07	1.98	2.23	0.44	2.01	2.02
Cs	0.44	0.20	0.36	0.14	1.03	1.20	1.06	0.14	0.25	0.27	0.22	0.19	0.38	0.07	0.06	0.72	0.08
Sc	47	49	47	50	46	49	49	50	50	53	41	33	35	35	45	47	50
Ta	0.19	0.20	0.35	0.16	0.12	0.37	0.36	0.22	0.07	0.21	0.16	0.95	0.09	0.11	0.03	0.18	0.20
U	0.20	0.13	0.13	0.71	0.11	0.40	0.13	0.31	0.36	0.09	0.07	0.29	0.35	0.63	0.04	0.04	0.08
La	4.13	3.33	5.94	3.01	2.52	6.91	7.58	4.10	1.79	3.98	2.99	11.71	6.79	8.94	0.89	2.95	3.71
Ce	10.70	10.26	16.28	9.14	7.51	20.06	20.51	12.14	5.64	10.87	8.33	27.82	13.87	18.39	2.28	8.68	9.47
Pr	1.73	1.74	2.66	1.59	1.25	3.28	3.25	2.00	1.03	1.75	1.35	3.81	1.93	2.40	0.38	1.51	1.47
Nd	9.56	9.90	14.46	9.32	7.14	18.19	17.90	11.48	6.29	9.62	7.45	18.21	9.36	11.29	2.13	8.81	8.16
Sm	3.63	3.85	5.47	3.71	2.82	6.87	6.86	4.41	2.77	3.64	2.96	5.33	2.97	3.17	0.86	3.50	3.24
Eu	1.36	1.39	1.91	1.43	1.08	2.51	2.27	1.47	1.08	1.31	1.13	1.83	0.99	0.86	0.42	1.36	1.23
Gd	4.89	5.24	7.21	5.26	3.76	8.86	9.08	6.01	4.03	4.58	4.08	5.84	3.40	3.27	1.27	4.74	4.45
Tb	0.94	1.00	1.35	0.98	0.71	1.73	1.72	1.13	0.80	0.85	0.77	0.98	0.60	0.54	0.25	0.91	0.90
Dy	6.12	6.49	8.77	6.51	4.72	11.30	11.42	7.50	5.45	5.61	5.12	6.05	3.93	3.33	1.73	6.09	6.09
Ho	1.31	1.38	1.86	1.39	0.99	2.45	2.50	1.60	1.18	1.19	1.13	1.21	0.83	0.70	0.38	1.29	1.30
Er	3.71	3.77	5.19	3.84	2.68	6.78	7.00	4.41	3.41	3.25	3.11	3.24	2.34	1.99	1.07	3.57	3.59
Tm	0.53	0.55	0.75	0.55	0.40	0.97	1.00	0.64	0.48	0.45	0.45	0.45	0.34	0.30	0.15	0.51	0.51
Yb	3.21	3.46	4.60	3.34	2.42	6.08	6.30	3.95	3.00	2.81	2.84	2.74	2.17	1.94	1.01	3.20	3.05
Lu	0.50	0.52	0.71	0.52	0.38	0.95	0.97	0.62	0.47	0.43	0.43	0.43	0.35	0.31	0.15	0.50	0.46

Note: LOI—Loss on ignition.

REE and MORB-normalized diagrams show that all the MORB-like samples, except the transitional E-MORB/WPB, are LREE depleted and have essentially flat MORB-normalized patterns (Fig. 10A–C) typical of normal MORB (N-MORB). The Th depletion shown by one of the amphibolites (Fig. 10B) is unusual; this sample plots outside the fields defined by terrestrial volcanic rocks on Figure 7B, suggesting Th may have been mobile during metamorphism.

Three of the pillow lava samples are unusually highly fractionated, as indicated by enrichment in incompatible elements (e.g., Zr, Y, P, REEs; Table 2); they have sufficiently high TiO_2 (>2 wt%) and FeO^T (>12 wt%) to be classified as Fe-Ti basalts (Sinton et al., 1983; Fig. 5). Such high degrees of TiO_2 and FeO^T enrichment do not occur in arc magmas, which are typically highly oxidizing; their high oxygen fugacities result in earlier crystallization of Fe-Ti oxide, which leads to lower TiO_2 and FeO^T and higher SiO_2 with continued fractionation (e.g., Harper, 2003).

The gabbro block is depleted in incompatible elements and has a positive Eu anomaly (Fig. 10C), suggesting that it is a cumulate. It has a low Ti/V ratio (12), a negative Ta anomaly, and enrichment in Th. These features, which are essentially unaffected by crystal accumulation, are characteristics of arc magmas. Thus, the gabbro is interpreted as a cumulate formed from an arc-affinity magma, probably IAT. In outcrop, the gabbro appears to be intrusive into a metavolcanic rock, but the contact between the block and the serpentinite matrix is not exposed; thus it is possible that the gabbro is intrusive into the mélange.

Two of the hypabyssal rocks in the mélange of the Elk River outlier are distinct in that they have calc-alkaline affinities (Fig. 7B). They are highly enriched in LREEs and Th, and have a large negative Ta anomaly (Fig. 10D) typical of calc-alkaline magmas erupted in arcs. Their patterns are remarkably similar to a felsic dike (01KK5al; Table 3) that cuts serpentinite in the mélange. The margins of these arc-affinity hypabyssal "blocks" are not exposed, so it is possible they are actually dikes, which, together with the felsic dike, might be related to the nearby ca. 145-Ma Pearse Peak pluton.

Dikes Cutting Galice Formation and Serpentinite

One sample from the dikes that cut the Galice Formation was analyzed (Table 3) and is presumably related to the Pearse Peak pluton, which also cuts the Galice Formation. The dike is a porphyritic hornblende-pyroxene microgabbro with 50.5 wt% SiO_2. Its LREEs are enriched, and this sample has a MORB-normalized pattern (Fig. 10E) characteristic of calc-alkaline magmas with strong Th enrichment and a prominent negative Ta anomaly. This sample plots in the calc-alkaline field on the Th/Yb versus Ta/Yb diagram (Fig. 7B), within the field defined by dikes and plutons that cut the Galice Formation elsewhere in the western Klamath terrane (Barnes et al., this volume, Chapter 17).

The dacite dike that cuts sheared serpentinite in the mélange also has calc-alkaline affinities (Figs. 7B and 10E). It may be related to the Pearse Peak pluton, although it plots outside the field defined by dikes and plutons cutting the Galice Formation elsewhere in the western Klamath terrane (Fig. 7E). It has a very similar MORB-normalized pattern to a hornblende-phyric hypabyssal rock shown on discriminant diagrams as a block in the mélange (cf. Fig. 10D and E). As discussed above, the margins of this "block" were not observed and it could actually be a dike.

TABLE 3. LATE DIKES

	Cuts	
Sample	Serpentinite 01KK5A	Galice Formation 02ER04B
SiO_2	66.25	50.48
TiO_2	0.52	1.43
Al_2O_3	16.61	14.91
FeO^T	4.22	11.86
MnO	0.08	0.23
MgO	1.88	7.17
CaO	3.60	10.46
Na_2O	6.52	2.28
K_2O	0.19	0.98
P_2O_5	0.14	0.20
Total	100.00	100.00
LOI	1.47	1.37
Ba	59	368
Rb	2	26
Sr	125	455
Y	24.4	28.2
Zr	113	83
Nb	2.8	3.5
Th	2.76	0.87
Pb	1.0	1.4
Ga	14	16
Zn	21	102
Cu	4	128
Ni	4	31
V	84	379
Cr	2	110
Hf	3.43	2.37
Cs	0.11	2.61
Sc	15	51
Ta	0.19	0.23
U	0.88	0.35
La	14.23	8.50
Ce	27.90	19.28
Pr	3.53	2.82
Nd	15.81	14.31
Sm	4.22	4.51
Eu	1.14	1.47
Gd	4.16	5.01
Tb	0.68	0.84
Dy	4.19	5.19
Ho	0.90	1.06
Er	2.57	2.81
Tm	0.39	0.40
Yb	2.53	2.48
Lu	0.41	0.38

Notes: LOS—Loss on ignition.

DISCUSSION

Dike Complex

The dike complex consists of virtually 100% parallel dikes with scarce diorite screens and amphibole-rich hydrothermal metamorphic assemblages, all of which are consistent with it being the lower part of a sheeted dike complex of an ophiolite. Because the dike complex is in fault contact with the Galice Formation, a reasonable assumption is that it represents a fragment of ophiolitic basement correlative to the Josephine ophiolite that underlies the Galice in much of the western Klamath terrane (Fig. 1). This assumption is supported by the occurrence of other rocks in the eastern part of the Elk outlier that may be part of an ophiolite basement, including pillow basalt (locally adjacent to chert) and the large Iron Mountain ultramafic body (Dott, 1966, 1971; Coleman, 1972). Coleman (1972) interpreted the ultramafic body to be a klippe on top of the Galice, which Blake et al. (1985a) subsequently assigned to the Snowcamp terrane (Figs. 1 and 2), but the outcrop pattern of contacts shown on the maps by Dott (1966, 1971) and Coleman (1972) imply the boundaries of the ultramafic body are steeply dipping.

As evident in Figures 5–8, the sheeted dike samples from the Elk River outlier are remarkably similar to the sheeted dike complex and lower pillow lavas of the type Josephine ophiolite in northwestern California (Harper, 1984, 2003, 2004). Although not shown on Figure 9, REE and MORB-normalized patterns are also virtually identical. Low-Ti boninitic rocks in both suites are variable in their trace- and rare-earth chemistry, with crossing REE and MORB-normalized patterns, and include some strongly LREE-depleted samples (Harper, 2004). The non-boninitic samples characteristically plot in the IAT field on the Y-Cr diagram, but in the MORB field on the Ti-V diagram; this feature, indicating transitional IAT-MORB magmatic affinity, is characteristic not only for the type area of the Josephine ophiolite (Harper, 1984, 2004), but for samples from the southernmost (Wyld and Wright, 1988) and northernmost (Yule, 1996) parts of the ophiolite. Boninitic rocks only make up ~10–15% of the type Josephine ophiolite (Harper, 2004), however, whereas half of the analyzed dike-complex samples in this study are boninitic. The higher proportion of boninitic rocks in the dike complex on the Elk River (50% of samples) could be an artifact of sampling, as three of the boninitic samples (00ER682A–C; Table 1) come from one large landslide block. None of the samples from the dike complex analyzed in this study are similar to the more MORB-like, and generally highly fractionated, upper pillow lavas in the Josephine ophiolite (Figs. 6 and 7), but dikes related to these upper pillow lavas are very rare in the sheeted dike complex of the Josephine ophiolite (Harper, 2004).

The fault-bounded nature and limited extent (~0.3 km^2) of the dike complex in the study area allows for other possible correlations. The dike complex may be a large block in the adjacent mélange, but this appears unlikely, as the dike complex samples are geochemically distinct from blocks in the serpentinite (Figs. 5–10). Another possibility is that the dike complex represents a fragment of ophiolitic rocks (Coast Range ophiolite?) of the Snowcamp terrane. According to Blake et al. (1985a), ultramafic rocks of the Snowcamp terrane structurally overlie the western Klamath terrane in the eastern part of the Elk outlier (whether these rocks are actually a klippe is uncertain, as discussed above), and a sliver of western Klamath terrane is reported by Blake et al. (1985b) to underlie the Coast Range ophiolite just south of the Klamath Mountains. If these relationships are correct, and if the dike complex correlates with ophiolites of the Snow Camp terrane, then the dike complex would have to be a small down-dropped block. Geochemically, the Coast Range ophiolite, including the remnants in the Snowcamp terrane, is very similar to the Josephine ophiolite (Kosanke, 2000; Metzger et al., 2002), including the local presence of late-stage MORB-affinity basalts, such as Fe-Ti basalts (Harper, 2004; Shervais et al., 2004). In addition, recent work on the Snowcamp terrane suggests that it may well be an outlier of the Josephine ophiolite (Harper et al., 2000; Schoonmaker et al., this volume).

Even though it is uncertain whether the dike complex is basement for the Galice Formation, the igneous geochemistry clearly indicates a supra-subduction zone origin. The presence of a MORB component in Group 1 dikes is most compatible with their origin in a spreading center or rifting arc rather than the roots of an unrifted arc. The presence of boninites in ophiolites is commonly taken as evidence for formation in a forearc, but boninites also erupt in back-arc basins, especially close to the arc axis (Deschamps and Lallemand, 2003; Harper, 2004).

Serpentinite-Matrix Mélange

One of the primary goals of this chapter is to determine whether the mafic blocks in the mélange are petrogenetically related to the dike complex. Figures 5–10 indicate that the geochemistry of the mélange blocks is distinct from that of the dike complex. The MORB-affinity rocks that comprise the majority of mélange blocks show no arc component on any of the diagrams, including the lack of any negative Ta anomaly (Fig. 10A–C). The hypabyssal rocks having arc affinities are also geochemically distinct from samples of the dike complex (cf. Figs. 10D and 9B,C), having calc-alkaline rather than IAT-MORB and boninitic affinities. The only mélange sample that may have magmatic affinities similar to those of the dike complex is the gabbro cumulate, based on its negative Ta anomaly (Fig. 10C).

The volcanic, amphibolite, and one of the hypabyssal blocks have clear N-MORB affinities with no arc component. N-MORB affinities strongly suggest formation at a spreading center. Furthermore, these mélange blocks lack the small negative Ta anomaly common in many back-arc basin basalts (e.g., Pearce et al., 1994; Hawkins, 1995). Thus, a mid-ocean ridge origin is likely for these rocks. The amphibolites may have formed in an oceanic fracture zone or along a ridge segment characterized by low magma supply and extensive faulting where amphibolites are known to occur (e.g., Silant'yev, 1979;

Robinson et al., 2000). Unusual highly fractionated Fe-Ti basalts, such as those found in two pillow-lava blocks in the mélange (Fig. 5), are most common near propagating tips of spreading centers on mid-ocean and back-arc basin spreading centers, although they also occur in proximity to some fracture zones (Sinton et al., 1983; Pearce et al., 1994). Shearing that created the mélange could have taken place in a fracture zone setting, as proposed for the lithologically similar Rattlesnake Creek terrane mélange (Wright and Wyld, 1994), although the presence of Colebrooke Schist blocks(?) would require later reactivation.

In the absence of similarities in magmatic affinities between mélange blocks and the dike complex, absence of dikes similar to the dike complex cutting the mélange, and absence of a depositional contact between the mélange and Lower Cretaceous Humbug Mountain Conglomerate, there are few age constraints on the age of mélange blocks or mélange formation. The poor exposure of the mélange and its contacts also create uncertainties as to whether the mélange underlies the Colebrooke Schist along its western contact and whether high-standing areas of Colebrooke Schist and sandstone ("mss" in Fig. 3) within the mapped area of the mélange are either fault or mélange blocks. The mélange is in fault contact with both the Pickett Peak terrane and the western Klamath terrane, and there is no definitive evidence indicating to which, if either, it belongs. If the dacite dike cutting serpentinite in the mélange is related to the Pearse Peak diorite, then the mélange would be part of the western Klamath terrane. However, if outcrops of Colebrooke Schist within the mélange are blocks in the mélange rather than small fault blocks, then the mélange formed, or was at least reactivated, after metamorphism of the Colebrooke Schist (ca. 135–125 Ma; Brown and Blake, 1987). In addition, if the sandstone within the mapped area of the mélange ("mss" in Fig. 3) is a block derived in part from the Myrtle Group, the formation or last reactivation of the mélange was after deposition of the Myrtle Group in Early Cretaceous time. The mélange may have formed during thrusting of the Colebrooke Schist beneath the western Klamath terrane, in which case it does not belong to either the western Klamath terrane or Pickett Peak terrane.

Serpentinite-matrix mélange occurs elsewhere in the western Klamath terrane, locally as basement beneath the Rogue Formation (Yule, 1996; Yule et al., this volume) and beneath the Devils Elbow remnant of the Josephine ophiolite (Wyld and Wright, 1988), both interpreted as rift-edge facies of the Josephine ophiolite. Volcanic blocks in these mélanges have MORB affinities (Wyld and Wright, 1988; Yule et al., 1994; Yule, 1996). The mélanges are very similar to serpentinite-matrix mélange of the Rattlesnake Creek terrane (Rattlesnake Creek terrane) that structurally overlies the western Klamath terrane (Fig. 1). The Rattlesnake Creek terrane contains mafic volcanic and amphibolite blocks having N-MORB, E-MORB, and WPB magmatic affinities (Fig. 5), in addition to cross-cutting arc-affinity intrusions, which now occur mostly as blocks in the mélange (reactivation of the mélange is indicated by local presence of younger blocks; Wright and Wyld, 1994). Furthermore, the Rattlesnake Creek terrane mélange and the Rattlesnake Creek terrane-like mélange in the western Klamath terrane are locally intruded and/or overlain by mafic complexes similar in age to the Josephine ophiolite and interpreted as rift-edge facies (Snoke, 1977; Saleeby et al., 1982; Wyld and Wright, 1988; Yule, 1996; Yule et al., this volume). Thus, the serpentinite-matrix mélange and dike complex in the study area together might comprise a rift-edge facies similar to that beneath the Rogue Formation, Devils Elbow remnant of the Josephine ophiolite, or Preston Peak ophiolite. The possibility that the analyzed gabbro is both intrusive into the mélange and related to the dike complex is most consistent with this interpretation.

The mafic volcanic blocks have similarities to the upper pillow lava unit in the Josephine ophiolite, which has dominantly MORB affinities and contains abundant Fe-Ti basalts (Harper, 2003; Fig. 5). In detail, however, the Josephine Fe-Ti basalts have a small arc component and were derived from more enriched mantle (higher Ta/Yb), as evident in Figure 7. The absence of transitional IAT-MORB or boninitic affinity blocks in the mélange is also inconsistent with derivation of blocks from the Josephine ophiolite.

Greenstone blocks, many pillowed, are common in the Colebrooke Schist and, like the mélange blocks in the study area, have MORB magmatic affinities and include some highly fractionated Fe-Ti basalts (Plake, 1989; Katrib, 2005; Fig. 5). Serpentinite is a very minor component of the Colebrooke Schist, however, but it is likely that the Colebrooke contributed blocks to serpentinite-matrix mélanges above and below the Colebrooke (Roure and Blanchet, 1983; Blake et al., 1985a; Katrib, 2005; Schoonmaker et al., this volume). Available geochemical data come from two large pillow basalt blocks from the serpentinite-matrix mélange above the Colebrooke Schist in the Snowcamp Mountain area. They have transitional IAT-MORB affinities (Katrib, 2005; Fig. 5) and are thus distinct from the pillow basalts in the mélange of the study area; in addition, no amphibolite blocks have been found. Nevertheless, this mélange contains some similar block types to that in the study area, including ultramafic cumulates and hornblende diorite, as well as blocks of Colebrooke Schist.

Northward Translation of the Elk Outlier

Jayko and Blake (1993) used geologic and paleomagnetic data to argue for northward translation of terranes in southwest Oregon during the latest Jurassic to mid-Eocene. They pointed out that dextral transpression is likely to have occurred during this time, based on reconstructions of North American–Farallon plate motions, and suggested the Elk outlier was transported north from the latitude of the southern Klamath Mountains, based primarily on the outlier's position far to the west of the rest of the western Klamath terrane, as well as west of the Coast Range ophiolite (Snowcamp terrane).

The dike complex associated with the Elk outlier lies north of the northernmost extent of the Josephine ophiolite (Fig. 1).

Thus, if the dike complex is in fact a remnant of the Josephine ophiolite, then its presence would be consistent with at least modest amounts of northward translation. However, if the dike complex and underlying serpentinite mélange together constitute a rift-edge facies, then no northward translation is required, because such facies occur east of the outlier beneath the Rogue and Galice Formations (Yule, 1996; Yule et al., this volume). Nevertheless, the serpentinite-matrix mélange and dike complex together are similar to the Devils Elbow remnant of the Josephine ophiolite in the southern Klamath Mountain (rift-edge facies), except for the apparent absence of boninitic rocks in the latter (Wyld and Wright, 1988). Consequently, the rift-edge-facies interpretation of the dike complex and mélange associated with the Elk outlier would still be consistent with northern translation. If the Elk outlier has not been translated northward, its current position implies that the western Klamath terrane extended far to the west of the Klamath Mountains province and, if the dike complex is a remnant of the Josephine ophiolite, that the Josephine ophiolite extended far to the northwest (Fig. 1).

Detrital modes of graywacke and conglomerates, and possibly ages of detrital zircon in graywackes, in the Galice Formation of the Elk outlier might provide a way of testing for lateral translation. Detrital modes of graywacke and conglomerate are quite different for the Galice Formation and correlative units in the Sierra Nevada foothills, and they may change from north to south in the Galice Formation itself (Seiders, 1991; MacDonald et al., this volume).

CONCLUSIONS

1. Rocks previously mapped as the diorite phase of the Pearse Peak pluton actually comprise a dike complex and serpentinite-matrix mélange.
2. The dike complex is most likely part of a sheeted dike complex, in which case, it represents an ophiolite remnant, possibly of the Josephine ophiolite. Samples from the dike complex have transitional IAT-MORB and boninitic affinities, suggesting a supra-subduction zone origin. The magmatic affinities are the same as those for sheeted dike complex of the Josephine ophiolite, but also similar to lavas of the Coast Range ophiolite. The fault-bounded nature of the dike complex, however, makes it uncertain that the dike complex is actually part of the Elk outlier of the western Klamath terrane.
3. Mafic blocks in the serpentinite-matrix mélange mostly have MORB affinities, suggesting the mélange is petrogenetically unrelated to the dike complex. The mélange may be older basement to the dike complex or part of the adjacent Pickett Peak terrane. The presence of amphibolite blocks, MORB-to-WPB affinities of volcanic and hypabyssal blocks, and presence of unusual Fe-Ti basalts is similar to older basement that locally occurs elsewhere in the western Klamath terrane. The mélange and dike complex together could comprise a rift-edge facies similar to the Devils Elbow remnant of the Josephine ophiolite.
4. The presence of a probable Josephine ophiolite remnant in the Elk outlier so far from the main mass of the Josephine ophiolite would imply that the ophiolite originally extended far to the north *or* that the Elk outlier was translated northward relative to the Klamath Mountains province.

ACKNOWLEDGMENTS

We are grateful to Nathan Christman, Eyrene Fisher, Khris Karambela, Dennis Leikam, Charles Preppernau, Eric Price, and Jimmy Wetenkamp for their assistance in the field and lab. We appreciate Michael Cox, Kelley Giaramita, and Samantha Schiller for camp management, food preparation, and companionship. We thank Margaret McHugh of the National Forest Service, Gold Beach, Oregon, for logistical support of fieldwork. We thank the many landowners who granted us permission to work on their property, and Moore Mill Company for providing vehicular access to their property. We are grateful to Bob Miller and an anonymous reviewer for their thoughtful reviews and helpful suggestions that improved the quality of the manuscript. Financial support to MJG was provided through Research, Scholarship, and Creative Activity Grants from California State University, Stanislaus. Financial support for GDH was provided by National Science Foundation grant EAR-9706798 and a State University of New York Albany Faculty Research Awards Program grant.

APPENDIX

TABLE A-1. PETROGRAPHY OF SELECTED MELANGE BLOCKS AND LATE DIKES

Sample	Rock type	Textures	Grain size (mm)	Minerals (Primary)	Minerals (Secondary)†	Comments
00ER74A	Pillow basalt	ig, so, q, rs	0.1	aug, Ftox	ab, cc, pp, qz, sp	Possible vesicles; altered glassy rim
01ER27A	Diabase/microgabbro	eg, so	1.5	aug, hb, Ftox	ab, pp(?)	Purple titaniferous aug
01ER30C	Microgabbro	ig	0.6	aug, Ftox	ab, ch, pr, sp, ur	
00ER76B	Gabbro/diorite	cum, hg	1.5		act, hb	Variable cataclasis
01ER14A	Igneous breccia	fgt			ep, pp, qz	Clasts (described below) in sand-sized matrix of aug, pl, Ftox
	Basaltic clast	gl, rs	0.1	aug, pl(?)	ur	mp of pl and ur
	Hypabyssal clast	so, tr	0.5	aug, Ftox, pl(?)	ur	
	Dacite(?) clast	po	0.1	K-sp, pl(?), qz		mp of Ksp and/or pl
01ER29A	Dacite breccia	fgt				Dacite clasts (descibed below) in sand-sized matrix of pl, qz
	Dacite clast	ftd, po	0.1	Ftox, pl(?), qz	cc	
01ER02	Clinopyroxenite	cum(?), hg(?)	4.0	aug		
01ER34	ol-clinopyroxenite		3.0	aug, se, pm after ol(?)	se	Some show complete replacement of aug by ur
02ER58	Peridotite	rm	1.5	aug, bs, Cr-sp	se	bs and Cr-sp aligned
02ER51	hyp-amphibolite	lep	0.2	aug, brn-hb, Ftox, hyp, pl, rt	ep, pr	Polygonal grain boundaries
01ER16A	gt-amphibolite	lep	1.0	gt, hb, pl, rt	ab, ch, ep, qz, sp	
00ER27	Metagreywacke	ps, sa, weak foliation	0.3	aug, Ftox, li, pl, qz	cc	Aligned white mica
01KK5A	Dacite	po	1.0	Ftox, pl(?), qz	ab, ch, ms	Undulose qz; cataclasis cuts serpentinite
02ER04B	Microgabbro	po	1.5–2.0	aug, hb	bt, ur	Grains up to 5 mm; cuts Galice Formation

Abbreviations: ab—albite; act—actinolite; aug—augite; brn—brown; bs—bastite (serpentinite pseudomorphs after orthopyroxene); bt—biotite; cc—calcite; ch—chlorite; Cr-sp—chrome spinel; eg—equigranular; ep—epidote-group mineral; fgt—fragmental; ftd—felted; Ftox—iron-titanium oxide; gl—altered glassy texture; gt—garnet; hb—hornblende; hg—hypidiomorphic granular; hyp—hypersthene; ig—intergranular; Ksp—K-felspar; lep—lepidoblastic; li—lithic fragments; mp—microphenocryst; ms—muscovite; ol—olivine; pl—plagioclase; pm—pseudomorphs; po—porphyritic; pp—pumpellyite; pr—prehnite; q—quench textures; qz—quartz; rm—ribbon and mesh texture in serpentinite (typical of replacement of olivine by lizardite and chrysotile; O'Hanley, 1996); rs—radial splays of augite; rt—rutile; se—serpentine; so—sub-ophitic; sp—sphene; tr—trachytic texture; ur—uralitic actinolite.

†Albite psedomorphs after plagioclase.

TABLE A-2. UTM COORDINATES OF GEOCHEMICAL SAMPLES

Sample	North (m)	East (m)	Sample	North (m)	East (m)
00ER21A	4729777	387442	00ER11B	4728533	386899
00ER41A	4729750	387488	01BMC1A	4728533	386899
00ER41B	4729750	387488	01ER19A	4728265	386398
01ER46A	4729750	387488	01ER41	4728751	385847
00ER682A	4729609	387516	02ER49B	4729282	386621
00ER682B	4729609	387516	01ERKK2	4728133	386384
00ER682C	4729609	387516	01ER27A	4729969	386216
02ER09B	4730147	386978	01ER30C	4729769	386145
02ER11A	4729525	387478	02ER16A	4728589	386125
00ER68A	4729745	387506	00ER76B	4730482	386301
00ER74A	4730208	386355	02ER51	4729279	386738
00ER74B	4730205	386345	01ER16A	4728495	386474
00ER74C	4730212	386395	01KK5A	4728586	386000
00ER93	4728940	385964	02ER04B	4725048	387735
00ER94B	4728801	386315			

Notes: Locations are in zone 10. UTM—Universal Transverse Mercator.

REFERENCES CITED

Aalto, K.R., 1968, The sedimentology of the Late Jurassic (Portlandian) Otter Point Formation of southwestern Oregon [M.S. thesis]: Madison, University of Wisconsin, 60 p.

Barnes, C.G., Snoke, A.W., Harper, G.D., Frost, C.D., McFadden, R.R., Bushey, J.C., and Barnes, M.A.W., 2006, this volume (Chapter 17), Arc plutonism following regional thrusting: Petrology and geochemistry of syn- and post-Nevadan plutons in the Siskiyou Mountains, Klamath Mountains province, California, *in* Snoke, A.W., and Barnes, C.G., eds., Geological studies in the Klamath Mountains province, California and Oregon: A volume in honor of William P. Irwin: Boulder, Colorado, Geological Society of America Special Paper 410, doi: 10.1130/2006.2410(17).

Blake, M.C., Jr., Engebretson, D.C., Jayko, A.S., and Jones, D.L., 1985a, Tectonostratigraphic terranes in southwest Oregon, *in* Howell, D.G., ed., Tectonostratigraphic terranes of the Circum-Pacific region: Houston, Circum-Pacific Council for Energy and Mineral Resources, Earth Sciences Series 1, p. 147–157.

Blake, M.C., Jr., Jayko, A.S., and McLaughlin, R.J., 1985b, Tectonostratigraphic terranes of the northern Coast Ranges, California, *in* Howell, D.G., ed., Tectonostratigraphic terranes of the circum-Pacific region: Houston, Circum-Pacific Council for Energy and Mineral Resources, Earth Sciences Series 1, p. 159–171.

Brown, E.H., and Blake, M.C., Jr., 1987, Correlation of Early Cretaceous blueschists in Washington, Oregon and Northern California: Tectonics, v. 6, p. 795–806.

Burchfiel, B.C., and Davis, G.A., 1981, Triassic and Jurassic tectonic evolution of the Klamath Mountain–Sierra Nevada geologic terrane, *in* Ernst, W.G., ed., The geotectonic development of California—Rubey volume 1: Englewood Cliffs, New Jersey, Prentice-Hall, p. 50–70.

Coleman, R.G., 1972. The Colebrooke Schist of southwestern Oregon and its relation to the tectonic evolution of the region: Washington, D.C., U.S. Geological Survey Bulletin 1339, 61 p.

Crawford, A.J., Falloon, T.J., and Green, D.H., 1989, Classification, petrogenesis and tectonic setting of boninites, *in* Crawford, A.J., ed., Boninites and related rocks: London, Unwin Hyman, p. 2–49.

Davis, G.A., 1969, Tectonic correlations, Klamath mountains and western Sierra Nevada, California: Geological Society of America Bulletin, v. 80, p. 1095–1108.

Deschamps, A., and Lallemand, S., 2003 Geodynamic setting of Izu-Bonin-Mariana boninites, *in* Larter, R.D., and Leat, P.T., eds., Intra-oceanic subduction systems; Tectonic and magmatic processes: Oxford, Blackwell Scientific Publications, Geological Society Special Publication 219, p. 163–185.

Dick, H.J.B., 1976, The origin and emplacement of the Josephine Peridotite of southwestern Oregon [Ph.D. dissertation]: New Haven, Yale University, 409 p.

Dickinson, W.R., Hopson, C.A., and Saleeby, J.B., 1996, Alternate origins of the Coast Range Ophiolite (California); Introduction and implications: GSA Today, v. 6, p. 1–10.

Diller, J.S., 1903, Description of the Port Orford Quadrangle, Oregon: Washington, D.C., U.S. Geological Survey, Geological Atlas Folio 89, 6 p.

Dott, R.H., Jr., 1965, Mesozoic-Cenozoic tectonic history of the southwestern Oregon coast in relation to Cordilleran orogenesis: Journal of Geophysical Research, v. 18, p. 4687–4707.

Dott, R.H., Jr., 1966, Late Jurassic unconformity exposed in southwestern Oregon: Ore Bin, v. 28, p. 85–97.

Dott, R.H., Jr., 1971, Geology of the southwestern Oregon coast west of the 124th meridan: Portland, Oregon Department of Geology and Mineral Industries Bulletin 69, 63 p.

Garcia, M.O., 1982, Petrology of the Rogue River island-arc complex, southwest Oregon: American Journal of Science, v. 282, p. 783–807.

Gorman, C.M., III, 1985, Geology, geochemistry, and geochronology of the Rattlesnake Creek Terrane, west-central Klamath Mountains, California [M.S. thesis]: Salt Lake City, University of Utah, 111 p.

Hacker, B.R., Donato, M.M., Barnes, C.G., McWilliams, M.O., and Ernst, W.G., 1995, Timescales of orogeny: Jurassic construction of the Klamath Mountains: Tectonics, v. 14, p. 677–703, doi: 10.1029/94TC02454.

Harper, G.D., 1980, Structure and petrology of the Josephine ophiolite and overlying metasedimentary rocks, northwestern California [Ph.D. dissertation]: Berkeley, University of California, Berkeley, 281 p.

Harper, G.D., 1984, The Josephine ophiolite, northwestern California: Geological Society of America Bulletin, v. 95, p. 1009–1026.

Harper, G.D., 1995, Pumpellyosite and prehnitite associated with epidosite in the Josephine ophiolite—Ca metasomatism during upwelling of hydrothermal fluids at a spreading axis, *in* Schiffman, P., and Day, H., eds., Low grade metamorphism of mafic rocks: Boulder, Colorado, Geological Society of America Special Paper 296, p. 101–122.

Harper, G.D., 2003, Fe-Ti basalts and propagating rift tectonics in the Josephine ophiolite: Geological Society of America Bulletin, v. 115, p. 771–787, doi: 10.1130/0016-7606(2003)115<0771:FBAPTI>2.0.CO;2.

Harper, G.D., 2004, Tectonic implications of boninite, arc tholeiite, and MORB magma types in the Josephine ophiolite, California-Oregon, *in* Dilek, Y., and Robinson, P.T., eds., Ophiolites in Earth history: Oxford, Blackwell Scientific Publications, Geological Society Special Publication 218, p. 207–230.

Harper, G.D., 2006, this volume, Structure of syn-Nevadan dikes and their relationship to deformation of the Galice Formation, western Klamath terrane, northwestern California, *in* Snoke, A.W., and Barnes, C.G., eds., Geological studies in the Klamath Mountains province, California and Oregon: A volume in honor of William P. Irwin: Boulder, Colorado, Geological Society of America Special Paper 410, doi: 10.1130/2006.2410(06).

Harper, G.D., Bowman, J.R., and Kuhns, R., 1988, Field, chemical, and isotopic aspects of submarine hydrothermal metamorphism of the Josephine ophiolite, Klamath Mountains, California-Oregon: Journal of Geophysical Research, v. 93, p. 4625–4657.

Harper, G.D., Grady, K., and Wakabayashi, J., 1990, A structural study of a metamorphic sole beneath the Josephine ophiolite, western Klamath terrane, California-Oregon, *in* Harwood, D.S., and Miller, M.M., eds., Paleozoic and early Mesozoic paleogeographic relations; Sierra Nevada, Klamath Mountains, and related terranes: Boulder, Colorado, Geological Society of America Special Paper 255, p. 379–396.

Harper, G.D., Saleeby, J.B., and Heizler, M., 1994, Formation and emplacement of the Josephine ophiolite and the age of the Nevadan Orogeny in the Klamath Mountains, California-Oregon: U/Pb zircon and $^{40}Ar/^{39}Ar$ geochronology: Journal of Geophysical Research, v. 99, p. 4293–4321, doi: 10.1029/93JB02061.

Harper, G.D., Grady, K., and Coulton, A.J., 1996, Origin of the amphibolite "sole" of the Josephine ophiolite; Emplacement of a cold ophiolite over a hot arc: Tectonics, v. 15, p. 296–313, doi: 10.1029/95TC02525.

Harper, G.D., Kosanke, S., Giaramita, M.J., Saleeby, J.B., and Heizler, M., 2000, Tectonostratigraphy and geochemistry of the Coast Range Ophiolite at Snowcamp Mountain, SW Oregon: Implications for the exotic vs. native origin of the Coast Range Ophiolite. Geological Society of America Abstracts with Programs, v. 32, p. A-17.

Hawkins, J.W., 1995, Evolution of the Lau Basin; Insights from ODP Leg 135, *in* Taylor, B., and Natland, J., eds., Active margins and marginal basins of the western Pacific: Washington, D.C., Geophysical Monograph 88, American Geophysical Union, p. 125–173.

Hopson, C.A., Mattinson, J.M., and Pessagno, E.A., Jr., 1981, Coast Range ophiolite, western California, *in*, Ernst, W.G., ed., The Geotectonic Development of California—Rubey Volume 1: Englewood Cliffs, New Jersey, Prentice-Hall, p. 418–510.

Irwin, W.P., 1966, Geology of the Klamath Mountains province, *in* Bailey, E.H., ed., Geology of northern California: San Francisco, California Division of Mines and Geology Bulletin 190, p. 19–38.

Irwin, W.P., 1981, Tectonic accretion of the Klamath Mountains, *in* Ernst, W.G, ed., The geotectonic development of California—Rubey volume 1: Englewood Cliffs, New Jersey, Prentice-Hall, p. 29–49.

Irwin, W.P., 1994, Geologic map of the Klamath Mountains, California and Oregon: Reston, Virginia, U.S. Geological Survey Miscellaneous Field Investigations, I-2148, scale 1:500,00, 2 sheets.

Jachens, R.C., Barnes, C.G., and Donato, M.M., 1986, Subsurface configuration of the Orleans Fault; Implications for deformation in the western Klamath Mountains, California: Geological Society of America Bulletin, v. 97, p. 388–395, doi: 10.1130/0016-7606(1986)97<388:SCOTOF>2.0.CO;2.

Jayko, A.S., and Blake, M.C., Jr., 1993, Northward displacements of forearc slivers in the Coast Ranges of California and southwest Oregon during the Late Mesozoic and Early Cenozoic, *in* Dunn, G.C., and McDougall, K.A., eds., Mesozoic paleogeography of the western United States II: Los Angeles, Pacific Section, Society of Economic Paleontologists and Mineralogists, v. 71, p. 19–36.

Johnson, D.M., Hooper, P.R., and Conrey, R.M., 1999, XRF analysis of rocks and minerals for major and trace elements on a single low dilution Li-tetraborate fused bead: Advances in X-ray Analysis, v. 41, p. 843–867.

Kaiser, W.G., 1962, The Late Mesozoic geology of the Pearse Peak diorite, southwest Oregon [M.S. thesis]: Madison, University of Wisconsin, 73 p.

Katrib, J., 2005, Source of meta-igneous blocks and structure of the Colebrooke Schist in the Snowcamp Peak area, Pickett Peak terrane, southwestern Oregon [M.S. thesis]: Albany, New York, University at Albany, 111 p.

Koch, J.G., 1966, Late Mesozoic stratigraphy and tectonic history, Port Orford–Gold Beach area, southwestern Oregon coast: Bulletin of the American Association of Petroleum Geologists, v. 50, p. 25–71.

Kosanke, S.B., 2000, The geology, geochronology, structure and geochemistry of the Wild Rogue Wilderness remnant of the Coast Range Ophiolite, southwest Oregon; Implications for the magmatic and tectonic evolution of the Coast Range Ophiolite [Ph.D. dissertation]: Albany, State University of New York, 774 p.

Le Maitre, R.W., Streckeisen, A., Zanettin, B., Le Bas, M.J., Bonin, B., Bateman, P., Bellieni, G., Dudek, A., Efremova, S., Keller, J., Lameyre, J., Sabine, P.A., Schmid, R., Sorensen, H., and Woolley, A.R., 2002, Igneous rocks; a classification and glossary of terms; Recommendations of the International Union of Geological Sciences Subcommission on the Systematics of Igneous Rocks (2nd edition): Cambridge, Cambridge University Press, 236 p.

MacDonald, J.H., Jr., Harper, G.D., and Zhu, B., 2006, this volume, Petrology, geochemistry, and provenance of the Galice Formation, Klamath Mountains, Oregon and California, *in* Snoke, A.W., and Barnes, C.G., eds., Geological studies in the Klamath Mountains province, California and Oregon: A volume in honor of William P. Irwin: Boulder, Colorado, Geological Society of America Special Paper 410, doi: 10.1130/2006.2410(04).

Meffre, S., Aitchison, J.C., and Crawford, A.J., 1996, Geochemical evolution and tectonic significance of boninites and tholeiites from the Koh ophiolite, New Caledonia: Tectonics, v. 15, p. 67–83, doi: 10.1029/95TC02316.

Metzger, E.P., Miller, R.B., and Harper, G.D., 2002, Geochemistry and tectonic setting of the ophiolitic Ingalls Complex, North Cascades, Washington; Implications for correlations of Jurassic Cordilleran ophiolites: Journal of Geology, v. 110, p. 543–560, doi: 10.1086/341759.

Murton, B.J., Peate, D.W., Arculus, R.J., Pearce, J.A., and van-der-Laan, S.R., 1992, Trace element geochemistry of volcanic rocks from Site 786; The Izu-Bonin forearc, *in* Dearmont, L.H., et al., eds., Proceedings of the Ocean Drilling Program, scientific results 125: College Station, Texas, Ocean Drilling Program, p. 211–235.

O'Hanley, D.S., 1996, Serpentinites; Records of tectonic and petrological history: Oxford, Oxford University Press, Oxford Monographs on Geology and Geophysics 34, 277 p.

Pearce, J.A., 1982, Trace element characteristics of lavas from destructive plate boundaries, *in* Thorpe, E.S., ed., Andesites: New York, John Wiley and Sons, p. 525–548.

Pearce, J.A., 1983, Role of the sub-continental lithosphere in magma genesis at active continental margins, *in* Hawksworth, C.J., and Norry, M.J., eds., Continental basalts and mantle xenoliths: Nantwich, UK, Shiva Publishing, p. 230–249.

Pearce, J.A., 1996, A user's guide to basalt discrimination diagrams, *in* Wyman, D.A., ed., Trace element geochemistry of volcanic rocks: Applications for massive sulphide exploration: St. John's, Newfoundland, Memorial University, Geological Association of Canada Short Course Notes, v. 12, p. 79–114.

Pearce, J.A., and Parkinson, I.J., 1993, Trace element models for mantle melting: application to volcanic arc petrogenesis, *in* Prichard, H.M., et al., eds., Magmatic processes and plate tectonics: Oxford, Blackwell Scientific Publications, Geological Society Special Publication 76, p. 373–403.

Pearce, J.A., and Peate, D.W., 1995, Tectonic implications of the composition of volcanic arc magmas: Annual Review of Earth and Planetary Sciences, v. 23, p. 251–286, doi: 10.1146/annurev.ea.23.050195.001343.

Pearce, J.A., Lippard, S.J., and Roberts, S., 1984, Characteristics and tectonic significance of suprasubduction zone (ssz) ophiolites, *in* Kokellar, B.P., and Howell, M.F., eds., Marginal basin geology: Oxford, Blackwell Scientific Publications, Geological Society Special Publication 16, p. 77–94.

Pearce, J.A., Ernewein, M., Bloomer, S.H., Parson, L.M., Murton, B.J., and Johnson, L.E., 1994, Geochemistry of Lau Basin volcanic rocks: influence of ridge segmentation and arc proximity, *in* Mellie, J.L., ed., Volcanism associated with extension at consuming plate margins: Oxford, Blackwell Scientific Publications, Geological Society Special Publication 81, p. 53–75.

Plake, T.D., 1989, Structure and petrology of the Colebrooke Schist, southwestern Oregon [M.S. thesis]: Bellingham, Western Washington University, 88 p.

Robinson, P.T., Dick, H.J.B., Natland, J.H., and the ODP Leg 176 Shipboard Party, 2000, Lower oceanic crust formed at an ultra-slow-spreading ridge; Ocean Drilling Program Hole 735B, Southwest Indian Ridge, *in* Dilek, Y., et al., eds., Ophiolites and oceanic crust; New insights from field studies and the Ocean Drilling Program: Boulder, Colorado, Geological Society of America Special Paper 349, p. 75–86.

Roure, F., and Blanchet, R., 1983, A geological transect between the Klamath Mountains and the Pacific Ocean (southwestern Oregon); A model for paleosubduction: Tectonophysics, v. 91, p. 53–72, doi: 10.1016/0040-1951(83)90057-4.

Saleeby, J.B., 1992, Petrotectonic and paleogeographic settings of U.S. Cordilleran ophiolites, *in* Burchfiel, B.C., et al., eds., The Cordilleran Orogen; Conterminous United States: Boulder, Colorado, Geological Society of America, The Geology of North America, v. G-3, p. 653–682.

Saleeby, J.B., Harper, G.D., Snoke, A.W., and Sharp, W. 1982, Time relations and structural-stratigraphic patterns in ophiolite accretion, west-central Klamath Mountains, California: Journal of Geophysical Research, v. 87, p. 3831–3848.

Saleeby, J.B., and Harper, G.D., 1993, Tectonic relations between the Galice Formation and the Condrey Mountain Schist, Klamath Mountains, northern California, *in* Dunn, G.C., and McDougall, K.A., eds., Mesozoic paleogeography of the western United States II: Los Angeles, Pacific Section, Society of Economic Paleontologists and Mineralogists, v. 71, p. 61–80.

Schoonmaker, A., Harper, G.D., Heizler, M.T., 2006, this volume, Tectonic emplacement of the Snowcamp remnant of the Coast Range ophiolite near Game Lake, southwestern Oregon, *in* Snoke, A.W., and Barnes, C.G., eds., Geological studies in the Klamath Mountains province, California and Oregon: A volume in honor of William P. Irwin: Boulder, Colorado, Geological Society of America Special Paper 410, doi: 10.1130/2006.2410(09).

Seiders, V.M., 1991, Conglomerate stratigraphy and tectonics in the Franciscan assemblage of northern California and implications for Cordilleran tectonics: Reston, Virginia, U.S. Geological Survey Open-File Report PFR 91-50, 10 p.

Shervais, J.W., 1982, Ti-V plots and the petrogenesis of modern and ophiolitic lavas: Earth and Planetary Science Letters, v. 59, p. 101–118, doi: 10.1016/0012-821X(82)90120-0.

Shervais, J.W., Kimbrough, D.L., Renne, P., Hanan, B.B , Murchey, B., Snow, C.A., Zoglman Schuman, M.M., and Beaman, J., 2004, Multi-stage origin of the Coast Range ophiolite, California: Implications for the life cycle of supra-subduction zone ophiolites: International Geology Review, v. 46, p. 289–315.

Silant'yev, S.A., 1979, Metabasites of the epidote-amphibolite metamorphic facies of the Atlantic Ocean floor: International Geology Review, v. 21, p. 1253–1259.

Sinton, J.M., Wilson, D.S., Christie, D.M., Hey, R.N., and Delaney, J.R., 1983, Petrologic consequences of rift propagation on ocean spreading ridges: Earth and Planetary Science Letters, v. 62, p. 193–207, doi: 10.1016/0012-821X(83)90083-3.

Snoke, A.W., 1977, A thrust plate of ophiolitic rocks in the Preston Peak area, Klamath Mountains, California: Geological Society of America Bulletin, v. 88, p. 1641–1659, doi: 10.1130/0016-7606(1977)88<1641:ATPOOR>2.0.CO;2.

Steiger, R.H., and Jager, E., 1977, Subcommission of geochronology: Convention on the use of decay constants in geo- and cosmochronology: Earth and Planetary Science Letters, v. 36, p. 359–362, doi: 10.1016/0012-821X(77)90060-7.

Sun, S.S., and McDonough, W.F., 1989, Chemical and isotopic systematics of oceanic basalts: Implications for mantle composition and processes, *in* Saunders, A.D., and Norry, M.J., eds., Magmatism in the ocean basins: Oxford, Blackwell Scientific Publications, Geological Society Special Publication 42, p. 313–345.

Walker, G.W., and MacLeod, N.S., 1991, Geologic map of Oregon: Reston, Virginia, U.S. Geological Survey, scale 1:500,000, 2 sheets.

Wells, F.G., and Peck, D.L., 1961, Geologic map of Oregon west of the 121st meridian: Washington, D.C., U.S. Geological Survey Miscellaneous Geologic Inventory Map I-325, scale 1:500,000.

Winchester, J.A., and Floyd, P.A., 1976, Geochemical magma type discrimination; Application to altered and metamorphosed basic igneous rocks: Earth and Planetary Science Letters, v. 28, p. 459–469, doi: 10.1016/0012-821X(76)90207-7.

Wright, J.E., and Fahan, M.R., 1988, An expanded view of Jurassic orogenesis in the western United States Cordillera: Middle Jurassic (pre-Nevadan) regional metamorphism and thrust faulting within an active arc environment, Klamath Mountains, California: Geological Society of America Bulletin, v. 100, p. 859–876, doi: 10.1130/0016-7606(1988)100 <0859:AEVOJO>2.3.CO;2.

Wright, J.E., and Wyld, S.J., 1994, The Rattlesnake Creek terrane, Klamath Mountains, California: An early Mesozoic volcanic arc and its basement of tectonically disrupted oceanic crust: Geological Society of America Bulletin, v. 106, p. 1033–1056, doi: 10.1130/0016-7606(1994)106<1033:TRCTKM>2.3.CO;2.

Wyld, S.J., and Wright, J.E., 1988, The Devils Elbow ophiolite remnant and overlying Galice Formation: New constraints on the Middle to Late Jurassic evolution of the Klamath Mountains, California: Geological Society of America Bulletin, v. 100, p. 29–44, doi: 10.1130/0016-7606(1988)100 0029:TDEORA 2.3.CO;2.

Yule, J.D., 1996, Geologic and tectonic evolution of Jurassic marginal ocean basin lithosphere, Klamath Mountains, Oregon [Ph.D. dissertation]: Pasadena, California Institute of Technology, 308 p.

Yule, J.D., Saleeby, J.B., and Barnes, C.G., 1994, Geochemistry of Rattlesnake Creek terrane fragments contained within the western Jurassic belt, Oregon Klamath Mountains: Geological Society of America Abstracts with Programs, v. 26, p. 106–107.

Yule, J.D., Saleeby, J.B., and Barnes, C.G., 2006, this volume, A rift-edge facies of the Late Jurassic Rogue–Chetco arc and Josephine ophiolite, Klamath Mountains, Oregon, *in* Snoke, A.W., and Barnes, C.G., eds., Geological studies in the Klamath Mountains province, California and Oregon: A volume in honor of William P. Irwin: Boulder, Colorado, Geological Society of America Special Paper 410, doi: 10.1130/2006.2410(03).

Manuscript Accepted by the Society 10 January 2006

Printed in the USA

Geological Society of America
Special Paper 410
2006

Tectonic emplacement of the Snowcamp remnant of the Coast Range ophiolite near Game Lake, southwestern Oregon

Adam Schoonmaker*
Gregory D. Harper
Department of Geological Sciences, State University of New York Albany, 1400 Washington Avenue, Albany, New York 12222, USA
Matthew T. Heizler
New Mexico Institute of Mining and Technology, 801 Leroy Place, Socorro, New Mexico 87801, USA

ABSTRACT

Field mapping and structural analysis indicate that the Snowcamp remnant of the Coast Range ophiolite, in the vicinity of Game Lake, southwest Oregon, was thrust to the northeast over mafic phyllonite and amphibolite correlated with the Chetco complex. The contact between the upper ophiolitic and lower mafic sections is marked by a ~5- to 10-m-thick serpentinite unit and paralleled by mylonitic foliations in the lower mafic section. Syndeformational tonalites intrude the lower plate. The fabric-forming metamorphic mineral assemblages (amphibolite facies) in the mafic section suggest a temperature range of ~550–600+ °C and medium pressure, whereas quantitative thermobarometry indicates 615–720 °C and 3.5–6 kb (11.5–20 km depth). A mylonitic greenschist-facies mineral assemblage in phyllonites below the thrust overprints the amphibolite-facies fabric, indicating that thrusting continued during cooling. The $^{40}Ar/^{39}Ar$ analyses indicate that thrusting and syntectonic intrusion occurred by at least 154.2 ± 2.0 and 149.1 ± 0.4 Ma, respectively.

The character of this thrust strongly resembles that of the Madstone Cabin thrust that emplaced the Josephine ophiolite over the Rogue–Chetco arc complex. The proposed correlation between the two faults implies that either the Snowcamp remnant of the Coast Range ophiolite has undergone a similar emplacement history as the Josephine ophiolite *or* that the ophiolitic rocks near Game Lake may be an outlier of the Josephine ophiolite juxtaposed by postemplacement faulting against a Coast Range remnant around Snowcamp Mountain to the south.

Keywords: Coast Range ophiolite, Snowcamp remnant, Josephine ophiolite, thermobarometry, $^{40}Ar/^{39}Ar$ geochronology, structural analysis, Chetco complex, Madstone Cabin thrust, Game Lake thrust, southwest Oregon

*Email: Adam.Schoonmaker@uvm.edu.

Schoonmaker, A., Harper, G.D., and Heizler, M.T., 2006, Tectonic emplacement of the Snowcamp remnant of the Coast Range ophiolite near Game Lake, southwestern Oregon, *in* Snoke, A.W., and Barnes, C.G., eds., Geological studies in the Klamath Mountains province, California and Oregon: A volume in honor of William P. Irwin: Geological Society of America Special Paper 410, p. 177–197, doi: 10.1130/2006.2410(09). For permission to copy, contact editing@geosociety.org.

INTRODUCTION

Significant features of western North America are the Middle Jurassic Josephine and Coast Range ophiolite belts. Although details surrounding the origin and history of the Josephine ophiolite are well understood (e.g., Harper and Wright, 1984; Saleeby, 1992; Harper et al., 1996), the tectonic origin and emplacement history of the Coast Range ophiolite remain controversial (e.g., Dickinson et al., 1996). The Josephine ophiolite was likely formed in a marginal basin formed during rifting of a composite island-arc complex and was subsequently thrust over the margin of western North America during the Nevadan orogeny (Saleeby et al., 1982; Harper and Wright, 1984; Wyld and Wright, 1988; Saleeby, 1992; Harper et al., 1994).

The tectonic setting for the origin of the Coast Range ophiolite, however, has remained controversial for some time (e.g., Moores, 1970; Hopson et al., 1981; Shervais and Kimbrough, 1985; Robertson, 1989; Dickinson et al., 1996). The controversy arises in part from conflicting evidence from the crustal components of the ophiolite, the character of the overlying sedimentary rocks, and differing interpretations of paleomagnetic data (see Stern and Bloomer, 1992, and Dickinson et al., 1996, for reviews). Proposed models fall into two broad categories of either a native or exotic origin for the ophiolite. Exotic settings for the origin of the Coast Range ophiolite, whether a mid-ocean ridge or intraoceanic arc, require postgenesis accretion to Late Jurassic North America, which may imply the presence of an inboard suture, volcanic arc, and possible accretionary prism. Evidence for or against such a suture is missing in California, due to the presence of the postaccretion Great Valley group that obscures the eastern contacts of the ophiolite remnants with the Late Jurassic continental margin of the Sierra Nevada foothills, although such a contact may exist for the Great Valley ophiolite in the western Sierra Nevada metamorphic belt (Godfrey and Dilek, 2000).

One possible exception occurs in the Game Lake area of southwest Oregon, where the Snowcamp remnant of the Coast Range ophiolite has been found in fault contact with intrusive rocks of the Chetco complex, part of the Josephine ophiolite-arc–rifted-arc system (Fig. 1). The nature of this thrust, here named the "Game Lake thrust," and its implications for the origin and tectonic history of the Coast Range ophiolite, are the subject of this chapter.

The results of work in the Game Lake area show that many significant similarities exist between the Game Lake thrust and Madstone Cabin thrust, the latter of which is exposed ~25 km to the east of the Game Lake area and separates the overlying Josephine ophiolite from the underlying Chetco complex. A correlation between the two thrusts has significant implications for the question of the original tectonic setting for the Coast Range ophiolite (i.e., they formed in similar tectonic environments), or alternately, it may indicate that at least part of the Snowcamp terrane is a remnant of the Josephine ophiolite, rather than the Coast Range ophiolite. To address the question of whether the two thrusts are correlative, the pertinent features of the Madstone Cabin thrust and associated units are described below, followed by a description of the rocks, age, and pressure and temperature conditions of the Game Lake thrust. The significance of similarities and differences between the two thrusts and their tectonic implications are evaluated.

JOSEPHINE OPHIOLITE AND MADSTONE CABIN THRUST

Tectonic Overview

The Josephine ophiolite is part of a series of east-dipping Paleozoic and Mesozoic thrust slices that generally young to the west. These slices represent a series of marine sedimentary rocks and mélanges, ophiolites, and arc terranes accreted during extended and ongoing east-dipping subduction along the evolving west coast of North America (Burchfiel and Davis, 1981; Irwin, 1981; Saleeby, 1992). The Josephine ophiolite and cover rocks make up the Jurassic section of the Smith River subterrane of the western Klamath terrane of Blake et al. (1985) and have been described in detail by a number of previous workers (Vail, 1977; Harper, 1980, 1984; Harper et al., 1994). Isotopic ages of the ophiolite include a 162 ± 1-Ma U-Pb zircon age (1-σ) and a 165 ± 3-Ma $^{40}Ar/^{39}Ar$ hornblende age reported by Harper et al. (1994; revised 2-σ zircon age of 162 +7/–2-Ma by Palfy et al., 2000), and a 164 ± 1 Ma U-Pb zircon age (1-σ) by Wright and Wyld (1986). Pessagno and Blome (1990) originally reported Callovian radiolarians from sedimentary rocks interbedded with pillow lavas and Oxfordian radiolarians from the hemipelagic sedimentary rocks overlying the crustal sequence, but this fauna may be older (Bajocian and Bathonian, respectively), according to Baumgartner et al. (1995). They are in turn conformably overlain by the Kimmeridgian, fossil-bearing, continentally-derived flysch of the Galice Formation (Harper, 1983). Based on the above isotopic dates, the Josephine ophiolite may be ca. 169–160 Ma, but the fossil calibration of Baumgartner et al. (1995) favors an age toward the older end of this range.

The Josephine ophiolite is interpreted to have formed in an intra-arc basin as arc magmatic activity shifted from the Middle Jurassic western Hayfork terrane, built on older accreted terranes, to the outboard Middle to Late Jurassic Chetco complex and correlative metavolcanic and volcaniclastic rocks of the Rogue Formation (Harper and Wright, 1984; Wright and Wyld, 1986; Wyld and Wright, 1988; Harper et al., 1994). The volcanic sequence of the inboard, Middle Jurassic arc complex has been dated at ca. 177–168 Ma, whereas the plutonic part of the complex, which cuts across a number of older terranes of the Klamath Mountains province, has been dated at 174–159 Ma (U-Pb zircon: Wright and Fahan, 1988). Hacker et al. (1995) dated the age of the western Hayfork arc at 177–169 Ma ($^{40}Ar/^{39}Ar$). The younger Chetco complex has yielded U-Pb zircon and $^{40}Ar/^{39}Ar$ (hornblende) ages of 160–156 Ma (Yule and Saleeby, 1993; Hacker et al., 1995; Yule, 1996; Yule et al., this volume).

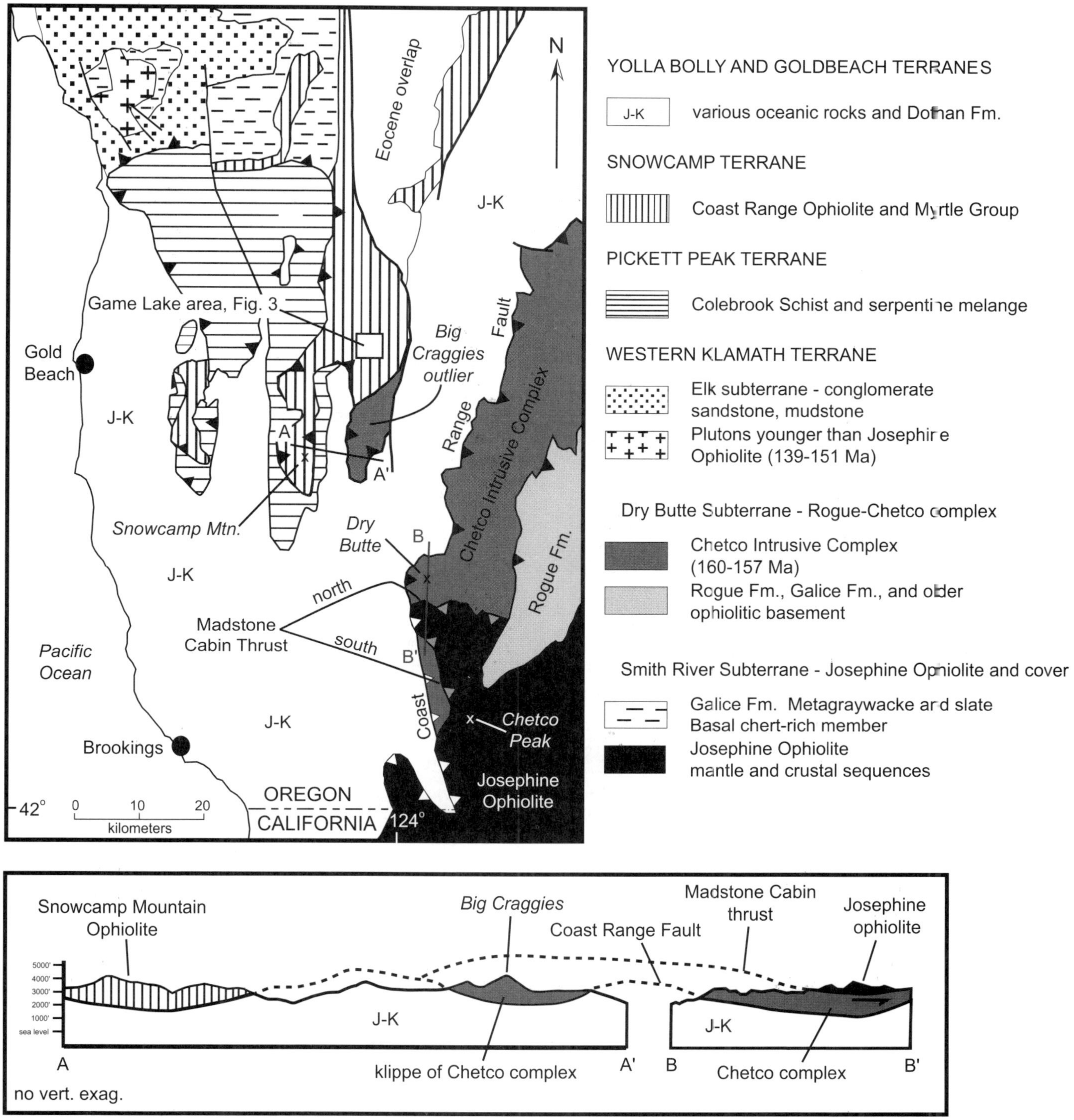

Figure 1. Regional geologic map and schematic cross-sections across the Snowcamp terrane (A–A′) and the Madstone Cabin thrust (B–B′).

Shortly after generation of the Josephine ophiolite, the arc-ophiolite–rifted-arc system collapsed and was imbricated by underthrusting during the Nevadan orogeny (Harper and Wright, 1984; Harper et al., 1994). The ophiolite and associated rocks of the Smith River subterrane were thrust beneath the older continental margin along the Orleans (Preston Peak) thrust, which has a minimum displacement of 40 km (Harper et al., 1994) to 110 km (Jachens et al., 1986). The outboard Middle to Late Jurassic arc, consisting of the Chetco complex (Illinois River complex of Yule, 1996, and Yule et al., this volume) and the metavolcanic/volcaniclastic Rogue Formation were thrust beneath the Smith River subterrane along the Madstone Cabin thrust (Ramp, 1975; Loney and Himmelberg, 1977; Harper et al., 1990, 1994) and are interpreted to be contemporaneous with thrusting along the Orleans roof fault (Harper et al., 1994). Collapse of the arc-ophiolite–rifted-arc system is consistent with deposition and subsequent deformation and metamorphism of the Upper Jurassic flysch of the Galice Formation, which is locally unconformably overlain by Early Cretaceous sedimentary rocks (Harper et al., 1994).

Madstone Cabin Thrust

The Madstone Cabin thrust is exposed in two areas of the western Klamath terrane of southwest Oregon: a northern exposure that includes the area near Vulcan Peak and Dry Butte and a southern reentrant near Chetco Peak (Fig. 1). It separates overlying peridotite of the Josephine ophiolite, which was thrust northwestward (north-northeastward in present position) over mafic units of the Chetco complex and possibly the Rogue Formation (Loney and Himmelberg, 1977; Harper et al., 1990; Saleeby and Harper, 1993). The presently indicated direction of transport direction (north-northeast), which differs from the inferred original transport direction (northwest), is interpreted to be the result of ~65° of regional clockwise rotation during post-Nevadan oroclinal bending (Saleeby and Harper, 1993). The post-Nevadan Valen Lake thrust (Coast Range fault) of Ramp (1975) cuts the Madstone Cabin thrust and emplaces the Cretaceous Dothan Formation (Yolla Bolly terrane of the Franciscan Complex) beneath the amalgamated western Klamath terrane (Fig. 1).

The Madstone Cabin thrust and/or associated lithologic units have been well studied by Himmelberg and Loney (1973), Ramp (1975), Dick (1976), Loney and Himmelberg (1976, 1977), Cannat and Boudier (1985), Grady (1990), Harper et al. (1990), Coulton (1995), and Harper et al. (1994, 1996) and is schematically illustrated in Figure 2. The base of the Josephine peridotite is marked by a 20- to 40-m-thick antigorite serpentinite mylonite that is structurally concordant with deformation fabrics in the amphibolite and mafic phyllonite units below the thrust (Grady, 1990). According to Harper et al. (1996), this occurrence of syndeformational antigorite intergrown with brucite indicates thrusting temperatures in the range of 300–400 °C (although antigorite + brucite may be stable up to ~450 °C; Evans, 2004), likely formed during the late stages of thrusting, is substantially lower than ~800+ °C interpreted for strained basal peridotite mylonites in other ophiolites that have been obducted onto continental margins (Harper et al., 1990).

Beneath the serpentinite mylonite lies up to ~300 m of amphibolite showing evidence of greenschist-facies retrograde metamorphism. In addition, local occurrences of mafic phyllonite occur along the trace of the Madstone Cabin thrust that Grady (1990) determined to be retrogressed amphibolite.

Harper et al. (1994) reported $^{40}Ar/^{39}Ar$ cooling ages of 152 ± 0.8 Ma and 151 ± 1.1 Ma for hornblende in amphibolite below the base of the Josephine ophiolite, similar to a 153 ± 3-Ma K/Ar hornblende age reported by Dick (1976). These ages are consistent with the 151.9 ± 1.0- and 152.8 ± 0.8-Ma (1-σ) $^{40}Ar/^{39}Ar$ hornblende cooling ages from amphibolite and 149.9 ± 0.3-Ma (1-σ) $^{40}Ar/^{39}Ar$ muscovite cooling age from a deformed pegmatite, intruding the amphibolite (Harper et al., 1994). Furthermore, a deformed syntectonic muscovite-bearing granite pegmatite dike that cuts the amphibolite yielded a zircon U-Pb igneous age of 151 ± 1 Ma (Harper et al., 1994). Harper et al. (1994) also reported that the dike gave a $^{40}Ar/^{39}Ar$ muscovite-cooling gradient of 150–146 Ma. Based on these ages, Harper et al. (1994) determined that movement along the Madstone Cabin thrust, which resulted in the development of the amphibolite and phyllonitic foliations and lineations, occurred ca. 155–145 Ma, with the greatest displacement having occurred by 150 Ma. The base of the amphibolite is both intruded by (Loney and Himmelberg, 1977) and grades into (Ramp, 1975; Dick,

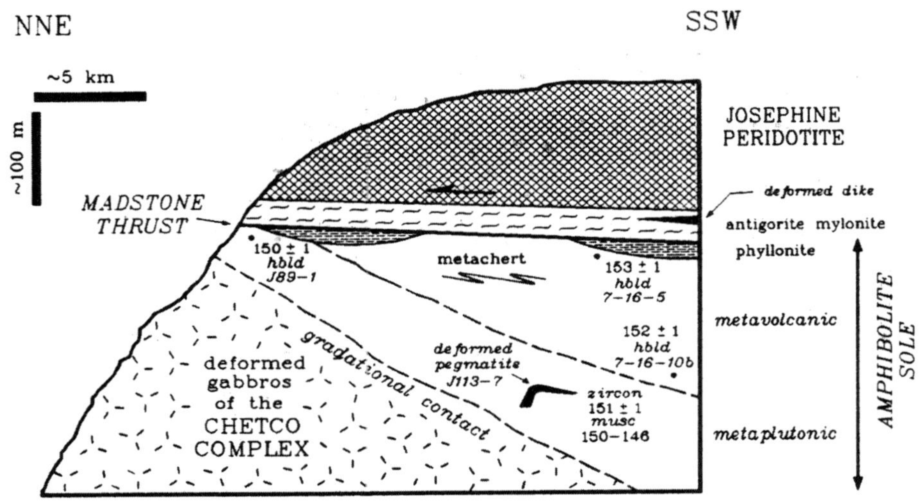

Figure 2. Schematic cross-section of the Madstone Cabin thrust. From Harper et al. (1996).

1976) hornblende metagabbro, suggesting that active arc magmatism was also occurring during the formation of the amphibolite fabric (Harper et al., 1990).

The timing and structural concordance of the amphibolite and phyllonite of the lower plate with the serpentinite mylonite that overprints low-T serpentinite of the upper plate led Harper et al. (1996) to conclude that the amphibolite-facies conditions may have prevailed in the lower plate during early stages of Nevadan shortening, as a result of ongoing gabbroic intrusion below the thrust, but culminated in emplacement of the cold (~350 °C) ophiolite over the hot, active arc complex at lower greenschist-grade conditions. This process resulted in the retrograde greenschist-facies metamorphism of the arc complex concurrent with prograde antigorite recrystallization in serpentinite along the base of the peridotite. Such a model differs from typical metamorphic soles, where heating of the underlying amphibolite comes from the overlying hot ophiolite.

COAST RANGE OPHIOLITE, SNOWCAMP TERRANE, AND GEOLOGY OF THE GAME LAKE AREA

Snowcamp Terrane and Its Correlation with the Coast Range Ophiolite

The Coast Range ophiolite has been described and hypotheses concerning its origin have been discussed by a number of workers (e.g., Hopson et al., 1981; Jayko et al., 1987; Robertson, 1989; Shervais, 1990; Stern and Bloomer, 1992; Dickinson et al., 1996; Shervais et al., 2004). Remnants of the Coast Range ophiolite are present throughout California and extend into southwest Oregon in the Snowcamp terrane (Blake, 1984; Blake et al., 1985; Saleeby, 1992). Published ages (K-Ar hornblende, U-Pb zircon) for the ophiolite, summarized in Shervais et al. (2004), range from 173 to 143 Ma. However, the lack of precision and reliability of some of the early, published ages led them to propose a range of 172- to 164–160 Ma for ophiolite formation (also 170–165 Ma in Hopson et al., 1996), which is slightly older but overlaps the age range of the Josephine ophiolite. The age of ophiolitic rocks in the Snowcamp terrane is constrained by U-Pb zircon dates of 169 ± 1 Ma (1-σ) from a plagiogranite dike in the Snowcamp Mountain area (Saleeby, 1984) and 164 ± 1 Ma (1-σ) in a plagiogranite from the Wild Rogue remnant northeast of the Game Lake area.

Sedimentary rocks of the Great Valley Group overlie remnants of the Coast Range ophiolite in California. In the Snowcamp terrane, there is the conglomeritic Myrtle Group (Riddle Formation) that is largely unmetamorphosed, but locally shows some evidence of metamorphism (Blake et al., 1985). The contact between conglomerate and underlying pillow lavas, northeast of Snowcamp Mountain, is obscured and is marked by a small ravine, suggesting a possible fault relationship, rather than a depositional one. Harper (2003) also reported chert and volcaniclastic sandstone overlying pillow lavas within the Snowcamp terrane. These relationships differ from those of the Josephine ophiolite, which is overlain by continentally derived flysch of the Galice Formation that was subsequently affected by Nevadan deformation. Furthermore, deposition of the Riddle and overlying Days Creek Formations continued during Nevadan time, which, in the sedimentary rocks overlying the Josephine ophiolite, is marked by an unconformity (Blake et al., 1985).

The remnants, in California, are dismembered and floored by the Franciscan Complex, which, in the Game Lake area, includes the Colebrook Schist, part of the Pickett Peak terrane.

Game Lake Area

The Game Lake area, of the Snowcamp terrane, lies ~7 km north of the Big Craggies and ~25 km northwest of the Madstone Cabin thrust near Vulcan Peak (Fig. 1). The region surrounding Game Lake was first studied and mapped at 1:125,000 by Coleman (1972) and later mapped at 1:62,500 by Page et al. (1981), who recognized that an outlier of the Chetco complex underlies Big Craggies proper. Blake et al. (1985) later incorporated their own data with that of previous workers to delineate the Snowcamp terrane, which includes the rocks in the Game Lake area. It is a generally north–south-striking group of fault-bounded rocks composed of a dismembered ophiolite and cover sequence disrupted by late faulting (Blake, 1984; Blake et al., 1985; Harper et al., 2000). Units include serpentinized peridotite, gabbro, a sheeted dike complex, and pillow basalts. The unmetamorphosed sedimentary cover conformably(?) overlies the ophiolite and is composed of the latest Jurassic to Early Cretaceous, flyschlike Riddle and Days Creek Formations of the Myrtle Group (Blake, 1984; Blake et al., 1985).

The Snowcamp terrane is mapped in fault contact (Page et al., 1981) with an outlier of metagabbros and amphibolite of the Chetco complex of the western Klamath terrane, mainly exposed in the Big Craggies Botanical Area west of Snowcamp Mountain (Fig. 1). Coleman (1972) reported a K/Ar age of 155 ± 5 Ma (recalculated using decay constants of Steiger and Jager, 1977) for hornblende from this area, similar to that for amphibolite of the Game Lake area (discussed below).

Lithic Units

Major contacts between the Snowcamp terrane and Chetco complex of the Big Craggies south of Game Lake on previously published maps (Coleman, 1972; Page et al., 1981) are inferred and not exposed. Coleman (1972) briefly mentioned amphibolitic rocks around Game Lake Peak as occurring as inclusions in serpentinite.

Less than a kilometer north of Game Lake, unsheared peridotite (mostly harzburgite with minor dunite) is exposed along an approximately east-to-west oriented ridge and structurally overlies mafic phyllonite along the shallowly north-dipping Game Lake thrust (Figs. 3 and 4). The peridotite and mafic phyllonite are separated by a thin layer (5–10 m) of coherently

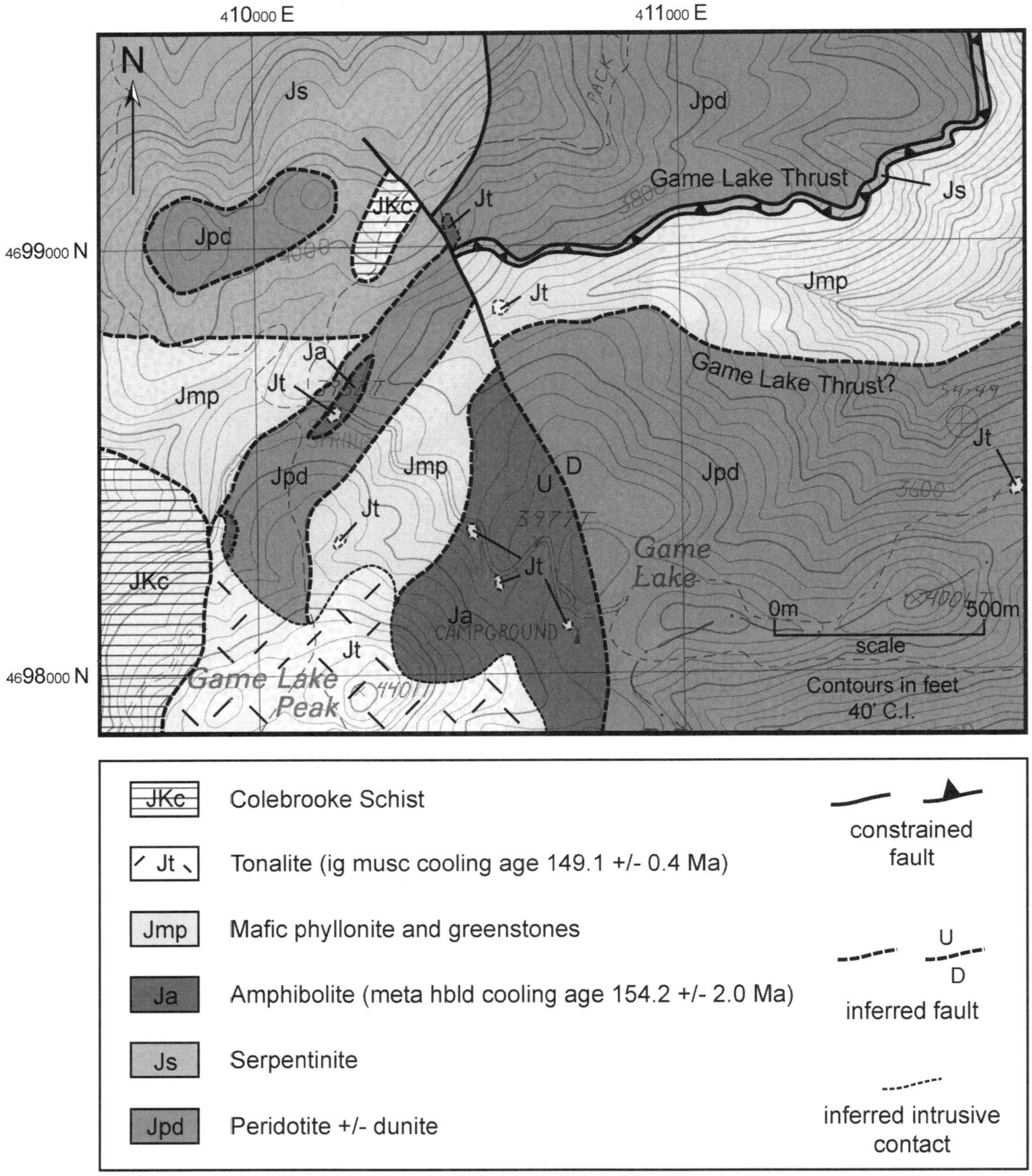

Figure 3. Geologic map of the Game Lake area.

foliated chrysotile- and lizardite-bearing serpentinite (determined by X-ray diffraction), which displays cleavage concordant with both the thrust surface and foliation in the underlying units (Fig. 4). The western limit of the peridotite is marked by a >200-m-wide band of incoherent and sheared serpentinite mélange that strikes north to south. The serpentinite mélange is characterized by a highly foliated serpentine matrix containing variably sized (up to ~100 m) blocks of amphibolite; mafic phyllonite; tonalite; and Colebrook Schist of the blueschist–greenschist-facies Pickett Peak terrane, which in the vicinity of Game Lake is typically a dark, fine-grained, highly foliated, slaty phyllite. These blocks commonly display varying degrees of cataclastic texture.

Underlying the peridotite and serpentinite are 100+ m of nearly continuously exposed and well-indurated mafic phyllonite and greenstone (Fig. 4). The mafic phyllonite displays a

Figure 4. View of the Game Lake thrust (looking approximately north).

mylonitic foliation of variable intensity that parallels the thrust surface. However, locally, the rocks resemble massive greenstone rather than phyllonite, due to postfoliation late-stage retrograde metamorphism that commonly overprints the tectonite fabric in thin-section. Few relict plagioclase, hornblende, epidote, and clinozoisite porphyroclasts of early epidote-amphibolite- to amphibolite-facies metamorphism are seen, largely replaced by fine-grained retrograde greenschist-facies mineral assemblages, including epidote, actinolite, chlorite, and albite (Fig. 5). Any early foliations associated with amphibolite-facies metamorphism are no longer evident, and highly altered relict plagioclase grains show poor fabric development. The existing tectonite fabric is defined by strongly oriented, fine-grained fibrous actinolite that has completely, or nearly completely, replaced hornblende porphyroclasts. Evidence for further post-tectonic metamorphism is shown by ubiquitous fine-grained epidote overgrowths that locally obscure the macroscopic fabric (Fig. 5), as well as coarse-grained prehnite found filling vugs and veins.

Structurally, below the mafic phyllonite is a thick sequence of gneissic amphibolite that shows variable amounts of retrograde metamorphism. Unaltered pleochroic blue-green, green, or yellow hornblende; untwinned plagioclase (An_{20-45}); epidote; quartz; and to a lesser extent, ilmenite and sphene define a well-developed mylonitic fabric and is indicative of amphibolite-facies metamorphism (Fig. 6). Scarce apatite and fine-grained biotite are also present. Porphyroclasts up to 3–5 mm in size of brown hornblende and plagioclase are present, which are interpreted to be the relict igneous phases of a gabbroic protolith. Porphyroclastic crystals show varying degrees of imbrication and rotation, dynamic grain-size reduction, and pressure shadow overgrowths. In those samples that show postdeformational alteration, fine-grained actinolite, epidote, and chlorite replace hornblende, and plagioclase crystals show varying degrees of saussuritization. Saussuritization is greatest in the calcic cores of large porphyroclastic plagioclase crystals, whereas the finer-grained, recrystallized, and untwinned plagioclase grains

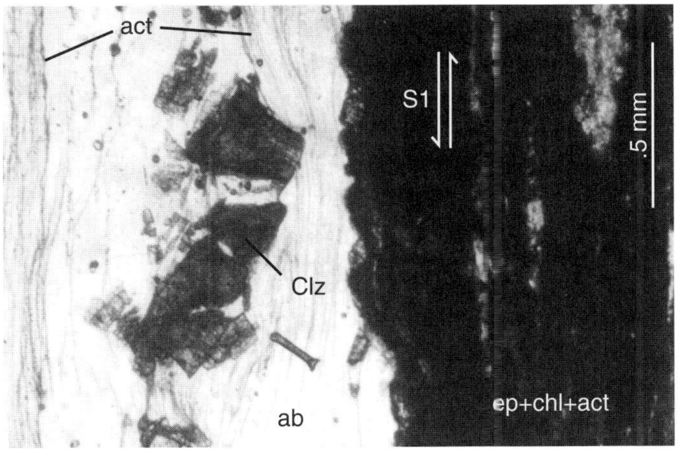

Figure 5. Plane-polarized light (2.5×) photomicrograph of mylonitic texture of mafic phyllonite (BC-307). Note imbricated and rotated porphyroclasts of clinozoisite (Clz) and fibrous actinolite (act). Light-colored albite-rich zone (ab) shows strong foliation (S1), whereas foliation in dark area is partly obscured by nonfoliated chlorite (chl) and epidote (ep).

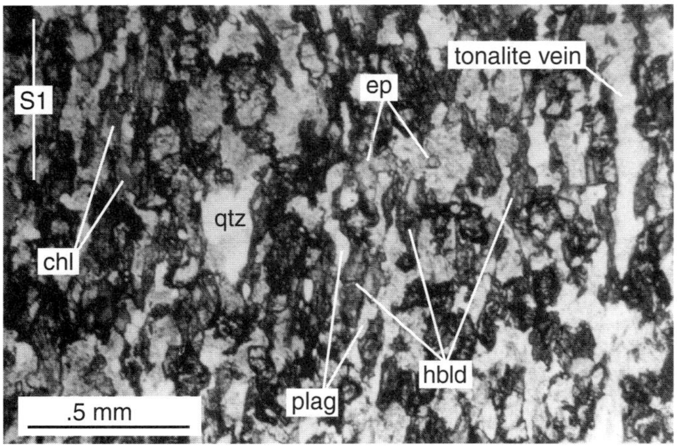

Figure 6. Plane-polarized light (2.5×) photomicrograph of amphibolite (BC-136). Foliation (S1) defined by hornblende (hbld) and plagioclase (plag) and, to a lesser extent, by epidote (ep) and quartz (qtz). Slight alteration of hornblende to chlorite (chl) is present. Tonalite vein is deformed parallel to foliation.

are less altered. Vein-filling epidote, albite, quartz, calcite, and prehnite are also locally present.

A large mass of metagabbro (foliated gabbro) occurs south of the Game Lake area underlying the Big Craggies, where amphibolite is also locally present. The metagabbro displays a variable but generally low degree of metamorphic fabric development that is protomylonitic (Fig. 7) but locally grades to a gneissic foliation typical of the amphibolites around Game Lake. Twinned hornblende and twinned and zoned plagioclase crystals in the protomylonitic metagabbro are significantly larger (up to 4–5 mm) than those of the amphibolite. Imbrication and apparent rotation of early-formed crystals is commonly seen at both macro- and microscopic scales. Dismembered lenses of greenschist up to ~1-m thick and several meters long are drawn out parallel to the mineral foliation.

Muscovite-garnet tonalite veins, dikes, and stocks intrude the mafic units below the Game Lake thrust. Intrusions cut across, or are strung out parallel to, mineral foliations in the adjacent amphibolite and mafic phyllonite. Locally, a strong internal foliation is developed in some intrusions, defined by aligned metamorphic hornblende and flattened or recrystallized plagioclase. More commonly, texturally igneous porphyroclasts of plagioclase and muscovite show partial recrystallization along weak foliations (Fig. 8). Most of the intrusions exhibit significant greenschist-facies metamorphism in the form of chlorite, epidote, and actinolite replacement of hornblende and saussuritization of plagioclase. The mutually crosscutting relationship between the tonalites and foliated country rocks suggests diachronous intrusive events during deformation.

Exposures of Colebrook Schist, part of the Franciscan Complex (Blake et al., 1985), occur as blocks within the serpentinite mélange and in the area intervening between the Game Lake area and Snowcamp Mountain, which lies ~8 km to the south-southwest, a small part of which appears in the southwest corner of Figure 3. The Colebrook Schist in the Game Lake area usually occurs as a dark, fine-grained, well-foliated phyllite or slate. It structurally underlies the mafic phyllonite, amphibolite, and peridotite. The contact between these units is not exposed, but outcrops between mafic phyllonite or amphibolite and Colebrook Schist are observed separated by only a few meters in some places. The distribution of the units suggests that this contact crosscuts the contacts between the peridotite, mafic phyllonite, and amphibolite (Fig. 3).

No sedimentary rocks overlying the ophiolitic rocks in the Game Lake area were observed, with the exception of a pebble conglomerate at Horse Sign Butte, ~1 km to the east-northeast (not shown in Fig. 3) of Game Lake. These have been correlated with the Myrtle Group (Page et al., 1981). The conglomerate is unmetamorphosed, but bedding dips steeply to the north. Its contact with underlying serpentinite is covered, although they are separated by only a couple of meters of cover in a small saddle of the ridgeline, suggesting a fault contact between the two.

Structural Geology

Structures exposed in the Game Lake area include amphibolite and phyllonite mylonitic foliations and associated lineations, at least two generations of folds, cataclastic fabrics, and veins. The amphibolite and mafic phyllonite are L-S tectonites that display a generally continuous foliation (S_1) defined by preferentially oriented hornblende and plagioclase (Fig. 6), and actinolite and chlorite (Fig. 5), respectively. In the mafic phyllonite directly below the Game Lake thrust, a strong mylonitic fabric is present that locally includes planar zones of ultramylonitic foliation, indicative of inhomogeneous strain local-

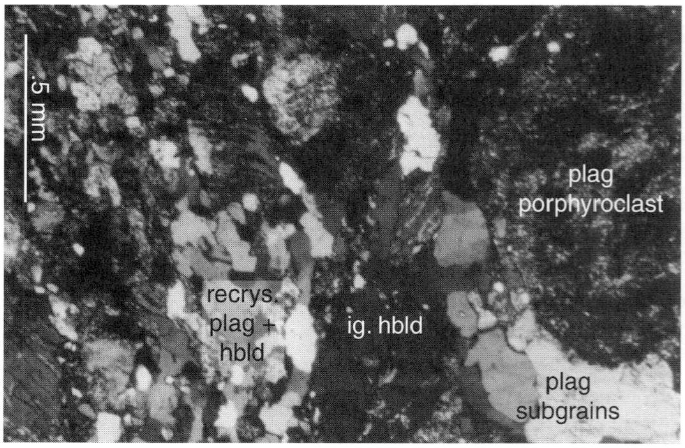

Figure 7. Crossed-nicols (2.5×) photomicrograph of metagabbro (BC-005) showing protomylonitic texture. Large igneous (ig) plagioclase (plag) porphyroclasts show significant saussuritization and subgrain development (lower right). Small, oriented, recrystallized grains of hornblende (hbld) and plagioclase define foliation.

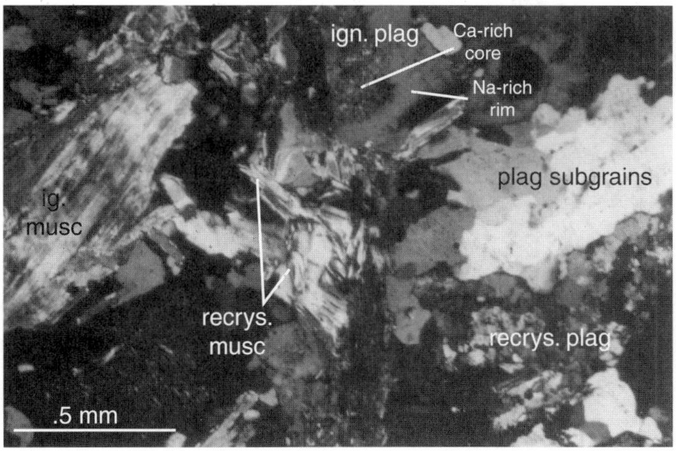

Figure 8. Crossed-nicols (2.5×) photomicrograph of tonalite (BC-133). Large igneous muscovite (ig. musc) is strained rimmed by fine-grained recrystallized muscovite (recrys. musc). Present are large igneous plagioclase (ign plag), small recrystallized plagioclase (recrys. plag), and subgrains.

Figure 9. Field photograph of amphibolite (BC-037) with folded tonalite dikelet that parallels S_1 amphibolitic foliation. Pencil is 14 cm long.

ized along discrete surfaces. Oriented hornblende crystals in amphibolites and actinolite and greenschist-facies pseudomorphs after hornblende in mafic phyllonites define variably developed mineral stretching lineations (L_1).

In most areas surrounding Game Lake, serpentinized peridotite and serpentinite mélange display noncoherent, anastomosing cleavage formed by brittle shearing, but the ~5- to 10-m-thick serpentinized base of the peridotite above the Game Lake thrust is coherently foliated parallel to the thrust surface that separates it from the underlying mafic phyllonite and the mylonitic foliation in the phyllonite. A generally weak, but locally continuous, foliation is present in tonalite intrusions. Mineral foliations in the metagabbros are generally protomylonitic and easily identifiable, but coherent mineral lineations are rarely evident.

Tight to isoclinal similar folds in the amphibolite have axial surfaces parallel to the foliation and display fold hinges (F_1) that approximate the mineral lineation direction (Fig. 9). Late, open folds (F_2: Fig. 10) and a rare spaced cleavage (S_2) that poorly approximates the axial surface to the open folds deform the structures (mylonitic foliations, tight to isoclinal folds) described above.

The dominant S_1 surfaces generally dip to the northeast and northwest. A stereonet plot of the poles to cleavage in the phyllonite and amphibolite around Game Lake (Fig. 11A) are scattered but define a π-circle, the pole of which plunges shallowly to the northeast. The scatter in the foliation pole data is inferred to be result of the late F_2 folding event (after Loney and Himmelberg, 1977). Similar orientations in the foliation and lineation data are seen in the structural data of Harper et al. (1990) from the northeast-dipping, southern exposure of the Madstone Cabin thrust near Chetco Peak (Figs. 1 and 11B) and from the south-dipping, northern exposure of the thrust (Loney and Himmelberg, 1977), except that orientations of structures in the north appear to have been rotated, relative to the structures of the Chetco Peak area, by ~40° around an axis that approximates the south-plunging, late, fold axis (Fig. 11C and D). The magnitude of this rotation approximates the amount of rotation inferred by Loney and Himmelberg (1977) to explain the degree of scatter in poles to the amphibolite foliation near the Madstone Cabin thrust.

Based on morphology and crosscutting relations, Loney and Himmelberg (1977) and Harper et al. (1990) described two generations of early folds that deform the amphibolite foliation: isoclinal (F_1 and F_{1a}, respectively) folds that are in turn deformed

Figure 10. Field photograph of open fold in amphibolite (BC-136). Dog's head is 14 cm wide.

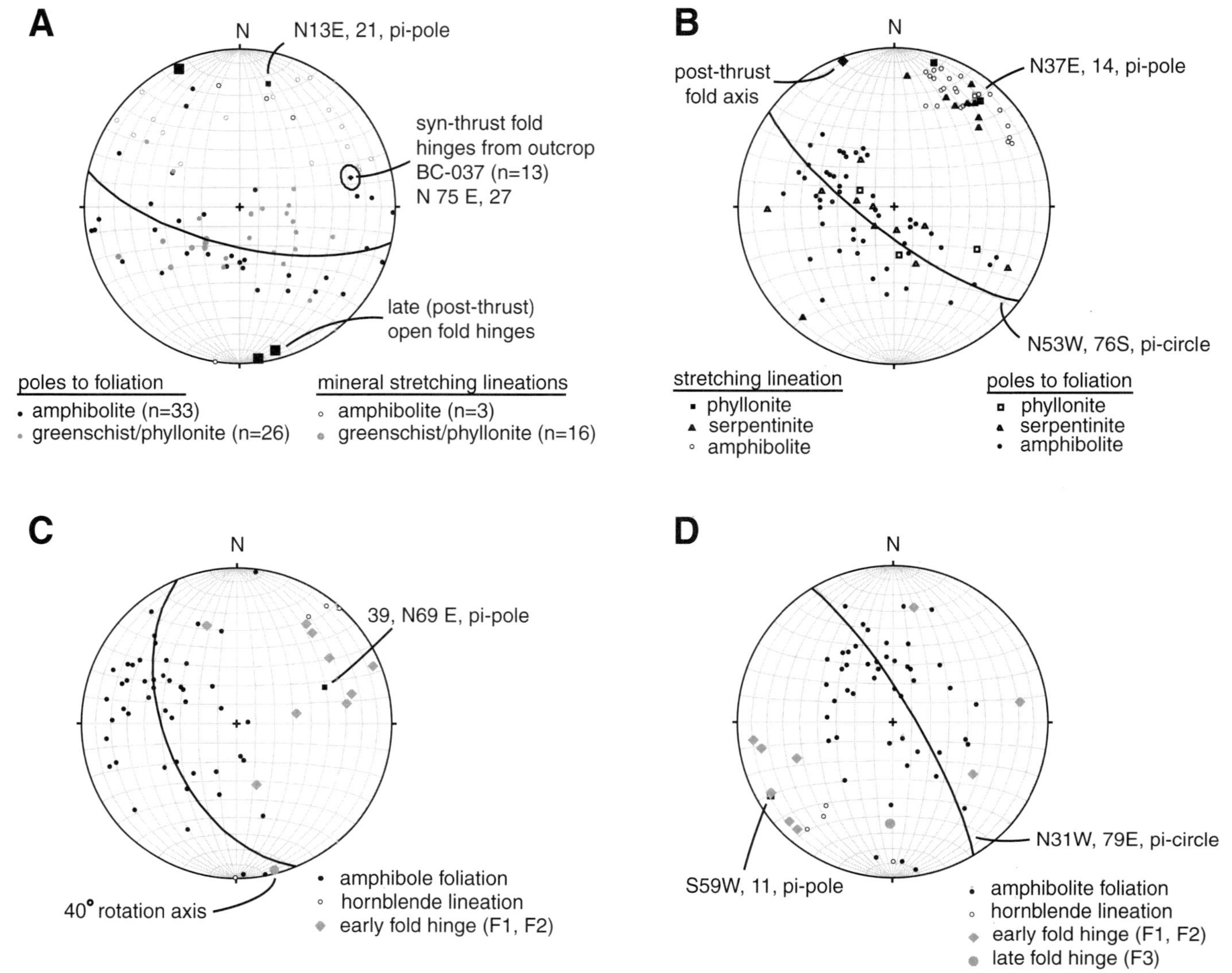

Figure 11. Structural data plotted on equal-area, lower-hemisphere projections from Game Lake and Madstone Cabin thrusts. (A) Structural data from Game lake area (this study). (B) Structural data from southern exposure of Madstone Cabin thrust (modified from Harper et al., 1990). (C) Rotated structural data from northern exposure of Madstone Cabin thrust (adapted from Loney and Himmelberg, 1977). (D) Unrotated structural data from northern exposure of Madstone Cabin thrust (modified from Loney and Himmelberg, 1977).

by tight (F_2 and F_{1a}, respectively) folds. These authors concluded on the basis of similar fold-hinge orientations, axial surface foliations, and mineral lineations, that they are related to a single inhomogeneous folding event. In the Game Lake area, both tight and isoclinal folds (collectively F_1 in this chapter) in the amphibolite foliation are observed, but no determination of timing between these folds is made in this chapter, as they were not observed in crosscutting relationship. However, in the hinges of isoclinally folded amphibolites, two generations of hornblende growth are observed. An early folded amphibolitic fabric in an isoclinal fold hinge is overprinted by oriented hornblende that parallels the axial surface. This configuration suggests either two episodes of hornblende growth or two deformation events during ongoing hornblende growth, the second of which is consistent with the conclusion of the previous authors that the two folding events are related to one inhomogeneous deformation. In the Game Lake area, micro- and macroscopic F_1 folds of the foliation and tonalite veins and dikelets occur (Fig. 9) that have axial surfaces subparallel to the local foliation surface.

Along the Madstone Cabin thrust, the orientations of mineral lineations in phyllonites are subparallel to early fold hinges (F_1 and F_2 of Loney and Himmelberg, 1977, and F_{1a} and F_{1b} of Harper et al., 1990) in phyllonites and amphibolites. They generally trend shallowly to the northeast (after 40° rotation of the data from Loney and Himmelberg, 1977) and cluster near the π-poles of the associated foliations (Fig. 12). This relationship suggests that fold axes and mineral lineation development were rotated toward parallelism with the bulk transport direction developed during thrusting (Harper et al., 1990). In the Game Lake area, a similar relationship is observed, with a significant difference. At Game Lake, two populations of mineral stretching lineations are present (Figs. 11A, 12A and B). One group clusters around a direction that is shallow to the northeast (Fig. 12B) and approximates the observed early fold hinges (F_1) and the π-pole to the mylonitic foliation. But, a second group clusters around a direction that plunges shallowly to the northwest (Fig. 12A). This second population likely does not represent a second lineation-producing event, but is part of the scatter of a single population resulting from observed late (F_2) conical folding, such that initial northeast-trending lineations are rotated around a small circle whose axis approximates the observed late fold axes seen in the area and whose apices are inferred to terminate to the south (Fig. 12C). This interpretation is supported by shear sense data (described below) that show that the inferred sense of motion associated with northwest-trending lineations (northwest-dipping foliations) are top to the southeast, whereas the shear sense in rocks associated with northeast-trending lineations (northeast-dipping foliations) are top to the northeast. However, this style of folding (conical) differs from the cylindrical folding described by Loney and Himmelberg (1977) and Harper et al. (1990) to explain the differing average orientations of the foliations and lineations between the northern and southern exposures of the Madstone Cabin thrust. One possible explanation for this discrepancy is that conical folds in the Game Lake area may represent the axial terminations of cylindrical folds represented along the Madstone Cabin thrust (Ramsay and Huber, 1987). It is not clear whether the late folding and rotation of thrust-related structures is directly related to oroclinal bending or an unrelated event.

Macro- and microscopic shear sense indicators are largely present in the mafic phyllonite and, to a lesser extent, in the amphibolite. Shear sense was determined on sections cut parallel to the mineral stretching lineation (L_1) and perpendicular to the dominant foliation (S_1). Indicators include asymmetric porphyroclast tails, imbricated porphyroclasts, asymmetrically folded veins, and C-S fabrics. Asymmetric porphyroclasts (plagioclase) are dominantly σ-type, although some δ-type porphyroclasts are present. Imbricated porphyroclasts (plagioclase and hornblende) are dominantly fractured in a normal (antithetic) sense to the overall shear sense. The assumed transport direction during thrusting is parallel to the average mineral stretching lineations, isoclinal fold axes, and π-pole to the dominant foliations, which trend to the northeast. Overall shear sense for the upper (ophiolite) block of the Game Lake thrust is top to the northeast (present-day direction), similar to that determined by Harper et al. (1990) for the Josephine ophiolite and the Madstone Cabin thrust. Asymmetrically folded veins and tonalite intrusions are generally consistent with other shear sense indicators, although in one exposure of the mafic phyllonite (BC-173), an opposite sense in an folded tonalite vein was determined than from the C-S fabric developed in the same section. This overall consistency suggests that the anomalous asymmetric fold exposed may be the short overturned limb, giving an erroneous sense of shear (Simpson, 1986). The C-S fabrics are visible in thin-section but are often difficult to interpret, due to the late retrograde metamorphism that has obscured many of the syndeformational structures at the microscopic level.

Geochronology

Two absolute ages were determined using $^{40}Ar/^{39}Ar$ methodology (Fig. 13; see the section on analytical methods). The release spectrum from metamorphic hornblende from an amphi-

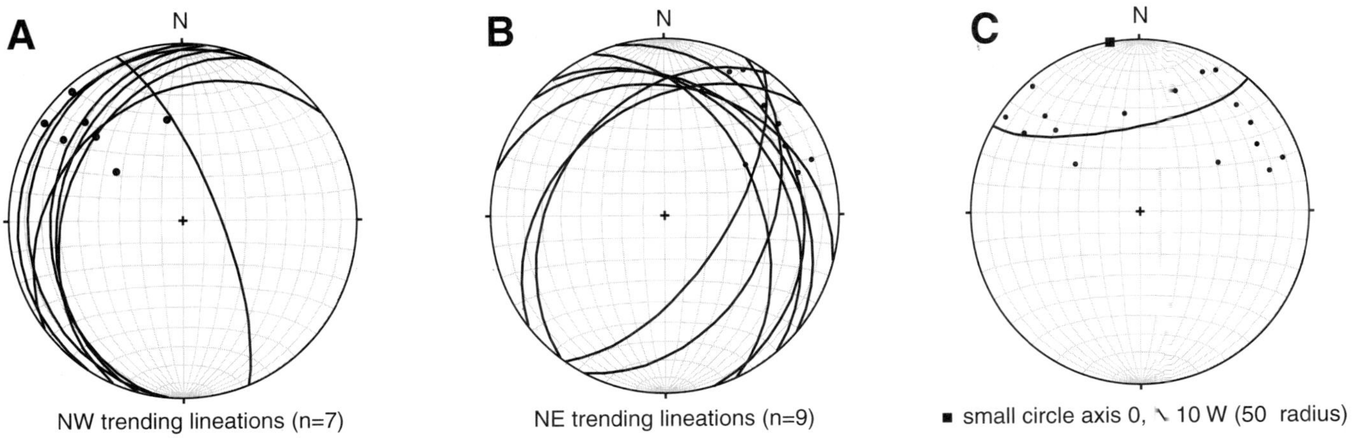

Figure 12. Lineation data plotted on equal-area, lower-hemisphere projections from Game Lake thrust. (A) Plot of northwest-trending lineations and associated foliations on which they occur. (B) Plot of northeast-trending lineations and associated foliations. (C) Plot of northwest- and northeast-trending lineations. These data scatter along a small circle, which suggest conical folding, with a pole (0°, N10°W, 50° radius) that approximates the average late fold axis.

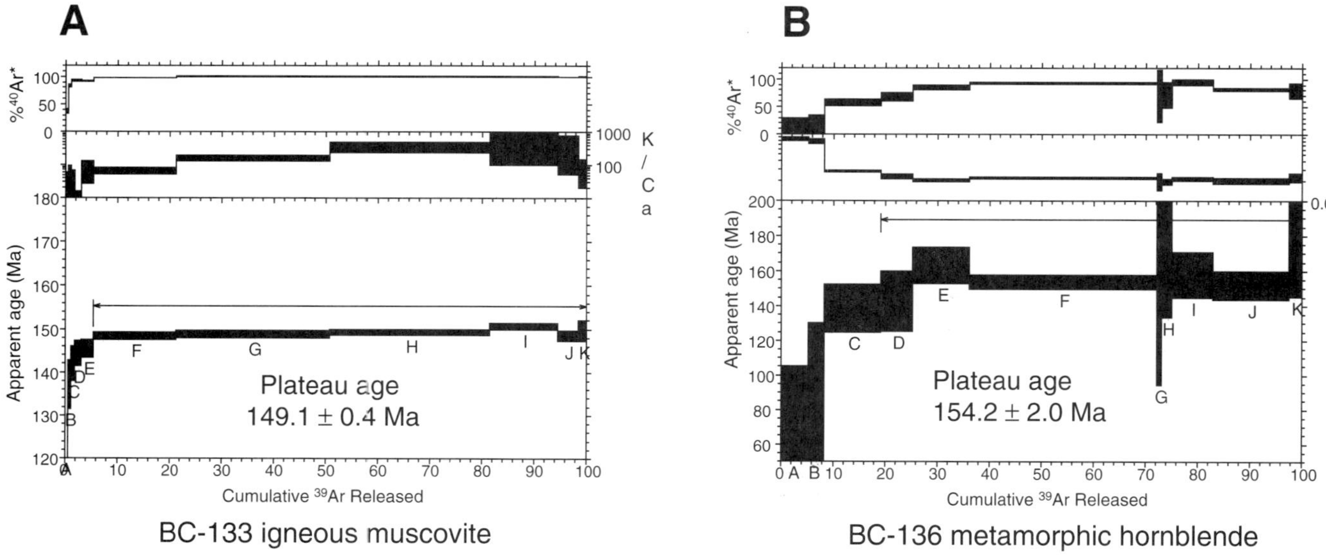

Figure 13. Release spectra for $^{40}Ar/^{39}Ar$. (A) Tonalite sample BC-133 (igneous muscovite). (B) Amphibolite sample BC-136 (metamorphic hornblende).

bolite (BC-136) gives a plateau age of 154.2 ± 2.0 Ma (1-σ). Igneous muscovite from a weakly deformed tonalite intrusion (BC-133) gives a plateau age of 149.1 ± 0.4 Ma (1-σ). The slightly younger age for igneous-textured muscovite within the tonalite likely reflects the lower closure temperature of muscovite (~350 °C) following its crystallization relative to that of metamorphic hornblende (~400–500 °C; McDougall and Harrison, 1999). Metamorphic resetting of older (pre-deformation) muscovite is considered unlikely, as the tonalite sample analyzed crosscuts the amphibolite foliation and is only weakly foliated.

Based on the closure temperature of hornblende, the hornblende cooling age probably represents the end of amphibolite-grade metamorphism during ongoing deformation. Deformation associated with thrusting is inferred to have continued during cooling to greenschist-grade conditions, based on the textural relationships (foliated versus nonfoliated greenschist-facies minerals in the mafic phyllonite and amphibolite and the poorly deformed nature of the tonalite intrusion itself) described in preceding sections. Thus, the muscovite cooling age represents a minimum age for the end of the thrusting.

Based on the ages given above, the end of amphibolite-facies metamorphism of amphibolite and intrusion of tonalite likely occurred during the Nevadan orogeny. The ages also correspond closely with the ca. 153- to 150-Ma $^{40}Ar/^{39}Ar$ age range obtained from amphibolites and 150- to 146-Ma $^{40}Ar/^{39}Ar$ age and 151 ± 1-Ma (1-σ) U-Pb age from syntectonic intrusions along the Madstone Cabin thrust (Harper et al., 1994), as well as with a K-Ar age of 155 ± 5 Ma from metagabbro in the Big Craggies area.

Pressure and Temperature Conditions during Deformation

Pressure and temperature conditions at the time of thrusting can be estimated using observed metamorphic assemblages and microprobe analyses in conjunction with published geothermobarometers that utilize mineral compositions that equilibrate during igneous crystallization or metamorphic recrystallization. In the Game Lake area, qualitative estimates of pressure and temperature can be made in the mafic phyllonites beneath the thrust, based on the relict amphibolite-facies mineral assemblages and the dominant retrograde greenschist-facies assemblage. In the amphibolites, which retain an unaltered early amphibolite-facies mineral assemblage, quantitative estimates of temperature during thrust-related metamorphism can be made using the amphibole-plagioclase thermometry of Holland and Blundy (1994). For syntectonic tonalites, quantitative estimates of pressure during intrusion and deformation are based on the phengite content of igneous and metamorphic muscovite, respectively, using the geobarometer of Massonne and Schreyer (1987).

Amphibolite-Facies Metamorphic Conditions. In the amphibolites, the well-preserved mineral assemblage associated with the deformation fabric consists of hornblende-plagioclase (An_{20-45})-epidote-quartz-sphene-ilmenite (±biotite and apatite; Fig. 6). Chlorite, actinolite, saussuritized plagioclase, and vein epidote and prehnite are also present, which show textural replacement of the primary fabric-forming minerals and are considered to be part of the retrograde mineral assemblage. The presence of hornblende and oligoclase/andesine and absence of primary chlorite suggest amphibolite-facies conditions in excess of ~550 °C (assuming medium pressures; Apted and Liou, 1983; Bucher and Frey, 2002). The presence of epidote in textural equilibrium with hornblende and calcium-bearing plagioclase indicates epidote-amphibolite subfacies and does not typically indicate temperatures in excess of ~600 °C (Bucher and Frey, 2002), although epidote may be stable at temperatures >600 °C under pressure conditions above ~4 kb (Apted and

Liou, 1983). Similarly, aluminous hornblende crystals in the amphibolite that have Al_2O_3 wt.% that range from ~10 to 13% suggest temperatures of ~550 °C or more (Apted and Liou, 1983).

The hornblende-plagioclase geothermometers of Holland and Blundy (1994) were applied to microprobe analyses of quartz-bearing amphibolite (samples BC-037c and BC-136 from the Game Lake area and SA-44 from the Madstone Cabin thrust; note that BC-136 is the same sample analyzed for $^{40}Ar/^{39}Ar$ geochronology; see the section on analytical methods). Several constraints on the ratio of elements present in various sites in hornblende and plagioclase and the temperature range over which the thermometers were valid are given by Holland and Blundy (1994). These constraints were met in all analyses of the Game Lake and Madstone Cabin thrust amphibolites, and ferric iron content was calculated using the method described by Holland and Blundy (1994). The effect of fluorine in the O(3)-site was not modeled by Holland and Blundy (1994) but is below the detection limit in all amphibole crystals analyzed. In most cases, visibly unaltered crystal interiors were analyzed, but two compositional cross-sections were constructed using a series of probe points from rim-to-core-to-rim to evaluate the possibility of postmetamorphism re-equilibration of the amphibole-plagioclase pair compositions. Anomalously low temperatures result from calculations using hornblende rims and suggests alteration occurred, most likely during greenschist-facies metamorphism.

The quantitative temperature estimation is pressure dependant and based on the partitioning of tetrahedral aluminum and sodium between hornblende and plagioclase. The thermometer was applied to metamorphic hornblende in textural equilibrium with coexisting plagioclase and represents a metamorphic equilibration temperature. Holland and Blundy (1994) presented two thermometers: one that can be applied to quartz-bearing hornblende and plagioclase assemblages (thermometer A) and a second that can be applied to assemblages with or without quartz (thermometer B). In the case of the samples analyzed, free quartz is present, so both thermometers may be applied. The results from thermometer A are presented in Figure 14. The slope of the data from thermometer A is less steep than thermometer B (not illustrated) and is therefore less pressure dependant, although the steeper slope of thermometer B results in a greater overall range of temperature (~600–750 °C for thermometer A and 535–825 °C for thermometer B). The temperature estimate is pressure dependant, and a pressure range of 3.5–6 kb is assumed, based on the results of phengite geobarometry (see below). This assumption may be invalid if rapid exhumation was occurring during the several million years between hornblende growth in the amphibolite and the later stages of tonalite intrusion (the phengites analyzed in geobarometric section utilize the later-stage, weakly deformed intrusions). If this is the case, then the temperature estimates given below may be underestimations. However, based on the metamorphic assemblage present in the amphibolites (discussed below), it seems likely that rapid exhumation did not occur.

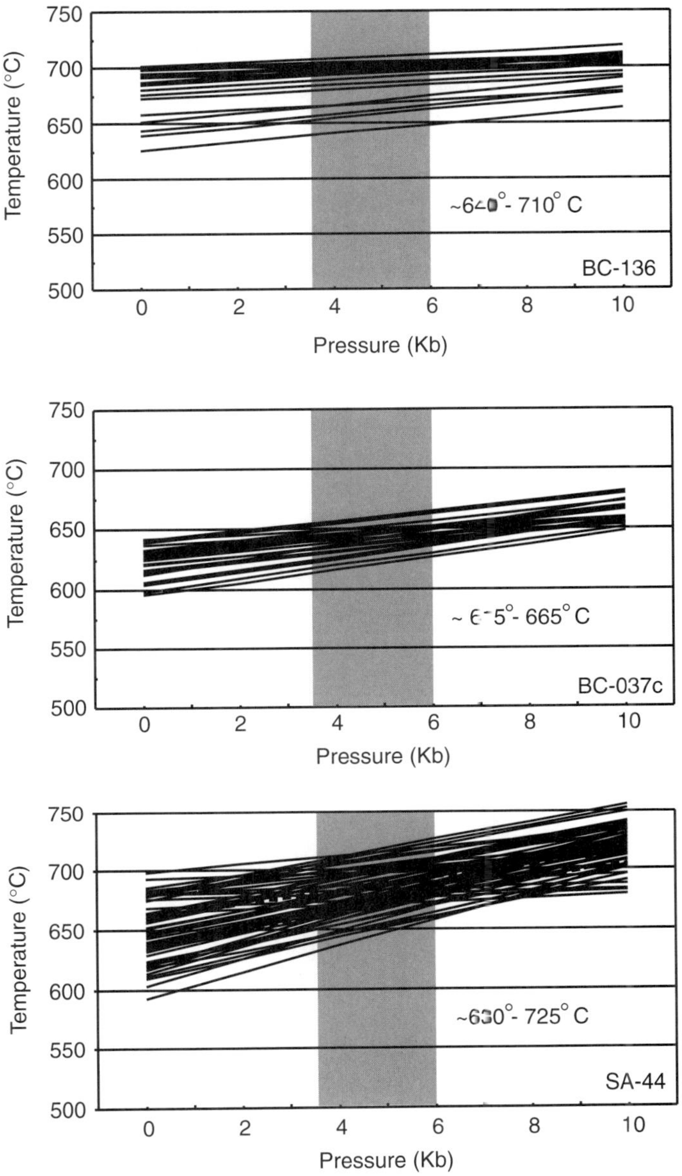

Figure 14. Results of application of geothermometer A (tremolite-richterite) of Holland and Blundy (1994) to coexisting hornblende-plagioclase pairs in amphibolites from Madstone Cabin thrust (SA-44) and Game Lake thrust (BC-136, BC-037c). Shaded areas represent pressure range derived from phengite barometer (see Fig. 15).

Thermometer A gives a constrained (by igneous phengite barometry) temperature range of ~615–720 °C. Although sample BC-037c shows a slightly lower range in temperature than that determined for BC-136 and SA-44 (Madstone Cabin thrust), and a slightly steeper slope for the Madstone thrust sample (SA-44) is apparent, the overlap in temperature ranges is nearly coincident, indicating that similar temperature conditions likely prevailed along the two thrusts.

The lower limit of the constrained range of geothermometer A results (~615 °C) is higher than the upper limit (~600 °C) of the qualitative estimation of the amphibolite-facies metamorphism based on the mineral assemblage. This lack of overlap suggests that pressure conditions above 4 kb may have prevailed during metamorphism and stabilized amphibolite-facies epidote above 600 °C.

The early amphibolite-facies mineral assemblage is rarely preserved in the mafic phyllonites and consists of rare porphyroclasts of hornblende, clinozoisite, and saussuritized plagioclase that are largely replaced by a retrograde greenschist-facies assemblage (Fig. 5). The paucity of equilibration textures and mineral associations from this relict assemblage precludes definitive statements regarding temperature and pressure conditions during the formation of this assemblage. However, the few relict porphyroclasts present suggest conditions consistent with those interpreted for the amphibolite.

Retrograde Greenschist-Facies Metamorphic Conditions. The mafic phyllonite is almost completely dominated by greenschist-grade mineral assemblages, which is in contrast with the amphibolite and tonalite that show a minor, though variable, greenschist-grade mineral assemblage component. These assemblages include actinolite, epidote, clinozoisite, chlorite, and plagioclase. Epidote, chlorite, prehnite, calcite, and quartz are also present as fracture and vein-filling minerals. Actinolite and chlorite are commonly observed in partial replacement of hornblende in amphibolites and tonalite.

In the mafic phyllonites, oriented actinolite alternating with plagioclase-rich domains are the dominant fabric-forming minerals (Fig. 5). The reactions that replace hornblende, calcic plagioclase, and quartz with actinolite, albite, chlorite, and epidote occur below ~500 °C (Apted and Liou, 1983; Best, 2003) and indicate cooling through greenschist conditions under hydrous conditions. Most mafic phyllonites also contain a fine-grained *nonfoliated* greenschist mineral assemblage that consists dominantly of epidote with some chlorite and actinolite that variably overprints the mylonitic fabrics (Fig. 5). In rocks in which this assemblage is well developed, the terms "greenschist," or even "greenstone," rather than "phyllonite," is more appropriate. These textures suggest that, along with the presence of undeformed veins containing epidote and chlorite, fabric development associated with deformation ceased while greenschist-facies metamorphic conditions still prevailed.

In the amphibolites, the retrograde assemblage is less well developed. Actinolite rims and chlorite embayments into hornblende are common, as is saussuritization of the cores of large, relict, zoned igneous plagioclase porphyroclasts. The alteration of igneous porphyroclasts contrasts with metamorphic oligoclase/andesine formed as part of the amphibolite-facies assemblage, which shows little or no visible alteration, although some albitization of their rims has occurred.

Tonalite Intrusion Conditions. An igneous mineral assemblage that includes coarse-grained to pegmatitic plagioclase, hornblende, muscovite, and quartz dominates tonalitic dikes and stocks. Minor orthoclase, biotite, and trace amounts of nearly consumed garnet are also present in some samples. Discordant intrusions that crosscut the amphibolite fabric, display weak, spaced, anastomosing foliations composed of fine-grained recrystallized plagioclase and muscovite (Fig. 8). Intrusions, strongly deformed parallel to the amphibolite fabric, are dominated by fabric-forming metamorphic hornblende and plagioclase, suggesting that intrusion occurred during amphibolite-facies metamorphism. Chlorite, actinolite, and epidote commonly replace hornblende, and igneous plagioclase cores are saussuritized but show no evidence of fabric development during retrograde metamorphism. Also present are nonfoliated veins that cut larger foliated plutons. Veins and fractures in the tonalitic rocks are commonly filled with epidote, chlorite, or plagioclase. The above relationships suggest diachronous intrusive events during cooling from amphibolite-facies through greenschist-facies metamorphic conditions.

The geobarometer of Massonne and Schreyer (1987) was used to estimate pressure conditions during igneous crystallization and subsequent metamorphic recrystallization. The phengite content of igneous and metamorphic muscovite (in the presence of the buffering assemblage of orthoclase, quartz, and phlogopite [biotite]) in two tonalite samples from the Game Lake area (BC-123 and BC-133) and from a similar intrusion along the Madstone Cabin thrust (J82-3) were determined. The phengite geobarometer is temperature dependant, but some constraints on the temperature ranges during igneous crystallization and subsequent metamorphic recrystallization can be imposed. The temperature of intrusion and igneous crystallization is taken to be ~700 °C, which is the approximate solidus temperature over a wide range of pressures of the muscovite-bearing assemblage (Massonne and Schreyer, 1987). However, little information on the reactions that occur in granitoid rocks is available (Bucher and Frey, 2002) to constrain the upper stability limit of muscovite during metamorphism, although muscovite in the presence of the buffering assemblage was stable up to ~700 °C in the experiments of Massonne and Schreyer (1987). A minimum temperature of metamorphism is assumed to be ~300 °C, the lower limit of greenschist metamorphism.

Average silica contents (on an 11 oxygen basis) for muscovite-bearing tonalites in the Game Lake area range from 3.05 to 3.10 atoms per formula unit (apfu), with a total range of 3.02–3.13 apfu (Fig. 15) for igneous muscovite, and is 3.06 apfu (total range of 3.04–3.10 apfu) for metamorphic muscovite. The average silica contents correspond to pressures that range from ~3.5 to ~6 kb (~11.5–20 km depth, assuming a density of 3.3 g/cm^3 for oceanic crust) for equilibration during igneous crystallization (at 700 °C) and 0 to ~4 kb (~13 km depth) for equilibration during metamorphic recrystallization (300–700 °C).

Sample J82-3, from the Madstone Cabin thrust, gives average silica contents of 3.07 apfu (total range, 3.05–3.1 apfu) for igneous muscovite and 3.09 apfu (3.02–3.13 apfu) for metamorphic muscovite. The average silica content corresponds to pressures of 3.5 to ~5 kb (11.5–16.5 km) for igneous muscovite and 0 to ~6 kb (~20 km) for metamorphic muscovite. Harper

A

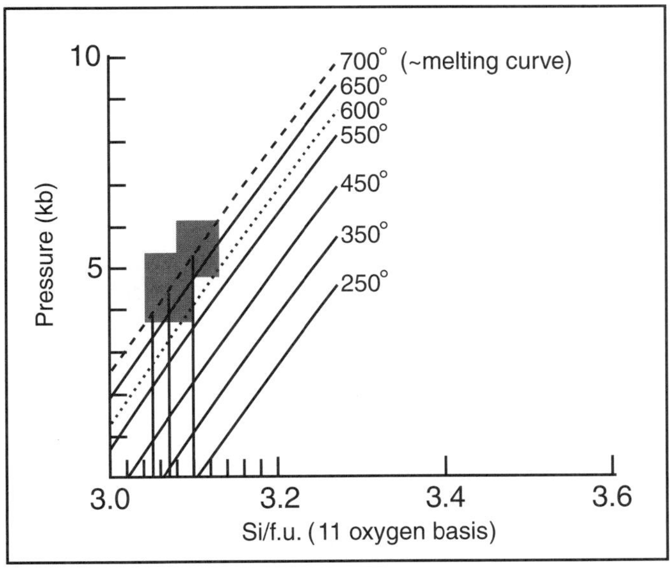

Average silica content of igneous muscovite/sample		
Madstone Cabin Thrust:		
Si = 3.07	J82-3 (ign.)	n = 13
Game Lake Thrust:		
Si = 3.10	BC-133 (ign.)	n = 5
Si = 3.05	BC-123 (ign.)	n = 7

B

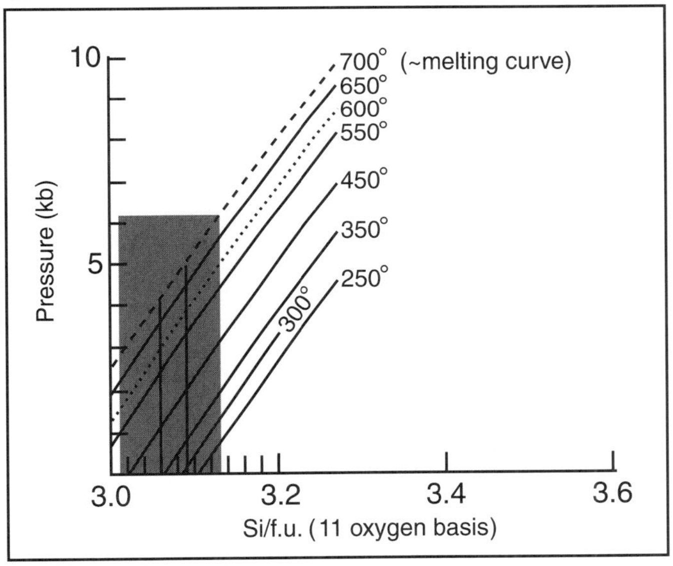

Average silica content of metamorphic muscovite/sample		
Madstone Cabin Thrust:		
Si = 3.09	J82-3 (meta.)	n = 11
Game Lake Thrust:		
Si = 3.06	BC-133 (meta.)	n = 6
Si = 3.06	BC-123 (meta.)	n = 6

Figure 15. Results of application of phengite geobarometer of Massonne and Schreyer (1987) to igneous (A) and metamorphic (B) muscovite in tonalites from Madstone Cabin thrust (J82-3) and Game Lake thrust (BC-123, BC-133). Vertical bars represent average silica content of igneous and metamorphic muscovite from each sample. Shaded area represents the range of all analyses for each sample using constraints given in text.

et al. (1990) also used the geothermobarometer of Massonne and Schreyer to assess the pressure conditions along the Madstone Cabin thrust. The reported silica content of sample J113-7 is 3.12 apfu, from which they determined a pressure of equilibration of 5.5 kb (~18 km) at 700 °C.

The similarity in the pressure estimates above suggests that pressure conditions during intrusion and deformation along both the Madstone Cabin and in Game Lake thrusts were comparable and that the upper oceanic block was ~1.5–20 km thick. Also, the overlap in silica contents between igneous and metamorphic muscovite from both areas suggests that metamorphism of muscovite may have occurred shortly after intrusion and igneous crystallization along both thrusts (assuming metamorphic muscovite stability up to ~700 °C).

SUMMARY AND DISCUSSION

The results of field mapping and structural analysis of the units exposed near Game Lake show that ultramafic rocks from the Snowcamp remnant of the Coast Range ophiolite were thrust northwestward over mafic phyllonites and amphibolites correlated with the Chetco complex along a low-angle fault (Game Lake thrust). Following Saleeby and Harper (1993), the present-day direction of thrusting (north-northeast) is inferred to be the result of ~65° of regional clockwise rotation resulting from post-Nevadan oroclinal bending. The mylonitic to ultramylonitic foliation in the mafic phyllonite directly below the thrust parallels the fault contact and is interpreted to be a thrust-related fabric. Similar foliation orientations present in amphibolites structurally beneath the phyllonites suggest that the amphibolites were also deformed and metamorphosed during thrusting. Sense-of-shear indicators indicate northeast-directed displacement of the upper ophiolitic plate. Late, possibly conical, folds likely deformed those shear-sense indicators that indicate southeast-directed displacement. Diachronous muscovite-bearing tonalite dikes and plutons intruded the lower plate before cooling of muscovite at ca. 149 Ma and probably during amphibolite-facies metamorphism that ended by ca. 154 Ma. Strongly foliated and deformed intrusions are inferred to have formed during relatively early stages of thrusting, whereas weakly foliated tonalites that cross-cut the amphibolite fabric are interpreted to have intruded during relatively late stages of thrusting.

The dominant fabric-forming mineral assemblage in the amphibolite indicates amphibolite-facies conditions (~550 °C to more than ~600 °C) were followed by retrograde greenschist-facies metamorphism with little to no associated fabric development. In contrast, a similar, but relict, amphibolite-facies assemblage is present in some phyllonite, overprinted by a stronger fabric-forming retrograde greenschist-facies (less than ~500 °C) assemblage. These relationships suggest that early thrusting occurred at amphibolite-facies metamorphic conditions. This process was followed by continued fabric development in phyllonites during the later stages of thrusting at lower greenschist-facies conditions and was localized directly below the Game Lake thrust. The geothermometer of Holland and Blundy (1994) gives a temperature range of ~615–720 °C for metamorphic hornblende and plagioclase pairs in amphibolite. The geobarometer of Massonne and Schreyer (1987) gives a pressure range of 3.5–6.0 kb (11.5–20 km depth) for igneous muscovite and 0–6 kb for metamorphic muscovite in tonalite. These ranges are consistent (although the upper end of the range may be too hot) with those required to form the observed metamorphic assemblages, assuming epidote stabilization above 600 °C (>4 kb). The similarity of the pressure-dependent silica contents of igneous and metamorphic muscovite suggests deformation occurred shortly after intrusion, which agrees with the timing of intrusion and deformation of the tonalites based on structural relationships. The absolute age of thrusting is taken to be within the range of ca. 154 ± 2.0 Ma for the amphibolite (i.e., prior to the end of amphibolite-facies metamorphism) to ca. 148 Ma based on the cooling age of muscovite, which has a closure temperature similar to the lower limit of greenschist-facies metamorphism in the phyllonite. Harper et al. (1996) presented

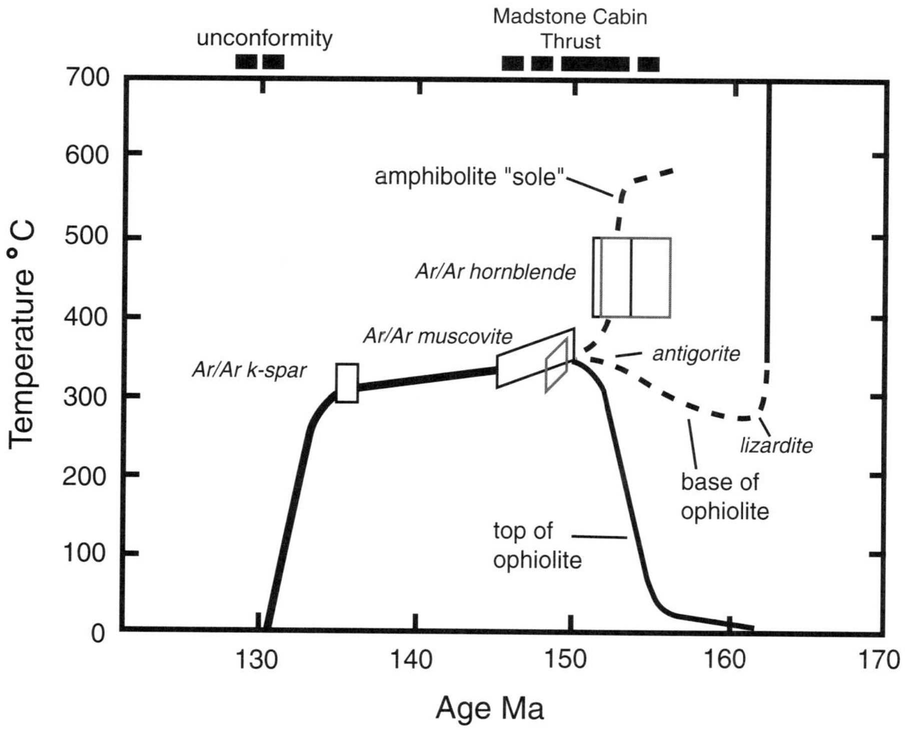

Figure 16. Temperature-time path for Madstone Cabin and Game Lake thrusts. Black lines and boxes (Madstone Cabin thrust) modified from Harper et al. (1996). Gray boxes represent temperature-time data ($^{40}Ar/^{39}Ar$ ages and closure temperature ranges) from Game Lake thrust.

a temperature-time path for the upper and lower plates of the Madstone Cabin thrust, upon which the equivalent data from the Game Lake thrust is superimposed (Fig. 16), and shows excellent correlation between the two thrusts.

The characteristics of the Game Lake thrust are very similar to those of the Madstone Cabin thrust (Table 1A), indicating that the two thrusts are correlative. Similarities in lithologic units, structural orientations and deformational styles, conditions of metamorphism, metamorphic histories, cooling ages, and syntectonic intrusions are shared by both thrusts. In addition, Harper et al. (1990) noted that "fresh" peridotite overlies serpentinite mylonite, which in turn overlies mafic phyllonite and amphibolite along the Madstone Cabin thrust. They hypothesized that fluids in the lower plate caused retrograde metamorphism of the mafic rocks, but were sealed from the upper plate by volume expansion during serpentinization along the thrust surface (see also Shervais et al., 2005). Alternately, the base of the peridotite may have been a sink for fluids rising up out of the lower block as serpentinization used up the available fluids, preventing flow much beyond the fault interface (J.W. Shervais, personal commun., 2004). Along the Game Lake thrust, a similar situation can be inferred where relatively unaltered peridotites are separated from highly retrograded mafic phyllonites by an ~5- to 10-m-thick, coherently foliated, serpentinite layer.

Two notable differences between the two thrusts exist (Table 1A). First, the serpentinite mylonite that forms the base of the Josephine Peridotite above the Madstone Cabin thrust contains antigorite (+brucite), whereas coherent serpentinite above the Game Lake thrust consists of the lower-temperature serpentine polymorphs lizardite and chrysotile. The antigorite along the Madstone Cabin thrust overprints lizardite and chrysotile, indicating that the upper plate underwent prograde metamorphism from less than ~350 °C up to 300–400 °C (Harper et al., 1996) or 450 °C (Evans, 2004). Second, the mafic phyllonite section at Game Lake is significantly thicker than observed along much of the Madstone Cabin thrust. Harper et al. (1990, 1996) suggested that the phyllonite was a product of strong retrograde greenschist-facies metamorphism of an amphibolite derived from a gabbro protolith, the result of the channeling of fluids along the thrust. In the Game Lake area, the mafic phyllonite also appears to be retrograded amphibolite, but the fine-grained recrystallized nature of the phyllonite differs significantly from the coarse-grained retrograded amphibolite, with no evidence for transitional units. This observation suggests that the phyllonite may have formed as a result of retrograde metamorphism accompanied by grain-size reduction of an amphibolite derived from a basaltic protolith. This process may also account for the difference in thickness of the phyllonites observed along the two thrusts.

The results of this study show that the Madstone Cabin and Game Lake thrusts are correlative, and therefore, the ultramafic rocks at Game Lake are correlative with the Josephine ophiolite. Several possibilities follow from these correlations:

TABLE 1A. SUMMARY OF CHARACTERISTICS OF GAME LAKE AND MADSTONE CABIN THRUSTS

	Game Lake thrust	Madstone Cabin thrust
Similarities		
Lithic units	Peridotite and dunite	Harzburgite and dunite
	Coherent basal serpentinite	Basal serpentinite mylonite
	Mafic phyllonite	Mafic phyllonite
	Amphibolite	Amphibolite
	Syntectonic muscovite-garnet tonalite	Syntectonic muscovite-garnet tonalite
Structures	NE-trending early, tight to isoclinal folds	NE-trending early, tight to isoclinal folds
	NE trend of π-pole to amphibolite and mafic phyllonite foliations	NE trend of π-pole to amphibolite and mafic phyllonite foliations
	NE- and NW-trending lineations	NE-trending lineation
	NE transport direction of upper block (modern coordinates)	NE transport direction of upper block (modern coordinates)
	Late open folds	Late open folds
Cooling ages	154.2 ± 2.0 Ma ($^{40}Ar/^{39}Ar$, hornblende, 1-σ)	150.3 ± 1.0 Ma ($^{40}Ar/^{39}Ar$, hornblende, 1-σ)
	149.1 ± 0.4 Ma ($^{40}Ar/^{39}Ar$, muscovite, 1-σ)	150–146 Ma ($^{40}Ar/^{39}Ar$, muscovite 1-σ)
Amphibolite	Prograde amphibolite facies	Prograde amphibolite facies
Syntectonic	550 to >600 °C, 3.5–6 kb	550 to >600 °C, 3.5–5 kb (this study)
Tonalite		400–540 °C, 5.5 kb
Metamorphic history	Retrograde greenschist facies	Retrograde greenschist facies
	0–4 kb	0–6 kb (this study)
Differences		
Basal serpentinite composition	Chrysotile and lizardite	Lizardite overprinted by antigorite – brucite
Thickness of mafic phyllonite	>100 m	Discontinuous

Sources: Madstone Cabin thrust data from Harper et al. (1990, 1994, 1996) and Loney and Himmelberg (1977). Game Lake thrust data from this study.

TABLE 1B. SUMMARY OF SIGNIFICANT FEATURES OF THE JOSEPHINE AND COAST RANGE OPHIOLITES

	Coast Range Ophiolite	Josephine Ophiolite
Overlying cover	Great Valley Group Continuous sedimentation during the Nevadan Orogeny Volcaniclasts and chert (Oxfordian–Tithonian)	Myrtle Group Unconformity (Nevadan Orogeny) Galice Flysch (Kimmeridgian) Hemipelagic sequence (Oxfordian or Bathonian)
Crustal sequence	Chert interlayered with basalt (Bathonian to Early Kimmeridgian) 172 to 164–160 Ma isotopic age Complete but dismembered ophiolite remnants Island arc geochemistry	Tuffaceous chert between volcanic flows (Late Callovian or Bajocian) 169 to 160 Ma isotopic age Complete intact ophiolite sequence Island arc and late mid-ocean ridge basalt geochemistry
Subjacent rocks	Franciscan complex	Rogue–Chetco arc complex

Sources: Data for Coast Range Ophiolite from Shervais (1990), Hopson et al. (1981), Hopson et al. (1996), and Shervais et al. (2004). Data for Josephine Ophiolite from Loney and Himmelberg (1977), Saleeby et al. (1982), Pessagno and Blome (1990), Harper et al. (1994), Palfy et al. (2000), and Harper (2003).

1. If the Snowcamp terrane is part of the Coast Range ophiolite, then the Coast Range and Josephine ophiolites are correlative, and the Coast Range ophiolite has followed a similar emplacement history as the Josephine ophiolite (i.e., not exotic to North America).
2. The Snowcamp ophiolite is not a remnant of the Coast Range ophiolite at all, but is an outlier of the Josephine ophiolite.
3. The peridotite at Game Lake is unrelated to the ophiolite remnant at Snowcamp Mountain proper, and a terrane boundary is required somewhere between the two areas.

In the Game Lake area, the Colebrook Schist structurally underlies both the mafic phyllonite and peridotite. Its outcrop distribution indicates that this contact crosscuts the Game Lake thrust and that the amalgamated ophiolite and arc rocks were emplaced over the Franciscan rocks. This relationship is, in part, similar to that seen with other Coast Range ophiolite remnants. Beyond the features of the Snowcamp terrane that suggest correlation with Coast Range ophiolite described in earlier sections (nonflysch sedimentation during the Nevadan orogeny, lack of deformation in cover sedimentary rocks, age; see Table 1B), other evidence to support possibility 1 includes geochemical evidence from parts of the Rogue River (Wild Rogue ophiolite) and Snowcamp Mountain that indicates some affinity with other Coast Range ophiolite remnants (see Harper et al., 2000; Kosanke, 2000), suggesting that the Snowcamp remnant may be transitional between the Josephine and Coast Range ophiolites. Pillow lavas in the western part of the Wild Rogue ophiolite have island-arc affinities, and overlying metavolcanics are high-Cr or calc-alkaline, similar to many Coast Range ophiolite remnants (Kosanke, 2000; Shervais et al., 2004). However, the eastern part of the Wild Rogue ophiolite contains sheeted dikes, rare in the Coast Range ophiolite but common in the Josephine ophiolite (Harper et al., 2002). Also, a large Nevadan-aged shear zone cuts the Wild Rogue ophiolite (Blosson Bar shear zone; Kosanke, 2000).

If possibility 2 is correct, then the Snowcamp ophiolite should be redefined as an outlier of the Josephine ophiolite, rather than of the Coast Range ophiolite. Such a lack of correlation would require that resolving the tectonic history of the Coast Range ophiolite, according to the relationships within the Snowcamp ophiolite, is unlikely to succeed. However, this does not *preclude* such a resolution, based on other, outside evidence, and similarities between the Coast Range and Snowcamp ophiolites may still be the result of similar tectonic histories.

Several pieces of evidence suggest that possibility 3 may be correct. A large, late, high-angle fault along the western boundary of the field area suggests that a terrane boundary may be present. In addition, the pressure determination based on Massonne and Schreyer (1987) indicates that the blocks overlying both the Madstone Cabin and Game Lake thrust were at least 11.5 km thick during thrusting. This contrasts with the thicknesses of other Coast Range remnants, which are unusually thin, particularly the plutonic igneous section, relative to typical ocean sections (Hopson et al., 1981). Finally, an ophiolite section, also part of the Snowcamp terrane, several kilometers to the north along the Rogue River shows evidence of being related geochemically, at least in part, to the Josephine ophiolite (Kosanke, 2000).

ACKNOWLEDGMENTS

We thank John W. Shervais and John Wakabayashi for their instructive and constructive reviews. Scott Soricelli was an invaluable assistant during the 2000 field season. The Department of Earth and Atmospheric Sciences at the State University of New York at Albany (especially Vince Idone and Sally Marsh), National Academy of Sciences through Sigma Xi (Grant-in-aid of Research), and the Geological Society of America (GSA Penrose Research Grant #6750-00) provided monetary funds to support fieldwork and microprobe analyses.

APPENDIX: ANALYTICAL METHODS

The $^{40}Ar/^{39}Ar$ analysis was conducted at New Mexico Geochronology Research Laboratory, Socorro. Uncertainties in ages are given at the 1-σ level. Plateau ages incorporate uncertainties in interfering reaction corrections and J factors. Ages determined relative to FC-1 Fish Canyon Tuff sanidine interlaboratory standard at 27.84 Ma. Decay constants and isotopic abundances after Steiger and Jager (1977).

Mineral analyses used for geothermobarometry were obtained using the JEOL 8900 Electron Microprobe at Binghamton University, Binghamton, New York (William Blackburn). Geothermometry was determined using the method of Holland and Blundy (1994). Ferric iron was recalculated according to Spear and Kimball (1984) and Robinson et al. (1982), with two additional constraints: (1) total M2 cations (Al[VI],Ti, and Fe^{+3}) cannot exceed 2 apfu; and (2) Ca in the M4 site cannot exceed 2.00 apfu. Flourine content was analyzed for and found to be below detection limits for all samples. Geobarometry was determined using the method described in Massonne and Schreyer (1987).

Structural data was analyzed using the program Stereonet v. 6.1.1 (Dr. Richard W. Allmendinger).

REFERENCES CITED

Apted, M.J., and Liou, J.G., 1983, Phase relations among greenschist, epidote-amphibolite, and amphibolite in a basaltic system: American Journal of Science, v. 283-A, p. 328–354.

Baumgartner, P.O., Bartolino, A., Carter, E.S., Conti, M., Cortese, G., Danelian, T., De Wever, P., Dumitrica, P., Dimitrica-Jud, R., Gorican, S., Geux, J., Hull, D.M., Kito, N., Marcucci, M., Matsuoka, A., Murchey, B., O'Dogherty, L., Savary, J., Vishnevskaya, V., Widz, D., and Yao, A., 1995, Middle Jurassic to Early Cretaceous radiolarian biochronology of Tethys based on unitarian associations, *in* Baumgartner, P.O., et al., eds., Middle Jurassic to Lower Cretaceous Radiolaria of Tethys: Occurrences, systematics, biochronology: Lausanne, Switzerland, Mémoires de Géologie (Lausanne): p. 1013–1048.

Best, M.G., 2003, Igneous and metamorphic petrology: Malden, Massachussetts, Blackwell Science Publications, 729 p.

Blake, M.C., Jr., 1984, Tectonostratigraphic terranes in southwestern Oregon, *in* Nilsen, T.H., ed., Geology of the Upper Cretaceous Hornbrook Formation, Oregon and California: Los Angeles, Pacific Section, Society of Economic Paleontologists and Mineralogists, v. 42, p. 159–165.

Blake, M.C., Jr., Engebretson, D.C., Jayko, A.S., and Jones, D.L., 1985, Tectonostratigraphic terranes in southwest Oregon, *in* Howell, D.G., ed., Tectonostratigraphic terranes of the Circum-Pacific region: Houston, Circum-Pacific Council for Energy and Mineral Resources, Earth Sciences Series 1, p. 147–157.

Bucher, K., and Frey, M., 2002, Petrogenesis of metamorphic rocks: Berlin, Springer, 341 p.

Burchfiel, B.C., and Davis, G.A., 1981, Triassic and Jurassic tectonic evolution of the Klamath Mountains–Sierra Nevada geologic terrane, *in* Ernst, W.G., ed., The geotectonic development of California—Rubey volume 1: Englewood Cliffs, New Jersey, Prentice Hall, p. 51–70.

Cannat, M., and Boudier, F., 1985, Structural study of intra-oceanic thrusting in the Klamath Mountains, northern California: Implications on accretion geometry: Tectonics, v. 4, p. 435–452 (and corrections v. 4, p. 598–601).

Coleman, R.G., 1972, The Colebrook Schist of southwestern Oregon and its relation to the tectonic evolution of the region: Washington, D.C., U.S. Geological Survey Bulletin 1339, 61 p.

Coulton, A.J., 1995, Fault-related oceanic and emplacement-age serpentinization in the Josephine ophiolite of NW California and SW Oregon [Ph.D. dissertation]: Albany, State University of New York, 251 p.

Dick, H.J.B., 1976, The origin and emplacement of the Josephine peridotite of southwestern Oregon [Ph.D. dissertation]: New Haven, Connecticut, Yale University, 409 p.

Dickinson, W.R., Hopson, C.A., and Saleeby, J.B., 1996, Alternate origins of the Coast Range ophiolite (California): Introduction and implications: GSA Today, v. 6, p. 2–10.

Evans, B., 2004, The serpentine multisystem revisited: Chrysotile is metastable: International Geology Review, v. 46, p. 479–506.

Godfrey, N.J., and Dilek, Y., 2000, Mesozoic assimilation of ocean crust and island arc into the North American continental margin in California and Nevada: Insights from geophysical data, *in* Dilek, Y., et al., eds., Ophiolites and ocean crust: New insights from field studies and the ocean drilling program: Boulder, Colorado, Geological Society of America Special Paper 349, p. 365–382.

Grady, K.A., 1990, Geology and structure of the rocks associated with the basal (Madstone) thrust of the Josephine ophiolite in southwestern Oregon: Evidence for a metamorphic sole [M.S. thesis]: Albany, State University of New York, 160 p.

Hacker, B.R., Donato, M.M., Barnes, C.G., McWilliams, M.O., and Ernst, W.G., 1995, Timescales of orogeny: Jurassic construction of the Klamath Mountains: Tectonics, v. 14, p. 677–703, doi: 10.1029/94TC02454.

Harper, G.D., 1980, Structure and petrology of the Josephine ophiolite and overlying sedimentary rocks, northern California [Ph.D. dissertation]: Berkeley, University of California, 260 p.

Harper, G.D., 1983, A depositional contact between the Galice Formation and a Late Jurassic ophiolite in northwestern California and southwestern Oregon: Oregon Geology, v. 45, p. 3–9.

Harper, G.D., 1984, The Josephine ophiolite, northwestern California: Geological Society of America Bulletin, v. 95, p. 1009–1026, doi: 10.1130/0016-7606(1984)95<1009:TJONC>2.0.CO;2.

Harper, G.D., 2003, Tectonic implications of boninite, arc tholeiite, and MORB magma types in the Josephine Ophiolite, California-Oregon, *in* Dilek, Y., and Robinson, P.T., eds., Ophiolites in Earth history: London, Blackwell Scientific Publications, Geological Society Special Publications 218, p. 207–230.

Harper, G.D., and Wright, J.E., 1984, Middle to late Jurassic tectonic evolution of the Klamath Mountains, California-Oregon: Tectonics, v. 3, p. 759–772.

Harper, G.D., Grady, K., and Wakabayashi, J., 1990, A structural study of a metamorphic sole beneath the Josephine ophiolite, western Klamath terrane, California-Oregon, *in* Harwood, D.S., and Miller, M.M., eds., Paleozoic and early Mesozoic paleogeographic relations; Sierra Nevada, Klamath Mountains, and related terranes: Boulder, Colorado, Geological Society of America Special Paper 255, p. 379–396.

Harper, G.D., Saleeby, J.B., and Heizler, M., 1994, Formation and emplacement of the Josephine ophiolite and the Nevadan orogeny in the Klamath Mountains, California-Oregon: U/Pb zircon and $^{40}Ar/^{39}Ar$ geochronology: Journal of Geophysical Research, v. 99, p. 4293–4321, doi: 10.1029/93JB02061.

Harper, G.D., Grady, K., and Coulton, A.J., 1996, Origin of the amphibolite "sole" of the Josephine ophiolite: Emplacement of a cold ophiolite over a hot arc: Tectonics, v. 15, p. 296–313, doi: 10.1029/95TC02525.

Harper, G.D., Kosanke, S., and Giaramita, M.J., 2000, Tectonostratigraphy and geochemistry of the Coast Range Ophiolite at Snowcamp Mountain, SW Oregon; Implications for the exotic vs. native origin of the Coast Range Ophiolite: Geological Society of America Abstracts with Programs, v. 32, p. 17.

Harper, G.D., Giaramita, M.J., and Kosanke, S.B., 2002, Josephine and Coast Range Ophiolites, Oregon and California, *in* Moore, G.W., ed., Field guide to the geologic processes in Cascadia: Portland, Oregon, 98th Annual Meeting of the Cordilleran Section of the Geological Society of America Guidebook, p. 1–22.

Himmelberg, G.R., and Loney, R.A., 1973, Petrology of the Vulcan Peak alpine-type peridotite, southwestern Oregon: Geological Society of America Bulletin, v. 84, p. 1585–1600, doi: 10.1130/0016-7606(1973)84<1585:POTVPA>2.0.CO;2.

Holland, T., and Blundy, J., 1994, Non-ideal interactions in calcic amphiboles and their bearing on amphibole-plagioclase thermometry: Contributions to Mineralogy and Petrology, v. 116, p. 433–447, doi: 10.1007/BF00310910.

Hopson, C.A., Mattinson, J.M., and Pessagno, E.A., Jr., 1981, Coast Range ophiolite, western California, *in* Ernst, W.G., ed., The geotectonic development of California—Rubey volume 1: Englewood Cliffs, New Jersey, Prentice Hall, p. 419–510.

Hopson, C.A., Pessagno, E.A., Jr., Mattinson, J.M., Luyendyk, B.P., Beebe, W., Hull, D.M., Munoz, I.M., and Blome, C.D., 1996, Coast Range Ophiolite as paleoequatorial mid-ocean lithosphere, *in* Dickinson, W.R., et al., Alternate origins of the Coast Range ophiolite (California): Introduction and implications: GSA Today, v. 6, no. 2, p. 5–7.

Irwin, W.P., 1981, Tectonic accretion of the Klamath Mountains, *in* Ernst, W.G., ed., The geotectonic development of California—Rubey volume 1: Englewood Cliffs, New Jersey, Prentice Hall, p. 30–49.

Jachens, R.C., Barnes, C.G., and Donato, M.M., 1986, Subsurface configuration of the Orleans fault: Implications for deformation in the western Klamath Mountains, California: Geological Society of America Bulletin, v. 97, p. 388–395, doi: 10.1130/0016-7606(1986)97<388:SCOTOF>2.0.CO;2.

Jayko, A.S., Blake, M.C., Jr., and Harms, T., 1987, Attenuation of the Coast Range ophiolite by extensional faulting, and nature of Coast Range "thrust," California: Tectonics, v. 6, p. 475–488.

Kosanke, S., 2000, The geology, geochronology, structure, and geochemistry of the Rogue Wilderness ophiolite, SW Oregon: Implications for the magmatic and tectonic evolution of the Coast Range Ophiolite [Ph.D. dissertation]: Albany, State University of New York, 755 p.

Loney, R.A., and Himmelberg, G.R., 1976, Structure of the Vulcan Peak alpine-type peridotite, southwestern Oregon: Geological Society of America Bulletin, v. 87, p. 259–274, doi: 10.1130/0016-7606(1976)87<259:SOTVPA>2.0.CO;2.

Loney, R.A., and Himmelberg, G.R., 1977, Geology of the gabbroic complex along the northern border of the Josephine peridotite, Vulcan Peak area, southwestern Oregon: Journal of Research of the U.S. Geological Survey, v. 5, no. 6, p. 761–781.

Massonne, H.-J., and Schreyer, W., 1987, Phengite geobarometry based on the limiting assemblage with K-feldspar, phlogopite, and quartz: Contributions to Mineralogy and Petrology, v. 96, p. 212–224, doi: 10.1007/BF00375235.

McDougall, I., and Harrison, T.M., 1999, Geochronology and thermochronology by the $^{40}Ar/^{39}Ar$ method: New York, Oxford University Press, 269 p.

Moores, E.M., 1970, Ultramafics and orogeny with models of the US Cordillera and Tethys: Nature, v. 228, p. 837–842, doi: 10.1038/228837a0.

Page, N.J., Gray, F., Cannon, J.K., Foose, M.P., Lipin, B., Moring, B.C., Nicholson, S.W., Sawlin, M.G., Till, A., and Ziemianski, W.P., 1981, Geologic map of the Kalmiopsis Wilderness area, Oregon: Reston, Virginia, U.S. Geological Survey Miscellaneous Field Studies Map MF-1240-A, scale 1:62,500, 1 sheet.

Palfy, J., Smith, P.L., and Mortensen, J.K., 2000, A U-Pb and $^{40}Ar/^{39}Ar$ time scale for the Jurassic: Canadian Journal of Earth Sciences, v. 37, p. 923–944, doi: 10.1139/cjes-37-6-923.

Pessagno, E.A., Jr., and Blome, C.D., 1990, Implication of the new Jurassic stratigraphic, geochronometric, and paleolatitudinal data from the western Klamath terrane (Smith River and Rogue Valley subterranes): Geology, v. 18, p. 665–668, doi: 10.1130/0091-7613(1990)018<0665:IONJSG>2.3.CO;2.

Ramp, L., 1975, Geology and mineral resources of the upper Chetco drainage area, Oregon: Portland, Oregon Department of Geology and Mineral Industry Bulletin 88, 47 p.

Ramsay, J.G., and Huber, M.I., 1987, The techniques of modern structural geology, Volume 2: Folds and fractures: London, Academic Press, p. 309–700.

Robertson, A.H.F., 1989, Palaeoceanography and tectonic setting of the Coast Range ophiolite, central California: Evidence from the extrusive rocks and the volcaniclastic sediment cover: Marine and Petroleum Geology, v. 6, p. 194–220, doi: 10.1016/0264-8172(89)90001-9.

Robinson, P., Schumacher, J.C., and Spear, F.S., 1982, Formulation of electron probe analyses, *in* Veblen, D.R., and Ribbe, P.H., eds., Amphiboles: Petrology and experimental phase relations: Washington, D.C., Mineralogical Society of America, Reviews in Mineralogy 9B, p. 6–9.

Saleeby, J.B., 1984, Pb/U zircon ages from the Rogue river area, western Jurassic belt, Klamath Mountains, Oregon: Geological Society of America Abstracts with Programs, v. 16, p. 331.

Saleeby, J.B., 1992, Petrotectonic and paleogeographic settings of U.S. Cordilleran ophiolites, *in* Burchfiel, B.C., et al., eds., The Cordilleran Orogen: Conterminous U.S.: Boulder, Colorado, Geological Society of America, The Geology of North America, v. G-3, p. 653–682.

Saleeby, J.B., and Harper, G.D., 1993, Tectonic relations between the Galice Formation and the Condrey Mountain Schist, Klamath Mountains, northern California, *in* Dunn, G.C., and McDougall, K.A., eds., Mesozoic paleogeography of the western United States II: Los Angeles, Pacific Section, Society of Economic Paleontologists and Mineralogists, v. 71, p. 61–80.

Saleeby, J.B., Harper, G.D., Snoke, A.W., and Sharp, W.D., 1982, Time relations and structural-stratigraphic patterns in ophiolite accretion, west central Klamath Mountains, California: Journal of Geophysical Research, v. 87, p. 3831–3848.

Shervais, J.W., 1990, Island arc and ocean crust ophiolites: Contrasts in the petrology, geochemistry and tectonic style of ophiolite assemblages in the California Coast Ranges, *in* Malpas, J., et al., eds., Ophiolites: Oceanic crustal analogues: Nicosia, Cyprus, Geological Survey Department, p. 507–520.

Shervais, J.W., and Kimbrough, D.L., 1985, Geochemical evidence for the tectonic setting of the Coast Range ophiolite: A composite island arc-oceanic crust terrane in western California: Geology, v. 13, p. 35–38, doi: 10.1130/0091-7613(1985)13<35:GEFTTS>2.0.CO;2.

Shervais, J.W., Kimbrough, D.L., Renne, P., Hanan, B.B., Murchey, B., Snow, C.A., Schuman, M.M.Z., and Beaman, J., 2004, Multi-stage origin of the Coast Range ophiolite, California: Implications for the life cycle of supra-subduction zone ophiolites: International Geology Review, v. 46, p. 2–28.

Shervais, J.W., Kolesar, P., and Andreason, K., 2005, A field and chemical study of serpentinization, Stonyford, California: Chemical flux and mass balance: International Geology Review, v. 47, p. 1–23.

Simpson, C., 1986, Determination of movement sense in mylonites: Journal of Geological Education, v. 34, p. 246–261.

Spear, F.S., and Kimball, C., 1984, RECAMP—A Fortran IV program for estimating Fe^{3+} contents in amphiboles: Computers & Geosciences, v. 10, p. 317–325, doi: 10.1016/0098-3004(84)90029-3.

Steiger, R.H., and Jager, E., 1977, Subcommission on geochronology: Convention on the use of decay constants in geo- and cosmochronology: Earth and Planetary Science Letters, v. 36, p. 359–362, doi: 10.1016/0012-821X(77)90060-7.

Stern, R.J., and Bloomer, S.H., 1992, Subduction zone infancy: Examples from the Eocene Izu-Bonin-Mariana and Jurassic California arcs: Geological

Society of America Bulletin, v. 104, p. 1621–1636, doi: 10.1130/0016-7606(1992)104<1621:SZIEFT>2.3.CO;2.

Vail, S.G., 1977, Geology and Geochemistry of the Oregon Mountain area, southwestern Oregon and northern California [Ph.D. dissertation]: Corvallis, Oregon State University, 159 p.

Wright, J.E., and Fahan, M.R., 1988, An expanded view of Jurassic orogenesis in the western United States Cordillera: Middle Jurassic (pre-Nevadan) regional metamorphism and thrust faulting within and active arc environment, Klamath Mountains, California: Geological Society of America Bulletin, v. 100, p. 859–876, doi: 10.1130/0016-7606(1988)100<0859:AEVOJO>2.3.CO;2.

Wright, J.E., and Wyld, S.J., 1986, Significance of xenocrystic Precambrian zircon contained within the southern continuation of the Josephine ophiolite: Devils Elbow ophiolite remnant, Klamath Mountains, northern California: Geology, v. 14, p. 671–674, doi: 10.1130/0091-7613(1986)14<671:SOXPZC>2.0.CO;2.

Wyld, S.J., and Wright, J.E., 1988, The Devils Elbow ophiolite remnant and overlying Galice Formation: New constraints on the Middle to Late Jurassic evolution of the Klamath Mountains, California: Geological Society of America Bulletin, v. 100, p. 29–44, doi: 10.1130/0016-7606(1988)100<0029:TDEORA>2.3.CO;2.

Yule, J.D., 1996, Geologic and tectonic evolution of the Jurassic marginal basin lithosphere, Klamath Mountains, Oregon [Ph.D. thesis]: Pasadena, California Institute of Technology, 308 p.

Yule, J.D., and Saleeby, J.B., 1993, Highly extended oceanic lithosphere: The basement and wallrocks for the late Jurassic Rogue–Chetco oceanic arc, Oregon Klamath Mountains: Geological Society of America Abstracts with Programs, v. 25, p. 169.

Yule, J.D., Saleeby, J.B., and Barnes, C.G., 2006, this volume, A rift-edge facies of the Late Jurassic Rogue–Chetco arc and Josephine ophiolite, Klamath Mountains, Oregon, *in* Snoke, A.W., and Barnes, C.G., eds., Geological studies in the Klamath Mountains province, California and Oregon: A volume in honor of William P. Irwin: Boulder, Colorado, Geological Society of America Special Paper 410, doi: 10.1130/2006.2410(03).

Manuscript Accepted by the Society 10 January 2006

Printed in the USA

Geological Society of America
Special Paper 410
2006

Petrology and geochemistry of the Middle Jurassic Ironside Mountain batholith: Evolution of potassic magmas in a primitive arc setting

Calvin G. Barnes*
Elizabeth V. Mars
Department of Geosciences, Texas Tech University, Lubbock, Texas 79409-1053, USA
Susan Swapp
Carol D. Frost
Department of Geology and Geophysics, Dept. 3006, 1000 E. University Avenue, University of Wyoming, Laramie, Wyoming 82071, USA

ABSTRACT

The Ironside Mountain batholith consists of the 170-Ma Ironside Mountain pluton, the quartz diorite of Happy Camp Mountain, the 167-Ma Denny intrusive complex, and the West China Peak intrusive complex. The contact aureole of the Ironside Mountain pluton is present in both the western and eastern Hayfork terranes, which indicates that emplacement followed regional thrusting along the Wilson Point thrust. The batholith marks the beginning of a 15-m.y.-long episode of arc magmatism. Al-in-hornblende barometry suggests emplacement at ~4 kb. The Denny and West China Peak complexes are predominantly gabbro to olivine pyroxenite characterized by calcic plagioclase. Along with the quartz diorite of Happy Camp Mountain, these plutons have calcic bulk compositions, abundant amphibole, and low FeO/(FeO + MgO). Their chemical and petrographic features suggest oxidized, H_2O-rich parental magmas. In contrast, the Ironside Mountain pluton has higher FeO/(FeO + MgO), shows K_2O enrichment with increasing SiO_2 to ultrahigh-K compositions, and is typified by 2- and 3-pyroxene assemblages in which hydrous mafic silicates are sparse or absent, except in the most evolved rocks. The elements Rb, Zr, and Ba all behave incompatibly. Initial Nd and Sr isotope ratios (170 Ma) are uniform at ε_{Nd} of ~5.2 and $^{87}Sr/^{86}Sr$ of ~0.7037. These data suggest that the pluton lacks significant crustal input, unlike younger Middle Jurassic plutons, and that it evolved primarily by fractional crystallization of a reduced, H_2O-poor arc tholeiite. Enrichment of K_2O is interpreted to result from differentiation at relatively high pressure (≥8 kb), under conditions in which augite fractionation predominated over olivine fractionation. High-P fractionation without appreciable crustal input is probably related to the thermal structure of the crust. Emplacement immediately after regional thrusting placed parental magmas into cool, nonreactive host rocks. Succeeding Middle Jurassic magmatism encountered progressively hotter crust, with consequent crustal melting and assimilation.

Keywords: Klamath Mountains, arc magmatism, Jurassic, potassic

*E-mail: Cal.Barnes@ttu.edu.

Barnes, C.G., Mars, E.V., Swapp, S., and Frost, C.D., 2006, Petrology and geochemistry of the Middle Jurassic Ironside Mountain batholith: Evolution of potassic magmas in a primitive arc setting, *in* Snoke, A.W., and Barnes, C.G., eds., Geological studies in the Klamath Mountains province, California and Oregon: A volume in honor of William P. Irwin: Geological Society of America Special Paper 410, p. 199–221, doi: 10.1130/2006.2410(10). For permission to copy, contact editing@geosociety.org.

INTRODUCTION

Mesozoic volcanic and plutonic rocks in the Klamath Mountain province have long been considered to be excellent examples of primitive arc magmatism (Moore, 1959; Hotz, 1971; Irwin, 1972). Most of these rocks are calcic or calc-alkaline (Hotz, 1971; Barnes et al., 1992). However, Lanphere et al. (1968) and Hotz (1971) recognized a group of plutons with unusual K_2O-rich compositions. In modern terminology, these are plutons with evolved rock compositions in the monzonitic and quartz monzonitic range, rather than in the tonalitic and granodioritic range. The Ironside Mountain batholith and the Wildwood pluton are the largest examples of these K_2O-rich plutons (Hotz, 1971), and the Ironside Mountain batholith is the subject of this chapter. It is the largest plutonic unit in the Klamath province, with a length of 72 km and a maximum width of 12 km.

The Ironside Mountain and Wildwood plutons have radiometric ages in the 169- to 170-Ma range (Wright and Fahan, 1988) and on the basis of these ages and earlier K-Ar and Rb-Sr geochronology (Lanphere et al., 1968), Irwin and Wooden (1999) defined a "Western Hayfork" magmatic episode that encompassed the previously mentioned plutons, the western Hayfork terrane, and a number of small plutons with similar ages throughout the province. One of the principal assumptions of this grouping is that Ironside Mountain–type plutons are the intrusive equivalents of the western Hayfork terrane volcanic and volcaniclastic strata. Geologic data presented here show that this interpretation must be reconsidered. Furthermore, we show that Ironside Mountain–type magmatism is unusual relative to other Klamath plutonism and that it occurred in a specific, post-contractional tectonic setting. In this chapter, we describe the Ironside Mountain batholith, its contact relations, mineral assemblages, and geochemical composition. We then propose a petrogenetic model for its origin.

METHODS

The first detailed field study of the Ironside Mountain batholith was part of the Ph.D. dissertation of D. Charlton (1979). He mapped the rugged central part of the batholith and its host rocks. He interpreted the main plutonic unit to be crudely zoned, and he also mapped several small, compositionally distinct intrusive bodies within or adjacent to the main unit. Our work extended Charlton's mapping to the north and south. Particularly in the south, much of the pluton is poorly exposed, such that exposures were limited to road cuts. However, glaciated exposures in the northeastern part of the pluton permitted inspection of larger, more continuous outcrops.

Bulk geochemical analysis was done at Texas Tech University by inductively coupled plasma atomic emission spectrometry for major elements, plus Sr, Zr, Y, Nb, Ba, Sc, Cu, Cr, and Zn. The element Rb was analyzed by flame emission. For samples whose sample number ends in "91," the rare-earth elements, Th, Ta, and Hf, were analyzed by instrumental neutron activation analysis at Oregon State University. Rare-earth element concentrations in samples whose numbers end in "88" were determined by inductively coupled plasma emission spectroscopy after separation by ion exchange methods (see Shannon et al., 1997). These data were supplemented by analyses from Lanphere et al. (1968) and Charlton (1979).

Mineral compositions were analyzed on an automated JEOL 8900 Superprobe at the University of Wyoming. Nominal instrument conditions were 15-kV accelerating potential and a beam current of 10–20 nA. Standards were natural and synthetic silicates and oxides; data were reduced using ZAF corrections.

The original Sr isotopic data were produced by R.W. Kistler at the U.S. Geological Survey. More recent data were collected at the University of Wyoming, where bulk rock Nd and Sr isotopic compositions were determined by thermal ionization mass spectrometry. The Sm, Nd, Rb, and Sr concentrations were determined by isotope dilution on aliquots of the same sample dissolved for isotope ratio measurements. See Table 7 for analytical details.

GEOLOGIC SETTING

Regional Geology

The Ironside Mountain batholith intruded a stack of imbricate tectonostratigraphic terranes separated by east-dipping thrust faults (e.g., Irwin, 1972; Charlton, 1979; Wright, 1982). From structurally lowest, these are the Rattlesnake Creek terrane, the western Hayfork terrane, and the eastern Hayfork terrane (terminology of Wright, 1982; Fig. 1). All of these units and the batholith were thrust over the Galice Formation during the Nevadan orogeny (e.g., Harper and Wright, 1984).

West of the Ironside Mountain batholith, the Galice Formation consists primarily of semipelitic slate and phyllite, with lesser amounts of chlorite-rich phyllite, greenschist, and minor greenstone (Young, 1978). To the north, the Galice Formation contains ca. 153-Ma detrital zircons (Miller et al., 2003) and is cut by dikes and plutons as old as 151 ± 2 Ma (Saleeby et al., 1982; Harper et al., 1994).

The Rattlesnake Creek terrane consists of a lower ophiolitic mélange that is locally unconformably overlain by coherent metasedimentary and metavolcanic sequences, some of which have volcanic arc affinities (Gray, 1986; Wright and Wyld, 1994). The mélange contains fossil-bearing blocks as young as Norian and the entire sequence is cut by plutons as young as ca. 193 Ma (Wright and Wyld, 1994).

The western Hayfork terrane structurally overlies the Rattlesnake Creek terrane along a regional thrust fault. The terrane consists primarily of clastic metasedimentary rocks that range from cobble-rich breccia (lahar deposits) through graded sandstones to fine-grained massive and well-layered argillite. Andesitic and basaltic lavas are uncommon (Irwin, 1972; Wright, 1982). The unit is generally considered to be a clastic apron deposited along the flanks of a volcanic arc. Donato et al.

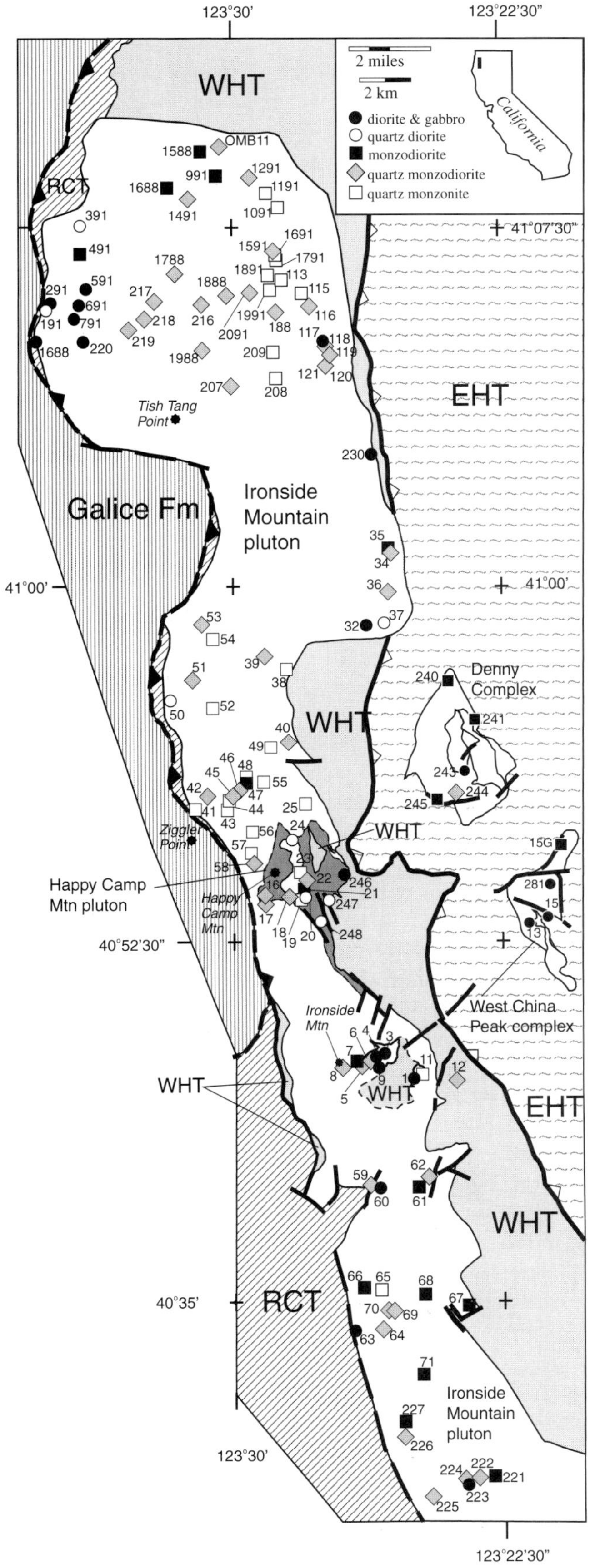

Figure 1. Simplified geologic map of the Ironside Mountain batholith. Bold lines are faults; unornamented lines are high angle faults; Orleans thrust fault shown with filled barbs; Wilson Point thrust fault shown with open barbs. Numbers adjacent to sample location symbols are sample numbers. EHT—eastern Hayfork terrane, RCT—Rattlesnake Creek terrane, WHT—western Hayfork terrane. Modified after Charlton (1979), Donato et al. (1983), and Mars, (1999).

(1996) showed that, at the latitude of the Oregon-California border, the western Hayfork terrane was deposited on Rattlesnake Creek terrane basement (also see Wright and Fahan, 1988). This stratigraphic link suggests that the thrust fault that separates the Rattlesnake Creek terrane and the western Hayfork terrane is of limited displacement (Wright and Fahan, 1988; Donato et al., 1996). The $^{40}Ar/^{39}Ar$ ages on hornblende (considered to be detrital volcanic hornblende) from the western Hayfork range from 168 ± 1.9 Ma to 173 ± 1 Ma (Wright and Fahan, 1988; Donato et al., 1996), and at least one 174-Ma pluton (Squaw Mountain pluton; U-Pb, zircon; Irwin and Wooden, 1999) intrudes the terrane.

The eastern Hayfork terrane overlies the western Hayfork terrane along the Wilson Point thrust (Wright, 1982; Fig. 1). The eastern Hayfork terrane is a chert-argillite mélange and broken formation; it has yielded late Triassic fossils (Wright, 1982; Ernst, 1998). The timing of displacement on the Wilson Point thrust is uncertain, and Wright and Fahan (1988) suggested that it postdated emplacement of the Ironside Mountain batholith. Below we show that the batholith intruded the Wilson Point thrust and therefore places an important constraint on the age of thrusting.

Magmatism with chemical similarities to the Ironside Mountain batholith ended at ca. 167 Ma with emplacement of the Denny complex (see below). Younger magmatic activity, herein referred to as "Wooley Creek belt" magmatism, began at ca. 165 Ma (Hacker et al., 1995; Allen and Barnes, this volume) and continued to ca. 156 Ma (Yule et al., this volume). Wooley Creek belt plutons are composite; they all display field, geochemical, and isotopic evidence for open-system behavior, such as magma mixing and mingling, in situ assimilation, and deep-seated crustal melting (Barnes et al., 1986, 1990, 1995; Gribble et al., 1990).

Intrusive Units and Definition of the Batholith

Irwin (1960) named the Ironside Mountain batholith for an elongate plutonic mass that extends 72 km from 40°32′N latitude to 41°10′N (Fig. 1). The main part of the Ironside Mountain batholith reaches 12 km wide in its northern part, where it is in intrusive contact with host rocks to the east and west. In his original definition, Irwin (1960) mapped an arm that extended southeast from the center of the main mass; however, this extension was shown by Charlton (1979) to be a separate pluton, the Denny zoned complex (Fig. 1). The U-Pb (zircon) dating of the Ironside Mountain batholith and Denny Zoned Complex (Wright and Fahan, 1988) yielded ages of 169–170 Ma and 167 Ma, respectively.

Charlton's (1979) detailed study of the central part of the Ironside Mountain batholith identified a number of intrusive units internal and external to the main mass of the batholith. The principal external units are the so-called "concentrically-zoned complexes" at Denny and West China Peak (Fig. 1), both of which are intrusive into the eastern Hayfork terrane. These plutons contain ultramafic to intermediate rocks that range from olivine pyroxenite to quartz diorite.

Within the main outcrop area of the batholith, Charlton defined two principal units. The "Ironside Mountain pluton" is the elongate, dioritic to quartz monzonitic body that underlies most of the batholith (Fig. 1). Charlton (1979) interpreted this part of the batholith to be gradationally zoned from diorite to quartz monzonite (but see below). Charlton (1979) also mapped a sequence of dioritic, quartz dioritic, and tonalitic rocks that underlie an ~8-km^2 part of the pluton (Fig. 1) and small areas east of Ironside Mountain. This "Happy Camp Mountain pluton" is intruded by dikes from the Ironside Mountain pluton but has not been radiometrically dated. Charlton (1979) also mapped a number of small dioritic and gabbroic bodies, some of which have appinitic texture.

Although our mapping suggests that rocks similar to those of the Happy Camp Mountain pluton are unusual elsewhere in the batholith, we retain the terminology introduced by Charlton with slight modification. We consider the "Ironside Mountain batholith" to consist of the Ironside Mountain pluton, as defined by Charlton (1979), the "quartz diorite of Happy Camp Mountain," the "Denny intrusive complex," and the "West China Peak intrusive complex." Definitive inclusion of the latter pluton in the Ironside Mountain batholith awaits radiometric dating. This chapter is primarily on the Ironside Mountain pluton, but data are also presented for the quartz diorite of Happy Camp Mountain, the Denny intrusive complex, and the West China Peak intrusive complex.

Contact Relationships

Most of the contacts between the Ironside Mountain pluton and Rattlesnake Creek terrane are sheared. Contact zones are typically marked by mylonitic or submylonitic fabrics and local development of S-C fabrics. Along the northwestern contact, metaserpentinite of the Rattlesnake Creek terrane is tectonically shuffled into the pluton. In some locations, the plutonic rocks have been rodingitized adjacent to the metaserpentinite. However, in the same region (Mill Creek drainage: ~123°34′W, 41°08′N), the grain size of the pluton decreases from medium- and coarse-grained in the interior to fine-grained within ~50 m of the contact. We interpret this grain size decrease to result from undercooling against Rattlesnake Creek terrane host rocks and therefore conclude that the original intrusive contact was intensely modified by later faulting, possibly during Nevadan-age thrusting.

Contacts between the Ironside Mountain pluton and the western Hayfork terrane are clearly intrusive in many locations. These intrusive relations are most commonly seen as development of pyroxene and hornblende hornfels in the western Hayfork terrane and xenoliths of western Hayfork terrane rocks near Ironside Mountain (Charlton, 1979; Fig. 1). Unusual quartz-rich sandstone of the western Hayfork terrane is in contact with the pluton northwest of the Denny complex (south of sample location 37; Fig. 1). Contact metamorphism of the sandstone resulted in a plagioclase + orthoclase + quartz + biotite + garnet + cordierite assemblage.

The Ironside Mountain pluton is in contact with the eastern Hayfork terrane for ~4 km near 123°25′W and 41°00′N. Along this contact, chert-argillite breccias of the eastern Hayfork terrane are metamorphosed to augite-bearing hornfels, and metaserpentinite blocks in the chert-argillite matrix contain augite neoblasts intergrown with olivine. These contact metamorphic effects overprint subgreenschist- to greenschist-facies regional assemblages (e.g., Wright, 1982) and show that the Ironside Mountain pluton intruded the eastern Hayfork terrane.

In summary, the Ironside Mountain batholith was emplaced into a stack of tectonostratigraphic terranes (Rattlesnake Creek, western Hayfork, and eastern Hayfork) that were juxtaposed prior to 170 Ma. The Ironside Mountain pluton intruded the Wilson Point thrust, and therefore places a minimum age on juxtaposition of the eastern and western Hayfork terranes. This assemblage of thrust slices plus the intrusive rocks of the batholith were thrust over the Galice Formation along the Orleans thrust during the Nevadan orogeny (155–150 Ma; Saleeby et al., 1982; Harper and Wright, 1984; Harper et al., 1994), so that parts of the Ironside Mountain batholith are in direct thrust contact over slates of the Galice Formation (Fig. 1).

PETROGRAPHY

Ironside Mountain Pluton

The principal rock types in the Ironside Mountain pluton range from diorite to quartz monzonite (Fig. 1). In most parts of the pluton, zoning is weak or absent. Charlton (1979) mapped a gradational contact in the Ironside Mountain area between western diorite and monzodiorite and eastern quartz monzonite; however, he did not give sample locations, so it is impossible to represent his modal data in Figure 1. In the northern part of the pluton, a crude gradation exists from predominantly dioritic rocks in the west through quartz monzodiorite and monzodiorite in the central part and quartz monzonite near the eastern contact (Fig. 1).

Mafic magmatic enclaves are sparse. They range from gabbro and diorite through monzodiorite. Most are fine grained and equigranular; a few have plagioclase phenocrysts as long as 5 mm. Unlike other dioritic plutons in the Klamath Mountains (e.g., Grayback: Barnes et al., 1995; Johnson and Barnes, this volume; Wooley Creek: Barnes, 1983; Bear Mountain: Snoke et al., 1981; Bushey et al., this volume; English Peak: Schmidt, 1994), the Ironside Mountain pluton lacks swarms of mafic magmatic enclaves and abundant synplutonic mafic dikes.

Texture varies little in the pluton and is predominantly hypidiomorphic granular, with euhedral to subhedral pyroxenes and lath-shaped to blocky plagioclase, anhedral biotite, and interstitial quartz. Alkali feldspar habits range from interstitial to poikilitic to blocky. Where present, hornblende is interstitial in the more mafic rocks and becomes progressively more prismatic in evolved compositions.

Gabbro, Diorite, Monzogabbro, and Monzodiorite. These rocks are characterized by augite, orthopyroxene, and commonly by inverted pigeonite (Fig. 2; Table 1). A few samples contain olivine rimmed by a vermicular intergrowth of orthopyroxene and Fe-Ti oxide. Others contain the vermicular intergrowth but lack olivine; such samples are inferred to have had olivine on the liquidus. The sequence of pyroxene crystallization was apparently augite and pigeonite (the two phases show mutual rimming relations) followed by orthopyroxene, which rims inverted pigeonite. In many samples, hornblende is absent and in others, biotite is rimmed by or intergrown with hornblende. Plagioclase ranges from An_{66} to about An_{48} in quartz-free samples and from An_{56} to An_{28} in quartz-bearing ones (Table 2). Apatite and Fe-Ti oxides are common accessory phases.

Quartz Monzodiorite and Quartz Diorite. These rocks show a transition from two- and three-pyroxene mafic assemblages to increasing proportions of biotite and hornblende. Alkali feldspar habits range from poikilitic to blocky, with alkali feldspar crystals euhedral against interstitial quartz. Plagioclase ranges from An_{50} to An_{30} (Table 2). Where it is in contact with alkali feldspar, it shows rounded (resorbed?) contacts. Accessory phases are apatite, Fe-Ti oxides, and typically zircon ± allanite.

Quartz Monzonite, and Syenite. The proportion of biotite and hornblende increases in quartz monzonite, but some samples still preserve three-pyroxene assemblages as inclusions in hornblende. More commonly, the pyroxenes are altered to actinolitic amphibole and cummingtonite. Plagioclase ranges from An_{45} to An_{22}. An unusual hornblende-bearing pyroxene syenite dike (92OMB116C) contains plagioclase that ranges from An_{64} to An_{14}; most crystals range from An_{47} to An_{28}. Some samples contain alkali feldspar phenocrysts as long as 2 cm, and poikilitic alkali feldspar with euhedral faces against quartz is also present. Accessory phases are the same as for quartz monzonite.

The sequence of crystallization, as inferred from rimming and possible reaction relations, is shown in Figure 3. Olivine appeared first, followed by augite and plagioclase. We infer augite stability before pigeonite because cumulate olivine pyroxenites in the batholith contain augite but lack pigeonite (see below). Hornblende stability is shown with a dashed line because it is absent in many samples. Likewise, we infer the plagioclase in quartz monzodiorite and quartz monzonite was not a solidus phase but was in reaction with the interstitial melt (e.g., Nekvasil and Lindsley, 1990).

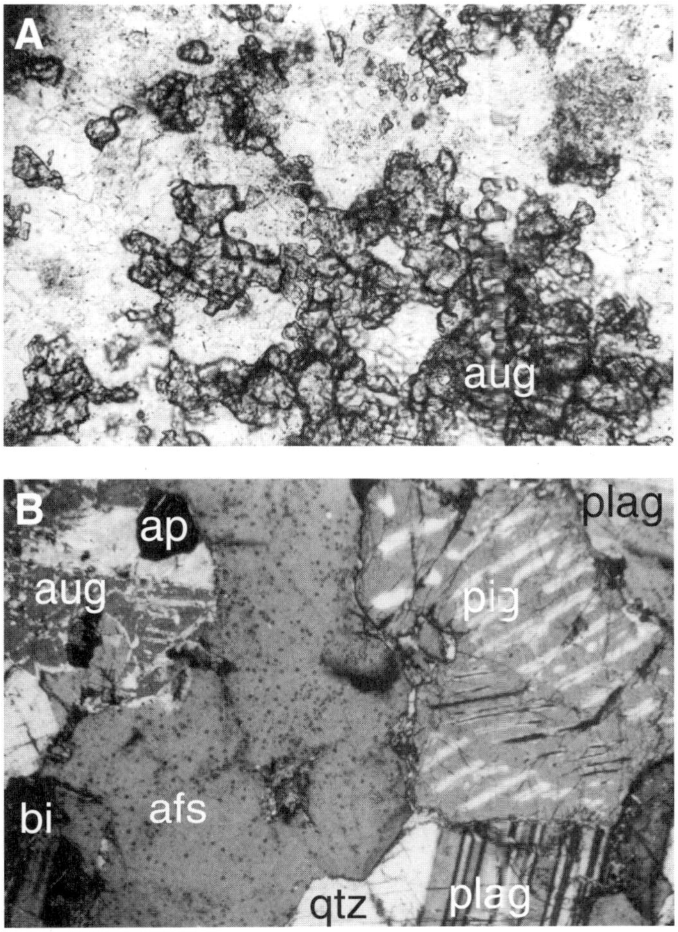

Figure 2. (A) Photomicrograph of pyroxene hornfels from the contact of the Ironside Mountain pluton and chert-argillite breccia of the eastern Hayfork terrane, sample 93OMB234. Field of view is 0.5 mm. (B) Photomicrograph of quartz-bearing biotite, three-pyroxene monzogabbro (sample OMB1688) showing augite (aug), inverted pigeonite (pig), plagioclase (plag), poikilitic orthoclase (afs), and interstitial biotite (bi) and quartz (qtz). The augite is intergrown with prismatic apatite (ap). Field of view is 1.8 mm.

Quartz Diorite of Happy Camp Mountain

This unit is typically medium-grained and hypidiomorphic granular. Augite and orthopyroxene are partly rimmed by olive to pale-green hornblende; brown biotite is interstitial. Lath-shaped to blocky plagioclase ranges from An_{51} to An_{36} (Table 2), and alkali feldspar is interstitial to blocky. The accessory phases are zircon, apatite, and minor Fe-Ti oxides.

TABLE 1. SAMPLE LOCATIONS AND ROCK TYPES

Sample	N latitude (° ′ ″)	W longitude (° ′ ″)	Rock type
Ironside Mountain pluton			
OMB11A	41 09 13.2	123 30 18.3	Biotite-bearing pyroxene monzodiorite
OMB11B	41 09 13.2	123 30 18.3	Dioritic dike
OMB1588A	41 09 06.4	123 30 50.2	Quartz- and biotite-bearing three-pyroxene monzodiorite
OMB1588B	41 09 06.4	123 30 50.2	Biotite-bearing three-pyroxene monzodiorite; mafic enclave
OMB1588C	41 09 06.4	123 30 50.2	Biotite-bearing three-pyroxene gabbro; mafic enclave
OMB1688	41 08 21.2	123 31 45.6	Quartz-bearing biotite three-pyroxene monzogabbro
OMB1788	41 06 31.7	123 34 14.0	Biotite three-pyroxene quartz monzodiorite
OMB1888	41 06 03.9	123 30 10.5	Biotite hornblende three-pyroxene quartz monzodiorite
OMB1988	41 04 56.0	123 30 50.7	Biotite three-pyroxene quartz monzodiorite
OMB491	41 06 55.1	123 34 14.0	Biotite three-pyroxene monzodiorite
OMB791	41 06 35.3	123 34 26.6	Altered biotite hornblende two-pyroxene diorite
OMB991	41 08 36.3	123 30 23.0	Quartz-bearing biotite three-pyroxene monzodiorite
OMB1191	41 08 13.3	123 29 00.5	Biotite hornblende three-pyroxene quartz monzonite
OMB1391	41 08 37.1	123 29 47.4	Hornblende lamprophyre dike
OMB1491	41 08 06.9	123 31 11.6	Biotite three-pyroxene quartz monzodiorite
OMB1591A	41 07 00.1	123 28 52.1	Biotite hornblende three-pyroxene quartz monzodiorite
OMB1591B	41 07 00.1	123 28 52.1	Three-pyroxene quartz monzodiorite
OMB1691	41 06 54.6	123 28 48.4	Biotite granite dike
OMB1891	41 06 29.3	123 28 59.4	Biotite hornblende two-pyroxene quartz monzonite
OMB2091	41.06 07.7	123 29 31.8	Olivine- and biotite-bearing gabbronorite
92OMB1	40 49 39.4	123 24 57.9	Augite hornblende appinite
92OMB3	40 50 08.3	123 25 50.5	Quartz-bearing biotite two-pyroxene diorite
92OMB4	40 50 04.7	123 26 03.5	Hornblende diorite
92OMB5	40 50 01.0	123 26 08.8	Hornblende-bearing biotite two-pyroxene quartz monzodiorite
92OMB7	40 50 04.7	123 26 32.7	Two-pyroxene diorite; mafic enclave
92OMB11	40 49 43.0	123 24 43.3	Two-pyroxene hornblende quartz monzodiorite
92OMB12	40 49 46.5	123 23 51.3	Augite hornblende quartz monzodiorite
92OMB22	40 53 35.5	123 27 58.6	Biotite two-pyroxene hornblende quartz monzodiorite
92OMB23	40 53 57.7	123 28 05.4	Biotite hornblende quartz monzodiorite
92OMB25	40 55 25.4	123 27 56.0	Pyroxene- and biotite-bearing hornblende quartz monzonite
92OMB32	40 58 54.5	123 25 56.3	Quartz- and K-feldspar-bearing hornblende biotite diorite
92OMB35	41 00 47.2	123 25 35.5	Biotite- and quartz-bearing hornblende three-pyroxene monzonite
92OMB36	40 59 53.7	123 25 35.4	Biotite hornblende granite
92OMB38	40 58 16.0	123 28 27.3	Augite-bearing biotite hornblende quartz monzonite
92OMB39	40 58 31.9	123 29 02.7	Biotite-bearing hornblende quartz monzodiorite
92OMB40	40 56 43.2	123 28 25.2	Biotite hornblende quartz monzodiorite
92OMB43	40 55 18.7	123 30 05.9	Biotite- and pyroxene-bearing hornblende quartz monzonite
92OMB44	40 55 30.2	123 30 05.2	Biotite-bearing pyroxene quartz monzonite
92OMB54	40 58 54.5	123 30 32.8	Biotite two-pyroxene hornblende quartz monzonite
92OMB55	40 55 50.4	123 29 04.8	Biotite two-pyroxene hornblende quartz monzonite
92OMB57	40 54 22.3	123 29 25.6	Biotite hornblende quartz monzonite
92OMB58	40 54 08.4	123 29 20.9	Hornblende biotite two-pyroxene quartz monzodiorite
92OMB61	40 47 23.7	123 24 49.6	Biotite-bearing pyroxene monzodiorite
92OMB67	40 44 54.9	123 23 25.7	Biotite hornblende two-pyroxene monzodiorite
92OMB70	40 44 48.5	123 25 38.0	Hornblende biotite granodiorite dike
92OMB113	41 06 23.5	123 28 39.0	Quartz-bearing two-pyroxene hornblende monzonite
92OMB207	41 04 10.8	123 30 01.0	Quartz-bearing biotite three-pyroxene monzodiorite
92OMB116A	41 05 51	123 27 48	Biotite two-pyroxene hornblende quartz monzodiorite
92OMB117	41 05 08	123 27 27	Biotite three-pyroxene diorite; mafic enclave
92OMB118	41 05 06	123 27 25	Biotite two-pyroxene quartz monzodiorite
92OMB121	41 04 37	123 27 21	Quartz-bearing biotite three-pyroxene monzodiorite
92OMB188	41 06 08	123 28 44	Biotite two-pyroxene quartz monzodiorite
92OMB220A	41 05 06	123 24 11	Olivine pyroxenite
92OMB220B	41 05 06	123 24 11	Gabbronorite
93OMB224	40 41 16	123 23 35	Biotite hornblende quartz monzodiorite
93OMB227	40 42 29	123 25 12	Biotite- and quartz-bearing two-pyroxene monzodiorite
Quartz diorite of Happy Camp Mountain			
92OMB19	40 53 22 8	123 28 04.9	Pyroxene-bearing biotite hornblende quartz monzonite
92OMB24	40 54 39.0	123 28 17.4	Biotite augite hornblende quartz diorite
93OMB247	40 53 22	123 26 50	Biotite two-pyroxene quartz diorite

TABLE 1. *Continued*

Sample	N latitude (° ′ ″)	W longitude (° ′ ″)	Rock type
West China Peak complex			
92OMB13	40 52 53	123 21 41	Quartz-bearing biotite augite diorite
92OMB14A	40 53 12	123 21 06	Olivine pyroxenite
92OMB15	40 53 02	123 21 09	Quartz-bearing hornblende biotite two-pyroxene diorite
92OMB15G	40 54 34	123 20 53	Biotite hornblende augite monzodiorite
Denny complex			
93OMB244	40 55 38	123 23 47	Hornblende biotite two-pyroxene quartz diorite
93OMB245	40 55 30	123 24 20	Two-pyroxene hornblende diorite

TABLE 2. RANGE OF PLAGIOCLASE COMPOSITIONS AND AVERAGE ALKALI FELDSPAR COMPOSITIONS

	Plagioclase maximum and minimum An			K-feldspar average composition (standard deviation)		
Sample	An	Ab	Or	An	Ab	Or
Ironside Mountain pluton						
OMB1688. Biotite three-pyroxene quartz-bearing monzodiorite						
max An	46.5	51.7	1.8	1.8	14.3	82.5
min An	27.2	70.1	2.7	(3.7)	(8.0)	(11.2)
OMB491. Biotite three-pyroxene monzodiorite						
max An	65.8	32.6	1.6			
min An	50.9	47.0	2.0			
OMB991. Biotite three-pyroxene quartz-bearing monzodiorite						
max An	55.7	42.9	1.4	2.9	17.2	78.7
min An	32.6	66.5	0.9	(4.4)	(7.5)	(11.5)
OMB1191. Biotite amphibole three-pyroxene quartz monzodiorite						
max An	48.7	50.0	1.3	1.6	15.1	82.6
min An	29.4	69.2	1.4	(3.5)	(9.4)	(12.4)
92.OMB116C. Pyroxene syenite						
max An	63.9	35.8	0.4	1.7	16.7	81.0
min An	14.2	83.6	2.3	(2.8)	(9.6)	(11.7)
93OMB221C. Biotite-bearing orthopyroxene augite monzodiorite						
max An	58.1	39.8	2.1	3.2	13.6	81.7
min An	47.7	49.6	2.8	(5.5)	(3.5)	(8.4)
93OMB224. Biotite amphibole quartz monzodiorite						
max An	45.1	53.5	1.4	0.5	15.9	82.1
min An	22.3	76.6	1.1	(0.7)	(4.9)	(8.4)
Quartz diorite of Happy Camp Mountain						
93OMB248. Biotite orthopyroxene augite diorite						
max An	50.9	47.6	1.5			
min An	35.8	62.8	1.5			
Denny complex						
93OMB242. Plagioclase-bearing olivine amphibole clinopyroxenite						
Average An	88.8	11.0	0.1			
93OMB243. Gabbronorite						
max An	72.9	26.6	0.6			
min An	50.0	48.7	1.3			
West China Peak dike in eastern Hayfork terrane						
93OMB278. Amphibolte augite diorite						
max An	88.90	10.89	0.21			
min An	25.05	73.83	1.12			

Notes: Ab = Na/(Ca + Na + K); An = Ca/(Ca + Na + K); Or = K/(Ca + Na + K).

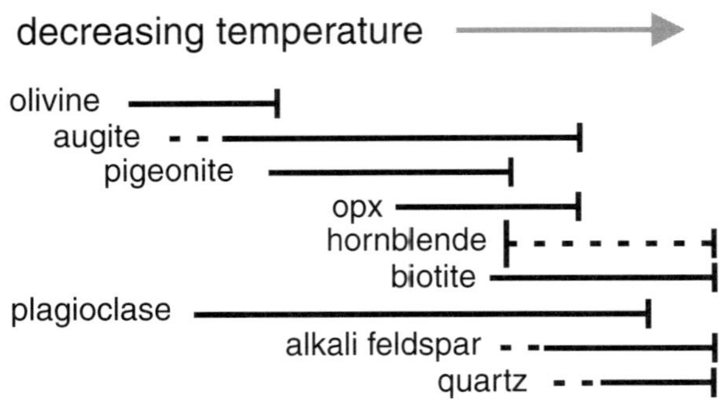

Figure 3. Inferred sequence of crystallization. Solid lines indicate mineral stability in all samples; broken lines indicate minerals present in some samples.

Denny Intrusive Complex

Pyroxenite in the Denny intrusive complex is medium to coarse grained, with subhedral to anhedral olivine (Fo_{72}; Table 3), subhedral augite, and interstitial calcic plagioclase (An_{88}; Table 2). Olivine and plagioclase are compositionally uniform within 1 mole %. Minor element concentrations in olivine (CaO, NiO, and MnO) are less than 0.06, 0.11, and 0.48, respectively. The remainder of the intrusion is predominantly gabbroic to quartz dioritic, with variable proportions of augite, orthopyroxene, hornblende, and biotite. Plagioclase from sample 93OMB243, an opx augite gabbro, is as calcic as An_{73}, and most plagioclase is more calcic than An_{60}; rim compositions are as sodic as An_{48} (Table 2). When compared to the Ironside Mountain pluton, the dioritic rocks of the Denny complex are distinct in their relative abundance of hornblende, the lack of inverted pigeonite, and lower abundances of alkali feldspar and Fe-Ti oxides.

West China Peak Intrusive Complex

Pyroxenite in the West China Peak intrusive complex is similar to that in the Denny intrusive complex. Olivine is slightly more magnesian than that in the Denny complex (Fo_{83} to Fo_{78}; Table 3), but is otherwise similar in composition. The remaining samples range from gabbro to monzodiorite; they are petrographically similar to dioritic and monzodioritic rocks of the Ironside Mountain pluton, including the presence of relict olivine in gabbroic and dioritic samples. A fine-grained hornblende augite gabbro dike adjacent to the West China Peak complex has plagioclase with An_{89} to An_{73} cores and An_{28} to An_{25} rims.

Mineral Compositions

The ranges of plagioclase compositions are given above for each rock type and in Table 2. Compositions of alkali feldspar and the mafic silicates are presented here.

Alkali Feldspar. Poikilitic and phenocrystic alkali feldspar is typically microperthitic and lacks polysynthetic twins. Average compositions were calculated from multiple line traverses across oikocrysts or phenocrysts (Table 2). Although the num-

TABLE 3. SUMMARY OF OLIVINE AND PYROXENE COMPOSITIONS

	Olivine	Clinopyroxene				Orthopyroxene			
Sample	Mg#	Mg#	Mg	Fe	Ca	Mg#	Mg	Fe	Ca
Ironside Mountain pluton									
OMB1688		0.638	0.358	0.203	0.439	0.522	0.509	0.467	0.024
OMB491		0.640	0.363	0.204	0.433	0.532	0.52	0.458	0.023
OMB991		0.621	0.351	0.214	0.436	0.492	0.479	0.494	0.028
OMB1191		0.579	0.323	0.235	0.443	0.437	0.424	0.546	0.03
92OMB116C		0.576	0.326	0.24	0.435	0.465	0.445	0.517	0.033
93OMB221		0.656	0.37	0.194	0.436				
Quartz diorite of Happy Camp Mountain									
93OMB248		0.633	0.351	0.203	0.446	0.482	0.471	0.505	0.024
Denny complex									
93OMB242	0.70–0.72	0.843	0.46	0.086	0.454				
93OMB243		0.754	0.424	0.139	0.437	0.615	0.601	0.376	0.022
West China Peak complex									
93OMB278		0.765	0.43	0.133	0.438				
93OMB280	0.79–0.82	0.864	0.463	0.073	0.464				

Note: Mg# = Mg/(Mg + Fe).

ber of points used to calculate these averages ranges from 24 to 129, standard deviations are still large. Despite these large standard deviations, the average compositions of alkali feldspars from a range of Ironside Mountain pluton rocks types (monzogabbro to syenite) are quite similar, with molar Or from 78.7 to 82.6, Ab from 14.3 to 17.2, and An from 0.5 to 3.2.

Pyroxene. All pyroxenes analyzed during this study are exsolved. In the Ironside Mountain pluton, augite contains orthopyroxene lamellae parallel to (001) and (010), which indicates exsolution of pigeonite followed by orthopyroxene. Inverted pigeonite contains augite lamellae parallel to (001) and (010), whereas primary orthopyroxene contains augite lamellae parallel to (010). In contrast, augite from the Denny and West China Peak complexes contains exsolution lamellae of orthopyroxene parallel to (010); orthopyroxene from the Denny complex exsolved augite on (010).

Because of the complex exsolution among pyroxenes in the batholith, detailed reconstruction of primary pyroxene compositions has not yet been undertaken. Initial compositional data of exsolved samples show lower values of Mg/(Mg + Fe) in the Ironside Mountain pluton compared to the Denny and West China Peak complexes (Table 3). For example, the Mg/(Mg + Fe) ratio in augite from the Ironside Mountain pluton ranges from 0.66 to 0.58. Augite in Denny and West China Peak gabbros and pyroxenites has Mg/(Mg + Fe) of 0.75–0.77 and 0.84–0.86, respectively. Orthopyroxene in the Ironside Mountain pluton ranges in Mg/(Mg + Fe) from 0.52 to 0.42, whereas orthopyroxene in Denny complex gabbro has Mg/(Mg + Fe) of 0.62.

Amphibole. Amphibole was analyzed from three samples of the Ironside Mountain pluton (Table 4). The amphibole shows a limited compositional range, with Mg/(Mg + Fe) from 0.47 to 0.36. Calcium (cations per formula unit) ranges from 1.9 to 1.7, Ti from 0.25 to 0.08, and total alkali content from 0.78 to 0.35 (most values >0.55). With few exceptions, Al content ranges from 1.67 to 1.42 per formula unit. According to the classification of Leake et al. (1997), these amphiboles straddle the boundary between ferropargasite and ferro-edenite (Fig. 4). Amphi-

TABLE 4. REPRESENTATIVE AMPHIBOLE AND BIOTITE ANALYSES

	Amphibole						Biotite					
	Ironside Mountain pluton		Denny Complex		West China Peak		Ironside Mountain pluton				HCM	WCP
Sample	OMB 1191	OMB 991	93OMB 224	93OMB 243	93OMB 278	93OMB 278	OMB 1688	OMB 491	OMB 991	OMB 1191	93OMB 248	93OMB 278
SiO_2	42.19	43.85	42.79	43.65	45.56	45.44	36.51	35.93	36.19	35.77	35.63	35.25
Al_2O_3	9.10	8.59	8.56	9.89	7.80	7.88	13.25	13.94	13.68	13.66	4.61	15.64
FeO	22.56	20.42	23.52	15.84	20.64	19.06	22.31	21.58	22.65	25.08	22.19	22.77
MgO	8.01	9.84	7.50	11.63	10.29	10.75	10.97	10.00	10.76	8.19	11.30	9.58
TiO_2	2.03	1.31	1.72	2.59	1.39	1.47	4.05	5.39	3.59	4.99	3.66	3.39
Cr_2O_3	0.00	0.00	0.00	0.03	0.00	0.02	0.01	0.03	0.00	0.00	0.03	0.00
MnO	0.39	0.33	0.70	0.22	0.42	0.37	0.17	0.07	0.11	0.22	0.19	0.19
CaO	12.14	11.96	10.59	11.23	10.86	10.86	0.00	0.00	0.00	0.03	0.09	0.00
Na_2O	1.55	1.46	1.47	1.56	1.24	1.38	0.04	0.02	0.13	0.19	0.02	0.07
K_2O	1.43	1.27	1.10	0.86	0.84	0.81	9.68	9.65	9.87	9.34	8.33	9.27
Cl	0.68	0.57	0.29	0.00	0.16	0.03	0.58	0.54	0.56	0.39	0.09	0.05
F	0.15	0.22	0.00	0.00	0.00	0.00	0.68	0.23	0.50	0.19	0.00	0.00
Total	100.03	99.58	98.15	97.49	99.17	98.04	97.84	97.15	97.72	97.88	96.11	96.21
	Cations per 23 oxygens						Cations per 22 oxygens					
Si	6.396	6.567	6.619	6.537	6.821	6.837	5.710	5.680	5.701	5.714	5.700	5.677
Al	1.627	1.517	1.560	1.747	1.377	1.397	2.443	2.598	2.541	2.571	2.754	2.968
Fe^{2+}	2.860	2.558	3.042	1.983	2.584	2.398	2.918	2.853	2.984	3.350	2.969	3.067
Mg	1.810	2.197	1.729	2.596	2.297	2.412	2.557	2.356	2.526	1.949	2.695	2.301
Ti	0.231	0.147	0.200	0.292	0.157	0.166	0.477	0.641	0.426	0.599	0.440	0.411
Cr	0.000	0.000	0.000	0.003	0.000	0.003	0.001	0.004	0.000	0.000	0.003	0.000
Mn	0.050	0.042	0.092	0.027	0.054	0.047	0.023	0.009	0.015	0.030	0.026	0.026
Ca	1.972	1.919	1.756	1.802	1.742	1.751	0.000	0.000	0.000	0.005	0.016	0.000
Na	0.457	0.423	0.440	0.453	0.360	0.402	0.013	0.007	0.040	0.060	0.008	0.023
K	0.277	0.242	0.216	0.164	0.161	0.155	1.931	1.946	1.984	1.904	1.699	1.905
Cl	0.174	0.144	0.075	0.001	0.041	0.008	0.154	0.143	0.148	0.107	0.025	0.014
F	0.074	0.105	0.000	0.000	0.000	0.000	0.337	0.117	0.251	0.097	0.000	0.000
Total	15.927	15.860	15.730	15.605	15.594	15.575	16.563	16.354	16.615	16.384	16.335	16.392
Mg#	0.388	0.462	0.362	0.567	0.471	0.501	0.467	0.452	0.458	0.368	0.476	0.429

Notes: HCM—quartz diorite of Happy Camp Mountain, WCP—West China Peak complex, Mg# = Mg/(Mg + Fe).

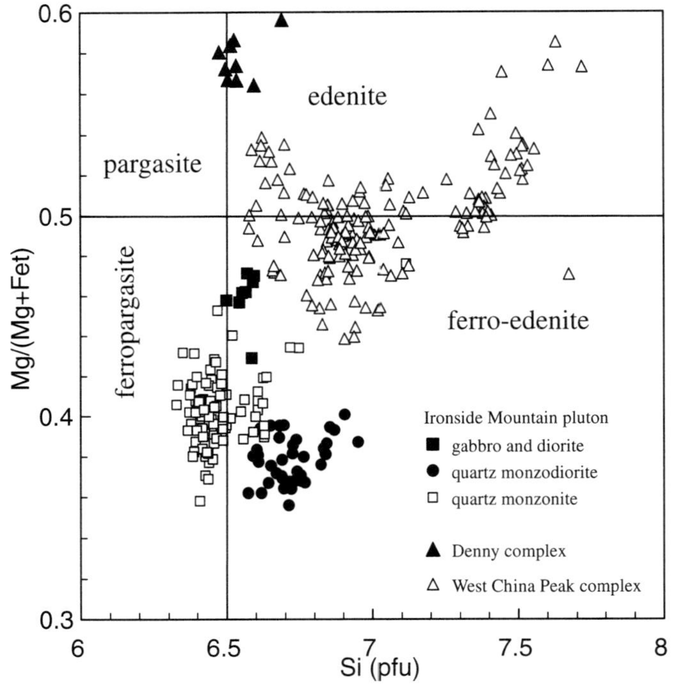

Figure 4. Amphibole classification. Fet indicates total Fe.

bole in a two-pyroxene gabbro from the Denny complex is edenitic (Fig. 4), and amphibole from a biotite hornblende quartz monzodiorite of the West China Peak complex is edenitic to ferro-edenitic. The fluorine content of Ironside Mountain pluton amphibole reaches 0.57 wt% and Cl contents reach 0.25 wt%. Halogen concentrations are below detection limits in amphiboles from other plutons in the batholith.

Biotite. As is typical of many plutonic rocks, biotite from individual samples of the Ironside Mountain pluton is homogeneous. It is characterized by relatively low Mg/(Mg + Fe) (Table 4): from 0.51 (monzogabbro) to 0.35 (quartz monzonite). The Ironside Mountain biotites also contain moderate halogen contents, with F from 0.21 to 0.58 wt% and Cl from 0.40 to 0.57 wt%. There is no obvious correlation between bulk composition and halogen content, although biotite from the most mafic samples shows the highest F and Cl contents. Biotite from the quartz diorite of Happy Camp Mountain has Mg/(Mg + Fe) of 0.47, similar to that of Ironside Mountain pluton samples, but has halogen contents at or below detection limits. Biotite from a West China Peak gabbro has similar Mg/(Mg + Fe) of 0.43 and undetectable halogen contents.

Estimates of P and H_2O Content

Hammarstrom and Zen (1986) and Hollister et al. (1987) showed that for samples with the assemblage calcic amphibole + plagioclase + biotite + K-feldspar + quartz + Fe-Ti oxide + titanite, the total Al content of amphibole crystallized near the solidus is a function of total pressure. Amphibole was analyzed from three samples of the Ironside Mountain pluton. All three samples contain the required mineral assemblage except for titanite, and all show petrographic features that suggest amphibole was stable at the solidus. The lack of titanite should not affect pressure estimates, because Ti content in the melt was buffered by ilmenite (e.g., Anderson and Smith, 1995). The calibration of Schmidt (1992) results in average pressure estimates of 4.1 ± 0.4, 4.4 ± 0.4, and 4.2 ± 0.2 kb for quartz monzonite, quartz monzodiorite, and quartz-bearing monzodiorite, respectively. These estimates are consistent with the presence of contact metamorphic cordierite + garnet in the aureole.

The highest An content in plagioclase from the Ironside Mountain pluton is in the An_{66} range. In contrast, intercumulus plagioclase in hornblende clinopyroxenite from the Denny complex reaches An_{88}. Gabbroic rocks from the Denny and West China Peak complexes have An_{73} and An_{89} plagioclase, respectively. Experimental data (Beard, 1986; Sisson and Grove, 1993) show that plagioclase crystallized from H_2O-rich magmas is calcic relative to plagioclase from similar, but H_2O-poor magmas. These data are consistent with the paucity of hydrous mafic minerals in the Ironside Mountain pluton compared to the Denny and West China Peak complexes, and they emphasize the H_2O-poor nature of the Ironside Mountain pluton.

GEOCHEMISTRY

Major and Trace Elements

Major element compositions of samples from the Ironside Mountain batholith are presented in Table 5 and illustrated in Figure 5. Data from Lanphere et al. (1968), Hotz (1971) and Charlton (1979) are also plotted. These plots are scaled to emphasize the trends of the main part of the batholith; thus, a fine-grained biotite granite dike from the northern part of the Ironside Mountain pluton (sample OMB1691; Table 5) is only plotted in Figure 5A–C, and olivine pyroxenites from the Denny and West China Peak intrusive complexes are not plotted. All but one sample from the Denny and West China Peak intrusive complexes contain less than 54 wt% SiO_2; the sample with higher SiO_2

Figure 5. Major element variation and geochemical classifications. (A) K_2O vs. SiO_2. Boundary lines from Le Maitre et al. (2002) and Rickwood (1989). Shaded field shows the range of compositions from the 165- to 156-Ma Wooley Creek belt plutons. (B) FeO/(FeO + MgO) vs. SiO_2, where FeO is total iron. Boundary between ferroan and magnesian fields from Frost et al. (2001). Shaded field shows the range of compositions from the 165- to 156-Ma Wooley Creek belt plutons; few compositions plot in the ferroan field. (C) Modified alkali-lime index versus SiO_2. See Frost et al. (2001). (D) TiO_2 vs. SiO_2. (E) CaO vs. SiO_2. Shaded field shows the range of compositions from the 165- to 156-Ma Wooley Creek belt plutons. (F) Na_2O vs. SiO_2. Data points outlined are central zone samples enriched in Na_2O and Sr relative to the rest of the Ironside Mountain pluton.

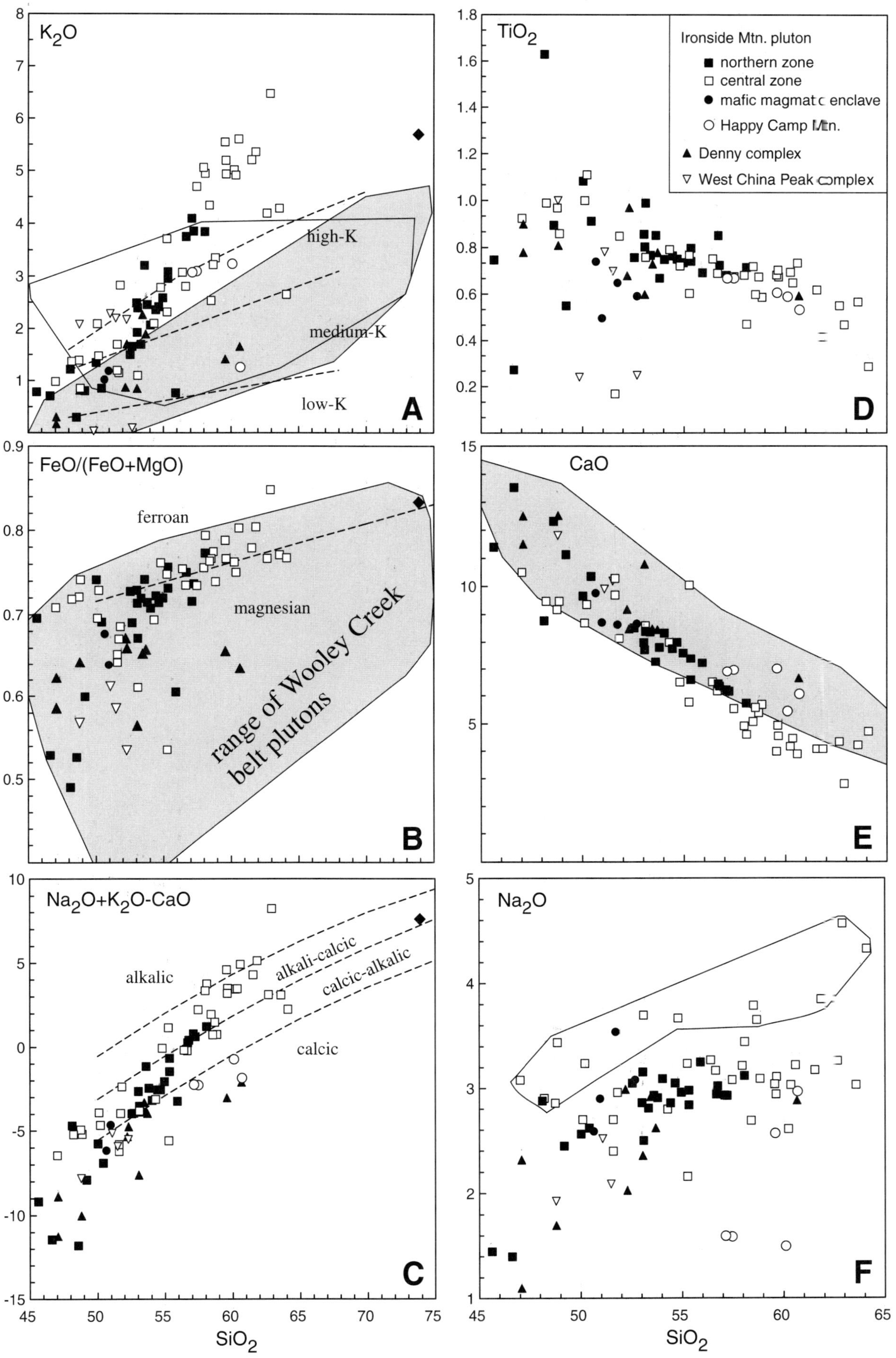

K2O
high-K
medium-K
low-K
A
TiO2
Ironside Mtn. pluton
northern zone
central zone
mafic magmatic enclave
Happy Camp Mtn.
Denny complex
West China Peak complex
D
FeO/(FeO+MgO)
ferroan
magnesian
range of Wooley Creek belt plutons
B
CaO
E
Na2O+K2O-CaO
alkalic
alkali-calcic
calcic-alkalic
calcic
C
Na2O
F
SiO2
SiO2

TABLE 5. REPRESENTATIVE MAJOR AND TRACE ELEMENT CONCENTRATIONS

	OMB 1588A	OMB 1588B	OMB 1688	OMB 1788	OMB 1888	OMB 491	OMB 991	OMB 1191	OMB 1491	OMB 1591A	OMB 1591B	OMB 1691
						Ironside Mountain pluton						
Major oxide concentrations (wt%)												
SiO_2	54.96	52.68	54.65	53.80	56.75	50.42	55.33	58.08	53.35	55.33	52.55	73.89
TiO_2	0.74	0.59	0.75	0.67	0.73	0.91	0.74	0.72	0.77	0.80	0.76	0.12
Al_2O_3	16.94	17.44	17.17	17.32	16.19	17.72	16.70	15.95	16.38	16.08	16.31	14.10
Fe_2O_3	2.60	2.04	2.61	3.01	1.97	2.69	1.79	2.10	3.64	2.23	2.72	0.13
FeO	6.72	7.65	6.99	6.25	6.64	8.12	7.26	6.57	6.95	6.99	8.05	1.09
MnO	0.18	0.22	0.20	0.19	0.18	0.20	0.19	0.17	0.22	0.18	0.25	0.02
MgO	3.52	4.25	3.73	3.57	3.03	4.70	3.25	2.48	3.98	2.89	3.92	0.24
CaO	7.60	8.67	8.00	7.81	6.35	10.37	7.39	5.74	8.38	6.58	8.52	1.07
Na_2O	2.96	3.08	3.05	2.91	3.02	2.62	2.98	3.12	2.81	2.84	3.05	2.97
K_2O	2.59	1.66	2.42	2.46	3.76	0.85	2.96	3.85	1.70	3.09	1.50	5.70
P_2O_5	0.44	0.28	0.41	0.34	0.41	0.13	0.40	0.42	0.50	0.47	0.44	0.05
LOI	1.83	1.73	1.06	1.72	1.31	0.97	0.99	1.35	1.28	1.16	1.39	0.99
Total	101.07	100.29	101.04	100.04	100.33	99.69	99.97	100.54	99.95	98.64	99.46	100.37
$Mg/(Mg + Fe^t)$	0.409	0.444	0.416	0.415	0.391	0.443	0.395	0.343	0.410	0.364	0.400	0.263
Trace element concentrations (ppm)												
Rb	73	47	62	65	104	19	74	106	48	86	36	183
Sr	654	695	699	626	591	580	644	472	640	516	517	377
Zr	87	48	91	41	109	35	142	218	31	141	69	93
Y	27.0	22.6	25.8	27.0	29.2	15.0	27.7	35.8	26.6	33.6	31.8	4.2
Nb	9	4	9	8	9	2	5	11	4	7	3	2
Ba	751	523	747	740	1357	424	1089	923	499	775	404	1427
Sc	24.0	28.8	24.7	25.4	21.2	33.9	25.1	22.0	30.2	26.1	34.8	1.2
V	215	206	227	222	170	380	203	160	285	209	242	11
Cr	16	22	18	15	19	42	36	20	27	13	25	b.d.
Ni	b.d.	b.d.	b.d.	b.d.	b.d.	b.d.	26	23	22	b.d.	b.d.	b.d.
Cu	120	53	57	75	88	60	81	114	174	147	88	7
Zn	94	94	98	96	97	93	86	87	105	94	99	15

Notes: Analyzed by inductively coupled plasma atomic emission spectrometry, Melanie Barnes, analyst. Abbreviations: b.d.—below detection limits, HCM—quartz diorite of Happy Camp Mountain, LOI—loss on ignition, n.d.—not analyzed.

content is quartz diorite. Samples from the quartz diorite of Happy Camp Mountain range from 57.1 to 61.7 wt% SiO_2 and have distinctly lower K_2O contents than most samples of the batholith at similar silica contents (Fig. 5A). The Ironside Mountain pluton ranges from 45.5 to 63.6 wt% SiO_2 (Fig. 5A). In these figures, the pluton is divided into northern and central parts, the boundary of which is arbitrarily placed at ~41°02′. The southern part of the pluton was not sampled.

The batholith is subalkaline according to the criteria of MacDonald and Katsura (1964). More detailed distinctions may be made with the geochemical classification of Frost et al. (2001), which shows that the trend of the Ironside Mountain pluton crosses the boundaries between calcic, calc-alkalic, and alkali-calcic rocks (Fig. 5C). In contrast, the other plutons in the batholith are nearly entirely calcic (Fig. 5C). In the Frost et al. (2001) classification scheme, many samples from the Ironside Mountain pluton plot along the boundary between ferroan and magnesian suites (Fig. 5B) and straddle the calc-alkaline/tholeiitic boundary of Middlemost (1975; not shown). According to the criteria of Miyashiro (1974), the pluton is tholeiitic, and its compositional trend is similar to the pigeonite series defined by Kuno (1968). The quartz diorite of Happy Camp Mountain and the Denny and West China Peak complexes are magnesian according to Frost et al. (2001). In Miyashiro's (1974) classification, the Denny and West China Peak compositions plot across the tholeiite/calc-alkaline boundary, but all samples of the quartz diorite of Happy Camp Mountain plot in the calc-alkaline field.

The range of K_2O contents in the Ironside Mountain pluton, which extends from medium to ultra-high K fields, clearly distinguishes it from other plutons in the province (Fig. 5A; Lanphere et al., 1968; Hotz, 1971; Barnes et al., 1992). At 65% SiO_2, few other plutons in the province contain 3 wt% K_2O and none contain 4 wt% (Barnes et al., 1992, 1995, 1996). The most evolved rocks of the pluton are also distinct from typical Klamath Mountain plutons (as characterized by the late Middle Jurassic plutons of the Wooley Creek belt; Barnes et al., 1992) in their relatively Fe-rich and Ca-poor nature (Fig. 5B and E).

OMB 1981	92OMB 3	92OMB 4	93OMB 221	92OMB 24	93OMB 247	93OMB 244	93OMB 245	93OMB 242	93OMB 280	92OMB 13	92OMB 15
Ironside Mountain pluton			**HCM**			**Denny complex**			**West China Peak**		
57.23	48.86	47.01	50.55	60.69	59.58	60.65	53.07	50.46	51.81	51.07	48.78
0.67	0.86	0.93	0.79	0.53	0.61	0.59	0.60	0.29	0.23	0.78	1.00
15.68	20.07	20.15	17.44	15.59	16.15	16.49	15.09	3.41	2.30	14.97	13.46
2.41	10.29	11.86	11.03	6.28	7.73	6.84	9.40	7.46	7.53	10.62	11.39
6.01	n.d.	n.d.	n.d.	n.d.	n.d.	n.d.	n.d.	n.d.	n.d.	n.d.	n.d.
0.17	0.18	0.21	0.19	0.12	0.14	0.13	0.18	0.16	0.15	0.20	0.21
2.92	3.22	4.39	4.9	3.42	3.64	3.54	6.50	16.81	17.15	6.05	7.78
6.16	9.48	10.51	10.25	6.07	7.00	6.64	10.80	20.72	19.59	9.92	11.81
2.93	3.44	3.08	2.97	2.97	2.57	2.88	2.36	0.24	0.23	2.52	1.93
3.86	0.86	0.98	1.37	1.26	1.42	1.66	0.85	0.04	0.03	2.29	2.09
0.36	0.96	0.51	0.35	0.14	0.24	0.16	0.21	0.01	0.02	0.42	0.73
1.47	1.50	1.57	0.26	2.05	1.45	0.75	0.95	0.68	0.71	0.50	1.12
99.87	99.72	101.20	100.10	99.12	100.52	100.35	100.01	100.28	99.75	99.34	100.09
0.389	0.383	0.423	0.468	0.519	0.483	0.506	0.578	0.817	0.819	0.530	0.575
114	18	17	24	32	34	36	16	b.d.	b.d.	53	58
556	1476	1066	892	437	625	557	594	105	62	736	962
178	17	44	43	103	96	87	46	6	2	70	45
27.2	21.6	34.0	21.7	19.8	21.1	19.0	20.2	11.4	8.4	24.1	29.9
8	3	3	3	6	4	3	2	b.d.	b.d.	5	7
799	820	367	654	639	555	830	436	23	20	688	1093
22.5	17.2	35.8	35.3	19.0	20.4	24.3	38.8	n.d.	103	35.3	35.5
180	180	255	310	164	208	134	265	n.d.	152	234	290
32	17	20	42	72	59	50	148	n.d.	1295	129	228
23	24	b.d.	b.d.	49	b.d.	27	25	n.d.	152	50	48
106	6	45	65	54	89	10	39	n.d.	20	39	115
96	90	88	85	64	91	83	81	n.d.	25	84	92

Like most plutons in the province, Na_2O contents show considerable scatter (Fig. 5F). However, much of the scatter can be resolved by recognition of two groups within the pluton. One group is characterized by higher Na_2O over a range of SiO_2 values; it is outlined in Figure 5F. These samples are also richer in Sr and Al_2O_3. Reasons for this compositional distinction are discussed below.

Variations in selected trace element concentrations are shown in Figure 6. The smaller number of data points compared to Figure 5 is because not all data taken from the literature included trace element values.

The concentration of Rb in the Ironside Mountain pluton shows continuous increase with increasing SiO_2 (Fig. 6A). The Zr and Ba contents also increase (Fig. 6D and E, respectively), but both show considerable scatter, particularly at SiO_2 contents higher than ~55 wt%. Furthermore, many samples with the highest Ba concentrations belong to the group of Na_2O-rich rocks illustrated in Figure 5F. If this group is excluded from the compositional trend, then Ba concentrations increase until SiO_2 concentrations reach ~56 wt%, after which little variation in Ba is observed.

Strontium, V, and Sc (Fig. 6B, C, and F) all decrease with increasing SiO_2, as do TiO_2 (Fig. 5D), P_2O_5, and Zn (not shown). Each element shows broad scatter at SiO_2 contents below ~54 wt%, and most show reasonably well-defined trends at higher silica contents. Although Sr concentrations scatter widely, if the group of Na_2O-rich samples discussed is excluded (data points outlined in Fig. 6B), Sr concentrations clearly decrease at SiO_2 contents above 54 wt%. Except for pyroxene-rich cumulates, Sc contents are less than 40 ppm and decrease among samples with SiO_2 >54 wt%. The element Cr (Table 5) is similarly enriched in the pyroxenites, but otherwise is less abundant than 80 ppm and is not correlated with SiO_2. Zr concentrations reach 120 ppm in some pyroxenites, are in the 90–100 ppm range in dioritic and monzodioritic rocks, and decrease to ~60 ppm in the most evolved samples (Table 5).

Figure 7A shows rare-earth element patterns for dioritic to quartz monzonitic rocks from the northern part of the Ironside

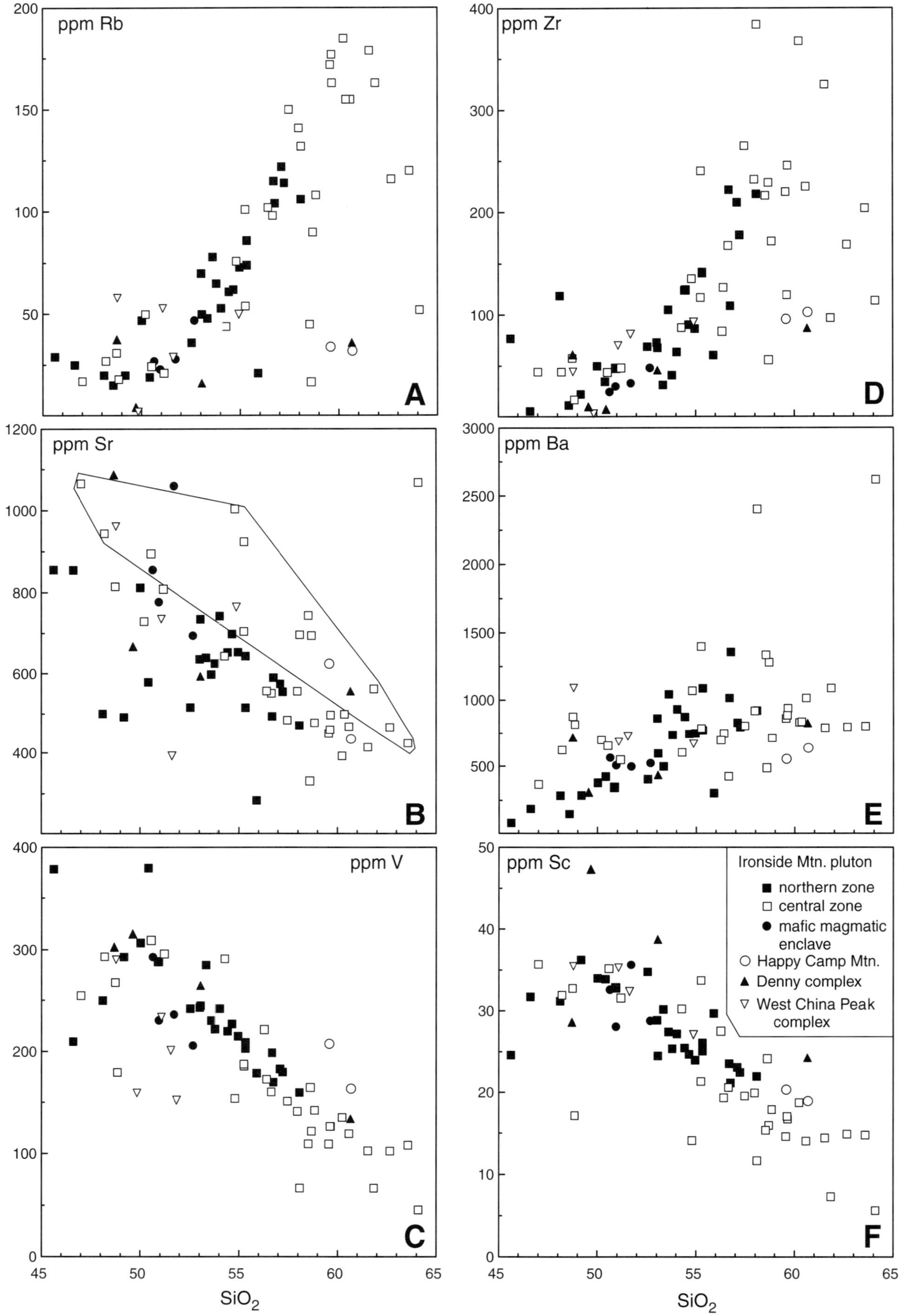
ppm Rb
A
ppm Sr
B
ppm V
C
ppm Zr
D
ppm Ba
E
ppm Sc
F
SiO2
SiO2
Ironside Mtn. pluton
northern zone
central zone
mafic magmatic enclave
Happy Camp Mtn.
Denny complex
West China Peak complex

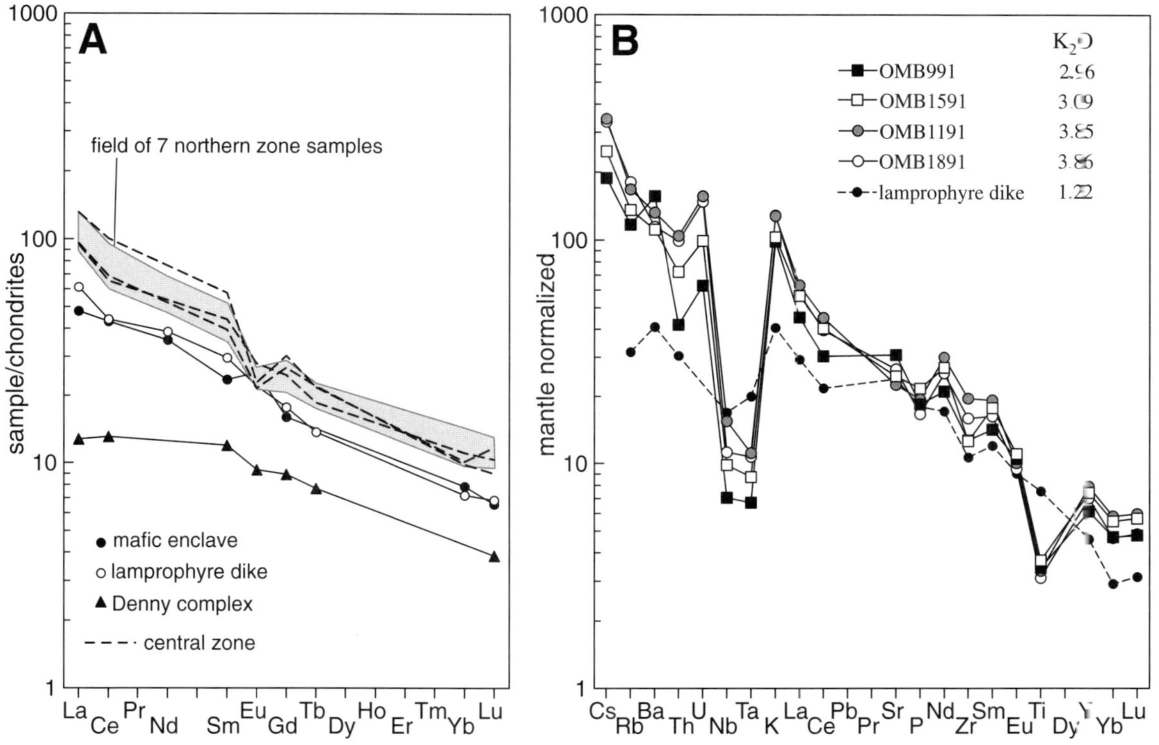

Figure 7. (A) Chondrite-normalized rare-earth element diagrams. The shaded field represents seven samples from the northern zone of the Ironside Mountain pluton; the dashed patterns are for samples from the central part of the pluton. The mafic enclave (OMB1588C) and lamprophyric dike are from the northern zone. (B) Mantle-normalized (Sun and McDonough, 1989) multi-element patterns. The key shows sample numbers and K_2O contents.

Mountain pluton (Table 6). A mafic magmatic enclave (sample OMB1588C) and a hornblende lamprophyre dike (OMB1391) are also plotted. Except for the enclave and dike, the rare-earth element patterns are quite similar, with moderate negative slopes and a negative Eu anomaly. Rare-earth element concentrations broadly increase as a function of increasing SiO_2 (or K_2O). The pattern of the mafic magmatic enclave is similar to that of the host pluton, but with lower rare-earth element abundances and a positive Eu anomaly. The pattern of the lamprophyric dike is similar to the enclave except that the dike sample lacks an Eu anomaly.

The mantle-normalized multi-element diagram (Fig. 7B, Table 6) also shows tightly clustered patterns for four samples of the northern Ironside Mountain pluton. Compared to their neighbor elements, U and K show positive anomalies and Nb, Ta, Ti, and P(?) show negative anomalies. The lamprophyric dike shows similar, but more subdued negative anomalies except for Ti (Fig. 7B).

←

Figure 6. Trace element variation as a function of SiO_2. In B, data points outlined are central zone samples enriched in Na_2O and Sr relative to the rest of the Ironside Mountain pluton. See text for discussion.

Isotopes

Initial $^{87}Sr/^{86}Sr$ values for eleven samples from the Ironside Mountain pluton (Table 7, calculated at 170 Ma) range from 0.70364 to 0.70397. These data are in agreement with initial $^{87}Sr/^{86}Sr$ values of 0.7036 reported by Masi et al. (1981). One

TABLE 6. TRACE ELEMENT COMPOSITIONS

Sample	OMB991	OMB1191	OMB1591A	OMB1891
La	30.93	43.11	38.53	39.53
Ce	53.55	79.53	71.16	70.15
Nd	28.36	40.4	36.33	34.14
Sm	6.293	8.505	7.825	7.204
Eu	1.76	1.683	1.851	1.589
Tb	0.833	1.051	0.952	0.8407
Yb	2.31	2.86	2.709	2.285
Lu	0.352	0.44	0.42	0.3569
Cs	1.48	2.71	1.94	2.63
Hf	3.53	4.58	3.69	4.05
U	1.31	3.27	2.08	3.1
Th	3.55	8.86	6.12	8.42
Ta	0.273	0.456	0.356	0.439

Notes: Analyses by instrumental neutron activation analysis at the University of Texas, El Paso. Concentrations in ppm.

TABLE 7. Nd AND Sr ISOTOPIC DATA FOR THE IRONSIDE MOUNTAIN BATHOLITH

Sample	Rb (ppm)	Sr (ppm)	$^{87}Rb/^{86}Sr$	$^{87}Sr/^{86}Sr$	Initial $^{87}Sr/^{86}Sr$	Sm (ppm)	Nd (ppm)	$^{147}Sm/^{144}Nd$	$^{143}Nd/^{144}Nd$	Initial $^{143}Nd/^{144}Nd$	Initial ε_{Nd}	Nd model age†
University of Wyoming*												
OMB1491W	32.11	509.37	0.1824	0.70438	0.703935	4.180	19.98	0.12652	0.512820	0.512679	5.08	580 Ma
OMB1788	49.87	543.68	0.2619	0.70444	0.703806	5.700	29.33	0.11753	0.5128245	0.512694	5.36	520 Ma
92OMB39W	93.96	637.20	0.4266	0.70494	0.703905	6.083	33.00	0.11147	0.512815	0.512691	5.30	500 Ma
92OMB13W	40.14	653.19	0.1778	0.70484	0.704412	5.393	27.59	0.11819	0.5127211	0.512590	3.33	690 Ma
U.S. Geological Survey§												
OMB1588A	73.8	644	0.329	0.70444	0.70364							
OMB1588C	34.2	859	0.115	0.70392	0.70364							
OMB1688	63.3	701	0.126	0.70427	0.70397							
OMB1888	97.5	589	0.48	0.70485	0.70369							
OMB1191	104	471	0.639	0.70525	0.70371							
OMB1591B	35.4	541	0.189	0.704213	0.70376							
OMB1691	179	371	1.397	0.707015	0.70364							
OM1891	111	581	0.553	0.705071	0.70373							

*Analytical details: ~80 to 100 mg of sample were dissolved in $HF-HNO_3$. After conversion to chlorides, one-third of the sample was spiked with ^{87}Rb, ^{84}Sr, ^{149}Sm, and ^{146}Nd. Elements Rb, Sr, and rare-earth elements were separated by conventional cation-exchange procedures; Sm and Nd were further separated in di-ethyl-hexyl orthophosphoric acid colums. All isotopic measurements were made on a VG Sector multi-collector mass spectrometer at the University of Wyoming. An average $^{87}Sr/^{86}Sr$ isotopic ratio of 0.710251 ± 20 (2σ) was measured for NBS 987 Sr, and an average $^{143}Nd/^{144}Nd$ ratio of 0.511846 ± 11 (2σ) was measured for the LaJolla Nd standard. Uncertainties in Sr isotopic ratio measurements are ±0.00002 and uncertainties in Nd isotopic ratio measurements are ±0.00001 (2σ). Blanks are <50 pg for Rb, Sr, Nd, and Sr, and no blank correction was made. Uncertainties in Rb, Sr, Nd, and Sm concentrations are ±2% of the measured value; uncertainties on initial ε_{Nd} = ±0.5 epsilon units. Initial Sr and Nd isotopic ratios and initial ε_{Nd} values are calculated for 170 Ma.

†Nd model ages are calculated based upon the depleted mantle model of Goldstein et al. (1984).

§See Kistler et al. (1986) for analytical details. Analyst: R.W. Kistler.

biotite augite quartz-bearing diorite from the West China Peak intrusive complex has initial $^{87}Sr/^{86}Sr$ (170 Ma) of 0.70441.

Initial ε_{Nd} values were determined for three Ironside Mountain pluton samples (Table 7); they range from 4.2 to 6.1 (calculated at 170 Ma), and the West China Peak diorite has ε_{Nd} of 5.4. Two whole-rock $\delta^{18}O$ values were reported for the Ironside Mountain pluton by Masi et al. (1981); they are 7.7 and 8.8‰.

When compared to isotopic data for slightly younger, but pre-Nevadan Klamath plutons (e.g., Wooley Creek belt plutons; Masi et al., 1981; Barnes et al., 1990, 1992, 1995; Gribble et al., 1990), the Ironside Mountain pluton is characterized by some of the lowest initial $^{87}Sr/^{86}Sr$, highest ε_{Nd}, and lowest $\delta^{18}O$ values. Only a few samples from the Grayback pluton (Barnes et al., 1995) have more primitive isotope values.

DISCUSSION

Compositional Variation in the Ironside Mountain Pluton

Magmatism responsible for the Ironside Mountain batholith is unusual in the Klamath Mountains in a number of ways. First, it occurred shortly after regional thrusting along the Wilson Point thrust. As such, we interpret Ironside Mountain magmatism to mark the beginning of a 15-m.y. episode of magmatic activity that encompassed the Ironside Mountain batholith and related plutons, the large plutons of the Wooley Creek belt, the Josephine ophiolite, and the magmatic arc characterized by the Rogue Formation and the Illinois River gabbro. This episode ended with the onset of Nevadan thrusting at ca. 155 Ma (Harper and Wright, 1984).

Second, relative to typical Klamath magmatism, magmas of the Ironside Mountain pluton evolved toward high K_2O contents and high FeO/(FeO + MgO). In the SiO_2 range of 50–60%, no other plutonic sequence in the province extends to ultra-high K_2O compositions and few (e.g., sparse tholeiitic enclaves and synplutonic dikes in the Grayback pluton: Barnes et al., 1995) have FeO/(FeO + MgO) values greater than 0.7. Furthermore, the most evolved rocks in the pluton have SiO_2 contents of ~64% SiO_2, whereas evolved rocks in the younger plutonic sequences reach 75% SiO_2 (e.g., Barnes et al., 1992, 1996).

Third, unlike nearly all other large Jurassic plutons in the province, the Ironside Mountain pluton lacks significant evidence of magma mixing or mingling. This observation is supported by the lack of reverse zoning of plagioclase, the absence of synplutonic mafic dikes, and the paucity of mafic enclaves.

Mafic rocks in the batholith are also characterized by mafic mineral assemblages indicative of low to very low H_2O contents (e.g., two- or three-pyroxene ± olivine); this assemblage changes gradually to biotite + hornblende in the most evolved rocks (Table 1). Clearly, both biotite and hornblende were only stable late in the crystallization sequence (Fig. 3). Although such assemblages constitute parts of larger plutons in the Wooley Creek belt (e.g., Barnes, 1983; Cotkin and Medaris, 1993) and the Bear Mountain intrusive complex (Snoke et al., 1981; Bushey et al., this volume), the amount of H_2O-poor magma represented by the Ironside Mountain pluton is many orders of magnitude larger.

Another striking characteristic of the Ironside Mountain pluton is the relatively coherent compositional trend. In Figures 5 and 6, this trend encompasses samples collected over an ~50-km-long strike length. Although some hint of zoning is present in the northern part of the pluton (Fig. 1), there is little evidence for compositional zoning elsewhere in the body, except that the most evolved samples were collected in the central part of the pluton. Even so, the central zone contains all of the lithologic types present in the pluton. This lack of obvious zoning could be the result of piecemeal emplacement of numerous magma batches, all of which cooled slowly enough to have similar textural features. This possibility cannot be disproved on geochemical grounds, but it is noteworthy that in areas with good exposure (e.g., east-central parts of the northern zone), internal intrusive contacts were not observed.

If the Ironside Mountain pluton represents magma emplacement over a relatively short time, which is consistent with field relationships, then three simple models may be considered. The first is that compositional variation in the pluton arose via crystal-liquid separation (fractional or equilibrium crystallization), with or without crystal accumulation and magma recharge. The second is that such variation results from variable degrees of partial melting of rather uniform source rocks, and the third is that the variation is the result of variable proportions of mixing between mafic and felsic end members.

The third possibility is unlikely for at least two reasons. First, initial $^{87}Sr/^{86}Sr$ values are uncorrelated with fractionation (e.g., Rb content; Table 7). In fact, the narrow range of initial $^{87}Sr/^{86}Sr$ and ε_{Nd} of the Ironside Mountain pluton is distinct compared to all other large Middle Jurassic plutons in the province, all of which have significant crustal components (e.g., Gribble et al., 1990; Barnes et al., 1992). Second, many binary element plots show curved trends, which is not compatible with simple mixing. In addition, textural and field evidence for mixing (e.g., mantled and disequilibrium crystals; Anderson, 1976) or mingling (e.g., mafic magmatic enclaves) are nearly absent in the pluton.

The first and second possibilities may be tested in binary plots in Figure 8A and B. These diagrams plot the ratio of strongly incompatible element/moderately incompatible element on the *y*-axis against a strongly incompatible element on the *x*-axis (Treuil et al., 1977). For fractional crystallization, the ratio plotted on the *y*-axis shows little variation (near zero slope), whereas for equilibrium partial melting, the ratio has a distinct positive slope. In Figure 8, Rb is used as the "perfectly" incompatible element and Zr and K_2O are used as incompatible elements with slightly larger bulk partition coefficients. The Rb/Zr (Fig. 8A) and Rb/K_2O (Fig. 8B) ratios show little variation; they are not consistent with an origin by partial melting.

We are left with the likelihood that compositional variation in the Ironside Mountain pluton resulted primarily from crystal-

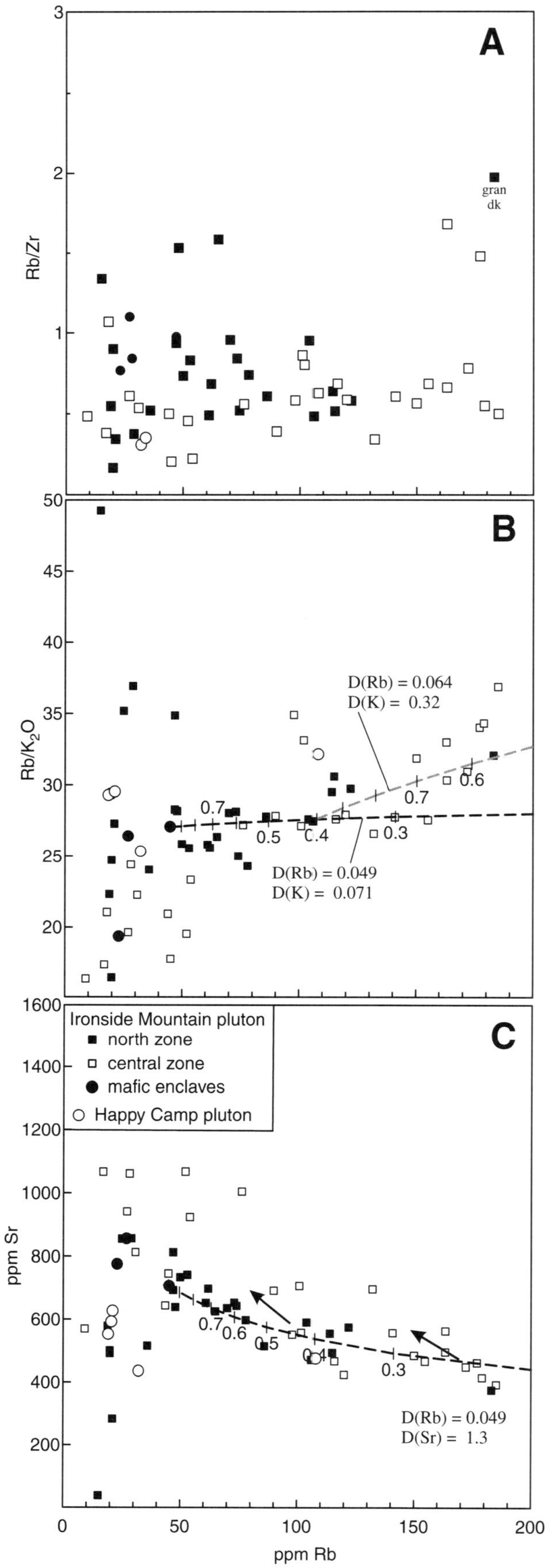

liquid separation ± crystal accumulation. In systems such as the Ironside Mountain pluton, the compositional effects of accumulation can be particularly difficult to determine, because the curvature of compositional trends is not large and there are few fine-grained rocks that might indicate a likely parental composition. The two extreme cases are that the composition of the parental magma was (1) among the most evolved ones observed, so that the compositional trend reflects mixtures of accumulated crystals and trapped liquid (e.g., Charlton, 1979), and (2) among the most mafic ones observed, so that most rock samples represent solidified liquids.

The first case is unlikely, because the mineral assemblage in the most evolved rocks in the pluton is not equivalent to the mineral assemblage in the most mafic rocks, and because plagioclase An contents decrease with increasing bulk rock SiO_2 (see the section on petrography). The second case is also unlikely for the following reasons. First, with few exceptions, major and trace element plots (Figs. 5 and 6) show the most scatter below ~53% SiO_2. At higher SiO_2 contents, most element plots show well-defined trends. This is particularly true for the compatible elements, such as P_2O_5, MnO, V, Zn, and TiO_2, and suggests that the low-SiO_2 rocks consist of variable proportions of accumulated plagioclase and mafic minerals. The plot of Sc versus SiO_2 (Fig. 6F) is especially instructive. The element Sc is strongly compatible in pyroxene and petrographic evidence shows that one or more pyroxenes crystallized at high temperature. Therefore, if the most mafic rocks in the pluton represent parental compositions, they should have the highest Sc contents, and Sc should decrease monotonically with increasing SiO_2. Instead, at low SiO_2 contents, Sc concentrations are approximately constant to ~53% SiO_2 (Fig. 6F). At higher SiO_2 contents, Sc abundances decrease. This pattern suggests that many rocks with <53% SiO_2 are partial cumulates, presumably of a combination of plagioclase and pyroxene, such that the abundance of cumulate plagioclase diluted Sc concentrations. Most of these samples lack quartz and either lack K-feldspar or contain interstitial (rather than poikilitic or subhedral) K-feldspar. In contrast, nearly all samples with ≥53% SiO_2 contain poikilitic

Figure 8. (A) Variation of Rb/Zr ratio as a function of Rb content. (B) Variation of Rb/K_2O as a function of Rb. (C) Variation of Sr as a function of Rb. In B and C, dashed lines represent fractionation models for the Ironside Mountain pluton, assuming a parental basaltic andesite with 45 ppm Rb, 710 ppm Sr, and 1.66 wt% K_2O (equivalent to ~53% SiO_2). This composition is similar to mafic enclave sample OMB1588B (Table 5). Bulk partition coefficients, D, are shown for each element. Tic marks and numbers along the curves represent the fraction of melt remaining. Fractionation of Fe-Ti oxides would not change the shape of the curves but would increase fractionation efficiency (i.e., the space between tic marks would increase). In B and C, the black dashed lines represent a fractionating assemblage of 58% plagioclase, 29% augite, and 13% low-Ca pyroxene. In B, the gray dashed line represents a fractionating assemblage of 55% plagioclase, 5% K-feldspar, 28% augite, and 12% low-Ca pyroxene. In C, arrows indicate the effect of plagioclase accumulation.

to subhedral K-feldspar and interstitial quartz. Thus, the lower SiO_2 samples are interpreted to have lost interstitial melt. In addition to accumulation in the low-SiO_2 samples, a group of samples from the central zone of the pluton also stands out in terms of its higher Na_2O and Sr abundances (Figs. 5 and 6). These higher concentrations are consistent with accumulation of intermediate plagioclase.

Figures 8B and C show simple fractional crystallization models for Rb, Sr, and K_2O in the Ironside Mountain pluton. A fractionating assemblage of plagioclase, augite, and low-Ca pyroxene from a parent with ~53% SiO_2 readily fits the general trend of Rb versus Sr (Fig. 8C). This assemblage also fits the Rb and K_2O data, except among the most differentiated samples (Fig. 8B). These evolved samples can be explained if K-feldspar is added to the fractionating assemblage at Rb contents of ~100 ppm (K_2O contents of ~3.3%). This minor change in assemblage would have no appreciable effect on the trend plotted in Figure 8C.

We conclude that compositional variation in the Ironside Mountain pluton resulted from a combination of fractional crystallization and local accumulation of plagioclase ± pyroxene. The data are consistent with a parental magma similar to mafic enclave OMB1588B (Table 5); a basaltic andesite with ~53% SiO_2, 17.4% Al_2O_3, and 4.3% MgO.

Parental Magma

Discussion of the origin of parental magma(s) for the Ironside Mountain pluton focuses on two themes, the subduction-related origin of the magma and the causes for the parental magma compositions. Figure 7 clearly shows all of the signatures expected of supra-subduction zone magmatism. The rocks are characterized by low abundances of Nb, Ta, and TiO_2 and by high abundances of large ion lithophile elements. The resultant trace element patterns are characteristic of arc magmas and imply a source region modified by transfer of slab-derived fluids and/or melts.

Understanding the origin of the parental magma is more problematic because the rocks do not fit precisely into common pigeonholes for arc magmas. For example, parts or all of the pluton are classified as ferroan or tholeiitic, which is consistent with the inferred low H_2O contents. However, arc tholeiite is generally poor in K_2O and Rb rather than K_2O-rich (e.g., Gill, 1981). In contrast, typical potassic arc magmas (e.g., shoshonites) are defined by magnesian compositions and low FeO/(FeO + MgO) (e.g., Müller and Groves, 1995).

The inferred parental magma has FeO/(FeO + MgO) values too high for the magma to be in equilibrium with a mantle assemblage. This observation is also consistent with its low Cr and Ni contents. In addition, Figures 5 and 6 show that total Fe, TiO_2, Cr, and V all decrease with increasing SiO_2. These trends suggest that, along with plagioclase and pyroxenes, Fe-Ti oxide phases were stable, which indicates that the magma underwent significant differentiation prior to emplacement. This deeper-level differentiation was probably responsible for Fe enrichment and probably involved fractionation of olivine, plagioclase, clinopyroxene, and Cr-spinel from a parental, mantle-derived, low fO_2, low H_2O arc tholeiite. Significant H_2O content would result in evolution along a calc-alkaline line of descent (e.g., Grove and Kinzler, 1986). Finally, low H_2O contents are consistent with the petrographic observation that plagioclase was in reaction with late-stage melts in the Ironside Mountain pluton (e.g., Nekvasil et al., 2000).

This two-stage differentiation scenario is schematically illustrated in Figure 9. If it is correct, then one explanation for the high K_2O content of the magmas arises from experimental studies of Meen (1990). He showed that under relatively low pressure conditions, olivine is the dominant fractionating mafic mineral and that its removal causes SiO_2 enrichment. At higher pressure (8–10 kb), augite fractionation is more important, with consequent increase in abundances of elements such as K_2O and Rb at nearly constant SiO_2 values. The implication of this two-stage differentiation processes is that the mantle-derived magmas must have ponded near the base of relatively thick crust (Fig. 9).

Evidence for a thick Klamath crust during Middle Jurassic time comes from an unusual set of 160-Ma andesitic dikes (Barnes and Allen, this volume). These dikes contain phenocrysts of garnet, plagioclase, Al-rich amphibole, quartz, and epidote. The mineral assemblage and mineral compositions are consistent with equilibration pressures in the 7- to 9-kb range, and the presence of quartz phenocrysts indicates that the assemblage formed in the crust, not in the mantle.

There is no evidence for regional contractional deformation between Ironside Mountain magmatism and emplacement of the garnet-bearing andesite dikes at 160 Ma. Therefore, it is unlikely that crustal thickness increased (except possibly by magmatic underplating) after Ironside Mountain time. Hence, initial high-P (Moho-level?) differentiation of the mantle-derived precursor to the Ironside Mountain parental magma is consistent with available data.

Why Do Wooley Creek Belt Plutons Differ from the Ironside Mountain Pluton?

In the transition from Ironside Mountain magmatism (170–168 Ma) to Wooley Creek belt magmatism (165–156 Ma), magma compositions changed from predominantly K-and Fe-rich, reducing, and H_2O-poor to almost entirely calc-alkaline to calcic, H_2O-rich, and oxidized. Hornblende became an important early phase in crystallization sequences, crustally derived magmas became important parts of the suite, and mixing of mantle and crustal magmas became commonplace (Barnes, 1983; Barnes et al., 1986, 1990, 1995; Gribble et al., 1990). Because Ironside Mountain magmatism followed regional thrusting (Wilson Point thrust) and because no regional contractional deformation separated Ironside Mountain and Wooley Creek belt activity, we surmise that the two magma suites ascended

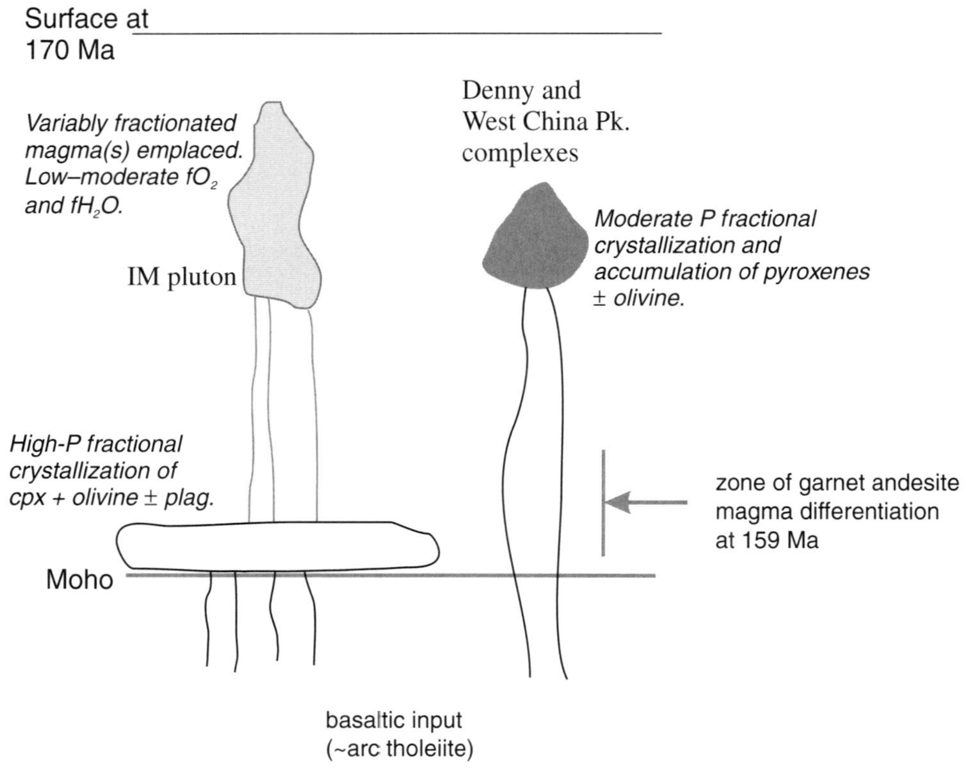

Figure 9. Schematic illustration of the differentiation stages of the Ironside Mountain batholith. Low-fO_2, low-H_2O magmas parental to the Ironside Mountain pluton ponded near the base of the crust. High pressure (7–9 kb) fractionation of augite ± olivine resulted in K_2O enrichment relative to SiO_2. Magmas parental to the Denny and West China Peak complexes were richer in H_2O, had higher fO_2, and underwent little or no differentiation in the deep crust.

through the same structural sequence. Why, then, are they so distinct?

In addition to the possibility of distinct parental magmas (and magma sources), many other explanations have been proposed for the transition from tholeiitic to calc-alkaline magmatism. They include magma mixing (Sakuyama, 1981), increase in H_2O content (Grove and Kinzler, 1986), and crustal contamination (Grove and Baker, 1984; Grove and Kinzler, 1986; but see Hunter, 1998). In the present case, there is evidence that all three explanations are possible: each Wooley Creek belt pluton shows evidence for mixing, predominantly high H_2O contents, and crustal contamination (e.g., Barnes et al., 1990; Gribble et al., 1990; Schmidt, 1994; Johnson and Barnes, this volume). Without discounting the idea of distinct magma source regions, we suggest that these differences also arose because of the changing thermal structure of Klamath Mountain crust from Middle to late Middle Jurassic time.

Because Ironside Mountain magmatism followed regional thrusting, it is likely that the crustal section through which early Ironside Mountain magmas passed was relatively cool and not preconditioned by earlier western Hayfork magmatism. This scenario is supported by the paucity of large western Hayfork-age plutons in the province. As a result, the Ironside Mountain magmas were emplaced into and at the base of a relatively cool crustal section. With succeeding magmatism, the crustal section became hotter and ultimately underwent partial melting during Wooley Creek time.

In this scenario, interaction between the cool crustal section and Ironside Mountain magmas would have been difficult, owing to the amounts of heat necessary to raise crustal temperatures to conditions where assimilation was possible (e.g., Spera and Bohrson, 2001). Instead, these early magmas would have evolved primarily by fractional crystallization ± magma mixing and mingling. However, as the magmatism continued, underplating and intraplating (e.g., Barnes et al., 2002) of mafic magma into the progressively hotter crust would promote melting, hybridization of mantle and crustal magmas, and long-lived magma systems capable of extended open-system behavior.

CONCLUSIONS

Diverse calcic to potassic plutons comprise the Middle Jurassic Ironside Mountain batholith. The Denny and West China Peak complexes range from pyroxenitic to gabbroic and dioritic. Their parental magmas were sufficiently H_2O-rich to support early amphibole stability. The calcic quartz diorite of Happy Camp Mountain was similarly H_2O-rich, but arose from a more evolved parent. The volumetrically important part of the batholith, the 70-km-long Ironside Mountain pluton, is unusual in the Klamath Mountains because it evolved to potassic (quartz monzonitic) compositions. It has petrographic, chemical, and isotopic signatures that suggest approximately closed-system differentiation of an H_2O-poor, low-fO_2 arc tholeiitic parent. Because K_2O enrichment (relative to SiO_2) is difficult at low pressures, we suggest that such enrichment occurred during early, high pressure fractional crystallization at or just above the base of the crust (≥7 kb).

Emplacement of the Ironside Mountain pluton followed regional thrust faulting. Presumably this contractional deformation reorganized the Klamath crustal sequence so that parental Ironside Mountain magmas entered cool crustal rocks that could not be readily melted or assimilated. With successive magma emplacement during the 165- to 156-Ma Wooley Creek magmatic event, crustal melting ensued and assimilation came to characterize this younger plutonic suite.

ACKNOWLEDGMENTS

We thank Ken Johnson and Greg Harper for helpful discussions and assistance in the field and, particularly, Melanie Barnes for field and laboratory assistance. We also thank Jonathan Miller and Tore Prestvik for their thoughtful reviews and James Hoover for assistance with instrumental neutron activation analyses (University of Texas at El Paso). Our research on the Ironside Mountain pluton, and Klamath magmatism in general, has greatly benefited from the long-term interest and input of Porter Irwin, Bob Coleman, Mary Donato, and Ron Kistler. We thank them all for their help and encouragement. Partial support was provided by National Science Foundation grants EAR-9902912, EAR-9117103, and EAR-8720141.

REFERENCES CITED

Allen, C.M., and Barnes, C.G., 2006, this volume, Ages and some cryptic sources of Mesozoic plutonic rocks in the Klamath Mountains, California and Oregon, *in* Snoke, A.W., and Barnes, C.G., eds., Geological studies in the Klamath Mountains province, California and Oregon: A volume in honor of William P. Irwin: Boulder, Colorado, Geological Society of America Special Paper 410, doi: 10.1130/2006.2410(11).

Anderson, A.T., 1976, Magma mixing: Petrological process and volcanological tool: Journal of Volcanology and Geothermal Research, v. 1, p. 3–33, doi: 10.1016/0377-0273(76)90016-0.

Anderson, J.L., and Smith, D.R., 1995, The effects of temperature and fO_2 on the Al-in-hornblende barometer: American Mineralogist, v. 80, p. 549–559.

Barnes, C.G., 1983, Petrology and upward zonation of the Wooley Creek batholith, Klamath Mountains, California: Journal of Petrology, v. 24, p. 495–537.

Barnes, C.G., and Allen, C.M., 2006, this volume, Depth of origin of late Middle Jurassic garnet andesite, southern Klamath Mountains, California, *in* Snoke, A.W., and Barnes, C.G., eds., Geological studies in the Klamath Mountains province, California and Oregon: A volume in honor of William P. Irwin: Boulder, Colorado, Geological Society of America Special Paper 410, doi: 10.1130/2006.2410(13).

Barnes, C.G., Allen, C.M., and Saleeby, J.B., 1986, Open- and closed-system characteristics of a tilted plutonic system, Klamath Mountains, California: Journal of Geophysical Research, v. 91, p. 6073–6090.

Barnes, C.G., Allen, C.M., Hoover, J.D., and Brigham, R.H., 1990, Magmatic components of a tilted plutonic system, Klamath Mountains, California, *in* Anderson, J.L., ed., The nature and origin of Cordilleran magmatism: Boulder, Colorado, Geological Society of America Memoir 174, p. 331–346.

Barnes, C.G., Petersen, S.W., Kistler, R.W., Prestvik, T., and Sundvoll, B., 1992, Tectonic implications of isotopic variation among Jurassic and early Cretaceous plutons, Klamath Mountains: Geological Society of America Bulletin, v. 104, p. 117–126, doi: 10.1130/0016-7606(1992)104<0117:TIOIVA>2.3.CO;2.

Barnes, C.G., Johnson, K., Barnes, M.A., Prestvik, T., Kistler, R.W., and Sundvoll, B., 1995, The Grayback pluton: Magmatism in a Jurassic back-arc environment, Klamath Mountains, Oregon: Journal of Petrology, v. 36, p. 397–415.

Barnes, C.G., Petersen, S.W., Kistler, R.W., Murray, R., and Kays, M.A., 1996, Source and tectonic implications of tonalite-trondhjemite magmatism in the Klamath Mountains: Contributions to Mineralogy and Petrology, v. 123, p. 40–60, doi: 10.1007/s004100050142.

Barnes, C.G., Yoshinobu, A., Prestvik, T., Nordgulen, Ø., Karlsson, H., and Sundvoll, B., 2002, Mafic magma intraplating: Anatexis and hybridization in arc crust, Bindal Batholith, Norway: Journal of Petrology, v. 43, p. 2171–2190, doi: 10.1093/petrology/43.12.2171.

Barnes, C.G., Snoke, A.W., Harper, G.D., Frost, C.D., McFadden, R.R., Bushey, J.C., and Barnes, M.A.W., 2006, this volume (Chapter 17), Arc plutonism following regional thrusting: Petrology and geochemistry of syn- and post-Nevadan plutons in the Siskiyou Mountains, Klamath Mountains province, California, *in* Snoke, A.W., and Barnes, C.G., eds., Geological studies in the Klamath Mountains province, California and Oregon: A volume in honor of William P. Irwin: Boulder, Colorado, Geological Society of America Special Paper 410, doi: 10.1130/2006.2410(17).

Beard, J.S., 1986, Characteristic mineralogy of arc-related cumulate gabbros: Implications for the tectonic setting of gabbroic plutons and for andesite genesis: Geology, v. 14, p. 848–851, doi: 10.1130/0091-7613(1986)14<848:CMOACG>2.0.CO;2.

Bushey, J.C., Snoke, A.W., Barnes, C.G., and Frost, C.D., 2006, this volume, Geology of the Bear Mountain intrusive complex, Klamath Mountains, California, *in* Snoke, A.W., and Barnes, C.G., eds., Geological studies in the Klamath Mountains province, California and Oregon: A volume in honor of William P. Irwin: Boulder, Colorado, Geological Society of America Special Paper 410, doi: 10.1130/2006.2410(14).

Charlton, D.W., 1979, Geology of part of the Ironside Mountian quadrangle, northern California, Klamath Mountains [Ph.D. dissertation]: Santa Barbara, University of California, 542 pp.

Cotkin, S.J., and Medaris, L.G., Jr., 1993, Evaluation of the crystallization conditions for the calcalkaline Russian Peak intrusive complex, Klamath Mountains, northern California: Journal of Petrology, v. 34, p. 543–571.

Donato, M.M., Barnes, C.G., and Gray, G.G., 1983, Geologic map of the Orleans Mountain roadless area, Humboldt and Siskiyou Counties, California: Reston, Virginia, U.S. Geological Survey Miscellaneous Field Investigations Map MF-1526-A, scale, 1:62,500, 1 oversized sheet.

Donato, M.M., Barnes, C.G., and Tomlinson, S.L., 1996, The enigmatic Applegate Group of southwestern Oregon: Age, correlation, and tectonic affinity: Oregon Geology, v. 58, p. 79–91.

Ernst, W.G., 1998, Geology of the Sawyers Bar area, Klamath Mountains, northern California: Sacramento, California Division of Mines and Geology Map Sheet 47, scale 1:48,000, 1 sheet, 59 p.

Frost, B.R., Barnes, C.G., Collins, W.J., Arculus, R.J., Ellis, D.J., and Frost, C.D., 2001, A geochemical classification for granitic rocks: Journal of Petrology, v. 42, p. 2033–2048.

Gill, J., 1981, Orogenic andesites and plate tectonics: New York, Springer-Verlag, 390 pp.

Goldstein, S.L., O'Nions, R.K., and Hamilton, P.J., 1984, A Sm-Nd isotopic study of atmospheric dusts and particulates from major river systems: Earth and Planetary Science Letters, v. 70, p. 221–236, doi: 10.1016/0012-821X(84)90007-4.

Gray, G.G., 1986, Native terranes of the central Klamath Mountains, California: Tectonics, v. 5, p. 1043–1054.

Gribble, R.F., Barnes, C.G., Donato, M.M., Hoover, J.D., and Kistler, R.W., 1990, Geochemistry and intrusive history of the Ashland pluton, Klamath Mountains, California and Oregon: Journal of Petrology, v. 31, p. 883–923.

Grove, T.L., and Baker, M.B., 1984, Phase equilibrium controls on the tholeiitic versus calc-alkaline differentiation trends: Journal of Geophysical Research, v. 89, p. 3252–3274.

Grove, T.L., and Kinzler, R.J., 1986, Petrogenesis of andesites: Annual Review of Earth and Planetary Sciences, v. 14, p. 417–454, doi: 10.1146/annurev.ea.14.050186.002221.

Hacker, B.R., Donato, M.M., Barnes, C.G., McWilliams, M.O., and Ernst, W.G., 1995, Timescales of orogeny: Jurassic construction of the Klamath Mountains: Tectonics, v. 14, p. 677–703, doi: 10.1029/94TC02454.

Hammarstrom, J.M., and Zen, E., 1986, Aluminum in hornblende: An empirical igneous geobarometer: American Mineralogist, v. 71, p. 1297–1313.

Harper, G.D., and Wright, J.E., 1984, Middle to Late Jurassic tectonic evolution of the Klamath Mountains, California-Oregon: Tectonics, v. 3, p. 759–772.

Harper, G.D., Saleeby, J.B., and Heizler, M., 1994, Formation and emplacement of the Josephine ophiolite and the Nevadan orogen in the Klamath Mountains, California-Oregon: U/Pb zircon and $^{40}Ar/^{39}Ar$ geochronology: Journal of Geophysical Research, v. 99, p. 4293–4321, doi: 10.1029/93JB02061.

Hollister, L.S., Grissom, G.C., Peters, E.K., Stowell, H.H., and Sisson, V.B., 1987, Confirmation of the empirical correlation of Al in hornblende with pressure of solidification of calc-alkaline plutons: American Mineralogist, v. 72, p. 231–239.

Hotz, P.E., 1971, Plutonic rocks of the Klamath Mountains, California and Oregon: Washington, D.C., U.S. Geological Survey Professional Paper 684-B, p. 1–20.

Hunter, A.G., 1998, Intracrustal controls on the coexistence of tholeiitic and calc-alkaline magma series at Aso Volcano, SW Japan: Journal of Petrology, v. 39, p. 1255–1284, doi: 10.1093/petrology/39.7.1255.

Irwin, W.P., 1960, Geologic reconnaissance of the northern Coast Ranges and Klamath Mountains, California, with a summary of the mineral resources: San Francisco, Bulletin of the California Division of Mines and Geology 179, 80 pp.

Irwin, W.P., 1972, Terranes of the western Paleozoic and Triassic belt in the southern Klamath Mountains, California: Washington, D.C., U.S. Geological Survey Professional Paper 800-C, p. 103–111.

Irwin, W., and Wooden, J., 1999, Plutons and accretionary episodes of the Klamath Mountains, California and Oregon: Reston, Virginia, U.S. Geological Survey Open-File Report 99-374, 1 sheet.

Johnson K., and Barnes, C.G., 2006, this volume, Magma mixing and mingling in the Grayback pluton, Klamath Mountains, Oregon, *in* Snoke, A.W., and Barnes, C.G., eds., Geological studies in the Klamath Mountains province, California and Oregon: A volume in honor of William P. Irwin: Boulder, Colorado, Geological Society of America Special Paper 410, doi: 10.1130/2006.2410(12).

Kistler, R.W., Chappell, B.W., Peck, D.L., and Bateman, P.C., 1986, Isotopic variation in the Tuolumne Intrusive Suite, central Sierra Nevada, California: Contributions to Mineralogy and Petrology, v. 94, p. 205–220, doi: 10.1007/BF00592937.

Kuno, H., 1968, Differentiation of basalt magmas, *in* Hess, H.H., and Poldevaart, A., eds., Basalts, vol. II: New York, Wiley Intersciences Publications, p. 624–688.

Lanphere, M.A., Irwin, W.P., and Hotz, P.E., 1968, Isotopic age of the Nevadan orogeny and older plutonic and metamorphic events in the Klamath Mountains, California: Geological Society of America Bulletin, v. 79, p. 1027–1052.

Leake, B., Woolley, A., Arps, C., Birch, W., Gilbert, M., Grice, J., Hawthorne, F., Kato, A., Kisch, H., Krivovichev, V., Linthout, K., Laird, J., Mandarino, J., Schumacher, J., Smith, D., Stephenson, N., Ungaretti, L., Whittaker, E., and Youzhi, G., 1997, Nomenclature of amphiboles: Report of the Subcommittee on Amphiboles of the International Mineralogical Association, Commission on New Minerals and Mineral Names: American Mineralogist, v. 82, p. 1019–1037.

Le Maitre, R.W., Streckeisen, A., Zanettin, B., Le Bas, M.J., Bonin, B., Bateman, P., Bellieni, G., Dudek, A., Efremova, S., Keller, J., Lameyre, J., Sabine, P.A., Schmid, R., Sorensen, H., and Woolley, A.R., 2002, Igneous Rocks: A classification and glossary of terms, recommendations of the International Union of Geological Sciences Subcommission on the Systematics of Igneous Rocks (2nd edition): Cambridge, Cambridge University Press, 252p.

MacDonald, G.A., and Katsura, T., 1964, Chemical composition of Hawaiian lavas: Journal of Petrology, v. 5, p. 83–133.

Mars, E.V., 1999, Petrology and geochemistry of the Ironside Mountain batholith, Klamath Mountains, northern California [MS thesis]: Lubbock, Texas Tech University, 71 pp.

Masi, U., O'Neil, J.R., and Kistler, R.W., 1981, Stable isotope systematics in Mesozoic granites of central and northern California and southwestern Oregon: Contributions to Mineralogy and Petrology, v. 76, p. 116–126, doi: 10.1007/BF00373691.

Meen, J.K., 1990, Elevation of potassium content of basaltic magma by fractional crystallization: The effect of pressure: Contributions to Mineralogy and Petrology, v. 104, p. 309–331, doi: 10.1007/BF00321487.

Middlemost, E.A.K., 1975, The basalt clan: Earth Science Reviews, v. 11, p. 337–364, doi: 10.1016/0012-8252(75)90039-2.

Miller, J., Miller, R., Wooden, J., and Harper, G., 2003, Geochronologic links between the Ingalls ophiolite, North Cascades, Washington and the Josephine ophiolite, Klamath Mts., Oregon and California: Geological Society of America Abstracts with Programs, v. 35, p. 113.

Miyashiro, A., 1974, Volcanic rock series in island arcs and active continental margins: American Journal of Science, v. 274, p. 321–355.

Moore, J.G., 1959, The quartz diorite boundary line in the western United States: Journal of Geology, v. 67, p. 198–210.

Müller, D., and Groves, D.I., 1995, Potassic igneous rocks and associated gold-copper mineralization: New York, Springer-Verlag, 252 pp.

Nekvasil, H., and Lindsley, D.H., 1990, Termination of the 2 feldspar + liquid curve in the system Ab-Or-An-H_2O at low H_2O contents: American Mineralogist, v. 75, p. 1071–1079.

Nekvasil, H., Simon, A., and Lindsley, D.H., 2000, Crystal fractionation and the evolution of intra-plate *hy*-normative igneous suites: Insights from the feldspars: Journal of Petrology, v. 41, p. 1743–1757, doi: 10.1093/petrology/41.12.1743.

Rickwood, P.C., 1989, Boundary lines within petrologic diagrams which use oxides of major and minor elements: Lithos, v. 22, p. 247–263, doi: 10.1016/0024-4937(89)90028-5.

Sakuyama, M., 1981, Petrological study of the Myoko and Kurohime volcanoes, Japan: Crystallization sequence and evidence for magma mixing: Journal of Petrology, v. 22, p. 553–583.

Saleeby, J.B., Harper, G.D., Snoke, A.W., and Sharp, W.D., 1982, Time relations and structural-stratigraphic patterns in ophiolite accretion, west-central Klamath Mountains, California: Journal of Geophysical Research, v. 87, no. B5, p. 3831–3848.

Schmidt, B.L., 1994, The petrology and geochemistry of the English Peak intrusive suite, Klamath Mountains, California [Ph.D. dissertation]: Lubbock, Texas Tech University, 216 p.

Schmidt, M.W., 1992, Amphibole composition in tonalite as a function of pressure: An experimental calibration of the Al-in-hornblende barometer: Contributions to Mineralogy and Petrology, v. 110, p. 304–310, doi: 10.1007/BF00310745.

Shannon, W.M., Barnes, C.G., and Bickford, M.E., 1997, Grenville magmatism in west Texas: Petrology and geochemistry of the Red Bluff granitic suite: Journal of Petrology, v. 38, p. 1279–1305, doi: 10.1093/petrology/38.10.1279.

Sisson, T.W., and Grove, T.L., 1993, Experimental investigations of the role of H_2O in calc-alkaline differentiation and subduction zone magmatism: Contributions to Mineralogy and Petrology, v. 113, p. 143–166, doi: 10.1007/BF00283225.

Snoke, A.W., Quick, J.E., and Bowman, H.R., 1981, Bear Mountain igneous complex, Klamath Mountains, California: An ultrabasic to silicic calc-alkaline suite: Journal of Petrology, v. 22, p. 501–552.

Spera, F., and Bohrson, W., 2001, Energy-constrained open-system magmatic processes I: General model and energy-constrained assimilation and fractional crystallization (EC-AFC) formulation: Journal of Petrology, v. 42, p. 999–1018, doi: 10.1093/petrology/42.5.999.

Sun, S.-s., and McDonough, W.F., 1989, Chemical and isotopic systematics of oceanic basalts: Implications for mantle composition and processes, *in* Saunders, A., and Norry, M., eds., Magmatism in the ocean basins: Oxford, Geological Society Special Publication, p. 313–345.

Treuil, M., Joron, J.L., and Jaffrezic, H., 1977, Trace element geochemistry of magmatic rock series of converging and diverging plate boundaries: Journal of Radioanalytical Chemistry, v. 38, p. 351–362.

Wright, J.E., 1982, Permo-Triassic accretionary subduction complex, southwestern Klamath Mountians, northern California: Journal of Geophysical Research, v. 87, p. 3805–3818.

Wright, J.E., and Fahan, M.R., 1988, An expanded view of Jurassic orogenesis in the western United States Cordillera: Middle Jurassic (pre-Nevadan) regional metamorphism and thrust faulting within an active arc environment, Klamath Mountains, California: Geological Society of America Bulletin, v. 100, p. 859–876, doi: 10.1130/0016-7606(1988)100<0859:AEVOJO>2.3.CO;2.

Wright, J.E., and Wyld, S.J., 1994, The Rattlesnake Creek terrane, Klamath Mountains, California: An early Mesozoic volcanic arc and its basement of tectonically disrupted oceanic crust: Geological Society of America Bulletin, v. 106, p. 1033–1056, doi: 10.1130/0016-7606(1994)106<1033:TRCTKM>2.3.CO;2.

Young, J.C., 1978, Geology of the Willow Creek quadrangle, Humboldt and Trinity Counties, California: San Francisco, California Division of Mines and Geology, Map Sheet 31, scale 1:62,500, 1 sheet, 16 p.

Yule, J.D., Saleeby, J.B., and Barnes, C.G., 2006, this volume, A rift-edge facies of the Late Jurassic Rogue–Chetco arc and Josephine ophiolite, Klamath Mountains, Oregon, *in* Snoke, A.W., and Barnes, C.G., eds., Geological studies in the Klamath Mountains province, California and Oregon: A volume in honor of William P. Irwin: Boulder, Colorado, Geological Society of America Special Paper 410, doi: 10.1130/2006.2410(03).

Manuscript Accepted by the Society 10 January 2006

Geological Society of America
Special Paper 410
2006

Ages and some cryptic sources of Mesozoic plutonic rocks in the Klamath Mountains, California and Oregon

Charlotte M. Allen*
Research School of Earth Sciences, Australia National University, Canberra, ACT 0200, Australia
Calvin G. Barnes
Department of Geosciences, Texas Tech University, Lubbock, Texas 79409-1053, USA

ABSTRACT

A growing body of evidence indicates that Middle Jurassic to Early Cretaceous plutons recorded changing sources during tectonic evolution of the Klamath Mountain province. The data set now includes U-Pb zircon ages and zircon trace element compositions determined by laser ablation inductively coupled plasma mass spectrometry (LA-ICP-MS). Thirteen rock samples were dated, and these data refine thermal ionization mass spectrometry (TIMS) data where inheritance was problematic, or provide new U-Pb ages. Individual plutonic suites, previously defined on the basis of crystallization age, isotope and elemental compositions, and petrogenetic style, show characteristic inherited zircon age ranges and zircon trace element patterns. Moreover, ages of inherited zircons in these suites are distinct and, in at least three suites, indicate the presence of cryptic (unexposed) source rocks. The zircon data complement oxygen, Nd, and Sr isotope whole-rock data that, when taken together, suggest a number of major changes in the crustal column with time. Middle Jurassic magmatism began with the oceanic(?) arc-related western Hayfork terrane comprising volcanic, volcaniclastic, and plutonic components. After regional thrusting on the ca. 170-Ma Wilson Point thrust, the Ironside Mountain batholith and Wooley Creek suite of plutons were emplaced. The former shows little evidence of interaction with the crust, but the latter contains Middle Jurassic inheritance and Sr, Nd, and oxygen isotope signatures suggestive of interaction with metasedimentary crustal rocks. Following Nevadan thrusting (ca. 153–150 Ma), emplacement of western Klamath suite plutons in the western Klamath Mountains province involved significant assimilation of Galice Formation metasedimentary rocks. This activity was followed by emplacement of tonalite-trondhjemite-granodiorite (ttg) plutons in the eastern Klamath Mountains province, which were derived by partial melting of metabasic rocks. Their zircon trace element signatures indicate diverse magma histories and, at least locally, multiple magma sources. Inherited zircons in ttg plutons suggest late Middle Jurassic to Late Jurassic sources, younger than the Josephine ophiolite. The youngest magmatism in the Klamath Mountains province consists of broadly granodioritic plutons, which, on

*E-mail: Charlotte.Allen@anu.edu.au.

Allen, C.M., and Barnes, C.G., 2006, Ages and some cryptic sources of Mesozoic plutonic rocks in the Klamath Mountains, California and Oregon, *in* Snoke, A.W., and Barnes, C.G., eds., Geological studies in the Klamath Mountains province, California and Oregon: A volume in honor of William P. Irwin: Geological Society of America Special Paper 410, p. 223–245, doi: 10.1130/2006.2410(11). For permission to copy, contact editing@geosociety.org.

the basis of limited data, show variable petrogenesis and zircon inheritance. At least one of these plutons (136-Ma Yellow Butte pluton) contains a ca. 150-Ma inheritance that indicates the presence of Late Jurassic crustal rocks beneath the eastern Klamath terrane.

Keywords: U-Pb geochronology, Klamath Mountains, trace elements, Jurassic, Cretaceous, radiogenic isotopes

INTRODUCTION

Since the publication of Lanphere et al. (1968), workers in the Klamath Mountains province have recognized a number of igneous and metamorphic events on the basis of radiometric dating. Dating of plutons was particularly important, because many are stitching plutons: they either crosscut or are cut by regional faults. Moreover, early workers, such as Lanphere et al. (1968) and especially Hotz (1971), recognized that in many cases, plutons within a distinct age group also had distinct compositions. Hotz's work served as the basis for a number of detailed studies of the petrology, geochemistry, age, and tectonic importance of Klamath plutons. The sophistication of the analytical and conceptual tools used in these studies has improved through time, such that the petrogenetic history of a number of Klamath Mountains province plutons is now well known. The most recent broadly based discussions of Klamath Mountains province plutons are those of Irwin (1985) and Irwin and Wooden (1999), who used existing and new geochronologic data to refine pluton age groups, and Barnes et al. (1992b), who showed that plutons of specific ages and in specific geographic locations are characterized by distinct elemental and isotopic compositions.

Detailed petrogenetic studies (Snoke et al., 1981; Barnes et al., 1990, 1992a, 1995b, 1996; Gribble et al., 1990; Bushey et al., this volume) showed that most Klamath Mountains province plutons are upper crustal manifestations of much larger, possibly crustal-scale magmatic systems. These conclusions arose in part from isotopic evidence (e.g., Barnes et al., 1990), but also from major and trace element modeling (e.g., Barnes et al., 1986, 1990, 1996). In fact, because middle crustal levels are rarely exposed in the province and deep crustal exposures are absent, our knowledge of middle to deep crustal levels has been obtained primarily from studying the crustal component of plutons (e.g., Barnes et al., 1992a). This type of "remote sensing" is difficult, because isotopic and trace element models are rarely strongly constrained.

With the advent of laser ablation inductively coupled plasma mass spectrometry (LA-ICP-MS) analysis, it is now possible to date 32-µm-diameter spots on single zircon crystals and simultaneously determine the trace element compositions of the crystal (e.g., Barnes and Allen, this volume). This type of dating, along with sensitive high-resolution ion microprobe (SHRIMP) and single crystal thermal ionization mass spectrometry (TIMS) methods (e.g., Chamberlain et al., this volume) has provided an entirely new perspective on magma history by identifying and dating zircons precipitated from their host melt, as well as those inherited from magma source rocks and incorporated from assimilated crust.

In this chapter, we present recent LA-ICP-MS zircon dating of thirteen Middle Jurassic to Early Cretaceous Klamath Mountains province plutons. We interpret the data to show that plutons in specific age and composition groups (Barnes et al., 1992b) also have distinctive zircon inheritance patterns. Although the data are preliminary in nature, we suggest that the combination of petrogenetic study, isotopic variation, and inherited zircon age patterns provides otherwise unattainable information about the substructure of the Klamath Mountains province, and that this substructure changed through time as the result of thrusting events in Middle Jurassic through Early Cretaceous time.

GEOLOGIC SETTING

Major terrane units

The Klamath Mountain province consists of a series of thrust-bounded tectonostratigraphic terranes (e.g., Irwin, 1960, 1972, 1981; Fig. 1). Most terranes originated in suprasubduction zone (SSZ) environments. The traditional view of terrane amalgamation and accretion in the Klamath Mountains province is by alternation of terrane formation as part of an SSZ complex followed by contractional deformation and collapse against the growing assembly of terranes. Because most of these terranes are described in detail elsewhere in this volume (Lindsley-Griffin et al.; Barrow and Metcalf; Scherer et al.; Snoke and Barnes), only short descriptions are presented here; terranes are described from east to west, from structurally highest to lowest (Fig. 1).

The eastern Klamath terrane consists of the Trinity, Redding, and Yreka subterranes. The Trinity subterrane consists of

→

Figure 1. Geologic map of the Klamath Mountain province after Irwin (1994) and Irwin and Wooden (1999) showing principal tectonostratigraphic terranes and plutons color coded according to age group. Numbers associated with each pluton are U-Pb (zircon) ages (Ma) unless noted as hornblende (h) or biotite (b) K-Ar ages or hornblende (H) $^{40}Ar/^{39}Ar$ ages. Data from this chapter and sources summarized in Irwin and Wooden (1999).

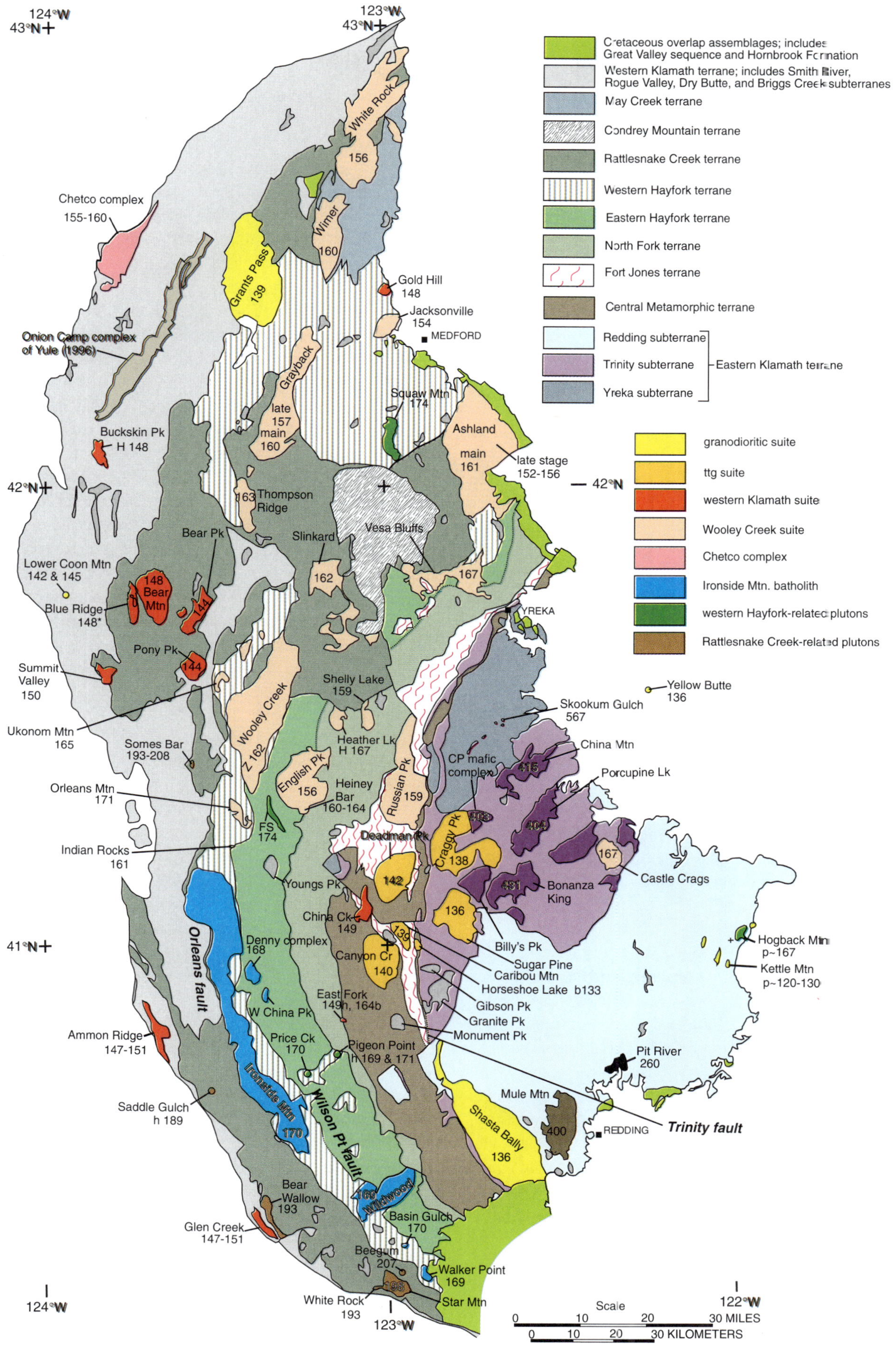

124°W
43°N
123°W
43°N
Cretaceous overlap assemblages; includes Great Valley sequence and Hornbrook Formation
Western Klamath terrane; includes Smith River, Rogue Valley, Dry Butte, and Briggs Creek subterranes
May Creek terrane
Condrey Mountain terrane
Rattlesnake Creek terrane
Western Hayfork terrane
Eastern Hayfork terrane
North Fork terrane
Fort Jones terrane
Central Metamorphic terrane
Redding subterrane
Trinity subterrane
Yreka subterrane
Eastern Klamath terrane
granodioritic suite
ttg suite
western Klamath suite
Wooley Creek suite
Chetco complex
Ironside Mtn. batholith
western Hayfork-related plutons
Rattlesnake Creek-related plutons
White Rock
156
Wimer
160
Grants Pass
139
Chetco complex
155-160
Gold Hill
148
Jacksonville
154
MEDFORD
Onion Camp complex of Yule (1996)
Grayback
late
157
main
160
Squaw Mtn
174
Ashland
main
161
late stage
152-156
Buckskin Pk
H 148
42°N
163 Thompson Ridge
Bear Pk
Slinkard
Vesa Bluffs
162
167
Lower Coon Mtn
142 & 145
148 Bear Mtn
144
Blue Ridge
148*
YREKA
Pony Pk
144
Summit Valley
150
Shelly Lake
159
Yellow Butte
136
Skookum Gulch
567
Ukonom Mtn
165
Somes Bar
193-208
Heather Lk
H 167
China Mtn
Wooley Creek
Z 162
English Pk
156
Heiney Bar
160-164
Russian Pk
159
CP mafic complex
415
Porcupine Lk
Orleans Mtn
171
FS
174
Deadman Pk
Craggy Pk
138
403
404
167
Indian Rocks
161
Youngs Pk
142
431
Bonanza King
Castle Crags
136
China Ck
149
Orleans fault
41°N
Denny complex
168
139
Canyon Cr
140
Billy's Pk
Sugar Pine
Caribou Mtn
Horseshoe Lake b133
Hogback Mtn
p~167
Kettle Mtn
p~120-130
W China Pk
East Fork
149h, 164b
Gibson Pk
Granite Pk
Monument Pk
Ammon Ridge
147-151
Price Ck
170
Pigeon Point
h 169 & 171
Pit River
260
Saddle Gulch
h 189
Ironside Mtn
170
Wilson Pt fault
Mule Mtn
400
REDDING
Trinity fault
Shasta Bally
136
Bear Wallow
193
169 Wildwood
Basin Gulch
170
Glen Creek
147-151
Beegum
207
Walker Point
169
White Rock
193
193
Star Mtn
124°W
123°W
122°W
Scale
0
10
20
30 MILES
0
10
20
30 KILOMETERS

Ordovician peridotite intruded by Silurian–Devonian gabbroic to ultramafic plutons and associated bimodal dike swarms (Wallin and Metcalf, 1998). The Yreka subterrane contains Silurian–Devonian mélange and volcaniclastic turbidites. This unit contains Precambrian detrital zircon ($^{207}Pb/^{206}Pb$ ages of 556–579 Ma; Wallin et al., 1995). The Redding subterrane is Lower Devonian to Middle Jurassic in age; it consists of interbedded marine clastic and volcaniclastic rocks (Miller and Harwood, 1990), volcanic rocks, and limestone.

The structurally lower Central Metamorphic terrane is in contact with the eastern Klamath terrane along the east-dipping Trinity fault (Fig. 1). The Central Metamorphic terrane consists of a metabasic unit (Salmon Hornblende Schist) with mid-ocean ridge basalt (MORB) affinities (Barrow and Metcalf, this volume) and the metasedimentary Grouse Ridge Formation. The $^{40}Ar/^{39}Ar$ ages of hornblende from the Salmon Hornblende Schist indicate Early Permian metamorphism (Barrow and Metcalf, this volume). To the west, the Central Metamorphic terrane is in fault contact with the Stuart Fork/Fort Jones terrane, a subduction complex that underwent ca. 220-Ma blueschist-facies metamorphism (Hotz et al., 1977). It consists primarily of semi-pelitic schist, with local metabasic rocks (Goodge, 1995).

The Sawyers Bar terrane (Ernst, 1999) represents a nearly complete SSZ complex associated with a west-facing arc. From west to east, it consists of an accretionary complex (the chert-argillite mélange and broken formation of the eastern Hayfork terrane; Wright, 1982; Goodge and Renne, 1993), ophiolitic fragments of the North Fork terrane (Irwin, 1972; Ando et al., 1983), primitive arc metabasalts (the Salmon River unit of Ernst, 1993), and back-arc–related metasedimentary rocks of the St. Clair Creek unit (Ernst, 1999). Fossils from these units range from Permian to earliest Jurassic age, suggesting either a very long-lived arc or, more probably, diverse, reworked fossil components in an Early to Middle Jurassic arc. Miller et al. (1988) reported 2.05-Ga detrital zircons from the eastern Hayfork terrane, which is consistent with paleogeographic links to a nearby but unspecified continent.

At its western margin, the Sawyers Bar terrane overlies the Middle Jurassic western Hayfork terrane along the Wilson Point thrust. The western Hayfork terrane in turn overlies the Rattlesnake Creek terrane. The Rattlesnake Creek terrane consists of a basal ophiolitic mélange overlain by a coherent sequence of metavolcanic and metasedimentary rocks interpreted by Wright and Wyld (1994) to be arc related. Fossil assemblages and radiometric ages of mélange blocks suggest protolith ages from Permian to Late Triassic and possibly Early Jurassic (Irwin and Blome, 2004, and on the CD-ROM accompanying this volume and in the GSA Data Repository[1]); the cover sequence ranges from Late Triassic to Early Jurassic age (Wright and Wyld, 1994).

[1]GSA Data Repository item 2006196, Fossil localities of Rattlesnake Creek, is available online at www.geosociety.org/pubs/ft2006.htm, or on request from editing@geosociety.org or Documents Secretary, GSA, P.O. Box 9140, Boulder, CO 80301, USA.

The western Hayfork terrane consists of arc-related metamorphosed volcanic, volcaniclastic, and argillitic rocks. The $^{40}Ar/^{39}Ar$ ages on volcanic and detrital hornblende range from 179 to 169 Ma (e.g., Wright and Fahan, 1988; Hacker et al., 1995). The contact between the Rattlesnake Creek terrane and western Hayfork terrane is a thrust fault (Salt Creek thrust); however, the presence of Rattlesnake Creek terrane detritus in basal western Hayfork terrane beds suggests that this fault has minor displacement (Donato et al., 1996) and that the Rattlesnake Creek terrane was the depositional basement of the western Hayfork terrane (Wright and Fahan, 1988).

The Rattlesnake Creek–western Hayfork terrane assembly overlies younger rocks to the west along the regionally extensive Orleans thrust (Fig. 1). These structurally lower rocks represent a late Middle Jurassic SSZ complex originally named the "western Klamath belt" (Irwin, 1960), now called the "western Klamath terrane." Detailed study of the northern part of this belt (e.g., Harper, 1980, 1984; Harper and Wright, 1984; Harper et al., 1988, 1994; Yule et al., this volume) shows that it originated by forearc(?) rifting in the Rattlesnake Creek terrane. This rifting led to the formation of the 161- to 165-Ma Josephine ophiolite (Wright and Wyld, 1986; Harper et al., 1994). As the rift evolved, arc magmatism shifted to an outboard position. This new arc sequence is preserved as the late Middle Jurassic Rogue Formation (Garcia, 1982) and Chetco plutonic complex (Yule et al., this volume; Fig. 1). Sediments were laid down on both the Josephine ophiolite and the Rogue Formation contemporaneous with magmatism (e.g., Pinto-Auso and Harper, 1985; Miller et al., 2003). The basal sediments are predominantly hemipelagic (Pinto-Auso and Harper, 1985); these give way to a thick flysch sequence (Galice Formation), whose provenance includes the active Rogue arc and metamorphic rocks of the older Klamath Mountains province terranes (Pinto-Auso and Harper, 1985). SHRIMP dating of detrital zircons from a basal Galice greywacke (Miller et al., 2003) showed a complex population of zircon ages, with prominent peaks at ca. 153 Ma, 185 Ma, 220 Ma, and 270 Ma, as well as two Proterozoic zircons.

Middle Jurassic to Early Cretaceous plutonism

Various subdivisions of Middle Jurassic to Early Cretaceous magmatism have been made on the basis of age (Irwin, 1985; Irwin and Wooden, 1999) or age and composition (Barnes et al., 1992b). In Table 1, we present a revision of the groups outlined by Barnes et al. (1992b) that shows U-Pb (zircon) ages of plutons and their compositional groups. This information is also summarized in Figure 2. Details of relationships between age and elemental and isotopic compositions are discussed below.

Arc magmatism in the western Hayfork terrane ended with terrane amalgamation along the Wilson Point thrust (Barnes et al., this volume, Chapter 10). This east-dipping fault places arc-related rocks of the western Hayfork terrane beneath the eastern Hayfork terrane. Thrusting was followed by emplace-

TABLE 1. SUMMARY OF U-Pb (ZIRCON) AGES, COMPOSITIONAL GROUPS, AND PETROGENETIIC PROCESSES

Pluton		U-Pb ages (Ma)	Characteristics*	Reference§
Post-western Klamath granodioritic suite				
YB	Yellow Butte	136.4 ± 0.9	Diorite, tonalite, granodiorite	[1]
SBP	Shasta Bally	136	Diorite, tonalite, granodiorite	[6]
GP	Grants Pass	139 ± 2	Diorite, tonalite, granodiorite	[4]
			Magma mingling, assimilation, crustal melting(?)	
Post-western Klamath tonalite-trondhjemite-granodiorite suite				
SP	Sugar Pine	136.3 ± 1.0	*Crustal melting, magma mingling*	[1]
CP	Craggy Peak	136.6 ± 1.4	*Crustal melting, magma mingling*	[1, 10]
CM	Caribou Mountain	138.1 ± 1.1	*Crustal melting, magma mingling*	[1, 10, 11]
CC	Canyon Creek	140.8 ± 1.5	*Crustal melting, magma mingling*	[1, 8, 12, 13]
DP	Deadman Peak	142.3 ± 1.1	Includes gabbroic cumulates	[1, 8, 10]
Probable western Klamath suite plutons				
ChCk	China Creek	149	Unknown	[5]
GC	Glen Creek	147–151	Unknown	[8]
AR	Ammon Ridge	147–151	Unknown	[8]
Western Klamath plutonic suite				
LCM	Lower Coon Mountain	145 +1/–2	Ultramafic, minor gabbro, diorite, quartz diorite (dated) and granodiorite	[4]
BP	Bear Peak	143 ± 1.3	Composite pluton; ultramafic to granodioritic; *mixing, assimilation, fractional crystallization, crustal melting*	[1, 14]
PP	Pony Peak	144.2 ± 2.5		[1]
		146 ± 3	Gabbro to tonalite; assimilation, magma *Mixing(?) crustal melting(?)*	[4]
BHP	Buckskin Peak	148	Unknown	[4]
BMIC	Bear Mtn intrusive complex	149 ± 0.6 to 144.9 ± 0.8	Ultramafic, gabbroic, dioritic	[3, 7, 15]
			Tonalitic; *magma mixing, fractional crystallization, assimilation, crustal melting*	
SV	Summit Valley	150	Ultramafic to tonalitic	[4]
Chetco/Illinois River plutonic rocks				
Chetco complex		155–160	Multiple plutons(?)	[9]
			Ultramafic, gabbro, diorite	
Wooley Creek suite (formerly Wooley Creek and Grayback belts)				
JP	Jacksonville	154	Quartz diorite–tonalite	[5]
WR	White Rock	156	Tonalite-trondhjemite	[9, 10]
RP	Russian Peak	159	Composite pluton; ultramafic to granodioritic	[8, 16]
SLP	Shelley Lake	159.1 ± 1.3	Dioritic to granodioritic	[1, 17]
IR	Indian Rocks dike	159.5 ± 2.0	Tonalite	[1, 18]
AP	Ashland pluton	152–156	Biotite granite	[9, 19]
		160–161	Composite pluton; dioritic to quartz monzonitic; *mixing, assimilation, crustal melting*	[9, 19]
EP	English Peak (main stage)	156 ± 0.7	Composite pluton; gabbroic to granodioritic; *magma mixing, assimilation*	[1, 20]
EP	English Peak (Heiney Bar)	160–164	Dioritic	[8]
GB	Grayback	157 ± 2	Composite pluton; tonalite and granodiorite	[9, 21]
		160 ± 2	Gabbro to tonalite; *mixing, assimilation*	
W	Wimer	160	Dioritic	[9]
TR	Thompson Ridge	163 ± 2	Dioritic	[9]
WC	Wooley Creek	161 +4/–2	Composite plutons; gabbro to granite; *mixing, assimilation, crustal melting, fractional crystallization*	[2]
SP	Slinkard	161 +4/–2	Composite; gabbro to tonalite; granite; *mixing, assimilation, crustal melting*	[2]

(*continued*)

TABLE 1. *Continued*

Pluton		U-Pb ages (Ma)	Characteristics*	Reference§
VB	Vesa Bluffs	166.9 ± 1.9	Composite pluton; gabbro to tonalite, late granite	[1]
CR	Castle Crags	166.8 ± 1.8	Composite pluton alkali-rich granodiorite, trondhjemite	[1]
Unassociated plutons				
UM	Ukonom Mountain	~165	Deformed quartz diorite, tonalite	[1]
OM	Orleans Mountain	170.7 ± 1.7	Deformed quartz diorite, tonalite	[1]
Ironside Mountain batholith (IMB)				
West China Peak complex		Undated	Ultramafic to dioritic	
Denny complex		168	Ultramafic to dioritic	[8]
Walker Point		169	Unknown	
Wildwood pluton		169	Similar to Ironside Mtn pluton	[8]
Price Creek		170	Unknown	
Basin Gulch		170	Unknown	
Ironside Mountain pluton		170	Monzodioritic; mildly alkaline, potassic; *primarily fractional crystallization.*	[8, 22]
Pre-Wilson Point thrust plutons (western Hayfork terrane association?)				
Squaw Mountain		174	Diorite, tonalite(?)	[8]
Forks of Salmon		174	Gabbro, diorite	[8]

*Rock types in roman; inferred mechanisms that produced compositional ranges in *italics*.

§References: 1. This chapter; 2. Barnes et al. (1986); 3. Chamberlain et al. (this volume); 4. Harper et al. (1994); 5. Irwin and Wooden (1999); 6. Lanphere and Jones (1978); 7. Saleeby et al. (1982); 8. Wright and Fahan (1988); 9. Yule et al. (1996); 10. Barnes et al. (1996); 11. Davis (1963); 12. Barker et al. (1979); 13. Peterman (1979); 14. McFadden et al. (this volume); 15. Snoke et al. (1981); Bushey et al. (this volume); Chamberlain et al. (this volume); 16. Cotkin and Medaris (1993); 17. Dorais (1983); 18. Barnes and Allen (this volume); 19. Gribble et al. (1990); 20. Schmidt (1994); 21. Barnes et al. (1995b); 22. Barnes et al. (this volume, Chapter 10).

ment of plutons of the Ironside Mountain batholith, parts of which contact metamorphosed upper and lower plates of the thrust. The timing of thrusting is narrowly constrained to be 170 Ma in age, because $^{40}Ar/^{39}Ar$ ages of hornblende in the western Hayfork terrane are as young as 169 Ma and because the Ironside Mountain pluton, which cuts the Wilson Point thrust, has a U-Pb zircon age of 170 Ma (Wright and Fahan, 1988). Barnes et al. (this volume, Chapter 10) interpreted thrusting to be of regional importance and suggested that it disrupted the thermal, as well as the tectonostratigraphic, structure of the crust.

The Ironside Mountain batholith (168–170 Ma) therefore represents the beginning of a magmatic cycle that ended at ca. 156 Ma as laid out below. The magmatic cycle comprises plutons of the Ironside Mountain batholith, primarily in the southwestern part of the Klamath Mountains province, the Wooley Creek suite of plutons (167–156 Ma) that intruded pre-late Middle Jurassic terranes (Fig. 1), the Josephine ophiolite, and the Rogue–Chetco arc complex (Table 1). This magmatic cycle ended at or before the beginning of contractional deformation and regional thrusting on the Orleans fault (Harper et al., 1994), during which the older, inboard terranes were juxtaposed over the terranes of the western Klamath terrane. This contractional event is referred to as the "Nevadan orogeny" (Blackwelder, 1914; Hinds, 1934; Lanphere et al., 1968).

Nevadan thrusting began after ca. 153 Ma, on the basis of ages of detrital zircons in the footwall Galice Formation (Miller et al., 2003), and was complete by 150 Ma, as shown by pinning plutons (Harper et al., 1994). During thrusting, pre-Nevadan plutons were truncated and are now rootless (Jachens et al., 1986). The "western Klamath suite" of plutons, which represents the oldest post-Nevadan magmatism (150–144 Ma), is primarily exposed in the western part of the province (Fig. 1). Most or all of these plutons were emplaced into, or passed through the Galice Formation (e.g., Harper et al., 1994; Bushey et al., this volume; McFadden et al., this volume).

At ca. 142 Ma, the locus of pluton emplacement shifted to the east, particularly in the southern Klamath Mountains province (Figs. 1 and 2). Barnes et al. (1992b) recognized two compositional groups in this age range: a set of tonalite-trondhjemite-granodiorite (ttg suite) plutons in the east-central Klamath Mountains province (Fig. 1) and three dioritic to granodioritic plutons (Grants Pass, Shasta Bally, and Yellow Butte: "granodioritic suite"). These two groups of "post-western Klamath suite" plutons record the youngest Mesozoic magmatism in the Klamath Mountains province, which ceased by ca. 136 Ma.

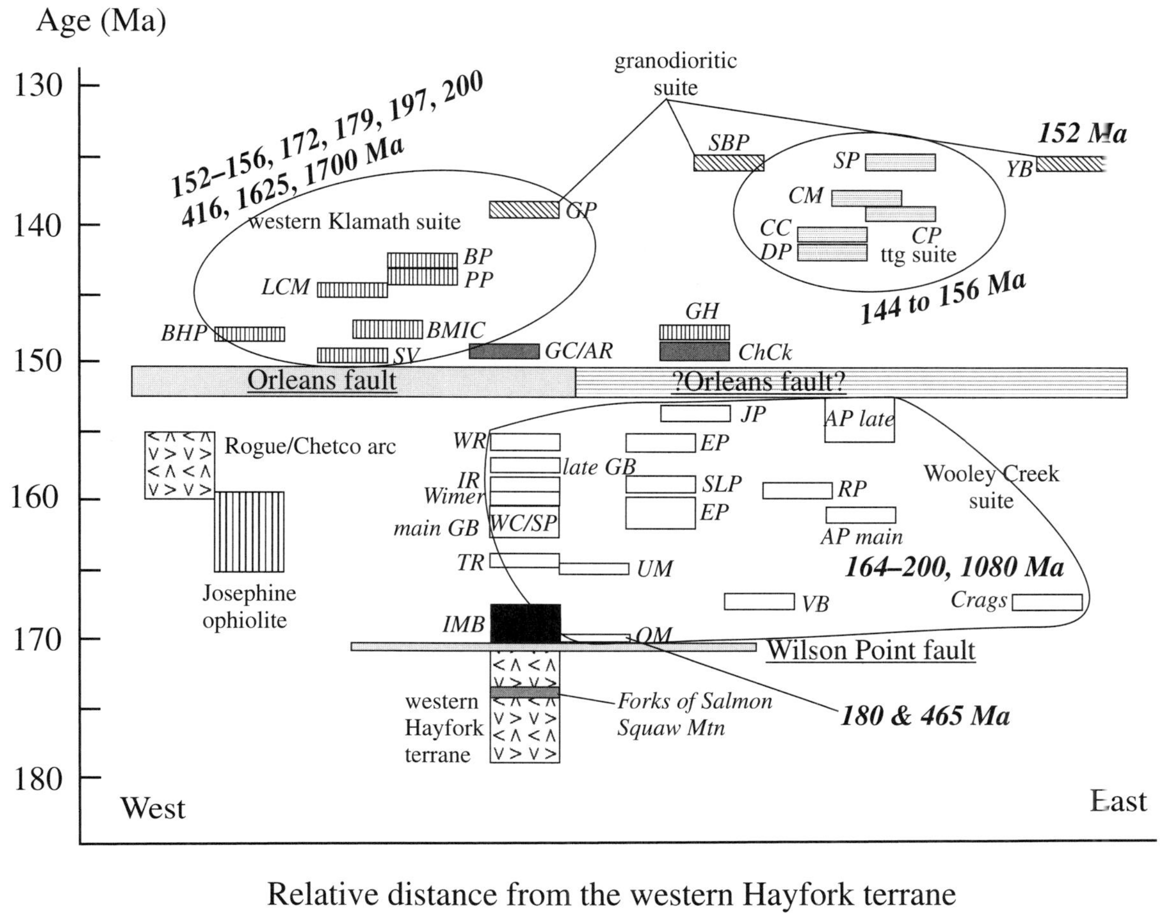

Figure 2. Chart showing the ages of Middle Jurassic–Early Cretaceous magmatic events in the Klamath Mountains and their relative position compared to the western Hayfork terrane. Long horizontal bars show how two regional thrusting episodes (Wilson Point and Orleans faults) separate broad magmatic events. Narrow bars show individual plutons. Circled groups of plutons appear to share similar zircon inheritance, which is shown in bold italic text. It is noteworthy that not all plutons in each circled group have been analyzed for zircon inheritance. Abbreviations of pluton names (see Fig. 1): AP—Ashland; BHP—Buckskin Peak; BMIC—Bear Mountain intrusive complex; BP—Bear Peak; CC—Canyon Creek; ChCk—China Creek; CM—Caribou Mountain; CP—Craggy Peak; Crags—Castle Crags; DP—Deadman Peak; EP—English Peak; GB—Grayback; GC/AR—Glen Creek and Ammon Ridge; GH—Gold Hill; GP—Grants Pass; IMB—Ironside Mountain batholith; IR—garnet andesite dikes; JP—Jacksonville; LCM, Lower Coon Mountain; OM—Orleans Mountain; PP—Pony Peak; RP—Russian Peak; SBP—Shasta Bally; SLP—Shelly Lake; SP—Sugarpine; SV—Summit Valley; TR—Thompson Ridge; UM—Ukonom Mountain; VB—Vesa Bluffs; WC/SP—Wooley Creek/Slinkard; WR—White Rock; YB—Yellow Butte.

METHODS

Analytical Techniques

The U-Pb zircon LA-ICP-MS ages were obtained for thirteen rock samples from Mesozoic plutons that lacked U-Pb ages or from ones for which conventional U-Pb analysis gave equivocal results (e.g., Canyon Creek pluton: Wright and Fahan, 1988). Sample information is listed in Table 2. Zircons were separated from rock samples varying in size from 3 to 20 kg at Texas Tech University, Lubbock. With the exception of the Yellow Butte pluton sample, which was processed on a Rodgers gold table, samples were crushed and then ground in a tungsten carbide swing mill, the heavy mineral fraction was concentrated with a gold pan, and then zircon was separated using standard heavy liquid separation and hand picking. Hand-picked zircons were placed into windows in the backing of double sticky tape on a glass base. After a sample was loaded, the window was covered and secured with tape before the next window was cut. Glass and zircon standards were mounted in a similar fashion. A cone was placed on the tape after the backing was removed, and epoxy resin was poured into the cone. The cast mount was polished to expose zircon interiors. Mounts were photographed at 5- or 10× magnification on a polarizing microscope, and cathodoluminescence

TABLE 2. SAMPLE DESCRIPTIONS, LOCATIONS, AND SUMMARY OF AGE DATA

Sample	Analytical session	Pluton	Label in Fig. 2	Age* (Ma)	1 s.e.	MSWD	Analysis used in age	Total analysis	Number of rejected analyses: Criterion A§	Criterion B†	Inheritance	Young outlier	Location**	Rock type
Post-western Klamath granodiorites														
FPY-23	030920	Yellow Butte	YB	136.4	0.9	0.85	45	49		3	1		^{5}62025, ^{45}99880	Biotite hornblende granodioritc
Post-western Klamath tonalite-trondhjemite-granodiorite suite														
KM-121	031208	Sugarpine	SP	136.3	1.0	1.13	29	30		1			^{4}16460, ^{45}51070	Pyroxene-bearing biotite hornblende quartz diorite
FPCP1	031208	Craggy Peak	CP	138.2	1.3	2.70	27	30		1	2		^{5}15989, ^{45}70067	Biotite hornblende tonalite
FPC-20	030131	Caribou Mountain	CM	139.2	1.9	2.29	27	28	1				^{5}03400, ^{45}43826	Biotite tonalite
FPCC1	031208	Canyon Creek	CC	140.0	1.3	1.90	27	27			3		^{4}98177, ^{45}26504	Hornblende-bearing biotite tonalite
FPD-188	031208	Deadman Peak	DP	142.3	1.1	1.71	30	30					^{5}01270, ^{45}57555	Biotite hornblende granodiorite
Western Klamath suite														
BP-288	040310	Bear Peak	BP	143.7	1.3	1.73	27	41		1	13		^{4}52067, ^{46}15849	Biotite granodiorite
PP-688	030920	Pony Peak	PP	144.2	2.5	1.79	11	25	5	3	6		^{4}53210, ^{46}06095	Biotite hornblende tonalite
Wooley Creek suite														
FPE-288	031208	English Peak	EP	156.2	1.0	0.99	25	25					^{4}79820, ^{45}74670	Pyroxene-bearing biotite hornblende quartz diorite
MMB-950	031224	Shelley Lake	SLP	159.1	1.3	1.21	21	30	2	1	6		^{4}96586, ^{45}94161	Hornblende biotite quartz
OMB-I	030131	Indian Rocks	IR	160.5	1.9	1.65	26	27				1	^{4}64693, ^{45}60477	Andesitic dike, phenocrysts of garnet, plagioclase, hornblende ± quartz
FPV-4	030131	Vesa Bluffs	VB	167.1	1.8	1.08	21	27			6		^{5}18480, ^{46}29325	Biotite tonalite
EK-188	030131	Castle Crags	Crags	167.5	2.0	1.36	21	27	3		3		^{5}55885, ^{45}56736	Biotite hornblende granodiorite
Unassociated plutons														
OMB-3090	031224	Orleans Mountain	OM	170.7	1.7	1.70	23	30	1	1	4	1	^{4}63475, ^{45}69170	Biotite hornblende quartz diorite
KM-52		Ukonom Mountain§§	—	165									^{4}59810, ^{46}03120	Biotite hornblende tonalite

Notes: MSWD—mean square of weighted deviates; s.e.—standard error.

*Age is common Pb-corrected $^{206}Pb/^{238}U$ age employing a ^{208}Pb-based common Pb calculation. See text for details.

§Criterion A is rejection when uncertainty of $^{206}Pb/^{238}U$ is more than twice that expected from counting statistics.

†Criterion B is rejection based on discordance, in this case, when $^{206}Pb/^{238}U$ and $^{207}Pb/^{235}U$ ages differ by more than 5%.

**UTM coordinates, 1927 NAD, zone 10, easting, northing.

§§See Appendix 1.

images were produced on a Hitachi S-225ON scanning electron microscope at 15 kV and 100–300× magnification.

Zircons were analyzed in five sessions in a 13-month period. Epoxy mounts containing multiple unknown populations, Temora 2 zircon standard (416.8 ± 1 Ma: Black et al., 2004), National Institute of Standards & Technology (NIST) 610 glass and 98-521 zircon standard were placed in a sample cell connecting a 193-nm wavelength Lambda Physik LPX 120I ArF excimer laser to an Agilent 7500S ICP-MS through a signal-smoothing manifold. The laser delivered 100-mJ constant energy at 5 Hz. Ablated holes were 32 µm in diameter. Masses used in geochronology (^{206}Pb, ^{207}Pb, ^{208}Pb, ^{232}Th, ^{235}U, and ^{238}U) were each analyzed for 0.04 s, and other elements of geochemical interest (^{29}Si, ^{31}P, ^{91}Zr, ^{177}Hf), and seven rare earth elements (REEs) were each collected for 0.01 s. Data were collected in peak jumping mode with a mass sweep time of 0.384 s. To avoid nonlinear detector response at very high count rates, ^{235}U was employed only in the event that U concentration was >2500 ppm; otherwise, ^{235}U was calculated from $^{238}U/137.88$. Because of Hg backgrounds ten times that of Pb and the isobaric interference of ^{204}Hg and ^{204}Pb, ^{204}Pb was not collected.

Backgrounds (laser off) were collected for ~25 s at the beginning of each analysis, after which the laser-on signal was collected for 35 s with a sample drilling rate of ~0.7 µm/s. The analysis cycle consisted of Temora 2 (zircon standard), NIST 610, 98-521, or R33 (secondary zircon standard) and then ten unknowns. Unknowns were analyzed in a round-robin fashion, so that the averages of the standards were applicable. Two standard materials were necessary because (1) precision for geochronology requires that down-hole fractionation effects on isotope ratios be modeled from zircon and (2) homogenous glass is required for calculating concentrations, especially Th/U (assumed to be 1.01870 in NIST 610), as standard zircons have uniform ages but are compositionally heterogeneous. Fractionation factors for the isotope ratios of interest (known ratio/measured ratio), $^{206}Pb/^{238}U$, and the like, derived from each mass sweep of the average Temora standard, were applied to the corresponding mass sweep in the unknown. Similarly, fractionation factors were derived from known/measured count ratios in NIST 610 for $^{232}Th/^{238}U$ and all elements ratioed to ^{29}Si. Elemental concentrations were derived by multiplying element/^{29}Si count ratios according to the NIST 610 fractionation factor, and by 32.77 wt% SiO_2, the assumed concentration in stoichiometric zircon. This technique of applying fractionation factors from the average standard on a mass sweep–by–mass sweep basis corrects for down-hole mass fractionation.

Uncertainties quoted for an individual age analysis (1 standard error [s.e.]; Table 3 and Appendix 1 [Appendices 1–3 are on the CD-ROM accompanying this volume and in the GSA Data Repository[2]]) include observed uncertainties plus a factor added in quadrature that is the standard deviation of the particular isotope ratio observed in average Temora 2. For instance, the standard deviation for $^{206}Pb/^{238}U$ observed in average Temora 2 in an analytical session is 1–2%. This added term accommodates the imperfections of a natural standard as well as instrument drift, and dominates the 1-s.e. uncertainties quoted for individual analyses. The success of this data reduction method is demonstrated by the 98-521 zircon data (Ballard et al., 2001; Table 3). Daily averages are in agreement with one another and with the laboratory's running average (42.6 ± 0.2 Ma). It is noteworthy that the mean square of the weighted deviates (MSWDs) for these populations from a well-behaved natural zircon standard are only somewhat greater than unity (1.2–1.6). Further details of the technique are presented in Harris et al. (2004).

Individual analyses were inspected for obvious age changes during the ablation, and for inclusions, particularly with reference to P and La contents. Ballard et al. (2002) have shown how inclusions express themselves, and how they can be avoided by visually selecting a portion (minimum of ten mass scans) of the ablation.

Common Pb corrections based on both ^{207}Pb and ^{208}Pb were calculated (see Appendix 1); the latter is preferred and is described here. Based on an initial calculation of $^{206}Pb/^{238}U$ age, knowing $^{232}Th/^{208}Pb$ at that age, and using a model common Pb composition of that age (Cumming and Richards, 1975), the amount of calculated excess (or deficit) ^{208}Pb was used to calculate the amount of common ^{206}Pb, common ^{207}Pb, and the total common Pb content. The appropriate amounts of common ^{206}Pb, and ^{207}Pb were then subtracted from (or added to) $^{206}Pb/^{238}U$, $^{207}Pb/^{235}U$, and $^{207}Pb/^{206}Pb$, and then the process was repeated at that new $^{206}Pb/^{238}U$ age, until these ratios stopped changing at the 0.1-Ma equivalent level. Because data on a set of zircons of a single age result in a cloud around concordia, disallowing small negative common Pb correction would bias the data set, and thus is allowed in age calculations. The resulting ratios are called the "^{208}Pb-corrected ages." The $^{206}Pb/^{238}U$ ages are quoted for analyses of grains younger than 1000 Ma, and $^{207}Pb/^{206}Pb$ ages for those older.

After compilation of analyses for each sample, two filters were applied to evaluate the quality of the selected portion and ablation. If observed uncertainty (s.e.) for the isotope ratio for that selected part of the analysis was more than twice that expected from counting statistics, the analysis was rejected, because such variation could result from a mixed-age analysis, inclusions, or the like. This circumstance was only encountered in 2.8% of 426 analyses of unknown zircons from rock samples (Table 2). Then concordancy was tested. For grains <1000 Ma, a grain was deemed concordant if the $^{206}Pb/^{238}U$ and $^{207}Pb/^{235}U$ ages, including uncertainty, agreed within 10%. Concordance in older grains was tested in a similar way but using $^{206}Pb/^{238}U$ and $^{207}Pb/^{206}Pb$. Discordant grains were detected in 2.6% of unknowns (Table 2).

The compiled population for a sample was inspected for

[2]GSA Data Repository item 2006198, Appendices 1–3, is available online at www.geosociety.org/pubs/ft2006.htm, or on request from editing@geosociety.org or Documents Secretary, GSA, P.O. Box 9140, Boulder, CO 80301, USA.

TABLE 3. ANALYSES OF IN-HOUSE ZIRCON STANDARD 98-521

						Number of rejected analyses	
Sample-date	Age* (Ma)	2 s.e.	MSWD	N	n	Discordance	Inheritance
030131	42.43	0.66	1.33	12	12		
030920	42.25	0.69	1.54	13	13		
031208	42.68	0.54	1.63	12	11	1	
031224	42.98	0.43	1.50	12	10	1	1(45.3 Ma)
040310	42.47	0.65	1.24	9	6	3	
Total	42.67	0.28	1.52	58	52	5	1

Notes: Running average exclusive of these data = 42.6 ± 0.2 Ma (n = 288). Sensitive high-resolution ion microprobe age = 43.1 ± 0.6 (n = 10; Ballard et al., 2001). MSWD—mean square of weighted deviates; N—total number of analyses; n—number of analyses used in age calculation; s.e.—standard error.

*Age is ^{208}Pb-corrected ^{206}Pb/^{238}U.

gross inheritance (ages many standard deviations older than the weighted mean, n = 24, 5.6%; Table 2 and Fig. 3). Once these were excluded, the remaining ablations were examined to determine if they represented a single age population by means of MSWD and by trends on cumulative probability plots (Fig. 4). If MSWD was greater than 2 and a distinct break in age distribution was obvious, slightly older zircons were interpreted to be inherited. Many examples occur in the data set, the most striking of which are the Vesa Bluffs and Orleans Mountain bodies. Zircons interpreted as inherited in this fashion make up 5.0% of the total analyses, for a total of 10.6% of all grains interpreted as inherited. In two instances, outliers younger than the main population were encountered and omitted from population age calculation. A weighted mean age for the rock was calculated from the remaining analyses and the weighted mean of the ^{208}Pb-corrected ^{206}Pb/^{238}U ages are reported in Table 2 and Figure 4. Uncertainties reported for populations are ±2 s.e.

Phosphorus was analyzed to detect inclusions, so that parts of the ablation that encountered inclusions could be avoided during age analysis. Hafnium and seven REEs were collected to examine variations in zircon composition. Lanthanum is at or below detection limits in zircon (<1 ppm). There is no rigorous secondary standard for these elements, because natural zircon standards are heterogeneous with respect to trace elements; precision is assumed to be ±2%. The REEs in zircon were used to calculate the ratios of Ce^{IV}/Ce^{III} (positive anomaly), and of Eu^{II}/Eu^{III} (negative anomaly), based on an average crust composition as a proxy for parent magma composition using the technique of Ballard et al. (2002). The average-crust-normalized REEs were plotted against ionic radius (Onuma plot) and a line fitted to Nd, Sm, Dy, and Lu to define an REE "baseline". The projected value at the ionic radii for Ce and Eu give the estimated Ce^{III} and Eu^{III} contents, respectively. In Figure 5, only trace element values for analyses used in the age calculation are shown. Moreover, if the fit to the baseline REEs (Nd, Sm, Dy, and Lu) was poor, no anomaly is plotted. The Eu ratio is better expressed as Eu_N/Eu'_N or the amount of measured Eu relative to that predicted from surrounding REEs (if $Eu^{III} = Eu_{tot}$) because zircon does not incorporate Eu^{II}. These REE anomalies reflect crystallization conditions: temperature, oxidation state, competing mineral species, and source magma composition.

Isotopic data for Nd, Sr, and oxygen are compiled in Appendix 3. Most data are from the literature, but data for the Forks of Salmon, Vesa Bluffs, Thompson Ridge, Pony Peak, Yellow Butte, Grants Pass, Gold Hill, Jacksonville, and Salmon Mountain plutons were previously unpublished. The Rb and Sr isotope data for these samples were determined by R.W. Kistler at

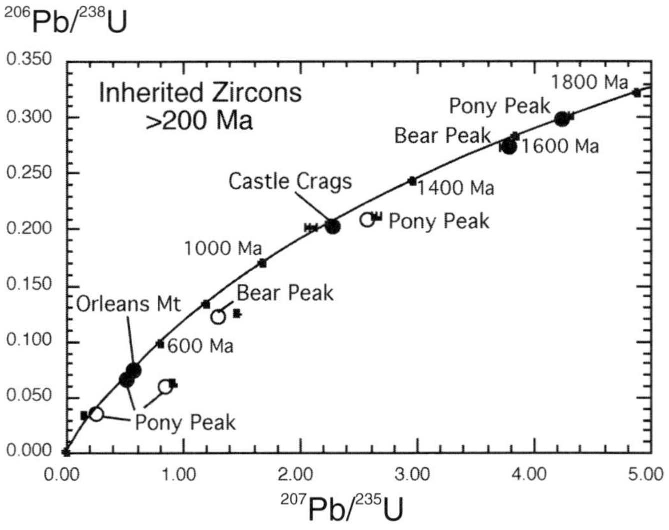

Figure 3. Concordia plot for obviously inherited, pre-Mesozoic zircons from the thirteen Mesozoic samples. Circles, filled and unfilled, are uncorrected U-Pb isotope data for concordant and discordant grains, respectively. Small crosses with error bars are ^{208}Pb-corrected data for all points. Note the prominence of Pony Peak and Bear Peak. Tick marks on concordia are at 200-Ma intervals.

the U.S. Geological Survey. Analytical methods are given in Barnes et al. (1992a). Oxygen isotope ratios were analyzed at Texas Tech University or the University of Chicago by H.R. Karlsson and J. Browning. Oxygen was liberated from silicates using the BrF_5 method of Clayton and Mayeda (1963) and converted to CO_2 by passage over a hot graphite rod. Oxygen isotopic ratios were obtained on a VG SIRA-12 dual-inlet mass spectrometer. All values are relative to Vienna Standard Mean Ocean Water. Silicate analyses are precise to ±0.2 per mil. The average and standard deviation obtained for NBS-28 is 9.45 ± 0.2 per mil (s.e. is ±0.02 per mil).

ZIRCON GEOCRHONOLGY, INHERITANCE, AND TRACE ELEMENT CHARACTERISTICS

Crystallization Ages of Plutons

The U-Pb data and interpreted ages produced in this study are presented in Appendix 1 and summarized in Table 2. Ages are presented roughly in order of decreasing age. When combined with the results of Yule (1996), Yule et al. (1996; this volume), and Irwin and Wooden (1999), no new age groups are recognized.

Zircons from the Vesa Bluffs pluton yielded an age of 167.1 ± 1.8 Ma, which is consistent with its placement in the Wooley Creek suite. Zircons from the Castle Crags pluton gave a similar age (167.5 ± 2.0 Ma), which suggests that it also belongs to the Wooley Creek suite. A sample from an unusual set of garnet-phyric andesitic dikes (Indian Rocks, sample OMBI) that crop out in the southern Klamath Mountains province yielded a 160.5 ± 1.9-Ma age. Detailed analysis of these dikes is given in a separate chapter (Barnes and Allen, this volume), but the age data clearly place it in the Wooley Creek suite. The Shelly Lake pluton contains a subtle inherited component; when this is excluded, an age of 159.1 ± 1.3 Ma is obtained. A sample of the main stage of the English Peak pluton yielded an age of 156.2 ± 1.0 Ma.

The Orleans Mountain pluton is one of four small quartz diorite to tonalite plutons in the west-central Klamath Mountains province (Fig. 1). These plutons are characterized by subsolidus foliation formed under upper greenschist-facies conditions. Zircons from the Orleans Mountain pluton gave an age of 170.7 ± 1.7 Ma, which is similar to a conventional 165-Ma TIMS age for the Ukonom Mountain pluton (Appendix 2). Although the age of the Orleans Mountain pluton overlaps with that of the Ironside Mountain suite, zircon inheritance (see below) and geochemical data suggest placement in the Wooley Creek suite.

The Pony Peak and Bear Peak plutons gave zircon ages of 144.2 ± 2.5 and 143.7 ± 1.3 Ma, respectively. These ages overlap with the youngest ages from the nearby Bear Mountain intrusive complex, which ranges in age from 149.9 ± 0.6 to 144.9 ± 0.8 Ma (Chamberlain et al., this volume). However, the compositional similarities between these two plutons and the Bear Mountain intrusive complex suggest that they are related. Elsewhere in this volume, Barnes et al. (Chapter 17) include the Bear Mountain, Summit Valley, Lower Coon Mountain, Pony Peak, and Bear Peak plutons in their so-called "western Klamath suite" of plutons, 151–144 Ma.

There are two distinct compositional groups of plutons younger than the western Klamath suite (post-western Klamath plutons). The ttg suite of plutons crops out in the eastern Klamath Mountains province (Fig. 1): the Sugar Pine, Craggy Peak, Caribou Mountain, Canyon Creek, and Deadman Peak. Our data significantly refine previously published ages for the ttg plutons. This refinement comes about because earlier multi-crystal U-Pb zircon dating (Wright and Fahan, 1988) found discordance between U-Pb and Pb-Pb ages, which we interpret as the effects of subtle inheritance (see below). The new data indicate 142.3- to 136.3-Ma emplacement ages (Table 1).

Among the post-western Klamath granodiorite suite plutons (Fig. 1), only the Yellow Butte pluton was dated in this study. The Yellow Butte pluton yielded a zircon age of 136.4 ± 0.9 Ma, which is in the age range of the other two plutons in the suite: 136 Ma, Shasta Bally pluton and 139 Ma Grants Pass pluton. Table 1 shows that on the basis of age, the ttg and granodioritic suites are indistinguishable; however major element compositions show the latter to be considerably more potassic (Barnes et al., 1992b).

Zircon Inheritance

The zircon inheritance of dated samples is described below in order of increasing age. Concordia and cumulative probability plots are provided in Figures 3 and 4. Two kinds of inheritance are discussed: obvious—that which is many standard deviations from the mean of the main population; and subtle—that which is defined on the basis of probability plots and population MSWD, as discussed below. Detailed discussion of this technique can be found above and in Campbell et al. (personal commun.)

Post-Western Klamath Plutons. The Yellow Butte pluton is the only member of the post-western Klamath granodiorite suite dated in this study (136.4 ± 0.9 Ma; MSWD = 0.85). It is a well-behaved sample and has the greatest number of analyses (48). It contains only a hint of inheritance—a small, 153-Ma core that was pierced at the end of an analysis that otherwise gave a 132-Ma age. This is only one of two instances in this study in which a core-rim pair with distinctly different ages was encountered in a single analysis.

Among the ttg suite plutons, the Sugarpine and Deadman Peak plutons contained well-behaved, inheritance-free zircon populations (136.3 ± 1.0 Ma, MSWD = 1.13 and 142.2 ± 1.1 Ma, MSWD = 1.71, respectively). In contrast, straightforward dating of the Craggy Peak, Caribou Mountain, and Canyon Creek plutons results in the same age within analytical uncertainty (138.2–140.0 Ma), all with high MSWD values (>3), but none with obvious pre-Jurassic inheritance. However, each sample

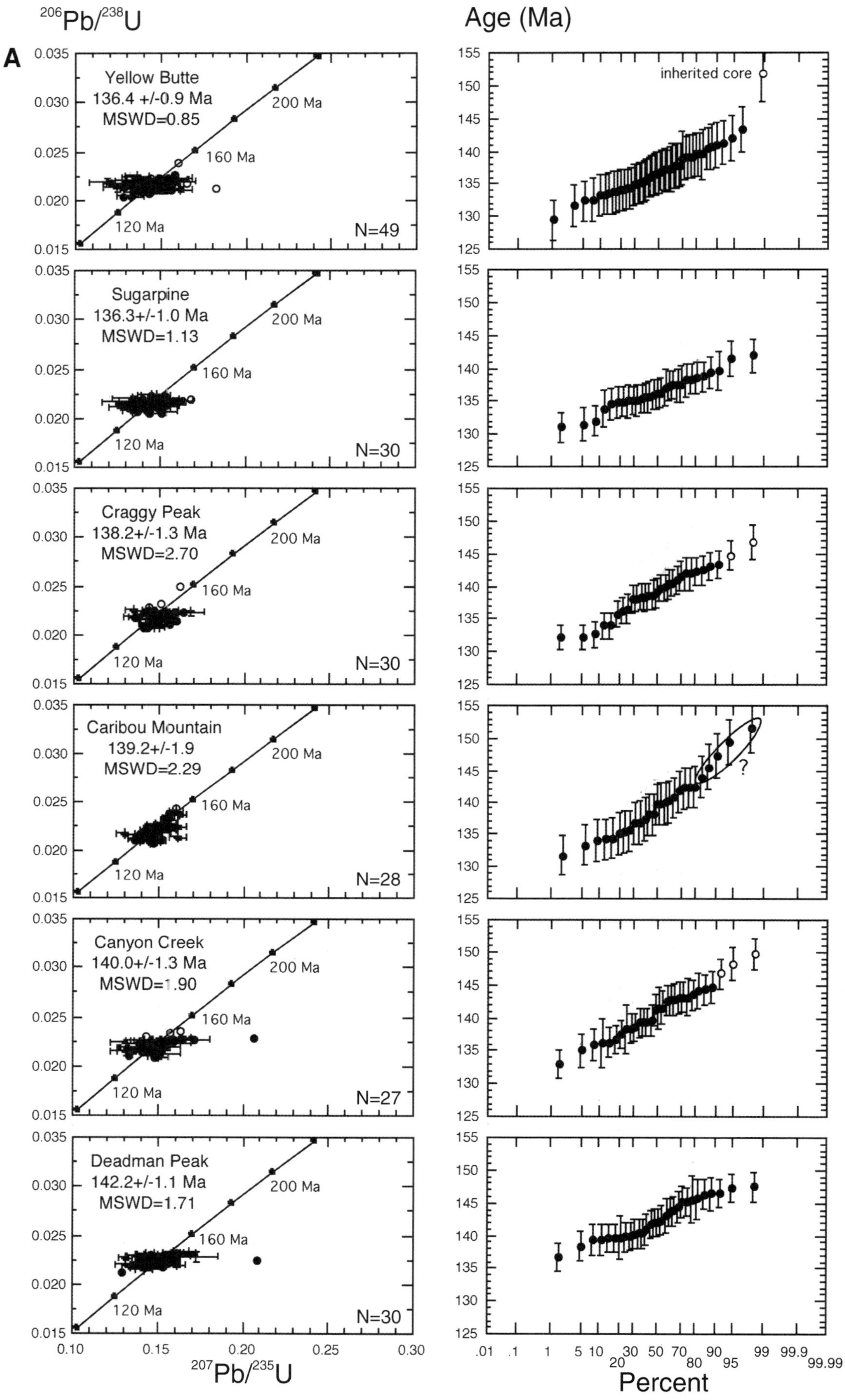
A
$^{206}Pb/^{238}U$
Age (Ma)
Yellow Butte
136.4 +/-0.9 Ma
MSWD=0.85
N=49
inherited core
Sugarpine
136.3+/-1.0 Ma
MSWD=1.13
N=30
Craggy Peak
138.2+/-1.3 Ma
MSWD=2.70
N=30
Caribou Mountain
139.2+/-1.9
MSWD=2.29
N=28
?
Canyon Creek
140.0+/-1.3 Ma
MSWD=1.90
N=27
Deadman Peak
142.2+/-1.1 Ma
MSWD=1.71
N=30
120 Ma
160 Ma
200 Ma
0.015
0.020
0.025
0.030
0.035
0.10
0.15
0.20
0.25
0.30
$^{207}Pb/^{235}U$
125
130
135
140
145
150
155
.01 .1 1 5 10 20 30 50 70 80 90 95 99 99.9 99.99
Percent

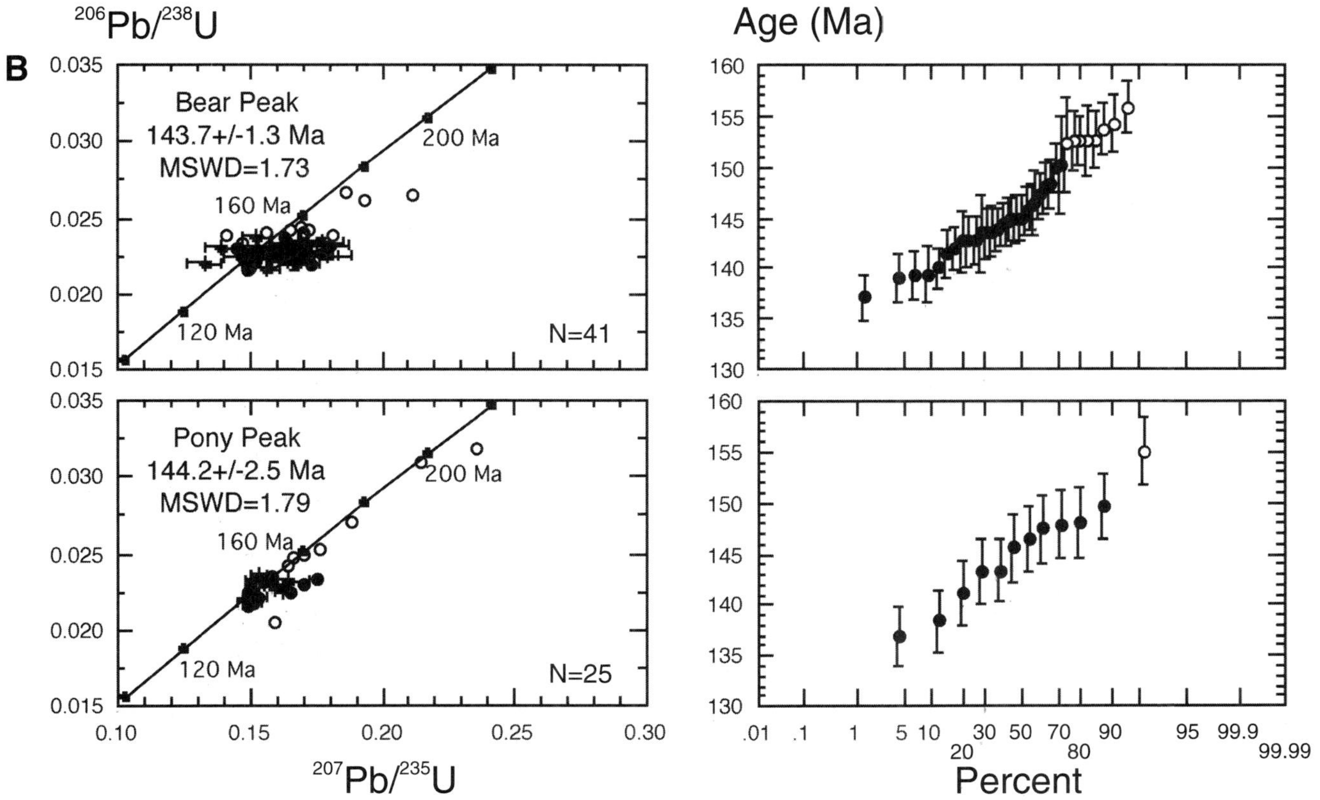

Figure 4. Concordia (left) and cumulative probability (right) diagrams for rocks dated in this study. In the concordia plots, filled and unfilled circles are uncorrected U-Pb isotope data. Unfilled circles represent data omitted from age calculation due to discordance, high internal noise, or inheritance. Crosses with error bars are the ^{208}Pb-corrected data corresponding to the filled circles. In the probability diagrams, ^{208}Pb-corrected ages are plotted. Points excluded as inherited *on the basis of these plots* are unfilled circles. All error bars are ±1-σ. Tick marks on concordia are at 20-Ma intervals. (A) Post-western Klamath-age plutons. (B) Dated western Klamath-age plutons. MSWD—mean square of weighted deviates.

contains zircons in the ca. 142- to 152-Ma age range. In the case of the Craggy Peak and Canyon Creek plutons, we interpret these zircons as subtle inheritance, ages of which are omitted from ages reported in Table 1 (see Fig. 4A). In the case of the Caribou Mountain pluton (Fig. 4) an age of 139.2 ± 1.9 Ma with MSWD of 2.29 results from the use of all data; however, an age of 137.5 ± 1.4 Ma with MSWD of 1.00 results if the five oldest grains are interpreted as subtle inheritance. The former, more conservative, interpretation is adopted here, because three of the five suspect grains were the last three analyses of the analytical session, leaving doubt *in this case only* that interpretation of subtle inheritance is actually unaccounted-for analytical drift.

Western Klamath Plutons. The Bear Peak and Pony Peak plutons contain both obvious and subtle inheritance. Two obviously inherited zircons were analyzed from the Bear Peak sample (Table 2, Fig. 3). In addition, Bear Peak provides a particularly good example of how probability plots are used to judge subtle inheritance in conjunction with MSWD. During the analytical session, secondary standard 98-521 gave MSWD of 1.52, which can be expected by a single-aged population (Table 3). If 152- to 156-Ma grains from Bear Peak are incorporated in the age calculation, the MSWD is 3.65. However, if grains above the 152-Ma inflection point in the probability plot (Fig. 4B) are not used, the calculated age is 143.7 ± 1.3 Ma with MSWD = 1.73, similar to MSWD for the standard.

Five of the nine pre-Mesozoic zircons analyzed (Fig. 3) are from the Pony Peak sample. One analysis gave a 138.5 ± 3.0-Ma rim and a 200.1 ± 4.4-Ma core. Pony Peak also contains subtle inheritance, with one 155-Ma grain judged to be inherited.

Wooley Creek Suite Plutons and Unassigned Orleans Mountain Pluton. The English Peak pluton and the garnet-phyric andesitic dike have no detected zircon inheritance of inheritance. The Shelly Lake pluton contains a subtle inherited component, 168 ± 2 Ma. The remaining Wooley Creek suite plutons, Vesa Bluffs, Castle Crags, and Orleans Mountain all display inheritance. Castle Crags contains obvious inherited zircons (189, 198, and 1230 Ma; Fig. 3) and the Orleans Mountain sample contained one 465-Ma grain. This group of zircons is the only one with significant common Pb; it appears to be well corrected using the 208Pb method described in the methods section

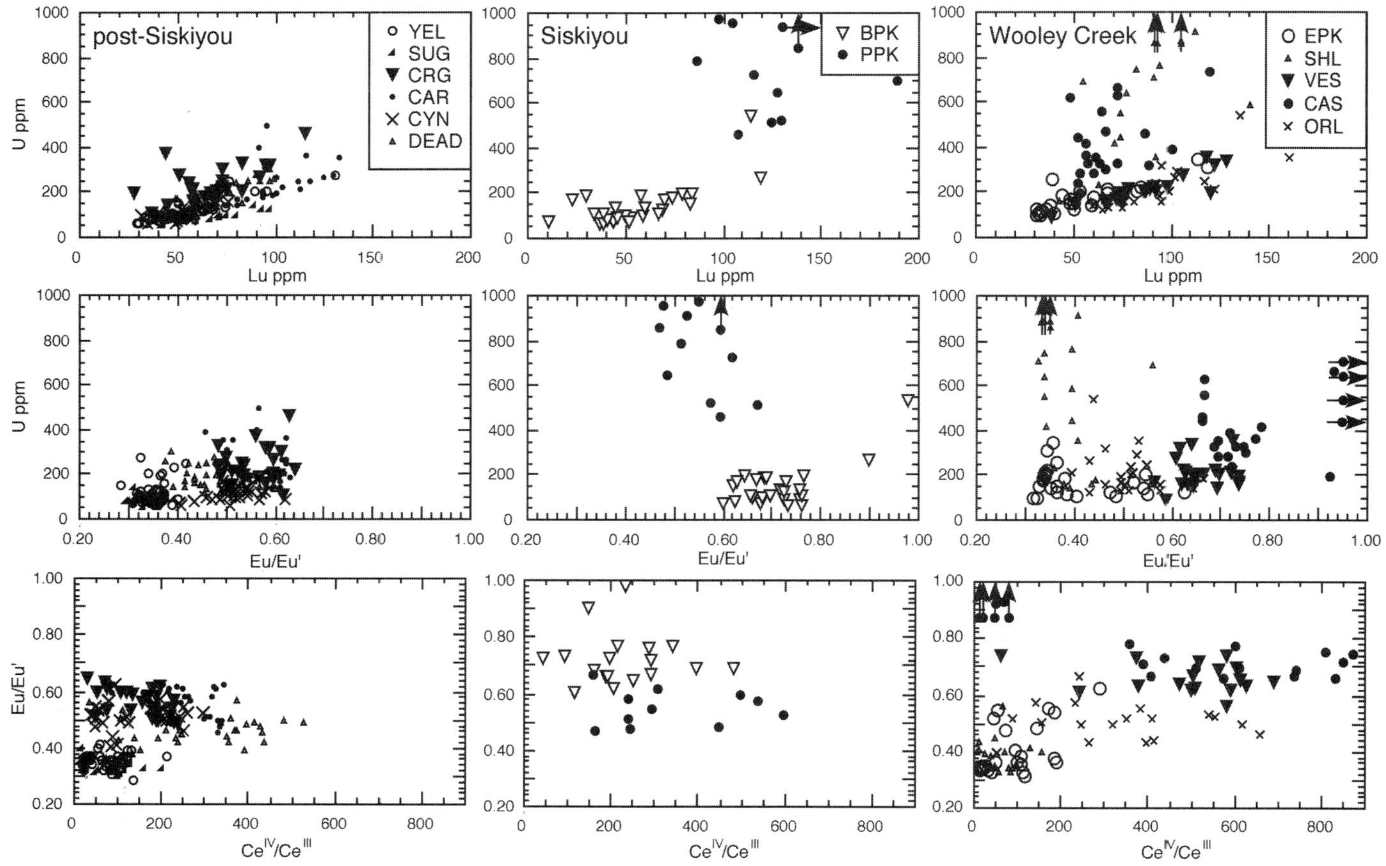

Figure 5. Trace element compositions for concordant, primary magmatic zircons. Zircon compositions excluded from these plots include inherited ones, those with high internal noise or discordance, and those that did not provide good line-fits ($r^2 < 0.99$) through elements used to define the baseline: Nd-Sm-Dy-Lu (see the section on analytical techniques). Sample abbreviations: BPK—Bear Peak; CAR—Caribou Mountain; CAS—Castle Crags; CRG—Craggy Peak; CYN—Canyon Creek; DEAD—Deadman Peak; EPK—English Peak; ORL—Orleans; PPK—Pony Peak; SHL—Shelly Lake; SUG—Sugarpine; VES—Vesa Bluffs; YEL—Yellow Butte.

above (see Fig. 4). Vesa Bluffs and Orleans Mountain otherwise display very similar age distributions, and together, they contain a distinct 182-Ma population (nine analyses). This similarity supports inclusion of the Orleans Mountain pluton in the Wooley Creek suite.

Discussion of Inherited Zircon

Prior to this work, inherited zircon in Klamath Mountains province plutons was rarely reported, and then only in cases where some of the inherited zircons were Precambrian (e.g., Harper et al., 1994). However, the laser ablation method can reliably identify subtle inheritance just a few m.y. older than the igneous age. Although inherited zircons were not found in all of the samples, each age and/or composition group appears to contain distinctive inheritance patterns.

In the Wooley Creek suite, all inherited zircons are 164–200 Ma in age except for two pre-Jurassic grains detected in the Orleans Mountain and Castle Crags plutons. These results were unexpected on the basis of previous conventional TIMS zircon studies on other Wooley Creek suite rocks, which resulted in concordant U-Pb ages (e.g., Barnes et al., 1986; Wright and Fahan, 1988). The youngest of these inherited zircons could be cannibalized Wooley Creek suite– or Ironside Mountain batholith–aged zircons (168 ± 2 Ma component in Shelly Lake, $n = 6$), whereas the oldest inherited Mesozoic grains (ca. 200 Ma) have ages that partly overlap those from the Rattlesnake Creek terrane (Wright and Wyld, 1994).

In contrast to the Wooley Creek suite, the two dated western Klamath suite plutons, the Pony Peak and Bear Peak, show much more complex inheritance patterns, consistent with previous TIMS U/Pb zircon age results of Harper et al. (1994), who reported discordant U-Pb data for Pony Peak and Lower Coon Mountain plutons. These plutons had lower intercepts of 146 ± 3 and 145 +1/–2 Ma, and upper intercepts of 1555 +598/–336 and 755 +158/–174 Ma, respectively. The upper intercepts are consistent with our results, showing the presence of Proterozoic zircons in both samples. Moreover, several samples from the

Bear Mountain intrusive complex contain inherited zircons (Chamberlain et al., this volume).

Among the three Early Cretaceous granodioritic plutons, only the Grants Pass pluton has been shown to have Proterozoic inheritance. It gave a lower intercept of 139 ± 2 Ma and an upper intercept of 1792 +168/–189 Ma (Harper et al., 1994). In contrast, our results from the 136.4 ± 0.9-Ma Yellow Butte pluton showed evidence of only one 153-Ma inherited core among forty-eight analyzed grains.

Some of the plutons in the Early Cretaceous ttg suite show inheritance in the 150- to 160-Ma range (Table 1). (It is possible that analysis of a larger zircon population from these plutons will identify inheritance in all of them.) This inheritance may explain the difficulty Wright and Fahan (1988) had in interpreting multi-grain TIMS results for these plutons. Moreover, it suggests either a common contaminant for the ttg magmas or a common source. Barnes et al. (1992b) suggested the latter and inferred the source to be metabasic rocks of the underthrust 165- to 161-Ma Josephine ophiolite. Although our results are consistent with a common source, they clearly show that such a source was younger than the Josephine ophiolite.

Trace Elements in Zircon

Some general comments about contrasts in zircon composition are appropriate here in terms of what is known about individual plutons and plutonic suites. We focus on the magnitude of ranges of trace element composition rather than on absolute concentrations, and in particular, on U and Lu concentrations and average crust-normalized Eu and Ce anomalies.

Uranium and Lu have high zircon/melt distribution coefficients that vary strongly and negatively with temperature, at least at 20 kb (D. Rubatto and J. Hermann, personal commun., 2005). For example, a synthetic hydrous granite melt composition yielded zircon/melt partition coefficients for U from about 15 to 180 between 1050 and 800 °C, respectively. Zircons in relatively low-temperature magmas have even higher partition coefficients for Lu, about 1–450 in the same experiments. We argue that linear trends in the top panels of Figure 5 result from fractional crystallization (i.e., increases in both U and Lu in the magma due to removal of major mineral phases) and from the decrease in temperature and consequent increase in zircon/melt distribution coefficient.

The rare earth *anomalies* in zircon are only partly related to fractional crystallization. The Eu/Eu′ in zircon reflects the supply of Eu^{III} available, relative to other REEs, when zircon crystallizes. This Eu^{III} supply is primarily controlled by plagioclase fractionation, either in the magma or in the source rock. Magma oxidation state also influences the Eu^{II}/Eu^{III} but to a lesser degree. Therefore, the Eu/Eu′ in zircon is essentially a passive indicator of magma composition and oxygen fugacity. In contrast, Ce/Ce′ is controlled by the strong affinity of zircon for any Ce^{IV} available in the magma (still a small percentage of total Ce). This is because Ce^{IV} and Zr^{IV} share not only the same charge but also similar ionic radii, so that zircon preferentially incorporates Ce^{IV} over Ce^{III}, particularly at lower temperatures (Rohrlach, 2002). Clearly, the Ce/Ce′ ratio is also dependent on oxidation state, which determines the amount of Ce^{IV} in the magma.

Post-western Klamath plutons all display strong positive correlations of Lu and U, with the Craggy Peak and Canyon Creek samples displaying the greatest dispersion at concentrations of U >300 ppm. The Sugarpine sample has the most restricted U range (maximum, 133 ppm), and the Sugarpine and Yellow Butte samples display the lowest and most restricted Eu anomalies (0.3–0.4). In contrast, the Deadman Peak sample has the greatest variation in Eu anomaly (0.38–0.60) and Ce anomaly (35–530). The relatively restricted range of U concentrations in Deadman Peak zircon is indicative of limited fractionation, which therefore suggests that the large range in Eu/Eu′ results from a range of parental magmas (i.e., the range of Eu anomalies arises from multiple magmas, probably due to magma mixing). The range of Ce anomalies in the Deadman Peak sample may be similarly interpreted as the result of crystallization from magmas with a range of oxidation conditions and temperatures. The high Ce anomalies in this sample represent higher oxidation states and/or lower temperatures during zircon crystallization than any other post-western Klamath body. The Canyon Creek zircons have maximum U concentration of only 160 ppm, but Eu/Eu′ ranges from 0.41 to 0.62. The Caribou Mountain sample is potentially the most fractionated (highest U) and it has a modest Eu anomaly range (0.46–0.62). With respect to Ce anomalies, the Sugarpine sample is the most restricted (17–203). When the REE anomalies are plotted against one another (Fig. 5), two groups emerge: those of restricted range (Sugarpine and Yellow Butte) and those with wide ranges, the extreme being Deadman Peak.

The two western Klamath-age plutons display distinct trace element features. Uranium and Lu are strongly correlated in the Bear Peak sample and U concentrations are, except for two analyses, <200 ppm. Except for the same two analyses, the Eu anomaly is limited to the 0.60–0.77 range, but the Ce anomaly range is large (60–465). The Pony Peak sample has higher U concentrations than any analyzed pluton except Shelly Lake (Fig. 5), but correlation with Lu is poor.

Wooley Creek group plutons display strong positive correlation of Lu and U (Fig. 5), but Shelly Lake and Castle Crags show more scatter than the others. Most grains have <800-ppm U except for Shelly Lake, which contains many relatively U rich zircons (to 1600 ppm). This finding is at odds with Eu and Ce anomalies from the Shelly Lake sample, which are the most limited of the suite (Fig. 5). The Castle Crags zircons show moderate U enrichment (to 600 ppm), have a moderate to high Eu/Eu′ range, and have a remarkably wide range of Ce anomaly values (12–910). Vesa Bluffs and Orleans Mountain samples are similar in their trace element ranges, and three of the five Wooley Creek suite plutons display very wide ranges in zircon Ce anomalies.

The use of zircon trace element data in petrogenetic studies is still a nascent field. However, it is already clear that many

Wooley Creek suite plutons can be distinguished from western Klamath and post-western Klamath plutons on the basis of higher U contents and larger Eu and Ce anomalies. We interpret zircons from the Yellow Butte, Sugarpine, and English Peak samples to reflect homogenous parental magmas that underwent little evolution in terms of fractional crystallization or changes in temperature and oxidation state. It is noteworthy that these conclusions refer to individual samples, not to their host pluton. For example, Schmidt (1994) showed that the English Peak pluton is composite and consists of a number of isotopically distinct pulses. In contrast to these zircons with simple, restricted trace element behaviors, zircon from the Shelly Lake pluton crystallized from strongly fractionated magma(s) in which plagioclase had not yet separated. Large Ce and Eu anomalies uncorrelated with large variation in U concentration probably reflect prolonged crystallization from a hybrid magma with multiple magma sources (i.e., magma mixing), instead of fractional crystallization (e.g., Deadman Peak).

There is no simple correlation of inheritance patterns with the trace element trends. However, the Yellow Butte, Suparpine, and English Peak samples all have limited ranges of U concentration and Eu and Ce anomalies and all lack inherited zircon. This pattern may be indicative of relatively homogeneous, metabasic source rocks.

Longevity of Individual Plutonic Systems

On the basis of traditional TIMS dating of multiple crystal populations, Yule (1996) showed that the Grayback and Ashland plutons of the Wooley Creek suite (Fig. 1, Table 1) were emplaced over several million years and can therefore be considered to be composite bodies. Similarly, the published dates for the English Peak pluton span 160–164 Ma (Wright and Fahan, 1988). The part of the pluton dated by Wright and Fahan (1988) is actually the oldest unit of the magma system: the Heiney Bar pluton of Seyfert (1964). Our sample is from the main stage of the pluton; it yielded an age of 156.2 ± 1.0 Ma. This range indicates that English Peak magmatic activity spanned at least 4 m.y. and perhaps 8 m.y. In contrast, samples from the composite Wooley Creek batholith–Slinkard pluton that range from quartz diorite to two-mica granite were dated by traditional multicrystal TIMS methods and yielded contemporaneous ages of ca. 161–162 Ma within analytical uncertainty (Barnes et al., 1986). Reasons why some Klamath Mountains province plutonic systems spanned several million years of activity and other batholith-scale plutons were emplaced in shorter time intervals are not yet clear.

ISOTOPIC VARIATION AND COMPOSITIONAL GROUPINGS

Figure 6 presents the available Sr, Nd, and oxygen isotope data for Middle Jurassic to Early Cretaceous Klamath Mountains province plutons (see Appendix 3). This data set fills in compositional and/or age groupings outlined by Barnes et al. (1992b) but does not significantly change them. The most general feature of the data set is the distinction between pre-Nevadan plutons, particularly the Wooley Creek suite, and post-western Klamath plutons (ttg and granodioritic suites). The former has generally higher initial $^{87}Sr/^{86}Sr$, with most samples >0.7038 and the latter has lower initial $^{87}Sr/^{86}Sr$, with all samples <0.7038 (Fig. 6B). Moreover, all but three samples from the Wooley Creek suite have $\delta^{18}O$ greater than +8‰ and several evolved samples have $\delta^{18}O$ above +10‰. In contrast, none of the ttg or granodiorite suite plutons have $\delta^{18}O$ greater than +10‰ and most of the data cluster between +6 and +8‰. Samples from the Ironside Mountain pluton display a limited range of both Nd and Sr isotope ratios (Barnes et al., this volume, Chapter 10; Fig. 6A), although a single sample from the undated West China Peak complex has Nd and Sr isotope ratios that overlap with the Wooley Creek suite.

These isotopic distinctions are illustrated as a function of age in Figure 7, which shows that the lowest $\delta^{18}O$ and initial $^{87}Sr/^{86}Sr$ values of the pre-Nevadan Ironside Mountain batholith and Wooley Creek suite plutons are higher than those of post-Nevadan western Klamath and ttg suite plutons. These plots also emphasize the wide compositional range of the Wooley Creek suite compared to the limited variability of post-Nevadan magmatism.

DISCUSSION AND PETROGENETIC IMPLICATIONS

Earlier in this chapter and elsewhere in the volume (Barnes et al., Chapters 10 and 17; Bushey et al.; McFadden et al.), the relationships between Middle Jurassic to Early Cretaceous magmatism and tectonism in the Klamath Mountains province are discussed. In this time frame, regional faulting permits division of Klamath Mountains province magmatism into three distinct intervals. The oldest is the western Hayfork terrane with coeval plutons (Forks of Salmon and Squaw Mountain). This period was followed, after thrusting on the Wilson Point thrust, by emplacement of the Ironside Mountain batholith and the Wooley Creek suite, with coeval outboard magmatic activity that formed the Josephine ophiolite and the Rogue–Checto arc complex. Following Nevadan thrusting (Orleans fault), the western Klamath suite was emplaced primarily in the western Klamath Mountains province. Post-western Klamath magmatism encompassed the ttg and granodiorite suites. With the exception of the young granodioritic plutons, petrogenetic analysis of each of these magmatic suites is in the literature; therefore, in the following section, we give an overview of the inferred petrogenesis of these magmatic units and discuss the implications of our new data.

Western Hayfork Arc

Geochemical data on volcanic clasts from the western Hayfork terrane (Barnes et al., 1995a), and on the coeval Forks of

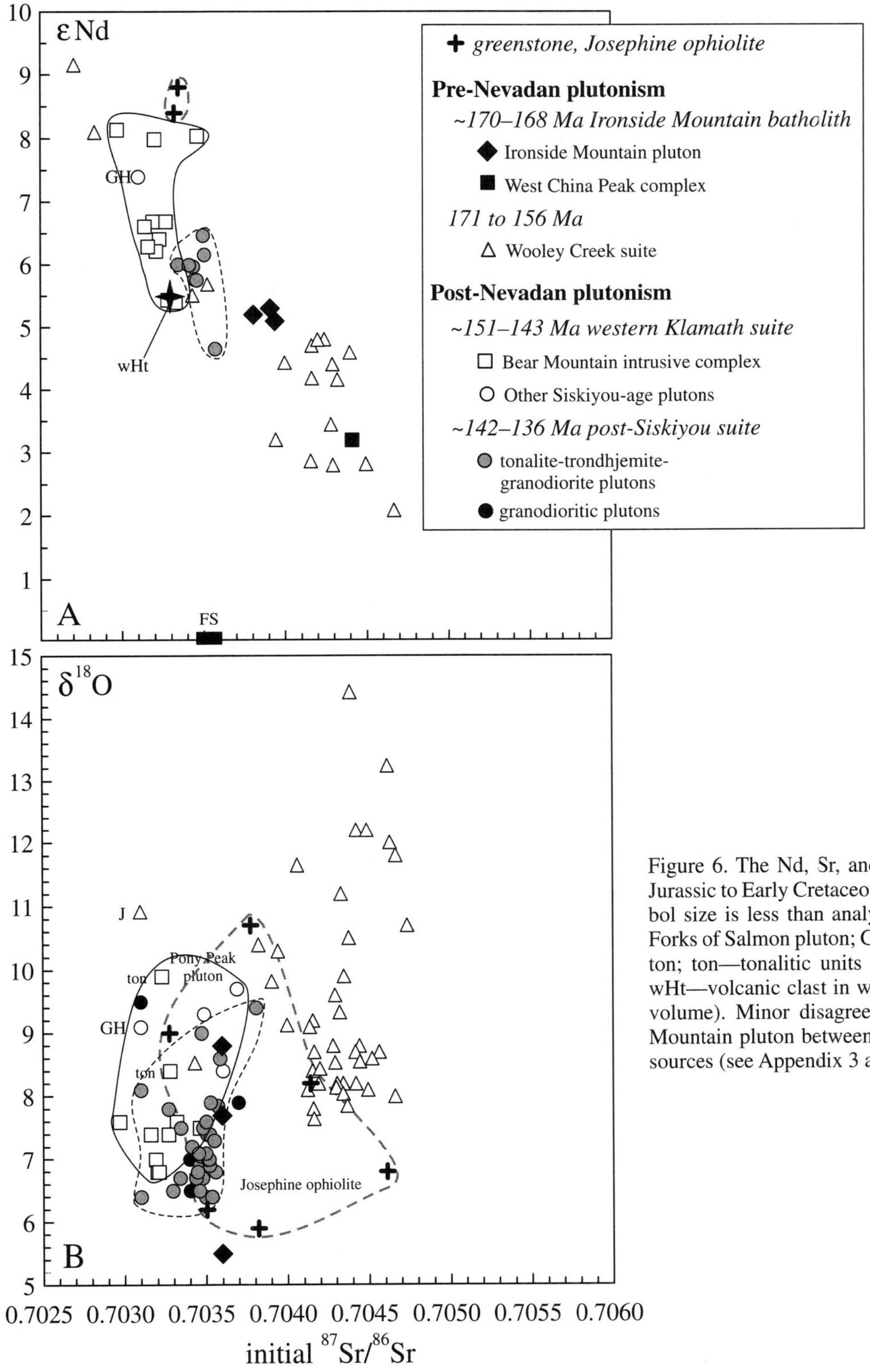

Figure 6. The Nd, Sr, and oxygen isotope compositions of Middle Jurassic to Early Cretaceous plutons in the Klamath Mountains. Symbol size is less than analytical uncertainty. FS—initial $^{87}Sr/^{86}Sr$ of Forks of Salmon pluton; GH—Gold Hill pluton; J—Jacksonville pluton; ton—tonalitic units in the Bear Mountain intrusive complex; wHt—volcanic clast in western Hayfork terrane (Bushey et al., this volume). Minor disagreement in initial $^{87}Sr/^{86}Sr$ for the Ironside Mountain pluton between panels A and B result from different data sources (see Appendix 3 and Barnes et al., this volume, Chapter 10).

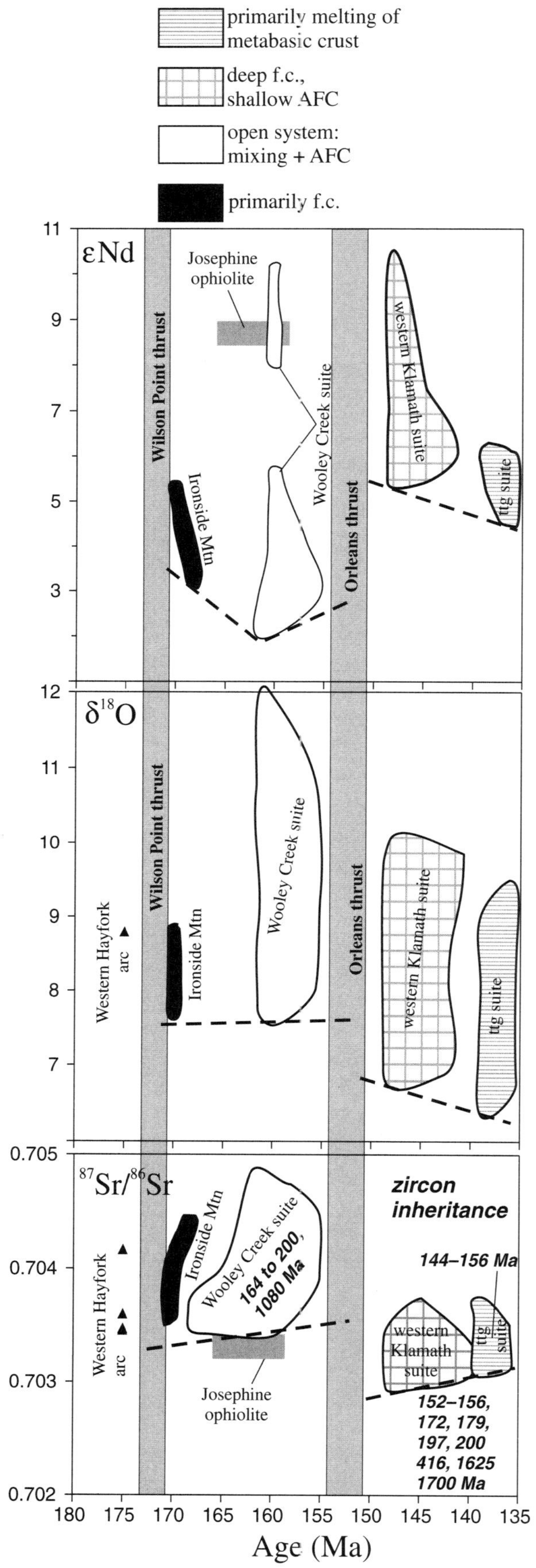

Figure 7. Variation of Nd, Sr, and oxygen isotope values and inherited zircon ages in Middle Jurassic–Early Cretaceous plutons as a function of time. Vertical bars show approximate times of regional thrust faulting. Dashed lines represent minimum isotope ratios displayed by plutons in individual time and/or composition groups. AFC—assimilation and fractional crystallization; f.c.—fractional crystallization; ttg—tonalite-trondhjemite-granodiorite.

Salmon pluton, indicate that the bulk of western Hayfork volcanism ranged from basalt to mafic andesite and was broadly calcic to calc-alkaline. Wright (1982) found similar results. Sparse isotopic data suggest ε_{Nd} and initial $^{87}Sr/^{86}Sr$ in the +5.5 and 0.7035 range, respectively (Appendix 3, Fig. 6). These data and the relatively flat REE patterns displayed by these volcanogenic rocks (Barnes et al., 1995a) are consistent with primitive arc magmatism. Too few data are available to evaluate possible effects of crustal contamination.

Ironside Mountain Batholith

The largest pluton in the batholith (Ironside Mountain pluton) ranges from calcic to alkalic and is characterized by unusual K_2O enrichment (Barnes et al., this volume, Chapter 10). The other plutons in the batholith are calcic and commonly contain abundant mafic to ultramafic cumulates (Charlton, 1979; Barnes et al., this volume, Chapter 10). Limited isotopic data for the Ironside Mountain batholith show it to have initial $^{87}Sr/^{86}Sr$ and ε_{Nd} slightly more primitive than those of most of the Wooley Creek suite (Fig. 6A). The enrichment in K_2O, curvilinear data arrays on trace and major element diagrams, and limited variation in isotopic compositions led Barnes et al. (this volume, Chapter 10) to propose that the Ironside Mountain pluton evolved by deep-seated fractional crystallization of a dry, low-fO_2 arc tholeiite, with little or no crustal assimilation. Other plutons in the batholith presumably had similar, but more H_2O-rich parental magmas.

Wooley Creek Suite

On the basis of the new zircon data, we suggest that the Wooley Creek suite consists of plutons as old as the 170.7-Ma Orleans Mountain pluton, and as young as late-stage granite in the Ashland pluton (152–156 Ma; Yule, 1996). The reason for including the Orleans Mountain pluton in this grouping is the similarity of zircon inheritance patterns in all of the plutons now included in the group. The Wooley Creek suite therefore overlaps the age range of the Ironside Mountain batholith; distinctions between the two groups are primarily geochemical (Barnes et al., 1992b, this volume, Chapter 17). All of the large Wooley Creek suite plutons are composite and some (e.g., Ashland) may span age ranges of as much as 9 m.y. The suite ranges from mafic and ultramafic cumulates to two-mica granites in the Slinkard and Vesa Bluffs plutons and shows a commensurate range of isotopic values, with both the highest and lowest ε_{Nd} and initial $^{87}Sr/^{86}Sr$ (Fig. 6). Barnes et al. (1990, 1992b, 1996) interpreted these characteristics to reflect contamination of mafic Wooley Creek suite magmas by assimilation of crustal rocks combined with mixing of additional basaltic input. They also suggested that many of the felsic rocks of the Wooley Creek suite were crustal melts, with or without later magma mixing (e.g., Barnes et al., 1996). A significant proportion of this assimilation and crustal melting was not at emplacement level, because even the most primitive rocks in the suite show oxygen isotope evidence for crustal contamination, and because many of the host rocks are compositionally inappropriate end-members (e.g., Barnes et al., 1990, 1992b, 1995b; Gribble et al., 1990). Instead, previous authors proposed that assimilation occurred in deeper crustal levels and required a metasedimentary contaminant that has not been identified in outcrop.

Zircon inheritance in these plutons ranges from 164 to 186 Ma. Possible contaminants at the level of emplacement (Rattlesnake Creek terrane, western Hayfork terrane, and eastern Hayfork terrane) do not encompass this age range (see above). In fact, ages from 180 to 186 Ma are not known in surface samples from the Klamath Mountains province. We therefore conclude that the isotopic and zircon inheritance data, and the presence of two-mica granites, are consistent with the presence of a probable Middle Jurassic metasedimentary terrane in the Klamath Mountains province subsurface during emplacement of the Wooley Creek suite.

Western Klamath Suite

The western Klamath plutonic suite was emplaced during and after Nevadan-age thrusting (151–143 Ma; Barnes et al., this volume, Chapter 17; Table 1). Rock types range from ultramafic to tonalitic and granodioritic, and most plutons are either calcic or calc-alkalic. These plutons are typically characterized by low K_2O contents compared to the Wooley Creek suite; however, the Buck Lake unit of the Bear Mountain pluton is relatively K_2O-rich (Bushey et al., this volume). Western Klamath suite plutons show a narrow range of initial $^{87}Sr/^{86}Sr$, from 0.7029 to 0.7037 and ε_{Nd} (+5.4 to +8.2), but $\delta^{18}O$ varies from +6.8‰ to +9.9‰. The ranges of isotope values for western Klamath suite plutons overlap slightly with those of the Wooley Creek suite (Fig. 6), but western Klamath suite plutons generally have lower initial $^{87}Sr/^{86}Sr$, higher ε_{Nd}, and generally lower $\delta^{18}O$ values. Barnes et al. (this volume, Chapter 17) suggested that much of the compositional variation within and among western Klamath suite plutons arises from fractional crystallization of at least two distinct mafic parents: H_2O-poor basalt to basaltic andesite and H_2O-rich basalt. However, major and trace element patterns and high $\delta^{18}O$ values of late-stage tonalitic and granodioritic units are consistent with an origin by crustal melting or substantial crustal assimilation.

The two western Klamath suite plutons dated in this work contain complex zircon inheritance age patterns, with ages from 154 Ma to Middle Proterozoic. Traditional TIMS dating has also shown the presence of Proterozoic inheritance in the Bear Mountain intrusive complex (Chamberlain et al., this volume), Pony Peak, and Lower Coon Mountain plutons, as well as the post-western Klamath Grants Pass pluton (Harper et al., 1994). In addition, zircons dated in this study have relatively broad ranges of Eu and Ce anomalies. This evidence indicates that either the parental magmas of these plutons arose from distinctive zircon-bearing source rocks or that they assimilated such rocks.

Miller et al. (2003) showed that graywacke collected near the base of the Galice Formation contains a detrital zircon assemblage nearly identical to that seen in the western Klamath suite plutons, including peaks at ca. 153, 185, 220, and 270 Ma. Miller et al. (2003) also reported the presence of Proterozoic zircons in this graywacke. The striking age correspondence of inherited zircons (Fig. 3) and exposed graywackes suggest that zircon inheritance in western Klamath suite plutons, and in the Grants Pass pluton, was derived from the Galice Formation. There are two ways in which Galice zircons could have been incorporated into these plutons. One is that the Galice Formation underwent partial melting and that these partial melts were emplaced with their load of detrital zircons. This idea is untenable for at least three reasons. First, partial melting of Galice Formation metasedimentary rocks should result in peraluminous magmas, which are not observed in abundance in the western Klamath suite. Second, coarse-grained mafic (gabbroic pegmatite) differentiates in the Bear Mountain pluton contain Proterozoic zircons (Chamberlain et al., this volume). Such gabbroic compositions cannot be obtained by partial melting of metagraywacke. Third, the metamorphic grade of the Galice Formation is greenschist facies or lower (Snoke, 1977). Regional geologic reconstructions and gravity data indicate that the Galice Formation dips shallowly beneath overlying Rattlesnake Creek terrane (Jachens et al., 1986) and, in the western Klamath Mountains province, did not reach depths sufficient to achieve anatexis.

An alternative way that Galice zircons could have been incorporated into western Klamath suite magmas is by assimilation of Galice rocks. Most parental western Klamath magmas were relatively hot (e.g., Bushey et al., this volume) and could have assimilated large proportions of the relatively soluble Galice metasedimentary rocks. Such assimilation could account for the large Eu and Ce anomaly ranges observed in the primary magmatic zircons—changes of temperature and oxidation state should be the main factors affecting these anomalies. More oxidizing and cooler magmas should result in higher $^{IV}Ce/^{III}Ce$. We therefore ascribe the inherited zircon of these plutons to assimilation during magma rise and emplacement.

Post-Western Klamath Plutonism

From 142 to 136 Ma, two compositional groups of plutons were emplaced (Table 1). The older group consists of ttg plutons of the central Klamath Mountains province. These plutons are characterized by low K_2O (Barnes et al., 1992b, 1995b) and are entirely calcic. Petrogenetic study (Barnes et al., 1996) indicated that these plutons are primarily partial melts of deep-seated metabasic crustal rocks. These authors suggested that the most likely source was the metabasic part of the Josephine ophiolite, which was thrust beneath older Klamath Mountains province terranes during the Nevadan orogeny. Ages of the few inherited zircons in the suite (146–156 Ma) are not entirely consistent with this idea, because the U-Pb ages of the Josephine ophiolite are too old (161–165 Ma). Furthermore, the source could not be from the flysch that overlies the Josephine Ophiolite (Galice Formation), because inherited zircon ages younger than 150 Ma must post-date Galice deposition (as well as for petrologic reasons cited above). Therefore, the zircon inheritance in ttg suite plutons is suggestive of the presence of an unknown zircon-bearing metabasic terrane in the Klamath Mountains province subsurface during Early Cretaceous time.

The granodioritic suite has somewhat higher K_2O contents and ranges from calcic to calc-alkalic. No detailed petrogenetic studies have been carried out, but preliminary data suggest that these plutons represent a combination of crustal melting and magma mixing.

Among plutons of the two post-western Klamath groups, zircon trace element geochemistry exemplifies at least two distinct types of magmatic process. The Sugarpine pluton represents an end-member derived from a mafic (zircon-free) source, presumably by partial melting. Zircon crystallized at a discrete point during magma evolution and is therefore low in U content and has limited ranges of Eu and Ce anomalies. The Deadman Peak sample represents another end-member, in which limited U enrichment in zircon suggests minor fractionation but large ranges of Eu and Ce anomalies suggest multiple sources, perhaps due to magma mixing. If so, these multiple sources were mafic and therefore lacked pre-existing (inherited) zircons.

CONCLUSIONS

Detailed studies cited here and presented elsewhere in this volume demonstrate Klamath Mountain pluton suites are diverse in element and isotopic compositions, mode of emplacement, and longevity. Our results extend suite distinctions to include characteristics of inherited zircons such as their abundance, age, and trace element characteristics. With the exception of the post-western Klamath ttg suite, too few plutons in each age and/or composition group have been analyzed to permit definitive statements. At present, our results suggest the following: Ironside Mountain suite magmas evolved primarily by deep-seated fractional crystallization; they show little evidence (in this case, isotopic) of crustal interaction. Wooley Creek suite magmas show ample evidence of interaction of mafic magmas with crustal rocks and/or melts. Their isotopic compositions and zircon inheritance suggest that the assimilated rocks were primarily Middle Jurassic sedimentary strata. Such a unit is not exposed in the province.

Western Klamath suite magmas also underwent deep-seated fractional crystallization. At least two parental magmas can be identified, one H_2O-rich and the other H_2O-poor. During magma storage (and fractionation) or during ascent to the level of emplacement, these magmas assimilated metasedimentary rocks of the Galice Formation and inherited their distinctive zircon population. Post-western Klamath ttg suite magmas arose by partial melting of metabasic rocks. The few inherited zircons in these units indicate widely variable magmatic processes and, at least locally, multiple, compositionally distinct source rocks. The metabasic source(s) were younger than the Josephine ophiolite, and thus represent yet another rock unit not exposed in the province. The youngest, granodioritic, plutonic suite is, as yet, poorly known from a petrogenetic standpoint. That the Grants Pass pluton contains a significant proportion of inherited zircon (including Proterozoic ones), whereas the Yellow Butte pluton yielded one 153-Ma inherited zircon core, suggests diverse magma histories and source and/or contaminant rocks. However, because the Yellow Butte pluton presumably intrudes Paleozoic rocks, this single inherited core indicates that Jurassic or younger rocks are present in the subsurface at the longitude of Mount Shasta.

ACKNOWLEDGMENTS

This work relies on assistance, advice, and discussion with many co-workers. In particular, we thank Ron Kistler, who graciously provided Sr isotope data over many years, Melanie Barnes, for her assistance in the field and laboratory, James Browning and H.R. Karlsson, who provided much of the oxygen isotope data, and R. Tosdal for dating the Ukonom Mountain zircons. We thank Porter Irwin, Allan Kays, Mary Donato, Greg Harper, Art Snoke, Gary Ernst, and Bob Coleman for their insight and encouragement, and Rob Gribble, Jimmy Boardman, Scott Petersen, Ken Johnson, Bennetta Schmidt, and Susan Tomlinson for their contributions in graduate school and beyond. John Hogan and Carol Frost provided helpful reviews. Funding for this research came from numerous National Science Foundation grants, the most recent of which are EAR–9902912 to CGB and EAR–9902807 to A.W. Snoke.

REFERENCES CITED

Ando, C.J., Irwin, W.P., Jones, D., and Saleeby, J., 1983, The ophiolitic North Fork terrane in the Salmon River region, central Klamath Mountains, California: Geological Society of America Bulletin, v. 94, p. 236–252, doi: 10.1130/0016-7606(1983)94<236:TONFTI>2.0.CO;2.

Ballard, J.R., Palin, J.M., Williams, I.S., and Campbell, I.H., 2001, Two ages of porphyry intrusion resolved for the super-giant Chuquicamata copper deposit of northern Chile by ELA-ICP-MS and SHRIMP: Geology, v. 29, p. 383–386, doi: 10.1130/0091-7613(2001)029<0383:TAOPIR>2.0.CO;2.

Ballard, J.R., Palin, J.M., and Campbell, I.H., 2002, Relative oxidation states of magmas inferred from Ce(IV)/Ce(III) in zircon: Application to porphyry copper deposits of northern Chile: Contributions to Mineralogy and Petrology, v. 144, p. 347–364.

Barker, F., Millard, H.T., Jr., and Knight, R.J., 1979, Reconnaissance geochemistry of Devonian island-arc volcanic and intrusive rocks, west Shasta district, California, *in* Barker, F., ed., Trondhjemites, dacites, and related rocks: Amsterdam, Elsevier, p. 531–545.

Barnes, C.G., and Allen, C.M., 2006, this volume, Depth of origin of late Middle Jurassic garnet andesite, southern Klamath Mountains, California, *in* Snoke, A.W., and Barnes, C.G., eds., Geological studies in the Klamath Mountains province, California and Oregon: A volume in honor of William P. Irwin: Boulder, Colorado, Geological Society of America Special Paper 410, doi: 10.1130/2006.2410(13).

Barnes, C.G., Allen, C.M., and Saleeby, J.B., 1986, Open- and closed-system characteristics of a tilted plutonic system, Klamath Mountains, California: Journal of Geophysical Research, v. 91, p. 6073–6090.

Barnes, C.G., Allen, C.M., Hoover, J.D., and Brigham, R.H., 1990, Magmatic components of a tilted plutonic system, Klamath Mountains, California, *in* Anderson, J.L., ed., The nature and origin of Cordilleran magmatism: Boulder, Colorado, Geological Society of America Memoir 174, p. 331–346.

Barnes, C.G., Barnes, M.A., and Kistler, R.W., 1992a, Petrology of the Caribou Mountain pluton, Klamath Mountains, California: Journal of Petrology, v. 33, p. 95–124.

Barnes, C.G., Petersen, S.W., Kistler, R.W., Prestvik, T., and Sundvoll, B., 1992b, Tectonic implications of isotopic variation among Jurassic and Early Cretaceous plutons, Klamath Mountains: Geological Society of America Bulletin, v. 104, p. 117–126, doi: 10.1130/0016-7606(1992)104 <0117:TIOIVA>2.3.CO;2.

Barnes, C.G., Donato, M.M., Barnes, M.A., Yule, J.D., Hacker, B.R., and Helper, M.A., 1995a, Geochemical compositions of metavolcanic and metasedimentary rocks, western Jurassic and western Paleozoic and Triassic belts, Klamath Mountains, Oregon and California: Reston, Virginia, U.S. Geological Survey Open-File Report 95-227.

Barnes, C.G., Johnson, K., Barnes, M.A., Prestvik, T., Kistler, R.W., and Sundvoll, B., 1995b, The Grayback pluton: Magmatism in a Jurassic back-arc environment, Klamath Mountains, Oregon: Journal of Petrology, v. 36, p. 397–415.

Barnes, C.G., Petersen, S.W., Kistler, R.W., Murray, R., and Kays, M.A., 1996, Source and tectonic implications of tonalite-trondhjemite magmatism in the Klamath Mountains: Contributions to Mineralogy and Petrology, v. 123, p. 40–60, doi: 10.1007/s004100050142.

Barnes, C.G., Mars, E.V., Swapp, S., and Frost, C.D., 2006, this volume (Chapter 10), Petrology and geochemistry of the Middle Jurassic Ironside Mountain batholith: Evolution of potassic magmas in a primitive arc setting, *in* Snoke, A.W., and Barnes, C.G., eds., Geological studies in the Klamath Mountains province, California and Oregon: A volume in honor of William P. Irwin: Boulder, Colorado, Geological Society of America Special Paper 410, doi: 10.1130/2006.2410(10).

Barnes, C.G., Snoke, A.W., Harper, G.D., Frost, C.D., McFadden, R.R., Bushey, J.C., and Barnes, M.A.W., 2006, this volume (Chapter 17), Arc plutonism following regional thrusting: Petrology and geochemistry of syn- and post-Nevadan plutons in the Siskiyou Mountains, Klamath Mountains province, California, *in* Snoke, A.W., and Barnes, C.G., eds., Geological studies in the Klamath Mountains province, California and Oregon: A volume in honor of William P. Irwin: Boulder, Colorado, Geological Society of America Special Paper 410, doi: 10.1130/2006.2410(17).

Barrow, W.M., and Metcalf, R.V., 2006, this volume, A reevaluation of the paleotectonic significance of the Paleozoic Central Metamorphic terrane, eastern Klamath Mountains, California: New constraints from trace element geochemistry and ^{40}Ar/^{39}Ar thermochronology, *in* Snoke, A.W., and Barnes, C.G., eds., Geological studies in the Klamath Mountains province, California and Oregon: A volume in honor of William P. Irwin: Boulder, Colorado, Geological Society of America Special Paper 410, doi: 10.1130/2006.2410(19).

Black, L.P., Kamo, S.L., Allen, C.M., Davis, D.W., Aleinikoff, J.N., Valley, J.W., Mundil, R., Campbell, I.H., Korsch, R.J., Williams, I.S., and Foudoulis, C., 2004, Improved ^{206}Pb/^{238}U microprobe geochronology by the monitoring of a trace-element-related matrix effect; SHRIMP, ID-TIMS, ELA-ICP-MS and oxygen isotope documentation for a series of zircon standards: Chemical Geology, v. 205, p. 115–140, doi: 10.1016/j.chemgeo.2004.01 .003.

Blackwelder, E., 1914, A summary of the orogenic epochs in the geologic history of North America: Journal of Geology, v. 22, p. 633–654.

Bushey, J.C., Snoke, A.W., Barnes, C.G., and Frost, C.D., 2006, this volume, Geology of the Bear Mountain intrusive complex, Klamath Mountains, California, *in* Snoke, A.W., and Barnes, C.G., eds. Geological studies in the Klamath Mountains province, California and Oregon: A volume in honor of William P. Irwin: Boulder, Colorado, Geological Society of America Special Paper 410, doi: 10.1130/2006.2410(14)

Chamberlain, K.R., Snoke, A.W., Barnes, C.G., and Bushey, J.C., 2006, this volume, New U-Pb radiometric dates of the Bear Mountain intrusive complex, Klamath Mountains, California, *in* Snoke, A.W., and Barnes, C.G., eds., Geological studies in the Klamath Mountains province, California and Oregon: A volume in honor of William P. Irwin: Boulder, Colorado, Geological Society of America Special Paper 410, doi: 10.1130/2006.2410(15).

Charlton, D.W., 1979, Geology of part of the Ironside Mountain quadrangle, northern California, Klamath Mountains [Ph.D. dissertation]: Santa Barbara, University of California, 542 p.

Clayton, R., and Mayeda, T., 1963, The use of bromine pentafluoride in the extraction of oxygen from oxides and silicates for isotopic analysis: Geochemica et Cosmochimica Acta, v. 27, p. 43–52, doi: 10.1016/0016-7037 (63)90071-1.

Cotkin, S.J., and Medaris, L.G., Jr., 1993, Evaluation of the crystallization conditions for the calcalkaline Russian Peak intrusive complex, Klamath Mountains, Northern California: Journal of Petrology, v. 34, p. 543–571.

Cumming, G.L., and Richards, J.R., 1975, Ore lead isotope ratios in a continuously changing Earth: Earth and Planetary Science Letters, v. 28, p. 155–171, doi: 10.1016/0012-821X(75)90223-X.

Donato, M.M., Barnes, C.G., and Tomlinson, S.L., 1996, The enigmatic Applegate Group of southwestern Oregon: Age, correlation, and tectonic affinity: Oregon Geology, v. 58, p. 79–91.

Dorais, M.J., 1983, The geology and petrology of the Shelly Lake pluton, central Klamath Mountains, California [MS thesis]: Eugene, University of Oregon, 147 p.

Ernst, W.G., 1993, Chemically distinct mafic dike/sill sequences of contrasting age ranges, Sawyers Bar area, central Klamath Mountains, northern California: Journal of Petrology, v. 34, p. 63–75.

Ernst, W.G., 1999, Mesozoic petrotectonic development of the Sawyers Bar suprasubduction-zone arc, central Klamath Mountains, northern California: Geological Society of America Bulletin, v. 111, p. 1217–1232, doi: 10.1130/0016-7606(1999)111<1217:MPDOTS>2.3.CO;2.

Garcia, M.O., 1982, Petrology of the Rogue River island-arc complex, southwest Oregon: American Journal of Science, v. 282, p. 783–807.

Goodge, J.W., 1995, Pre-Middle Jurassic accretionary metamorphism in the southern Klamath Mountains of northern California, USA: Journal of Metamorphic Geology, v. 13, p. 93–110.

Goodge, J.W., and Renne, P.R., 1993, Mid-Paleozoic olistoliths in eastern Hayfork terrane mélange, Klamath Mountains: Implications for late Paleozoic–early Mesozoic Cordilleran forearc development: Tectonics, v. 12, p. 279–289.

Gribble, R.F., Barnes, C.G., Donato, M.M., Hoover, J.D., and Kistler, R.W., 1990, Geochemistry and intrusive history of the Ashland pluton, Klamath Mountains, California and Oregon: Journal of Petrology, v. 31, p. 883–923.

Hacker, B.R., Donato, M.M., Barnes, C.G., McWilliams, M.O., and Ernst, W.G., 1995, Timescales of orogeny: Jurassic construction of the Klamath Mountains: Tectonics, v. 14, p. 677–703, doi: 10.1029/94TC02454.

Harper, G.D., 1980, The Josephine ophiolite—Remains of a Late Jurassic marginal basin in northwestern California: Geology, v. 8, p. 333–337, doi: 10.1130/0091-7613(1980)8<333:TJOOAL>2.0.CO;2.

Harper, G.D., 1984, The Josephine ophiolite, northwestern California: Geological Society of America Bulletin, v. 95, p. 1009–1026, doi: 10.1130/0016-7606(1984)95<1009:TJONC>2.0.CO;2.

Harper, G.D., and Wright, J.E., 1984, Middle to Late Jurassic tectonic evolution of the Klamath Mountains, California-Oregon: Tectonics, v. 3, p. 759–772.

Harper, G.D., Bowman, J.R., and Kuhns, R., 1988, A field, chemical, and stable isotope study of subseafloor metamorphism of the Josephine ophiolite, California-Oregon: Journal of Geophysical Research, v. 93, p. 4625–4656.

Harper, G.D., Saleeby, J.B., and Heizler, M., 1994, Formation and emplacement of the Josephine ophiolite and the Nevadan orogeny in the Klamath Mountains, California-Oregon: U/Pb zircon and $^{40}Ar/^{39}Ar$ geochronology: Journal of Geophysical Research, v. 99, p. 4293–4321, doi: 10.1029/93JB02061.

Harris, A.C., Allen, C.M., Bryan, S.E., Campbell, I.H., Holcombe, R.J., and Palin, J.M., 2004, ELA-ICP-MS U-Pb zircon geochronology of regional volcanism hosting the Bajo de la Alumbrera Cu-Au deposit: Implications for porphyry-related mineralization: Mineralium Deposita, v. 39, p. 46–67, doi: 10.1007/s00126-003-0381-0.

Hinds, N.E.A., 1934, The Jurassic age of the last granotoid intrusives in the Klamath Mountains and Sierra Nevada, California: American Journal of Science, 5th series, v. 27, p. 182–192.

Hotz, P.E., 1971, Plutonic rocks of the Klamath Mountains, California and Oregon: Washington, D.C., U.S. Geological Survey Professional Paper 684-B, p. B1–B20.

Hotz, P.E., Lanphere, M.A., and Swanson, D.A., 1977, Triassic blueschist from northern California and north-central Oregon: Geology, v. 5, p. 659–663, doi: 10.1130/0091-7613(1977)5<659:TBFNCA>2.0.CO;2.

Irwin, W.P., 1960, Geological reconnaissance of the northern Coast Ranges and Klamath Mountains, California, with a summary of the mineral resources: San Francisco, California Division of Mines Bulletin 179, 80 p.

Irwin, W.P., 1972, Terranes of the western Paleozoic and Triassic belt in the southern Klamath Mountains, California: Washington, D.C., U.S. Geological Survey Professional Paper 800C, p. C101–C111.

Irwin, W.P., 1981, Tectonic accretion of the Klamath Mountains, *in* Ernst, W.G., ed., The geotectonic development of California—Rubey volume 1: Englewood Cliffs, New Jersey, Prentice-Hall, p. 29–49.

Irwin, W.P., 1985, Age and tectonics of plutonic belts in accreted terranes of the Klamath Mountains, California and Oregon, *in* Howell, D.G., ed., Tectonostratigraphic terranes of the Circum-Pacific region: Houston, Circum-Pacific Council for Energy and Mineral Resources, p. 187–199.

Irwin, W.P., 1994, Geologic map of the Klamath Mountains, California and Oregon: Reston, Virginia, U.S. Geological Survey Miscellaneous Investigations Series Map I-2148, scale 1:500,000, 2 sheets.

Irwin, W.P., and Blome, C.D., 2004, Fossil localities of the Rattlesnake Creek, western and eastern Hayfork, and North Fork terranes of the Klamath Mountains: Reston, Virginia, U.S. Geological Survey Open-File Report 2004-1094, 1 sheet, 50 p.

Irwin, W.P., and Wooden, J.L., 1999, Plutons and accretionary episodes of the Klamath Mountains, California and Oregon: Reston, Virginia, U.S. Geological Society Open-File Report 99-374, 1 sheet.

Jachens, R.C., Barnes, C.G., and Donato, M.M., 1986, Subsurface configuration of the Orleans fault: Implications for deformation in the western Klamath Mountains, California: Geological Society of America Bulletin, v. 97, p. 388–395, doi: 10.1130/0016-7606(1986)97<388:SCOTOF>2.0.CO;2.

Lanphere, M.A., and Jones, D.L., 1978, Cretaceous time scale for North America, *in* Cohee, G., et al., eds., Contributions to the geologic time scale: Tulsa, Oklahoma, American Association of Petroleum Geologists, p. 259–268.

Lanphere, M.A., Irwin, W.P., and Hotz, P.E., 1968, Isotopic age of the Nevadan orogeny and older plutonic and metamorphic events in the Klamath Mountains, California: Geological Society of America Bulletin, v. 79, p. 1027–1057.

Lindsley-Griffin, N., Griffin, J.R., Farmer, J.D., Sivers, E.A., Bruckno, B., and Tozer, M.K., 2006, this volume, Ediacaran cyclomedusoids and the paleogeographic setting of the Neoproterozoic–early Paleozoic Yreka and Trinity terranes, eastern Klamath Mountains, California, *in* Snoke, A.W., and Barnes, C.G., eds., Geological studies in the Klamath Mountains province, California and Oregon: A volume in honor of William P. Irwin: Boulder, Colorado, Geological Society of America Special Paper 410, doi: 10.1130/2006.2410(20).

McFadden, R.R., Snoke, A.W., and Barnes, C.G., 2006, this volume, Structural and petrologic evolution of the Bear Peak intrusive complex, Klamath Mountains, California, *in* Snoke, A.W., and Barnes, C.G., eds., Geological studies in the Klamath Mountains province, California and Oregon: A volume in honor of William P. Irwin: Boulder, Colorado, Geological Society of America Special Paper 410, doi: 10.1130/2006.2410(16).

Miller, J., Miller, R., Wooden, J., and Harper, G., 2003, Geochronological links between the Ingalls ophiolite, North Cascades, Washington and the Josephine ophiolite, Klamath Mts., Oregon and California: Geological Society of America Abstracts with Programs, v. 35, p. 113.

Miller, M.M., Saleeby, J.B., Harper, G.D., and Wright, J.E., 1988, Jurassic orogenesis in the Klamath Mountains: Episodic tectonism and arc superposition within an evolving convergent margin: Geological Society of America Abstracts with Programs, v. 20, p. A274.

Miller, M.M., and Harwood, D.S., 1990, Paleogeographic setting of upper Paleozoic rocks in the northern Sierra and eastern Klamath terranes, northern California, *in* Harwood, D.S., and Miller, M.M., eds., Paleozoic and early Mesozoic paleogeographic relations; Sierra Nevada, Klamath Mountains, and related terranes: Boulder, Colorado, Geological Society of America Special Paper 255, p. 175–192.

Peterman, Z., 1979, Strontium isotope geochemistry of Late Archean to Late Cretaceous tonalites and trondhjemites, *in* Barker, F., ed., Trondhjemites, dacites, and related rocks: Developments in Petrology 6: Amsterdam, Elsevier, p. 133–147.

Pinto-Auso, M., and Harper, G.D., 1985, Sedimentation, metallogenesis, and tectonic origin of the basal Galice Formation overlying the Josephine ophiolite, northwestern California: Journal of Geology, v. 93, p. 713–725.

Rohrlach, B.D., 2002, Tectonic evolution, petrochemistry, geochronology and palaeohydrology of the Tampakan Porphyry and high sulphidation epithermal Cu-Au deposit Mindanao, Philippines [Ph.D. dissertation]: Canberra, Australian National University, 499 p.

Saleeby, J.B., Harper, G.D., Snoke, A.W., and Sharp, W.D., 1982, Time relations and structural-stratigraphic patterns in ophiolite accretion, west-central Klamath Mountains, California: Journal of Geophysical Research B, v. 87, p. 3831–3848.

Scherer, H.H., Snow, C.A., and Ernst, W.G., 2006, this volume, Geologic-petrochemical comparison of early Mesozoic mafic arc terranes: Western Paleozoic and Triassic belt, Klamath Mountains, and Jura–Triassic arc belt, Sierran Foothills, *in* Snoke, A.W., and Barnes, C.G., eds., Geological studies in the Klamath Mountains province, California and Oregon: A volume in honor of William P. Irwin: Boulder, Colorado, Geological Society of America Special Paper 410, doi: 10.1130/2006.2410(18).

Schmidt, B.L., 1994, The petrology and geochemistry of the English Peak intrusive suite, Klamath Mountains, California [Ph.D. dissertation]: Lubbock, Texas Tech University, 216 p.

Seyfert, C.K., 1964, Geology of the Sawyers Bar area, Klamath Mountains, northern California [Ph.D. dissertation]: Stanford, California, Stanford University, 186 p.

Snoke, A.W., 1977, A thrust plate of ophiolitic rocks in the Preston Peak area, Klamath Mountains, California: Geological Society of America Bulletin, v. 88, p. 1641–1659, doi: 10.1130/0016-7606(1977)88<1641:ATPOOR>2.0.CO;2.

Snoke, A.W., and Barnes, C.G., 2006, this volume, The development of tectonic concepts for the Klamath Mountains province, California and Oregon, *in* Snoke, A.W., and Barnes, C.G., eds., Geological studies in the Klamath Mountains province, California and Oregon: A volume in honor of William P. Irwin: Boulder, Colorado, Geological Society of America Special Paper 410, doi: 10.1130/2006.2410(01).

Snoke, A.W., Quick, J.E., and Bowman, H.R., 1981, Bear Mountain igneous complex, Klamath Mountains, California: An ultrabasic to silicic calc-alkaline suite: Journal of Petrology, v. 22, p. 501–552.

Wallin, E.T., and Metcalf, R.V., 1998, Supra-subduction zone ophiolite formed in an extensional forearc: Trinity terrane, Klamath Mountains, California: Journal of Geology, v. 106, p. 591–608.

Wallin, E.T., Coleman, D.S., Lindsley-Griffin, N., and Potter, A.W., 1995, Silurian plutonism in the Trinity terrane (Neoproterozoic and Ordovician), Klamath Mountains, California, USA: Tectonics, v. 14, p. 1007–1013, doi: 10.1029/95TC01447.

Wright, J.E., 1982, Permo-Triassic accretionary subduction complex, southwestern Klamath Mountains, California: Journal of Geophysical Research, v. 87, p. 3805–3818.

Wright, J.E., and Fahan, M.R., 1988, An expanded view of Jurassic orogenesis in the western United States Cordillera: Middle Jurassic (pre-Nevadan) regional metamorphism and thrust faulting within an active arc environment, Klamath Mountains, California: Geological Society of America Bulletin, v. 100, p. 859–876, doi: 10.1130/0016-7606(1988)100<0859:AEVOJO>2.3.CO;2.

Wright, J.E., and Wyld, S.J., 1986, Significance of xenocrystic Precambrian zircon contained within the southern continuation of the Josephine ophiolite: Devils Elbow ophiolite remnant, Klamath Mountains, northern California: Geology, v. 14, p. 671–674.

Wright, J.E., and Wyld, S.J., 1994, The Rattlesnake Creek terrane, Klamath Mountains, California: An early Mesozoic volcanic arc and its basement of tectonically disrupted oceanic crust: Geological Society of America Bulletin, v. 106, p. 1033–1056, doi: 10.1130/0016-7606(1994)106<1033:TRCTKM>2.3.CO;2.

Yule, J.D., 1996, Geologic and tectonic evolution of Jurassic marginal ocean basin lithosphere, Klamath Mountains, Oregon [Ph.D. dissertation]: Pasadena, California Institute of Technology, 308 p.

Yule, J.D., Saleeby, J.B., and Barnes, C.G., 1996, The 175–135 Ma plutons of the Klamath Mountains province reinterpreted as a single oceanic arc batholithic system: Geological Society of America Abstracts with Programs, v. 28, p. 128.

Yule, J.D., Saleeby, J.B., and Barnes, C.G., 2006, this volume, A rift-edge facies of the Late Jurassic Rogue–Chetco arc and Josephine ophiolite, Klamath Mountains, Oregon, *in* Snoke, A.W., and Barnes, C.G., eds., Geological studies in the Klamath Mountains province, California and Oregon: A volume in honor of William P. Irwin: Boulder, Colorado, Geological Society of America Special Paper 410, doi: 10.1130/2006.2410(03).

Manuscript Accepted by the Society 10 January 2006

Printed in the USA

Geological Society of America
Special Paper 410
2006

Magma mixing and mingling in the Grayback pluton, Klamath Mountains, Oregon

Kenneth Johnson*
Department of Natural Sciences, University of Houston–Downtown, 1 Main Street, Houston, Texas 77002, USA
Calvin G. Barnes
Department of Geosciences, Texas Tech University, Lubbock, Texas 79409, USA

ABSTRACT

The 160-Ma Grayback pluton provides ample evidence for the role of mantle-derived magmas in the crystallization history of the magma chamber. Mafic magmas tapped several distinct mantle sources, and evidence for magma mixing is manifest in (1) widespread mafic enclave swarms in which basaltic magma mingled with basaltic to andesitic magma, (2) late-stage granodioritic dikes containing pillow-shaped enclaves of basalt and hybrid andesite, and (3) late-stage tonalitic dikes including microdioritic enclaves contained within a dark-colored hybrid matrix. These magma-mingling zones reflect large-scale magma mixing processes in the Grayback system. Much of the compositional variation among Grayback main-stage samples was apparently produced by magma mixing between mafic and intermediate magmas similar to those exposed in some of the mafic enclave swarms. Mixing between basaltic and granitic end-members at deeper levels in the Grayback magma chamber produced hybrid andesite. Enclaves of basalt and hybrid andesite are observed in late-stage granodioritic dikes that cut the mafic enclave swarms. Mingling between partially crystalline basaltic and tonalitic mushes, resulting in a mega- and microscopically complex hybrid, is observed in the late-stage tonalitic dikes. Few Grayback main-stage samples were apparently produced in this manner.

The Grayback pluton is one of several similar-aged plutons in the Wooley Creek suite, in which mafic magma played a prominent role in petrogenesis. The Grayback system is unusual in that mafic magmas from several obviously different sources were present throughout its crystallization history. This circumstance probably resulted from its position in an extensional, back-arc setting.

Keywords: magma mixing, magma mingling, Klamath Mountains, hybridization, enclave

*E-mail: JohnsonK@uhd.edu.

Johnson, K., and Barnes, C.G., 2006, Magma mixing and mingling in the Grayback pluton, Klamath Mountains, Oregon, *in* Snoke, A.W., and Barnes, C.G., eds., Geological studies in the Klamath Mountains province, California and Oregon: A volume in honor of William P. Irwin: Geological Society of America Special Paper 410, p. 247–267, doi: 10.1130/2006.2410(12). For permission to copy, contact editing@geosociety.org.

INTRODUCTION

Most Jurassic plutons in the Klamath Mountains province occupy "belts" or suites that are defined by relatively narrow age ranges, geographic settings, and exposed rock types (Hotz, 1971; Irwin, 1985; Irwin and Wooden, 1999). More recent work has shown that plutons within belts share broadly similar petrogenetic histories. This commonality is particularly the case for plutons within the Ironside Mountain batholith (Barnes et al., this volume, Chapter 10), the Wooley Creek suite (Barnes, 1983; Barnes et al., 1986, 1995; Gribble et al., 1990), which now includes several plutons from the Grayback belt of Irwin (1985), and for Late Jurassic to Early Cretaceous tonalitic and trondhjemitic intrusions in the eastern Klamath Mountains (Barnes et al., 1996; Allen and Barnes, this volume). Plutons within the Wooley Creek suite evolved largely as open systems, involving assimilation of crustal rocks, recharge of mantle-derived magma, and/or mixing between mantle and crustal melts. Detailed study of such open systems provides information about the source(s) of magmas, the nature of the crust through which they ascended, and the timing of magma mixing/mingling events.

Magma mixing is commonly invoked to explain compositional variations in plutonic and volcanic suites from a variety of tectonic environments. Evidence for the interaction of magmas at the outcrop scale is found in banded pumice and mafic magmatic enclaves. Mafic magmatic enclaves (hereafter referred to as "enclaves") are common and often widespread features of both volcanic and plutonic rocks, and there is general consensus that they represent blobs of mafic magma that were rapidly undercooled when brought into contact with cooler, more felsic magma. Striking examples of basaltic "pillows" in felsic rocks have been presented that leave little doubt, at least at the outcrop scale, as to their magmatic origin (e.g., Wiebe, 1991; Wiebe et al., 2001). Often the petrogenetic significance of enclaves is uncertain, other than to indicate that mafic magma was present during the crystallization history of the pluton. If the presence of enclaves in mingling zones can be used to demonstrate that magma mixing was important in the compositional evolution of a magma, then such features may provide information about the physical and chemical processes involved in mixing.

The late Middle Jurassic Grayback pluton in southwestern Oregon is a large, reversely zoned member of the Wooley Creek plutonic suite. Magma-mingling zones are extensively exposed throughout the pluton. They feature a wide variety of intermingled rock types and provide evidence for mingling throughout the pluton's history. The observed mixing and mingling relationships indicate that mafic mantle-derived magmas were not only present, but played an important role in the evolution of the Grayback pluton. The Grayback pluton is not unique in this respect; all plutons in the Wooley Creek belt exhibit characteristics of open-system behavior and the involvement of mafic magmas during their crystallization histories (Barnes et al., 1986, 1987; Gribble et al., 1990). However, the Grayback pluton is particularly well suited for such study, because magma-mixing processes were arrested in situ in a variety of mixing stages and involved a range of end-member compositions. In this chapter, we describe the range of mingling zones in the Grayback pluton and demonstrate their utility as small-scale analogues to pluton-wide processes. We then discuss the tectonic conditions under which the Grayback system evolved, conditions that may have allowed for widespread mixing and mingling behavior. In the following text, "mixing" refers to the process of blending two or more magmas together to form reasonably uniform hybrid magma, and "mingling" refers to incomplete mixing in which end-members are physically distinct, typically resulting in enclave-bearing rocks or mingling zones.

REGIONAL GEOLOGY

The Klamath Mountains of northwestern California and southwestern Oregon comprise an arcuate assemblage of four lithotectonic terranes, which represent fragments of oceanic crust and island arcs that were accreted to the western margin of North America during Late Paleozoic through Middle Mesozoic time (Irwin, 1981). These units are, from east to west and from structurally highest to lowest, the eastern Klamath, central metamorphic, western Paleozoic and Triassic, and the western Jurassic terranes or belts (Fig. 1). The belts are separated by east-dipping thrust faults, so that structurally higher rocks were thrust over generally younger rocks to the west. Thrusting of the western Paleozoic and Triassic over the western Jurassic belt along the Orleans fault was responsible for some of the deformation associated with the Nevadan orogeny at ca. 152–150 Ma (Saleeby et al., 1982; Harper and Wright, 1984).

Calc-alkaline and calcic magmatism occurred during Jurassic and Early Cretaceous times. Barnes et al. (1992a) showed that pre-Nevadan plutons are chemically and isotopically distinct from post-Nevadan ones, reflecting a change from largely mantle-derived magmas to those formed by crustal melting in response to underthrusting of ophiolitic rocks beneath older terranes to the east (Barnes et al., 1992b, 1996). A sample from the Grayback pluton main stage yielded a zircon U-Pb age of 160 ± 2 Ma (Yule et al., this volume), whereas late-stage samples yielded a zircon U-Pb age of 157 ± 2 Ma (Yule et al., this volume) and a hornblende $^{40}Ar/^{39}Ar$ age of 157.3 ± 1.4 Ma (Hacker et al., 1995). These ages place Grayback magmatism within the age range of plutons belonging to the 165- to 156-Ma "Wooley Creek suite" that includes the Wooley Creek batholith and Slinkard pluton (Barnes et al., 1986), the Russian Peak pluton (Cotkin and Medaris, 1993), the Ashland pluton (Gribble et al., 1990), and the English Peak pluton (Schmidt, 1994). To the west of the Wooley Creek belt are the 165- to 159-Ma Josephine ophiolite (Wyld and Wright, 1988; Harper et al., 1994) and 161- to 156-Ma Rogue–Chetco volcanic arc complex (Garcia, 1982; Yule et al., this volume).

The Grayback pluton intruded rocks of the Rattlesnake Creek and western Hayfork terranes, the westernmost of four

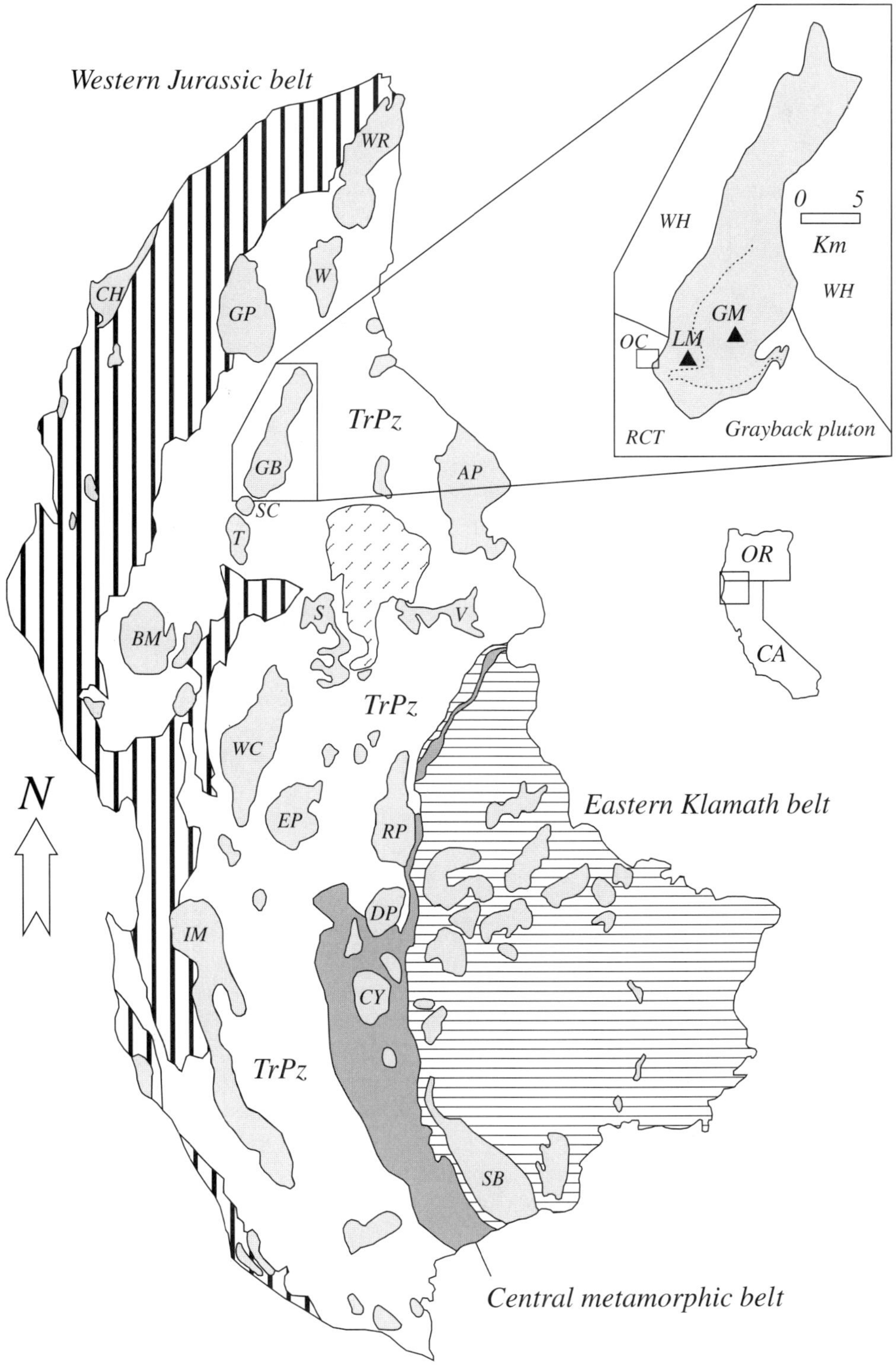

Figure 1. Geologic map of the Klamath Mountains showing boundaries between lithotectonic belts and locations of Jurassic and Cretaceous plutons. Inset shows locations of Lake Mountain (LM) and Grayback Mountain (GM) within the Grayback pluton. Square labeled "OC" represents the approximate locality of Oregon Caves National Monument. Thin dotted line within the Grayback pluton separates quartz-poor lithologies in the interior from quartz-rich ones along the margin (Barnes et al., 1995). AP—Ashland pluton; BM—Bear Mountain intrusive complex; CH—Chetco intrusive complex; CY—Canyon Creek pluton; DP—Deadman Peak pluton; EP—English Peak pluton; GB—Grayback pluton; GP—Grants Pass pluton; IM—Ironside Mountain batholith; RCT—Rattlesnake Creek terrane; RP—Russian Peak pluton; S—Slinkard pluton; SB—Shasta Bally pluton; SC—Sucker Creek pluton; T—Thompson Ridge pluton; TrPz—western Paleozoic and Triassic composite belt; V—Vesa Bluffs pluton; W—Weimer pluton; WC—Wooley Creek batholith; WH—western Hayfork terrane; WR—White Rock pluton. Simplified from Irwin (1985).

terranes of the western Paleozoic and Triassic belt (Irwin, 1972; Donato et al., 1996; Ernst, 1998). The Grayback pluton is the largest of a linear, north–northeast-trending chain of four coeval plutons (ca. 161–160 Ma; Yule et al., this volume) that consists of the Weimer, Grayback, Sucker Creek, and Thompson Ridge plutons. The host rocks that separate these plutons are essentially dike swarms with as much as 50% dike rock (Barnes et al., 1995; Donato et al., 1996). The presence of mafic dike swarms along the southern edge of the Grayback pluton, evidence for contemporaneous normal faulting, and its setting inboard of the coeval Rogue–Chetco arc led Barnes et al. (1995) to conclude that it was emplaced in a back-arc setting.

SAMPLE PREPARATION AND ANALYSIS

Samples were crushed and powdered in tungsten carbide and alumina shatterboxes. All samples were analyzed by inductively coupled plasma atomic emission spectrometry (ICP-AES) at Texas Tech University following methods described in Shannon (1994). Rare earth elements (REEs) were separated using the cation exchange elution method of Shannon (1994) and then analyzed by ICP-AES. Rubidium abundances were determined by flame atomic emission.

Mineral compositions were determined with a JEOL JXA-733 electron probe microanalyzer at Southern Methodist University, Dallas, Texas. Operating conditions were 15-kV accelerating potential, 20-nA beam current, and a 10-µm spot, defocused to 20 µm for plagioclase. Online corrections were made with the Bence-Albee procedure, and a suite of natural and synthetic minerals and glasses was used as standards.

GEOLOGY OF THE MAGMA-MINGLING ZONES

The main stage of the Grayback pluton consists of tonalite, quartz diorite, gabbro, and norite that were intruded by syn-plutonic noritic and gabbroic magmas, some of which resulted in widely dispersed enclaves throughout the pluton (referred to as "main-stage enclaves"). Late-stage activity was characterized by the intrusion of tonalitic and granodioritic dikes, many of which were enclave bearing. Detailed discussions of the geology and petrology of the Grayback pluton were presented in Godchaux (1969) and Barnes et al. (1995).

This study focuses on three types of magma-mingling zones in the Grayback pluton: (1) large, widespread swarms of mafic enclaves in a mafic to intermediate host; (2) late-stage granodioritic dikes containing pillow-shaped basaltic and andesitic enclaves; and (3) late-stage tonalitic dikes containing microdioritic enclaves enclosed in an intermediate matrix.

Mafic Enclave Swarms

Field Observations. These swarms consist of microgabbroic enclaves in gabbroic to dioritic host rocks. They are exposed on Lake Mountain and the ridge extending north from Grayback Mountain (Fig. 1). Some can be traced laterally for several hundred meters. They are complex intrusive zones within the main stage of the pluton and are intruded by coarse-grained hornblende gabbro and enclave-bearing felsic dikes.

Individual enclaves range from <1.0 cm to >2.5 m in longest dimension. Enclaves are generally lens-shaped and oriented parallel or subparallel to foliation in the host (Fig. 2A). In some places, enclaves constitute >90% of outcrop area and are separated by a thin septum (<0.5 cm) of host rock. The enclave boundaries are sharp and well defined, but lack the fine-grained "chilled" margins suggestive of rapid undercooling of the enclave magma.

On Grayback Mountain, enclaves are generally flattened at the margins of the mingling zones with aspect ratios approaching 35:1, whereas in the center of the zone, aspect ratios are estimated to be <10:1. The enclaves are plastically deformed around brittle xenoliths, resulting in pressure shadows in which prismatic hornblende is randomly oriented. Elsewhere in the mingling zones, hornblende is parallel to the foliation.

Veins of the host rock, generally <0.5 cm in width, are common in most enclaves. However, most do not penetrate the entire enclave, but pinch out. These veins are generally parallel to the foliation of the host from which they originate and to the direction of enclave elongation (Fig. 2B). Aggregates of coarse-grained plagioclase and poikilitic hornblende, as much as 1.5 cm in length, are common throughout the finer-grained enclaves. The long axes of these aggregates are parallel to the foliation in the host and to the veins of host within the enclaves. They are commonly collinear with pinched out veins (Fig. 2C and D).

Petrography. The host rocks to these enclave swarms are hornblende norite (on Grayback Mountain) and biotite-hornblende quartz gabbro (on Lake Mountain). Host rocks consist of euhedral to subhedral, reversely to normally zoned plagioclase (An_{48-63}) + magnesio-hornblende (Mg/[Mg + Fe] = 0.52–0.62) + Fe-Ti oxides ± orthopyroxene ($En_{53}Fs_{47}Wo_0$ to $En_{46}Fs_{52}Wo_2$) ± biotite ± quartz (Johnson, 1991). Titanite and zircon are minor accessories, whereas apatite is common. Hornblende is prismatic or poikilitic, with inclusions of plagioclase, apatite, and oxides. Minor late-stage cummingtonite is associated with hornblende.

Enclaves are micronorite (Grayback Mountain) and biotite hornblende quartz microgabbro (Lake Mountain). They have the same mineral assemblages as their respective hosts. Plagioclase compositions in the enclaves reach An_{69}, whereas hornblende and orthopyroxene compositions are similar to those in host rocks (Johnson, 1991). Plagioclase xenocrysts from the host are strongly zoned and are partially resorbed and embayed by quartz. Hornblende in the quartz microgabbro is prismatic, but it forms rims around pyroxene grains in the micronoritic enclaves. Aspect ratios of apatite in the enclaves rarely exceed 5:1 and are very similar to those in the host.

In thin section, enclaves typically exhibit a well-developed lamination defined by the parallel alignment of plagioclase laths (Fig. 2E), although enclaves with little or no lamination were observed. Lamination in the enclaves is generally parallel to the enclave/host interface, and may be deflected around mineral aggregates and xenocrysts from the host (Fig. 2F).

Late-Stage Granodioritic Dikes

Field Observations. Granodioritic dikes containing basaltic and andesitic "pillows" were studied at several localities including Lake Mountain summit, the U.S. Forest Service Boundary Trail south of Grayback Mountain, and along logging roads east of Oregon Caves National Monument in the southern portion of the pluton.

Basaltic enclaves are fine-grained, dark green to gray mafic pillowlike blobs that range from a few centimeters to slightly less than one meter in diameter. The basalt-granodiorite inter-

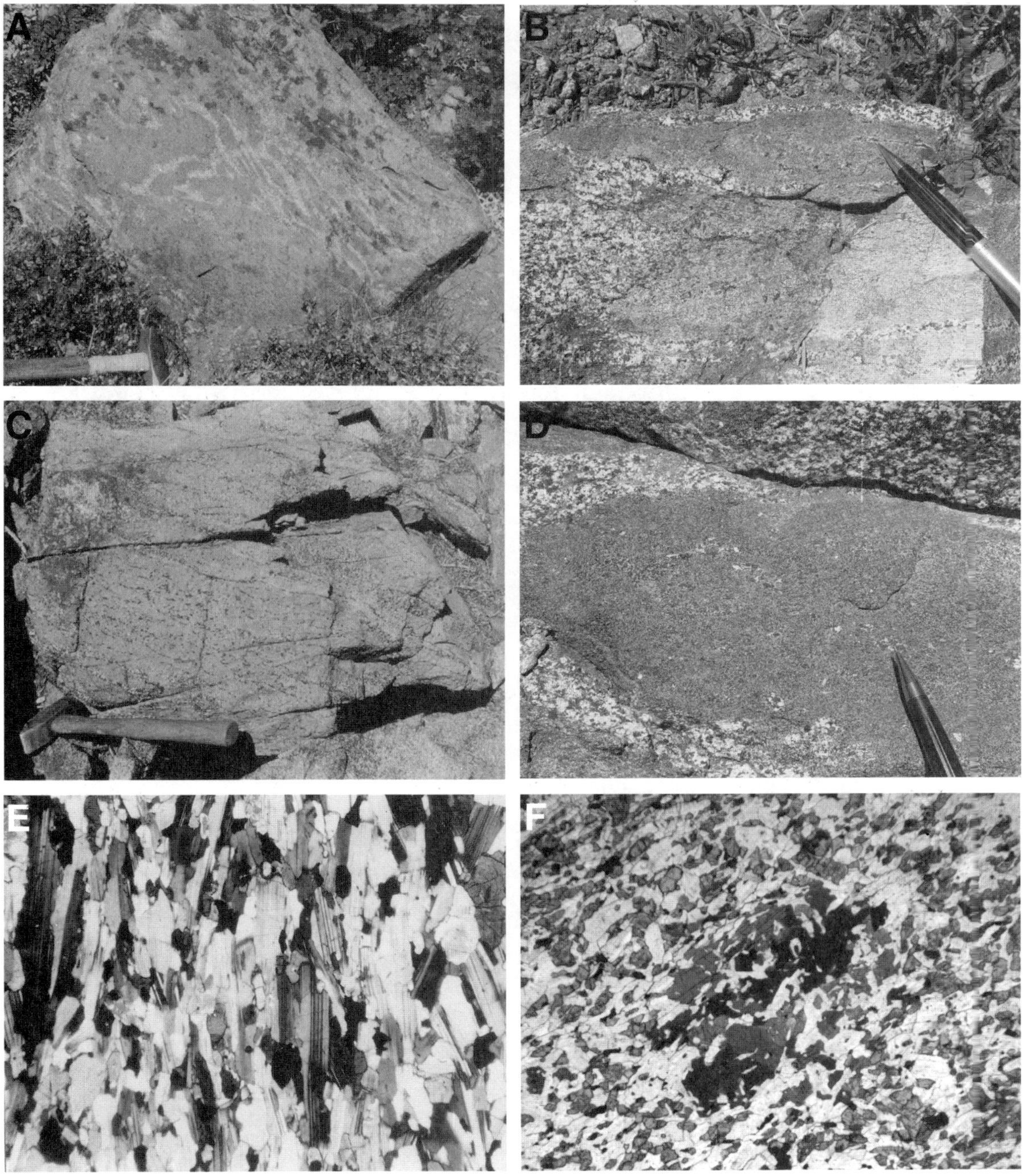

Figure 2. Photographs and photomicrographs of enclaves in the mafic enclave swarms. (A) Tightly packed enclaves of all sizes showing plastic behavior and high degree of elongation. (B) Coarse-grained vein of host material in fine-grained enclave. Vein terminates near tip of pencil. (C) Trails of hornblende xenocrysts in pyroxene-rich enclaves. Xenocryst trails are oriented parallel to lamination in the enclaves and foliation in the host. (D) Remnants of host vein material, now manifest as plagioclase and hornblende xenocrysts, within pyroxene-rich enclave. (E) Pronounced lamination in pyroxene-rich enclave defined by orientation of plagioclase laths (field of view is 6.0 mm). (F) Hornblende xenocryst in pyroxene-rich enclave. Plagioclase laths (white) and pyroxene grains (gray) are deflected around xenocryst (mostly extinct; field of view is 6.0 mm).

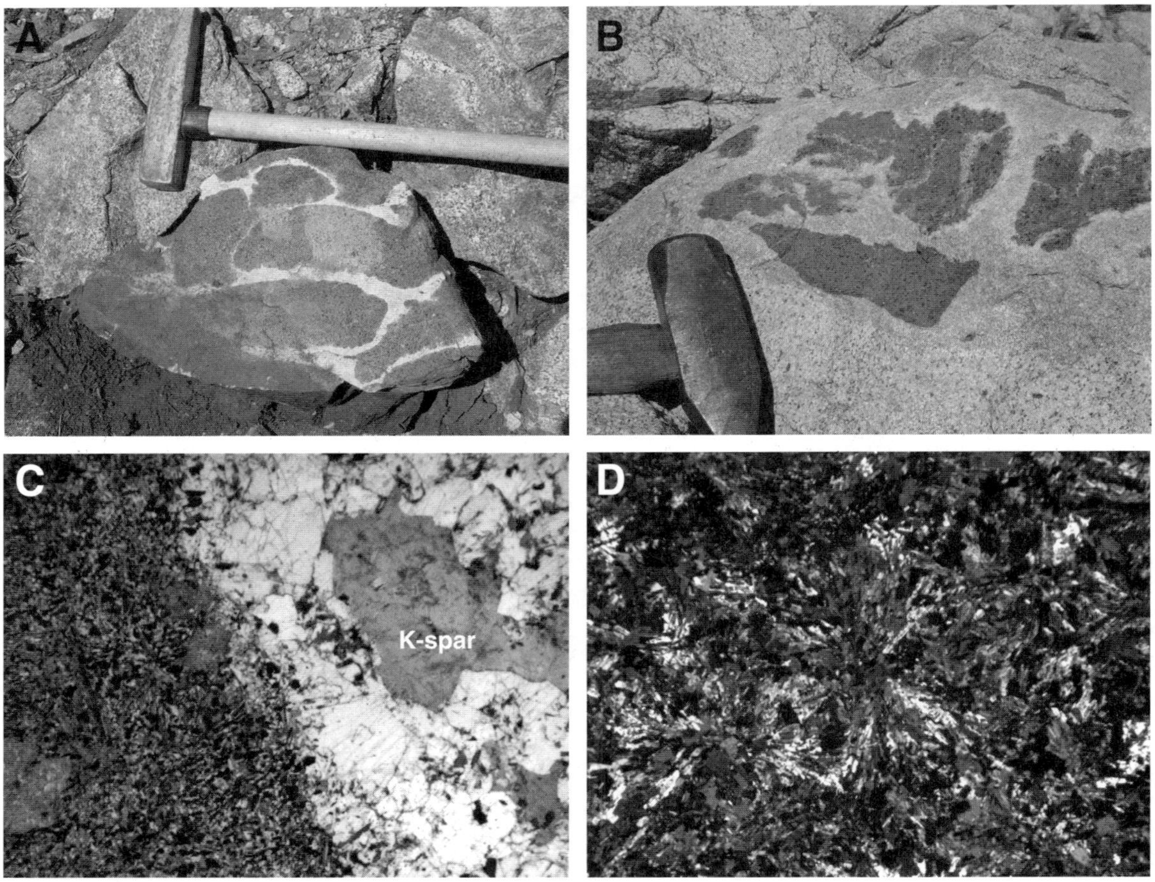

Figure 3. Photographs and photomicrographs of enclaves in late-stage granodioritic dikes. (A) Basaltic pillows in float block of granodiorite. Note the close-packed nature of the enclaves. (B) Thin trondhjemitic haloes surrounding basaltic enclaves. Haloes are lacking in K-feldspar and biotite. (C) Sharp interface between basaltic enclave (left) and granodioritic host (right; field of view is 6.0 mm). (D) Radiating plagioclase laths in hybrid andesitic enclave (field of view is 6.0 mm).

face is generally very sharp and convex toward the host (Fig. 3A), and the enclaves lack a preferred orientation. The enclaves are commonly enveloped by trondhjemitic zones (Fig. 3B) ~0.75 cm wide that lack K-feldspar and biotite. On fresh enclave surfaces, clots of randomly oriented plagioclase phenocrysts and clots of actinolitic hornblende 2–3 mm in length are visible.

Andesitic enclaves are less abundant than basaltic ones. They are fine-grained, pale-green, and intermediate in color between the basaltic pillows and granodioritic host. These enclaves are coarser grained than the basaltic enclaves, and plagioclase phenocrysts are not visible on fresh surfaces. In addition, their margins are more diffuse than those of the basaltic enclaves, and the granodiorite adjacent to the enclave/host contact is somewhat darker than elsewhere.

Petrography. Biotite granodiorite has medium-grained hypidiomorphic granular texture and consists of plagioclase + K-feldspar + quartz + biotite + Fe-Ti oxides. Apatite, titanite, allanite, and zircon are accessories. Inclusions of biotite, apatite, Fe-Ti oxides, and minor amphibole (hornblende?) are common in plagioclase. Quartz and K-feldspar have interstitial and poikilitic habits. Apatite forms euhedral acicular prisms with aspect ratios of 20:1 or greater. Amphibole occurs as small rounded blebs up to 0.1 mm in diameter.

Basaltic enclaves are porphyritic and consist of plagioclase + biotite + actinolitic hornblende + Fe oxides and accessory titanite and apatite. Euhedral to subhedral plagioclase phenocrysts reach 3.0 mm in length. Hornblende and biotite are abundant groundmass phases. Aggregates of actinolitic amphibole (up to 1.0 mm in diameter) are rimmed by biotite, and probably represent relict pyroxene (e.g., Castro et al., 1990). Apatite occurs as acicular prisms; some are hollow, and aspect ratios exceeding 30:1 are common. Enclave margins adjacent to trondhjemitic zones in the host are sharp (Fig. 3C) and locally enriched in biotite.

Andesitic enclaves contain the same mineral assemblage as the host granodiorite. Long prismatic crystals of plagioclase form radiating clusters up to 4.0 mm in diameter (Fig. 3D). Inclusions of biotite, actinolite, apatite, and Fe-Ti oxides in plagioclase are common. Large poikilitic grains of K-feldspar up to 4.0 mm in diameter enclose plagioclase, biotite, quartz, titanite, and actinolite. Quartz is also interstitial and titanite is poikilitic. Actinolite (0.01–0.30 mm in length) occurs as small stubby blebs, tabular grains, and needles.

Late-Stage Tonalitic Dikes

Field Observations. Late-stage tonalitic dikes commonly contain mingling zones of variably hybridized enclaves. The best-exposed occurrence is on the summit of Lake Mountain. Sizes of individual enclaves range from <1 cm to >20 cm in longest dimension. Enclaves are fine- to medium-grained and black to dark gray and set in a medium- to coarse-grained, dark- to medium-gray matrix that grades into the pale gray of the tonalitic host (Fig. 4A). Contacts between the matrix and enclaves are commonly diffuse, and veins of host material in the enclaves are rare. Enclaves are typically oblate with tapered ends and aspect ratios <2:1. The enclaves are crudely oriented parallel to the gradational contact between the hybrid matrix and the tonalitic dike (Fig. 4A).

Enclaves are generally isolated, separated from one another by as much as 10–15 cm of matrix. They exhibit a wide range of grain sizes, as does the matrix, thereby making it difficult to discern the outlines of smaller enclaves. In fact, the matrix is littered with small, barely recognizable enclaves, which made sampling for chemical analysis difficult (Fig. 4B).

Petrography. The biotite tonalite consists of plagioclase (An_{28-37}) + quartz + biotite + Fe oxides ± hornblende ± K-feldspar. Accessory phases are apatite, as inclusions in plagioclase, and zircon. Euhedral to subhedral plagioclase (up to 3.0 mm in length) exhibits weak normal zonation. Quartz and minor K-feldspar are interstitial. Minor hornblende occurs as inclusions in biotite.

Microdioritic enclaves consist of plagioclase (An_{45-49}) + hornblende + biotite + Fe-Ti oxides + clinopyroxene + orthopyroxene ± quartz. Apatite is common, with aspect ratios of 5:1–10:1. Enclaves commonly exhibit a strong foliation defined by the alignment of euhedral to subhedral hornblende and plagioclase. Biotite is poikilitic and quartz is interstitial. Pyroxene occurs as cores in hornblende and as large clots rimmed with hornblende.

Rocks from the dark "hybrid" matrix appear to contain components of both enclave and host, resulting in a strongly bimodal distribution of grain size (Fig. 4C). Strongly zoned plagioclase grains with resorbed rims and cores are common, although unresorbed grains are also present. Pyroxene is generally rimmed with hornblende. Apatite exhibits two habits: long, acicular needles and short, stubby prisms. Relatively siliceous hybrid compositions contain more biotite, plagioclase, and quartz, and less pyroxene and hornblende.

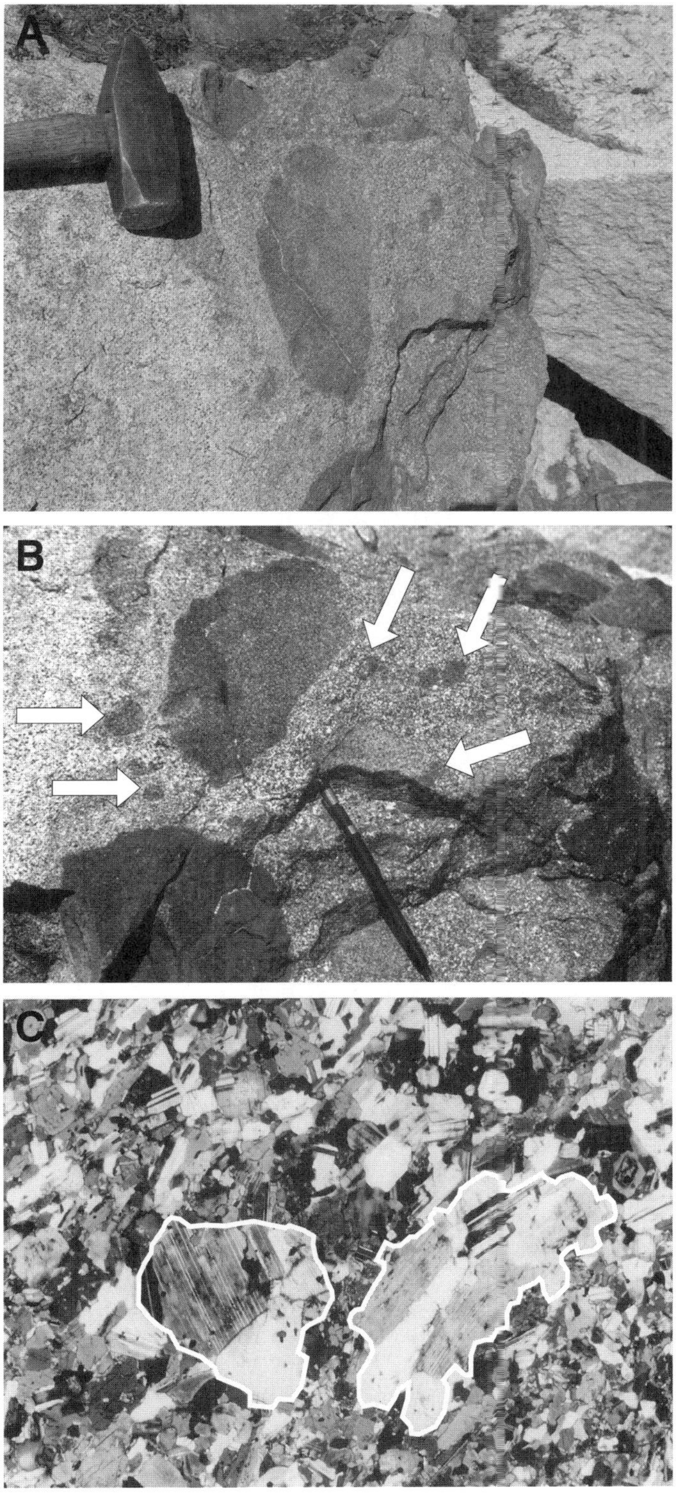

Figure 4. Photographs and photomicrographs of enclaves in late-stage tonalitic dikes. (A) Dioritic enclaves in dark colored hybrid matrix, which grades into lighter-colored tonalite (at hammer head). (B) Enclaves within hybrid matrix of the tonalitic dike. Arrows point to small, inconspicuous enclaves, which contribute to the bimodal grain size in the hybrid matrix. (C) Coarse-grained crystal aggregates (outlined) within the hybrid matrix of a tonalitic dike (field of view is 6.0 mm).

WHOLE-ROCK GEOCHEMISTRY

Mafic Enclave Swarms

Micronoritic enclaves from the Grayback Mountain enclave swarm most closely resemble high-Al island arc tholeiites, on the basis of their low K_2O and high CaO/Al_2O_3 and FeO*/MgO (Table 1, Fig. 5). There is little difference between enclave and host compositions, with the exception of slightly higher K_2O, Cr, Zr, and Rb, and slightly lower TiO_2, Fe (total Fe as FeO), and Ni abundances in the host (Fig. 6). In addition, host material (GM-25-8) adjacent to a large enclave contains higher Na_2O, V, Nb, Y, Zr, and Ba, and lower MgO than host material isolated from enclaves (Table 1).

TABLE 1. MAJOR AND TRACE ELEMENT COMPOSITIONS OF REPRESENTATIVE SAMPLES FROM THE GRAYBACK PLUTON MAGMA MINGLING ZONES

Sample	SiO_2	TiO_2	Al_2O_3	Fe*	MnO	MgO	CaO	Na_2O	K_2O	P_2O_5	LOI	Total	Sc
Mafic Enclave Swarms													
GM-2A	48.90	0.92	20.49	10.84	0.19	3.83	9.95	3.26	0.25	0.44	0.46	99.52	23.5
GM-1A	49.96	1.03	18.18	12.58	0.24	5.01	10.65	2.58	0.21	0.32	0.18	100.90	34.3
GM-1B	50.84	1.02	18.69	11.85	0.21	4.17	9.70	2.93	0.26	0.44	0.67	100.78	25.4
GM-24	50.08	0.55	19.26	9.05	0.17	5.09	12.01	2.27	0.45	0.25	1.24	100.41	38.0
GM-25-1	48.06	0.94	18.82	12.42	0.21	5.26	10.27	2.69	0.35	0.30	0.58	99.89	43.5
GM-25-2	48.99	0.89	18.51	12.71	0.22	5.07	10.06	2.57	0.31	0.30	0.61	100.23	36.1
GM-25-3	49.34	0.88	18.61	12.60	0.23	5.03	10.37	2.33	0.28	0.30	0.63	100.59	35.8
GM-25-4	49.26	0.87	18.30	12.84	0.23	5.13	10.42	2.36	0.29	0.30	0.60	100.59	35.7
GM-25-5	49.77	0.86	18.32	12.90	0.23	4.94	10.11	2.30	0.29	0.31	0.68	100.70	34.4
GM-25-6	48.25	0.84	18.60	12.51	0.23	5.16	11.17	2.40	0.37	0.27	0.58	100.38	39.9
GM-25-7	48.55	0.90	18.65	14.08	0.24	5.13	9.65	2.32	0.29	0.31	0.61	100.72	38.0
GM-25-8	49.97	0.91	19.66	10.56	0.15	3.80	9.86	3.19	0.40	0.37	0.73	99.60	38.4
LM-11	53.23	0.82	18.65	10.50	0.21	3.41	8.70	3.10	1.28	0.37	n.d.	100.27	n.d.
LM-7-1	57.04	0.73	19.36	6.63	0.09	2.59	7.39	3.03	1.10	0.25	1.92	100.12	4.8
LM-7-2	52.07	0.67	14.61	11.13	0.22	7.44	10.71	1.94	0.43	0.19	0.53	99.94	49.1
LM-7-3	52.23	0.65	15.49	10.46	0.21	7.16	10.83	2.06	0.41	0.19	0.42	100.11	43.2
LM-7-4	52.52	0.63	15.43	10.11	0.21	6.99	11.13	2.30	0.43	0.18	0.45	100.38	45.3
LM-7-5	52.08	0.67	15.42	10.60	0.21	6.96	10.61	2.33	0.50	0.19	0.51	100.08	50.0
LM-7-6	50.25	0.81	15.87	11.62	0.23	6.56	10.23	2.38	0.67	0.21	0.58	99.41	50.0
LM-7-7	57.67	0.76	18.71	6.59	0.08	2.57	6.73	2.70	1.52	0.24	1.58	99.15	4.8
Late-stage granodioritic dikes													
LM-1	54.85	1.73	17.27	8.03	0.17	4.45	7.60	3.91	1.71	0.36	1.11	101.18	22.6
LM-2	73.75	0.27	13.81	2.68	0.07	0.51	1.76	3.16	3.94	0.06	0.60	100.61	n.d.
LM-3	73.03	0.32	14.52	2.11	0.04	0.77	2.29	3.71	3.03	0.09	0.69	100.60	4.0
LM-18A	51.69	1.84	17.68	8.90	0.17	5.96	7.85	3.68	1.83	0.41	0.87	100.88	23.5
LM-18B	71.07	0.41	14.64	2.87	0.05	1.01	2.46	3.86	2.67	0.11	0.73	99.89	5.6
LM-29A	51.35	1.78	17.37	8.61	0.15	5.82	8.06	3.71	1.82	0.37	1.01	100.05	23.3
LM-29B	69.14	0.62	14.56	3.51	0.06	1.94	3.08	3.76	2.96	0.15	0.91	100.68	8.1
LM-29C	59.56	1.25	15.60	6.39	0.12	4.14	5.42	3.74	2.62	0.27	0.77	99.87	15.8
LM-29D	59.68	1.22	15.67	6.32	0.11	4.15	5.63	3.99	2.00	0.25	0.90	99.95	15.4
GBP-152B	71.99	0.36	14.88	2.30	0.04	0.78	2.41	3.67	3.36	0.09	0.00	99.88	4.5
Late-stage tonalitic dikes													
LM-15A	47.97	1.01	15.19	12.29	0.25	8.32	11.46	1.44	0.66	0.24	1.25	100.08	58.3
LM-16B	51.21	0.96	16.21	10.90	0.19	6.25	9.68	3.26	0.68	0.27	0.67	100.28	44.0
LM-16C	50.08	0.97	16.02	11.11	0.19	6.88	10.68	2.80	0.49	0.26	0.46	99.94	46.9
LM-19	50.88	1.20	17.29	10.67	0.21	5.57	9.76	3.06	0.57	0.35	0.54	100.10	40.3
LM-26A1	52.47	0.95	15.66	10.73	0.20	6.05	9.20	3.24	0.75	0.22	0.67	100.14	43.1
LM-26A2	54.48	1.05	17.35	9.40	0.18	4.62	8.16	3.28	1.11	0.29	0.90	100.82	35.2
LM-26A3	63.35	0.73	16.72	5.99	0.13	2.32	4.97	3.60	1.23	0.18	0.86	100.08	15.5
LM-27A	54.15	1.06	17.31	9.64	0.19	4.80	8.33	3.17	1.00	0.31	1.31	101.28	33.6
LM-27B2	50.70	1.05	16.80	11.41	0.21	6.19	10.02	3.12	0.59	0.27	0.50	100.86	46.3
LM-28	67.65	0.51	16.04	4.17	0.08	1.22	4.07	4.06	1.47	0.14	0.62	100.02	8.2
LM-10	62.88	0.62	17.68	5.50	0.09	1.19	4.41	4.58	1.35	0.16	n.d.	98.46	9.9

Notes: Major element oxides are expressed in wt%; trace elements in ppm. Total Fe as Fe_2O_3. Samples starting with "GM" are from Grayback Mountain; those beginning with "LM" are from Lake Mountain. Abbreviation: n.d.—not determined.

Unlike enclave-host relationships exposed on Grayback Mountain, enclaves in the Lake Mountain enclave swarm are more chemically distinct from their host (Table 1, Fig. 6). Lake Mountain enclaves have lower SiO_2, TiO_2, Al_2O_3, Na_2O, K_2O, P_2O_5, Zr, Sr, Ba, Rb, and FeO*/MgO, and higher MgO, CaO, Sc, V, Cr, Ni, Y, and CaO/Al_2O_3 than their host rocks (Fig. 6). Enclave compositions are similar to calc-alkaline basalt and basaltic andesite on the basis of their higher K_2O and lower FeO*/MgO, relative to tholeiitic compositions. Enclave and host samples have light REE (LREE)-enriched chondrite-normalized patterns (Fig. 7). The host sample (LM-11) has a straight REE pattern, whereas the enclave (LM-7–4) has higher ΣREE and exhibits a negative Eu anomaly (Eu/Eu* = 0.73). Both enclave and host have similar initial $^{87}Sr/^{86}Sr$ (0.7042) and

V	Cr	Ni	Cu	Zn	Nb	Y	Zr	Ba	Sr	Rb	Rock
228	27	11	91	92	16	19.7	22	248	620	3	Enclave
293	8	12	55	111	24	13.6	10	219	513	2	Enclave
247	37	10	90	107	19	19.9	22	225	575	4	Enclave
223	77	24	283	75	16	17.5	58	259	497	7	Host
279	12	8	117	102	23	29.7	37	251	482	3	Enclave margin
290	12	12	109	112	23	20.2	21	232	480	3	Enclave core
292	10	13	172	119	26	19.9	21	226	483	3	Enclave core
282	28	115	164	109	26	20.2	30	226	480	3	Enclave core
293	9	10	191	108	23	19.2	40	226	482	3	Enclave core
294	11	10	135	117	27	23.8	19	232	487	3	Enclave core
300	16	36	111	130	28	24.6	15	225	492	3	Enclave margin
268	30	14	125	89	22	31.1	73	317	542	5	Host adjacent to enclosure
n.d.	n.d.	n.d.	n.d.	n.d.	n.d.	28.8	17	546	475	33	Host
174	27	10	119	66	12	4.6	107	602	470	24	Host adjacent to enclosure
257	140	32	98	100	24	36.9	43	177	304	5	Enclave margin
241	127	30	130	86	20	31.2	55	176	315	4	Enclave core
224	124	29	136	87	19	33.5	62	174	325	4	Enclave core
222	111	26	152	92	19	35.4	57	197	334	6	Enclave core
223	96	28	86	93	18	35.6	40	284	340	9	Enclave margin
122	29	14	54	60	7	4.3	151	1131	470	38	Host adjacent to enclosure
170	112	47	28	83	33	37.3	169	218	339	62	Enclave-basalt
n.d.	n.d.	n.d.	n.d.	n.d.	n.d.	n.d.	n.d.	1342	154	75	Host
24	12	5	4	33	5	10.3	141	970	241	63	Host
184	156	83	26	74	24	28.6	172	184	355	62	Enclave-basalt
32	18	11	20	73	24	27.2	154	189	371	55	Enclave-basalt
174	158	86	6	32	12	21.3	133	774	292	48	Host
57	47	31	6	57	18	27.0	135	457	282	55	Enclave-andesite
122	114	75	7	56	19	23.2	150	320	284	45	Enclave-andesite
25	10	7	5	39	19	15.0	155	1108	235	50	Host
305	135	11	43	122	1	31.0	50	239	395	15	Enclave
243	98	7	35	113	2	29.0	64	323	399	11	Enclave
259	119	10	113	118	2	28.3	49	226	429	7	Enclave
225	64	4	19	119	4	28.9	73	243	501	10	Enclave
225	107	6	24	121	4	28.7	56	361	390	15	Enclave
188	45	3	17	101	4	28.5	112	521	453	29	Matrix
88	18	4	8	73	3	7.9	189	769	432	39	Host
193	49	6	17	108	3	24.0	87	505	468	26	Matrix
237	79	15	12	111	5	40.0	28	269	435	10	Enclave
53	9	4	4	49	3	10.7	229	1088	366	59	Host
51	8	6	7	66	11	13.4	282	895	366	31	Host

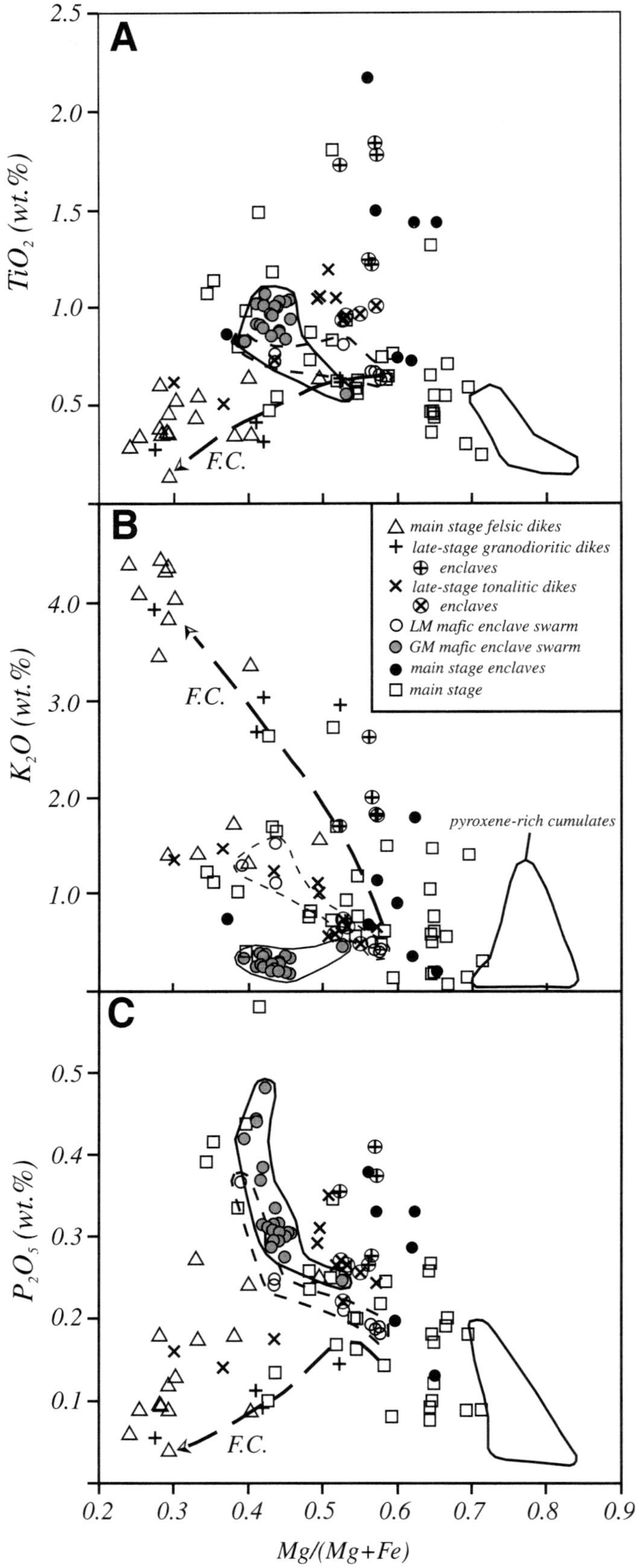

Figure 5. Variation diagrams showing compositions from the main-stage, mafic enclave swarms, and late-stage tonalitic and granodioritic dikes, including enclave and hybrid compositions. Samples within the field outlined by the solid line are from the Grayback Mountain mafic enclave swarm, whereas those within the field outlined by the dashed line represent the Lake Mountain mafic enclave swarm. Thick dashed line labeled "F.C." represents the evolution of mafic magmas by fractional crystallization (see Barnes et al., 1995). Outlined fields at Mg/(Mg + Fe) >0.7 represent the compositions of pyroxene-rich cumulates.

ε_{Nd} (+5.5 and +4.8, respectively; Barnes et al., 1995), which are comparable to rocks of the main stage of the Grayback pluton.

Late-Stage Granodioritic Dikes

Basaltic enclaves in the late-stage granodioritic dikes are high-Al and calc-alkaline in composition. They have similar SiO_2 contents to micronoritic enclaves in the mafic enclave swarms, but have higher TiO_2, Na_2O, K_2O, Zr, and Rb, and lower FeO, MnO, CaO, Sc, V, and CaO/Al_2O_3 (Table 1, Fig. 5). One enclave sample is characterized by a linear chondrite-normalized REE pattern that lacks a Eu anomaly, by high ε_{Nd} (+5.5), and by low initial $^{87}Sr/^{86}Sr$ (0.7027; Barnes et al., 1995). Apparently unmodified host granodiorite has SiO_2 >70 wt%, K_2O to nearly 4 wt%, and Ba abundances of 800–1342 ppm (Fig. 5). In addition, a granodiorite sample has low heavy REE (HREE) abundances (Yb_N = 5.5), and a slight negative Eu anomaly (Eu/Eu* = 0.76; Fig. 7). Andesitic enclaves are compositionally intermediate between basaltic enclaves and granodioritic host samples (Table 1); indeed, andesitic samples plot along a straight line between the basaltic and granodioritic end-members for most elements (Fig. 8).

Additional samples from the margin of an andesitic enclave and from dark-colored host granodiorite adjacent to the enclave were analyzed (Table 1). Compared to unmodified granodiorite, the granodiorite adjacent to the enclave is higher in TiO_2, FeO, MgO, CaO, Sc, V, Cr, Ni, and Y, and lower in SiO_2 and Sr. The sample from the margin of the andesitic enclave is depleted in Na_2O and Zr, but enriched in K_2O, Y, Ba, and Rb, relative to the enclave core.

Late-Stage Tonalitic Dikes

Tonalitic rocks from late-stage dikes are distinct from the main-stage tonalites. Barnes et al. (1995) characterized the late-stage rocks as high-Al tonalites, whereas main-stage tonalites are low-Al (c.f. Barker, 1979). In addition to its higher Al_2O_3 contents, the late-stage tonalite is characterized by higher SiO_2, FeO*/MgO, and Zr, and chondrite-normalized REE patterns that show greater fractionation of HREE from LREE and positive Eu anomalies.

Microdioritic enclaves in the late-stage tonalitic dikes have similar SiO_2 contents to basaltic enclaves in the late-stage granodioritic dikes, but have higher FeO, MgO, CaO, Sc, V, Zn, Ba,

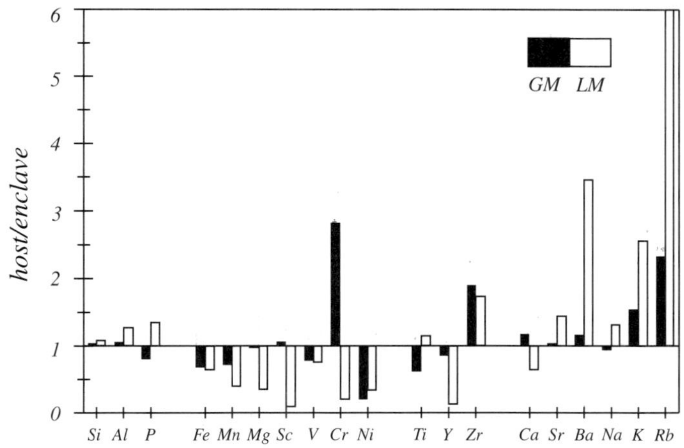

Figure 6. Enrichment-depletion diagram comparing representative enclave and host compositions from the mafic enclave swarms on Grayback (GM) and Lake (LM) Mountains. Analyses are from cores of large (9–12 cm thick) enclaves and from host rock material not directly adjacent to enclaves.

Sr, and CaO/Al_2O_3, and lower TiO_2, Al_2O_3, Na_2O, K_2O, P_2O_5, Nb, Zr, and Rb (Table 1, Figs. 5 and 8). Concentrations of most elements in these microdioritic enclaves are similar to or intermediate between micronoritic and microgabbroic enclaves in the mafic enclave swarms, but microdiorite has higher TiO_2, Na_2O, K_2O, Ba, and Rb, and lower CaO, Ni, and Nb. A microdioritic enclave from a late-stage tonalitic dike has a LREE-

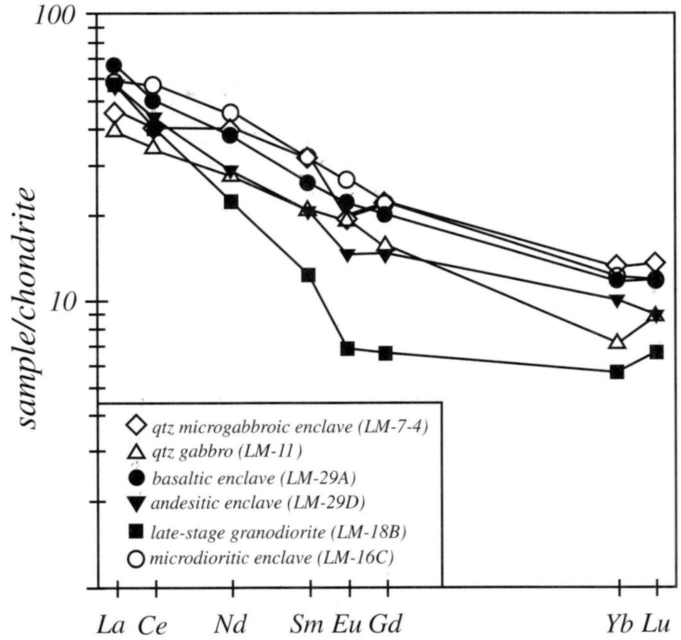

Figure 7. Chondrite-normalized rare earth element diagrams for selected samples from magma-mingling zones within the Grayback pluton. Chondrite values are from Haskin et al. (1968). Samples LM-18B, LM-29A, and LM-29D are from the same granodioritic dike.

enriched chondrite-normalized pattern, and lacks a Eu anomaly (Eu/Eu* = 0.99; Fig. 7). This same enclave yielded an initial $^{87}Sr/^{86}Sr$ value similar to those of a microgabbroic enclave and its host from the mafic enclave swarm on Lake Mountain (0.7042), but has a much higher ε_{Nd} (+10.3; Barnes et al., 1995).

The dark matrix that envelops enclaves in the late-stage tonalitic dikes has a wide range of compositions, from ~51 to 63 wt% SiO_2 (Table 1). There is considerable overlap of matrix compositions and those of enclave and tonalitic samples, which reflects the difficulty in distinguishing the different lithologies (e.g., micro-enclaves in similarly dark matrix rock; Fig. 4B). Matrix compositions plot along straight lines between enclave and tonalitic end-members for most elements, with the marked exception of TiO_2 and P_2O_5, which are generally higher in matrix compositions (Fig. 8).

INTERACTIONS BETWEEN MAGMAS IN THE MINGLING ZONES

Enclaves in the Grayback pluton mingling zones possess characteristics that indicate they were liquid or only partially crystallized when brought into contact with their host magmas. For example, their fine grain size, evidence for plastic deformation, and the acicular nature of apatite crystals preclude an autolithic (e.g., early cumulate) origin and suggest that the enclave magmas were undercooled by contact with the host magma. Below, we discuss the nature of the interactions between magmas in the various magma-mingling zones exposed in the Grayback pluton and the implications of these mingling zones for magma mixing processes in the petrogenesis of the Grayback system.

Mafic Enclave Swarms

The mafic enclave swarms on Grayback and Lake Mountains represent relatively early mingling episodes in the Grayback pluton. Moreover, they represent mingling between two or more mafic and/or intermediate magmas, as opposed to mingling between mafic and felsic compositions, as in the late-stage dikes. The mingling zone on Grayback Mountain is unusual in that there is little compositional difference between enclave and host rocks.

One of the puzzling features of the mafic enclave swarms is the aggregates of coarse-grained poikilitic hornblende and plagioclase within finer-grained pyroxene-rich enclaves (Fig. 2C and F). We interpret these as relict veins of host material caught between enclaves during emplacement. The residual liquid in the veins was either squeezed out or incorporated into the enclave magma during localized contraction. In this scenario, the coarser hornblende and plagioclase would be xenocrystic to the enclaves, akin to the quartz ocelli frequently described in mafic enclaves in quartz-rich host rocks (e.g., Vernon, 1983). The preservation of quartz ocelli in mafic enclaves is typically cited as evidence that the enclaves represent hybrid compositions, which is taken to imply that magma mixing occurred prior to commingling and enclave formation.

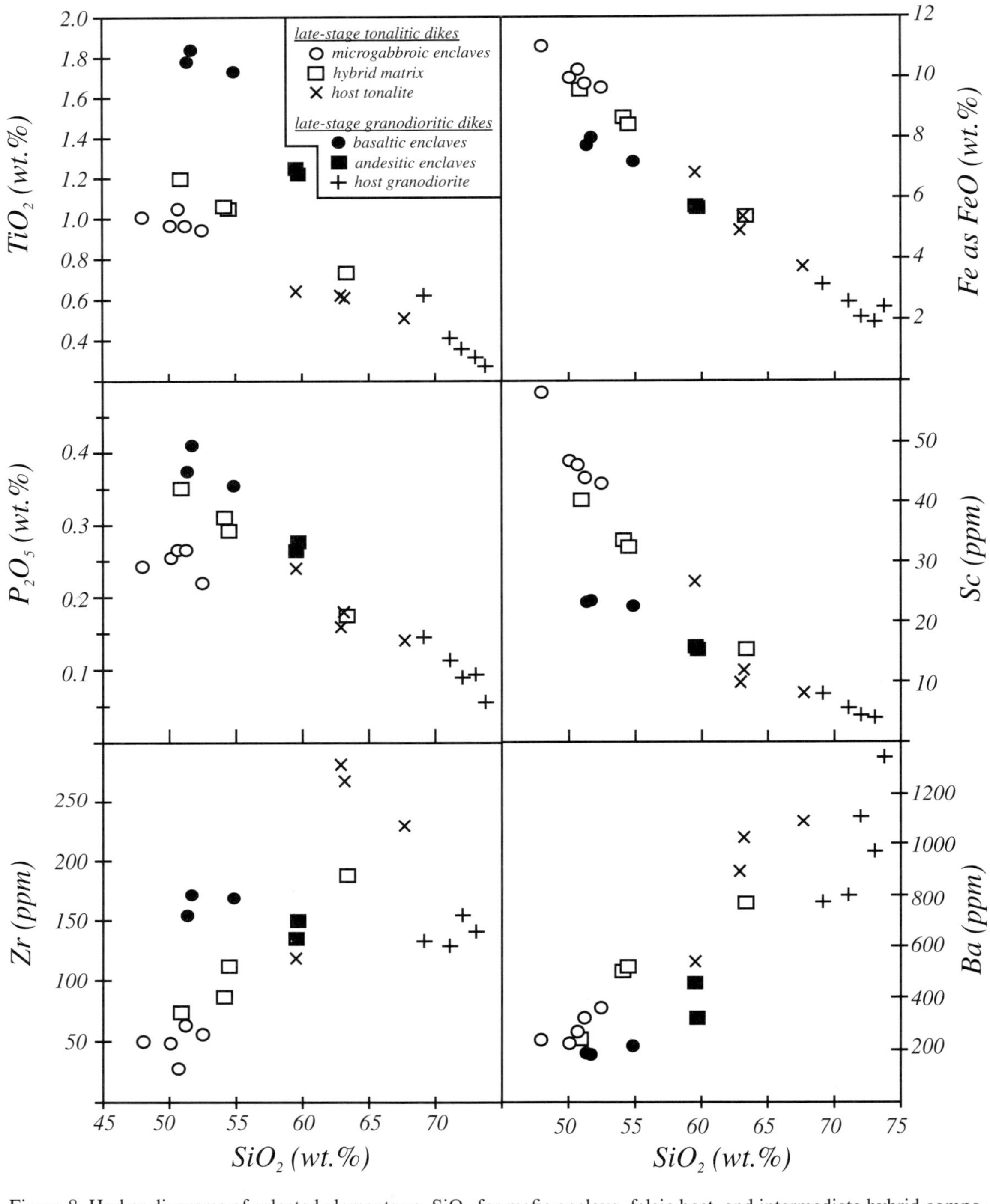

Figure 8. Harker diagrams of selected elements vs. SiO_2 for mafic enclave, felsic host, and intermediate hybrid compositions in the late-stage granodioritic and tonalitic dikes.

Some veins of the host rock contain fragments of enclave material that were apparently separated from the enclave during vein formation. Thus, field evidence indicates that hybridization of the magmas occurred in two ways: (1) incorporation of host magma into enclave magma by collapse of veins, and (2) fragmentation of enclave material into the host magma. This mechanism of hybridization might explain why there is less compositional distinction between enclave and host rocks on Grayback Mountain, where the degree of localized flattening and thus incorporation of host magma into the enclaves, was apparently greater than on Lake Mountain.

It could be argued that enclave and host rocks in the Lake Mountain mingling zone represent the mafic and more evolved components, respectively, of a single magma undergoing fractional crystallization. However, the host has higher TiO_2, Al_2O_3, P_2O_5, Sr, and Zr, and lower ΣREE than the enclaves, which is not the expected result of fractional crystallization of the observed phases. Therefore, we conclude that enclave and host compositions in the Lake Mountain mafic enclave swarm represent magmas from two calc-alkaline sources.

Although their outcrops are similar in appearance, it is clear that rocks comprising the mafic enclave swarm on Grayback Mountain are very different in composition from those exposed on Lake Mountain (Fig. 5). In contrast to Grayback Mountain samples, those from the Lake Mountain mafic enclave swarm are similar in composition to rocks comprising the noncumulate main stage of the Grayback pluton.

Late-Stage Granodioritic Dikes

The presence of both basaltic and andesitic enclaves in the late-stage granodioritic dikes suggests either that the enclave magmas were derived from two distinct sources or that the andesitic enclaves represent a hybridized mixture of basaltic and granodioritic magmas. We tested the mixing scenario, using element ratios according to the method of Langmuir et al. (1978). If a suite of rocks is related by binary mixing, then the samples should define a hyperbolic curve in ratio-ratio plots, define straight lines in companion plots, and remain in the same relative order in both ratio-ratio and companion plots. This approach, using Sc, Cr, K, and Ba abundances (Fig. 9A), produced consistent results that suggest binary mixing of basaltic and granodioritic end-members formed the andesitic magma. Major element least-squares mass balance calculations indicate that a mixture of 58% basalt and 42% granodiorite would produce a hybrid composition similar to that of the andesite (Table 2). Mixing calculations using the REE abundances of the same basaltic and granodioritic end-members produced hybrid compositions with chondrite-normalized REE patterns similar to the andesitic enclaves (Fig. 9B).

The presence of basaltic and hybrid andesitic enclaves in the late-stage granodioritic dikes is suggestive of a multi-stage mixing/mingling process (Koyaguchi, 1986). In the first stage, basaltic magma intruded the silicic magma, producing a layer of hybrid andesite at their interface. In the second stage, dike formation and magma ascent disrupted the basaltic and andesitic layers, resulting in their incorporation as enclaves in the silicic host magma. A post-emplacement stage involved chemical exchange between the magmas. The abundance of biotite and actinolite in the basaltic enclaves and the biotite- and K-feldspar–poor trondhjemitic halo that envelops several enclaves is suggestive of migration of H_2O and K^+ from the granodioritic to basaltic magmas (Watson and Jurewicz, 1984).

Major element mass balance calculations indicate that the dark-colored granodioritic host adjacent to some andesitic enclaves represents a mixture of ~18% andesitic and 82% granodioritic magmas (Table 2). The andesitic magma would have had a lower temperature and a resultant higher viscosity, relative to the basaltic magma, which would have prevented complete mechanical mixing of the andesitic and granodioritic magmas. We suggest instead that the dark-colored granodiorite formed by late-stage diffusive exchange between andesitic and granodioritic magmas during and/or after emplacement of the dike. This post-emplacement exchange was more pronounced than that between basaltic enclaves and host, because the basaltic magma was supercooled to a greater degree, which would have hindered chemical diffusion.

Late-Stage Tonalitic Dikes

As in the case of the late-stage granodioritic dikes, ratio-ratio and companion diagrams using enclave, host, and intermediate samples from late-stage tonalitic dikes (in this case, using Sc, Cr, Zr, and Ba abundances) suggest that magma mixing played a dominant role in the formation of intermediate compositions (Fig. 10). Minor deviations from linearity in the companion diagrams may result from fractional crystallization of augite in the enclave magma (see also Smith et al., 1990; Wiebe et al., 2002), resulting in enclave compositions with lower Sc/Zr values than those predicted for mixing alone (Fig. 10).

Major element mass-balance calculations suggest that binary mixing between dioritic and tonalitic magmas can produce almost the entire range of intermediate compositions observed in the late-stage tonalitic dikes. Most calculations yielded acceptable results (sum of the squares of residuals [Σr^2]) less than unity; Table 2). However, high negative Al_2O_3 residuals and high positive MgO and CaO residuals in some calculations (Table 2) suggest accumulation of plagioclase in, and possibly removal of augite from, the hybrid magma.

Anomalously high TiO_2 and P_2O_5 abundances in hybrid rocks have been noted in mingling zones elsewhere (Wood, 1978; Vogel, 1982; Marshall and Sparks, 1984; D'Lemos, 1996; Kuritani et al., 2003) and have been attributed to pre-mixing fractional crystallization in the mafic magma, mixing between compositionally heterogeneous magmas, post-mixing TiO_2 and P_2O_5 enrichment in the hybrid, and anomalous diffusion of Ti and P complexes. Enrichment of TiO_2 in the matrix compositions is inconsistent with the removal of augite from the hybrid magma (see above), and mixing between heterogeneous magmas is not suggested by the uniform TiO_2 and P_2O_5 in enclave and host samples. Alternatively, nonlinear relationships might reflect differences in diffusion coefficients between the elements of interest. However, differences in diffusion coefficients are unlikely to affect element correlations at the scale of hand samples used in this study (Perugini and Poli, 2004). In light of the fractional crystallization scenario described above, we suggest that the anomalously high TiO_2 and P_2O_5 contents in the matrix samples (Fig. 8) are a reflection of pre-mixing removal of augite (and apatite) from the enclave magma.

Discussion of Rheologic Models

Introduction. Petrographic observations and whole-rock chemical compositions can be used to estimate the rheologic

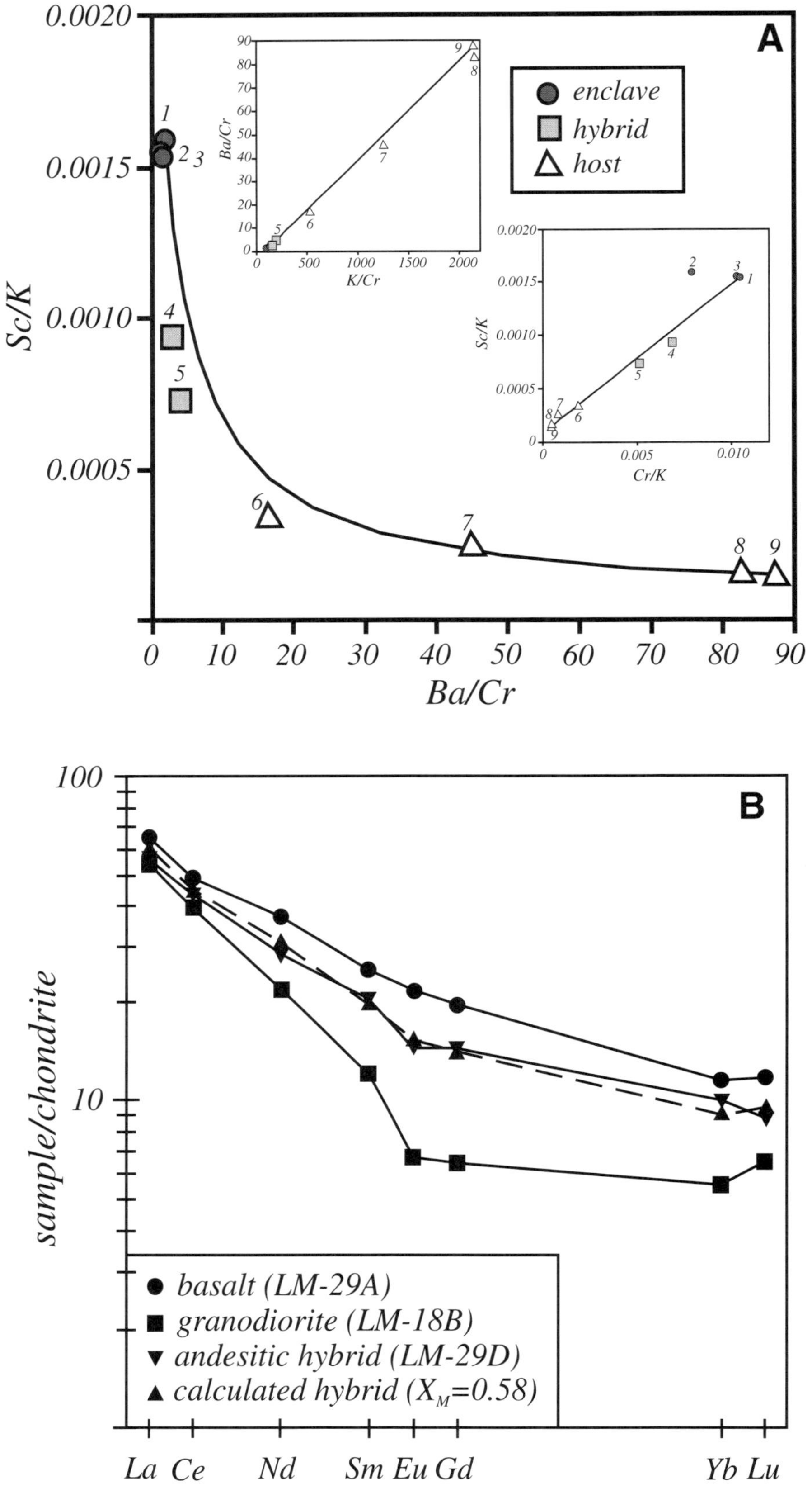

Figure 9. (A) Ratio-ratio and companion (insets) diagrams for basaltic and andesitic enclave and host compositions from a late-stage granodioritic dike. The solid lines represent the calculated hybrid compositions produced by mixing of basaltic and granodioritic magmas. The numbered samples in the diagram of Sc/K vs. Ba/Cr generally retain their relative order in the companion diagrams, suggesting that the intermediate compositions are hybrids of the two end-members (Langmuir et al., 1978). (B) Chondrite-normalized rare-earth element diagram comparing the composition of an andesitic enclave (LM-29D) with the calculated mixture (dashed line) of 58% basaltic enclave (LM-29A) and 42% granodioritic host (LM-18B) magmas, as suggested by major element mass-balance calculations.

TABLE 2. RESULTS OF MAJOR ELEMENT MIXING CALCULATIONS FOR SAMPLES FROM THE LATE-STAGE GRANODIORITIC AND TONALITIC DIKES

		Residuals													
Model	Mafic	SiO_2	TiO_2	Al_2O_3	FeO	MnO	MgO	CaO	Na_2O	K_2O	P_2O_5	Felsic	Mix	X_M	ssr
1A	LM-29A	–0.05	–0.01	0.55	–0.11	–0.00	–0.35	0.08	–0.22	0.18	–0.01	LM-18B	LM-29D	0.58	0.53
1B	LM-29D	–0.10	–0.07	0.26	–0.02	0.01	–0.37	–0.06	0.12	–0.41	–0.00	LM-18B	LM-29B	0.18	0.41
2A	LM-27B2	0.22	–0.09	–0.50	–0.07	–0.03	0.10	0.53	0.23	0.03	–0.00	LM-28	LM-26A3	0.24	0.66
2B	LM-16C	0.22	–0.12	–0.68	–0.19	–0.03	0.21	0.63	0.17	0.01	–0.01	LM-28	LM-26A3	0.23	1.05
2C	LM-27B2	0.30	–0.13	–0.73	0.24	0.00	0.37	0.43	0.07	–0.31	–0.06	LM-28	LM-26A2	0.76	1.13
2D	LM-16C	0.19	–0.03	–0.19	–0.28	–0.00	0.21	0.50	–0.37	–0.12	–0.02	LM-28	LM-[illegible]6B	0.93	0.59

Notes: The terms "Mafic," "Felsic," and "Mix" refer to the mafic, felsic, and observed hybrid compositions, respectively, used in the calculations. Residuals are the differences between the calculated mixture and the observed mixture. Abbreviations: X_M—proportion of mafic and member used in the calculations; ssr—sum of the squares of the residuals.

conditions under which magma mixing and mingling are possible (Sparks and Marshall, 1986; Frost and Mahood, 1987). Complete mixing between two magmas of contrasting compositions requires the equilibrium temperature of the mixture to be greater than the solidus temperature of the mafic end-member. The equilibrium temperature depends on the initial temperatures and mass fractions of the end-member magmas, the heat capacities of liquid and solid components, and the heats of fusion of the solids (Frost and Mahood, 1987).

Viscosity (η)-composition relationships from the mingling zones in the Grayback pluton were determined using the algorithms compiled by Frost and Lindsay (1988). Resultant viscosity curves are shown in Figures 11–14, with respect to the mass fraction of the mafic end member in each mixture (X_M). Intersections of viscosity curves of coexisting magmas are circled, and represent the conditions at which both magmas possess equal viscosities ($\acute{\eta}$, the "equilibrium" viscosity), and are thus most likely to mix. However, it should be noted that hybridization can occur between magmas with viscosity differences as great as 10^2 poise (Frost and Mahood, 1987) and at X_M greater than that at $\acute{\eta}$ (Sparks and Marshall, 1986). Terminations of the mafic end-member curves represent the η and X_M values where

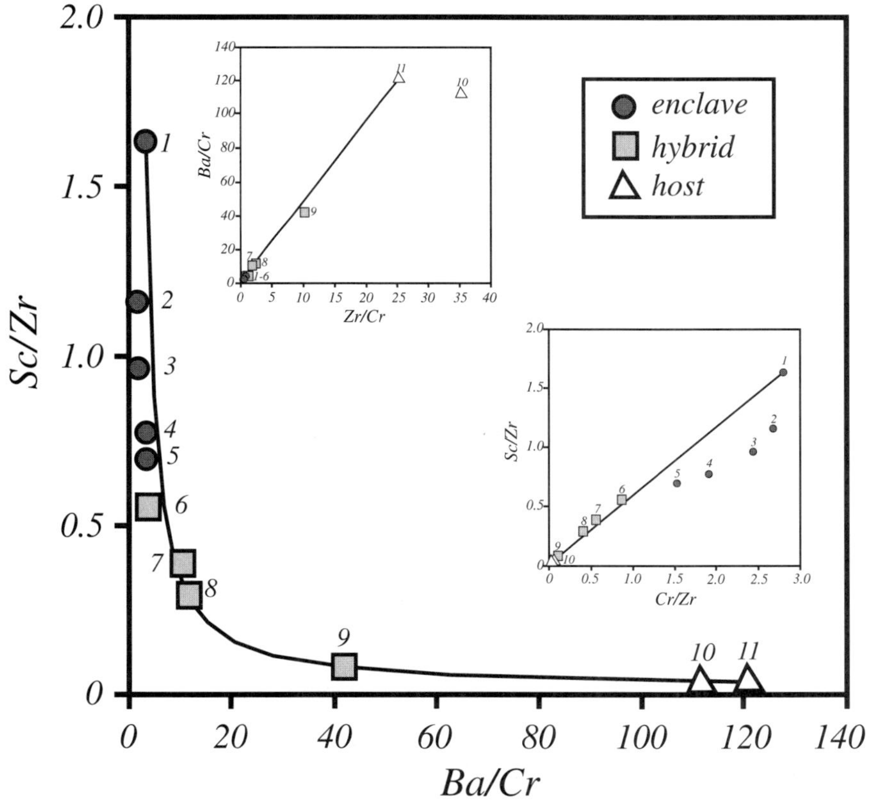

Figure 10. Ratio-ratio and companion (insets) diagrams for microdioritic enclave hybrid matrix, and host tonalite compositions in a late-stage tonalitic dike. The solid lines represent the calculated hybrid compositions produced by mixing of dioritic and tonalitic magmas. As in Figure 9, the numbered samples in the ratio-ratio diagram generally retain their relative order in the accompanying companion diagrams, suggesting a mixing origin for intermediate compositions. Deviation of hybrid compositions from the calculated mixing line in the companion plot of Sc/Zr vs. Cr/Zr probably reflects fractional crystallization of augite (see text).

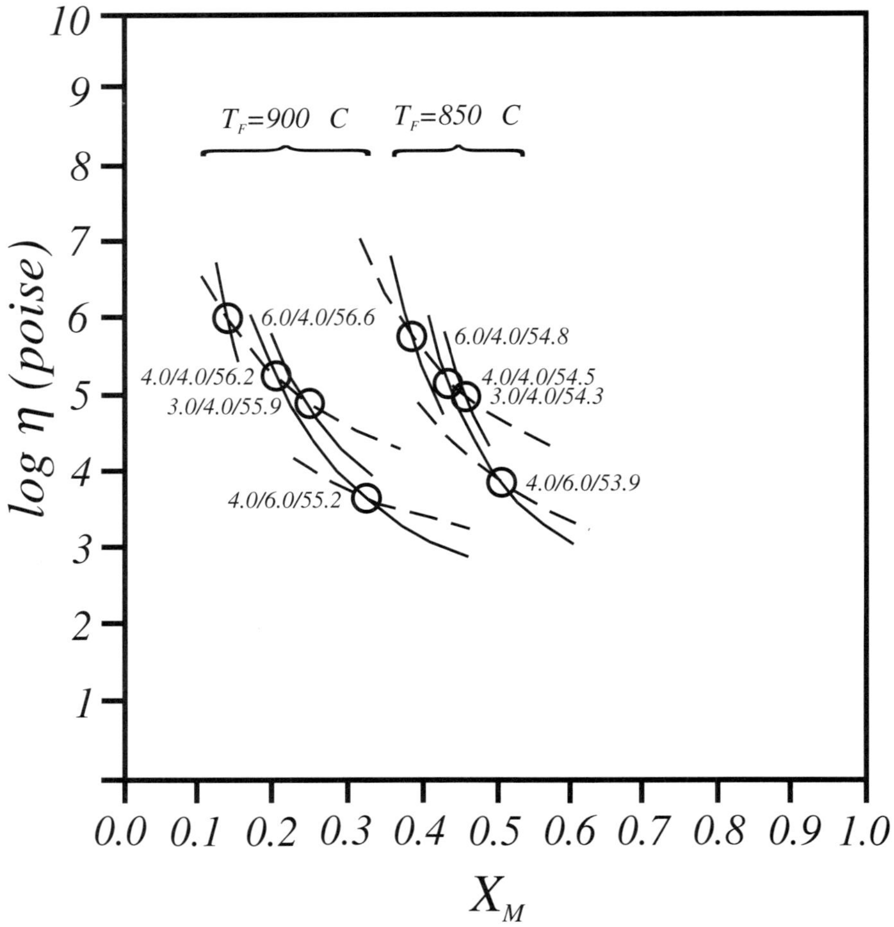

Figure 11. Viscosity-composition diagram for enclave-host pairs in the Lake Mountain mafic enclave swarm, showing the variations in viscosity and composition of hybrid magmas as a function of initial temperatures and H_2O contents of the enclave (solid curves) and host (dashed curves) magmas. Only the intersections of enclave-host pairs are shown and are labeled X/Y/Z, where X is the initial H_2O content of the enclave magma, Y is the initial H_2O content of the host magma, and Z is the SiO_2 content (in wt%) of the resultant hybrid. Two sets of intersections are shown for initial host magma temperatures of 850 °C and 900 °C.

the mafic magma is >60% crystallized, and thus essentially acts as a solid (Frost and Mahood, 1987).

Lake Mountain Mafic Enclave Swarm. One of the puzzling aspects of the mafic enclave swarms is the preservation of enclave and host, despite their compositional similarities. If these rocks are reflective of magma mingling in the main stage of the pluton, then it can be assumed that main-stage mixing occurred earlier in the Grayback system and that conditions changed enough to prohibit thorough mixing of these magmas within the enclave swarm. The rheological properties of enclave and host compositions were investigated in an effort to determine why the magmas did not mechanically mix within the mafic enclave swarm at the present level of exposure.

The hornblende-rich nature of enclaves and host rocks in the Lake Mountain enclave swarm suggests that both enclave and host magmas had initial H_2O contents >3.0 wt% (e.g., Sisson and Grove, 1993). For an enclave magma temperature of 1050 °C, mixing to produce an intersection of mafic and felsic curves at $\acute{\eta}$ requires a host magma temperature <950 °C. Temperatures >950 °C result in enclave magma with a lower viscosity than the host magma at all values of X_M. A relatively small decrease in host magma temperature (50 °C) results in a large increase in X_M ($\Delta X_M > 0.2$) required for mixing at $\acute{\eta}$ (Fig. 11). For mixing at equal viscosities, an increase in enclave magma H_2O content results in a slight increase in $\acute{\eta}$ and a slight decrease in X_M (Fig. 11), whereas increasing the host magma H_2O content results in slightly decreasing $\acute{\eta}$ and increasing X_M.

Complete mixing between the enclave and host magmas would have required at least 15% of the enclave end-member. That complete mixing did not occur within the mafic enclave swarm may reflect a mass decrease in the enclave end member to below 15%, or a shift of $\acute{\eta}$ to higher values of X_M as a result of lower magma temperatures or loss of H_2O from the enclave magma (Fig. 11).

Late-Stage Granodioritic Dikes. Figure 12 shows viscosity-composition curves that describe mixing between basaltic and granodioritic magmas from the late-stage granodioritic dikes. The presence of plagioclase phenocrysts and the lack of early hornblende suggest that the basaltic magma had less than ~3.0 wt% H_2O (e.g., Eggler, 1972). Rare hornblende inclusions in plagioclase and the lack of evidence for vapor saturation constrain the H_2O content in the granodioritic magma to between ~4.0 and 6.5 wt% (Naney, 1983). Mixing of anhydrous basaltic magma with granodioritic magma having 4.0 wt% H_2O would

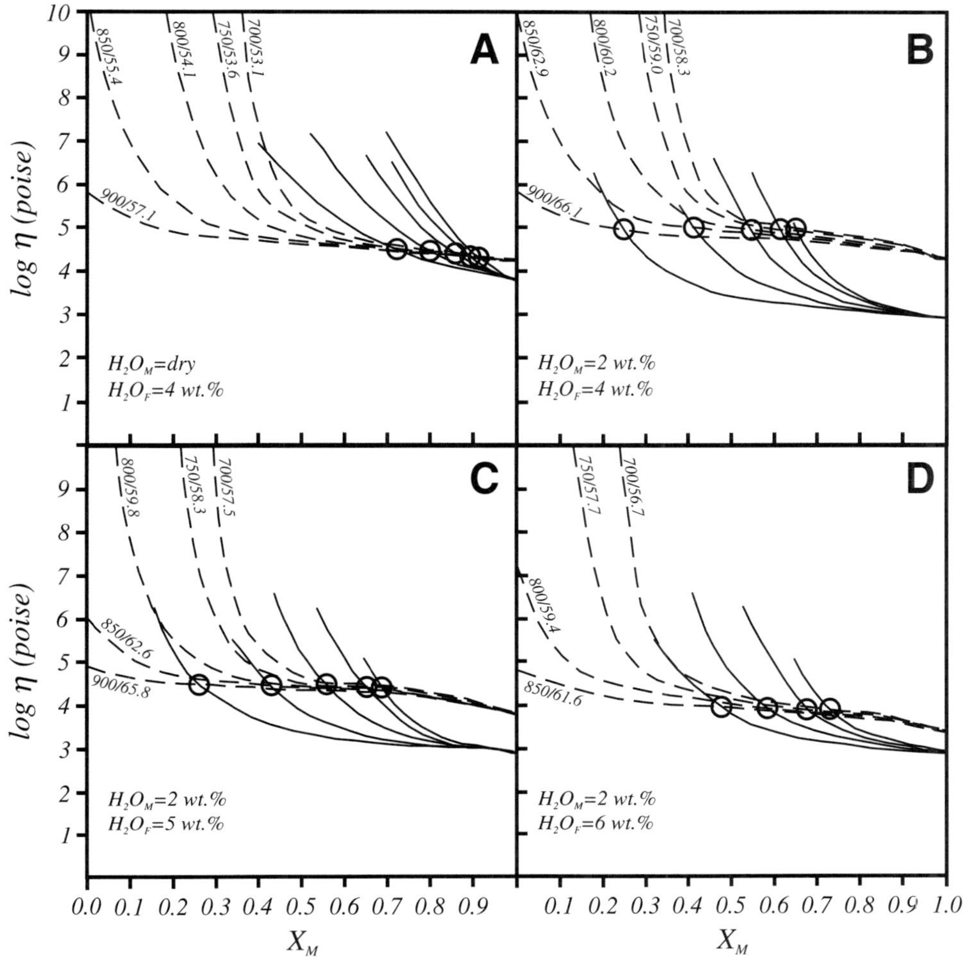

Figure 12. Viscosity-composition diagram for enclave (basalt)-host pairs in a late-stage granodioritic dike, showing the variations in viscosity and composition of hybrid magmas as a function of initial temperatures and H_2O contents of the enclave (solid curves) and host (dashed curves) magmas. Each panel in the figure represents different conditions of initial H_2O contents in the enclave (H_2O_M) and host (H_2O_F) magmas. The dashed curves in each panel are labeled X/Y, where X is the initial temperature (in °C) of the host magma and Y is the SiO_2 content (in wt%) of the resultant hybrid.

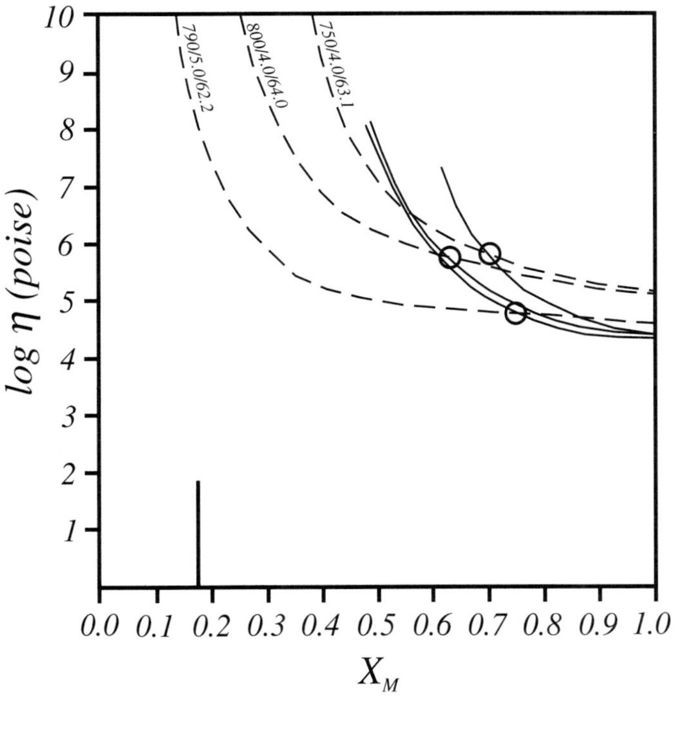

result in a relatively narrow range of hybrid compositions (~53–57 wt% SiO_2), requiring X_M >0.7 (Fig. 12A). In contrast, mixing involving hydrous basaltic magma (2.0 wt% H_2O) results in a wider spectrum of hybrid compositions (~56–66 wt% SiO_2) requiring proportionally less mafic magma (X_M = 0.25–0.73; Fig. 12B–D), depending on the initial temperature and H_2O content of the host magma. The equilibrium viscosity $\acute{\eta}$ decreases from 10^5 to 10^4 poise with increasing H_2O content in the host magma (Fig. 12B–D). At X_M = 0.53, the proportion of enclave magma estimated by major and trace element mass

Figure 13. Viscosity-composition diagram for enclave (andesite)-host pairs in the same late-stage granodiorite (as in Fig. 12), showing variations in viscosity and composition of the hybrid, which is assumed to be the dark-colored granodiorite adjacent to several andesitic enclaves, as a function of initial temperatures and H_2O contents of the enclave (solid curves) and host (dashed curves) magmas. Dashed curves are labeled X/Y/Z, where X is the initial temperature of the host magma, Y is the initial H_2O content of the host magma, and Z is the SiO_2 content (in wt%) of the resultant hybrid. The vertical line at X_M = 0.18 represents the proportion of enclave magma required by major element mass balance to produce the dark-colored granodiorite (see text).

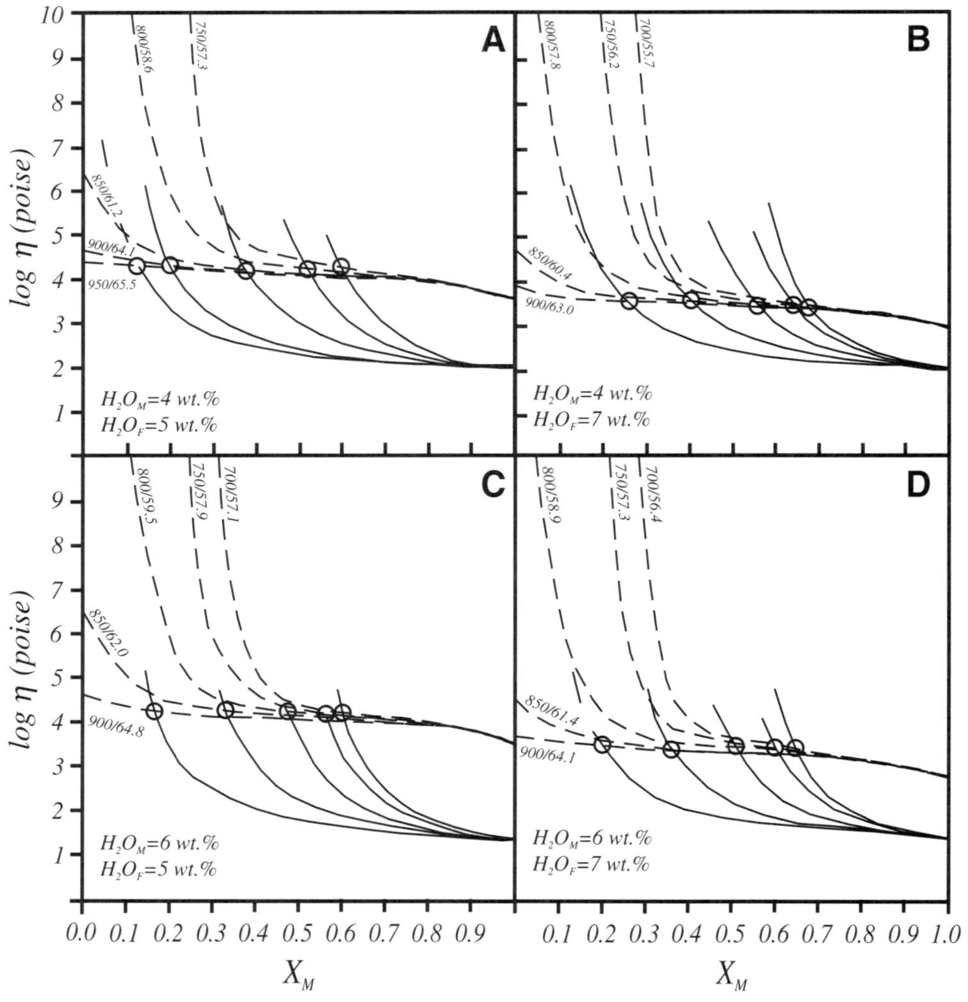

Figure 14. Viscosity-composition diagram for enclave (diorite)-host pairs in a late-stage tonalitic dike, showing the variations in viscosity and composition of hybrid magmas as a function of initial temperatures and H_2O contents of the enclave (solid curves) and host (dashed curves) magmas. Each panel in the figure represents different conditions of initial H_2O contents in the enclave (H_2O_M) and host (H_2O_F) magmas. The dashed curves in each panel are labeled X/Y, where X is the initial temperature (in °C) of the host magma and Y is the SiO_2 content (in wt%) of the resultant hybrid.

balance calculations (see discussion above), mixing at $\acute{\eta}$ requires host magma temperatures of 775–800 °C for initial H_2O contents of 4–6 wt%.

Additional calculations were performed to evaluate the conditions and extent of mixing between the hybrid andesitic and host granodioritic magmas, the results of which are shown in Figure 13. The initial temperature and H_2O content of the andesitic end-member were taken from results of the basalt-granodiorite mixing calculations (see above). Intersections between curves representing the andesitic and granodioritic end-members ($\acute{\eta}$) occur at higher viscosities ($10^{4.6}$–$10^{5.8}$ poise; Fig. 13) than those between basaltic and granodioritic magmas (<10^5 poise; Fig. 12). In addition, mixing between andesitic and granodioritic magmas at $\acute{\eta}$ could only occur at X_M >0.60, which is significantly greater than the X_M~ 0.18 predicted by major element mass balance. These results indicate that conditions suitable for complete binary mixing of andesitic and granodioritic magmas in the late-stage granodioritic dikes were never achieved and that the dark-colored granodiorite adjacent to the andesitic enclaves was instead produced by diffusion of chemical components across the enclave-host contact.

Late-Stage Tonalitic Dikes. Mixing results for the late-stage tonalitic dikes are more difficult to assess, because the hybrid compositions span a wide range in compositions. The hornblende-rich nature of the microdioritic enclaves suggests relatively high H_2O contents of 3–6 wt% (Sisson and Grove, 1993); the low An content of plagioclase suggests, however, that the H_2O content was at the low end of the range (Sisson and Grove, 1993). The lack of orthopyroxene and the abundance of early biotite in the tonalitic rocks suggest that the felsic magma was H_2O-rich, but not H_2O-saturated (5–7 wt% H_2O; Scaillet and Evans, 1999). Calculations were performed using a range of host magma H_2O contents (5.0–7.0 wt%) and initial temperatures (700–950 °C) and using an enclave magma H_2O content of 4.0 wt% (Fig. 14). An increase in host magma H_2O content from 5.0 to 7.0 wt% causes a decrease in $\acute{\eta}$ from $10^{4.2}$ to $10^{3.5}$ poise; however, an increase in the enclave magma H_2O content from 4.0 to 6.0 wt% over the same range of host magma H_2O contents (5.0–7.0 wt%) and initial temperatures (750–900 °C) does not affect $\acute{\eta}$ ($10^{4.3}$–$10^{3.5}$ poise). Ideal mixing of enclave and host magmas at $\acute{\eta}$ cannot account for the range in hybrid compositions observed in the late-stage tonalitic dikes, particularly those having X_M >0.7 (SiO_2 <55 wt%), which is probably a result of accumulation (see above).

The coarser grain size of the microdioritic enclaves, compared to the basaltic enclaves in the granodioritic dikes, suggests that the dioritic magma did not experience the same degree of undercooling or that it was largely crystalline when it came into contact with the tonalitic magma. The former suggests a smaller temperature contrast between mafic and felsic magmas, whereas the latter might indicate a high contrast in initial viscosities. A decrease in the temperature contrast results in mixing at progressively lower X_M (Fig. 14). However, a decrease in the initial viscosity contrast between the two magmas results in ideal mixing at slightly greater values for X_M, but still well below the values suggested by mass-balance calculations (Fig. 14). The bimodal grain size of the hybrid rocks is characteristic of mixing between crystal-rich mushes (e.g., D'Lemos, 1996); therefore, it is unlikely that the hybrid compositions were produced by mixing of liquid magmas having equal viscosities (see Fig. 14).

DISCUSSION

Barnes et al. (1995) documented a great number of mafic rock types in the Grayback pluton. The lithologic heterogeneity observed, even at the hand sample scale, attests to the presence of coexisting magmas throughout main-stage magmatism. The main stage of the pluton comprises gabbro, norite, quartz diorite, and lesser amounts of pyroxenite and tonalite. Coeval calc-alkaline and evolved tholeiitic basaltic magmas point to a heterogeneous mantle source. Barnes et al. (1995) demonstrated that fractional crystallization of plagioclase + magnetite + olivine ± clinopyroxene ± orthopyroxene ± hornblende ± apatite from calc-alkaline basaltic magma of the main stage could produce felsic compositions similar to those observed in the late-stage granodioritic and granitic dikes. Such a process could explain the abundance of cumulate pyroxenites and melagabbros with Mg/(Mg + Fe) >0.58 (see Barnes et al., 1995).

Fractional crystallization resulted in an initial increase in P_2O_5, followed by an abrupt decrease at the onset of apatite crystallization (Fig. 5C). In addition, TiO_2 (Fig. 5A), Al_2O_3, and Sr abundances steadily decreased owing to the removal of Fe-Ti oxides, clinopyroxene, and plagioclase. Few main-stage samples follow the fractional crystallization trend observed in Figure 5, but instead exhibit increasing TiO_2, Al_2O_3, P_2O_5, and Sr with decreasing Mg/(Mg + Fe). Likewise, few samples follow the mixing trends exhibited by enclave-host pairs in the late-stage granodioritic and tonalitic dikes. It is difficult to discern a clear liquid line of descent with respect to most elements, but most of the noncumulate main-stage samples plot near the field of enclave and host compositions represented by the Lake Mountain mafic enclave swarm (Fig. 5). The compositional similarities between the Lake Mountain mafic enclave swarm and Grayback main-stage samples suggest that mixing between calc-alkaline basaltic and andesitic magmas occurred throughout much of the Grayback history, probably at a greater depth than represented by the current level of exposure. The enclave and host magmas observed in the Lake Mountain mafic enclave swarm were prevented from mixing, possibly as a result of devolatilization or a temperature decrease (and hence, a higher crystallinity). A similar scenario could be envisioned for tholeiitic compositions in the mafic enclave swarm on Grayback Mountain, which tapped a distinctly different mantle source (Barnes et al., 1995).

Some basaltic magma evolved to granitic compositions through fractional crystallization. Minor assimilation of metasedimentary rocks gave rise to the elevated $^{87}Sr/^{86}Sr$ and $\delta^{18}O$ contents of the granites (Barnes et al., 1992a). Continued influx of basaltic magma and subsequent mixing with the silicic magma produced hybrid andesite, presumably at a deeper level in the Grayback system. Intrusion of the granitic magma to the present level of exposure entrained small volumes of basaltic and andesitic magmas that quickly solidified in their new environment, producing the mafic and intermediate enclaves. Although the late-stage granodioritic dikes lack evidence for extensive mixing between mafic and felsic magmas in situ, the presence of hybrid andesitic enclaves indicates that such mixing did occur in the Grayback system, requiring close to 60% of the basaltic end-member for complete mixing to take place. That the hybrid andesite compositions are unlike any main-stage samples suggests either that the andesite produced in this manner was volumetrically minor or that large volumes did not ascend to the present level of exposure.

Heat from mafic magma partially melted mafic metavolcanic rocks residing deep in the Klamath crust, generating small volumes of high-Al tonalitic magma. Major and trace element signatures of the tonalitic rocks (high Al_2O_3 and Na_2O, and low K_2O, Y, and HREE; Drummond and Defant, 1990) suggest that partial melting occurred in equilibrium with a garnet- and hornblende-bearing residual assemblage at pressures exceeding 10 kb (Wolf and Wyllie, 1994; Rapp and Watson, 1995). Injection of calc-alkaline basaltic magma into the high-Al tonalitic magma resulted in limited mixing and the magma mingling relationships observed in the late-stage tonalitic dikes. Results of this study indicate that this basalt-tonalite mixing did not play a significant role in the compositional variation of the Grayback main stage.

Other plutons in the Wooley Creek belt experienced open-system behavior similar to the Grayback pluton. In the Wooley Creek batholith, basalt mixed with fractionated dacite in the upper part of the magma chamber to produce hybrid andesite (Barnes et al., 1986, 1987). Open-system behavior in the Ashland pluton (Gribble et al., 1990) led to mixing between alkalic basalt and tonalite and resultant quartz monzodioritic magma, which evolved by coupled fractionation and assimilation to more felsic compositions. Schmidt (1994) determined that mixing between andesitic magma and anatectic granodiorite and granite was important during later stages of the composite English Peak pluton. In addition, mafic enclaves throughout the Russian Peak pluton indicate not only the presence of mafic magma in the system but, coupled with ranges in Sr and O isotope abundances (Cotkin and Medaris, 1993), hint at the possible role of magma mixing in its history. The Grayback pluton, however, is unusual amongst these plutons in that mafic magmas from multiple mantle sources were prevalent *throughout* its history. It is likely that ascent of mantle-derived magmas into the Grayback system was facilitated by extension in the back-arc of the contemporaneous Rogue–Chetco arc complex (Harper et al., 1994). This setting, directly inboard of the arc, may have experienced greater extension than further inboard or to the south, positions currently occupied by the other Wooley Creek belt plutons.

CONCLUSIONS

A diverse range of magma-mingling zones in the Grayback pluton provides valuable information about the physicochemical interactions of magmas, and documents the coexistence of several mantle-derived magmas throughout the evolution of the Grayback system. Comparisons between rock compositions preserved in the magma-mingling zones and those from the main stage of the pluton suggest that most of the compositional variation within the Grayback pluton resulted from mixing of calc-alkaline basaltic and andesitic magmas. Late-stage activity resulted in the emplacement of enclave-bearing granodioritic and tonalitic dikes, which exhibit different styles of interaction between mafic and felsic end-members.

Grayback magmatism was coeval with other plutons of the Wooley Creek belt, which all show evidence for a significant role for mafic magmas in their evolution. Increased interaction of mantle-derived magmas and crustal rocks in Wooley Creek belt plutonic systems, compared with earlier episodes of Jurassic magmatism, was facilitated by extension in the back-arc region of a contemporaneous volcanic arc. Emplacement of these mafic magmas into already-warmed crust allowed for assimilation of crustal rocks and/or mixing with anatectic melts.

ACKNOWLEDGMENTS

We are pleased to have the opportunity to contribute to this volume honoring Porter Irwin's pioneering and influential work in the Klamath Mountains. We thank Bennetta Schmidt and Melanie Barnes for help in the field, and Bill Shannon and Melanie for their invaluable assistance in the laboratory. We appreciate the thorough and constructive reviews of Nancy McMillan and Daniel Lux. Funding was provided by a Penrose Bequest of the Geological Society of America (KJ) and National Science Foundation grants EAR-8720141 and EAR-9117103 (CGB).

REFERENCES CITED

Allen, C.M., and Barnes, C.G., 2006, this volume, Ages and some cryptic sources of Mesozoic plutonic rocks in the Klamath Mountains, California and Oregon, *in* Snoke, A.W., and Barnes, C.G., eds., Geological studies in the Klamath Mountains province, California and Oregon: A volume in honor of William P. Irwin: Boulder, Colorado, Geological Society of America Special Paper 410, doi: 10.1130/2006.2410(11).

Barker, F., 1979, Trondhjemite: Definition, environment, and hypotheses of origin, *in* Barker, F., ed., Trondhjemites, dacites, and related rocks: Amsterdam, Elsevier, p. 1–12.

Barnes, C.G., 1983, Petrology and upward zonation of the Wooley Creek batholith, Klamath Mountains, California: Journal of Petrology, v. 24, p. 495–537.

Barnes, C.G., Allen, C.M., and Saleeby, J.B., 1986, Open- and closed-system characteristics of a tilted plutonic system, Klamath Mountains, California: Journal of Geophysical Research, v. 91, p. 6073–6090.

Barnes, C.G., Allen, C.M., and Brigham, R.H., 1987, Isotopic heterogeneity in a tilted plutonic system, Klamath Mountains, California: Geology, v. 15, p. 523–527, doi: 10.1130/0091-7613(1987)15<523:IHIATP>2.0.CO;2.

Barnes, C.G., Petersen, S.W., Kistler, R.W., Prestvik, T., and Sundvoll, B., 1992a, Tectonic implications of isotopic variation among Jurassic and Early Cretaceous plutons, Klamath Mountains: Geological Society of America Bulletin, v. 104, p. 117–126, doi: 10.1130/0016-7606(1992)104<0117:TIOIVA>2.3.CO;2.

Barnes, C.G., Barnes, M.A., and Kistler, R.W., 1992b, Petrology of the Caribou Mountain pluton, Klamath Mountains, California: Journal of Petrology, v. 33, p. 95–124.

Barnes, C.G., Johnson, K., Barnes, M.A., Prestvik, T., Kistler, R.W., and Sundvoll, B., 1995, The Grayback pluton: Magmatism in a Jurassic back-arc environment, Klamath Mountains, Oregon: Journal of Petrology, v. 36, p. 397–416.

Barnes, C.G., Petersen, S.W., Kistler, R.W., Murray, R., and Kays, M.A., 1996, Source and tectonic implications of tonalite-trondhjemite magmatism in the Klamath Mountains: Contributions to Mineralogy and Petrology, v. 123, p. 40–60, doi: 10.1007/s004100050142.

Barnes, C.G., Mars, E.V., Swapp, S., and Frost, C.D., 2006, this volume (Chapter 10), Petrology and geochemistry of the Middle Jurassic Ironside Mountain batholith: Evolution of potassic magmas in a primitive arc setting, *in* Snoke, A.W., and Barnes, C.G., eds., Geological studies in the Klamath Mountains province, California and Oregon: A volume in honor of William P. Irwin: Boulder, Colorado, Geological Society of America Special Paper 410, doi: 10.1130/2006.2410(10).

Castro, A., de la Rosa, J.D., and Stephens, W.E., 1990, Magma mixing in the subvolcanic environment: Petrology of the Gerena interaction zone near Seville, Spain: Contributions to Mineralogy and Petrology, v. 106, p. 9–26, doi: 10.1007/BF00306405.

Cotkin, S.J., and Medaris, L.G., Jr., 1993, Evaluation of the crystallization conditions for the calc-alkaline Russian Peak Intrusive Complex, Klamath Mountains, northern California: Journal of Petrology, v. 34, p. 543–571.

D'Lemos, R.S., 1996, Mixing between granitic and dioritic crystal mushes, Guernsey, Channel Islands, UK: Lithos, v. 38, p. 233–257, doi: 10.1016/0024-4937(96)00007-2.

Donato, M.M., Barnes, C.G., and Tomlinson, S.L., 1996, The enigmatic Applegate Group of southwestern Oregon: Age, correlation, and tectonic affinity: Oregon Geology, v. 58, p. 79–91.

Drummond, M.S., and Defant, M.J., 1990, A model for trondhjemite-tonalite-dacite genesis and crustal growth via slab melting: Archean to modern comparisons: Journal of Geophysical Research, v. 95, p. 21,503–21,521.

Eggler, D.H., 1972, Amphibole stability in H_2O-undersaturated calc-alkaline melts: Earth and Planetary Science Letters, v. 15, p. 28–34, doi: 10.1016/0012-821X(72)90025-8.

Ernst, W.G., 1998, Geology of the Sawyers Bar area, Klamath Mountains, northern California: Sacramento, California Division of Mines and Geology Map Sheet 47, scale 1:48,000, 59 p.

Frost, T.P., and Lindsay, J.R., 1988, MAGMIX: A BASIC program to calculate viscosities of interacting magmas of differing composition, temperature, and water content: Computers & Geosciences, v. 14, p. 213–228, doi: 10.1016/0098-3004(88)90005-2.

Frost, T.P., and Mahood, G.A., 1987, Field, chemical, and physical constraints on mafic-felsic magma interaction in the Lamarck Granodiorite, Sierra Nevada, California: Geological Society of America Bulletin, v. 99, p. 272–291, doi: 10.1130/0016-7606(1987)99<272:FCAPCO>2.0.CO;2.

Garcia, M.O., 1982, Petrology of the Rogue River island-arc complex, southwest Oregon: American Journal of Science, v. 282, p. 783–807.

Godchaux, M.M., 1969, Petrology of the Grayback Igneous Complex and contact aureole, Klamath Mountains, southwestern Oregon [Ph.D. dissertation]: Eugene, University of Oregon, 223 p.

Gribble, R.F., Barnes, C.G., Donato, M.M., Hoover, J.D., and Kistler, R.W., 1990, Geochemistry and intrusive history of the Ashland pluton, Klamath Mountains, California and Oregon: Journal of Petrology, v. 31, p. 883–923.

Hacker, B.R., Donato, M.M., Barnes, C.G., McWilliams, M.O., and Ernst, W.G., 1995, Timescales of orogeny: Jurassic construction of the Klamath Mountains: Tectonics, v. 14, p. 677–703, doi: 10.1029/94TC02454.

Harper, G.D., and Wright, J.E., 1984, Middle to Late Jurassic tectonic evolution of the Klamath Mountains, California-Oregon: Tectonics, v. 3, p. 759–772.

Harper, G.D., Saleeby, J.B., and Heizler, M., 1994, Formation and emplacement of the Josephine ophiolite and the Nevadan orogeny in the Klamath Mountains, California-Oregon: U/Pb zircon and $^{40}Ar/^{39}Ar$ geochronology: Journal of Geophysical Research, v. 99, p. 4293–4321, doi: 10.1029/93JB02061.

Haskin, L.A., Haskin, M.A., Frey, F.A., and Wildeman, T.R., 1968, Relative and absolute abundances of the rare earths, *in* Ahrens, L.H., ed., Origin and distribution of the elements: Oxford, Pergamon Press, p. 889–912.

Hotz, P.E., 1971, Plutonic rocks of the Klamath Mountains, California and Oregon: Washington, D.C., U.S. Geological Survey Professional Paper 684-B, 20 p.

Irwin, W.P., 1972, Terranes of the western Paleozoic and Triassic belt in the southern Klamath Mountains, California: Washington, D.C., U.S. Geological Survey Professional Paper 800-C, p. 103–111.

Irwin, W.P., 1981, Tectonic accretion of the Klamath Mountains, *in* Ernst, W.G., ed., The geotectonic development of California—Rubey volume 1: Englewood Cliffs, New Jersey, Prentice-Hall, p. 29–49.

Irwin, W.P., 1985, Age and tectonics of plutonic belts in accreted terranes of the Klamath Mountains, California and Oregon, *in* Howell, D.G., ed., Tectonostratigraphic terranes of the Circum-Pacific Region: Houston, Circum-Pacific Council for Energy and Mineral Resources, Earth Science Series 1, p. 187–199.

Irwin, W.P., and Wooden, J., 1999, Plutons and accretionary episodes of the Klamath Mountains, California and Oregon: Reston, Virginia, U.S. Geological Survey Open-File Report 99-374, 1 sheet.

Johnson, K., 1991, Magma mixing and mingling in the Grayback pluton, Klamath Mountains, southwest Oregon [M.S. thesis]: Lubbock, Texas Tech University, 122 p.

Koyaguchi, T., 1986, Evidence for two-stage mixing in magmatic inclusions and rhyolitic lava domes on Nijima Island, Japan: Journal of Volcanology and Geothermal Research, v. 29, p. 71–98, doi: 10.1016/0377-0273(86)90040-5.

Kuritani, T., Yokoyama, T., Kobayashi, K., and Nakamura, E., 2003, Shift and rotation of composition trends by magma mixing: 1983 eruption of Miyake-jima volcano, Japan: Journal of Petrology, v. 44, p. 1895–1916, doi: 10.1093/petrology/egg063.

Langmuir, C.H., Vocke, R.D., Hanson, G.N., and Hart, S.R., 1978, A general mixing equation with applications to Icelandic basalts: Earth and Planetary Science Letters, v. 37, p. 380–392, doi: 10.1016/0012-821X(78)90053-5.

Marshall, L.A., and Sparks, R.S.J., 1984, Origin of some mixed-magma and net-veined ring intrusions: Journal of the Geological Society (London), v. 141, p. 171–182.

Naney, M.T., 1983, Phase equilibria of rock-forming ferromagnesian silicates in granitic systems: American Journal of Science, v. 283, p. 993–1033.

Perugini, D., and Poli, G., 2004, Analysis and numerical simulation of chaotic advection and chemical diffusion during magma mixing: Petrologic implications: Lithos, v. 78, p. 43–66, doi: 10.1016/j.lithos.2004.04.039.

Rapp, R.P., and Watson, E.B., 1995, Dehydration melting of metabasalt at 8–32 kbar: Implications for continental growth and crust-mantle recycling: Journal of Petrology, v. 36, p. 891–931.

Saleeby, J.B., Harper, G.D., Snoke, A.W., and Sharp, W.D., 1982, Time relations and structural-stratigraphic patterns in ophiolite accretion, west-central Klamath Mountains, California: Journal of Geophysical Research B, v. 87, p. 3831–3848.

Scaillet, B., and Evans, B.W., 1999, The 15 June 1991 eruption of Mount Pinatubo. I. Phase equilibria and pre-eruption P–T–fO_2–fH_2O conditions of the dacite magma: Journal of Petrology, v. 40, p. 381–411, doi: 10.1093/petrology/40.3.381.

Schmidt, B.L., 1994, The petrology and geochemistry of the English Peak intrusive suite, Klamath Mountains, California [Ph.D. dissertation]: Lubbock, Texas Tech University, 216 p.

Shannon, W.M., 1994, Petrogenesis of Middle Proterozoic alkaline granitic rocks, Franklin Mountains, west Texas [Ph.D. dissertation]: Lubbock, Texas Tech University, 240 p.

Sisson, T.W., and Grove, T.L., 1993, Experimental investigations of the role of H_2O in calc-alkaline differentiation and subduction zone magmatism: Contributions to Mineralogy and Petrology, v. 113, p. 143–166, doi: 10.1007/BF00283225.

Smith, E.I., Feuerbach, D.L., Naumann, T.R., and Mills, J.G., 1990, Mid-Miocene volcanic and plutonic rocks in the Lake Mead area of Nevada and Arizona: Production of intermediate igneous rocks in an extensional environment, *in* Anderson, J.L., ed., The nature and origin of Cordilleran magmatism: Boulder, Colorado, Geological Society of America Memoir 174, p. 169–194.

Sparks, R.S.J., and Marshall, L.A., 1986, Thermal and mechanical constraints on mixing between mafic and silicic magmas: Journal of Volcanology and Geothermal Research, v. 29, p. 99–124, doi: 10.1016/0377-0273(86)90041-7.

Vernon, R.H., 1983, Restite, xenoliths, and microgranitoid enclaves in granites: Journal and Proceedings, Royal Society of New South Wales, v. 166, p. 77–103.

Vogel, T.A., 1982, Magma mixing in the acidic-basic complex of Ardnamurchan: Implications on the evolution of shallow magma chambers: Contributions to Mineralogy and Petrology, v. 79, p. 411–423, doi: 10.1007/BF01132071.

Watson, E.B., and Jurewicz, S.R., 1984, Behavior of alkalies during diffusive interaction of granitic xenoliths with basaltic magma: Journal of Geology, v. 92, p. 121–131.

Wiebe, R.A., 1991, Commingling of contrasted magmas and generation of mafic enclaves in granitic rocks, *in* Didier, J., and Barbarin, B., eds., Enclaves and granite petrology: Amsterdam, Elsevier, Developments in Petrology 13, p. 393–402.

Wiebe, R.A., Frey, H., and Hawkins, D.P., 2001, Basaltic pillow mounds in the Vinalhaven intrusion, Maine: Journal of Volcanology and Geothermal Research, v. 107, p. 171–184, doi: 10.1016/S0377-0273(00)00253-5.

Wiebe, R.A., Blair, K.D., Hawkins, D.P., and Sabine, C.P., 2002, Mafic injections, in situ hybridization, and crystal accumulation in the Pyramid Peak granite, California: Geological Society of America Bulletin, v. 114, p. 909–920, doi: 10.1130/0016-7606(2002)114<0909:MIISHA>2.0.CO;2.

Wolf, M.B., and Wyllie, P.J., 1994, Dehydration-melting of amphibolite at 10 kbar: The effects of temperature and time: Contributions to Mineralogy and Petrology, v. 115, p. 369–383, doi: 10.1007/BF00320972.

Wood, D.A., 1978, Major and trace element variations in the Tertiary lavas of eastern Iceland and their significance with respect to the Iceland geochemical anomaly: Journal of Petrology, v. 19, p. 393–436.

Wyld, S.J., and Wright, J.E., 1988, The Devils Elbow ophiolite remnant and overlying Galice Formation: New constraints on the Middle to Late Jurassic evolution of the Klamath Mountains, California: Geological Society of America Bulletin, v. 100, p. 29–44, doi: 10.1130/0016-7606(1988)100<0029:TDEORA>2.3.CO;2.

Yule, J.D., Saleeby, J.B., and Barnes, C.G., 2006, this volume, A rift-edge facies of the Late Jurassic Rogue–Chetco arc and Josephine ophiolite, Klamath Mountains, Oregon, *in* Snoke, A.W., and Barnes, C.G., eds., Geological studies in the Klamath Mountains province, California and Oregon: A volume in honor of William P. Irwin: Boulder, Colorado, Geological Society of America Special Paper 410, doi: 10.1130/2006.2410(03).

MANUSCRIPT ACCEPTED BY THE SOCIETY 10 JANUARY 2005

Printed in the USA

Geological Society of America
Special Paper 410
2006

Depth of origin of late Middle Jurassic garnet andesite, southern Klamath Mountains, California

Calvin G. Barnes*
Department of Geosciences, Texas Tech University, Lubbock, Texas 79409-1053, USA
Charlotte M. Allen
Research School of Earth Sciences, Australia National University, Canberra, ACT 0200, Australia

ABSTRACT

A distinctive set of andesitic dikes crops out in the Klamath Mountains province south of ~41°15′. These dikes are characterized by phenocrysts of Al-rich amphibole, Ca-rich almandine-pyrope garnet, plagioclase, and, commonly, quartz, set in a fine-grained groundmass of plagioclase, quartz, alkali feldspar, and amphibole. Conspicuous microphenocryst/accessory phases are allanite/epidote, zircon, and apatite. A variety of thermobarometric methods plus comparison with published experimental data indicates the phenocryst assemblage was stable at pressure >7 kb and at temperatures between 850 °C and the solidus. The best estimate of pressure of phenocryst equilibration is 8–9 kb and ~800 °C, which corresponds to depths of 25–30 km. Emplacement pressure was probably 3–4 kb, and the lack of low-pressure equilibration features suggests that the dike magmas rose and cooled quickly. Rare earth element (REE) patterns for zircon are distinct from regional zircon compositions (samples in Klamath River sand) but are consistent with crystallization in equilibrium with garnet. Moreover, the zircons have strongly depleted light REE patterns, which is consistent with co-precipitation of allanite/epidote. A U-Pb (zircon) age of 160.5 ± 1.9 Ma was determined for one of the dike samples. This age is coeval with emplacement of the voluminous Wooley Creek plutonic suite in the central and northern parts of the province (north of latitude 41°15′). It is not clear why sparse dikes with high-pressure assemblages crop out in the southern Klamath Mountains province and voluminous coeval plutons are present in the central and northern parts. This distribution may reflect distinct tectonic regimes in the north (extension) versus the south (contraction), differences in melt productivity, or differences in the composition of deep crustal rocks.

Keywords: andesite, high pressure, garnet, crustal thickness, rare earth elements, REE

*Email: al.barnes@ttu.edu.

Barnes, C.G., and Allen, C.M., 2006, Depth of origin of late Middle Jurassic garnet andesite, southern Klamath Mountains, California, *in* Snoke A.W., and Barnes, C.G., eds., Geological studies in the Klamath Mountains province, California and Oregon: A volume in honor of William P. Irwin: Geological Society of America Special Paper 410, p. 269–286, doi: 10.1130/2006.2410(13). For permission to copy, contact editing@geosociety.org.

INTRODUCTION

The Klamath Mountains province is considered to be an example of crustal accretion in which arc-related terranes were juxtaposed along shallow thrust faults during episodic contractional deformation. The typically low metamorphic grade of these terranes (e.g., Coleman et al., 1988; Ernst, 1998) and the shallow dip of terrane-bounding thrust faults has led most workers to think of the province as one in which thin-skinned deformation dominated and high-pressure metamorphism associated with relatively thick crust was absent. In fact, aside from subduction complexes metamorphosed under blueschist-facies conditions (e.g., Stuart Fork formation [= Fort Jones terrane]: Hotz et al., 1977; Goodge, 1990), the only metamorphic rocks reported from the province that reached equilibrium pressures of 7 kb are parts of the Rattlesnake Creek terrane exposed along the California-Oregon border (Grover, 1984; Lieberman and Rice, 1986).

Nevertheless, isotopic studies (Barnes et al., 1990, 1992, 1995; Gribble et al., 1990) showed that magmas parental to many late Middle Jurassic plutons underwent intense, deep-seated interaction with diverse crustal rocks and that crustal anatexis was common during this time. Such deep-seated magma differentiation is widely invoked to explain isotopic and element variations in many arc magmatic systems (DePaolo, 1981; Kistler et al., 1986; Ague and Brimhall, 1988), yet direct evidence for such activity is uncommon except where arc basement is exposed (e.g., DeBari, 1997).

In the southern Klamath Mountains, a distinctive suite of porphyritic garnet andesite dikes was emplaced at ca. 160 Ma. Emplacement was at relatively low-pressure conditions; however, the phenocryst suite of the dikes (garnet + amphibole + plagioclase ± quartz ± epidote) indicates equilibration at high pressure. In this chapter, we describe examples of the dikes, their phenocryst assemblages, and geochemical compositions. We then discuss their importance in terms of differentiation of arc magmas in the province.

GEOLOGIC SETTING

Garnet andesite dikes crop out as far north as Indian Rocks[1] (latitude 41°11′48″; Table 1) and at least as far south as 40°53′41″. The dikes intrude the eastern Hayfork terrane, the western Hayfork terrane, and the Ironside Mountain batholith. They range in width from <1 m to at least 30 m. The intrusion at Indian Rocks (Fig. 1) consists of a narrow (1–2 m) dike that can be traced for ~650 m and then apparently widens to form a broader intrusive mass. The depth of emplacement is difficult to determine, because contact metamorphic aureoles were not observed. Loose constraints are provided by the emplacement conditions of nearby plutons. The Ironside Mountain pluton (Fig. 1A), which was intruded by garnet andesite dikes, was emplaced at ~4 kb (Barnes et al., this volume, Chapter 10). Contact metamorphic conditions at the southern end of the 161-Ma Wooley Creek batholith (15 km north of Indian Rocks; Fig. 1A) were at a pressure of ~3.5 kb (Barnes et al., 1986).

[1]Indian Rocks is sacred ground for the local Native Americans and should be treated with respect.

GEOCHRONOLOGY

Zircons separated from sample OMB-I, collected near Indian Rocks, were dated by laser-ablation inductively coupled plasma mass spectrometry (LA-ICP-MS) at the Research School of Earth Sciences (RSES), Australia National University (ANU), Canberra. The method is similar to that reported in Harris et al. (2004). An excimer laser was used to ablate zircons in a sample cell connected to an Agilent 7500 quadrupole ICP-MS via a smoothing device. The in-house zircon standard (98-521) run at the same time (January 2003) yielded 42.7 ± 0.6 Ma, with MSWD of 1.32. This value is consistent with a running average age of 42.6 ± 0.2 Ma for three hundred individual analyses on twenty analytical days in a year. The uncertainty on individual grains includes a term in quadrature that expresses the uncertainty in the primary zircon standard, TEMORA 2, which is taken to be concordant and 416.75 ± 0.24 Ma in age (Black et al., 2003). Concentration data are based on National Institute of Standards and Technology (NIST) 610 silicate glass standard assuming stoichiometric SiO_2 (32.77 wt%) as the internal standard. Europium and Ce anomalies are calculated using the method of Ballard et al. (2002) and are relative to average crust. The data are tabulated in Appendix 1 of Allen and Barnes (this volume).

Twenty-seven zircons from OMB-I were analyzed, and all were concordant ($^{206}Pb/^{238}U$ and $^{207}Pb/^{235}U$ ages agree within 5%; Fig. 2), and the measured uncertainty in $^{206}Pb/^{238}U$ for a given analysis was always less than twice the theoretical or expected uncertainty. One 149-Ma grain was excluded from the age calculation, because it was distinct from the rest of the population. The remaining twenty-six grains yield a ^{208}Pb-based, common Pb-corrected $^{206}Pb/^{238}U$ age of 160.5 ± 1.9 Ma with MSWD of 1.72; an MSWD only slightly greater than that observed in the in-house standard. The age using no common Pb correction is identical, and the MSWD is only marginally larger (1.74). There is a suggestion that two distinct age populations are present. These populations are divided at 162 Ma on a probability plot (inset, Fig. 2) but more data would be needed to prove that the older grains represent subtle inheritance. Removing these eight older analyses does not shift the age outside the uncertainty quoted above.

PETROGRAPHY

All samples are porphyritic and all contain phenocrysts of plagioclase, amphibole, and garnet (Table 1). Quartz phenocrysts are present in some samples (Table 1, Fig. 3A). The microphenocryst assemblage includes garnet, apatite (to 1 mm

TABLE 1. MODAL DATA

Sample	OMB-I Range	OMB-I Average	s.d.	92OMB 31A	92OMB 31B	92OMB 114	92OMB 281
Phenocrysts and microphenocrysts							
Plagioclase	19.5–27.8	23.2	3.44	30.4	26.8	24.8	17.1
Amphibole	18.2–22.5	20.0	1.84	14.4	14.4	12.7	17
Garnet	tr–0.4	0.2	0.17	tr	tr	tr	0.2
Quartz	0.7–0.9	0.8	0.09	n.d.	n.d.	n.d.	0.1
Opaque minerals	0.2–0.5	0.4	0.12	0.4	0.3	0.9	tr
Epidote	tr–0.2	0.1	0.08	n.d.	tr	tr	tr
Apatite	0.2–0.4	0.3	0.09	0.1	tr	0.1	0.3
Groundmass	52.0–55.2	53.7	1.31	54.7	58.5	61.5	64.9
Prehnite	0.5–2.0	1.4	0.63	n.d.	n.d.	n.d.	0.4
Total		100.0		100.0	100.0	100.0	100.0
Number of points		3000		1000	1000	1000	1000

Notes: s.d.—standard deviation, tr—trace amount, n.d.—not detected.

long), allanite/epidote, and ilmenite (Fig. 3B). The intergranular groundmass of these dikes is typically fine grained to very fine grained and consists of plagioclase, amphibole, and quartz ± K-feldspar ± Fe-Ti oxides. Accessory zircon is present in the groundmass and as inclusions in amphibole.

The dike collected at location 92OMB31B contains medium-grained, hypidiomorphic granular, cognate dioritic enclaves, which consist of euhedral to subhedral amphibole and altered intergranular to poikilitic plagioclase. Amphibole crystals range from robust to acicular prisms and show no preferred orientation; they display the same color zoning as amphibole phenocrysts in the host rock (see below) and rarely include garnet. Apatite and Fe-Ti oxides are accessory phases. Some interstices in the sample are filled with prehnite ± elongate pale amphibole ± twinned calcite.

Plagioclase phenocrysts reach at least 1 cm in length. They are typically euhedral, but some are subhedral or anhedral. Plagioclase shows seriate distribution in some samples but is unimodal in others. In all samples, plagioclase zoning is weak, fine-scale, and oscillatory, with a narrow range of compositions from An_{51} to An_{43} (Table 2). Inclusions in plagioclase are amphibole oriented parallel to zoning boundaries, ilmenite, and, rarely, garnet.

Amphibole phenocrysts are euhedral to subhedral, reach at least 5 mm in length, and typically show seriate distribution. Most crystals show oscillatory color zoning (Fig. 3C); colors range from brownish olive to pale blue-green, to pale green/tan. Some crystals have pale to colorless rims suggestive of late-stage alteration. Garnet, ilmenite, apatite, and quartz (in sample OMB-I) are inclusions. Amphibole ranges from pargasite to ferropargasite. Typical Mg/(Mg + Fe) values range from 0.41 to 0.66 (Table 3). Al contents are high; most range from 2.3 to 2.9 cations per formula unit. Compared to amphibole from the older Ironside Mountain pluton, amphibole from the garnet andesite dikes has low K (typically 0.06–0.13 cations per formula unit) and halogen contents are very low or below detection limits.

Garnet phenocrysts vary from euhedral to anhedral and reach at least 5 mm in diameter. They are burgundy red in hand sample and pale pink in thin section. Garnets are variously intergrown with apatite, ilmenite, amphibole (Fig. 3A), and color-zoned euhedral epidote (Fig. 3B). Some crystals contain inclusions of hollow, acicular apatite (Fig. 3B), and some contain zircon inclusions. The garnets are primarily mixtures of almandine (53–60%), grossular (17–28%), and pyrope (13–18%) components, with minor spessartine component (3–6%; Table 4). If ferric iron is calculated on the basis of charge balance, the average andradite component is less than 2%.

Allanite/epidote phenocrysts show a narrow range of compositions (Table 5), with pistacite components ranging from 0.17 to 0.20. This range is lower than is normally assumed for igneous epidote (e.g., Brandon et al., 1996). It is not clear whether this is an effect of the bulk composition of the host magma, a function of temperature of crystallization, or related to the fO_2 of the magma. Brandon et al. (1996) also suggested that TiO_2 contents in igneous epidote are <0.2 wt%, and all but one epidote from the garnet andesites are in this range (Table 5).

Quartz phenocrysts are present in two samples (Table 1); maximum diameter is at least 1.2 mm. The phenocrysts vary from euhedral to anhedral; rounded crystal faces and resorption channels are common. Zircon and ilmenite occur as inclusions in quartz.

On the basis of inclusion relationships, the order of crystallization of phenocryst phases is interpreted to be garnet, plagioclase + hornblende, and quartz. Zircon and apatite appear as inclusions in all phenocryst phases, and allanite/epidote is included in garnet. Therefore, these accessory minerals are interpreted to be early crystallizing phases.

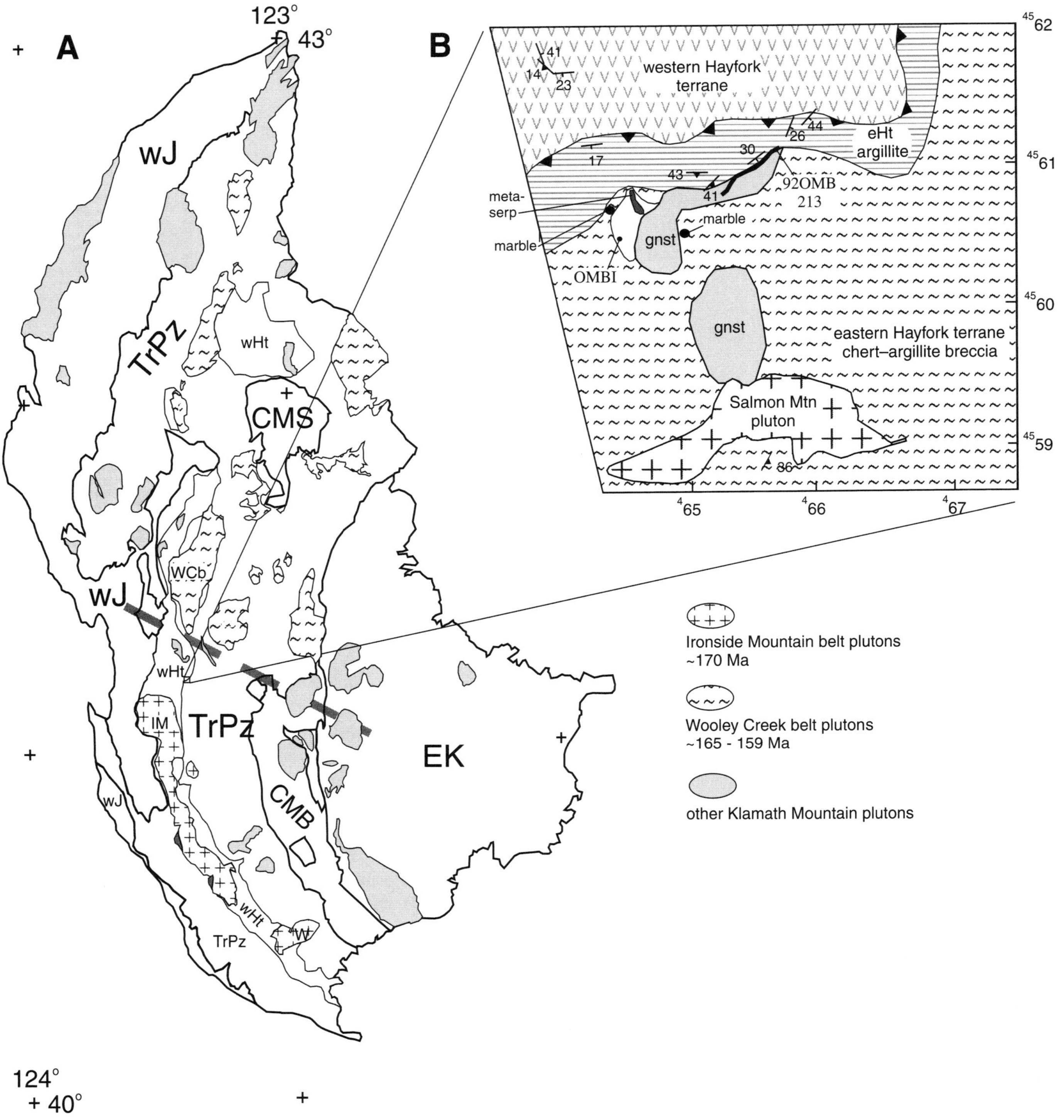

Figure 1. (A) Simplified geologic map of tectonostratigraphic terranes and plutons in the Klamath Mountains, northern California and southern Oregon (after Irwin and Wooden, 1999). Terranes and composite terranes: wJ—western Jurassic; TrPz—western Paleozoic and Triassic; wHt—western Hayfork terrane (part of the TrPz); CMB—central metamorphic belt; CMS—Condrey Mountain Schist; EK—eastern Klamath. The dashed gray line separates the northern part of the province, where 165- to 159-Ma plutons of the Wooley Creek belt crop out from the southern part of the province. South of this line, ca. 160 Ma plutonic rocks are limited to garnet andesite dikes. WCb is the Wooley Creek batholith; IM is the 170-Ma Ironside Mountain pluton. (B) Geologic map of Indian Rocks area with UTM coordinate marks. The contact between the western and eastern Hayfork terranes is thought to be a thrust fault with eastern Hayfork terrane (eHt) in the hanging wall on the basis of regional mapping and the shape of the contact. The garnet andesite intrusion at Indian Rocks consists of a plug-like mass (unpatterned area; location OMB-I) and a south-dipping dike (heavy black line; location 92OMB213) that intrude chert, argillite, marble, metaserpentinite (metaserp), and greenstone (gnst) of the eastern Hayfork terrane. The marble, metaserpentinite, and greenstone are thought to be blocks in a chert-argillite mélange matrix (e.g., Wright, 1982). The Salmon Mountain pluton is primarily diorite and quartz diorite (Boardman, 1985).

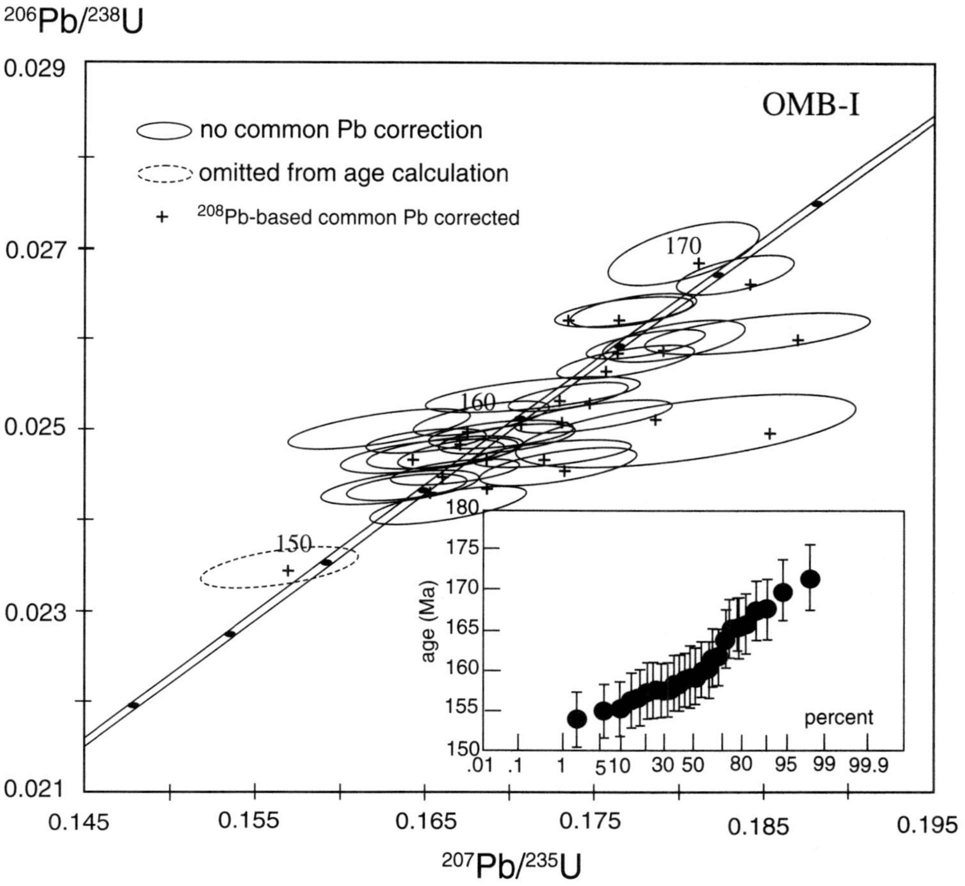

Figure 2. Concordia plot for zircons from sample OMB-I Ellipses are 1-σ uncertainties. The inset is a probability plot that shows the possibility of two age populations. See text for discussion.

GEOCHEMISTRY

Whole-Rock Compositions

The garnet andesite dike rocks show a limited range of major element compositions. The SiO_2 contents range from 59 to 64.5% and K_2O contents are, with one exception, less than 1.6% (Table 6, Fig. 4A). These values are in contrast with synplutonic and roof-zone dikes of the nearly coeval (161 Ma) Wooley Creek batholith, which display a wide compositional range and K_2O contents to nearly 4% (Fig. 4A). At equivalent SiO_2 contents, the garnet andesite dikes are also characterized by higher Al_2O_3 (17.5–20%; Fig. 4B) and Na_2O contents, and lower MgO, FeO_t, Y, and V contents than the dikes associated with the Wooley Creek batholith. The dike samples range from metaluminous to mildly peraluminous, with A/CNK from 0.90 to 1.04.

Rare earth element (REE) plots for two garnet andesite dikes are shown in Figure 4C (Table 7). The patterns show moderate light REE enrichment compared to the heavy rare earths. A small positive Eu anomaly is present in both samples, with Eu/Eu* of 1.04 and 1.06 for OMB-I and 92OMB-114, respectively. Multi-element diagrams for the same two samples (Fig. 4D) show patterns characteristic of arc magmatism, with relative depletion of Nb and Ti and enrichment of K and Sr. Although many of these dikes intruded the same crustal section as the somewhat older Ironside Mountain pluton (Barnes et al., this volume, Chapter 10), they do not share similar trace element contents or patterns (Fig. 4C and D).

Strontium isotope data for four analyzed samples (Table 8) show little variation, with an average initial $^{87}Sr/^{86}Sr$ value (calculated at 160.5 Ma) of 0.70470 ± 0.00028. This range of initial $^{87}Sr/^{86}Sr$ is typical of coeval plutons from the Wooley Creek belt but is higher than the Ironside Mountain pluton, whose initial $^{87}Sr/^{86}Sr$ ranges from 0.7036 to 0.7039 (Barnes et al., 1992; Allen and Barnes, this volume).

MINERAL TRACE ELEMENT COMPOSITIONS

LA-ICP-MS analyses of zircon and other major phases (garnet, plagioclase, hornblende, allanite/epidote, and apatite) were undertaken primarily to examine the magmatic REE and phosphorus budgets and to investigate whether zircon, which appears in textural equilibrium with garnet, carries a distinct REE pattern that earmarks these zircons as being from a garnet-bearing (i.e., relatively high pressure) assemblage (e.g., Day et al.,

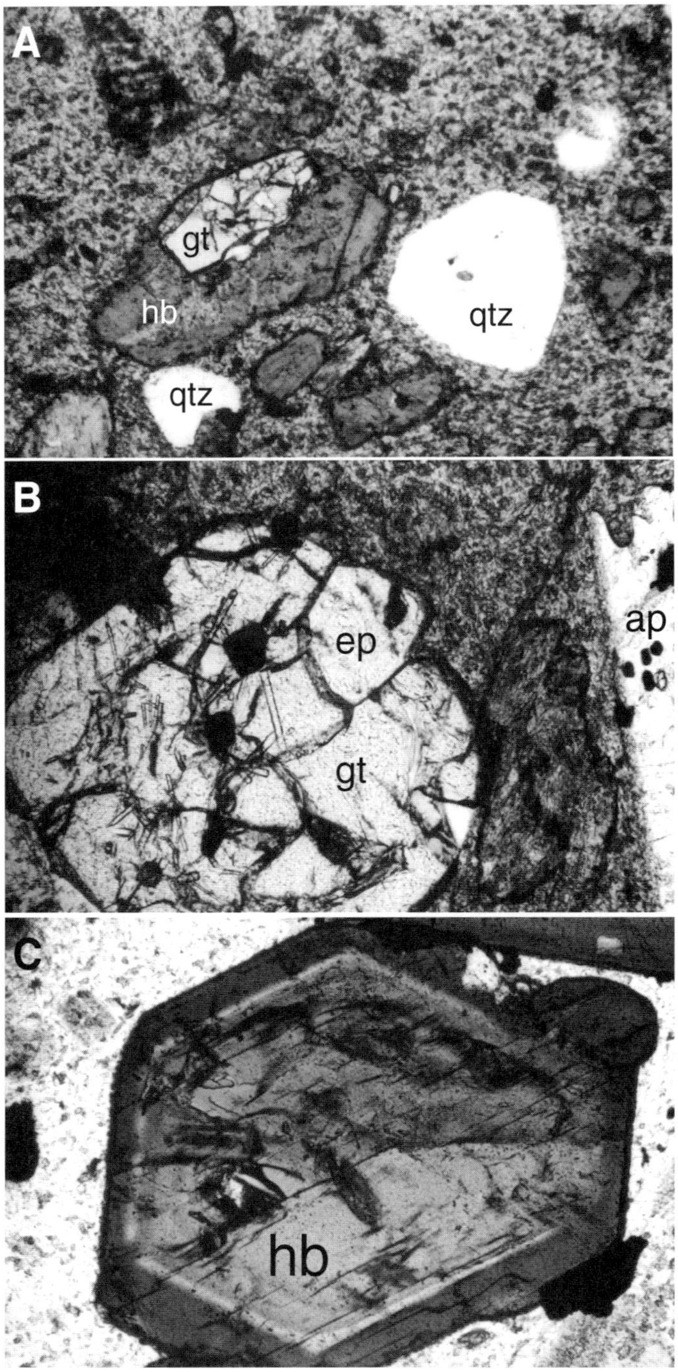

Figure 3. (A) Photomicrograph of amphibole (hb), garnet (gt), and quartz (qtz) from sample OMB-I. Note that garnet is intergrown with amphibole and amphibole is intergrown with quartz. Field of view is 1.8 mm. (B) Photomicrograph of garnet phenocrysts from sample OMB-I. The garnet is intergrown with epidote (ep; outlined by visible dark boundaries). Garnet also contains inclusions of magnetite with cubic outlines and acicular apatite. The phenocrysts on the right side of the photo are apatite (ap). Field of view is 1 mm. (C) Photomicrograph of a color-zoned amphibole (hb) phenocryst from sample 92OMB31B.

TABLE 2. REPRESENTATIVE PLAGIOCLASE COMPOSITIONS

Sample Spot	OMB-IA Core	OMB-IA Rim	OMB-IC Rim	OMB-ID Rim	OMB-114 Core
Wt% oxide					
SiO_2	55.66	56.91	57.88	55.90	55.88
TiO_2	0.02	0.00	0.00	0.05	0.00
Al_2O_3	28.34	28.23	26.83	28.44	28.18
FeO	0.04	0.02	0.03	0.07	0.10
CaO	9.50	9.06	8.66	9.89	9.74
Na_2O	5.65	5.80	6.37	5.53	5.51
K_2O	0.15	0.17	0.21	0.17	0.14
Total	99.36	100.20	99.98	100.05	99.55
Cations per 8 oxygens					
Si	2.512	2.540	2.589	2.506	2.518
Ti	0.000	0.000	0.000	0.001	0.000
Al	1.507	1.484	1.414	1.502	1.497
Fe	0.001	0.000	0.000	0.001	0.002
Ca	0.458	0.432	0.415	0.475	0.469
Na	0.494	0.501	0.551	0.480	0.480
K	0.008	0.009	0.012	0.009	0.006
Total	4.979	4.966	4.981	4.974	4.973
Ca/(Ca + Na)	0.48	0.46	0.43	0.50	0.49

1992). ICP-MS analyses of trace elements for the six phases employed the same run table, and except for zircon, minerals were analyzed in polished thin section. The diameter of the ablation spot ranged from 32 to 65 µm. All data were reduced relative to surrounding pairs of analyses of NIST 612 (Table 9) and relative accuracy and precision are demonstrated by repeat analysis of basalt glass standard BCR-2 g (Table 9). Detection limit calculation is after Longerich et al. (1996). The major element used for the internal standard varied, and was either an assumed stoichiometric abundance (SiO_2 in zircon and CaO in apatite), or was CaO as measured by electron microprobe at RSES, ANU.

All analyzed zircons (Table 10) have positive Ce and negative Eu anomalies (Fig. 5), but heavy REE slopes are variable, with chondrite-normalized Lu_N/Dy_N ranging from 1.4 to 3.4. Zircon is the most U-rich phase in the rock (150–825 ppm, average 469) and Th ranges from 19 to 115 ppm with a 53-ppm average. Of the six core-rim pairs analyzed, three show no distinct difference in REE patterns and three consistently show rims with flatter heavy REE patterns, and higher Lu concentrations. An example pair is given in Figure 5.

Three garnets were analyzed; they have total REE of 170–470 ppm, except for the one illustrated in Figure 3B (40–80 ppm). These values are of the same order as zircon (100–360 ppm). However, individual garnets show distinct trace element features and zoning (Table 11). Values of Lu_N/Dy_N (chondrite-normalized using the values of McDonough and Sun, 1995) vary from 4.1 to 1.2, and the garnet illustrated in Figure 3B has Lu_N/Dy_N of ~3 in the core, ~1 in the rim, and lacks the negative Eu anomaly obvious for the other garnet grains. Thorium is below or near detection limits (~0.02 ppm) in all samples.

TABLE 3. REPRESENTATIVE AMPHIBOLE COMPOSITIONS

Sample	OMBI	OMBI	OMB ID	OMB 31A	OMB 31A	OMB 114	OMB 114
Grain, spot	1, rim	2, rim	2, core	2, rim	2, rim	1, rim	4, rim
Wt% oxides							
SiO_2	42.93	43.70	42.65	40.10	40.08	41.51	39.94
Al_2O_3	15.88	13.71	13.86	16.02	15.66	14.50	17.04
FeO	15.23	15.63	15.59	13.27	14.28	17.63	18.16
MgO	9.47	9.81	9.97	11.12	10.51	9.07	7.43
TiO_2	1.01	1.17	1.09	1.45	1.30	0.70	0.51
MnO	0.29	0.28	0.70	0.23	0.27	0.52	0.52
CaO	10.27	10.47	10.38	11.65	11.69	10.45	10.68
Na_2O	1.64	1.58	1.43	2.30	2.29	1.54	1.68
K_2O	0.64	0.45	0.46	0.48	0.49	0.37	0.64
Cl	0.04	0.02	0.02	0.02	0.02	0.02	0.02
F	0.21	0.14	0.05	0.05	0.09	0.37	0.00
Total	97.68	96.95	96.10	96.59	96.53	96.13	96.60
Cations per 23 oxygens							
Si	6.332	6.501	6.419	6.001	6.031	6.284	6.085
Al	2.761	2.404	2.458	2.826	2.777	2.587	3.060
Fe	1.879	1.945	1.962	1.661	1.797	2.232	2.314
Mg	2.081	2.175	2.236	2.480	2.357	2.046	1.687
Ti	0.112	0.131	0.123	0.163	0.147	0.080	0.058
Mn	0.036	0.035	0.089	0.029	0.034	0.067	0.067
Ca	1.623	1.669	1.674	1.868	1.885	1.695	1.743
Na	0.469	0.456	0.417	0.667	0.668	0.452	0.496
K	0.120	0.085	0.088	0.092	0.094	0.071	0.124
Total	15.413	15.401	15.467	15.788	15.790	15.514	15.635
Cl	0.010	0.005	0.005	0.005	0.005	0.005	0.005
F	0.098	0.066	0.024	0.024	0.043	0.177	0.000
Mg/(Mg + Fe)	0.526	0.528	0.478	0.599	0.567	0.478	0.422

TABLE 4. REPRESENTATIVE GARNET COMPOSITIONS

Sample	OMBIA	OMBIA	OMBIA	OMBIC	OMB119	OMB119
Spot	Near rim	Core	Rim	Rim	Core	Rim
Wt% oxide						
SiO_2	37.33	37.51	37.54	37.8	38.17	38.29
Al_2O_3	20.16	20.45	20.46	20.5	20.71	20.83
TiO_2	0.58	0.41	0.43	0.44	0.26	0.27
FeO	25.68	27.1	27.25	25.76	25.09	24.91
MnO	2.36	3.08	2.95	2.94	3.23	3.3
MgO	3.21	4.7	4.47	4.3	4.66	4.48
CaO	10.23	6.11	5.93	7.34	7.22	7.24
Total	99.55	99.36	99.03	99.08	99.34	99.32
Cations per 12 oxygens; Fe^{3+} estimated by charge balance						
Si	2.967	2.981	2.995	3.005	3.017	3.025
Al	1.887	1.915	1.922	1.920	1.929	1.939
Ti	0.033	0.023	0.023	0.025	0.014	0.016
Fe^{3+}	0.113	0.080	0.053	0.041	0.030	0.006
Fe^{2+}	1.592	1.720	1.764	1.671	1.628	1.641
Mn	0.158	0.205	0.199	0.196	0.215	0.220
Mg	0.380	0.555	0.532	0.509	0.547	0.526
Ca	0.870	0.520	0.506	0.625	0.610	0.612
Total	8.000	7.999	7.994	7.990	7.990	7.986
Garnet components using estimated Fe^{3+} values						
Pyrope	0.127	0.185	0.177	0.170	0.182	0.175
Almandine	0.531	0.573	0.588	0.557	0.543	0.547
Grossular	0.234	0.133	0.142	0.188	0.188	0.201
Andradite	0.056	0.040	0.027	0.020	0.015	0.003
Spessartine	0.053	0.068	0.066	0.065	0.072	0.073
Garnet components assuming all Fe as Fe^{2+}						
Almandine	0.548	0.585	0.595	0.563	0.547	0.548
Pyrope	0.122	0.180	0.174	0.167	0.181	0.175
Grossular	0.279	0.169	0.166	0.205	0.201	0.204
Spessartine	0.051	0.067	0.065	0.064	0.071	0.073

TABLE 5. REPRESENTATIVE ALLANITE/EPIDOTE ANALYSES

Sample	OMBIA (1)	OMBID (2)	OMBIE (3)	92OMB-119 (4)	92OMB-119 (5)
Wt% oxide					
SiO_2	36.34	36.49	36.33	36.55	35.62
Al_2O_3	25.36	24.74	25.37	24.65	23.63
TiO_2	0.30	0.15	0.31	0.16	0.18
Fe_2O_3	8.88	8.00	8.97	8.71	9.15
MnO	0.24	0.24	0.25	0.24	0.27
MgO	0.31	0.46	0.33	0.41	0.52
CaO	21.55	19.80	21.42	20.27	19.06
Total*	92.98	89.88	92.97	90.99	88.43
Cations per 12.5 oxygens					
Si	2.994	3.080	2.993	3.060	3.070
Al	2.462	2.461	2.462	2.433	2.400
Ti	0.019	0.010	0.019	0.010	0.012
Fe^{3+}	0.550	0.508	0.557	0.549	0.594
Mn	0.017	0.017	0.017	0.017	0.020
Mg	0.038	0.058	0.040	0.051	0.067
Ca	1.902	1.791	1.890	1.818	1.760
Total	7.982	7.925	7.979	7.939	7.922
Ps†	0.183	0.171	0.184	0.184	0.198

Notes: (1)—Partial inclusion in garnet (Fig. 3B); (2)—single analysis of epidote phenocryst; (3)—average phenocryst composition (n = 5); (4)—average phenocryst composition; (5)—average phenocryst composition.

*Total assuming all Fe as Fe^{3+}. Low totals probably due to unanalyzed rare earth elements and yttrium.

†Ps—pistacite component = $Fe^{3+}/(Fe^{3+} + Al)$.

Uranium is below detection limits in the garnet shown in Figure 3B, but averages 0.04 ppm in the other grains. Phosphorus contents range from 130 to 280 ppm.

Allanite/epidote yielded a consistent REE pattern for the four analyzed grains (Table 12). Total REE content varies from 10,500 to 30,500 ppm, about one hundred times greater than that of garnet or zircon. The ratios Lu_N/La_N and Lu_N/Dy_N average 0.005 and 0.115, respectively, and there is a negligible to small positive Eu anomaly. Allanite/epidote is the most Th-rich phase in the rock (220–500 ppm); it also carries ~60–90 ppm U. Phosphorus contents range from 2430 to 5000 ppm.

Plagioclase is a REE-, U-, and Th-poor phase (Table 13). Light REE patterns do not show core-rim zoning, and average La content is 1.5 ppm. Heavy REEs are below detection limits (~0.05 ppm), as are U and Th (~0.01 ppm, each). These plagioclase crystals carry 27.4 ppm P on average.

Six hornblende crystals were analyzed (Table 14). Four grains are euhedral phenocrysts, and four are microscopic decorations (inclusions) in plagioclase phenocrysts. These two types have different REE patterns. The hornblende inclusions have, with one exception, steeper heavy REE patterns (Lu_N/Dy_N ranges from 0.17 to 0.19) than do the euhedral phenocrysts (0.20–0.69), but all grains have negative Eu anomalies. Moreover, in all three core-rim pairs from the euhedral crystals, the

TABLE 6. MAJOR AND TRACE ELEMENT ABUNDANCES

Sample	OMB-I	92OMB 114	92OMB 119	92OMB 179	92OMB 183	92OMB 213	92OMB 31A	92OMB 31B
Major element abundances in wt%								
SiO_2	59.94	60.78	62.85	61.60	64.49	60.47	61.29	51.85
TiO_2	0.56	0.44	0.36	0.40	0.50	0.60	0.39	0.60
Al_2O_3	17.91	19.92	18.57	19.43	18.46	18.31	18.77	18.34
Fe_2O_3	0.76	4.80	3.71	4.27	4.42	5.31	4.50	6.85
FeO	4.14	n.d.	n.d.	n.d.	n.d.	n.d.	n.d.	n.d.
MnO	0.12	0.12	0.11	0.12	0.09	0.12	0.12	0.16
MgO	2.74	1.65	1.37	1.62	1.65	2.41	2.57	5.67
CaO	6.36	6.53	5.05	6.56	1.74	5.50	3.54	9.48
Na_2O	4.25	3.94	4.25	4.07	4.94	3.70	8.79	4.69
K_2O	1.22	1.12	1.60	0.68	1.38	1.35	0.58	1.08
P_2O_5	0.13	0.15	0.14	0.13	0.13	0.13	0.11	0.10
LOI	1.94	2.03	2.22	2.03	3.03	2.13	1.18	2.04
Total	100.05	101.49	100.23	100.92	100.82	100.03	101.83	100.85
Mg/(Mg + Fe^t)	0.50	0.40	0.42	0.43	0.42	0.47	0.53	0.62
Trace element abundances in ppm								
Rb	32	41	54	18	43	44	14	22
Sr	452	568	417	549	395	474	382	460
Zr	78	86	86	96	85	84	69	59
Y	15	13	9	15	9	14	14	20
Nb	9	3	7	b.d.	6	10	8	b.d.
Ba	621	562	727	484	1491	515	189	644
Sc	13.4	6.7	5.3	7.2	8.1	12.9	10.5	26.7
V	81	51	36	51	35	77	72	160
Cr	29	6	5	7	16	22	20	39
Cu	7	6	5	6	5	bd	10	29
Zn	70	80	64	64	76	69	54	70

Notes: Sample locations (N latitude, W longitude): OMB-I (41°11′48″, 123°25′20″); 92.OMB31 (40°58′53″, 123°25′59″); 92.OMB119 (41°05′51″, 123°28′35″); 92.OMB179 (41°14′06″, 123°27′10″); 92.OMB183 (41°05′27″, 123°26′19″); 93.OMB281 (40°53′41″, 123°21′08″). Abbreviations: b.d.—Below detection limits, LOI—loss on ignition, n.d.—not determined.

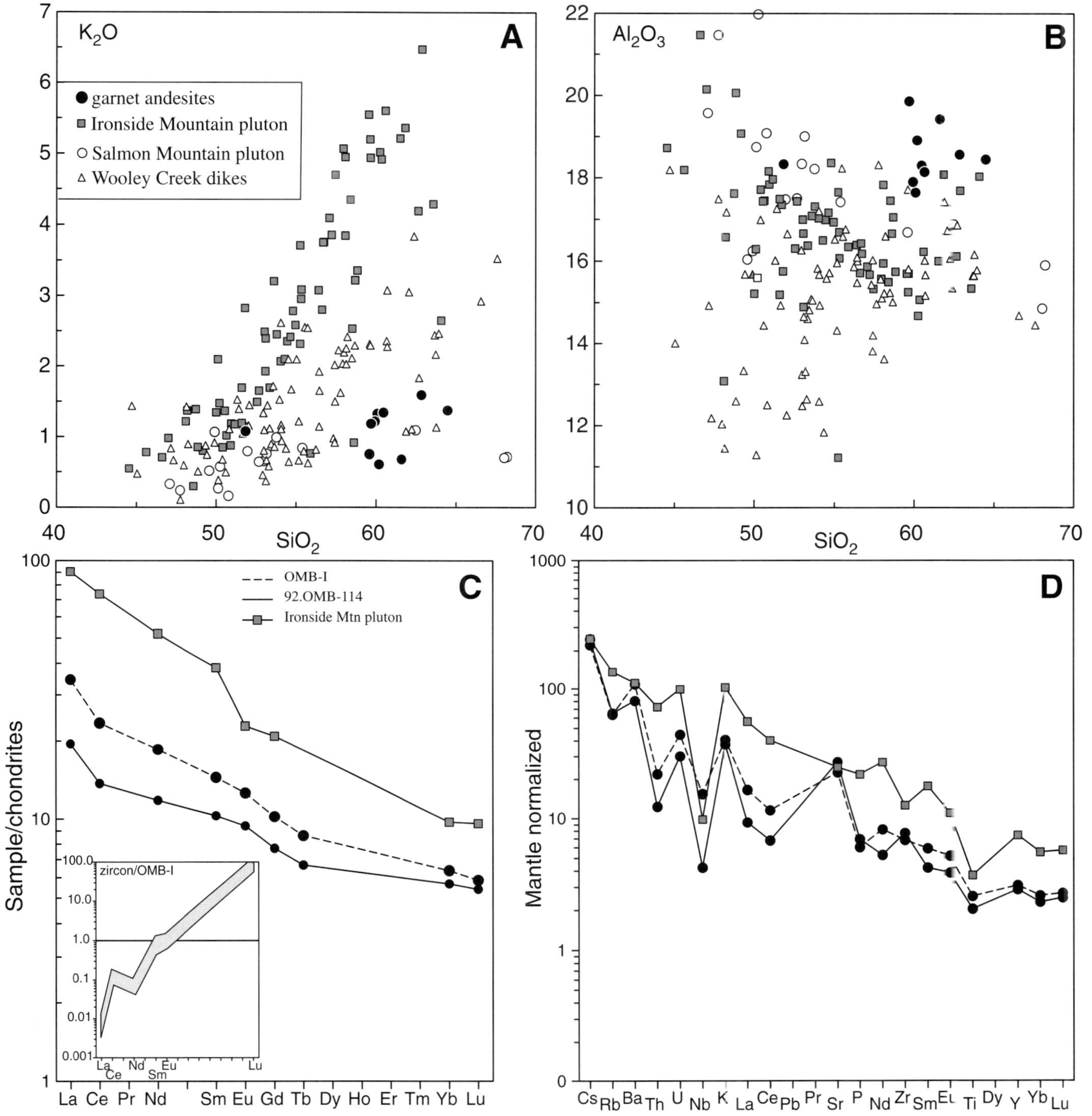

Figure 4. Major and trace element compositions and trace element patterns. Variation of (A) K_2O and (B) Al_2O_3 as a function of SiO_2. Compositions of the 170-Ma Ironside Mountain pluton and dikes associated with the ca. 161-Ma Wooley Creek batholith are shown for comparison. Note the relatively low K_2O and high Al_2O_3 contents of the garnet andesites. (C) Rare earth element patterns and (D) mantle-normalized (Sun and McDonough, 1989) multielement diagram of two garnet andesite dikes. Data for a monzogabbro from the Ironside Mountain pluton is shown for comparison. The inset in C shows rare earth element variation in dated zircon crystals normalized to the composition of the host rock, sample OMB-I.

TABLE 7. TRACE ELEMENT CONCENTRATIONS

Sample	OMB-I	92OMB114
La	11.33	6.41
Ce	20.52	11.97
Nd	11.06	7.072
Sm	2.60	1.861
Eu	0.865	0.647
Tb	0.40	0.31
Yb	1.27	1.13
Lu	0.198	0.183
Cs	1.74	1.93
Hf	2.41	2.39
U	0.93	0.63
Th	1.85	1.05
Ta	0.40	0.18

Notes: Concentrations in ppm. Analysis by instrumental neutron activation.

cores have steeper heavy REE patterns than the rims. In other words, Lu contents are 1.7–3.3 times higher in the rims than in the cores. Total REE contents are 45–96 ppm. Thorium and U contents are low, averaging 0.04 for both elements for all hornblende analyses.

Apatite total REE varies from 960 to 1470 ppm for the three grains analyzed (Table 15), and this phase is second only to allanite/epidote in terms of total REE content. The apatite shows consistent REE patterns, all with negative Eu anomalies, with some fanning of the heavy REE. Values for Lu_N/Dy_N range from 0.11 to 0.22. Average Th contents are 3.7 ppm, and U contents are 5.2 ppm.

Zircons from Garnet Andesite Compared to Other Zircons from the Klamath Mountains

Compositions of zircons from the garent andesite samples were compared to compositions of detrital zircons from the Klamath River, which we interpret as representative of zircons produced during Mesozoic magmatism in the Klamath Mountains (Allen et al., 2002; Fig. 5). Zircons from the garnet andesite dikes have average Th/U values of 0.07–0.16, P contents of 100–238 ppm (with one exception), and normalized Eu/Eu* from 0.5 to 0.7 (see Appendix 1 in Allen and Barnes, this volume, and Table 10). In contrast, 130- to 180-Ma zircons sampled from the Klamath River (Allen et al., 2002) typically have higher Th/U (0.3–0.8), higher P contents (typically 200–500 ppm), and lower normalized Eu/Eu* (generally <0.5). These data suggest that in garnet andesite melts, Th and P were depleted prior to zircon growth, presumably by crystallization of allanite/epidote and apatite, respectively. This scenario is consistent with the presence of epidote and apatite microphenocrysts, inclusion relationships, and with the differences in apatite and zircon saturation temperatures (see below). The relatively high Eu/Eu* values (small anomaly size) suggest that plagioclase fractionation was not as important in the garnet andesite magmas as it was in more typical Klamath magma systems (Fig. 5). The average garnet andesite zircons have a shallower normalized Dy to Lu slope than does the average Klamath River detrital zircon; nonetheless, the most striking contrast of the garnet andesite zircons compared to those sampled in the Klamath River is the very low overall REE content, particularly the light REEs. Average Ce and Lu contents of OMB-1 zircons are 2.3 and 17.5 ppm, respectively, whereas in Klamath River zircons >150 Ma they are 19 and 77 ppm. Thus, the Indian Rocks zircons are about six to eight times poorer in total REE contents than those from other Mesozoic Klamath plutons.

TABLE 8. STRONTIUM ISOTOPE DATA

Sample	Rb (ppm)	Sr (ppm)	$^{87}Rb/^{86}Sr$	$^{87}Sr/^{86}Sr$	$^{87}Sr/^{86}Sr$ 159 Ma
OMB-IZ	30.0	457	0.190	0.705293	0.70486
92OMB31A	11.1	393	0.082	0.705068	0.70488
92OMB31B	19.0	468	0.118	0.705127	0.70486
92OMB114	41.8	543	0.222	0.704715	0.70421

DISCUSSION

Although the garnet andesite dikes were emplaced at relatively low pressure (3–4 kb; based on pressure estimates from nearby plutons), a phenocryst assemblage of Ca-rich garnet, aluminous amphibole, plagioclase, and allanite/epidote is suggestive of relatively high-pressure crystallization (e.g., Eggler and Burnham, 1973; Allen and Boettcher, 1983; Zen and Hammarstrom, 1984; Evans and Vance, 1987; Dawes and Evans, 1991; Day et al., 1992; Green, 1992). In this section, we present the results of thermobarometric calculations, compare the dike mineralogy to experimental studies, and discuss implications of the results.

Inclusion and intergrowth relationships between garnet and amphibole phenocrysts in the garnet andesites indicate that these minerals grew in equilibrium with each other. The exchange equilibria between garnet and calcic amphibole have been experimentally calibrated in terms of pressure for Mg and Fe exchange reactions (Kohn and Spear, 1990). Results of both calculations are shown in Figure 6. They indicate equilibrium pressures of at least 6.9 kb, and as much as 9.5 kb.

Hammarstrom and Zen (1986) showed that the Al content in calcic amphibole is, in part, a function of total pressure, and Hollister et al. (1987) described possible controlling equilibria. According to Hollister et al. (1987), application of the Al-in-hornblende barometer requires some control of temperature, as well as an equilibrium assemblage that includes plagioclase, quartz, titanite or an Fe-Ti oxide, biotite, and K-feldspar. Anderson and Smith (1995) suggested that the pressure estimates are insensitive to the presence of K-feldspar. The garnet andesite

TABLE 9. PRIMARY AND SECONDARY GLASS STANDARD COMPOSITIONS

Analysis	P	Sr	Y	Zr	Nb	Ba	La	Ce	Pr	Nd	Sm	Eu	Gd	Tb	Dy	Ho	Er	Tm	Yb	Lu	Hf	Ta	Pb	Th	U
NIST 612*	55.2	76.4	37.8	43.3	38.1	38.5	35.5	38.2	37.1	35.0	36.4	34.6	37.1	36.3	35.7	37.9	38.0	37.7	39.5	37.6	34.8	39.8	38.7	37.3	36.9
Bcr2g-1†	2222	327	31.8	193	11.9	660	24.4	52	6.4	27.6	6.1	1.9	6.4	1.0	6.1	1.3	3.5	0.5	3.2	0.5	4.3	0.7	11.1	5.5	1.6
Bcr2g-2†	2204	328	31.3	193	11.7	656	23.9	52	6.4	27.3	6.1	1.9	6.5	1.0	6.2	1.2	3.5	0.5	3.2	0.5	4.3	0.8	11.0	5.4	1.6
Bcr2g-3†	2291	326	31.2	193	11.6	667	24.6	53	6.6	28.8	6.3	1.9	6.3	1.0	6.2	1.2	3.6	0.5	3.3	0.5	4.4	0.7	11.5	5.5	1.7
Average	2239	326.9	31.5	193	11.8	661	24.3	52.2	6.5	27.9	6.1	1.9	6.4	1.0	6.2	1.2	3.5	0.5	3.3	0.5	4.3	0.7	11.2	5.5	1.6
Standard deviation	45	1.2	0.3	0.2	0.1	6	0.3	0.5	0.1	0.8	0.1	0.0	0.1	0.0	0.0	0.0	0.1	0.0	0.1	0.0	0.0	0.0	0.2	0.1	0.0
Accepted§	1615	335	39.4	210	13.1	672	24.4	51.9	6.48	28.4	6.58	1.98	6.67	1.06	6.33	1.32	3.73	n.d.	3.34	0.5	4.9	0.81	10.3	6.03	1.62

Notes: Concentrations in ppm. Abbreviation: n.d.—not determined.

*National Institute of Standards and Technology (NIST) 612 values after Pearce et al. (1997) but include in-house refinements based on round-robin analyses of well-known standard glasses.

†These concentrations are based the assumption of 7.12 wt% CaO in BCR-2g.

§Solution inductively coupled plasma mass spectrometry (ICP-MS) values from Norman et al. (1998) except P, by electron probe. Norman et al. (1998) reported Zr = 192 ppm by particle-induced x-ray emission and Zr = 194 ppm by laser ablation ICP-MS analysis.

TABLE 10. ZIRCON TRACE ELEMENT COMPOSITIONS IN PPM

Analysis	P	Sr	Y	ZrO_2 (wt%)	Nb	Ba	La	Ce	Pr	Nd	Sm	Eu	Gd	Tb	Dy	Ho	Er	Tm	Yb	Lu	Hf	Ta	Pb*	Th	U
z-1r	255	0.40	486	60.30	0.5	0.2	b.d.	1.5	0.03	0.5	1.6	0.8	10.8	3.3	40	14.2	64	13.5	125	24.6	7130	0.2	0.34	19	152
z-1c	188	0.14	416	64.86	0.5	b.d.	0.01	1.3	0.04	0.5	1.6	0.7	10.1	3.2	37	12.2	53	10.8	98	18.9	8272	0.2	0.32	17	167
z-2r	175	0.18	366	70.09	0.6	0.2	0.03	1.3	0.03	0.5	1.9	0.8	12.3	3.6	37	10.8	42	7.8	63	11.3	8348	0.2	0.41	28	231
z-2c	149	0.37	386	63.97	0.7	0.5	0.03	1.0	0.02	0.6	1.8	0.7	13.1	4.0	39	11.4	43	8.0	68	11.9	9343	0.2	0.33	21	258
z-4	175	0.18	507	63.71	1.4	0.1	b.d.	1.8	0.04	0.6	2.0	0.8	13.9	4.4	47	14.8	62	12.5	111	20.6	8154	0.4	0.51	32	297
z-5	158	0.47	469	65.65	2.1	0.9	0.03	2.0	0.04	0.8	2.6	1.0	18.4	5.0	49	14.0	54	10.2	87	15.6	8655	0.6	1.13	64	569
z-6c	236	0.21	662	64.15	0.8	0.2	b.d.	1.7	0.04	1.0	3.4	1.5	24.0	7.1	69	19.4	69	12.0	95	16.3	8381	0.2	0.60	38	361
z-6r	120	0.10	193	64.41	1.3	b.d.	b.d.	1.1	0.03	0.3	1.3	0.6	8.8	2.3	22	5.7	20	3.5	29	4.9	8871	0.4	0.80	44	528
z-7	223	0.40	495	61.56	1.9	0.7	0.03	3.6	0.02	1.0	2.8	1.3	17.9	5.3	52	14.4	52	9.1	72	11.4	8566	0.6	1.41	89	599
z-8r	154	0.16	230	64.57	1.6	0.1	0.04	1.7	0.02	0.4	1.0	0.5	6.6	2.0	21	6.8	29	5.9	54	10.1	8996	0.6	0.96	53	564
z-8c	140	0.23	265	67.33	1.5	0.1	b.d.	1.6	0.02	0.6	1.8	0.7	12.3	3.2	30	8.0	28	4.9	39	6.4	8858	0.5	1.06	62	600
z-10	181	0.17	634	70.62	1.8	0.0	b.d.	2.2	0.13	0.7	2.0	1.0	16.6	5.3	60	19.0	78	15.5	135	24.5	8336	0.5	0.81	48	398
z-12r	406	0.52	296	69.98	6.0	0.1	0.18	2.8	0.03	0.9	1.5	0.6	8.1	2.5	28	8.9	39	8.4	79	14.2	9750	0.9	1.32	90	646
z-12c	141	0.13	527	70.64	1.8	b.d.	b.d.	1.6	0.03	0.6	2.9	1.4	23.3	6.7	62	16.0	54	9.0	69	11.0	9879	0.6	0.69	42	575
z-14	153	0.12	300	67.46	2.2	0.1	0.02	2.1	0.04	0.7	1.7	0.7	10.8	3.2	31	9.1	35	6.9	62	10.9	9838	0.9	1.46	91	885
z-15	238	0.18	672	66.23	1.7	0.0	b.d.	3.4	0.02	0.8	2.0	1.1	15.2	5.2	61	20.2	88	17.7	156	28.5	8126	0.5	1.00	62	408
z-16r	154	0.13	372	67.32	2.2	b.d.	b.d.	1.5	0.03	0.3	1.1	0.5	8.0	2.6	32	11.1	50	10.8	102	20.2	8821	0.7	0.51	31	410
z-16c	180	0.13	404	63.86	1.5	0.0	b.d.	2.5	0.01	0.7	1.8	0.8	13.2	4.0	41	11.9	46	8.8	74	12.9	8816	0.5	0.93	60	438
z-19	160	0.12	324	65.65	2.8	b.d.	b.d.	2.2	0.03	0.6	1.6	0.8	11.8	3.6	38	11.3	45	8.9	78	13.9	9139	0.9	1.91	115	823

Notes: Analyses of zircon mineral separates mounted in epoxy and polished. Concentrations are based the assumption of 32.77 wt% SiO_2. All values except those for ZrO_2 are in ppm. Abbreviations: b.d.—below detection limits, c—core, r—rim.

*This estimation of Pb content for zircon and other minerals is derived from assuming minerals have the same Pb isotope content as National Institute of Standards and Technology (NIST) 612 (i.e., that ^{208}Pb makes up 53% of total Pb).

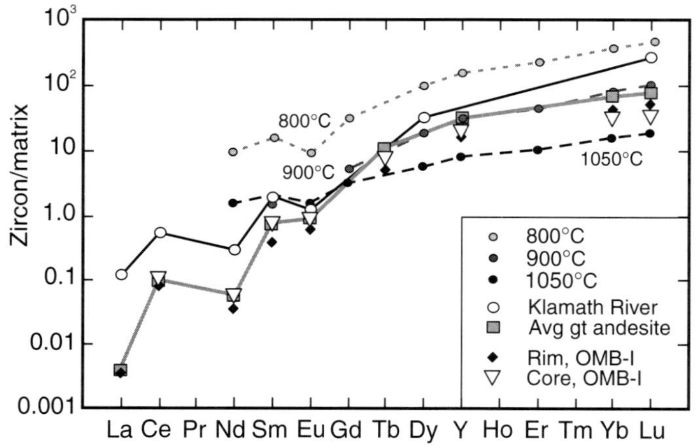

Figure 5. Rare-earth element ratios of Zr in zircon/Zr in rock, melt, or matrix (distribution coefficient). Temperature labels in centigrade are for experiments of D. Rubatto and J. Hermann (personal commun., 2005). Compositions of detrital zircons from the Klamath River are normalized to average crust. The average composition of zircon from garnet andesite sample OMB-I has a pattern very similar to the 900 °C/20-kb experiment, although it contains less Sm. The OMB-I pattern is also distinct from the average Klamath River zircon, which is thought to represent zircons from garnet-free Mesozoic plutons from the region. The core-rim zoning of heavy REE from a single OMB-I zircon grain (symbols without connecting lines) is consistent with falling temperature.

dikes lack K-feldspar and biotite phenocrysts, and several lack quartz phenocrysts (Table 1), and so do not contain the required equilibrium assemblage. Nevertheless, we calculated pressure on the basis of Schmidt's (1992) experimental calibration for the purposes of comparison with the garnet-amphibole geobarometer. Amphibole rim compositions yielded pressure estimates from 12.8 to 8.2 kb, which is similar to the range of all amphibole compositions, which gave estimates from 13.1 to 8.0 kb.

Anderson and Smith (1995) presented an iterative procedure that solved simultaneous equations for pressure dependence and temperature dependence (Holland and Blundy, 1994; edenite-richterite reaction). Application of the Anderson and Smith (1995) method to three garnet andesite samples (Fig. 6) yields pressure and temperature estimates (as ranges): OMB-I: 8.5–8.9 kb, 709–729 °C; 92OMB114: 7.4–7.5 kb, 768 °C; 92OMB31B: 6.7–7.1 kb, 792–804 °C (shaded box in Fig. 6). Uncertainties on pressure estimates are in the range of 0.8–1.6 kb, on temperature estimates from 14 to 40 °C. The discrepancy in pressure and temperature between OMB-I and the other two samples may be related to buffering assemblage: OMB-I contains quartz phenocrysts but the other two samples do not (Table 1).

Temperature estimates were also obtained using apatite (Harrison and Watson, 1984) and zircon (Watson and Harrison, 1983) saturation thermometry. Apatite saturation temperatures range from 814 to 872 °C, whereas zircon saturation temperatures are in the 703- to 732-°C range. Interpretation of mineral saturation temperatures can be problematic, owing to crystal accumulation and inheritance (e.g., Miller et al., 1992, 2003) and disequilibrium crystallization, particularly of apatite (e.g., Bacon, 1989). Nevertheless, because the rocks are at most weakly peraluminous, and because apatite phenocrysts and microphenocrysts are present in the dikes, a case may be made for apatite saturation in the melt. This condition argues for a minimum temperature of apatite stability of ~810 °C. In contrast, zircon saturation temperatures suggest that zircon was not stable until near-solidus temperatures. It is noteworthy that the relatively low heavy REE contents of zircons in these rocks are consistent with coprecipitation of zircon and REE-richer phases (allanite/epidote and apatite) as well as garnet (see above and Fig. 4C). This in turn suggests considerably higher-termperature zircon saturation than was calculated.

A number of experimental studies have established the stability field of garnet in andesitic and dacitic compositions (e.g., Eggler and Burnham, 1973; Allen et al., 1975; Green, 1992). These studies consistently show that for moderate to high H_2O contents, igneous garnet is stable at pressures greater than 7–8 kb and that this minimum pressure is weakly correlated with temperature. In most experiments on andesitic compositions, augite and/or orthopyroxene are part of the high-pressure assemblage, along with garnet, plagioclase, and amphibole. Quartz stability is reached between 800 and 850 °C, with higher stability temperatures at lower H_2O contents (Green, 1992). Experiments on a dacitic composition (Green, 1992; Fig. 5) reproduce the phenocryst assemblage of sample OMB-I (garnet, amphibole, plagioclase, quartz) at pressures >7 kb and temperatures <860 °C with 3 wt% H_2O, and pressure >8.5 kb and temperatures between 800 and 850 °C with 5 wt% H_2O. This range of pressures and temperatures overlaps the values derived from our thermobarometric estimates. We conclude that the phenocryst assemblages in the garnet andesite dikes reflect initial high-pressure crystallization, probably between 8 and 9 kb. These pressures suggest crystallization depths between 25 and 30 km.

Recent experiments to explore garnet-zircon equilibria in hydrous, peraluminous granitic melts bear on our interpretation of the garnet andesite dike rocks (D. Rubatto and J. Hermann, personal commun., 2005), although these experiments were conducted at 20 kb, clearly at much higher pressures than our estimates for phenocryst growth in the garnet andesite magmas. The mineral assemblage in the experiments is garnet + zircon + allanite + orthopyroxene (opx) + kyanite (900 and 1050 °C). At 800 °C, opx and kyanite were absent and biotite + staurolite were present, and at 750 °C, garnet, opx, and kyanite were absent but staurolite + biotite + phengite were present. Therefore, the experimental data permit comparison of zircon REE patterns from the garnet andesites to those from garnet-present experiments (Fig. 6). For the experimental conditions reported in D. Rubatto and J. Hermann (personal commun., 2005), the zircon saturation thermometer of Harrison and Watson (1983) is correct in shape but yields temperatures ~80 degrees too low. If this

TABLE 11. GARNET TRACE ELEMENT COMPOSITIONS

Analysis	CaO	P	Sr	Y	Zr	Nb	Ba	La	Ce	Pr	Nd	Sm	Eu	Gd	Tb	Dy	Ho	Er	Tm	Yb	Lu	Hf	Ta	Pb	Th	U
bgc1	6.49	203	0.11	518	67	0.46	0.01	0.01	0.18	0.14	2.6	4.5	2.1	19	5.7	61	18	70	12	101	17	0.7	0.04	b.d.	b.d.	0.03
bgc2	6.49	215	0.11	541	73	0.54	b.d.	b.d.	0.16	0.13	2.5	4.7	2.2	20	5.8	65	19	73	13	106	18	0.7	0.04	b.d.	b.d.	0.05
bgr1	6.81	281	0.19	712	87	3.95	b.d.	b.d.	0.19	0.15	2.8	5.3	2.4	23	7.0	80	26	106	21	179	33	0.9	0.17	0.09	0.02	0.05
bgr2	6.24	253	0.16	600	77	0.72	0.11	0.01	0.19	0.14	2.9	4.5	2.2	20	6.3	70	21	87	16	131	23	0.8	0.07	b.d.	b.d.	0.05
bgc3	6.21	175	0.11	281	53	0.21	0.14	0.02	0.17	0.14	2.9	5.7	1.9	20	4.8	40	10	35	6	45	7	0.7	0.02	b.d.	b.d.	0.05
bgr3	7.08	140	0.06	288	44	0.18	b.d.	0.03	0.10	0.08	1.5	4.6	1.6	22	5.6	47	11	32	5	36	6	0.5	0.02	b.d.	b.d.	b.d.
agr1	8.86	130	0.06	77	57	0.09	0.01	b.d.	b.d.	b.d.	0.1	0.3	0.3	2	0.9	9	3	9	1	10	2	0.6	0.01	0.12	b.d.	b.d.
agc2	10.10	266	0.15	149	88	0.38	0.03	0.03	0.10	0.02	0.2	0.5	0.4	3	1.1	14	5	19	4	30	5	0.9	0.04	0.06	b.d.	0.02
agr2	8.24	185	0.10	81	67	0.19	0.06	0.02	0.03	b.d.	0.1	0.2	0.3	2	0.8	9	3	10	2	12	2	0.7	0.02	0.05	0.01	0.01

Notes: Concentrations are based on CaO wt% as determined by electron microprobe. All values are in ppm. Abbreviations: b.d.—below detection limits, c—core, r—rim.

TABLE 12. EPIDOTE/ALLANITE TRACE ELEMENT COMPOSITIONS

Analysis	CaO	P	Sr	Y	Zr	Nb	Ba	La	Ce	Pr	Nd	Sm	Eu	Gd	Tb	Dy	Ho	Er	Tm	Yb	Lu	Hf	Ta	Pb	Th	U
1aec1	22.6	5002	393	510	117	0.08	25	2555	5637	676	2666	501	165	347	37	154	19.2	36	3.5	18.2	2.1	3.0	0.01	8.2	324	82
1aer1	22.0	3714	412	438	116	0.05	20	2086	4641	549	2189	413	145	289	31	133	16.6	30	2.9	14.0	1.6	2.9	0.01	6.3	217	74
1ae3	21.5	4118	542	382	139	0.07	60	6794	14109	1661	6221	834	188	441	38	135	15.5	27	2.5	11.9	1.4	3.7	b.d.	22.1	938	64
1ae4	21.4	2438	621	390	133	0.34	23	2292	4670	560	2250	411	134	282	29	120	14.7	26	2.4	10.8	1.2	3.9	0.02	22.2	386	91
1ae5	21.7	3561	529	439	162	0.04	37	4080	8603	1014	3958	625	159	378	36	140	17.0	31	2.9	14.6	1.6	4.1	b.d.	13.7	605	84

Notes: Concentrations are based on CaO wt% as determined by electron microprobe. All values in ppm. Abbreviations: b.d.—below detection limits, c—core, r—rim.

TABLE 13. PLAGIOCLASE TRACE ELEMENT COMPOSITIONS

Analysis	CaO	P	Sr	Y	Zr	Nb	Ba	La	Ce	Pr	Nd	Sm	Eu	Pb
bp2-1	8.2	38	540	0.03	0.02	b.d.	61	1.9	3.1	0.31	0.8	0.06	0.6	3.6
bp2-2	8.3	26	509	0.02	0.01	0.00	63	1.4	2.2	0.22	0.7	0.10	0.7	3.4
bp2-3	8.3	27	483	0.02	b.d.	b.d.	72	1.2	1.9	0.19	0.7	0.06	0.8	3.9
bp2-4	8.3	26	499	0.02	b.d.	b.d.	58	2.0	3.0	0.27	0.9	0.07	0.7	3.2
bp2-5	8.4	32	488	0.05	0.06	0.01	55	1.4	2.3	0.23	0.7	0.08	0.5	2.6
bp2-6	8.4	30	509	0.02	0.01	b.d.	45	1.3	2.0	0.19	0.7	0.06	0.5	2.6
bp2-7	8.2	26	541	0.02	b.d.	b.d.	74	1.4	2.1	0.18	0.7	0.08	0.7	3.9
bp2-8	8.3	22	499	0.03	0.03	0.00	75	1.2	1.8	0.18	0.6	0.10	0.7	4.0
bp2-9	8.3	29	563	0.02	b.d.	b.d.	82	1.7	2.6	0.23	0.8	0.09	0.7	4.3
bp1-1c	8.2	21	768	0.03	b.d.	b.d.	93	1.3	1.9	0.19	0.5	b.d.	0.7	4.7
bp1-1r	8.3	25	674	0.04	b.d.	b.d.	90	1.4	2.2	0.21	0.7	b.d.	0.7	4.3

Notes: Concentrations are based on CaO wt% as determined by electron microprobe. All values in ppm. Those elements given in Table 10 but not listed here are below detection limits. Abbreviations: b.d.—below detection limits, c—core, r—rim.

TABLE 14. HORNBLENDE TRACE ELEMENT COMPOSITIONS

Analysis	CaO	P	Sr	Y	Zr	Nb	Ba	La	Ce	Pr	Nd	Sm	Eu	Gd	Tb	Dy	Ho	Er	Tm	Yb	Lu	Hf	Ta	Pb	Th	U
p2h1	10.7	44	23	19	21	2.9	37	1.5	9.2	2.5	18.1	9.0	1.6	10.2	1.3	5.8	0.7	1.4	0.1	0.8	0.1	1.4	0.07	0.96	0.02	0.04
p2h2	10.5	55	25	18	22	2.2	32	2.3	13.9	3.5	23.6	8.7	1.7	8.6	1.1	5.1	0.7	1.5	0.1	0.8	0.1	1.3	0.04	0.73	0.02	0.03
p2h3	10.5	51	23	15	25	2.4	35	1.6	9.9	2.6	19.0	8.4	1.5	9.2	1.0	4.5	0.6	1.2	0.1	0.7	0.1	1.3	0.07	0.82	0.03	0.03
p2h4	10.8	60	38	19	26	1.8	45	1.5	8.0	1.9	13.7	5.9	1.4	6.6	0.9	4.8	0.8	1.7	0.2	1.3	0.2	1.2	0.06	0.97	0.03	0.03
bhr1	10.5	64	36	41	25	7.2	41	3.3	18.3	4.4	28.6	9.4	2.0	9.6	1.4	8.3	1.6	4.4	0.6	3.8	0.5	1.3	0.11	0.93	0.04	0.04
mhr1	10.9	42	22	27	27	6.8	27	1.2	6.8	1.7	11.7	5.0	1.2	5.8	0.9	5.5	1.0	2.9	0.4	2.6	0.4	1.3	0.15	3.78	0.15	0.08
mhc1	10.8	496	33	22	39	2.9	47	1.7	9.4	2.4	18.0	7.8	1.4	8.2	1.1	5.4	0.9	2.0	0.3	1.7	0.2	2.2	0.09	1.91	0.03	0.07
ahr1	10.7	65	26	27	21	2.5	38	2.1	12.3	3.1	21.6	7.6	1.5	8.2	1.1	6.2	1.0	2.6	0.3	2.1	0.2	1.2	0.06	0.88	0.03	0.03
ahc1	11.2	55	25	15	23	2.4	42	1.1	6.4	1.7	13.0	6.4	1.4	7.2	0.9	4.4	0.6	1.3	0.1	0.8	0.1	1.3	0.07	0.88	0.02	0.04
ahr2	10.8	88	34	29	23	6.1	45	2.1	12.3	3.1	20.9	7.7	1.7	8.1	1.1	6.6	1.1	3.0	0.4	2.4	0.3	1.2	0.09	0.97	0.03	0.05
ahc2	11.0	55	24	16	19	2.2	39	1.4	9.0	2.4	17.4	7.6	1.5	8.4	1.0	4.5	0.7	1.3	0.1	0.7	0.1	1.1	0.05	0.85	0.02	0.04

Notes: Concentrations are based on CaO wt% as determined by electron microprobe. All values in ppm. Abbreviations: c—core, p—inclusion in plagioclase, r—rim.

TABLE 15. APATITE TRACE ELEMENT COMPOSITIONS

Analysis	Sr	Y	Zr	Nb	Ba	La	Ce	Pr	Nd	Sm	Eu	Gd	Tb	Dy	Ho	Er	Tm	Yb	Lu	Pb	Th	U
Ma-1	255	80	2.4	0.8	3.8	154	481	75	352	63	8.9	47.3	4.7	20.3	3.0	6.1	0.7	3.3	0.4	2.2	1.8	3.7
Ma-2	292	94	0.8	0.2	20.9	145	439	68	333	64	9.5	52.7	5.4	24.5	3.5	7.4	0.8	4.2	0.5	1.5	3.6	5.7
Ma-3	249	63	0.8	0.0	2.1	116	361	58	290	54	7.1	40.2	3.8	16.2	2.4	4.7	0.5	2.5	0.3	1.1	1.2	2.4
Ma-4	286	96	0.9	0.2	2.4	147	440	69	341	65	9.7	53.8	5.4	24.5	3.6	7.5	0.8	4.2	0.6	1.3	3.5	5.6
Aa-1	249	62	1.8	0.7	2.8	178	599	84	398	71	11.6	51.1	4.7	18.4	2.3	4.3	0.4	1.9	0.2	1.2	4.8	5.6
Aa-2	255	83	2.4	0.5	3.1	182	610	85	400	77	12.5	58.6	5.6	23.0	3.1	6.3	0.6	3.2	0.4	1.5	7.1	7.7
Aa-3	239	66	0.8	0.2	2.4	143	453	72	361	73	9.4	56.6	5.1	19.5	2.5	4.6	0.4	1.9	0.2	1.8	3.8	5.9

Notes: Concentrations assume CaO = 54.0 wt%. All values in ppm. Hafnium and Ta were below detection limits.

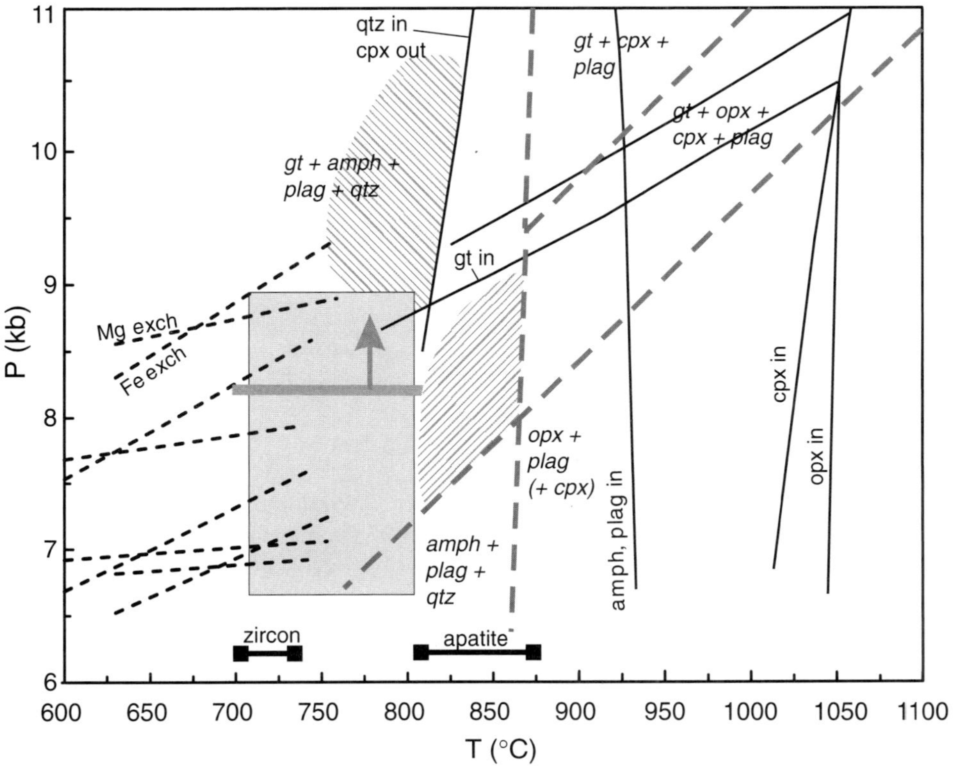

Figure 6. Summary of pressure and temperature estimates. The black short-dashed lines are Mg-Fe-exchange equilibrium boundaries for garnet-amphibole equilibria (Kohn and Spear, 1990) in the analyzed samples. The shaded box represents the range of pressure and temperature estimates from combined Al-in-hornblende barometry and plagioclase-hornblende thermometry (Anderson and Smith, 1995). The shaded line with arrow is the lower limit Al-in-hornblende pressure estimate using the Schmidt (1992) calibration, and the two thick black lines are the temperature ranges of zircon and apatite saturation thermometers (see text). The long-dashed gray and solid black lines bound stability fields determined by Green (1992) for a garnet-bearing dacite at 3 wt% and 5 wt% H_2O, respectively. Italicized labels give melt-present assemblages for 3 wt% H_2O. Boundaries for 5 wt% H_2O conditions are labeled according to the down-temperature mineral stability. Hachured areas show experimental conditions consistent with stability of the observed phenocryst assemblages of the garnet andesites. Abbreviations: amph—amphibole; cpx—clinopyroxene; gt—garnet; opx—orthopyroxene; plag—plagioclase; qtz—quartz.

discrepancy is true for zircons in the garnet andesites, then zircon saturation might have been as high as ~800 °C, which is consistent with our observations of inclusion relationships.

In Figure 5, the REE contents of experimentally-grown zircons are plotted normalized to compositions of their host melts, and the compositions of zircons from sample OMB-I are normalized to bulk rock compositions. Because it is impossible to determine a rock composition for detrital zircons, the pattern for the Klamath River zircons is based on zircon/average crust.

It is apparent that the average OMB-I zircon pattern is a close match to the 900-°C experiment, although the natural zircons have lower average Sm contents. However, the experiments of D. Rubatto and J. Hermann (personal commun., 2005) predict zircon saturation at 800 °C. Therefore, the shape of the OMB-I zircon REE pattern is consistent with equilibrium with garnet, but the match between calculated saturation temperature and the "matching" experimental REE pattern is from 100 to nearly 200 °C too low. The cause of this discrepancy (equilibrium assemblage, pressure of crystallization, nucleation delay, etc.) is uncertain.

Comparison of REE patterns of zircon in equilibrium with garnet in the experiments and in OMB-I shows that heavy REE are similar, but are also not readily distinguished from the heavy REE patterns of zircons from garnet-absent assemblages (e.g., Klamath River zircons). The more conspicuous difference in the case of OMB-I zircons is the strong light REE depletion. This depletion we interpret to be caused by equilibrium with allanite/epidote, which contains higher total REE and which controls the zircon light REE pattern. Thus, at least in this sample, the earmark of zircon that crystallized from a high-pressure assemblage is as much the light REE depletion (because of the coexistence with allanite/epidote) as it is the flatter heavy REE pattern (due to the presence of garnet).

One of the striking petrographic features of these dikes is the lack of low-pressure equilibration, particularly of garnet but also of epidote (cf., Day et al., 1992; Brandon et al., 1996). This absence suggests that the dike magmas moved from high pressure to the level of emplacement rapidly and that they cooled quickly. An explanation may be found in the probable H_2O contents of the magmas. All experimental studies that reproduce the equilibrium assemblage required significant H_2O contents (>3 wt%; typically, 5 wt%). The presence of quartz phenocrysts in some garnet andesite samples indicates magmatic temperature <850 °C, which implies even greater H_2O contents dissolved in the melt phase. We speculate that final emplacement of the garnet andesite dikes was driven by exsolution of H_2O, which would have caused rapid emplacement and pressure quench solidification of interstitial melt. Abundant deuteric H_2O might also explain the presence of late-stage prehnite in many of these samples (Table 1).

Emplacement of the ca. 170-Ma Ironside Mountain batholith (Barnes et al., this volume, Chapter 10) began a period of arc magmatism in the central Klamath Mountains that extended at least to 156 Ma (Yule, 1996) and ended with the onset of Nevadan thrusting (e.g., Harper et al., 1994). Most Ironside Mountain magmas lack isotopic evidence for significant interaction with crustal rocks (Barnes et al., this volume, Chapter 10), but younger magmatic rocks (e.g., Wooley Creek belt plutons; ca. 163–156 Ma) show abundant Sr, Nd, and oxygen isotopic evidence for crustal anatexis and assimilation of crustal material into mantle-derived basaltic magmas (Barnes et al., 1990, 1995; Gribble et al., 1990; Allen and Barnes, this volume). These authors showed that although some crustal assimilation occurred at the level of emplacement, much of the assimilation responsible for the isotopic signatures required a deep-seated source or sources. Barnes et al. (this volume, Chapter 10) suggested that because Ironside Mountain magmatism immediately followed regional thrusting, the crust was too cool to melt or to significantly contaminate mantle-derived magmas. However, as the arc evolved, the influx of mafic magmas into the deep crust heated the adjacent rocks, caused local partial melting, and permitted large-scale contamination. The garnet andesite dikes support the idea that during late Middle Jurassic time (ca. 160 Ma) magmas were stored and evolved in the deep crustal environment.

It has long been recognized (Irwin, personal commun.) that late Middle Jurassic plutons are widespread in the central and northern Klamath Mountains but are virtually absent in the southern part of the province. Interestingly, the outcrop zone of the coeval garnet andesite dikes seems to be the converse of this pattern, with these dikes present in the southern Klamath Mountains but absent in the central and north. Hacker et al. (1995) explained the distribution of large late Middle Jurassic plutons as the result of extensional deformation associated with plutonism (central and north) and a relative lack of extension to the south. In this scenario, deep-seated magmas in the southern Klamath province would have risen into the shallow crust only when they evolved to low-density, presumably H_2O-rich compositions. If this scenario was the case, then only evolved, H_2O-rich magmas such as the garnet andesite magmas would have reached the upper crust of the southern Klamath Mountains during this time.

CONCLUSIONS

Distinctive ca. 160-Ma garnet andesite dikes are widespread in the southern Klamath Mountains province. They contain a phenocryst assemblage of Ca-rich garnet + plagioclase (andesine) + Al-rich calcic amphibole ± quartz ± epidote. Experimental studies on similar rocks and thermobarometric calculations indicate that this assemblage was stable at pressures >7 kb and probably in the 8–9 kb range. Zircons in apparent equilibrium with garnet have heavy REE patterns that match experiments in which zircon and garnet are in equilibrium with melt (D. Rubatto and J. Hermann, personal commun., 2005), but the difference between the patterns for zircons in garnet-present and garnet-absent assemblages is subtle. The overall low REE abundances, and the extreme light REE depletion is thought to result from co-precipitation of allanite/epidote and apatite; these are much more conspicuous features of the zircon geochemistry.

Although the dikes are coeval with voluminous late Middle Jurassic magmatism in the central and northern parts of the province, no large plutons of that age intruded the southern Klamath Mountains. This paucity may reflect distinct tectonic realms, with extensional deformation in the central and northern areas and neutral or contractional deformation in the south. In this case, the H_2O-rich garnet andesite magmas rose, due to buoyancy contrasts as H_2O exsolved at near-solidus temperatures.

ACKNOWLEDGMENTS

We thank Porter Irwin for initial discussions concerning these interesting rocks and Melanie Barnes for assistance in the field and laboratory. Susan Swapp graciously provided a program for garnet-amphibole barometry. We gratefully acknowledge Ron Kistler for providing Sr isotope analyses and for many helpful conversations about granite petrogenesis, and Helen Lang and Craig White for their reviews. This work received partial support from National Science Foundation grant EAR-9117103.

REFERENCES CITED

Ague, J.J., and Brimhall, G.H., 1988, Magmatic arc asymmetry and distribution of anomalous plutonic belts in the batholiths of California: Effects of assimilation, crustal thickness, and depth of crystallization: Geological Society of America Bulletin, v. 100, p. 912–927, doi: 10.1130/0016-7606(1988)100<0912:MAAADO>2.3.CO;2.

Allen, C.M., and Barnes, C.G., 2006, this volume, Ages and some cryptic sources of Mesozoic plutonic rocks in the Klamath Mountains, California and Oregon, *in* Snoke, A.W., and Barnes, C.G., eds., Geological studies in the Klamath Mountains province, California and Oregon: A volume in honor of William P. Irwin: Boulder, Colorado, Geological Society of America Special Paper 410, doi: 10.1130/2006.2410(11).

Allen, C.M., Barnes, C.G., and Campbell, I.H., 2002, U-Th-U ages from Klamath River detrital zircons using LA-ICP-MS: Geological Society of America Abstracts with Programs, v. 34, p. 434.

Allen, J.C., and Boettcher, A.L., 1983, The stability of amphibole in andesite and basalt at high pressures: American Mineralogist, v. 68, p. 307–314.

Allen, J.C., Boettcher, A.L., and Marland, G., 1975, amphiboles in andesite and basalt: I. Stability as a function of P-T-fO_2: American Mineralogist, v. 60, p. 1069–1085.

Anderson, J.L., and Smith, D.R., 1995, The effects of temperature and fO_2 on the Al-in-hornblende barometer: American Mineralogist, v. 80, p. 549–559.

Bacon, C.R., 1989, Crystallization of accessory phases in magmas by local saturation adjacent to phenocrysts: Geochimica et Cosmochimica Acta, v. 53, p. 1055–1066, doi: 10.1016/0016-7037(89)90210-X.

Ballard, J.R., Palin, J.M., and Campbell, I.H., 2002, Relative oxidation states of magmas inferred from Ce(IV)/Ce(III) in zircon: Application to porphyry copper deposits of northern Chile: Contributions to Mineralogy and Petrology, v. 144, p. 347–364.

Barnes, C.G., Allen, C.M., and Saleeby, J.B., 1986, Open- and closed-system characteristics of a tilted plutonic system, Klamath Mountains, California: Journal of Geophysical Research, v. 91, p. 6073–6090.

Barnes, C.G., Allen, C.M., Hoover, J.D., and Brigham, R.H., 1990, Magmatic components of a tilted plutonic system, Klamath Mountains, California, *in* Anderson, J.L., ed., The nature and origin of Cordilleran magmatism: Boulder, Colorado, Geological Society of America Memoir 174, p. 331–346.

Barnes, C.G., Petersen, S.W., Kistler, R.W., Prestvik, T., and Sundvoll, B., 1992, Tectonic implications of isotopic variation among Jurassic and early Cretaceous plutons, Klamath Mountains: Geological Society of America Bulletin, v. 104, p. 117–126, doi: 10.1130/0016-7606(1992)104<0117:TIOIVA>2.3.CO;2.

Barnes, C.G., Johnson, K., Barnes, M.A., Prestvik, T., Kistler, R.W., and Sundvoll, B., 1995, The Grayback pluton: Magmatism in a Jurassic back-arc environment, Klamath Mountains, Oregon: Journal of Petrology, v. 36, p. 397–415.

Barnes, C.G., Mars, E.V., Swapp, S., and Frost, C.D., 2006, this volume (Chapter 10), Petrology and geochemistry of the Middle Jurassic Ironside Mountain batholith: Evolution of potassic magmas in a primitive arc setting, *in* Snoke, A.W., and Barnes, C.G., eds., Geological studies in the Klamath Mountains province, California and Oregon: A volume in honor of William P. Irwin: Boulder, Colorado, Geological Society of America Special Paper 410, doi: 10.1130/2006.2410(10).

Black, L.P., Kamo, S.L., Allen, C.M., Aleinikoff, J.N., Davis, D.W., Korsch, R.J., and Foudoulis, C., 2003, TEMORA 1: A new zircon standard for Phanerozoic U-Pb geochronology: Chemical Geology, v. 200, p. 155–170, doi: 10.1016/S0009-2541(03)00165-7.

Boardman, J.J., 1985, Petrology of the Salmon Mountain stock, Klamath Mountains, California [M.S. thesis]: Lubbock, Texas Tech University, 78 p.

Brandon, A.D., Creaser, R.A., and Chacko, T., 1996, Constraints on rates of granitic magma transport from epidote dissolution kinetics: Science, v. 271, p. 1845–1848.

Coleman, R.G., Mortimer, N., Donato, M.M., Manning, C.E., and Hill, L.B., 1988, Tectonic and regional metamorphic framework of the Klamath Mountains and adjacent Coast Ranges, California and Oregon, *in* Ernst, W.G., ed., Metamorphism and crustal evolution of the western United States: Englewood Cliffs, New Jersey, Prentice-Hall, p. 1061–1097.

Dawes, R.L., and Evans, B.W., 1991, Mineralogy and geothermobarometry of magmatic epidote-bearing dikes, Front Range, Colorado: Geological Society of America Bulletin, v. 103, p. 1017–1031, doi: 10.1130/0016-7606(1991)103<1017:MAGOME>2.3.CO;2.

Day, R.A., Green, T.H., and Smith, I.E.M., 1992, The origin and significance of garnet phenocrysts and garnet-bearing xenoliths in Miocene calc-alkaline volcanics from Northland, New Zealand: Journal of Petrology, v. 33, p. 125–161.

DeBari, S., 1997, Evolution of magmas in continental and oceanic arcs: The role of the lower crust: Canadian Mineralogist, v. 35, p. 501–519.

DePaolo, D., 1981, Trace element and isotopic effects of combined wallrock assimilation and fractional crystallization: Earth and Planetary Science Letters, v. 53, p. 189–202, doi: 10.1016/0012-821X(81)90153-9.

Eggler, D.H., and Burnham, C.W., 1973, Crystallization and fractionation trends in the system andesite-H_2O-CO_2-O_2 at pressures to 10 kb: Geological Society of America Bulletin, v. 84, p. 2517–2532, doi: 10.1130/0016-7606(1973)84<2517:CAFTIT>2.0.CO;2.

Ernst, W.G., 1998, Geology of the Sawyers Bar area, Klamath Mountains, northern California: Sacramento, California Division of Mines and Geology Map Sheet 47, scale 1:48,000, 59 p. pamphlet.

Evans, B.W., and Vance, J.A., 1987, Epidote phenocrysts in dacitic dikes, Boulder County, Colorado: Contributions to Mineralogy and Petrology, v. 96, p. 178–185, doi: 10.1007/BF00375231.

Goodge, J.W., 1990, Tectonic evolution of a coherent Late Triassic subduction complex, Stuart Fork terrane, Klamath Mountains, northern California: Geological Society of America Bulletin, v. 102, p. 86–101, doi: 10.1130/0016-7606(1990)102<0086:TEOACL>2.3.CO;2.

Green, T.H., 1992, Experimental phase equilibrium studies of garnet-bearing I-type volcanics and high-level intrusives from Northland, New Zealand: Transactions of the Royal Society of Edinburgh: Earth Sciences, v. 83, p. 429–438.

Gribble, R.F., Barnes, C.G., Donato, M.M., Hoover, J.D., and Kistler, R.W., 1990, Geochemistry and intrusive history of the Ashland pluton, Klamath Mountains, California and Oregon: Journal of Petrology, v. 31, p. 883–923.

Grover, T.W., 1984, Progressive metamorphism west of the Condrey Mountain dome, north-central Klamath Mountains, northern California [M.S. thesis]: Eugene, University of Oregon, 129 p.

Hacker, B.R., Donato, M.M., Barnes, C.G., McWilliams, M.O., and Ernst, W.G., 1995, Timescales of orogeny: Jurassic construction of the Klamath Mountains: Tectonics, v. 14, p. 677–703, doi: 10.1029/94TC02454.

Hammarstrom, J.M., and Zen, E., 1986, Aluminum in hornblende: An empirical igneous geobarometer: American Mineralogist, v. 71, p. 1297–1313.

Harper, G.D., Saleeby, J.B., and Heizler, M., 1994, Formation and emplacement of the Josephine ophiolite and the Nevadan orogen in the Klamath Mountains, California-Oregon: U/Pb zircon and $^{40}Ar/^{39}Ar$ geochronology: Journal of Geophysical Research, v. 99, p. 4293–4321, doi: 10.1029/93JB02061.

Harris, A.C., Allen, C.M., Bryan, S.E., Campbell, I.H., Holcombe, R.J., and Palin, J.M., 2004, ELA-ICP-MS U-Pb zircon geochronology of regional volcanism hosting the Bajo de la Alumbrera Cu-Au deposit: Implications for porphyry-related mineralization: Mineralium Deposita, v. 39, p. 46–67, doi: 10.1007/s00126-003-0381-0.

Harrison, T.M., and Watson, E.B., 1983, Kinetics of zircon dissolution and zirconium diffusion in granitic melts of variable water content: Contributions to Mineralogy and Petrology, v. 84, p. 66–72.

Harrison, T.M., and Watson, E.B., 1984, The behavior of apatite during crustal anatexis: Equilibrium and kinetic considerations: Geochimica et Cosmochimica Acta, v. 48, p. 1467–1477, doi: 10.1016/0016-7037(84)90403-4.

Holland, T., and Blundy, J., 1994, Non-ideal interactions in calcic amphiboles and their bearing on amphibole-plagioclase thermometry: Contributions to Mineralogy and Petrology, v. 116, p. 433–447, doi: 10.1007/BF00310910.

Hollister, L.S., Grissom, G.C., Peters, E.K., Stowell, H.H., and Sisson, V.B., 1987, Confirmation of the empirical correlation of Al in hornblende with pressure of solidification of calc-alkaline plutons: American Mineralogist, v. 72, p. 231–239.

Hotz, P.E., Lanphere, M.A., and Swanson, D.A., 1977, Triassic blueschist from northern California and north-central Oregon: Geology, v. 5, p. 659–663, doi: 10.1130/0091-7613(1977)5<659:TBFNCA>2.0.CO;2.

Irwin, W.P., and Wooden, J.L., 1999, Plutons and accretionary episodes of the Klamath Mountains, California and Oregon: Reston, Virginia, U.S. Geological Survey Open-File Report 99-374, 1 sheet.

Kistler, R.W., Chappell, B.W., Peck, D.L., and Bateman, P.C., 1986, Isotopic variation in the Tuolumne intrusive suite, central Sierra Nevada, California: Contributions to Mineralogy and Petrology, v. 94, p. 205–220, doi: 10.1007/BF00592937.

Kohn, M.J., and Spear, F.S., 1990, Two new geobarometers for garnet amphibolites, with applications to southeastern Vermont: American Mineralogist, v. 75, p. 89–96.

Lieberman, J.E., and Rice, J.M., 1986, Petrology of marble and peridotite in the Seiad ultramafic complex, northern California, USA: Journal of Metamorphic Geology, v. 4, p. 179–199.

Longerich, H.P., Jackson, S.E., and Gunter, D., 1996, Laser ablation inductively coupled plasma mass spectrometric transient signal data acquisition and analyte concentration calculation: Journal of Analytical Atomic Spectrometry, v. 11, p. 899–904, doi: 10.1039/ja9961100899.

McDonough, W.F., and Sun, S.-s., 1995, The composition of the Earth: Chemical Geology, v. 120, p. 223–253, doi: 10.1016/0009-2541(94)00140-4.

Miller, C.F., Hanchar, J.M., Wooden, J.L., Bennett, V.C. Harrison, T.M., Wark, D.A., and Foster, D.A., 1992, Source region of granite batholith: Evidence from lower crustal xenoliths and inherited accessory minerals: Transactions of the Royal Society of Edinburgh: Earth Sciences, v. 83, p. 49–62.

Miller, C.F., McDowell, S.M., and Mapes, R.W., 2003, Hot and cold granites? Implications of zircon saturation temperatures and preservation of inheritance: Geology, v. 31, p. 529–532, doi: 10.1130/0091-7613(2003)031<0529:HACGIO>2.0.CO;2.

Norman, M.D., Griffin, W.L., Pearsom, N.J., Garcia, M.O., and O'Reilly, S.Y., 1998, Quantitative analysis of trace element abundances in glasses and minerals: A comparison of laser ablation inductively coupled plasma mass spectrometry, solution inductively coupled plasma mass spectrometry, proton microprobe and electron microprobe data: Journal of Analytical Atomic Spectrometry, v. 13, p. 477–482, doi: 10.1039/a707972i.

Pearce, N.J.G., Perkins, W.T., Westgate, J.A., Gorton, M.P., Jackson, S.E., Neal, C.R., and Chenery, S.P., 1997, A compilation of new and published major and trace element data for NIST SRM 610 and NIST SRM 612 glass reference materials: Geostandards Newsletter, v. 21, p. 115–144.

Schmidt, M.W., 1992, Amphibole composition in tonalite as a function of pressure: An experimental calibration of the Al-in-hornblende barometer: Contributions to Mineralogy and Petrology, v. 110, p. 304–310, doi: 10.1007/BF00310745.

Sun, S.-s., and McDonough, W.F., 1989, Chemical and isotopic systematics of oceanic basalts: Implications for mantle composition and processes, *in* Saunders, A.D., and Norry, M.J., eds., Magmatism in the ocean basins: London: Geological Society Special Publication, pp. 313–345.

Watson, E.B., and Harrison, T.M., 1983, Zircon saturation revisited: Temperature and composition effects in a variety of crustal magma types: Earth and Planetary Science Letters, v. 64, p. 295–304, doi: 10.1016/0012-821X(83)90211-X.

Wright, J.E., 1982, Permo-Triassic accretionary subduction complex, southwestern Klamath Mountains, northern California: Journal of Geophysical Research, v. 87, p. 3805–3818.

Yule, J.D., 1996, Geologic and tectonic evolution of Jurassic marginal ocean basin lithosphere, Klamath Mountains, Oregon [Ph.D. dissertation]: Pasadena, California Institute of Technology, 308 p.

Zen, E.-A., and Hammarstrom, J.M., 1984, Magmatic epidote and its petrologic significance: Geology, v. 12, p. 515–518, doi: 10.1130/0091-7613(1984)12<515:MEAIPS>2.0.CO;2.

Manuscript Accepted by the Society 10 January 2006

Printed in the USA

Geological Society of America
Special Paper 410
2006

Geology of the Bear Mountain intrusive complex, Klamath Mountains, California

Jonathan C. Bushey
Arthur W. Snoke*
Department of Geology and Geophysics, Department 3006,
1000 East University Avenue, University of Wyoming, Laramie, Wyoming 82071, USA
Calvin G. Barnes
Department of Geosciences, Texas Tech University, Lubbock, Texas 79409-1053, USA
Carol D. Frost
Department of Geology and Geophysics, Department 3006,
1000 East University Avenue, University of Wyoming, Laramie, Wyoming 82071, USA

ABSTRACT

The Bear Mountain intrusive complex, Klamath Mountains, California, is a multiphase, ultramafic to silicic plutonic suite emplaced into the Rattlesnake Creek terrane during the Late Jurassic (ca. 151–147 Ma). The intrusive complex includes five plutonic units: (1) elongated, flanking bodies of ultramafic to gabbroic rocks (Blue Ridge, Clear Creek, and Cedar Creek intrusions); (2) biotite + two-pyroxene diorite/monzodiorite of the Buck Lake plutonic unit; (3) biotite-bearing hornblende ± pyroxene gabbro/diorite of the Punchbowl plutonic unit; (4) biotite + hornblende ± pyroxene (± quartz) diorite of the Doe Flat plutonic unit; and (5) minor biotite ± hornblende quartz diorite to tonalite/granodiorite of the Wilderness Falls plutonic unit. Units 2, 3, and 4 constitute the Bear Mountain pluton. The petrogenesis of these units involved emplacement of distinct batches of magma into the same magmatic center during a complex, multistage intrusion history. Inclusions of older units are found within younger units, and dikes related to younger units intrude older units. Mineral assemblages and Al-in-hornblende barometry indicate early crystallization of pyroxene at >1100 °C and late crystallization of hornblende and biotite at >700 °C between 3 and 5 kb. A dynamothermal aureole extends for ~600–1000 m from the external intrusive contact of the Bear Mountain pluton. Metabasic rocks within this dynamothermal aureole have been strongly deformed and recrystallized to well-foliated and lineated hornblende schist or fine-grained amphibolite and exhibit a "pluton-down" sense of shear (i.e., the Bear Mountain pluton). Adjacent to the intrusive complex, contact-metamorphic conditions were hornblende-hornfels facies and locally reached pyroxene-hornfels facies. At some localities in the inner aureole, anatexis occurred, forming migmatitic amphibolite gneiss. The effects of thermal metamorphism apparently extend >2 km from the intrusive contact and are evidenced by various mineral

*E-mail: snoke@uwyo.edu.

Bushey, J.C., Snoke, A.W., Barnes, C.G., and Frost, C.D., 2006, Geology of the Bear Mountain intrusive complex, Klamath Mountains, California, *in* Snoke, A.W., and Barnes, C.G., eds., Geological studies in the Klamath Mountains province, California and Oregon: A volume in honor of William P. Irwin: Geological Society of America Special Paper 410, p. 287–315, doi: 10.1130/2006.2410(14). For permission to copy, contact editing@geosociety.org.

assemblages in metaserpentinite. A concentric, margin-parallel magmatic foliation is found throughout the main body of the Bear Mountain pluton. Map-scale evidence suggests emplacement of the Bear Mountain pluton as a series of sheetlike intrusive bodies that subsequently sagged into an oval-shaped pluton due to negative buoyancy and overall isostatic downward displacement. The three-dimensional shape of the complex is poorly known, but it may be funnel-shaped with a dense root of ultramafic rocks. The ultramafic rocks of the Bear Mountain intrusive complex are primarily cumulates from H_2O- and MgO-rich basaltic parental magmas. Differentiation of these magmas gave rise to H_2O- and Al_2O_3-rich basalt/basaltic andesite from which rocks of the Punchbowl unit accumulated. Further differentiation of this magma type resulted in the development of the parental magma for the Doe Flat unit. Magmas parental to the Buck Lake unit were similar to the Punchbowl parent, except that Buck Lake magmas were H_2O poor. All of these magmas have trace-element, Sr, Nd, and oxygen isotope values characteristic of a supra-subduction zone tectonic setting. The late-stage tonalite/granodiorite magmas have geochemical signatures consistent with an origin by partial melting of metabasic crustal rocks with residual garnet; therefore, they cannot be explained by local, in situ anatexis of host rock metabasites.

Keywords: Jurassic, oceanic-arc pluton, ultramafic rocks, monzodioritic rocks, tonalite/granodiorite, metaserpentinite, migmatitic amphibolite, Rattlesnake Creek terrane, Siskiyou Mountains

INTRODUCTION

The Klamath Mountains province of northwestern California and southwestern Oregon is composed of imbricate, regional thrust sheets of oceanic rocks (Fig. 1). These thrust sheets include ancient island-arc sequences, ophiolitic assemblages, and accretionary complexes. The age of emplacement of the thrust sheets is roughly oldest in the east and youngest in the west. Thus, the Klamath Mountains province is a classic example of continental growth through the progressive lateral accretion of oceanic rocks along an active continental margin. The thrust sheets are intruded by a diverse suite of plutonic rocks that range in age from early Paleozoic to Early Cretaceous and in composition from ultrabasic to silicic. The relationship between regional contraction, extension, and plutonism is still incompletely understood, despite numerous studies (e.g., Saleeby et al., 1982; Harper and Wright, 1984; Wright and Fahan, 1988; Hacker et al., 1995; Irwin and Wooden, 1999). Furthermore, there are substantial compositional and isotopic variations of the plutons (Barnes et al., 1992). For example, some Middle Jurassic plutons show little evidence in contamination by crustal rocks, whereas Early Cretaceous plutons exhibit abundant evidence for widespread contamination during magma genesis (Barnes et al., 2003; Barnes et al., this volume, Chapter 17).

The purpose of this study was a detailed, field-based analysis of the emplacement and petrogenetic history of an intrusive complex comprised of mantle-derived mafic magmas that intruded accreted terranes of oceanic rocks immediately after regional contraction. The plutonic complex studied in detail was the Bear Mountain intrusive complex, originally described by Snoke et al. (1981). New U-Pb zircon radiometric data (Chamberlain et al., this volume) and regional geologic relationships (Snoke, 1977; Saleeby et al., 1982; Harper et al., 1994) indicate that the Bear Mountain intrusive complex was emplaced immediately after regional thrust faulting along the Orleans fault system associated with the Late Jurassic Nevadan orogeny. This study bears on the important problem of the petrogenesis of oceanic-arc plutonic complexes and their relationship to regional contraction associated with ongoing subduction-related processes at an evolving continental margin.

GEOLOGIC SETTING

The Bear Mountain intrusive complex intruded the Rattlesnake Creek terrane in Late Jurassic time (Saleeby et al., 1982; Chamberlain et al., this volume). The Rattlesnake Creek terrane includes a basal unit of serpentinite-matrix mélange (Wright and Wyld, 1994), which is inferred to be unconformably overlain by a cover sequence consisting of mafic metavolcanic rocks (pillow lavas and volcaniclastic rocks) and lesser amounts of intercalated siliceous meta-argillite and metachert. The cover sequence of the Rattlesnake Creek terrane in the Preston Peak area (Bear Basin Road sequence of Snoke, 1977; Fig. 2) is

Figure 1. Generalized geologic map of the Klamath Mountains province. BMic = Bear Mountain intrusive complex; BPic = Bear Peak intrusive complex (see McFadden et al., this volume); O. ft. = Orleans fault. The Preston Peak fault of Snoke (1977) is the local name for the regional Orleans (thrust) fault. Modified from Irwin (1994).

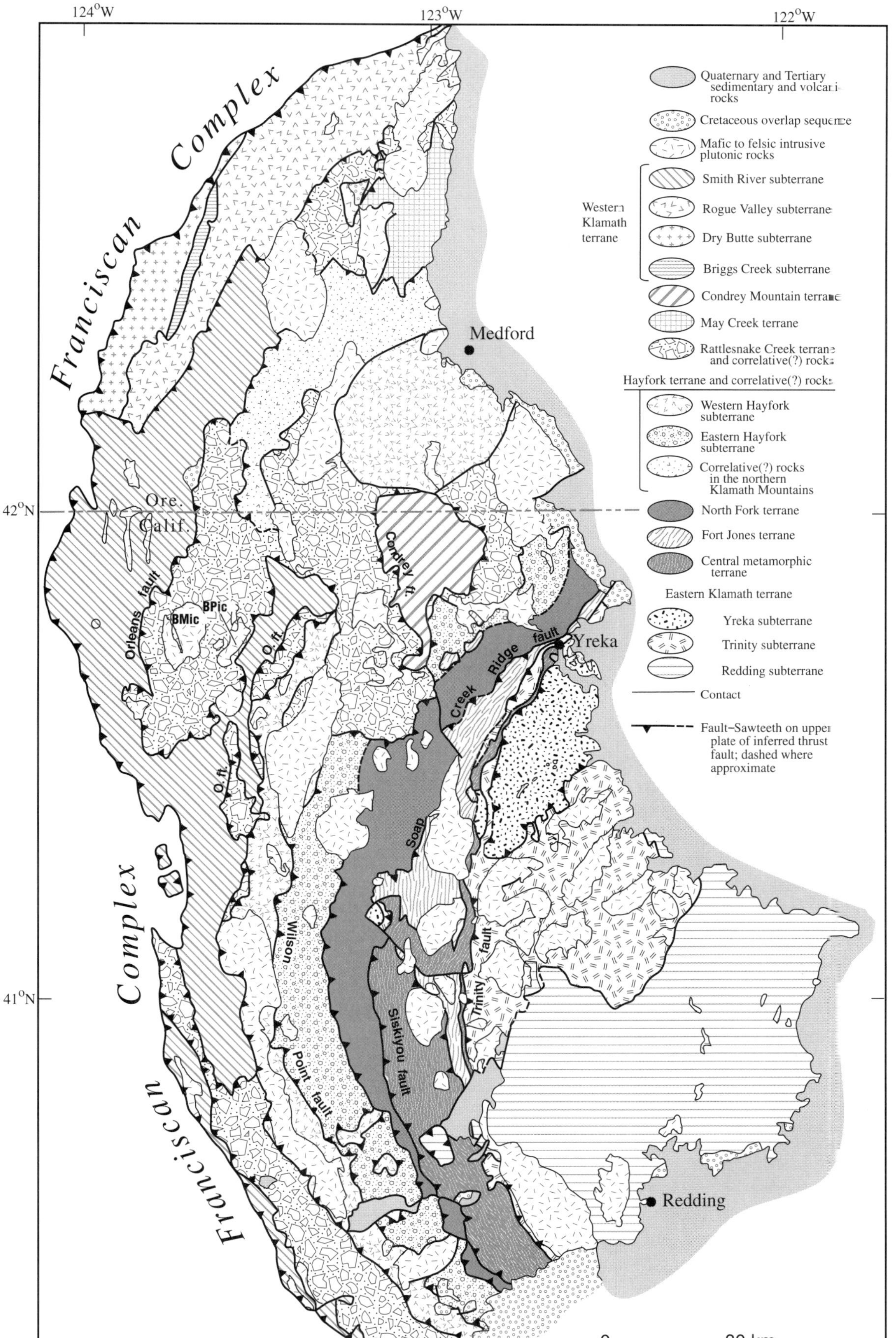

124°W
123°W
122°W
42°N
41°N
Franciscan Complex
Franciscan Complex
Ore.
Calif.
Medford
Yreka
Redding
Orleans fault
BMic
BPic
O. ft.
O. ft.
Condrey ft.
Creek Ridge fault
Soap
Wilson Point fault
Siskiyou fault
Trinity fault
0
30 km
Quaternary and Tertiary sedimentary and volcanic rocks
Cretaceous overlap sequence
Mafic to felsic intrusive plutonic rocks
Western Klamath terrane
Smith River subterrane
Rogue Valley subterrane
Dry Butte subterrane
Briggs Creek subterrane
Condrey Mountain terrane
May Creek terrane
Rattlesnake Creek terrane and correlative(?) rocks
Hayfork terrane and correlative(?) rocks
Western Hayfork subterrane
Eastern Hayfork subterrane
Correlative(?) rocks in the northern Klamath Mountains
North Fork terrane
Fort Jones terrane
Central metamorphic terrane
Eastern Klamath terrane
Yreka subterrane
Trinity subterrane
Redding subterrane
Contact
Fault–Sawteeth on upper plate of inferred thrust fault; dashed where approximate

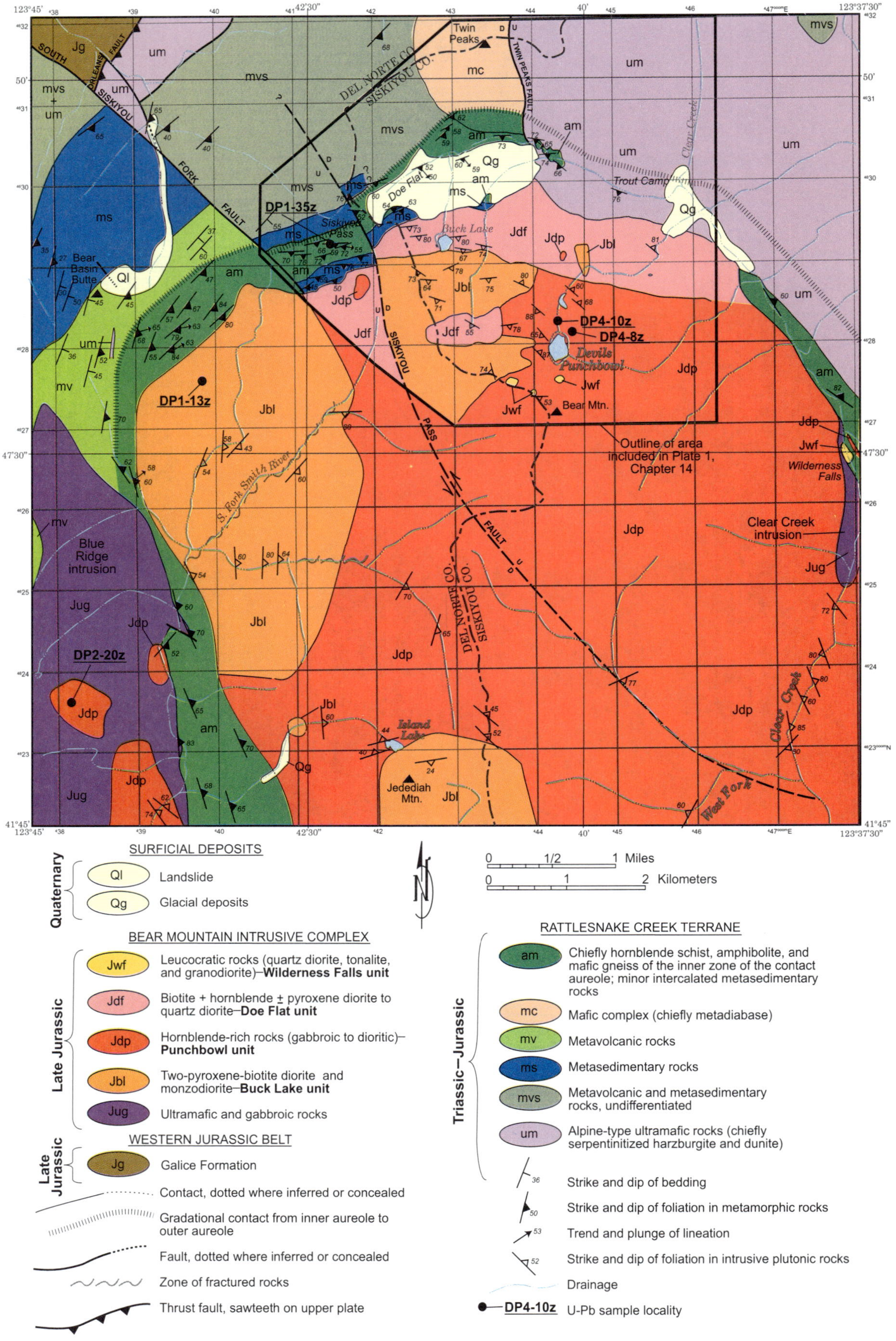
123°45'
123°37'30"
41°45'
42'30"
40'
50'
47'30"
SOUTH FORK FAULT
ORLEANS FAULT
SISKIYOU
TWIN PEAKS FAULT
DEL NORTE CO.
SISKIYOU CO.
SISKIYOU PASS FAULT
Twin Peaks
Doe Flat
Siskiyou Pass
Buck Lake
Trout Camp
Clear Creek
Bear Basin Butte
Devils Punchbowl
Bear Mtn.
S. Fork Smith River
Island Lake
Jedediah Mtn.
West Fork
Wilderness Falls
Clear Creek intrusion
Blue Ridge intrusion
DP1-35z
DP1-13z
DP4-10z
DP4-8z
DP2-20z
Outline of area included in Plate 1, Chapter 14
0 1/2 1 Miles
0 1 2 Kilometers
Quaternary
SURFICIAL DEPOSITS
Ql Landslide
Qg Glacial deposits
Late Jurassic
BEAR MOUNTAIN INTRUSIVE COMPLEX
Jwf Leucocratic rocks (quartz diorite, tonalite, and granodiorite)–Wilderness Falls unit
Jdf Biotite + hornblende ± pyroxene diorite to quartz diorite–Doe Flat unit
Jdp Hornblende-rich rocks (gabbroic to dioritic)–Punchbowl unit
Jbl Two-pyroxene-biotite diorite and monzodiorite–Buck Lake unit
Jug Ultramafic and gabbroic rocks
Late Jurassic
WESTERN JURASSIC BELT
Jg Galice Formation
Contact, dotted where inferred or concealed
Gradational contact from inner aureole to outer aureole
Fault, dotted where inferred or concealed
Zone of fractured rocks
Thrust fault, sawteeth on upper plate
Triassic–Jurassic
RATTLESNAKE CREEK TERRANE
am Chiefly hornblende schist, amphibolite, and mafic gneiss of the inner zone of the contact aureole; minor intercalated metasedimentary rocks
mc Mafic complex (chiefly metadiabase)
mv Metavolcanic rocks
ms Metasedimentary rocks
mvs Metavolcanic and metasedimentary rocks, undifferentiated
um Alpine-type ultramafic rocks (chiefly serpentinitized harzburgite and dunite)
Strike and dip of bedding
Strike and dip of foliation in metamorphic rocks
Trend and plunge of lineation
Strike and dip of foliation in intrusive plutonic rocks
Drainage
DP4-10z U-Pb sample locality

considered to be Late Triassic–Early Jurassic in age. This stratigraphic age is supported by Early Jurassic radiolarians collected in a similar sequence of rocks exposed ~20 mi (~32 km) to the north-northeast (sample no. I-91-77, locality 35 of Fig. 1 in Irwin et al., 1978; Saleeby et al., 1982), although the exact stratigraphic and structural relationships between these rocks and the rocks of the Rattlesnake Creek terrane exposed in the map area are uncertain. The rocks of the Rattlesnake Creek terrane are intruded by the mafic complex of the Preston Peak ophiolite (Snoke, 1977), which was dated at ca. 164 Ma by U-Pb (zircon; originally reported as ca. 160 Ma by Saleeby et al., 1982, but revised in Saleeby and Harper, 1993).

Two plutons (Blue Ridge intrusion and Bear Mountain pluton) as well as a number of smaller bodies that range in composition from ultramafic to tonalitic comprise the Bear Mountain intrusive complex (Fig. 2). The age of the Bear Mountain intrusive complex was originally estimated on the basis of two U-Pb (zircon) ages of the Bear Mountain pluton to be ca. 153–149 Ma (Saleeby et al., 1982). However, Chamberlain et al. (this volume) show that the Bear Mountain pluton, a related tonalitic body, and a late mafic dike were emplaced between ca. 151 and 145 Ma. These new geochronologic data indicate that the Bear Mountain intrusive complex intruded the Rattlesnake Creek terrane after Nevadan-age thrusting (Orleans [thrust] fault or Preston Peak fault of Snoke, 1977). Basic field relationships originally noted in Snoke (1977) and further substantiated in this study also support this conclusion. Snoke (1977) mapped the northwest–southeast-striking South Siskiyou Fork fault as a cross fault locally truncating his Preston Peak fault (i.e., Orleans thrust fault). The footwall of the Preston Peak fault is the Upper Jurassic Galice Formation of the western Klamath terrane. Snoke (1977, his Fig. 3) did not extend the South Siskiyou Fork fault beyond the Preston Peak fault; therefore, this fault is best interpreted as a tear fault and thus part of the Preston Peak (Orleans) thrust fault system. When the South Siskiyou Fork fault is traced to the southeast toward the Bear Mountain intrusive complex, it is significant that the external contact of the intrusive complex is *not* offset by this major fault (Fig. 2). Thus, these field relationships strongly support the relative chronology that the Bear Mountain intrusive complex was emplaced after regional contraction (i.e., development of the Orleans thrust fault system).

Given the above field relationships and the geochronologic data, the Bear Mountain intrusive complex must have intruded the Upper Galice Formation, which is a widespread unit in the footwall of the Orleans thrust fault (Jachens et al., 1986). Several studies have indicated that rocks of the Galice Formation contain detrital Precambrian components (Miller and Saleeby, 1995; Miller et al., 2003; Frost et al., this volume). Consequently, a focus of this study was to determine whether these ancient components are also contained in the magmas of the Bear Mountain intrusive complex (see the section on isotopes below).

Contact metamorphism of the wall rocks (Rattlesnake Creek terrane) locally reached pyroxene-hornfels facies, although hornblende-hornfels conditions were more typical. Snoke (1972) originally documented and discussed the contact aureole surrounding the Bear Mountain intrusive complex. In regard to the aureole adjacent to the Bear Mountain pluton, he delineated two zones (inner and outer) based on megascopic textural changes. Inner zone rocks are typically characterized by a penetrative dynamothermal foliation, and in metabasaltic rocks, hornblende is visible in hand sample. Locally, particularly in the septum between the Bear Mountain and Blue Ridge plutons, metabasic rocks in the dynamothermal aureole underwent anatexis to form migmatitic amphibolite gneiss (Snoke, 1972; Obenshain, 1981; Snoke et al., 1981; Snoke and Barnes, 2002).

Figure 2. Geologic map of the Bear Mountain intrusive complex. Modified from Snoke et al. (1981); Bushey (2003); A.W. Snoke (unpublished data).

ROCK UNITS

Rattlesnake Creek Terrane

Ultramafic Rocks. The ultramafic rocks are dominantly fine-grained, black to green or blue-green metaserpentinite. The metaserpentinites are massive and generally exhibit no foliation outside the dynamothermal inner aureole, although a high-temperature mantle plastic flow fabric is locally preserved in some exposures of partially serpentinized peridotite of the Preston Peak ophiolite (Snoke, 1977). Tectonic blocks of metabasalt as well as other rock types occur in the metaserpentinite. Metamorphosed mafic dikes, interpreted as related to the mafic complex of the Preston Peak ophiolite (Snoke, 1977), locally intrude the metaserpentinite. Other unmetamorphosed mafic dikes that intrude the metaserpentinite are probably related to the Bear Mountain intrusive complex.

Within the inner zone of the contact aureole of the Bear Mountain pluton, the metaserpentinites commonly show a weak to moderate foliation. Medium- to coarse-grained, sheaflike aggregates of acicular tremolite, and possibly anthophyllite, are locally apparent on weathered surfaces; but in many cases, these porphyroblasts are extensively replaced by serpentine-group minerals. At some localities, the tremolite-anthophyllite aggregates define a mineral lineation in the foliation plane.

The bulk of the metaserpentinite consists of serpentine-group minerals (commonly comprising >50 modal %). However, virtually all the samples collected contain some relict olivine grains. Tremolite and talc are commonly also part of the metamorphic mineral assemblage as well as Fe-Cr spinel, chlorite, opaque oxides, sulfides, and carbonate minerals.

Within the inner zone of the contact aureole, the metaserpentinites contain the subassemblage olivine ± tremolite ± anthophyllite ± talc ± orthopyroxene (Table 1). Nickel- or magnesium-rich chlorite occurs in some samples. These assemblages are indicative of pyroxene-hornfels to hornblende-

TABLE 1. SUMMARY OF CONTACT-METAMORPHIC MINERAL ASSEMBLAGES IN METASERPENTINITE ADJACENT TO THE BEAR MOUNTAIN PLUTON

Mineral subassemblages in the inner aureole (all rocks also contain serpentine-group minerals + opaque iron oxides ± sulfide minerals ± carbonate minerals):

olivine + orthopyroxene + magnesium-rich chlorite + nickel-rich chlorite
olivine + orthopyroxene + tremolite + chlorite
olivine + tremolite + anthophyllite + chlorite
olivine + tremolite + chlorite + talc

Mineral subassemblages in the outer aureole (all rocks also contain serpentine-group minerals + opaque iron oxides ± sulfide minerals ± carbonate minerals):

olivine + talc + tremolite + chlorite + Fe-Cr spinel
olivine + talc + chlorite + Fe-Cr spinel

Primary mineral assemblage beyond the outer aureole in serpentinized peridotite (all rocks also contain serpentine-group minerals + opaque iron oxides ± sulfide minerazls ± carbonate minerals):

olivine + enstatite + diopsidic augite + Fe-Cr spinel

hornfels facies contact-metamorphic conditions (e.g., Trommsdorff and Evans, 1972, 1974; Springer, 1974; Arai, 1975; Frost, 1975). However, in many of the inner zone metaserpentinites, the contact-metamorphic mineral assemblage has undergone re-serpentinization, and thus only scattered relict mineral phases and textural traces identify these rocks as contact-metamorphosed metaserpentinites. In Table 1, the contact-metamorphic mineral assemblages in metaserpentinite adjacent to the northern margin of the Bear Mountain intrusive complex are summarized.

Metabasaltic Tectonic Inclusion and Mafic Dikes in Serpentinite-Matrix Mélange. Within the inner zone of the contact aureole, a metabasaltic tectonic block (Plate 1, on the CD-ROM accompanying this volume and in the GSA Data Repository[1]) and scattered mafic dikes are metamorphosed to well-foliated, fine-grained amphibolite or hornblende schist. Margins of the mafic dikes as well as the metabasaltic tectonic inclusion are altered to tough, dense, pale gray-green metarodingite along contacts with metaserpentinite.

The metabasaltic tectonic inclusion in metaserpentinite is a fine-grained hornblende schist consisting of hornblende + plagioclase + Fe-Ti oxides + titanite ± quartz ± epidote ± clinopyroxene. Compositional layering and strongly oriented acicular hornblende grains define these mafic rocks as L-S tectonites. Metarodingite margins of the metabasaltic tectonic block consist of fine-grained diopside, epidote, titanite, and minor plagioclase.

Metabasaltic Rocks of the Cover Sequence. The metabasaltic rocks of the inferred "cover sequence" of the Rattlesnake Creek terrane are dominantly fine-grained and locally pillowed. These metabasaltic rocks are commonly grayish-green, reflecting the regional greenschist-facies metamorphism experienced by this part of the Rattlesnake Creek terrane (Coleman et al., 1988). Relict igneous texture, manifested by tabular plagioclase feldspar, is apparent in many samples located beyond the dynamothermal effects of the emplacement of the Bear Mountain pluton. Within the inner zone of the contact aureole, the metabasaltic rocks have been penetratively deformed and metamorphosed to a fine- to fine–medium-grained amphibolite or hornblende schist. These rocks have a prominent foliation due to hornblende, plagioclase, and sometimes biotite-rich folia. Plagioclase-rich leucosomes form 2–25 vol% of the rock and when sufficiently abundant give these mafic rocks a migmatitic character (Obenshain, 1981; Snoke and Barnes, 2002). A well-developed, steeply plunging hornblende mineral lineation is commonly visible on foliation planes. The foliation is parallel to the margin of the Bear Mountain pluton (Fig. 2) and is less well developed with distance from the intrusive contact. Beyond the inner zone, the penetrative foliation virtually disappears, and the rocks are noticeably more massive and commonly hornfelsic (Snoke, 1972).

The relict igneous textures of the metabasaltic rocks are especially evident under the microscope. Most plagioclase occurs as fine- to medium-grained, tabular crystals and/or as patches of fine-grained, recrystallized, anhedral grains in the shape of relict tabular to blocky phenocrysts. Fine-grained hornblende, commonly rimmed with actinolite, also exhibits the original, blocky crystal shape (after igneous pyroxene) characteristic of the original igneous protolith. Very rarely, primary pyroxene remains in the cores of these pseudomorphs. Regional metamorphism resulted in growth of fine-grained prismatic actinolite and anhedral to subhedral epidote and the replacement of original calcic plagioclase by albite.

Within the inner zone of the contact aureole, plagioclase and hornblende are completely or almost completely recrystallized, with only a few relict igneous habits remaining. The rock is predominantly fine grained except for these relict phenocrysts. Metamorphic layering is conspicuous in thin section, with plagioclase-, hornblende-, and minor biotite-rich layers defining the foliation along with the overall strong acicular to prismatic grain-shape fabric of hornblende. Quartz is commonly present in leucocratic segregation layers or as monomineralic lenses/layers.

Metasedimentary Rocks of the Cover Sequence. This rock unit is chiefly composed of fine-grained, tan to dark gray metasiltstone, siliceous meta-argillite, and metachert. Locally, the unit includes interlayered metavolcaniclastic rocks, meta-

[1]GSA Data Repository item 2006199, Plate 1, Geologic map of northern Bear Mountain intrusive complex, Klamath Mountains, is available online at www.geosociety.org/pubs/ft2006.htm, or on request from editing@geosociety.org or Documents Secretary, GSA, P.O. Box 9140, Boulder, CO 80301, USA.

basaltic rocks, and lenses of serpentinite. The unit exhibits a moderate to strong foliation.

Within the inner zone of the contact aureole of the Bear Mountain pluton, the unit is composed of brown to purplish metachert, fine-grained quartzose schist and brownish to purplish dark-gray biotite + hornblende ± quartz schist with scarce white calc-silicate lenses (deformed, original concretions[?]).

Biotite and granoblastic quartz + plagioclase are typical in siliceous meta-argillite beyond the inner zone of the contact aureole, indicating the broad thermal effects of the Bear Mountain intrusive complex. Foliation, if present, is defined by fine-grained, oriented biotite. Some outer-zone metasedimentary rocks are spotted slates or fine-grained schists. The original porphyroblasts in these rocks are typically pseudomorphosed to an aggregate of phyllosilicate minerals, Fe-Ti oxides, and quartz, although cordierite was identified in a few samples (see Snoke, 1972).

Mafic Complex of the Preston Peak Ophiolite (after Snoke, 1977). The mafic complex is dominantly fine- to fine–medium-grained gray to gray-green metadiabase and metabasalt (Snoke, 1977). It commonly has a salt-and-pepper texture defined by tabular, altered plagioclase and amphibole grains. A crosscutting network of white veinlets consisting of clinozoisite, calcite, and prehnite is characteristic.

Within the inner zone of the contact aureole of the Bear Mountain pluton, rocks of the mafic complex are light to dark gray, fine- to fine–medium-grained amphibolite similar to metabasaltic rocks of the cover sequence of the Rattlesnake Creek terrane. They also have foliation or cleavage subparallel to the margin of the Bear Mountain pluton.

In thin section, the original igneous texture is well preserved. Most of the plagioclase occurs as fine- to medium-grained, tabular crystals that are locally pseudomorphosed to fine-grained aggregates of anhedral grains. Fine- to medium-grained, green to olive-green hornblende is commonly rimmed with prismatic to acicular actinolite. Actinolite, chlorite, and epidote are the products of regional greenschist-facies metamorphism that affected these rocks prior to the emplacement of the Bear Mountain intrusive complex.

The mafic complex only extends partially into the inner zone of the contact aureole of the Bear Mountain intrusive complex. Thus, hornblende and plagioclase of the mafic complex in the inner zone of the aureole did not completely recrystallize during dynamothermal metamorphism. There are varying amounts of apparent original, relict igneous plagioclase and hornblende crystals in rocks of the mafic complex within the inner aureole; foliation in these rocks is weak to moderate.

Bear Mountain Intrusive Complex

Introduction. The oldest intrusive units of the Bear Mountain intrusive complex are the Blue Ridge, Clear Creek, and Cedar Creek ultramafic-gabbroic intrusions, and related ultramafic to gabbroic dikes. This highly mafic magmatism was followed by intrusion of the Bear Mountain pluton (Figs. 2 and 3, Plate 1) that chiefly consists of the Buck Lake, Punchbowl, and Doe Flat plutonic units. The youngest intrusive rocks in the Bear Mountain intrusive complex are numerous, but volumetrically minor, felsic dikes, sheets, and small irregular masses that range from quartz diorite to granite. They are typified by the

Figure 3. Aerial view of the northern end of the Bear Mountain pluton. Bear Mountain (6411 feet [1955 m]) is the prominent peak above the Devil's Punchbowl cirque at the north end of a long ridge (left-center of photograph).

Wilderness Falls body (Snoke, 1972; Fig. 2). The youngest intrusive rocks are a suite of dikes that range in composition from mafic to intermediate and are both aphyric and phyric (augite, hornblende, and/or plagioclase).

The ultramafic-gabbroic bodies are interpreted as the oldest plutons of the Bear Mountain intrusive complex, based on the following field relationships: (1) clinopyroxenitic dikes derived from the Blue Ridge intrusion are deformed and contact-metamorphosed in the aureole of the Bear Mountain pluton (Snoke et al., 1981, their Fig. 3); (2) xenoliths of ultramafic to gabbroic rocks occur in the Punchbowl unit (Fig. 4A); and (3) small stocks of hornblende gabbro/diorite to tonalite intrude the Blue Ridge intrusion (Fig. 2), one of which yielded the oldest U-Pb radiometric date from the Bear Mountain intrusive complex (ca. 151 Ma: Chamberlain et al., this volume).

Within the Bear Mountain pluton, the sequence of emplacement was determined by crosscutting relationships (e.g., inclusions of older intrusive units in younger ones and dikes of younger units in older ones). However, exposures that provide information about the relative chronology of emplacement of the plutonic units are scarce. Evidence that the Doe Flat unit is the youngest intrusive stage is clearest. Xenolithic blocks of Buck Lake and Punchbowl affinity are enclosed in Doe Flat rocks north of Devils Punchbowl and west of the Siskiyou Pass fault (Fig. 2, Plate 1). Also, a band of distinctive low color index (~15–25), foliated biotite + hornblende ± quartz diorite is present along the southern margin of the Doe Flat unit. This important field characteristic is useful in mapping the boundary with adjacent units to the south, where the Doe Flat unit truncates the Buck Lake–Punchbowl contact at a high angle (Plate 1).

The Buck Lake unit is cut by dikes of the Punchbowl unit. Moreover, Snoke et al. (1981) documented locally extensive hydration of the H_2O-poor primary igneous assemblage in the unit (see below). This hydration is particularly apparent near contacts with adjacent plutonic units, which suggests a source of H_2O from the younger, more hydrous magmas.

The Wilderness Falls unit intrudes the Punchbowl unit south and southwest of Devils Punchbowl (Fig. 4B) and also intrudes the Doe Flat unit (Plate 1). It is therefore the youngest unit in the Bear Mountain pluton.

Perhaps the most striking aspect of the Bear Mountain intrusive complex is its outcrop-scale heterogeneity, defined by variations in grain size, mineral assemblage and modes, and texture. In areas of good exposure, it is apparent that much of this variation is due to piecemeal (incremental) emplacement of magmas. In many cases, rocks with homogeneous mineral assemblage, grain size, and texture occur as steeply dipping sheet-like bodies, which suggests emplacement as numerous dikes (Petford et al., 1993).

The most prevalent and recognizable structure in the Bear Mountain pluton is magmatic to hypersolidus foliation. On weathered surfaces, the foliation appears as parallel alignment and elongation of mafic mineral aggregates (pyroxene, biotite, and/or hornblende, depending on the unit) within a field of white plagioclase. In thin section, subhedral tabular plagioclase is aligned parallel to subparallel to the mafic aggregates. In rocks where quartz is abundant, as in some Doe Flat and Wilderness Falls rocks, this mineral occurs in ribbon-shaped aggregates, indicating the importance of late solid-state deformation during part of the development of the fabric in the Bear Mountain pluton.

Lineations are locally present in the Doe Flat and Wilderness Falls units. These rocks are generally L>S tectonites. The lineation, like the foliation, is seen in outcrop due to the alignment of biotite and/or hornblende aggregates within a field of white plagioclase.

Blue Ridge Ultramafic to Gabbroic Intrusion. The Blue Ridge intrusion includes a broad compositional range: dunite, wehrlite, olivine clinopyroxenite, hornblende-olivine clinopyroxenite, magnetite-hornblende clinopyroxenite, and various gabbroic rocks (Snoke et al., 1981). Dunite is scarce and occurs as streaks, lenses, and irregular masses in wehrlite. Igneous layering is scarce in the ultramafic rocks, and evidence of mechanical disruption of original mineralogical layering is apparent in some outcrops (e.g., dunite-wehrlite breccia, Fig. 4D). The gabbroic rocks are massive to foliated, and the nature of the foliation varies from magmatic mineralogical layering (Fig. 4C) to high-temperature, solid-state penetrative foliation (see petrographic description below). Hornblende-calcic plagioclase pegmatite (Fig. 4E) is widely distributed in the clinopyroxene-rich ultramafic rocks, especially in the Blue Ridge intrusion.

The dunite is nearly pure forsteritic olivine, but diopsidic augite and chromian spinel are common accessory minerals. Secondary serpentine-group minerals and magnetite are ubiquitous. The dunite is allotriomorphic-granular, with grain sizes ranging from 1 to 3 mm. The clinopyroxene-rich ultramafic rocks are allotriomorphic- to hypidiomorphic-granular rocks and consist essentially of diopsidic augite, olivine, and hornblende in varying proportions (Fig. 5A). The gabbroic rocks are unfoliated to well foliated and exhibit a broad range in color index and modal proportions (Snoke et al., 1981, their Table 3). Mineralogical layering, interpreted as original cumulate layering, is present in some exposures of gabbroic rocks (Fig. 4C). In some gabbroic rocks, the foliation may be largely related to magmatic flowage, whereas in other gabbroic rocks, solid-state deformational features are conspicuous (e.g., widespread dynamic recrystallization of plagioclase grains and crystal-plastic strain effects in olivine and pyroxene grains).

Buck Lake Plutonic Unit. The Buck Lake unit is predominantly medium- to coarse-grained, hornblende-bearing to hornblende + biotite + two-pyroxene ± quartz diorite/monzodiorite. The color index ranges from ~15 to 55 but averages ~30 (Snoke et al., 1981, their Table 4). Scattered enclaves of fine-grained, plagioclase + pyroxene-phyric porphyry occur in this unit.

The diorite and monzodiorite of the Buck Lake unit range from fine to coarse grained, but are commonly medium to coarse grained. Subhedral tabular plagioclase is the dominant mineral phase, and it commonly defines a prominent magmatic foliation

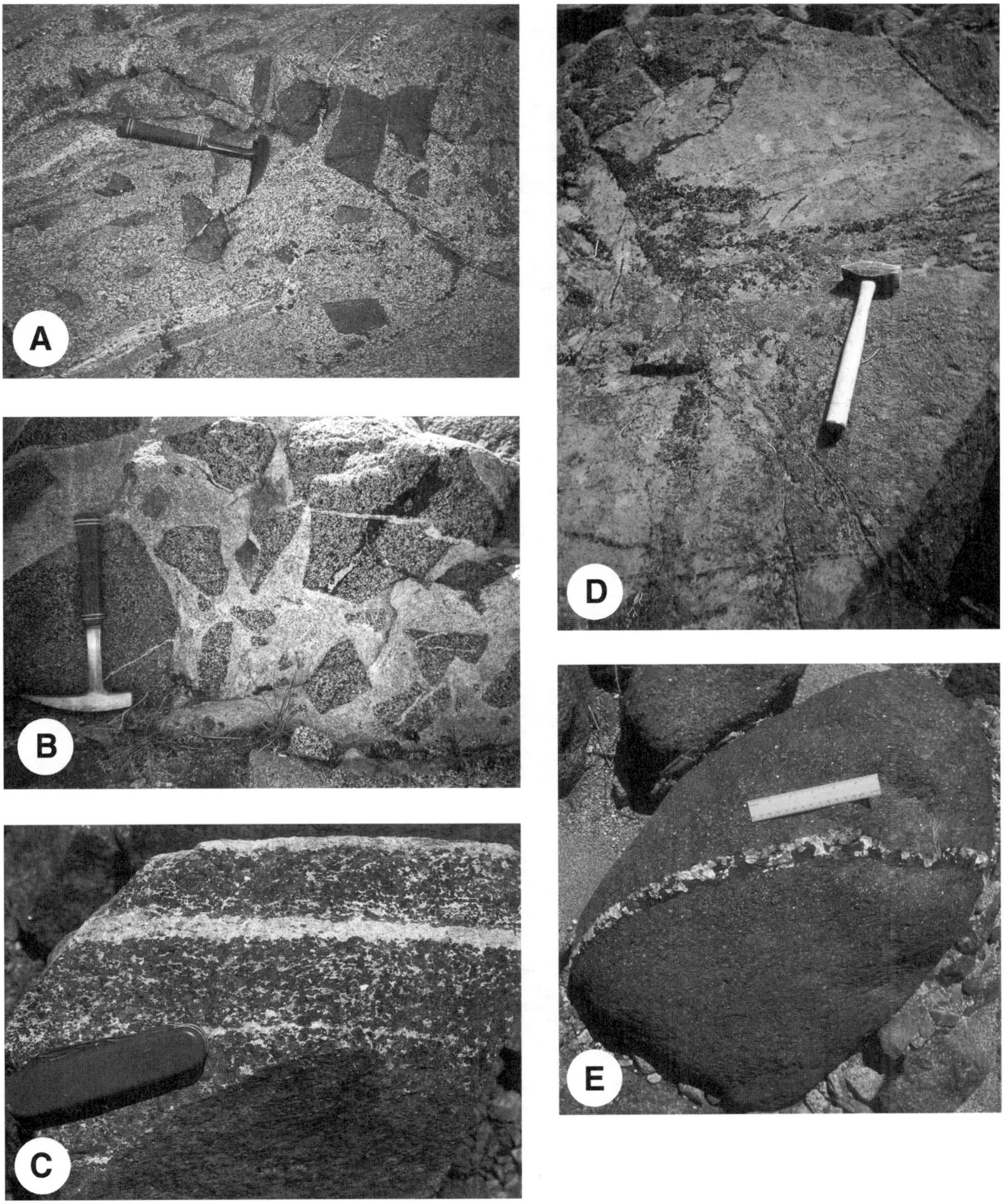

Figure 4. Photographs of various field characteristics and relationships in the Bear Mountain intrusive complex. (A) Angular clinopyroxenitic xenoliths in pyroxene-hornblende gabbro of the Punchbowl unit, Clear Creek area. (B) Angular xenoliths of pyroxene-hornblende gabbro of the Punchbowl unit in leucocratic biotite granodiorite (Wilderness Falls unit), on ridge southeast of Devils Punchbowl. (C) Mineralogical layering (original cumulate layering) in olivine gabbro with anorthositic layers, Blue Ridge intrusion. (D) Dunite-wehrlite breccia near Elk Lake, Blue Ridge intrusion. (E) Hornblende-calcic plagioclase pegmatite dikelet in wehrlite, South Fork of the Smith River.

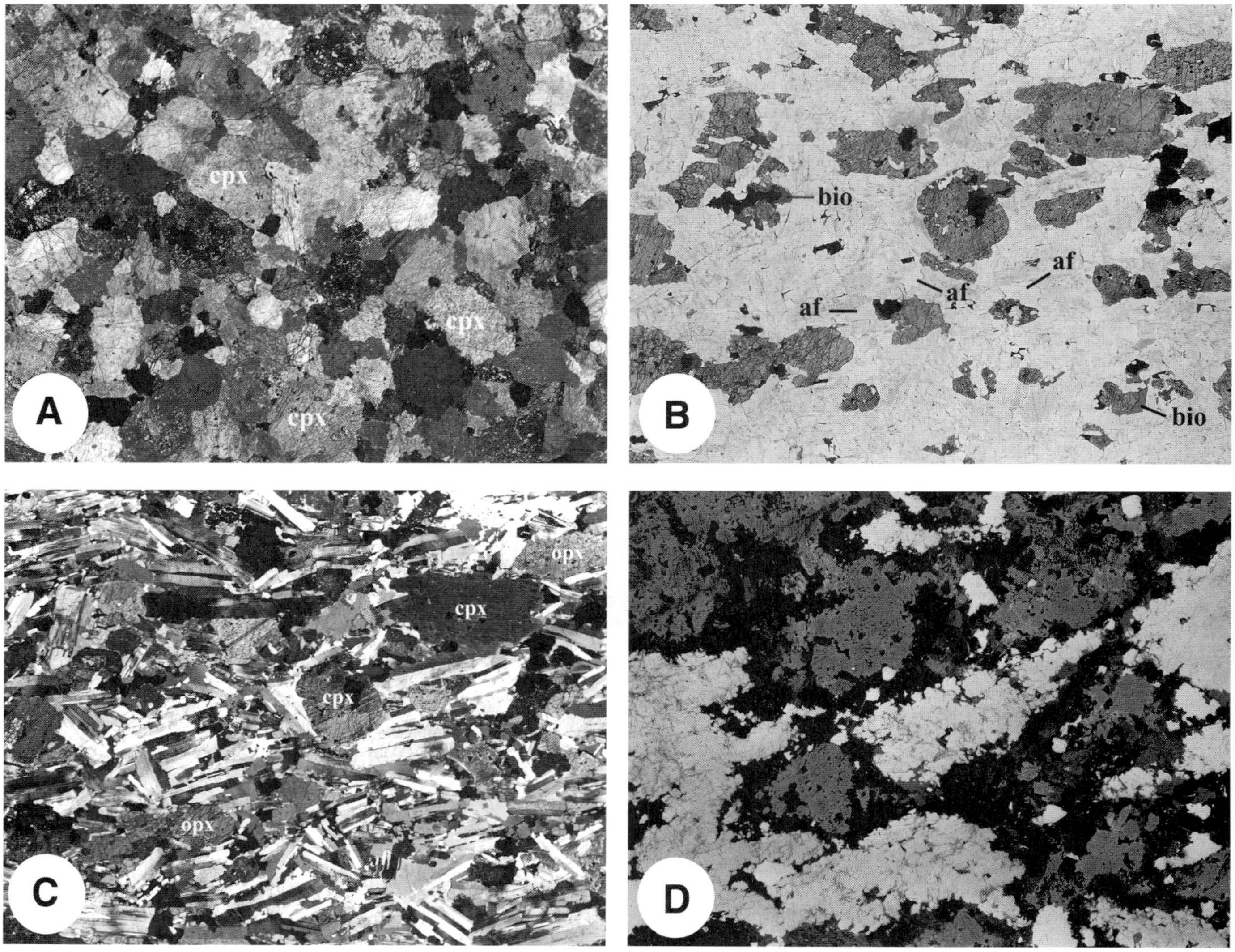

Figure 5. Photomicrographs of various rocks from the Bear Mountain intrusive complex. (A) Hornblende-olivine clinopyroxenite from the Blue Ridge intrusion. Diopsidic augite grains (cpx) are partially replaced by late magmatic brown hornblende. Cross-polarized light. Frame width is ~25 mm. (B) Coarse-grained Buck Lake biotite–two-pyroxene monzodiorite; af = alkali feldspar; bio = biotite. Plain light. Frame width is ~25 mm; (C) Same view as B but in cross-polarized light. Note the strong plagioclase fabric, whereas biotite and alkali feldspar form scattered, post-fabric irregular patches in this rock (labeled in B); cpx = clinopyroxene; opx = orthopyroxene. (D) Coarse-grained hornblende-pyroxene gabbro of the Punchbowl unit, illustrating the replacement of original pyroxene by hornblende and the contemporaneous reaction with adjacent calcic plagioclase. Note the restriction of the Fe-Ti oxides to inclusions in mafic grains and symplectite of pyroxene–Fe-Ti oxides in the upper right corner of the photograph. Plain light. Frame width is ~25 mm.

(Fig. 5B and C). Aggregates of pyroxenes and biotite are elongated subparallel to the plagioclase foliation.

The dominant mafic minerals of the unit are pyroxenes, with augite more abundant than orthopyroxene (Fig. 5B and C). Exsolution lamellae parallel to (100) occur in orthopyroxene and augite. Some orthopyroxene grains exhibit a second set of lamellae parallel to (001). These lamellae are coarse and widely spaced. In crystals twinned on (100), a herringbone pattern results (Fig. 6). These textural relationships indicate the former presence of pigeonite that subsequently inverted to orthopyroxene (Hess, 1941). The former presence of pigeonite in plutonic rocks from the Klamath Mountains is unusual, and the only other known examples are from the Ironside Mountain batholith (Barnes et al., this volume, Chapter 10). Pyroxenes occur as individual grains or in multigrain aggregates. In both textural occurrences, biotite commonly partially surrounds the pyroxene(s), but biotite also occurs as individual grains distributed throughout the rock. Varying amounts of green, olive, and brown hornblende (Z tint) are found in almost all samples. The hornblende typically occurs as thin to thick reaction rims partially

Figure 6. Inverted pigeonite in a coarse-grained biotite–two-pyroxene monzodiorite from the Buck Lake unit. The broad lamellae are parallel to (001) and consist of augite exsolved prior to inversion. The herringbone pattern is a reflection of (100) twinning in the original pigeonite grain. Thinner lamellae parallel to (100) are not detectable in the photograph. Cross-polarized light.

surrounding pyroxene. Quartz and alkali feldspar are in the Buck Lake unit. Quartz is present in nearly every sample, and its origin is at least in part related to the late-magmatic replacement of orthopyroxene by biotite and quartz (Snoke, 1972). Alkali feldspar forms large poikilitic grains surrounding and preferentially replacing plagioclase. Myrmekite is everywhere associated with the patchy alkali feldspar, typically along boundaries with plagioclase grains.

The plagioclase + pyroxene-phyric porphyry is basically similar to other Buck Lake unit rocks. However, the porphyry typically has distinct phenocrysts of plagioclase and pyroxene, has a higher color index, and contains more Fe-Ti oxides.

Punchbowl Plutonic Unit. The Punchbowl unit is predominantly medium-grained, weakly to well foliated, biotite-bearing pyroxene + hornblende gabbro and diorite. Hornblende is the dominant mafic mineral, with distinctly subordinate biotite and pyroxene. The rocks vary from hornblende gabbro/diorite to biotite + pyroxene + hornblende gabbro/diorite and include pyroxene + hornblende gabbro and biotite + hornblende diorite. Also included in the Punchbowl unit is a coarse-grained, biotite-bearing to biotite + pyroxene + hornblende gabbro/diorite. The color index in the unit ranges from ~20 to 45, with many samples ~40. In outcrop, the rocks tend to show a moderate to strong foliation, due to alignment of elongated aggregates of hornblende (±biotite ± pyroxene). Pyroxene grains commonly form the cores of these aggregates.

Various enclaves are common in the Punchbowl unit. They include (1) biotite + pyroxene diorite of the Buck Lake unit, (2) plagioclase + hornblende-phyric porphyry (similar to the plagioclase + pyroxene-phyric porphyry of the Buck Lake unit), and (3) clinopyroxenite and/or hornblendite cumulates and their altered equivalents (Fig. 4A). Numerous fine-grained biotite + hornblende (±quartz) (micro)diorite and hornblende ± pyroxene (micro)gabbro dikes occur in the Punchbowl plutonic suite, and locally, these rock types are intermingled. These mingled dikes may intrude subparallel to foliation, truncate foliation, or both.

Tabular, subhedral plagioclase is the dominant phase in the Punchbowl unit. These tabular grains are commonly aligned and define a magmatic foliation along with subparallel, elongated aggregates of mafic minerals. Plagioclase grain boundaries with hornblende are commonly irregular, suggesting a late-magmatic reaction that probably also involved the replacement of original pyroxene by hornblende (Fig. 5D). Fine-grained (~0.1 mm), anhedral plagioclase grains (neoblasts) commonly occur along the boundaries of the much larger tabular plagioclase and indicate high-temperature grain-boundary migration, which perhaps developed during the transition from hypersolidus to subsolidus conditions.

The dominant mafic mineral is hornblende. It is dark green, olive-green, or brown (Z tint). Pyroxene and/or cummingtonite are commonly found in the cores of coarse hornblende grains. The modal amount of biotite is highly variable but is always subordinate to hornblende. The biotite is associated with hornblende. Quartz was not identified in any sample, although trace amounts of alkali feldspar are locally present.

Doe Flat Plutonic Unit. This unit forms the northern zone of the Bear Mountain pluton (Plate 1). The Doe Flat unit contains xenoliths of the Buck Lake and Punchbowl units, but it is intruded by small bodies of the Wilderness Falls unit.

The Doe Flat unit is chiefly medium-grained, pyroxene-bearing, biotite + hornblende quartz diorite that is moderately to well foliated and locally lineated. In outcrop, the rocks commonly exhibit a moderate to strong foliation defined by aligned plagioclase grains and aggregates of mafic minerals. The lineated rocks are mylonitic and record strong crystal-plastic deformation. The color index of the unit ranges from ~15 to 45 but averages ~30. Hornblende is the dominant mafic mineral, although biotite is also abundant.

The Doe Flat unit contains scarce enclaves of plagioclase + hornblende-phyric porphyry similar to inclusions in the Punchbowl suite. There are also scarce hornblendite enclaves, but fewer than in the Punchbowl unit, and no clinopyroxenitic inclusions were found. Fine-grained biotite + hornblende (micro) diorite and hornblende ± pyroxene (micro)gabbro dikes, like those found in the Punchbowl suite, are also found in the Doe Flat unit, but they also are less common.

The microscopic textures characteristic of rocks of the Doe Flat unit are similar to those seen in the Buck Lake and Punchbowl units, especially in regard to plagioclase. Subhedral, tabular plagioclase commonly defines a magmatic foliation and is the dominant mineral phase of the unit. Oscillatory zoning is common. Some plagioclase grains are kinked, and grain boundaries vary from straight to serrated. The dominant mafic mineral is green or olive-green hornblende (Z tint). Pyroxene cores occur in some of the coarser hornblende grains. Colorless patches of cummingtonite are associated with some of the hornblende grains. Biotite is commonly abundant, and this characteristic is useful in distinguishing the Doe Flat from the Punchbowl units. Biotite is commonly found associated with individual hornblende grains or surrounding multigrain aggregates of hornblende. Quartz is also characteristic of the Doe Flat unit and is another distinguishing feature from the Punchbowl unit.

Wilderness Falls Plutonic Unit. The Wilderness Falls unit forms numerous small intrusive bodies (dikes and sheets) in the older plutonic units. The name applied to this unit is derived from the Wilderness Falls locality along Clear Creek (Fig. 2) that Snoke (1972) originally mapped and Snoke et al. (1981) included in their "leucocratic rocks."

The rocks are typically fine grained, moderately to well foliated, and locally lineated. The unit generally consists of dark biotite-hornblende quartz diorite to tonalite (color index = 30–35) and light biotite ± muscovite tonalite (color index = 10–20). In general, dikes of the biotite tonalite intrude rocks of the biotite-hornblende quartz diorite, but at the Wilderness Falls locality, they are mingled.

Grain size is generally fine, but medium-grained and even pegmatitic variants are present. Plagioclase is generally blocky and subhedral and exhibits oscillatory zoning. In the dark quartz dioritic rocks, biotite is fine grained, subhedral to euhedral, reddish brown (Z tint), and aligned subparallel with hornblende, thereby defining a magmatic foliation. In the light tonalitic rocks, if a foliation exists, it is typically defined by biotite and to a lesser extent by aligned plagioclase. In the light tonalitic rocks, muscovite and alkali feldspar are common subordinate phases. Evidence of solid-state deformation is present in some samples of the Wilderness Falls plutonic unit and is manifested as recrystallized quartz ribbons with subgrain development.

Mafic to Intermediate Dikes. A variety of mafic to intermediate dikes are found throughout the Bear Mountain intrusive complex. The most common dikes are aphyric and equigranular and are either hornblende microgabbros or microdiorites, depending on plagioclase composition. Porphyritic dikes are less common in the Bear Mountain intrusive complex and include clinopyroxene-, hornblende-, and/or plagioclase-phyric dikes.

Enclaves. Enclaves of various compositions are found throughout the Bear Mountain intrusive complex, including olivine clinopyroxenite, olivine melagabbro, hornblendite, hornblende gabbro, mafic hornfels, and biotite + hornblende ± pyroxene ± quartz schist. Sizes of the enclaves range from a few centimeters to tens of meters. Many of the enclaves sampled retained at least some original igneous textures when viewed in thin section, although all the samples were extensively recrystallized.

Clinopyroxenite enclaves were only found in the Punchbowl unit (Fig. 4A), although hornblendite enclaves found in the Doe Flat unit are probably hydrated equivalents of these cumulates. The mafic hornfels and biotite + hornblende ± pyroxene ± quartz schist are found in the three major plutonic units—Buck Lake, Punchbowl, and Doe Flat—and are believed to be inclusions of metabasalt, metavolcaniclastic, and/or possibly metasedimentary wall rocks.

GEOCHEMISTRY

Methods

Major element oxide contents along with Sr, Zr, Y, Ba, Sc, V, Cr, Ni, Cu, and Zn were analyzed by inductively coupled plasma atomic emission spectrometry, and Rb was determined by flame emission. Analytical techniques are described in Shannon et al. (1997). The rare earth elements (REEs), Ta, Nb, Cs, Hf, and Th, were determined by inductively coupled plasma mass spectrometry at Washington State University, Pullman. Additional major- and trace-element data are from Snoke et al. (1981). Oxygen was liberated from silicates using the BrF_5 method of Clayton and Mayeda (1963) and converted to CO_2 by passage over a hot graphite rod. Oxygen isotopic ratios were obtained on a VG SIRA-12 dual-inlet mass spectrometer. All values are reported relative to Vienna–Standard Mean Ocean Water. Silicate analyses are precise to ±0.2 per mil. The average and standard deviation obtained for NBS-28 is 9.45 ± 0.2 per mil (standard error is ±0.02 per mil). Neodymium and Sr isotopic compositions were determined at the University of Wyoming by thermal ionization mass spectrometry. Concentrations of Sm, Nd, Rb, and Sr were determined by isotope dilution on aliquots of the same sample dissolved for isotope ratio measurements. Further analytical details are given in the footnotes to Table 2.

TABLE 2. Nd AND Sr ISOTOPIC DATA FOR THE BEAR MOUNTAIN INTRUSIVE COMPLEX AND ITS AUREOLE

Sample	Unit	Rb (ppm)	Sr (ppm)	$^{87}Rb/^{86}Sr$	$^{87}Sr/^{86}Sr$	$^{87}Sr/^{86}Sr$ (147 Ma)	Sm (ppm)	Nd (ppm)	$^{147}Sm/^{144}Nd$	$^{143}Nd/^{144}Nd$	$^{143}Nd/^{144}Nd$ (147 Ma)	Initial ε_{Nd} (147 Ma)	Nd model age (Ma)
Bear Mountain Intrusive Complex													
PPS-39	BM mzdi	51.35	744.5	0.1995	0.703624	0.70321	5.284	24.55	0.13015	0.512914	0.512789	6.64	434
PPS-115	BM ton	25.14	435.1	0.1672	0.703590	0.70324	1.467	9.82	0.09038	0.512860	0.512773	6.32	361
PPS-213	BM mzdi	15.57	617.0	0.0730	0.703361	0.70321	3.631	15.54	0.14128	0.512992	0.512856	7.95	335
PPS-244	BR um+	9.32	161.8	0.1667	0.703821	0.70347	2.159	9.006	0.14501	0.512998	0.512859	8.00	341
PPS-410	BR um+	1.75	358.1	0.0142	0.703174	0.70314	1.458	4.210	0.20939	0.512969	0.512768	6.23	—
PPS-423	BM ton	19.42	833.3	0.0674	0.703114	0.70297	3.028	18.61	0.09839	0.512957	0.512862	8.06	258
PPS-531	BM ton	41.45	503.8	0.2380	0.703796	0.70330	3.471	18.09	0.11607	0.512836	0.512724	5.38	493
PPS-534	BM mzdi	39.87	480.9	0.2398	0.703840	0.70334	5.243	23.57	0.13453	0.512867	0.512738	5.64	546
PPS-567	BM hb gb	13.76	526.9	0.0756	0.703438	0.70328	4.004	17.08	0.14180	0.512926	0.512790	6.65	478
PPS-572	BM mzdi	29.38	661.3	0.1285	0.703438	0.70317	5.489	25.58	0.12978	0.512893	0.512768	6.24	469
DP4-3	BM hb gb	1.43	781.1	0.0053	0.703212	0.70320	3.627	15.606	0.14057	0.512909	0.512774	6.34	506
DP4-6	BM hb gb	2.90	1048.0	0.0080	0.703346	0.70333	2.133	9.701	0.13298	0.512852	0.512724	5.38	565
DP4-7	BM hb gb	17.71	534.7	0.0958	0.703339	0.70314	3.469	16.302	0.12868	0.512911	0.512787	6.61	431
Rattlesnake Creek terrane migmatitic amphibolite in aureole													
PP538A(F)*	tonalite	6.00	621.5	0.0279	0.703180	0.70312	1.15	5.46	0.12724	0.512940	0.512818	7.20	373
PP538A (M)	amphibolite	3.35	217.9	0.0444	0.703171	0.70308	3.73	11.23	0.20067	0.513123	0.512930	9.39	333
DP1 21C host	amphibolite	11.19	174.8	0.1851	0.704027	0.70364	4.21	16.16	0.15753	0.512952	0.512800	6.86	573
DP1 21C Leucosome	tonalite	16.56	235.9	0.2031	0.704038	0.70361	2.49	10.46	0.14363	0.512918	0.512780	6.46	510
PP79-556	tonalite	5.25	793.0	0.0191	0.703375	0.70333	3.22	18.05	0.10775	0.512939	0.512836	7.55	306

Notes: About 80 to 100 mg of sample were dissolved in $HF\text{-}HNO_3$. After conversion to chlorides, one-third of the sample was spiked with ^{87}Rb, ^{84}Sr, ^{149}Sm, and ^{146}Nd. The Rb, Sr, and rare-earth elements were separated by conventional cation-exchange procedures; Sm and Nd were further separated in di-ethyl-hexyl orthophosphoric acid columns. All isotopic measurements were made on a VG Sector multicollector mass spectrometer at the University of Wyoming, Laramie. An average $^{87}Sr/^{86}Sr$ isotopic ratio of 0.710251 ± 20 (2σ) was measured for NBS 987 Sr, and an average $^{143}Nd/^{144}Nd$ ratio of 0.511846 ± 11 (2σ) was measured for the La Jolla Nd standard. Uncertainties in Sr isotopic ratio measurements are ±0.00002 and uncertainties in Nd isotopic ratio measurements are ±0.00001 (2σ). Blanks are 50 pg for Rb, Sr, Nd, and Sm, and no blank correction was made. Uncertainties in Rb, Sr, Nd, and Sm concentrations are ±2% of the measured value; uncertainties on initial ε_{Nd} are ±0.3 epsilon units. The Nd model ages are calculated based on the depleted mantle model of Goldstein et al. (1984). Abbreviations: hb gb—hornblende ± pyroxene ± biotite gabbro/diorite; mzdi—biotite two-pyroxene monzodiorite/diorite; ton—biotite tonalite; umt—Blue Ridge intrusion ultramafic and gabbroic rocks.

*The Rb concentration for PP538A(F) by inductively coupled plasma atomic emission spectrometry at Texas Tech University, Lubbock.

Results

Major and Trace Elements. Analyzed samples of the Blue Ridge intrusion and related bodies consist chiefly of ultramafic rocks (dunite, wehrlite, hornblende + olivine clinopyroxenite) and olivine gabbro. Ultramafic samples are characterized by high MgO and low Al_2O_3 contents (Table 3, Fig. 7C and D). Rb contents are ≤40 ppm (most <4 ppm) and Zr contents are <22 ppm. In contrast, Sc ranges from 41 to 91 ppm, Cr from 727 to 2118 ppm, and Ni from 185 to 590 ppm (Table 4, Fig. 8). These compositions support the conclusions of Snoke et al. (1981) that the Blue Ridge ultramafic rocks are cumulates of clinopyroxene ± olivine. The FeO*/(FeO* + MgO) ratios of the ultramafic rocks range from 0.49 to 0.24 (Fig. 7A). Gabbroic rocks from the intrusion have FeO*/(FeO* + MgO) values at the high end of the range of the ultramafic rocks, from 0.48 to 0.40. The gabbros also show a wide range of Al_2O_3 contents, from 8.6 to 23.1 wt% and low K_2O (≤0.20 wt%, Fig. 7D and E).

In the Punchbowl unit, trace elements Sr, Sc, and Zr vary widely and typically show little or no correlation with SiO_2 (Fig. 8). Barium abundances show a weak positive correlation with SiO_2 (Fig. 8). In the Doe Flat unit, the average Sr abundance is lower than in the Punchbowl unit (631 versus 878 ppm), yet the range of Sr in the Punchbowl unit almost completely overlaps that of the Doe Flat unit. Among Doe Flat samples, there are weak positive correlations of Ba and Zr with SiO_2 and negative correlation of Sc with SiO_2 (Fig. 8). In the Buck Lake unit, Sr and Sc show weak or very weak correlations with SiO_2 and Ba and Zr are positively correlated. The Buck Lake unit has the highest Ba and Zr contents of all units in the Bear Mountain pluton.

Most late-stage dikes collected within the Bear Mountain pluton plot in the same clusters as samples from the three intrusive units (Figs. 7 and 8). One sample has higher MgO and lower FeO*/(FeO* + MgO), similar to gabbroic rocks from the Blue Ridge intrusion. Another sample (PPS-115) is a magnesian, calc-alkaline granitic porphyry with very low contents of Sr and Ba (both <100 ppm; Fig. 8).

The Wilderness Falls tonalite and late-stage felsic rocks that intrude the Blue Ridge intrusion plot on the ferroan-magnesian boundary; are broadly calcic, rather than calc-alkaline (Fig. 7); and are mildly peraluminous or metaluminous. All but one sample have K_2O <2.0 wt%. Trace-element concentrations in these samples vary widely (Fig. 8) and commonly show little correlation with SiO_2.

The host rocks to the Bear Mountain intrusive complex contain two distinct types of mafic dike rock that can be distinguished on the basis of phenocryst assemblages. Gabbroic dikes contain pyroxene (or amphibole after pyroxene). Some are pyroxene-phyric, others are equigranular. Dioritic dikes are intermediate in composition and typically plagioclase-phyric. Pyroxene phenocrysts are sparse in the dioritic dikes, but hornblende phenocrysts are common.

The gabbroic dikes have low FeO*/(FeO* + MgO) values and low SiO_2 contents (46–50 wt%, with one exception at 54 wt%; Fig. 7). Their MgO contents range from 8.5 to 15 wt%, and most have between 10 and 15 wt% Al_2O_3 (Fig. 7). Their Cr contents range from 340 to 1755 ppm and average 634 ppm; Sc ranges from 32 to 55 ppm and averages 44 ppm (Table 4). Compared to the Bear Mountain pluton, they have lower Sr and higher Sc contents, but their Ba and Zr contents overlap with those of Bear Mountain samples.

The dioritic dikes have a similar range in SiO_2 contents as the gabbroic dikes but have higher FeO*/(FeO* + MgO), lower MgO, and higher Al_2O_3 and K_2O (Fig. 7). Compared to the gabbroic dikes, the dioritic dikes have lower Sc (average 24 ppm) and much lower Cr (average 29 ppm). The dioritic dikes consistently plot near samples from the Bear Mountain pluton. In most instances, the dikes have compositions intermediate between those of the Punchbowl and Doe Flat units. This transitional position is particularly apparent in terms of MgO, Ba, and Zr (Figs. 7C and 8C and D).

Rare-Earth Elements. Chondrite-normalized REE patterns for clinopyroxenitic and gabbroic rocks from the Blue Ridge intrusion show distinct concave-downward patterns and broad overlap among clinopyroxenite and gabbro (Fig. 9A). Most clinopyroxenites show small negative Eu anomalies, whereas all but one of the gabbros have positive Eu anomalies. These patterns are consistent with accumulation of pyroxene ± olivine to form the ultramafic rocks and variable proportions of pyroxene and plagioclase to form the gabbros.

The three units of the Bear Mountain pluton have similar light REE-enriched patterns (Fig. 9B). Most samples also show apparent, slight negative Ce anomalies (Fig. 9B). Samples from the Punchbowl unit show the widest range of abundances, and two of these samples have positive Eu anomalies, indicative of plagioclase accumulation. The single analyzed sample from the Doe Flat unit also has a small positive Eu anomaly, suggesting plagioclase accumulation. The remaining Punchbowl samples and all samples of the Buck Lake unit have slight negative Eu anomalies (Fig. 9B).

Figure 9C shows REE patterns for the Wilderness Falls tonalite, a tonalitic intrusion into the Blue Ridge pluton, and two late-stage granitic dikes. All of these samples show reversed J-shaped patterns, with steep light and middle REE slopes that flatten in the heavy REEs. Europium anomalies are either lacking or are slightly positive.

The REE patterns of mafic dikes from the host rocks of the Bear Mountain intrusive complex are shown in Figure 9D. The gabbroic and dioritic dikes have nearly parallel patterns with moderate negative slopes, a slight dip at Ce, and weak, slightly negative Eu anomalies. The gabbroic dikes have lower REE abundances by a factor of ~1.5–2.0.

Isotopes. Table 2 presents Nd and Sr isotope values for the Bear Mountain intrusive complex. Initial $^{87}Sr/^{86}Sr$ and ε_{Nd} values were calculated for 147 Ma. The entire suite of samples from the Blue Ridge and Bear Mountain plutons shows little variation in initial $^{87}Sr/^{86}Sr$ (0.70315–0.70334) and ε_{Nd} (5.4–8.0; Fig. 10). The $\delta^{18}O$ values also show a narrow range,

TABLE 3. REPRESENTATIVE MAJOR ELEMENT DATA

	Doe Flat unit				Punchbowl unit			Buck Lake unit									Related dike rocks		Wilderness Falls		
Sample*	JB-35a	JB-35b	PPS-567	Ave. (*n* = 10)	DP4-3	DP4-6	Ave. (*n* = 19)	JB-373	PP-39	PPS-213	PPS-534	PPS-572	DPI-13Z	JB-28	JB-270	Ave. (*n* = 16)	PPS-115	DPI-21A	JB-396A	PPS-531	DP4-11
SiO_2	51.54	55.24	55.00	53.15	46.57	48.14	46.93	45.76	52.08	52.30	54.84	52.62	52.34	53.90	49.55	52.79	72.40	68.37	66.99	61.66	68.44
TiO_2	1.27	0.79	0.81	0.96	1.45	0.63	1.24	1.59	1.04	1.05	1.14	1.16	1.38	0.88	1.42	1.08	0.18	0.40	0.37	0.56	0.24
Al_2O_3	18.36	19.58	17.98	19.14	20.08	27.09	20.75	15.14	17.98	17.14	16.30	17.15	19.08	19.94	21.07	18.61	14.50	15.36	16.42	15.16	16.60
Fe_2O_3	9.62	6.34	8.53	—	12.12	5.09	—	15.58	9.64	9.63	9.45	9.23	8.44	7.53	9.29	—	1.69	3.05	3.89	6.07	3.83
FeO	n.d.	n.d.	n.d.	7.82	n.d.	n.d.	9.47	n.d.	n.d.	n.d.	n.d.	n.d.	n.d.	n.d.	n.d.	7.91	n.d.,	n.d.	n.d.	n.d.	n.d.
MnO	0.14	0.13	0.18	0.16	0.17	0.07	0.16	0.21	0.17	0.17	0.16	0.17	0.13	0.15	0.15	0.15	0.06	0.04	0.09	0.12	0.07
MgO	4.47	2.90	3.59	3.62	4.16	1.72	4.59	7.56	3.69	4.67	3.71	4.24	2.64	3.04	3.16	3.58	0.49	0.72	1.43	4.57	0.89
CaO	8.35	7.57	7.81	8.43	10.52	13.00	11.15	10.78	7.78	8.89	7.10	8.80	9.31	8.19	10.57	8.55	2.59	2.45	4.28	6.20	4.01
Na_2O	5.04	5.78	3.68	3.85	3.60	3.27	2.94	1.87	3.66	3.42	3.55	3.37	4.06	3.72	3.68	3.73	5.08	3.32	3.71	3.06	3.88
K_2O	0.27	0.32	1.05	0.80	0.43	0.27	0.56	0.50	2.51	1.42	2.11	1.98	0.88	1.74	0.43	1.56	2.06	4.80	1.86	1.60	1.69
P_2O_5	0.30	0.24	0.24	0.25	0.28	0.37	0.33	0.19	0.54	0.29	0.35	0.36	0.57	0.25	0.69	0.34	0.06	0.12	0.16	0.19	0.14
LOI	0.51	0.73	0.89	0.65	0.75	0.62	0.89	0.66	0.57	0.17	0.70	0.74	1.07	0.79	0.04	0.54	0.59	0.51	0.75	0.74	0.82
Total	99.86	99.60	99.75		100.13	100.27		99.84	99.65	99.16	99.40	99.81	99.89	100.13	100.06		99.71	99.14	99.95	99.93	100.62
Mg#	0.48	0.48	0.45		0.40	0.40		0.49	0.43	0.49	0.44	0.48	0.38	0.44	0.40		0.37	0.32	0.42	0.60	0.32
ASI	0.80	0.85	0.86		0.81	0.95		0.67	0.83	0.76	0.81	0.75	0.82	0.89	0.87		0.96	1.04	1.06	0.86	1.09

	Blue Ridge intrusion					Gabbroic dikes									Dioritic dikes				
Sample	PPS-412	PPS-417	PPS-427	PPS-529	PPS-410	HB4-3	HB4-6	PPS-244	PPS-576	DP1-35	DPI-51	DP2-9	DP1-35Z	Ave. (*n* = 10)	DP1-9B	DP1-9C	DP2-2	DP1-18Z	Ave. (*n* = 8)
SiO_2	47.58	47.66	51.17	40.62	47.17	46.92	49.08	48.32	49.49	48.84	46.25	47.80	53.26	48.61	49.50	49.03	53.89	50.86	50.53
TiO_2	0.28	0.35	0.39	0.45	0.67	0.32	0.51	0.75	0.77	0.70	1.06	0.71	0.77	0.87	1.30	1.86	0.94	1.19	1.19
Al_2O_3	1.81	2.38	2.33	3.18	10.33	16.10	20.50	13.04	11.80	12.15	11.91	12.30	14.94	12.63	18.55	20.34	18.36	17.75	18.25
Fe_2O_3	10.23	8.07	6.67	17.16	11.86	9.30	6.91	10.33	9.90	9.81	11.75	10.80	8.54	10.47	10.28	9.92	8.43	10.32	10.09
FeO	n.d.	n.d.	n.d.	n.d.	n.d.	n.d.	n.d.	n.d.	n.d.	n.d.	n.d.	n.d.	n.d.	n.d.	n.d.	n.d.	n.d.	n.d.	n.d.
MnO	0.19	0.17	0.14	0.27	0.23	0.16	0.13	0.19	0.18	0.16	0.21	0.18	0.13	0.18	0.16	0.17	0.15	0.18	0.17
MgO	23.50	21.05	17.05	27.90	12.89	11.51	6.38	11.72	11.70	15.26	13.65	14.83	8.27	11.87	4.54	3.15	3.55	3.95	4.25
CaO	13.45	18.01	21.07	6.89	13.95	13.86	14.08	9.46	11.39	8.34	11.48	9.31	7.97	10.43	8.20	7.26	8.13	8.52	8.95
Na_2O	0.15	0.26	0.27	0.38	1.19	1.20	2.24	2.89	1.86	1.65	2.05	1.61	2.73	2.03	4.11	4.90	3.78	3.79	3.70
K_2O	0.02	0.04	0.03	0.12	0.10	0.11	0.13	1.07	0.82	0.29	0.46	0.48	0.82	0.64	0.83	0.82	1.25	1.36	1.14
P_2O_5	0.00	0.02	0.01	0.02	0.01	0.01	0.01	0.15	0.14	0.12	0.16	0.10	0.12	0.14	0.29	0.65	0.25	0.30	0.30
LOI	2.07	1.48	0.53	2.39	1.08	0.21	0.48	1.63	1.35	2.54	1.41	1.97	2.40	1.79	2.27	1.86	1.18	1.53	1.49
Total	99.29	99.50	99.65	99.39	99.46	99.69	100.48	99.56	99.39	99.85	100.39	100.09	99.94	99.88	100.06	99.96	99.89	99.75	100.06
Mg#	0.82	0.84	0.84	0.76	0.68	0.71	0.65	0.69	0.70	0.75	0.70	0.73	0.66	0.69	0.47	0.39	0.45	0.43	0.45
ASI	0.07	0.07	0.06	0.24	0.38	0.59	0.70	0.57	0.49	0.68	0.49	0.62	0.76		0.85	0.99	0.84	0.79	

Notes: Analysis by inductively coupled plasma atomic emission spectrometry at Texas Tech University, Lubbock. See Shannon et al. (1997) for analytical details. ASI—alumina saturation index; LOI—loss on ignition; n.d.—not determined.

*Rock types: JB-35a, fine-grained biotite-bearing hornblende diorite dike; JB-35b, biotite hornblende diorite; PPS-567, hornblende biotite quartz diorite; DP4-3, pyroxene-bearing hornblende diorite; DP4-6, hornblende diorite; JB-373, biotite two-pyroxene hornblende quartz-bearing diorite; PP-39, biotite two-pyroxene monzodiorite; PPS-213, biotite two-pyroxene diorite; PPS-534, biotite hornblende pyroxene diorite; PPS-572, biotite two-pyroxene monzodiorite; DP1-13Z, biotite two-pyroxene monzodiorite; JB-28, biotite two-pyroxene monzodiorite; JB-270, two-pyroxene biotite hornblende diorite; PPS-115, plagioclase-phyric porphyry in Buck Lake unit; DP1-21A, biotite granite in host-rock hornfels; JB-396A, biotite tonalite; PPS-531, biotite tonalite (dark phase); DP4-11, biotite tonalite.

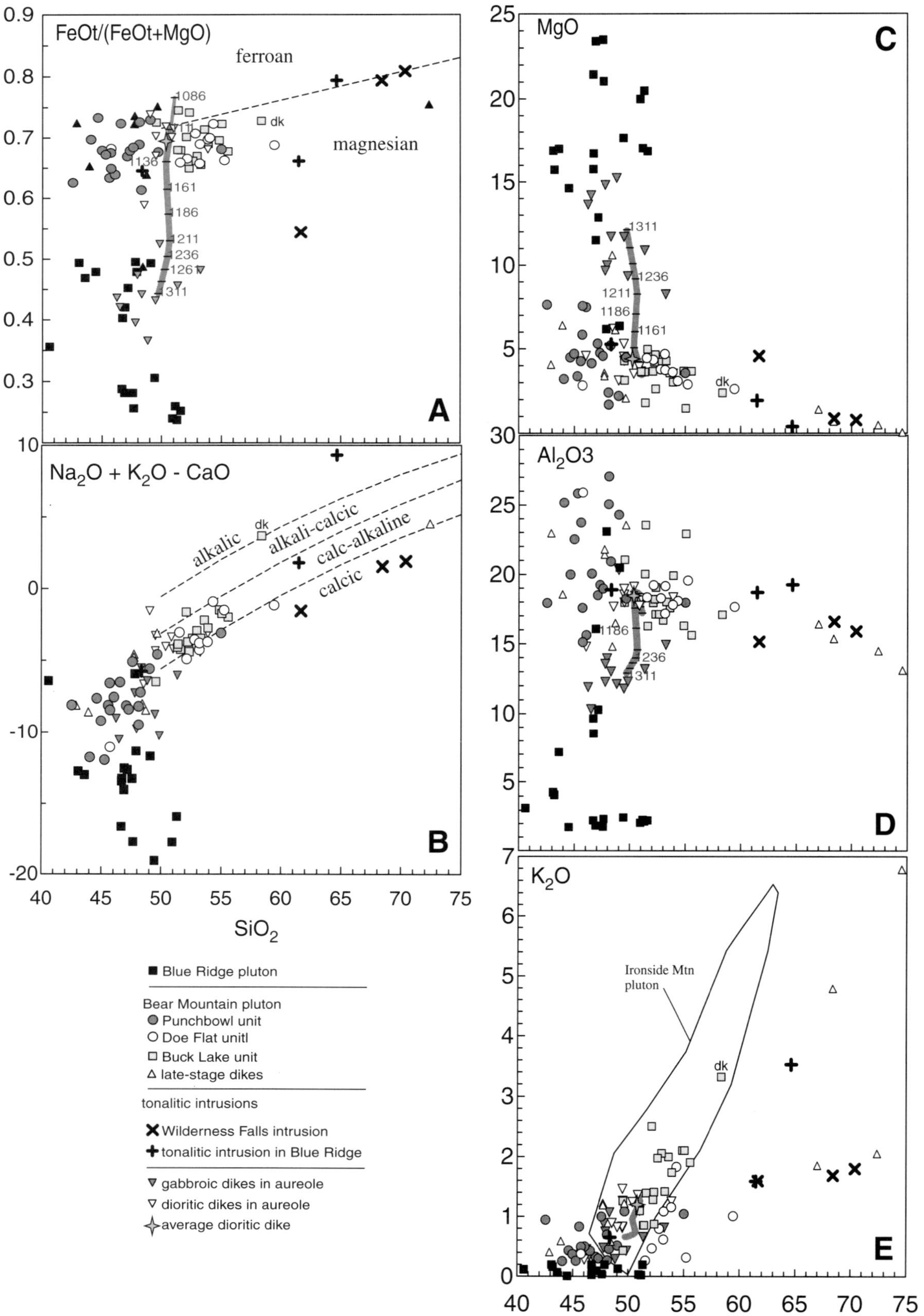

FeOt/(FeOt+MgO)
ferroan
magnesian
A
Na2O + K2O - CaO
alkalic
alkali-calcic
calc-alkaline
calcic
dk
B
SiO2
MgO
C
Al2O3
D
K2O
Ironside Mtn pluton
E
SiO2
Blue Ridge pluton
Bear Mountain pluton
Punchbowl unit
Doe Flat unitl
Buck Lake unit
late-stage dikes
tonalitic intrusions
Wilderness Falls intrusion
tonalitic intrusion in Blue Ridge
gabbroic dikes in aureole
dioritic dikes in aureole
average dioritic dike

from +5.9 to +7.6‰. There is no apparent correlation between composition and radiogenic isotope ratios. For example, neither Sr contents and initial $^{87}Sr/^{86}Sr$ nor Nd contents and ε_{Nd} are correlated. For Bear Mountain pluton samples, there is possibly a weak correlation between $\delta^{18}O$ and initial $^{87}Sr/^{86}Sr$ (Fig. 10), but $\delta^{18}O$ is not correlated with major or trace element compositions.

A tonalite from the Wilderness Falls intrusion has higher $\delta^{18}O$ and low ε_{Nd} compared to the rest of the suite. In contrast, tonalite that intrudes the Blue Ridge pluton has the highest ε_{Nd} and lowest initial $^{87}Sr/^{86}Sr$ measured, and $\delta^{18}O$ for this sample overlaps the range of the Bear Mountain pluton (Fig. 10). Figure 10 also shows isotopic compositions of four migmatitic amphibolite host rock samples and of one gabbroic dike that intrudes amphibolitic host rocks. Figure 10 shows the narrow range of Bear Mountain intrusive complex plutonic rocks compared to local metasedimentary rocks from the Rattlesnake Creek terrane and Galice Formation (Frost et al., this volume).

DISCUSSION

Parental Magmas

As illustrated in Figures 7C–E and 8, compositional trends within individual intrusive units are essentially lacking. The possible exception is the Buck Lake unit, in which incompatible elements (e.g., K_2O, Rb, Ba, Zr) increase with increasing SiO_2. Otherwise, each unit is best described as having broadly scattered compositions. In the Blue Ridge intrusion, this lack of a well-defined differentiation trend is clearly associated with the fact that most, if not all, rocks in the pluton are cumulates. The REE patterns (Fig. 9) illustrate the influence of cumulate clinopyroxene ± plagioclase in all samples. In the intrusive units of the Bear Mountain pluton, the wide ranges in MgO, Al_2O_3, and K_2O (Fig. 7C–E) are also consistent with variable proportions of cumulus pyroxene, hornblende, and plagioclase. These effects of crystal accumulation complicate petrogenetic interpretation, because the identification of possible parental magmas is difficult.

Presumably the magmas of the Blue Ridge intrusion were basaltic, because they were capable of producing voluminous olivine and olivine + clinopyroxene cumulates (Snoke et al., 1981). Furthermore, they must have contained sufficient H_2O to permit widespread late-stage amphibole stability and contained sufficient K_2O to stabilize late-stage phlogopitic biotite.

In contrast, magmas of the Punchbowl unit of the Bear Mountain pluton were probably H_2O-rich evolved basalt or basaltic andesite, due to the presence of cumulate clinopyroxene, hornblende, and plagioclase. A similar magma might be parental to the Buck Lake unit; however, such a magma would have to be poorer in H_2O, because amphibole and biotite are present only as late-stage, noncumulate phases. The Buck Lake parent may have had higher K_2O contents than the Punchbowl parent.

All the parental magmas of the Bear Mountain intrusive complex were isotopically primitive, such that the rocks of the complex have among the highest initial ε_{Nd} and lowest $\delta^{18}O$ and initial $^{87}Sr/^{86}Sr$ of the igneous rocks of the Klamath Mountains province (Fig. 10A and B). Their isotopic compositions probably were similar to those of the Rattlesnake Creek terrane metabasalts, which would be expected, because both are mantle-derived basaltic magmas. The isotopic compositions of the Bear Mountain rocks therefore do not rule out assimilation of metabasalt during the evolution of the complex. However, based on their very different isotopic compositions, assimilation of Galice and Rattlesnake Creek terrane argillaceous rocks by Bear Mountain magmas cannot have contributed a significant volume to the pluton (Fig. 10C).

Barnes et al. (this volume, Chapter 17) show that the gabbroic and dioritic dikes that intrude the Bear Mountain intrusive complex host rocks were coeval with emplacement of the two large plutons. This timing means that one or both sets of dikes may be parental to the adjacent plutons. Specifically, Figure 7A shows that the gabbroic dikes have FeO*/(FeO* + MgO) values that overlap those of the gabbroic samples from the Blue Ridge intrusion. In addition, these dikes have the high Cr, Ni, and Sc contents necessary to explain the high values of these elements in Blue Ridge cumulates. Similarly, the dioritic dikes plot essentially in the middle of the data field from the Bear Mountain pluton—specifically, between data for the Punchbowl and Doe Flat units (Figs. 7 and 8). Furthermore, REE patterns of the dioritic dikes are almost nearly parallel to Bear Mountain samples that lack positive Eu anomalies (Fig. 9B).

The possibility that the gabbroic dikes were parental to the Blue Ridge intrusion was modeled using the PELE program (Boudreau, 1999), a simplified version of MELTS. The starting composition was an average of the gabbroic dikes (Table 5). Other input parameters were 3.5 kb pressure, fO_2 of quartz-fayalite-magnetite buffer + 1, and 1.0 wt% H_2O, and fractional crystallization was assumed. Calculated compositional trends are plotted in Figure 7.

The model predicts a liquidus temperature of 1311 °C, with olivine (Fo_{89}) as the liquidus phase. Clinopyroxene was calculated to join olivine on the liquidus at 1211 °C, at which point the olivine is Fo_{85}. At 1136 °C, plagioclase (An_{79}) also joins the fractionating assemblage. At this temperature, olivine is Fo_{78}. Olivine leaves the liquidus between 1085 and 1060 °C.

Figure 7. (A, B) Classification—$FeO^t/(FeO^t + MgO)$ vs. wt% SiO_2 diagram and modified alkali-lime index ($Na_2O + K_2O - CaO$) vs. wt% SiO_2 diagram—according to Frost et al. (2001). (C–E) Variation of MgO, Al_2O_3, and K_2O as a function of SiO_2 content. The broad gray curves are calculated liquid line of descent of a Mg-rich gabbroic dike composition (see text). Tic marks along these curves are 25 °C temperature intervals. In (E) the outline represents the main compositional trend of the ca. 170-Ma Ironside Mountain pluton (Barnes et al., this volume, Chapter 10).

TABLE 4. REPRESENTATIVE TRACE ELEMENT DATA

	Doe Flat unit		Punchbowl unit				Buck Lake unit							Related dike rocks					Wilderness Falls	
Sample	JB-35a	JB-35b	PPS-567	DP4-3	DP4-6	JB-373	PP-39	PPS-213	PPS-534	PPS-572	DP1-13Z	JB-28	JB-270	PPS-115	DP1-21A	JB-291B	JB-389	JB-396A	PPS-531	DP4-11
Analyzed by ICP-AES at Texas Tech University, Lubbock																				
Rb	4	3	18	4	4	10	46	25	44	36	17	34	5	32	94	23	10	47	46	46
Sr	544	588	584	887	1070	557	692	707	486	643	853	785	885	498	346	692	414	569	516	592
Zr	150	237	139	34	35	57	191	106	252	227	66	127	30	103	264	155	81	144	103	173
Y	23.6	24.9	25.0	17.5	9.9	16.8	25.6	19.5	26.1	25.3	23.3	19.7	22.0	6.9	10.6	24.5	17.7	7.2	12.0	6.9
Ba	366	371	272	218	140	159	593	305	431	440	334	390	270	753	1425	343	165	741	529	777
Sc	26.2	16.3	17.9	25.6	8.4	41.4	20.2	23.9	23.3	26.0	19.9	17.1	22.9	2.1	4.1	13.9	35.4	4.9	15.4	1.8
V	207	155	151	333	103	507	189	249	194	185	147	161	245	12	29	187	256	35	116	11
Cr	16	8	6	22	b.d.	64	18	25	26	38	8	12	16	b.d.	b.d.	b.d.	459	23	179	9
Ni	n.d.	n.d.	b.d.	n.d.	n.d.	38	8	13	9	10	n.d.	n.d.	12	b.d.	b.d.	b.d.	137	b.d.	50	b.d.
Cu	3	2	23	274	43	111	109	63	43	78	167	74	156	1	14	29	55	9	14	7
Zn	39	33	87	204	63	122	112	94	108	100	86	78	103	15	24	123	82	80	79	86
Be	1.1	1.2	1.0	1.1	0.3	0.9	0.9	0.9	1.0	1.2	0.9	1	0.9	1.5	0.9	1.0	0.8	1.2	1.4	3.1
Analyzed by ICP-MS at Washington State University, Pullman																				
La	n.d.	n.d.	14.09	12.11	9.11	n.d.	19.64	13.15	19.64	20.79	n.d.	n.d.	n.d.	17.14	38.57	n.d.	n.d.	n.d.	17.84	n.d.
Ce	n.d.	n.d.	30.04	26.04	17.93	n.d.	40.69	27.39	40.66	42.95	n.d.	n.d.	n.d.	27.65	62.81	n.d.	n.d.	n.d.	34.30	n.d.
Pr	n.d.	n.d.	3.99	3.58	2.29	n.d.	5.32	3.65	5.27	5.61	n.d.	n.d.	n.d.	2.79	6.44	n.d.	n.d.	n.d.	4.12	n.d.
Nd	n.d.	n.d.	18.24	16.69	10.53	n.d.	23.92	16.98	23.59	24.89	n.d.	n.d.	n.d.	9.52	23.06	n.d.	n.d.	n.d.	16.90	n.d.
Sm	n.d.	n.d.	4.95	4.44	2.59	n.d.	6.10	4.53	6.02	6.42	n.d.	n.d.	n.d.	1.75	3.99	n.d.	n.d.	n.d.	3.78	n.d.
Eu	n.d.	n.d.	1.42	2.00	1.27	n.d.	1.67	1.36	1.52	1.68	n.d.	n.d.	n.d.	0.53	1.13	n.d.	n.d.	n.d.	1.08	n.d.
Gd	n.d.	n.d.	4.83	4.25	2.42	n.d.	5.68	4.38	5.71	5.79	n.d.	n.d.	n.d.	1.31	3.02	n.d.	n.d.	n.d.	3.07	n.d.
Tb	n.d.	n.d.	0.82	0.65	0.36	n.d.	0.87	0.68	0.89	0.91	n.d.	n.d.	n.d.	0.20	0.40	n.d.	n.d.	n.d.	0.45	n.d.
Dy	n.d.	n.d.	5.04	3.83	2.10	n.d.	4.98	4.07	5.37	5.36	n.d.	n.d.	n.d.	1.22	2.29	n.d.	n.d.	n.d.	2.57	n.d.
Ho	n.d.	n.d.	1.03	0.74	0.40	n.d.	1.00	0.83	1.07	1.04	n.d.	n.d.	n.d.	0.24	0.46	n.d.	n.d.	n.d.	0.49	n.d.
Er	n.d.	n.d.	2.86	1.94	1.01	n.d.	2.72	2.19	2.88	2.80	n.d.	n.d.	n.d.	0.74	1.22	n.d.	n.d.	n.d.	1.24	n.d.
Tm	n.d.	n.d.	0.41	0.27	0.14	n.d.	0.37	0.31	0.41	0.39	n.d.	n.d.	n.d.	0.12	0.18	n.d.	n.d.	n.d.	0.18	n.d.
Yb	n.d.	n.d.	2.46	1.60	0.80	n.d.	2.23	1.87	2.48	2.36	n.d.	n.d.	n.d.	0.81	1.18	n.d.	n.d.	n.d.	1.08	n.d.
Lu	n.d.	n.d.	0.39	0.24	0.13	n.d.	0.35	0.30	0.39	0.37	n.d.	n.d.	n.d.	0.15	0.21	n.d.	n.d.	n.d.	0.17	n.d.
Ba	n.d.	n.d.	271.2	219.4	134.6	n.d.	569.6	313.3	422.1	448.2	n.d.	n.d.	n.d.	717.7	1444.5	n.d.	n.d.	n.d.	518.3	n.d.
Th	n.d.	n.d.	1.20	0.17	0.80	n.d.	3.23	1.65	4.16	3.47	n.d.	n.d.	n.d.	3.61	7.51	n.d.	n.d.	n.d.	3.56	n.d.
Nb	n.d.	n.d.	5.65	3.33	1.90	n.d	6.18	4.46	7.48	7.00	n.d	n.d.	n.d.	6.50	5.12	n.d.	n.d.	n.d.	5.81	n.d.
Y	n.d.	n.d.	27.61	19.13	10.75	n.d	26.02	20.98	28.80	28.10	n.d	n.d.	n.d.	7.40	11.95	n.d.	n.d.	n.d.	13.36	n.d.
Hf	n.d.	n.d.	3.22	0.96	1.00	n.d	3.77	2.28	3.85	3.92	n.d	n.d.	n.d.	2.70	6.44	n.d.	n.d.	n.d.	2.80	n.d.
Ta	n.d.	n.d.	0.34	0.18	0.14	n.d	0.44	0.31	0.53	0.47	n.d	n.d.	n.d.	0.57	0.27	n.d.	n.d.	n.d.	0.49	n.d.
U	n.d.	n.d.	0.41	0.07	0.25	n.d	1.14	0.59	1.52	1.20	n.d	n.d.	n.d.	1.32	1.59	n.d.	n.d.	n.d.	1.55	n.d.
Pb	n.d.	n.d.	4.37	3.02	4.01	n.d	7.41	5.23	8.09	7.79	n.d	n.d.	n.d.	15.33	7.64	n.d.	n.d.	n.d.	7.31	n.d.
Rb	n.d.	n.d.	17.53	2.00	3.39	n.d	42.62	23.02	44.12	37.46	n.d	n.d.	n.d.	30.50	87.89	n.d.	n.d.	n.d.	46.97	n.d.
Cs	n.d.	n.d.	0.86	0.62	0.64	n.d	1.91	1.37	2.26	1.76	n.d	n.d.	n.d.	1.44	2.28	n.d.	n.d.	n.d.	2.15	n.d.
Sr	n.d.	n.d.	579.4	913.4	1160.3	n.d	676.33	704.9	493.3	645.2	n.d	n.d.	n.d.	474.5	344.8	n.d.	n.d.	n.d.	535.2	n.d.
Sc	n.d.	n.d.	19.83	26.09	8.64	n.d	22.14	24.99	26.22	29.64	n.d	n.d.	n.d.	2.26	4.72	n.d.	n.d.	n.d.	17.05	n.d.
Zr	n.d.	n.d.	115.2	26.40	36.13	n.d	132.1	74.71	131.69	144.16	n.d	n.d.	n.d.	90.39	240.38	n.d.	n.d.	n.d.	98.58	n.d.

Sample	Blue Ridge intrusion PPS-412	PPS-417	PPS-427	PPS-529	PPS-410	HB4-3	HB4-6	Gabbroic dikes PPS-244	PPS-576	DP1-35	DPI-51	DP2-9	DP1-35Z	Ave. (*n* = 10)	Dioritic dikes DP1-9B	DP1-9C	DP2-2	DP1-18Z	Ave. (*n* = 8)
Analyzed by ICP-AES at Texas Tech University, Lubbock																			
Rb	3	3	2	4	4	3	3	23	17	5	9	10	18	12	14	16	32	22	21
Sr	29	56	45	96	362	540	727	240	569	285	262	245	480	298	564	708	612	672	688
Zr	11	16	14	20	16	11	12	58	62	64	77	55	80	71	123	315	97	127	117
Y	5.3	6.1	6.5	8.7	10.1	6.6	5.7	13.3	14.0	12.9	16.3	12.5	15.9	15	22.3	46.9	19.3	20.0	21
Ba	14	16	11	31	47	44	69	174	184	71	323	152	251	186	330	311	321	297	289
Sc	75.2	79.1	96.2	40.7	65.4	41.6	34.3	39.9	47.3	38.9	45.1	38.6	32.2	44	28.2	25.1	18.3	20.7	24
V	131	115	157	109	260	165	206	208	201	197	262	204	190	234	269	115	200	264	278
Cr	1622	1801	1921	727	433	371	163	779	942	1662*	605	1755	576	634	39	7	8	23	32
Ni	391	314	185	590	92	n.d.	n.d.	229	154	n.d.	266	n.d.	n.d.	162	n.d.	n.d.	n.d.	n.d.	24
Cu	73	18	31	8	94	71	51	76	13	52	69	52	119	58	207	124	57	133	120
Zn	41	25	15	114	58	92	80	71	69	114	63	89	72	81	165	196	152	85	119
Be	0.1	n.d.	0.1	0.1	0.2	b.d.	b.d.	0.3	0.4	b.d.	0.7	b.d.	0.2	b.d.	0.8	0.9	0.9	3.2	1
Analyzed by ICP-MS at Washington State University, Pullman																			
La	n.d.	1.23	1.01	n.d.	1.39	n.d.	1.49	7.62	6.18	6.77	n.d.	5.35	n.d.	n.d.	15.24	n.d.	12.07	n.d.	n.d.
Ce	n.d.	3.47	3.20	n.d.	3.91	n.d.	3.27	15.77	14.61	13.35	n.d.	11.74	n.d.	n.d.	32.27	n.d.	25.23	n.d.	n.d.
Pr	n.d.	0.59	0.58	n.d.	0.71	n.d.	0.50	2.19	2.10	1.75	n.d.	1.63	n.d.	n.d.	4.21	n.d.	3.41	n.d.	n.d.
Nd	n.d.	3.32	3.51	n.d.	4.27	n.d.	2.73	10.44	10.09	7.88	n.d.	7.92	n.d.	n.d.	19.20	n.d.	15.51	n.d.	n.d.
Sm	n.d.	1.22	1.33	n.d.	1.74	n.d.	0.98	2.96	3.03	2.28	n.d.	2.39	n.d.	n.d.	5.11	n.d.	4.26	n.d.	n.d.
Eu	n.d.	0.40	0.43	n.d.	0.68	n.d.	0.58	0.97	0.98	0.80	n.d.	0.79	n.d.	n.d.	1.59	n.d.	1.30	n.d.	n.d.
Gd	n.d.	1.40	1.53	n.d.	2.07	n.d.	1.17	2.91	3.04	2.41	n.d.	2.56	n.d.	n.d.	5.03	n.d.	4.08	n.d.	n.d.
Tb	n.d.	0.23	0.25	n.d.	0.36	n.d.	0.19	0.47	0.49	0.44	n.d.	0.43	n.d.	n.d.	0.80	n.d.	0.66	n.d.	n.d.
Dy	n.d.	1.39	1.52	n.d.	2.26	n.d.	1.19	2.82	2.93	2.75	n.d.	2.57	n.d.	n.d.	4.74	n.d.	4.07	n.d.	n.d.
Ho	n.d.	0.26	0.29	n.d.	0.45	n.d.	0.24	0.56	0.59	0.57	n.d.	0.53	n.d.	n.d.	0.96	n.d.	0.83	n.d.	n.d.
Er	n.d.	0.67	0.76	n.d.	1.18	n.d.	0.63	1.50	1.64	1.60	n.d.	1.41	n.d.	n.d.	2.66	n.d.	2.26	n.d.	n.d.
Tm	n.d.	0.09	0.10	n.d.	0.16	n.d.	0.09	0.21	0.22	0.22	n.d.	0.21	n.d.	n.d.	0.36	n.d.	0.32	n.d.	n.d.
Yb	n.d.	0.54	0.58	n.d.	0.97	n.d.	0.52	1.27	1.32	1.38	n.d.	1.29	n.d.	n.d.	2.23	n.d.	2.03	n.d.	n.d.
Lu	n.d.	0.08	0.09	n.d.	0.15	n.d.	0.08	0.21	0.21	0.22	n.d.	0.19	n.d.	n.d.	0.36	n.d.	0.32	n.d.	n.d.
Ba	n.d.	12.94	9.66	n.d.	41.85	n.d.	65.37	169.8	183.4	73.3	n.d.	153.1	n.d.	n.d.	333.5	n.d.	345.4	n.d.	n.d.
Th	n.d.	0.09	0.06	n.d.	0.03	n.d.	0.06	1.04	0.54	1.24	n.d.	0.88	n.d.	n.d.	1.62	n.d.	2.26	n.d.	n.d.
Nb	n.d.	0.41	0.22	n.d.	0.26	n.d.	0.17	1.57	2.35	4.28	n.d.	2.95	n.d.	n.d.	7.12	n.d.	4.72	n.d.	n.d.
Y	n.d.	6.58	7.21	n.d.	11.12	n.d.	5.94	14.49	15.50	14.89	n.d.	13.77	n.d.	n.d.	25.32	n.d.	22.04	n.d.	n.d.
Hf	n.d.	0.48	0.43	n.d.	0.55	n.d.	0.31	1.62	1.74	1.63	n.d.	1.45	n.d.	n.d.	3.16	n.d.	2.45	n.d.	n.d.
Ta	n.d.	0.03	0.02	n.d.	0.02	n.d.	0.01	0.13	0.16	0.28	n.d.	0.21	n.d.	n.d.	0.47	n.d.	0.32	n.d.	n.d.
U	n.d.	0.05	0.03	n.d.	0.02	n.d.	0.03	0.38	0.27	0.45	n.d.	0.32	n.d.	n.d.	0.58	n.d.	0.84	n.d.	n.d.
Pb	n.d.	0.92	0.85	n.d.	1.17	n.d.	1.20	2.49	2.27	1.44	n.d.	2.23	n.d.	n.d.	3.30	n.d.	4.97	n.d.	n.d.
Rb	n.d.	1	1	n.d.	1	n.d.	2	23	16	4	n.d.	8	n.d.	n.d.	12	n.d.	34	n.d.	n.d.
Cs	n.d.	0.21	0.29	n.d.	0.33	n.d.	0.55	0.74	0.58	0.67	n.d.	0.77	n.d.	n.d.	0.84	n.d.	1.43	n.d.	n.d.
Sr	n.d.	56.28	46.79	n.d.	352.87	n.d.	708.82	235.68	580.17	297.35	n.d.	252.80	n.d.	n.d.	564.42	n.d.	627.46	n.d.	n.d.
Sc	n.d.	74	93	n.d.	67	n.d.	34	39	51	37	n.d.	30	n.d.	n.d.	31	n.d.	21	n.d.	n.d.
Zr	n.d.	12.27	10.55	n.d.	11.92	n.d.	7.98	52.43	56.26	58.89	n.d.	46.77	n.d.	n.d.	111.20	n.d.	82.25	n.d.	n.d.

Notes: All abundances in ppm. Abbreviations: b.d.—below detection limits; ICP-AE—inductively coupled plasma atomic emission spectrometry; ICP-MS—inductively coupled plasma mass spectrometry; n.d.—not determined.

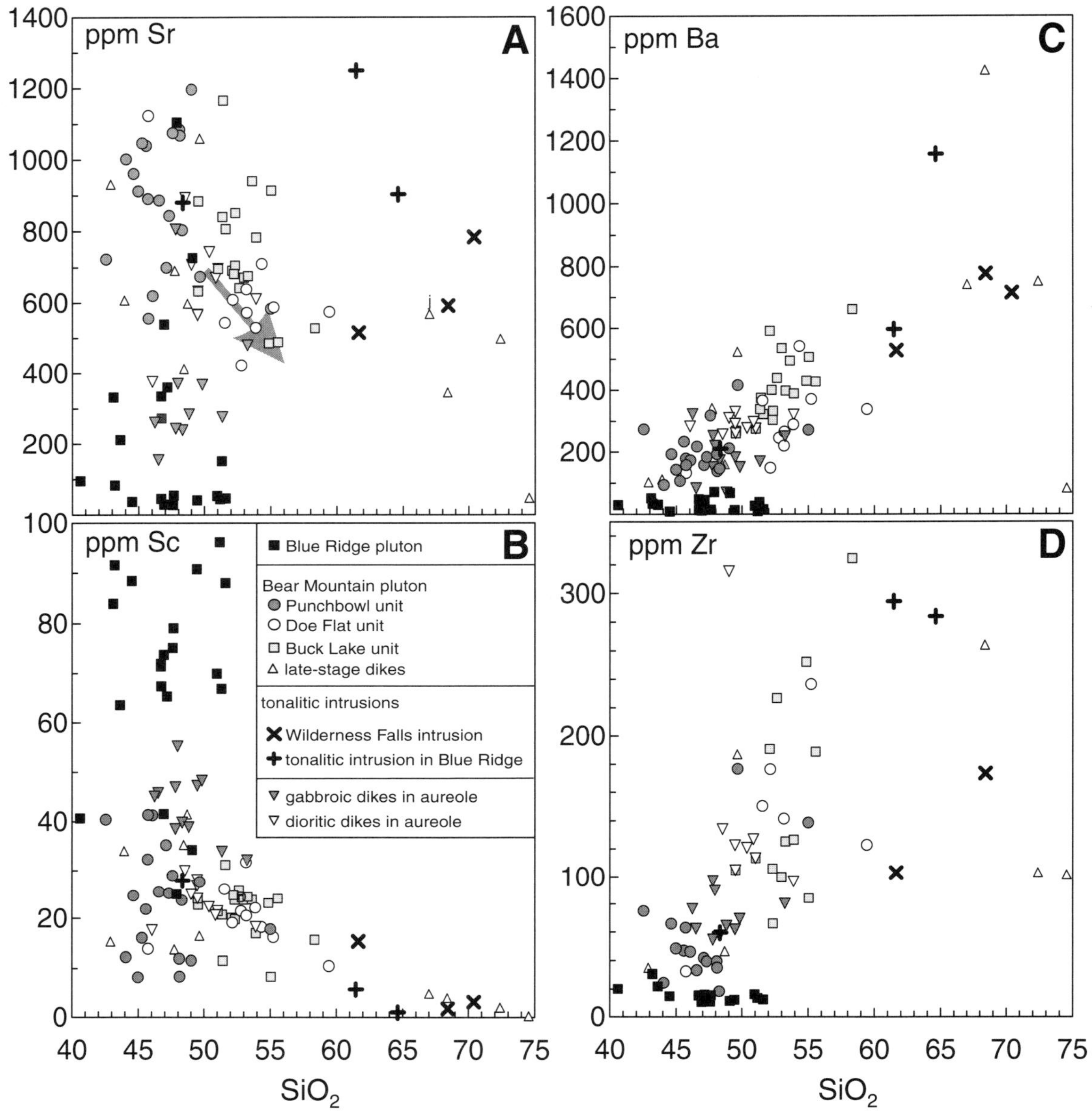

Figure 8. Variation of trace-element contents as a function of SiO_2. The arrow in A indicates the inferred change in Sr content during fractionation of dioritic magma parental to Punchbowl unit cumulates and Doe Flat unit differentiates.

These results are consistent with the observed crystallization order and overall mineral compositions in the Blue Ridge intrusion. Olivine in Blue Ridge dunites reaches Fo_{83} and is as magnesian as Fo_{74} in olivine gabbro (Snoke et al., 1981). Furthermore, the large temperature gaps between the liquidus and appearance of clinopyroxene (~100 °C) and between clinopyroxene and plagioclase stability (~75 °C) can explain the abundance of dunite and olivine clinopyroxenite in the Blue Ridge intrusion compared to olivine gabbro (Snoke et al., 1981). The first plagioclase to appear is somewhat less calcic than is seen in Blue Ridge gabbros (An_{79} in the model versus An_{85} in olivine melagabbro); however, such differences are dependent on the choice of parental magma and on initial (assumed) H_2O contents (e.g., Beard, 1986).

Not only does the calculated mineral assemblage fit the observed Blue Ridge assemblages reasonably well, but the calculated residual liquid at ~1136 °C is similar to the average composition of the dioritic dikes (Table 3 and Fig. 7). Specifically, the model matches SiO_2, Al_2O_3, and CaO abundances quite well, and it also predicts an increase in the FeO*/(FeO* + MgO) value at nearly constant SiO_2 content (Fig. 7). This increase in FeO*/(FeO* + MgO) is particularly interesting, be-

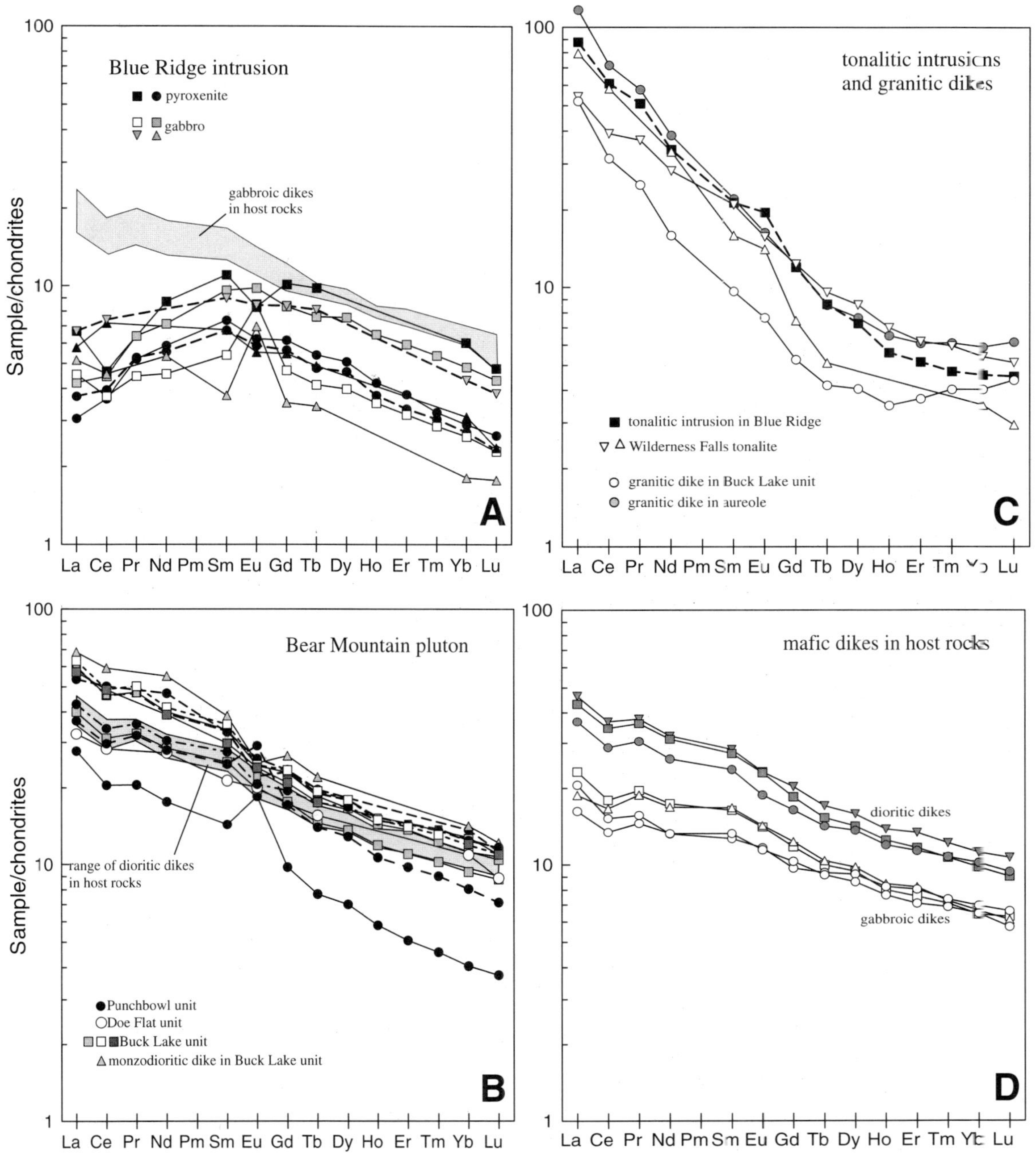

Figure 9. Rare-earth-element plots.

cause few plutonic rocks in the Klamath Mountains province have such high values, whereas many plutonic rocks have the lower FeO*/(FeO* + MgO) values displayed by the Blue Ridge gabbros (e.g., Barnes et al., 1990, 1995). That the differentiation model used to explain the cumulate assemblages in the Blue Ridge intrusion also reproduces the composition of the dioritic dikes is probably not fortuitous. Instead, we suggest that the dioritic dikes are the products of fractional crystallization of olivine and clinopyroxene from parental magmas similar to the gabbroic dikes.

Fractionation of the average gabbroic dike composition to make the average dioritic dike composition represents ~35–40% crystallization. Therefore, abundances of perfectly incompatible element contents should increase by a factor of ~1.6, which is the case for Rb, Ba, and the heavy REEs; however, the light REEs and Zr are enriched by a factor of ~2 (e.g., Figs. 8 and 9).

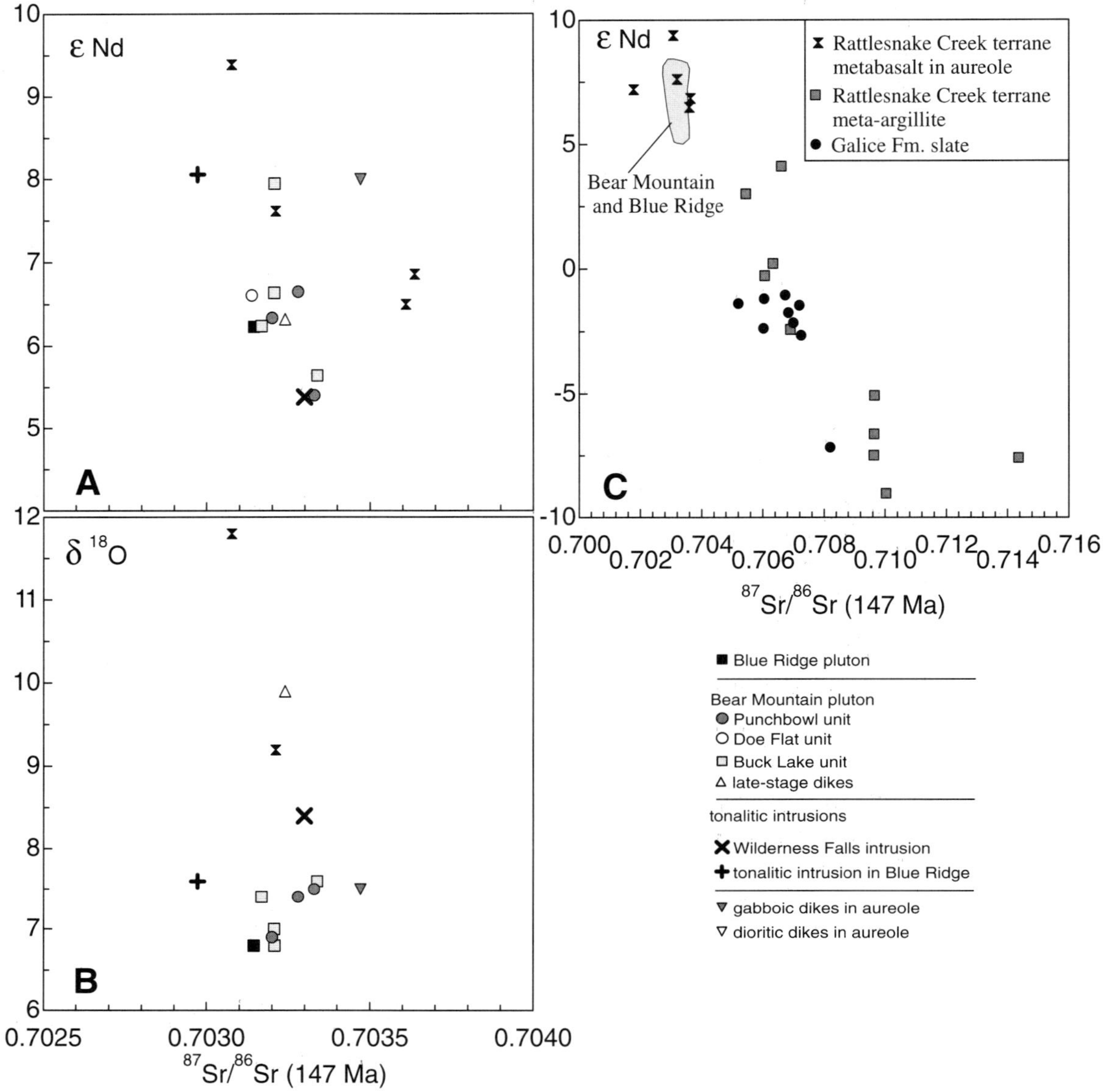

Figure 10. Isotope compositions.

These discrepancies are, at least in part, a function of the choice of parental compositions and the use of average values to represent the two dike populations. We conclude that magmas similar to the gabbroic dikes were parental to cumulates of the Blue Ridge intrusion *and* to the dioritic dike magmas.

Because the dioritic dike compositions plot between the lower-SiO_2 Punchbowl unit and higher-SiO_2 Doe Flat unit (Figs. 7 and 8), the possibility exists that the Punchbowl unit, Doe Flat unit, and dioritic dikes are related. We suggest that the Punchbowl rocks are cumulates from a dioritic dikelike parent and that the Doe Flat rocks crystallized from the evolved liquid produced by this process. It is likely that many Doe Flat samples are also cumulates (e.g., the positive Eu anomaly in the Doe Flat sample; Fig. 9), and had variable proportions of intercumulus melt. This scenario is consistent with the mafic mineral assemblages of the two units (dominated by hornblende with minor early pyroxene), with the lower Sr content of the Doe Flat compared to the Punchbowl samples (due to plagioclase accumulation in the Punchbowl magmas), and with the higher contents of incompatible elements, such as Zr and Ba, in Doe Flat samples (Fig. 8). Unfortunately, realistic thermodynamic models of this fractionation scheme are not possible because of the importance of hornblende, which has not been incorporated into either MELTS or PELE.

On the basis of the similarities in major element contents (Fig. 7) and REE patterns (Fig. 9) between the dioritic dikes and

TABLE 5. INPUT FOR FRACTIONATION MODEL

SiO_2	49.68*
TiO_2	0.89
Al_2O_3	12.91
Fe_2O_3	2.09
FeO	7.75
MgO	12.13
CaO	10.66
Na_2O	2.07
K_2O	0.65
P_2O_5	0.14
H_2O	1.02
fO_2	QFM + 1
T step	25 °C
P	3.5 kb

Note: QFM—quartz-fayalite-magnetite buffer.
*Values in wt%.

Buck Lake unit, we suggest that the dioritic magmas are logical parental magmas, with the exception of H_2O content. The paucity of cumulate hornblende in the Buck Lake rocks (some samples lack hornblende entirely, see above and Snoke et al., 1981), indicates that the Buck Lake parental magma contained much less H_2O than the Punchbowl–Doe Flat parent.

Enrichments of K_2O, Rb, and Ba in the Buck Lake rocks are distinctive, both in the Bear Mountain intrusive complex and in the Klamath Mountains province. Barnes et al. (this volume, Chapter 10) illustrate similar but more pronounced alkali enrichment in the ca. 170-Ma Ironside Mountain pluton. Their petrogenetic model suggests that alkali enrichment was a function of moderate-pressure fractional crystallization in which clinopyroxene fractionation predominated over olivine fractionation (see Meen, 1990). If this model is applicable to the Buck Lake unit, it suggests that H_2O-poor magma similar in other respects to the Punchbowl–Doe Flat dioritic parent fractionated deeper in the crust and then rose to the level of emplacement already enriched in alkalis.

An alternative hypothesis for compositional variation in the Punchbowl and Doe Flat units calls on mafic magmas parental to the Punchbowl rocks mixing with tonalitic magmas similar to those that intrude the Blue Ridge intrusion. For example, Figure 8B–D shows broadly linear variations among these units. However, if such mixing were important, the REE patterns of evolved Punchbowl and Doe Flat samples would show significant rotation, with higher light REE and lower heavy REE. This pattern is clearly not the case (Fig. 9B). Therefore, we interpret mixing of mafic and tonalitic magmas in the Punchbowl and Doe Flat units to be negligible. However, mingling and mixing of mafic magmas with Wilderness Falls tonalite is likely but requires additional data for thorough evaluation.

The Wilderness Falls intrusion and tonalitic rocks intrusive into the Blue Ridge pluton show a wide range of major and trace element compositions. They are generally calcic or calc-alkaline and have moderate K_2O contents (Fig. 7) and Sr contents between 500 and 1250 ppm (Fig. 8A). Although they share similar REE patterns (Fig. 9C), they are markedly distinct in Zr contents, with 100–180 ppm Zr in Wilderness Falls samples and ~290 ppm Zr in Blue Ridge tonalitic rocks.

The tonalite REE patterns are characteristic of control by garnet ± amphibole, either as fractionating phases or as a residual minerals during partial melting (e.g., Arth and Hanson, 1972; Barker and Arth, 1976; Drummond and Defant, 1990). Fractionation of a parental magma similar to any Blue Ridge or Bear Mountain magma is unlikely because of the comparatively high Sr contents of the tonalites. The arrow in Figure 8A shows the inferred fractionation trend responsible for the Punchbowl–Doe Flat rocks. Further fractionation should result in lower Sr contents (<500 ppm) rather than the higher ones seen in the tonalitic units. Fractionation from the parental magma of the Buck Lake unit is also unlikely, because fractionated daughter magmas should have K_2O contents >2.5 wt%, whereas all but one tonalitic samples has K_2O <2 wt% (Fig. 7E). These values suggest that if fractional crystallization was responsible for the tonalitic magmas, the parental composition was unlike the inferred magmas for the Bear Mountain and Blue Ridge plutons.

The alternative, partial melting, suggests metabasaltic source rocks. Partial melting of such a source with residual garnet could explain the REE patterns and also the high Sr concentrations (e.g., Drummond and Defant, 1990; Barnes et al., 1996). This metabasaltic source cannot be the migmatitic host rocks to the plutons (Plate 1), because garnet is lacking in the Bear Mountain intrusive complex aureole. The source could be structurally deeper metabasaltic rocks of the Josephine ophiolite, or even more deeply seated (unknown) metabasites. In either case, garnet stability requires partial melting at pressures greater than ~10 kb (Rapp et al., 1991). In turn, this condition implies either the presence of greatly thickened crust during Bear Mountain intrusive complex magmatism or partial melting of a newly subducting slab and transport of such a slab melt through the mantle wedge with little modification. We prefer the former interpretation; that is, deep-crustal melting.

In summary, we suggest that at least three types of mafic magma were parental to the Bear Mountain intrusive complex. The most primitive is magnesian basalt represented by the gabbroic dikes. Such magmas underwent fractional crystallization, leaving olivine- and clinopyroxene-rich cumulates, as in the Blue Ridge intrusion. The residual magmas were high-Al, relatively low-Mg basalts similar to the dioritic dikes. Water-rich magmas of this type were parental to the Punchbowl and Doe Flat units. In this case, the Punchbowl rocks represent cumulates of plagioclase + hornblende ± pyroxene, and Doe Flat rocks crystallized from the evolved liquid. Magmas parental to the Buck Lake unit were probably similar to the dioritic dikes but had lower H_2O contents. It is possible that the K_2O enrichment seen in the Buck Lake unit is due to differentiation at higher pressure than that of emplacement. The late-stage tonalitic intrusions at Wilderness Falls and in the Blue Ridge pluton are interpreted as the products of partial melting of metabasaltic rocks in deep crust that was thickened during the Late Jurassic Nevadan orogeny. A local, contact anatectic source is unlikely.

Arc Affinity

If the gabbroic and dioritic dikes are indeed similar to Bear Mountain intrusive complex parental magmas, then their normalized, multi-element patterns should be indicative of their source region. Figure 11 shows primitive-mantle-normalized diagrams for the gabbroic and dioritic dikes. The gabbroic and dioritic dikes have similar patterns, with depletions of Nb, Ta, Zr, and Ti and enrichments in K and Sr. Such patterns are characteristic of a subduction-modified source and indicate that the mantle beneath the Bear Mountain intrusive complex was either part of an active supra-subduction zone system or had been modified by subduction metasomatism prior to Nevadan thrusting.

EMPLACEMENT HISTORY

Structural Evidence for Emplacement History

The deflection of the adjacent host rocks around the margins of the Bear Mountain intrusive complex (Fig. 2) indicates that the forceful shouldering aside of wall rocks was an important emplacement mechanism for this plutonic complex. This interpretation is also supported by (1) the concentric igneous foliation pattern within the Bear Mountain pluton (Fig. 2, Plate 1); (2) widespread, small-scale, isoclinal folding in the rocks of the inner aureole; and (3) a steeply plunging mineral lineation in the rocks of the inner aureole, suggesting subvertical elongation during deformation of the wall rocks (e.g., see Compton, 1955, his Fig. 15).

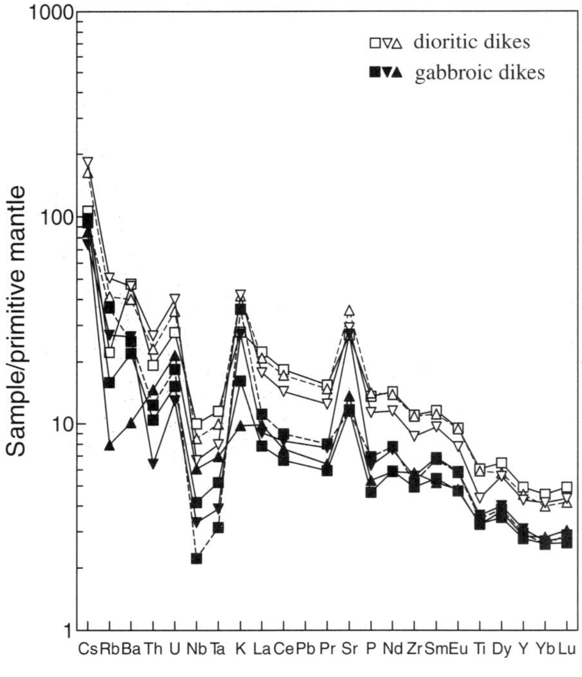

Figure 11. Multi-element diagram normalized to primitive mantle (Sun and McDonough, 1989).

The alignment of the Twin Peaks fault with the Bear Mountain pluton (Fig. 2, Plate 1) suggests that emplacement may have been also facilitated by this north–south-striking, steeply dipping, reverse fault (i.e., part of the Orleans fault system). The lack of displacement of the intrusive contact across the projected strike of the South Siskiyou Fork fault suggests that this element of the Orleans fault system also may have played a minor role in the emplacement of the Bear Mountain pluton (i.e., northwest-bulging of this intrusive body; Fig. 2).

The rocks of the inner aureole are characterized by a steeply dipping, penetrative foliation that commonly contains a steeply plunging mineral lineation (e.g., hornblende in hornblende schist or fine-grained amphibolite). Leucosome veinlets, interpreted as evidence of partial melting in the inner aureole (Snoke and Barnes, 2002), are common near the intrusive contact with the Bear Mountain pluton; some of these veinlets are distinctly asymmetric when viewed in the plane parallel to the lineation but perpendicular to foliation (Fig. 12A). This asymmetry consistently indicates a "pluton-down" sense of shear. Some porphyroblasts in inner-aureole hornblende schists or fine-grained amphibolites are asymmetric, typically forming σ-type porphyroblasts (Fig. 12B and C), and also indicate a "pluton-down" sense of shear. These data suggest that the Bear Mountain pluton foundered late in its emplacement history (e.g., Glazner, 1994; Glazner and Miller, 1997; also see Snoke et al., 2001, for evidence of late-stage foundering or sinking of a similar oceanic-arc pluton on Tobago, West Indies).

Another important aspect of the emplacement history of the Bear Mountain intrusive complex is that the Bear Mountain pluton is a heterogeneous, composite intrusive body consisting of multiple, mappable plutonic units that in turn consist of numerous intrusive sheets. Thus, the Bear Mountain intrusive complex in general—as well as the delineated plutonic units (Plate 1)—grew in a piecemeal (incremental) fashion and consequently represents the emplacement of oceanic-arc magmas over a minimum duration of 1.5 m.y. and possibly as long as 6 m.y. (Chamberlain et al., this volume). The sheetlike nature of the Bear Mountain intrusive complex is indicated by (1) the flanking ultramafic-mafic intrusions (Blue Ridge, Clear Creek, and Cedar Creek) that dip inward toward the center of the intrusive complex (Snoke et al., 1981; Fig. 2) and (2) the outcrop pattern of the Doe Flat plutonic unit (Plate 1). We envision that the assembly of the Bear Mountain intrusive complex was by the intrusion of numerous sheetlike bodies of oceanic-arc magma ranging from early-stage basaltic magmas to late-stage felsic magmas. The early basaltic magmas differentiated and formed extensive ultramafic to mafic cumulates, which played an instrumental role in the late sagging or foundering of the intrusive

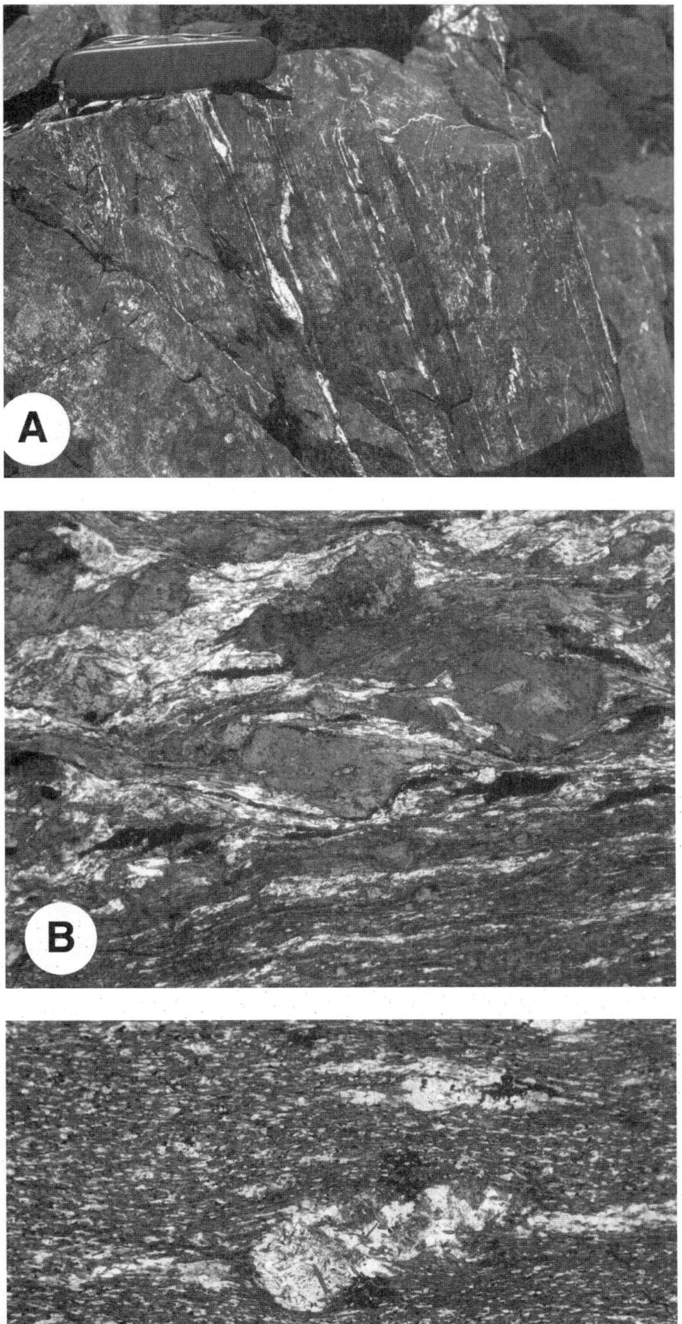

Figure 12. Field and photomicrographs demonstrating sense-of-shear in the dynamothermal aureole of the Bear Mountain intrusive complex. (A) Asymmetric lenticular leucosome veinlet in migmatitic amphibolite of the inner aureole of the Bear Mountain intrusive complex. Sense-of-shear is pluton-down. (B) Asymmetric σ-type hornblende porphyroblast in fine-grained amphibolite (hornblende schist) from the inner aureole of the Bear Mountain intrusive complex. Sense-of-shear is dextral. Plane-polarized light. Frame width is 4.3 mm. (C) Asymmetric σ-type plagioclase porphyroblast in fine-grained amphibolite (hornblende schist) from the inner aureole of the Bear Mountain intrusive complex. Sense-of-shear is dextral. Plane-polarized light. Frame width is 4.3 mm.

complex, yielding a "pluton-down" sense of shear in inner aureole wall rocks (Fig. 12). The Bear Mountain intrusive complex is roughly oval shaped (Wagner and Saucedo, 1987; Fig. 1), and its three-dimensional form possibly may be funnel shaped with a dense root of ultramafic rocks; this possibility is supported by the high positive Bouguer gravity anomaly associated with the Bear Mountain intrusive complex (Kim and Blank, 1973).

Temperature and Depth of Crystallization

Contact-Metamorphic Mineral Assemblages in Host Rocks. Mineral assemblages in the contact aureole of the Bear Mountain intrusive complex provide some temperature information regarding the conditions of magma emplacement. Near the intrusive contact, metaserpentinites contain the subassemblage olivine + orthopyroxene + nickel- or magnesium-rich chlorite (Table 1), indicative of metamorphic temperatures between 700 and 800 °C (Trommsdorff and Evans, 1972; Arai, 1975). Metarodingites within the dynamothermal aureole contain diopside + epidote + titanite ± glossular ± plagioclase, indicative of the high-temperature zones IV and V of Frost (1975). The thermal effects within the ultramafic unit extend for over 2 km from the intrusive contact (Table 1). At this distance from the contact, olivine, talc, and tremolite, which are stable in the ~500–600 °C temperature range (Trommsdorff and Evans, 1972; Arai, 1975), are present (Table 1). Metabasaltic rocks of the cover sequence of the Rattlesnake Creek terrane near the contact with the Bear Mountain intrusive complex typically contain plagioclase with dark green to olive-green or even brown hornblende with scarce diopsidic augite. This assemblage indicates high-temperature amphibolite (or hornblende-hornfels) to pyroxene-hornfels facies, equating to ~650–700 °C. Two-pyroxene mafic gneiss locally occurs in the host-rock septum that separates the Bear Mountain pluton and Blue Ridge intrusion (Snoke, 1972; Fig. 2). Tonalitic leucosome veinlets occur in migmatitic amphibolite gneiss of the inner aureole near the intrusive contact with the Bear Mountain pluton. Experimental partial melting studies of metabasaltic rocks indicate that amphibolite melting can occur at temperatures as low as 700 °C at H_2O-saturated conditions and higher than ~900 °C for dehydration melting (e.g., Rapp et al., 1991; Winther and Newton, 1991; Wolf and Wyllie, 1991). Outside the dynamothermal inner zone of the aureole, the prograde metamorphic mineral assemblage in the metabasaltic rocks and mafic complex of the Preston Peak ophiolite is sodic plagioclase + actinolitic amphibole ± epidote ±

titanite/leucoxene. This assemblage indicates upper greenschist-facies metamorphic conditions (including epidote-amphibolite facies), equating to ~450–500 °C.

Pressure constraints in the contact aureole are almost nonexistent compared to temperature constraints. In pressure-temperature space, the reaction curves for ultramafic rocks are nearly vertical, leaving pressure unconstrained. Most assemblages within the cover sequence lack pressure-dependent mineral assemblages. A spotted, slaty argillite was found in the metasedimentary rocks outside the inner zone of the aureole, and Snoke (1972) also reported similar rocks in his study of the outer aureole of the Bear Mountain intrusive complex. The shape of the porphyroblasts in hand sample is indicative of either andalusite or cordierite (Pattison and Tracy, 1991). However, under the microscope, these porphyroblasts are completely replaced by a micromosaic of phyllosilicates and quartz. If the porphyroblasts were originally andalusite, this paragenesis would indicate a pressure of ~4 kb at 500 °C, 4.5 kb at 550 °C, and as low as ~2 kb at 700 °C (Pattison and Tracy, 1991). Snoke et al. (1981) reported cordierite in two metasiltstones, one from inside and one from outside the inner zone of the aureole of the Bear Mountain intrusive complex. Within the inner zone, where the temperature was 600–700+ °C, cordierite would be stable up to ~3–4.5 kb, depending on X_{Mg}. Outside the dynamothermal aureole, where the temperature was ~500 °C, cordierite would only be stable below ~2 kb (Pattison and Tracy, 1991). Thus, contact metamorphic mineral assemblages suggest emplacement at >700 °C and from 2 to 4.5 kb.

Quartz-Ulvöspinel-Ilmenite-Fayalite (QUILF) Thermometry of the Buck Lake Unit. The Buck Lake unit contains biotite, orthopyroxene, clinopyroxene, inverted pigeonite, and Fe-Ti oxides. This assemblage is appropriate for use of the Ca-QUILF system (Lindsley and Frost, 1992) to estimate magmatic temperature. The first observation is that the presence of inverted pigeonite is indicative of high magmatic temperatures. Using a pressure estimate of 3 kb (which is reasonable on the basis of contact-metamorphic assemblages) and XFe in orthopyroxene of 0.45 (Snoke et al., 1981), Lindsley and Frost (1992) estimated clinopyroxene crystallization in excess of 1100 °C. They also showed that exsolution in the original pigeonite grains took place at ~1100 °C and that the orthopyroxene formed at ~1000 °C. All of these results are consistent with the calculated temperature at which Buck Lake–like magma was multiply saturated (~1136 °C; see discussion above on parental magmas).

Hornblende-Plagioclase Thermobarometry. Plagioclase and hornblende compositions from a number of samples of the Bear Mountain intrusive complex and dikes that intrude the aureole were used to estimate emplacement pressures (Hammarstrom and Zen, 1986; Anderson and Smith, 1995). Samples PPS-567 and DP4-7 are biotite-hornblende quartz diorite and were collected from the Doe Flat unit. Sample PPS-531 is a hornblende-bearing, biotite tonalite from the "dark phase" of the Wilderness Falls intrusion, and sample DP2-21 is a pyroxene- and quartz-bearing hornblende diorite that intruded the Blue Ridge pluton. These samples contain the assemblage plagioclase + hornblende + biotite + quartz + titanite + Fe-Ti oxides, and so are appropriate for use in the Al-in-hornblende barometery.

Among the dikes that intrude the dynamothermal contact-metamorphic aureole, DP2-10 is a foliated biotite-hornblende quartz diorite, DP1-9C is a pyroxene-hornblende quartz diorite, and DP1-9B and DP1-14 are hornfelsic dioritic dikes that preserve primary plagioclase and hornblende phenocrysts. These samples lack the complete equilibrium assemblage; in particular, primary titanite is typically absent and quartz is a groundmass phase.

The temperature-corrected barometer of Anderson and Smith (1995) uses the hornblende-plagioclase thermometers of Holland and Blundy (1994). These calculations were done using a spreadsheet provided by J.L. Anderson (University of Southern California, Los Angeles). Care was taken to analyze rim compositions and, where possible, to choose analytical spots near coexisting quartz (cf., Hollister et al., 1987). Possible sources of

TABLE 6. REPRESENTATIVE RESULTS OF Al-IN-HORNBLENDE BAROMETRY AND HORNBLENDE-PLAGIOCLASE THERMOMETRY

Sample*	PPS-567	DP4-7	PPS-531	DP2-21	DP2-10
P (kb), rims at plagioclase	4.1 ± 1.3	5.7 ± 1.2	5.2 ± 1.4	3.5 ± 0.4	3.6 ± 0.8
T (°C)	740 ± 17	730 ± 15	660 ± 25	750 ± 32	700 ± 23
P (kb), rims at quartz	3.8 ± 0.6	5.0 ± 0.8	5.4 ± 1.2		
T (°C)	730 ± 15	720 ± 18	660 ± 21		

*Rock types: PPS-567—biotite + hornblende quartz diorite, Doe Flat unit; DP4-7—biotite + hornblende quartz diorite, Doe Flat unit; PPS-531—hornblende-bearing, biotite tonalite, Wilderness Falls unit; DP2-21—pyroxene-bearing, hornblende (quartz-bearing) diorite, part of composite pluton that intrudes the Blue Ridge intrusion; DP2-10—foliated biotite + hornblende quartz diorite (leucosome segregation [dike] in migmatitic amphibolite), Bear Mountain pluton aureole, South Fork of the Smith River area.

error in these calculations include analytical uncertainty, post-crystallization reequilibration, nonequilibrium conditions between the plagioclase and hornblende, and uncertainty of the thermometer and barometer. Representative calculated temperatures and pressures are shown in Table 6. The results may be summarized as ~5.3 kb and 725 °C for the Doe Flat unit, ~3.5 kb and 760 °C for a quartz dioritic intrusion in Blue Ridge pluton, ~3.3 kb and 735 °C for intermediate dikes in the aureole, and ~5.1 kb and 700 °C for the Wilderness Falls unit.

With the exception of samples from the Doe Flat and Wilderness Falls units, the results are in the 3.1- to 3.8-kb range, which is consistent with pressures estimated on the basis of contact-metamorphic assemblages. Estimated temperatures are generally at or near solidus conditions. Reasons for the two discrepant (~5 kb) results are difficult to evaluate. It is possible that the hornblende was not in equilibrium with the analyzed plagioclase. Alternatively, the hornblende compositions may reflect higher pressures of crystallization. This alternative is consistent with field evidence cited above for late-stage "pluton-down" motion of the Bear Mountain pluton and with the location of the Doe Flat and Wilderness Falls bodies at the margin of the pluton. One might envision descent of the central mass of the pluton and consequent upward migration of deep-seated fractionated magma, which was injected along the original pluton–host-rock contact and crystallized to form all or part of the Doe Flat and Wilderness Falls bodies.

CONCLUSIONS

1. The Bear Mountain intrusive complex was assembled in a piecemeal (incremental) fashion but is divisible into broad, mappable units: the Blue Ridge, Clear Creek, and Cedar Creek ultramafic-gabbroic intrusions; Buck Lake plutonic unit; Punchbowl plutonic unit; Doe Flat plutonic unit; Wilderness Falls plutonic unit; and late-stage mafic dikes.
2. The Bear Mountain intrusive complex was emplaced after thrusting related to the Late Jurassic Nevadan orogeny based on the truncation of the Twin Peaks and South Siskiyou Fork faults (part of the Nevadan Orleans fault system) by the northern margin of the Bear Mountain intrusive complex (also see Chamberlain et al., this volume).
3. The initial parental magma for the Bear Mountain intrusive complex was an arc-related, H_2O- and MgO-rich basalt, which in turn differentiated to form extensive ultramafic-mafic cumulates and H_2O- and Al_2O_3-rich basalt/basaltic andesite, from which the Punchbowl unit accumulated. Further differentiation yielded the Doe Flat unit. Magmas of the Buck Lake unit were similar to the Punchbowl magma type but were H_2O poor. The late-stage quartz diorite/tonalite/granodiorite magmas were generated by the partial melting of metabasic crustal rocks.
4. The Bear Mountain intrusive complex experienced late-stage sagging or foundering due to the negative buoyancy of cumulates that formed the dense root of this intrusive complex, and the metamorphosed dynamothermal wall rocks that surround the Bear Mountain pluton consistently indicate a "pluton-down" sense of shear.

ACKNOWLEDGMENTS

This research was funded by National Science Foundation grant EAR-9902807 to AWS and EAR-9902912 to CGB. Discussions in the field and office with Rory McFadden were helpful in interpreting the evolution of the Bear Mountain intrusive complex. The help and advice of Susan Swapp in obtaining and interpreting mineral analyses with the University of Wyoming, Laramie, automated JEOL JXA-8900 electron probe microanalyzer are greatly appreciated. Melanie Barnes (Texas Tech University, Lubbock) supervised major and trace element analysis. We thank technical reviewers James S. Beard, Todd C. Feeley, and Aaron S. Yoshinobu for their many helpful comments and suggestions, which improved our final manuscript.

REFERENCES CITED

Anderson, J.L., and Smith, D.R., 1995, The effects of temperature and fO_2 on the Al-in-hornblende barometer: American Mineralogist, v. 80, p. 549–559.

Arai, S., 1975, Contact metamorphosed dunite-harzburgite complex in the Chugoku district, western Japan: Contributions to Mineralogy and Petrology, v. 52, p. 1–16, doi: 10.1007/BF00377998.

Arth, J.G., and Hanson, G.N., 1972, Quartz diorites derived by partial melting of eclogite or amphibolite at mantle depths: Contributions to Mineralogy and Petrology, v. 37, p. 161–174, doi: 10.1007/BF00371074.

Barker, F., and Arth, J.G., 1976, Generation of trondhjemitic-tonalitic liquids and Archean bimodal trondhjemitic-basalt suites: Geology, v. 4, p. 596–600, doi: 10.1130/0091-7613(1976)4<596:GOTLAA>2.0.CO;2.

Barnes, C.G., Allen, C.M., Hoover, J.D., and Brigham, R.H., 1990, Magmatic components of a tilted plutonic system, Klamath Mountains, California, *in* Anderson, J.L., ed., The nature and origin of Cordilleran magmatism: Boulder, Colorado, Geological Society of America Memoir 174, p. 331–346.

Barnes, C.G., Scott, W.P., Kistler, R.W., Prestvik, T., and Sundvoll, B., 1992, Tectonic implications of isotopic variation among Jurassic and Early Cretaceous plutons, Klamath Mountains: Geological Society of America Bulletin, v. 104, p. 117–126, doi: 10.1130/0016-7606(1992)104<0117:TIOIVA>2.3.CO;2.

Barnes, C.G., Johnson, K., Barnes, M.A., Prestvik, T., Kistler, R.W., and Sundvoll, B., 1995, The Grayback pluton: Magmatism in a Jurassic back-arc environment, Klamath Mountains, Oregon: Journal of Petrology, v. 36, p. 397–415.

Barnes, C.G., Petersen, S.W., Kistler, R.W., Murray, R., and Kays, M.A., 1996, Source and tectonic implications of tonalite-trondhjemite magmatism in the Klamath Mountains: Contributions to Mineralogy and Petrology, v. 123, p. 40–60, doi: 10.1007/s004100050142.

Barnes, C.G., Allen, C.M., and Snoke, A.W., 2003, Episodes of Middle to Late Jurassic arc development in the Klamath Mountains, California and Oregon: EOS, Transactions of the American Geophysical Union, v. 84, no. 46, Fall Meeting suppl., abstract V31C-0946, p. F1545–F1546.

Barnes, C.G., Mars, E.V., Swapp, S., and Frost, C.D., 2006, this volume (Chapter 10), Petrology and geochemistry of the Middle Jurassic Ironside Mountain batholith: Evolution of potassic magmas in a primitive arc setting, *in* Snoke, A.W., and Barnes, C.G., eds., Geological studies in the Klamath Mountains province, California and Oregon: A volume in honor of William P. Irwin: Boulder, Colorado, Geological Society of America Special Paper 410, doi: 10.1130/2006.2410(10).

Barnes, C.G., Snoke, A.W., Harper, G.D., Frost, C.D., McFadden, R.R., Bushey, J.C., and Barnes, M.A.W., 2006, this volume (Chapter 17), Arc plutonism following regional thrusting: Petrology and geochemistry of syn- and post-Nevadan plutons in the Siskiyou Mountains, Klamath Mountains province, California, *in* Snoke, A.W., and Barnes, C.G., eds., Geological studies in the Klamath Mountains province, California and Oregon: A volume in honor of William P. Irwin: Boulder, Colorado, Geological Society of America Special Paper 410, doi: 10.1130/2006.2410(17).

Beard, J.S., 1986, Characteristic mineralogy of arc-related cumulate gabbros: Implications for the tectonic setting of gabbroic plutons and for andesite genesis: Geology, v. 14, p. 848–851, doi: 10.1130/0091-7613(1986)14<848:CMOACG>2.0.CO;2.

Boudreau, A.E., 1999, PELE—A version of the MELTS software program for the PC platform: Computers & Geosciences, v. 25, p. 201–203, doi: 10.1016/S0098-3004(98)00117-4.

Bushey, J.C., 2003, Petrology and structural geology of the northern part of the Bear Mountain intrusive complex and contact aureole, Klamath Mountains, Del Norte and Siskiyou Counties, California [M.S. thesis]: Laramie, University of Wyoming, 166 p.

Chamberlain, K.R., Snoke, A.W., Barnes, C.G., and Bushey, J.C., 2006, this volume, New U-Pb radiometric dates of the Bear Mountain intrusive complex, Klamath Mountains, California, *in* Snoke, A.W., and Barnes, C.G., eds., Geological studies in the Klamath Mountains province, California and Oregon: A volume in honor of William P. Irwin: Boulder, Colorado, Geological Society of America Special Paper 410, doi: 10.1130/2006.2410(15).

Clayton, R., and Mayeda, T., 1963, The use of bromine pentafluoride in the extraction of oxygen from oxides and silicates for isotopic analysis: Geochimica et Cosmochimica Acta, v. 27, p. 43–52, doi: 10.1016/0016-7037(63)90071-1.

Coleman, R.G., Manning, C.E., Mortimer, N., Donato, M.M., and Hill, L.B., 1988, Tectonic and regional metamorphic framework of the Klamath Mountains and adjacent Coast Ranges, California and Oregon, *in* Ernst, W.G., ed., Metamorphism and crustal evolution of the western United States—Rubey volume VII: Englewood Cliffs, New Jersey, Prentice Hall, p. 1061–1097.

Compton, R.R., 1955, Trondhjemite batholith near Bidwell Bar, California: Geological Society of America Bulletin, v. 66, p. 9–44.

Drummond, M.S., and Defant, M.J., 1990, A model for trondhjemite-tonalite-dacite genesis and crustal growth via slab melting: Archean to modern comparisons: Journal of Geophysical Research, v. 95, p. 21,503–21,521.

Frost, B.R., 1975, Contact metamorphism of serpentinite, chloritic blackwall and rodingite at Paddy-Go-Easy Pass, central Cascades, Washington: Journal of Petrology, v. 16, p. 272–313.

Frost, B.R., Barnes, C.G., Collins, W.J., Arculus, R.J., Ellis, D.J., and Frost, C.D., 2001, A geochemical classification for granitic rocks: Journal of Petrology, v. 42, p. 2033–2048, doi: 10.1093/petrology/42.11.2033.

Frost, C.D., Barnes, C.G., and Snoke, A.W., 2006, this volume, Nd and Sr isotopic data from argillaceous rocks of the Galice Formation and Rattlesnake Creek terrane, Klamath Mountains: Evidence for the input of Precambrian sources, *in* Snoke, A.W., and Barnes, C.G., eds., Geological studies in the Klamath Mountains province, California and Oregon: A volume in honor of William P. Irwin: Boulder, Colorado, Geological Society of America Special Paper 410, doi: 10.1130/2006.2410(05).

Glazner, A.F., 1994, Foundering of mafic plutons and density stratification of continental crust: Geology, v. 22, p. 435–438, doi: 10.1130/0091-7613(1994)022<0435:FOMPAD>2.3.CO;2.

Glazner, A.F., and Miller, D.M., 1997, Late-stage sinking of plutons: Geology, v. 25, p. 1099–1102, doi: 10.1130/0091-7613(1997)025<1099:LSSOP>2.3.CO;2.

Goldstein, S.L., O'Nions, R.K., and Hamilton, P.J., 1984, A Sm-Nd study of atmospheric dusts and particulates from major river systems: Earth and Planetary Science Letters, v. 70, p. 221–236.

Hacker, B.R., Donato, M.M., Barnes, C.G., McWilliams, M.O., and Ernst, W.G., 1995, Timescales of orogeny: Jurassic construction of the Klamath Mountains: Tectonics, v. 14, p. 677–703, doi: 10.1029/94TC02454.

Hammarstrom, J.M., and Zen, E., 1986, Aluminum in hornblende: An empirical igneous geobarometer: American Mineralogist, v. 71, p. 1297–1313.

Harper, G.D., and Wright, J.E., 1984, Middle to Late Jurassic evolution of the Klamath Mountains, California-Oregon: Tectonics, v. 3, p. 759–772.

Harper, G.D., Saleeby, J.B., and Heizler, M., 1994, Formation and emplacement of the Josephine ophiolite and the Nevadan orogeny in the Klamath Mountains, California-Oregon: U/Pb zircon and ^{40}Ar/^{39}Ar geochronology: Journal of Geophysical Research, v. 99, p. 4293–4321, doi: 10.1029/93JB02061.

Hess, H.H., 1941, Pyroxenes of common mafic magmas: American Mineralogist, v. 26, p. 515–535, 573–594.

Holland, T., and Blundy, J., 1994, Non-ideal interactions in calcic amphiboles and their bearing on amphibole-plagioclase thermometry: Contributions to Minerology and Petrology, v. 116, p. 433–447, doi: 10.1007/BF00310910.

Hollister, L.S., Grissom, G.C., Peters, E.K., Stowell, H.H., and Sisson, V.B., 1987, Confirmation of the empirical correlation of Al in hornblende with pressure of solidification of calc-alkaline plutons: American Mineralogist, v. 72, p. 231–239.

Irwin, W.P., 1994, Geologic map of the Klamath Mountains, California and Oregon: Reston, Virginia, U.S. Geological Survey Miscellaneous Investigations Map I-2148, scale 1:500,000, 2 sheets.

Irwin, W.P., and Wooden, J.L., 1999, Plutons and accretionary episodes of the Klamath Mountains, California and Oregon: Reston, Virginia, U.S. Geological Survey Open-File Report 99-374. (http://geopubs.wr.usgs.gov/open-file/of99-374/).

Irwin, W.P., Jones, D.L., and Kaplan, T.A., 1978, Radiolarians from pre-Nevadan rocks of the Klamath Mountains, California and Oregon, *in* Howell, D.G., and McDougall, K.A., eds., Mesozoic paleogeography of the western United States I: Los Angeles, Pacific Section, Society of Economic Paleontologists and Mineralogists, p. 303–310.

Jachens, R.C., Barnes, C.G., and Donato, M.M., 1986, Subsurface configuration of the Orleans fault: Implications for deformation in the western Klamath Mountains, California: Geological Society of America Bulletin, v. 97, p. 388–395, doi: 10.1130/0016-7606(1986)97<388:SCOTOF>2.0.CO;2.

Kim, C.K., and Blank, H.R., Jr., 1973, Bouguer gravity map of California—Weed Sheet: San Francisco, California Division of Mines and Geology, scale 1:250,000, 2 sheets, 8 p. pamphlet.

Lindsley, D.H., and Frost, B.R., 1992, Equilibria among Fe-Ti oxides, pyroxenes, olivine, and quartz: Part I. Theory: American Minerologist, v. 77, p. 987–1003.

McFadden, R.R., Snoke, A.W., and Barnes, C.G., 2006, this volume, Structural and petrologic evolution of the Bear Peak intrusive complex, Klamath Mountains, California, *in* Snoke, A.W., and Barnes, C.G., eds., Geological studies in the Klamath Mountains province, California and Oregon: A volume in honor of William P. Irwin: Boulder, Colorado, Geological Society of America Special Paper 410, doi: 10.1130/2006.2410(16).

Meen, J.K., 1990, Elevation of potassium content of basaltic magma by fractional crystallization: the effect of pressure: Contributions to Mineralogy and Petrology, v. 104, p. 309–331, doi: 10.1007/BF00321487.

Miller, J., Miller, R., Wooden, J., and Harper, G., 2003, Geochronologic links between the Ingalls ophiolite, North Cascades, Washington and the Josephine ophiolite, Klamath Mts., Oregon and California: Geological Society of America Abstracts with Programs, v. 35, p. 113.

Miller, M.M., and Saleeby, J.B., 1995, U-Pb geochronology of detrital zircon from Upper Jurassic synorogenic turbidites, Galice Formation, and related rocks, western Klamath Mountains: Correlation and Klamath Mountains provenance: Journal of Geophysical Research, v. 100, p. 18,045–18,058, doi: 10.1029/95JB00761.

Obenshain, K.R., 1981, Amphibolite anatexis within a contact aureole [M.S. thesis]: Athens, University of Georgia, 85 p.

Pattison, D.R.M., and Tracy, R.J., 1991, Phase equilibria and thermobarometry of metapelites: Reviews in Mineralogy, v. 26, p. 105–207.

Petford, N., Kerr, R.C., and Lister, J.R., 1993, Dike transport of granitoid magmas: Geology, v. 21, p. 845–848, doi: 10.1130/0091-7613(1993)021<0845:DTOGM>2.3.CO;2.

Rapp, R.P., Watson, E.B., and Miller, C.F., 1991, Partial melting of amphibolite/eclogite and the origin of Archean trondhjemites and tonalities: Precambrian Research, v. 51, p. 1–25, doi: 10.1016/0301-9268(91)90092-O.

Saleeby, J.B., and Harper, G.D., 1993, Tectonic relations between the Galice Formation and the Condrey Mountain Schist, Klamath Mountains, northern California, *in* Dunne, G.C., and McDougall, K.A., eds., Mesozoic paleogeography of the western United States II: Los Angeles, Pacific Section, Society of Economic Paleontologists and Mineralogists, v. 71, p. 61–80.

Saleeby, J.B., Harper, G.D., Snoke, A.W., and Sharp, W.D., 1982, Time relations and structural-stratigraphic patterns in ophiolite accretion, west central Klamath Mountains, California: Journal of Geophysical Research, v. 87, p. 3831–3848.

Shannon, W.M., Barnes, C.G., and Bickford, M.E., 1997, Grenville magmatism in West Texas: Petrology and geochemistry of the Red Bluff granitic suite: Journal of Petrology, v. 38, p. 1279–1305, doi: 10.1093/petrology/38.10.1279.

Snoke, A.W., 1972, Petrology and structure of the Preston Peak area, Del Norte and Siskiyou counties, California [Ph.D. dissertation]: Stanford, California, Stanford University, 274 p.

Snoke, A.W., 1977, A thrust plate of ophiolitic rocks in the Preston Peak area, Klamath Mountains, California: Geological Society of America Bulletin, v. 88, p. 1641–1659, doi: 10.1130/0016-7606(1977)88<1641:ATPOOR>2.0.CO;2.

Snoke, A.W., and Barnes, C.G., 2002, Anatexis and dynamothermal metamorphism during the emplacement of a syntectonic oceanic arc-root pluton, Klamath Mountains, California: Geological Society of America Abstracts with Programs, v. 34, p. A-22.

Snoke, A.W., Quick, J.E., and Bowman, H., 1981, Bear Mountain igneous complex, Klamath Mountains, California: An ultrabasic to silicic calc-alkaline suite: Journal of Petrology, v. 22, p. 501–552.

Snoke, A.W., Rowe, D.W., Yule, J.D., and Wadge, G., 2001, Petrologic and structural history of Tobago, West Indies: A fragment of the accreted Mesozoic oceanic arc of the southern Caribbean: Boulder, Colorado, Geological Society of America Special Paper 354, 54 p. with 4 plates.

Springer, R.K., 1974, Contact metamorphosed ultramafic rocks in the western Sierra Nevada Foothills, California: Journal of Petrology, v. 15, p. 160–195.

Sun, S.-s., and McDonough, W.F., 1989, Chemical and isotopic systematics of oceanic basalts: Implications for mantle composition and processes, *in* Saunders, A.D., and Norry, M.J., eds., Magmatism in the ocean basins: London, Geological Society Special Publication 42, p. 313–345.

Trommsdorff, V., and Evans, B.M., 1972, Progressive metamorphism of antigorite schist in the Bergell tonalite aureole (Italy): American Journal of Science, v. 272, p. 423–437.

Trommsdorff, V., and Evans, B.W., 1974, Alpine metamorphism of peridotitic rocks: Schweizerische Mineralogische und Petrographische Mitteilungen, v. 54, p. 333–352.

Wagner, D.L., and Saucedo, G.J., 1987, Geologic map of the Weed quadrangle: Sacramento, California Division of Mines and Geology Regional Geologic Map Series, Map 4A, scale 1:250,000, 4 sheets, 15 p. pamphlet.

Winther, K.T., and Newton, R.C., 1991, Experimental melting of hydrous low-K tholeiite: Evidence on the origin of Archaean cratons: Bulletin of the Geological Society of Denmark, v. 39, p. 213–228.

Wolf, M.B., and Wyllie, P.J., 1991, Dehydration-melting of solid amphibolite at 10 kbar: Textural development, liquid interconnectivity and applications to the segregation of magmas: Mineralogy and Petrology, v. 44, p. 151–179, doi: 10.1007/BF01166961.

Wright, J.E., and Fahan, M.R., 1988, An expanded view of Jurassic orogenesis in the western United States Cordillera: Middle Jurassic (pre-Nevadan) regional metamorphism and thrust faulting within an active arc environment, Klamath Mountains, California: Geological Society of America Bulletin, v. 100, p. 859–876, doi: 10.1130/0016-7606(1988)100<0859:AEVOJO>2.3.CO;2.

Wright, J.E., and Wyld, S.J., 1994, The Rattlesnake Creek terrane, Klamath Mountains, California: An early Mesozoic volcanic arc and its basement of tectonically disrupted oceanic crust: Geological Society of America Bulletin, v. 106, p. 1033–1056, doi: 10.1130/0016-7606(1994)106<1033:TRCTKM>2.3.CO;2.

Manuscript Accepted by the Society 10 January 2006

Printed in the USA

Geological Society of America
Special Paper 410
2006

New U-Pb radiometric dates of the Bear Mountain intrusive complex, Klamath Mountains, California

Kevin R. Chamberlain*
Arthur W. Snoke
Department of Geology and Geophysics, Department 3006,
1000 East University Avenue, University of Wyoming, Laramie, Wyoming 82071, USA
Calvin G. Barnes
Department of Geosciences, Texas Tech University, Lubbock, Texas 79409-1053, USA
Jonathan C. Bushey
Department of Geology and Geophysics, Department 3006,
1000 East University Avenue, University of Wyoming, Laramie, Wyoming 82071, USA

ABSTRACT

New, high-precision U-Pb titanite (sphene) and zircon dates from five samples of the Bear Mountain intrusive complex establish the timing and duration of magmatism. The oldest, magmatic date (150.5 ± 0.6 Ma) comes from dark-colored titanite from a biotite-hornblende tonalite that is part of a composite pluton that intrudes the Blue Ridge ultramafic-gabbroic intrusion. Pale titanite and zircon from this sample yielded a distinctly younger date of 149.3 ± 0.3 Ma. A similar pattern of mineral dates is also apparent in two samples of the areally extensive Punchbowl unit of the Bear Mountain pluton. Dark-colored titanite in one of these samples yielded a date of 149.5 ± 0.6 Ma, whereas the dates of pale titanite and zircon are 147.4 ± 0.3 Ma. The second sample of the Punchbowl unit only contained a single morphology of pale titanite, which yielded the same date as zircon (148.2 ± 0.3 Ma). The U-Pb zircon date of the Buck Lake unit of the Bear Mountain pluton, 148.2 ± 0.2 Ma, supports field evidence that the Buck Lake unit was emplaced synchronously with the Punchbowl unit. A lower age limit on magmatism in the Bear Mountain intrusive complex comes from a 145.4 ± 0.4-Ma zircon date from a late crosscutting mafic dike. All samples exhibit slight inheritance in the zircon data, with 152- to 150-Ma minimum ages. The mafic dike contains inherited components that are at least 264 Ma and possibly Paleoproterozoic in age.

The new dates constrain magmatism in the Bear Mountain intrusive complex to the period from 151 to 147 Ma, with a minimum duration of 1.5 m.y. and a maximum of 6 m.y. The dates establish that the emplacement and crystallization of the Bear Mountain intrusive complex post-dated regional thrust faulting (Orleans fault) associated with the Nevadan orogeny, including the South Siskiyou Fork fault, which is interpreted as an oblique-slip tear fault associated with the Orleans (thrust) fault system.

*E-mail: kchamber@uwyo.edu.

Chamberlain, K.R., Snoke, A.W., Barnes, C.G., and Bushey, J.C., 2006, New U-Pb radiometric dates of the Bear Mountain intrusive complex, Klamath Mountains, California, *in* Snoke, A.W., and Barnes, C.G., eds., Geological studies in the Klamath Mountains province, California and Oregon: A volume in honor of William P. Irwin: Geological Society of America Special Paper 410, p. 317–332, doi: 10.1130/2006.2410(15). For permission to copy, contact editing@geosociety.org.

The pattern of mineral dates from the composite pluton intruded into the Blue Ridge intrusion as well as the areally more extensive Punchbowl unit indicate that crystallization of these bodies occurred over 1.5–2 m.y., due to either insulating effects of the intrusive complex and/or magma recharge.

Keywords: Klamath Mountains, northwestern California, Jurassic, U-Pb geochronology, titanite, zircon, igneous rocks, oceanic-arc pluton, thrust faulting, Nevadan orogeny

INTRODUCTION

The tectonic development of the Klamath Mountains province involved multiple contractional events, which led to the progressive accretion of predominantly oceanic terranes onto the western North American continental margin (Irwin, 1972, 1981; Harper and Wright, 1984; Harper et al., 1994; Irwin and Wooden, 1999; Fig. 1). Many of these tectonostratigraphic terranes represent parts of oceanic to near-continental arc sequences subsequently intruded by plutonic magmas that ranged from ultramafic to silicic. Some of these plutons are "pinning plutons" (after the terminology of Howell et al., 1985) that intrude across regional, terrane-bounding faults, whereas other plutons are truncated by regional faults. Thus, precise dating of these plutons provides important constraints on the timing and duration of tectonic activity within the Klamath Mountains province (e.g., Lanphere et al., 1968; Davis et al., 1978; Irwin, 1985; Wright and Fahan, 1988; Irwin and Wooden, 1999).

One of the most important and widespread contractional events in the Klamath Mountains province was thrusting of terranes of the western Paleozoic and Triassic belt (Irwin, 1972) over terranes of the western Jurassic belt (Blake, 1984; Blake et al., 1985). This event typically emplaced higher-grade rocks of the hanging-wall allochthon onto lower-grade rocks of the footwall along the east-rooted, west-directed Orleans fault (Hershey, 1906, 1911), a regional thrust fault with a minimum horizontal displacement of ~70 km (Jachens et al., 1986). In the central Klamath Mountains, the age of thrusting is constrained by ca. 153-Ma detrital zircon ages in footwall rocks of the Galice Formation (Miller et al., 2003) and the ca. 150-Ma Summit Valley "pinning pluton" (Harper et al., 1994). This orogenic event (commonly referred to as the "Nevadan orogeny"; Blackwelder, 1914; Taliaferro, 1942; Snoke and Barnes, this volume) also involved polyphase folding of the metasedimentary rocks of the footwall (Upper Jurassic Galice Formation; e.g., Snoke, 1977; Harper and Wright, 1984; Harper et al., 1994; Gray, this volume).

The Bear Mountain intrusive complex consists of two large plutons, a number of smaller intrusive bodies, and associated dikes (Snoke et al., 1981; Bushey et al., this volume). The large proportion of ultramafic and gabbroic to dioritic rocks that comprise the Bear Mountain intrusive complex suggests that this intrusive complex is the underpinnings of a primitive oceanic arc (Snoke et al., 1981, 1982). Initial U-Pb zircon radiometric dating of the Bear Mountain intrusive complex (Saleeby et al., 1982) yielded Late Jurassic (i.e., Nevadan) ages. Specifically, Saleeby et al. (1982) reported ages as old as ca. 153 Ma and as young as ca. 149 Ma. As the timing of thrusting on the Orleans fault was refined, it became apparent that these U-Pb zircon dates suggested a syn-thrusting emplacement history for the Bear Mountain intrusive complex. This chapter reports the results of high-precision dating of zircon and titanite (sphene) from several intrusive units in the Bear Mountain intrusive complex. The data are interpreted to suggest that the intrusive complex was emplaced immediately after thrusting ended, although field and petrographic evidence indicates that deformation continued during the emplacement of the intrusive complex. These new data also indicate a protracted (several m.y.) magmatic history of the Bear Mountain intrusive complex, which has important implications for the longevity of magmatism in arc-related settings.

GEOLOGIC SETTING

At the present level of exposure, the Bear Mountain intrusive complex intrudes the Rattlesnake Creek terrane, including the late Middle Jurassic (ca. 164 Ma) mafic complex of the Preston Peak ophiolite (Snoke, 1977; Saleeby et al., 1982; Harper et al., 1994). Furthermore, the Bear Mountain pluton truncates the Twin Peaks fault, a north–south-striking fault that separates the rocks of the mafic complex on the west side of the fault (footwall) from serpentinized peridotite of the basement complex of the Rattlesnake Creek terrane on the east side (hanging wall) of the fault (Fig. 2). This fault is interpreted as a high-angle, reverse fault related to the Orleans fault system (Preston Peak fault [thrust] of Snoke, 1977), the terrane-bounding fault that separates the Rattlesnake Creek terrane from the western Klamath terrane, including the Upper Jurassic Galice Formation (Fig. 1). In the same light, the Bear Mountain intrusive complex truncates the northwest–southeast-striking, South Siskiyou Fork fault (Snoke, 1977; Fig. 2) that is interpreted as a tear fault

Figure 1. Generalized geologic map of the Klamath Mountains province. BMic—Bear Mountain intrusive complex; BPic—Bear Peak intrusive complex (see McFadden et al., this volume); O. ft.—Orleans fault. The Preston Peak fault of Snoke (1977) is the local name for the regional Orleans (thrust) fault. Modified from Irwin (1994).

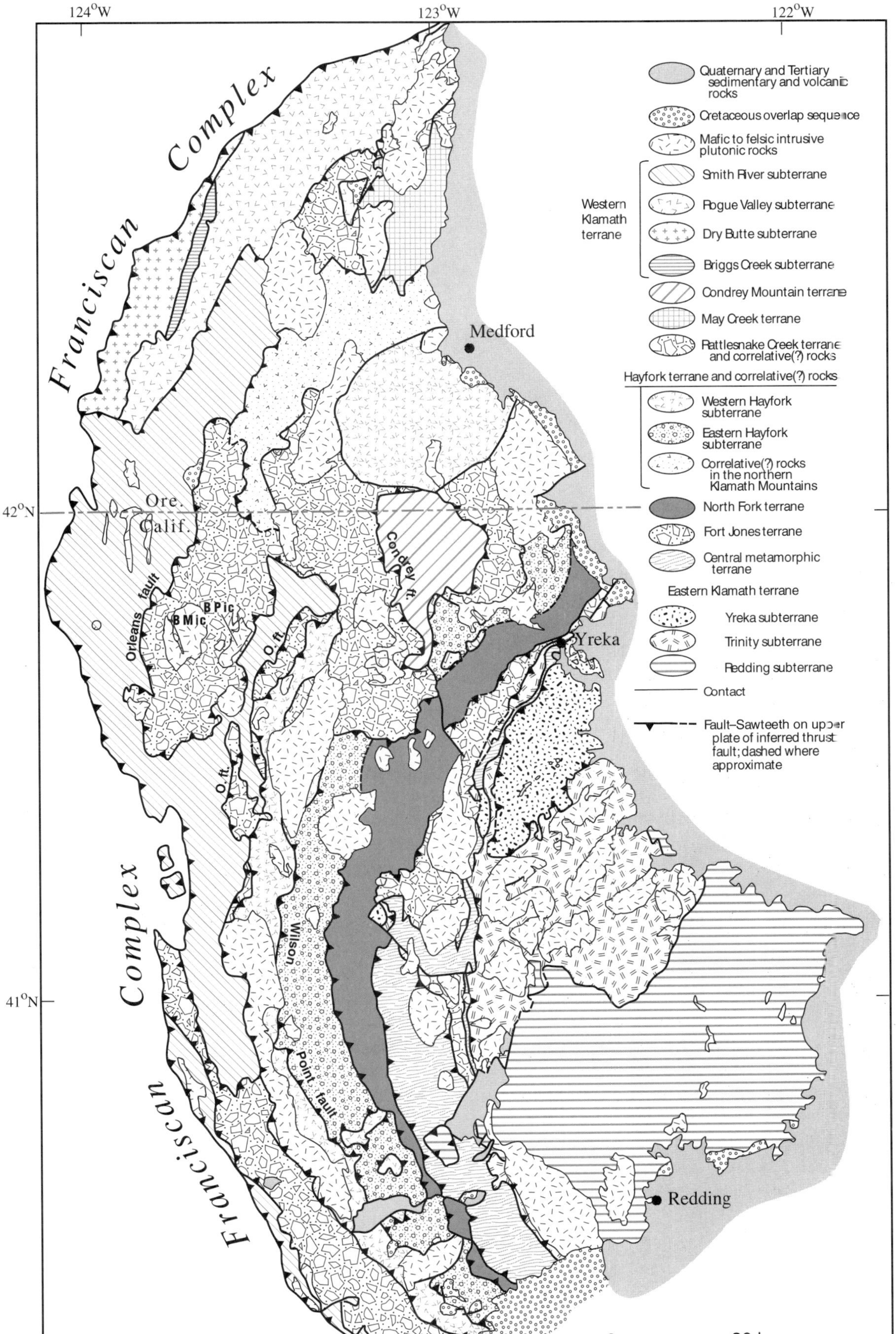
124°W
123°W
122°W
42°N
41°N
Franciscan Complex
Franciscan Complex
Medford
Yreka
Redding
Ore.
Calif.
Orleans fault
BPic
BMic
O. ft.
O. ft.
Condrey ft.
Wilson Point fault
0
30 km
Quaternary and Tertiary sedimentary and volcanic rocks
Cretaceous overlap sequence
Mafic to felsic intrusive plutonic rocks
Western Klamath terrane
Smith River subterrane
Rogue Valley subterrane
Dry Butte subterrane
Briggs Creek subterrane
Condrey Mountain terrane
May Creek terrane
Rattlesnake Creek terrane and correlative(?) rocks
Hayfork terrane and correlative(?) rocks
Western Hayfork subterrane
Eastern Hayfork subterrane
Correlative(?) rocks in the northern Klamath Mountains
North Fork terrane
Fort Jones terrane
Central metamorphic terrane
Eastern Klamath terrane
Yreka subterrane
Trinity subterrane
Redding subterrane
Contact
Fault–Sawteeth on upper plate of inferred thrust fault; dashed where approximate

associated with the Orleans fault (see Bushey et al., this volume, for a more detailed discussion of the field relationships). These field relationships thus suggest that the Bear Mountain intrusive complex was emplaced after tectonic juxtaposition of the Rattlesnake Creek terrane above the western Klamath terrane. The age of the Upper Jurassic Galice Formation, which is widespread and the stratigraphically highest unit in the western Klamath terrane, is based on the presence of the pelecypod *Buchia concentrica* (Sowerby), which ranges from the late Oxfordian to middle Kimmeridgian (Imlay, 1959) and the presence of ca. 153-Ma detrital zircon grains (Miller et al., 2003).

The contact-metamorphic effects in the country rocks surrounding the Bear Mountain intrusive complex are widespread and locally intense (Snoke et al., 1981; Bushey et al., this volume). Rocks of the Rattlesnake Creek terrane commonly reached hornblende-hornfels–facies conditions in the inner aureole of the Bear Mountain intrusive complex and locally reached pyroxene-hornfels–facies conditions (two-pyroxene–bearing mafic rocks; Fig. 2). The rocks of the inner aureole are chiefly hornblende schists or fine-grained amphibolite with a pervasive foliation and commonly a steeply plunging mineral lineation, defined by hornblende grains. Subordinate metasedimentary rocks (chiefly siliceous meta-argillite) are intercalated with the metabasic rocks of the inner aureole. Some of the metabasic rocks of the Rattlesnake Creek terrane were partially melted in the inner aureole, yielding migmatitic amphibolites with tonalitic leucosomes (Snoke and Barnes, 2002).

The Bear Mountain intrusive complex was initially subdivided into five broad lithologic units based chiefly on petrography (Snoke et al., 1981): (1) ultramafic and gabbroic rocks of the Blue Ridge intrusion, other small ultramafic intrusive bodies, and related dikes; (2) biotite–two-pyroxene monzodiorite/diorite; (3) heterogeneous hornblende-rich gabbro and diorite and related dikes; (4) leucocratic rocks, chiefly biotite ± hornblende tonalite and granodiorite; and (5) late dikes (mafic to felsic). Field relationships, including the contact metamorphism of clinopyroxenitic dikes of unit 1 by unit 2, indicate that the Blue Ridge intrusion and other satellitic ultramafic-gabbroic bodies are the oldest units of the intrusive complex. Leucocratic rocks intrude all other units except the late dikes; however, it was not clear whether these leucocratic rocks represented a single stage of magmatism or encompassed the entire Bear Mountain intrusive complex time interval. Relative relationships between units 2 and 3 were much more equivocal, but Snoke et al. (1981) concluded that unit 2 was older than unit 3. Recent geologic mapping by Bushey et al. (this volume) has revised these subdivisions, particularly in regard to units 2 and 3. These authors designate three plutonic units in the northern part of the Bear Mountain pluton: Buck Lake, Punchbowl, and Doe Flat. The Buck Lake unit is basically equivalent to unit 2 of Snoke et al. (1981), whereas the Punchbowl and Doe Flat units are unit 3 of Snoke et al. (1981). Bushey et al. (this volume) cite specific field relationships, including crosscutting dikes and inclusions of one unit within another, that indicate that the Doe Flat unit is younger than either the Buck Lake or Punchbowl units, whereas field relationships between the latter two units suggest a rough synchronicity. The new geochronological data presented in this chapter support this basic chronology.

URANIUM-LEAD GEOCHRONOLOGY

Five samples were collected from the Bear Mountain intrusive complex to determine the crystallization ages of different intrusive phases and overall duration of magmatism in the intrusive complex. The samples include a biotite-hornblende tonalite (DP2-20z) that is part of a small, composite pluton emplaced into the Blue Ridge intrusion (Fig. 2); two samples from the Punchbowl unit (Bushey et al., this volume), a predominantly hornblende-pyroxene gabbro-diorite (DP4-8z and 10z); a biotite–two-pyroxene diorite from the Buck Lake unit (DP1-13z) previously dated by Saleeby et al. (1982); and a late, undeformed crosscutting mafic dike (DP1-35z) that provides a lower age limit on magmatic activity during the development of the Bear Mountain intrusive complex.

Zircons were recovered and analyzed from each of these samples, and three of them also contained magmatic titanite (sphene). The titanite data are particularly useful, for titanite typically exhibits neither significant Pb loss nor inherited xenocrystic cores (e.g., Frost et al., 2001), both of which can add significant complexity in the interpretation of zircon data. The majority of the zircon analyses come from air-abraded grains, but some unabraded grains were also analyzed to test for metamorphic overgrowths and to monitor the overall effect of air abrasion on discordance. Two zircon fractions were analyzed by the annealing and partial dissolution method of Mattinson (2005), termed "chemical abrasion thermal ionization mass spectrometry," or CA-TIMS. This method is touted to reduce Pb loss effects in young zircons more effectively than simple mechanical abrasion. Sample locations are shown in Figure 2; data are presented in Tables 1 and 2. Laboratory procedures are described in Table 1. Concordia coordinates and radiogenic ages of individual analyses were calculated by using model mantle Pb isotopic values for 150 Ma (Zartman and Doe, 1981) for initial Pb and assigning variation in these values that approximates the range of modern-day measured mid-ocean ridge basalt (MORB). Model values were used rather than measured isotopic compositions of coexisting feldspars, because the samples had little to no alkali feldspar and measurements of plagioclase are typically slightly radiogenic, due to in situ decay of uranium. For Jurassic rocks, $^{206}Pb/^{238}U$ dates are generally the most reliable, and using a slightly radiogenic plagioclase $^{206}Pb/^{204}Pb$ value would lead to an age bias. These considerations only

Figure 2. Geologic map of the Bear Mountain intrusive complex. Sample locations for geochronology are indicated. Modified from Snoke et al. (1981); Bushey et al. (this volume); A.W. Snoke (unpublished data).

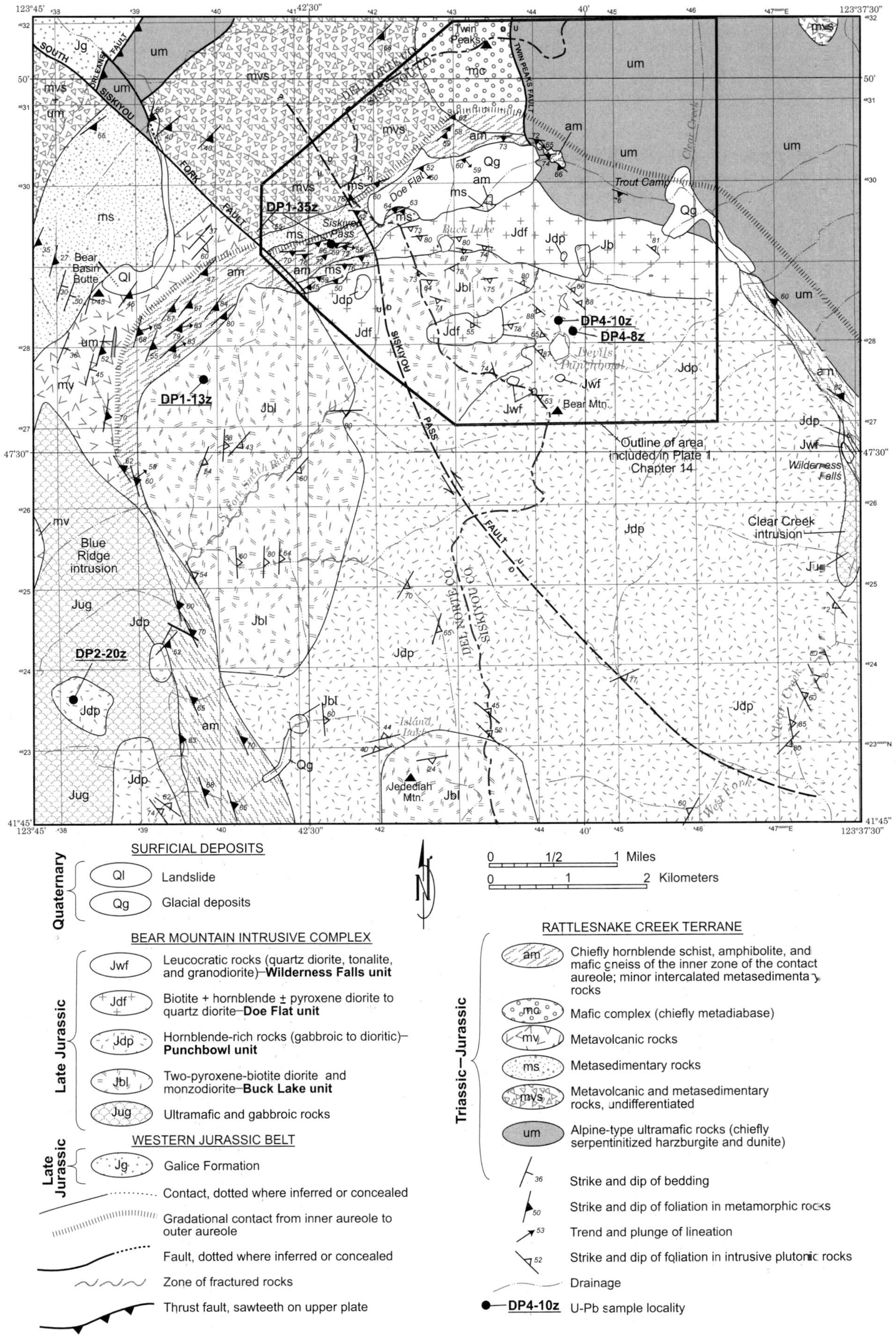
123°45'
42'30"
40'
123°37'30"
50'
47'30"
41°45'
SOUTH FORK FAULT
ORLEANS FAULT
SISKIYOU
DEL NORTE CO.
SISKIYOU CO.
TWIN PEAKS FAULT
SISKIYOU PASS FAULT
Twin Peaks
Clear Creek
Trout Camp
Doe Flat
Siskiyou Pass
Buck Lake
Bear Basin Butte
DP1-35z
DP1-13z
DP2-20z
DP4-10z
DP4-8z
Devils Punchbowl
Bear Mtn.
Outline of area included in Plate 1, Chapter 14
Wilderness Falls
Clear Creek intrusion
Blue Ridge intrusion
Island Lake
Jedediah Mtn.
West Fork
0 1/2 1 Miles
0 1 2 Kilometers
Quaternary
SURFICIAL DEPOSITS
Ql Landslide
Qg Glacial deposits
BEAR MOUNTAIN INTRUSIVE COMPLEX
Late Jurassic
Jwf Leucocratic rocks (quartz diorite, tonalite, and granodiorite)—Wilderness Falls unit
Jdf Biotite + hornblende ± pyroxene diorite to quartz diorite—Doe Flat unit
Jdp Hornblende-rich rocks (gabbroic to dioritic)—Punchbowl unit
Jbl Two-pyroxene-biotite diorite and monzodiorite—Buck Lake unit
Jug Ultramafic and gabbroic rocks
WESTERN JURASSIC BELT
Late Jurassic
Jg Galice Formation
Contact, dotted where inferred or concealed
Gradational contact from inner aureole to outer aureole
Fault, dotted where inferred or concealed
Zone of fractured rocks
Thrust fault, sawteeth on upper plate
RATTLESNAKE CREEK TERRANE
Triassic–Jurassic
am Chiefly hornblende schist, amphibolite, and mafic gneiss of the inner zone of the contact aureole; minor intercalated metasedimentary rocks
mc Mafic complex (chiefly metadiabase)
mv Metavolcanic rocks
ms Metasedimentary rocks
mvs Metavolcanic and metasedimentary rocks, undifferentiated
um Alpine-type ultramafic rocks (chiefly serpentinitized harzburgite and dunite)
Strike and dip of bedding
Strike and dip of foliation in metamorphic rocks
Trend and plunge of lineation
Strike and dip of foliation in intrusive plutonic rocks
Drainage
DP4-10z U-Pb sample locality

TABLE 1. DATA FOR U-Pb

					Corrected atomic ratios§				
Sample	Weight (mg)	U (ppm)	RadPb (ppm)	ComPb† (ppm)	$^{206}Pb/^{204}Pb$	%err	$^{238}U/^{204}Pb$	%err	$^{206}Pb/^{208}Pb$
DP2-20z bio-hbl tonalite (UTM S10 0438055E 4623651N)									
sphene							150.5 ± 0.61 Ma (MSWD = 0.0514) (3 pt.*		
sph nm1.0a lt 0.25mm	2.200	29	1.12	0.31	153.19	(1.1)	5774	(1.2)	3.03
sph nm1.0a dk 0.40mm*	2.250	77	5.37	2.01	74.74	(0.5)	2391	(0.7)	0.94
sph nm1.0a yel 0.25mm p1*	2.020	49	1.86	0.41	193.94	(1.0)	7438	(1.1)	2.55
sph nm1.0a yel 0.25mm p2*	1.950	49	1.82	0.36	217.32	(1.5)	8423	(1.7)	2.55
zircon							149.33 ± 0.29 Ma (MSWD = 1.3) (4 pt.* zr		
zr d-2 c'less euh*	0.046	56	2.51	1.10	92.20	(14)	3176	(18)	5.08
zr d-1 yel euh sm	0.028	118	3.03	0.22	776.59	(144)	33281	(147)	6.27
zr d-1 yel euh lg 7g	0.056	69	2.51	0.87	131.37	(16)	4942	(18)	6.32
zr d-2 yel euh sm aa*	0.129	32	1.02	0.21	245.20	(27)	9682	(30)	5.11
zr d-2 yel euh lg 24g aa*	0.167	83	2.01	0.02	5281	(207)	225237	(208)	6.72
zr d-2 yel euh lg 29g CA*	0.251	49	1.17	0.01	10336	(111)	440203	(112)	6.97
DP4-8z hbl-px gabbro (UTM S10 0444377E 4628108N)							149.5 ± 0.64 Ma 6/8 age one sphene pt.;		
zr d-1 pk anh sA	0.309	140	3.3	0.07	2968	(38)	128170	(38)	8.56
zr nm1 pk anh sB	0.144	144	3.5	0.10	2188	(56)	91454	(57)	10.38
zr d-1 yel anh sC aa*	0.334	221	5.2	0.00	508189	(3820)	21899000	(3820)	7.41
zr d-1 lt br anh sD aa*	0.446	83	1.9	0.01	15318	(228)	659810	(228)	11.13
zr d-1 yel anh sE aa*	0.520	294	6.9	0.06	7458	(27)	321820	(27)	7.84
zr d-1 yel anh sF aa	0.232	151	3.5	0.03	6252	(100)	273960	(100)	8.12
zr d-1 br anh sG aa*	0.844	387	9.4	0.12	4673	(7.9)	200780	(7.9)	6.70
sph nm1.0a dk 0.35mm 11g	1.410	88	4.5	0.64	218.72	(0.9)	8563	(0.9)	0.91
sph nm1.0a lt 0.20mm*	1.900	45	2.2	0.95	86.39	(0.4)	2940	(0.4)	2.62
DP4-10z hbl-px gabbro late pod (UTM S10 0444315E 4628198N)							148.24 ± 0.30 Ma (MSWD=0.94) (4 pt.*		
zr nm2 lt br anh sA	0.181	206	4.86	0.01	35299	(529)	1529150	(530)	7.33
zr nm2 lt br anh sB	0.013	1001	24.6	0.30	4864	(102)	205887	(102)	6.73
zr nm2 lt br anh 8g aa2*	0.743	24	0.62	0.03	1091	(32)	46122	(33)	6.34
zr nm2 lt br anh 24g aa*	0.102	143	4.26	0.57	380.20	(13)	15569	(14)	4.26
sph nm1.0a yel 0.35mm*	1.110	11	0.52	0.22	88.96	(2.6)	3057	(3.3)	3.02
sph nm1.0a yel 0.35mm p2*	4.440	64	2.28	0.49	202.96	(0.3)	7994	(0.6)	2.78
DP1-13z bio-2 px monzodiorite (UTM S10 0439716E 4627659N)							148.17 ± 0.21 Ma (MSWD = 0.92) (5 pt.*		
zr nm1 yel euh 5g	0.055	190	4.60	0.10	2688	(81)	115490	(81)	8.21
zr nm1 yel euh 6g aa*	0.055	183	4.51	0.18	1500	(92)	63892	(92)	7.27
zr nm2 yel euh 7g	0.055	220	5.37	0.11	2822	(143)	119910	(143)	7.31
zr d-1 yel anh incl. 36g aa*	0.186	112	2.73	0.07	2456	(64)	105040	(64)	6.97
zr d-1 yel euh 17g aa*	0.074	64	1.60	0.07	1414	(79)	60085	(79)	6.81
zr nm1 yel euh 25g aa2*	0.186	100	2.40	0.01	20555	(603)	881850	(603)	7.00
zr nm1 yel euh 7g CA*	0.278	94	2.25	0.01	17713	(91)	760950	(91)	7.06
DP1-35z late mafic dike (UTM S10 0441513E 4629247N)							≥145.43 ± 0.42 Ma (MSWD = 0.43) (2 pt.*		
zr nm2 yel euh 36g	0.052	537	12.61	0.56	1351	(31)	59839	(31)	8.21
zr nm2 yel euh 13g aa*	0.015	261	6.46	0.44	865	(121)	37196	(121)	7.76
zr nm3 yel euh 18g	0.051	266	6.52	0.12	3234	(144)	135130	(144)	7.85
zr d-1 yel euh 70g aa	0.074	194	4.64	0.16	1748	(66)	74909	(66)	9.04
zr nm2 yel euh incl. 65g aa	0.089	242	5.88	0.07	5403	(133)	224960	(133)	8.82
zr nm3 yel euh 13g aa*	0.002	1092	29.55	4.72	345	(34)	14336	(34)	9.05
zr nm3 yel euh 12g aa2	0.005	711	18.19	2.02	518	(33)	21839	(33)	9.50

Notes: d, nm represent angles of diamagnetic and paramagnetic susceptibility on a barrier style Frantz separator at 1.4 amps; nm1.0a represent lower amperage paramagnetic splits. Mineral dissolution and chemistry were adapted from methods developed by Krogh (1973), Parrish et al. (1987), and Mattinson (2005). Aliquots of dissolved sample were spiked with a mixed $^{208}Pb/^{235}U$ tracer. The Pb and U samples were loaded onto single rhenium filaments with silica gel and graphite, respectively; isotopic compositions were measured in multicollector, static mode on a VG Sector 54 mass spectrometer at the University of Wyoming, Laramie, with ^{204}Pb in Daly-photomultiplier collector if the $^{206}Pb/^{204}Pb$ value was greater than 200, and all other isotopes in Faraday collectors. Mass discrimination factors of 0.108 to 0.060 ± 0.06 %/amu were determined for Pb based on silica gel batch and replicate analyses of National Institute of Standards and Technology (NIST) SRM 981; U mass discrimination was consistently 0.0 ± 0.06%/amu. Procedural blanks for zircon decreased from 10 pg to less than 5 pg Pb during the course of the study. Uranium blanks were consistently less than 1 pg. Sphene blanks were 20 pg Pb and less than 1 pg U. Concordia coordinates, U-Pb dates, $^{206}Pb/^{204}Pb$ vs. $^{238}U/^{204}Pb$ isochrons, weighted averages, and uncertainties were calculated using PBDAT and ISOPLOT programs (Ludwig 1988, 1991); initial Pb isotopic compositions estimated by the Zartman and Doe (1981) model mantle for 150 Ma. The decay constants used by PBDAT are those recommended by the IUGS Subcommission on Geochronology (Steiger and Jäger, 1977): 0.155125×10^{-9}/yr for ^{238}U, 0.98485×10^{-9}/yr for ^{235}U and present-day $^{238}U/^{235}U = 137.88$. The $^{208}Pb/^{235}U$ tracer was calibrated against MIT2 gravimetric standard and yielded a $^{206}Pb/^{238}U$ date of 419.26 ± 0.64 Ma for zircon standard R33 ($^{207}Pb/^{235}U$ date of 420.11 ± 0.59 Ma), in good agreement with the ROM date for R33 (Black et al., 2004). The $^{206}Pb/^{238}U$ dates corrected for ^{230}Th disequilibrium used the equations of Schärer (1984). Ratios of Th/U for the minerals are based on radiogenic ^{208}Pb and growth from 150

Corrected atomic ratios (cont.)						Ages# (Ma)					Rho	Rho
$^{206}Pb/^{238}U$	%err	$^{207}Pb/^{235}U$	%err	$^{207}Pb/^{206}Pb$	%err	$^{206}Pb/^{238}U$	abs err	$^{206}Pb_{Th}/^{238}U$	$^{207}Pb/^{235}U$	$^{207}Pb/^{206}Pb$	6/8-7/5‡	a-m††
sph 6/4 vs. 238/4 isochron date)												
0.02345	(0.49)	0.1595	(1.07)	0.0493	(0.80)	149.41	(0.73)	149.48	150.2	163	0.72	0.97
0.02381	(1.02)	0.1603	(3.26)	0.0488	(2.76)	151.71	(1.53)	151.69	151.0	139	0.62	0.71
0.02368	(0.45)	0.1615	(0.73)	0.0495	(0.44)	150.88	(0.67)	150.94	152.0	170	0.83	0.95
0.02369	(0.44)	*no measurement of ^{207}Pb*				150.92	(0.65)	150.98				0.98
weighted average 206/238 date)												
0.02343	(0.84)	0.1666	(3.20)	0.0516	(2.77)	149.28	(1.24)	149.36	156.5	267	0.60	1.00
0.02280	(0.44)	0.1565	(1.59)	0.0498	(1.35)	145.33	(0.64)	145.41	147.6	185	0.63	1.00
0.02298	(0.64)	0.1614	(2.35)	0.0509	(2.04)	146.46	(0.92)	146.54	151.9	238	0.59	1.00
0.02349	(0.44)	0.1607	(1.34)	0.0496	(1.12)	149.66	(0.65)	149.75	151.3	177	0.62	1.00
0.02337	(0.32)	0.1586	(0.45)	0.0492	(0.29)	148.91	(0.47)	148.99	149.5	159	0.78	1.00
0.02344	(0.31)	0.1586	(0.33)	0.0491	(0.09)	149.36	(0.46)	149.45	149.5	152	0.96	1.00
147.43 ± 0.33 Ma (MSWD = 1.42) (5 pt.* sph and zr 6/4 vs. 238/4 isochron date)												
0.02302	(0.32)	0.1559	(0.34)	0.0491	(0.12)	146.72	(0.46)	146.81	147.1	153	0.93	1.00
0.02373	(0.32)	0.1612	(0.40)	0.0493	(0.22)	151.17	(0.47)	151.27	151.7	161	0.83	1.00
0.02321	(0.32)	0.1571	(0.34)	0.0491	(0.11)	147.88	(0.46)	147.97	148.1	152	0.95	1.00
0.02319	(0.31)	0.1572	(0.35)	0.0492	(0.13)	147.77	(0.46)	147.87	148.2	155	0.92	1.00
0.02312	(0.32)	0.1566	(0.32)	0.0491	(0.06)	147.33	(0.46)	147.42	147.8	155	0.98	1.00
0.02276	(0.32)	0.1542	(0.40)	0.0491	(0.23)	145.05	(0.45)	145.14	145.6	154	0.82	1.00
0.02319	(0.32)	0.1571	(0.34)	0.0492	(0.10)	147.76	(0.47)	147.84	148.2	155	0.96	1.00
0.02346	(0.43)	0.1590	(0.76)	0.0491	(0.52)	149.50	(0.64)	149.48	149.8	155	0.75	0.94
0.02333	(0.81)	0.1604	(1.57)	0.0499	(0.97)	148.64	(1.18)	148.70	151.0	189	0.86	0.80
sph and zr weighted average 206/238 date)												
0.02307	(0.32)	0.1580	(0.34)	0.0497	(0.12)	147.04	(0.46)	147.13	148.9	179	0.94	1.00
0.02354	(0.31)	0.1603	(0.38)	0.0494	(0.19)	149.99	(0.47)	150.08	150.9	166	0.87	1.00
0.02326	(0.32)	0.1597	(0.45)	0.0498	(0.28)	148.21	(0.47)	148.30	150.4	186	0.78	1.00
0.02328	(0.35)	0.1595	(0.64)	0.0497	(0.48)	148.33	(0.51)	148.41	150.3	181	0.68	1.00
0.02328	(0.78)	0.1626	(1.96)	0.0507	(1.50)	148.33	(1.15)	148.40	153.0	226	0.72	0.99
0.02316	(0.49)	0.1574	(0.71)	0.0493	(0.38)	147.62	(0.72)	147.68	148.4	162	0.87	0.68
aa & CA zr weighted average 206/238 date)												
0.02312	(0.32)	0.1565	(0.41)	0.0491	(0.24)	147.35	(0.47)	147.44	147.7	153	0.83	1.00
0.02319	(0.33)	0.1574	(0.62)	0.0492	(0.46)	147.80	(0.48)	147.89	148.5	159	0.67	1.00
0.02339	(0.32)	0.1592	(0.55)	0.0494	(0.40)	149.03	(0.48)	149.12	150.0	165	0.70	1.00
0.02322	(0.32)	0.1576	(0.43)	0.0492	(0.26)	147.95	(0.47)	148.04	148.6	159	0.80	1.00
0.02323	(0.34)	0.1576	(0.59)	0.0492	(0.43)	148.05	(0.50)	148.14	148.6	157	0.69	1.00
0.02329	(0.32)	0.1577	(0.40)	0.0491	(0.21)	148.40	(0.47)	148.49	148.7	154	0.85	1.00
0.02325	(0.31)	0.1572	(0.32)	0.0490	(0.06)	148.19	(0.46)	148.28	148.3	150	0.98	1.00
aa zr weighted average 206/238 date)												
0.02229	(0.32)	0.1514	(0.40)	0.0493	(0.21)	142.08	(0.45)	142.17	143.2	161	0.84	1.00
0.02278	(0.39)	0.1562	(1.22)	0.0497	(1.02)	145.21	(0.56)	145.30	147.4	182	0.62	1.00
0.02380	(0.33)	0.1630	(0.49)	0.0497	(0.33)	151.62	(0.49)	151.71	153.3	179	0.75	1.00
0.02310	(0.36)	0.1571	(0.49)	0.0493	(0.30)	147.22	(0.52)	147.31	148.2	163	0.79	1.00
0.02394	(0.32)	0.1700	(0.38)	0.0515	(0.18)	152.50	(0.47)	152.60	159.5	264	0.87	1.00
0.02283	(0.44)	0.1560	(1.15)	0.0496	(0.96)	145.49	(0.63)	145.59	147.2	174	0.60	1.00
0.02292	(0.40)	0.1558	(0.78)	0.0493	(0.60)	146.06	(0.58)	146.15	147.0	163	0.65	1.00

Ma. Abbreviations: aa—air abraded; anh—anhedral; CA—chemically abraded; dk—dark yellow; euh—euhedral; g—number of grains; incl.—with inclusions; lt—pale yellow to colorless; lt br—light brown; mm—shortest diameter in millimeters; MSWD—mean square of weighted deviates; p—pick number; s—single grain; sm—small; sph—sphene; yel—yellow to honey colored; zr—zircon.

*Used in age calculation.

†ComPb: common Pb corrected for laboratory blank of 10 ± 6 pg Pb for zircon and 20 ± 6 pg Pb for sphene analyses. Isotopic composition of blank was 19.09 ± 0.2, 15.652 ± 0.1, 38.81 ± 0.2 for 6/4, 7/4, and 8/4, respectively.

§$^{206}Pb/^{204}Pb$, $^{238}U/^{204}Pb$ corrected for blank and mass discrimination, all others corrected for blank, mass discrimination and initial Pb, using model mantle Pb isotopic values for 150 Ma ($^{206}Pb/^{204}Pb$, $^{207}Pb/^{204}Pb$, $^{208}Pb/^{204}Pb$ =17.80, 15.46, 37.50, respectively; Zartman and Doe, 1981). Values in parentheses are 2σ errors in percent, except for $^{206}Pb/^{238}U$ ages, which are absolute in m.y.

#Ages: $^{206}Pb_{Th}/^{238}U$ date corrected for initial ^{230}Th disequilibria, using estimated $[Th/U]_{magma}$ = 2.85, based on averages of measured Bear Mountain intrusive complex samples. $[Th/U]_{mineral}$ calculated from radiogenic ^{208}Pb and growth from 150 Ma. $\lambda(^{232}Th) = 4.9475 \times 10^{-11}$/yr; $\lambda(^{230}Th) = 9.22 \times 10^{-6}$/yr.

‡Rho 6/8-7/5: $^{206}Pb/^{238}U$ vs. $^{207}Pb/^{235}U$ error correlation coefficient.

§§Rho a-m: $^{206}Pb/^{204}Pb$ vs. $^{238}U/^{204}Pb$ error correlation coefficient.

TABLE 2. VARIATIONS IN $^{206}Pb/^{238}U$ DATES WITH CHOICE OF INITIAL Pb ISOTOPIC COMPOSITION

		Dates (Ma)			
	$^{206}Pb/^{204}Pb$ corrected	$^{206}Pb/^{238}U$ ZD man	abs err	$^{206}Pb/^{238}U$ SK 150	abs err
DP2-20z bio-hbl tonalite					
sphene					
sph nm1.0a lt 0.25mm	153	149.41	(0.73)	148.67	(0.72)
sph nm1.0a dk 0.40mm*	75	151.71	(1.53)	149.93	(1.53)
sph nm1.0a yel 0.25mm p1*	194	150.88	(0.67)	150.31	(0.66)
sph nm1.0a yel 0.25mm p2*	217	150.92	(0.65)	150.41	(0.65)
zircon					
zr d-2 c'less euh 0.05mm*	92	149.28	(1.24)	147.94	(1.17)
zr d-1 yel euh sm 0.03mm	777	145.33	(0.64)	145.20	(0.56)
zr d-1 yel euh lg 7g 0.1mm	131	146.46	(0.92)	145.60	(0.88)
zr d-2 yel euh sm 0.03mm aa*	245	149.66	(0.65)	149.22	(0.61)
zr d-2 yel euh lg 24g 0.2mm aa*	5281	148.91	(0.47)	148.89	(0.47)
zr d-2 yel euh lg 29g 0.15mm CA*	10336	149.36	(0.46)	149.35	(0.46)
DP4-8z hbl-px gabbro					
zr d-1 pk anh sA	2968	146.72	(0.46)	146.69	(0.46)
zr nm1 pk anh sB	2188	151.17	(0.47)	151.13	(0.47)
zr d-1 yel anh sC aa*	508189	147.88	(0.46)	147.88	(0.46)
zr d-1 lt br anh sD aa*	15318	147.77	(0.46)	147.77	(0.46)
zr d-1 yel anh sE aa*	7458	147.33	(0.46)	147.32	(0.46)
zr d-1 yel anh sF aa	6252	145.05	(0.45)	145.04	(0.45)
zr d-1 br anh sG aa*	4673	147.76	(0.47)	147.73	(0.47)
sph nm1.0a dk 0.35mm 11g	219	149.50	(0.64)	149.01	(0.64)
sph nm1.0a lt 0.20mm*	86	148.64	(1.18)	147.20	(1.18)
DP4-10z hbl-px gabbro late pod					
zr nm2 lt br anh sA	35299	147.04	(0.46)	147.04	(0.46)
zr nm2 lt br anh sB	4864	149.99	(0.47)	149.97	(0.46)
zr nm2 lt br anh 8g aa2*	1091	148.21	(0.47)	148.12	(0.47)
zr nm2 lt br anh 24g aa*	380	148.33	(0.51)	148.06	(0.51)
sph nm1.0a yel 0.35mm*	89	148.33	(1.15)	146.94	(1.15)
sph nm1.0a yel 0.35mm p2*	203	147.62	(0.72)	147.08	(0.71)

Notes: $^{206}Pb/^{204}Pb$ corrected for blank and mass discrimination; $^{206}Pb/^{238}U$ dates calculated with initial Pb isotopic values (6/4, 7/4, 8/4) of the Zartman and Doe (1981) mantle (ZD man) at 150 Ma (17.80 ± 0.5, 15.46 ± 0.05, 37.50 ± 0.2) and the Stacey and Kramers (1975) model (SK 150) at 150 Ma (18.474 ± 0.5, 15.618 ± 0.05, 38.356 ± 0.2). Uncertainties for model isotopic values are based on observed ranges for present-day mid-ocean ridge basalt. See Table 1 for other abbreviations.

*Used in age calculation.

affect a few analyses; many of the zircon and titanite analyses are insensitive to the choice of initial Pb isotopic composition, as they have high ratios of radiogenic to common Pb. Correction of $^{206}Pb/^{238}U$ dates for possible Th/U disequilibrium followed Schärer (1984). The disequilibrium corrected dates are used for all $^{206}Pb/^{238}U$ weighted average dates; they overlap in error with the uncorrected values (Table 1).

Biotite-Hornblende Tonalite (DP2-20z)

This sample is part of an intermediate to felsic, composite pluton intruded into the Blue Ridge ultramafic-gabbroic intrusion (Fig. 2). Field relationships indicate that the Blue Ridge intrusion is the earliest stage of the Bear Mountain intrusive complex (Snoke et al., 1981). Because zircon has not been observed in the Blue Ridge intrusion, this composite, crosscutting pluton provides the best means of constraining the age of the host ultramafic-gabbroic intrusion. The older part of the composite pluton is medium-grained, augite-hornblende quartz-bearing diorite with accessory titanite. This phase of the composite pluton contains scarce olivine xenocrysts, presumably derived from the host Blue Ridge intrusion. The dated sample intrudes the quartz-bearing diorite. It is a medium-grained, biotite-hornblende tonalite. Zircon is present as <10-µm inclusions in hornblende, biotite, and plagioclase, and as intergranular crystals as much as 100-µm long. Titanite commonly forms subhedral decorations on hornblende. Some grains show rational faces against adjacent crystals, whereas other grains are ragged, especially against plagioclase. Some late-stage titanite is intergrown with Fe-Ti oxide minerals ± chlorite. The tonalite is chemically similar to other tonalitic plutons in the area (e.g., Pony Peak pluton; Barnes et al., this volume, Chapter 17).

Zircons from this sample appear to be a single morphological population of colorless to pale yellow euhedral grains that are interpreted to be magmatic. There is a wide range in size of the isolated zircon grains, from 350 µm to less than 50 µm in length. Fractions of small and large-sized grains were picked separately to test for any age bias. Zircons appear to be free of cores under visible light. Titanite grains are large (250–400 µm in diameter), euhedral to anhedral grains that range in color from dark to pale yellow. Pale titanite overgrowths occur occasionally on darker cores. Dark titanite commonly indicates a higher temperature growth than pale titanite (e.g., Frost et al., 2001; Aleinikoff et al., 2002). Thus, end-member representatives were selected and analyzed separately to determine whether there was an age progression in titanite growth.

The U-Pb data from titanite and zircon plot in a broad array on a concordia plot (Fig. 3A), with $^{206}Pb/^{238}U$ dates that range from 151.7 to 145.3 Ma (Table 1). There is a general pattern in the ages, with the dark titanite yielding the oldest ages (ca. 152 Ma), pale titanite and abraded (mechanical and chemical) zircon the next oldest (ca. 149.3 Ma), and unabraded zircons the youngest (ca. 149.3–145.3 Ma). The titanite analyses and some of the zircon analyses have $^{206}Pb/^{204}Pb$ of 200 or less. In these cases, the calculated $^{206}Pb/^{238}U$ dates are moderately dependent on the choice of initial Pb isotopic composition (e.g., Verts et al., 1996). For example, the difference in calculated $^{206}Pb/^{238}U$ date between using 150-Ma Zartman and Doe (1981) model mantle values and Stacey and Kramers (1975) model Pb values is ca. 0.5 m.y. for an analysis with a $^{206}Pb/^{204}Pb$ of 200 (Table 2, Fig. 4A). The values in Table 2 and Figure 4A include error propagation of variations in these model isotopic values that approximate the measured variations in MORB. Inclusion of these large variations produces a strong dependence between age uncertainty and measured $^{206}Pb/^{204}Pb$ and a generous estimation of the age uncertainty. The zircon weighted mean date is immobile with this treatment, as it is largely controlled by data with $^{206}Pb/^{204}Pb$ values greater than 200. $^{206}Pb/^{238}U$ dates from yellow titanite are older than the zircon mean date with either reduction. An initial $^{206}Pb/^{204}Pb$ composition that is more radiogenic than the Stacey and Kramers (1975) model could bring the titanite dates into concordance with the zircon date, but it would also produce more scatter in the titanite dates and push the dark titanite date to younger values than the pale population. We favor the Zartman and Doe (1981) model mantle values for the initial Pb isotopic compositions, due to the arc setting of the intrusive complex and the relatively primitive chemical compositions of the Bear Mountain intrusive complex.

A $^{238}U/^{204}Pb$ versus $^{206}Pb/^{204}Pb$ isochron plot avoids this complexity, as no initial Pb correction is needed to plot the analyses. The three yellow to dark titanite analyses yield a good linear fit on a $^{238}U/^{204}Pb$ versus $^{206}Pb/^{204}Pb$ isochron plot (mean square weighted deviation [MSWD] = 0.0514; Fig. 3B). The isochron date of 150.5 ± 0.61 Ma is within error of the weighted, radiogenic $^{206}Pb/^{238}U$ date of these three analyses (150.97 ± 0.44 Ma; Fig. 3A) when reduced with model mantle

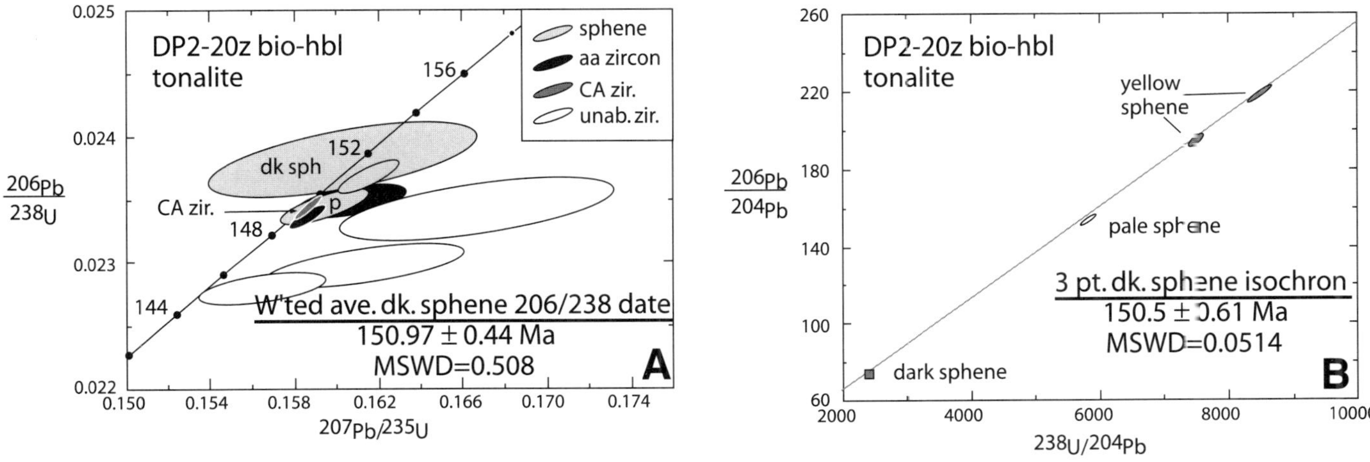

Figure 3. Concordia plot and $^{238}U/^{204}Pb$ vs. $^{206}Pb/^{204}Pb$ isochron plot of data from biotite-hornblende tonalite of the Blue Ridge intrusion (DP2-20z). (A) Data were reduced with model mantle Pb isotopic compositions for 150 Ma (Zartman and Doe, 1981) for the concordia plot. Data from one yellow titanite (sphene) analysis are not shown, as ^{207}Pb was not measured. However, the calculated $^{206}Pb/^{238}U$ date from this analysis is included in the weighted average. Data from dark (dk) and pale (p) titanite fractions are indicated. (B) Data used for the isochron calculation are indicated by filled symbols. The data from the pale titanite do not plot on this isochron and are consistent with microtextural evidence that the pale titanite grew after the darker grains. Agreement between the isochron date of the three dark to yellow titanite analyses (150.5 ± 0.61 Ma) and their weighted radiogenic $^{206}Pb/^{238}U$ date (150.97 ± 0.44 Ma) is interpreted to suggest that 150-Ma model mantle Pb isotopic compositions are appropriate for the initial Pb of these minerals.

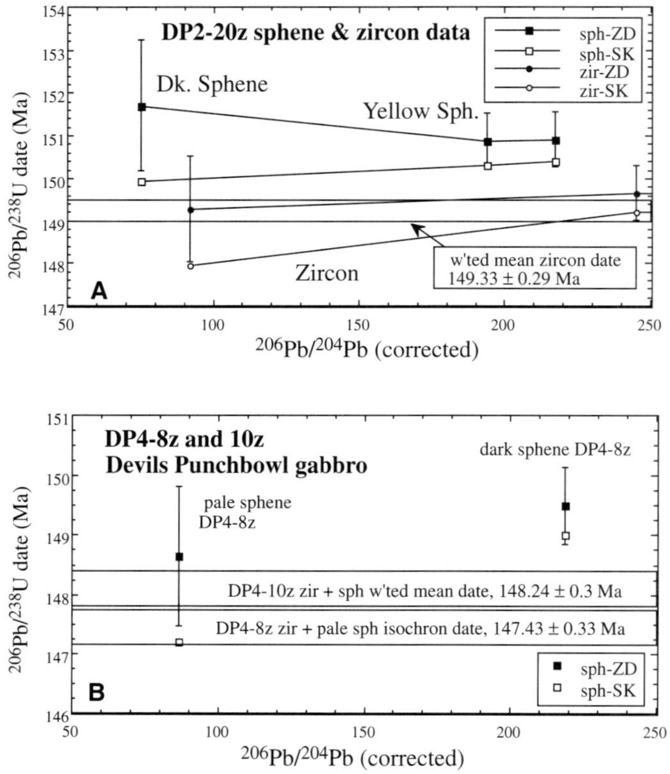

Figure 4. Plots of radiogenic $^{206}Pb/^{238}U$ dates vs. blank-corrected $^{206}Pb/^{204}Pb$ to demonstrate the effect of choice of initial Pb isotopic composition. Data were reduced with 150-Ma values from either the Zartman and Doe (1981; ZD—filled symbols in figure) model mantle ($^{206}Pb/^{204}Pb$ of 17.80 ± 0.5) or Stacey and Kramers (1975; SK—open symbols in figure) crustal model ($^{206}Pb/^{204}Pb$ of 18.474 ± 0.5). The magnitude of the effect of this choice and the size of the age uncertainties both decrease with increasing blank-corrected $^{206}Pb/^{204}Pb$. Uncertainties in model $^{206}Pb/^{204}Pb$ values are estimated to include the range of measurements for modern mid-ocean ridge basalt (Zartman and Doe, 1981). Uncertainties on SK reduced dates are not shown, but are the same magnitudes as for ZD reductions. (A) Dates from yellow titanite (sph) from DP2-20z are older and outside error of the weighted mean age of zircon (zir) analyses for both choices of initial Pb $^{206}Pb/^{204}Pb$. (B) The $^{206}Pb/^{238}U$ date of dark titanite from DP4-8z is outside errors of both Punchbowl zircon dates for the Zartman and Doe (1981) reduction, but overlaps the zircon date of DP4-10z when reduced with Stacey and Kramers (1975) $^{206}Pb/^{204}Pb$ value.

Pb isotopic values for 150 Ma (Zartman and Doe, 1981). This result strengthens our interpretation that model mantle Pb isotopic compositions are appropriate as estimates of initial Pb for these analyses. The pale titanite results do not plot on the isochron from the other three titanite analyses (Fig. 3B) and have a distinctly younger $^{206}Pb/^{238}U$ date (149.4 ± 0.73 Ma; Table 1, Fig. 3A). This result is consistent with the observation of pale overgrowths on darker titanite and suggests to us that either the pluton crystallized slowly as it cooled over a 1- to 2-m.y. period, or a slightly younger magmatic event triggered an additional pulse of titanite growth in this body (see below).

Variation in radiogenic $^{206}Pb/^{208}Pb$ (Table 1) also supports evidence for either a protracted period of titanite growth or multiple events. The dark titanite analysis has the lowest $^{206}Pb/^{208}Pb$ (~0.9), pale titanite has the highest (~3.0), and the yellow titanite analyses are intermediate (~2.5). We interpret these results to indicate that the U/Th composition of the melt evolved by preferential incorporation of Th into accessory phases and that this evolution took 1–2 m.y. in this pluton, based on the ages of the end-member titanites. Variations in U/Th between minerals and parental magma can lead to age biases in young rocks due to deficits or excesses of radiogenic ^{206}Pb (Schärer, 1984). To test for this possibility, we calculated $^{206}Pb/^{238}U$ dates corrected for Th/U disequilibrium (Table 1) by using radiogenic ^{208}Pb as a measure of Th in the minerals and a magma Th/U of 2.85 based on measured compositions from the Bear Mountain intrusive complex (Bushey et al., this volume). The effect of this correction is slight, with changes of only 0.02–0.08 m.y. for the titanite dates (Table 1), and it fails to narrow the age differences between dark titanite, pale titanite, and zircon over a wide range of magmatic Th/U values (Fig. 5).

We interpret the $^{238}U/^{204}Pb$ versus $^{206}Pb/^{204}Pb$ isochron date of 150.5 ± 0.61 Ma from dark and yellow titanite as the best estimate for the age of initial crystallization of this pluton. On this basis, the zircon data either reflect slightly later crystallization in the pluton (149.3 ± 0.3 Ma) or some degree of Pb loss (0.6–3.5%) despite rigorous air abrasion of grains for two of the analyses and chemical abrasion for another. Calculated zircon saturation temperature (Watson and Harrison, 1983) for this sample is 842 °C, based on chemical compositions reported in Bushey et al. (this volume). This temperature is consistent with relatively late zircon crystallization in a dry, tonalitic cumulate. It is also possible that the dark titanite grew in the earlier quartz-bearing dioritic magma. If so, then the dark titanite could represent xenocrystic material from the older host rock. In this case, however, the dark titanite would have had to retain its older U-Pb date for several 100 k.y. as the later magma evolved to tonalite and precipitated zircon and pale titanite. Closure temperatures of at least 700 °C have been reported for titanite in metamorphic settings (Scott and St-Onge, 1995).

Pyroxene-Hornblende Gabbro, Punchbowl Unit (DP4-8z and DP4-10z)

This unit is part of a large, composite pluton that is exposed over ~40 km². It consists of pyroxene-hornblende gabbro that locally grades into anorthositic gabbro as well as biotite-hornblende (quartz-bearing) diorite (Snoke et al., 1981; Bushey et al., this volume). Modal proportions vary considerably, even at the outcrop scale. Most samples have major and trace element compositions suggestive of accumulation from a high-Al basaltic magma (Bushey et al., this volume; Barnes et al., this volume, Chapter 17). Two samples were collected for geochronology: a coarse-grained, slightly pegmatitic sample with hornblende, plagioclase, biotite, and minor quartz (DP4-8z) and a late-stage,

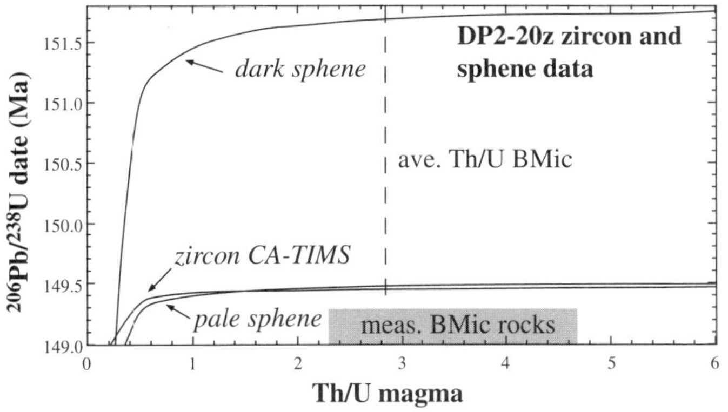

Figure 5. Plot of $^{206}Pb/^{238}U$ mineral dates from DP2-20z corrected for Th/U disequilibrium for a range of magma Th/U values. Dark titanite (sphene) and chemical abrasion thermal ionization mass spectrometry zircon analyses have the most variation in Th/U of any samples (indicated by radiogenic $^{206}Pb/^{208}Pb$ values of 0.94 and 6.97 respectively). Extreme Th/U magma values of <0.4 are required to bring the dates of these two minerals into concordance. Measured Th/U of representative rocks of the Bear Mountain intrusive complex (BMic) range from 2.29 to 4.72 and average 2.85 (Bushey et al., this volume).

magmatic pod (DP4-10z) with distinctive, concentric layers of plagioclase and hornblende, weakly developed harrisitic texture, and a coarse-grained interior (3-cm crystals; Fig. 6). The hornblende is prismatic, with interstitial to poikilitic plagioclase (~An_{60}). Accessory phases are Fe-Ti oxides, interstitial alkali feldspar, and subhedral to anhedral titanite up to 1.5 mm in length. Ragged fractures in the hornblende show replacement by epidote ± fine-grained titanite ± chlorite, and the plagioclase shows deformation that ranges from bent crystals to subgrain development. This pod crosscuts magmatic foliation in the pyroxene-hornblende gabbro and is interpreted as a pocket of late, H_2O-rich liquid in the crystallizing pluton. Both samples yielded zircon and titanite. Zircon from both samples are large (~750 × 450 µm, typically) and anhedral. The titanite from DP4-8z ranges in color from dark and reddish to pale yellow, although no core-rim relationships were detected. The titanite from DP4-10z is more uniformly honey colored.

The zircon data from DP4-8z, the texturally earlier sample of the Punchbowl unit, display a range in $^{206}Pb/^{238}U$ dates from 151.2 to 145.1 Ma (Table 1, Fig. 7A). All of these analyses are from single grains, and five of the seven were air abraded. Four of the analyses of air-abraded zircon plot in a tight cluster with overlapping $^{206}Pb/^{238}U$ dates and indicate a weighted average of 147.7 ± 0.23 Ma. Analyses of dark titanite and one single zircon have older $^{206}Pb/^{238}U$ dates, ca. 149.5 and ca. 151 Ma, respectively. We interpret the date from the older zircon to reflect slight inheritance, because it is so much older than those from all the other analyses. The two zircon analyses that are younger than the cluster of four are interpreted to reflect Pb loss. The $^{206}Pb/^{238}U$ date from the cluster of four zircons is interpreted as the crystallization age of zircon in this sample, although it is possible that there is some Pb loss even in these analyses. However, the data from the pale titanite analysis lie on a linear regression of these four zircon analyses on a $^{238}U/^{204}Pb$ versus $^{206}Pb/^{204}Pb$ isochron plot (Fig. 7C). Including the titanite data yields an isochron date of 147.43 ± 0.33 Ma with a good linear fit (MSWD = 1.42). This date is indistinguishable from the weighted average $^{206}Pb/^{238}U$ date of the zircon alone and is interpreted to indicate that the pale titanite grew at the same time as these zircon grains. The data from dark titanite is off the linear regression of the pale titanite and zircon analyses (Fig. 7C) and has a $^{206}Pb/^{238}U$ date that is nearly 2 m.y. older (149.5 ± 0.64 Ma). The radiogenic $^{206}Pb/^{208}Pb$ for the two titanite analyses are distinct as well (0.9 versus 2.6) and support our interpretation that titanite grew over a 2-m.y. period. The pattern of mineral dates is very similar in this sample to the pattern in DP2-20z discussed above, and we interpret them to record a progression of crystallization of the pluton. Initial crystallization occurred at 149.5 ± 0.64 Ma, based on data from dark titanite, and was followed 1–2 m.y. later by crystallization of zircon and pale titanite.

The U-Pb data from DP4-10z, the late magmatic pod in the Punchbowl unit, is slightly less complicated because there is only one morphology of titanite, and the data from two titanite analyses are internally consistent. Titanite data yield a weighted $^{206}Pb/^{238}U$ date of 147.9 ± 0.6 Ma and have similar radiogenic $^{206}Pb/^{208}Pb$ values of ~3.02–2.78 (Table 1). Data from two fractions of small, air-abraded zircons overlap this date and produce a four-point, titanite and zircon, $^{206}Pb/^{238}U$-weighted date of 148.24 ± 0.30 Ma (MSWD = 0.94; Fig. 7B, Table 1). A linear regression of these same four data points on a $^{238}U/^{204}Pb$ versus $^{206}Pb/^{204}Pb$ isochron plot yields nearly the same result, 148.12 ± 0.41 Ma, with a good linear fit (MSWD = 1.49; Fig. 7D). Data from two large, single zircon grains yield both older and younger $^{206}Pb/^{238}U$ dates compared to this cluster (ca. 150

Figure 6. Photograph of late-stage magmatic pod (DP4-10z) from the Punchbowl unit. Note lens cap for scale. Concentric layers of plagioclase (beneath lens cap) and hornblende surround coarse-grained interior composed of 3-cm length crystals.

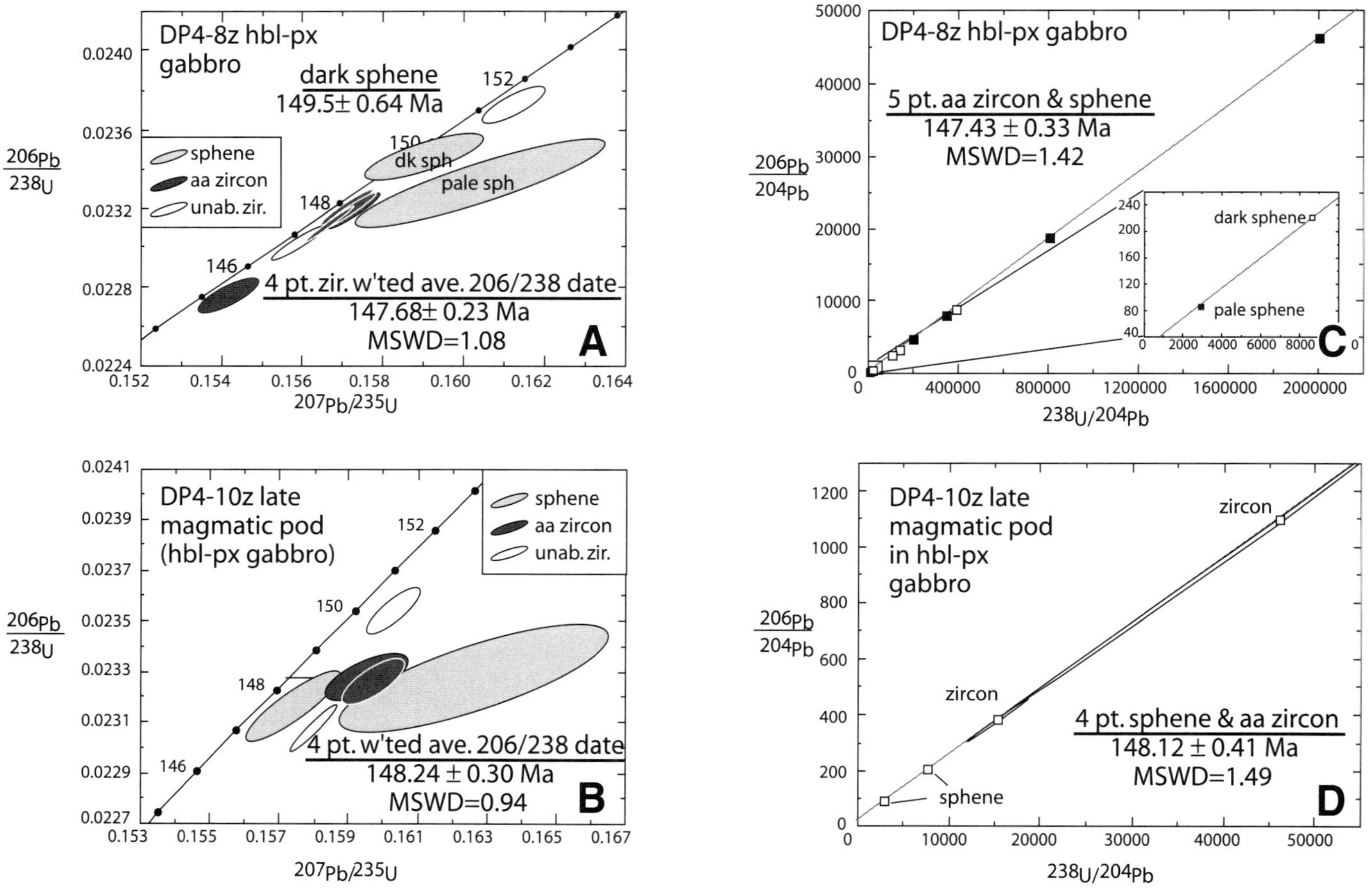

Figure 7. (A, B) Concordia plots and (C, D) $^{238}U/^{204}Pb$ vs. $^{206}Pb/^{204}Pb$ isochron plots of data from two samples of the Punchbowl unit. DP4-8z is from a coarse-grained, pyroxene-hornblende gabbro that is fairly typical of the intrusion. DP4-10z is from a distinct, late-stage magmatic pod, shown in Figure 6. Isochron date from four air-abraded zircon fractions and pale titanite data from DP4-8z (147.43 ± 0.33 Ma; filled symbols in C) agrees with the radiogenic $^{206}Pb/^{238}U$ date of the four air-abraded zircon fractions alone (147.68 ± 0.23 Ma in A) and is interpreted to indicate that pale titanite grew at the same time as zircon in this sample. Data from dark titanite plot above this isochron and reduce to an older radiogenic $^{206}Pb/^{238}U$ date of 149.5 ± 0.64 Ma. The late, magmatic pod (DP4-10z) yielded only one morphology of titanite. Zircon and titanite grew at the same time in this sample (148.24 ± 0.30 Ma), slightly earlier than the growth of pale titanite and zircon in DP4-8z. Taken together, these data document 1–2 m.y. of crystallization in the Punchbowl unit, either due to multiple pulses of magma injection or protracted crystallization. Both samples also contain evidence for slight inheritance of zircon that is at least 151 Ma in age.

and ca. 147 Ma, respectively; Table 1) and are interpreted to reflect slight inheritance and Pb loss. Our best estimate for the timing of crystallization of this late magmatic pod comes from the weighted average $^{206}Pb/^{238}U$ date of the two titanite and two air-abraded zircon analyses, 148.24 ± 0.30 Ma. This date is barely within error to slightly older than the date of crystallization of pale titanite and zircon from the texturally earlier Punchbowl sample, DP4-8z, and 1–2-m.y. later than the dark titanite date from that sample (Fig. 4B). The $^{206}Pb/^{238}U$ dates of titanite from DP4-8z and DP4-10z are somewhat dependent on choice of initial Pb isotopic composition (Table 2, Fig. 4B), but we interpret the agreement between the $^{238}U/^{204}Pb$ versus $^{206}Pb/^{204}Pb$ isochron dates and the radiogenic $^{206}Pb/^{238}U$ dates to support our choice of the Zartman and Doe (1981) mantle value for initial $^{206}Pb/^{204}Pb$. The model $^{207}Pb/^{204}Pb$ values may not be appropriate, however, as both zircon and titanite data from DP4-8z and DP4-10z are slightly discordant, and the degree of discordance increases with proportion of common Pb (inverse of $^{206}Pb/^{204}Pb$). A different choice of initial $^{207}Pb/^{204}Pb$ could bring all these values into concordance.

Taken together, the data from these two samples of the Punchbowl unit are interpreted to record a protracted period of crystallization. Dark titanite commonly has higher Fe and Al content (e.g., Aleinikoff et al., 2002) and grows at higher temperatures than pale titanite (Frost et al., 2001). The $^{206}Pb/^{238}U$ date of the dark titanite from DP4-8z is the earliest date (149.5 ± 0.6 Ma) and is interpreted to record the initial stages of crystallization of the pluton. Pale titanite and zircon from both samples record crystallization 1–2 m.y. after the dark titanite and document continued crystallization of an evolving magma, perhaps at lower temperatures. Calculated zircon saturation temperatures for these samples are ~640–650 °C (Watson and Harrison, 1983; Bushey et al., this volume), reflecting the cumulate nature of this unit and near-solidus crystallization of zircon.

Biotite–Two-Pyroxene Quartz-Bearing Diorite, Buck Lake Unit (DP1-13z)

This is a medium- to coarse-grained plutonic unit with tabular plagioclase, ortho- and clinopyroxene, and oikocrystic biotite. The analyzed sample contains scarce alkali feldspar, but other samples from the outcrop have poikilitic alkali feldspar and are monzodioritic. The Buck Lake unit is distinct from the Punchbowl unit in the relative abundance of alkali feldspar and paucity of primary hornblende (Snoke et al., 1981; Bushey et al., this volume). Field relationships between the Buck Lake and Punchbowl unit have been interpreted to suggest synchronicity (Snoke et al., 1981; Bushey et al., this volume). Zircon grains are euhedral, prismatic, 100 × 250 µm, and honey colored. No titanite was recovered from this sample. The U-Pb data from seven multigrain analyses plot in a relatively tight cluster with $^{206}Pb/^{238}U$ dates that range from 149.1 ± 0.5 to 147.4 ± 0.5 Ma (Fig. 8). The spread in dates is interpreted to reflect a combination of inheritance and Pb loss. Our best estimate for the crystallization age of the unit comes from the weighted $^{206}Pb/^{238}U$ date of four air-abraded zircon fractions and one chemically abraded fraction: 148.2 ± 0.2 Ma (Fig. 8, Table 1). This date is within error of the last stages of crystallization of the Punchbowl unit and 2–3 m.y. later than the initial crystallization date of the tonalite (DP2-20z). Saleeby et al. (1982) reported a date of 153 ± 2 Ma for the monzodiorite unit based on a single 10-mg fraction of zircon. Their analysis had a relatively low $^{206}Pb/^{204}Pb$ of 700, and the absolute date is relatively sensitive to choice of initial Pb isotopic composition. The $^{206}Pb/^{204}Pb$ ratios of our data are significantly higher (1400–20,000), and the calculated dates are less sensitive to choice of initial isotopic composition. It is also possible that the Saleeby et al. (1982) analysis included a significant inherited component, as there is evidence for the presence of inheritance in our data, and they dissolved about one-hundred times as many grains for their analysis as we did.

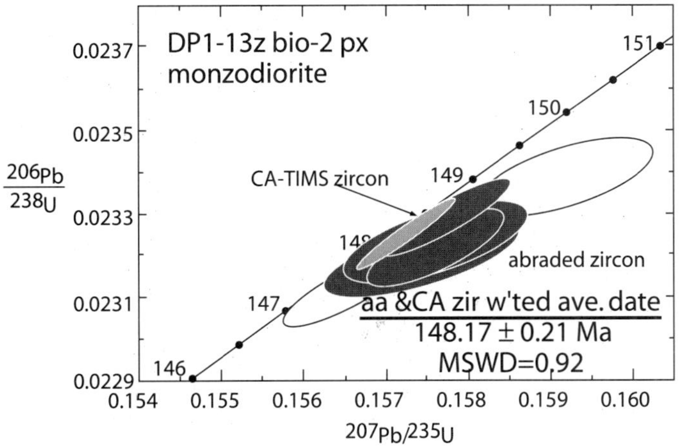

Figure 8. Concordia plot of zircon U-Pb data from biotite–two-pyroxene (bio-2 px) monzodiorite of Buck Lake unit (DP1-13z). Data from air-abraded grains (aa; black-filled ellipses) and annealed and chemically abraded grains (chemical abrasion thermal ionization mass spectrometry; gray-filled ellipse) have a weighted mean $^{206}Pb/^{238}U$ date, 148.17 ± 0.21 Ma, within error of the date of the Punchbowl unit.

Late Mafic Dike (DP1-35z)

This sample is from a quartz-bearing pod in a 2-m-thick, medium- to fine-grained, hornblende quartz diorite dike. The dike crosscuts foliation in the inner aureole of the Bear Mountain intrusive complex, lacks penetrative deformation, and its age thus places a lower limit on the timing of magmatism of the intrusive complex. The sample is an unusual 0.25-m-diameter portion of the dike, with visible free quartz, ~3 mm in size, and a matrix that is finer than the rest of the dike. The pod can be due to either late-stage differentiation or partial digestion of a country rock. We prefer the former interpretation, because quartz is intergrown with all of the primary phases in the sample. The zircons are relatively small (~150 µm or less in length), euhedral, yellow grains. Visible inclusions exist in many of the grains. The U-Pb data from seven multigrain fractions show the widest dispersion of any of the samples in this study, with $^{206}Pb/^{238}U$ dates from ca. 153 to 142 Ma. We interpret the data to imply that most of the zircon grew during crystallization of the dike and is not derived from significantly older country rock. Data from one fraction of inclusion-rich grains that were strongly air-abraded are distinctly discordant and have a $^{207}Pb/^{206}Pb$ age of 264 Ma. These results establish that inherited zircon components exist and that they are at least 260 Ma and may be as old as Paleoproterozoic (Fig. 9). Our best estimate of the crystallization age of the zircon comes from the weighted $^{206}Pb/^{238}U$ average date from two fractions of air-abraded zircon (145.4 ± 0.4 Ma; Fig. 9). Data from three other analyses are interpreted to reflect some inherited components, and one fraction of unabraded grains displays Pb loss. The date of ca. 145 Ma is clearly later than that for all other samples (Fig. 10), consistent with the field relationships, and establishes a maximum duration of magmatism in the Bear Mountain complex of 4–6 m.y. Moreover, this age is similar to U-Pb ages determined for the nearby Bear Peak and Pony Peak plutons (Allen and Barnes, this volume).

DISCUSSION

The new geochronological data presented in this chapter are complex and fairly difficult to interpret as precise crystallization ages, because each sample produced a 3- to 10-m.y. range of dates. The spread in dates reflects minor inheritance and Pb loss, but also adds insight into duration of crystallization processes. Even in the most conservative interpretation, the data establish that the Bear Mountain intrusive complex was emplaced be-

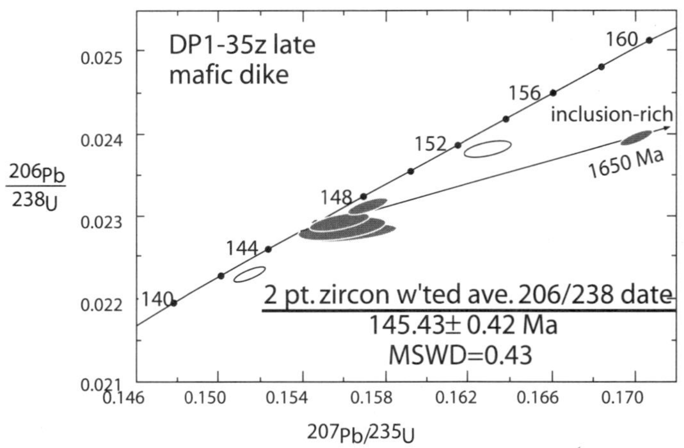

Figure 9. Concordia plot of zircon U-Pb data from late mafic dike, DP1–35z. Data from air-abraded, inclusion-rich grains yield strong evidence for inherited components, at least as old as 264 Ma ($^{207}Pb/^{206}Pb$ date) and possibly Paleoproterozoic. The weighted mean $^{206}Pb/^{238}U$ date, 145.43 ± 0.42 Ma, of two analyses of inclusion-free, air-abraded grains is distinctly later than the dates of the other samples analyzed.

tween ca. 151 and 147 Ma, with a minimum duration of 1.5 m.y. (Fig. 10). If the late mafic dike is considered to be part of the complex, then the duration of magmatism could have been as much as 6 m.y. The new data also support the relative chronology determined from field relationships, although they clearly argue against using petrographic characteristics as indicative of a general age progression during magma genesis. For example, leucocratic rocks commonly intrude the mafic rocks of the Bear Mountain intrusive complex. The sample of biotite-hornblende tonalite (DP2-20z) that yielded the oldest ages in this new geochronological study is from a satellitic, composite pluton that intrudes the Blue Ridge ultramafic-gabbroic intrusion. The absolute age of this leucocratic rock, however, is older than dated samples from either the Punchbowl or Buck Lake plutonic units.

The age pattern from old, dark titanite to younger, pale titanite and zircon in the Blue Ridge and Punchbowl samples has

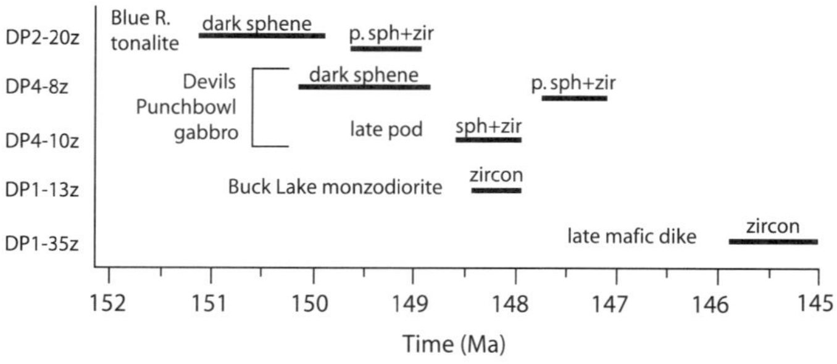

Figure 10. Summary plot of mineral dates from Bear Mountain intrusive complex samples in this study. Abbreviations: p. sph—pale sphene; R.—Ridge; zir—zircon.

intriguing implications for crystallization processes and mineral isotopic systematics. One possible interpretation of this pattern is that the dark titanite grains are xenocrystic and inherited from an earlier body. We are not aware of a previously documented occurrence of inherited titanite in magmatic rocks, so this could be the first recorded instance of the phenomenon. This interpretation would also imply U-Pb closure temperatures in titanite that are higher than generally accepted (600–700 °C; e.g., Frost et al., 2001; Aleinikoff et al., 2002) for the older age to survive during generation of the new pluton. An alternative interpretation of the age pattern for titanite and zircon is that the data have captured a crystallization history of 1.5–2 m.y. in each pluton. The tonalite intruded into the Blue Ridge intrusion (DP2-20z) is a differentiate in the core of a composite pluton (Fig. 2), so it is conceivable that cooling and crystallization was relatively slow (1–2 m.y.). The Punchbowl unit is a complex body produced by multiple intrusive sheets (hundreds to thousands of them), so a model of protracted crystallization in this body is certainly possible. It is important to note, however, that the Punchbowl unit exhibits more or less uniform magmatic foliation, and there must have been sufficient interstitial melt to permit communication of strain throughout the composite pluton as it cooled. We envision that later pulses of magmatism either slowed the cooling and crystallization process by injecting additional heat, or actually spurred pulses of crystal growth through renewed fluids and heat. A combination of these processes is also possible. At this stage, we favor the protracted crystallization interpretation, although we cannot completely rule out xenocrystic titanite.

Coleman et al. (2004) and Glazner et al. (2004) invoked incremental assembly by successive sheets to explain the nearly 4-m.y. range of zircon dates for the Half Dome granodiorite of the Tuolumne intrusive suite. Our data from the Bear Mountain intrusive complex differ from their results in that we have documented protracted crystallization histories (1–2 m.y.) within single samples. The zircon dates from the Half Dome granodiorite young toward the center (Coleman et al. 2004), consistent with protracted crystallization, although the authors do not favor this interpretation. Both studies demonstrate that simple thermal models of cooling and crystallization may be too simplistic, and that detailed, high-precision geochronology in plutonic complexes can elucidate fine-scale features of crystallization.

The new dates on the Bear Mountain intrusive complex place constraints on the timing of deformation in the region. The South Siskiyou Fork fault (Snoke, 1977; Snoke et al., 1981; Fig. 2) is truncated by the Punchbowl unit of the Bear Mountain intrusive complex and thereby predates 149.5 ± 0.6 Ma. This northwest–southeast-striking, high-angle fault is interpreted as a tear fault associated with the Preston Peak (thrust) fault (Snoke, 1977), a local equivalent of the regional Orleans fault system. Upper Jurassic Galice Formation is part of the footwall block of the Orleans fault and contains ca. 153-Ma detrital zircon grains (Miller et al., 2003). Thus, the South Siskiyou Fork fault must have formed between 153 and 149 Ma. This relationship indicates that the Bear Mountain intrusive complex was emplaced subsequent to the development of the Orleans (Preston Peak) fault and that the emplacement of the intrusive complex post-dated regional thrust faulting associated with the Late Jurassic Nevadan orogeny (Snoke and Barnes, this volume).

Combining our results with new dates on other plutons in the western Klamath Mountains (e.g., Allen and Barnes, this volume) suggests that the Bear Mountain intrusive complex is part of a long-lived magmatic (arc) pulse that began as early as ca. 151 Ma, probably during Nevadan-age thrusting, and extended at least to ca. 144 Ma (e.g., age of the Pony Peak pluton). In fact, magmatism in the Klamath Mountains continued, with no younger thrusting events, until ca. 136 Ma. However, there seems to be a distinctive set of mainly dioritic plutons that crop out in the western Klamath Mountains that intrude either the Galice Formation (of the western Klamath terrane) or Rattlesnake Creek terrane. Every one of these plutons that has been dated has an age between 150 and 144 Ma (Irwin and Wooden, 1999; Allen and Barnes, this volume). In addition, the Bear Mountain intrusive complex and other "western Klamath" plutons share a number of geochemical similarities, which supports the idea that they are part of a distinct episode of arc magmatism (Barnes et al., this volume, Chapter 17).

CONCLUSIONS

The U-Pb data from titanite and zircon have established that the Bear Mountain intrusive complex was emplaced between 151 and 147 Ma, after the Late Jurassic Nevadan orogeny. Emplacement and crystallization of the complex took at least 1.5 m.y. and may have lasted 6 m.y. Based on age ranges of titanite and zircon, crystallization of individual plutons may have taken 1.5–2 m.y. The timing of deformation of the South Siskiyou Fork fault, an oblique-slip tear fault associated with the Preston Peak (thrust) fault (Orleans fault system), is constrained to the period from ca. 153 to 149 Ma. The Bear Mountain intrusive complex is part of long-lived pulse of arc magmatism that lasted from 151 Ma to at least 144 Ma.

ACKNOWLEDGMENTS

This research was funded by National Science Foundation grants EAR-9902807 to AWS and EAR-9902912 to CGB. Jessica Anderson, Diana Strickland, and Peter Valley assisted with sample preparation and mass spectrometry. We thank Mark D. Schmitz and James M. Mattinson for their thorough and helpful reviews.

REFERENCES CITED

Aleinikoff, J.N., Wintsch, R.P., Fanning, C.M., and Dorais, M.J., 2002, U-Pb geochronology of zircon and polygenetic titanite from the Glastonbury Complex, Connecticut, USA: An integrated SEM, EMPA, TIMS, and SHRIMP study: Chemical Geology, v. 188, p. 125–147, doi: 10.1016/S0009-2541(02)00076-1.

Allen, C.M., and Barnes, C.G., 2006, this volume, Ages and some cryptic sources of Mesozoic plutonic rocks in the Klamath Mountains, California and Oregon, *in* Snoke, A.W., and Barnes, C.G., eds., Geological studies in the Klamath Mountains province, California and Oregon: A volume in honor of William P. Irwin: Boulder, Colorado, Geological Society of America Special Paper 410, doi: 10.1130/2006.2410(11).

Barnes, C.G., Snoke, A.W., Harper, G.D., Frost, C.D., McFadden, R.R., Bushey, J.C., and Barnes, M.A.W., 2006, this volume (Chapter 17), Arc plutonism following regional thrusting: Petrology and geochemistry of syn- and post-Nevadan plutons in the Siskiyou Mountains, Klamath Mountains province, California, *in* Snoke, A.W., and Barnes, C.G., eds., Geological studies in the Klamath Mountains province, California and Oregon: A volume in honor of William P. Irwin: Boulder, Colorado, Geological Society of America Special Paper 410, doi: 10.1130/2006.2410(17)

Black, L.P., Kamo, S.L., Allen, C.M., Davis, D.W., Aleinikoff, J.N., Valley, J.W., Mundil, R., Campbell, I.H., Korsch, R.J., Williams, I.S., and Foudoulis, C., 2004, Improved $^{206}Pb/^{238}U$ microprobe geochronology by the monitoring of a trace-element-related matrix effect; SHRIMP, ID-TIMS, ELA-ICP-MS and oxygen isotope documentation for a series of zircon standards: Chemical Geology, v. 205, p. 115–140, doi: 10.1016/j.chemgeo.2004.01.003.

Blackwelder, E., 1914, A summary of the orogenic epochs in the geologic history of North America: Journal of Geology, v. 22, p. 633–654.

Blake, M.C., Jr., 1984, Tectonostratigraphic terranes in southwestern Oregon, *in* Nilsen, T.H., ed., Geology of the Upper Cretaceous Hornbrook Formation, Oregon and California: Los Angeles, Pacific Section, Society of Economic Paleontologists and Mineralogists, p. 159–165.

Blake, M.C., Jr., Engebretson, D.C., Jayko, A.S., and Jones, D.L., 1985, Tectonostratigraphic terranes in southwest Oregon, *in* Howell, D.G., ed., Tectonostratigraphic terranes of the Circum-Pacific region: Houston, Circum-Pacific Council for Energy and Mineral Resources, Earth Sciences Series 1, p. 147–157.

Bushey, J.C., Snoke, A.W., Barnes, C.G., and Frost, C.D., 2006, this volume, Geology of the Bear Mountain intrusive complex, Klamath Mountains, California, *in* Snoke, A.W., and Barnes, C.G., eds., Geological studies in the Klamath Mountains province, California and Oregon: A volume in honor of William P. Irwin: Boulder, Colorado, Geological Society of America Special Paper 410, doi: 10.1130/2006.2410(14).

Coleman, D.S., Gray, W., and Glazner, A.F., 2004, Rethinking the emplacement and evolution of zoned plutons: Geochronologic evidence for incremental assembly of the Tuolumne intrusive suite, California: Geology, v. 32, p. 433–436, doi: 10.1130/G20220.1.

Davis, G.A., Monger, J.W.H., and Burchfiel, B.C., 1978, Mesozoic construction of the Cordilleran "collage," central British Columbia to central California, *in* Howell, D.G., and McDougall, K.A., eds., Mesozoic paleogeography of the western United States, Pacific Coast Paleogeography II: Los Angeles, Pacific Section, Society of Economic Paleontologists and Mineralogists, p. 1–32.

Frost, B.R., Chamberlain, K.R., and Schumacher, J., 2001, Sphene (titanite): Phase relations and role as a geochronometer: Chemical Geology, v. 172, p. 131–148, doi: 10.1016/S0009-2541(00)00240-0.

Glazner, A.F., Bartley, J.M., Coleman, D.S., Gray, W., and Taylor, R.Z., 2004, Are plutons assembled over millions of years by amalgamation from small magma chambers? GSA Today, v. 14, no. 4/5, p. 4–11, doi: 10.1130/1052-5173(2004)014<0004:APAOMO>2.0.CO;2.

Gray, G.G., 2006, this volume, Structural and tectonic evolution of the western Jurassic belt along the Klamath River corridor, Klamath Mountains, California, *in* Snoke, A.W., and Barnes, C.G., eds., Geological studies in the Klamath Mountains province, California and Oregon: A volume in honor of William P. Irwin: Boulder, Colorado, Geological Society of America Special Paper 410, doi: 10.1130/2006.2410(07).

Harper, G.D., and Wright, J.E., 1984, Middle to Late Jurassic tectonic evolution of the Klamath Mountains, California-Oregon: Tectonics, v. 3, p. 759–772.

Harper, G.D., Saleeby, J.B., and Heizler, M., 1994, Formation and emplacement of the Josephine ophiolite and the Nevadan orogeny in the Klamath Mountains, California-Oregon: U/Pb zircon and $^{40}Ar/^{39}Ar$ geochronology: Journal of Geophysical Research, v. 99, p. 4293–4321, doi: 10.1029/93JB02061.

Hershey, O.H., 1906, Some western Klamath stratigraphy: American Journal of Science, 4th series, v. 21, p. 58–66.

Hershey, O.H., 1911, Del Norte County [California] geology: Mining and Scientific Press, v. 102, p. 468.

Howell, D.G., Jones, D.L., and Schermer, E.L., 1985, Tectonostratigraphic terranes of the Circum-Pacific region, *in* Howell, D.G., ed., Tectonostratigraphic terranes of the Circum-Pacific region: Houston, Circum-Pacific Council for Energy and Mineral Resources, Earth Sciences Series 1, p. 3–30.

Imlay, R.W., 1959, Succession and speciation of the pelecypod *Aucella:* Washington, D.C., U.S. Geological Survey Professional Paper 314-G, p. 155–169, plates 16–19.

Irwin, W.P., 1972, Terranes of the western Paleozoic and Triassic belt in the southern Klamath Mountains, California: Washington, D.C., U.S. Geological Survey Professional Paper 800-C, p. C103–C111.

Irwin, W.P., 1981, Tectonic accretion of the Klamath Mountains, *in* Ernst, W.G., ed., The geotectonic development of California—Rubey volume 1: Englewood Cliffs, New Jersey, Prentice-Hall, p. 29–70.

Irwin, W.P., 1985, Age and tectonics of plutonic belts in accreted terranes of the Klamath Mountains, California and Oregon, *in* Howell, D.G., ed., Tectonostratigraphic terranes of the circum-Pacific region: Houston, Circum-Pacific Council for Energy and Mineral Resources, Earth Sciences Series 1, p. 187–199.

Irwin, W.P., 1994, Geologic map of the Klamath Mountains, California: Reston, Virginia, U.S. Geological Survey Miscellaneous Investigations Series Map I–2148, scale 1:500,000, 2 sheets.

Irwin, W.P., and Wooden, J.L., 1999, Plutons and accretionary episodes of the Klamath Mountains, California and Oregon: Reston, Virginia, U.S. Geological Survey Open-File Report 99-374, 1 sheet.

Jachens, R.C., Barnes, C.G., and Donato, M.M., 1986, Subsurface configuration of the Orleans fault: Implications for deformation in the western Klamath Mountains, California: Geological Society of America Bulletin, v. 97, p. 388–395, doi: 10.1130/0016-7606(1986)97<388:SCOTOF>2.0.CO;2.

Krogh, T.E., 1973, A low-contamination method for hydrothermal decomposition of zircon and extraction of U and Pb for isotopic age determinations: Geochimica et Cosmochimica Acta, v. 37, p. 485–494, doi: 10.1016/0016-7037(73)90213-5.

Lanphere, M.A., Irwin, W.P., and Hotz, P.E., 1968, Isotopic age of the Nevadan orogeny and older plutonic and metamorphic events in the Klamath Mountains, California: Geological Society of America Bulletin, v. 79, p. 1027–1052.

Ludwig, K.R., 1988, PBDAT for MS-DOS, a computer program for IBM-PC compatibles for processing raw Pb-U-Th isotope data, version 1.24: Reston, Virginia, U.S. Geological Survey Open-File Report 88-542, 32 p.

Ludwig, K.R., 1991, ISOPLOT for MS-DOS, a plotting and regression program for radiogenic-isotope data, for IBM-PC compatible computers, version 2.75: Reston, Virginia, U.S. Geological Survey Open-File Report 91-445, 45 p.

Mattinson, J.M., 2005, Zircon U-Pb chemical abrasion ("CA-TIMS") method: Combined annealing and multi-step partial dissolution analysis for improved precision and accuracy of zircon ages: Chemical Geology, v. 220, p. 47–66, doi: 10.1016/j.chemgeo.2005.03.011.

McFadden, R.R., Snoke, A.W., and Barnes, C.G., 2006, this volume, Structural and petrologic evolution of the Bear Peak intrusive complex, Klamath Mountains, California, *in* Snoke, A.W., and Barnes, C.G., eds., Geological studies in the Klamath Mountains province, California and Oregon: A volume in honor of William P. Irwin: Boulder, Colorado, Geological Society of America Special Paper 410, doi: 10.1130/2006.2410(16).

Miller, J., Miller, R., Wooden, J., and Harper, G., 2003, Geochronologic links between the Ingalls ophiolite, North Cascades, Washington and the Josephine ophiolite, Klamath Mts., Oregon and California: Geological Society of America Abstracts with Programs, v. 35, p. 113.

Parrish, R.R., Roddick, J.C., Loveridge, W.D., and Sullivan, R.D., 1987, Uranium-lead analytical techniques at the geochronology laboratory, Geological Survey of Canada, *in* Radiogenic age and isotopic studies, Report 1: Ottawa, Geological Survey of Canada Paper 87-2, p. 3–7.

Saleeby, J.B., Harper, G.D., Snoke, A.W., and Sharp, W.D., 1982, Time relations and structural-stratigraphic patterns in ophiolite accretion, west-central Klamath Mountains, California: Journal of Geophysical Research, v. 87, p. 3831–3848.

Schärer, U., 1984, The effect of initial ^{230}Th disequilibrium on young U-Pb ages: The Makalu case, Himalaya: Earth and Planetary Science Letters, v. 67, p. 191–204, doi: 10.1016/0012-821X(84)90114-6.

Scott, D.J., and St-Onge, M.R., 1995, Constraints on Pb closure temperature in titanite based on rocks from the Ungava orogen, Canada: Implication for U-Pb geochronology and P-T-t path determinations: Geology, v. 23, p. 1123–1126, doi: 10.1130/0091-7613(1995)023<1123:COPCTI>2.3.CO;2.

Snoke, A.W., 1977, A thrust plate of ophiolitic rocks in the Preston Peak area, Klamath Mountains, California: Geological Society of America Bulletin, v. 88, p. 1641–1659, doi: 10.1130/0016-7606(1977)88<1641:ATPOOR>2.0.CO;2.

Snoke, A.W., and Barnes, C.G., 2002, Anatexis and dynamothermal metamorphism during the emplacement of a syntectonic oceanic arc-root pluton, Klamath Mountains, California: Geological Society of America Abstracts with Programs, v. 34, no. 5, p. A-22.

Snoke, A.W., and Barnes, C.G., 2006, this volume, The development of tectonic concepts for the Klamath Mountains province, California and Oregon, *in* Snoke, A.W., and Barnes, C.G., eds., Geological studies in the Klamath Mountains province, California and Oregon: A volume in honor of William P. Irwin: Boulder, Colorado, Geological Society of America Special Paper 410, doi: 10.1130/2006.2410(01).

Snoke, A.W., Quick, J.E., and Bowman, H.R., 1981, Bear Mountain igneous complex, Klamath Mountains, California: An ultrabasic to silicic calc-alkaline suite: Journal of Petrology, v. 22, p. 501–552.

Snoke, A.W., Sharp, W.D., Wright, J.E., and Saleeby, J.B., 1982, Significance of mid-Mesozoic peridotitic to dioritic intrusive complexes, Klamath Mountains—western Sierra Nevada, California: Geology, v. 10, p. 160–166, doi: 10.1130/0091-7613(1982)10<160:SOMPTD>2.0.CO;2.

Stacey, J.S., and Kramers, J.D., 1975, Approximation of terrestrial lead isotope evolution by a two-stage model: Earth and Planetary Science Letters, v. 26, p. 207–221, doi: 10.1016/0012-821X(75)90088-6.

Steiger, R.H., and Jäger, E., 1977, Subcommission on geochronology: Convention on the use of decay constants in geo- and cosmochronology: Earth and Planetary Science Letters, v. 36, p. 359–362, doi: 10.1016/0012-821X(77)90060-7.

Taliaferro, N.L., 1942, Geologic history and correlation of the Jurassic of southwestern Oregon and California: Geological Society of America Bulletin, v. 53, p. 71–112.

Verts, L.A., Chamberlain, K.R., and Frost, C.D., 1996, U-Pb sphene dating of metamorphism: The importance of sphene growth in the contact aureole of the Red Mountain pluton, Laramie Mountains, Wyoming: Contributions to Mineralogy and Petrology, v. 125, p. 186–199, doi: 10.1007/s004100050215.

Watson, E.B., and Harrison, T.M., 1983, Zircon saturation revisited: Temperature and composition effects in a variety of crustal magma types: Earth and Planetary Science Letters, v. 64, p. 295–304, doi: 10.1016/0012-821X(83)90211-X.

Wright, J.E., and Fahan, M.R., 1988, An expanded view of Jurassic orogenesis in the western United States Cordillera: Middle Jurassic (pre-Nevadan) regional metamorphism and thrust faulting within an active arc environment, Klamath Mountains, California: Geological Society of America Bulletin, v. 100, p. 859–876, doi: 10.1130/0016-7606(1988)100<0859:AEVOJO>2.3.CO;2.

Zartman, R.E., and Doe, B.R., 1981, Plumbotectonics—The model: Tectonophysics, v. 75, p. 135–162, doi: 10.1016/0040-1951(81)90213-4.

Manuscript Accepted by the Society 10 January 2006

Printed in the USA

Geological Society of America
Special Paper 410
2006

Structural and petrologic evolution of the Bear Peak intrusive complex, Klamath Mountains, California

Rory R. McFadden
Arthur W. Snoke*
Department of Geology and Geophysics, Department 3006,
1000 East University Avenue, University of Wyoming, Laramie, Wyoming 82071, USA
Calvin G. Barnes
Department of Geosciences, Texas Tech University, Lubbock, Texas 79409-1053, USA

ABSTRACT

The Bear Peak intrusive complex is a Late Jurassic (ca. 144 Ma) composite plutonic suite that ranges in composition from ultramafic to silicic. Clinopyroxene- and hornblende-rich ultramafic cumulate rocks form an intrusion breccia that is complexly intruded by multiple generations of crosscutting gabbroic to dioritic dikes. The bulk of the intrusive complex consists of mappable gabbroic to quartz dioritic to tonalitic/granodioritic units.

The Bear Peak intrusive complex was emplaced into rocks of the Rattlesnake Creek terrane, producing a dynamothermal contact aureole. Contact metamorphism was chiefly at hornblende-hornfels-facies conditions and grades into regional greenschist-facies metamorphism. Andalusite, cordierite, and chloritoid form small porphyroblasts in some of the more aluminous metasedimentary rocks, indicating low-pressure contact metamorphism (<4 kb). Al-in-hornblende geobarometry in quartz dioritic to tonalitic rocks also suggests pressure conditions of ~4 kb. Pseudomorphs of original chiastolite porphyroblasts developed during contact metamorphism of pelitic horizons in the Upper Jurassic Galice Formation, which lies in the footwall of the regional Orleans thrust fault, indicate that the Bear Peak intrusive complex was emplaced after regional contraction related to the Nevadan orogeny.

The Bear Peak intrusive complex is an example of the extended compositional range characteristic of some oceanic-arc plutonic suites and demonstrates how multiple, chiefly magmatic processes, can yield a broad range of rock compositions within a single intrusive complex. Mafic magmatic enclaves are common in most of the plutonic units of the Bear Peak intrusive complex, and distinctive migmatitic amphibolite enclaves indicate that magma temperatures were sufficient to facilitate dehydration-melting of metabasic rocks. The distribution of host-rock enclaves and screens suggest that much of the gabbroic to quartz dioritic parts of the Bear Peak intrusive complex were emplaced as magmatic sheets that coalesced into mappable, relatively homogeneous units that grew by piecemeal intrusion. Ultramafic-mafic cumulates and hornblende gabbro crystallized from a high-Mg, low-Al basaltic parent, whereas

*E-mail: snoke@uwyo.edu.

McFadden, R.R., Snoke, A.W., and Barnes, C.G., 2006, Structural and petrologic evolution of the Bear Peak intrusive complex, Klamath Mountains, California, *in* Snoke, A.W., and Barnes, C.G., eds., Geological studies in the Klamath Mountains province, California and Oregon: A volume in honor of William P. Irwin: Geological Society of America Special Paper 410, p. 333–355, doi: 10.1130/2006.2410(16). For permission to copy, contact editing@geosociety.org.

high-Al, low-Mg contents in quartz dioritic rocks suggest an evolved basaltic or basaltic andesite parent. The biotite tonalite/granodiorite rocks have high Sr values (>700 ppm), large Sr/Y and Ba/Y ratios, and reverse J-shaped rare-earth-element (REE) patterns. These features are characteristic of partial melting of metabasic rocks in which amphibole ± garnet are residual phases. Thus, major, trace, and REE compositions indicate at least two batches of magma were involved in the petrogenesis of the Bear Peak intrusive complex. Complex field relationships and geochemical data suggest that multiple magmas passed through the cumulates and presumably fed structurally higher mafic units in the complex.

Keywords: Jurassic, oceanic-arc pluton, ultramafic rocks, enclaves, contact metamorphism, migmatitic amphibolite, Rattlesnake Creek terrane, Galice Formation, Siskiyou Mountains

INTRODUCTION

The emplacement of juvenile magmas into accreted, oceanic terranes is an important mechanism for the growth of crust along active continental margins. Although this igneous process and the accretion of the wall rocks (oceanic terranes) are broadly related to the subduction of oceanic lithosphere, detailed knowledge of the interplay between magmatism, tectonism, and recycling of accreted crust is still poorly understood.

The ca. 144-Ma Bear Peak intrusive complex is one of a suite of upper mesozonal plutonic complexes that were emplaced after Late Jurassic regional thrust faulting (i.e., Orleans thrust fault; Barnes et al., this volume, Chapter 17). This thrusting has been interpreted as an important structural feature related to the collapse of a Late Jurassic back-arc basin (Snoke, 1977; Saleeby et al., 1982; Harper and Wright, 1984). Bear Peak magmatism thus represents a post-contractional event. Our new geologic mapping of the Bear Peak intrusive complex indicates that it was constructed in piecemeal fashion by the intrusion of numerous dikes or magmatic sheets that coalesced into a plutonic complex. Furthermore, the extended compositional range from ultramafic to silicic indicates that this plutonic complex is characteristic of some oceanic-arc plutonic suites (Snoke et al., 1981; Pearcy et al., 1990; DeBari, 1997; Snoke et al., 2001). Deep incision of the Bear Peak intrusive complex provides ~1300 m of topographic (structural) relief, so that the combination of detailed geologic mapping, petrography, and geochemical studies reveal the types of magmatic processes that can yield a broad range in rock composition in a single plutonic complex.

REGIONAL GEOLOGIC SETTING

The Klamath Mountains province is composed of arcuate, west-directed regional thrust sheets of oceanic rocks that form distinct tectonostratigraphic terranes (Fig. 1). The province is part of an extensive, diverse megabelt of accreted oceanic terranes that forms much of the western half of the North American Cordilleran orogen (Dickinson, 2004). These tectonostratigraphic terranes were accreted onto the North American continental margin from early Paleozoic through Late Jurassic time. The regional thrust sheets were intruded by ultramafic to felsic plutons, which range in age from Silurian to Early Cretaceous. These plutons expose the roots of several long-lived, composite oceanic-arc systems.

Regional contraction, intra-arc extension, and plutonism were important deformational events during the development of the Klamath Mountains. However, relationships among these processes in the province are still debated (e.g., Saleeby et al., 1982; Harper and Wright, 1984; Wright and Fahan, 1988; Harper et al., 1994; Hacker et al., 1995; Irwin and Wooden, 1999; Barnes et al., 2003). Our study focuses on a Late Jurassic, composite, oceanic-arc pluton; thus, relationships to regional events center on tectonic activity during Jurassic and Early Cretaceous time, when arc magmatism was closely associated with multiple contractional deformations as well as an intervening regional extensional history (i.e., development of the Josephine and related ophiolites).

FIELD RELATIONS

Introduction

The Bear Peak intrusive complex is a Late Jurassic (ca. 144 Ma), composite plutonic suite that intruded the Rattlesnake Creek terrane (Wright and Wyld, 1994) and Upper Jurassic Galice Formation, the hanging wall and footwall of the Orleans thrust fault, respectively (Fig. 1). This relationship indicates that the Bear Peak intrusive complex was emplaced after regional thrust faulting along the Orleans fault.

A dynamothermal contact aureole in rocks of the Rattlesnake Creek terrane and Galice Formation is indicated by solid-state plastic deformation of the wall rocks as well as an increase

Figure 1. Generalized geologic map of the Klamath Mountains province. BMic—Bear Mountain intrusive complex, BPic—Bear Peak intrusive complex, O. ft.—Orleans fault. Modified from Irwin (1994).

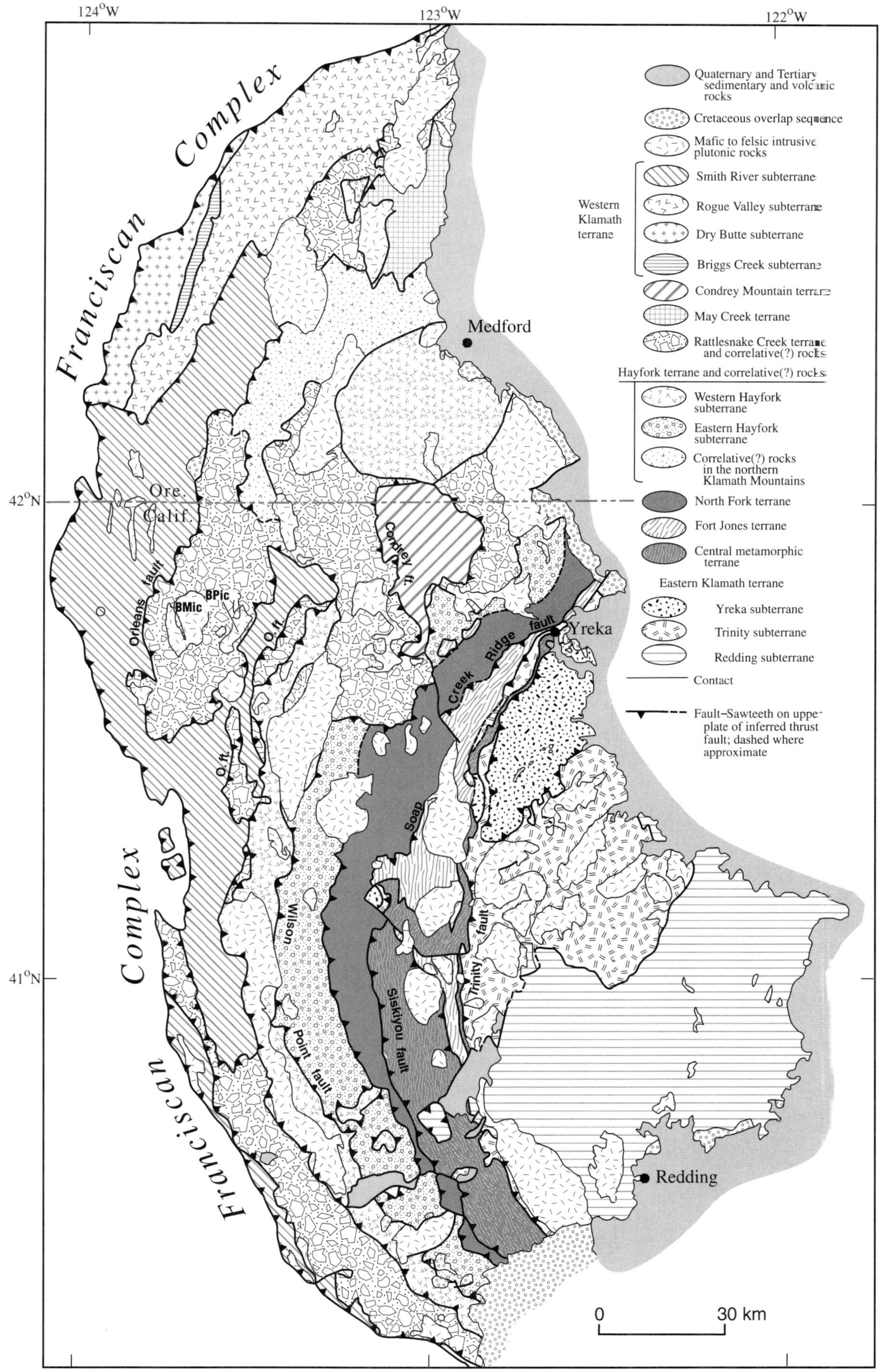
124°W
123°W
122°W
42°N
41°N
Franciscan Complex
Complex
Franciscan
Medford
Yreka
Redding
Ore.
Calif.
Orleans fault
BMic
BPic
O. ft.
O. ft.
Condrey ft.
Creek Ridge fault
Soap
Wilson Point fault
Siskiyou fault
Trinity fault
0
30 km
Quaternary and Tertiary sedimentary and volcanic rocks
Cretaceous overlap sequence
Mafic to felsic intrusive plutonic rocks
Western Klamath terrane
Smith River subterrane
Rogue Valley subterrane
Dry Butte subterrane
Briggs Creek subterrane
Condrey Mountain terrane
May Creek terrane
Rattlesnake Creek terrane and correlative(?) rocks
Hayfork terrane and correlative(?) rocks
Western Hayfork subterrane
Eastern Hayfork subterrane
Correlative(?) rocks in the northern Klamath Mountains
North Fork terrane
Fort Jones terrane
Central metamorphic terrane
Eastern Klamath terrane
Yreka subterrane
Trinity subterrane
Redding subterrane
Contact
Fault–Sawteeth on upper plate of inferred thrust fault; dashed where approximate

in metamorphic grade toward the intrusive complex. Contact metamorphism was chiefly under hornblende-hornfels–facies conditions and grades into the regional greenschist-facies metamorphism characteristic of this part of the Klamath Mountains (Coleman et al., 1988).

The Bear Peak intrusive complex is a heterogeneous body consisting of distinct intrusions that vary in mineral composition and texture. There are abundant mafic dikes that crosscut the intrusive units, and the units commonly contain mafic magmatic enclaves. Also, wall-rock xenoliths and migmatitic amphibolite enclaves are conspicuous throughout the complex (Fig. 2; Plate 1 on the CD-ROM accompanying this volume and in the GSA Data Repository[1]). Five lithologic units comprise the Bear Peak intrusive suite and are displayed in Figure 2. From oldest to youngest they are: (1) ultramafic-mafic complex, (2) hornblende gabbro, (3) biotite-hornblende quartz diorite, (4) biotite tonalite/granodiorite, and (5) synplutonic mafic dikes.

Ultramafic-Mafic Complex

The ultramafic-mafic complex forms a centrally located, topographically low mass within the Bear Peak intrusive complex (Fig. 2). This unit varies in composition from coarse-grained hornblende (±olivine) clinopyroxenite to hornblendite and hornblende melagabbro. The typical mineral assemblage is diopsidic augite + hornblende ± olivine ± orthopyroxene. Hornblende and clinopyroxene oikocrysts that can reach ~6 cm are locally common. The hornblendite commonly contains relict enclaves of hornblende (±olivine) clinopyroxenite (Fig. 3A). The hornblendite grades into hornblende melagabbro with an increase in the abundance of interstitial plagioclase. An important aspect of the ultramafic-mafic complex is that these cumulate rocks were intruded by numerous gabbroic to dioritic dikes (Figs. 3B–D and 4). The complexity of these intrusive relationships has yielded a map-scale intrusion breccia, consisting of ultramafic-mafic rocks as the host unit and crisscrossed by numerous mafic to intermediate dikes.

Hornblende (±olivine) clinopyroxenite consists chiefly of subhedral to anhedral diopsidic augite that is commonly partially to completely replaced by olive-green (Z tint) hornblende. Olivine (~Fo_{85}–Fo_{75}) is commonly partially replaced by a mixture of serpentine-group minerals and magnetite. Orthopyroxene grains are typically scarce, anhedral, and partially replaced by hornblende. Poikilitic hornblende grains in the hornblendite commonly contain many relict clinopyroxene grains. The hornblende (±olivine) clinopyroxenite does not contain intercumulus plagioclase, whereas hornblendite has scarce intercumulus (i.e., interstitial) plagioclase (~An_{65}). Hornblende melagabbro is texturally similar to the hornblendite, except that it contains more interstitial plagioclase.

[1]GSA Data Repository item 2006200, Plate 1, Geologic map of Bear Peak intrusive complex, Siskiyou County, is available online at www.geosociety.org/pubs/ft2006.htm, or on request from editing@geosociety.org or Documents Secretary, GSA, P.O. Box 9140, Boulder, CO 80301, USA.

Hornblende Gabbro

The hornblende gabbro unit crosscuts the ultramafic-mafic rocks and contains enclaves of clinopyroxenite and hornblendite. The unit is texturally and mineralogically heterogeneous. Grain size varies from fine-grained, equigranular to pegmatitic. Rock types that are part of this unit include massive; fine-, medium-, to coarse-grained hornblende gabbro; porphyritic hornblende-clinopyroxene gabbro; pegmatitic hornblende gabbro; and scarce medium-grained pyroxene-hornblende gabbro. Hornblende gabbro is the most common rock type. It is equigranular, characterized by subhedral to euhedral hornblende grains, and has a high color index (>60). Leucocratic, plagioclase-rich segregations and dikelets are common throughout the hornblende gabbro unit. The pegmatitic hornblende gabbro and porphyritic hornblende-clinopyroxene gabbro are found as dikes and plugs throughout the intrusive complex.

Intermediate to calcic plagioclase and abundant olive-green to brown (Z tint) hornblende are the main mineral phases. Plagioclase exhibits weak normal zoning, ranging from ~An_{52} to An_{65}. Clinopyroxene, biotite, and quartz are minor constituents in some samples. Titanite, Fe-Ti oxides, and apatite are common accessory minerals in these gabbroic rocks. The porphyritic hornblende-clinopyroxene gabbro as well as pegmatitic hornblende gabbro contain large grains of prismatic hornblende that can reach ~4 cm in length. The groundmass in the porphyritic gabbro consists of hornblende, clinopyroxene, plagioclase, quartz, and titanite. The most widespread deuteric alteration is the replacement of hornblende by epidote + chlorite + Fe-Ti oxides. The sericitization of plagioclase and replacement of biotite by chlorite ± white mica are other common postmagmatic alterations in the gabbroic rocks.

Biotite-Hornblende Quartz Diorite

Biotite-hornblende quartz diorite is medium-grained and equigranular with a color index between 30 and 50. It is the most extensive map-scale unit of the intrusive suite. Magmatic to hypersolidus foliation defined by plagioclase and hornblende is common and dips steeply toward the interior of the pluton. Fabric development ranges from massive (nonfoliated) to strongly foliated, whereas lineation is not present. Mafic dikes intrude the quartz diorite and mingling relationships are widespread. Mafic biotite-hornblende (quartz) diorite magmatic enclaves are also common throughout the quartz diorite unit. Melting and injection relationships are conspicuous where migmatitic amphibolite enclaves are incorporated into the quartz diorite.

The major igneous minerals in the quartz diorite unit are plagioclase, hornblende, biotite, quartz, and Fe-Ti oxides. Titanite and apatite are common accessory minerals. Trace amounts of alkali feldspar occur in some samples. Hornblende grains are typically prismatic and dark green (Z tint). Biotite is brown, shows no preferred orientation, and commonly forms patchy grains that have partially replaced hornblende. Augite armored by rims of hornblende locally occurs. The Fe-Ti oxides are typically closely associated with the mafic minerals, especially hornblende. Interstitial quartz is anhedral and weakly strained. Local subgrain development in quartz indicates intracrystalline strain at subsolidus conditions. Various deuteric alterations are conspicuous throughout the quartz diorite unit, including the replacement of hornblende by epidote + chlorite and the sericitization of plagioclase. Biotite is also commonly replaced by white mica ± chlorite ± epidote.

Biotite Tonalite/Granodiorite

The tonalite/granodiorite unit consists of equigranular, medium-grained biotite ± hornblende tonalite and granodiorite with trace to ~7 modal% alkali feldspar. It is the least extensive map-scale unit of the Bear Peak intrusive complex. Biotite tonalite/granodiorite is generally homogeneous and weathers yellow; the color index ranges from 10 to 30. The unit is generally massive, but a magmatic fabric defined by oriented biotite plates and tabular plagioclase is evident at some localities. Enclaves of mafic microdiorite locally occur near contacts with the quartz diorite unit. Mafic dikes, fine-grained mafic magmatic enclaves, and migmatitic amphibolite enclaves also occur in this unit.

The common igneous minerals are subhedral biotite, blocky oscillatory-zoned plagioclase (~An_{20-35}), anhedral quartz, and sparse alkali feldspar. Subgrain development and grain-boundary suturing in quartz indicate intracrystalline strain at subsolidus conditions. Apatite, titanite, and Fe-Ti oxides are common accessory phases. Acicular to prismatic deep blue-green hornblende (Z tint) occurs in some samples. Deuteric alteration of biotite to chlorite ± white mica ± epidote and of plagioclase to sericite is widespread in the tonalite/granodiorite unit.

Mafic Dikes and Mafic Magmatic Enclaves

Northwest-striking mafic dikes commonly intrude the quartz diorite unit and locally intrude the hornblende gabbro and biotite tonalite/granodiorite units (Fig. 2, Plate 1). Conspicuous mingling zones are found along the margins of these mafic dikes where enclaves have pillow shapes, and lobate bulb-and-cusp contacts are conspicuous. Mutual back-veining is apparent along many contacts. Flow fabrics in the dikes, where present, are parallel to contacts with the host. Locally, mafic dikes have chilled margins and exhibit abrupt textural variations from fine-grained to very coarse over a meter scale. The fragmentation of once-continuous dikes occurs where they have been dismembered and mechanically incorporated into the magma, leading to trains of subangular to rounded enclaves (Fig. 5). The mafic magmatic enclaves are commonly lobate and flattened in the plane of foliation. They are petrographically similar to the mafic dikes.

Hornblende in the mafic dikes is olive green (Z tint) and is typically acicular to prismatic in habit. Plagioclase is tabular and sometimes aligned by magmatic flow, whereas quartz, a minor component in these rocks, is anhedral and exhibits undulatory extinction, indicating modest solid-state strain. Patchy brown biotite is common in the mafic dikes, and Fe-Ti oxides, titanite, and apatite are the accessory minerals. Hornblende is locally altered to a colorless amphibole, and patchy biotite has also locally overgrown hornblende. Epidote ± chlorite are deuteric alteration products of hornblende and, white mica + chlorite replace biotite.

Red Hill Creek Pluton

The intrusive rocks that comprise the Red Hill Creek pluton are massive, generally equigranular, coarse-grained biotite tonalite, trondhjemite, and granodiorite. This relatively small intrusion is shown as an undifferentiated body on the Bear Peak intrusive complex map (Plate 1), although future, more detailed work would probably yield distinct mappable units. The Red Hill Creek rocks consist chiefly of plagioclase, quartz, and biotite, whereas alkali feldspar is a greatly subordinate varietal mineral. Quartz accounts for ~25–35 modal% of the granitic rocks of the Red Hill Creek pluton; color index ranges from 10 to 20. The Red Hill Creek pluton is significantly more quartzofeldspathic than the adjacent Bear Peak intrusive complex. North–northwest-striking dikes, which intrude the granitic rocks, include fine-grained mafic rocks (microdiorite to microgabbro), fine- to medium-grained biotite + hornblende gabbro, and aplite/pegmatite. Xenoliths of clinopyroxenite occur in the Red Hill Creek pluton. Also, dikes and small masses of fine-grained gabbro, characterized by scattered hornblende phenocrysts and a groundmass consisting hornblende + plagioclase + clinopyroxene, intrude the pluton.

Plagioclase forms blocky subhedral grains that exhibit conspicuous oscillatory zoning (An_{15}–An_{45}). Quartz grains are anhedral, with subgrain development and evidence of grain-boundary suturing (migration). Dark brown to reddish-brown (Z tint) biotite occurs as subhedral, tabular grains. Alkali feldspar forms scarce interstitial grains. Blue-green hornblende occurs in a few samples. Apatite, titanite, zircon, and Fe-Ti oxides are accessory phases. White mica and epidote are common deuteric alterations of biotite, and sericitization of plagioclase grains is common.

Host Rocks

Galice Formation. The Galice Formation consists of light- to medium-gray metagraywacke interlayered with black slate. The metagraywacke is predominantly an impure chert/cherty argillite-clast lithic wacke with quartz and subordinate plagioclase grains. Bedding is locally preserved, but transposed by a pervasive cleavage (S_1) related to near-isoclinal folding and synchronous pressure solution. The Galice Formation is contact metamorphosed near the northeast contact of the Bear Peak intrusive complex. This contact metamorphism is manifested by

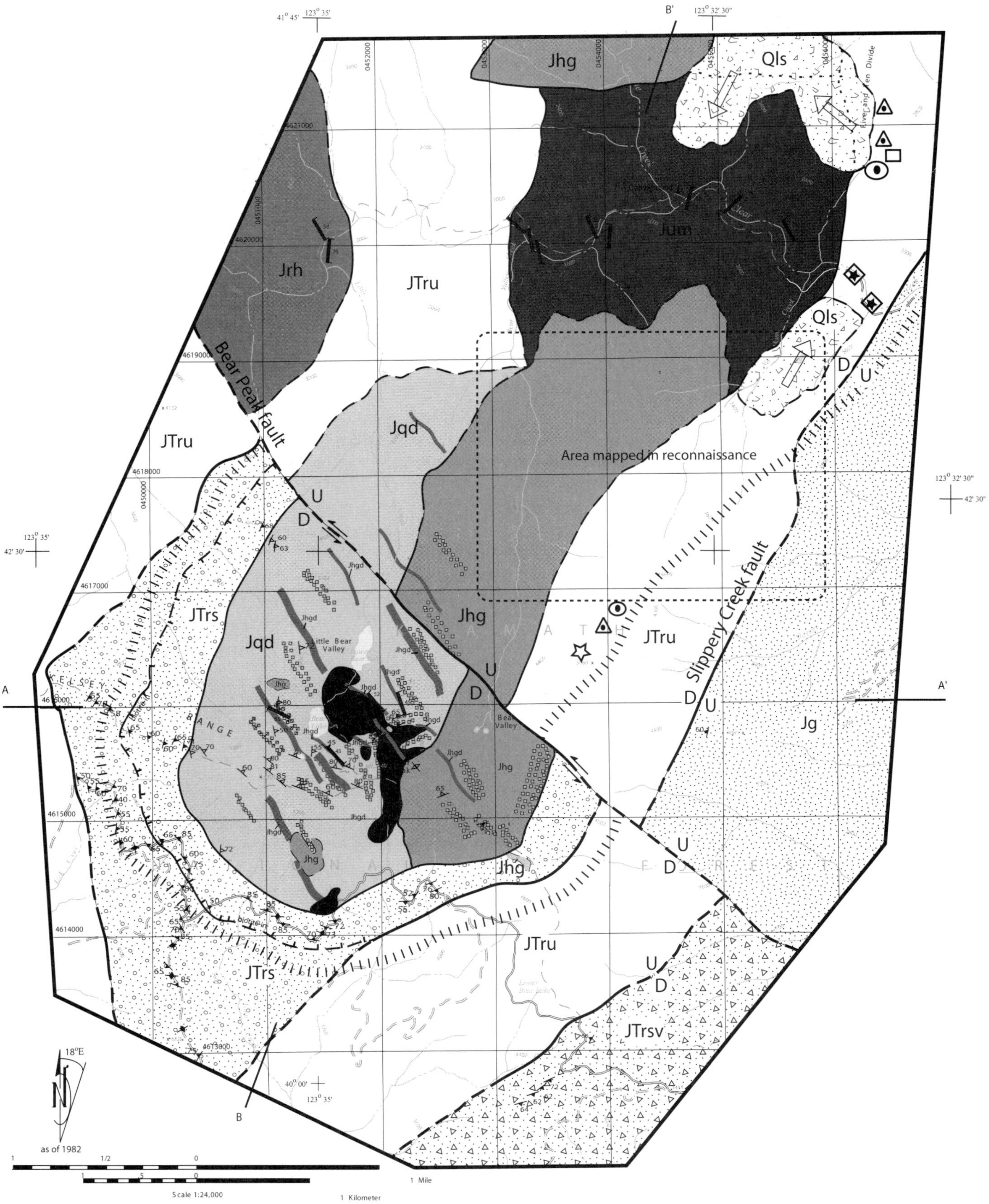

Topographic base derived from U.S.G.S. Bear Peak, California 7.5 minute quadrangle

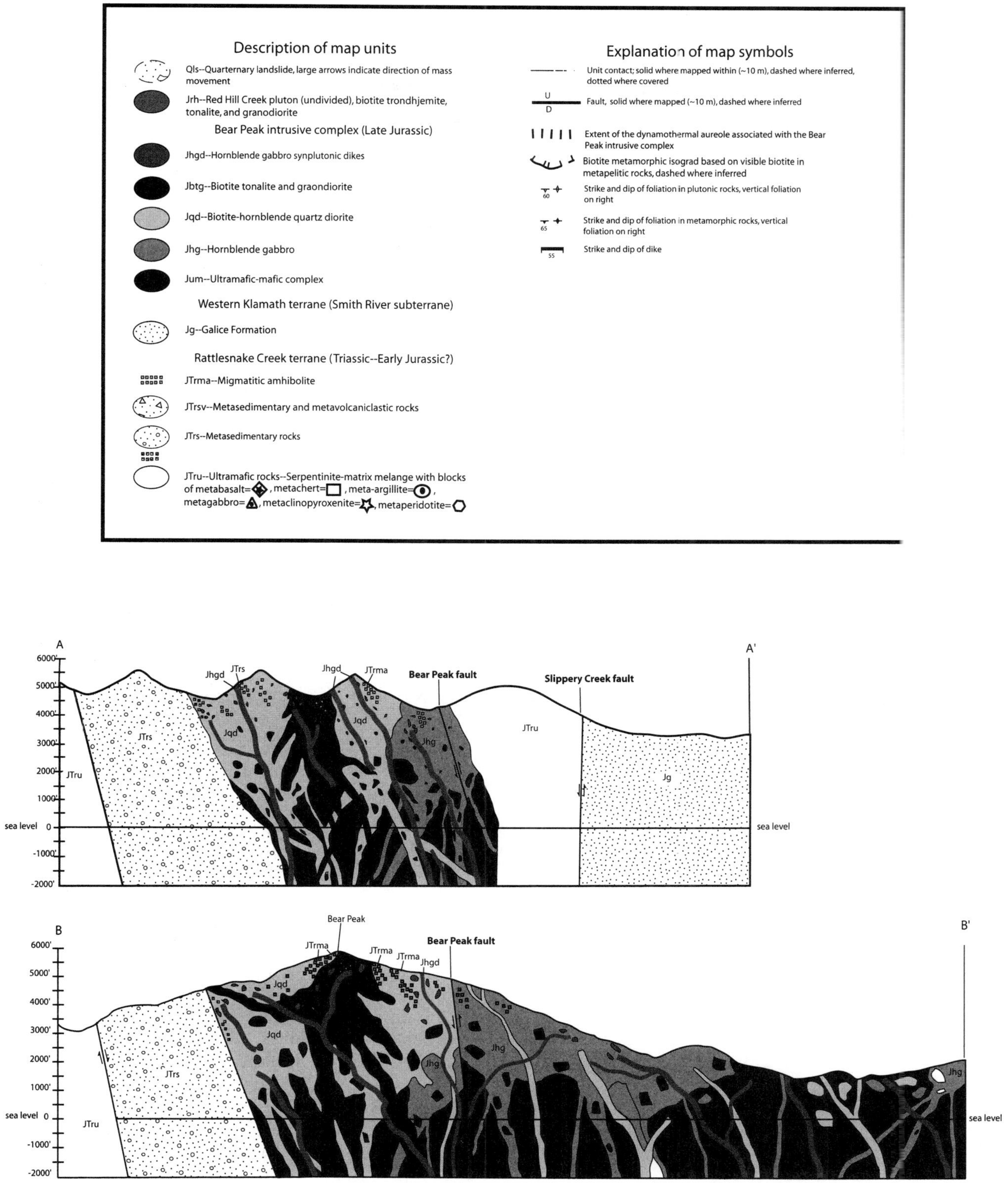

Figure 2. (A) Generalized geologic map of the Bear Peak intrusive complex, Siskiyou County, California. (B) Cross-sections of Bear Peak intrusive complex, Siskiyou County, California. Unit descriptions are given in the figure. Refer to Plate 1 (on the CD-ROM accompanying this volume and in the GSA Data Repository [see footnote 1]) for a colored geologic map of the Bear Peak intrusive complex on a topographic base with cross-sections.

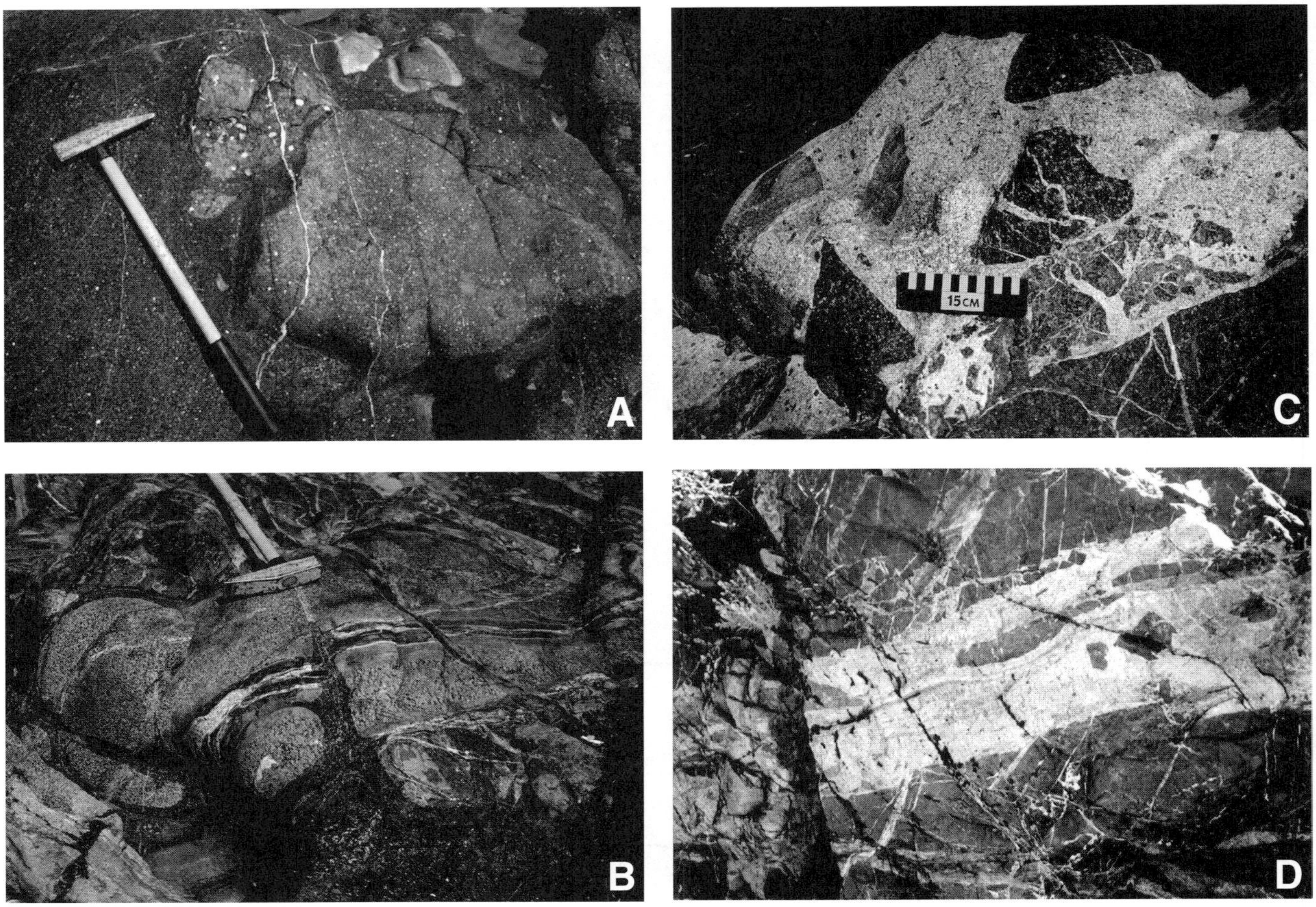

Figure 3. Field photographs of the ultramafic-mafic complex, showing the complexity and heterogeneity of the unit. (A) Subangular hornblende clinopyroxenite enclave within hornblendite. The clasts near the top of the photograph are xenoliths of metaserpentinite that has been altered chiefly to talc + Fe-Ti oxides + chlorite; relict Fe-Cr spinel grains also occur in these rocks. (B) Crosscutting relationships in the intrusion breccia of the ultramafic-mafic complex. (C) Intrusion breccia engulfing angular enclaves of hornblendite. (D) Dioritic dike with angular enclaves of hornblendite. The dike also shows the development of veins within the hornblendite rocks.

chloritoid and scarce pseudomorphs of porphyroblastic chiastolite, developed in pelitic horizons of the Galice Formation. The presence of contact-metamorphic assemblages localized near the Bear Peak intrusive complex contact indicates that the pluton intruded lower-plate rocks structurally below the Orleans thrust fault.

Foliation in contact-metamorphosed metapelitic rocks of the Galice Formation is defined by strongly aligned, fine-grained biotite and graphite. These rocks also include fine-grained, granoblastic aggregates of quartz. Chloritoid porphyroblasts have grown across the foliation, suggesting a post-kinematic development. Prismatic pseudomorphs of chiastolite porphyroblasts are aggregates of phyllosilicates and carbonaceous material. The biotite-graphite fabric wraps around these porphyroblasts.

Interlayered Metasedimentary and Metavolcaniclastic Rocks of the Rattlesnake Creek Terrane. The metasedimentary and metavolcaniclastic unit crops out in the southeast corner of the map area (Fig. 2, Plate 1). These rocks are interpreted as part of a coherent cover sequence of the Rattlesnake Creek terrane that originally was deposited unconformably upon serpentinite-matrix mélange (Wright and Wyld, 1994). The metasedimentary rocks consist of metachert, siliceous meta-argillite, and slaty meta-argillite. The metavolcaniclastic rocks range from felsic to mafic in composition. The metasedimentary and metavolcaniclastic rocks were regionally metamorphosed at greenschist-facies conditions. The metasedimentary rocks consist chiefly of tabular brown biotite, anhedral quartz, and subordinate plagioclase. The metavolcaniclastic rocks consist chiefly of calcic

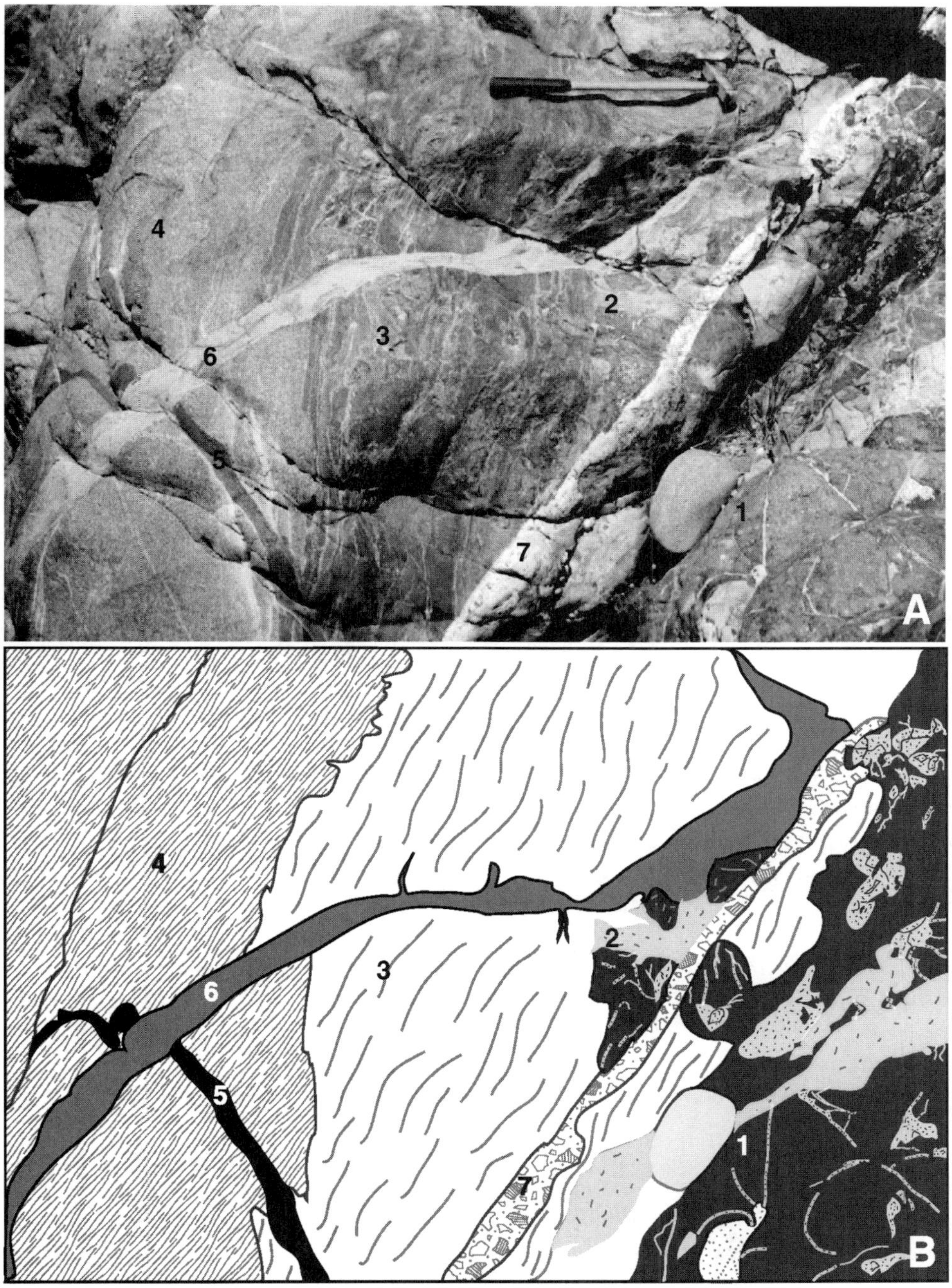

Figure 4. (A) Photograph of crosscutting relationships that are conspicuous in the ultramafic-mafic complex of Bear Peak intrusive complex. The order of intrusive units is based on crosscutting relationships and is shown from oldest to youngest (earliest unit, clinopyroxenite not shown): (1) hornblendite, (2) fine-grained diorite forming intrusion breccia, (3) hornblende gabbro displaying mingling and mixing relationships with foliated hornblende diorite, (4) foliated hornblende diorite, (5) late hornblende diorite dike, (6) late fine-grained hornblende gabbro dike, and (7) pegmatitic hornblende diorite dike. (B) Sketch of the intrusive relationships drawn from photograph in A.

amphibole (actinolite?), brown biotite, clinozoisite, plagioclase, and quartz.

Metasedimentary Rocks of the Rattlesnake Creek Terrane. This unit is in contact with the southern part of the Bear Peak intrusive complex (Fig. 2, Plate 1). The unit consists of interlayered metachert, meta-argillite, siliceous meta-argillite, and slaty meta-argillite. In the contact aureole, metamorphic grade is hornblende-hornfels facies characterized by megascopic biotite, a hornfelsic texture, crystal-plastic deformation, and local partial melting of pelitic rocks. Scarce porphyroblasts of chloritoid, cordierite, andalusite, garnet, and muscovite are also present. This rock unit may be a large composite block within the mélange of the Rattlesnake Creek terrane; or alternatively, it may be part of the originally overlying "cover sequence," subsequently down-faulted into the mélange.

The mineral assemblages of the metasedimentary rocks consist chiefly of fine-grained, anhedral quartz, strongly aligned biotite grains, and anhedral Fe-Ti oxides. Biotite-rich domains

Figure 5. Photograph of mafic magmatic enclaves (dismembered dike) in biotite-hornblende quartz diorite, Upper Bear Lake area. Such synplutonic dikes are evidence of coexisting gabbroic and dioritic magmas during the development of the Bear Peak intrusive complex.

wrap around the garnet porphyroblasts, suggesting synkinematic growth. Chloritoid and muscovite porphyroblasts cut across the foliation indicating post-kinematic growth. Probable original cordierite and/or andalusite porphyroblasts are now chiefly pseudomorphosed to elliptical aggregates of phyllosilicate minerals and quartz.

Ultramafic Rocks of the Rattlesnake Creek Terrane. The ultramafic host rocks are the matrix of the serpentinite-matrix mélange. The mélange contains blocks of metabasalt, metachert, meta-argillite, metagabbro, and metaclinopyroxenite (Plate 1). The metaserpentinite rocks are fine-grained, range in color from dark green to blue black to black, and consist chiefly of an aggregate of serpentine-group minerals, talc, tremolite, and opaque oxides. The metaserpentinite is commonly massive and forms angular outcrops that weather brown to deep orange.

Xenoliths of metaserpentinite wall rock are found in the ultramafic-mafic complex (Figs. 3A and 6). The fragmentation of these inclusions may be due to a combination of thermal stresses and dehydration reactions. The presence of ultramafic xenoliths in the Bear Peak intrusive complex suggests that wall-rock assimilation (e.g., Kelemen and Ghiorso, 1986) may have played a role in the development of the ultramafic-mafic complex.

Serpentine-group minerals in the metaserpentinites are texturally variable and occur as interlocking aggregates, as pseudomorphs after olivine and/or pyroxene, and in veinlets. A colorless, bladed serpentine-group mineral is especially common in the metaserpentinites. It is probably antigorite and may reflect thermal metamorphism related to the Bear Peak intrusive complex. Tremolite forms colorless, acicular to prismatic grains and is commonly associated with talc. Forsteritic olivine, developed during contact metamorphism, was found in some metaserpentinites exposed in Clear Creek between the intrusive contacts of the Bear Peak intrusive complex and Red Hill Creek pluton. Various opaque oxides (probably magnetite and Fe-Cr spinel) as well as magnesite are also common in the metaserpentinites. The typical prograde contact metamorphic mineral assemblage is antigorite + talc + opaque oxides ± tremolite ± olivine ± chlorite ± magnesite. The xenoliths found in the ultramafic-mafic complex consist of talc + magnetite + Fe-Cr spinel + chlorite.

Migmatitic Amphibolite and Metasedimentary Enclaves

The enclaves of migmatitic amphibolite and metasedimentary rocks are abundant, recognizable units in the Bear Peak intrusive complex. They are oriented subparallel to the magmatic foliation in the host plutonic rocks and strike northwest–southeast (subparallel to the synplutonic mafic dikes). These enclaves appear to separate distinct, sheetlike bodies of plutonic rock and thereby provide evidence for an important mode of magmatic emplacement during the development of this plutonic complex (Fig. 2, Plate 1; see discussion below). Near the eastern host-rock–intrusive complex contact, migmatitic amphibolite and hornblende gabbro form a sheeted margin.

The migmatitic amphibolite enclaves are metabasites that preserve relict lithic clasts, heterogeneous grain size in the matrix, and relict pyroxene and plagioclase. These rocks are described as "migmatitic" because they consist of variable proportions of amphibolitic melanosome pervasively injected by leucosome veinlets. The leucosomes consist of quartz and plagioclase with minor biotite, hornblende, and alkali feldspar,

Figure 6. Field photographs depicting the ultramafic xenoliths in the ultramafic-mafic complex. (A) The rock above the hammer is an ultramafic xenolith that has been boudinaged. A thin reaction zone consisting chiefly of tremolite occurs along the contact with hornblendite. The mineral assemblage in the xenolith is talc + Fe-Ti oxides + Fe-Cr spinel + chlorite. (B) Fragmented ultramafic xenoliths (~3 cm). The reaction zone developed around the xenoliths is a narrow rim of tremolite in turn surrounded by hornblendite.

whereas the melanosome (including the relict lithic clasts) consists chiefly of hornblende and plagioclase with minor biotite and clinozoisite/epidote.

Metapelitic and siliceous argillite enclaves are found near the pluton-host rock contact toward Red Hill (Fig. 2, Plate 1). The melanosome mineral assemblage in metapelitic rocks is biotite + quartz + cordierite, whereas in siliceous argillite, it is biotite + quartz. Both metasedimentary rock types are injected by leucosome with the mineral assemblage quartz + plagioclase with minor biotite.

Late Crosscutting Faults

The Bear Peak fault is a left-lateral, oblique-slip fault with a northwest-strike and unknown displacement. The northeast block is the upthrown side. There is noticeable map-scale offset of the intrusive contacts of the Bear Peak intrusive complex, with many intrusive and host rock units truncated by the fault.

The Slippery Creek fault is a high-angle, northeast-striking normal fault. This fault juxtaposes Rattlesnake Creek terrane rocks in the hanging wall of the Orleans fault on the west with footwall rocks of the Galice Formation on the east. The Slippery Creek fault is a late normal fault that cuts the regional Orleans thrust fault.

GEOCHEMISTRY

Samples from the Bear Peak intrusive complex have a wide variation in SiO_2 content, ranging from 42 to 72 wt%. Based on the classification of Frost et al. (2001), the sample suite is chiefly calcic to calc-alkaline, magnesian, and metaluminous, although some tonalitic and granodioritic samples are mildly peraluminous (Fig. 7). Also, several samples of quartz diorite straddle the magnesian-ferroan boundary.

The ultramafic-mafic rocks are characterized by high MgO and low Al_2O_3 contents. They show considerable scatter in minor elements (not shown) that may reflect chemical interaction with the mafic and/or intermediate dikes, or possibly these characteristics reflect multiple processes (i.e., crystal accumulation and/or assimilation of metaserpentinite). It is evident that Cr and Ni are abundant in clinopyroxenite and hornblende melagabbro samples (Fig. 8D and E). The Bear Peak intrusive complex ultramafic-mafic rocks are similar to other ultramafic-mafic complexes in the Klamath Mountains, such as the Blue Ridge intrusion (e.g., Snoke et al., 1981). The hornblende gabbro, quartz diorite, and tonalite/granodiorite units have chemical trends that contrast with one another (Fig. 8, Table 1). The amount of Al_2O_3 increases with increasing SiO_2 in the high-MgO hornblende gabbros, whereas Al_2O_3 decreases with increasing SiO_2 in the biotite tonalite/granodiorite unit. The low-MgO quartz dioritic rocks show approximately constant high Al_2O_3 contents of ~18–19 wt%. Trends similar to those for Al_2O_3 can be seen in Sr abundances, but the data are more scattered (Fig. 8C). The quartz diorite and biotite tonalite/granodiorite samples show low concentrations of Rb, Ni, and Cr that are characteristic of calc-alkaline rocks. However, the gabbroic samples, interpreted as cumulates, have extremely high Cr and Ni abundances. The biotite tonalite/granodiorite rocks (62–72 SiO_2 wt%) are characterized by large Sr/Y and Ba/Y ratios and high Sr (>700 ppm) values.

A sample of hornblende melagabbro (1188) from the ultramafic-mafic complex shows a concave downward rare-earth-element (REE) pattern with an overall negative slope (Fig. 9, Table 2). The pattern reaches maximum normalized REE abundances at Sm, and heavier REE (HREE) abundances are higher than all other analyzed samples (Fig. 9). In contrast, the light REEs (LREEs) have a normalized La/Sm ratio <1. This pattern is typical of rocks whose REE abundances are controlled by accumulation of clinopyroxene and/or hornblende and accessory minerals. The biotite-hornblende quartz diorite and hornblende gabbro samples display parallel, negative slopes (normalized La/Yb ratios range from 5 to 8), and HREE concentrations are about ten times those of chondrites. Europium anomalies are absent or slightly positive (Fig. 9). The biotite tonalite/granodiorite rocks exhibit distinctive steep negative slopes (Ce/Yb = 16.70 and 22.98) and some have nearly flat chondrite-normalized HREE (reverse J-shape). Such patterns are common among high-Al_2O_3 tonalite and trondhjemite rocks (Arth and Hanson, 1972; Barker and Arth, 1976; Arth et al., 1978; Barnes et al., 1996).

CONDITIONS OF EMPLACEMENT

Age of Emplacement

Although a contact between the Bear Peak intrusive complex and Upper Jurassic Galice Formation is not exposed, contact-metamorphic textures and assemblages in the Galice Formation within ~600 m of the contact argue for an intrusive relationship. If so, the pluton intruded both the hanging wall and footwall of the Orleans thrust fault. This assertion is consistent with recent U-Pb (zircon) dating of a late-stage granodiorite (Allen and Barnes, this volume), which yielded an age of 143.7 ± 1.3 Ma. As the Orleans thrust is pinned by the ca. 150-Ma Summit Valley pluton (Harper et al., 1994), the Bear Peak intrusive complex must have passed through and thermally metamorphosed the Galice Formation. Furthermore, Allen and Barnes (this volume) reported distinctive zircon inheritance that is similar to detrital zircon from basal metagraywacke of the Galice Formation (Miller and Saleeby, 1995; Miller et al., 2003). These data led Barnes et al. (this volume, Chapter 17) to place the Bear Peak intrusive complex in their "western Klamath plutonic suite" (151–144 Ma).

Estimates of Contact-Metamorphic Conditions

Adjacent to the ultramafic-mafic complex, the host rocks are mainly metaserpentinite. The peak metamorphic assemblage is olivine + tremolite + talc + magnetite + Fe-Cr spinel that suggests temperatures between 550 and 600 °C at pressures of ~3 kb (Tracy and Frost, 1991). The metamorphic grade reached hornblende-hornfels–facies metamorphism in the semipelitic host rocks, and the lack of clinopyroxene in the semipelitic rocks is consistent with temperatures below 675 °C (Russ-Nabelek, 1989). Spotted slaty argillite in the Rattlesnake Creek terrane contains porphyroblasts of andalusite and chloritoid, as well as inferred cordierite porphyroblasts, now pseudomorphs consisting of an aggregate of phyllosilicate minerals, quartz, and opaque oxides. Hornfelsic metapelitic rocks in the Galice Formation contain chloritoid porphyroblasts and scarce chiastolite pseudomorphs, now chiefly altered to phyllosilicate minerals. Pattison and Tracy (1991) summarized experimental data, indicating that andalusite is stable at ~4.5 kb (450 MPa) at 550 °C, and at 650 °C, andalusite has a maximum pressure stability of ~3.5 kb (350 MPa). Therefore, the contact-metamorphic assemblages in the Bear Peak aureole indicate emplacement pressures of ~3–4.5 kb and maximum aureole temperatures of ~550–675 °C.

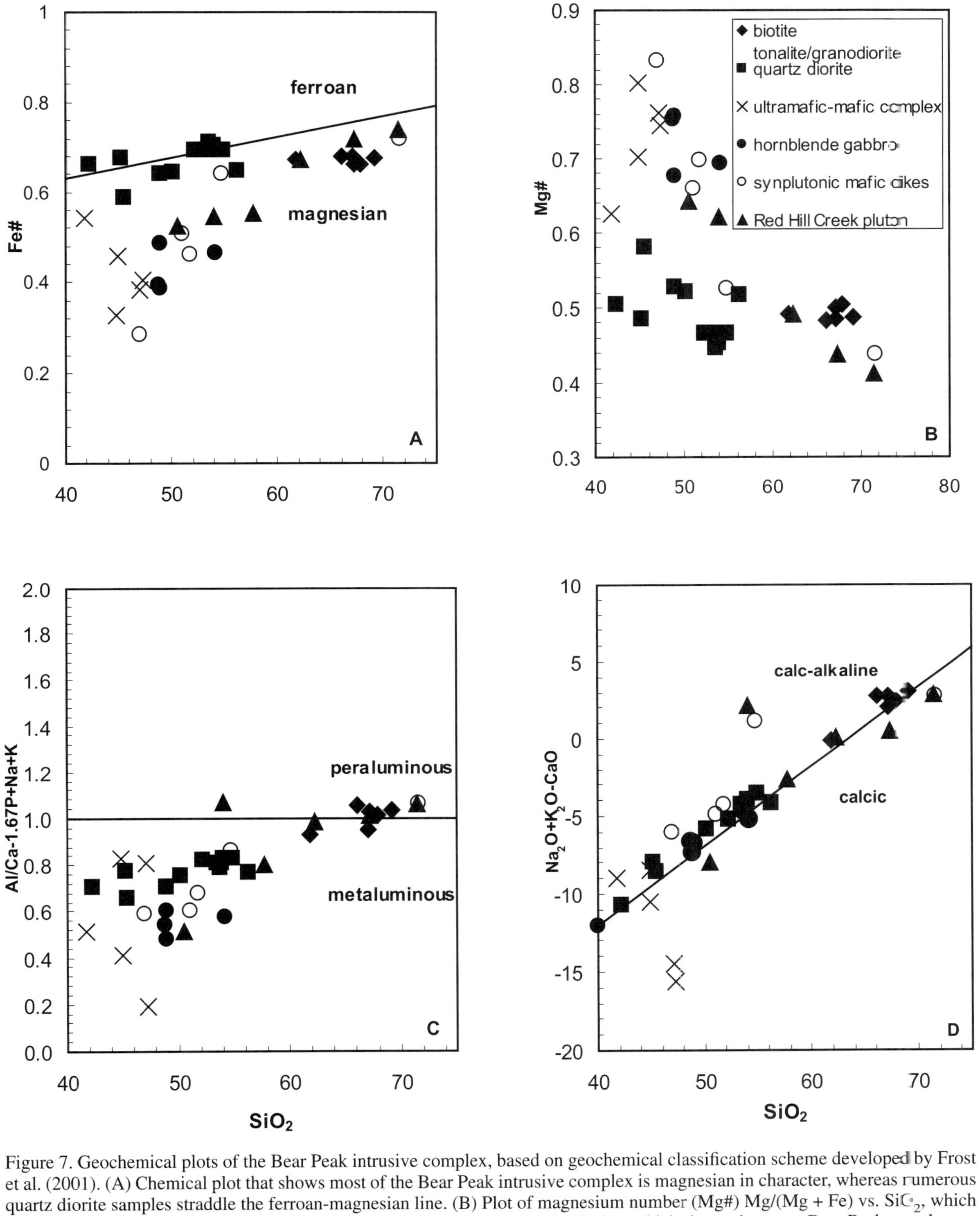

Figure 7. Geochemical plots of the Bear Peak intrusive complex, based on geochemical classification scheme developed by Frost et al. (2001). (A) Chemical plot that shows most of the Bear Peak intrusive complex is magnesian in character, whereas numerous quartz diorite samples straddle the ferroan-magnesian line. (B) Plot of magnesium number (Mg#) Mg/(Mg + Fe) vs. SiO_2, which displays two distinct groups of samples. (C) Plot of the aluminum-saturation index, which shows that most Bear Peak samples are metaluminous, except for some samples from the biotite tonalite/granodiorite unit and Red Hill Creek pluton. (D) Plot of the modified alkali-lime index, which shows the Bear Peak intrusive complex samples straddle the calcic–calc-alkalic line.

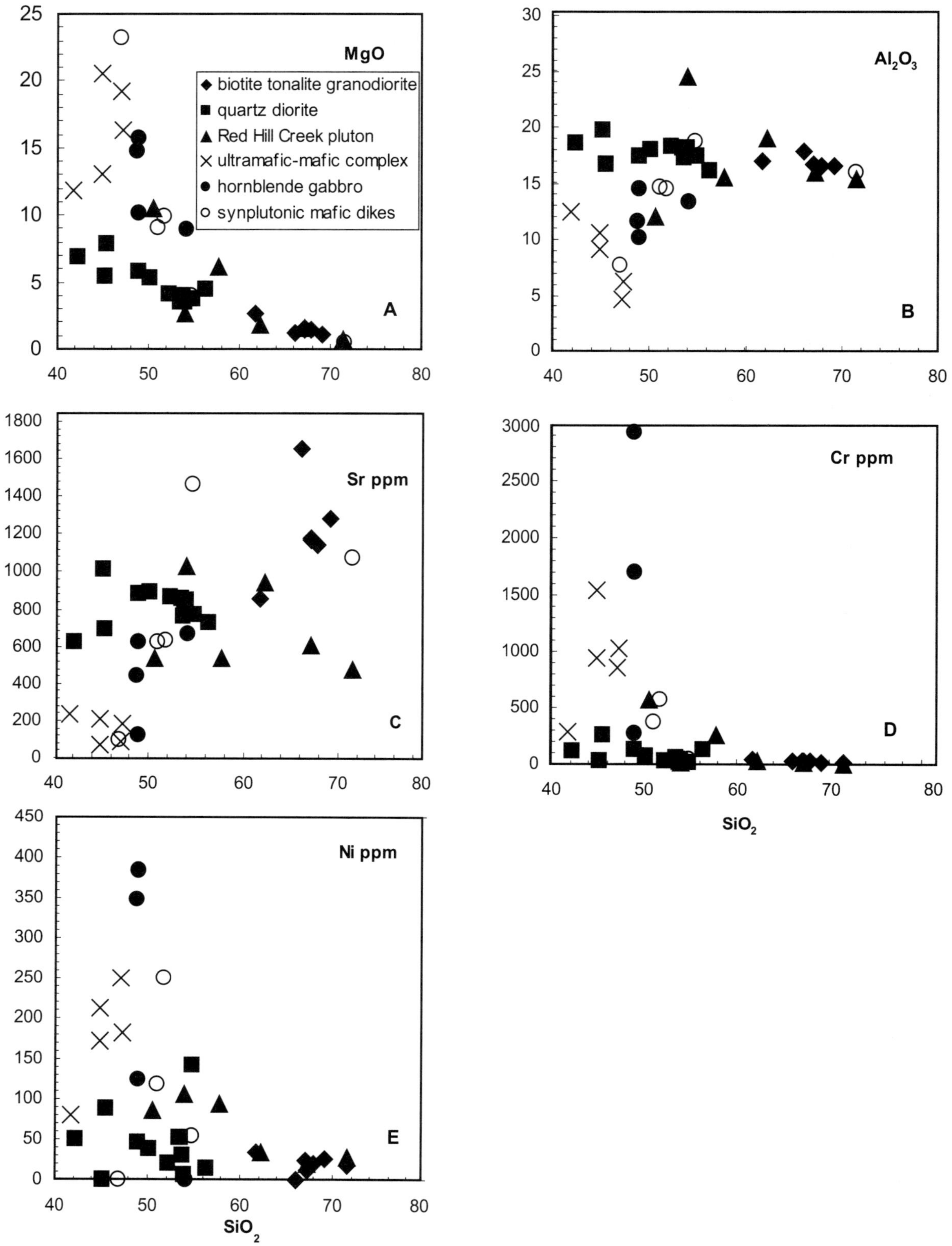

Figure 8. Major and trace element variation diagrams for the Bear Peak intrusive complex. (A) Plot of MgO vs. SiO_2 exhibits two distinct trends: one high-MgO trend for the hornblende gabbro unit and a low-MgO trend for the quartz diorite unit. (B) The Al_2O_3 vs. SiO_2 plot shows an increase in Al_2O_3 for the hornblende gabbros and constant high Al_2O_3 for the quartz diorites. (C) The Sr vs. SiO_2 plot shows very high concentrations for the biotite tonalite/granodiorite unit that may suggest melting from a crustal source region. (D) Plot of Cr vs. SiO_2 displays high Cr values for the ultramafic-mafic rocks as well as some hornblende gabbros. (E) Concentrations of Ni are also high in the ultramafic-mafic complex and hornblende gabbros.

TABLE 1. REPRESENTATIVE WHOLE-ROCK GEOCHEMICAL DATA FOR ROCKS FROM THE BEAR PEAK INTRUSIVE COMPLEX

	BP4	BP988A	BP350A	BP382	BP18	BP253	BP279	BP291	BP357	BP348A	BP340A	BP430B
Sample	Ultramafic-mafic rocks		Hornblende gabbro		Biotite-hornblende quartz diorite		Biotite tonalite/ granodiorite		Synplutonic mafic dikes		Red Hill Creek pluton	
SiO_2	47.07	47.24	48.96	48.89	50.15	54.89	69.14	67.21	51.04	54.75	71.48	50.55
TiO_2	0.49	0.67	1.15	0.88	0.98	0.85	0.28	0.32	0.93	0.74	0.20	0.85
Al_2O_3	4.70	6.23	14.44	10.16	18.03	17.42	16.59	16.49	14.60	18.77	15.42	12.00
FeO_t	11.90	11.08	9.70	9.92	9.60	8.36	2.37	3.04	9.25	7.06	2.20	11.59
MnO	0.19	0.15	0.14	0.22	0.19	0.17	0.06	0.08	0.19	0.14	0.08	0.25
MgO	19.15	16.29	10.19	15.69	5.29	3.69	1.14	1.54	9.00	3.94	0.78	10.52
CaO	15.18	16.74	10.35	9.93	10.05	8.22	3.44	3.99	9.55	6.02	3.30	10.73
Na_2O	0.62	0.99	2.65	0.99	3.18	3.29	5.03	4.49	3.62	6.65	3.83	2.16
K_2O	0.08	0.13	0.95	1.62	1.08	1.31	1.52	1.55	1.08	0.48	2.28	0.59
P_2O_5	0.02	0.07	0.09	0.07	0.33	0.34	0.07	0.09	0.08	0.24	0.09	0.20
LOI	0.92	1.17	1.31	1.22	1.05	1.09	0.79	1.09	1.09	0.97	0.72	0.70
Total	100.33	100.76	99.93	99.59	99.94	99.62	100.44	99.90	100.44	99.77	100.38	100.14
Mg#	0.76	0.74	0.68	0.74	0.52	0.47	0.49	0.50	0.66	0.53	0.41	0.64
Fe#	0.38	0.40	0.49	0.39	0.64	0.69	0.68	0.66	0.51	0.64	0.74	0.52
MALI	–14.48	–15.62	–6.76	–7.32	–5.79	–3.61	3.11	2.05	–4.86	1.11	2.81	–7.98
ASI	0.81	0.19	0.60	0.48	0.76	0.83	1.04	1.02	0.60	0.86	1.06	0.52
Rb	2.00	22.00	14.00	17.00	25.00	3.00	33.00	36.00	37.00	51.00	23.00	13.00
Sr	182.00	425.31	669.39	672.11	805.52	86.94	1292.38	1182.42	614.78	1081.10	846.00	765.00
Zr	28.40	59.07	n.d.	56.31	97.98	n.d.	119.90	95.75	115.00	188.00	107.00	149.00
Y	11.30	14.53	11.68	14.67	17.47	8.14	3.46	5.93	8.22	11.86	19.70	22.00
Nb	11.00	n.d.	14.11	n.d.	n.d.	13.39	n.d.	n.d.	n.d	n.d.	13.00	16.00
Ba	54.70	475.10	271.76	298.78	458.02	14.18	986.00	835.16	637.80	959.89	415.00	401.00
Sc	85.50	40.96	38.30	66.03	21.09	75.30	3.89	7.43	6.08	6.25	19.40	28.40
V	280.00	242.97	194.00	299.23	199.01	191.00	46.63	58.68	54.45	50.39	191.00	233.00
Cr	1025.00	2958.00	407.00	269.71	18.42	860.00	20.55	21.84	8.32	13.88	32.00	41.00
Ni	182.00	347.14	94.51	124.00	29.24	250.00	25.01	11.45	19.05	105.36	7.00	53.00
Cu	202.00	70.53	9.24	35.22	11.63	288.00	23.18	24.56	12.39	4.00	13.00	11.00
Zn	n.d.	60.21	72.72	62.88	116.00	49.90	42.41	41.76	70.50	36.67	n.d.	n.d.
Be	n.d.	0.48	1.07	0.62	0.90	0.30	1.20	1.20	0.89	1.62	n.d.	n.d.

Notes: Major elements in wt%, trace elements in ppm. The classification scheme used is based on Frost et al. (2001). ASI—Aluminum saturation index, which classifies granitic rocks as peraluminous or metaluminous; Fe#—Fe/(Fe + Mg); LOI—loss on ignition; MALI—modified alkali-lime index, which indicates whether a granitic rock is calcic, calc-alkalic, or alkalic; Mg#—(Mg/40.3)/[(Mg/40.3) + (2 × Fe/159.7)]; n.d.—not detected.

Geothermobarometry

To estimate temperature and pressure of final mineral equilibration, rim compositions of hornblende and plagioclase grains were determined on the University of Wyoming, Laramie, automated JEOL JXA-8900 electron probe microanalyzer. The Al-in-hornblende geobarometer correlates Al^{tot} content of magmatic hornblende linearly with crystallization pressure of intrusion (Hammarstrom and Zen, 1986). In amphibole-bearing tonalitic to granodioritic rocks, the magmatic system can be described by ten components: SiO_2–TiO_2–Al_2O_3–Fe_2O_3–FeO–MgO–CaO–Na_2O–K_2O–H_2O. Commonly, tonalite and granodiorite are nine-phase systems: hornblende + biotite + plagioclase + potassic feldspar + quartz + titanite + Fe-Ti oxide + melt + fluid phase. Such systems are trivariant, with three degrees of freedom, which leads to three intensive variables (Hollister et al., 1987). The fO_2 may be buffered by a second Fe-Ti oxide or epidote. If temperature lies close to the wet solidus (nearly isothermal), then pressure is the only variable (Hollister et al., 1987; Schmidt, 1992). In cases where the solidus temperature is poorly known, the barometer is sensitive to both fO_2 and temperature (Anderson and Smith, 1995). Therefore, it is important to correct for the temperature of equilibration of the nine-phase assemblage and to analyze rocks that fall within the experimental calibrations, which is the case for most of the Bear Peak intrusive complex samples. All samples have the appropriate assem-

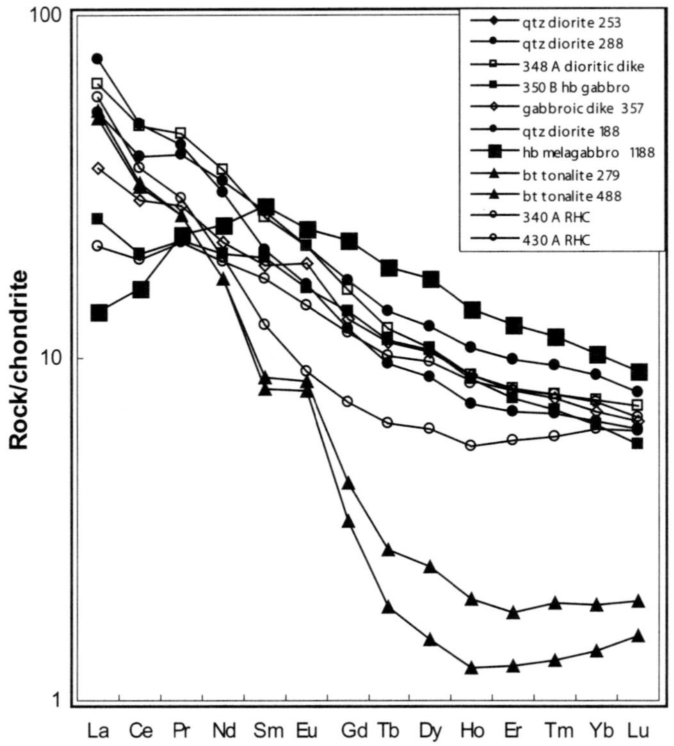

Figure 9. Rare-earth-element (REE) diagram of the Bear Peak intrusive complex rocks. The REE plot of the hornblende melagabbro sample (1188) shows a concave-downward trend with an overall negative slope. There is also a depletion of heavy and light REEs. The quartz diorite and hornblende gabbro display a negative slope with light REE enrichment. The biotite tonalitic/granodioritic rocks are strongly fractionated; have small, positive Eu anomalies; and exhibit a flat heavy-REE pattern. These trends show a reverse J-shaped pattern characteristic of high-Al_2O_3 tonalitic/trondhjemitic rocks (Barnes et al., 1996).

blage for Al-in-hornblende barometry: hornblende + plagioclase + biotite + quartz + potassic feldspar + titanite + Fe-Ti oxides (Hammarstrom and Zen, 1986; Hollister et al., 1987).

We used the approach of Anderson and Smith (1995), in which temperature is calculated from the two plagioclase-hornblende thermometers (Thb1 and Thb2) developed by Holland and Blundy (1994), and pressure estimates are corrected using the estimated temperature by iteration. Temperature estimates based on Thb2 are preferred, due to the possibility that quartz crystallized after the equilibration of plagioclase and hornblende (Anderson and Smith, 1995).

Table 3 shows the rock types and estimated pressure and temperature for the four samples. Quartz diorite samples, BP19b and BP251, yielded hornblende–plagioclase equilibrium temperatures from 750 to 770 °C (±20 °C) and pressures from 3.7 to 3.9 kb (±0.7 kb). Samples from the biotite tonalite/granodiorite unit, BP174 and BP288, have estimated temperatures from 690 to 720 °C (±30 °C) and pressures of 4.5–4.7 kb (±0.5 kb).

Pressure estimates derived from the two rock types overlap (within analytical uncertainty), although the overlap is slight. This near lack of agreement may be the result of lack of equilibrium among the analyzed phases; however, care was taken to analyze minerals in apparent textural equilibrium (McFadden, 2004). Different pressure estimates may also be related to differences in bulk composition or the lack of quartz as a SiO_2-buffering phase when amphibole and plagioclase were in equilibrium. If the precision of the pressure estimates are assumed to be correct, then the data suggest that late-stage parts of the Bear Peak intrusive complex crystallized at higher pressure than the early units. If this scenario is true, the higher pressure estimated for the younger stage of the pluton could result

TABLE 2. REPRESENTATIVE RARE-EARTH-ELEMENT DATA FROM THE BEAR PEAK INTRUSIVE COMPLEX

Sample	BP253 qd	BP279 btg	BP288 qd	BP340A RHC	BP348A dike	BP350B hg	BP357 dike	BP430A RHC	BP1188 umc	BP188 qd	BP488 btg
La	17.08	16.49	24.52	18.98	20.81	8.30	11.73	6.93	4.48	17.23	17.37
Ce	32.50	27.78	42.10	31.59	41.81	17.69	25.31	16.95	13.91	34.07	28.40
Pr	4.06	2.90	4.66	3.27	5.05	2.46	3.10	2.43	2.55	4.38	2.91
Nd	17.71	10.23	18.07	11.70	21.05	12.04	13.02	11.34	14.67	19.56	10.11
Sm	4.38	1.59	3.71	2.26	4.64	3.55	3.36	3.08	5.03	4.84	1.48
Eu	1.29	0.59	1.12	0.63	1.45	1.10	1.30	0.97	1.62	1.46	0.55
Gd	3.91	1.08	3.02	1.86	3.89	3.38	3.22	2.94	5.46	4.18	0.83
Tb	0.59	0.13	0.45	0.30	0.57	0.53	0.52	0.48	0.86	0.64	0.09
Dy	3.57	0.74	2.62	1.86	3.18	3.14	3.12	2.92	5.07	3.70	0.45
Ho	0.69	0.14	0.51	0.39	0.62	0.61	0.62	0.60	0.97	0.74	0.09
Er	1.86	0.36	1.39	1.15	1.63	1.53	1.62	1.61	2.50	1.97	0.25
Tm	0.27	0.06	0.21	0.18	0.23	0.21	0.23	0.23	0.35	0.29	0.04
Yb	1.69	0.38	1.31	1.25	1.50	1.28	1.40	1.49	2.06	1.78	0.28
Lu	0.26	0.07	0.21	0.21	0.25	0.19	0.22	0.23	0.31	0.27	0.05

Notes: All values in ppm. Abbreviations: btg—biotite tonalite/grandiorite; hg—hornblende gabbro; qd—quartz diorite; RHC—Red Hill Creek pluton; umc—ultramafic-mafic complex.

TABLE 3. CALCULATED CRYSTALLIZATION TEMPERATURE AND PRESSURE FOR THE BEAR PEAK INTRUSION COMPLEX USING ALUMINUM-IN-HORNBLENDE GEOTHERMOBAROMETRY

Rock unit	Sample	T_B (°C)	Pressure at T_B (kb)	T_A (°C)	Pressure at T_A (kb)
Quartz diorite	BP19b	771.1 ± 20.6	3.71 ± 0.70	805.2 ± 24.3	2.84 ± 0.78
Quartz diorite	BP251	757.2 ± 19.3	3.91 ± 0.54	785.6 ± 21.0	3.25 ± 0.57
Biotite tonalite	BP174	691.7 ± 31.1	4.48 ± 0.55	722.8 ± 27.3	4.05 ± 0.59
Biotite granodiorite	BP288	720.7 ± 24.4	4.76 ± 0.43	745.3 ± 21.4	4.32 ± 0.37

Note: See text for definitions of T_A and T_B.

from foundering of the intrusive complex during emplacement, to magma loading at higher structural levels, or to hornblende crystallization at higher pressure and a lack of equilibration at the pressure of emplacement.

DISCUSSION

Development of the Bear Peak Intrusive Complex

Abundant enclaves, distinct intrusions that vary in mineral composition and texture, and the overall heterogeneity of the Bear Peak intrusive complex imply growth by piecemeal intrusion. Also, wall-rock xenoliths, migmatitic amphibolite enclaves (see below), and solid-state foliation in the wall rocks suggest that the Bear Peak intrusive complex formed by coalescence of multiple magmatic sheets (Fig. 10). The magmatic fabric intensity and enclave density is variable across the complex, further supporting sheetlike assembly of the pluton. Field evidence from the ultramafic-mafic complex and the calcic to calc-alkalic units indicates a multistage emplacement history. Emplacement of the Bear Peak intrusive complex originated with intrusion of basaltic magma, from which olivine + diopsidic augite + hornblende accumulated to form the ultramafic-mafic complex. The cumulate rocks were subsequently intruded by numerous gabbroic to dioritic dikes (Figs. 3 and 4), forming an intrusion breccia. The dikes incorporated angular/subangular to subrounded enclaves of clinopyroxenite and hornblendite, thereby indicating that the cumulate rocks were sufficiently solidified to undergo brittle deformation. The ultramafic-mafic complex also incorporated and perhaps assimilated meta-serpentinite xenoliths.

Back-veining by the intrusive units, bulb-and-cusp contacts, conformity of flow fabrics in the mafic dikes and intrusive rocks, and progressive fragmentation indicate that the mafic dikes and intrusive units were essentially contemporaneous. The mingled margins suggest that hot, mafic magmas were undercooled against a cooler but still mobile host. The presence of synplutonic mafic dikes and associated mafic magmatic enclaves is commonly attributed to mingling of mafic and intermediate to felsic magmas at the level of emplacement (Barnes, 1983; Barbarin, 1991; Bussel, 1991), whereas dispersed mafic magmatic enclaves are attributed to mingling below the level of exposure (Vernon, 1983; Barnes et al., 1992; McNulty et al. 1996). The presence of both synplutonic dikes and dispersed mafic magmatic enclaves suggests that magma mingling occurred prior to, during(?), and after emplacement of the Bear Peak intrusive complex magmas. On the basis of geochemical analysis (Barnes et al., this volume, Chapter 17), it is also possible to infer that significant magma mixing affected some Bear Peak magmas.

During the emplacement of the Bear Peak intrusive complex, a dynamothermal aureole developed in the wall rocks through solid-state deformation. Finally, after the Bear Peak intrusive complex had solidified, oblique-slip displacement along the Bear Peak fault and minor displacement along the Slippery Creek fault occurred.

Migmatitic Amphibolite Enclaves

The migmatitic amphibolite enclaves in the Bear Peak intrusive complex are of problematic origin for at least two reasons: (1) lithologically similar rocks are not exposed along the contacts of the intrusive complex and (2) the enclaves show evidence for partial melting, yet none of the exposed aureole rocks underwent significant anatexis. An exception to item 1 is the presence of mafic hornfels (i.e., amphibolitic rocks) adjacent to the eastern margin of the Bear Peak intrusive complex in the Clear Creek area (Fig. 2, Plate 1). However, these rocks are relatively small blocks in serpentinite-matrix mélange of the Rattlesnake Creek terrane and therefore are not areally extensive. The evidence for partial melting (item 2) comes from the presence of crosscutting leucosomes, which suggest a mobile, leucocratic magma capable of dilational intrusion. The tonalitic compositions of the leucosomes is consistent with hornblende-dehydration melting of mafic to intermediate amphibolitic protoliths (Drummond and Defant, 1990; Beard and Lofgren, 1991; Rapp et al., 1991; Rushmer, 1991; Wolf and Wyllie, 1994).

At least some of the migmatitic enclaves had a volcaniclastic protolith, because fragmental textures and structures are preserved in less-strained parts of the amphibolite. Local presence of relict clinopyroxene and plagioclase phenocrysts also suggests that the protolith had heterogeneous grain size, which is typical of volcaniclastic rocks in the province. In contrast,

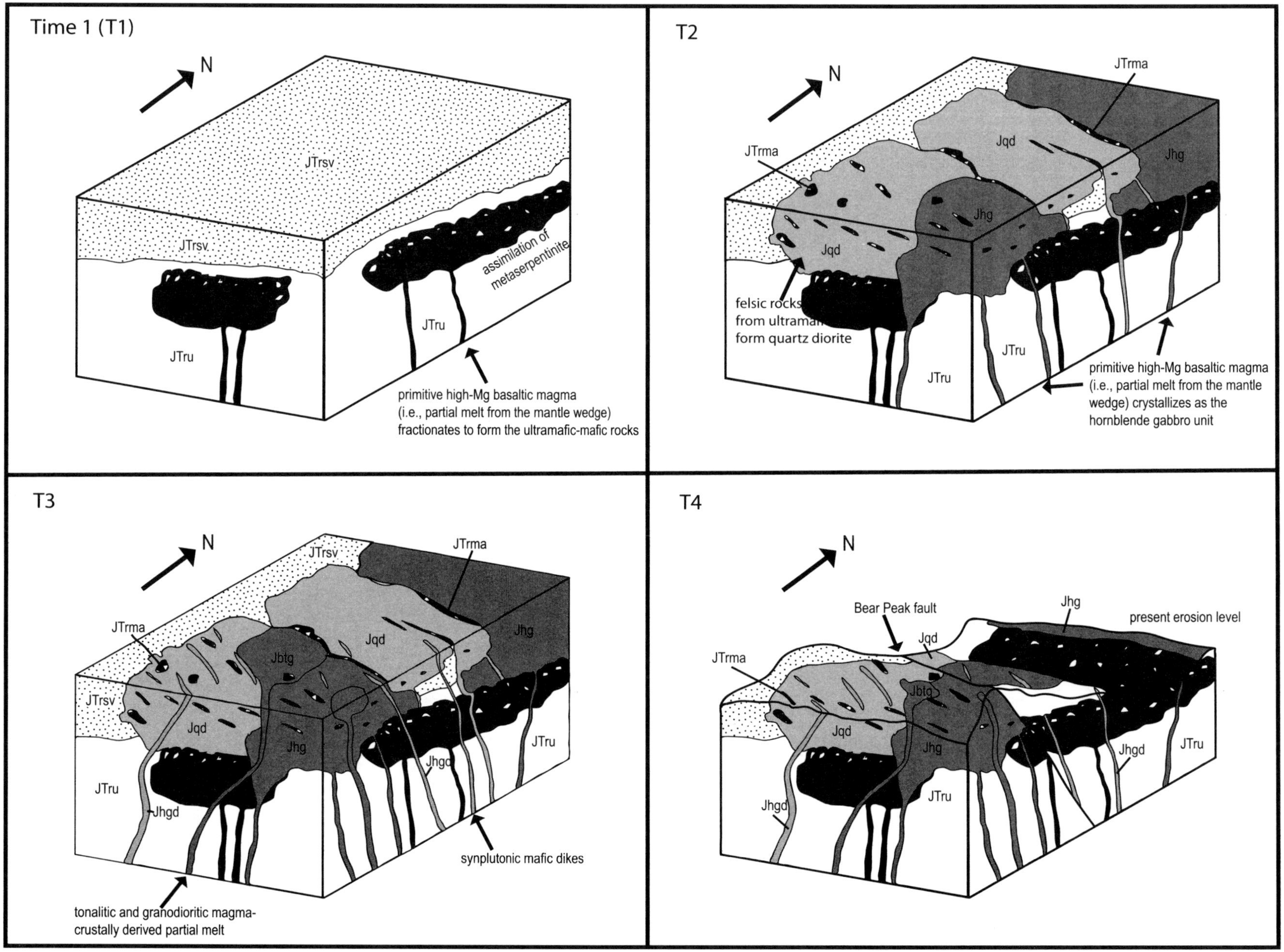

Figure 10. Petrogenetic model, which summarizes the evolution of the Bear Peak intrusive complex. Jhgd—synplutonic mafic dikes; Jbtg—biotite tonalite/granodiorite; Jqd—biotite-hornblende quartz diorite; Jhg—hornblende gabbro; Jum—ultramafic-mafic rocks; Jtrma—migmatitic amphibolite enclaves; JTrsv—metasedimentary and metavolcaniclastic rocks of the Rattlesnake Creek terrane cover sequence; Jtru—ultramafic rocks (serpentinite-matrix mélange) of the Rattlesnake Creek terrane.

many of the migmatitic enclaves consist chiefly of leucosome and melanosome, and relict textures of the original protolith are absent. The origin of these rocks becomes less problematic if one takes a regional view. Mafic to intermediate volcaniclastic and volcanic rocks form part of the Rattlesnake Creek terrane cover sequence between Bear Peak and Pony Peak (Figs. 1 and 2, Plate 1) and are common in the klippe of Rattlesnake Creek terrane between Dillon Creek and Orleans (i.e., Prospect Hill area; Gray, 1985).

Dehydration-melting of amphibolitic rocks requires temperatures greater than 850–900 °C at 3 kb (Beard and Lofgren, 1991), although melting in the presence of interstitial H_2O (water-saturated conditions) can occur at lower temperatures, depending on the ambient pressure (Beard and Lofgren, 1991). The compositions of typical magmas in the Bear Peak intrusive complex suggest that thermal conditions were appropriate to achieve the partial melting of the basic to intermediate enclaves. As an example, liquidus temperatures were calculated for two Bear Peak intrusive complex samples using the program PELE (Boudreau, 1999). A basaltic dike (BP258) with 2.0 wt% H_2O has a liquidus temperature of 1311 °C at 4 kb. Quartz diorite sample BP288 with 1 wt% H_2O yielded a calculated liquidus temperature of 1174 °C. The latter sample would be ~70% crystalline at 1000 °C; therefore, one can infer that most magmas emplaced in the Bear Peak intrusive complex were hotter than 1000 °C. Clearly, thermal conditions for partial melting of the migmatitic enclave protoliths were possible in the pluton.

Nevertheless, questions remain: how did the enclaves come to be in the intrusive complex and what was their source? At least three hypotheses of origin are possible for the enclaves (Fig. 11): (1) sheets of magma coalesced and surrounded screens of mafic volcaniclastic rock, leading to in situ partial melting; (2) amphibolitic or metabasic xenoliths were torn from a deep conduit and entrained in Bear Peak magmas; and (3) mafic volcaniclastic rocks from the Rattlesnake Creek terrane cover sequence (possible roof of the Bear Peak intrusive complex) were stoped into the intrusive complex and sank to their present level.

Data in support of the in situ screen hypothesis (1) include the northwest-southeast orientation of the migmatitic amphibolite enclaves; the wrapping of magmatic fabric around the enclaves; and the septa-like nature of extensive, albeit discontinuous, enclaves separating intrusive units. This interpretation is consistent with the observation that, near contacts with metasedimentary host rocks, metasedimentary enclaves are completely surrounded by the host pluton. However, the in situ screen hypothesis is not consistent with the lack of mafic volcaniclastic rocks in the aureole. In the classic case of in situ screen formation (e.g., Pitcher, 1970; Pitcher and Berger, 1972), a ghost stratigraphy defined by enclaves is in geologic continuity with rock units and contacts in the aureole. Thus, in the Bear Peak intrusive complex, development of in situ screens is only viable if one assumes that a package of volcaniclastic rocks in the Rattlesnake Creek terrane was completely engulfed by magma.

Hypothesis 2 involves entrainment and upward transport of mafic rocks torn from magmatic conduit(s). Regional geologic relationships suggest that such rocks would be from either the Galice Formation or Josephine ophiolite. However, the Galice Formation in the area around the Bear Peak intrusive complex lacks volcaniclastic rocks, and such rocks are essentially absent in the Josephine ophiolite (Harper, 1984, 2003; Harper et al., 1994). Moreover, the probable low viscosity and yield strength of parental Bear Peak magmas make entrainment and upward transport of amphibolitic rafts of the dimensions observed unlikely.

The stoping hypothesis (3) requires that the cover sequence of the Rattlesnake Creek terrane consisted, in part, of volcaniclastic rocks similar to the sequence exposed southeast of the complex. Had significant stoping of the typical Rattlesnake Creek terrane cover sequence occurred, one might expect a greater proportion of metasedimentary enclaves in the pluton, as well as randomly oriented enclaves. The paucity of metasedimentary enclaves could result from a cover (roof) sequence that was chiefly volcaniclastic rocks or from complete melting and incorporation of metasedimentary enclaves into the host magma. This latter idea is unlikely, owing to the quartz-rich, and therefore refractory, nature of many of the host rocks. Finally, the present orientation of the enclaves may be related to post-stoping rotation by late-stage magmatic deformation, which is consistent with the orientation of magmatic foliations, subparallel to the orientation of the enclave blocks.

We conclude that the heterogeneity of the enclaves, their textural characteristics, their subparallel orientations, and the presence of similar protolith rocks within a few km of the pluton are consistent with the development of in situ screens and/or stoping.

Interpretation of Geochemical Data

Major, trace, and REE abundances (Tables 1 and 2, Figs. 7 and 8) suggest that the Bear Peak intrusive complex had at least two types of mafic parental magma that resulted in three of the plutonic units: (1) ultramafic-mafic rocks, (2) hornblende gabbro, and (3) biotite-hornblende quartz diorite.

A basaltic parental magma is probable for the ultramafic-mafic rocks, based on development of magnesian olivine (Fo_{85}–Fo_{75}), diopsidic augite, Fe-Cr spinel, and calcic plagioclase (~An_{65}). Most samples of this unit are probably cumulate, which would explain their high CaO and MgO and low Al_2O_3. Accumulation of olivine ± augite ± hornblende is consistent with the high Cr and Ni contents, as is the concave-downward REE pattern (Fig. 9, sample 1188). Early stability of clinopyroxene without plagioclase plus the abundance of cumulate hornblende indicate that the parental magma was H_2O-rich. This property in turn suggests that this MgO- and H_2O-rich parental magma had a source in a mantle wedge above a subducting slab.

The gabbroic unit probably also had basaltic parental magmas, which is supported by the overlap in compositions of fine-

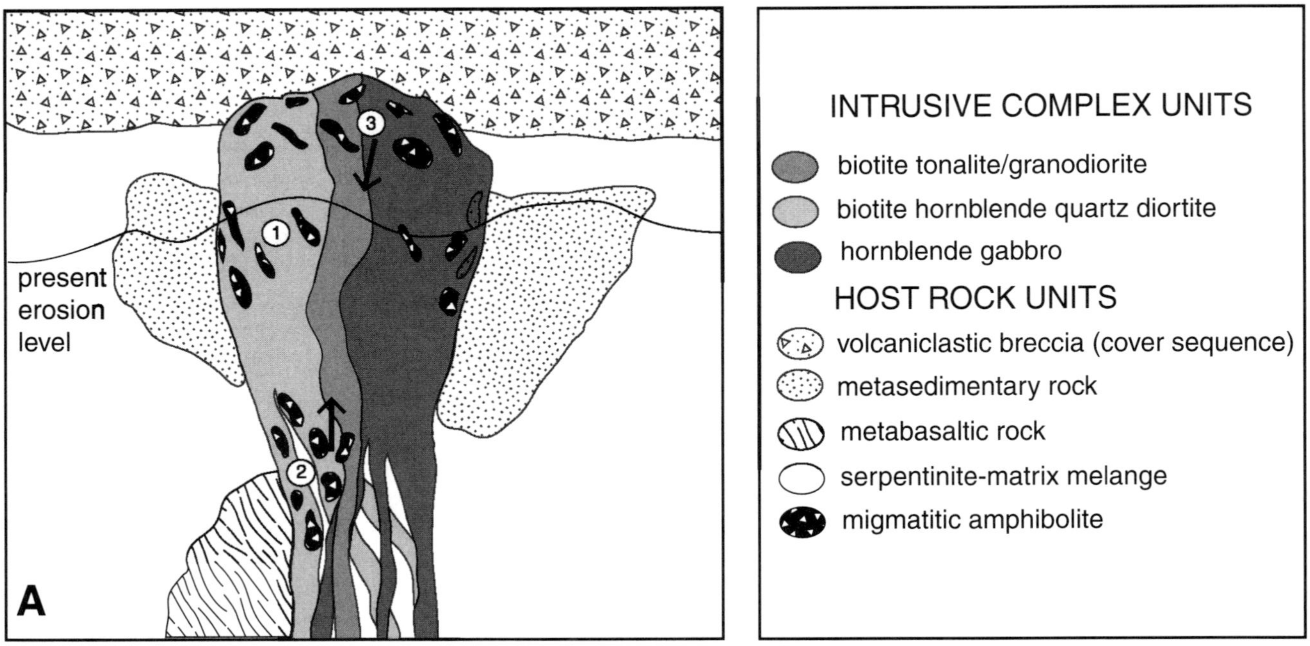

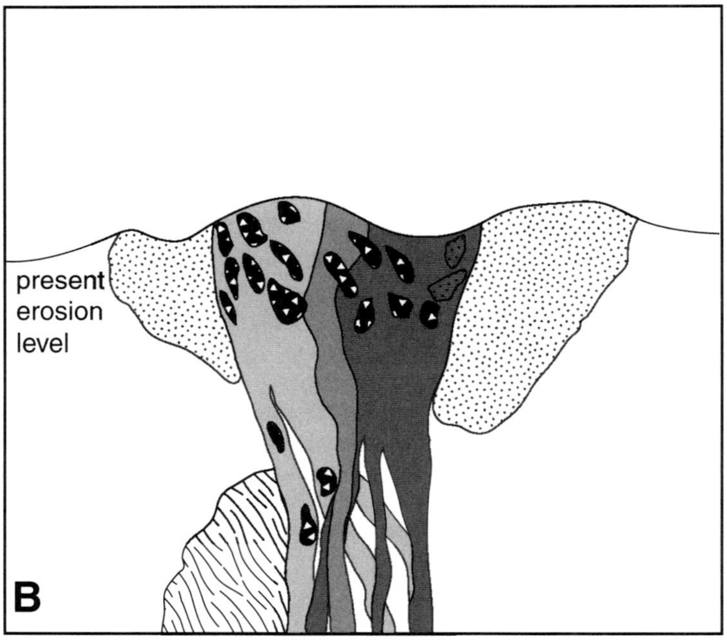

Figure 11. Schematic illustration of three possible hypotheses for the occurrence of enclaves of migmatitic amphibolite in the Bear Peak intrusive complex. (A) Illustration representing the Bear Peak intrusive complex during emplacement. Hypothesis 1 proposes that the Bear Peak intrusive complex intruded as sheets of magma that coalesced and surrounded screens of metavolcaniclastic rock that in turn led to in situ dehydration-melting of hornblende. Hypothesis 2 proposes the entrainment of amphibolitic to metabasaltic host-rock xenoliths (i.e., enclaves) ripped from a deep conduit. Hypothesis 3 proposes metavolcaniclastic rocks from the cover sequence of the Rattlesnake Creek terrane (Bear Peak intrusive complex roof rocks) were stoped into the intrusive complex and subsequently sank and rotated to their present level and position. Arrows near hypotheses 2 and 3 indicate the inferred direction of transport for the host-rock enclaves. See text for discussion of the three hypotheses. (B) Illustration representing Bear Peak intrusive complex at its current level of erosion.

grained basaltic dikes and the gabbroic unit (Figs. 7 and 8). The parental magmas to the gabbroic rocks were probably also H_2O-rich, which would explain the early stability of hornblende, the presence of hornblende phenocrysts locally, the extremely variable grain sizes (fine-grained to pegmatitic), and the presence of plagioclase-rich dikelets and veinlets (thought to represent low-temperature melt fractions). Together, the coherent compositional trends shown by the ultramafic-mafic and gabbroic units (Figs. 7 and 8) are consistent with a single parental magma type for both. Such a model has been proposed for similar rocks in the region (Bear Mountain intrusive complex: Snoke et al., 1981; Barnes et al., this volume, Chapter 17) and elsewhere (DeBari, 1997).

The nature of magmas parental to the quartz dioritic unit is less certain. Their phase assembly is consistent with evolved basaltic or andesitic magmas. On the basis of Al_2O_3 and MgO variation (Fig. 8), we surmise that plagioclase was an early stable phase in the parental magmas (along with hornblende), which in turn suggests crystallization from H_2O-bearing magmas, but ones with less H_2O than magmas parental to the ultramafic-mafic and gabbroic units. In a companion paper, Barnes et al. (this volume, Chapter 17) further investigated the origin of these magmas. They concluded that the parental magmas arose by partial melting of a hydrated mantle wedge and then evolved by fractional crystallization prior to emplacement in the Bear Peak intrusive complex.

Several of the tonalitic and granodioritic rocks have distinctive features, such as relatively high Mg/(Mg + Fe), high Sr values (>700 ppm), large Sr/Y and Ba/Y ratios, and reverse J-shaped REE patterns (Figs. 7–9). Such features are characteristic of partial melting of metabasic rocks in which amphibole ± garnet are residual phases (e.g., Martin, 1987; Drummond and Defant, 1990; Beard and Lofgren, 1991; Rapp et al., 1991; Rushmer, 1991; Wolf and Wyllie, 1994; Barnes et al., 1996; Wyllie et al., 1997). Although it is tempting to relate the tonalitic and granodioritic magmas to partial melting of migmatitic enclaves, we view this possibility as unlikely for at least two reasons. First, the chemistry of the tonalitic and granodioritic rocks indicates residual garnet in the source, yet the migmatitic enclaves lack garnet. Second, the dated granodiorite sample contains inherited zircon that is virtually identical to detrital zircon in the Galice Formation (Miller and Saleeby, 1995; Miller et al., 2003; Allen and Barnes, this volume). This observation requires that the magma parental to the dated sample, and presumably to other tonalite-granodiorite samples, assimilated Galice Formation metasedimentary rocks, and therefore originated well below the level of emplacement.

One of the characteristic features of the Bear Peak intrusive complex is the presence of partly disrupted synplutonic mafic dikes. Their crosscutting relationships, fragmentation and incorporation into host units, and formation of adjacent mafic magmatic enclaves are typical of mechanical mingling while the host magmas were still hot. Moreover, the presence of dispersed mafic magmatic enclaves suggests that mingling also occurred below the level of emplacement.

The synplutonic dikes appear to be compositionally similar to the MgO-rich gabbroic unit (Tables 1 and 2, Figs. 7–9). That these dikes intrude all stages of the Bear Peak intrusive complex indicates that the MgO-rich parental magma type was being generated in the underlying mantle wedge throughout the history of the Bear Peak intrusive complex.

CONCLUSIONS

The 144-Ma Bear Peak intrusive complex is a composite plutonic suite that ranges in composition from ultramafic to silicic, indicative of the extended compositional range characteristic of some oceanic-arc plutonic suites (e.g., Snoke et al., 2001). Contact-metamorphic assemblages and Al-in-hornblende barometry indicate crystallization at 3.5–4.7 kb, thus implying a mesozonal emplacement (12–20 km). Contact-metamorphic temperatures of 650–770 °C suggest high magmatic temperatures or significant flow of magma past the host rocks.

The intrusive complex grew by the multistage emplacement of magmas, commonly as dikes and large-scale, sheetlike bodies. This process is evident in the early development of the Bear Peak intrusive complex, as manifested by the intrusion of multiple gabbroic to dioritic dikes in the central (structurally low) ultramafic-mafic complex. The piecemeal growth of the Bear Peak intrusive complex continued throughout its development, as indicated by the presence of synplutonic mafic dikes and mafic magmatic enclaves in the major intrusive units of the complex. However, some space for the magmas also was made by stoping, as indicated by enclaves and screens of migmatitic amphibolite and xenoliths of quartzofeldspathic and ultramafic rocks. The in situ partial melting of metabasic and quartzofeldspathic rocks indicate the high temperatures of the intruded magmas of the Bear Peak intrusive complex, and this conclusion is also consistent with the estimated contact-metamorphic temperatures. Thus, the distribution of xenoliths and wall-rock screens as well as the mineralogical and textural heterogeneity of the intrusive plutonic rocks are the key field relations that suggest that the gabbroic to dioritic to tonalitic units of the Bear Peak intrusive complex were chiefly emplaced as multiple magmatic sheets. These sheets eventually coalesced into distinct, mappable plutonic rock units.

Geochemical features of the earliest two stages and the synplutonic dikes show that MgO- and H_2O-rich basaltic magma was present in the system from its inception to its end. Such magmas probably originated in the underlying mantle wedge, and some must have ponded in the crust. Such ponded magmas may have given rise to the MgO-poor parent for the quartz diorite unit. Furthermore, ponding in, and assimilation of, the underlying Galice Formation could explain the evolved nature of, and inherited zircon in, the late-stage tonalites and granodiorites. With evidence for a mantle source, deep-crustal ponding, mid-crustal interaction and assimilation, and upper-crustal stoping, the Bear Peak intrusive complex is a clear manifestation of a crustal magma column and cannot be viewed as a small, isolated pluton.

ACKNOWLEDGMENTS

This research was funded by National Science Foundation grant EAR-9902807 to AWS and EAR-9902912 to CGB. The Gregg Ranch Foundation also partially supported the fieldwork of RRM. Discussions in the field and office with J.C. Bushey were helpful in interpreting the evolution of the Bear Peak intrusive complex. The help and advice of Susan Swapp in obtaining and interpreting mineral analyses with the University of Wyoming, Laramie, automated JEOL JXA-8900 electron probe microanalyzer are greatly appreciated. Drew S. Coleman and Susan M. DeBari provided us with many comments and suggestions that were helpful in preparing the final version of this chapter.

REFERENCES CITED

Allen, C.M., and Barnes, C.G., 2006, this volume, Ages and some cryptic sources of Mesozoic plutonic rocks in the Klamath Mountains, California and Oregon, *in* Snoke, A.W., and Barnes, C.G., eds., Geological studies in the Klamath Mountains province, California and Oregon: A volume in honor of William P. Irwin: Boulder, Colorado, Geological Society of America Special Paper 410, doi: 10.1130/2006.2410(11).

Anderson, J.L., and Smith, D.R., 1995, The effects of temperature and fO_2 on the Al-in-hornblende barometer: American Mineralogist, v. 80, p. 549–559.

Arth, J.G., and Hanson, G.N., 1972, Quartz diorites derived by partial melting of eclogite or amphibolite at mantle depths: Contributions to Mineralogy and Petrology, v. 37, p. 161–174, doi: 10.1007/BF00371074.

Arth, J.G., Barker, F., Peterman, Z.E., and Friedman, I., 1978, Geochemistry of the gabbro–diorite–tonalite–trondhjemite suite of southwest Finland and its implications for the origin of tonalitic and trondhjemitic magmas: Journal of Petrology, v. 19, p. 289–316.

Barbarin, B., 1991, Enclaves in the Mesozoic calc-alkaline granitoids of the Sierra Nevada batholith, California, *in* Didier, J., and Barbarin, B., eds., Enclaves and granite petrology: Amsterdam, Elsevier, Developments in Petrology 13, p. 135–153.

Barker, F., and Arth, J.G., 1976, Generation of trondhjemitic-tonalitic liquids and Archean bimodal trondhjemitic-basalt suites: Geology, v. 4, p. 596–600, doi: 10.1130/0091-7613(1976)4<596:GOTLAA>2.0.CO;2.

Barnes, C.G., 1983, Petrology and upward zonation of the Wooley Creek batholith, Klamath Mountains, California: Journal of Petrology, v. 24, p. 495–537.

Barnes, C.G., Petersen, S.W., Kistler, R.W., Prestvik, T., and Sundvoll, B., 1992, Tectonic implications of isotopic variation among Jurassic and Early Cretaceous plutons, Klamath Mountains: Geological Society of America Bulletin, v. 104, p. 117–126, doi: 10.1130/0016-7606(1992)104<0117:TIOIVA>2.3.CO;2.

Barnes, C.G., Petersen, S.W., Kistler, R.W., Murray, R., and Kays, M.A., 1996, Source and tectonic implications of tonalite-trondhjemite magmatism in the Klamath Mountains: Contributions to Mineralogy and Petrology, v. 123, p. 40–60, doi: 10.1007/s004100050142.

Barnes, C.G., Allen, C.M., and Snoke, A.W., 2003, Episodes of Middle to Late Jurassic arc development in the Klamath Mountains, California and Oregon: EOS, Transactions of the American Geophysical Union, v. 84, no. 46, Fall Meeting suppl. Abstracts.

Barnes, C.G., Snoke, A.W., Harper, G.D., Frost, C.D., McFadden, R.R., Bushey, J.C., and Barnes, M.A.W., 2006, this volume (Chapter 17), Arc plutonism following regional thrusting: Petrology and geochemistry of syn- and post-Nevadan plutons in the Siskiyou Mountains, Klamath Mountains province, California, *in* Snoke, A.W., and Barnes, C.G., eds., Geological studies in the Klamath Mountains province, California and Oregon: A volume in honor of William P. Irwin: Boulder, Colorado, Geological Society of America Special Paper 410, doi: 10.1130/2006.2410(17).

Beard, J.S., and Lofgren, G.E., 1991, Dehydration melting and water-saturated melting of basaltic and andesitic greenstones and amphibolites at 1, 3, and 6.9 kb: Journal of Petrology, v. 32, p. 365–401.

Boudreau, A.E., 1999, PELE—A version of the MELTS software program for the PC platform: Computers & Geosciences, v. 25, p. 201–203, doi: 10.1016/S0098-3004(98)00117-4.

Bussel, M.A., 1991, Enclaves in the Mesozoic and Cenozoic granitoids of the Peruvian Coastal batholith, *in* Didier, J., and Barbarin, B., eds., Enclaves and granite petrology: Amsterdam, Elsevier, Developments in Petrology 13, p. 155–165.

Coleman, R.G., Manning, C.E., Mortimer, N., Donato, M.M., and Hill, L.B., 1988, Tectonic and regional metamorphic framework of the Klamath Mountains and adjacent Coast Ranges, California and Oregon, *in* Ernst, W.G., ed., Metamorphism and crustal evolution of the western United States—Rubey volume VII: Englewood Cliffs, New Jersey, Prentice Hall, p. 1061–1097.

DeBari, S.M., 1997, Evolution of magmas in continental and oceanic arcs: The role of the lower crust: Canadian Mineralogist, v. 35, p. 501–519.

Dickinson, W.R., 2004, Evolution of the North American Cordillera: Annual Review of Earth and Planetary Sciences, v. 32, p. 13–45, doi: 10.1146/annurev.earth.32.101802.120257.

Drummond, M.S., and Defant, M.J., 1990, A model for trondhjemite-tonalite-dacite genesis and crustal growth via slab melting: Archean to modern comparisons: Journal of Geophysical Research, v. 95, p. 21,503–21,521.

Frost, B.R., Barnes, C.G., Collins, W.J., Arculus, R.J., Ellis, D.J., and Frost, C.D., 2001, A Geochemical classification for granitic rocks: Journal of Petrology, v. 42, p. 2033–2048, doi: 10.1093/petrology/42.11.2033.

Gray, G.G., 1985, Structural, geochronologic, and depositional history of the western Klamath Mountains, California and Oregon: Implications for the early to middle Mesozoic tectonic evolution of the western North American Cordillera [Ph.D. dissertation]: Austin, University of Texas, 161 p.

Hacker, B.R., Donato, M.M., Barnes, C.G., McWilliams, M.O., and Ernst, W.G., 1995, Timescales of orogeny: Jurassic construction of the Klamath Mountains: Tectonics, v. 14, p. 677–703, doi: 10.1029/94TC02454.

Hammarstrom, J.A., and Zen, E., 1986, Aluminum in hornblende: An empirical igneous geobarometer: American Mineralogist, v. 71, p. 1297–1313.

Harper, G.D., 1984, The Josephine ophiolite, northwestern California: Geological Society of America Bulletin, v. 95, p. 1009–1026, doi: 10.1130/0016-7606(1984)95<1009:TJONC>2.0.CO;2.

Harper, G.D., 2003, Fe-Ti basalts and propagating-rift tectonics in the Josephine ophiolite: Geological Society of America Bulletin, v. 115, p. 771–787, doi: 10.1130/0016-7606(2003)115<0771:FBAPTI>2.0.CO;2.

Harper, G.D., and Wright, J.E., 1984, Middle to Late Jurassic tectonic evolution of the Klamath Mountains, California-Oregon: Tectonics, v. 3, p. 759–772.

Harper, G.D., Saleeby, J.B., and Heizler, M., 1994, Formation and emplacement of the Josephine ophiolite and the Nevadan orogeny in the Klamath Mountains, California-Oregon: U/Pb zircon and $^{40}Ar/^{39}Ar$ geochronology: Journal of Geophysical Research, v. 99, p. 4293–4321, doi: 10.1029/93JB02061.

Holland, T., and Blundy, J., 1994, Non-ideal interactions in calcic amphiboles and their bearing on amphibole-plagioclase thermometry: Contributions to Mineralogy and Petrology, v. 116, p. 433–447, doi: 10.1007/BF00310910.

Hollister, L.S., Grissom, G.C., Peters, E.K., Stowell, H.H., and Sisson, V.B., 1987, Confirmation of the empirical correlation of Al in hornblende with pressure of solidification of calc-alkaline plutons: American Mineralogist, v. 65, p. 231–239.

Irwin, W.P., 1994, Geologic map of the Klamath Mountains, California and Oregon: Reston, Virginia, U.S. Geological Survey Miscellaneous Investigations Series Map I-2148, scale 1:500,000, 2 sheets.

Irwin, W.P., and Wooden, J.L., 1999, Plutons and accretionary episodes of the Klamath Mountains, California and Oregon: Reston, Virginia, U.S. Geological Survey Open-File Report 99–374. (http://geopubs.wr.usgs.gov/open-file/of99-374/).

Kelemen, P.B., and Ghiorso, M.S., 1986, Assimilation of peridotite in zoned calc-alkaline plutonic complexes: Evidence from the Big Jim Complex, Washington Cascades: Contributions to Mineralogy and Petrology, v. 94, p. 12–28, doi: 10.1007/BF00371222.

Martin, H., 1987, Petrogenesis of Archean trondhjemites, tonalites, and granodiorites from eastern Finland: Major and trace element geochemistry: Journal of Petrology, v. 28, p. 921–953.

McFadden, R.R., 2004, Bear Peak intrusive complex, Klamath Mountains, California: Construction and petrogenesis of an ultramafic to silicic plutonic suite [M.S. thesis]: Laramie, University of Wyoming, 111 p.

McNulty, B.A., Tong, W., and Tobisch, O.T., 1996, Assembly of a dike-fed magma chamber: The Jackass Lakes pluton, central Sierra Nevada, California: Geological Society of America Bulletin, v. 108, p. 926–940, doi: 10.1130/0016-7606(1996)108<0926:AOADFM>2.3.CO;2.

Miller, J., Miller, R., Wooden, J., and Harper, G., 2003, Geochronological links between the Ingalls ophiolite, North Cascades, Washington and the Josephine ophiolite, Klamath Mts., Oregon and California: Geological Society of America Abstracts with Programs, v. 35, p. 113.

Miller, M.M., and Saleeby, J.B., 1995, U-Pb geochronology of detrital zircon from Upper Jurassic synorogenic turbidites, Galice Formation, and related rocks, western Klamath Mountains: Correlation and Klamath Mountains provenance: Journal of Geophysical Research, v. 100, p. 18,045–18,058, doi: 10.1029/95JB00761.

Pattison, D.R.M., and Tracy, R.J., 1991, Phase equilibria and thermobarometry of metapelites: Reviews in Mineralogy, v. 26, p. 105–206.

Pearcy, L.G., DeBari, S.M., and Sleep, N.H., 1990, Mass balance calculations for two sections of island arc crust and implications for the formation of continents: Earth and Planetary Science Letters, v. 96, p. 427–442, doi: 10.1016/0012-821X(90)90018-S.

Pitcher, W.S., 1970, Ghost stratigraphy in intrusive granites: A review, *in* Newall, G., and Rast, N., eds., Mechanisms of igneous intrusion: Liverpool, Gallery Press, p. 123–140.

Pitcher, W.S., and Berger, A.R., 1972, The geology of Donegal: A study of granite emplacement and unroofing: New York, Wiley-Interscience, Regional Geology Series, 435 p.

Rapp, R.P., Watson, E.B., and Miller, C.F., 1991, Partial melting of amphibolite/eclogite and the origin of Archean trondhjemites and tonalites: Precambrian Research, v. 51, p. 1–25, doi: 10.1016/0301-9268(91)90092-O.

Rushmer, T., 1991, Partial melting of two amphibolites: Contrasting experimental results under fluid-absent conditions: Contributions to Mineralogy and Petrology, v. 107, p. 41–59.

Russ-Nabelek, C., 1989, Isochemical contact metamorphism of mafic schist, Laramie Anorthosite complex, Wyoming: Amphibole compositions and reactions: American Mineralogist, v. 74, p. 530–548.

Saleeby, J.B., Harper, G.D., Snoke, A.W., and Sharp, W.D., 1982, Time relations and structural stratigraphic patterns in ophiolite accretion, west central Klamath Mountains, California: Journal of Geophysical Research, v. 87, p. 3831–3848.

Schmidt, M.W., 1992, Amphibole composition in tonalite as a function of pressure: An experimental calibration of the Al-in-hornblende barometer: Contributions to Mineralogy and Petrology, v. 110, p. 304–310, doi: 10.1007/BF00310745.

Snoke, A.W., 1977, A thrust plate of ophiolitic rocks in the Preston Peak area, Klamath Mountains, California: Geological Society of America Bulletin, v. 88, p. 1641–1659, doi: 10.1130/0016-7606(1977 88<1641:ATPOOR>2.0.CO;2.

Snoke, A.W., Quick, J.E., and Bowman, H.R., 1981, Bear Mountain igneous complex, Klamath Mountains, California: An ultrabasic to silicic calc-alkaline suite: Journal of Petrology, v. 22, p. 501–552.

Snoke, A.W., Rowe, D.W., Yule, J.D., and Wadge, G., 2001, Petrologic and structural history of Tobago, West Indies: A fragment of the accreted Mesozoic oceanic arc of the southern Caribbean: Boulder, Colorado, Geological Society of America Special Paper 354, 54 p.

Tracy, R.J., and Frost, B.R., 1991, Phase equilibria and thermobarometry of calcareous, ultramafic and mafic rocks, and iron formations: Reviews in Mineralogy, v. 26, p. 207–289.

Vernon, R.H., 1983, Restite, xenoliths, and microgranitoid enclaves in granites: Royal Society of New South Wales Journal, v. 116, p. 77–103.

Wolf, M.B., and Wyllie, P.J., 1994, Dehydration-melting of amphibolite at 10 kbar: The effects of temperature and time: Contributions to Mineralogy and Petrology, v. 115, p. 369–383, doi: 10.1007/BF00320972.

Wright, J.E., and Fahan, M.R., 1988, An expanded view of Jurassic orogenesis in the western United States Cordillera: Middle Jurassic (pre-Nevadan) regional metamorphism and thrust faulting within an active arc environment, Klamath Mountains, California: Geological Society of America Bulletin, v. 100, p. 859–876, doi: 10.1130/0016-7606(1988)100<0859:AEVOJO>2.3.CO;2.

Wright, J.E., and Wyld, S.J., 1994, The Rattlesnake Creek terrane, Klamath Mountains, California: An early Mesozoic volcanic arc and its basement of tectonically disrupted oceanic crust: Geological Society of America Bulletin, v. 106, p. 1033–1056, doi: 10.1130/0016-7606(1994)106<1033:TRCTKM>2.3.CO;2.

Wyllie, P.J., Wolf, M.B., and van der Laan, S.R., 1997, Conditions for formation of tonalites and trondhjemites: Magmatic sources and products, *in* De Wit, M.J., and Ashwal, L.D., eds., Greenstone belts: Oxford, Clarendon Press, p. 256–266.

Manuscript Accepted by the Society 10 January 2006

Printed in the USA

Geological Society of America
Special Paper 410
2006

Arc plutonism following regional thrusting: Petrology and geochemistry of syn- and post-Nevadan plutons in the Siskiyou Mountains, Klamath Mountains province, California

Calvin G. Barnes*
Department of Geosciences, Texas Tech University, Lubbock, Texas 79409-1053, USA
Arthur W. Snoke
*Department of Geology and Geophysics, Department 3006,
1000 East University Avenue, University of Wyoming, Laramie, Wyoming 82071, USA*
Gregory D. Harper
*Department of Earth and Atmospheric Sciences,
State University of New York, Albany, New York 12222, USA*
Carol D. Frost
Rory R. McFadden
Jonathan C. Bushey
*Department of Geology and Geophysics, Department 3006,
1000 East University Avenue, University of Wyoming, Laramie, Wyoming 82071, USA*
Melanie A.W. Barnes
Department of Geosciences, Texas Tech University, Lubbock, Texas 79409-1053, USA

ABSTRACT

A group of plutons were emplaced in the western Klamath Mountains province during the waning stages of the Late Jurassic Nevadan orogeny. Published U-Pb (zircon) ages indicate that the "western Klamath plutonic suite" was emplaced in the age range of 151–144 Ma. Crosscutting relationships, development of contact metamorphic aureoles, and the presence of distinctive inherited zircon populations indicate that the magmas intruded the footwall and hanging-wall rocks of the principal Nevadan thrust fault. The plutons are chiefly gabbroic to dioritic in composition, but commonly include ultramafic rocks and contain smaller volumes of tonalite and granodiorite. Hornblende is the most common mafic phase, except for some ultramafic rocks in which clinopyroxene ± olivine are locally distinctive, the two-pyroxene dioritic to monzodioritic rocks of the Buck Lake unit of the Bear Mountain pluton, and the most felsic rocks in which biotite is the most abundant mafic phase.

Compositions of fine-grained mafic dikes suggest the presence of two principal parental, H_2O-rich magmas: primitive basalt and evolved basalt/basaltic andesite. The former was parental to the ultramafic rocks of this suite. It was also parental to the basalt/basaltic andesite magmas by deep-seated fractional crystallization pro-

*E-mail: Cal.Barnes@ttu.edu.

Barnes, C.G., Snoke, A.W., Harper, G.D., Frost, C.D., McFadden, R.R., Bushey, J.C., and Barnes, M.A.W., 2006, Arc plutonism following regional thrusting: Petrology and geochemistry of syn- and post-Nevadan plutons in the Siskiyou Mountains, Klamath Mountains province, California, *in* Snoke, A.W., and Barnes, C.G., eds., Geological studies in the Klamath Mountains province, California and Oregon: A volume in honor of William P. Irwin: Geological Society of America Special Paper 410, p. 357–376, doi: 10.1130/2006.2410(17). For permission to copy, contact editing@geosociety.org.

cesses. The latter magmas were parental to the gabbroic and dioritic units. Many of the felsic rocks show evidence of origins by partial melting of metabasaltic crustal rocks, particularly their low heavy rare-earth element concentrations and high Sr/Y ratios. Mixing of crustal melts with primitive basaltic magmas was locally important (e.g., Pony Peak pluton). The mafic parental magmas show trace element features typical of an origin by partial melting of a subduction-modified mantle wedge. It is unclear whether subduction was coeval with western Klamath magmatism or whether the subduction signature developed as the result of Middle Jurassic subduction.

Keywords: arc magmatism, Nevadan orogeny, differentiation, crustal melting, Jurassic, western Klamath plutonic suite

INTRODUCTION

During Jurassic and Early Cretaceous time, arc magmatism in the Klamath Mountains province was episodically punctuated by contractional deformation (Wright and Fahan, 1988; Hacker and Ernst, 1993; Hacker et al., 1995). These contractional events are represented by regional thrust faults that separate distinct, commonly arc-related, tectonostratigraphic terranes (Irwin, 1960, 1972, 1981, 1994; Davis, 1968). Episodic contraction is apparently lacking in many modern intraoceanic arcs, in which arc crust appears to be in nearly continuous extension (Hawkins et al., 1984; Hamilton, 1988) due to slab rollback. Dewey (1980) discussed a plate-tectonic model for convergent margins where increased coupling, in part related to the age of the lithosphere being subducted, could lead to contraction within a continental magmatic arc (his Peruvian-type). Busby (2004) argued that during Late Cretaceous time, the continental arc of Baja California, Mexico, was a "compressional arc" (see her Fig. 22).

In the Klamath Mountains province, one of the most intensively studied contractional events is the Late Jurassic Nevadan orogeny (Saleeby et al., 1982; Harper and Wright, 1984; Harper et al., 1994), during which tectonostratigraphic terranes of the western Paleozoic and Triassic belt and structurally higher units were thrust over a supra-subduction zone complex that included the Rogue–Chetco arc, Josephine ophiolite, and overlying flysch of the Upper Jurassic Galice Formation. Thrusting began at ca. 155 Ma and ended by 150 Ma, according to Saleeby et al. (1982), Harper and Wright (1984), and Harper et al. (1994), although Nevadan deformation continued until at least 146 Ma (Harper et al., 1994). Recent high-precision radiometric dating has shown that arc-style magmatism was active in both the hanging wall and footwall of the key Nevadan thrust fault (Orleans fault of Hershey, 1906, 1911; Preston Peak fault [thrust] of Snoke, 1977; Fig. 1) until 156 Ma (Yule, 1996), and detrital zircons in the footwall flysch (Galice Formation; Fig. 1) are as young as ca. 153 Ma (Miller et al., 2003). These data support the ideas of Harper and Wright (1984) and Harper et al. (1994) that Nevadan thrusting and deposition of the Galice flysch were, at least in part, synchronous. Moreover, the work of Miller et al. (2003) indicates that somewhere in the source region of the footwall Galice Formation, magmatism lasted until ca. 153 Ma.

Magmatism in the footwall of the Orleans thrust may have been continuous during thrusting, because Saleeby and Harper (1993) and Harper et al. (1994) reported deformed dikes with ages as old as ca. 151 Ma, and the Orleans thrust is intruded by a number of plutons with ages as old as ca. 150 Ma (Harper et al., 1994). Furthermore, the studies of Snoke et al. (1981), Harper et al. (1994), Allen and Barnes (this volume), Bushey et al. (this volume), and McFadden et al. (this volume) indicate that large volumes of mafic to felsic magma were emplaced into the footwall and hanging wall of the Orleans thrust from 150 to 144 Ma.

The combination of new (Allen and Barnes, this volume; Chamberlain et al., this volume) and existing geochronologic data (e.g., Harper et al., 1994; Irwin and Wooden, 1999) combined with new field, geochemical, and isotopic data on these syn- and immediately post-Nevadan plutonic rocks provides a means of characterizing the nature of magmatic activity before, during, and after contractional deformation. In this chapter, we emphasize the compositions and origins of syn- and post-Nevadan magmatism, herein referred to as the "western Klamath plutonic suite." Questions that we address include the origins and evolution of parental magmas; interactions with, and effects on, the crust through which these magmas passed; and the significance of arc magmatism that closely followed an episode of thrust faulting during a regional contractional orogeny.

Figure 1. Generalized geologic map of the north-central Klamath Mountains, California, showing most of the plutons of the western Klamath plutonic suite: Bear Mountain intrusive complex (Blue Ridge and Bear Mountain plutons), Bear Peak pluton, Red Hill Creek pluton (R.H. pluton), Pony Peak pluton, and Summit Valley pluton. The syn-Nevadan dikes and sills are not shown on this geologic map. The locations of the Lower Coon Mountain intrusion, Ammon Ridge pluton, and Glen Creek gabbro-ultramafic complex are shown in Figure 1 in Allen and Barnes (this volume). The basis for this map is Wagner and Saucedo (1987).

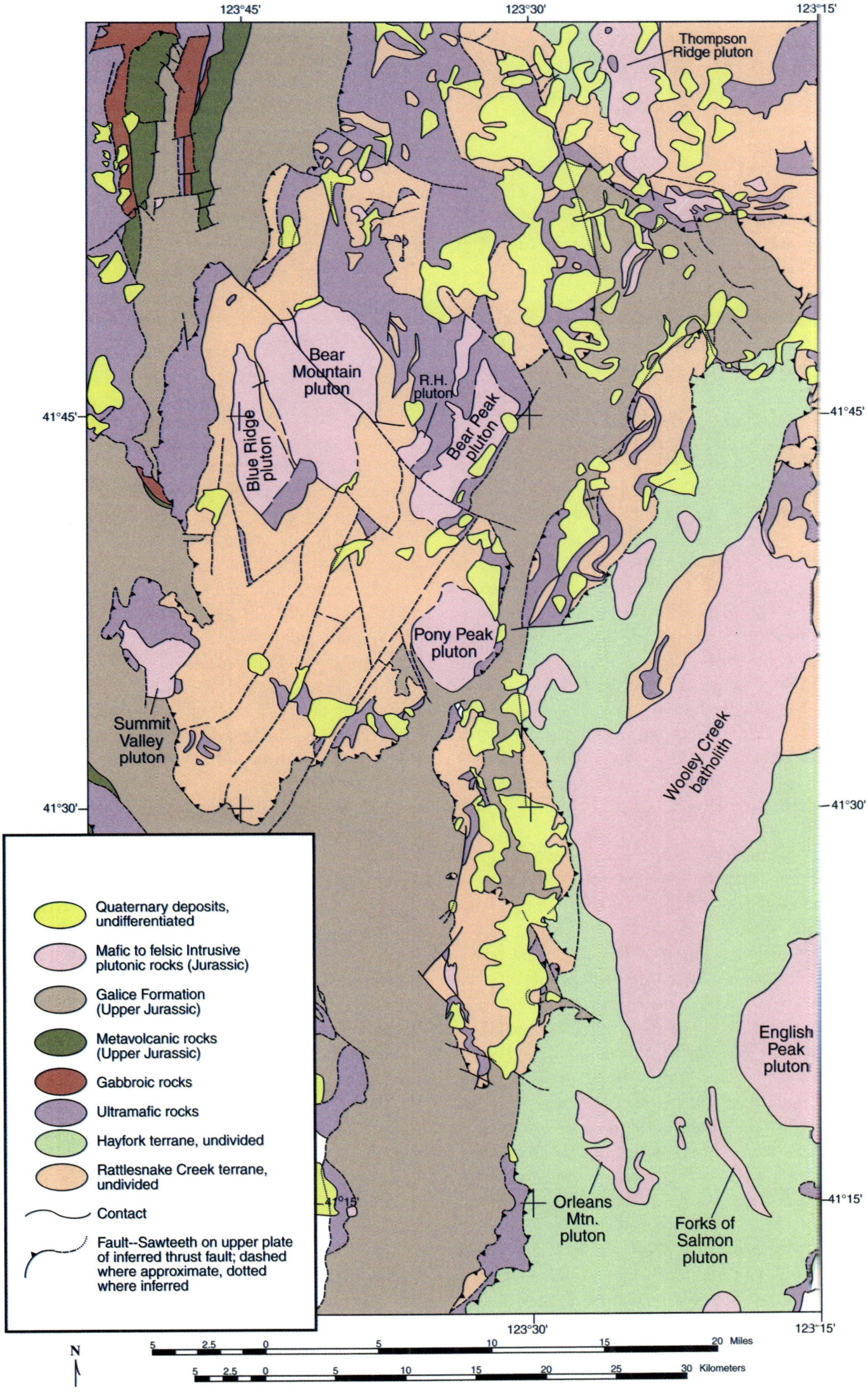

123°45'
123°30'
123°15'
41°45'
41°30'
41°15'
Thompson Ridge pluton
Bear Mountain pluton
R.H. pluton
Bear Peak pluton
Blue Ridge pluton
Pony Peak pluton
Summit Valley pluton
Wooley Creek batholith
English Peak pluton
Orleans Mtn. pluton
Forks of Salmon pluton
Quaternary deposits, undifferentiated
Mafic to felsic Intrusive plutonic rocks (Jurassic)
Galice Formation (Upper Jurassic)
Metavolcanic rocks (Upper Jurassic)
Gabbroic rocks
Ultramafic rocks
Hayfork terrane, undivided
Rattlesnake Creek terrane, undivided
Contact
Fault--Sawteeth on upper plate of inferred thrust fault; dashed where approximate, dotted where inferred
N
5 2.5 0 5 10 15 20 Miles
5 2.5 0 5 10 15 20 25 30 Kilometers

GEOLOGIC SETTING

Regional Setting

In the western Klamath Mountains, the Nevadan orogeny is associated with juxtaposition of terranes of the western Paleozoic and Triassic belt (Irwin, 1960) over rocks of the western Jurassic belt along the Orleans fault (Hershey, 1906, 1911; Preston Peak fault [thrust] of Snoke, 1977; Fig. 1). Hanging-wall rocks immediately above the Orleans thrust belong to the Rattlesnake Creek terrane (Fig. 1), which consists of an ophiolitic mélange overlain by various cover sequences of Upper Triassic to Lower Jurassic arc volcanic and volcaniclastic rocks (Gray, 1986; Wright and Wyld, 1994) and argillaceous to siliceous metasedimentary rocks (Frost et al., this volume). The Rattlesnake Creek terrane, in turn, is overlain by the western Hayfork terrane (cf. Wright, 1982; Fig. 1). This contact is a low-angle fault that has been interpreted as a minor thrust that modified a depositional contact (Wright and Fahan, 1988; Donato et al., 1996).

In the area of study (Fig. 1), the footwall consists of the basal Josephine ophiolite and overlying (meta)sedimentary rocks of the Upper Jurassic Galice Formation. The latter consist of interbedded metawacke, arenaceous phyllite, slate, and scarce metaconglomerate (e.g., Snoke, 1977; Harper, 1984; Pinto-Auso and Harper, 1985). To the north of the area, the footwall also encompasses late Middle Jurassic arc rocks of the Rogue Formation (Garcia, 1982) and their presumed intrusive equivalents (Chetco intrusive complex; Jorgenson, 1970; Hotz, 1971; Yule, 1996). The Rogue Formation was deposited on the Josephine ophiolite, whereas the Chetco intrusive complex was intruded into it.

In the western Klamath Mountains, we have identified a suite of magmatic rocks that were synchronous with or immediately post-dated Nevadan thrusting. These rocks either intrude the Galice Formation or show geologic or geochonrologic evidence of post-dating Galice deposition. They are characteristically hornblende-rich, have associated ultramafic cumulates, and show evidence for an Mg-rich parental magma. Syn-thrusting magmatism is represented by a suite of variably deformed dikes and sills (Harper et al., 1994; Harper, this volume; see below). Post-thrusting plutonism encompasses ultramafic to tonalitic plutons that were emplaced along the length of the western part of the province (Irwin and Wooden, 1999). In the study area, these plutons are Summit Valley, Pony Peak, Bear Peak, and Red Mountain Creek, and the cluster of plutons of the Bear Mountain intrusive complex. We additionally discuss the Lower Coon Mountain pluton (Harper et al., 1994), Buckskin Peak, Ammon Ridge, and Glen Creek plutons (Young, 1978; Irwin and Wooden, 1999; see Fig. 1 *in* Allen and Barnes, this volume).

Two younger plutonic suites were emplaced after western Klamath magmatism. One consists of a group of tonalite-trondhjemite-granodiorite plutons (142–136 Ma; "ttg suite") and the other of a group of granodioritic plutons (139–136 Ma; "granodiorite suite"). These younger suites are discussed in Allen and Barnes (this volume).

Magmatism during and after Nevadan Thrusting

Syn-Nevadan Dikes and Sills. A suite of distinctive porphyritic dikes and sills intrudes footwall rocks beneath the Orleans thrust (Josephine ophiolite and Galice Formation; see Harper et al., 1994; Harper, this volume). These rocks have U-Pb (zircon) ages and deformational features that indicate emplacement during and after Nevadan thrusting (Saleeby and Harper, 1993; Harper et al., 1994). The intrusions range in thickness from ~12 cm to 2 m. Most are mafic to intermediate in composition and range in color from dark gray to medium grayish green. They are distinguished in the field by the presence of hornblende phenocrysts and/or acicular hornblende in the groundmass. These traits are in contrast to those of dikes of the Josephine ophiolite, which are generally aphyric and lack igneous hornblende.

The oldest dated sample is broadly granodioritic; it yielded an age of 151 ± 1.5 Ma (Harper et al., 1994). A basaltic dike yielded an age of 150.5 ± 1.5 Ma, and a dike with basaltic andesite composition was dated at 146 ± 1 Ma (Harper et al., 1994).

Summit Valley Pluton. This 150 ± 1-Ma pluton (Harper et al., 1994) is the oldest known pluton to intrude the Orleans (Preston Peak) thrust; there are well-developed contact-metamorphic assemblages in both hanging-wall and footwall rocks (Norman, 1984). The pluton is heterogeneous and shows complex crosscutting intrusive relationships. It is predominantly hornblende gabbro, but ranges from hornblendite and scarce olivine clinopyroxenite to hornblende diorite (Norman, 1984). Hypersolidus and subsolidus deformation fabrics are common; some are cut by dikes that themselves are deformed (Griesau, 1992). High-temperature shear zones reach ~300 m in thickness and occur within variably deformed gabbro. These high-temperature, banded mylonites consist of granoblastic-polygonal brown hornblende, diopsidic augite, and plagioclase. Epidote-prehnite veins with slickenfibers are common and are interpreted to have formed during cooling of the pluton to the ambient temperature of the host rocks, which were undergoing lower greenschist-facies regional metamorphism (Griesau, 1992; Griesau and Harper, 1992).

Bear Mountain Intrusive Complex. The Bear Mountain intrusive complex consists of several plutons described in detail by Snoke et al. (1981) and Bushey et al. (this volume). The complex consists of two large bodies, the Blue Ridge and Bear Mountain plutons, and a number of small, related intrusions (Bushey et al., this volume; Fig. 1).

The Blue Ridge pluton consists chiefly of wehrlite and hornblende ± olivine clinopyroxenite, with subordinate dunite and various gabbroic rocks, including hornblende gabbro pegmatite. This pluton is undated, but it is intruded by a biotite-hornblende quartz diorite to tonalite pluton with a U-Pb (zircon) age of 150.5 ± 0.61 Ma (Chamberlain et al., this volume).

The Bear Mountain pluton consists of at least three major intrusive units. These are detailed in Bushey et al. (this volume) and are named the "Punchbowl plutonic unit" (pyroxene-hornblende gabbro and diorite), "Buck Lake plutonic unit" (biotite–two-pyroxene ± hornblende diorite to monzodiorite), and "Doe Flat plutonic unit" (biotite-hornblende quartz diorite to tonalite). The U-Pb (zircon) ages of the Punchbowl and Buck Lake units are identical with analytical uncertainty (ca. ± 0.5 Ma) at 148 Ma (Chamberlain et al., this volume). Snoke (1972) and Snoke et al. (1981) showed that the Bear Mountain pluton intrudes a high-angle fault that offsets the Orleans (Preston Peak) thrust. Therefore, the pluton must post-date the thrust. This conclusion is consistent with the 147- to 148-Ma U-Pb (zircon) ages.

A number of small tonalitic to granitic intrusive bodies cuts all three of the main intrusive units (Bushey et al., this volume). The largest of these bodies is called the "Wilderness Falls plutonic unit," after good exposures at Wilderness Falls along Clear Creek (Snoke et al., 1981). This body is petrographically similar to tonalite intruded into the Blue Ridge pluton, but must be younger, because it intrudes ca. 148-Ma rocks of the Bear Mountain pluton.

Two sets of mafic dikes are present in the aureole between the Blue Ridge and Bear Mountain plutons (Snoke and Barnes, 2002; Bushey et al., this volume). One set, referred to as the "gabbroic dikes," is Mg-rich and typically contains augite phenocrysts or their reaction products. The second set, the "dioritic dikes," is mafic but poorer in MgO and contains hornblende and plagioclase phenocrysts. Field relations (Snoke and Barnes, 2002; Bushey et al., this volume) show that these dikes were intruded and deformed during development of the dynamothermal aureole of the Bear Mountain pluton and are therefore coeval with the Bear Mountain intrusive complex.

Bear Peak and Red Hill Creek Plutons. The Bear Peak pluton (Fig. 1) is an elongate pluton with a central, structurally deep zone of hornblende clinopyroxenite and gabbroic to dioritic rocks (McFadden et al., this volume). Structurally higher parts of the pluton consist of gabbroic through granodioritic rocks that were emplaced as a series of sheetlike intrusions. The pluton clearly intrudes and contains xenoliths of the hanging-wall Rattlesnake Creek terrane, but a contact with the footwall Galice Formation is not exposed at the present level of erosion. However, samples of the Galice Formation collected within ~600 m of the pluton have undergone contact metamorphism (e.g., these samples contain biotite and scarce chiastolite porphyroblasts). This relationship, along with the 143.7 ± 1.3-Ma U-Pb (zircon) age from the Bear Peak pluton (Allen and Barnes, this volume), indicates that this pluton is a member of the western Klamath plutonic suite. Allen and Barnes (this volume) reported a population of inherited zircon with an age spectrum similar to that of detrital zircons from the Galice Formation (Miller et al., 2003).

The small Red Hill Creek pluton crops out west of the Bear Peak pluton (Fig. 1). It is predominantly tonalite and granodiorite and is intruded by fine-grained microgabbroic dikes. Evidence for magma mingling (e.g., mafic magmatic enclaves, synplutonic dikes) is abundant in both Bear Peak and Red Hill Creek plutons.

Pony Peak Pluton. The Pony Peak pluton (Fig. 1) ranges from hornblende (±pyroxene) gabbro to biotite hornblende tonalite and granodiorite. It has a dynamothermal contact aureole in the footwall Galice Formation (Harper et al., 1994), and semipelitic rocks in the Galice Formation contain andalusite porphyroblasts. Contact effects in the hanging-wall Rattlesnake Creek terrane are primarily limited to hornfels development. Two U-Pb (zircon) ages of 146 ± 3 and 144.2 ± 2.5 Ma were determined by Harper et al. (1994) and Allen and Barnes (this volume), respectively. The multicrystal dating of Harper et al. (1994) resulted in a Middle Proterozoic upper intercept and laser-ablation inductively coupled plasma mass spectrometry dating (Allen and Barnes, this volume) confirmed the presence of a suite of inherited zircons similar to detrital zircons from the Galice Formation.

Evidence for magma mingling in the Pony Peak pluton is primarily in the form of mafic magmatic enclaves. Synplutonic dikes, such as seen in the Bear Peak pluton, were not observed, and the proportion of isolated enclaves and enclave swarms is smaller than in the Bear Peak pluton.

Lower Coon Mountain, Buckskin Peak, Ammon Ridge, and Glen Creek Plutons. At least two other plutons exposed in the north-central Klamath Mountains appear to be part of the western Klamath plutonic suite. These are the Lower Coon Mountain and Buckskin Peak plutons. Moreover, at least two plutons in the southern Klamath Mountains should probably be considered members of this plutonic suite: the Ammon Ridge pluton (Young, 1978) and Glen Creek gabbro-ultramafic complex (Irwin et al., 1974; Irwin and Wooden, 1999; see Fig. 1 in Allen and Barnes, this volume).

The Lower Coon Mountain pluton is a subhorizontal, ≥300-m-thick, sill-like body that intrudes the Galice Formation (Harper, 1980; Gray and Page, 1985). It consists of >90% ultramafic rocks, mostly olivine ± plagioclase clinopyroxenite, along with less abundant olivine gabbronorite, and hornblende diorite. Scarce granitic dikes intrude the Galice Formation along the margins of the pluton and have yielded U-Pb zircon ages of ca. 142 and ca. 145 Ma (with Proterozoic upper intercepts; Harper et al., 1994). Ultramafic rocks are exposed from the base to the roof of the pluton, suggesting that it has a cumulate origin. No igneous layering, foliation, or solid-state deformation fabrics were observed. The narrow contact aureole consists of pelitic hornfels with relict andalusite (now muscovite) that overprints the pervasive slaty cleavage of the Galice Formation. The plutonic rocks are partially altered to low-grade metamorphic mineral assemblages. The ca. 149-Ma Buckskin Peak pluton intrudes the Josephine peridotite (Irwin and Wooden, 1999; Fig. 1 *in* Allen and Barnes, this volume). It consists of biotite + hornblende quartz diorite (Harper et al., 1994), but has otherwise received little study. The Ammon Ridge pluton was mapped and described by Young (1978), and Wright and Fahan (1988) reported Late Jurassic U-Pb zircon ages (147–151 Ma). Young (1978) described the pluton as chiefly consisting of hornblende ± biotite ± augite diorite with variable amounts of quartz. Young (1978) also described a distinctive, chiastolite-bearing dynamothermal contact aureole in the Upper Jurassic Galice Formation. The Glen Creek gabbro-ultramafic complex was mapped by Irwin et al. (1974), who described it as consisting of medium-grained to pegmatitic hornblende gabbro with subordinate hornblendite and peridotite. He mapped a hornfelsic contact aureole in the Galice Formation. Wright and Fahan (1988) reported 147- to 151-Ma U-Pb ages from this pluton. Thus, the ages and intrusive relationships of the Ammon Ridge and Glen Creek plutons are consistent with other members of the western Klamath plutonic suite. Moreover, the presence of hornblende-bearing cumulates and of hornblende as a near-liquidus phase in gabbroic to dioritic rocks supports inclusion in the western Klamath suite.

Emplacement Conditions

The presence of andalusite (chiastolite) in the aureoles of the Pony Peak (Harper et al., 1994), Bear Peak (McFadden et al., this volume), Lower Coon Mountain, and Ammon Ridge (Young, 1978) plutons, and of cordierite in the aureole of the Bear Mountain pluton (Snoke et al., 1981) indicate relatively shallow emplacement (pressure ≤ 4 kb, depth ≤ 14 km; Pattison, 1992). Use of Al-in-hornblende barometry (Anderson and Smith, 1995, calibration) yielded 3.7–4.2 kb for the Pony Peak pluton and 2.8–4.8 kb for the Bear Peak pluton. Calculated Al-in-hornblende pressures on the Bear Mountain pluton resulted in a range of pressure estimates from 3.5 to 5.0 kb, and pressure estimates from syn-emplacement dikes in the host rocks were 3.1–3.7 kb (Bushey et al., this volume). In summary, aureole assemblages and Al-in-hornblende barometry indicate emplacement of the western Klamath magmas in the upper crust at depths ≤15 km.

GEOCHEMISTRY

Introduction

The narrow time interval of the western Klamath plutonic suite, the predominance of hornblende-bearing mafic to intermediate compositions, and the distribution of these plutons in the western part of the province suggest that they belong to a single magmatic suite. In the following section, the geochemical compositions of the rocks are characterized and compared, and the compositions of dikes associated with the Bear Mountain intrusive complex are used to infer parental compositions. The data show that most western Klamath plutons had two H_2O-rich parental magmas: one a primitive basalt and the other an evolved basalt or basaltic andesite. H_2O-poor parental magma was much less common.

Petrographic summaries of the western Klamath suite plutons are given in Appendix A1, geochemical data are compiled in Appendix A2, and analytical methods are given in Appendix A3, all on the CD-ROM accompanying this volume or in the GSA Data Repository.[1]

Major and Trace Element Compositions

The suite of samples shows a wide range of SiO_2 contents, ranging from <40 to 74% (Fig. 2). However, several of the western Klamath suite plutons have a less extended SiO_2 range. For example, few samples of the Bear Mountain pluton have SiO_2 contents >55 wt%, silica contents in the Blue Ridge ultramafic-mafic pluton are <52 wt%, and Summit Valley pluton samples are <58 wt%. In contrast, the Bear Peak and Pony Peak plutons and syn-Nevadan dikes range in SiO_2 contents from as low as 47 wt% to as high as 74 wt%. Most samples are metaluminous, but a small group of tonalitic and granodioritic rocks from the Wilderness Falls, Bear Peak, and Pony Peak intrusions is mildly peraluminous.

In the classification of Frost et al. (2001), most samples plot in the magnesian field (Fig. 2A), although samples from the Bear Mountain pluton straddle the magnesian-ferroan boundary. Similarly, most Bear Mountain pluton samples are calc-alkaline (Fig. 2B) but show scatter on either side of the calc-alkaline field. Most Bear Peak samples and late-stage tonalitic intrusions associated with the Bear Mountain intrusive complex and Summit Valley plutons are calcic, and Pony Peak samples straddle the calcic–calc-alkaline boundary (Fig. 2B). Samples of the syn-Nevadan dikes and sills scatter between the calcic and alkalic fields in Figure 2B. This scatter is very probably the result of alkali metasomatism (Appendix A1).

Because the plutonic rocks discussed in this chapter are similar in age and geologic setting, one might expect them to have similar magmatic histories. Here we use the largest and best documented of these intrusions, the Bear Mountain intrusive complex (Snoke et al., 1981; Bushey et al., this volume), as a model to which the other plutonic rocks may be compared.

Bear Mountain Intrusive Complex. The Bear Mountain intrusive complex was derived primarily from basaltic parental magmas (Snoke et al., 1981). At least three basaltic parents were present, two of which are represented by mafic dikes in the Bear Mountain aureole. The petrogenetic relationships determined by Bushey et al. (this volume) are summarized here and as insets in Figure 3A and B. The most primitive of the parental magmas are represented by the compositions of the Mg-rich gabbroic dikes (black star in Fig. 3A and B). The Blue Ridge pluton is interpreted as a cumulate from such magmas, with cumulus olivine ± augite ± plagioclase. Fractionation of gabbroic dike-type magma gave rise to the second parental magma (gray star in

[1]GSA Data Repository item 2006201, Appendices 1–3, is available online at www.geosociety.org/pubs/ft2006.htm, or on request from editing@geosociety.org or Documents Secretary, GSA, P.O. Box 9140, Boulder, CO 80301, USA.

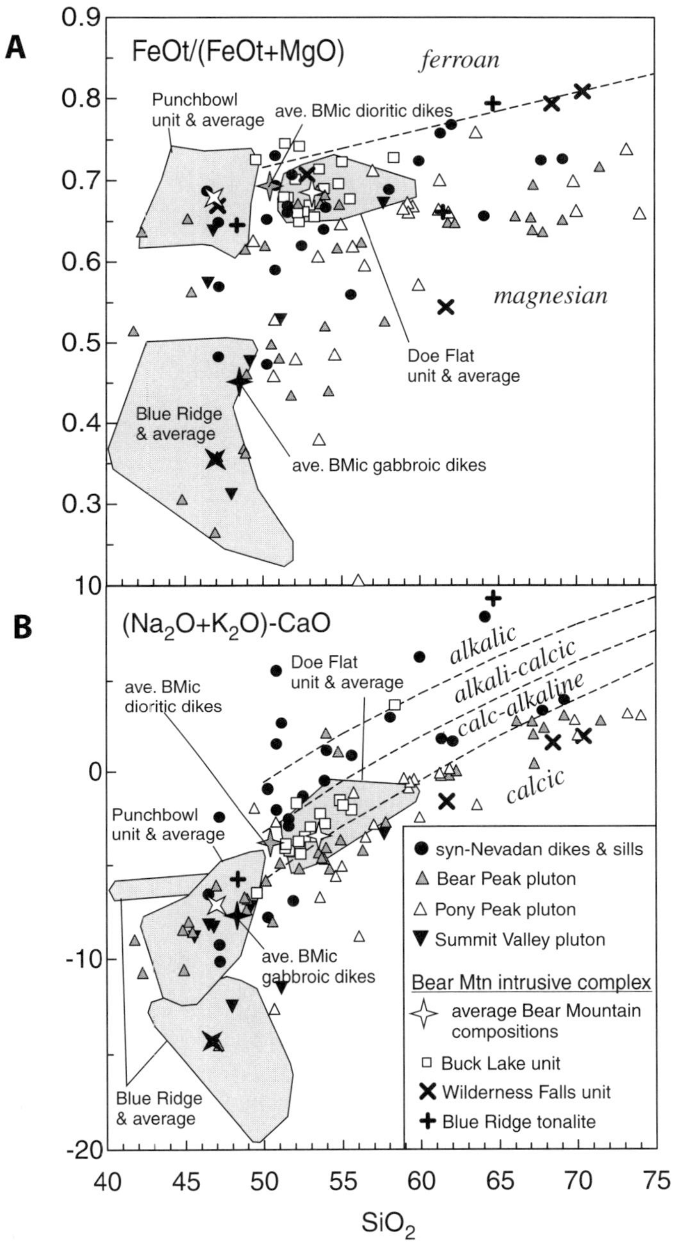

Figure 2. Geochemical classification according to Frost et al. (2001). (A) Most samples are magnesian, but parts of the Bear Mountain intrusive complex plot in the ferroan field. (B) Similarly, most samples are either calcic or calc-alkaline. Fields are shown for the Blue Ridge pluton and Punchbowl and Doe Flat units of the Bear Mountain pluton. Star symbols are average compositions of the Blue Ridge pluton (diagonal black), Punchbowl (diagonal white), and Doe Flat (white) units of the Bear Mountain pluton, and gabbroic (black) and dioritic (gray) dikes associated with the Bear Mountain intrusive complex. Bmic—Bear Mountain intrusive complex.

Fig. 3A and B), the plagioclase + hornblende-phyric dioritic dikes. Crystal-liquid separation of this parental magma type yielded cumulates (i.e., the gabbro/diorite Punchbowl unit of the Bear Mountain pluton) and fractionated magma (the Doe Flat unit; Fig. 3A and B). The third parental magma that of the more potassic Buck Lake unit, was compositionally similar to the dioritic dike magmas but had higher K_2O and lower H_2O contents (Bushey et al., this volume).

The close petrogenetic relationship between gabbroic and dioritic dikes is illustrated by their having very similar rare-earth-element (REE) patterns (Fig. 4A). Both groups of dikes have moderate negative slopes, little or no Eu anomaly, and a slight hump from Pr to Sm. These patterns are mirrored in the REE patterns of most samples from the Bear Mountain pluton (Fig. 4B), which is consistent with magmas similar to the gabbroic and dioritic dikes as parental magmas. The positive and negative Eu anomalies seen in samples of the Bear Mountain pluton indicate plagioclase accumulation or fractionation, respectively.

Late-stage tonalitic and granitic intrusions associated with the Bear Mountain intrusive complex are characterized by high Sr contents (500 to ~1250 ppm; Fig. 5). Their REE patterns (Fig. 4C) are distinct from those of the main plutonic units, in that they have steeper slopes, generally higher light REE (LREE) abundances, and heavy REE (HREE) contents that are less than seven times chondritic abundances. Furthermore, the HREE parts of the patterns are nearly flat. The high Sr contents, LREE enrichment, steep REE patterns, and low HREE contents are all characteristic of an origin by partial melting of metabasic rocks in which garnet ± amphibole were residual phases (e.g., Martin, 1987; Drummond and Defant, 1990; Barnes et al., 1996).

Syn-Nevadan Dikes and Sills. Detailed petrologic discussion of this suite of samples is complicated by their nearly pervasive alteration and by their lack of a single parental magma composition. Nevertheless, these rock compositions tend to plot in a trend that is coincident with the evolutionary path inferred for the Bear Mountain intrusive complex (particularly Fig. 3A and C), which implies evolution primarily by fractional crystallization. In addition, the majority of the dikes and sills have REE patterns similar to those of Bear Mountain intrusive complex gabbroic and dioritic dikes (Fig. 4D). The most mafic samples straddle the Bear Mountain intrusive complex gabbroic dike patterns, and LREE contents for the suite of samples are broadly correlated to SiO_2. The REE patterns of the majority of samples (ruled pattern in Fig. 4D) differ slightly from the Bear Mountain intrusive complex dioritic dikes in the somewhat steeper patterns of the latter group (Fig. 4D).

Two of the most evolved samples of the syn-Nevadan dikes and sills (quartz diorite and tonalite) have crossing REE patterns, HREE abundances greater than fifteen times that of chondrites (Fig. 4D), and Sr contents <600 ppm (Fig. 5). These features are also consistent with a fractional crystallization trend, as inferred from compositional variation in the Bear Mountain intrusive complex, and particularly with fractionation of calcic amphibole (e.g., Arth et al., 1978).

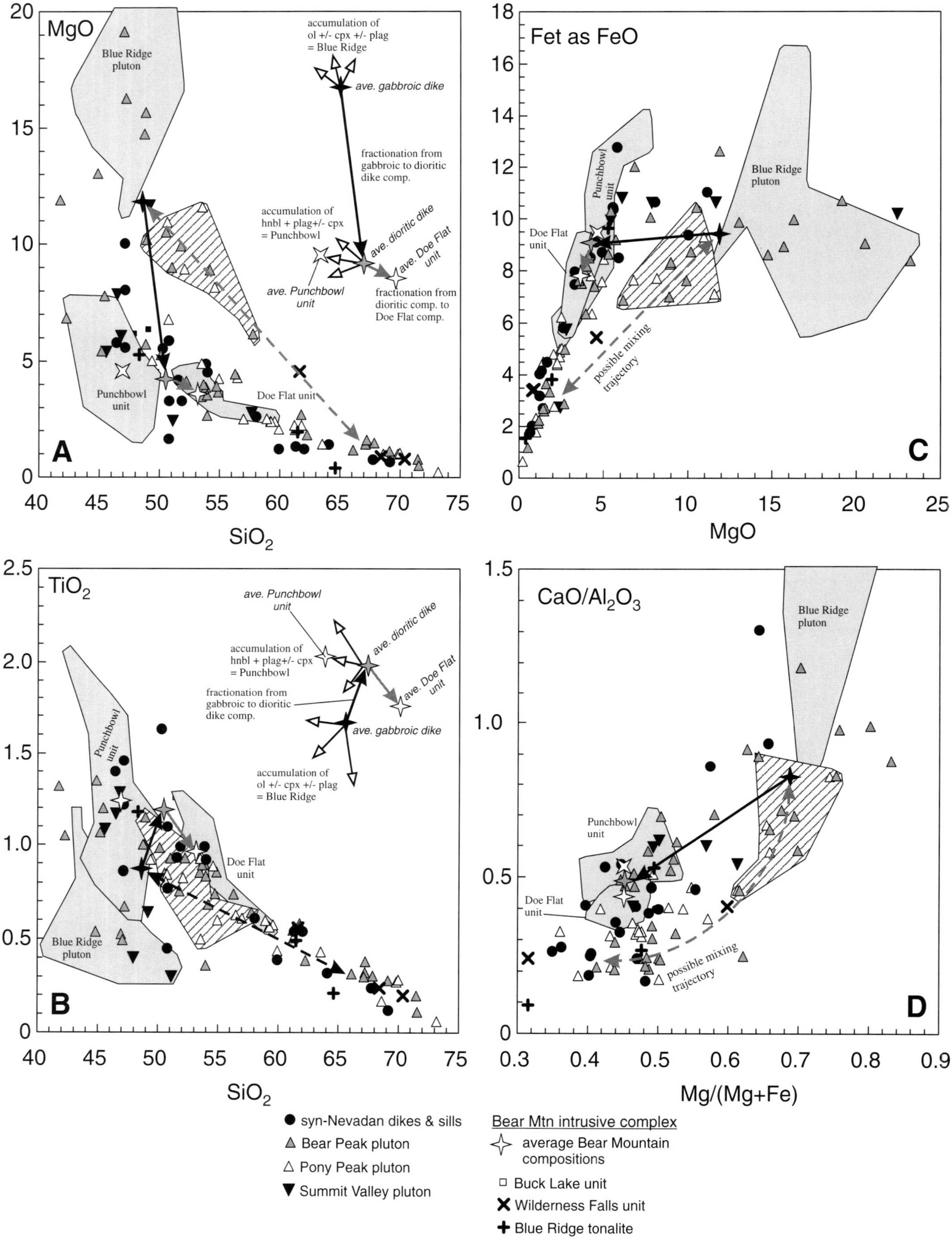
MgO
Blue Ridge pluton
accumulation of ol +/- cpx +/- plag = Blue Ridge
ave. gabbroic dike
fractionation from gabbroic to dioritic dike comp.
accumulation of hnbl + plag+/- cpx = Punchbowl
ave. dioritic dike
ave. Doe Flat unit
ave. Punchbowl unit
fractionation from dioritic comp. to Doe Flat comp.
Punchbowl unit
Doe Flat unit
A
SiO2
Fet as FeO
Punchbowl unit
Blue Ridge pluton
Doe Flat unit
possible mixing trajectory
C
MgO
TiO2
ave. Punchbowl unit
ave. dioritic dike
accumulation of hnbl + plag+/- cpx = Punchbowl
fractionation from gabbroic to dioritic dike comp.
ave. Doe Flat unit
ave. gabbroic dike
accumulation of ol +/- cpx +/- plag = Blue Ridge
Punchbowl unit
Doe Flat unit
Blue Ridge pluton
B
SiO2
CaO/Al2O3
Blue Ridge pluton
Punchbowl unit
Doe Flat unit
possible mixing trajectory
D
Mg/(Mg+Fe)
syn-Nevadan dikes & sills
Bear Peak pluton
Pony Peak pluton
Summit Valley pluton
Bear Mtn intrusive complex
average Bear Mountain compositions
Buck Lake unit
Wilderness Falls unit
Blue Ridge tonalite

Summit Valley Pluton. The Summit Valley pluton is, like the Bear Mountain intrusive complex, predominantly mafic to intermediate (Appendix A1). Compositional variation appears to follow an evolutionary path similar to that of the Bear Mountain intrusive complex gabbroic and dioritic dikes (Fig. 3A and C). With the exception of two samples, an ultramafic cumulate (Norman, 1984) and a hornblende diorite, Summit Valley samples have Al_2O_3 contents >18.5% (Appendix A2). This is similar to values seen in the Punchbowl unit of the Bear Mountain intrusive complex and suggests that plagioclase is probably a cumulate phase in many Summit Valley rocks. This inference is supported by the high Sr contents of these rocks: 582–1151 ppm (Fig. 5A; Appendix A2).

Bear Peak and Pony Peak Plutons. The geochemistry of the Bear Peak and Pony Peak plutons is discussed together, because both plutons have similar extended compositional ranges, with SiO_2 contents as high as 74 wt% (Fig. 3), and because they contain rocks that diverge from the Bear Mountain intrusive complex fractional crystallization trend. In Figure 3A, it is apparent that cumulate rocks from the structurally deepest, central part of the Bear Peak pluton have compositions similar to the Blue Ridge pluton. Because of their high contents of MgO and FeOt, we infer a similar origin for these rocks: accumulation of augite ± hornblende ± olivine. In addition, some of the Mg-rich aspects of the Bear Peak cumulates may have resulted from assimilation of metaserpentinite host rocks (McFadden et al., this volume). Figure 3 (especially Fig. 3A and C) shows that some samples from the Bear Peak and Pony Peak plutons follow the Bear Mountain intrusive complex fractional crystallization trend. However, another group of samples, shown with a diagonally ruled field in Figure 3, contains significantly higher MgO contents than do Bear Mountain intrusive complex samples with the same SiO_2 values. These samples are typically equigranular, medium- and fine-grained rocks whose compositions are probably close to melt compositions. In Figure 3, this field encompasses, or extends away from, the average composition of Bear Mountain intrusive complex gabbroic dikes.

Figure 3. Major element variation diagrams (wt%). Variation of (A) MgO, (B) TiO_2, and (C) total iron as FeO as functions of SiO_2 content. (D) Variation of CaO/Al_2O_3 as a function of Mg/(Mg + Fe). Oxides were chosen to illustrate the influence of cumulate assemblages on possible liquid lines of descent and magma mixing. Fields are shown for the Blue Ridge pluton and Punchbowl and Doe Flat units of the Bear Mountain pluton. Data for the Buck Lake unit of the Bear Mountain pluton are not shown; see Bushey et al. (this volume). Star symbols are average compositions of gabbroic (black) and dioritic dikes (gray) associated with the Bear Mountain intrusive complex, and Doe Flat (white) and Punchbowl (diagonal white) units. Solid arrows with filled heads show the inferred fractional crystallization trends of the various Bear Mountain intrusive complex units. In A and B, the inset diagrams show this path, along with the effects of crystal accumulation (arrows with unfilled heads) in the Blue Ridge pluton and Punchbowl unit. The field with a ruled pattern represents rocks from the Bear Peak and Pony Peak plutons with high MgO compared to the Bear Mountain intrusive complex fractional crystallization path. The double-headed, dashed arrow is the inferred mixing trajectory for mixtures of Mg-rich parental magmas with crustally derived tonalitic and granodioritic magmas in the Bear Peak and Pony Peak plutons. Abbreviations: cpx—clinopyroxene, hnbl—hornblende, plag—plagioclase.

Bear Peak and Pony Peak samples with SiO_2 >64 wt% have major element concentrations similar to those of the syn-Nevadan dikes and sills and tonalitic units of the Bear Mountain intrusive complex (Fig. 3). Trace element abundances and REE patterns (Figs. 4 and 5) show that this overlap does not imply similar origins. For example, compared to the syn-Nevadan dikes and sills, most evolved Bear Peak and Pony Peak samples have higher Sr contents (Fig. 5A), higher Ba/Y and Sr/Y ratios (Fig. 5D and E), and steeper REE patterns (Figs. 4 and 5C). Moreover, the REE patterns of evolved Bear Peak and Pony Peak samples generally have flat chondrite-normalized HREE abundances (reverse J-shape). Two granodiorites from the Bear Peak pluton have especially low HREE values (<2.5 times chondrites; Fig. 4E) and slight positive Eu anomalies.

When compared to the Bear Mountain intrusive complex, intermediate Bear Peak samples are distinct, because they show decreasing REE abundances with differentiation, whereas Bear Mountain intrusive complex rocks show overall increase of REE abundances (Figs. 4A and 5B). The same is true for the middle and heavy REEs in Pony Peak samples (Fig. 4E and F). These decreases may result from fractionation of REE-rich accessory minerals. However, if fractional crystallization controlled differentiation of Bear Peak and Pony Peak magmas with SiO_2 >55%, then one would expect plagioclase removal to cause a significant decrease in Sr contents. This process may be the case for some samples with Sr <700 ppm, but it cannot account for the samples with Sr >700–800 ppm (Fig. 5). Furthermore, fractional crystallization in the Bear Mountain intrusive complex system resulted in a decrease in Mg/(Mg + Fe) from ~0.68 to ~0.45 (Figs. 3 and 5), with an increase in Sr from ~300 ppm to ~700 ppm, followed by a decrease in Sr abundance. (Note that the higher Sr contents in the Bear Mountain intrusive complex Punchbowl unit are probably the result of plagioclase accumulation.) The Sr-rich samples from Bear Peak and Pony Peak are distinct in having higher SiO_2 and also higher Mg/(Mg + Fe) than the probable parental magma to the Punchbowl unit. These elevated values suggest that a fractional crystallization origin for these rocks is unlikely.

The steep REE slopes, high Sr contents, plus high Sr/Y and Ba/Y ratios of Bear Peak and Pony Peak samples with SiO_2 greater than ~65 wt% are characteristic of magmas derived by partial melting of metabasic rocks in which garnet ± amphibole ± clinopyroxene are residual phases (Martin, 1987; Drummond and Defant, 1990). In addition, plagioclase may not be a residual phase during partial melting of metabasic rocks. In such cases, positive Eu anomalies such as those seen in the Bear Peak granodiorites are possible, because residual phases such as clinopyroxene have negative Eu anomalies. Therefore, we suggest that a likely origin for Bear Peak and Pony Peak samples

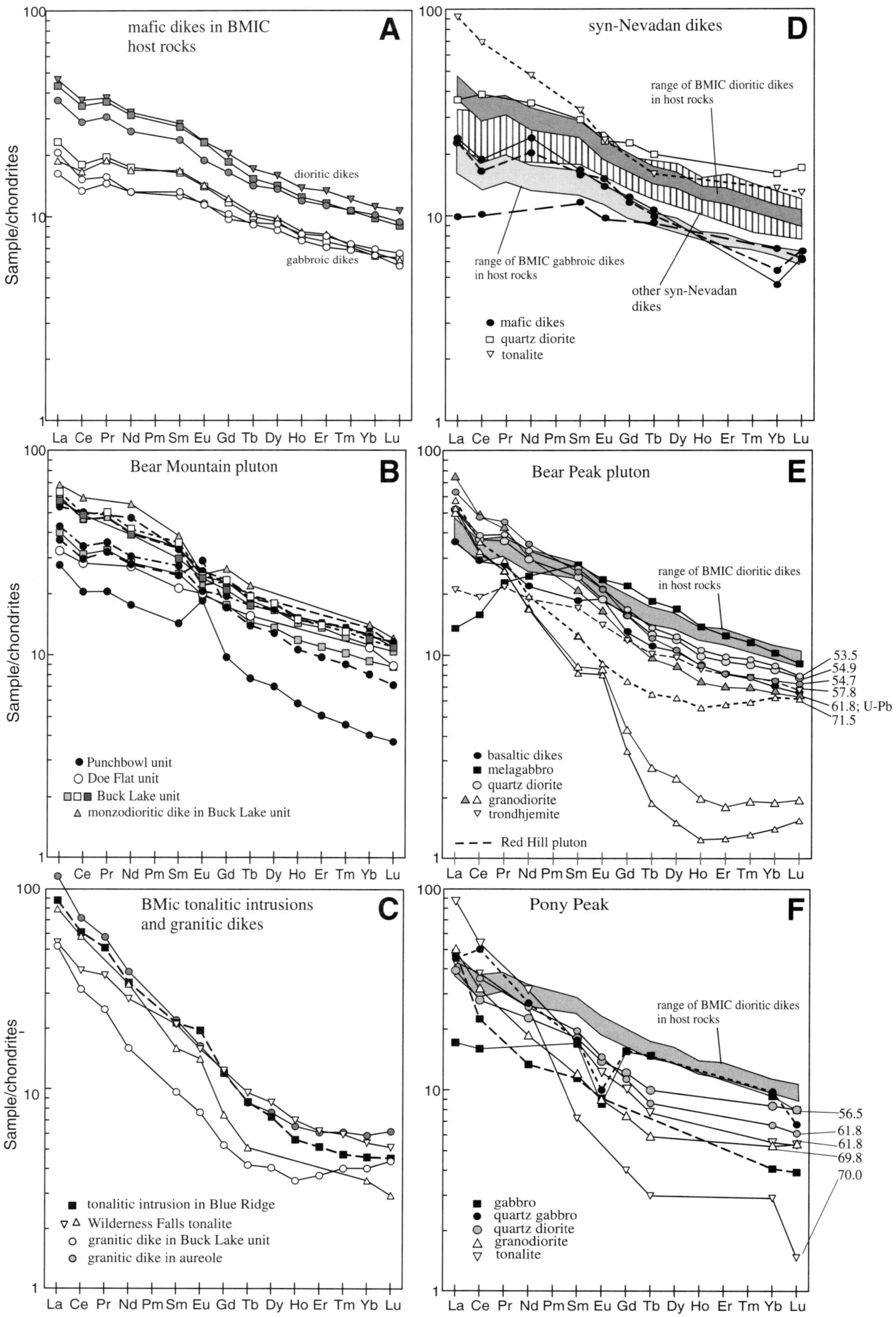

A
mafic dikes in BMIC
host rocks
dioritic dikes
gabbroic dikes
Sample/chondrites
D
syn-Nevadan dikes
range of BMIC dioritic dikes
in host rocks
range of BMIC gabbroic dikes
in host rocks
other syn-Nevadan
dikes
mafic dikes
quartz diorite
tonalite
B
Bear Mountain pluton
Punchbowl unit
Doe Flat unit
Buck Lake unit
monzodioritic dike in Buck Lake unit
E
Bear Peak pluton
range of BMIC dioritic dikes
in host rocks
53.5
54.9
54.7
57.8
61.8; U-Pb
71.5
basaltic dikes
melagabbro
quartz diorite
granodiorite
trondhjemite
Red Hill pluton
C
BMic tonalitic intrusions
and granitic dikes
tonalitic intrusion in Blue Ridge
Wilderness Falls tonalite
granitic dike in Buck Lake unit
granitic dike in aureole
F
Pony Peak
range of BMIC dioritic dikes
in host rocks
56.5
61.8
61.8
69.8
70.0
gabbro
quartz gabbro
quartz diorite
granodiorite
tonalite
La Ce Pr Nd Pm Sm Eu Gd Tb Dy Ho Er Tm Yb Lu
100
10
1

with high Sr/Y and Ba/Y and steep, reverse J-shaped REE patterns is by partial melting of metabasites, presumably at garnet-amphibolite to eclogite facies.

The previous discussion explains the existence of both basaltic (Bear Mountain intrusive complex gabbroic dikes) and felsic (granodioritic to tonalitic) magmas beneath the Bear Peak and Pony Peak plutons. Figure 3 shows that the MgO-rich samples with SiO_2 between 50 and 58 wt% could result from mixing of two such magmas. This scenario is consistent with the inverse correlation between HREEs and SiO_2 in these plutons (Fig. 4E and F). A similar explanation (magma mixing) is possible for MgO-poor samples with SiO_2 from 56 to 64 wt%. In this case, the mafic end member would be similar to evolved basaltic magma (diorite dike magma of the Bear Mountain intrusive complex) and the felsic end-member would be similar to granodioritic and tonalitic crustal magmas. As with the MgO-rich intermediate rocks, mixing would explain the inverse correlation between HREEs and SiO_2.

Isotopic Compositions

With the exception of the Bear Mountain intrusive complex, few isotopic data are available for the western Klamath plutonic rocks (Allen and Barnes, this volume; Fig. 6). Bushey et al. (this volume) showed that the Bear Mountain intrusive complex has low initial $^{87}Sr/^{86}Sr$ (0.70297–0.70347, with an average value of 0.70325 ± 0.00009 for the Bear Mountain pluton) and high ε_{Nd} (5.4–8.1; average of 6.4 ± 0.1 for the Bear Mountain pluton). All initial values are calculated at 147 Ma. Oxygen isotope ratios vary from $\delta^{18}O$ as low as +5.9‰ in clinopyroxenite to as high as +9.9‰ in a late-stage granitic dike (Bushey et al., this volume). The average $\delta^{18}O$ for seven samples of the Bear Mountain pluton is +7.2 ± 0.04‰, and for three samples of the Blue Ridge pluton, it is +6.6 ± 0.5‰. Initial Sr isotope data (at 145 Ma) are also available for five samples from the Pony Peak pluton (Allen and Barnes, this volume). They average 0.70356 ± 0.00008. Three of these samples were analyzed for oxygen isotope ratios, with values of $\delta^{18}O$ from +8.4 to +9.7‰.

Although initial $^{87}Sr/^{86}Sr$ values for the Bear Mountain and Pony Peak plutons are very similar, they are statistically distinct (Fig. 6). This distinction is also apparent in their $\delta^{18}O$ values. The Bear Mountain pluton has average $\delta^{18}O$ slightly higher than accepted "mantle values" (~5.5–6.2‰; Eiler et al., 1995), whereas $\delta^{18}O$ in samples from the Pony Peak pluton is at least 1.5‰ higher than the mantle range. Although the problems inherent in interpreting $\delta^{18}O$ values are well known (particularly deuteric equilibration and isotopic exchange with circulating fluids), the differences in $\delta^{18}O$ between the two plutons are suggestive that the Pony Peak pluton has a larger proportion of crustal material than does the Bear Mountain pluton.

Trace-Element Discrimination Diagrams

Because western Klamath magmatism overlapped with and immediately followed Nevadan contractional deformation, the tectonic setting of this magmatism is of interest. Figure 7A is a tectonic discrimination diagram developed by Pearce and Peate (1995) that uses ratios of three elements, Th, Ta, and Yb, which are incompatible during fractional crystallization of basaltic magmas. All mafic and intermediate samples of the western Klamath plutons plot in the calc-alkaline zone of the volcanic arc field in Figure 7A, and the mafic dikes associated with the Bear Mountain intrusive complex plot in the center of the data cluster. All samples that plot in or near the shoshonite field are evolved rocks, primarily tonalite, granodiorite, and granite. Such rock types are inappropriate for use in this type of discrimination diagram and are plotted here to show the effects of HREE depletion on the element ratios.

Figure 7B shows chondrite-normalized multi-element plots of the gabbroic and dioritic dikes associated with the Bear Mountain intrusive complex. The patterns show distinct depletion of Nb, Ta, and Ti and enrichments of K and Sr; all characteristics of arc-related magmas.

DISCUSSION

The origin and compositional evolution of western Klamath suite plutons provide a snapshot of nascent epicontinental arc magmatism. The oldest magmatism in this event, the earliest syn-Nevadan sills and dikes, apparently overlapped waning stages of Nevadan thrusting. Moreover, emplacement of pluton-scale magmas occurred within 1 or 2 m.y. of the end of thrusting, at a time when Nevadan deformation was still active (e.g., high-temperature mylonites in the Summit Valley pluton). Minor mylonitic deformation is present in plutons as young as the 144-Ma Pony Peak pluton (Harper et al., 1994). Therefore, on the basis of age data, crosscutting relationships, and zircon inheritance, we view the syn-Nevadan dikes and sills, Bear Mountain intrusive complex, Summit Valley, Bear Peak, Pony Peak, and Lower Coon Mountain intrusions as part of a single magmatic episode. It is probable that the Buckskin Peak, Ammon Ridge, and Glen Creek plutons are also part of this episode, but this suggestion must be tested with geochemical and petrologic study.

The most obvious feature of western Klamath magmatism, relative to pre-Nevadan magmatism and to younger ttg

Figure 4. Chondrite-normalized rare-earth-element plots. (A) Gabbroic and dioritic dikes in the host rocks of the Bear Mountain intrusive complex. These dikes are thought to represent parental magmas to the Bear Mountain intrusive complex. (B) Bear Mountain pluton. (C) Tonalitic intrusions and granitic dikes of the Bear Mountain intrusive complex. (D) Syn-Nevadan dikes. The ranges of gabbroic and dioritic Bear Mountain intrusive complex dikes are shown as shaded fields. (E) Bear Peak pluton. (F) Pony Peak pluton. In E and F, the field of Bear Mountain intrusive complex dioritic dikes is shown. Numbers on the right side of panels E and F indicate the SiO_2 contents of the samples. Bmic—Bear Mountain intrusive complex.

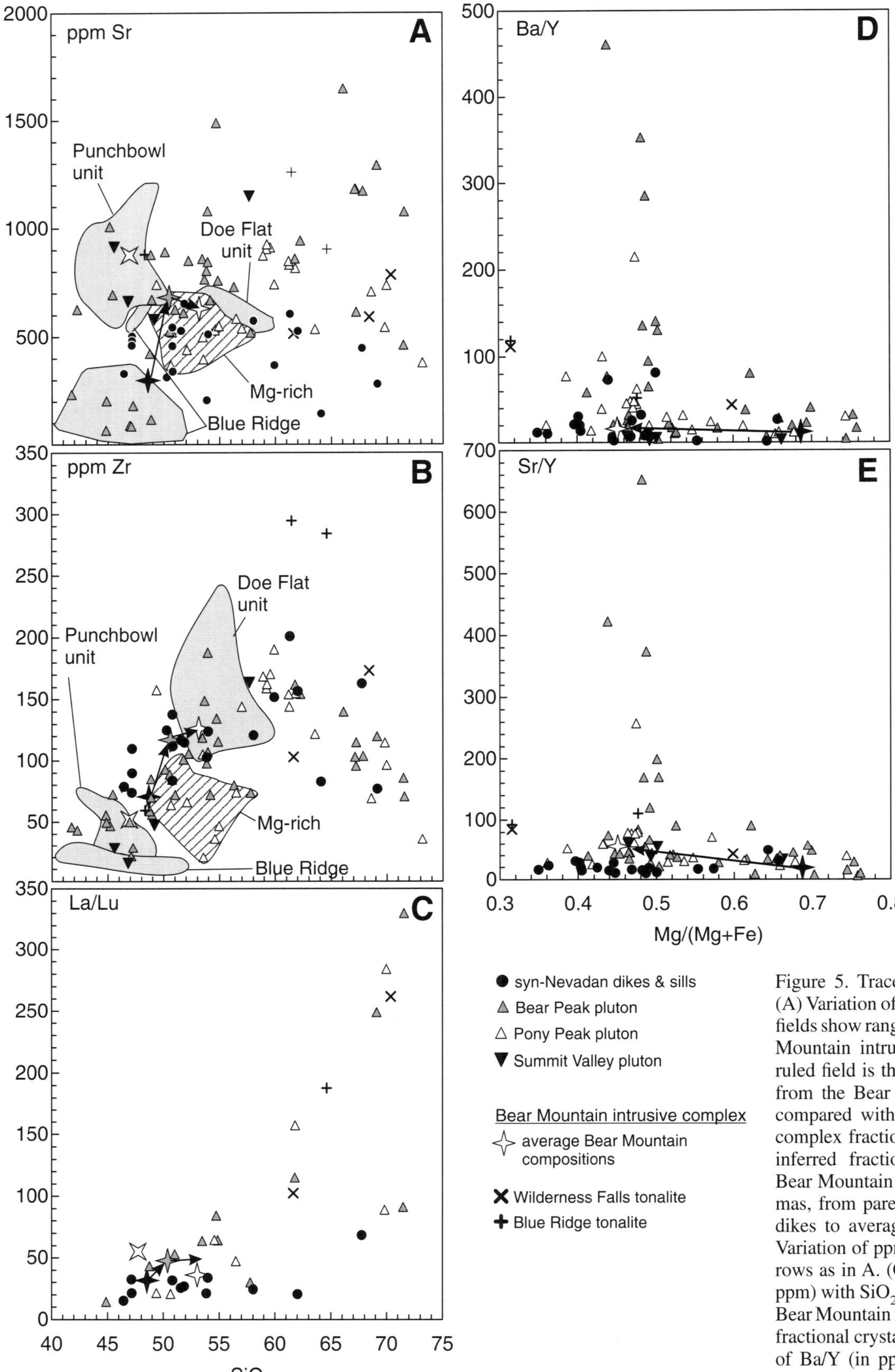

Figure 5. Trace element variation diagrams. (A) Variation of ppm Sr with SiO_2. The shaded fields show ranges of intrusive units in the Bear Mountain intrusive complex; the diagonally ruled field is the range of high MgO samples from the Bear Peak and Pony Peak plutons compared with the Bear Mountain intrusive complex fractionation path. Arrows show the inferred fractional crystallization trend for Bear Mountain intrusive complex mafic magmas, from parental gabbroic dikes to dioritic dikes to average Doe Flat composition. (B) Variation of ppm Zr with SiO_2. Fields and arrows as in A. (C) Variation in La/Lu ratio (in ppm) with SiO_2. The arrows show the inferred Bear Mountain intrusive complex trends due to fractional crystallization as in A. (D) Variation of Ba/Y (in ppm) with Mg/(Mg + Fe). The arrow shows the inferred Bear Mountain intrusive complex trends due to fractional crystallization. (E) Variation of Sr/Y (in ppm) with Mg/(Mg + Fe). The arrow shows the inferred Bear Mountain intrusive complex trends due to fractional crystallization.

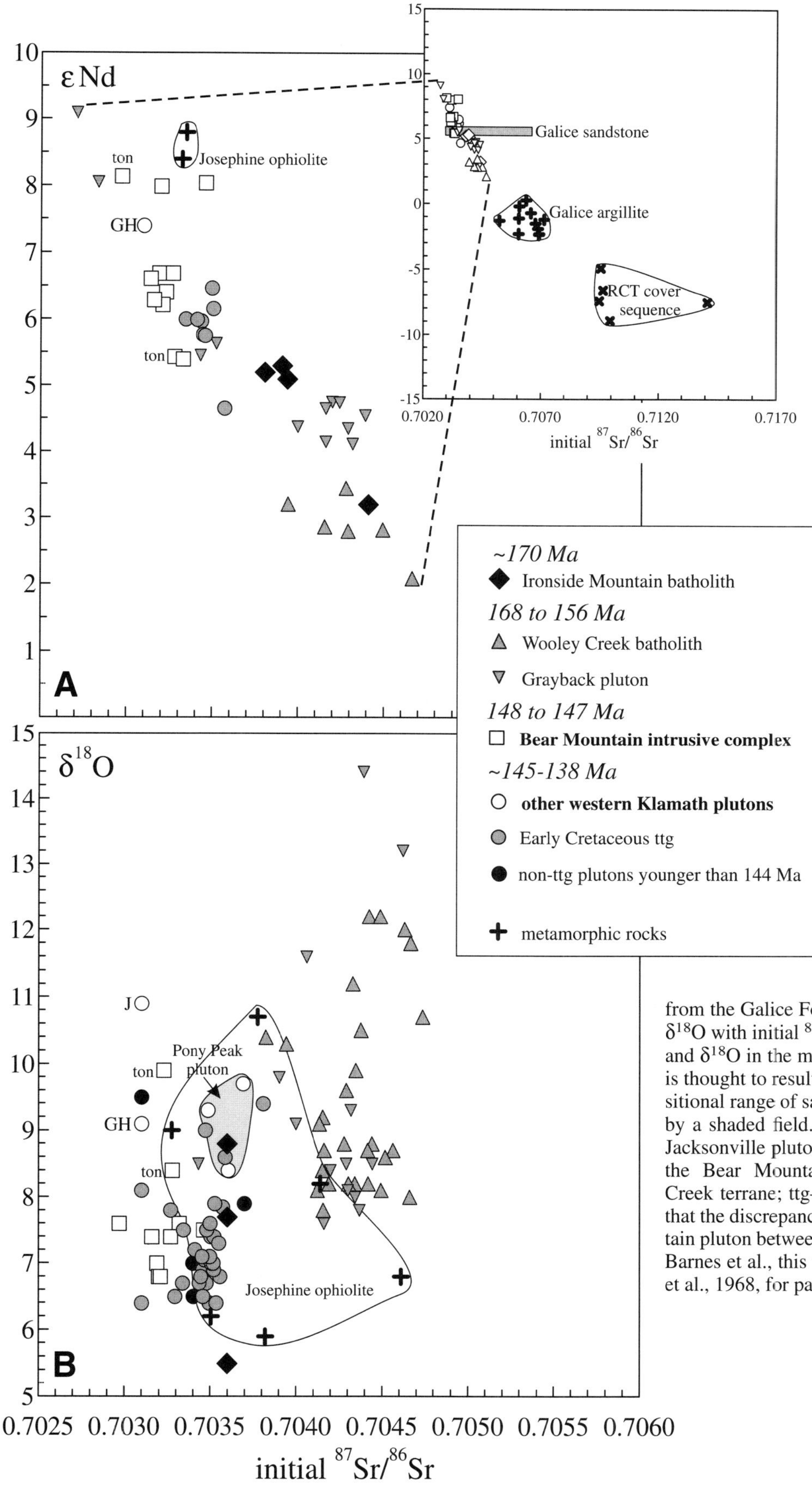

Figure 6. Isotopic compositions, showing the range of Bear Mountain intrusive complex and Pony Peak plutons, other Middle Jurassic to Early Cretaceous plutonic rocks, and metamorphic rocks from the western Klamath Mountains. (A) Variation of ε_{Nd} with initial $^{87}Sr/^{86}Sr$ (calculated at the age of the specific pluton). The Bear Mountain intrusive complex is noteworthy for its comparatively high ε_{Nd} and low initial $^{87}Sr/^{86}Sr$. Two metabasic samples from the Josephine ophiolite overlap the initial $^{87}Sr/^{86}Sr$ of the Bear Mountain intrusive complex but have slightly higher ε_{Nd}. The inset shows the compositional ranges of argillaceous rocks from the Galice Formation and Rattlesnake Creek terrane (Frost et al., this volume) and the range of initial $^{87}Sr/^{86}Sr$ of metasandstones from the Galice Formation (Coleman, 1972). (B) Variation of $\delta^{18}O$ with initial $^{87}Sr/^{86}Sr$. The large range of initial $^{87}Sr/^{86}Sr$ and $\delta^{18}O$ in the metabasic rocks from the Josephine ophiolite is thought to result from hydrothermal alteration. The compositional range of samples from the Pony Peak pluton is shown by a shaded field. GH—Gold Hill pluton (ca. 145 Ma); J—Jacksonville pluton (ca. 154 Ma); ton—tonalitic intrusions of the Bear Mountain intrusive complex; RCt—Rattlesnake Creek terrane; ttg—tonalite-trondhjemite-granodiorite. (Note that the discrepancy in initial $^{87}Sr/^{86}Sr$ for the Ironside Mountain pluton between panels A and B is due to different sources: Barnes et al., this volume, Chapter 10, for panel A; Lanphere et al., 1968, for panel B.)

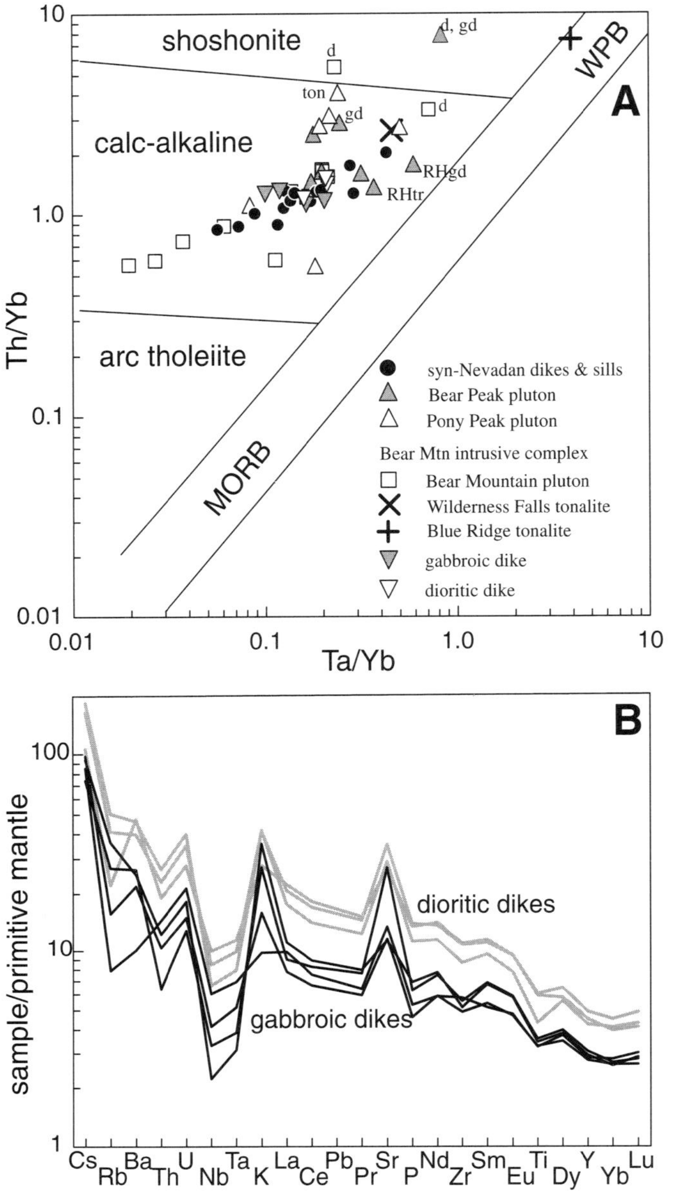

Figure 7. (A) Tectonic discrimination diagram (Pearce and Peate, 1995). All mafic samples plot in the calc-alkaline field. Samples that plot outside this field are tonalitic or granodioritic and contain a large component of crustal material (see text). They are, therefore, not appropriate for projection into this system and are shown as examples of the effects of crustal interaction. Abbreviations: d—dike sample; gd—granodiorite; MORB—mid-ocean ridge basalt; RH—Red Hill intrusion (part of the Bear Peak plutonic sequence); ton—tonalite; tr—trondhjemite; WPB—within-plate basalt. (B) Primitive mantle-normalized multielement plot of the gabbroic and dioritic dikes associated with the Bear Mountain intrusive complex.

magmatism, is the predominance of primitive, hydrous basaltic parental magmas. REE and major element data suggest that these parental magmas were similar throughout the western Klamath suite, although certainly not identical to one another. They were typically MgO-rich, enriched in LREE, and had olivine + augite ± plagioclase as near-liquidus phases. Most were H_2O-rich, as shown by early stability of hornblende, but the Buck Lake unit of the Bear Mountain intrusive complex requires an H_2O-poor mafic parent.

The western Klamath suite plutons are also similar in their petrogenetic histories. In the oldest members of the suite (syn-Nevadan dikes, Summit Valley pluton, and most of the Bear Mountain intrusive complex), compositional variation was primarily due to crystal-liquid fractionation, either fractional crystallization or crystal accumulation. Despite the clear importance of crystal-liquid separation, deep-seated interaction with crustal rocks is apparent in all plutons of the suite. In the Bear Mountain intrusive complex, this interaction takes the form of late-stage, predominantly tonalitic intrusions, such as the dated tonalite of the Blue Ridge pluton and the Wilderness Falls unit of the Bear Mountain pluton; whereas in the Summit Valley pluton, the presence of inherited zircon as old as Proterozoic (Harper et al., 1994) indicates some contribution from (meta) sedimentary rocks. Younger members of the suite (e.g., Bear Peak, Pony Peak, Lower Coon Mountain plutons) all contain felsic rocks with significant inherited zircon populations (Harper et al., 1994; Allen and Barnes, this volume). In the Bear Peak and Pony Peak plutons, REE patterns indicate residual garnet in crustal source rocks and therefore deep-seated melting. Moreover, the inverse correlation of HREE and SiO_2 contents in intermediate Pony Peak and Bear Peak samples suggests mixing of HREE-richer, mantle-derived magmas with HREE-depleted crustal magmas.

Petrogenesis

Our petrogenetic interpretations are shown schematically in Figure 8. This figure is derived from data presented here and from other chapters in this volume (Allen and Barnes; Bushey et al.; McFadden et al.). In Figure 8, the emplacement depth of the plutons is constrained by pressure estimates cited above. Magma differentiation at deeper levels is less well constrained, as discussed below.

Primitive, hydrous, basaltic magmas were emplaced early in the western Klamath magmatic event as gabbroic dikes associated with the Bear Mountain intrusive complex, as the parental magma to the Blue Ridge intrusion, and presumably as parental magma to at least some of the syn-Nevadan dikes (Fig. 8). Similar magmas were parental to the central cumulate zone of the Bear Peak pluton. Some of this primitive magma ponded deep in the crust (Fig. 8). The depth(s) of ponding is not known; however, for reasons discussed below, one possible location is within and/or near the top of the mafic part of the Josephine ophiolite. The ponded magmas underwent fractional crystallization along Fe-enrichment trends to produce evolved basaltic to andesitic magmas, which were emplaced to form dioritic dikes associated with the Bear Mountain intrusive complex, parental magmas to the Punchbowl unit of the Bear Mountain pluton, and hornblende gabbro to biotite-hornblende quartz

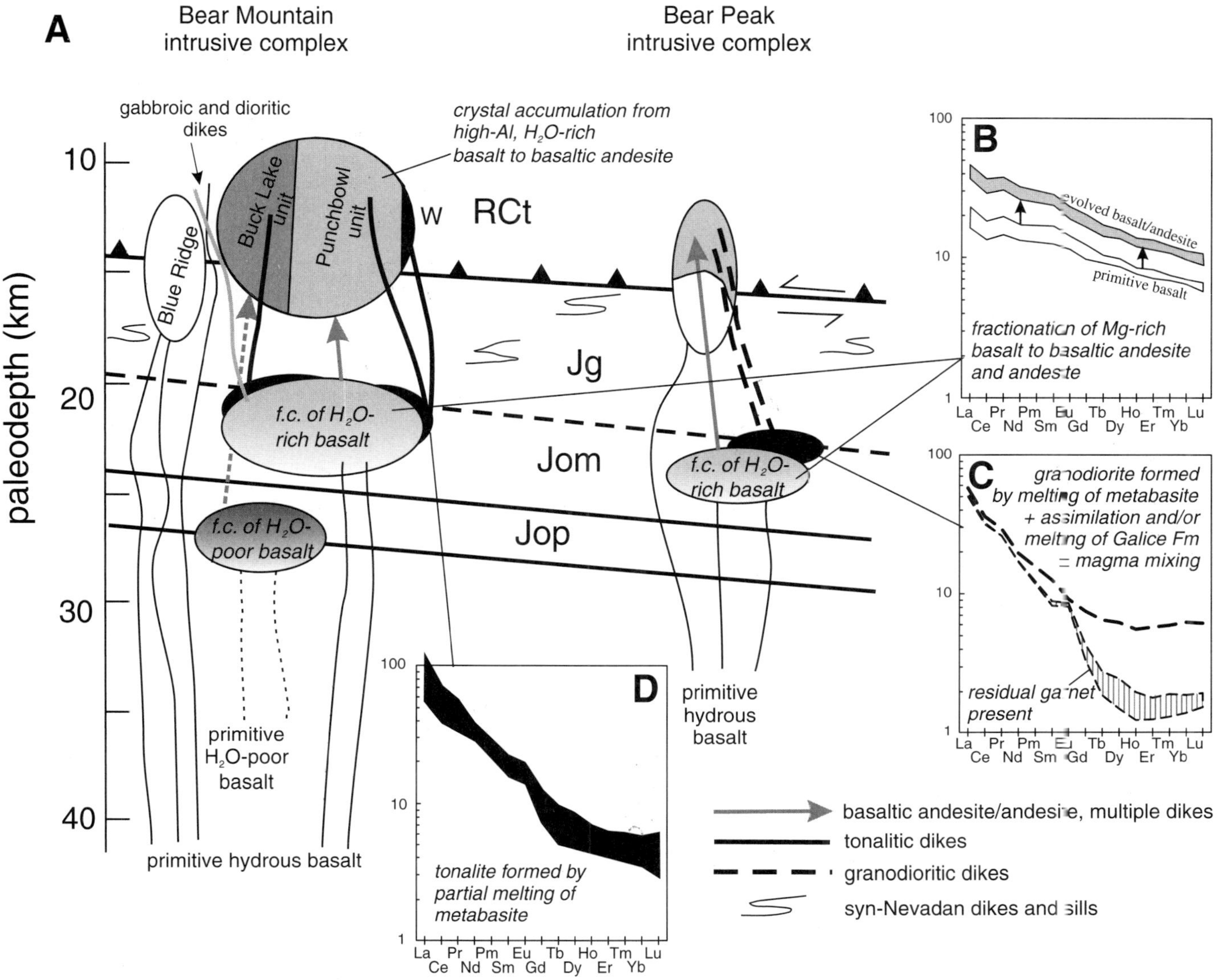

Figure 8. (A) Schematic cross-section to illustrate depth and types of western Klamath magma differentiation processes. Abbreviations: f.c.—fractional crystallization; Jg—Jurassic Galice Formation; Jom and Jop—the crustal section and peridotite, respectively, of the Josephine ophiolite; RCt—Rattlesnake Creek terrane. (B,C) Rare-earth-element patterns associated with various petrogenetic processes, including deep-seated fractional crystallization of parental high-Mg basaltic magmas, partial melting of metabasic rocks to form tonalite of Wilderness Falls, and hybridization of magmas with Galice Formation metasedimentary rocks/melts to form granodioritic rocks of the Bear Peak complex. Paleodepth estimates from Bushey et al. and McFadden et al. (this volume); Thickness estimates of Josephine ophiolite and Galice Formation from Harper (1980), Jachens et al. (1986), and Evans (1987).

diorite units in the Bear Peak pluton. This differentiation process is characterized by the uniform increase in REEs from parental basalt to evolved andesite (Fig. 8). Some mafic magma in the Bear Mountain intrusive complex was also poor in H_2O. This magma was similar to the primitive hydrous basalt except in terms of H_2O content. It differentiated to H_2O-poor andesitic magma that was emplaced as the Buck Lake unit of the Bear Mountain pluton (Fig. 8).

Tonalitic to granodioritic members of the suite (e.g., Wilderness Falls intrusion, much of the Pony Peak pluton, dikes in Bear Peak pluton) have trace element signatures consistent with an origin by partial melting of metabasic rocks (see above and Fig. 8). Thorough evaluation of the source of the crustal magmas is problematic, primarily because of a paucity of isotope data. Possible sources include the Upper Jurassic Galice Formation, mafic rocks of the Josephine ophiolite, or possible (cryptic) underlying terranes. The Galice Formation as a primary source of crustal magmas is unlikely for at least two reasons. First, much of the Galice Formation consists of intercalated meta-arenite and meta-argillite, and metawacke is common. Partial melting of such sedimentary protoliths results in peraluminous to strongly peraluminous magmas (e.g., Vielzeuf and Montel, 1994). However, even the most felsic rocks in the Bear Peak and Pony Peak plutons have an alumina saturation index <1.02. Second, the average $^{87}Sr/^{86}Sr$ of Galice Formation metawackes (at 147 Ma) is 0.7050 (range, 0.70296–0.70585; Coleman, 1972; Fig. 5) and the average of Galice Formation argillaceous rocks is 0.70646 (Frost et al., this volume). In contrast, the initial $^{87}Sr/^{86}Sr$ of the Pony Peak pluton is nearly uniform (0.70366–0.70379) in samples that range in SiO_2 content from 50.7 to 70 wt%. These values are significantly lower than those of the Galice rocks.

The $^{87}Sr/^{86}Sr$ values of mafic rocks from the Josephine ophiolite (at 147 Ma) have an average of 0.70373 ± 0.00044 (Shaw and Wasserburg, 1984; Barnes et al., 1996; Fig. 6), which is within the range of the Pony Peak samples. Therefore, on the basis of Sr isotope compositions, metabasic rocks of the Josephine ophiolite are possible sources of the evolved tonalitic and granodioritic rocks of the Pony Peak (and Bear Peak?) plutons. Mixing of such crust-derived magmas with mafic magmas would cause little or no change in initial $^{87}Sr/^{86}Sr$ but would result in higher $\delta^{18}O$ values than expected from fractional crystallization of a mantle-derived basaltic magma.

Figure 8 illustrates the simplest explanation for these signatures, in which primitive hydrous basalt ponded in the mafic section of the Josephine ophiolite. Heat released during cooling and crystallization of these magmas caused partial melting of the adjacent mafic rocks to form tonalitic magmas. These magmas were then emplaced as small tonalitic plutons, such as the Wilderness Falls intrusion and similar dikes. In some cases, particularly in the Pony Peak and Bear Peak systems, these crust-derived magmas mixed with basaltic magmas, giving rise to their extended compositional ranges and Mg-rich rocks that characterize both plutons (Fig. 3). The principal objection to this model is that the REE patterns of these tonalitic rocks suggest a pressure of origin of at least 8–10 kb. Simple cross-sections of the region do not place the mafic section of the Josephine ophiolite at such depths (25–35 km; Fig. 8). If further study shows the inferred depth of the Josephine mafic section to be correct, then the source of western Klamath crustal magmas is presumably in a deeper, cryptic mafic terrane.

Evolved rocks from the Pony Peak and Bear Peak plutons contain inherited zircons (Allen and Barnes, this volume), and these zircons show an age spectrum essentially identical to that of the detrital zircons in the Galice Formation (Miller et al., 2003). Chamberlain et al. (this volume) also found evidence for inherited Proterozoic zircon in units of the Bear Mountain intrusive complex, and Harper et al. (1994) reported Proterozoic inheritance in the Summit Valley pluton and granodiorites in the margin of the Lower Coon Mountain pluton. It is clear that these inherited zircons did not originate from the Josephine ophiolite, which formed in a narrow time span (166–162 Ma) and whose tectonic setting was oceanic (Pessagno, this volume). Moreover, the zircons were not inherited from rocks of the Rattlesnake Creek terrane, because inherited zircons are as young as 153 Ma, but the age of the Rattlesnake Creek terrane is Triassic to no younger than Middle Jurassic. It is conceivable that the inherited zircons came from a cryptic subduction-related accretionary wedge assemblage beneath the Josephine ophiolite. We consider this possibility unlikely, because such a source would need a detrital zircon population identical to that of the Galice Formation and yet should produce metaluminous magmas with $^{87}Sr/^{86}Sr$ ratios <0.7038 at 148 Ma.

We interpret the inherited zircons to be derived from the Galice Formation (Allen and Barnes, this volume; Fig. 8). If such inherited, Galice-like zircons were present only in plutons with evidence for hybridization of mantle-derived and of crustal magmas (Bear Peak and Pony Peak plutons), one could conclude that crustal melting involved metabasic rocks of the Josephine ophiolite *and* Galice metasedimentary rocks. (This possibility is shown schematically in Fig. 8.) However, the presence of inherited zircon in plutons that lack any evidence for mixing of crustal melts (Bear Mountain intrusive complex, Summit Valley?) suggests that some zircons were entrained by assimilation of Galice metasedimentary rocks during transport and emplacement. Such assimilation would have little influence on the initial $^{87}Sr/^{86}Sr$ ratios in these plutons because (1) the $^{87}Sr/^{86}Sr$ values of the Galice metasedimentary rocks are not drastically higher than those of the plutons and (2) because the magmas were generally Sr-rich (e.g., 647 ppm average in Pony Peak), whereas Galice rocks have lower Sr contents (average 136 ppm; Coleman, 1972; Frost et al., this volume).

Tectonic Setting

The trace element plots in Figure 7, the high concentrations of MgO, Ni, and Cr (Fig. 3, Appendix A2) in the primitive basaltic rocks, and evidence for high H_2O contents in these primitive magmas (e.g., the nearly ubiquitous presence of hornblende) are characteristic of mantle-derived magmas in an arc-related tectonic setting. Primitive H_2O-rich magmas are also characteristic of many modern arcs, the closest of which is the High Cascade arc (e.g., Wagner et al., 1995; Bacon et al., 1997; Grove et al., 2002). Therefore, the most probable source of these parental basaltic magmas is mantle modified by arc-related metasomatism. However, western Klamath magmatism was contemporaneous with and immediately followed contractional deformation (Nevadan orogeny). This orogenic event has commonly been considered to accompany collapse of an arc–back-arc assemblage (Rogue–Chetco arc, Josephine ophiolite, and Galice back-arc basin). It is not clear whether subduction was active during this deformation or the source of the magmas was metasomatized during a previous time of subduction.

Comparison with Other Klamath Plutonic Suites

On the basis of our data, it appears that early in the western Klamath plutonic event, large magma batches evolved primarily by fractional crystallization, with late assimilation of host rocks during transport and/or emplacement. Crustal melting was minor and probably restricted to the aureoles of the plutons, perhaps because the middle and lower crust was cool (Snoke et al., 1981; Bushey et al., this volume). In contrast, younger plutons show clear evidence for crustal melting and for incorporation of crust-derived magmas by magma mixing. This evidence suggests that with time, deep-seated crustal rocks were sufficiently hot to undergo partial melting and that these magmas were large enough in volume to interact and mix with mafic, mantle-derived magmas. Presumably, the increase in crustal magmas with time resulted from heating of the crust by basaltic magma trapped in middle or lower crustal depths. In other words, the western Klamath magmas changed from being primarily mantle-derived, with fractional crystallization driving magma evolution, to a combined derivation from mantle and crustal sources, with fractional crystallization, magma mixing, and assimilation driving magma evolution. Similar sequences characterize many High Cascade volcanoes, for example, Crater Lake (Bacon, 1983; Druitt and Bacon, 1989), Mount Shasta (Baker et al., 1994; Grove et al., 2002), and Medicine Lake Highlands (Grove and Donnelly-Nolan, 1986; Grove et al., 1997).

A similar, ~15 m.y.-long sequence of magmatic events began in Middle Jurassic time (Barnes et al., this volume, Chapter 10). In this case, the Ironside Mountain pluton and related rocks were emplaced soon after the end of thrusting along the Wilson Point thrust. These plutons resulted primarily from mantle melting and subsequent fractional crystallization (and crystal accumulation). As with the Bear Mountain intrusive complex, they were characterized by evolution to relatively low values of Mg/(Mg + Fe), but generally had lower H_2O and higher K_2O contents than did typical western Klamath magmas. Younger plutons (Wooley Creek suite) had higher H_2O contents, commonly had granitic units of crustal origin, and showed abundant evidence for mixing and assimilation of crustal magmas and rocks (Allen and Barnes, this volume, and references therein).

There are broad similarities between this Middle Jurassic magmatism and the western Klamath episode. Early western Klamath magmas evolved primarily by fractional crystallization, whereas later ones had larger crustal components (Pony Peak and Bear Peak). One can stretch this comparison to a regional scale by noting that no tectonic break exists between western Klamath magmatism and the slightly younger, but compositionally distinct, Late Jurassic to Early Cretaceous ttg plutons of the eastern part of the Klamath Mountains province (Fig. 1 in Allen and Barnes, this volume). These ttg plutons are primarily due to melting of metabasic crustal rocks (Barnes et al., 1996). Thus, if the western Klamath and ttg events are part of a single magmatic episode, then the change from early, predominantly mantle-derived magmatism to later, predominantly crustal magmatism is quite pronounced. Moreover, the time span of such a post-Nevadan magmatic episode (150–136 Ma) is essentially identical to that of the Middle Jurassic event that followed Wilson Point thrusting (170–156 Ma). Evidently, the Klamath Mountains province provides two excellent examples of the birth, growth, and demise of epicontinental arc systems that each encompass ~15 m.y. of geologic time.

CONCLUSIONS

The western Klamath plutonic suite represents magmatism synchronous with and immediately following regional contractional deformation. The entire suite results from emplacement of primitive, mantle-derived, primarily H_2O-rich basalt at various levels in the crust. Differentiation of these basalts yielded ultramafic cumulates (Blue Ridge and Lower Coon Mountain plutons and the ultramafic part of the Bear Peak pluton) and differentiated basaltic magma. The latter was parental to gabbroic cumulates (e.g., Punchbowl unit of the Bear Mountain pluton) and dioritic differentiates (e.g., Doe Flat unit of the Bear Mountain pluton). Heat from ponded basaltic magmas caused local crustal melting to form tonalitic magmas. Some of these crustal melts were emplaced with little modification, whereas others mixed with evolved basaltic magmas (Pony Peak and Bear Peak plutons). Assimilation of metasedimentary crustal rocks was widespread, especially in the youngest members of the suite (e.g., Pony Peak, Bear Peak plutons). In sum, magmatism of the western Klamath suite involved most of the crustal section.

ACKNOWLEDGMENTS

Magmatism in the Klamath Mountains province has been an important focus of research ever since Porter Irwin mapped batholith-scale plutons (Irwin, 1960), and Davis (1963) and Lipman (1963) subsequently provided detailed descriptions of some of them. The relationships between magmatism and contractional deformation were recognized by Irwin and other workers (e.g., Davis et al., 1965; Lanphere et al., 1968; Hotz, 1971; Irwin, 1985; Irwin and Wooden, 1999) and re-emphasized by Bob Coleman and his co-workers in the 1980s (e.g., Coleman et al., 1988). We thank them all for their contributions, and we particularly thank Porter Irwin, Bob Coleman, Mary Donato, and Ron Kistler for their interest, input, and encouragement. We thank Howard Day and Todd Feeley for their thorough, helpful, and timely reviews. Field and analytical work for this project received support from numerous National Science Foundation grants to CGB, GDH, and AWS, the latest of which are EAR-9902912 to CGB and EAR-9902807 to AWS. The Gregg Ranch Foundation provided field funds to RRM for his work on the Bear Peak pluton.

REFERENCES CITED

Allen, C.M., and Barnes, C.G., 2006, this volume, Ages and some cryptic sources of Mesozoic plutonic rocks in the Klamath Mountains, California and Oregon, *in* Snoke, A.W., and Barnes, C.G., eds., Geological studies in the Klamath Mountains province, California and Oregon: A volume in honor of William P. Irwin: Boulder, Colorado, Geological Society of America Special Paper 410, doi: 10.1130/2006.2410(11).

Anderson, J.L., and Smith, D.R., 1995, The effects of temperature and fO_2 on the Al-in-hornblende barometer: American Mineralogist, v. 80, p. 549–559.

Arth, J.G., Barker, F., Peterman, Z.E., and Friedman, I., 1978, Geochemistry of the gabbro-diorite-tonalite-trondhjemite suite of southwest Finland and its implications for the origin of tonalitic and trondhjemitic magmas: Journal of Petrology, v. 19, p. 289–316.

Bacon, C., 1983, Eruptive history of Mount Mazama and Crater Lake caldera, Cascade Range, U.S.A: Journal of Volcanology and Geothermal Research, v. 18, p. 57–115, doi: 10.1016/0377-0273(83)90004-5.

Bacon, C., Bruggman, P., Christiansen, R., Clynne, M., Donnelly-Nolan, J., and Hildreth, W., 1997, Primitive magmas at five Cascade volcanic fields: Melts from hot, heterogeneous sub-arc mantle: Canadian Mineralogist, v. 35, p. 397–423.

Baker, M.B., Grove, T.L., and Price, R., 1994, Primitive basalts and andesites from the Mt. Shasta region, N. California: Products of varying melt fraction and water content: Contributions to Mineralogy and Petrology, v. 118, p. 111–129, doi: 10.1007/BF01052863.

Barnes, C.G., Petersen, S.W., Kistler, R.W., Murray, R., and Kays, M.A., 1996, Source and tectonic implications of tonalite-trondhjemite magmatism in the Klamath Mountains: Contributions to Mineralogy and Petrology, v. 123, p. 40–60, doi: 10.1007/s004100050142.

Barnes, C.G., Mars, E.V., Swapp, S., and Frost, C.D., 2006, this volume (Chapter 10), Petrology and geochemistry of the Middle Jurassic Ironside Mountain batholith: Evolution of potassic magmas in a primitive arc setting, *in* Snoke, A.W., and Barnes, C.G., eds., Geological studies in the Klamath Mountains province, California and Oregon: A volume in honor of William P. Irwin: Boulder, Colorado, Geological Society of America Special Paper 410, doi: 10.1130/2006.2410(10).

Busby, C., 2004, Continental growth at convergent margins facing large ocean basins: A case study from Mexico convergent-margin basins of Baja California, Mexico: Tectonophysics, v. 392, p. 241–277, doi: 10.1016/j.tecto.2004.04.017.

Bushey, J.C., Snoke, A.W., Barnes, C.G., and Frost, C.D., 2006, this volume, Geology of the Bear Mountain intrusive complex, Klamath Mountains, California, *in* Snoke, A.W., and Barnes, C.G., eds., Geological studies in the Klamath Mountains province, California and Oregon: A volume in honor of William P. Irwin: Boulder, Colorado, Geological Society of America Special Paper 410, doi: 10.1130/2006.2410(14).

Chamberlain, K.R., Snoke, A.W., Barnes, C.G., and Bushey, J.C., 2006, this volume, New U-Pb radiometric dates of the Bear Mountain intrusive complex, Klamath Mountains, California, *in* Snoke, A.W., and Barnes, C.G., eds., Geological studies in the Klamath Mountains province, California and Oregon: A volume in honor of William P. Irwin: Boulder, Colorado, Geological Society of America Special Paper 410, doi: 10.1130/2006.2410(15).

Coleman, R.G., 1972, The Colebrook schist of southwestern Oregon and its relation to the tectonic evolution of the region: Washington, D.C., U.S. Geological Survey Bulletin 1339, 61 p.

Coleman, R.G., Manning, C.E., Mortimer, N., Donato, M.M., and Hill, L.B., 1988, Tectonic and regional metamorphic framework of the Klamath Mountains and adjacent Coast Ranges, California and Oregon, *in* Ernst, W.G., ed., Metamorphism and crustal evolution of the western United States—Rubey volume VII: Englewood Cliffs, New Jersey, Prentice-Hall, p. 1061–1097.

Davis, G.A., 1963, Structure and mode of emplacement of Caribou Mountain pluton, Klamath Mountains, California: Geological Society of America Bulletin, v. 74, p. 331–348.

Davis, G.A., 1968, Westward thrust faulting in the south-central Klamath Mountains, California: Geological Society of America Bulletin, v. 79, p. 911–934.

Davis, G.A., Holdaway, M.J., Lipman, P.W., and Romey, W.D., 1965, Structure, metamorphism, and plutonism in the south-central Klamath Mountains, California: Geological Society of America Bulletin, v. 76, p. 933–966.

Dewey, J.F., 1980, Episodicity, sequence, and style at convergent plate boundaries, *in* Strangeway, D.W., ed., The continental crust and its mineral deposits: Ottawa, Ontario, Geological Association of Canada Special Paper 20, p. 553–573.

Donato, M.M., Barnes, C.G., and Tomlinson, S.L., 1996, The enigmatic Applegate Group of southwestern Oregon: Age, correlation, and tectonic affinity: Oregon Geology, v. 58, p. 79–91.

Druitt, T.H., and Bacon, C.R., 1989, Petrology of the zoned calcalkaline magma chamber of Mount Mazama, Crater Lake, Oregon: Contributions to Mineralogy and Petrology, v. 101, p. 245–259.

Drummond, M.S., and Defant, M.J., 1990, A model for trondhjemite-tonalite-dacite genesis and crustal growth via slab melting: Archean to modern comparisons: Journal of Geophysical Research, v. 95, p. 21,503–21,521.

Eiler, J.M., Farley, K.A., Valley, J.W., Stolper, E.M., Hauri, E.H., and Craig, H., 1995, Oxygen isotope evidence against bulk recycled sediment in the mantle sources of Pitcairn island lavas: Nature, v. 377, p. 138–141, doi: 10.1038/377138a0.

Evans, J.G., 1987, Deformation of the Josephine Peridotite, California and Oregon: Reston, Virginia, U.S. Geological Survey Professional Paper 1378, 45 p.

Frost, B.R., Barnes, C.G., Collins, W.J., Arculus, R.J., Ellis, D.J., and Frost, C.D., 2001, A geochemical classification for granitic rocks: Journal of Petrology, v. 42, p. 2033–2048.

Frost, C.D., Barnes, C.G., and Snoke, A.W., 2006, this volume, Nd and Sr data from argillaceous rocks of the Galice Formation and Rattlesnake Creek terrane: Evidence for the input of Precambrian sources, *in* Snoke, A.W., and Barnes, C.G., eds., Geological studies in the Klamath Mountains province, California and Oregon: A volume in honor of William P. Irwin: Boulder, Colorado, Geological Society of America Special Paper 410, doi: 10.1130/2006.2410(5).

Garcia, M.O., 1982, Petrology of the Rogue River island-arc complex, southwest Oregon: American Journal of Science, v. 282, p. 783–807.

Gray, F., and Page, N.J., 1985, Geologic map of the Lower Coon Mountain pluton, Del Norte County, California: Reston, Virginia, U.S. Geological Survey Open-File Report 85-148, scale 1:24,000, 1 sheet.

Gray, G.G., 1986, Native terranes of the central Klamath Mountains, California: Tectonics, v. 5, p. 1043–1054.

Griesau, N., 1992, A kinematic study of the Summit Valley plutonic complex [M.S. thesis]: Albany, State University of New York, 133 p.

Griesau, N., and Harper, G.D., 1992, Deformation of a Late Jurassic plutonic complex intruding the Orleans thrust, west-central Klamath Mountains, California: Geological Society of America Abstracts with Programs, v. 24, p. 28.

Grove, T.L., and Donnelly-Nolan, J.M., 1986, The evolution of young silicic lavas at Medicine Lake Volcano, California: Implications for the origin of compositional gaps in calc-alkaline series lavas: Contributions to Mineralogy and Petrology, v. 92, p. 281–302, doi: 10.1007/BF00572157.

Grove, T., Donnelly-Nolan, J., and Housh, T., 1997, Magmatic processes that generated the rhyolite of Glass Mountain, Medicine Lake volcano, N California: Contributions to Mineralogy and Petrology, v. 127, p. 205–223, doi: 10.1007/s004100050276.

Grove, T., Parman, S., Bowring, S., Price, R., and Baker, M., 2002, The role of an H_2O-rich fluid component in the generation of primitive basaltic andesites and andesites from the Mt. Shasta region, N California: Contributions to Mineralogy and Petrology, v. 142, p. 375–396.

Hacker, B.R., and Ernst, W.G., 1993, Jurassic orogeny in the Klamath Mountains: A geochronological analysis, *in* Dunne, G.C., and McDougall, K.A., eds., Mesozoic paleogeography of the western United States II: Los Angeles, Pacific Section, Society of Economic Paleontologists and Mineralogists, v. 71, p. 37–60.

Hacker, B.R., Donato, M.M., Barnes, C.G., McWilliams, M.O., and Ernst, W.G., 1995, Timescales of orogeny: Jurassic construction of the Klamath Mountains: Tectonics, v. 14, p. 677–703, doi: 10.1029/94TC02454.

Hamilton, W.B., 1988, Plate tectonics and island arcs: Geological Society of America Bulletin, v. 100, p. 1503–1527, doi: 10.1130/0016-7606(1988)100<1503:PTAIA>2.3.CO;2.

Harper, G.D., 1980, The Josephine ophiolite—Remains of a Late Jurassic marginal basin in northwestern California: Geology, v. 8, p. 333–337, doi: 10.1130/0091-7613(1980)8<333:TJOOAL>2.0.CO;2.

Harper, G.D., 1984, The Josephine ophiolite, northwestern California: Geological Society of America Bulletin, v. 95, p. 1009–1026, doi: 10.1130/0016-7606(1984)95<1009:TJONC>2.0.CO;2.

Harper, G.D., 2006, this volume, Structure of syn-Nevadan dikes and their relationship to deformation of the Galice Formation, western Klamath terrane, northwestern California, *in* Snoke, A.W., and Barnes, C.G., eds., Geological studies in the Klamath Mountains province, California and Oregon: A volume in honor of William P. Irwin: Boulder, Colorado, Geological Society of America Special Paper 410, doi: 10.1130/2006.2410(6).

Harper, G.D., and Wright, J.E., 1984, Middle to Late Jurassic tectonic evolution of the Klamath Mountains, California-Oregon: Tectonics, v. 3, p. 759–772.

Harper, G.D., Saleeby, J.B., and Heizler, M., 1994, Formation and emplacement of the Josephine ophiolite and the Nevadan orogeny in the Klamath Mountains, California-Oregon: U/Pb zircon and $^{40}Ar/^{39}Ar$ geochronology: Journal of Geophysical Research, v. 99, p. 4293–4321, doi: 10.1029/93JB02061.

Hawkins, J.W., Bloomer, S.H., Evans, C.A., and Melchoir, J.T., 1984, Evolution of intra-oceanic arc-trench systems: Tectonophysics, v. 102, p. 175–205, doi: 10.1016/0040-1951(84)90013-1.

Hershey, O.H., 1906, Some western Klamath stratigraphy: American Journal of Science, 4th series, v. 21, p. 58–66.

Hershey, O.H., 1911, Del Norte County [California] geology: Mining and Scientific Press, v. 102, p. 468.

Hotz, P.E., 1971, Plutonic rocks of the Klamath Mountains, California and Oregon: Washington, D.C., U.S. Geological Survey Professional Paper 684-B, p. B1–B20.

Irwin, W.P., 1960, Geological reconnaissance of the northern Coast Ranges and Klamath Mountains, California, with a summary of the mineral resources: San Francisco, California Division of Mines Bulletin 179, 80 p.

Irwin, W.P., 1972, Terranes of the western Paleozoic and Triassic belt in the southern Klamath Mountains, California: Washington, D.C., U.S. Geological Survey Professional Paper 800C, p. C101–C111.

Irwin, W.P., 1981, Tectonic accretion of the Klamath Mountains, *in* Ernst, W.G., ed., The geotectonic development of California—Rubey volume 1: Englewood Cliffs, New Jersey, Prentice-Hall, p. 29–49.

Irwin, W.P., 1985, Age and tectonics of plutonic belts in accreted terranes of the Klamath Mountains, California and Oregon, *in* Howell, D.G., ed., Tectonostratigraphic terranes of the Circum-Pacific region: Houston, Circum-Pacific Council for Energy and Mineral Resources, Earth Sciences Series 1, p. 187–199.

Irwin, W.P., 1994, Geologic map of the Klamath Mountains, California and Oregon: Reston, Virginia, U.S. Geological Survey Miscellaneous Investigations Series Map I-2148, scale 1:500,000, 2 sheets.

Irwin, W.P., and Wooden, J.L., 1999, Plutons and accretionary episodes of the Klamath Mountains, California and Oregon: Reston, Virginia, U.S. Geological Survey Open-File Report 99-374, 1 sheet.

Irwin, W.P., Wolfe, E.W., Blake, M.C., Jr., Cunningham, C.G., Jr., 1974, Geologic map of the Pickett Peak quadrangle, Trinity County, California: Reston, Virginia, U.S. Geological Survey Map GQ-1111, one sheet, 1:62,500.

Jachens, R.C., Barnes, C.G., and Donato, M.M., 1986, Subsurface configuration of the Orleans fault: Implications for deformation in the western Klamath Mountains, California: Geological Society of America Bulletin, v. 97, p. 388–395, doi: 10.1130/0016-7606(1986)97<388:SCOTOF>2.0.CO;2.

Jorgenson, D.B., 1970, Petrology and origin of the Illinois River gabbro, a part of the Josephine peridotite-gabbro complex, Klamath Mountains [Ph.D. dissertation]: Santa Barbara, University of California, 195 p.

Lanphere, M.A., Irwin, W.P., and Hotz, P.E., 1968, Isotopic age of the Nevadan orogeny and older plutonic and metamorphic events in the Klamath Mountains, California: Geological Society of America Bulletin, v. 79, p. 1027–1052.

Lipman, P.W., 1963, Gibson Peak pluton: A discordant composite intrusion in the southeastern Trinity Alps, northern California: Geological Society of America Bulletin, v. 74, p. 1259–1280.

Martin, H., 1987, Petrogenesis of Archaean trondhjemites tonalites and granodiorites from eastern Finland: Major and trace element geochemistry: Journal of Petrology, v. 28, p. 921–953.

McFadden, R.R., Snoke, A.W., and Barnes, C.G., 2006, this volume, Structural and petrologic evolution of the Bear Peak intrusive complex, Klamath Mountains, California, *in* Snoke, A.W., and Barnes, C.G., eds., Geological studies in the Klamath Mountains province, California and Oregon: A volume in honor of William P. Irwin: Boulder, Colorado, Geological Society of America Special Paper 410, doi: 10.1130/2006.2410(16).

Miller, J., Miller, R., Wooden, J., and Harper, G., 2003, Geochronological links between the Ingalls ophiolite, North Cascades, Washington and the Josephine ophiolite, Klamath Mts., Oregon and California: Geological Society of America Abstracts with Programs, v. 35, p. 113.

Norman, E.A.S., 1984, The structure and petrology of the Summit Valley area, Klamath Mountains, California [M.S. thesis]: Salt Lake City, University of Utah, 148 p.

Pattison, D.R.M., 1992, Stability of andalusite and sillimanite and the Al_2SiO_5 triple point: Constraints from the Ballachulish aureole, Scotland: Journal of Geology, v. 100, p. 423–446.

Pearce, J.A., and Peate, D.W., 1995, Tectonic implications of the composition of volcanic arc magmas: Annual Review of Earth and Planetary Sciences, v. 23, p. 251–285, doi: 10.1146/annurev.ea.23.050195.001343.

Pessagno, E.A., Jr., 2006, this volume, Faunal evidence for the tectonic transport of Jurassic terranes in Oregon, California, and Mexico, *in* Snoke, A.W., and Barnes, C.G., eds., Geological studies in the Klamath Mountains province, California and Oregon: A volume in honor of William P. Irwin: Boulder, Colorado, Geological Society of America Special Paper 410, doi: 10.1130/2006.2410(02).

Pinto-Auso, M., and Harper, G.D., 1985, Sedimentation, metallogenesis, and tectonic origin of the basal Galice Formation overlying the Josephine ophiolite, northwestern California: Journal of Geology, v. 93, p. 713–725.

Saleeby, J.B., and Harper, G.D., 1993, Tectonic relations between the Galice Formation and the Condrey Mountain Schist, Klamath Mountains, northern California, *in* Dunn, G.C., and McDougall, K.A., eds., Mesozoic paleogeography of the western United States II: Los Angeles, Pacific Section, Society of Economic Paleontologists and Mineralogists, v. 71, p. 61–80.

Saleeby, J.B., Harper, G.D., Snoke, A.W., and Sharp, W.D., 1982, Time relations and structural stratigraphic patterns in ophiolite accretion, west central Klamath Mountains, California: Journal of Geophysical Research, v. 87, p. 3831–3848.

Shaw, H.F., and Wasserburg, G.J., 1984, Isotopic constraints on the origin of Appalachian mafic complexes: American Journal of Science, v. 284, p. 319–329.

Snoke, A.W., 1972, Petrology and structure of the Preston Peak area, Del Norte and Siskiyou Counties, California [Ph.D. dissertation]: Stanford, California, Stanford University, xix + 274 p., 3 plates.

Snoke, A.W., 1977, A thrust plate of ophiolitic rocks in the Preston Peak area, Klamath Mountains, California: Geological Society of America Bulletin, v. 88, p. 1641–1659, doi: 10.1130/0016-7606(1977)88<1641:ATPOOR>2.0.CO;2.

Snoke, A.W., Quick, J.E., and Bowman, H.R., 1981, Bear Mountain igneous complex, Klamath Mountains, California: An ultrabasic to silicic calc-alkaline suite: Journal of Petrology, v. 22, p. 501–552.

Vielzeuf, D., and Montel, J.M., 1994, Partial melting of metagreywackes. Part I. Fluid-absent experiments and phase relationships: Contributions to Mineralogy and Petrology, v. 117, p. 375–393, doi: 10.1007/BF00307272.

Wagner, D.L., and Saucedo, G.J., compilers, 1987, Geologic map of the Weed quadrangle, California: San Francisco, California Division of Mines and Geology Regional Geologic Map Series Map 4A, scale 1:250,000.

Wagner, T.P., Donnelly-Nolan, J.M., and Grove, T.L., 1995, Evidence of hydrous differentiation and crystal accumulation in the low-MgO, high-Al_2O_3 Lake

Basalt from Medicine Lake volcano, California: Contributions to Mineralogy and Petrology, v. 121, p. 201–216, doi: 10.1007/s004100050099.

Wright, J.E., 1982, Permo-Triassic accretionary subduction complex, southwestern Klamath Mountains, California: Journal of Geophysical Research, v. 87, p. 3805–3818.

Wright, J.E., and Fahan, M.R., 1988, An expanded view of Jurassic orogenesis in the western United States Cordillera: Middle Jurassic (pre-Nevadan) regional metamorphism and thrust faulting within an active arc environment, Klamath Mountains, California: Geological Society of America Bulletin, v. 100, p. 859–876, doi: 10.1130/0016-7606(1988)100<0859:AEVOJO>2.3.CO;2.

Wright, J.E., and Wyld, S.J., 1994, The Rattlesnake Creek terrane, Klamath Mountains, California: An early Mesozoic volcanic arc and its basement of tectonically disrupted oceanic crust: Geological Society of America Bulletin, v. 106, p. 1033–1056, doi: 10.1130/0016-7606(1994)106<1033:TRCTKM>2.3.CO;2.

Young, J.C., 1978, Geology of the Willow Creek quadrangle, Humboldt and Trinity counties, California: San Francisco, California Division of Mines and Geology Map Sheet 31, scale 1:62,500, 1 sheet and 16 p. pamphlet.

Yule, J.D., 1996, Geologic and tectonic evolution of Jurassic marginal ocean basin lithosphere, Klamath Mountains, Oregon [Ph.D. dissertation]: Pasadena, California Institute of Technology, 308 p.

Manuscript Accepted by the Society 10 January 2006

Printed in the USA

Geological Society of America
Special Paper 410
2006

Geologic-petrochemical comparison of early Mesozoic mafic arc terranes: Western Paleozoic and Triassic belt, Klamath Mountains, and Jura–Triassic arc belt, Sierran Foothills

Hannah H. Scherer
Cameron A. Snow
W.G. Ernst*
Department of Geological and Environmental Sciences, Stanford University, Stanford, California 94305-2115, USA

ABSTRACT

New field mapping and bulk-rock geochemical investigations in the southern Klamath Mountains and central Sierran Foothills combined with previous structural, petrochemical, and geochronologic studies allow the distinction between three Triassic–Jurassic basaltic arcs built along the continental edge versus two roughly coeval basaltic complexes that formed farther off the Californian margin. The three Klamath Mountains arcs are: (1) The Hayfork Summit–Salmon River segment of the southern North Fork terrane, formed offshore as a sequence of interlayered chert, volcaniclastic strata and shale, and ocean island basalt (OIB), deposited on a mélanged and serpentinized basement containing blocks of 310- to 265-Ma mid-ocean ridge basalt (MORB). (2) Northward, the Sawyers Bar sector of the central North Fork terrane formed closer to the continental margin; this mafic arc originated at ca. 200–170 Ma as a stack of interdigitated island-arc tholeiites (IAT) and minor OIBs interstratified with, and largely overlying, distal turbidites derived from eastern Klamath terranes. (3) The currently farther outboard Rattlesnake Creek terrane consists of continent-sourced, Lower Jurassic metasedimentary quartzose strata interbedded with island-arc volcanic rocks; this near-shore section was laid down on older ophiolitic basement consisting of tectonized serpentinite and MORB blocks. The remaining two arcs are in the Sierran Foothills: (4) The offshore Peñon Blanco arc consists of cherty and volcaniclastic sedimentary strata interlayered with 200-Ma mafic volcanic-plutonic arc rocks, all resting on a 300-Ma ophiolitic basement; suturing against the structurally higher Mariposa Formation took place after deformation of the latter at ca. 150 Ma. (5) The Slate Creek complex, and possibly the Lake Combie, Owl Gulch, and Sullivan Creek entities, formed along the margin of North America; superjacent units consist chiefly of 207- to 170-Ma volcaniclastic, sedimentary, and arc volcanic rocks deposited on an ophiolitic mélange basement. Metamorphic belts of the central Klamath Mountains and Sierran Foothills evidently contain both near-shore and offshore oceanic arcs.

Keywords: mafic igneous complexes, ophiolite, terranes, paleogeography, petrochemistry, late Paleozoic, Triassic, Jurassic, southern Klamath Mountains, western Sierra Nevada metamorphic belt

*E-mail: ernst@geo.stanford.edu.

Scherer H.H., Snow, C.A., and Ernst, W.G., 2006, Geologic-petrochemical comparison of early Mesozoic mafic arc terranes: Western Paleozoic and Triassic belt, Klamath Mountains, and Jura–Triassic arc belt, Sierran Foothills, *in* Snoke, A.W., and Barnes, C.G., eds., Geological studies in the Klamath Mountains province, California and Oregon: A volume in honor of William P. Irwin: Geological Society of America Special Paper 410, p. 377–392, doi: 10.1130/2006.2410(18). For permission to copy, contact editing@geosociety.org.

INTRODUCTION

Active continental margins are the locus of numerous tectonic processes that modify the composition, shape, and size of continents. These processes include the development of subduction complexes; arc magmatism; ophiolite genesis and emplacement; accretion of microcontinents, arcs, and seamounts; rifting; and reorganization of the crust by strike-slip faulting. The rock record of the Mesozoic North American Cordillera bears evidence of all these processes. In addition to continental arc volcanism (e.g., the voluminous Sierra Nevada batholith and related plutons in the Klamath Mountains), a large amount of material has been added to the craton through the accretion of mafic igneous and metasedimentary terranes. The Klamath Mountains and Sierran Foothills (western Sierra Nevada metamorphic belt) represent geologically complex products of such processes. Since invocation of the terrane concept by Irwin (1972), geologists working in these regions have examined and reexamined the origin and autochthonicity of these terranes. Due to the deformation that accompanies the accretionary process, it is generally not obvious whether a terrane is far-traveled or formed essentially in situ. In addition, diverse units of similar ophiolitic character (i.e., associations of variably serpentinized peridotite, gabbroic plutons, diabase, basalt, and oceanic sedimentary rocks) may be inappropriately grouped together as a single terrane.

The goal of this chapter is to examine the geologic and paleogeographic nature of mafic igneous complexes of late Paleozoic to early Mesozoic age in northern and central California. We recognize five assemblages of rather similar ages and lithologies. The Klamath terranes are: (1) the Hayfork Summit area of the southern North Fork terrane, (2) Sawyers Bar area of the central North Fork terrane, and (3) Rattlesnake Creek terrane. In the Sierran Foothills, we consider (4) the Peñon Blanco arc and (5) Slate Creek complex of the Jura–Triassic arc belt (Fig. 1). We hope to understand where these entities originated (i.e., are they native or exotic?) and how they are related to one another (i.e., did they form as one system or do they represent disparate, unrelated terranes?). To address these questions, we present relevant data from each of these complexes, including both previously published data and our new work. We have conducted detailed geologic mapping and have studied the petrology and bulk-rock and mineral geochemistry from the southern North Fork terrane and the Peñon Blanco arc (Fig. 1). This ongoing research provides new constraints regarding the plate-tectonic evolution of the Cordilleran margin during early Mesozoic time. Results arising from our geologic and bulk-rock petrochemical work are presented here as a modest tribute to Porter Irwin, whose lifetime of geologic research in northern and central California has provided both an inspiration and a solid scientific foundation for all future studies.

REGIONAL GEOLOGIC SETTING

Products of both continental arc volcanism-plutonism and accretion of oceanic and continental-derived materials are recorded in the terranes of the Klamath Mountains of northern California and southwestern Oregon (Irwin, 1960, 1972; Burchfiel and Davis, 1981) and the Sierran Foothills belt of central California (Saleeby, 1981; Schweickert, 1981; Sharp, 1988; Edelman and Sharp, 1989; Dilek et al., 1990). Some of the mafic-ultramafic lithostratigraphic entities are probably far-traveled and exotic to the North American margin, whereas others may have formed on or near the continental margin. Klamath Mountain and Sierran Foothill belts exhibit similar structural styles; major lithotectonic terranes are juxtaposed along a series of mainly north–south-striking Late Jurassic and older faults (typically, east-dipping, west-directed thrust faults). Reflecting progressive accretion to the North American continental margin, depositional and metamorphic ages of the oldest rocks in each terrane generally decrease westward, and stratified units are chiefly right-side-up, with tops facing east (Ernst, 1983). In both regions, oceanic rocks of late Paleozoic and early Mesozoic age occupy medial portions of the orogens. In the Klamath Mountains, these rocks are included in the western Paleozoic and Triassic belt; in the Sierran Foothills, they constitute the Jura–Triassic arc belt. We summarize geologic relationships among the terranes in each of these two belts and then examine the geochemistry of the mafic igneous complexes within them.

WESTERN PALEOZOIC AND TRIASSIC BELT

Arcuate, east-dipping allochthons are recognized based on distinctive lithotectonic units in the medial Klamath Mountains. Structurally beneath the Devonian central metamorphic belt and Late Triassic (metamorphic ages) Stuart Fork terrane (Lanphere et al., 1968; Hotz et al., 1977; Hacker and Goodge, 1990), Irwin (1972) distinguished three north–south-striking lithostratigraphic segments of the western Paleozoic and Triassic belt—the North Fork, Hayfork, and Rattlesnake Creek terranes (Fig. 1). In fact, Irwin proposed the terrane concept to explain their spatial relationships. Subsequent workers have modified this architecture to fit new data. Notably, the division of the Hayfork into the eastern and western Hayfork terranes by Wright (1982) and the definition of the Sawyers Bar composite terrane, combining the North Fork and eastern Hayfork lithotectonic entities (Ernst, 1991, 1998; Hacker et al., 1993), have contributed to the current divisions of the western Paleozoic and Triassic belt.

Recrystallization, contraction, and thickening accompanied relatively shallow-level tectonic insertion (Hacker et al., 1992)

Figure 1. Generalized geologic correlation map of the Klamath Mountains and the western Sierran Foothills. The Sawyers Bar, Salmon River, Hayfork Summit, and Peñon Blanco map areas are indicated. The Red Ant Schist, possibly correlative with the Stuart Fork terrane, is not shown separately but is included in the Calaveras complex in this compilation. The northwest-striking Browns Meadow fault (not labeled) transects the overlapping map boundaries between the Sawyers Bar (Ernst, 1998) and Salmon River (Ando, 1979) areas. Largely after Irwin (1981, 2003), Ando et al. (1983), Sharp (1988), Edelman and Sharp (1989), Ernst (1999), and Snow and Scherer (2006).

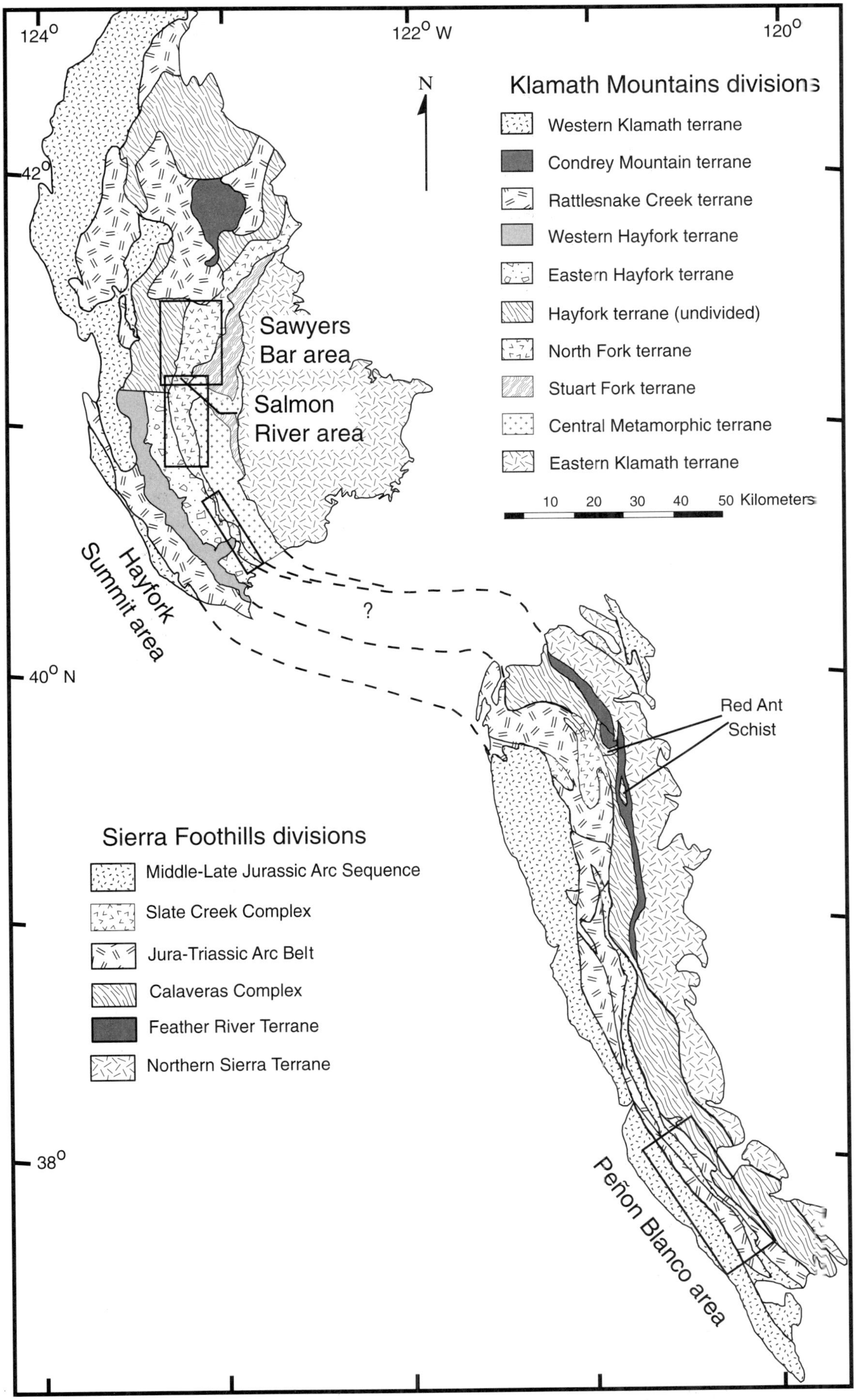

124°
122° W
120°
42°
40° N
38°
N
Klamath Mountains divisions
Western Klamath terrane
Condrey Mountain terrane
Rattlesnake Creek terrane
Western Hayfork terrane
Eastern Hayfork terrane
Hayfork terrane (undivided)
North Fork terrane
Stuart Fork terrane
Central Metamorphic terrane
Eastern Klamath terrane
10 20 30 40 50 Kilometers
Sawyers Bar area
Salmon River area
Hayfork Summit area
?
Red Ant Schist
Sierra Foothills divisions
Middle-Late Jurassic Arc Sequence
Slate Creek Complex
Jura-Triassic Arc Belt
Calaveras Complex
Feather River Terrane
Northern Sierra Terrane
Peñon Blanco area

of the greenschist-facies North Fork terrane outboard and structurally beneath the amphibolitic central metamorphic belt (Peacock and Norris, 1989) and the Stuart Fork blueschist terrane (Goodge, 1989) along the Siskiyou and related faults. These faults represent a major structural boundary with significant displacement, as evidenced by the contrasting metamorphic grades across them. Displacement along the Siskiyou system is constrained to post-Early Jurassic time by fossil ages in the North Fork terrane (Irwin et al., 1983) and to pre-late Middle Jurassic time by intrusion of the 159-Ma, crosscutting Russian Peak pluton (Wright and Fahan, 1988; Barnes et al., 1992, 1996), and the ca. 164-Ma East Fork stock (Irwin and Wooden, 1999). The tectonic significance of the contact between the eastern Hayfork subduction mélange, and the inboard, ophiolitic North Fork is less clear. Differential displacement between the two terranes took place along the Twin Sisters high-angle reverse fault, but offset appears to have been modest (Miller and Ernst, 1998), and deposition in both terranes was nearly contemporaneous (Wright, 1982). This scenario is supported by the recognition that sedimentary rocks of the North Fork and eastern Hayfork terranes are similar in both fossil ages—late Paleozoic to Jurassic for the North Fork, late Paleozoic to Triassic and possibly Jurassic for the eastern Hayfork (Irwin and Blome, 2004)—and mafic igneous and sedimentary rocks (Ando et al., 1983; Ernst, 1998). The eastern Hayfork mélange also contains mafic blocks similar to North Fork ocean island basalt (OIB) and island-arc tholeiite (IAT), suggesting its formation proximal to the North Fork terrane (Wright, 1982; Ernst, 1999). Slip along the Twin Sisters fault took place prior to emplacement of the 160- to 164-Ma English Peak pluton (Wright and Fahan, 1988; Irwin and Wooden, 1999).

The western Hayfork terrane lies structurally below the eastern Hayfork terrane and is separated from it by the Wilson Point thrust (Wright, 1982). Displacement along this major thrust fault occurred prior to intrusion of the 170-Ma Ironside Mountain crosscutting pluton (Wright and Fahan, 1988; Barnes et al., this volume, Chapter 10). Rocks of the western Hayfork terrane represent a Middle Jurassic (179–169 Ma) volcanic arc that received sediment from a terrigenous source. These arc rocks are in fault contact with the older Rattlesnake Creek terrane, but offset probably was minimal, and the Rattlesnake Creek terrane served as basement to the western Hayfork terrane (Wright and Fahan, 1988; Donato et al., 1996). These are the youngest volcanic rocks in the western Paleozoic and Triassic belt, and they are contemporaneous with mafic hypabyssal dikes as young as ca. 170 Ma that intruded the Sawyers Bar terrane (Hacker et al., 1993).

North Fork Terrane: Hayfork Summit Area

General Relationships. The North Fork terrane, the easternmost belt of the western Paleozoic and Triassic belt, as originally defined in the southern Klamath Mountains by Irwin 1972), is an ophiolitic assemblage composed of serpentinized ultramafic, metagabbroic, metabasaltic, metavolcaniclastic, and metasedimentary rocks. Later work on the North Fork terrane in the Salmon River area (Fig. 1) by Ando (1979) and Ando et al. (1983) identified a tectonically disrupted ophiolitic basement overlain by a superjacent series, consisting of chert, argillite, massive limestone, and alkalic volcanic rocks. These authors reported a discordant U-Pb zircon age of 310–265 Ma from a sample of plagiogranite associated with the basement complex, which they interpreted to represent the time of crystallization for the mafic-ultramafic unit. In keeping with Irwin's original hypothesis, they interpreted this complex as a tectonized mid-ocean ridge basalt (MORB) ophiolite. In addition, Ando et al. (1983) proposed that the OIB superjacent rocks represent a seamount built on this oceanic crust. New geochemical data from the Hayfork Summit area (Fig. 1) tend to support this conclusion.

Geologic Mapping and Petrology. Detailed mapping in the feebly recrystallized Hayfork Summit area indicates that lithologic units of the North Fork terrane described by Ando (1979) are also present in the southernmost Klamath Mountains. The chief metasedimentary lithologies are metachert, argillite, and volcanogenic metasedimentary rocks, constituting ~45, 35, and 20%, respectively, of the strata present in the map area. Metachert is interlayered with both argillite and volcanogenic metasedimentary rocks. Where it occurs with argillite, outcrops range from rhythmically bedded, tan to gray-green metachert (average 3- to 5-cm beds) with minor (~3 mm) shale partings, to dark gray, well-foliated argillite. Volcanogenic metasedimentary strata are distinguished from pelitic rocks by their distinctive bright green color; coarse-grained beds contain clasts of mafic volcanic rocks, plagioclase crystals, and chert fragments. Other metasedimentary rocks include marble and a shale-matrix, chert-rich breccia that grades to pure metashale. Marble commonly occurs as blocks (0.5–150 cm in diameter) incorporated in the weakly recrystallized volcanic flow breccias. The abundance of volcanic debris and apparent lack of continental detritus probably reflects proximity of the Hayfork Summit section to an offshore mafic volcanic edifice.

Alkalic mafic metavolcanic rocks, the Hayfork Summit metabasalts, are interstratified with the metasedimentary units. These bright green to dark gray-green metabasalts are massive to locally pillowed, amygdaloidal, and vesicular. Primary igneous phases include clinopyroxene, hornblende, and plagioclase. Pyroxene-bearing metabasalts and spatially associated hypabyssal rocks contain sub- to euhedral, glomeroporphyritic and interstitial clinopyroxene. The average composition of relict igneous clinopyroxene is $Wo_{42}En_{46}Fs_{12}$, with minor zoning from magnesian cores ($Wo_{45}En_{45}Fs_{10}$) to more ferrous rims ($Wo_{41}En_{38}Fs_{21}$) measured in a few samples. Clinopyroxene-free samples show relict euhedral plagioclase laths in the groundmass, now altered to albite, as well as plagioclase phenocrysts. No other definitively igneous phases are present in these units. Metamorphic phases include albite (An_{03}), chlorite, opaques (including pyrite), titanite, epidote, calcite, and pumpellyite. Plagioclase-rich units contain more neoblastic epidote than do the pyroxene-bearing rocks, reflecting differences in original bulk-rock compositions.

Serpentinite-matrix mélange is also present in the Hayfork Summit area. The mafic-ultramafic mélange is in tectonic contact with the metasedimentary and alkalic volcanic country rocks. The orientation of this contact is highly variable, but locally parallels the regional northwest–southeast foliation. Blocks in the mélange consist of variably metasomatized greenstone, metadiorite, and metagabbro and range from 1 to 200 m in diameter. Fine-grained greenstone is the most intensely altered lithology; alteration is greatest at the edges of blocks, where they are in sheared contact with host serpentinite. The serpentinite lacks evidence of contact metamorphism, indicating tectonic rather than plutonic emplacement. Metadiorite is made up of subhedral plagioclase; euhedral, green-brown, prismatic hornblende; anhedral quartz; minor epidote; and opaque oxides. Metagabbroic rocks consist of subhedral to euhedral plagioclase ($Ab_{87}Or_{10}An_{03}$), granular to interstitial clinopyroxene ($Wo_{39}En_{50}Fs_{11}$), titanite displaying exsolution lamellae of magnetite, and tremolitic amphibole.

Geologic units cropping out in the Hayfork Summit area are also present ~10 km to the northwest. Vesicular metabasalt is interbedded with metachert and metashale and is considered to be part of the Hayfork Summit metavolcanic and metasedimentary package. Greenstone and metagabbro similar to blocks in the serpentinite matrix mélange are larger (~500 m) fault-bounded bodies, and serpentinite matrix mélange containing metasomatized greenstone blocks is also present. We interpret this association to be a less-tectonized equivalent of the Hayfork Summit serpentinite matrix mélange. Samples collected from these units for geochemical analyses are treated as part of the mafic-ultramafic suite, with which they are correlated.

Bulk-Rock Geochemistry. Thirteen new bulk-rock X-ray fluorescence spectrometry (XRF) and inductively coupled plasma mass spectrometry (ICP-MS) analyses are listed in Table 1. The analytical data demonstrate that the two geologically distinct igneous units mapped in the Hayfork Summit area (Hayfork Summit metabasalts and mafic igneous rocks in serpentinite mélange) are also chemically distinct. Geochemical discrimination diagrams (Fig. 2) and elevated TiO_2, FeO*, and P_2O_5 contents indicate that Hayfork Summit metabasalts have OIB affinities, whereas mélange blocks in serpentinite tend to plot in the MORB field. Trace element abundances for Hayfork Summit metabasalts exhibit OIB patterns, decreasing steadily from light rare-earth elements (LREEs) to heavy REEs (HREEs) and lacking significant negative Nb and Ta anomalies (Fig. 3A). Metabasalt, metagabbro, and metadiorite blocks in the serpentinite matrix mélange have nearly flat trace element abundance patterns, similar to MORB (Fig. 3B); most analyzed samples are enriched in Ba, K, and Sr relative to normal MORB (N-MORB), probably reflecting alteration due to serpentinization, but display HREE patterns comparable to ocean-floor basalts.

Paleogeography. Paleomagnetic data from Permian red cherts in the North Fork terrane suggest that these basement rocks were located at least 1000 km south of the eastern Klamath terrane in Permian time (Mankinen et al., 1996). Based on paleogeographic implications of fossils in the eastern Klamath and eastern Hayfork terranes, Miller (1987) concluded that Permian Tethyan fusulinid-bearing limestone blocks in the eastern Hayfork metasedimentary mélange were derived from seamounts. Permian fusulinids are also present in limestone in the southern North Fork terrane (Irwin, 1972). These fossil-bearing carbonate units occur as blocks in mafic volcanic flow-breccias of OIB chemistry in the Hayfork Summit area. In the Salmon River area, Ando et al. (1983) described a large limestone layer that was deposited on alkalic volcanic rocks with a basal conglomerate of mafic volcanic clasts in the limestone. No fossil evidence is available for this unit due to intense recrystallization, but its clear association with alkalic volcanic rocks suggested to Ando et al. (as well as to us) that it is part of a seamount assemblage. The above considerations point to the likelihood that the southern North Fork terrane originated as an oceanic entity that was later incorporated in the convergent margin assemblage forming outboard of the North American craton. Interestingly, ongoing U-Pb studies of clastic zircons suggest that eastern Hayfork metasedimentary rocks in the Hayfork Summit area were derived from old continental crust.

Conclusion. North Fork units in the Hayfork Summit and Salmon River areas of the southern Klamath Mountains evidently formed as an interlayered, late Paleozoic–early Mesozoic complex of volcaniclastic shale, deep-water chert, limestone, and OIB resting on a late Paleozoic basement complex of MORB and serpentinized peridotite. Mafic igneous rocks of the North Fork terrane in this area lack a geochemical arc signature, suggesting their probable origin as an offshore seamount.

North Fork Terrane: Sawyers Bar Area

General Relationships. Geologic and geochemical contrasts between the North Fork terrane in the southern versus the central Klamath Mountains (Fig. 1) support their recognition as separate lithotectonic entities. Compared with the Hayfork Summit–Salmon River complex, these differences for the Sawyers Bar terrane include the presence of an unambiguous geochemical arc signature in mafic volcanic rocks, the occurrence of continental detritus in the associated distal turbidites, and the fact that the metabasaltic pile mainly overlies the metasedimentary section. In addition, limestone lenses and serpentinite bodies are nearly confined to the southernmost edge of the Sawyers Bar map area.

Mapping and geochemical analysis in the Sawyers Bar area led to division of the central North Fork belt into three interstratified, weakly metamorphosed units (Ernst, 1998): St. Clair Creek argillites and interlayered chert; minor North Fork sensu stricto OIBs and flow breccias; and voluminous Salmon River massive to pillowed IAT basalts, comagmatic hypabyssal dikes, and gabbroic plutons. Detailed $^{40}Ar/^{39}Ar$ geochronologic studies by Hacker et al. (1993, 1995) on relict hornblende from

TABLE 1. BULK-ROCK ANALYSES OF SAMPLES FROM THE HAYFORK SUMMIT AREA, SOUTHERN KLAMATH MOUNTAINS

	Blocks in serpentinite matrix melange (basalt, diorite, and gabbro)							Hayfork Summit volcanic rocks					
Sample	3-20	3-23	3-28B	3-134D	3-134G	3-138	3-143B	3-35	3-41C	3-51C	3-88B	3-101B	3-136
SiO_2	47.86	52.78	37.35	48.14	47.92	47.87	47.76	47.03	42.81	45.61	53.32	45.15	52.79
TiO_2	0.98	1.14	0.81	1.43	0.33	0.94	0.14	2.36	3.57	1.84	1.66	2.35	2.03
Al_2O_3	15.61	15.46	17.18	16.22	16.72	16.88	16.07	13.16	15.21	12.70	18.01	13.96	17.93
FeO*	9.25	8.29	7.05	12.13	6.28	8.28	4.65	9.87	11.52	10.70	6.98	10.72	8.52
MnO	0.16	0.13	0.17	0.13	0.13	0.15	0.09	0.16	0.07	0.15	0.22	0.21	0.24
MgO	7.21	6.70	7.98	5.94	9.14	7.62	13.20	7.57	7.96	12.12	4.26	10.74	3.31
CaO	11.65	7.01	25.29	9.34	13.42	11.78	12.05	11.62	4.85	8.74	3.05	8.28	3.12
Na_2O	2.88	5.27	0.00	3.58	2.63	2.44	1.03	3.11	4.66	2.20	6.61	2.73	6.94
K_2O	0.83	0.55	0.00	0.03	0.14	0.79	0.01	0.35	0.99	0.40	1.24	1.41	1.28
P_2O_5	0.08	0.12	0.08	0.11	0.03	0.09	0.01	0.33	0.56	0.23	1.02	0.35	1.02
LOI	3.28	2.44	3.98	2.86	3.06	2.96	4.84	4.26	7.66	5.00	3.46	3.63	2.65
Total	99.79	99.88	99.89	99.91	99.81	99.81	99.84	99.82	99.84	99.69	99.82	99.72	99.84
P	364.5	503.7	349.3	473.4	116.7	410.6	37.0	1434.0	2426.9	1021.5	4433.4	1520.2	4459.9
K	6923.3	4563.6	0.0	278.7	1160.8	6589.2	71.8	2913.8	8251.1	3322.6	10,272.7	11,714.6	10,646.0
Rb	19.1	9.1	0.2	0.3	1.6	7.1	0.4	5.1	17.9	4.0	10.9	18.3	16.1
Sr	775.5	226.3	10.3	247.2	682.4	542.1	48.2	247.1	74.1	288.3	389.5	240.2	343.9
Y	22.7	24.1	18.4	13.9	13.4	19.8	6.2	23.3	25.2	19.7	52.7	23.4	44.1
Zr	64.8	78.5	64.4	36.8	32.4	69.1	16.8	173.4	282.9	129.6	476.7	167.3	410.2
Nb	1.6	3.8	2.6	0.9	0.5	3.2	0.1	27.3	48.3	16.8	75.4	24.7	93.6
Cs	0.8	1.0	0.4	0.0	0.2	0.4	0.0	1.2	1.3	0.6	0.8	0.5	0.1
Ba	363.9	40.2	7.9	22.3	68.9	422.8	8.3	156.7	75.9	938.4	301.8	237.6	186.9
Hf	1.6	2.0	1.3	0.7	0.5	1.5	0.1	4.1	6.5	3.1	11.1	4.0	9.7
Ta	0.1	0.3	0.2	0.1	0.0	0.2	0.0	1.8	3.2	1.1	4.8	1.6	5.8
Pb	0.2	0.4	0.3	0.2	0.2	0.2	0.1	1.6	2.2	1.0	6.2	42.6	2.7
Th	0.2	0.3	0.2	0.0	0.1	0.3	0.0	2.5	4.1	1.5	7.8	2.3	9.0
U	0.0	0.1	0.1	0.0	0.0	0.1	0.0	0.7	0.8	0.4	2.2	0.6	2.1
Sc	40.3	44.7	31.7	41.7	39.8	34.1	33.6	40.4	29.6	34.1	10.8	36.7	9.5
Ti	5867.4	6821.7	4866.3	8580.0	1976.6	5643.4	851.9	14,139.4	21,378.8	11,022.5	9933.8	14,074.5	12,177.6
V	255.3	277.6	201.1	417.6	139.2	211.5	91.6	258.7	292.1	235.8	12.5	271.7	63.6
Cr	186.3	179.1	271.3	0.0	718.4	316.2	950.0	409.5	218.4	731.6	8.8	811.6	8.0
Ni	76.3	77.5	178.7	17.0	152.0	146.4	318.4	149.4	167.2	322.9	7.1	542.6	5.0
La	2.4	4.3	2.8	0.8	1.0	3.4	0.3	23.4	32.7	14.6	71.2	19.6	83.7
Ce	6.3	10.4	6.7	2.1	2.8	8.2	0.7	49.4	65.7	32.1	158.2	42.6	162.8
Pr	1.0	1.5	1.0	0.4	0.5	1.2	0.1	6.1	7.7	4.1	19.6	5.4	18.4
Nd	5.8	7.9	5.2	2.5	2.9	6.2	0.7	26.4	31.1	18.5	82.0	24.2	73.5
Sm	2.3	2.7	1.7	1.1	1.2	2.2	0.3	6.1	7.0	4.7	17.1	6.1	15.1
Eu	0.9	1.0	0.6	0.8	1.0	0.9	0.2	2.0	1.9	1.6	5.6	1.9	4.8
Gd	3.2	3.6	2.3	1.7	1.7	2.9	0.5	5.8	6.2	4.6	14.0	5.6	12.6
Tb	0.6	0.7	0.4	0.3	0.3	0.5	0.1	0.9	1.0	0.7	2.1	0.8	1.8
Dy	4.2	4.3	2.9	2.4	2.3	3.5	0.7	4.7	5.3	4.0	11.1	4.8	9.8
Ho	0.9	0.9	0.6	0.5	0.5	0.7	0.2	0.9	1.0	0.7	2.0	0.9	1.7
Er	2.5	2.5	1.7	1.4	1.4	2.1	0.4	2.3	2.3	1.8	5.1	2.2	4.3
Tm	0.4	0.4	0.2	0.2	0.2	0.3	0.1	0.3	0.3	0.2	0.7	0.3	0.6
Yb	2.3	2.2	1.5	1.3	1.2	1.9	0.4	1.7	1.8	1.4	4.0	1.7	3.4
Lu	0.37	0.35	0.24	0.21	0.18	0.29	0.06	0.24	0.26	0.20	0.61	0.25	0.51

Note: Samples were analyzed by X-ray fluorescence spectrometry and inductively coupled plasma mass spectrometry at the Washington State University GeoAnalytical Lab, Pullman. Trace elements in ppm. Abbreviation: LOI—loss on ignition.

*Fe as total iron oxide.

North Fork gabbros and diabases from the Sawyers Bar area suggest a period of igneous crystallization for both OIB and IAT units of the North Fork terrane from 200 to 170 Ma. Serpentinized peridotite occurs mostly along the southern extremity of the western terrane-bounding fault. Intimate interlayering of metabasalts and metasedimentary rocks and the orientation of pillow tops in the central North Fork belt suggest in situ, immature mafic island-arc development, as also reflected in the volumetric predominance of IAT bulk-rock geochemistry (Fig. 2B and C) documented by Ernst et al. (1991). Distal turbiditic argillite chiefly underlies the metabasaltic pile in the Sawyers Bar area, and the presence of quartz, plagioclase, rare potassium feldspar, mica schist, and Permian McCloud limestone detrital grains (Wright, 1982; Ernst, 1998) suggests that a source of the micrograywacke debris included Klamath terranes sited farther east. Our in-progress U-Pb studies of clastic zircons suggest that

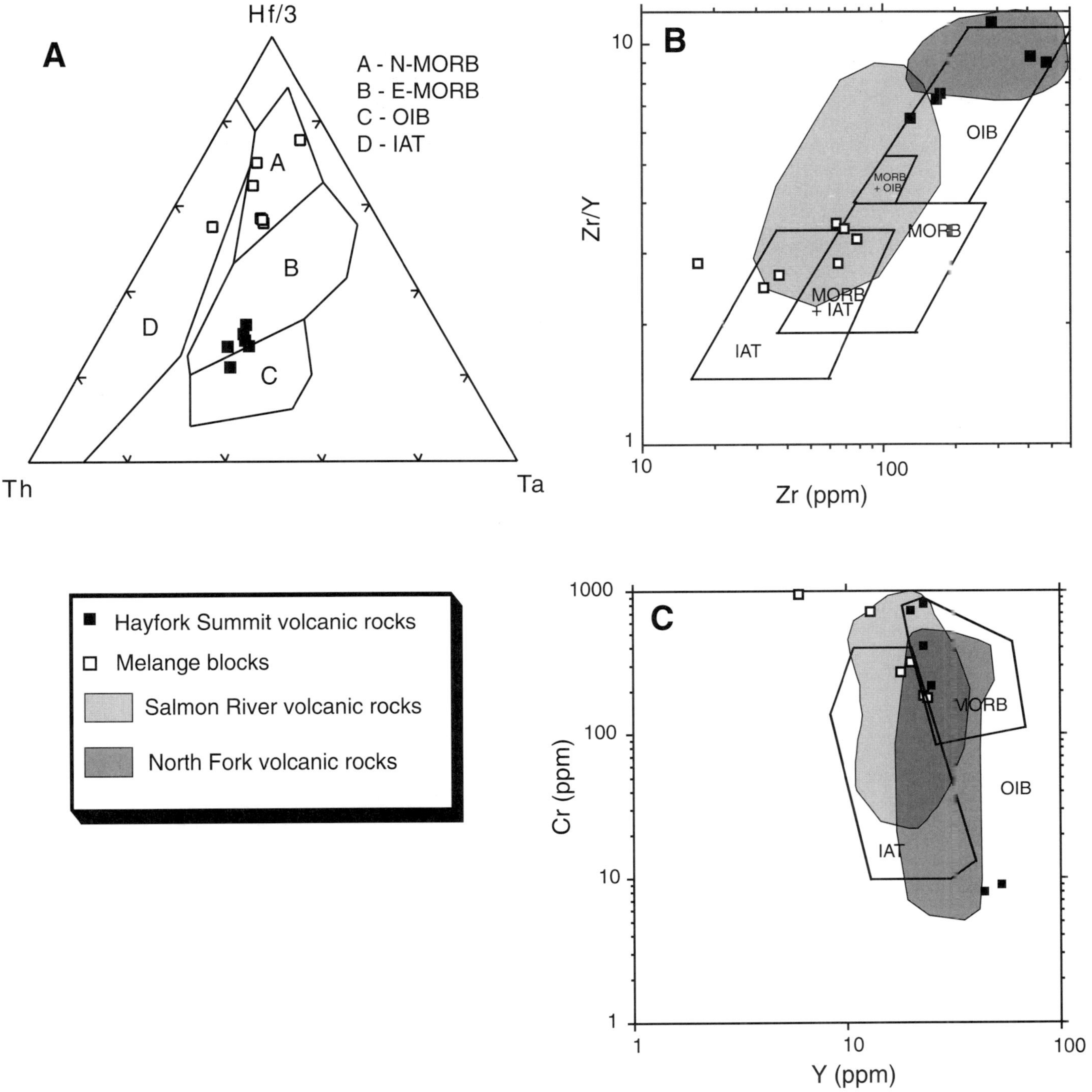

Figure 2. Bulk-rock geochemical discrimination diagrams illustrating analyzed compositional ranges for mafic igneous rocks in the Hayfork Summit and Sawyers Bar areas, North Fork terrane, Klamath Mountains. Classification schemes are after (A) Wood (1980), (B) Pearce and Norry (1979), (C) Pearce (1982). For Sawyers Bar mafic igneous rocks, the term "North Fork volcanic rocks" refers to the North Fork sensu stricto ocean island basalt unit, and the term "Salmon River volcanic rocks" refers to the island-arc tholeiite unit (Ernst, 1999). Abbreviations: E-MORB—enriched mid-ocean ridge basalt; IAT—island-arc tholeiite; N-MORB—normal mid-ocean ridge basalt; OIB—ocean island basalt.

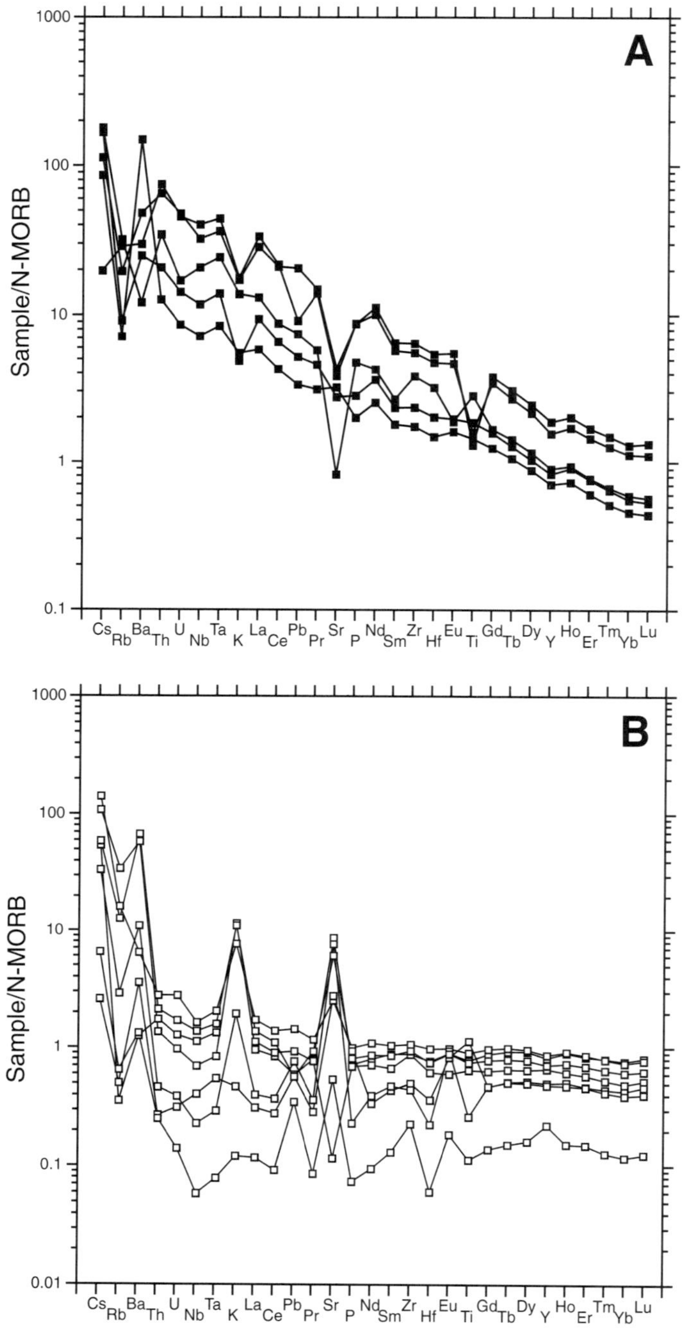

Figure 3. Bulk-rock spider diagrams showing element patterns normalized to normal mid-ocean ridge basalt (N-MORB; Sun and McDonough, 1989) for mafic igneous rocks in the Hayfork Summit area, North Fork terrane, Klamath Mountains. (A) Ocean-island metabasalts associated with metasedimentary strata. (B) Basalt and gabbro blocks from within the serpentinite matrix mélange.

North Fork distal turbidites in the Sawyers Bar area were derived from eastern Klamath terranes and more ancient continental crust.

Conclusion. The above geochemical and stratigraphic relationships led Ernst (1999) to postulate that the Sawyers Bar volcanic and sedimentary section formed mainly during Early to Middle Jurassic time in a suprasubduction-zone environment near or along the continental margin. In contrast, for rocks directly to the south, across the northwest extension of the Browns Meadow fault (Fig. 1), Ando et al. (1983) concluded that the ophiolitic section formed offshore.

Rattlesnake Creek Terrane

General Relationships. Irwin (1972) described the Rattlesnake Creek terrane, the westernmost belt of the western Paleozoic and Triassic, as a highly disrupted ophiolitic lithotectonic entity; later, it was studied in detail by Wright and Wyld (1994). These authors identified a serpentinite matrix mélange basement enclosing blocks of amphibolite, limestone, chert, basalt, greenstone, gabbro, and peridotite. Feebly metamorphosed basalt, amphibolite, and greenstone in this mélange exhibit N-MORB to enriched MORB (E-MORB) and within-plate (OIB) bulk-rock trace element geochemistry. The overlying sequence consists of recrystallized mafic igneous rocks and interlayered sedimentary rocks. The geochemical and petrologic characteristics of lavas of the cover sequence indicate the presence of a lower, primitive IAT suite and an upper, more evolved, volcanic arc suite. Associated metasedimentary rocks include chert, argillite, and coarse epiclastic rocks. The abundance of quartzose sedimentary and metasedimentary grains in the clastic rocks led Wright and Wyld (1994) to conclude that the cover sequence of the Rattlesnake Creek terrane formed proximal to a continental source of detritus. They showed that both the basement mélange and the superjacent section were intruded by 207- to 193-Ma gabbroic and dioritic plutons. Outboard of the Rattlesnake Creek terrane lies the structurally lower, Upper Jurassic Galice Formation, clastic debris of which was derived from the nearby continental margin (Miller and Saleeby, 1995).

Conclusion. The serpentinite mélange basement appears to represent an old fracture zone that formed in the oceanic lithosphere (Wright and Wyld, 1994). In contrast, these authors regarded the superjacent metasedimentary units as derived from the nearby continental margin in Triassic–Early Jurassic time; associated volcanic units have island-arc signatures, and are probably comagmatic with contemporaneous plutons elsewhere in the Klamath Mountains.

JURA–TRIASSIC ARC BELT

Introduction

The Jura-Triassic arc belt of the Sierran Foothills, located west of the Calaveras complex and east of the Great Valley sequence, comprises an ophiolitic Paleozoic serpentinite mélange and serpentinized ultramafic substrate, overlain by arc volcanic and volcaniclastic rocks (Snow and Scherer, 2006). The timing of magmatism appears to have been nearly coeval for most of the superjacent volcanic and volcaniclastic rocks, based on geochronology and on lithologic similarities, and geologic relations where radiometric dates are not available. However, multiple lines of evidence suggest that, although volcanism was roughly contemporaneous in several Sierran Foothills arc complexes, the paleogeographic origins and tectonic histories were different, judging by our new data from the Peñon Blanco arc combined with recent work on the Slate Creek complex (Day and Bickford, 2004). Below, we summarize evidence from these two petrotectonic assemblages to show that they are probably of contrasting origins, and that continental growth occurred by dissimilar processes—arc collision and near-margin volcanism, respectively.

Peñon Blanco Arc

General Relationships. The Peñon Blanco arc comprises a late Paleozoic mafic-ultramafic basement, the 300-Ma Tuolumne ophiolite mélange of Saleeby (1982), the overlying Upper Triassic–lowermost Jurassic mafic volcanic Jasper Point and Peñon Blanco formations, and the Don Pedro mafic intrusive complex. Previous workers interpreted this package of rocks as either a polygenetic ophiolite complex built in a transtensional setting at the continental margin (Saleeby, 1982) or an island arc of oceanic character (Bogen, 1984). Our studies in this area focused on illuminating the relationship of the Peñon Blanco arc to adjacent terranes; determining the nature of the contact with the tectonically higher, Upper Jurassic Mariposa Formation; and characterizing the mineralogy, metamorphism, and bulk-rock geochemistry of the mafic metavolcanic units.

Geologic Mapping and Petrology. The Jasper Point Formation comprises up to 900 m of aphyric to plagioclase-phyric, massive and pillowed metabasalts, capped by ~50 m of varicolored radiolarian metachert. Jasper Point mafic volcanic rocks are metamorphosed to lower actinolite-pumpellyite facies. Original igneous textures are predominantly overprinted by later recrystallization. Metamorphic assemblages are variable, but consist principally of chlorite, quartz, epidote, and actinolite, plus minor pumpellyite, albite, and/or white mica. Chlorite, quartz, and calcite are common in vesicles and veins. The contact between metachert and metabasalt of the Jasper Point Formation is sharp and was interpreted as conformable by Bogen (1984). Jasper Point metacherts have yielded latest Triassic to earliest Jurassic fossil ages (Bogen, 1985). Our mapping in the area indicates that weakly recrystallized uppermost metacherts and lowermost metavolcanic rocks of the Peñon Blanco Formation are intercalated; hence, there was no significant time break between chert deposition and the initiation of Peñon Blanco volcanism.

The Peñon Blanco Formation comprises ~700 m of basaltic metatuff, up to 3.5 km of massive to bedded augite and plagioclase-porphyritic metavolcanic rocks, 1–3.5 km of tuff breccia containing xenocrysts of augite and plagioclase, and hornblende-bearing metavolcanic and metavolcaniclastic rocks (Saleeby, 1982; Bogen, 1985; Sharp, 1988). Previous workers (e.g., Bogen, 1984) suggested that rhyolite exposed on the western limb of the Cotton Creek anticline could be included in the Peñon Blanco Formation. However, our new mapping shows that a sliver of Mariposa Formation lies between the Peñon Blanco metabasalts and the rhyolite, suggesting that the latter is more likely a part of the younger Gopher Ridge volcanic pile.

Recrystallized Peñon Blanco flow breccias are dominantly porphyritic-glomeroporphyritic and are characterized by primary igneous assemblages of (1) plagioclase, augite, and hornblende; (2) plagioclase and augite; and (3) plagioclase only; all phenocrysts are set in a fine-grained, crystalline to devitrified glassy groundmass. Plagioclase laths, which define a flow banding in many samples, are typically euhedral-subhedral, unzoned, and are largely altered to albite ($An_{\sim 09}$) Unzoned augite ($Wo_{45-50}En_{33-39}Fs_{10-22}$) grains are euhedral to subhedral, up to 6 mm in diameter, and typically occur as glomeroporphyritic clots. Breccia clasts in these flows are mineralogically similar to one another and to the matrix phase assemblage. Rare volcaniclastic fragments have mineral assemblages similar to those described below. Metamorphic grade is generally lower or not as well expressed in the Peñon Blanco flow breccias as in rocks of the Jasper Point Formation; it is typified by the subgreenschist-facies assemblage chlorite and epidote, with accessory albite, actinolite, pumpellyite, and/or quartz.

Metavolcaniclastic rocks of the Peñon Blanco Formation typically contain xenocrystic plagioclase and augite as well as fragments of plagioclase porphyry, augite porphyry, and volcaniclastic rip-up fragments. Zoning is characteristic in large igneous clinopyroxene crystals ($Wo_{47-50}En_{34-37}Fs_{13-19}$) and, in xenocrysts, is truncated abruptly at grain boundaries. Most of the plagioclase is altered to albite. Original clastic sedimentary units were locally derived and consisted exclusively of mafic arc detritus. Grain fragments are angular, indicating that transport distances were short. The metamorphic overprint assemblage consists of chlorite and epidote, plus minor quartz, albite, actinolite, and/or pumpellyite.

The Don Pedro intrusive complex consists predominantly of clinopyroxene and hornblende-bearing diorites, with subordinate wehrlite, clinopyroxenite, and gabbro. Saleeby (1982) obtained U-Pb ages of ca. 200 Ma from three Don Pedro diorite samples. New field mapping shows that Don Pedro intrusive units cut the Jasper Point metavolcanic rocks and lowermost Peñon Blanco Formation, but in turn are transected by diabase dikes that presumably fed the uppermost Peñon Blanco basalt flows. Therefore, we conclude that emplacement of the Don Pedro intrusive complex was coeval with the upper portion of the Peñon Blanco Formation. Don Pedro diorite is characterized by primary igneous assemblages of (1) plagioclase and clinopyroxene; (2) plagioclase, clinopyroxene, and hornblende; (3) plagioclase and hornblende; and (4) plagioclase plus quartz. Plagioclase is largely transformed to albite (An_{00-02}), and many of the hornblende and pyroxene crystals are strongly altered. Metamorphism is typically subgreenschist facies and consists of the assemblage chlorite and actinolite, with minor albite, quartz, ilmenite, saussurite, and/or serpentine-group minerals.

TABLE 2. BULK-ROCK ANALYSES OF SAMPLES FROM THE PENON BLANCO AREA, SIERRAN FOOTHILLS

	Peñon Blanco volcanic rocks										
Sample	PB-74B	PB-74A	PB-73A	PB-72B	PB-72A	PB-69B	PB-60C	PB-57D	PB-56A	PB-52A	PB-51E
SiO_2	52.36	52.40	50.33	49.52	55.52	56.11	60.15	53.69	48.52	52.28	54.31
TiO_2	0.36	0.36	0.38	0.36	0.31	0.83	0.57	0.39	0.49	0.36	0.40
Al_2O_3	15.80	15.61	14.40	15.42	14.34	19.20	15.04	15.15	18.18	14.52	15.31
FeO*	8.08	8.19	9.66	8.84	7.31	4.47	6.89	8.00	9.44	8.57	7.96
MnO	0.16	0.16	0.18	0.17	0.14	0.06	0.16	0.15	0.16	0.16	0.15
MgO	7.61	8.32	9.43	8.34	6.29	2.79	4.21	6.44	7.51	9.41	7.32
CaO	7.84	6.97	8.12	10.00	9.19	6.22	6.15	10.00	7.86	8.62	7.94
Na_2O	4.18	4.22	2.27	2.77	3.46	5.82	4.16	3.16	3.51	2.74	4.21
K_2O	0.10	0.23	0.76	0.32	0.03	1.20	0.22	0.14	0.50	0.24	0.02
P_2O_5	0.06	0.06	0.06	0.09	0.08	0.84	0.15	0.06	0.08	0.05	0.08
LOI	3.35	3.35	4.28	4.09	3.24	2.35	2.21	2.75	3.64	2.93	2.24
Total	99.88	99.88	99.88	99.90	99.91	99.89	99.90	99.91	99.89	99.88	99.92
P	240	259	253	372	349	3647	642	246	363	207	328
K	802	1946	6347	2655	257	9938	1867	1127	4174	1977	129
Rb	1.0	2.8	7.6	4.1	0.4	8.9	3.3	2.5	6.3	3.6	0.3
Sr	161.8	134.1	111.6	158.3	125.9	299.1	272.0	107.9	150.1	145.0	52.2
Y	12	13	14	15	14	50	18	13	16	14	16
Zr	33	31	33	35	31	169	74	32	39	32	34
Nb	0.39	0.38	0.36	0.78	0.61	6.18	2.31	0.46	0.65	0.29	0.81
Cs	0.20	0.37	0.40	0.42	0.14	0.44	0.19	0.45	0.60	0.43	0.12
Ba	38	61	136	62	29	157	62	35	78	37	11
Hf	0.65	0.61	0.67	0.68	0.54	4.28	1.74	0.62	0.84	0.57	0.70
Ta	0.02	0.03	0.02	0.05	0.04	0.38	0.16	0.03	0.04	0.02	0.05
Pb	2.85	1.65	1.44	1.72	1.88	1.29	1.64	2.61	1.70	1.72	1.39
Th	0.11	0.11	0.10	0.16	0.15	2.80	0.84	0.11	0.15	0.09	0.14
U	0.08	0.09	0.09	0.12	0.11	1.27	0.56	0.08	0.14	0.07	0.13
Sc	40	43	55	52	46	18	28	46	50	48	48
Ti	2133	2131	2302	2131	1836	5001	3429	2312	2931	2149	2393
V	254	252	296	306	289	152	233	291	371	266	260
Cr	368	421	260	140	122	4	31	124	86	346	207
Ni	103	105	61	29	28	11	17	35	31	104	48
La	1.25	1.47	1.20	2.13	2.07	29.78	7.54	1.30	1.50	1.17	1.75
Ce	2.85	3.23	2.57	4.21	3.90	56.32	15.87	2.84	3.55	2.68	3.97
Pr	0.48	0.53	0.44	0.64	0.58	7.40	2.20	0.46	0.61	0.47	0.63
Nd	2.62	2.98	2.46	3.35	3.08	33.30	10.29	2.56	3.48	2.77	3.57
Sm	1.00	1.11	1.06	1.23	1.12	8.18	2.80	1.08	1.40	1.11	1.31
Eu	0.42	0.46	0.39	0.45	0.39	2.13	0.89	0.41	0.47	0.47	0.48
Gd	1.46	1.50	1.55	1.73	1.54	8.16	2.79	1.48	1.88	1.62	1.79
Tb	0.28	0.30	0.31	0.33	0.31	1.32	0.48	0.29	0.37	0.30	0.34
Dy	1.96	2.03	2.14	2.30	2.08	8.02	2.95	2.02	2.52	1.98	2.33
Ho	0.42	0.44	0.47	0.51	0.47	1.72	0.64	0.45	0.56	0.44	0.52
Er	1.28	1.26	1.35	1.49	1.34	4.89	1.78	1.29	1.62	1.27	1.51
Tm	0.19	0.20	0.21	0.22	0.20	0.74	0.26	0.20	0.25	0.19	0.23
Yb	1.24	1.26	1.36	1.45	1.31	4.62	1.73	1.29	1.64	1.23	1.43
Lu	0.21	0.20	0.23	0.24	0.22	0.76	0.28	0.21	0.27	0.20	0.24

Notes: Samples were analyzed by X-ray fluorescence spectrometry (XRF) and inductively coupled plasma mass spectrometry (ICP-MS) at the Washington State University GeoAnalytical Lab, Pullman. Trace elements in ppm. Abbreviation: LOI—loss on ignition.

Bulk-Rock Geochemistry. Twenty-four mafic volcanic samples from the Jasper Point and Peñon Blanco formations were analyzed by XRF and ICP-MS methods. These constitute the first large set of bulk-rock geochemical data for any metamorphic complex in the central Sierran Foothills. Analytical results are presented in Table 2. The new bulk-rock data, summarized below, demonstrate that two geochemically distinct mafic volcanic suites are present in the Peñon Blanco arc: oceanic tholeiites in the Jasper Point Formation, and IATs in the Peñon Blanco Formation.

Jasper Point volcanic rocks are characterized by low SiO_2 (49–52 wt%), and high TiO_2 (>1.0 wt%) and FeO* (total iron as FeO ≥10 wt%). The Ta-Hf-Th and Cr-Y plots suggest an N-MORB–like affinity (Fig. 4A and C). The N-MORB–

Peñon Blanco volcanic rocks (cont.)							Jasper Point volcanic rocks						
PB-06B	PB-71B	PB-68B	PB-57A	PB-47B	PB-34D	PB-09A	PB-76A	PB-67E	PB-67A	PB-46A	PB-45C	PB-40C
51.42	51.74	55.04	51.52	55.92	56.17	52.31	49.51	50.31	50.66	48.60	52.77	50.14
0.40	0.54	0.42	0.44	0.33	0.32	0.34	1.23	1.07	1.03	0.89	1.51	1.01
15.29	16.11	16.27	15.12	15.51	16.53	16.50	13.38	14.19	13.86	15.12	13.19	13.41
9.35	8.72	7.84	9.56	7.84	7.46	8.73	14.18	10.80	10.40	9.78	12.82	12.11
0.19	0.16	0.15	0.17	0.15	0.16	0.18	0.20	0.19	0.19	0.19	0.28	0.20
8.25	6.99	5.79	7.78	5.85	5.41	7.03	7.09	7.09	7.35	8.99	6.71	8.10
9.53	8.90	8.50	8.91	7.79	8.31	9.32	7.94	9.71	10.11	11.92	8.45	10.00
2.16	4.20	2.40	2.84	2.95	3.04	2.56	3.38	2.91	2.67	1.79	2.73	2.52
0.43	0.19	1.09	0.22	0.84	0.30	0.31	0.24	0.90	1.04	0.66	0.18	0.06
0.05	0.06	0.12	0.06	0.09	0.07	0.07	0.12	0.09	0.07	0.07	0.11	0.07
2.82	2.27	2.24	3.29	2.61	2.13	2.53	2.62	2.62	2.53	1.89	1.14	2.30
99.88	99.89	99.86	99.91	99.90	99.90	99.88	99.90	99.90	99.90	99.88	99.89	99.91
231	264	526	241.5	383.6	325.4	303.2	526.5	401.6	324.5	302.1	495.4	306.9
3602	1555	9073	1858.9	6989.3	2455.3	2599.9	1964.6	7429.4	8593.7	5479.3	534.5	531.3
6.2	2.2	13.5	2.8	13.3	2.9	3.3	5.3	15.6	16.3	9.9	2.0	0.6
251.8	363.0	355.2	103.6	159.2	247.8	251.4	39.7	93.1	70.3	65.2	78.2	52.7
12	14	13	12.7	11.7	11.8	11.7	35.7	28.4	28.2	23.4	40.8	28.6
33	37	51	37.2	41.5	39.0	38.2	71.4	61.5	59.9	53.8	85.5	61.1
0.63	0.99	1.87	0.8	0.8	0.8	0.7	0.7	0.5	0.4	0.7	1.2	0.6
0.22	0.14	0.40	0.6	0.9	0.2	0.2	0.3	0.4	0.6	0.4	0.2	0.0
77	52	179	37.7	143.7	70.4	74.3	26.9	70.8	106.7	17.7	23.5	4.4
0.61	0.73	1.15	0.7	0.8	0.8	0.8	1.8	1.5	1.5	1.2	2.3	1.4
0.04	0.06	0.12	0.0	0.1	0.0	0.0	0.1	0.0	0.0	0.1	0.1	0.0
1.89	2.18	2.45	1.6	1.9	2.1	2.3	0.3	0.3	0.2	0.4	0.5	0.3
0.12	0.26	0.45	0.1	0.3	0.2	0.2	0.1	0.1	0.0	0.1	0.1	0.1
0.07	0.18	0.20	0.1	0.2	0.1	0.1	0.2	0.1	0.0	0.1	0.0	0.0
52	49	40	47.7	42.8	36.9	48.1	51.6	55.9	53.1	50.5	56.6	54.9
2391	3248	2543	2659.2	2007.1	1940.2	2056.7	7392.5	6444.4	6179.5	5309.3	9024.3	6067.2
307	323	295	303.7	270.7	243.6	277.9	398.3	336.5	326.6	289.9	455.9	349.1
209	71	109	123.3	89.7	97.9	139.9	105.9	96.6	105.7	386.9	56.9	142.2
53	25	37	34.1	31.4	30.2	39.1	60.8	59.4	56.3	96.2	57.0	71.0
1.36	3.05	5.15	1.5	2.9	2.4	2.3	1.8	1.2	1.1	1.2	1.9	1.2
3.04	6.26	10.47	3.2	5.7	4.7	4.4	5.1	4.0	3.7	3.7	6.3	3.8
0.49	0.95	1.45	0.5	0.8	0.7	0.6	1.0	0.8	0.7	0.7	1.2	0.7
2.82	4.67	6.84	3.0	3.9	3.4	3.2	6.3	5.0	4.6	4.3	7.4	4.5
1.09	1.56	1.91	1.1	1.2	1.1	1.1	2.9	2.4	2.2	1.9	3.3	2.2
0.48	0.60	0.70	0.5	0.5	0.4	0.4	1.0	1.0	0.9	0.8	1.2	0.8
1.52	1.93	2.06	1.6	1.5	1.5	1.5	4.3	3.6	3.5	2.9	5.0	3.3
0.28	0.35	0.35	0.3	0.3	0.3	0.3	0.9	0.7	0.7	0.6	1.0	0.7
1.96	2.30	2.23	2.1	1.8	1.9	1.9	6.3	5.1	4.9	4.1	7.1	4.7
0.43	0.49	0.47	0.5	0.4	0.4	0.4	1.4	1.1	1.1	0.9	1.5	1.0
1.20	1.35	1.33	1.3	1.1	1.2	1.2	3.9	3.1	3.0	2.6	4.5	3.0
0.18	0.20	0.20	0.2	0.2	0.2	0.2	0.6	0.5	0.4	0.4	0.7	0.4
1.19	1.28	1.26	1.3	1.1	1.2	1.2	3.6	2.9	2.8	2.4	4.1	2.8
0.19	0.20	0.21	0.2	0.2	0.2	0.2	0.6	0.5	0.5	0.4	0.7	0.4

normalized trace-element spider diagrams show only a slight upward trend in the large ion lithophile elements (LILE) relative to N-MORB, and flat to slightly positive slopes for high field strength elements (HFSE) and HREE, which also support this interpretation (Fig. 5B). High values for LILE in these patterns probably reflect element mobility during low-grade metamorphism. As evident from Ta-Hf-Th and Cr-Y plots (Fig. 4A and C), Peñon Blanco volcanic and volcaniclastic rocks are characterized by high Th and low Y contents, indicative of an oceanic island-arc origin. This interpretation is supported by spider diagram patterns for the Peñon Blanco Formation. The flow breccias and volcaniclastic rocks are compositionally indistinguishable from one another. Both types are enriched in LILE and depleted in HFSE + HREE relative to N-MORB (Fig. 5A).

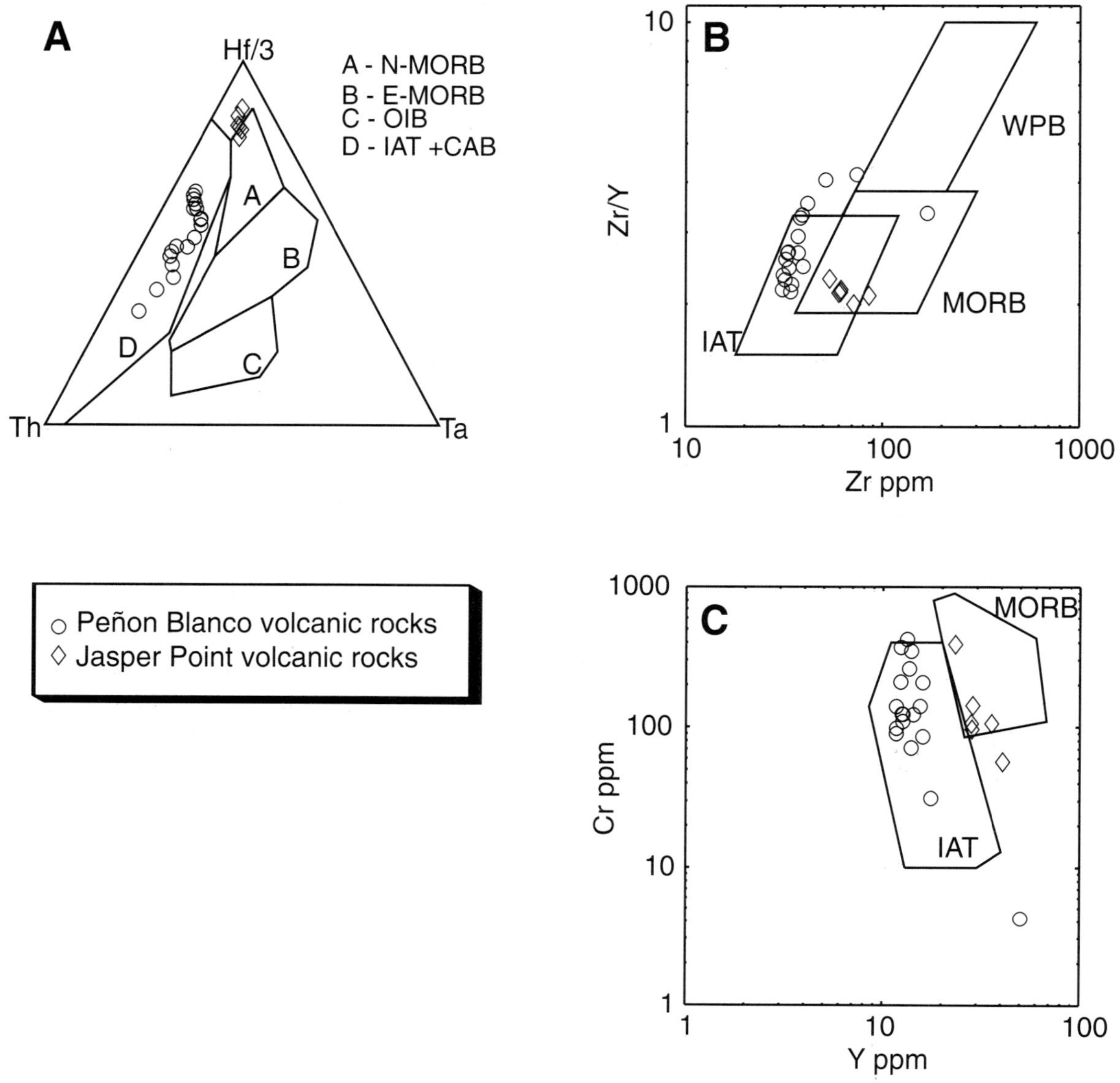

Figure 4. Bulk-rock geochemical discrimination diagrams illustrating analyzed compositional ranges for mafic igneous rocks in the Peñon Blanco area, Jura–Triassic arc, central Sierran Foothills. Classification schemes are after (A) Wood (1980), (B) Pearce and Norry (1979), and (C) Pearce (1982). Abbreviations: CAB—calc-alkaline basalt; E-MORB—enriched mid-ocean ridge basalt; IAT—island-arc tholeiite; N-MORB—normal mid-ocean ridge basalt; OIB—ocean-island basalt; WPB—within-plate basalt.

Paleogeography. Our new mapping, combined with previous geochronology, suggests that the Peñon Blanco offshore arc did not arrive at the margin until the Late Jurassic. This conclusion follows from the observation that the penetrative foliation of the Upper Jurassic Mariposa Formation is not present in the Peñon Blanco and Japer Point formations. Although this absence could be attributed to contrasts in lithologic competence, we do not regard that to be the case, because numerous intercalated mafic volcanic layers in the Mariposa Formation are strongly deformed. Evidently, the Mariposa section was penetratively deformed prior to its juxtaposition against rocks of the Peñon Blanco arc. We suspect that this foliation developed at shallow levels of the subduction zone while the Mariposa Formation was being offloaded along the accretionary margin and before arrival of the oceanic Peñon Blanco arc. Schweickert et al. (1984) concluded that this foliation must have developed at ca. 155–151 Ma, based on fossil ages and regional crosscutting relationships (151-Ma Guadalupe igneous complex). Apparently the Peñon Blanco arc collided with the margin shortly thereafter—much later than other terranes in the Jura–Triassic arc belt. Our ongoing U-Pb investigations of clastic zircons show that the provenance of the Mariposa Formation involved detritus

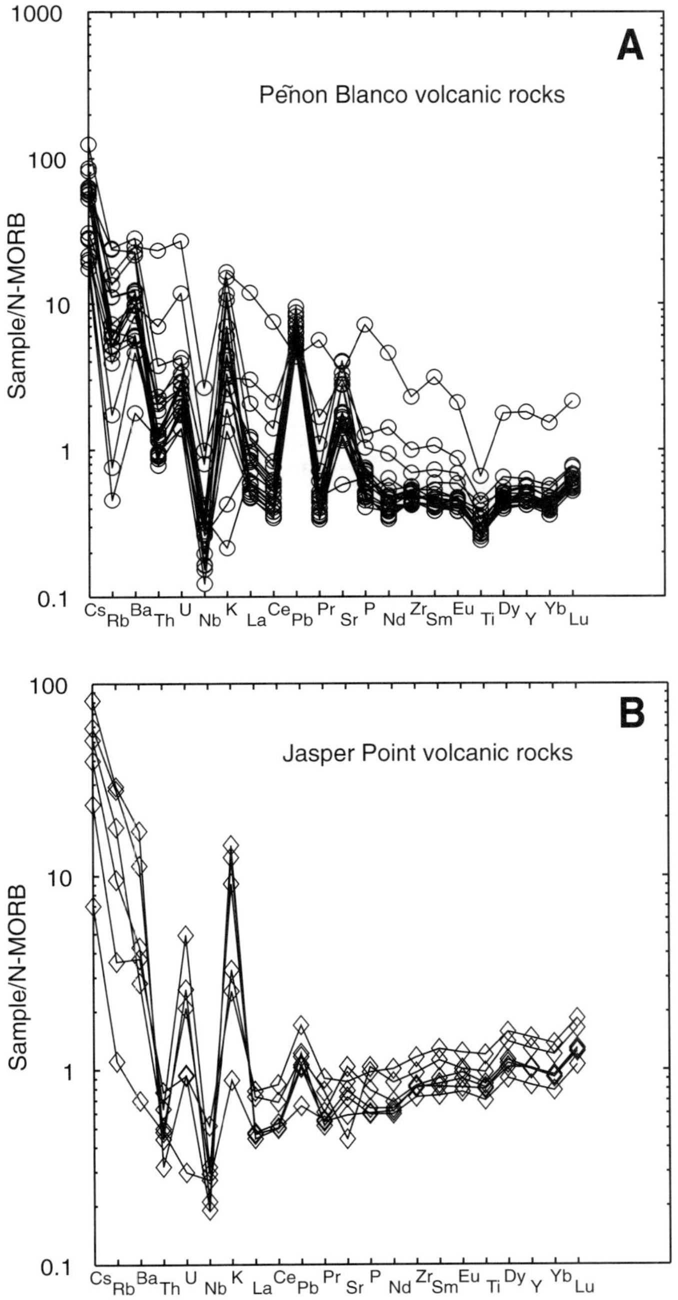

Figure 5. Bulk-rock spider diagrams showing element patterns normalized to normal mid-ocean ridge basalt (N-MORB; Sun and McDonough, 1989) for mafic igneous rocks in the Peñon Blanco area, Jura–Triassic arc, central Sierran Foothills.

ranging in age from Late Archean to Late Jurassic, reflecting formation at the continental margin.

Conclusion. The Peñon Blanco arc formed offshore as Lower Jurassic island-arc volcanic rocks intercalated with flow breccias and arc-derived clastic units above a late Paleozoic serpentinite-oceanic tholeiite basement; the mafic arc collided with the Sierran margin in Late Jurassic time.

Slate Creek Complex

General Relationships. The Slate Creek complex is an ophiolitic arc remnant comprising serpentinized ultramafic rocks, gabbroic-tonalitic intrusions, and volcanogenic sedimentary rocks characterized by major pyroclastic and volcaniclastic components, built on a basement of ophiolitic mélange (i.e., the Fiddle Creek complex of Day et al., 1985; Edelman et al., 1989; Day and Bickford, 2004). The superjacent section is polygenetic in origin, with volcanic-plutonic activity dated at 200 and 170 Ma (Fagan et al., 2001). Volcanic rocks in this complex underwent subgreenschist-greenschist–facies recrystallization, producing the metamorphic assemblage quartz, albite, chlorite, epidote, titanite, plus minor white mica, amphibole, and pumpellyite (Fagan and Day, 1997). Recent geochronology studies by Day and Bickford (2004) have firmly established the age of plutonic rocks in the Slate Creek complex at 207–202 Ma. Their work also showed Precambrian inheritance in zircons from Slate Creek tonalities, indicating that the plutons intruded through North American crust, and therefore that the terrane must have formed relatively in situ.

This complex appears to be comparable to others in the Sierran Foothills (e.g., the Lake Combie complex, the Owl Gulch volcanic rocks), and for our study, these were considered to be of similar paleogeographic origins. Also of analogous setting and origin is the ca. 187-Ma Sullivan Creek terrane, which is situated ~250 km to the southeast of the Slate Creek complex (Herzig and Sharp, 1992), and lies east of the Peñon Blanco arc. Herzig and Sharp (1992) mapped several sedimentary units interbedded with the arc volcanic rocks, which suggests formation proximal to, or on the margin, similar to the Slate Creek complex and related rocks.

Conclusion. These relationships suggest that at least two Early Jurassic arcs are preserved in the Sierran Foothills: (1) the near-shore Slate Creek complex (plus Lake Combie, Owl Gulch, and Sullivan Creek units) and (2) the offshore Peñon Blanco oceanic arc.

SUMMARY AND CONCLUSIONS

Key features for each of the terranes discussed above are listed in Table 3. The North Fork terrane in the southern Klamath Mountains is sufficiently dissimilar in structure and in petrochemistry from the North Fork terrane in the Sawyers Bar area to be considered as having a somewhat different origin. The most likely setting for construction of the Hayfork Summit–Salmon River section was as a Permian seamount, as originally suggested by Ando et al. (1983). Northward, this chiefly OIB-containing complex is juxtaposed against an active-margin, dominantly IAT arc produced during 200–170 Ma. We speculate that the northwest-striking Browns Meadow fault, located where the Sawyers Bar and Salmon River map areas overlap (Fig. 1), may represent a cryptic suture between the suprasubduction-zone North Fork terrane on the north and the offshore ophiolitic

TABLE 3. SUMMARY OF DATA FOR SELECTED ISLAND-ARC TERRANES

Offshore North Fork terrane: Hayfork Summit + Salmon River areas
1. MORB blocks spatially associated with serpentinite-matrix mélange; 310- to 265-Ma igneous crystallization age of basement rocks (Salmon River area).
2. No continental detritus in cherty argillites interbedded with mafic volcanic rocks (OIB).
3. OIB volcanic rocks lack a geochemical arc signature.
4. Permian fusulinids in limestone blocks in volcaniclastic rocks, Permian–Jurassic fossil ages in interlayered metacherts (Salmon River area).

Offshore(?) eastern Hayfork terrane: Hayfork Summit area
5. Detrital zircons in sedimentary strata indicate a Proterozoic continental source.

Near-shore North Fork terrane: Sawyers Bar area
1. Continent-derived distal turbidites (micrograywackes) interbedded with but chiefly underlie mafic volcanic rocks (IAT and volumetrically minor OIB). Detrital zircons suggest derivation from eastern Klamath terranes and old continental crust.
2. Clear geochemical arc signature of voluminous IAT units erupted contemporaneously with much less abundant OIB lavas.
3. Crystallization ages of 200–170 Ma for both series of mafic arc volcanic rocks.

Near-shore Rattlesnake Creek terrane
1. Basement is serpentinite-matrix mélange, including tectonic blocks of MORB.
2. Superjacent section consists of quartzose terrigenous sedimentary rocks interbedded with mafic volcanic arc rocks (IAT).
3. Lower volcanic arc section intruded by 207- to 193-Ma plutons.

Near-shore Galice Formation
4. Synorogenic turbidites of the western Klamath terrane derived from nearby arc margin.

Offshore Peñon Blanco arc
1. Ophiolitic basement of age 300 Ma, associated with mafic tectonic blocks of Triassic age.
2. Superjacent MORB lavas overlain by mafic arc volcanic rocks.
3. Only traces of ancient terrigenous debris; sedimentary rocks are mainly chert interbedded with base of the mafic volcanic section.
4. Latest Triassic–earliest Jurassic fossil ages in Jasper Point metacherts; 200-Ma arc volcanism and plutonism.

Near-shore Mariposa Formation
1. Detrital zircon ages of the Middle–Late Jurassic arc sequence range from Late Archean to Late Jurassic, indicating continental provenance.

Near-shore Slate Creek complex
1. Ophiolitic mélange basement.
2. Mafic plutons contain Precambrian inherited zircons, indicating intrusion in a continental margin setting.
3. Mafic volcanic and plutonic arc activity at 207–200 and 170 Ma.

Abbreviations: IAT—island-arc tholeiite; MORB—Mid-ocean ridge basalt; OIB—ocean island basalt.

terrane of the southern Klamath Mountains. If differential slip occurred, it must have been completed before intrusion of the small East Fork pluton at ca. 164 Ma (Irwin and Wooden, 1999). Structural relationships are currently being tested by ongoing ion microprobe U-Pb analyses of igneous and detrital zircons in North Fork and eastern Hayfork rocks of the central and southern Klamath Mountains.

Clearly, a period of arc volcanism occurred ca. 200 Ma in a variety of Klamath and Sierran Foothill paleogeographic settings. The Rattlesnake Creek arc was receiving coarse-grained, terrigenous detritus, indicating that it was near the edge of the continent at this time (Wright and Wyld, 1994). Island-arc volcanic rocks of the North Fork terrane were deposited on, and are interbedded with, Sawyers Bar micrograywackes derived from the eastern Klamath terrane; thus, this section was tied to the Late Triassic–Early Jurassic continental margin (Ernst, 1999). The 207- to 170-Ma Slate Creek complex and correlative terranes in the Sierran Foothills lack diagnostic sedimentary ties to the craton, but mafic plutons contain Precambrian inherited zircons, indicating eruption through North American crust (Day and Bickford, 2004). In contrast to these near-shore terranes, new data presented here indicate that the Peñon Blanco arc developed offshore in an oceanic setting and lacked ties to the North American craton until its collision and suturing in Late Jurassic time. In addition, Permian and earliest Mesozoic(?) oceanic rocks of the North Fork terrane in the southern Klamath Mountains probably originated in an offshore setting, then amalgamated with the Sawyers Bar terrane during Middle Jurassic time.

ACKNOWLEDGMENTS

Field mapping and the acquisition of geochemical data were supported by Stanford University and the U.S. Geological Survey's EDMAP program (grants 04HQAG0105 and 05HQAG0075). Bob Coleman, Brad Hacker, and Jason Saleeby provided constructive reviews of a first-draft manuscript. Howard Day, Art Snoke, and Cal Barnes provided further helpful feedback during the GSA reviewing process. We thank these scientists and institutions for support.

REFERENCES CITED

Ando, C.J., 1979, Structural and petrologic analysis of the North Fork terrane central Klamath Mountains, California [Ph. D. thesis]: Los Angeles, University of Southern California, 197 p.

Ando, C.J., Irwin, W.P., Jones, D.L., and Saleeby, J.B., 1983, The ophiolitic North Fork terrane in the Salmon River region, central Klamath Mountains, California: Geological Society of America Bulletin, v. 94, p. 236–252, doi: 10.1130/0016-7606(1983)94<236:TONFTI>2.0.CO;2.

Barnes, C.G., Peterson, S.W., Kistler, R.W., Prestvik, T., and Sundvoll, B., 1992, Tectonic implications of isotopic variation among Jurassic and Early Cretaceous plutons, Klamath Mountains: Geological Society of America Bulletin, v. 104, p. 117–126, doi: 10.1130/0016-7606(1992)104<0117:TIOIVA>2.3.CO;2.

Barnes, C.G., Peterson, S.W., Kistler, R.W., Murray, R., and Kays, M.A., 1996, Source and tectonic implications of tonalite-trondhjemite magmatism in the Klamath Mountains: Contributions to Mineralogy and Petrology, v. 123, p. 40–60, doi: 10.1007/s004100050142.

Barnes, C.G., Mars, E.V., Swapp, S., and Frost, C.D., 2006, this volume (Chapter 10), Petrology and geochemistry of the Middle Jurassic Ironside Mountain batholith: Evolution of potassic magmas in a primitive arc setting, *in* Snoke, A.W., and Barnes, C.G., eds., Geological studies in the Klamath Mountains province, California and Oregon: A volume in honor of William P. Irwin: Boulder, Colorado, Geological Society of America Special Paper 410, doi: 10.1130/2006.2410(10).

Bogen, N.L., 1984, Stratigraphy and sedimentary petrology of the Upper Jurassic Mariposa Formation, western Sierra Nevada, California, *in* Crouch, J.K., and Bachman, S.B., eds., Tectonics and sedimentation along the California margin: Los Angeles, Pacific Section, Society of Economic Paleontologists and Mineralogists, v. 38, p. 119–134.

Bogen, N.L., 1985, Stratigraphic and sedimentologic evidence of a submarine island-arc volcano in the lower Mesozoic Peñon Blanco and Jasper Point Formations, Mariposa Co., California: Geological Society of America Bulletin, v. 96, p. 1322–1331, doi: 10.1130/0016-7606(1985)96<1322:SASEOA>2.0.CO;2.

Burchfiel, B.C., and Davis, G.A., 1981, Triassic and Jurassic tectonic evolution of the Klamath Mountains–Sierra Nevada geologic terrane, *in* Ernst, W.G., ed., The geotectonic development of California: Englewood Cliffs, New Jersey, Prentice-Hall, p. 50–70.

Day, H.W., and Bickford, M.E., 2004, Tectonic setting of the Jurassic Smartville and Slate Creek complexes, northern Sierra Nevada, California: Geological Society of America Bulletin, v. 116, p. 1515–1528, doi: 10.1130/B25416.1.

Day, H.W., Moores, E.M., and Tuminas, A.C., 1985, Structure and tectonics of the northern Sierra Nevada: Geological Society of America Bulletin, v. 96, p. 436–450, doi: 10.1130/0016-7606(1985)96<436:SATOTN>2.0.CO;2.

Dilek, Y., Thy, P., Moores, E.M., and Grundvig, S., 1990, Late Paleozoic–early Mesozoic oceanic basement of a Jurassic arc terrane in the northwestern Sierra Nevada, California, *in* Harwood, D.S., and Miller, M.M., eds., Paleozoic and early Mesozoic paleogeographic relations; Sierra Nevada, Klamath Mountains, and related terranes: Boulder, Colorado, Geological Society of America Special Paper 255, p. 351–369.

Donato, M.M., Barnes, C.G., and Tomlinson, S.L., 1996, The enigmatic Applegate Group of southwestern Oregon; Age, correlation, and tectonic affinity: Oregon Geology, v. 58, p. 79–91.

Edelman, S.H., and Sharp, W.D., 1989, Terranes, early faults, and pre-Late Jurassic amalgamation of the western Sierra Nevada metamorphic belt, California: Geological Society of America Bulletin, v. 101, p. 1420–1433, doi: 10.1130/0016-7606(1989)101<1420:TEFAPL>2.3.CO;2.

Edelman, S.H., Day, H.W., Moores, E.M., Zigan, S.M., Murphy, T.P., and Hacker, B.R., 1989, Structure across a Mesozoic ocean-continent suture zone in the northern Sierra Nevada, California: Boulder, Colorado, Geological Society of America Special Paper 224, 56 p.

Ernst, W.G., 1983, Phanerozoic continental accretion and the metamorphic evolution of northern and central California: Tectonophysics, v. 100, p. 287–320, doi: 10.1016/0040-1951(83)90192-0.

Ernst, W.G., 1991, Petrological setting and inferred plate tectonic history of the Sawyers Bar terrane, central Klamath Mountains, northern California: Canadian Mineralogist, v. 29, p. 1051–1068.

Ernst, W.G., 1998, Geologic map of the Sawyers Bar area: Geochemical/petrotectonic evolution of an oceanic island arc, central Klamath Mountains, California: Sacramento, California Division of Mines and Geology Map Sheet 47, scale 1:48,000, 1 sheet with 59 p. pamphlet.

Ernst, W.G., 1999, Mesozoic petrotectonic development of the Sawyers Bar suprasubduction-zone arc, central Klamath Mountains, northern California: Geological Society of America Bulletin, v. 111, p. 1217–1232, doi: 10.1130/0016-7606(1999)111<1217:MPDOTS>2.3.CO;2.

Ernst, W.G., Hacker, B.R., Barton, M.D., and Sen, G., 1991, Igneous petrogenesis of magnesian metavolcanic rocks from the central Klamath Mountains, northern California: Geological Society of America Bulletin, v. 103, p. 56–72, doi: 10.1130/0016-7606(1991)103<0056:IPOMMR>2.3.CO;2.

Fagan, T.J., and Day, H.W., 1997, Formation of amphibole after clinopyroxene by dehydration reactions: Implications for pseudomorphic replacement and mass fluxes: Geology, v. 25, p. 395–398, doi: 10.1130/0091-7613(1997)025<0395:FOAACB>2.3.CO;2.

Fagan, T.J., Day, H.W., and Hacker, B.R., 2001, Timing of arc construction and metamorphism in the Slate Creek Complex, northern Sierra Nevada, California: Geological Society of America Bulletin, v. 113, p. 1105–1118, doi: 10.1130/0016-7606(2001)113<1105:TOACAM>2.0.CO;2.

Goodge, J.W., 1989, Polyphase metamorphic evolution of a Late Triassic subduction complex, Klamath Mountains, northern California: American Journal of Science, v. 289, p. 874–943.

Hacker, B.R., and Goodge, J.W., 1990, Comparison of early Mesozoic high-pressure rocks in the Klamath Mountains and Sierra Nevada, *in* Harwood, D.S., and Miller, M.M., eds., Paleozoic and early Mesozoic paleogeographic relations; Sierra Nevada, Klamath Mountains, and related terranes: Boulder, Colorado, Geological Society of America Special Paper 255, p. 277–295.

Hacker, B.R., Ernst, W.G., and Barton, M.D., 1992, Metamorphism, geochemistry, and origin of magnesian volcanic rocks, Klamath Mountains, California: Journal of Metamorphic Geology, v. 10, p. 55–69.

Hacker, B.R., Ernst, W.G., and McWilliams, M.O., 1993, Genesis and evolution of a Permian-Jurassic magmatic arc/accretionary wedge, and reevaluation of terranes in the central Klamath Mountains: Tectonics, v. 12, p. 387–409.

Hacker, B.R., Donato, M.M., Barnes, C.G., McWilliams, M.O., and Ernst, W.G., 1995, Timescales of orogeny: Jurassic construction of the Klamath Mountains: Tectonics, v. 14, p. 667–703.

Herzig, C.T., and Sharp, W.D., 1992, The Sullivan Creek terrane: A composite Jurassic arc assemblage, western Sierra Nevada metamorphic belt, California: Geological Society of America Bulletin, v. 104, p. 1292–1300, doi: 10.1130/0016-7606(1992)104<1292:TSCTAC>2.3.CO;2.

Hotz, P.E., Lanphere, M.A., and Swanson, D.A., 1977, Triassic blueschists from northern California and north-central Oregon: Geology, v. 5, p. 659–663, doi: 10.1130/0091-7613(1977)5<659:TBFNCA>2.0.CO;2.

Irwin, W.P., 1960, Geologic reconnaissance of the northern Coast Ranges and the southern Klamath Mountains, California, with a summary of the mineral resources: Sacramento, California Division of Mines Bulletin 179, 80 p.

Irwin, W.P., 1972, Terranes of the western Paleozoic and Triassic belt in the southern Klamath Mountains, California: Washington, D.C., U.S. Geological Survey Professional Paper 800-C, p. C103–C111.

Irwin, W.P., 1981, Tectonic accretion of the Klamath Mountains, *in* Ernst, W.G., ed., The geotectonic development of California—Rubey volume I: Englewood Cliffs, New Jersey, Prentice-Hall, p. 29–49.

Irwin, W.P., 2003, Correlation of the Klamath Mountains and the Sierra Nevada: Reston, Virginia, U.S. Geological Survey Open-File Report 02–490, 2 sheets.

Irwin, W.P., and Blome, C.D., 2004, Fossil localities of the Rattlesnake Creek, western and eastern Hayfork, and North Fork terranes of the Klamath Mountains: Reston, Virginia, U.S. Geological Survey Open-File Report 2004-1094, 45 p.

Irwin, W.P., and Wooden, J.L., 1999, Plutons and accretionary episodes of the Klamath Mountains, California and Oregon: Reston, Virginia, U.S. Geological Survey Open-File Report 99-374, 1 sheet.

Irwin, W.P., Wardlaw, B.R., and Kaplan, T.A., 1983, Conodonts of the western Paleozoic and Triassic belt, Klamath Mountains, California and Oregon: Journal of Paleontology, v. 57, p. 1030–1039.

Lanphere, M.A., Irwin, W.P., and Hotz, P.E., 1968, Isotopic age of the Nevadan orogeny and older plutonic and metamorphic events in the Klamath Mountains, California: Geological Society of America Bulletin, v. 79, p. 1027–1052.

Mankinen, E.A., Irwin, W.P., and Blome, C.D., 1996, Far-traveled Permian chert of the North Fork terrane, Klamath Mountains, California: Tectonics, v. 15, p. 314–328, doi: 10.1029/95TC03054.

Miller, D.E., and Ernst, W.G., 1998, Deformation and metamorphism of the Marble Mountain and Pony Camp areas, western Triassic and Paleozoic belt, central Klamath Mountains, northwestern California: International Geology Review, v. 39, p. 411–451.

Miller, M.M., 1987, Dispersed remnants of a northeast Pacific fringing arc: Upper Paleozoic terranes of Permian McCloud faunal affinity: Tectonics, v. 6, p. 807–830.

Miller, M.M., and Saleeby, J.B., 1995, U-Pb geochronology of detrital zircon from Upper Jurassic synorogenic turbidites, Galice Formation, and related rocks, western Klamath Mountains: Correlation and Klamath provenance: Journal of Geophysical Research, v. 100, p. 18,045–18,058, doi: 10.1029/95JB00761.

Peacock, S.M., and Norris, P.J., 1989, Metamorphic evolution of the central metamorphic belt, Klamath province, California: An inverted metamorphic gradient beneath the Trinity peridotite: Journal of Metamorphic Geology, v. 7, p. 191–209.

Pearce, J.A., 1982, Trace element characteristics of lavas from destructive plate boundaries, *in* Thorpe, R.S., ed., Andesites: New York, John Wiley & Sons, p. 525–548.

Pearce, J.A., and Norry, M.J., 1979, Petrogenetic implications of Ti, Zr, Y, and Nb variations in volcanic rocks: Contributions to Mineralogy and Petrology, v. 69, p. 33–47, doi: 10.1007/BF00375192.

Saleeby, J.B., 1981, Ocean-floor accretion and volcano-plutonic arc evolution of the Mesozoic Sierra Nevada, California, *in* Ernst, W.G., ed., The geotectonic development of California—Rubey volume 1: Englewood Cliffs, New Jersey, Prentice-Hall, p. 132–181.

Saleeby, J.B., 1982, Polygenetic ophiolite belt of the California Sierra Nevada: Geochronological and tectonostratigraphic development: Journal of Geophysical Research, v. 87, p. 1803–1824.

Schweickert, R.A., 1981, Tectonic evolution of the Sierra Nevada Range, *in* Ernst, W.G., ed., The geotectonic development of California—Rubey volume 1: Englewood Cliffs, New Jersey, Prentice-Hall, p. 87–131.

Schweickert, R.A., Bogen, N.L., Girty, G.H., Hanson, R.E., and Merguerian, C., 1984, Timing and structural expression of the Nevadan orogeny, Sierra Nevada, California: Geological Society of America Bulletin, v. 95, p. 967–979, doi: 10.1130/0016-7606(1984)95<967:TASEOT>2.0.CO;2.

Sharp, W.D., 1988, Pre-Cretaceous crustal evolution of the Sierra Nevada region, *in* Ernst, W.G., ed., Metamorphism and crustal evolution of the western United States—Rubey volume VII: Englewood Cliffs, New Jersey, Prentice-Hall, p. 824–864.

Snow, C.A., and Scherer, H.H., 2006, Terranes of the western Sierra Nevada Foothills metamorphic belt, California: A critical review: International Geology Review, v. 48, p. 46–62.

Sun, S.-s., and McDonough, W.F., 1989, Chemical and isotopic systematics of oceanic basalts: Implications for mantle composition and processes, *in* Saunders, A.D., and Norry, M.J., eds., Magmatism in the ocean basins: London, Geological Society of London Special Publication 42, p. 313–345.

Wood, D.A., 1980, The application of a Th-Hf-Ta diagram to problems of tectonomagmatic classification and to establishing the nature of crustal contamination of basaltic lavas of the British Tertiary volcanic province: Earth and Planetary Science Letters, v. 50, p. 11–30, doi: 10.1016/0012-821X(80)90116-8.

Wright, J.E., 1982, Permo-Triassic accretionary subduction complex, southwestern Klamath Mountains, northern California: Journal of Geophysical Research, v. 87, p. 3805–3818.

Wright, J.E., and Fahan, M.R., 1988, An expanded view of Jurassic orogenesis in the western United States Cordillera: Middle Jurassic (pre-Nevadan) regional metamorphism and thrust faulting within an active arc environment, Klamath Mountains, California: Geological Society of America Bulletin, v. 100, p. 859–876, doi: 10.1130/0016-7606(1988)100<0859:AEVOJO>2.3.CO;2.

Wright, J.E., and Wyld, S.J., 1994, The Rattlesnake Creek terrane, Klamath Mountains, California; An early Mesozoic volcanic arc and its basement of tectonically disrupted oceanic crust: Geological Society of America Bulletin, v. 106, p. 1033–1056, doi: 10.1130/0016-7606(1994)106<1033:TRCTKM>2.3.CO;2.

Manuscript Accepted by the Society 10 January 2006

Printed in the USA

Geological Society of America
Special Paper 410
2006

A reevaluation of the paleotectonic significance of the Paleozoic Central Metamorphic terrane, eastern Klamath Mountains, California: New constraints from trace element geochemistry and $^{40}Ar/^{39}Ar$ thermochronology

Wendy M. Barrow*
Rodney V. Metcalf
Department of Geoscience, University of Nevada Las Vegas, 4505 Maryland Parkway, Las Vegas, Nevada 89154-4010, USA

ABSTRACT

Interpretation of the Central Metamorphic terrane metabasites as the lower-plate of a Devonian intraoceanic subduction zone is a basic tenet of paleotectonic models for the eastern Klamath province. Specifically, such models hold that the (1) east-dipping Central Metamorphic terrane represents a subducted piece of oceanic lithosphere accreted to the overlying Trinity subterrane; (2) Trinity and Redding subterranes represent mantle and magmatic-arc crust of the overriding lithospheric plate; and (3) east-dipping Trinity thrust, separating the Central Metamorphic terrane and Trinity subterrane, represents the paleo-plate boundary. The subduction-zone model predicts that mid-ocean ridge basalt (MORB)-type oceanic basalts formed the protolith of Central Metamorphic terrane metabasites. We present new major (X-ray fluorescence spectrometry) and trace (inductively coupled plasma-mass spectrometry) element data for twenty Central Metamorphic terrane metabasites; these geochemical data demonstrate that (1) seafloor hydrothermal alteration had little or no effect on protolith compositions; (2) original igneous fractionation trends are preserved by immobile, incompatible trace elements; and (3) immobile trace elements preserve a distinct normal-MORB (N-MORB)-type composition. Fluid-mobile elements show varying degrees of enrichment relative to MORB, suggesting post-magmatic metasomatism. Based on these results, we conclude that the Central Metamorphic terrane represents oceanic lithosphere accreted to the upper-plate of an intraoceanic subduction-zone. Two new Early Permian $^{40}Ar/^{39}Ar$ ages of 274 ± 2 Ma are reported for Central Metamorphic terrane amphibolite near the Trinity fault are considerably younger than the previously reported Devonian age. These ages reflect Early Permian uplift and cooling of the Central Metamorphic terrane during a proposed episode of supra-subduction extension that reactivated the Devonian subduction-margin (Trinity thrust).

Keywords: Klamath Mountains, Devonian, Permian, convergent-margins, metabasite, ophiolite, geochemistry, thermochronology

*E-mail: wendyj@unlv.nevada.edu.

Barrow, W.M., and Metcalf, R.V., 2006, A reevaluation of the paleotectonic significance of the Paleozoic Central Metamorphic terrane, eastern Klamath Mountains, California: New constraints from trace element geochemistry and $^{40}Ar/^{39}Ar$ thermochronology, *in* Snoke, A.W., and Barnes, C.G., eds., Geological studies in the Klamath Mountains province, California and Oregon: A volume in honor of William P. Irwin: Geological Society of America Special Paper 410, p. 393–410, doi: 10.1130/2006.2410(19). For permission to copy, contact editing@geosociety.org.

INTRODUCTION

The Klamath Mountains province of northern California and southwest Oregon records over 300 m.y. of convergent-margin crustal growth. As such, the Klamath Mountains provide an excellent opportunity to test competing models for crustal growth in a long-lived orogen. One model holds that, in general, crustal growth in intraoceanic convergent-margins is mainly a result of the collision of exotic terranes with the continental margin (e.g., Jones et al., 1978; Speed, 1979; Speed et al., 1988). A second model focuses on long-lived subduction zones to generate crustal material in the over-riding plate by alternating extension and contraction (e.g., Burchfiel and Davis, 1975; Wright and Fahan, 1988; Miller, 1989; Goodge, 1990; Burchfiel and Royden, 1991; Hamilton, 1995; Hacker et al., 1995; Wallin and Metcalf, 1998; Collins, 2002). Crustal growth models for convergent-margins and paleo-plate reconstructions depend on the accurate identification of major paleo-plate boundaries. Such boundaries may be characterized by lower-plate rocks with mid-ocean ridge affinities and upper-plate rocks with subduction affinities. Subsequent deformation and metamorphism of these terranes can make it difficult to recognize such boundaries and reconstruct tectonic histories. Trace element studies may be useful in evaluating the tectonic-magmatic setting of such metamorphic terranes, provided that deformation, metamorphism, and/or hydrothermal activities have not substantially altered protolith compositions (Pearce, 1996).

The Klamath Mountains record crustal growth via a sequence of mid-Paleozoic to Cretaceous accretionary events (Burchfiel and Davis, 1975; Irwin, 1989). The oldest units are exposed in the Central Metamorphic terrane and Eastern Klamath terrane (Fig. 1) and may record the inception of Cordilleran convergent-margin tectonics (Burchfiel and Davis, 1975; Peacock, 1987a; Irwin, 1989; Wallin and Metcalf, 1998; Irwin and Wooden, 1999). It has been a long-standing view that these two terranes are remnants of an east-dipping Devonian subduction margin, with the Eastern Klamath terrane representing the over-riding plate and the Central Metamorphic terrane representing the subducted plate (Fig. 1). The Trinity thrust fault, which separates Eastern Klamath and Central Metamorphic terranes, may record a paleo-plate boundary and represent one of several major structures in the Klamath Mountains province. Recent work confirms the supra-subduction zone chemistry of the Eastern Klamath terrane magmatic rocks (Metcalf et al., 2000). Mineral assemblages in the Central Metamorphic terrane are consistent with a protolith of oceanic crust (Holdaway, 1965; Burchfiel and Davis, 1975; Peacock and Norris, 1989), but this interpretation has not been rigorously tested using geochemical techniques. An important aspect of the subduction margin model is the correlation of ages among key lithologies of the upper (Eastern Klamath terrane) and lower plates (Central Metamorphic terrane). Silurian and Devonian ages (mostly U/Pb zircon) are well documented for subduction-related elements of the upper plate, including subduction-zone magmatic rocks and trench/slope turbidites (Wallin and Gehrels, 1995; Wallin and Metcalf 1998). Metamorphic rocks of the Central Metamorphic terrane are generally considered to be Devonian in age (Lanphere et al., 1968), based on less reliable Rb-Sr isochron ages for a Central Metamorphic terrane metasedimentary unit.

The subduction-zone model predicts that Central Metamorphic terrane metabasites formed from a mid-ocean ridge basalt (MORB) protolith (i.e., typical oceanic lithosphere). Alternatively, potential Central Metamorphic terrane metabasite protoliths include within-plate, ocean island basalts (OIB) and volcanic-arc basalts (VAB), both of which are broadly consistent with accretionary models for the Klamath province. A primary goal of this study is the application of whole-rock trace element geochemistry to evaluate the tectonic-magmatic setting of the Central Metamorphic terrane protoliths. Specifically, these data are used to test whether the Central Metamorphic terrane protolith was MORB, as predicted by the subduction margin model. As part of this evaluation, trace element data are also used to assess element mobility during hydrothermal activity and/or metamorphism of the original igneous protolith. A secondary goal of this study is to provide the first modern $^{40}Ar/^{39}Ar$ thermochronology data for rocks of the Central Metamorphic terrane. The subduction margin model is discussed in light of these new data, in particular, whether the terranes of the eastern Klamath Mountains represent a major paleo-plate boundary.

GEOLOGICAL SETTING

The Klamath Mountains of northern California consist of an arcuate sequence of east-dipping lithotectonic units separated by east-dipping thrust faults (Irwin, 1960; Davis et al., 1965; Irwin, 1994; Fig. 1). This study focuses on lithotectonic units of the eastern Klamath Mountains, specifically, the Central Metamorphic and Eastern Klamath terranes, which are separated by the Trinity thrust fault (Fig. 1).

Eastern Klamath Terrane

The Eastern Klamath terrane forms the hanging wall of the Trinity thrust fault and is divided into the Trinity, Yreka, and Redding subterranes (Fig. 1).

Trinity Subterrane. The Trinity subterrane consists predominately of two lithotectonic assemblages, Ordovician peridotite and Silurian-Devonian mafic intrusive complexes (Jacobsen et al., 1984; Wallin et al., 1995; Wallin and Metcalf, 1998). The Ordovician Trinity peridotite is a moderately to highly depleted peridotite comprising mainly plagioclase lherzolite, clinopyroxene-poor lherzolite, harzburgite, and dunite (Davis et al., 1965; Quick, 1981; Peacock, 1987a; Wallin and Metcalf, 1998; Metcalf et al., 2000). One peridotite sample yielded an Ordovician (472 ± 32 Ma) Sm/Nd mineral–whole-rock isochron age (Jacobsen et al., 1984), interpreted as its crystallization age. The Trinity peridotite is intruded by Silurian-Devonian mafic

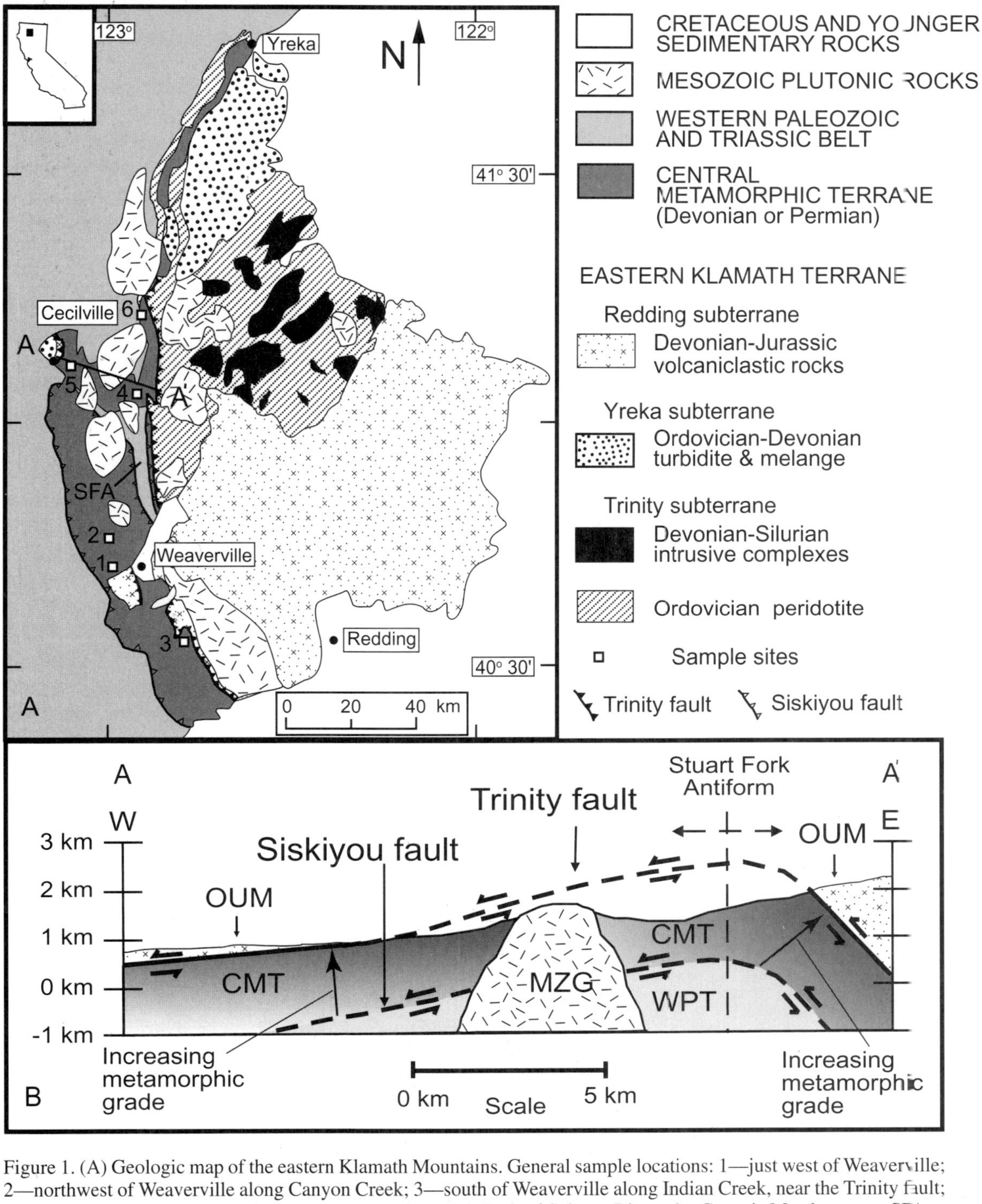

Figure 1. (A) Geologic map of the eastern Klamath Mountains. General sample locations: 1—just west of Weaverville; 2—northwest of Weaverville along Canyon Creek; 3—south of Weaverville along Indian Creek, near the Trinity fault; 4—Coffee Creek area; 5—Cecilville area, along the South Fork of Salmon River; 6—Carter's Meadow area; SFA—Stuart Fork antiform. Modified from Irwin (1994) and Wallin and Metcalf (1998). (B) Schematic cross-section of the eastern Klamath Mountains. The Trinity fault dips steeply on the east (main trace) and is subhorizontal on the west side beneath a klippe of Trinity subterrane peridotite. Units on the cross-section: CMT—Central Metamorphic terrane; MZG—Mesozoic Deadman Peak pluton; OUM—Trinity subterrane peridotite; WPT—western Paleozoic and Triassic belt.

igneous complexes that consist of pyroxenite, gabbro plutons, and mafic-felsic bimodal dike swarms (Metcalf et al., 2000). Gabbroic rocks yielded U/Pb zircon ages ranging from ca. 398 to 431 Ma (Wallin et al., 1995; Wallin and Metcalf, 1998 and personal commun.).

The Trinity peridotite has undergone at least two deformation episodes. An early high-temperature S_1 mantle fabric found throughout the Trinity subterrane likely formed during Ordovician rise of a mantle diapir (Quick, 1981; Cannat and Boudier, 1985). A younger S_2 foliation is defined by smears of serpentine, chlorite, and magnetite near the Trinity thrust (Lipman, 1964; Cannat and Boudier, 1985). The S_2 foliation is parallel to the Trinity thrust fault at the base of the peridotite as well as to foliation and lineation in underlying Central Metamorphic terrane amphibolite (Lipman, 1964).

Yreka Subterrane. The Yreka subterrane, which structurally overlies the Trinity subterrane, is a faulted package of turbidite, metagreywacke, and mélange (Potter et al., 1977; Lindsley-Griffin et al., 1991; Wallin and Trabert, 1994; Wallin and Metcalf, 1998). Cashman (1980) reported a Devonian age of metamorphism for the Yreka subterrane based on a hornblende K/Ar age of 389 ± 12 Ma from a thin band of amphibolite within the Duzel Formation. Detrital zircons in the basal sedimentary rocks of the Yreka subterrane yielded U/Pb ages of 410–420 Ma (Wallin and Gehrels, 1995), which, with the Devonian age of metamorphism, constrains the age of the Yreka subterrane to Late Silurian to Early Devonian (Wallin and Metcalf, 1998; Metcalf et al., 2000). The Yreka subterrane has been interpreted as trench-slope deposits formed at a paleo-plate margin (Wallin and Trabert, 1994; Wallin and Metcalf, 1998).

Redding Subterrane. The Redding subterrane consists of subduction-related mid-Paleozoic volcanic rocks overlain by a sequence of intercalated volcanic and marine sedimentary rocks, ranging in age from Mississippian to Jurassic (Kinkel et al., 1956; Sanborn, 1960; Irwin, 1977, 1989, 1994; Miller, 1989; Wallin and Metcalf, 1998). Most important to this study are the basal metavolcanic successions, specifically, the basal Copley Greenstone and the overlying Balaklala Rhyolite. The ages of the Copley and Balaklala volcanic units are constrained by the 400 ± 3-Ma (zircon U/Pb) Mule Mountain stock that intrudes both units (Albers et al., 1981), along with Devonian fish plates in sedimentary rocks interlayered with the Balaklala Rhyolite (Boucot et al., 1974). Trace element and isotopic geochemistry suggest correlation of the Copley Greenstone (at the base of the Redding subterrane) with the Silurian-Devonian mafic intrusive complex of the Trinity subterrane (Wallin and Metcalf, 1998; Metcalf et al., 1999, 2000). The volcanic and plutonic suites are therefore thought to represent parts of a cogenetic subduction-related complex (Wallin and Metcalf, 1998; Metcalf et al., 1999, 2000).

Central Metamorphic Terrane

The Central Metamorphic terrane is a fault-bounded package of metabasites (Salmon Formation) and metasedimentary rocks (Grouse Ridge Formation) with a common metamorphic history (Davis, 1968). The Trinity fault separates the Central Metamorphic terrane from the overlying Eastern Klamath terrane. The Siskiyou thrust fault separates the Central Metamorphic terrane from the underlying Stuart Fork Formation, St. Claire Creek, and North Fork units of the western Paleozoic and Triassic belt (Davis et al., 1965; Davis, 1968; Hacker et al., 1993; Fig. 1). Geophysical data indicate that the Central Metamorphic terrane extends to the east beneath the Trinity and Redding subterranes (Fuis et al., 1987). The Grouse Ridge Formation comprises metacarbonate, calc-silicate rocks and metabasite that structurally overlie the Salmon Formation, the latter consisting entirely of metabasite (Holdaway, 1965; Lanphere et al., 1968; Peacock and Norris, 1989). The protoliths of the Central Metamorphic terrane are generally consistent with oceanic crust (Salmon Formation) and pelagic sediments (Grouse Ridge Formation; Holdaway, 1965; Burchfiel and Davis, 1975; Peacock and Norris, 1989). The Central Metamorphic terrane has a reported inverted metamorphic field gradient (Fig. 1B) from greenschist at the base (above Siskiyou thrust fault) through epidote-amphibolite to amphibolite facies adjacent to the Trinity fault at the top of the section (Davis et al., 1965; Norris, 1983; Peacock and Norris, 1989). Peacock and Norris (1989) estimated the peak metamorphic temperatures of the Central Metamorphic terrane near the Trinity fault to be 650 ± 50 °C and peak temperatures at the base of the Central Metamorphic terrane to be 500 ± 50 °C. Pressure estimates for the Central Metamorphic terrane are broadly constrained at 5 ± 3 kb (Peacock and Norris, 1989). Lanphere et al. (1968) reported a range of hornblende K/Ar ages for the Salmon Formation from 186 to 286 Ma and muscovite K/Ar ages for the Grouse Ridge Formation from 262 to 329 Ma. Two Rb/Sr whole-rock ages for the Grouse Ridge Formation of 371 ± 11 and 374 ± 10 Ma have been considered the most reliable dates for the Central Metamorphic terrane (Lanphere et al., 1968). Lanphere et al. (1968) suggested that the K/Ar ages reflect minimum metamorphic cooling ages for the Central Metamorphic terrane.

At least two separate deformation events are recorded in the Central Metamorphic terrane, an earlier fabric-forming event and later map-scale folding (Davis et al., 1965; Cannat and Boudier, 1985). The main fabric in the Grouse Ridge and Salmon formations is an S_1 foliation defined by mineral alignment and compositional layering. The S_1 is folded about small thin-section scale folds that appear to be related to a large map-scale F_2 fold, an asymmetrical antiform (Stuart Fork antiform) trending generally north-south and plunging gently to the south (Davis et al., 1965; Fig. 1A). The Late Triassic Stuart Fork Formation (western Paleozoic and Triassic belt; Hotz, et al., 1977; Goodge, 1989) cores the Stuart Fork antiform and thus was folded along with the Central Metamorphic terrane and the Siskiyou thrust fault (Davis et al., 1965; Goodge, 1989, 1990). Timing of the large-scale folding is not well constrained, but it is most probably Middle Jurassic in age (Hacker et al., 1993; 1995).

Trinity Thrust Fault

The Trinity thrust fault is a moderately east dipping (~45°) structure at the main fault trace but is subhorizontal (may be dipping slightly to the west) beneath two klippen of Eastern Klamath terrane lithologies to the west of the main fault trace (Fig. 1, near the towns of Cecilville and Weaverville). At both locations, Trinity peridotite forms the hanging wall at the fault contact. The Trinity thrust was folded with the large-scale folding that occurred during the Late Jurassic about a subhorizontal north–south axis. At the main trace, the fault is a shear zone varying from a few meters to tens of meters in width. The shear zone is highly deformed, consisting mainly of serpenitized peridotite with clasts and blocks of Salmon Formation, Grouse Ridge Formation, and Trinity peridotite ranging from pebble to boulder size.

Irwin and Lipman (1962) suggested, based on stratigraphic discordance, that the contact between the Central Metamorphic terrane and the Eastern Klamath terrane is a low-angle thrust fault (Fig. 1). Davis (1968) agreed that the contact is a thrust fault, based on the arcuate map trace (concave to the east; Fig. 1) and a metamorphic discontinuity across the contact. Although the Central Metamorphic terrane below the Trinity thrust records a greenschist-facies through amphibolite-facies inverted metamorphic field gradient (Davis et al., 1965; Norris, 1983; Peacock and Norris, 1989; Fig. 1B), the mantle and magmatic rocks of the Trinity subterrane directly above the fault show little regional metamorphic overprint (Davis, 1968; Metcalf et al., 2000). Lipman (1964) and Peacock (1987a) noted that serpentinization of the Trinity peridotite and the metamorphic foliations within both Trinity peridotite (S_2) and the underlying Central Metamorphic terrane (S_1) are generally parallel to the Trinity fault. These fabrics suggest a common deformation event for both the Central Metamorphic terrane and Trinity peridotite and are thought to result from emplacement of the peridotite over the Central Metamorphic terrane along the Trinity thrust fault (Lipman, 1964; Cannat and Boudier, 1985).

METHODS

We collected major and trace element data on twenty Salmon Formation metabasite samples from the Central Metamorphic terrane and $^{40}Ar/^{39}Ar$ thermochronology data for hornblende separates from two metabasite samples. The Salmon metabasite samples used for geochemistry were collected in five general areas throughout the Central Metamorphic terrane: (1) just west of Weaverville (sample 4); (2) northwest of Weaverville along Canyon Creek (samples 5–8); (3) south of Weaverville along Indian Creek, near the Trinity fault (samples 9–12); (4) Coffee Creek area (samples 15–21); and (5) Cecilville area, along the South Fork of Salmon River (samples 29–32; Fig. 1A). The samples used for the $^{40}Ar/^{39}Ar$ dating were collected near Carter's Meadow (sample 41) and Coffee Creek (sample 21; Fig. 1A). All metabasite samples were examined petrographically to establish their suitability for chemical analysis (lacking hydrothermal alteration, mineralogy consistent with a basalt protolith) and thermochronology.

Geochemical Analysis

Major elements were analyzed by X-ray fluorescence spectrometry (XRF) and trace elements were analyzed by inductively coupled plasma mass spectrometry (ICP-MS) in the GeoAnalytical Laboratory at Washington State University, Pullman. Details of the analytical methods can be found on the Washington State University Web page (www.wsu.edu/~geology/Pages/Services/Geolab.html). Samples for XRF were prepared using tungsten carbide crushing equipment. Major elements were analyzed on a Rigaku automated XRF using fused beads (2:1 flux to sample ratio) and calibrated using U.S. Geological Survey (USGS) rock standards. Sample totals in Table 1 represent anhydrous values. Water contents were determined by loss on ignition (LOI) on a separate aliquot of sample and are reported in Table 1. Generally low LOI values (0.36–3.22 wt%, mean = 1.09 wt%; Table 1) suggest that water is largely bound in hydrous minerals (epidote and/or amphibole).

Samples for ICP-MS were prepared using steel crushing equipment to avoid tungsten carbide contamination, which interferes with analysis of Ta and Nb. Powders were digested using a combined fusion/dissolution procedure that ensures the complete dissolution of refractory phases such as zircon. Solutions were analyzed on a Sciex Elan model 250 ICP-MS; Washington State University calibrates using three in-house USGS rock standards that are calibrated against seventeen international standards.

Argon Geochronology

Hornblende mineral separates were prepared from two samples of Salmon Formation, an amphibolite (sample 21) and an epidote amphibolite (sample 41), by magnetic separation and handpicking. Samples were analyzed at the Nevada Isotope Geochronology Lab, University of Nevada, Las Vegas (http://www.unlv.edu/labs/nigl/home.htm), by the $^{40}Ar/^{39}Ar$ furnace step-heating method. Analytical data are reported at the confidence level of 1-σ (standard deviation). Plateau steps overlap in age at +2-σ analytical error and plateau ages are weighted by the inverse of the variance. Total gas ages are calculated by weighting the amount of ^{39}Ar released. A more detailed discussion of the laboratory's procedures is reported by Reiners et al. (2004).

PETROGRAPHY

The analyzed samples have mineral assemblages that range from greenschist, epidote amphibolite, to amphibolite metamorphic facies. Samples collected in the structurally lower portion of the Central Metamorphic terrane have greenschist-facies assemblages, whereas samples collected at structurally higher

TABLE 1. MAJOR ELEMENT WHOLE-ROCK CHEMICAL ANALYSES

Sample	4*	5†	6†	7†	8†	9§	10§	11§	12§	15#	16#	17#	18#	19#	20#	21#	29**	30**	31**	32**
SiO_2	49.73	49.76	49.45	49.57	49.93	48.15	47.76	48.03	50.92	49.05	50.38	52.34	49.94	51.95	50.73	51.09	48.24	49.27	49.75	54.02
Al_2O_3	14.04	13.64	13.08	13.28	13.87	13.23	12.82	12.8	12.85	13.52	13.83	13.17	13.55	13.45	14.41	12.83	15	16.77	14.5	15.53
TiO_2	1.199	1.847	2.469	2.039	1.9	2.433	2.925	2.843	2.632	2.219	1.884	1.85	1.931	1.653	1.372	2.164	1.842	1.078	1.235	1.058
FeO††	9.46	12.89	14.05	13.5	12.03	14.93	16.14	15.37	13.84	13.34	11.65	12.27	12.86	11.03	10.39	13.08	11.98	9.97	10.36	8.78
MnO	0.15	0.249	0.234	0.226	0.212	0.25	0.325	0.27	0.238	0.238	0.21	0.215	0.222	0.198	0.202	0.25	0.227	0.178	0.215	0.179
CaO	7.93	10.03	10.6	10.87	10.55	11.05	9.99	10.03	10.09	9.84	9.44	8.07	9.66	10.49	12.16	8.01	11.73	11.8	12.47	5.76
MgO	11.88	7.49	6.66	6.86	7.4	6.44	6.45	7.1	5.56	7.55	9.69	7.34	7.86	7.08	8.01	7.3	6.89	7.21	7.11	6.85
K_2O	1.03	0.13	0.14	0.15	0.16	0.32	0.08	0.08	0.25	0.11	0.07	0.14	0.11	0.07	0.08	0.09	0.3	0.31	0.26	1.88
Na_2O	3.86	3.27	2.23	2.35	3.09	2.4	2.17	2.4	2.54	3.28	2.32	3.75	3.18	3.36	2.34	4.06	2.68	3.22	3.64	4.97
P_2O_5	0.332	0.141	0.248	0.156	0.199	0.216	0.265	0.243	0.274	0.206	0.151	0.141	0.159	0.14	0.109	0.16	0.153	0.085	0.099	0.243
Total	99.61	99.44	99.16	99.00	99.34	99.42	98.93	99.16	99.20	99.35	99.62	99.28	99.47	99.43	99.80	99.03	99.05	99.89	99.64	99.27
LOI (%)	2.01	0.41	0.73	0.80	0.81	0.45	0.36	0.37	0.68	1.00	3.22	1.86	0.82	0.89	1.04	0.57	0.74	1.10	2.10	1.81

Notes: Values in wt%; LOI—loss on ignition.

*Sample collected just west of Weaverville and the LaGrange fault at site 1 ~ N40°44′, W123° (Fig. 1).

†Sample collected northwest of Weaverville along Canyon Creek at site 2 ~ N40°46′, W123°03′ (Fig. 1).

§Sample collected south of Weaverville along Indian Creek near the Trinity fault at site 3 ~ N40°36′, W122°48′ (Fig. 1).

#Sample collected in the Coffee Creek area at site 4 ~ N41°04′, W122°57′ and N41°05′, W122°57′ (Fig. 1).

**Sample collected in the Cecilville area along the South Fork of Salmon River at site 5 ~ N41°09′, W123°06′ (Fig. 1).

††Major elements are normalized on a volatile-free basis, with total Fe expressed as FeO.

levels have epidote amphibolite- and amphibolite-facies assemblages. These observations are consistent with an inverted metamorphic gradient reported by Davis et al. (1965), Norris (1983), and Peacock and Norris (1989).

Five of the samples exhibit greenschist-facies mineral assemblages and are composed mainly of chlorite (~25%), actinolite (~40%), plagioclase (~20%), epidote (~13%), and quartz (~3%) ± calcite ± apatite ± oxides. Four of the greenschist samples are very fine grained; the fifth (sample 31) is slightly coarser. Three of the very fine-grained samples have anastomosing spaced cleavage defined by actinolite and epidote. Two samples have compositional banding consisting of alternating layers of actinolite and quartz + plagioclase that is parallel to the mineral alignment foliation. In all samples, quartz is dynamically recrystallized and actinolite shows a strong preferred orientation that defines the S_1 foliation.

Five of the samples exhibit epidote amphibolite-facies mineral assemblages and are composed of hornblende (~50%), actinolite (~3%), plagioclase (~15%), epidote (~15%), and quartz (~10%) ± oxides ± titanite ± apatite. Crystal sizes in the epidote amphibolite samples vary from fine to medium grained (hornblende up to 3.0 mm in length and quartz up to 1.0 mm across). Most samples have a very pronounced deformation fabric, with compositional banding and strong preferred mineral orientation exhibited by the hornblende and epidote. The samples also have small shear bands filled with quartz that are at a moderate angle (~45°) to the main foliation and compositional layering. One sample, 9, does not have compositional banding or a preferred orientation of the crystals. The sample is medium grained, with dynamically recrystallized quartz and some large annealed quartz crystals. Small veins of epidote and quartz that crosscut the foliation at a high angle were observed in three samples. All five epidote amphibolite samples show little or no alteration to greenschist or lower-grade assemblages.

Ten samples exhibit amphibolite-facies mineral assemblages consisting of hornblende (~50%), plagioclase (~12%), and quartz (~15%) ± oxides ± epidote ± titanite ± apatite; eight samples have little to no secondary epidote and chlorite. Six out of the ten amphibolites have compositional banding of amphibole alternating with quartz and plagioclase. Three samples also have small shear bands filled with quartz that are at a moderate angle (~45°) to the main foliation and compositional layering. The crystal size ranges from fine (~0.01 mm across) to medium grained, with hornblende lengths up to 2.0 mm. Two samples, 4 and 32, show abundant alteration of hornblende to chlorite with much higher secondary chlorite percentages (24 and 15%). As discussed below, these two samples are compositionally distinct from the remainder of the data set.

GEOCHEMISTRY

The primary objective of this project was to establish the tectonic-magmatic setting of the Central Metamorphic terrane metabasite protoliths using empirically derived basalt discrimination diagrams. Because post-magmatic element mobility may cause significant compositional changes that could result in an incorrect interpretation, the data set was first evaluated to determine whether the chemical data preserve original igneous compositions or, alternatively, reflect post-magmatic alteration. The potential sources for such element mobility are (1) seafloor hydrothermal alteration and/or (2) fluid flux during dynamothermal metamorphism (e.g., conversion into schist and gneiss).

Evaluation of Element Mobility

Seafloor hydrothermal alteration occurs along mid-ocean ridges where seawater circulates into hot, newly formed oceanic crust, resulting in changes in mineralogy and chemical composition. In a study of fresh and hydrothermally altered basalts dredged from the Mid-Atlantic Ridge, Humphris and Thompson (1978) documented an inverse correlation between CaO and MgO. Both empirical and experimental data confirm that during hydrothermal alteration, CaO is leached from basalt and MgO is taken up on an approximately 1:1 molar basis, and that these compositional shifts increase with increasing water/rock ratio (Humphris and Thompson, 1978; Mottl, 1983). When plotted on a CaO versus MgO diagram (Humphris and Thompson, 1978; Mottl, 1983; Fig. 2), the majority of Central Metamorphic ter-

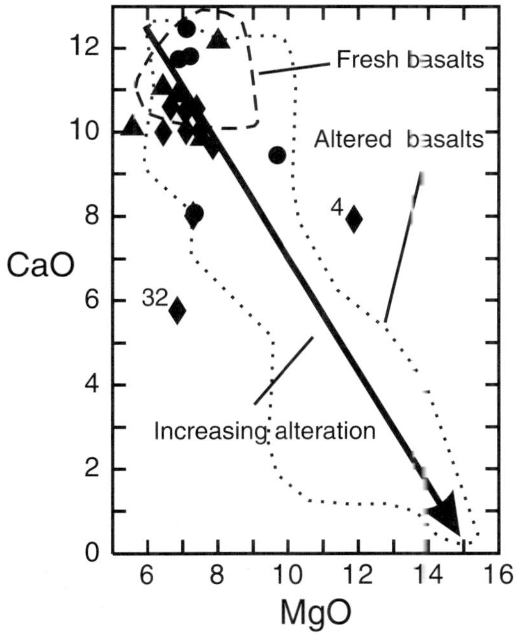

Figure 2. CaO wt% vs. MgO wt% diagram. Fifteen of the twenty Salmon Formation metabasites compare closely with fresh and slightly altered basalts from Humphris and Thompson (1978) and Mottl (1983) studies of fresh (dashed envelope) and hydrothermally altered basalt (dotted envelope). The arrow shows variation of CaO/MgO ratio with increasing water/rock ratio. The symbols for the greenschist samples are circles, epidote-amphibolite samples are triangles, and amphibolite samples are diamonds.

rane metabasalts compare closely with fresh basalts, whereas the rest are similar to slightly altered basalts. None of the Central Metamorphic terrane samples compare with the highly altered basalts (Fig. 2). These results suggest that seafloor hydrothermal alteration did not significantly affect the composition of the Central Metamorphic terrane protoliths. Samples 4 and 32 have the lowest CaO/MgO ratios in the data set, suggesting slightly more hydrothermal alteration (Fig. 2).

Incompatible trace element data (Table 2) were used to evaluate post-magmatic element mobility due to fluid flux during dynamothermal metamorphism by attempting to identify coherent magmatic fractionation trends in the data set. Incompatible elements preferentially partition into the melt during igneous fractional crystallization; consequently, incompatible element concentrations increase with increasing differentiation. Plots of two incompatible trace elements, therefore, should exhibit data arrays with positive slopes if post-magmatic alteration has not significantly affected whole-rock compositions. The approach used here follows Donato's (1991) study of the May Creek amphibolites, in which various incompatible elements were plotted against the highly immobile and incompatible element Zr. The majority of the samples (all but samples 4 and 32) show positive linear trends on plots of Zr versus Ti, Ce, Hf, Ta, Yb, and Y (Fig. 3A–F). Such trends indicate that original igneous abundances were preserved with little alteration of the immobile trace element compositions (Donato, 1991). When plotted against Zr, the more mobile trace elements show much more scatter, suggesting that post-magmatic alteration affected these elements (e.g., Rb, Cs; Fig. 3G and H). Samples 4 and 32 plot off the trends on most of the diagrams, including those for the immobile elements (Fig. 3). This discrepancy may indicate either post-magmatic alteration affecting both mobile and immobile elements, or more likely, a basaltic protolith from a different magmatic suite; therefore, samples 4 and 32 are discussed separately. Post-magmatic alteration had a minimal effect on immobile trace element abundances for the Central Metamorphic terrane metabasites. Thus, basalt discrimination diagrams that focus on the immobile trace elements should be useful in evaluating the tectonic-magmatic setting of Central Metamorphic terrane protoliths.

Protolith Discrimination

Major element compositions (Table 1) for Central Metamorphic terrane metabasites are consistent with a basaltic protolith (Fig. 4) as expected from metamorphic mineralogy. All samples exhibit a limited range of mafic compositions (SiO_2, 48–54 wt%, and MgO, 5.6–11.9 wt%; Fig. 4). Major element data are consistent with a subalkaline, tholeiitic basalt protolith similar to a mid-ocean ridge protolith, although this interpretation relies on the geochemically mobile alkali elements (Fig. 4). Metabasite TiO_2 contents (1.1 – 2.9 wt% mean = 1.92 wt%) are similar to MORB and higher than typically observed in primitive arc basalts.

Modern normal MORB (N-MORB), enriched MORB (E-MORB), OIB, and VAB exhibit significant and systematic differences in trace elements that are relatable to mantle source compositions (Figs. 5A–C and 6A). Compositions of OIB, VAB, and E-MORB are typically discussed relative to N-MORB. Within-plate basalts (i.e., OIB) are enriched with respect to MORB in Th, Nb, Ta, Ce, and light rare-earth elements (LREEs) and show a significant enrichment of Zr, Hf, and Ti (Pearce, 1996; Figs. 5A and 6A). This enrichment is consistent with a deeper mantle plume source, in contrast to the depleted upper mantle source for MORB (Morgan, 1972; Wilson, 1989; Pearce, 1996). Compared to N-MORB, VAB has a significant negative Nb (Ta) anomaly with respect to Th and Ce (Fig. 5B). E-MORB is enriched in the more incompatible elements in comparison to N-MORB, but not as enriched as OIB (Fig. 5A and C). E-MORB is also depleted in the least incompatible elements in comparison to N-MORB (Fig. 5C).

The Central Metamorphic terrane trace element data (Table 2) were evaluated using a variety of empirical basalt discrimination diagrams (Figs. 5–7). On N-MORB-normalized multielement diagrams (Fig. 5), Central Metamorphic terrane metabasites plot close to N-MORB values for the high-field-strength elements (HFSE) and lack significant negative Nb anomalies typical of VAB magmatic rocks. Large-ion lithophile elements (LILE), such as Cs, Ba, Pb, K, and Rb, however, show varying degrees of enrichment on the N-MORB-normalized plots (Fig. 5) that likely reflect post-magmatic element mobility, as discussed above (Fig. 3G and H). All metabasites, except samples 4 and 32, show N-MORB-like patterns on chondrite-normalized REE diagrams, specifically, values about ten times those of chondrite, LREE depletion, and flat heavy REE patterns (Fig. 6).

On plots that focus on the least mobile elements (e.g., Hf, Ta, Cr, Nb, Y, Yb, Ti, Th, V), the Central Metamorphic terrane metabasites show a clear N-MORB affinity (Fig. 7). On both the Ti-V diagram of Shervais (1982; Fig. 7A) and the Pearce (1982; Pearce et al., 1995) Cr-Y diagram (Fig. 7D) the Central Metamorphic terrane metabasites plot clearly in the MORB field. On the Th/Yb versus Nb/Yb ratio plots (Pearce, 1982; Pearce et al., 1995; Fig. 7C), the Central Metamorphic terrane samples plot near average N-MORB, indicating a depleted mantle source with no subduction-zone component. Similar results are seen in the Wood (1980) Th-Ta-Hf diagram (Fig. 7B). These latter two diagrams illustrate what Pearce et al. (1984) refers to as a coupling of Ta (and Nb) and Th, indicating N-MORB as opposed to VAB. The diagrams clearly show an N-MORB signature for the majority of the metabasite samples.

Two amphibolite samples (4 and 32) consistently plot apart from the majority of the data set. These samples have major element compositions consistent with an alkaline protolith (Fig. 4) and have HFSE and REE patterns more similar to E-MORB or OIB (Figs. 5 and 6). Both samples exhibit LILE enrichments typical of subduction zone magmatic rocks (Figs. 5 and 7B and C). In particular, the Th/Yb versus Nb/Yb plot (Fig. 7C)

TABLE 2. WHOLE-ROCK TRACE ELEMENTS ANALYSES

Sample	4*	5†	6†	7†	8†	9§	10§	11§	12§	15#	16#	17#	18#	19#	20#	21#	29**	30**	31**	32**
Ni	349	53	46	42	54	45	42	44	38	60	57	53	49	56	74	47	80	75	56	17
Cr	606	87	156	89	221	116	98	115	82	198	172	140	109	213	315	122	288	340	183	155
V	227	393	464	420	363	464	513	528	465	438	379	385	390	348	328	412	388	276	322	214
Ga	20	19	23	18	16	22	19	21	18	19	14	20	18	16	20	17	19	16	17	16
Cu	59	15	70	41	49	56	44	59	66	12	116	26	3	3	16	17	70	66	84	26
Zn	85	76	108	102	91	120	154	126	114	114	118	108	85	51	72	94	102	80	84	89
La	18.3	3.25	4.95	3.77	4	5.84	4.86	5.98	5.45	4.28	3.55	3.7	3.59	3.99	3.69	3.95	3.03	2.04	2.13	12.8
Ce	38.43	10.33	15.36	11.94	12.64	18.08	13.82	18.23	16.53	13.66	11.13	11.43	11.36	12.4	11.13	12.19	9.08	5.62	6.52	25.32
Pr	4.91	1.81	2.62	2.07	2.16	3.03	2.44	3.03	2.81	2.32	1.96	1.99	1.91	2.1	1.93	2.1	1.63	1.03	1.17	3.3
Nd	21.85	10.44	14.94	12.09	12.51	17.33	13.87	17.16	16.05	13.69	11.41	11.55	11.08	12.25	11.13	12.42	9.52	6.1	6.85	14.93
Sm	5.25	4.27	5.99	4.78	4.94	6.91	5.58	6.72	6.38	5.46	4.63	4.69	4.43	5	4.53	4.88	3.94	2.53	2.86	4.09
Eu	1.58	1.5	2.01	1.68	1.79	2.27	1.93	2.24	2.14	1.87	1.6	1.68	1.59	1.74	1.64	1.68	1.41	1.02	1.13	1.25
Gd	4.53	5.92	8.07	6.74	6.75	9.27	7.66	9.1	8.58	7.39	6.46	6.43	6.09	6.8	6.18	6.71	5.41	3.59	4.04	4.27
Tb	0.64	1.13	1.54	1.29	1.3	1.76	1.45	1.72	1.66	1.41	1.25	1.23	1.16	1.31	1.19	1.3	1.06	0.7	0.79	0.69
Dy	3.48	7.57	10.17	8.58	8.49	11.68	9.64	11.57	10.93	9.35	8.37	8.28	7.83	8.83	8.01	8.7	7.14	4.7	5.34	4.32
Ho	0.65	1.62	2.18	1.87	1.81	2.48	2.06	2.49	2.35	2.01	1.83	1.78	1.66	1.91	1.72	1.88	1.55	1.02	1.14	0.88
Er	1.68	4.58	6.05	5.26	4.97	6.86	5.65	6.87	6.43	5.63	5.22	5.02	4.67	5.42	4.82	5.28	4.36	2.84	3.17	2.35
Tm	0.22	0.67	0.88	0.79	0.72	1.02	0.82	1	0.94	0.81	0.75	0.72	0.67	0.78	0.71	0.76	0.62	0.41	0.46	0.34
Yb	1.32	4.14	5.44	4.91	4.36	6.25	5.1	6.26	5.89	5.09	4.73	4.47	4.21	4.83	4.33	4.81	3.91	2.52	2.82	2.11
Lu	0.21	0.66	0.86	0.78	0.68	0.99	0.8	0.98	0.91	0.81	0.74	0.71	0.66	0.76	0.69	0.74	0.62	0.4	0.44	0.33
Ba	209	13	18	14	14	12	75	12	20	30	24	26	14	13	16	14	93	96	68	565
Th	1.94	0.15	0.24	0.15	0.22	0.27	0.22	0.29	0.27	0.21	0.15	0.16	0.2	0.2	0.18	0.17	0.16	0.1	0.08	2.16
Nb	8.59	1.77	3.11	2.09	2.12	3.29	2.74	3.37	3.09	2.31	1.68	1.89	2.03	2.32	2.29	1.88	1.54	1.28	0.96	7.42
Y	16.72	42.01	56.54	48.55	47.03	65.33	54.71	66.69	62.62	51.61	47.51	45.5	42.74	48.95	44.46	48.49	40.57	26.76	29.7	23.15
Hf	2.93	2.92	4.21	3.38	3.45	4.99	3.97	4.9	4.6	3.72	3.12	3.2	3.07	3.45	3.09	3.5	2.73	1.66	1.83	2.53
Ta	0.57	0.13	0.22	0.15	0.16	0.24	0.2	0.25	0.24	0.17	0.13	0.14	0.15	0.17	0.16	0.14	0.11	0.09	0.07	0.48
U	0.71	0.06	0.09	0.06	0.09	0.1	0.13	0.12	0.13	0.09	0.27	0.06	0.06	0.07	0.06	0.06	0.08	0.09	0.08	0.73
Pb	3.64	1.18	1.39	1.39	1.27	1.14	1.56	1.32	1.28	1.27	2.08	2.11	0.85	0.98	0.88	0.87	1.88	3.16	1.03	5.75
Rb	9.3	1	1.8	1.7	2.1	0.8	5	0.9	4.9	1.3	1.1	2.2	1.3	0.6	1.3	1	5.2	3.6	3.1	43.9
Cs	1.7	1.15	1.1	1.11	1.23	1.19	1.52	1.15	1.58	1.18	1.29	1.21	1.15	1.13	1.2	1.16	1.53	1.51	1.09	1.45
Sr	185	103	125	126	95	87	138	95	123	83	71	97	89	162	117	66	165	212	86	176
Sc	22.8	47.4	48.1	47.7	49.7	53.6	55.2	63.2	58.7	45.6	48.6	46.1	46.3	47.6	48.7	45.9	45.8	45.9	47.1	38.9
Zr	106	97	144	114	116	172	135	175	163	122	101	106	101	113	102	116	91	54	60	91

Note: Values in ppm.

*Sample collected just west of Weaverville and the LaGrange fault at site 1: ~N40°44′, W123° (Fig. 1).

†Sample collected northwest of Weaverville along Canyon Creek at site 2: ~N40°46′, W123°03′ (Fig. 1).

§Sample collected south of Weaverville along Indian Creek near the Trinity fault at site 3: ~N40°36′, W122°48′ (Fig. 1).

#Sample collected in the Coffee Creek area at site 4: ~N41°04′, W122°57′ and N41°05′, W122°57′ (Fig. 1).

**Sample collected in the Cecilville area along the South Fork of Salmon River at site 5 ~ N41°09′, W123°06′ (Fig. 1).

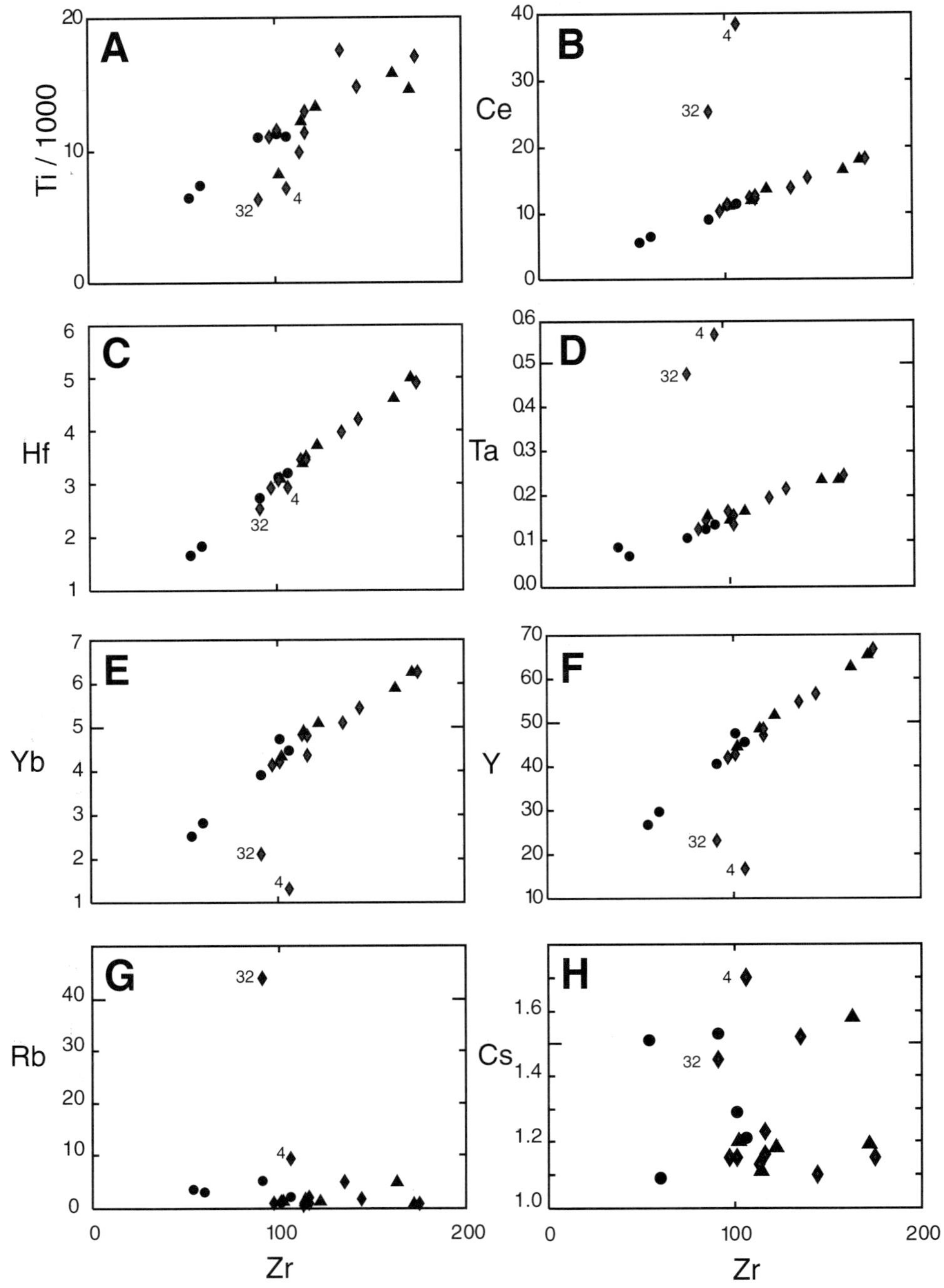

Figure 3. Plots of various incompatible elements vs. Zr. The symbols are as in Figure 2. Trace elements are reported in ppm.

suggests a protolith derived from an enriched source with a minor subduction-zone component.

The geochemical data clearly point to an N-MORB-type basalt as the protolith for the Central Metamorphic terrane metabasites. Although modern N-MORB-type basalts are erupted primarily at mid-ocean ridge spreading centers, basalts with N-MORB-like compositions are also recognized as eruptive products in modern backarc basins (e.g., the Lau basin; Hawkins, 1995). Basalts collected from modern backarc basins exhibit a range of VAB to MORB compositions, suggesting a heterogeneous mantle source and/or a mixing of VAB-type and MORB-type magmas. Backarc basalts with arc-like compositions are mostly low-K tholeiites more depleted in HFSE than NMORB. Typically, MORB-like backarc basalts carry some evidence of a subduction component, specifically, Nb-Ta depletion relative to Th and Ce and enrichments in light REE and LILE. Correlation of LILE and Zr concentrations in backarc basalts suggest LILE enrichments are attributable to source compositions rather than seafloor alteration (Hawkins, 1995). Only in mature backarc basins are compositions likely to be indistinguishable from MORB-type basalts.

The data presented here favor a mid-ocean ridge rather than a backarc basin setting for the generation of the N-MORB-like metabasite protolith. The LILE enrichments seen in the Central

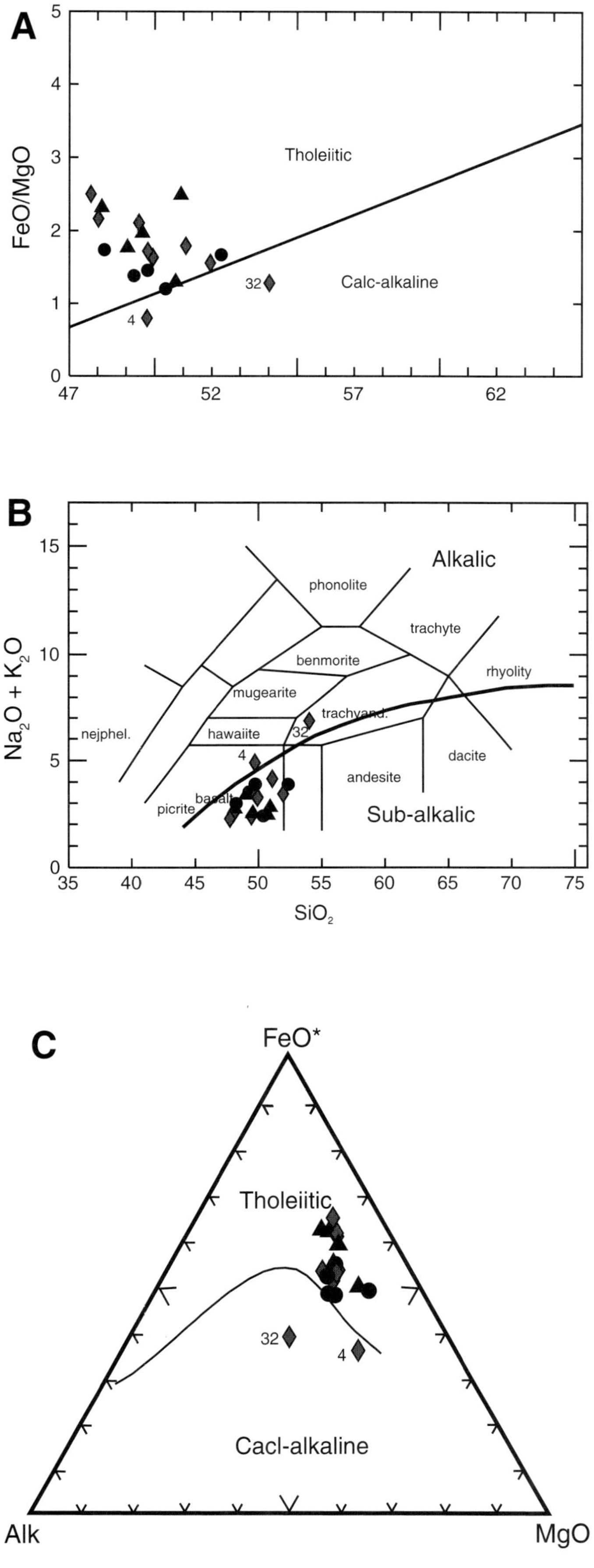

Metamorphic terrane data are not a feature of the original magmatic composition of the protolith, but rather are a consequence of post-magmatic addition of the most mobile elements. The only samples in the data set that exhibit VAB-like compositions are the two anomalous samples (4 and 32), but these are far more enriched than depleted arc tholeiites that typify backarc basins. Based on the preponderance and narrow range of N-MORB compositions, we argue that the Central Metamorphic terrane metabasite protolith was oceanic crust generated at a mid-ocean ridge spreading center.

ARGON GEOCHRONOLOGY

Hornblende separates from two samples of Central Metamorphic terrane Salmon metabasite (21 and 41) collected near the Trinity fault yielded similar Early Permian $^{40}Ar/^{39}Ar$ cooling ages of 274 ± 2 Ma (Fig. 8, Table 3). These ages are considerably younger than the previously reported Devonian age for the Central Metamorphic terrane metasediments (ca. 374-Ma Rb-Sr isochron muscovite–whole-rock ages; Lanphere et al. 1968). Sample 21 was collected in the Coffee Creek area (Fig. 1A) and did not give a plateau age, but does have a weighted mean age for heating steps 6 through 17 of 274.52 ± 1.62 Ma and a total gas age of 257.40 ± 1.37 Ma (Fig. 8, Table 3). Sample 41 was collected in the Carter Meadow area (Fig. 1A) and yielded a pseudo plateau age (47% of the gas released instead of >50%) of 274.10 ± 1.83 Ma for heating steps 8 through 12 and a total gas age of 260.65 ± 1.50 Ma (Fig. 8, Table 3). As discussed below, these results suggest cooling of Central Metamorphic terrane metabasites below ~500 °C at ca. 274 Ma.

DISCUSSION

It has been a long-standing view that the terranes of the eastern Klamath Mountains record elements of an east-dipping Devonian subduction zone (Burchfiel and Davis, 1975; Peacock, 1987a; Irwin, 1989; Peacock and Norris, 1989; Wallin and Metcalf, 1998; Irwin and Wooden, 1999; Metcalf et al., 2000). In this model, the Eastern Klamath terrane represents the upper-plate—including Silurian-Devonian arc-related magmatic rocks (volcanic rocks and plutonic rocks of the Redding and Trinity subterranes, respectively)—mantle wedge material (Ordovician peridotite of the Trinity subterrane) and trench-slope deposits (Silurian-Devonian mélange and turbidities of Yreka subterrane). Trinity-Redding magmatic rocks are broadly cogenetic and were generated by partial melting of a depleted, residual MORB mantle source enriched by subduction-derived fluids (Metcalf et al., 2000). The Eastern Klamath terrane records mid-Paleozoic initiation of subduction in a trench-proximal forearc setting

Figure 4. Major element variation diagrams: (A) Miyashiro (1978); (B) Cox et al. (1979; line separating alkaline and subalkaline is from Miyashiro, 1978); (C) alkali ($Na_2O + K_2O$)-FeO-MgO diagram. The symbols are as in Figure 2.

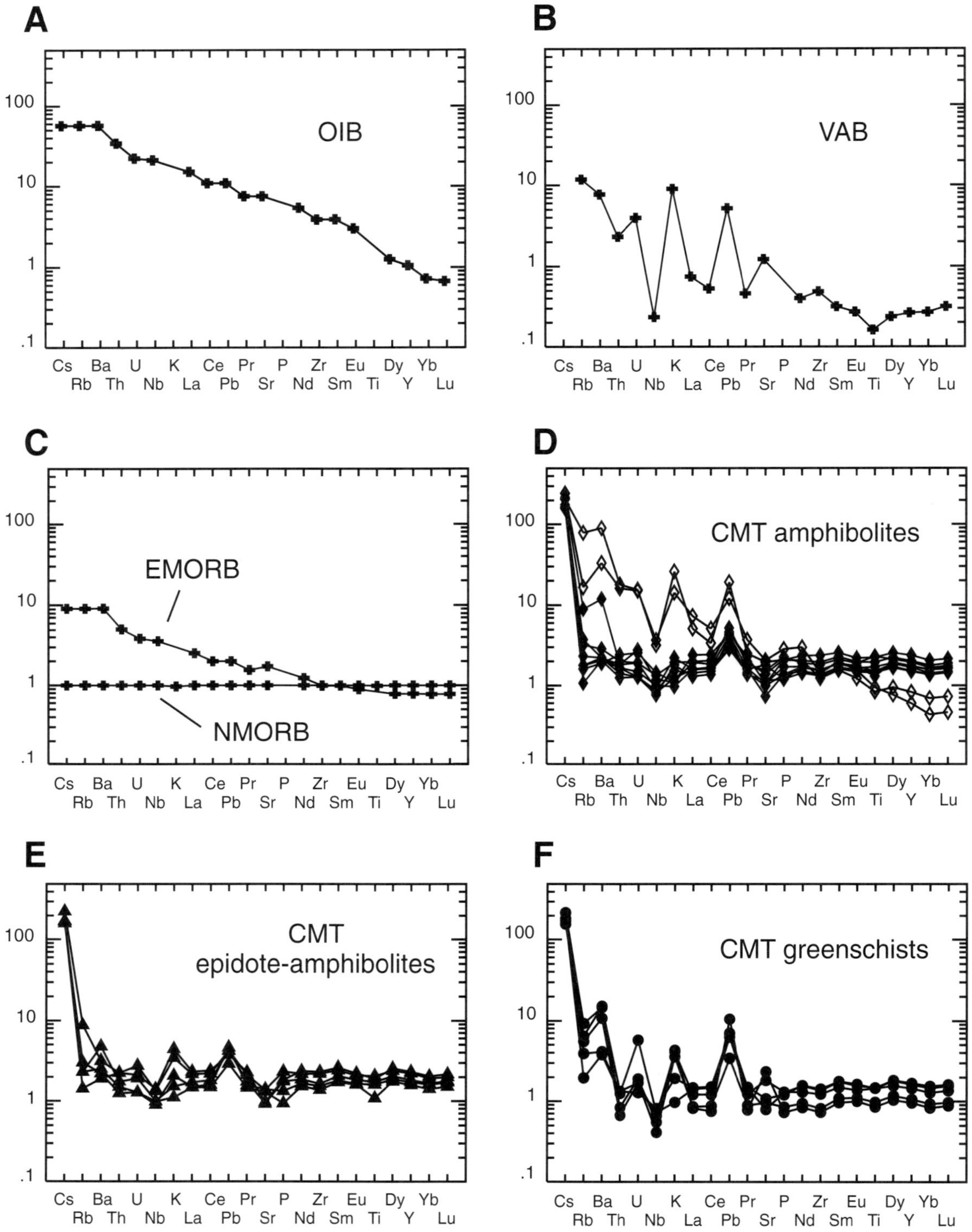

Figure 5. Trace element multielement diagrams showing patterns of: (A) ocean island basalt (OIB; Sun and McDonough, 1989); (B) volcanic-arc basalt (VAB; Pearce et al., 1995); (E) enriched (E-MORB) and normal mid-ocean ridge basalts (N-MORB; Sun and McDonough, 1989); (D) Central Metamorphic terrane (CMT) amphibolites (open diamonds are amphibolite samples 4 and 32); (E) Central Metamorphic terrane epidote-amphibolites; (F) Central Metamorphic terrane greenschists. Data are normalized to N-MORB using values from Sun and McDonough (1989). The symbols are the same as in Figure 2.

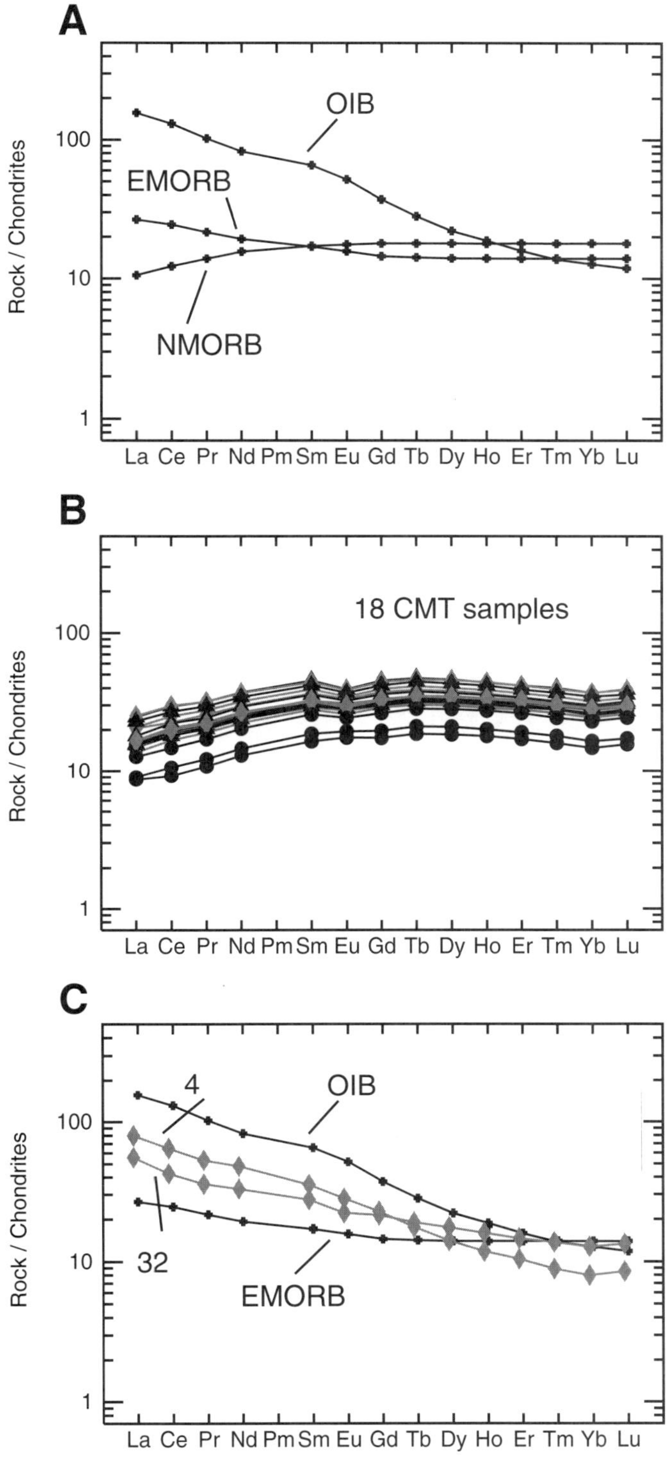

Figure 6. Chondrite-normalized rare earth elements diagrams showing patterns of: (A) enriched (E-MORB) and normal mid-ocean ridge basalt (N-MORB) and ocean island basalt (OIB; Sun and McDonough, 1989); (B) Central Metamorphic terrane (CMT) metabasites samples with N-MORB-like signatures; (C) samples 4 and 32 compared with E-MORB and OIB. The symbols are the same as in Figure 2. Normalizing values are from Sun and McDonough (1989).

comparable to Eocene forearc rocks formed during initiation of the Izu-Bonin-Mariana arc system (Wallin and Metcalf, 1998; Metcalf et al. 2000). The Devonian subduction model views the Central Metamorphic terrane as a remnant of subducted oceanic lithosphere that was accreted to the base of the overlying Trinity peridotite during subduction (Fig. 1B), with dynamothermal metamorphism of the Central Metamorphic terrane protolith driven by thrusting of the cold footwall oceanic plate beneath hot mantle of the overriding plate (Davis et al., 1965; Norris, 1983; Peacock and Norris, 1989). Our new geochemical data lend support to the Devonian subduction zone model; however, the new $^{40}Ar/^{39}Ar$ hornblende ages require some modification of the long-standing model, as discussed below.

Oceanic lithosphere subducted at modern convergent margins is generated predominately at mid-ocean spreading centers; thus, the subduction zone model for the eastern Klamath Mountains predicts a mid-ocean ridge protolith for the Central Metamorphic terrane. It has been recognized for decades that the metamorphic lithologies of the Central Metamorphic terrane are consistent with basaltic oceanic crust and overlying pelagic sediments (Davis, 1968). The Central Metamorphic terrane metabasite mineralogy, however, is equally consistent with other protoliths that fit broadly into the accretionary history of the Klamath Mountains province (e.g., VAB, OIB). Our geochemical results provide strong evidence that the protolith for the Central Metamorphic terrane metabasites was N-MORB-type basalt. These geochemical data demonstrate that (1) seafloor hydrothermal alteration had little or no effect on protolith compositions; (2) original igneous fractionation trends are preserved by immobile, incompatible trace elements; and (3) immobile trace elements preserve a distinct N-MORB-type composition. Observed enrichments in the fluid mobile LILE trace elements (e.g., K, Rb, Cs, Sr, Pb) are unrelated to original igneous compositions and result from post-magmatic metasomatism. The LILE metasomatism may have occurred during subduction-related accretion of the Central Metamorphic terrane protolith to the overlying Trinity peridotite, with LILE transported via slab-derived fluids generated at deeper levels. Alternatively, LILE metasomatism may have been derived from fluids expelled from below during emplacement of an underlying Triassic accretionary complex, represented by mélange of Stuart Fork Formation. Supporting the latter interpretation is the observation that greenschist samples from the base of the Central Metamorphic terrane show slightly greater LILE compared with epidote-amphibolite and amphibolite samples collected at structurally higher levels (Fig. 5). We interpret the protolith of the Central Metamorphic terrane metabasites as N-MORB-type oceanic crust generated at a mid-ocean ridge spreading center, consistent with subducted oceanic lithosphere, as predicted by the eastern Klamath subduction model.

The subduction zone model has been supported by several lines of evidence that point to synchronicity during Devonian time among supra-subduction zone activity in the Eastern Klamath terrane, thrusting of the Central Metamorphic terrane

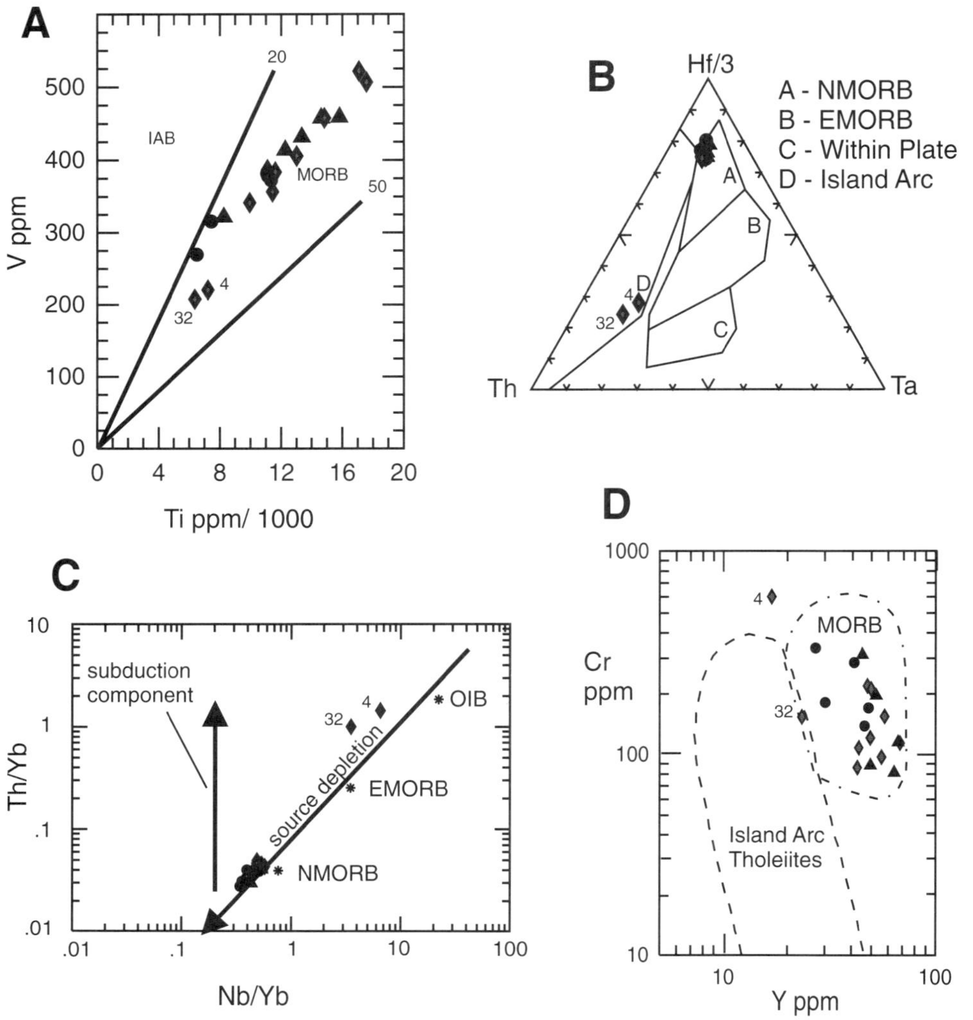

Figure 7. Empirically derived basalt discrimination diagrams. Anomalous samples 4 and 32 are labeled. (A) Ti vs. V (Shervais, 1982); (B) Th-Hf-Ta (Wood, 1980); (C) Nb/Yb vs. Th/Yb (modified from Pearce et al., 1995); (D) Cr vs. Y (Pearce, 1982; Pearce et al., 1995). The symbols are as in Figure 2. E-MORB—enriched mid-ocean ridge basalt; N-MORB—normal mid-ocean ridge basalt; OIB—ocean island basalt.

beneath the Eastern Klamath terrane, and dynamothermal metamorphism of the Central Metamorphic terrane. The Silurian-Devonian age of cogenetic magmatic activity in the Trinity and Redding subterranes and contemporaneous trench-slope deposition in the Yreka subterrane is well documented (Boucot et al., 1974; Cashman, 1980; Albers et al., 1981; Wallin and Gehrels, 1995; Wallin et al., 1995; Wallin and Metcalf, 1998; Metcalf et al., 2000). Metamorphic foliation within the Central Metamorphic terrane (S_1) and the overlying Trinity peridotite (S_2) are generally parallel with the Trinity fault, suggesting that foliation development was related to thrusting (Cannat and Boudier, 1985; Peacock, 1987a). The apparent inverted metamorphic gradient preserved in the Central Metamorphic terrane is the primary evidence supporting a hot plate over cold plate thermal model (Davis et al., 1965; Norris, 1983; Peacock and Norris, 1989) and suggests a dynamothermal metamorphism synchronous with thrusting. Two mineral–whole-rock Rb/Sr isochron ages of ca. 374 Ma for metasedimentary rocks of the Grouse Ridge Formation collected near the Trinity thrust fault (Lanphere et al., 1968) are the basis for assigning a Devonian age to Central Metamorphic terrane metamorphism. Our two new Early Permian (ca. 274 Ma) $^{40}Ar/^{39}Ar$ hornblende ages, however, necessitate some reevaluation of this aspect of the long-standing subduction model. Both metabasites dated in this study were collected in the upper structural levels of the Central Metamorphic terrane near the Trinity thrust fault, where Peacock and Norris (1989) estimated metamorphic temperatures on the order of 650 ± 50 °C. Given that hornblende Ar closure temperatures are on the order of 500 °C, the Early Permian dates reported here should be viewed as cooling ages, not peak metamorphic ages. The question becomes, are Early Permian Central Metamorphic terrane cooling ages compatible, in any way, with the Devonian subduction model?

Peacock (1987b, 1996) used heat flow models to investigate the thermal conditions produced by subduction of cold oceanic lithosphere beneath a hot mantle wedge providing a basis for evaluating the new Central Metamorphic terrane $^{40}Ar/^{39}Ar$ hornblende ages. These heat flow models show that under likely boundary conditions, a greenschist-amphibolite facies series cannot be generated under the steady-state conditions of an established subduction zone. Rather, under typical steady-state conditions, subducted oceanic lithosphere should pass through blueschist- to eclogite-facies P-T conditions (Peacock, 1996).

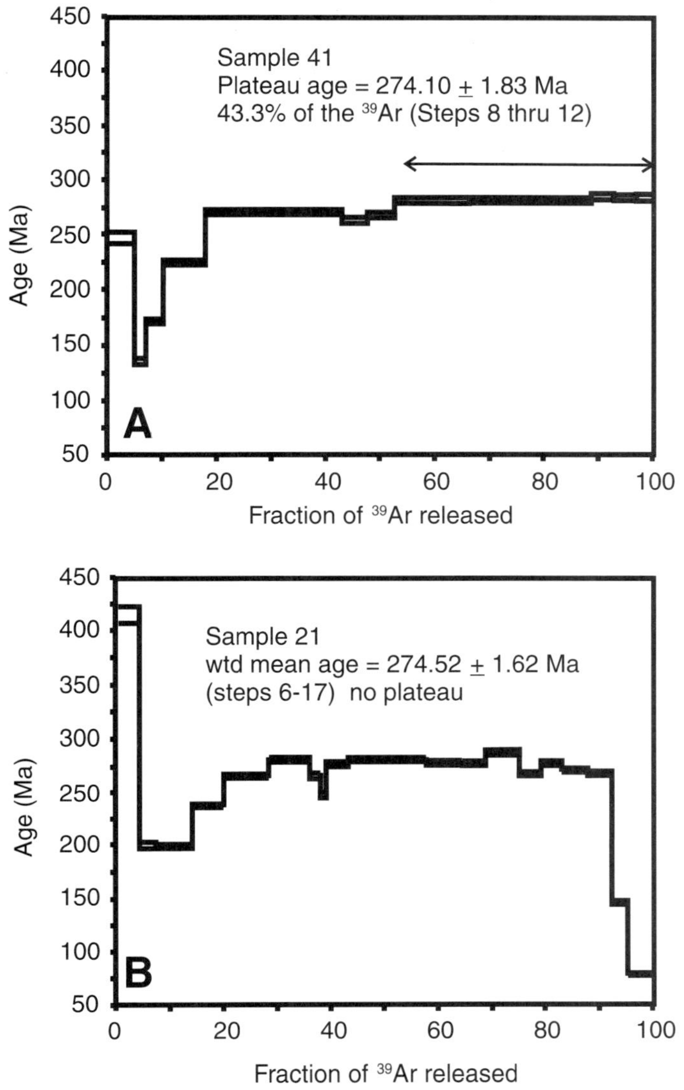

Figure 8. Spectrums of $^{40}Ar/^{39}Ar$: (A) sample 41; (B) sample 21.

Peacock (1987b) produced models specifically aimed at simulating the inverted geotherms necessary to produce inverted metamorphic gradients in the Central Metamorphic terrane and other suspected paleo-subduction margins. These models showed that during the early stages of subduction, prior to steady state, an amphibolite- to greenschist-facies inverted gradient could be generated while subducted oceanic material was being progressively accreted to the upper plate (Peacock, 1987b). Continued subduction would refrigerate the mantle wedge and the accreted oceanic material within a few million years of subduction initiation (Peacock, 1987b), blocking Ar diffusion in hornblende. Termination of subduction as a dead plate (as opposed to a subducted spreading ridge/slab window) would allow normal oceanic geotherms to reestablish in ca. 10 m.y. (Peacock, 1987b). Under such conditions, temperatures sufficient to reset Ar systematics in hornblende (~500 °C) would correspond to depths of 35–40 km (Peacock, 1987b), potentially exceeding the upper pressure limits of amphibolite-facies conditions. Thus, it seems unlikely that the Central Metamorphic terrane $^{40}Ar/^{39}Ar$ hornblende cooling ages can be related in any direct way to a Devonian subduction margin.

Several features of the Central Metamorphic terrane are reminiscent of metamorphic core complexes (Coney, 1980) and lead us to hypothesize that an Early Permian extensional exhumation event produced the ca. 274-Ma $^{40}Ar/^{39}Ar$ hornblende ages. Although the Central Metamorphic terrane records metamorphism at ~650 ± 50 °C and 4–8 kb adjacent to the Trinity fault (Peacock and Norris, 1989), petrologic data suggest that Silurian-Devonian gabbroic plutons in the Trinity subterrane were emplaced into relative cool peridotite (<500 °C) at pressures of less than ~3.3 kb and perhaps <2 kb (Schwindinger and Anderson, 1987; Barrow and Metcalf, 2004). Although Silurian-Devonian igneous units in the Trinity and Redding subterranes exhibit lower greenschist-facies alteration, the effects are not ubiquitous; most rocks retain primary igneous textures and, in many cases, primary mineral assemblages (Quick, 1981; Metcalf et al. 2000). Thus, the Trinity fault represents a metamorphic discontinuity separating higher-grade schist in the footwall from low-grade to unmetamorphosed rocks in the hanging wall. In addition, the Trinity fault is a serpentinite shear zone that overprints high-temperature (olivine) fabrics in the hanging wall peridotite and records hydration of peridotite at temperatures (475–570 °C) below those recorded by the footwall schist and (Peacock, 1987b), suggesting that the most recent motion on the fault postdated peak metamorphism in the footwall and ductile deformation in the hanging wall. The geometry of the Trinity fault, with a moderately east-dipping (~45°) main fault trace and subhorizontal segments beneath two Eastern Klamath terrane klippen to the west (Fig. 1), is similar to that of faults as-

TABLE 3. SUMMARY OF $^{40}Ar/^{39}Ar$ HORNBLENDE DATA

Sample	J*	Weight (mg)	Isochron (Ma)	Total gas age (Ma)	Plateau† age (Ma)	Weighted mean§ age (Ma)	T steps used (°C)	Percentage ^{39}Ar released
41	0.001521	13.51	None	260.7 ± 0.6	274.1 ± 1.83	274.1 ± 1.5	1160–1400	43.3
21	0.001642	18.08	None	257 ± 1.37	None	274 ± 1.62	1050–1280	64.2

Notes: Isotope beams in mV. Error in age includes 0.5% J error, all errors 1σ.
*J is the irradiation parameter.
†Pseudo plateau ages are based on temperature steps and fraction of ^{39}Ar listed.
§Weighted mean ages are based on temperature steps and fraction of ^{39}Ar listed.

sociated with major extensional core complexes. Early interpretation of the Trinity fault as a thrust was based largely on circumstantial evidence, such as stratigraphic discordance and metamorphic discontinuity across the fault and its arcuate map trace (Irwin and Lipman, 1962; Davis, 1968). Although a top-to-the-west (thrust) shear sense was reported by Cannat and Boudier (1985) for the Trinity fault, their work was based primarily on high-temperature fabrics (preferred crystallographic orientation in olivine) measured in Trinity peridotite (S_2) adjacent to the main trace of the fault. No shear sense indicators have been reported from lower-temperature serpentinite shear zone that represents the main trace of the Trinity fault.

We propose a modification to the eastern Klamath subduction model to explain the Early Permian ages. As with the long-standing model, we envision a history of Devonian subduction-related metamorphism and accretion of the ocean floor, Central Metamorphic terrane protolith (N-MORB) to upper-plate mantle peridotite. The inverted metamorphic gradient, high-temperature top-to-the-west shear indicators in the hanging wall peridotite, juxtaposition of N-MORB (Central Metamorphic terrane) and supra-subduction (Eastern Klamath terrane) lithologies across the Trinity fault, and perhaps the ca. 374-Ma Rb-Sr ages, record this portion of the history. Within a few million years of subduction initiation (based on Late Silurian–Early Devonian constraints from Eastern Klamath terrane magmatic rocks), dynamothermal metamorphism, accretion, and refrigeration of the Central Metamorphic terrane to temperatures below 500 °C was completed. Such refrigeration would be required to preserve the inverted metamorphic gradient (Peacock, 1987b). At that point, the accreted Central Metamorphic terrane material was at a greater paleo-depth than its current position relative to the adjacent Eastern Klamath terrane hanging wall. According to our hypothesis, during the Early Permian, the Trinity thrust fault was reactivated as an extensional structure. We suggest that during Permian time, the Central Metamorphic terrane and Eastern Klamath terrane resided in a supra-subduction setting, as evidenced by Permian volcanic/volcaniclastic units (Nosomi Formation; Dekkas andesite) and the Early Permian (261 Ma, U/Pb zircon) Pit River stock (Fraticelli et al., 1985; Charvet et al., 1990; Irwin, 1994) present in the Redding subterrane. Geochemical data for these Permian rocks point to subduction zone magmas derived from a fertile (enriched) mantle source (Charvet et al., 1990). We suggest that extension in the upper plate, probably in response to trench roll-back, reactivated the Trinity thrust fault as an extensional structure. The convection of fertile mantle into the mantle wedge may have produced a thermal pulse that outgassed the hornblende in the Central Metamorphic terrane at depth during the early stages of extension. The Early Permian $^{40}Ar/^{39}Ar$ hornblende ages likely record cooling of the Central Metamorphic terrane during exhumation. Such an extensional model is consistent with the observation that the upper plates of many modern oceanic subduction zones are in a state of extension (Hamilton, 1995) and with models that view the history of long-lived convergent margins as one of alternating periods of extension and contraction (Collins, 2002).

Our geochemical work demonstrates that the Trinity fault juxtaposes supra-subduction zone elements in the hanging wall and metamorphosed mid-ocean ridge crust in the footwall. Such juxtaposition demonstrates that the Trinity fault is a major structure within the architecture of the Klamath Mountains province and likely represents a paleo-plate boundary. Our new $^{40}Ar/^{39}Ar$ hornblende cooling ages point to a more complicated thermal history for the Central Metamorphic terrane than previously recognized and raises questions regarding timing and origin of peak metamorphism. The Early Permian extensional hypothesis provides an explanation for the hornblende cooling ages and presents some intriguing possibilities regarding the long-term evolution of the Klamath Mountains convergent margin. The extensional hypothesis, however, complicates the hot plate over cold plate model for the production of the apparent inverted metamorphic gradient in the Central Metamorphic terrane. The Early Permian extensional hypothesis and its implications needs to be more fully tested and is a focus of our continuing research.

CONCLUSIONS

1. Central Metamorphic terrane samples show little to no evidence of major element chemical changes due to seafloor metasomatism. Post-magmatic alteration of the Central Metamorphic terrane samples appears to have had minimal impact on the immobile trace element compositions. Immobile, incompatible trace elements appear to preserve original igneous fractionation trends.
2. A majority of Central Metamorphic terrane metabasites show a clear MORB geochemical signature on empirically derived basalt discrimination diagrams using relatively immobile elements (e.g., Th, Hf, Ta, Cr, Y, Yb, Ti, V).
3. Results support an interpretation that the Central Metamorphic terrane protolith was N-MORB-type oceanic lithosphere generated at a mid-ocean ridge spreading center and then accreted to the base of the over-riding plate at an intra-oceanic convergent margin.
4. The Trinity fault represents a major paleo-plate boundary, separating the N-MORB-like Central Metamorphic terrane and the overlying, arclike Eastern Klamath terrane.
5. Major paleo-plate boundaries may be recognized where N-MORB-type metabasites underlie an upper-plate sequence with magmatic rocks carrying arclike trace element compositions.
6. Two new $^{40}Ar/^{39}Ar$ hornblende ages record a previously unrecognized Early Permian (ca. 274 Ma) thermal event in the eastern Klamath Mountains. This event may record reactivation of the Trinity fault as an extensional structure in response to upper plate extension in a Late Paleozoic subduction zone.

ACKNOWLEDGMENTS

This research was supported by National Science Foundation Grant EAR 96-28552 to RVM and E.T. Wallin and grants to WMB from the Geological Society of America, the University of Nevada, Las Vegas (UNLV), Department of Geoscience, and the UNLV Graduate and Professional Student Association. We are grateful to Terry Spell and Kathy Zanetti for access to the UNLV Argon Laboratory and for their assistance in collecting and interpreting the data. WMB thanks Phyllis Barrow for help and encouragement. We thank Wanda Taylor, Terry Spell, and Michael Wells for their comments on an earlier version of the manuscript. The chapter benefited from careful reviews provided by Cal Barnes, John Goodge, and Bradley Hacker.

REFERENCES CITED

Albers, J.P., Kistler, R.W., and Kwak, L., 1981, The Mule Mountain stock, an early Middle Devonian pluton in northern California: Isochron West, v. 31, p. 17.

Barrow, W.M., and Metcalf, R.V., 2004, The mid-Paleozoic Trinity thrust fault of the eastern Klamath Mountains, northern California: Reactivated in the Permian as an extensional structure: Geological Society of America Abstracts with Programs, v. 36, p. 148.

Boucot, A.J., Dunkle, D.H., Potter, A.W., Savage, N.M., and Rohr, D.M., 1974, Middle Devonian orogeny in western North America? A fish and other fossils: Journal of Geology, v. 82, p. 691–708.

Burchfiel, B.C., and Davis, G.A., 1975, Nature and controls of Cordilleran orogenesis, western United States: Extensions of an earlier synthesis: Journal of Science, v. 275-A, p. 363–396.

Burchfiel, B.C., and Royden, L.H., 1991, Antler orogeny: A Mediterranean-type orogeny: Geology, v. 19, p. 66–69, doi: 10.1130/0091-7613(1991)019<0066:AOAMTO>2.3.CO;2.

Cannat, M., and Boudier, F., 1985, Structural study of intra-oceanic thrusting in the Klamath Mountains, northern California: Implications on accretion geometry: Tectonics, v. 4, p. 435–452.

Cashman, S.M., 1980, Devonian metamorphic event in the northeastern Klamath Mountains, California: Geological Society of America Bulletin, v. 91, p. 453–459, doi: 10.1130/0016-7606(1980)91<453:DMEITN>2.0.CO;2.

Charvet, J., Lapierre, H., Rouer, O., Coulon, C., Campos, C., Martin, P., and Lécuyer, C., 1990, Tectono-magmatic evolution of Paleozoic and early Mesozoic rocks in the eastern Klamath Mountains, California, and the Blue River Mountains, eastern Oregon–western Idaho, *in* Harwood, D.S., and Miller, M.M., eds., Paleozoic and early Mesozoic paleogeographic relations; Sierra Nevada, Klamath Mountains and related terranes: Boulder, Colorado, Geological Society of America Special Paper 255, p. 255–276.

Collins, W.J., 2002, Hot orogens, tectonic switching, and creation of continental crust: Geology, v. 30, p. 535–538.

Coney, P.J., 1980, Cordilleran metamorphic core complexes—An overview, *in* Crittenden, M.D., Jr., et al., eds., Cordilleran metamorphic core complexes: Boulder, Colorado, Geological Society of America Memoir 153, p. 7–34.

Cox, K.G., Bell, J.D., and Pankhurst, R.J., 1979, The interpretation of igneous rocks: London, Allen and Unwin, 450 p.

Davis, G.A., 1968, Westward thrust faulting in the south-central Klamath Mountains, California: Geological Society of America Bulletin, v. 79, p. 911–934.

Davis, G.A., Holdaway, M.J., and Lipman, P.W., 1965, Structure, metamorphism and plutonism in the south-central Klamath Mountains, California: Geological Society of America Bulletin, v. 76, p. 933–966.

Donato, M.M., 1991, Geochemical recognition of a captured back-arc basin metabasaltic complex, southwestern Oregon: Journal of Geology, v. 99, p. 711–728.

Fraticelli, L.A., Albers, J.P., and Zartman, R.E., 1985, The Permian Pit River stock of the McCloud plutonic belt, Eastern Klamath terrane, northern California: Isochron West, v. 44, p. 6–8.

Fuis, G.S., Zucca, J.J., Mooney, W.D., and Milkereit, B., 1987, A geologic interpretation of seismic-refraction results in northeastern California: Geological Society of America Bulletin, v. 98, p. 53–65, doi: 10.1130/0016-7606(1987)98<53:AGIOSR>2.0.CO;2.

Goodge, J.W., 1989, Polyphase metamorphic evolution of a Late Triassic subduction complex, Klamath Mountains, Northern California: American Journal of Science, v. 289, p. 874–943.

Goodge, J.W., 1990, Tectonic evolution of a coherent Late Triassic subduction complex, Stuart Fork Terrane, Klamath Mountains, Northern California: Geological Society of America Bulletin, v. 102, p. 86–101, doi: 10.1130/0016-7606(1990)102<0086:TEOACL>2.3.CO;2.

Hacker, B.R., Ernst, W.G., and McWilliams, M.O., 1993, Genesis and evolution of a Permian-Jurassic magmatic arc/accretionary wedge, and reevaluation of terranes in the central Klamath Mountains: Tectonics, v. 12, p. 387–409.

Hacker, B.R., Donato, M.M., Barnes, C.G., McWilliams, M.O., and Ernst, W.G., 1995, Timescales of orogeny: Jurassic construction of the Klamath Mountains: Tectonics, v. 14, p. 677–703, doi: 10.1029/94TC02454.

Hamilton, W.B., 1995, Subduction systems and magmatism, *in* Smellie, J.L., eds., Volcanism associated with extension at consuming margins: Boulder, Colorado, Geological Society Special Publication 81, p. 3–28.

Hawkins, J.W., Jr., 1995, The geology of the Lau basin, *in* Taylor, B., ed., Backarc basins: Tectonics and magmatism: New York, Plenum Press, p. 63–138.

Holdaway, M.J., 1965, Basic regional metamorphic rocks in part of the Klamath Mountains, northern California: American Mineralogist, v. 50, p. 953–977.

Hotz, P.E., Lanphere, M.A., and Swanson, D.A., 1977, Triassic blueschist from northern California and north-central Oregon: Geology, v. 5, p. 659–663, doi: 10.1130/0091-7613(1977)5<659:TBFNCA>2.0.CO;2.

Humphris, S.E., and Thompson, G., 1978, Hydrothermal alteration of oceanic basalts by seawater: Geochimica et Cosmichimica Acta, v. 42, p. 107–125, doi: 10.1016/0016-7037(78)90221-1.

Irwin, W.P., 1960, Geological reconnaissance of the northern Coastal Ranges and Klamath Mountains, California, with a summary of the mineral resources: Sacramento, California Division of Mines and Geology Bulletin 149, 80 p.

Irwin, W.P., 1977, Review of Paleozoic rocks of the Klamath Mountains, *in* Stewart, J.H., et al., eds, Paleogeography of the western United States, Pacific Coast Paleogeography Symposium 1: Los Angeles, Pacific Section, Society of Economic Paleontologists and Mineralogists, p. 441–454.

Irwin, W. P., 1989, Terranes of the Klamath Mountains, California and Oregon, *in* Blake, M.C., and Harwood, D.S., eds., Sedimentation and tectonics of western North America 2: Tectonic evolution of northern California: Washington, D.C., 28th International Geologic Congress Field Trip Guidebook T108, p. 19–32.

Irwin, W.P., 1994, Geologic map of the Klamath Mountains, California and Oregon: Reston, Virginia, U.S. Geological Survey Miscellaneous Investigation Series Map I-2148, scale 1:500,000, 2 sheets.

Irwin, W.P., and Lipman, P.W., 1962, A regional ultramafic sheet in eastern Klamath Mountains, California: Washington, D.C., U.S. Geological Survey Professional Paper 450-C, p. C18–C21.

Irwin, W.P., and Wooden, J.L., 1999, Plutons and accretionary episodes of the Klamath Mountains, California and Oregon: Reston, Virginia, U.S. Geological Survey Open-File Report 99-374.

Jacobsen, S.B., Quick, J.E., and Wasserburg, G.J., 1984, A Nd and Sr isotopic study of the Trinity peridotite—Implications for mantle evolution: Earth and Planetary Science Letters, v. 68, p. 361–378, doi: 10.1016/0012-821X(84)90122-5.

Jones, D.L., Blake, M.C., Jr., Bailey, E.H., and McLaughlin, R.J., 1978, Distribution and character of upper Mesozoic subduction complexes along the west coast of North America, *in* Burns, K.L., and Rutland, R.W.R., eds., Structural characteristics of tectonic zones: Tectonophysics, v. 47, p. 207–222, doi: 10.1016/0040-1951(78)90031-8.

Kinkel, A.R., Hall, W.E., and Albers, J.P., 1956, Geology and base-metal deposits of West Shasta copper-zinc district, Shasta County, California: Washington, D.C., U.S. Geological Survey Professional Paper 285, 156 p.

Lanphere, M.A., Irwin, W.P., and Holtz, P.E., 1968, Isotopic age of the Nevadan orogeny and older plutonic and metamorphic events in the Klamath Mountains, California: Geological Society of America Bulletin, v. 79, p. 1027–1052.

Lindsley-Griffin, N., Griffin, J.R., and Wallin, E.T., 1991, Redefinition of the Gazelle Formation of the Yreka terrane, Klamath Mountains, California: Paleogeographic implications, *in* Cooper, J.D., and Stevens, S.H., eds., Paleozoic Paleogeography of the western United States II: Los Angeles, Pacific Section, Society of Economic Paleontologists and Mineralogists, p. 609–624.

Lipman, P.W., 1964, Structure origin of an ultramafic pluton in the Klamath Mountains, California: American Journal of Science, v. 262, p. 199–222.

Metcalf, R.V., Willse, K.R., and Wallin, E.T., 1999, Trace element and Nd-Sr-Pb isotopic data from the Silurian Bonanza King mafic intrusive complex, Trinity terrane, CA: Evidence of supra-subduction zone magmatism in a proto-arc paleotectonic setting: Geological Society of America Abstracts with Programs, v. 31, p. A-79.

Metcalf, R.V., Wallin, E.T., Willse, K.R., and Muller, E.R., 2000, Geology and geochemistry of the ophiolitic Trinity terrane, California: Evidence of mid-Paleozoic ultradepleted supra-subduction zone magmatism in a proto-arc setting, *in* Dilek, Y., et al., eds., Ophiolites and oceanic crust: New insights from field studies and the Ocean Drilling Program: Boulder, Colorado, Geological Society of America Special Paper 349, p. 403–418.

Miller, M.M., 1989, Intra-arc sedimentation and tectonism: Late Paleozoic evolution of the eastern Klamath terrane, California: Geological Society of America Bulletin, v. 101, p. 170–187, doi: 10.1130/0016-7606(1989)101<0170:IASATL>2.3.CO;2.

Miyashiro, A., 1978, Nature of alkalic volcanic rock series: Contributions to Mineralogy and Petrology, v. 66, p. 91–104, doi: 10.1007/BF00376089.

Morgan, W.J., 1972, Plate motions and deep mantle convection: Boulder, Colorado, Geological Society of America Memoir 132, p. 7–22.

Mottl, M.J., 1983, Metabasalts, axial hot springs, and the structure of hydrothermal systems at mid-ocean ridges: Geological Society of America Bulletin, v. 94, p. 161–180, doi: 10.1130/0016-7606(1983)94<161:MAHSAT>2.0.CO;2.

Norris, P.J., 1983, Metamorphism at the base of the Trinity peridotite, Coffee Creek and the southern Trinity Alps area, northern California [Ph.D. thesis]: Los Angeles, University of California, 109 p.

Peacock, S.M., 1987a, Serpentinization and infiltration metasomatism in the Trinity peridotite, Klamath province, northern California: Implications for subduction zones: Contributions to Mineralogy and Petrology, v. 95, p. 55–70, doi: 10.1007/BF00518030.

Peacock, S.M., 1987b, Creation and preservation of subduction-related inverted metamorphic gradients: Journal of Geophysical Research, v. 92, p. 12,763–12,781.

Peacock, S.M., 1996, Thermal and petrologic structure of subduction zones, *in* Bebout, G.E., et al., eds., Subduction top to bottom: Washington, D.C., American Geophysical Union Geophysical Monograph 96, p. 119–133.

Peacock, S.M., and Norris, P.J., 1989, Metamorphic evolution of the central metamorphic belt, Klamath Province, California: An inverted metamorphic gradient beneath the Trinity peridotite: Journal of Metamorphic Geology, v. 7, p. 191–209.

Pearce, J.A., 1982, Trace element characteristics of lavas from destructive plate Boundaries, *in* Thorpe, R.S., ed., Andesites: Orogenic andesites and related rocks: Chichester, Wiley, p. 525–548.

Pearce, J.A., 1996, A user's guide to basalt discrimination diagrams, *in* Wyman, D. A., ed., Trace element geochemistry of volcanic rocks: Applications for massive sulphide exploration: St. John's, Newfoundland, Geological Association of Canada Short Course Notes 12, p. 79–113.

Pearce, J.A., Lipman, S.J., and Roberts, S., 1984, Characteristics and tectonic significance of supra-subduction zone ophiolites, *in* Kokelaar, B.P., and Howels, M.F., eds., Marginal basin geology: London, Geological Society Special Publication 16, p. 77–94.

Pearce, J.A., Keskin, M., Serri, G., and Innocenti, F., 1995, Tectonic significance of volcanism related to arc-continent collisions: EOS, Transactions, American Geophysical Union, v. 76, no. 46, suppl., p. 602–607.

Potter, A.W., Hotz, P.E., and Rohr, D.M., 1977, Stratigraphy and inferred tectonic framework of the lower Paleozoic rocks in the eastern Klamath Mountains, northern California, *in* Stewart, J.H., et al., eds., Paleozoic paleogeography of the western United States: Los Angeles, Pacific Section, Society of Economic Paleontologists and Mineralogists, p. 421–440.

Quick, J.E., 1981, Petrology and petrogenesis of the Trinity peridotite, an upper mantle diapir in the eastern Klamath Mountains, northern California: Journal of Geophysical Research, v. 86, p. 1836–1837.

Reiners, P.W., Spell, T.L., Nicolescu, S., and Zanetti, K.A., 2004, Zircon (U-Th)/He thermochronometry: He diffusion and intercalibration with ^{40}Ar/^{39}Ar dating: Geochimica et Cosmochimica Acta, v. 68, p. 1857–1887, doi: 10.1016/j.gca.2003.10.021.

Sanborn, A.F., 1960, Geology and paleontology of a part of the Big Bend quadrangle, Shasta County, California: Sacramento, California Division of Mines Special Report 63, 26 p.

Schwindinger, K.R., and Anderson, A.T., Jr., 1987, Probable low-pressure intrusion of gabbro into serpentinized peridotite, northern California: Geological Society of America Bulletin, v. 98, p. 364–372, doi: 10.1130/0016-7606(1987)98<364:PLIOGI>2.0.CO;2.

Shervais, J.W., 1982, Ti-V plots and the petrogenesis of modern and ophiolitic lavas: Earth and Planetary Science Letters, v. 59, p. 101–118, doi: 10.1016/0012-821X(82)90120-0.

Speed, R.C., 1979, Collided Paleozoic microplate in the western United States: Journal of Geology, v. 87, p. 279–292.

Speed, R.C., Elison, M.W., and Heck, F.R., 1988, Phanerozoic tectonic evolution of the Great Basin, *in* Ernst, W.G., ed., Metamorphism and crustal evolution of the western United States—Rubey Volume VII: Englewood Cliffs, New Jersey, Prentice-Hall, p. 572–605.

Sun, S.-s., and McDonough, W.F., 1989, Chemical and isotopic systematics of oceanic basalts: Implications for mantle composition and processes, *in* Saunders, A.D., and Norris, M.J., eds., Magmatism in the ocean basins: London, Geological Society London Special Publication 42, p. 313–325.

Wallin, E.T., and Gehrels, G.E., 1995, Sedimentary record of Silurian volcanism reveals denudation of the Trinity terrane and the timing of silicicastic sedimentation in the Yreka terrane, northern California: Geological Society of America Abstracts with Programs, v. 27, p. 83.

Wallin, E.T., and Metcalf, R.V., 1998, Supra-subduction zone ophiolite formed in an extensional forearc: Trinity terrane, Klamath Mountains: California: Journal of Geology, v. 106, p. 591–608.

Wallin, E.T., and Trabert, D.W., 1994, Eruption-controlled epiclastic sedimentation in a Devonian trench-slope basin: Evidence from sandstone petrofacies, Klamath Mountains: California: Journal of Sedimentary Research, v. A64, p. 373–385.

Wallin, E.T., Coleman, D.S., Lindsley-Griffin, N., and Potter, A.W., 1995, Silurian plutonism in the Trinity terrane (Neoproterozoic and Ordovician), Klamath Mountains, California, USA: Tectonics, v. 14, p. 1007–1013, doi: 10.1029/95TC01447.

Wilson, M., 1989, Igneous petrogenesis: London, Unwin Hyman, 466 p.

Wood, D.A., 1980, The application of a Th-Hf-Ta diagram to problems of tectonomagmatic classification and to establishing the nature of crustal contamination of basaltic lavas of the British Tertiary volcanic province: Earth and Planetary Science Letters, v. 50, p. 11–30, doi: 10.1016/0012-821X(80)90116-8.

Wright, J.E., and Fahan, M.R., 1988, An expanded view of Jurassic orogenesis in the western United States Cordillera: Middle Jurassic (pre-Nevadan) regional metamorphism and thrust faulting within an active arc environment, Klamath Mountains, California: Geological Society of America Bulletin, v. 100, p. 859–876, doi: 10.1130/0016-7606(1988)100<0859:AEVOJO>2.3.CO;2.

Manuscript Accepted by the Society 10 January 2006

Printed in the USA

Geological Society of America
Special Paper 410
2006

Ediacaran cyclomedusoids and the paleogeographic setting of the Neoproterozoic–early Paleozoic Yreka and Trinity terranes, eastern Klamath Mountains, California

Nancy Lindsley-Griffin*
John R. Griffin
Department of Geosciences, University of Nebraska, 214 Bessey Hall, Lincoln, Nebraska, 68588, USA
Jack D. Farmer
Department of Geological Sciences, PSF Room 686, Arizona State University, Tempe, Arizona 85287, USA
Eric A. Sivers*
Brian Bruckno*
Mary K. Tozer*
Department of Geosciences, University of Nebraska, 214 Bessey Hall, Lincoln, Nebraska, 68588, USA

ABSTRACT

Newly recognized fossil cyclomedusoids from the Yreka terrane include *Ediacaria* sp. and *Beltanella* sp. They are typical of the Ediacaran fossil assemblage, range from 640 to 575 Ma, and thus are latest Neoproterozoic (Vendian) in age. The Yreka terrane structurally overlies the Trinity terrane, which also includes Vendian rocks. The Yreka terrane is a polygenetic stack of sedimentary and metasedimentary thrust sheets consisting of the Vendian Antelope Mountain Quartzite, Siluro-Devonian turbidites (trench fill), lower Paleozoic mélanges (accretionary complexes), and the Lower Devonian Gazelle Formation (trench-slope basin deposits). The Trinity terrane is a polygenetic mafic-ultramafic complex consisting of multiple mantle tectonite blocks and two ophiolitic crustal sequences, one Vendian and one Siluro-Devonian. Multistage textures and structures within the Trinity terrane indicate Vendian or Cambrian ductile deformation in the mantle blocks, followed by pre-Early Ordovician amalgamation, then regional uplift and brittle deformation. The Siluro-Devonian crustal sequence developed on this polygenetic composite basement in a supra-subduction zone setting.

The Trinity and Yreka terranes formed close together, with some Yreka terrane components receiving Trinity terrane detritus. The Lower Devonian Gregg Ranch Complex was the active accretionary wedge on which the Gazelle Formation trench-

*E-mail: Lindsley-Griffin—nlg@unl.edu. Current addresses: Sivers—Montana Department of Environmental Quality, 662 North Warren Street, Helena, Montana 59601, USA; Bruckno—District Geologist, Department of Transportation, Staunton, Virginia 24401, USA; Tozer—8330 Fairfield Road, Brooklyn Park, Minnesota 55444, USA.

Lindsley-Griffin, N., Griffin, J.R., Farmer, J.D., Sivers, E.A., Bruckno, B., and Tozer, M.K., 2006, Ediacaran cyclomedusoids and the paleogeographic setting of the Neoproterozoic–early Paleozoic Yreka and Trinity terranes, eastern Klamath Mountains, California, *in* Snoke, A.W., and Barnes, C.G., eds., Geological studies in the Klamath Mountains province, California and Oregon: A volume in honor of William P. Irwin: Geological Society of America Special Paper 410, p. 411–431, doi: 10.1130/2006.2410(20). For permission to copy, contact editing@geosociety.org.

slope basin formed, accompanied by minor near-trench volcanism. Kinematic analysis of the Gregg Ranch Complex indicates convergence directed from the Yreka terrane toward the Trinity terrane, combined with a strike-slip component, probably during Early Devonian collision of outboard terranes. The Yreka terrane–Trinity terrane composite terrane was then stitched together by Middle Devonian dike swarms that fed overlapping lava flows. Because of the close spatial and temporal proximity between the Yreka and Trinity terranes, paleopoles from the Trinity terrane can be used to suggest paleolatitudes where Yreka terrane biota may have originated, and biogeography of Yreka terrane fossils limits the paleogeographic setting of both terranes.

Keywords: Ediacara, cyclomedusoid, Yreka terrane, Trinity terrane

INTRODUCTION

This chapter reports newly recognized Vendian cyclomedusoid fossils recovered from the Antelope Mountain Quartzite in the Yreka terrane and discusses contact relationships of the Yreka terrane with the structurally lower Trinity terrane (Fig. 1). We include new data on the Vendian fossils and Antelope Mountain Quartzite contact relationships, based on work by Lindsley-Griffin, Griffin, and Farmer. Data on contact relationships between the Yreka and Trinity terranes and the mechanisms and timing of their juxtaposition are from field and laboratory studies presented in M.S. theses by Tozer (1994), Bruckno (1997), and Sivers (2002). Names for individual Klamath terranes are consistent with recent U.S. Geological Survey usage (Irwin and Mankinen, 1998; Irwin, 2003), except in place of "Eastern Klamath terrane," we prefer the names of its individual components: Yreka terrane, Trinity terrane, and Redding terrane (Fig. 1).

Regional Setting

The Klamath superterrane of northern California and southern Oregon (Fig. 1) comprises a series of westward-younging arcuate belts that formed on oceanic crust at an unknown distance from Laurentia and amalgamated with older terranes in a series of collisions from mid-Paleozoic to mid-Mesozoic, as detailed by Irwin (e.g., 1973, 1977a,b, 1981, 1985; Irwin and Mankinen, 1998).

Yreka and Trinity Terranes. The oldest rocks in the Klamath superterrane crop out in the Yreka and Trinity terranes (Fig. 1), both of which contain rocks ranging in age from late Neoproterozoic to Early Devonian. Close spatial links between these two terranes are inferred from petrologic, sedimentological, and structural evidence. The Trinity terrane consists of ultramafic and mafic rocks that crop out over an area of ~1200 km^2; it forms the structural basement of the Yreka and Redding terranes (Fig. 1). The Yreka terrane consists of sedimentary and metasedimentary rocks that crop out over ~800 km^2 and were thrust over the Trinity terrane. The Yreka and Trinity terranes are stitched together by basaltic dike swarms that fed overlying basaltic lava flows (Lindsley-Griffin, 1991, 1994; Wallin et al., 1991). These lava flows have yielded a Middle Devonian paleopole (Mankinen et al., 2002) and overlie strata containing Early (to Middle?) Devonian fossils (Lindsley-Griffin et al., 1991), thus juxtaposition of the two terranes was complete by early Middle Devonian.

Geophysical evidence for the subsurface relationships of these two terranes indicates that a dense, relatively thin subhorizontal sheet of mafic-ultramafic rocks extends northwestward from the vicinity of Gazelle (Fig. 1) beneath the Yreka terrane, based on a positive gravity anomaly over the Yreka terrane (LaFehr, 1966; Griscom, 1977). An aeromagnetic traverse across the two terranes yielded a positive anomaly over the Trinity terrane that decreases over the Yreka terrane (Irwin and Bath, 1962). Griscom (1977) suggested that this decrease indicates that Trinity terrane basement extends beneath only the eastern part of the Yreka terrane rather than all the way to its western edge. The Trinity terrane projects eastward beneath the Redding terrane (Fig. 1) and Cenozoic volcanic rocks of the southern Cascade Range (Blakely et al., 1985, 1997; Fuis et al., 1987). Zucca et al. (1986), who incorporated seismic, gravity, and aeromagnetic data into their model, estimated the Trinity terrane to be 6.5–8 km thick. Zucca et al. (1986) also suggested that less dense rocks underlie the Trinity terrane, possibly consisting of imbricated thin sheets of terranes cropping out in the central and western Klamath Mountains.

Redding Terrane. The Middle Devonian–Early Jurassic Redding terrane has also been called the "Redding section" (e.g., Irwin, 1981) and the "eastern Klamath terrane" (e.g., Miller, 1987, 1989). The Redding terrane structurally overlies the southeastern edge of the Trinity-Yreka composite terrane (Fig. 1). Redding terrane exposures extend over a region of ~3800 km^2 and consist of marine metasedimentary strata derived in part from a series of Paleozoic and Mesozoic volcanic arcs (Irwin, 1981; Miller, 1987, 1989).

YREKA TERRANE

The Yreka terrane (Fig. 2) consists of stacked thrust sheets of marine sedimentary and metasedimentary rocks ranging in age from Vendian through Early Devonian. Each of these

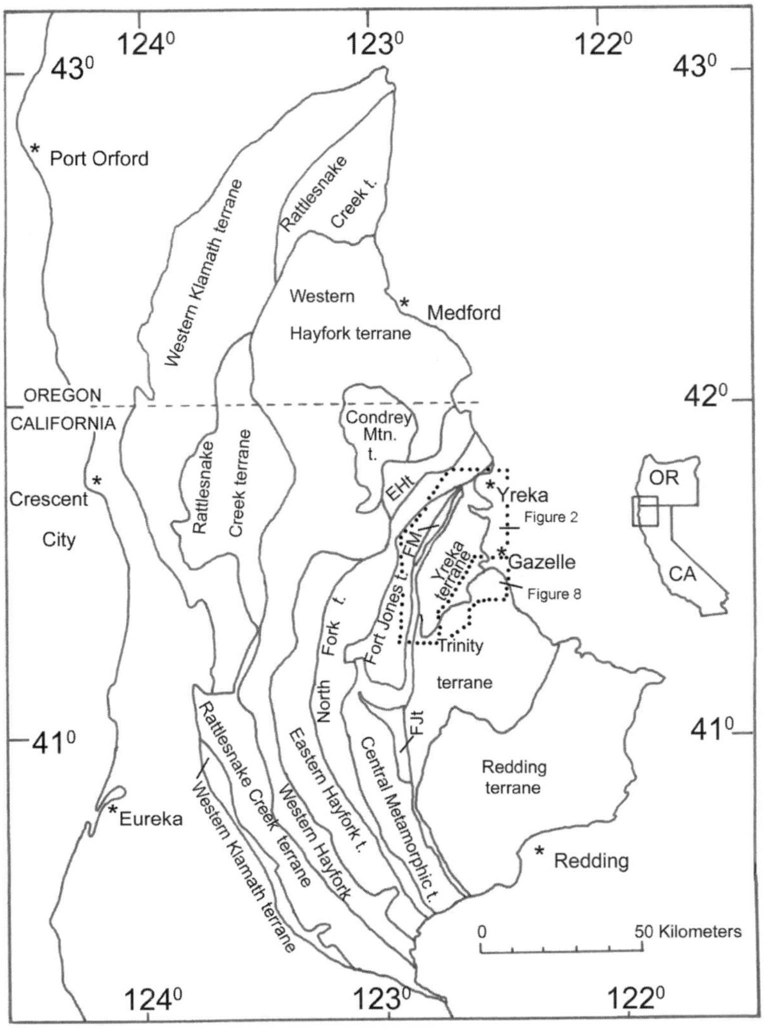

Figure 1. Principle terranes of the Klamath Mountains. Dotted lines outline areas shown in Figures 2 and 8. Adapted from Wallin et al. (1995) and Mankinen et al. (1996, 2002).

thrust sheets exhibits a distinct combination of lithology, internal structure, and metamorphic grade. Wells et al. (1959) recognized two stratigraphic units in Yreka terrane rocks, the metamorphic Ordovician(?) Duzel Formation and the unmetamorphosed Silurian(?) Gazelle Formation. Hotz (1977, 1978) renamed the Duzel Formation as "Duzel Phyllite," and defined three other formal units: Sissel Gulch Graywacke, Moffett Creek Formation, and Antelope Mountain Quartzite (Fig. 2). Subsequent student thesis work and detailed mapping has been summarized in review papers (e.g., Potter et al., 1977, 1990; Lindsley-Griffin and Griffin, 1983; Lindsley-Griffin et al., 1991; Wallin et al., 2000). The Duzel Phyllite, Sissel Gulch Graywacke, and Moffett Creek Formation are now considered to be Silurian or Siluro-Devonian (Wallin et al., 1992; Wallin and Gehrels, 1995), and the Gazelle Formation to be Early Devonian (Lindsley-Griffin et al., 1991). The recognition of fossil cyclomedusoids typical of the worldwide Ediacaran biota confirms that the Antelope Mountain Quartzite is Vendian in age.

Vendian Antelope Mountain Quartzite

The Antelope Mountain Quartzite occupies a separate thrust sheet at the northeast edge of the Yreka terrane (Fig. 2). Field mapping by Lindsley-Griffin and Griffin shows that the lower contact of the Antelope Mountain Quartzite is a fault. Bedding in the quartzite is discordant against structurally lower units everywhere along the contact, and cataclasized or mylonitized quartzite crops out locally just above the contact. Bedding and foliation in units below the contact are also discordant, striking into the contact. We interpret the mapped lower contact as a folded subhorizontal thrust fault locally offset by minor block faults. Structural cross-sections (Lindsley-Griffin and Griffin, 1983) and our recent field work place it structurally above thrust sheets of Duzel Phyllite and mélange of the Schulmeyer Gulch Complex. Other workers have also interpreted this contact as a low-angle fault (Klanderman, 1978; Cashman, 1980).

Cyclomedusoid fossils. Over 250 discoidal structures (Figs.

A

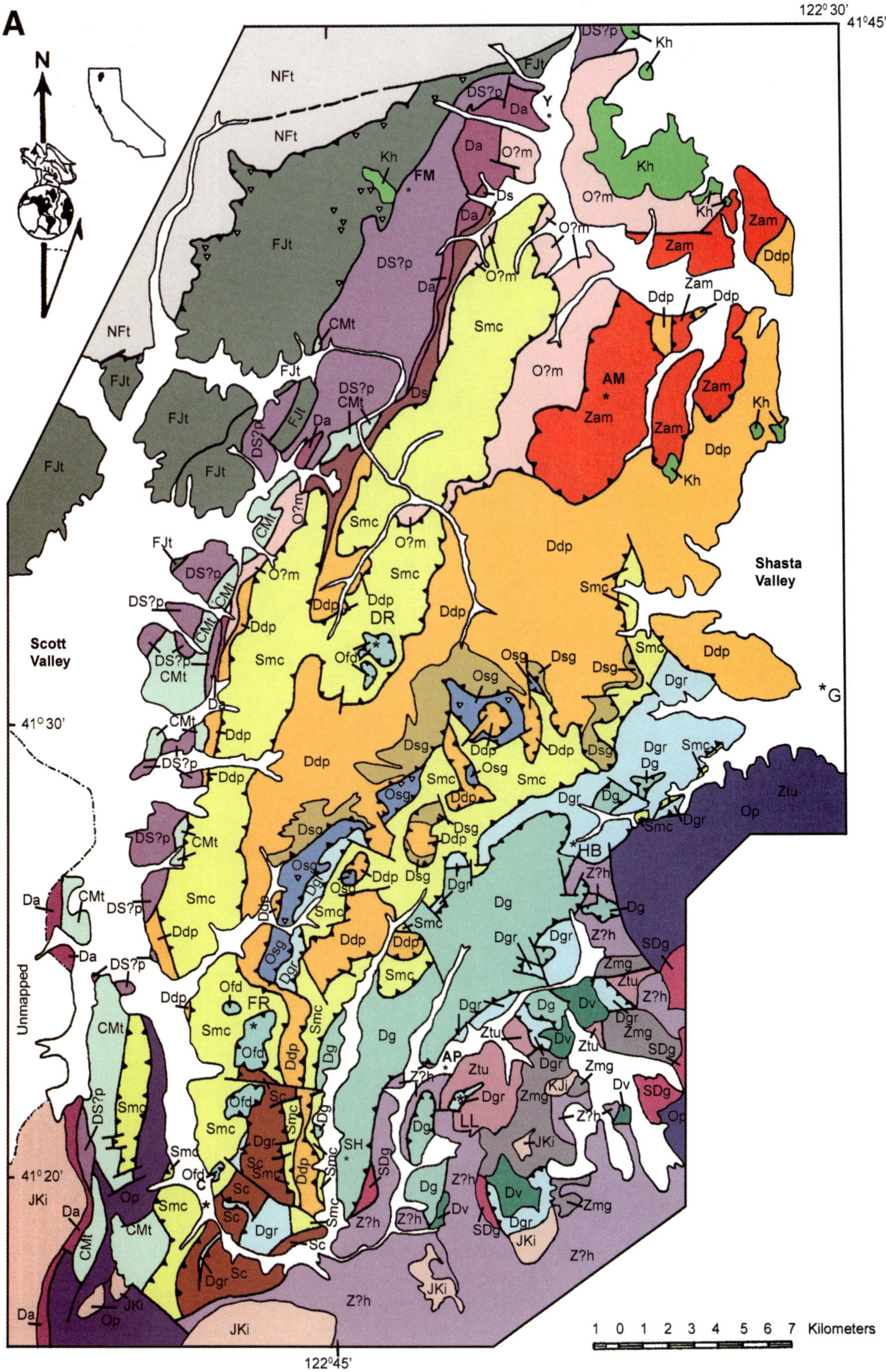

N
122° 30'
41°45'
Shasta Valley
Scott Valley
41° 30'
41° 20'
Unmapped
122°45'
1 0 1 2 3 4 5 6 7 Kilometers
FM
Y
AM
DR
FR
HB
AP
LL
SH
C
G
NFt
FJt
CMt
DS?p
Da
Ds
O?m
Kh
Zam
Ddp
Smc
Ofd
Osg
Dsg
Dgr
Dg
Ztu
Op
Z?h
SDg
Zmg
Dv
KJi
JKi
Sc

3 and 4), herein interpreted to be fossil medusoids, have been recovered from five localities within the Antelope Mountain Quartzite (Lindsley-Griffin et al., 1989, 2002, 2003). These forms are preserved as epireliefs on the upper surfaces of fine-grained metasiltstone and argillite beds (Fig. 3A–E). The assemblage is dominated by forms that are allied to the *Cyclomedusa* complex (Sun, 1986), a group of genera of fossil medusoids that are common components of the late Proterozoic (Vendian) Ediacaran fossil assemblage (Glaessner, 1984). The cyclomedusoid fossils are clustered on exposed bedding plane surfaces near the stratigraphic top of the formation in groups spaced 1–6 cm apart (Fig. 4A). The cyclomedusoids occur within intervals consisting of rhythmic siliceous mudstone layers 2–12 cm thick. These layers have phyllitic bases with mica grains several mm in diameter (Fig. 3E), which grade rapidly up into mudstone and then into very fine argillite at their tops. The in situ specimens in Figure 4A are on the argillite bed top, where the overlying bed has been eroded. In Figures 3E and 4B, a thin veneer of phyllitic metasiltstone, representing the base of the next overlying bed, covers and partially conceals the specimens.

Assessing biogenicity. Some assemblages of cyclomedusoid-like structures have been interpreted as pseudofossils (e.g., Sun, 1986), more specifically, as inorganic fluid escape structures formed by sediment dewatering during the compaction of water-saturated sediments (e.g., Lowe, 1975). Thus, it is vitally important to examine such structures for evidence of sediment fluidization. To evaluate the pseudofossil hypothesis, we prepared thin sections of the core regions of representative specimens from the Antelope Mountain assemblage and examined

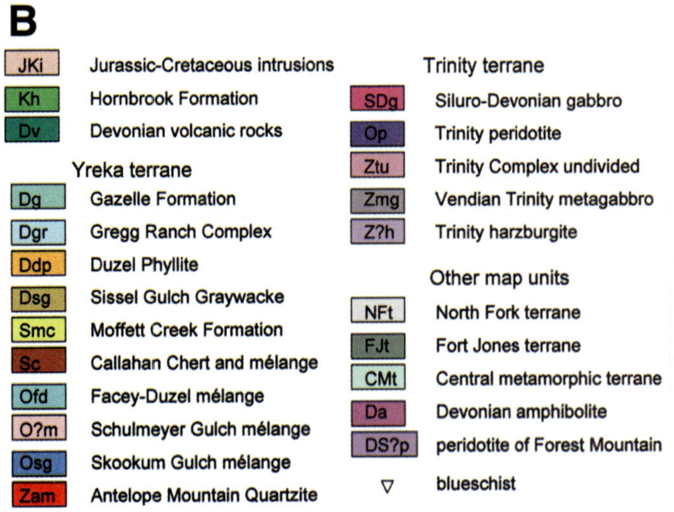

Figure 2. Geologic map of the Yreka terrane. Location symbols: AM—Antelope Mountain summit; AP—AP Ranch; C—Callahan; DR—Duzel Rock; FM—Forest Mountain summit; FR—Facey Rock; G—Gazelle; HB—Horseshoe Bend; LL—Lovers Leap; SH—Schneider Hill; Y—Yreka; no pattern or symbol—undivided Quaternary and Tertiary units. Adapted from Hotz (1977, 1978), Rohr (1978), and Lindsley-Griffin and Griffin (1983).

them under the petrographic microscope for evidence of structures that would indicate dewatering during sediment compaction. Thin-section analysis revealed that the discoidal structures consist of fine-grained, dark reddish-brown mudstone that contains small, isolated patches of microquartz (Fig. 3F). The structures reside in a matrix of thinly laminated, reddish metasiltstone and argillite, cemented by iron oxide. Thin sections also reveal pervasive post-depositional fracturing of both the structures and surrounding matrix, followed by the emplacement of late quartz veins. The larger veins crosscut and stand out in relief on the surface of most of the discoidal structures (Fig. 3). We interpret the quartz veins as resulting from brittle fracturing that followed the low-grade metamorphism and ductile folding that altered the original circular shape of the structures (Fig. 5).

With the exception of a few small micro-faults seen along the base of the structures, sedimentary laminae immediately beneath the medusoids conform to the base of the structures (Fig. 3F), indicating that the structures were more indurated than the matrix during compaction. The lack of disrupted laminae or other evidence of fluidization (e.g., crosscutting mud-filled veins, disturbed laminae) argues against an inorganic dewatering origin for the structures. On this basis, we interpret them as primary biogenic features.

Taxonomic affinities. Ediacaran medusoid taxa (and most notably, species of Cyclomedusa) have been broadly recognized in rocks of Neoproterozoic age, in part because of conspicuous taphonomic variations (Glaessner, 1984). *Cyclomedusa davidii* (Sprigg, 1947), the type species for the genus and the most common of the Ediacaran medusoids, is a circular to elliptical form, with a sub-umbrellar surface that is divided by fine radial grooves and a variable number of concentric furrows that further subdivide the disc into distinct regions. However, radial ornamentation is not observed in all species of this genus (Glaessner, 1984; Farmer et al., 1992), and it is likely that preservation of this feature depends on taphonomic (preservational) history. Consistent with previous reports of cyclomedusoid occurrences from Baltica (e.g., Farmer et al., 1992) and northwestern Laurentia (e.g., Narbonne and Hofmann, 1987; Narbonne and Aitken, 1990), radial ornamentation is rare among the Antelope Mountain medusoids. Most of these specimens more closely resemble the cyclomedusoid genus *Ediacaria,* which is characterized by a basic tripartite organization in which the ex-umbrellar surface is subdivided into three approximately equal regions by concentric furrows, along with a raised marginal rim of variable width (Figs. 3A and B and 5A). The central subdivision may be either raised or depressed (Fig. 3C–E), although the former is most common. Radial furrows may be present in *Ediacaria,* but only on the outer sub-umbrellar surface (Farmer et al., 1992, p. 189). However, no radial furrows have been observed in any of the Antelope Mountain specimens, perhaps reflecting a taphonomic bias toward ex-umbrellar preservation on upper bedding plane surfaces. Other specimens resemble the cyclomedusoid genus *Beltanella* Sprigg

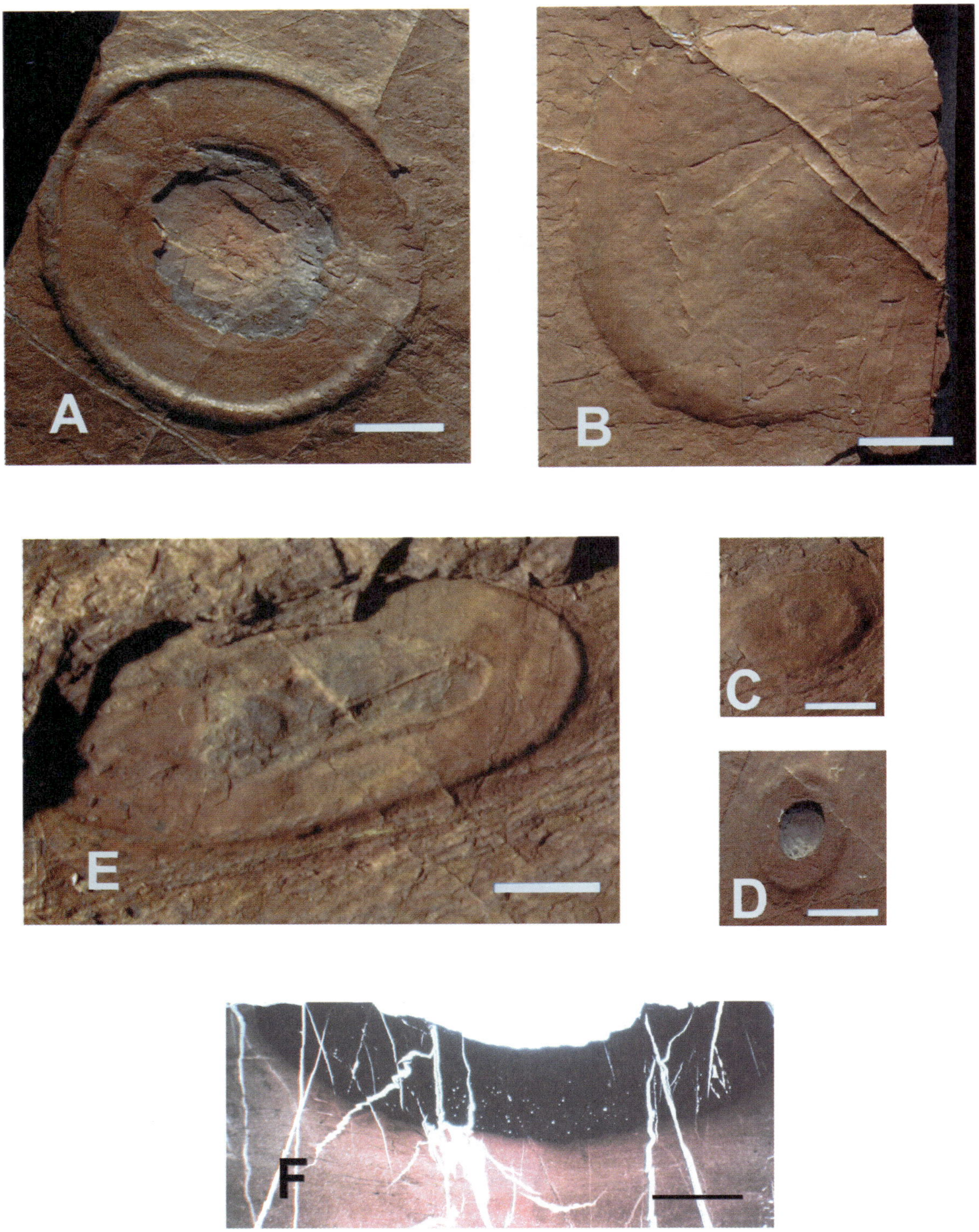

Figure 3. Ediacaran cyclomedusoids of the Antelope Mountain Quartzite. All examples are crossed by fractures filled with secondary quartz veins. (A) *Ediacaria* sp. Epirelief, showing tripartite organization with the ex-umbrellar surface divided into approximately equal regions by one or more concentric furrows. The central division is depressed in this specimen, but raised in others. The well-developed ridge around the margin is common in many specimens. Scale bar = 1.0 cm. (B) *Beltanella* sp. Epirelief showing a smooth, undivided ex-umbrellar surface with a slightly raised margin. A central papilla, common in this form genus, is absent in most of the Antelope Mountain specimens. Scale bar = 1.9 cm. (C) Small *Ediacaria* sp. showing tripartite organization with raised central division. Scale bar = 1.0 cm. (D) Small *Ediacaria* sp. showing tripartite organization with central depression. Scale bar = 1.0 cm. (E) *Ediacaria* sp. showing modified elliptical form resulting from secondary ductile deformation. Coarser-grained phyllite of the overlying layer partially conceals this in situ specimen. Scale bar = 1.0 cm. (F) Photomicrograph of a thin section of *Ediacaria* sp. Fossil consists of fine-grained, dark mudstone with small patches of microquartz and rests on lighter colored, iron-oxide cemented siltstone. Thin laminations in the siltstone conform with the base of the medusoid, except for small micro-faults along the lower contact of the fossil. Scale bar = 0.5 cm.

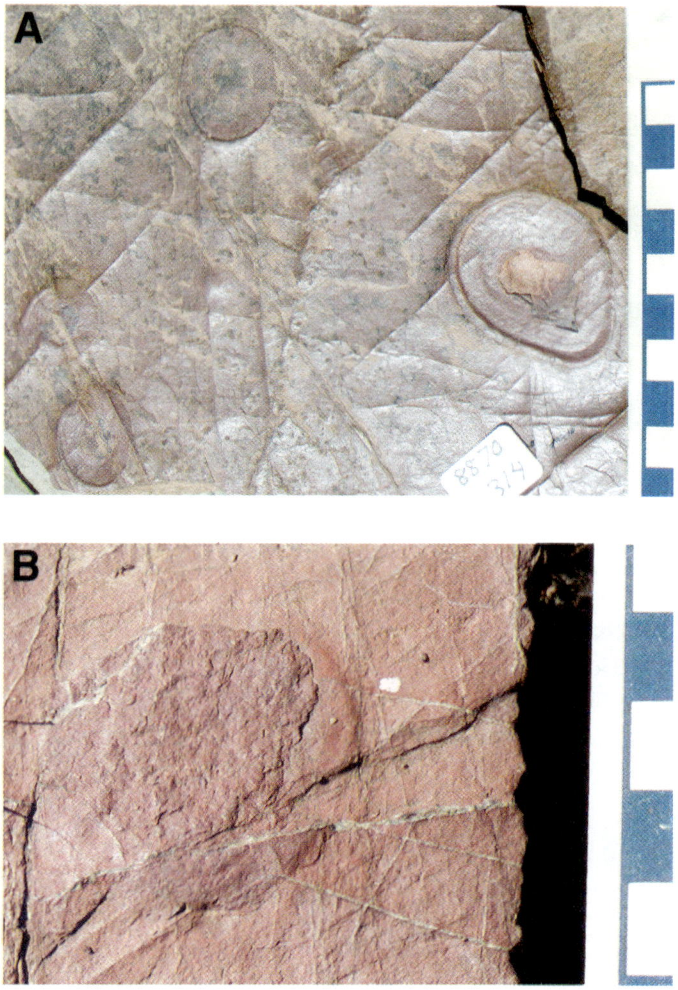

Figure 4. Cyclomedusoid occurrence. Examples are crossed by fractures filled with resistant quartz veins. (A) Four in situ epireliefs ranging from 2.5 to nearly 5 cm across, spaced 1–6 cm apart. Fossil impressions are on the argillitic top of a siliceous mudstone bed, revealed by erosion of the overlying bed. Scale in centimeters. (B) In situ epirelief on argillitic top of a siliceous mudstone bed, overlain and partially concealed by phyllitic metasiltstone base of the overlying layer. Scale in centimeters.

1947 (Farmer et al., 1992, and references therein). This genus exhibits a smooth ex-umbrellar surface and raised marginal rim but lacks a central subdivision. In some of the Antelope Mountain *Beltanella,* a central papilla is present (Fig. 3B). This structure has been observed in other occurrences of *Beltanella* and is interpreted as a seafloor attachment (Narbonne and Hofmann, 1987; Farmer et al., 1992).

Significance of shape variation. The Antelope Mountain fossil assemblage shows clear morphological gradations between the generic types described above (Fig. 5A). We interpret these variations as reflecting taphonomic differences that are entirely consistent with the range of taphonomically-induced variation observed in modern medusoid taxa (Norris, 1989). The Antelope Mountain specimens appear to fall well within the accepted range of variation for previously described form genera comprising the Cyclomedusa complex (see discussion by Farmer et al., 1992).

Overall shape variation in the Antelope Mountain Quartzite cyclomedusoid assemblage was further quantified by measuring the long and short axes of seventy-one specimens. A cross-plot of long and short axes (Fig. 5B) shows that the form varies from nearly circular to highly ellipsoidal (compare Figs. 3A and E). Basic characteristics of the medusoid Cnidaria include a radially-symmetrical, circular form. As seen in Figure 5B, departures from a circular shape are common among the Antelope Mountain cyclomedusoids. Consistent with the tectonic history

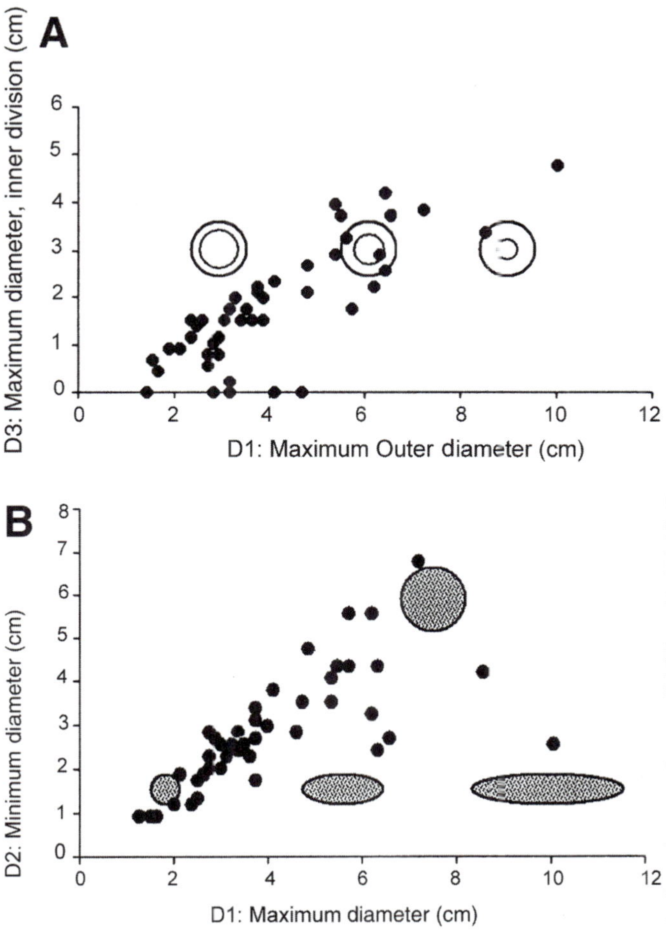

Figure 5. Shape analysis of cyclomedusoid fossils. (A) Maximum outer diameter plotted against maximum diameter of inner division, showing the gradational nature of the internal subdivisions of the ex-umbrellar surface in the Antelope Mountain forms. *Beltanella*-like forms are shown on the right side of the plot and *Ediacaria*-type forms on the left side. (B) Maximum outer diameter vs. minimum outer diameter, showing the variation in overall shape of the discoidal fossils from circular, regarded as the primary shape, to ellipsoidal forms, which reflect post-depositional deformation.

of the region discussed below, we attribute these departures from the primary circular form to later structural deformation.

Stratigraphy and Structure. The Antelope Mountain Quartzite includes three major lithologies: quartz arenite with minor pebble conglomerates, phyllite, and "bedded chert" (Hotz, 1977; Klanderman, 1978). Field work by Lindsley-Griffin and Griffin in 2003 demonstrates that most of the "cherts" are actually siliceous mudstones and argillites. Quartzite and phyllite each comprise ~40% of the unit, with siliceous mudstone and chert comprising ~20% of the unit. Only the resistant quartzites and cherts form readily observable outcrops, but on Antelope Mountain (Fig. 2), logging roads expose sections of pelitic rocks up to 100 m thick. Thick beds of quartz arenites alternating with phyllites (meta-pelites) dominate the lower part of the formation. Rhythmically bedded siliceous mudstone, argillite, and chert dominate the upper part of the formation, with minor interbeds of quartz arenite and phyllite.

The arenites are texturally immature and consist of 60–90% detrital quartz, both monocrystalline and polycrystalline, with ~2% detrital feldspar and 10–25% matrix (Klanderman, 1978; Bond and DeVay, 1980). The quartz granule and pebble conglomerate interbeds are similar in composition to the arenites but contain up to 12% of feldspar, chert, phyllite, and mudstone clasts (Klanderman, 1978). Plagioclase is more abundant than potassium feldspar in the arenites (Bond and DeVay, 1980). Wallin et al. (2000) analyzed forty-six highly diverse detrital zircon grains from the Antelope Mountain Quartzite and found concordant to slightly discordant U-Pb ages ranging from 3.2 to 1.32 Ga. Of these grains, 58% fall between 1.9 and 1.8 Ga (Paleoproterozoic). Wallin et al. (2000) reported three other groups of detrital zircon ages: (1) 13% of the grains range from 3.20 to 2.62 Ga (Archean); (2) 20% of the grains range from 2.53 to 2.08 Ga (Paleoproterozoic), with a small subset clustering around 2.23 Ga; and (3) 9% of the grains cluster around 1.33 Ga (Mesoproterozoic).

Metamorphic grade of the Antelope Mountain Quartzite is very low. The argillites exhibit no metamorphic grain growth beyond a faint sheen parallel to bedding. However, phyllites interbedded with the argillites exhibit a weak foliation, defined by flattened, recrystallized grains (Figs. 3E and 5B). Noto (2000) reported recrystallized quartz grains and secondary silica cement in the arenites. The grade is probably chlorite zone, low greenschist facies, based on thin-section descriptions of sericite and chlorite in the arenite matrix (Hotz, 1977; Klanderman, 1978).

Antelope Mountain Quartzite strata crop out in two main areas with different structural styles. The predominantly thin-bedded strata in the northernmost exposures (Fig. 2) form broad asymmetrical folds, whose subhorizontal F_1 fold axes warp gently about a subhorizontal F_2 fold axis nearly perpendicular to F_1. The subvertical network of quartz veins crosscuts the folds (Figs. 3 and 5). The gently dipping bedding of the northeastern exposures is right-side-up, based on graded bedding and ripple marks. The strata on Antelope Mountain and adjoining ridges (Fig. 2) dip more steeply, and F_1 folds are tighter. Small normal and reverse faults are common, but internal shearing is minimal, and thick packets of pelitic rocks exhibit their original undeformed bedding.

Provenance and Environment of Deposition. Bond and DeVay (1980) concluded that their provenance study suggested a source terrane of metamorphic and potassic plutonic rocks. They ruled out largely volcanic sources, such as a volcanic arc, as well as mature quartz sandstones of the craton, because of the relatively high proportion of polycrystalline to monocrystalline quartz grains. Their results did not support a source in the Laurentian craton. The textural immaturity of the quartz arenites and granule-pebble conglomerates (Klanderman, 1978) suggests a proximal source.

Bond and DeVay (1980) interpreted their data on the arenites as supporting deposition in a submarine fan near a passive continental margin. Hotz (1977) interpreted the Antelope Mountain Quartzite as a submarine fan complex, but the lack of Bouma sequence sedimentary structures in the well-sorted arenites suggests deposition by nonturbid gravity flows (Klanderman, 1978) or by some other process. Wallin et al. (2000) interpreted the arenite as shallow marine on the basis of symmetrical ripple marks whose crests are <5 cm apart, typical of water only centimeters deep, whereas Gehrels et al. (2000) preferred a submarine fan origin for the unit. Klanderman (1978) interpreted some pelitic intervals as hemipelagic deposits, which we consider likely for the chert intervals in the upper part of the formation. However, the thin rhythmic graded beds of the fossiliferous argillite-phyllite intervals (Figs. 3E and 5B) suggest sorting during turbid flow. We have found asymmetrical current ripple marks in quartzite immediately below the fossiliferous zones. These combined observations suggest an environment for the fossiliferous intervals of irregular fine-grained siliciclastic gravity flows interspersed with hemipelagic sedimentation. Over time, siliciclastic input decreased to almost zero and the environment became enriched in siliceous ooze and fine clays, allowing the cherts and siliceous mudstones to dominate the upper part of the formation.

Age of the Antelope Mountain Quartzite. Previous workers assumed a Paleozoic age for the Antelope Mountain Quartzite (e.g., Wells et al., 1959; Hotz, 1977; Bond and DeVay, 1980; Wallin et al., 2000). Although detrital zircons in the Antelope Mountain arenites may have been recycled, they originally were derived from Archean and Proterozoic continental sources 1.32 Ga or older (Wallin et al., 2000). However, the minimum age for the unit has been unconstrained until now. Radiometric age bracketing of the Ediacaran fossil assemblage, including the cyclomedusoid taxa discussed here, places it within the range 640–575 Ma, worldwide (Glaessner, 1979). Thus, the presence of cyclomedusoids in the upper part of the Antelope Mountain Quartzite indicates a formation age of late Precambrian (Vendian).

Turbidite Units of the Yreka Terrane

Three turbidite units occupying different Yreka terrane thrust sheets—the Moffett Creek Formation, Duzel Phyllite, and Sissel Gulch Graywacke—exhibit such strong similarities in composition and detrital zircon population that they likely originated in a single depositional system during the Late Silurian and Early Devonian (Wallin, 1989; Wallin et al., 1992; Gin and Wallin, 1994; Wallin and Gehrels, 1995).

Moffett Creek Formation. The Moffett Creek Formation consists of thick beds of massive grayish-brown calcareous metasiltstone and metasandstone with abundant white mica on weathered surfaces (Hotz, 1977). Sandstones range from quartz arenites with minor plagioclase and rare potassium feldspar (Hotz, 1977) to feldspathic wackes (Lindsley-Griffin, 1982). Sedimentary structures such as graded bedding, cross-laminae, sole markings, and rhythmically repeated sand-shale couplets attest to the turbidite origin of the formation (Lindsley-Griffin, 1982). Composite lithic grains of quartzite, volcanic rock, and metachert suggest sources of cratonic, volcanic edifice, and recycled subduction complex (Hotz, 1977; Wallin, 1989); the probable source of most of the detritus is a Siluro-Devonian volcanic carapace over the nearby Trinity terrane (Gin and Wallin, 1994). Detrital zircons (Wallin, 1989; Wallin et al., 1992) are Late Silurian; thus, the maximum age of the unit is Silurian. Light reflecting from bedding surfaces indicates a weak foliation of chlorite and white mica, typical of chlorite-zone greenschist metamorphism. Slaty cleavage subparallel to bedding is developed locally; both cleavage and bedding are folded into gentle open symmetrical folds. The Moffett Creek Formation has been interpreted as broken formation (Hotz, 1977), mélange (Potter et al., 1977), olistostromal turbidites (Lindsley-Griffin, 1982), and stratally continuous (Gin and Wallin, 1994), although its lack of exotic lithologies argues against applying the term "mélange." Sedimentary breccias and ductilely folded intervals confined between undeformed continuous layers attest to the olistostromal origin of parts of the Moffett Creek Formation.

Duzel Phyllite and Sissel Gulch Graywacke. The Duzel Phyllite and Sissel Gulch Graywacke interfinger with each other west of Gazelle (Fig. 2), with the Sissel Gulch Graywacke grading upward into Duzel Phyllite (Hotz, 1977). The Duzel Phyllite has uniformly thin (<2 cm) bedding of greenish chloritic phyllite and calcareous metasiltstone (Hotz, 1977). The metasiltstones consist of ~45% monocrystalline and polycrystalline quartz, 45% matrix (clay, chlorite, hematite, mica), 9% calcite cement, and 1% plagioclase (Klanderman, 1978). The Sissel Gulch Graywacke has thick, massive graywacke beds 15–100 cm thick, with thin (<3 cm) phyllitic partings (Hotz, 1977). The fine-grained, poorly sorted graywacke consists of abundant angular to subangular quartz, ~10% fresh plagioclase, no potassium feldspar, and many rock fragments (siliceous and mafic metavolcanic rock, metachert); minor constituents include titanite, apatite, epidote, and chlorite (Hotz, 1977). Composition of the Sissel Gulch Graywacke suggests that it was derived by erosion of a Late Silurian volcanic carapace over the Trinity terrane (Wallin and Gehrels, 1995).

The Duzel Phyllite is isoclinally folded with near-vertical axial surfaces and subvertical to steeply dipping bedding in most places. The F_2 deformation folded the F_1 axial surfaces into open asymmetrical folds with near-vertical axial surfaces roughly perpendicular to F_1 axial surfaces. Isolated highly ductile folds contained within isoclinal fold limbs that affect only a few beds are probably F_0 slump folds. Most of the Sissel Gulch Graywacke is too massive to readily reveal this folding style, although near the gradational contact with Duzel Phyllite, the Sissel-type beds are tightly folded. Quartzo-feldspathic segregation bands 1–10 cm thick are abundant in the Duzel Phyllite. Wallin and Gehrels (1995) report detrital zircons with concordant U-Pb ages of 420–410 Ma for the Sissel Gulch Graywacke; thus, the maximum age of deposition for the Sissel Gulch Graywacke and Duzel Phyllite is Late Silurian.

Gazelle Formation. The Gazelle Formation occupies a separate thrust sheet along the southeast edge of the Yreka terrane (Fig. 2). It consists of well-bedded and laterally continuous, thinly laminated, bluish-gray radiolarian-bearing shale and siliceous mudstone with minor interbeds of black, organic-rich calcarenite, and dark gray radiolarian chert. The calcarenite interbeds have yielded Lower Devonian (upper Siegenian to middle Emsian) conodonts (Savage, 1977; Lindsley-Griffin et al., 1991). The formation displays a general coarsening-upward trend, with volcaniclastic sandstone and conglomerate becoming more abundant up-section (Lindsley-Griffin et al., 1991; Wallin and Trabert, 1994). The Gazelle Formation is at least 1.25 km thick and is folded into an elongate syncline 22×8 km^2 (Wallin and Trabert, 1994). Wallin and Trabert (1994) described three distinct Gazelle Formation petrofacies: (1) chert-mudrock (94% polycrystalline rock fragments, 3% plagioclase, 3% monocrystalline quartz), (2) volcaniclastic (56% volcanic rock fragments, 32% plagioclase, 12% monocrystalline quartz), and (3) mixed. The mixed petrofacies consists of both chert-mudrock and volcanic components and evidences multiple sources to the basin. No potassium feldspar was observed in any of the petrofacies. The coarser siliciclastic sediments exhibit structures characteristic of sediment gravity flows, such as scoured beds; asymmetric ripple sets; rhythmically bedded, thinly laminated shales; interbedded siltstones and fine sandstones; and interbedded coarse turbidite sand bodies (Wallin and Trabert, 1994). The rhythmically bedded shales and siliceous mudstones exhibit dewatering veins and contain abundant radiolaria outlined by pyrite framboids and kerogen, suggesting organic-rich hemipelagic sedimentation in a marine-upwelling regime analogous to modern west-facing convergent margins, such as the Peru Trench (Lindsley-Griffin and Fisher, 1989; Lindsley-Griffin et al., 1991).

Informally Named Yreka Terrane Mélanges

The Yreka terrane includes four informally named mélanges and the formally named Gregg Ranch Complex. Each occupies its own thrust sheet and structural level and comprises different lithologic components, metamorphic grades, and structural histories (Hotz, 1977; Lindsley-Griffin and Griffin, 1983; Lindsley-Griffin et al., 1991).

Schulmeyer Gulch Mélange. Although Hotz (1977, 1978) termed this map unit the "Schulmeyer Gulch sequence," it is a metamorphosed tectonic mélange that consists of native and exotic blocks in semi-penetratively sheared matrix, lacks stratal continuity, and does not obey normal rules of stratigraphy (e.g., Hsu, 1968; Silver and Beutner, 1980). The Schulmeyer Gulch mélange occupies the structurally lowest thrust sheet in the northern Yreka terrane (Fig. 2), beneath thrust sheets of Antelope Mountain Quartzite, Moffett Creek Formation, and Duzel Phyllite. It is characterized by discontinuous ductilely stretched metasiltstone layers and tectonically sliced and flattened blocks entrained along semi-penetrative scaly foliation. Locally, the Schulmeyer Gulch matrix is folded into decimeter-scale isoclinal folds that are not reflected within the blocks. The Schulmeyer Gulch mélange includes the following blocks encased in intensely deformed, scaly-foliated phyllitic to schistose matrix: (1) marble of recrystallized impure carbonate with thin phyllitic folia, the largest over 600 m wide and 3 km long; (2) chert recrystallized to cryptocrystalline quartz (Klanderman, 1978, reported radiolarian molds in one chert block); (3) recrystallized quartz wacke and metasiltstone; (4) greenschist (sericite + chlorite ± epidote ± actinolite) with crinkly foliation; (5) hydrothermal quartz, massive multiply veined; (6) phyllite masses 0.5–2.0 m across cut by multiple generations of hydrothermal quartz veins that replace up to 80% of the original rock volume. These are distinguished from the phyllitic matrix because shear zones bound them and the quartz veins do not cross into the matrix. The metamorphosed Schulmeyer Gulch mélange is shown in Figure 2 as Ordovician(?) by analogy with the Skookum Gulch mélange, which is the only other metamorphosed mélange in the Yreka terrane. However, the Schulmeyer Gulch mélange is undated and could be older than Ordovician.

Skookum Gulch Mélange. The Upper Ordovician "schist of Skookum Gulch" comprises a thrust sheet at an intermediate structural level, exposed in structural windows between Facey Rock and Duzel Rock in the south-central part of the Yreka terrane (Fig. 2). It is one of the oldest blueschist mélanges in western North America. Discontinuous lenticular blocks are surrounded by phyllitic to schistose matrix metamorphosed to blueschist and greenschist; blueschist and greenschist lithologies are intricately interleaved at all scales down to thin-section scale. Blocks in the Skookum Gulch mélange include lawsonite- and glaucophane-bearing blueschist, serpentinite, plutonic and metabasaltic rocks, greenschist, metaquartzite and metagraywacke, calcareous and dolomitic marble, and metachert (Hotz, 1977; Cotkin, 1987; Haessig, 1988, 1989; Cotkin et al., 1992). Cotkin et al. (1992) analyzed glaucophane schist and chlorite schist by X-ray fluorescence for major elements and selected trace elements. Data for these metabasalt blocks suggest that they most likely formed at a spreading center, such as a mid-ocean ridge or back-arc basin (Cotkin et al., 1992). Cotkin et al. (1992) considered most of the schists in the Skookum Gulch mélange to have formed from oceanic sediments but interpreted the tourmaline-bearing quartz-albite schists as indicating continental provenance.

Haessig (1989) suggested that different block types were metamorphosed under a wide range of pressure-temperature conditions before being faulted into the matrix, whereas the phyllitic to schistose matrix experienced a very early blueschist-facies metamorphism, followed by prograde metamorphism to low greenschist facies. However, Cotkin et al. (1992) mapped a penetrative matrix foliation that they interpreted as forming under blueschist conditions, and considered the entire package of rocks to have been subducted, with blueschist matrix forming only where the protolith matrix was sufficiently basaltic to yield lawsonite or glaucophane. Haessig (1989) and Cotkin et al. (1992) described multiple generations of ductile and brittle deformational structures consistent with formation in a subduction zone. Haessig (1989) concluded that foliation and block alignment define a dominant northeast–southwest tectonic fabric, although some larger blocks are discordant to this direction. Haessig (1989) also concluded that the blueschist blocks experienced a more complex polyphase deformation than did the mélange matrix. Blueschist metamorphism occurred in Late Ordovician, based on K-Ar dates on white mica of 439 ± 13 and 451 ± 14 Ma (Potter et al., 1981), as well as a Rb-Sr age on phengite of 447 ± 9 Ma and a $^{40}Ar/^{39}Ar$ plateau age on phengite of 451 ± 4 Ma (Cotkin et al., 1992). Haessig (1989) reported detritus of quartz-albite schist and graphitic schist with deformation fabrics like those of the mélange matrix in Late Silurian conglomerates of the Horseshoe Gulch assemblage; she interpreted these as derived from the Skookum Gulch mélange, providing evidence that it was rapidly uplifted and rapidly eroded.

Vendian plutonic rocks comprise some of the tectonic blocks in the Skookum Gulch mélange. Wallin et al. (1988) obtained U-Pb zircon upper intercept ages of 565 ± 5 Ma from Skookum Butte, a large tonalite block, and upper intercept ages of ca. 575–565 Ma from a tonalite block at Trail Gulch. Zircons from the Trail Gulch block consist of two distinctly different fractions: one being "light pink, subhedral, transparent, and inclusion free" whereas the other is "subhedral to anhedral and slightly metamict" (Wallin, 1990, p. 1058). The light pink fraction yielded a $^{207}Pb/^{206}Pb$ age of 567 Ma, which Wallin interpreted as the crystallization age of the Trail Gulch block; the metamict fraction yielded a $^{207}Pb/^{206}Pb$ age of 1033 Ma, which he interpreted as the minimum crystallization age for an inherited component.

Vendian plutonic blocks in the Skookum Gulch mélange are typically cataclastically fractured to microbrecciated, but are locally undeformed. They exhibit both pressure solution and thin millimeter- to centimeter-scale cataclastic zones (Wallin et al., 1988). In the International Union of Geological Sciences classification scheme (Streckeisen, 1976), one of the samples described by Hotz (1977) is a tonalite; the other plots on the boundary between monzodiorite and quartz diorite/gabbro. We stained one sample from Skookum Butte for feldspars and found only plagioclase. Because these Skookum Gulch plutonic rocks are similar in composition and age to the Vendian (571–565 Ma) tonalitic rocks of the Trinity terrane, some authors (e.g., Hotz, 1977; Haessig et al., 1987) have suggested that they were derived from the Trinity Complex, but no direct evidence supports this hypothesis. The data suggest at least two sources for the plutonic blocks within the Skookum Gulch mélange, one containing inherited continental zircon and one probably oceanic.

Mélange of Callahan Area. Rocks exposed in the town of Callahan and the hills around it at the southern end of the Yreka terrane (Fig. 2) include blocks of Callahan Chert and other lithologies encased in a matrix of sheared siliceous mudstone. Romey first recognized the Callahan Chert as a distinct unit (Romey, 1962; Davis et al., 1965). Locally massive and recrystallized, the thinly laminated and well-bedded portions of the chert contain recrystallized radiolarian fossils as well as numerous dewatering veins and ductile, soft-sediment slump folds. Other blocks within the Callahan mélange include fossiliferous limestone, cherty limestone, and recrystallized limestone; lithic wacke and graywacke; red mudstone; and limestone cobble conglomerate with a variety of clast compositions, including basalt pillows and pillow fragments (Rohr, 1978). Late Silurian brachiopod fossils from a limestone block (Rohr, 1978) suggest a maximum age for the mélange, but radiolarians in the Callahan Chert are undated.

Although Rohr (1978) mapped the "mudstone of Thompson Gulch" as being interleaved with the chert by block faulting, the semi-penetrative scaly foliation in the mudstone wraps around large prominent blocks of Callahan Chert, which are entrained along the foliation. Tectonic slicing has broken some of the smaller exotic blocks into breccias, and block boundaries cut off primary bedding within the blocks. Thus, this assemblage shows structures typical of polygenetic sedimentary-tectonic mélanges. The Callahan mélange occupies a thrust sheet above the Moffett Creek Formation and below the thrust sheet of Facey Rock and Duzel Rock (Fig. 2). We consider it a separate mélange rather than part of the Gregg Ranch Complex, because it occupies a different structural level, contains a distinct population of block lithologies, and its mudstone matrix is more siliceous and more organic-rich than matrix of the Gregg Ranch Complex. Rohr (1978) reported turbidite-like interbeds of sandstone and granule conglomerate preserved locally in the mudstone; thus, parts of it may be a broken formation rather than a mélange. According to Rohr (1978), these quartz-poor interbeds are rich in plagioclase, altered mafic minerals, and lithic fragments, including mafic volcanic rocks, chert, and radiolarian-bearing mudstone, suggesting a recycled accretionary complex source. Metamorphic grade of the assemblage is chlorite zone, greenschist facies.

Mélange of Facey Rock and Duzel Rock. Facey Rock and Duzel Rock are large massive limestone blocks capping two of the highest points in the central and south-central Yreka terrane (Fig. 2); because they occupy the same structural level, they probably represent erosional outliers of a once-continuous thrust sheet. Both large limestone masses overlie an assemblage of blocks of mixed lithologies with limited stratal continuity, surrounded by pelitic matrix with locally developed semi-penetrative scaly foliation. Foliation is most intense immediately above the basal thrusts, suggesting that much of the disruption may be due to faulting. Facey Rock consists of Middle Ordovician massive limestone thrust over radiolarian chert and Late Ordovician platy-laminated limestone (Porter, 1973). The ages are based on conodonts from the limestone (Porter, 1973; Rohr, 1978). Lithologies under the Facey Rock limestone are poorly exposed but include sheared pelite, which we interpret as mélange matrix. Duzel Rock (Fig. 2) consists of two separate large blocks of massive, extensively recrystallized limestone resting on platy limestone (Nachman, 1977). The platy limestone exhibits slump folds and in places is sliced into smaller blocks strung out along the scaly foliation. No fossils have been recovered from the Duzel Rock limestones, but by analogy with Facey Rock, these also might be Ordovician. Lithologies that underlie the Duzel Rock limestones include smaller limestone blocks; black, blue, and green bedded chert; red and black mudstone; and both brecciated and pillowed basalt. Trace element analysis of the basalt reported by Nachman (1977) yielded Y/Nb ratios of less than 2, and Ti-Zr-Y ratios appropriate for within-plate basalt. Nachman (1977) interpreted these results as indicating the basalts at Duzel Rock formed at an oceanic hot-spot.

Gregg Ranch Complex

Mélange rocks of the Gregg Ranch Complex crop out along the boundary between the Yreka and Trinity terranes (Fig. 2), where they occupy one or more thrust sheets below the Gazelle Formation but structurally above the Trinity Complex. Lindsley-Griffin et al. (1991) defined the Gregg Ranch Complex to encompass all the relatively unmetamorphosed, lithologically diverse, stratigraphically discontinuous, structurally disrupted rock units that were originally included in the Gazelle and Duzel Formations by Wells et al. (1959). We recognize at least nine mélange units, based on distinct populations of block lithologies and faunas (Lindsley-Griffin, 1982; Lindsley-Griffin et al., 1991) and kinematic analysis (Sivers, 2002). We correlate the Horseshoe Gulch mélange unit, which overlies the Skookum Gulch mélange, with the Gregg Ranch Complex on the basis of its age, block composition, and structural level.

Fossils obtained from blocks within the Gregg Ranch Complex range in age from Middle Ordovician (Llanvirnian–Ashgillian) to Early Devonian (Emsian; Lindsley-Griffin et al., 1991). Tectonic deformation and emplacement of the mélange was Early or early Middle Devonian, as evidenced by the undeformed Middle Devonian lavas that overlie the mélange and undeformed feeder dikes that crosscut the mélange (Lindsley-Griffin, 1982, 1991). These Middle Devonian lavas have been correlated with the Middle Devonian Copley Greenstone and Balaklala Rhyolite, which comprise the base of the Redding terrane (Brouxel et al., 1988; Danielson, 1988; Wallin et al., 1991; Mankinen et al., 2002).

Blocks in the Gregg Ranch Complex range in size from millimeters up to the 100–300 m-wide limestones that cap Lovers Leap (Fig. 2); these may have originated as olistoliths (Lindsley-

Griffin, 1982). Block lithologies (Fig. 6) include shallow marine limestone and limestone cobble conglomerate, graywacke and graywacke cobble conglomerate, red and green mudstone, volcaniclastic siltstone and sandstone, bedded chert and chert breccia, lithic breccia and conglomerate, metabasalt, actinolite schist, marble, tonalite, gneissic metagabbro, peridotite, serpentinite, and metaperidotite. Some conglomeratic blocks that display nonfoliated internal sedimentary fabrics with partially rounded to well-rounded clasts probably represent sedimentary debris flows. Clast lithologies within these conglomerate blocks include mafic and ultramafic rocks identical to Trinity Complex lithologies as well as graywacke, shale, chert, fossiliferous limestone, metaquartzite, and phyllite. A graywacke cobble conglomerate at Lovers Leap (Fig. 2) contains tonalitic clasts of age 440 Ma (Early Silurian), based on a zircon concordia intercept of 455 Ma (Mattinson and Hopson, 1972), which Wallin et al. (1988) recalculated using the decay constants of Steiger and Jäger (1977).

Matrix of the Gregg Ranch Complex typically is greenish mudstone but locally grades into greenish-brown siltstone and fine sandstone; it is distinguished from the bluish shales of the Gazelle Formation by color and by scaly foliation (Fig. 6C). The mélange matrix is poorly exposed and outcrops are rare, but centimeter-scale lenticular chips of the matrix litter the subdued

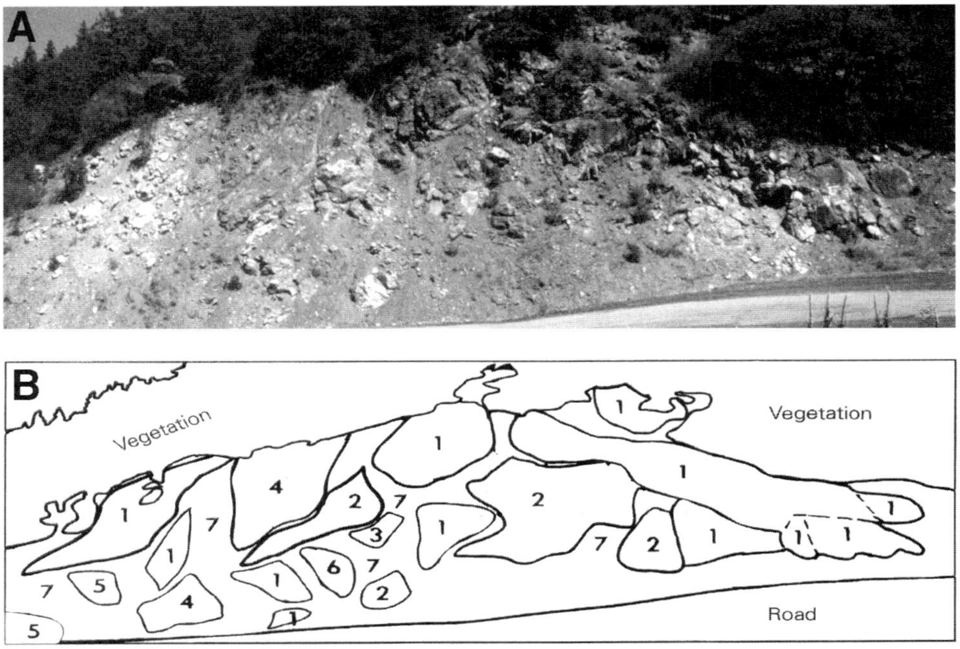

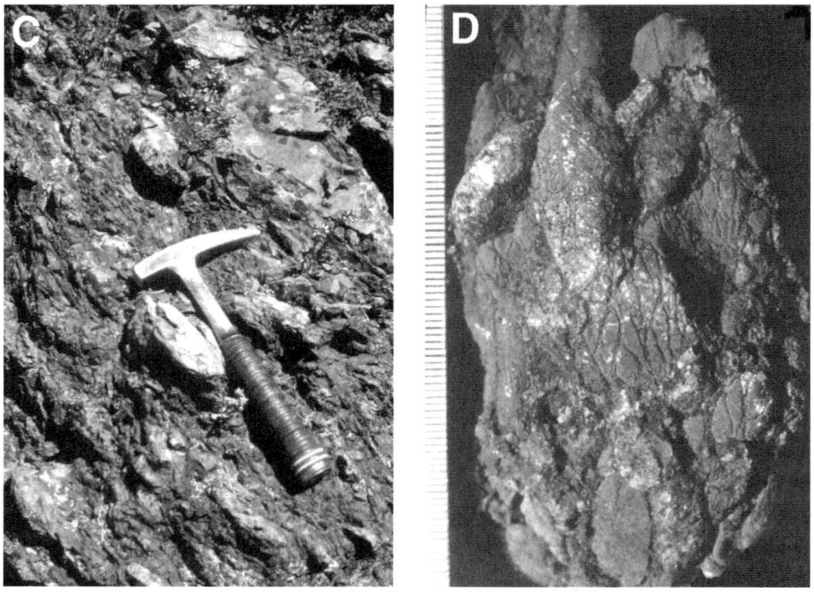

Figure 6. Mélange of the Gregg Ranch Complex. (A) Road-cut exposure with a high proportion of blocks to matrix at Horseshoe Bend (Fig. 2). Maximum height of outcrop is ~12 m. (B) Sketch of the outcrop in panel A: 1—light to dark gray tectonic breccia of recrystallized dolomicrite with calcite-spar and quartz veins, locally iron-oxide stained and pyritiferous; 2—dark gray tectonic breccia of organic-rich micrite; 3—dark yellowish-brown altered limonitic micrite and calcite-spar breccia; 4—slightly sheared greenish-gray pelite; 5—moderately sheared greenish to grayish-brown pelite; 6—moderately sheared dark greenish-black pelite; 7—intensely sheared greenish-gray pelitic matrix with scaly foliation. (C) Basalt phacoids in scaly foliated matrix. Small shaped fragments of basalt transported along the foliation form block train lineations. Hammer handle is 30 cm long. (D) Flattened and stretched limestone cobbles in graywacke matrix. Plane of flattening defines a foliation; long axes of clasts define a stretched-clast lineation. Sample is from a tectonic block of limestone cobble conglomerate at Lovers Leap (Fig. 2). Scale divisions along the left side of the photo are millimeters.

topographic surfaces. In streambeds and road cuts (Fig. 6A and B), scaly-foliated mélange matrix surrounding meter-scale blocks is readily visible. Many blocks are sliced and stretched into lenticular shapes (Fig. 6C and D). The anastomosing scaly foliation surfaces commonly offset or destroy bedding and transport clast fragments parallel to the surfaces. The proportion of matrix to blocks varies widely. The large outcrop of Gregg Ranch Complex exposed in a road cut at Horseshoe Bend on upper Willow Creek (Fig. 6A and B) displays a high proportion of blocks to matrix, with a fabric style in which the scaly foliation locally wraps around block margins. Blocks in this outcrop are meter to decimeter scale and consist of several different types of carbonate plus carbonate breccia, chert breccia, graywacke, and pelite. Large blocks do not exhibit internal foliation, so exposures smaller than a block may appear to be undeformed. For comparison, Figure 6C displays an outcrop with a high proportion of matrix to blocks, with a fabric style characterized by flattening and stretching of the blocks parallel to the scaly foliation. The ends of many blocks stretch out into elongated points that trail off into the foliation (Fig. 6C and D).

Kinematic Analysis of the Gregg Ranch Complex. In addition to the nearly ubiquitous scaly foliation, mélange of the Gregg Ranch Complex exhibits a number of kinematic structures such as subparallel long axes of tectonically sliced fragments, S- and Z-folds, and boudinaged tectonic slices entrained along S_1 scaly foliation surfaces. Figure 7 summarizes poles to 964 scaly foliations measured within the Gregg Ranch Complex between Gazelle and Callahan (Fig. 2). The S_1 scaly foliation predominantly strikes N16°E, with a near-vertical mean dip; the most common dip direction is ~85° west-northwest. The multiple maxima revealed by Figure 7 could represent (1) three separate foliation orientations; (2) two foliations, with one folded about a near-vertical east-plunging axis during north-south compression; or (3) two conjugate foliations that developed ~30 to 45° apart. However, we consider these explanations unsatisfactory.

Figure 8 shows kinematic data for each mélange unit plotted separately, keyed to a detailed geologic map. Data within each mélange unit are more tightly focused and dominated by only one orientation. In three localities (Fig. 8, stereograms 4, 6, and 11) the average scaly foliation is N16°E5°SE; in three other localities (Fig. 8, stereograms 3, 10, and 12), the foliation is N22°W5°SW. In one locality (Fig. 8, stereogram 5), a weak girdle suggests that the foliation is gently warped about a southwest-plunging fold axis. Lineations in the Gregg Ranch Complex are formed by (1) long axes of stretched clasts (Fig. 6D), (2) stretched tectonic blocks (Fig. 6C), (3) long axes of lenticular scaly folia (Fig. 6C), and (4) tectonically sliced fragments entrained along scaly foliation. Three mélange units exhibit relatively well-developed lineations (Fig. 8, stereograms 1, 2, 8, and 9). Most are roughly perpendicular to the contoured maxima for poles to scaly foliation, as one would predict.

Elliptical limestone clasts in mélange blocks of conglomerate at Lovers Leap (Figs. 6D and 9A) define a lineation (Fig. 8, stereogram 1) nearly perpendicular to the subvertical scaly foliation (Fig. 8, stereogram 3). Where scaly foliation is oriented at a high angle to the clast long axes, clasts are tectonically sliced and transported along the foliation, causing their apparent long axes to rotate into near-alignment with the foliation and producing an elongation lineation (Fig. 9B). This lineation is unique to mélange matrix immediately below the large kilometer-scale limestone blocks of the Lovers Leap mélange unit and different from that of matrix away from the large blocks, which is 5°S74°W (Fig. 8, stereogram 2). We interpret this 5°S74°W lineation as an early tectonic feature that was later crosscut and partially offset by the roughly north-south scaly foliation. The elongation lineation can be explained if the original clasts were right-laterally sheared parallel to the north-striking foliation (Fig. 9B), rotating the block long axes into a west-northwest–east-southeast trend.

Figure 7. Poles to 964 scaly foliation measurements in the Gregg Ranch Complex on lower-hemisphere, equal-area stereogram; contours at 2, 3.5, 5, 6, and 7%.

Contact Relations of the Gregg Ranch Complex. The Gregg Ranch Complex is overlain by the Gazelle Formation along a partially depositional, partially tectonized surface (Lindsley-Griffin and Fisher, 1989); it is structurally underlain by the Trinity terrane. Ophiolitic mélange blocks just above the contact are bounded by faults that can be traced downward into the Trinity terrane, suggesting that tectonic slicing has transported blocks of Trinity rocks upwards and encased them within mélange matrix. In other locations, Gregg Ranch complex is absent along the top of the Trinity terrane, with Gazelle Formation resting directly on the Trinity terrane along a discordant, sheared surface.

TRINITY TERRANE

Although a detailed discussion of the Trinity terrane (also called the "Trinity Complex") is beyond the scope of this chapter,

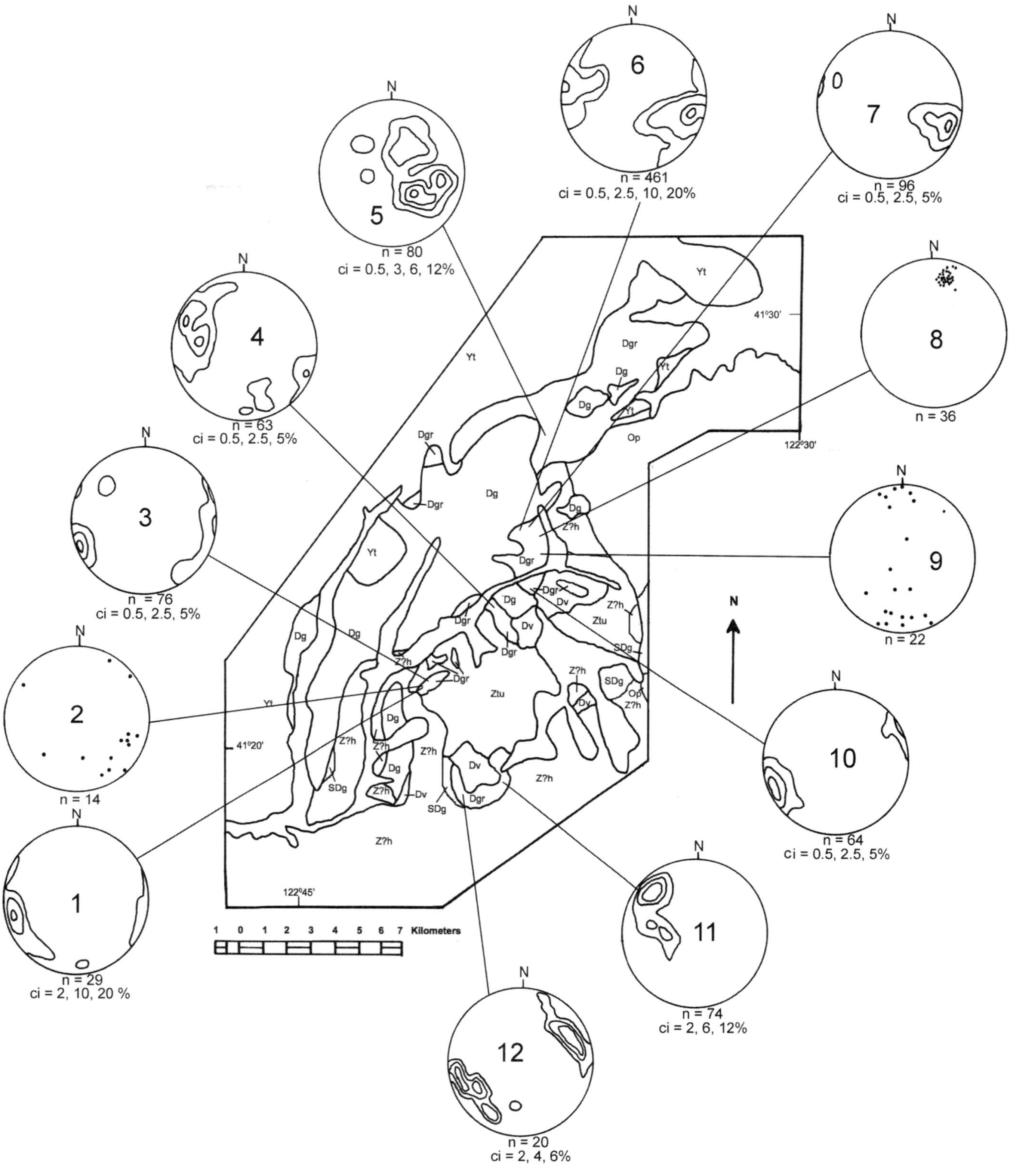

Figure 8. Kinematic map for the Gregg Ranch Complex. Stereograms are lower hemisphere equal-area projections, contoured where n >25. Abbreviations: ci—contour interval; n—number of data points. Numbered stereograms represent Lovers Leap area: 1—clast long axis lineations, 2—elongation lineations, 3—poles to scaly foliation; Gregg Ranch area: 4—poles to scaly foliation; upper Willow Creek area: 5—poles to scaly foliation; Mountain House area: 6—poles to scaly foliation in northwest, 7—poles to scaly foliation in southeast, 8—lineations on scaly foliation, 9—clast long axis lineations; Houston Creek area: 10—poles to scaly foliation; Cooper Peak: 11—poles to scaly foliation, east side; 12—poles to scaly foliation, west side. See legend for Figure 2. After Lindsley-Griffin et al. (1991) and Sivers (2002).

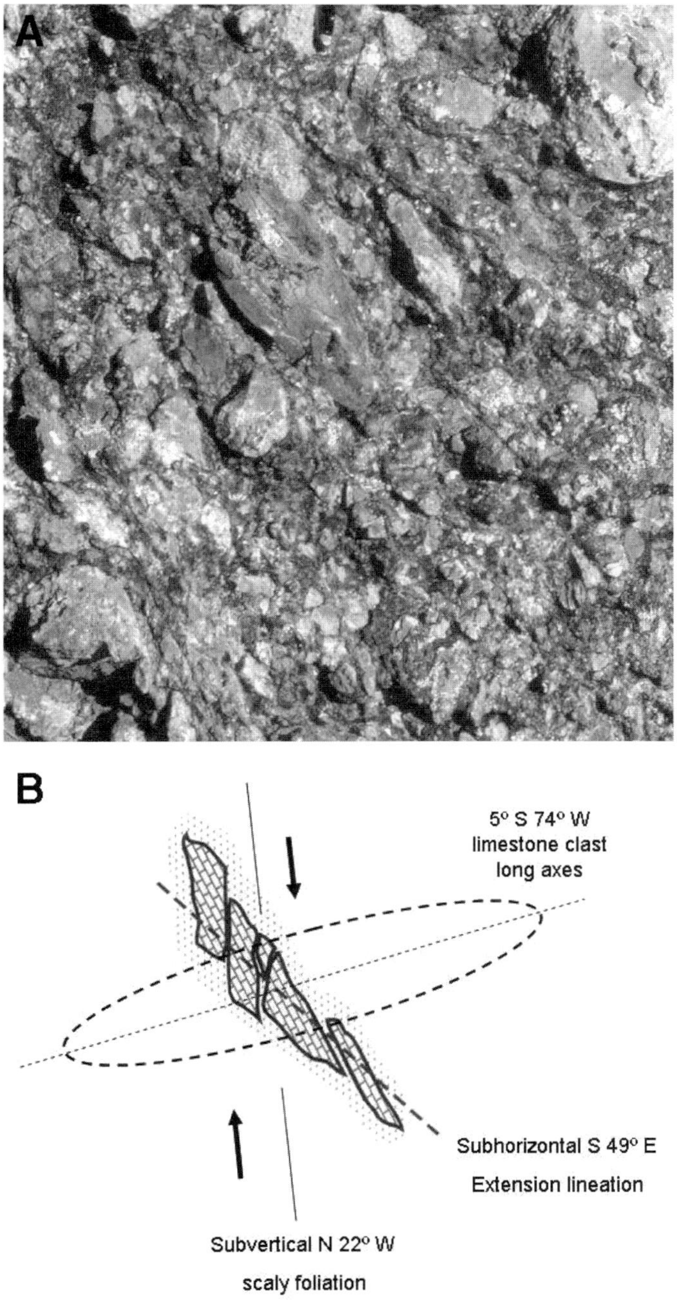

Figure 9. Elongation lineations in mélange of the Gregg Ranch Complex at Lovers Leap. (A) Conglomerate of limestone cobbles in a graywacke matrix, probably originating as a submarine debris flow, strongly overprinted by subvertical scaly foliation. (B) Sketch of relationship between original limestone clast long axes and the subhorizontal elongation lineation formed by tectonic slicing and transport of the clasts along scaly foliation surfaces. Arrows indicate sense of shear.

an overview of its main features is necessary to evaluate the tectonic history of the Yreka terrane. The Trinity terrane is polygenetic and formed over a period of 180 m.y. (Lindsley-Griffin, 1991; Wallin et al., 1991, 1995). It includes at least two distinct blocks of ultramafic mantle tectonites, one consisting of serpentinized harzburgite (olivine + orthopyroxene) and the other consisting of locally feldspathic harzburgite, lherzolite (olivine + clinopyroxene + orthopyroxene), and dunite (olivine). The latter block was termed the "Trinity peridotite" by Quick (1981, 1982). These two ultramafic blocks have different structural histories (Quick, 1981, 1982; Lindsley-Griffin, 1991) and are juxtaposed along ductile shear zones marked by mylonitized mafic and ultramafic rocks (Lindsley-Griffin, 1991; Wallin et al., 1991; Tozer, 1994).

Two ophiolitic crustal sequences of plutonic and extrusive rocks have been recognized within the Trinity terrane, a Vendian sequence associated with the harzburgite (Lindsley-Griffin, 1991; Wallin et al., 1991) and a Siluro-Devonian sequence that developed after the two ultramafic blocks were juxtaposed (Wallin et al., 1995; Wallin and Metcalf, 1998). The Vendian ophiolitic crustal sequence, along the northwestern edge of the Trinity terrane (Fig. 2), comprises deformed and locally gneissic metagabbro, oceanic plagiogranite, diabasic dikes and sills, and basaltic breccia and pillow lava (Lindsley-Griffin, 1982, 1991; Tozer, 1994; Bruckno, 1997). The plagiogranite has yielded euhedral to subhedral zircons with concordia upper intercept ages ranging from 550 ± 4 to 567 ± 4 Ma (Wallin et al., 1988), and the metagabbro has yielded zircon fractions with scattered, slightly discordant $^{207}Pb/^{206}Pb$ ages of 556 ± 2 to 579 ± 7 Ma (Wallin et al., 1995). These ages are Vendian, according to the revised geologic time scale of Gradstein et al. (2004), which places the boundary at 545 Ma. The Siluro-Devonian ophiolitic crustal sequence consists of voluminous, undeformed pegmatitic gabbro and pyroxenite plutons that intruded through the older peridotite (Schwindinger and Anderson, 1987; Cannat and Lécuyer, 1991; Lindsley-Griffin, 1991; Peterson et al., 1991; Wallin et al., 1991, 1995) and associated dikes and lavas (Wallin and Metcalf, 1998; Metcalf et al., 2000). The pegmatitic gabbro has yielded $^{207}Pb/^{206}Pb$ ages of 412 ± 10, 414 ± 8, 415 ± 2, and 419 ± 15 Ma (Wallin et al., 1988, 1995) and a concordant U-Pb zircon age of 415 ± 3 Ma (Wallin et al., 1991). This range of ages is now considered late Silurian (Ludlow–Pridoli) to Early Devonian (Lochkovian; Gradstein et al., 2004). Isolated tonalitic stocks and dikes dated as Early Ordovician (Jacobsen et al., 1984; Wallin et al., 1988, 1991; Lindsley-Griffin, 1991) intrude the Vendian ophiolitic crustal sequence and both ultramafic blocks.

The Vendian ophiolitic crustal sequence is internally folded and faulted, but its ophiolitic stratigraphy of metagabbro-plagiogranite-basalt is preserved within fault blocks. Although it lacks a sheeted dike complex, sills and feeder dikes along the base of the basalt crosscut the upper plagiogranite but not the metagabbro or the harzburgite, suggesting that it had been a conformable sequence with the harzburgite before being dismembered by faulting. These four components comprise the Vendian Trinity ophiolite (Lindsley-Griffin, 1991, 1994).

The Vendian Trinity ophiolite was juxtaposed against the "Trinity peridotite" of Quick (1981, 1982) before the Siluro-Devonian crustal sequence developed over both blocks. Dikes and stocks of the Siluro-Devonian crustal sequence crosscut all four of the Vendian ophiolite units as well as the Ordovician dikes. The China Mountain pluton, which intrudes and crosscuts a ductile shear zone between the blocks, is part of the Siluro-Devonian sequence (Wallin et al., 1995; Wallin and Metcalf, 1998); thus, the ductile deformation which juxtaposed the Vendian Trinity ophiolite with the "Trinity peridotite" was pre-Late Silurian. Although we have observed no dated Ordovician intrusions cutting the ductile shear zone, their undeformed character suggests that ductile deformation probably ended by Early Ordovician.

CONTACT BETWEEN TRINITY AND YREKA TERRANES

Mapping along the contact between the Trinity terrane and the Yreka terrane (Figs. 2 and 8) shows that the uppermost unit of the Trinity terrane varies along strike and that some faults that cut the Trinity terrane do not offset overlying rocks of the Yreka terrane. These observations indicate that the Trinity terrane was folded, faulted, and partially denuded prior to the juxtaposition of the two terranes. Yreka terrane units along the contact include Gazelle Formation and the structurally lower Gregg Ranch Complex. Mélange of the Gregg Ranch Complex varies greatly in thickness and is discontinuous, so that Gazelle Formation directly overlies the Trinity Complex in locations. Where mélange of the Gregg Ranch Complex rests on Trinity Complex plagiogranite, the contact zone exhibits a structural fabric subparallel to the contact. Lenticular fragments of Trinity plagiogranite are entrained within the mélange matrix, and foliation in the mélange matrix follows the contours of the competent plagiogranite boundaries. Thus, tectonic slicing incorporated fragments of the plagiogranite into the mélange during mélange formation. The contact between upper Gregg Ranch mélange and basal strata of the Gazelle Formation is exposed in road cuts at Horseshoe Bend (Fig. 6). In this location, the contact is a zone several tens of meters thick, within which deformation gradually decreases from semi-penetrative shearing in the mélange upward into continuous sedimentary strata of the Gazelle Formation. In the lower part of the contact zone, greenish-gray mélange matrix wraps around tectonic slices of deformed Gazelle-like black shale that are crossed by anastomosing shear zones. In the upper part of the contact zone, fault-bounded packets of black, well-bedded siliceous mudstone and shale of the Gazelle Formation are separated from exotic blocks of chert and limestone by greenish-gray scaly mélange matrix. We map the base of the Gazelle Formation as the lowest strata above exotic blocks, although these strata are typically boudinaged or sheared. Deformation in the Gazelle strata decreases upward, dying out completely within a few tens of meters.

DEVONIAN VOLCANIC ROCKS

We recognize two distinct sequences of Devonian volcanic and hypabyssal rocks in the area, a Lower Devonian assemblage that is part of the Yreka terrane and a Middle Devonian assemblage that postdates amalgamation of the Yreka and Trinity terranes.

Lower Devonian Volcanic Rocks

The Lower Devonian sequence includes mafic lavas, diabase feeder dikes, and the Schneider Hill tonalite stock. This sequence crops out along 6.5 km of the boundary between the Gazelle Formation and the Gregg Ranch Complex. At AP Ranch (Fig. 2), the lava rests on mélange of the Gregg Ranch Complex but underlies and partly interfingers with the base of the Lower Devonian Gazelle Formation. Along the southwest edge of Schneider Hill (Fig. 2), mafic pillow lava overlies Trinity terrane peridotite and Siluro-Devonian pegmatitic gabbro as well as the Schneider Hill tonalite stock (Fig. 2), which has been dated as ca. 400 Ma (Early Devonian; E.T. Wallin, personal commun., 1992). Fine-grained diabase dikes that cross lower pillow basalt flows can be traced upward to the bases of higher flows, with which they appear to merge, suggesting they were the conduits that fed those flows. Pegmatitic gabbro and tonalite xenoliths within the diabase dikes suggest in situ development of the pillow basalts over a gabbro-tonalite basement. Locally, the diabase feeder dikes coarsen in grain size and become feldspar-rich, merging into the tonalite stock with no chilled contacts. Based on these relationships, we consider these volcanic rocks to be the same age as the Schneider Hill stock (Early Devonian). The sequence consists of green to olive-green pillow lava with phenocrysts of feldspar and hornblende; the hornblende may be hydrothermal rather than primary. Pillow lavas are interbedded with pillow breccia and massive flows. The pillow lavas exhibit interpillow chert, and the pillow breccias have devitrified hyaloclastite separating the fragments. Composition of the Schneider Hill stock varies from hornblende-rich tonalite and hornblende diabase to quartz-rich tonalite and quartz aplite.

Middle Devonian Volcanic Rocks

The post-amalgamation Middle Devonian volcanic rock sequence consists of lavas and the dikes that fed them. The lavas consist of altered basaltic pillow lava, pillow breccia, hyaloclastite, and massive lava. Both the lava and the dikes consist of porphyritic basalt with 0.5-cm phenocrysts of green diopside and plagioclase altered to pink hydrogarnet. These diopside and hydrogarnet phenocrysts are seen only in the Middle Devonian lavas and help to distinguish them from Lower Devonian and Vendian pillow lavas of the area. Because these deposits have irregular basal contacts, they appear to have erupted onto uneven surfaces. Mankinen et al. (2002) were unable to demonstrate any post-emplacement tilting. The dikes, locally sheeted, intrude Trinity terrane rocks, Gregg Ranch mélange, and Gazelle Formation and can be traced through these units upward into undeformed lavas. Thus, they were erupted after juxtaposition of the Trinity and Yreka terranes, and their maximum age must be post-Emsian, because the Gazelle Formation is Emsian (Lindsley-Griffin et al., 1991). Geochemical data suggest that the Middle Devonian volcanic rocks may be cogenetic with the basal volcanic rocks of the Redding terrane to the south (Lapierre et al., 1985; Brouxel et al., 1988; Danielson, 1988; Wallin et al., 1991; Bruckno, 1997). In the Redding terrane, the Copley Greenstone and Balaklala Rhyolite interfinger with, and grade rapidly upward into, marine shales and graywackes of the Kennett Formation. A placoderm fishplate (Boucot et al., 1974) dates the Balaklala Rhyolite as Middle Devonian, and brachiopods (Boucot and Potter, 1977) date the Kennett Formation as Middle Devonian. The Balaklala Rhyolite is intruded by the Mule Mountain stock, which has been dated at 400 Ma (Albers et al., 1981; Kistler et al., 1985). Thus, the paleolatitude of 31° provides a paleogeographic constraint for the Yreka-Trinity composite terrane at the inception of the Copley-Balaklala volcanism that developed on it.

PALEOGEOGRAPHY AND TECTONIC SETTING

Relationship between Yreka and Trinity Terranes

The temporal link between the Yreka and Trinity terranes is well established; both contain rocks of Ordovician, Silurian, and Early Devonian age. Our recognition of Ediacaran cyclomedusoids (range, 640–575 Ma) in the Antelope Mountain Quartzite confirms that Yreka terrane history extends back into the late Neoproterozoic, overlapping the age of the oldest Trinity terrane rocks (580–550 Ma). Many authors have linked the Yreka terrane spatially to the Trinity terrane, based on detritus they assumed was derived from the Trinity terrane or its volcanic carapace (e.g., Mattinson and Hopson, 1972; Hotz, 1977; Haessig et al., 1987; Wallin et al., 1988; Gin and Wallin, 1994; Wallin and Gehrels, 1995). Even if the Yreka and Trinity terranes did not begin forming in the same location, they were juxtaposed by thrusting in the Early Devonian (Lindsley-Griffin et al., 1991; Wallin and Trabert, 1994) and since then have shared the same history. If we assume that the Yreka and Trinity terranes were spatially close during their entire history, then the Vendian paleolatitude of 7.1 ± 3.2° determined by Mankinen et al. (2002) can be used to approximate where both terranes might have been at that time, and biogeographical affinities of fossils from the Yreka terrane can be used as proxies for the paleogeographic setting of both terranes. The Middle Devonian paleolatitude of 31.1 ± 5.0° on lavas that overlap the Yreka-Trinity composite terrane (Mankinen et al., 2002) indicates that both terranes had migrated ~24° north or south of their starting position by ca. 390 Ma.

Source of Continental Detritus

Bond and DeVay (1980) concluded that the Antelope Mountain Quartzite was probably derived from a continental source other than the Laurentian craton but did not suggest where. The Antelope Mountain Quartzite contains Archean, Paleoproterozoic, and Mesoproterozoic detrital zircons with continental signatures that Wallin et al. (2000) suggested were derived from northern British Columbia, even though it lacks source rocks for the 1.3-Ga detrital zircons they found. They suggested that the Yreka terrane had been tectonically transported some 1000 km southward during late Paleozoic time. However that hypothesis is unnecessary, now that the Antelope Mountain Quartzite is known to be Vendian. Shortly before the unit was deposited, northern British Columbia was part of the Precambrian supercontinent of Rodinia (cf. Dalziel, 1997; Unrug, 1997; Karlstrom et al., 1999). It is more likely that the Antelope Mountain Quartzite and its detrital zircons were derived from relatively proximal continental fragments produced by the rifting of Rodinia. The inherited continental signature exhibited by zircons in the Trail Gulch tonalite block (Wallin et al., 1991) and the tourmaline-bearing quartz-albite schists (Cotkin et al., 1992) of the Skookum Gulch mélange may also have been derived from one of these fragments.

The timing of Rodinia's breakup is poorly constrained; estimates range from ca. 700 Ma (Dalziel, 1997; Unrug, 1997) to ca. 550 Ma (Dalziel, 1991; Moores, 1991; Veevers et al., 1997). According to Murphy and Nance (2004), Rodinia began rifting apart ca. 760 Ma, breaking into small continental fragments that drifted away from a short-lived southern hemisphere supercontinent termed "Pannotia" (Dalziel, 1997). Although the exact sequence and timing of breakup is controversial, these continental fragments probably included Australia, Antarctica, Laurentia, Baltica, and Siberia (e.g., Dalziel, 1997, Lawver et al., 2002). Pannotia began to break up ca. 550 Ma (Murphy and Nance, 2004). Because the oldest Trinity terrane rocks are 580–550 Ma in age, they formed after Rodinian rifting but before Pannotian rifting, ~7° north or south latitude, peripheral to Pannotia and smaller continental fragments.

The biogeographic affinities of Vendian cyclomedusoids suggest which continental fragments might have provided detritus to the Antelope Mountain Quartzite. Waggoner (1999) used parsimony analysis of endemism and phenetic clustering to analyze similarities and differences of Ediacaran biota and divided them into three biogeographic groups; one that includes Australia, northern Laurentia, Siberia, and Baltica; one spanning southern Gondwanan continents; and one very small Newfoundland group. Because cyclomedusoids from the Antelope Mountain Quartzite most closely resemble those from Finnmark (Farmer et al., 1992), a part of Baltica, and the MacKenzie Mountains in Canada (Narbonne and Hofmann, 1987; Narbonne and Aitken, 1990), part of northwestern Laurentia, we consider them to belong to Waggoner's (1999) Australian group. A number of published reconstructions of Rodinia show these elements clustered together in a near-equatorial position, with future ocean basins located ~7° north latitude (c.f. Rogers, 1996; Dalziel, 1997 Veevers et al., 1997; Waggoner, 1999).

Source of Oceanic Elements

Most of the Yreka terrane is oceanic in character: turbidites derived from one or more Siluro-Devonian or Devonian volcanic arcs: a blueschist-greenschist Ordovician mélange and an undated greenschist mélange, and at least three unmetamorphosed mélanges of early Paleozoic age. The evidence suggests an environment of one or more intraoceanic active margins that were relatively isolated from continental detritus. Kinematic analysis of the Gregg Ranch Complex, showing that different mélange units exhibit different orientations of scaly foliation, suggests that relative plate motions changed during convergence. Interaction of several small plates is suggested by different ages, block populations, and faunal characters of Yreka terrane mélanges.

Limestone blocks in the mélanges have yielded Early Ordovician through Early Devonian fossils, including brachiopods, gastropods, trilobites, rugose and tabulate corals, sphinctozoan sponges, and conodonts (Potter, 1990a,b; Potter et al., 1990). The shallow-water character of these fossils suggests that the limestones formed as fringing reefs around one or more chains of oceanic islands and atolls. Biogeographic affinities of Yreka terrane fossils (Potter et al., 1990; Soja and Antoshkina, 1997; Blodgett et al., 2002) suggest that such island chains may have been located near the continents of Laurentia, Siberia, Australia, Baltica, and China and possibly were contiguous with island chains that became Uralian and Alaskan accreted terranes. In addition, species that occur only in the Yreka terrane or in the Yreka terrane plus a few other localities (Potter et al., 1990) suggest relative isolation. For example, sphinctozoan sponges are found only in Australia, the Yreka terrane, Alaskan accreted terranes, and the Yukon (Rigby et al., 1988; Potter et al., 1990), suggesting that sponges might have been able to migrate between these places but were highly isolated from other regions.

CONCLUSIONS

The Yreka and Trinity terranes include rocks of the same age and probably were relatively close during their entire history, from Vendian through Early Devonian. Disc-shaped fossils in argillites of the Antelope Mountain Quartzite (Yreka terrane) are identified as representing the cyclomedusoid genera *Ediacaria* and *Beltanella* Sprigg 1947. The fossils are typical of the worldwide Ediacaran biota with a range of 640–575 Ma; thus, the Antelope Mountain Quartzite is Vendian in age. The fossils lack radial furrows and most closely resemble cyclomedusoids from Baltica (Farmer et al., 1992) and northwestern Laurentia (Narbonne and Hofmann, 1987; Narbonne and Aitken, 1990), suggesting that the formation was deposited within Waggoner's (1999) Australian biogeographic province.

The Yreka and Trinity terranes began forming around 7° latitude (Mankinen et al., 2002), most likely in the northern hemisphere, after the continental fragments of Baltica, Australia, Siberia, South China, and Laurentia rifted away from Rodinia. Detritus in the Antelope Mountain Quartzite, shown by Bond and DeVay (1980) to have been derived from a continent other than Laurentia, likely was derived from one or more of these fragments. As these continents drifted apart, the two terranes retained some connection with them but remained in an active oceanic setting that probably consisted of interconnected island chains, allowing limited mixing of faunal elements but also some isolation throughout the early Paleozoic. Early Devonian thrusting juxtaposed the terranes, interleaving the Antelope Mountain Quartzite with thrust sheets of mélanges and Siluro-Devonian turbidites. During this juxtaposition, the Lower Devonian Gazelle Formation was deposited in a trench-slope basin developed on actively forming mélange of the Gregg Ranch Complex (Lindsley-Griffin et al., 1991; Wallin and Trabert, 1994). Kinematic analysis of the Gregg Ranch Complex reveals that the average orientation of scaly foliations in different mélange units is either N16°E5°SE or N22°W5°SW. We interpret this orientation as indicating changes in relative plate motions during the convergence.

ACKNOWLEDGMENTS

W.P. Irwin encouraged many students of Klamath geology named herein. The support of the three student coauthors by the University of Nebraska, Lincoln (UNL), Department of Geosciences, and the contributions of thesis committee members Michael Blum and S.B. Treves are gratefully acknowledged. The Gregg Ranch Foundation, UNL Geosciences Department, Texaco Inc., and the Lincoln Gem and Mineral Club provided financial assistance to the student coauthors. Jeremy Brozak, UNL, helped prepare some of the figures. We thank field assistants Marc Fisher and Connie Gibb for their efforts, which contributed to this manuscript. We especially thank our reviewers, J.W. Hawkins, R.V. Metcalf, J.B. Wright, and the editors of this volume, C.G. Barnes and A.W. Snoke, for their many helpful suggestions.

REFERENCES CITED

Albers, J.P., Kistler, R.W., and Kwak, L., 1981, The Mule Mountain stock, an early Middle Devonian pluton in northern California: Isochron West, v. 31, p. 17.

Blakely, R.J., Jachens, R.C., Simpson, R.W., and Couch, R.W., 1985, Tectonic setting of the southern Cascade Range as interpreted from its magnetic and gravity fields: Geological Society of America Bulletin, v. 96, p. 43–48, doi: 10.1130/0016-7606(1985)96<43:TSOTSC>2.0.CO;2.

Blakely, R.J., Christiansen, R.L., Guffanti, M., Wells, R.E., Donnelly-Nolan, J.M., Muffler, L.J.P., Clynne, M.A., and Smith, J.G., 1997, Gravity anomalies, Quaternary vents, and Quaternary faults in the southern Cascade Range, Oregon and California: Implications for arc and backarc evolution: Journal of Geophysical Research, v. 102, p. 22,513–22,527.

Blodgett, R.B., Rohr, D.M., and Boucot, A.J., 2002, Paleozoic links among some Alaskan accreted terranes and Siberia based on megafossils, *in* Miller, E.L., et al., eds., Tectonic evolution of the Bering Shelf–Chukchi Sea–Arctic Margin and adjacent landmasses: Boulder, Colorado, Geological Society of America Special Paper 360, p. 273–290.

Bond, G.C., and DeVay, J.C., 1980, Pre-Upper Devonian quartzose sandstone in the Shoo Fly Formation, northern California: Petrology, provenance, and implications for regional tectonics: Journal of Geology, v. 88, p. 285–308.

Boucot, A.J., and Potter, A.W., 1977, Middle Devonian orogeny and biogeographical relations in areas along the North American Pacific Rim, *in* Murphy, M.A., et al., eds., Western North America: Devonian: Riverside, University of California–Riverside Campus Museum Contribution 4, p. 210–219.

Boucot, A.J., Dunkle, D.H., Potter, A., Savage, N.M., and Rohr, D., 1974, Middle Devonian orogeny in western North America? A fish and other fossils: Journal of Geology, v. 82, p. 691–708.

Brouxel, M., Lapierre, H., Michard, A., and Albarède, F., 1988, Geochemical study of an early Paleozoic island-arc–back-arc basin system. Part 2: Eastern Klamath, early to middle Paleozoic island-arc volcanic rocks (northern California): Geological Society of America Bulletin, v. 100, p. 1120–1130, doi: 10.1130/0016-7606(1988)100<1120:GSOAEP>2.3.CO;2.

Bruckno, B., 1997, Geochemistry of the Trinity ophiolite and the Trinity complex [M.S. thesis]: Lincoln, University of Nebraska, 97 p.

Cannat, M., and Lécuyer, C., 1991, Ephemeral magma chambers in the Trinity peridotite, northern California: Tectonophysics, v. 186, p. 313–328, doi: 10.1016/0040-1951(91)90366-Z.

Cashman, S.M., 1980, Devonian metamorphic event in the northeastern Klamath Mountains, California: Geological Society of America Bulletin, v. 91, part 1, p. 453–459, doi: 10.1130/0016-7606(1980)91<453:DMEITN>2.0.CO;2.

Cotkin, S.J., 1987, Conditions of metamorphism in an Early Paleozoic blueschist: Schist of Skookum Gulch, eastern Klamath Mountains, California: Contributions to Mineralogy and Petrology, v. 96, p. 192–200, doi: 10.1007/BF00375233.

Cotkin, S.J., Cotkin, M.L., and Armstrong, R.L., 1992, Early Paleozoic blueschist from the schist of Skookum Gulch, eastern Klamath Mountains, northern California: Journal of Geology, v. 100, p. 323–338.

Dalziel, I.W.D., 1991, Pacific margins of Laurentia and East Antarctica–Australia as a conjugate rift pair: Evidence and implications for an Eocambrian supercontinent: Geology, v. 19, p. 598–601, doi: 10.1130/0091-7613(1991)019<0598:PMOLAE>2.3.CO;2.

Dalziel, I.W.D., 1997, Neoproterozoic-Paleozoic geography and tectonics: Review, hypothesis, environmental speculation: Geological Society of America Bulletin, v. 109, p. 16–42, doi: 10.1130/0016-7606(1997)109<0016:ONPGAT>2.3.CO;2.

Danielson, J., 1988, Lithology and geochemistry of the French Gulch inlier, Klamath Mountains, northern California [M.S. thesis]: Chico, California State University, 230 p.

Davis, G.A., Holdaway, M.J., Lipman, P.W., and Romey, W.D., 1965, Structure, metamorphism, and plutonism in the south-central Klamath Mountains, California: Geological Society of America Bulletin, v. 76, p. 933–966.

Farmer, J., Vidal, G., Moczydłowska, M., Strauss, H., Ahlberg, P., and Siedlecka, A., 1992, Ediacaran fossils from the Innerelv Member (late Proterozoic) of the Tanafjorden area, northeastern Finnmark: Geological Magazine, v. 129, p. 181–195.

Fuis, G.S., Zucca, J.J., Mooney, W.D., and Milkereit, B., 1987, A geologic interpretation of seismic-refraction results in northeastern California: Geological Society of America Bulletin, v. 98, p. 53–65, doi: 10.1130/0016-7606(1987)98<53:AGIOSR>2.0.CO;2.

Gehrels, G.E., Dickinson, W.R., Darby, B.J., Harding, J.P., Manuszak, J.D., Riley, B.C.D., Spurlin, M.S., Finney, S.C., Girty, G.H., Harwood, D.S., Miller, M.M., Satterfield, J.I., Smith, M.T., Snyder, W.S., Wallin, E.T., and Wyld, S.J., 2000, Tectonic implications of detrital zircon data from Paleozoic and Triassic strata in western Nevada and northern California, *in* Soreghan, M.J., and Gehrels, G.E., eds., Paleozoic and Triassic paleogeography and tectonics of western Nevada and northern California: Boulder, Colorado, Geological Society of America Special Paper 347, p. 133–150.

Gin, G.M., and Wallin, E.T., 1994, A new interpretation for the Moffett Creek Formation, eastern Klamath Mountains, California: Geological Society of America Abstracts with Programs, v. 26, p. 54.

Glaessner, M.F., 1979, Precambrian, *in* Robison, R.A., and Teichert, C., eds., Treatise on invertebrate paleontology, Part A, Biogeography and biostratigraphy: Boulder, Colorado, Geological Society of America and University of Kansas Press, p. A79–A118.

Glaessner, M.F., 1984, The dawn of animal life: A biohistorical study: Cambridge, Cambridge University Press, 244 p.

Gradstein, F.M., Ogg, J.G., and Smith, A.G., eds., 2004, A geologic time scale 2004: Cambridge, Cambridge University Press, 384 p.

Griscom, A., 1977, Aeromagnetic and gravity interpretation of the Trinity ophiolite complex, northern California: Geological Society of America Abstracts with Programs, v. 9, p. 426–427.

Haessig, P.A., 1988, Structure and petrography of the schist of Skookum Gulch, Callahan-Yreka area, eastern Klamath Mountains, northern California [M.S. thesis]: Corvallis, Oregon State University, 169 p.

Haessig, P.A., 1989, Field relations and structure of the schist of Skookum Gulch, an Early Paleozoic mélange, Callahan-Yreka area, eastern Klamath Mountains: Geological Society of America Abstracts with Programs, v. 21, p. 88.

Haessig, P.A., Potter, A.W., Wallin, E.T., and Mattinson, J.M., 1987, Early Cambrian tonalite-plagiogranite from the Yreka-Callahan-Gazelle area, eastern Klamath Mountains, northern California: Geological Society of America Abstracts with Programs, v. 19, p. 689.

Hotz, P.E., 1977, Geology of the Yreka Quadrangle, Siskiyou County, California: Reston, Virginia, U.S. Geological Survey Bulletin 1436, 72 p.

Hotz, P.E., 1978, Geologic map of the Yreka Quadrangle and parts of the Fort Jones, Etna, and China Mountain Quadrangles, California: Reston, Virginia, U.S. Geological Survey Open-File Report 78-12, scale 1:62,500, 1 sheet.

Hsu, K.J., 1968, Principles of mélanges and their bearing on the Franciscan-Knoxville paradox: Geological Society of America Bulletin, v. 79, p. 1063–1074.

Irwin, W.P., 1973, Sequential minimum ages of oceanic crust in accreted tectonic plates of northern California and southern Oregon: Geological Society of America Abstracts with Programs, v. 5, p. 62–63.

Irwin, W.P., 1977a, Ophiolitic terranes of California, Oregon, and Nevada, *in* Coleman, R.G., and Irwin, W.P., eds., North American ophiolites: Oregon Department of Geology and Minerals Industries Bulletin v. 95, p. 75–92.

Irwin, W.P., 1977b, Review of Paleozoic rocks of the Klamath Mountains, *in* Stewart, J.H., et al., eds., Paleozoic paleogeography of the western United States, Pacific Coast Paleogeography Symposium I: Los Angeles, Pacific Section, Society of Economic Paleontologists and Mineralogists, p. 441–454.

Irwin, W.P., 1981, Tectonic accretion of the Klamath Mountains, *in* Ernst, W.G., ed., The geotectonic development of California—Rubey volume I: Englewood Cliffs, New Jersey, Prentice-Hall, p. 29–49.

Irwin, W.P., 1985, Age and tectonics of plutonic belts in accreted terranes of the Klamath Mountains, California and Oregon, *in* Howell, D.G., ed., Tectonostratigraphic terranes of the Circum-Pacific region: Houston, Circum-Pacific Council for Energy and Mineral Resources, Earth Science Series 1, p. 187–199.

Irwin, W.P., compiler, 2003, Correlation of the Klamath Mountains and Sierra Nevada: Reston, Virginia, U.S. Geological Survey Open-File Report 02-490, scale 1:1,000,000, 2 sheets.

Irwin, W.P., and Bath, G.D., 1962, Magnetic anomalies and ultramafic rock in northern California: Washington, D.C., U.S. Geological Survey Professional Paper 450-B, p. 65–67.

Irwin, W.P., and Mankinen, E.A., 1998, Rotational and accretionary evolution of the Klamath Mountains, California and Oregon, from Devonian to present time: Reston, Virginia, U.S. Geological Survey Open-File Report 98–114, 1 sheet.

Jacobsen, S.B., Quick, J.E., and Wasserburg, G.E., 1984, A Nd and Sr isotopic study of the Trinity peridotite: Implications for mantle evolution: Earth and Planetary Science Letters, v. 68, p. 361–378, doi: 10.1016/0012-821X(84)90122-5.

Karlstrom, K.E., Harlan, S.S., Williams, M.L., McLellan, J., Geissman, J.W., and Ahall, K.-L., 1999, Refining Rodinia: Geologic evidence for the Australia–western U.S. connection in the Proterozoic: GSA Today, v. 9, p. 1–7.

Kistler, R.W., Mckee, E.H., Futa, K., Peterman, Z.E., and Zartman, R.E., 1985, A reconnaissance Rb-Sr, Sm-Nd, U-Pb, and K-Ar study of some host rocks and ore minerals in the West Shasta Cu-Zn district, California: Economic Geology and the Bulletin of the Society of Economic Geologists, v. 80, p. 2128–2135.

Klanderman, D.S., 1978, Stratigraphy, structure, and depositional environments of the Antelope Mountain Quartzite, Yreka Quadrangle, California [M.S. Thesis]: Corvallis, Oregon State University, 115 p.

LaFehr, T.R., 1966, Gravity in the eastern Klamath Mountains, California: Geological Society of America Bulletin, v. 77, p. 1177–1190.

Lapierre, H., Brouxel, M., Albarède, F., and Cabanis, B., 1985, The Paleozoic and Mesozoic volcanism in the eastern Klamath Mountains (Northern California USA)—Part I: The Early Devonian intra-oceanic island-arc series: Terra Cognita—The Earth, v. 5, p. 320.

Lawver, L.A., Dalziel, I.W.D., Gahagan, L.M., Martin, K.M., and Campbell, D., 2002, PLATES 2002 Atlas of Plate Reconstructions (750 Ma to present day): PLATES Project: Austin, University of Texas Institute of Geophysics (http://www.ig.utexas.edu/ research/projects/plates/plates/htm).

Lindsley-Griffin, N., 1982, Structure, stratigraphy, petrology, and regional relationships of the Trinity ophiolite, eastern Klamath Mountains, California [Ph.D. dissertation]: Davis, University of California, 453 p.

Lindsley-Griffin, N., 1991, The Trinity complex: A polygenetic ophiolitic assemblage, *in* Cooper, J.D., and Stevens, C.H., eds., Paleozoic paleogeography of the United States II: Los Angeles, Pacific Section, Society of Economic Paleontologists and Mineralogists, v. 67, p. 589–607.

Lindsley-Griffin, N., 1994, The Cambrian Trinity ophiolite and related rocks of the lower Paleozoic Trinity Complex, northern California, U.S.A., *in* Ishiwatari, A., et al., eds., Ophiolites of the Circum-Pacific Region: Proceedings of the 29th International Geologic Conference, Kyoto, Japan, August 1992, Part D: Amsterdam, VSP International Science Publishers, p. 47–68.

Lindsley-Griffin, N., and Fisher, M., 1989, Significance of the tectonic contact between lower Paleozoic mélange and shales of the Gazelle Formation, Yreka terrane, Klamath Mountains, California: Geological Society of America Abstracts with Programs, v. 21, p. 107.

Lindsley-Griffin, N., and Griffin, J.R., 1983, The Trinity terrane: An early Paleozoic microplate assemblage, *in* Stevens, C.H., ed., Pre-Jurassic rocks in western North American suspect terranes: Los Angeles, Pacific Section, Society of Economic Paleontologists and Mineralogists, p. 63–75.

Lindsley-Griffin, N., Griffin, J.R., and De La Fuente, J., 1989, Unidentified fossil-like objects (UFOs) from the Antelope Mountain Quartzite, lower Paleozoic, Yreka terrane, Klamath Mountains, California: Geological Society of America Abstracts with Programs, v. 21, p. 107.

Lindsley-Griffin, N., Griffin, J.R., and Wallin, E.T., 1991, Redefinition of the Gazelle Formation of the Yreka terrane, Klamath Mountains, California: Paleogeographic implications, *in* Cooper, J.D., and Stevens, C.H., eds., Paleozoic paleogeography of the western United States II: Los Angeles, Pacific Section, Society of Economic Paleontologists and Mineralogists, v. 67, p. 609–624.

Lindsley-Griffin, N., Farmer, J.D., and Griffin, J.R., 2002, Ediacaran cyclomedusids from an accreted terrane in the eastern Klamath Mountains: CA: Geological Society of America Abstracts with Programs, v. 34, p. A272.

Lindsley-Griffin, N., Griffin, J.R., and Farmer, J.D., 2003, Significance of Ediacaran cyclomedusids and other Pacific rim biota in the Yreka terrane, eastern Klamath Mountains, California: Geological Society of America Abstracts with Programs, v. 35, p. A14.

Lowe, D.R., 1975, Water escape structures in coarse-grained sediments: Sedimentology, v. 22, p. 157–204.

Mankinen, E.A., Irwin, W.P., and Blome, C.D., 1996, Far-traveled Permian chert of the North Fork terrane, Klamath Mountains, California: Tectonics, v. 15, p. 314–328, doi: 10.1029/95TC03054.

Mankinen, E.A., Lindsley-Griffin, N., and Griffin, J.R., 2002, Concordant paleolatitudes for Neoproterozoic ophiolitic rocks of the Trinity Complex, Klamath Mountains, California: Journal of Geophysical Research, v. 107, p. 2254, doi: 10.1029/2001JB001623.

Mattinson, J.M., and Hopson, C.A., 1972, Paleozoic ophiolitic complexes in Washington and northern California: Carnegie Institution Yearbook, v. 71, p. 578–583.

Metcalf, R.V., Wallin, E.T., Willse, D.R., and Muller, E.R., 2000, Geology and geochemistry of the ophiolitic Trinity terrane, California: Evidence of middle Paleozoic depleted supra-subduction zone magmatism in a proto-arc setting, *in* Dilek, Y., et al., eds., Ophiolites and oceanic crust: New insights from field studies and the Ocean Drilling Program: Boulder, Colorado, Geological Society of America Special Paper 349, p. 403–418.

Miller, M.M., 1987, Dispersed remnants of a northeast Pacific fringing arc: Upper Paleozoic terranes of Permian McCloud faunal affinity, western United States: Tectonics, v. 26, p. 807–830.

Miller, M.M., 1989, Intra-arc sedimentation and tectonism: Late Paleozoic evolution of the eastern Klamath terrane, California: Geological Society of America Bulletin, v. 101, p. 170–187, doi: 10.1130/0016-7606(1989)101 <0170:IASATL>2.3.CO;2.

Moores, E.M., 1991, Southwest U.S.–East Antarctic (SWEAT) connection: A hypothesis: Geology, v. 19, p. 425–428, doi: 10.1130/0091-7613(1991)019 <0425:SUSEAS>2.3.CO;2.

Murphy, J.B., and Nance, R.D., 2004, How do supercontinents assemble? American Scientist, v. 92, p. 324–333, doi: 10.1511/2004.4.324.

Nachman, D.A., 1977, Geology of the Duzel Rock area, Yreka quadrangle, California [M.S. thesis]: Corvallis, Oregon State University, 153 p.

Narbonne, G.M., and Aitken, J.D., 1990, Ediacaran fossils from the Sewki Brook area, Mackenzie Mountains, northwestern Canada: Palaeontology, v. 33, p. 945–980.

Narbonne, G.M., and Hofmann, H.J., 1987, Ediacaran biota of the Wernecke Mountains, Yukon, Canada: Palaeontology, v. 30, p. 647–676.

Norris, R.D., 1989, Cnidarian taphonomy and affinities of the Ediacaran biota: Lethaia, v. 22, p. 381–395.

Noto, R.C., 2000, Structure and stratigraphy of the Antelope Mountain Quartzite, Klamath Mountains, northern California [M.S. thesis]: Las Vegas, University of Nevada, 150 p.

Peterson, S.W., Barnes, C.G., and Hoover, J.D., 1991, The Billy's Peak mafic complex of the Trinity sheet, California: Roots of a Paleozoic island arc, *in* Cooper, J.D., and Stevens, C.H., eds., Paleozoic paleogeography of the western United States II: Los Angeles, Pacific Section, Society of Economic Paleontologists and Mineralogists, v. 67, p. 625–633.

Porter, R., 1973, Geology of the Facey Rock area, Etna quadrangle, California [M.S. thesis]: Corvallis, Oregon State University, 87 p.

Potter, A.W., 1990a, Middle and Late Ordovician brachiopods from the eastern Klamath Mountains, northern California, Part I: Palaeontographica abteilung A, v. 212, p. 31–158.

Potter, A.W., 1990b, Middle and Late Ordovician brachiopods from the eastern Klamath Mountains, northern California, Part II: Palaeontographica abteilung A, v. 213, p.1–114.

Potter, A.W., Hotz, P.E., and Rohr, D.M., 1977, Stratigraphy and inferred tectonic framework of lower Paleozoic rocks in the eastern Klamath Mountains, northern California, *in* Stewart, J.H., et al., eds., Paleozoic paleogeography of the Western United States I: Los Angeles, Pacific Section, Society of Economic Paleontologists and Mineralogists, p. 421–440.

Potter, A.W., Hotz, P.E., and Lanphere, M., 1981, Evidence of Ordovician-Silurian subduction and Silurian or older igneous units of possible magmatic arc origin, eastern Klamath Mountains, northern California: Geological Society of America Abstracts with Programs, v. 13, p. 101.

Potter, A.W., Boucot, A.J., Bergström, S.W., Blodgett, R.B., Dean, W.T., Flory, R.A., Ormiston, A.R., Pedder, A.E.H., Rigby, J.K., Rohr, D.M., and Savage, N.M., 1990, Early Paleozoic stratigraphic, paleogeographic, and biogeographic relations of the eastern Klamath belt, northern California, *in* Harwood, D.S., and Miller, M.M., eds., Paleozoic and early Mesozoic paleogeographic relations: Sierra Nevada, Klamath Mountains, and related terranes: Geological Society of America Special Paper 255, p. 57–74.

Quick, J.E., 1981, Petrology and petrogenesis of the Trinity peridotite, an upper mantle diapir in the eastern Klamath Mountains, northern California: Journal of Geophysical Research, v. 86, p. 11,837–11,863.

Quick, J.E., 1982, The origin and significance of large, tabular dunite bodies in the Trinity peridotite, northern California: Contributions to Mineralogy and Petrology, v. 78, p. 413–422, doi: 10.1007/BF00375203.

Rigby, J.K., Potter, A.W., and Blodgett, R.B., 1988, Ordovician sphinctozoan sponges of Alaska and Yukon Territory: Journal of Paleontology, v. 62, p. 731–746.

Rogers, J.J.W., 1996, A history of continents in the past three billion years: Journal of Geology, v. 104, p. 91–107.

Rohr, D.M., 1978, Stratigraphy, structure, and early Paleozoic gastropoda of the Callahan area, Klamath Mountains, California [Ph.D. thesis]: Corvallis, Oregon State University, 316 p.

Romey, W.D., 1962, Geology of a part of the Etna quadrangle, Siskiyou County, California [Ph.D. thesis]: Berkeley, University of California, 93 p.

Savage, N.M., 1977, Lower Devonian conodonts from the Gazelle Formation, Klamath Mountains, northern California: Journal of Paleontology, v. 51, p. 57–62.

Schwindinger, K.R., and Anderson, A.T., Jr., 1987, Probable low-pressure intrusion of gabbro into serpentinized peridotite, northern California: Geological Society of America Bulletin, v. 98, p. 364–372, doi: 10.1130/0016-7606(1987)98<364:PLIOGI>2.0.CO;2.

Silver, E.A., and Beutner, E.C., 1980, Mélanges—Penrose conference report: Geology, v. 8, p. 32–34, doi: 10.1130/0091-7613(1980)8<32:M>2.0.CO;2.

Sivers, E.A., 2002, Kinematic analysis of the Gregg Ranch Complex mélange and implications for the history of the eastern Klamath Mountains, California [M.S. thesis]: Lincoln, University of Nebraska, 78 p.

Soja, C.M., and Antoshkina, A.I., 1997, Coeval development of Silurian stromatolite reefs in Alaska and the Ural Mountains: Implications for paleogeography of the Alexander terrane: Geology, v. 25, p. 539–542, doi: 10.1130/0091-7613(1997)025<0539:CDOSSR>2.3.CO;2.

Sprigg, R.C., 1947, Early Cambrian (?) jellyfishes from the Flinders Ranges, south Australia: Royal Society of South Australia, v. 71, p. 212–224.

Streckeisen, A.L., 1976, To each plutonic rock its proper name: Earth-Science Reviews, v. 12, p. 1–33, doi: 10.1016/0012-8252(76)90052-0.

Sun, W., 1986, Precambrian medusoids: The *Cyclomedusa* plexus and *Cyclomedusa*-like pseudofossils: Precambrian Research, v. 31, p. 325–360, doi: 10.1016/0301-9268(86)90039-2.

Tozer, M.K., 1994, Tectonic and structural implications of the Cambrian Trinity plagiogranite, eastern Klamath Mountains, northern California, U.S.A. [M.S. thesis]: Lincoln, University of Nebraska, 105 p.

Unrug, R., 1997, Rodinia to Gondwana: The geodynamic map of Gondwana supercontinent assembly: GSA Today, v. 7, p. 1–6.

Veevers, J.J., Walter, M.R., and Scheibner, E., 1997, Neoproterozoic tectonics of Australia-Antarctica and Laurentia and the 560 birth of the Pacific Ocean reflect the 400 m.y. Pangaean Supercycle: Journal of Geology, v. 105, p. 225–242.

Waggoner, B., 1999, Biogeographic analyses of the Ediacara biota: A conflict with paleotectonic reconstructions: Paleobiology, v. 25, p. 440–458.

Wallin, E.T., 1989, Provenance of lower Paleozoic sandstones in the eastern Klamath Mountains and the Roberts Mountains allochthon, California and Nevada [Ph.D. Thesis]: Lawrence, University of Kansas, 152 p.

Wallin, E.T., 1990, Petrogenetic and tectonic significance of xenocrystic Precambrian zircon in Lower Cambrian tonalite, eastern Klamath Mountains, California: Geology, v. 18, p. 1057–1059. Correction: Geology, v. 19, p. 192.

Wallin, E.T., and Gehrels, G.E., 1995, Sedimentary record of Silurian volcanism reveals denudation of the Trinity terrane and the timing of siliciclastic sedimentation in the Yreka terrane, northern California: Geological Society of America Abstracts with Programs, v. 27, p. 83.

Wallin, E.T., and Metcalf, R.V., 1998, Supra-subduction zone ophiolite formed in an extensional forearc: Trinity terrane, California: Journal of Geology, v. 106, p. 591–608.

Wallin, E.T., and Trabert, D.W., 1994, Eruption-controlled epiclastic sedimentation in Devonian trench-slope basin: Evidence from sandstone petrofacies, Klamath Mountains, California: Journal of Sedimentary Research, v. 64, p. 373–384.

Wallin, E.T., Mattinson, J.M., and Potter, A.W., 1988, Early Paleozoic magmatic events in the eastern Klamath Mountains, northern California: Geology, v. 16, p. 144–148, doi: 10.1130/0091-7613(1988)016<0144:EPMEIT>2.3.CO;2.

Wallin, E.T., Lindsley-Griffin, N., and Griffin, J.R., 1991, Overview of early Paleozoic magmatism in the eastern Klamath Mountains, California: An isotopic perspective, *in* Cooper, J.D., and Stevens, C.H., eds., Paleozoic paleogeography of the western United States II: Los Angeles, Pacific Section, Society of Economic Paleontologists and Mineralogists, v. 67, p. 581–588.

Wallin, E.T., Lindsley-Griffin, N., Griffin, J.R., Trabert, D.W., and Tozer, M.K., 1992, Early Paleozoic tectonics and sedimentation, Yreka and Trinity terranes, northern California: Boulder, Colorado, Geological Society of America Cordilleran Section Field Trip Guidebook, Trip No. 7, 38 p.

Wallin, E.T., Coleman, D.W., Lindsley-Griffin, N., and Potter, A.W., 1995, Silurian plutonism in the Trinity terrane (Neoproterozoic and Ordovician), Klamath Mountains, California, United States: Tectonics, v. 14, p. 1007–1013, doi: 10.1029/95TC01447.

Wallin, E.T., Noto, R.C., and Gehrels, G.E., 2000, Provenance of the Antelope Mountain Quartzite, Yreka terrane, California: Evidence for large-scale late Paleozoic sinistral displacement along the North American Cordilleran margin and implications for the mid-Paleozoic fringing-arc model, *in* Soreghan, M.J., and Gehrels, G.E., eds., Paleozoic and Triassic paleogeography and tectonics of western Nevada and northern California: Boulder, Colorado, Geological Society of America Special Paper 347, p.119–131.

Wells, F.G., Walker, G.W., and Merriam, C.W., 1959, Upper Ordovician (?) and Upper Silurian formations of the northern Klamath Mountains, California: Geological Society of America Bulletin, v. 70, p. 645–649.

Zucca, J.J., Fuis, G.S., Milkereit, B., Mooney, W.D., and Catchings, R.D., 1986, Crustal structure of northeastern California: Journal of Geophysical Research, v. 91, p. 7359–7382.

Manuscript Accepted by the Society 10 January 2006

Geological Society of America
Special Paper 410
2006

Cataclastic textures in La Grange fault rocks, Klamath Mountains, California

Susan M. Cashman*
Department of Geology, Humboldt State University, 1 Harpst Street, Arcata, California 95521, USA
Katharine V. Cashman*
Department of Geological Sciences, University of Oregon, 1272, Eugene, Oregon 97403, USA

ABSTRACT

A well-exposed, 1-km-long section of the La Grange fault, a major detachment fault in the southern Klamath Mountains, California, is examined for the dual purposes of analyzing the processes that operated during faulting and evaluating the tectonic significance of this fault. Black foliated fault-related rocks form a 25-cm-thick layer that caps the fault surface and consists of finely interlayered ultracataclasite and cataclasite. Features such as parallel slickenside striations on ultracataclasite layers, veins perpendicular to slickenside striations, clasts of ultracataclasite in cataclasite, and clasts of vein material in cataclasite record prolonged, brittle, extensional deformation along the La Grange fault. This faulting resulted in both south-southeastward transport of hanging-wall rocks and uplift and exhumation of footwall rocks. The presence of Yreka terrane units in the Oregon Mountain klippe suggests on the order of 60 km of southward displacement of the hanging wall of the La Grange fault. This faulting post-dates the assembly of accreted terranes at a long-lived accretionary margin, for which the Klamath Mountains province is renowned, and the overprint of extensional faulting both influences the map pattern and explains anomalies in extent and distribution of some Klamath terranes.

At the La Grange fault, the lithologic contrast between amphibolite in the footwall and siltstone, sandstone, chert, and mica schist in the hanging wall permits assessment of the relative contributions of footwall and hanging-wall rocks to the ultracataclasite. The ultracataclasite is composed of <10% single mineral grains ≤100 μm diameter in an ultra-fine-grained (<<1 μm) matrix. Single mineral grains are predominantly quartz, but also include calcite, pyrite or pyrrhotite, sphene, rutile, apatite, zircon, and barite. Matrix composition of the ultracataclasite is distinctly different from that of larger grains. Comparison of footwall and hanging-wall rock compositions indicates that most of the larger grains in the ultracataclasite are single crystals of mechanically resistant minerals, and most of these originated in the hanging wall. The less resistant grains in the cataclastic rocks (calcite and barite) are minerals that occur in veins; these were most likely introduced into the fault-related rocks relatively late in the faulting process. The preferential preservation of specific miner-

*E-mails: smc1@humboldt.edu; cashman@uoregon.edu.

Cashman, S.M., and Cashman, K.V., 2006, Cataclastic textures in La Grange fault rocks, Klamath Mountains, California, *in* Snoke, A.W., and Barnes, C.G., eds., Geological studies in the Klamath Mountains province, California and Oregon: A volume in honor of William P. Irwin: Geological Society of America Special Paper 410, p. 433–450, doi: 10.1130/2006.2410(21). For permission to copy, contact editing@geosociety.org.

als as larger grains within the ultracataclasite shows that the mechanical properties of individual minerals play an important role in comminution processes.

Micro-scale textures provide information about the processes that have operated during faulting and about the conditions and duration of faulting. The ultracataclasite is composed of rounded to subangular grains (1–100 μm diameter) in an extremely fine-grained (<<1 μm) matrix. The cataclasite contains ultracataclasite clasts (up to 500 μm) and angular to subrounded individual mineral grains (5–100 μm) in a fine-grained matrix. Texture appears to vary with depth in the fault-related rocks: the abundance of larger single-mineral grains increases upward, with larger grains occupying 3.3 ± 1.1 area% at 2 cm, 2.8 ± 1.1% at 17 cm, and 7.0 ± 1.3% at 24 cm above the base of the foliated ultracataclasite. Extremely small grain size in the ultracataclasite records extreme grain crushing, milling, and sustained cataclastic deformation. This extensive comminution is consistent with a very high degree of strain localization along the fault.

Keywords: cataclastic rocks, microstructure, Klamath Mountains, detachment fault, brittle deformation

INTRODUCTION

The purpose of this chapter is twofold. The intent is first to present new microstructural and compositional information about fault-related rocks on the surface of the La Grange fault and to investigate the processes of fault-rock development in this long-lived extensional fault zone. The second purpose is to consider implications of the structural development of the La Grange fault for the tectonic development of the eastern Klamath Mountains.

Fault-related rocks commonly exhibit a wide range of mesoscopic to microscopic features that attest to their complex evolution during multiple episodes of slip localized along persistent discontinuities (e.g., Davis, 1980; Chester et al., 1993; Evans and Chester, 1995; Snoke et al., 1998; Davis et al., 2004). For this reason, attempts to quantify brittle fault-rock microstructure have been largely confined to field studies of fault zone material formed at relatively shallow depth (within several kilometers of the surface) or relatively small displacement, such as fault gouge (e.g., Sammis et al., 1986, 1987) or deformation bands (e.g., Aydin and Johnson, 1983; Antonellini and Aydin, 1994; Cashman and Cashman, 2000), and to experimental studies of simple analog (e.g., Marone and Scholz, 1989; Mair and Marone, 1999) or natural samples (e.g., Mair et al., 2000; Hadizadeh and Johnson, 2003). A common feature of these studies has been a focus on quartz-rich samples deformed at relatively low confining pressures. In such systems, individual quartz grains are commonly angular, and particle size distributions tend to be fractal (self-similar), with fractal dimension D ~2.6 (Sammis et al., 1986), although extensive comminution may produce core zones with less angular grains and D >2.6 (Hadizadeh et al., 2004).

Here we extend analysis of brittle fault-rock microstructures to a study of ultracataclasite formation along the La Grange fault, a major detachment fault in the Klamath Mountains, California. Hydraulic mining of upper plate gold-bearing gravel in the late nineteenth and early twentieth centuries exposed a 1-km-long, planar, south-southwest- to south-southeast-dipping fault surface capped by up to 30 cm of black, fine-grained, strongly foliated, fault-related rocks. The lateral extent and accessibility of the hydraulic mine exposure, the distinctive compositions of hanging-wall versus footwall protoliths, and the large normal displacement across the La Grange fault make this fault an excellent locality for microstructural study of a regional-scale extensional fault.

Observations from this spectacularly well-exposed fault surface may help to resolve questions about fault zone development, such as the relative contributions of contrasting hanging-wall and footwall lithologies to the fault rocks, the evolution of particle sizes and shapes within the fault zone, and the role of fluids in fault zone development. In addition, a better understanding of the mechanics of faulting at the La Grange fault aids interpretation of the nature and importance of post-Nevadan extensional faulting in the Klamath Mountains province.

Previous Work

The La Grange fault was recognized by Irwin (1963, 1966) as one of a relatively small number of post-Nevadan faults in the Klamath Mountains province. He mapped the La Grange fault as a northeast-striking, shallowly southeast-dipping fault that places Devonian Abrams Mica Schist, Mississippian Bragdon Formation(?), and mid-Tertiary Weaverville Formation rocks in the hanging-wall block over Devonian Salmon Hornblende Schist and earliest Paleozoic Trinity sheet mafic/ultramafic rocks of the footwall (Irwin, 1963; Fig. 1). Irwin described the La Grange fault as one of the most important Tertiary faults in the province, noting both its structural position as the northernmost

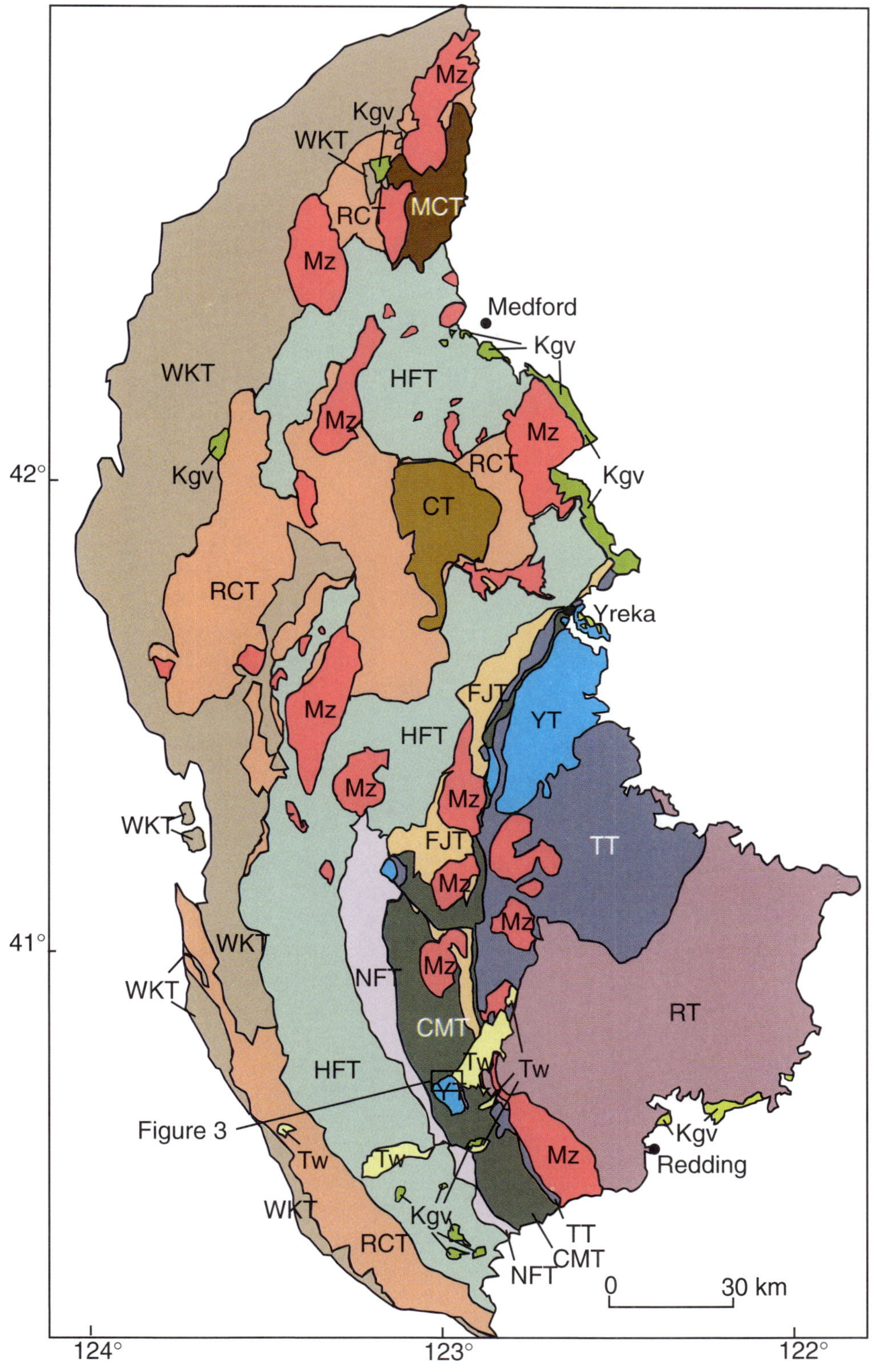

Figure 1. Simplified geologic map of terranes and lithotectonic units of the Klamath Mountains province. CMT—Central Metamorphic terrane; CT—Condrey Mountain terrane; FJT—Fort Jones terrane; HFT—Hayfork terrane; Kgv—Cretaceous Great Valley Group; MCT—May Creek terrane; Mz—Mesozoic plutons; NFT—North Fork terrane; RCT—Rattlesnake Creek terrane; RT—Redding terrane; TT—Trinity terrane; Tw—Tertiary Weaverville Formation; WKT—Western Klamath terrane; YT—Yreka subterrane. Modified from Irwin (1994).

Figure 2. View to the southwest of the La Grange fault exposure at the La Grange mine. A 20- to 30-cm-thick layer of foliated black cataclasite and ultracataclasite overlies sheared and vein-filled foliated greenstone.

of a zone of northeast-striking Cenozoic normal faults and the distinctive "mylonite" capping the exposed fault surface (Irwin, 1966; Fig. 2).

Following the recognition of metamorphic core complexes and associated tectonically detached and distended unmetamorphosed cover rocks throughout the North American Cordillera in the 1970s (e.g., Crittenden et al., 1980), Schweickert and Irwin (1989) proposed that the La Grange fault and Trinity terrane mafic/ultramafic rocks to the north are structurally analogous to the detachment faults and metamorphic core complexes found in eastern parts of the Cordillera. Supporting evidence for this correlation included the topographically high position of the structurally low Trinity terrane, the shallow (20–35°) dip of the La Grange fault away from the Trinity sheet, down-dip slickenside striations, "mylonitized" mafic rocks capping the footwall, younger-on-older faulting, and large normal displacement suggested by omission of several kilometers of structural thickness of bedrock units across the fault. Schweickert and Irwin's (1989) work marked a dramatic departure from the perception of the Klamath Mountains province as essentially inactive following the accretion of Klamath terranes in Paleozoic to mid-Jurassic time, and it initiated a flurry of related investigations. Phillips (1989), Phillips and Aalto (1989), and LeRoy (1993) documented lithofacies distribution and paleocurrent directions in Oligocene to lower Miocene Weaverville Formation rocks of the Weaverville and Hyampom basins, Cashman and Cashman (1996, 2003) analyzed microstructures of the fine-grained fault rocks on the La Grange fault surface, and Cashman and Elder (1993, 1996, 2002) mapped the regional distribution of the La Grange fault and related detachment faults and identified patterns of epithermal mineralization associated with the faulting.

Structural Setting

The La Grange fault strikes northeast, across the arcuate tectonostratigraphic terranes and the shallowly east-dipping terrane-bounding faults that form the framework of the Klamath Mountains province (Fig. 1). At the La Grange mine exposure, the fault places Tertiary Weaverville Formation and Paleozoic sedimentary and ultramafic rocks of the Oregon Mountain klippe over Devonian amphibolite. Footwall rocks at the La Grange mine are exclusively amphibolite of the Salmon Hornblende Schist (Central Metamorphic terrane). However, along the strike of the fault several tens of kilometers northeast of the La Grange mine, mafic and ultramafic rocks of the Trinity sheet (Eastern Klamath terrane) occupy the footwall (Schweickert and Irwin, 1989; Irwin, 1994; Cashman and Elder, 2002).

The Oregon Mountain klippe of Paleozoic sedimentary and metamorphic rocks (Eastern Klamath terrane and Central Metamorphic terrane) and unconformably overlying, poorly consolidated sediments of the middle Tertiary Weaverville Formation occupy the hanging wall of the fault (Fig. 3). A thin strip of serpentinized ultramafic rocks forms the eastern margin of the Oregon Mountain klippe (Irwin, 1963; Crawford, 2004). In addition, at the northeastern corner of the exposed La Grange fault surface, Irwin (1963) mapped a small exposure of Devonian Abrams Mica Schist (Central Metamorphic terrane) structurally overlying the fault surface. Cores of landslide material (several

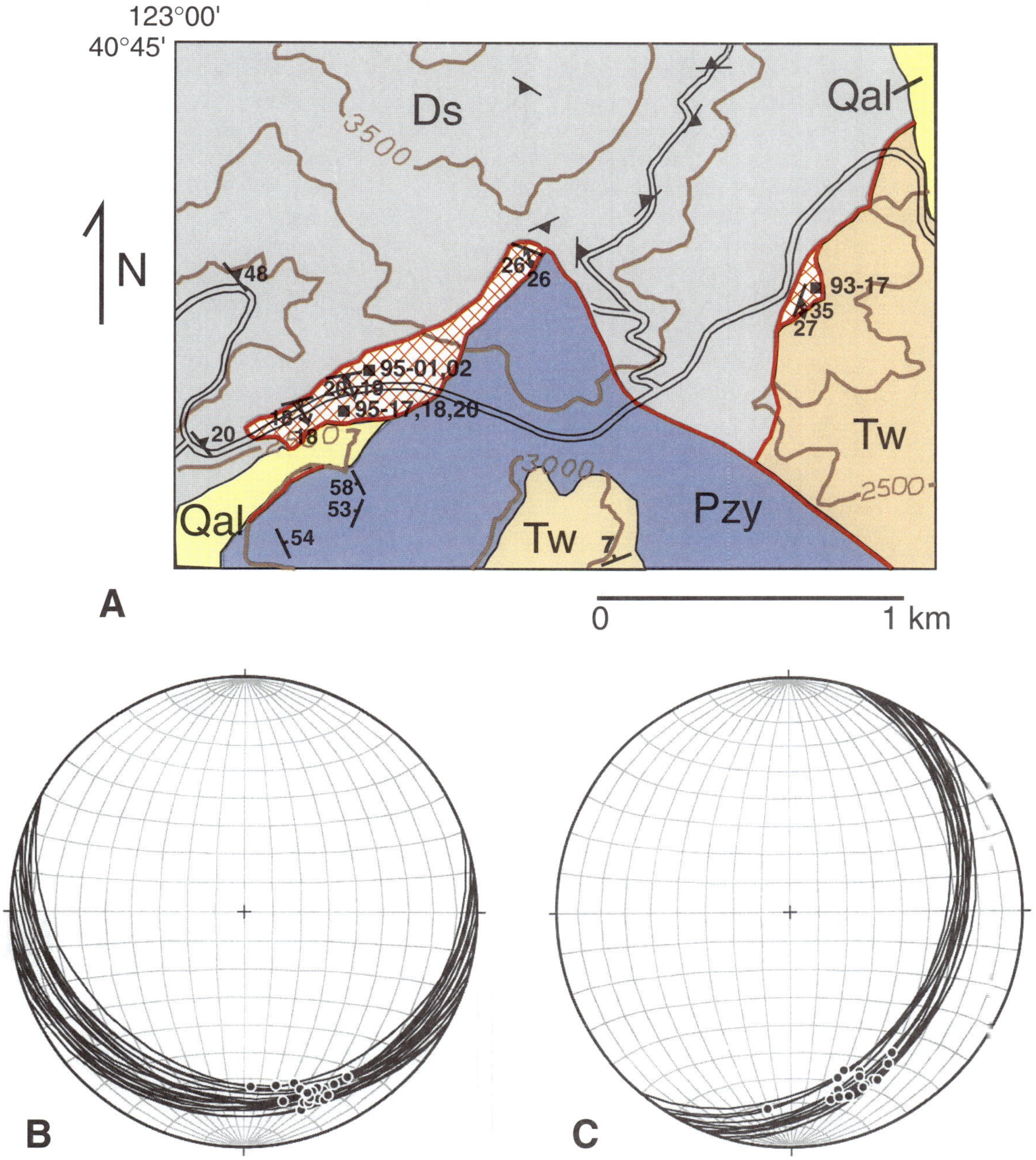

Figure 3. (A) Geologic map of the La Grange mine area. Filled circles indicate sample locations; cross-hatching shows the La Grange fault surface. Ds—Devonian Salmon Hornblende Schist; Pzy—Yreka subterrane units and minor associated Abrams Mica Schist; Qal—Quaternary alluvium; Tw—Tertiary Weaverville Formation. From Irwin (1963) and Cashman and Elder (unpublished mapping, 1993). (B) Equal-area stereonet plot of foliation (lines) and slickenside striations (filled circles), La Grange fault at the La Grange Mine, *N* = 20. (C) Equal-area stereonet plot of foliation (lines) and slickenside striations (filled circles), La Grange fault at exposure east of Oregon Mountain summit, *N* = 16.

hundred meters thick) from rocks overlying the La Grange fault were collected by CalTrans crews at the eastern end of the La Grange mine exposure in 1996. The cores contain mica schist, confirming Irwin's (1963) identification of Abrams Mica Schist in the hanging wall of the fault.

Sedimentary rocks in the hanging wall (the "Oregon Mountain outlier" of Irwin, 1966) were originally identified as "Bragdon Formation(?)" by Irwin (1963, 1966), who tentatively assigned these rocks to the Devonian-Jurassic Redding terrane of the eastern Klamath Mountains. Following identification and description of new stratigraphic units in the Yreka section of the eastern Klamath Mountains (e.g., Potter et al., 1977), hanging-wall rocks at the La Grange mine locality were reinterpreted to be correlative with part of the Ordovician–Devonian Yreka subterrane by Cashman and Elder (1993, 2002). These rocks consist largely of calcareous siltstone and sandstone with detrital mica grains, chert pebble conglomerate, mudstone, and chert of the Moffett Creek Formation. Lesser amounts of slate, chert, sheared chert and argillite, and small lenses of Silurian limestone (Silurian limestone in the Oregon Mountain area identified by W.P. Irwin, personal commun., 1992) are also found in the hanging wall at Oregon Mountain. These rocks are tentatively correlated with the Devonian–Silurian Gazelle Formation and Gregg Ranch Complex of Lindsley-Griffin (1991).

Mid-Tertiary sedimentary rocks ("stratified gravel": McDonald, 1909) of the Weaverville Formation were unconformably overlying hanging-wall bedrock units at the La Grange mine (McDonald, 1909; Irwin, 1963, 1966; Schweickert and Irwin, 1989) but have been largely removed by hydraulic mining. In a contemporary sketch of the mine (McDonald, 1909), Weaverville strata were shown to dip shallowly toward the fault surface. This structural geometry was recognized by Schweickert and Irwin (1989) as being similar to that of Cenozoic detachment faults that bound metamorphic core complexes elsewhere in the Cordillera. A dip of 7° northwest (toward the La Grange fault surface) in Weaverville sandstone 0.5 km south of Oregon Mountain Pass supports this structural interpretation (Cashman and Elder, unpublished mapping, 1993; Fig. 3A). Evidence that the La Grange fault was active during deposition of the Weaverville Formation is provided by fluvial conglomerates and debris flow deposits that coarsen and thicken toward the La Grange fault (Phillips, 1989; Phillips and Aalto, 1989).

The strongly foliated, ultra-fine-grained fault rocks capping the La Grange fault at the La Grange mine were originally identified as mylonite (Irwin, 1966) and as mafic mylonite (Schweickert and Irwin, 1989). Physical conditions required for mylonite formation—a minimum depth of 10–15 km for quartz-rich rocks in areas with a normal geothermal gradient—influenced tectonic interpretations of the La Grange fault (Schweickert and Irwin, 1989). However, subsequent microstructural analysis using high magnification scanning electron microscope (SEM) images shows textural evidence for extensive comminution and cataclastic deformation but no evidence of plastic deformation, thus prompting reclassification of the fault rocks as ultracataclasite rather than mylonite (Cashman and Cashman, 1996, 2003; Cashman and Elder, 2002). Reinterpretation of La Grange fault microstructures as resulting solely from brittle processes alters the tectonic interpretation of this regional-scale structure by decreasing the depth required for initiation of faulting.

PETROGRAPHY OF FOOTWALL ROCKS AND HANGING-WALL ROCKS

Detailed petrographic descriptions of bedrock units are beyond the scope of this chapter. However, a limited discussion of the minerals present in footwall and hanging-wall units is important, because it is from these rocks that La Grange fault rocks were derived.

The Salmon Hornblende Schist forms the footwall of the La Grange fault at the La Grange mine (Fig. 3A). This unit is well-foliated hornblende schist that commonly contains thin (0.5–2.0 cm) quartzofeldspathic lenses (Davis et al., 1965). The average composition of Salmon Hornblende Schist is hornblende, 72%; epidote, 14%; albite or oligoclase, 7%; quartz, 3%; opaque minerals, 2%; chlorite, 1%; actinolite, 0.5%; sphene, 0.5%; and trace amounts of rutile, hematite, garnet, clinopyroxene, and apatite (Davis et al., 1965; Table 1). Within several hundred meters of the La Grange fault surface, the hornblende schist becomes increasingly fine grained and veined. In the 10–15 m directly beneath the fault surface, footwall rocks consist of veined, altered greenstone with foliation roughly parallel to the overlying La Grange fault (Fig. 2).

Bedrock units in the hanging wall of La Grange fault at the La Grange mine and in the Oregon Mountain area south of the mine include sedimentary units of the Yreka subterrane (Moffett Creek Formation, Gazelle Formation, and Gregg Ranch Complex) and minor metamorphic rocks of the Abrams Mica Schist (Cashman and Elder, 1993). Point counts of three samples of Moffett Creek Formation sandstone, collected several hundred meters structurally above the La Grange fault, show that detrital grains are predominantly quartz, plagioclase, and mica, with minor chert, volcanic lithic grains, K-feldspar, and pyrite. Authigenic calcite is also a volumetrically important constituent of the Moffett Creek Formation (Table 1). Quantitative compositional data are not available for other, more fine-grained, sedimentary rocks in the hanging wall, but these rocks include chert pebble conglomerate, mudstone, chert, slate, and argillite.

The Abrams Mica Schist (Grouse Ridge Formation of Davis et al., 1965) consists of several different lithologies, including mica-quartz schist, micaceous quartzite, micaceous marble, calc-schist, and amphibolite. Mica-quartz schist is the only one of these rock types found in the hanging wall of the La Grange Fault at Oregon Mountain. Mica schist crops out above the fault surface at the northeastern edge of the La Grange mine exposure and several kilometers to the south at the southeastern edge of the Oregon Mountain klippe (Irwin, 1963). The average composition of mica-quartz schist and micaceous quartzite in the Abrams Mica Schist is quartz, 62%; plagioclase, 15%; muscovite, 10%; biotite, 5%; garnet, 2%; idocrase, 2%; epidote, 1%; and trace amounts of chlorite, calcite, sphene, rutile, and ore minerals (Davis et al., 1965).

DESCRIPTION OF FAULT ROCKS

Fault Surface

The La Grange fault is named for its exposure at the La Grange mine, a large hydraulic mine 8 km west of Weaverville, California. The mine was worked from 1851 to 1918 and produced $8,000,000 in gold from Weaverville Formation aurifer-

TABLE 1. CHEMICAL ANALYSES OF FAULT ZONE ROCKS AND FOOTWALL ROCKS, LA GRANGE FAULT

Sample	SHS1	SHS2	SHSg	93-17	95-17	95-20	SHS mean	std dev	ultc mean	std dev
SiO_2	46.94	49.18	48.55	56.37	55.25	50.69	48.06	1.59	54.10	3.01
TiO_2	2.05	1.94	1.08	1.24	0.85	0.94	1.99	0.08	1.01	0.21
Al_2O_3	12.36	12.21	14.65	10.60	11.50	11.19	12.28	0.11	11.10	0.45
FeO	14.70	13.48	7.62	9.83	7.54	8.69	14.09	0.86	8.68	1.14
MnO	0.24	0.23	0.17	0.29	0.17	0.29	0.24	0.01	0.25	0.07
MgO	8.88	7.94	9.23	10.37	10.60	13.36	8.41	0.67	11.45	1.66
CaO	11.38	11.17	12.72	9.50	11.07	13.14	11.28	0.15	11.24	1.83
Na_2O	2.69	3.14	3.58	0.86	1.60	0.78	2.92	0.32	1.08	0.45
K_2O	0.35	0.34	0.38	0.33	0.63	0.46	0.34	0.01	0.47	0.15
P_2O_5	0.14	0.21	0.24	0.17	0.15	0.17	0.17	0.05	0.16	0.02
S	0.00	0.00	0.00	0.08	0.16	0.06	0.00	0.00	0.10	0.05
F	0.06	0.05	0.06	0.05	0.04	0.05	0.06	0.01	0.04	0.01
H_2O	5.75	5.13	16.82	12.27	17.28	13.52				
Cr	154.7	146	310.8	516.5	545.3	807.6	150.35	6.2	623.1	160.4
Ni	12.4	14.1	114.6	317.6	350	545.6	13.25	1.2	404.4	123.4

Notes: Analyses by X-ray fluorescence using rhodium target tube and ground pellets. Eight U.S. Geological Survey rock standards were used for instrument calibration. Major element analyses are in wt% oxides and are normalized to 100% excluding water. Trace element analyses (Cr, Ni) are in ppm. SHS—Salmon Hornblende Schist; std dev—standard deviation; ultc—ultracataclasite.

ous gravel (O'Brien, 1965). Mining of upper plate gravel exposed a planar, south-southwest- to south-southeast-dipping fault surface 1 km in length on the west side of Oregon Mountain pass and a smaller southeast-dipping exposure immediately east of Oregon Mountain pass (Irwin, 1963; Cashman and Elder, 2002; Fig. 3). At both exposures, the fault is capped by up to 30 cm of black, fine-grained, strongly foliated, fault-related rocks (Figs. 2, 4, and 5). Fault-related rocks were originally termed "mylonite" (Irwin, 1966; Schweickert and Irwin, 1989), but were subsequently identified as ultracataclasite and cataclasite on the basis of SEM images (Cashman and Cashman, 1996). Contemporary mining reports describe the fault as "a great

Figure 4. Photomicrograph of foliated cataclasite/ultracataclasite, viewed perpendicular to foliation and parallel to slickenside striations, cross-polarized light. Texture records multiple episodes of deformation; the large clast of foliated ultracataclasite in center of image was broken off of an earlier ultracataclasite layer, cut by microfaults, and extended. Scale bar = 0.5 mm.

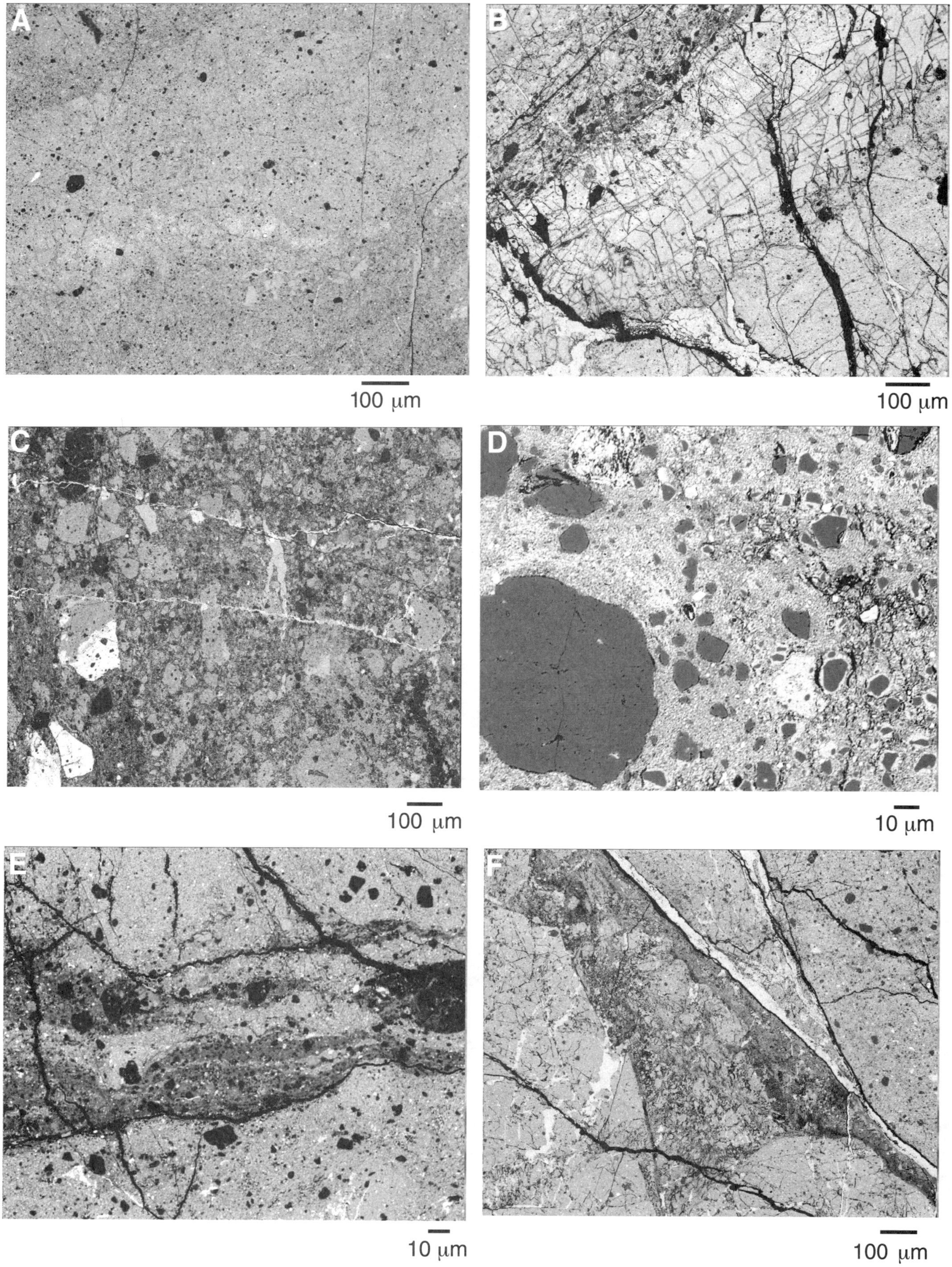
A
B
100 μm
100 μm
C
D
100 μm
10 μm
E
F
10 μm
100 μm

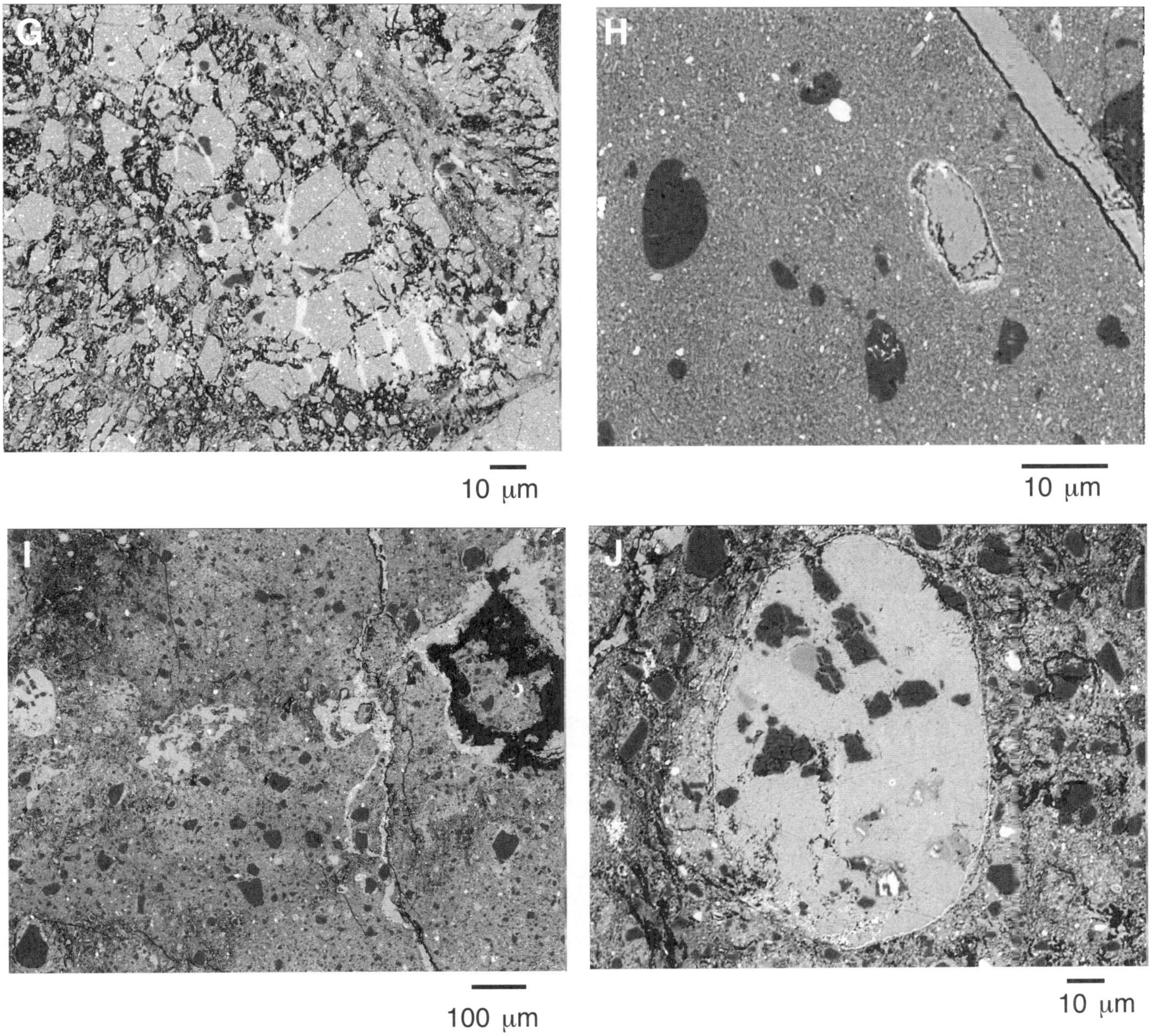

Figure 5. (A) Scanning electron microscope (SEM) backscatter image 95-18a100c, ultracataclasite. Sample collected 17 cm above the base of ultracataclasite layer, in southwest part of La Grange mine exposure. Black grains are quartz. Grains are extremely small, sub-rounded to well rounded, roughly equant, and show no evidence of recrystallization. (B) SEM image 95-20a100a. Sample collected 2 cm above base of ultracataclasite layer in southwest part of La Grange mine exposure. Foliation in upper-left corner of image is defined by a coarser-grained quartz-rich layer overlying a light-colored ultraclatclasite layer. (C) SEM image 95-01 501b, cataclasite. Common clast compositions are quartz (black) and ultracataclasite (light gray); veinlets are calcite. (D) SEM image 93-17a500a, ultracataclasite. Sample collected from fault exposure east of Oregon Mountain summit. Dark gray grains are quartz; many have lighter-colored rims composed of Si, Al, Mg, Ca, Fe, and Na. White grain at right center is rutile. (E) SEM image 95-18a500b, ultracataclasite. Very small-scale foliation shown by compositional variation in quartz-rich (dark) and quartz-poor (light) layers. Quartz grains (black) are more abundant in the quartz-rich layers. (F) SEM image 95-17f100a. Wedge-shaped cataclasite zone adjacent to a vein-filled fault contains discontinuous calcite veinlets (white) and fractured clasts. (G) SEM image, detail of 95-17f100a (see panel F). Fractured clasts are composed of ultracataclasite that formed in earlier faulting events. Broken ultracataclasite clast in center of image contains quartz grains (black) and calcite veinlets (white) formed prior to fragmentation of the clast. (H) SEM image 95-17 1600b. Ultracataclasite cut by calcite veinlet (light gray, upper right). Ultracataclasite contains well-rounded quartz grains (black) and a calcite clast (light gray) that appears to be a vein fragment. (I) SEM image 93-17a100b. Calcite (light gray) filling a fracture and intruding irregular spaces in ultracataclasite. (J) SEM image, detail of 93-17a100b (see panel I). Rounded grain is calcite and contains angular fragments of quartz (dark gray) and K-feldspar (medium gray).

smooth, slickensided fault plane trending N70°E and dipping 22°S" capped by "a parting of sticky clay, a few feet thick" along the fault plane, and record a zone of "disturbed stratification" in the stratified gravel directly overlying the fault surface (McDonald, 1909, p. 52). Hydraulic mining has removed all trace of these materials. In parts of the fault surface, the black fault-related rocks have also been removed by hydraulic mining, subsequent erosion, or a combination of the two.

The basal contact of the ultracataclasite is an abrupt, planar boundary (Fig. 2). At the southwestern edge of the mine exposure, the contact between black fault-related rocks and underlying sheared greenstone/amphibolite parallels both foliation in the fault rocks and the slightly undulatory foliation in sheared greenstone beneath.

The fault surface orientation varies between N80°E 18°S and N74°W 26°S at the main exposure at the La Grange mine, and is N25°E 29°SE at the smaller exposure immediately east of Oregon Mountain summit ("triangle exposure"; Fig. 3A). Foliation, defined by thin (<2 mm) black to charcoal gray, slickensided laminae, parallels the dip of the fault surface at both locations (Fig. 3B and C).

Prominent, closely spaced slickenside striations occur both on the exposed fault surface and on numerous discrete fault-parallel surfaces within the ultracataclasite. Slickenside striations consistently trend S21–22°E and plunge 18–26°SE, in spite of variations in fault plane orientation (Fig. 3B and C).

Steeply south-dipping, discrete small-scale faults, commonly showing several millimeters to several centimeters of down-to-the-south normal displacement, crosscut the ultracataclasite (Fig. 2). These cannot be traced below the base of the ultracataclasite, but die out at the contact between ultracataclasite and underlying foliated greenstone. Many small faults and fractures in the ultracataclasite host thin (commonly <2 mm) veins of calcite.

Fault Rocks

Examination of foliated ultracataclasite with a petrographic microscope and SEM shows two distinct, interlayered, fault-related rock types that can be distinguished on the basis of both grain size and grain composition (Cashman and Cashman, 1996, 2003; Figs. 4 and 5). Volumetrically dominant ultracataclasite layers ranging from several micrometers to several millimeters in thickness are interstratified with thin (<1 mm), generally foliation-parallel cataclasite zones. The ultracataclasite is composed of rounded to subangular grains (1–100 μm) in an extremely fine-grained <<1 μm matrix (Fig. 5A and B). Larger grains are commonly roughly equant and do not display preferred grain orientations. These grains constitute from 3 to 19% of samples studied; for most samples, larger grains constitute 7% or less of sample area.

Interlayered with the ultracataclasite are cataclasite zones composed of ultracataclasite clasts (up to 500 μm in diameter) and angular to subrounded individual mineral grains (5–100 μm) embedded in an ultracataclasite matrix (Fig. 5C–E). The cataclasite contains angular to sub-rounded grains that are equant to elongate and do not exhibit preferred orientation. Locally, cataclasite zones crosscut foliation in the ultracataclasite (Fig. 5F). Both the crosscutting relationship and the presence of ultracataclasite clasts in cataclasite show that the cataclasite formed later than the ultracataclasite.

Composition of Ultracataclasite and Cataclasite

Chemical analyses of cataclastic rocks show that these rocks have compositions distinctly different from those of hornblende schist in the footwall of the fault (Table 1). Both the altered hornblende schist ("greenstone") 1.0 m beneath the ultracataclasite and the ultracataclasite itself have high water content relative to the hornblende schist (12–17% compared to 5–6%). Normalized (anhydrous) compositions show that the ultracataclasite contains significantly higher SiO_2 and MgO, and lower FeO, Al_2O_3, TiO_2 and Na_2O than does the hornblende schist. These differences most likely represent incorporation into the ultracataclasite of material derived from Si-rich sedimentary rocks in the hanging wall of the fault. Greenstone within 1.0 m of the base of the cataclasite has substantially lower FeO and higher Al_2O_3, CaO, and Na_2O than does the hornblende schist from which it was derived. These differences likely reflect alteration of hornblende schist by fluids trapped beneath the ultra-fine-grained, impermeable, cataclastic rocks on the fault surface. Trace element compositions of ultracataclasite samples provide additional hints about possible sources for material incorporated in the ultracataclasite. In particular, concentrations of Cr and Ni in the ultracataclasite are substantially higher than those in the footwall rocks or in sedimentary rocks of the hanging wall; these constituents most likely derive from ultramafic rocks in the footwall north and east of the La Grange mine exposure.

An energy dispersive system (EDS) attached to the SEM was used to obtain qualitative compositional information for individual larger grains in the cataclasite and ultracataclasite. In spite of the broad range of rock compositions present in the footwall and hanging wall of the La Grange fault, only a limited suite of minerals occurs as large individual grains in the ultracataclasite and cataclasite on the fault surface. Single-mineral grains are most commonly (>95%) quartz. Rutile, apatite, pyrite or pyrrhotite, sphene, K-feldspar, and, locally, calcite or (rare) barite are also identifiable (Table 2). Ultracataclasite fragments are common in the cataclasite. Single mineral grains of quartz, calcite and, less commonly, barite, also occur as larger grains in coarser-grained cataclasite zones. The cataclasite layers show that the ultracataclasite has been rebroken into larger fragments relatively late in the development of the fault zone (Fig. 4).

A series of samples was collected from the base to the top of the 30-cm-thick layer of foliated black fault rocks at the southwest end of the mine exposure to investigate possible changes in texture and composition with depth in the layer. Sam-

TABLE 2. MINERALOGIC COMPOSITION OF FOOTWALL ROCKS, HANGING-WALL ROCKS, AND RESIDUAL GRAINS IN ULTRACATACLASITE

Mineral	Footwall: Salmon Hornblende Schist* (%)	Fault rocks: Larger clasts in ultracataclasite†	Hanging wall: Moffett Creek Formation‡ (%)	Hanging wall: Abrams Mica Schist# (%)
Hornblende	72			
Epidote	14			1
Plagioclase	7		13	15
Quartz	3	Very common	38	62
Pyrite and other "ore minerals"	2	Present	0.5	Trace
Chlorite	1			0.5
Actinolite	0.5			
Sphene	0.5	Present	‡	Trace
Rutile	Trace	Present		Trace
Hematite	Trace			Trace
Garnet	Trace			2
Clinopyroxene	Trace			
Apatite	Trace	Present	‡	
Mica			11	15
K-feldspar		Present	2	
Lithic grains			1.5	
Calcite		Locally common	27	1
Idocrase				2

*Salmon Hornblende Schist, Davis et al. (1965).
†Fault rocks; analyses derived from this work.
‡Moffett Creek Formation in Oregon Mountain area, S.M. Cashman (personal observation, 1993); minerals reported in Moffett Creek Schist in the Yreka area by Hotz (1977) but not identified in samples of Moffett Creek Formation from Oregon Mountain klippe.
#Abrams Mica Schist, "Grouse Ridge Formation, Mica-quartz schists and micaceous quartzite" (Table 1 *in* Davis et al., 1965).

ples cohesive enough to be examined as thin sections were not recovered from all levels, but useable samples were collected 2, 17, and 24 cm above the base of the foliated cataclastic rocks. All samples are composed primarily of ultracataclasite and all have fine (ten to several hundred micrometers thick) more quartz-clast–rich laminae. Texture appears to vary with depth in these samples: the abundance of larger grains increases upward. Larger grains constitute 3.3 ± 1.1% of the thin section at 2 cm, 2.8 ± 1.1% at 17 cm, and 7.0 ± 1.3% at 24 cm above the base of the foliated ultracataclasite. Major and trace element compositions of two of the samples also suggest variation in composition with depth. The sample near the base of the foliated ultracataclasite is enriched in MgO, CaO, FeO, MnO, Cr, and Ni and depleted in SiO_2, TiO_2, Al_2O_3, and Na_2O relative to the sample near the top of the ultracataclasite (Table 1).

A sample collected in a smaller "triangle" exposure of the La Grange fault, east of Oregon Mountain summit (Fig. 3A), has a distinctive texture. First, individual fragments show a wider size range than is observed in other samples, ranging from 5 µm to >100 µm in diameter. Second, roughly half of these grains have thin (1–2 µm) SEM-bright rims (Fig. 5D). Qualitative compositional data from the EDS are displayed as element maps in Figure 6. The maps show that Si, Al, Mg, Ca, Fe, and a small amount of Na are present in grain rims and that the rim material is compositionally indistinguishable from the matrix of the ultracataclasite. The nature of, and formation mechanism for, grain rims in this sample are not understood. Rims may be composed of pseudotachylite, or they may be composed of ultra-fine-grained particles accreted to the margins of larger quartz grains.

In summary, striking compositional differences between the ultracataclasite and the hornblende schist of the footwall, and from the base to the top of the ultracataclasite itself, record the contributions of both Si-rich hanging-wall rocks and more mafic footwall rocks to the ultracataclasite. The abundance of trace elements such as Cr and Ni in the ultracataclasite (Table 1) suggests that rocks not present at the La Grange mine exposure may also have contributed material to cataclastic rocks on the fault surface. Sample 95-20 from the base of the foliated ultracataclasite has higher values of Cr and Ni than does sample 95-17 from near the top of the ultracataclasite, suggesting that these elements were derived from footwall rocks, possibly ultramafic rocks of the Trinity terrane.

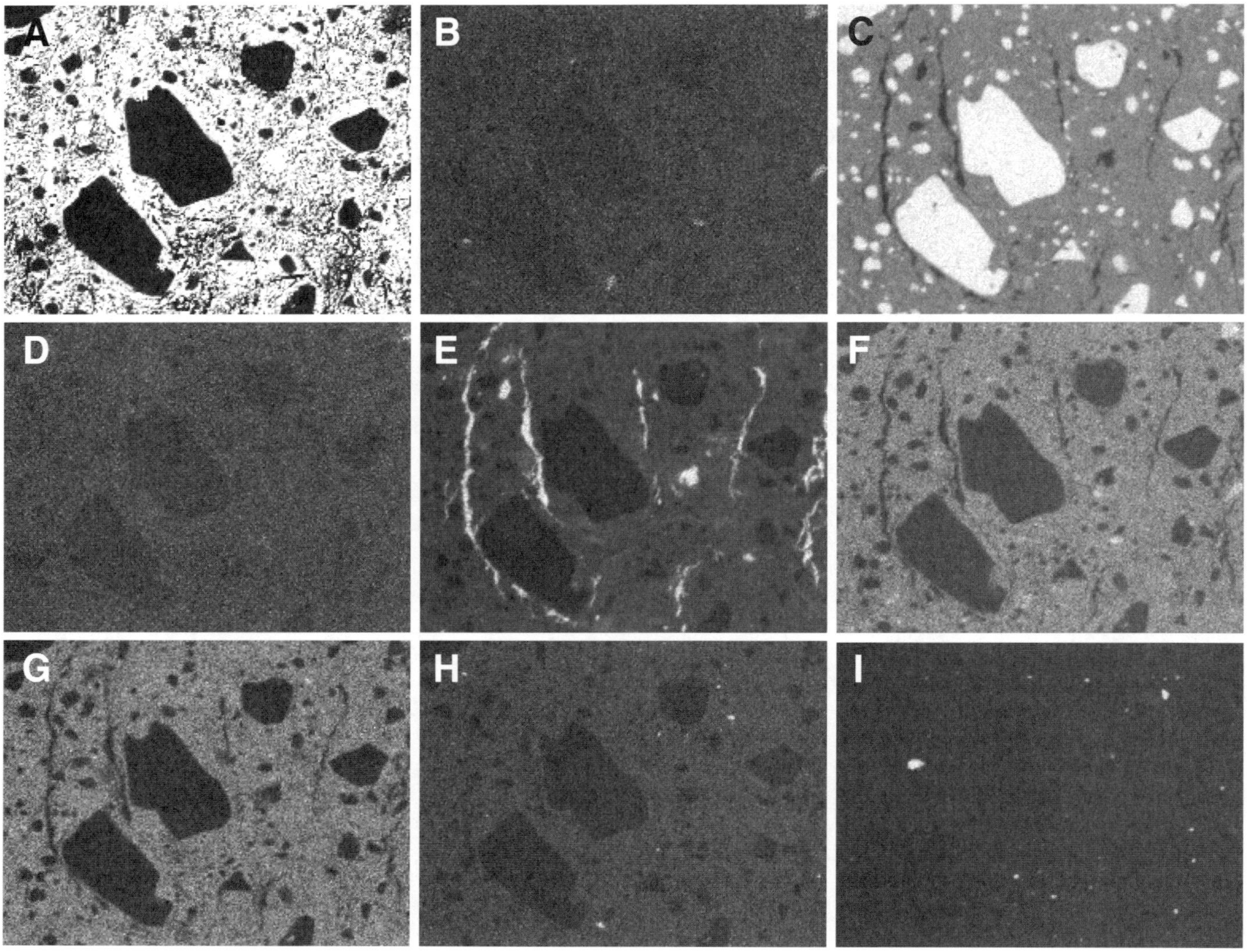

Figure 6. Scanning electron microscope (SEM) backscatter image and element maps of sample 93-17. Width of images is 120 µm. On element maps, lighter colors represent higher concentrations of the mapped element. (A) SEM backscatter image. Most of the larger grains have bright rims. (B) Na element map. (C) Si; this map shows abundant quartz grains. (D) K. (E) Ca; this map shows many tiny, suparallel calcite veins. (F) Al. (G) Mg. (H) Fe. (I) Ti; this map shows some grains of rutile. Grain rims do not stand out on element maps. Both rims and matrix contain Al, Si, Mg, K, Fe, and Na.

Microstructure of Ultracataclasite and Cataclasite

Foliation in the ultracataclasite is defined by relatively minor differences in rock texture and composition, primarily in the size and abundance of the largest grains. SEM and petrographic observation show thin ultracataclasite layers ranging from several to tens of millimeters in thickness (Figs. 4 and 5A, B, and E). Steeply-dipping micro-scale faults, commonly showing normal displacement, are locally abundant, as are pervasive sets of small (5–10 mm wide) calcite veins. Thin ultracataclasite layers are generally continuous and planar at the scale of a petrographic thin section. However, pinch-and-swell structure is developed locally in cataclasite and ultracataclasite layers (Fig. 5E). Occasional compositionally distinct (more quartz-poor or quartz-rich) and texturally distinct (more coarse- or fine-grained) lenses and stringers are interlayered with the foliation (Fig. 5B and E).

The extremely small grain size of the foliated ultracataclasite is interpreted as recording extreme grain crushing, milling, and sustained cataclastic deformation. In addition, textures preserved within individual larger grains record multiple episodes of grain breakage and vein injection. Fragments of calcite veins occur as grains in ultracataclasite (Fig. 5H); some calcite vein fragments in ultracataclasite include even smaller angular grains of quartz and K-feldspar (Fig. 5I and J).

Cataclasite zones are developed both parallel to foliation

and, less commonly, cutting across foliation in the ultracataclasite. Individual grains in cataclasite preserve evidence of multiple episodes of deformation that have included grain crushing, milling, vein formation, and repeated cataclastic deformation. Discontinuous veinlets within ultracataclasite clasts (Fig. 5F and G) and isolated calcite fragments within ultracataclasite (Fig. 5H) attest to a complex history of veining interspersed with cataclastic deformation in the fault zone.

Measurements of particle size distributions have been used to infer mechanisms of grain breakage in fault-related rocks (e.g., Sammis et al., 1986). Although the fine-grained nature of the ultracataclasite matrix precluded detailed grain size measurements, examination of back-scattered electron (BSE) images of a single ultracataclasite sample taken at increasing magnification (Fig. 7A) shows that the ratio of large to small grains changes with scale. These images indicate that the particle size distribution is not self-similar, in contrast to recent studies of both natural and experimental fault gouge (e.g., Sammis et al., 1986; Marone and Scholz, 1989; Blenkinsop and Fernandez, 2000). However, a similar series of SEM images of cataclasite (Fig. 7B), taken at the same range of magnifications as those of the ultracataclasite (Fig. 7A), shows that the cataclasite images are virtually indistinguishable from one another in terms of particle sizes. Each image has approximately the same ratio of large to smaller grains as any other image in the series, indicating that particle size distributions in the cataclasite are self-similar for magnifications of 100–3300×. Thus, it appears that whereas cataclasite formation appears to follow general models for grain size reduction developed from experimental and observational data on fault gouge, the extreme grain size reduction observed in the ultracataclasite requires more extensive crushing of larger grains.

Minor Faults and Veins

Foliated ultracataclasite and cataclasite layers on the surface of the La Grange fault are displaced by many small, parallel, closely-spaced, steeply south-dipping faults (Fig. 2). These minor faults strike roughly east-west, parallel to the La Grange fault, and have small (<1 cm) normal dip separations. Some minor faults are filled by veins.

Thin, planar veins are a prominent feature of the footwall both at the La Grange mine and at the "triangle" exposure east of Oregon Mountain Pass. Most veins have steep dips, perpendicular to foliation in the ultracataclasite, and die out into more irregular and widely dispersed veins in the underlying greenstone. Calcite is the most common vein mineral, followed by quartz, pyrite, and heulandite.

Several lines of evidence suggest that the vein formation accompanied slip on the La Grange fault. First, veins are more abundant and show more systematic orientations in the footwall than in the hanging wall, so veins do not appear to have formed along inactive structures. Second, sets of veins are systematically oriented relative to the fault surface and to the slickenside striations (Fig. 8A). At the southwest part of the main La Grange Mine exposure, two sets of veins 1–10 mm wide and 1–10 m long are oriented perpendicular to foliation and strike ~120° apart, with the obtuse angle between veins bisected by slickenside striations on the fault surface (Fig. 8A and B). At the fault exposure east of Oregon Mountain summit, veins 0.5–2 mm wide and 0.5–2 m long form a single set perpendicular to foliation and perpendicular to slickenside striations (Fig. 8C). Finally, kinematic indicators in fault zone rocks (slickenside striations, brittley extended clasts in the cataclasite, and asymmetric folds in greenstone directly underlying the fault surface) record extension in the same direction as that required for the opening of the vein-filled fractures (Fig. 8).

As vein orientations indicate that vein-filled minor faults in La Grange fault cataclastic rocks formed during extensional faulting, vein texture and composition provide clues to conditions under which faulting occurred. Calcite-heulandite veins at the southwest end of the La Grange mine exposure are the most useful in this regard. Heulandite crystals occur as small (0.1 mm long) radiating clusters on the outer edges of vein walls; these veins were subsequently filled by coarse-grained calcite. The presence of heulandite indicates that these veins formed at 240 ± 40 °C and 0–2 kb (Frey et al., 1991), which correlates to a maximum depth of ~7 km. Heulandite crystal clusters appear to have grown in the absence of neighbors, suggesting that crystallization occurred in the lower-pressure, shallower part of the heulandite stability range. Characteristics of both the veins and the minor faults they occupy are consistent with the interpretation that faults and vein minerals formed during the late stages of regional extensional faulting, during progressive uplift and unroofing of the fault surface.

DISCUSSION

Microscopic textures of fault rocks from the La Grange mine record pronounced and protracted strain localization during brittle faulting on the La Grange fault. Intense comminution and sustained cataclastic deformation were concentrated within the thin (30 cm) ultracataclasite layer on the exposed fault surface. Extensive comminution is consistent with very-high-strain localization, as implied by the tectonic juxtaposition of bedrock units with dramatically contrasting ages, metamorphic grades, and tectonostratigraphic affiliations across this fault. Textural and compositional observations from these rocks can be used to evaluate models for fragmentation processes that operate during the development of a major extensional fault zone.

Origin of Ultracataclasite

The lithologic contrast between amphibolite in the footwall and sandstone, chert, and mica schist in the hanging wall permits assessment of the relative contributions of footwall and hanging wall rocks to the ultracataclasite. Table 2 presents mineralogic compositions of footwall rocks, hanging-wall rocks,

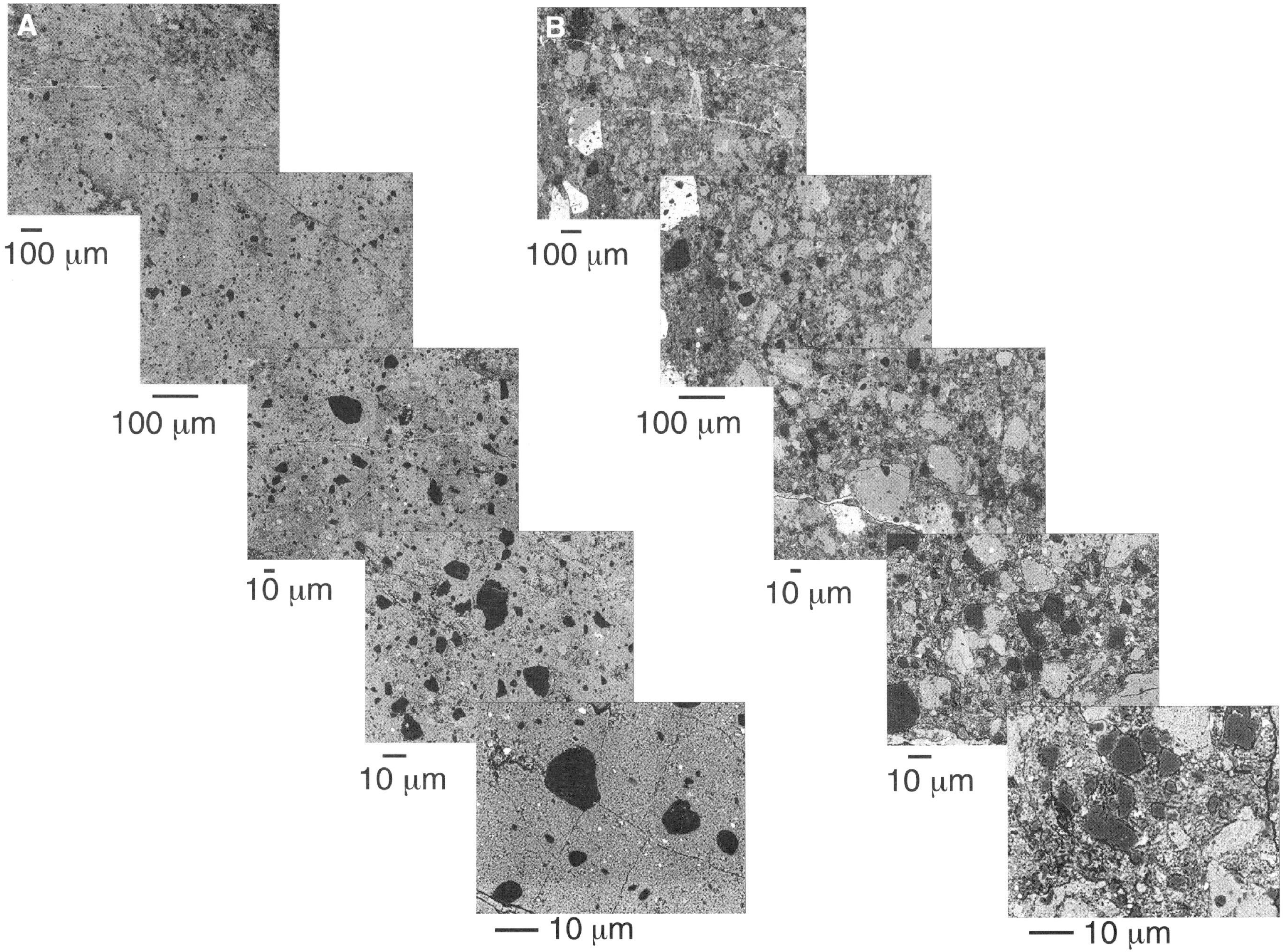

Figure 7. (A) Scanning electron microscope (SEM) images of ultracataclasite sample 95-02 taken at sequentially higher magnifications (100×, 200×, etc.) show that particle size distribution in this sample is not self-similar. The size of the largest grains in the image increases at higher magnification. (B) SEM images of sample 95-01, a cataclasite horizon within the ultracataclasite, taken at sequentially higher magnifications. Each image has approximately the same ratio of larger to smaller grains as any other image in the series; this sample displays scale-independent (self-similar) particle size distribution.

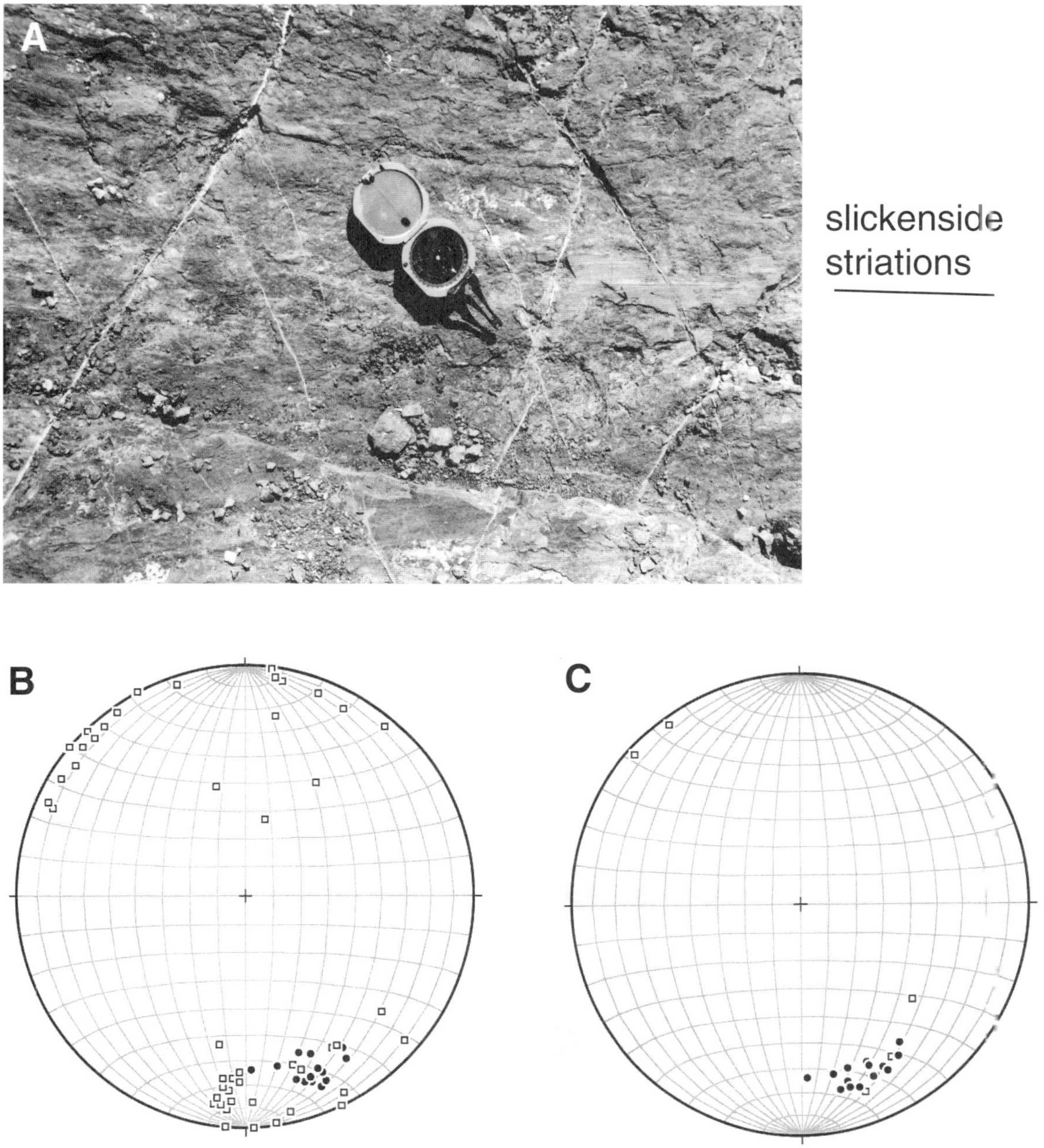

Figure 8. (A) Photograph of surface of the La Grange fault at the La Grange mine exposure, showing slickenside striations that bisect the obtuse angle between the two prominent sets of veins. (B) Equal-area stereonet slickenside striations (filled circles, $N = 20$) and poles to veins (open squares, $N = 47$), La Grange fault at La Grange Mine. (C) Equal-area stereonet plot of slickenside striations (filled circles, $N = 16$) and pole s to veins (open squares, $N = 5$), La Grange fault at exposure east of Oregon Mountain summit.

and residual larger single-mineral grains in the ultracataclasite. Larger single-mineral grains in ultracataclasite are predominantly quartz but also include calcite, pyrite, sphene, rutile, apatite, and zircon. Comparison of minerals present in the footwall and hanging wall indicates that most of the larger grains in the ultracataclasite appear to have originated in the hanging wall (Table 1). This interpretation is supported by the upward increase in both SiO_2 content and percentage of large quartz grains in three samples collected from different levels in the ultracataclasite.

In contrast, EDS analysis of the ultracataclasite matrix indicates that elements not found in single-mineral grains (e.g., Al, Mg, Na, K) are common in the matrix. Rocks in the footwall of the fault, and/or less resistant minerals from the hanging wall of the fault, must be the source of ultracataclasite matrix material.

Bulk rock compositions of ultracataclasite samples provide additional evidence: sample 95-20, from the base of the ultracataclasite, has more FeO, MnO, MgO, and CaO and less SiO_2, TiO_2, Al_2O_3, and Na_2O than does sample 95-17 from the top of the ultracataclasite (Table 1). Thus, compositional variations within the ultracataclasite match compositional differences between amphibolite in the footwall and sedimentary and metasedimentary rocks in the hanging wall.

Ultracataclasite compositions indicate that both hanging-wall and footwall rocks contributed to ultracataclasite formation and that individual resistant grains derived primarily from the hanging wall were physically mixed into a matrix derived, in part, from the footwall. Geochemical evidence from the La Grange fault indicates that mechanical mixing of rocks from the hanging wall and footwall was the primary mechanism for the formation of La Grange fault ultracataclasite. This finding corroborates the conclusion of Evans and Chester (1995) that mechanical mixing of rocks bounding the fault zone was the primary mechanism for formation of cataclastic rocks on the San Gabriel fault.

Lithologic Control of Fragmentation

The distribution and composition of La Grange fault ultracataclasite attest to the importance of parent-rock lithology in the mechanical development of fault-related rocks. First, cataclasite and ultracataclasite are found only on that portion of the fault for which amphibolite (Salmon Hornblende Schist) forms the footwall. Foliated cataclasite and ultracataclasite are not present on the fault surface east of the La Grange mine exposure, where the footwall is composed of ultramafic rocks (Cashman and Elder, 2002).

Second, the larger grains in the ultracataclasite are single crystals of mechanically resistant minerals, primarily quartz but also including rutile, apatite, pyrite or pyrrhotite, and sphene. Exceptions to this generalization are nonresistant calcite (common) and barite (rare) grains. Calcite and barite occur in veins and are likely preserved as larger grains in the cataclastic rocks because they were introduced into the fault rocks at a relatively late stage of the faulting process.

Third, matrix composition is distinctly different from that of larger single grains in the ultracataclasite. The compositional contrast between single-mineral grains and ultracataclasite matrix at the La Grange fault raises the question of the applicability of experimental and theoretical models based on behavior of a single grain composition (e.g., quartz) to the more complex behavior of heterogeneous fault zones.

Particle Size Distribution in Ultracataclasite

Preferential preservation of specific minerals as larger grains within the ultracataclasite attests to the importance of mechanical properties of individual minerals in comminution processes, as discussed by Rawling and Goodwin (2003) with regard to faulted, poorly lithified sediments from the Rio Grande rift, New Mexico. Their observations show that both the style of breakage and the resulting grain size distribution are dependant on relative grain strength; volcanic lithic grains failed by distributed microcracking and by ductile deformation (where altered), feldspars failed by transgranular fracturing, and quartz failed by flaking of small chips around grain margins. Preferential preservation of quartz and other mechanically resistant minerals as larger subangular to subrounded grains in La Grange fault ultracataclasite suggests that the pattern of differential fragmentation of different minerals observed by Rawling and Goodwin (2003) in poorly lithified sediments may also apply to more highly deformed cataclastic rocks. Furthermore, we suggest that the absence of self-similar particle size distributions in La Grange ultracataclasite (Fig. 7A) may reflect the contrasting mechanical properties of individual minerals. Therefore, the microstructure and particle size distribution produced by extensive cataclastic deformation may be as much a function of rock composition as they are of intensity of deformation.

Role of Fluid Flow in Faulting

Both the high water content of ultracataclasite samples (Table 1) and the presence of veins at all scales in these rocks (Figs. 4–6 and 8) attest to the importance of fluids in the formation of cataclastic rocks on the La Grange fault. Fragments of calcite veins are present as clasts in ultracataclasite (Fig. 5H), individual veins occur in ultracataclasite clasts within cataclasite layers (Fig. 5G), and tiny discontinuous calcite veins are pervasive in parts of the ultracataclasite matrix (Fig. 6E).

Several microstructural features in La Grange fault rocks suggest that fault slip was episodic and accompanied by fluid flow during the period of fault-related rock formation. The presence of ultracataclasite clasts in cataclasite (Fig. 5C, F, and G) requires extensive comminution followed by time sufficient for lithification of the ultracataclasite. Incorporation of vein calcite into both cataclasite and ultracataclasite (Fig. 5F–J) suggests that fluid flow through the fault zone was pervasive, was repeated many times, and may have contributed to lithification.

Although multiple episodes of veining accompanied formation of fault-related rocks, La Grange fault ultracataclasite lacks clear geochemical evidence for metasomatism. Geochemical studies of fault zone rocks have shown fault-related rocks to be depleted in Si, Al, K, Na, and Ca relative to immobile Ti (Goddard and Evans, 1995), or to be depleted in SiO_2, Na_2O, and K_2O but enriched in TiO_2, CaO, P_2O_5, and H_2O relative to protoliths (Evans and Chester, 1995). La Grange fault ultracataclasite has substantially higher water content than found in footwall rocks but does not show the element enrichment or depletion trends reported for other faults. At the La Grange fault, as at two study sites on the San Gabriel Fault (Evans and Chester, 1995), mechanical mixing of protoliths from both sides of the fault appears to be more important than chemical interaction with fluids in controlling the geochemistry of fault-related rocks.

Regional Implications of Extensional Faulting in the Klamath Mountains Province

Post-Nevadan extension on low-angle detachment faults such as the La Grange fault played an important role in framing the familiar map distribution of Klamath terranes. Regional detachment faulting, initially recognized by Schweickert and Irwin (1989), post-dates the assembly of the accreted terranes of the Klamath Mountains province (e.g., Irwin, 1966, 1972, 1977, 1981). Its overprint explains anomalies in the extent and distribution of Klamath terranes: klippen of Yreka terrane rocks at Cecilville and at Oregon Mountain, and the jagged northeast-striking northern margin of the Redding terrane (previously noted as a structural anomaly by Renne and Scott (1988; Fig. 1).

Extensional faulting in the Klamath Mountains province has also influenced the distribution of economic mineral deposits. Low-angle normal faulting at the La Grange mine locality created a topographic depression in which gold-bearing Tertiary gravels were preserved (Schweickert and Irwin, 1989). Extensional faulting also controlled the distribution of late Cretaceous muscovite-bearing felsic dikes and of epithermal mineralization and associated gold and/or copper massive sulfide deposits in the Klamath Mountains province (Cashman and Elder, 2002).

CONCLUSIONS

Mesoscale and microscale structures in fault-related rocks at the La Grange mine record prolonged, brittle, extensional deformation on the La Grange fault. Strongly foliated, fault-related rocks capping the fault surface at the La Grange mine are not ultramylonite, as originally described by Irwin (1966), but ultracataclasite, a term that had not been coined when Porter Irwin first described this fault. Extremely small grain size in the foliated ultracataclasite records extreme grain crushing, milling, and sustained cataclastic deformation.

Chemical composition of foliated ultracataclasite reveals that these fault-related rocks were derived in part from units currently present in the hanging wall and footwall of the fault, in part from vein minerals, and in part from rocks that are no longer present in the hanging wall or footwall (e.g., ultramafic rocks) but that contributed material to the fault zone earlier in the evolution of the fault. The compositional contrast between large single-mineral grains and matrix in the ultracataclasite highlights the importance of parent-rock lithology to the grain size distribution as well as to the composition of fault-related rocks. The prevalence of vein minerals as single crystal clasts in ultracataclasite, as veins preserved in ultracataclasite clasts within cataclasite zones, and in intact veins documents a prolonged history of vein formation alternating with slip events on the La Grange fault.

Kinematic indicators in fault zone rocks, including slickenside striations, brittle extended clasts, asymmetric folds in greenstone directly underlying the fault surface, and vein-filled fractures in foliated cataclasite and ultracataclasite, record south-southeast extension on the La Grange fault. Prolonged extensional faulting resulted in south-southeastward transport of hanging-wall rocks coupled with uplift and exhumation of footwall rocks. The presence of Yreka terrane units at Oregon Mountain, in the hanging wall of the La Grange fault, records on the order of 60 km of southward displacement of hanging-wall units. Truncation of the northern margin of Redding terrane rocks by the La Grange fault both highlights the unique character of this terrane-bounding fault and marks the change from accretionary tectonics to extensional tectonics in the history of the Klamath Mountains province.

ACKNOWLEDGMENTS

We thank Shasta Trinity National Forest for providing housing during the mapping phase of this project, and Scott North both for preparation of difficult thin sections and for X-ray fluorescence analysis of samples. We are grateful to D. Elder for many constructive discussions of the role of the La Grange fault in the tectonic evolution of the Klamath Mountains, and to L. Goodwin and J. Evans for thoughtful reviews of an earlier draft of this chapter.

REFERENCES CITED

Antonellini, M., and Aydin, A., 1994, Effect of faulting on fluid flow in porous sandstones: Petrophysical properties: American Association of Petroleum Geologists Bulletin, v. 78, p. 355–377.

Aydin, A., and Johnson, A.M., 1983, Analysis of faulting in porous sandstones: Journal of Structural Geology, v. 5, p. 19–31, doi: 10.1016/0191-8141(83)90004-4.

Blenkinsop, T.G., and Fernandez, T.R.C., 2000, Fractal characterization of particle size distributions in chromitites from the Great Dyke, Zimbabwe: Pure and Applied Geophysics, v. 157, p. 505–521.

Cashman, S.M., and Cashman, K.V., 1996, Evidence for episodic slip preserved in textures of fine-grained fault rocks, La Grange fault, California: Geological Society of America Abstracts with Programs, v. 28, p. 244–245.

Cashman, S.M., and Cashman, K.V., 2000, Cataclasis and deformation band formation in unconsolidated marine terrace sand, Humboldt County, California: Geology, v. 28, p. 111–114, doi: 10.1130/0091-7613(2000)028<0111:CADBFI>2.3.CO;2.

Cashman, S.M., and Cashman, K.V., 2003, Insights into development of a major extensional fault based on textural and compositional evidence from La Grange fault, CA, ultracataclasite: Geological Society of America Abstracts with Programs, v. 35, p. 180.

Cashman, S.M., and Elder, D.R., 1993, Tertiary detachment faulting and associated mineralization on the La Grange fault, Klamath Mountains, California (abs.): EOS, Transactions, American Geophysical Union, v. 74, p. 577–578.

Cashman, S.M., and Elder, D.R., 1996, The La Grange fault revisited: Evidence for regionally extensive detachment faulting, Klamath Mountains, California: Geological Society of America Abstracts with Programs, v. 28, p. 55.

Cashman, S.M., and Elder, D.R., 2002, Post-Nevadan detachment faulting in the Klamath Mountains, California: Geological Society of America Bulletin, v. 114, p. 1520–1534, doi: 10.1130/0016-7606(2002)114<1520:PNDFIT>2.0.CO;2.

Chester, F.M., Evans, J.P., and Biegel, R.L., 1993, Internal structure and weakening mechanisms of the San Andreas Fault: Journal of Geophysical Research, v. 98, p. 771–786.

Crawford, R., 2004, Petrology and microstructures of ultramafic rocks near Weaverville, California [B.A. thesis]: Arcata, California, Humboldt State University, 38 p.

Crittenden, M.D., Coney, P.J., and Davis, G.H., 1980, Cordilleran metamorphic core complexes: Boulder, Colorado, Geological Society of America Memoir 153, 490 p.

Davis, G.A., Holdaway, M.J., Lipman, P.W., and Romey, W.D., 1965, Structure, metamorphism, and plutonism in the south-central Klamath Mountains, California: Geological Society of America Bulletin, v. 76, p. 933–966.

Davis, G.H., 1980, Structural characteristics of metamorphic core complexes, southern Arizona, *in* Crittenden, M.D. Jr., et al., eds., Cordilleran metamorphic core complexes: Boulder, Colorado, Geological Society of America Memoir 153, p. 35–77.

Davis, G.H., Constentius, K.N., Dickinson, W.R., Rodriguez, E.P., and Cox, L.J., 2004, Fault and fault-rock characteristics associated with Cenozoic extension and core-complex evolution in the Catalina-Rincon region, southern Arizona: Geological Society of America Bulletin, v. 116, p. 128–141, doi: 10.1130/B25260.1.

Evans, J.P., and Chester, F.M., 1995, Fluid-rock interaction in faults of the San Andreas system: Inferences from San Gabriel fault rock geochemistry and microstructures: Journal of Geophysical Research, v. 100, p. 13,007–13,020, doi: 10.1029/94JB02625.

Frey, M., de Capitani, C., and Liou, J.G., 1991, A new petrogenetic grid for low-grade metabasites: Journal of Metamorphic Geology, v. 9, p. 497–509.

Goddard, J.V., and Evans, J.P., 1995, Chemical changes and fluid-rock interaction in faults of crystalline thrust sheets, northwestern Wyoming, U.S.A: Journal of Structural Geology, v. 17, p. 533–547, doi: 10.1016/0191-8141(94)00068-B.

Hadizadeh, J., and Johnson, W.K., 2003, Estimating local strain due to comminution in experimental cataclastic textures: Journal of Structural Geology, v. 25, p. 1973–1979, doi: 10.1016/S0191-8141(03)00016-6.

Hadizadeh, J., Konkachbaev, A., and Goldsby, D., 2004, Comminution characteristics of quartz in natural and experimental faults; Implications for brittle shear localization: EOS, Transactions, American Geophysical Union, v. 85, p. 422.

Hotz, P.E., 1977, Geology of the Yreka Quadrangle, Siskiyou County, California: Reston, Virginia, U.S. Geological Survey Bulletin 1436, 72 p.

Irwin, W.P., 1963, Preliminary geologic map of the Weaverville quadrangle, California: Washington, D.C., U.S. Geological Survey Miscellaneous Field Studies Map MF-275, scale 1:62,500, 1 sheet.

Irwin, W.P., 1966, Geology of the Klamath Mountains province, *in* Bailey, E.H., ed., Geology of Northern California: San Francisco, California Division of Mines and Geology Bulletin 190, p. 19–38.

Irwin, W.P., 1972, Terranes of the western Paleozoic and Triassic belt in the southern Klamath Mountains, California, *in* Geological Survey Research, 1972: Washington, D.C., U.S. Geological Survey Professional Paper 800-C, p. C103–C111.

Irwin, W.P., 1977, Review of Paleozoic rocks of the Klamath Mountains, *in* Stewart, J.H., et al., eds., Paleozoic paleogeography of the western United States I: Los Angeles, Pacific Section, Society of Economic Paleontologists and Mineralogists, p. 441–454.

Irwin, W.P., 1981, Tectonic accretion of the Klamath Mountains, *in* Ernst, W.G., ed., The geotectonic development of California—Rubey volume I: Englewood Cliffs, New Jersey, Prentice-Hall, p. 29–49.

Irwin, W.P., 1994, Geologic map of the Klamath Mountains: Reston, Virginia, U.S. Geological Survey Miscellaneous Investigations Series Map I-2148, scale 1:500,000, 1 sheet.

LeRoy, T.H., 1993, General reconnaissance of the northeast side of Hyampom Valley: Evidence for a Late Cenozoic structural basin, Southern Klamath Mountains, California [B.S. thesis]: Arcata, California, Humboldt State University, 27 p.

Lindsley-Griffin, N., 1991, The Trinity Complex: A polygenetic ophiolitic assemblage, *in* Cooper, J.D., and Stevens, C.H., eds., Paleozoic paleogeography of the western United States II: Los Angeles, Pacific Section, Society of Economic Paleontologists and Mineralogists, v. 67, p. 589–607.

Mair, K., and Marone, C., 1999, Friction of simulated fault gouge for a wide range of velocities and normal stresses: Journal of Geophysical Research, v. 104, p. 28,899–28,914, doi: 10.1029/1999JB900279.

Mair, K., Main, I., and Elphick, S., 2000, Sequential growth of deformation bands in the laboratory: Journal of Structural Geology, v. 22, p. 25–42, doi: 10.1016/S0191-8141(99)00124-8.

Marone, C., and Scholz, C.H., 1989, Particle size distribution and microstructures within simulated fault gouge: Journal of Structural Geology, v. 11, p. 799–814, doi: 10.1016/0191-8141(89)90099-0.

McDonald, D.F., 1909, The Weaverville–Trinity Center gold gravels, Trinity County, California: Washington, D.C., U.S. Geological Survey Bulletin 430, p. 48–58.

O'Brien, J.C., 1965, Mines and Mineral Resources of Trinity County: San Francisco, California Division of Mines and Geology County Report 4, 125 p.

Phillips, P.A., 1989, The geology of the Weaverville Formation in the vicinity of Weaverville, California [M.S. thesis]: Arcata, California, Humboldt State University, 174 p.

Phillips, P.A., and Aalto, K.R., 1989, The Weaverville Formation: Implications for tectonically controlled basin development in the central Klamath Mountains, *in* Colburn, I., et al., eds., Conglomerates in basin analysis: Los Angeles, Pacific Section, Society of Economic Paleontologists and Mineralogists, v. 62, p. 81–98.

Potter, A.W., Hotz, P.E., and Rohr, D.M., 1977, Stratigraphy and inferred tectonic framework of lower Paleozoic rocks in the eastern Klamath Mountains, northern California, *in* Stewart, J.H., et al., eds., Paleozoic paleogeography of the western United States I: Los Angeles, Pacific Section, Society of Economic Paleontologists and Mineralogists, p. 421–440.

Rawling, G.C., and Goodwin, L.B., 2003, Cataclasis and particulate flow in faulted, poorly lithified sediments: Journal of Structural Geology, v. 25, p. 317–331, doi: 10.1016/S0191-8141(02)00041-X.

Renne, P.R., and Scott, G.R., 1988, Structural chronology, oroclinal deformation, and the tectonic evolution of the southeastern Klamath Mountains, California: Tectonics, v. 7, p. 1223–1242.

Sammis, C.G., Osborne, R.H., Anderson, J.L., Mavowe, E., and White, P., 1986, Self-similar cataclasis in the formation of fault gouge: Pure and Applied Geophysics, v. 124, p. 53–77, doi: 10.1007/BF00875719.

Sammis, C.G., King, G., and Biegel, R., 1987, The kinematics of gouge deformation: Pure and Applied Geophysics, v. 125, p. 777–812, doi: 10.1007/BF00878033.

Schweickert, R.A., and Irwin, W.P., 1989, Extensional faulting in southern Klamath Mountains, California: Tectonics, v. 8, p. 135–149.

Snoke, A.W., Tullis, J., and Todd, V.R., eds., 1998, Fault-related rocks: A photographic atlas: Princeton, New Jersey, Princeton University Press, 613 p.

Manuscript Accepted by the Society 10 January 2006

Printed in the USA

Geological Society of America
Special Paper 410
2006

The Klamath peneplain: A review of J.S. Diller's classic erosion surface

K. R. Aalto*
Department of Geology, Humboldt State University, 1 Harpst Street, Arcata, California 95521, USA

ABSTRACT

Over one hundred years ago, J.S. Diller (1902) published his classic work, *Topographic Development of the Klamath Mountains,* in which he proposed that a regional lowland erosion surface, the "Klamath peneplain," formed in northwestern California during the late Miocene. W.P. Irwin's (1997) *Preliminary Map of Selected Post-Nevadan Geologic Features of the Klamath Mountains and Adjacent Areas* clearly delineates the extent of this erosion surface and provides detailed mapping and age determinations for overlap deposits in both the Klamath and Coast Range provinces. Among these deposits, basal late Miocene Wildcat Group, Saint George Formation, and Wimer Formation sediments are littoral. Rapid submergence of northwestern California to bathyal depths occurred, following initiation of Wildcat–Saint George–Wimer sedimentation. By 5 Ma, the Klamath peneplain was at sea level, based upon Wimer Formation overlap. A $^{40}Ar/^{39}Ar$ laser probe analysis of detrital micas indicates that Wildcat Group marine sediments are, in part, derived from the Idaho batholith. Sediment progradation and shoaling over a wide coastal shelf, onset of deposition of locally derived fluvial sediments in the Coast Ranges, and regional uplift during the middle Pleistocene may relate to the decreasing age of the subducting oceanic plate, decreasing dip angle of subduction, translation of the Sierra Nevada microplate northwestward as it impinges upon the Klamath block, and/or migration of the Blanco fracture zone northward along the continental margin.

Keywords: Klamath peneplain, J.S. Diller, W.P. Irwin, uplift, provenance, neotectonics

INTRODUCTION

Over one hundred years ago, J.S. Diller (1902, p. 15) published his classic work, *Topographic Development of the Klamath Mountains,* in which he proposed that a regional lowland erosion surface, the "Klamath peneplain," formed in the Klamath Mountains (Fig. 1) during the late Miocene:

The Klamath Mountains are characterized in many places by flattish or gently rounded summits and an approximate accordance in the altitudes of even-crested ridges, giving to these highlands the aspect of a dissected plateau. . . . The original more or less irregular plateau surface, of which we now have only remnants, for convenience of reference may be for the present assumed to be due to a form of planation, and designated the Klamath peneplain.

He further defined the "Bellspring peneplain" as having formed west of Klamath basement rocks (Fig. 2). This erosion surface is underlain by tilted Neogene sediments of the Coast Ranges and formed shortly following their deposition. The Bellspring

*E-mail: kra1@humboldt.edu.

Aalto, K.R., 2006, The Klamath peneplain: A review of J.S. Diller's classic erosion surface, *in* Snoke, A.W., and Barnes, C.G., eds., Geological studies in the Klamath Mountains province, California and Oregon: A volume in honor of William P. Irwin: Geological Society of America Special Paper 410, p. 451–463, doi: 10.1130/2006.2410(22). For permission to copy, contact editing@geosociety.org.

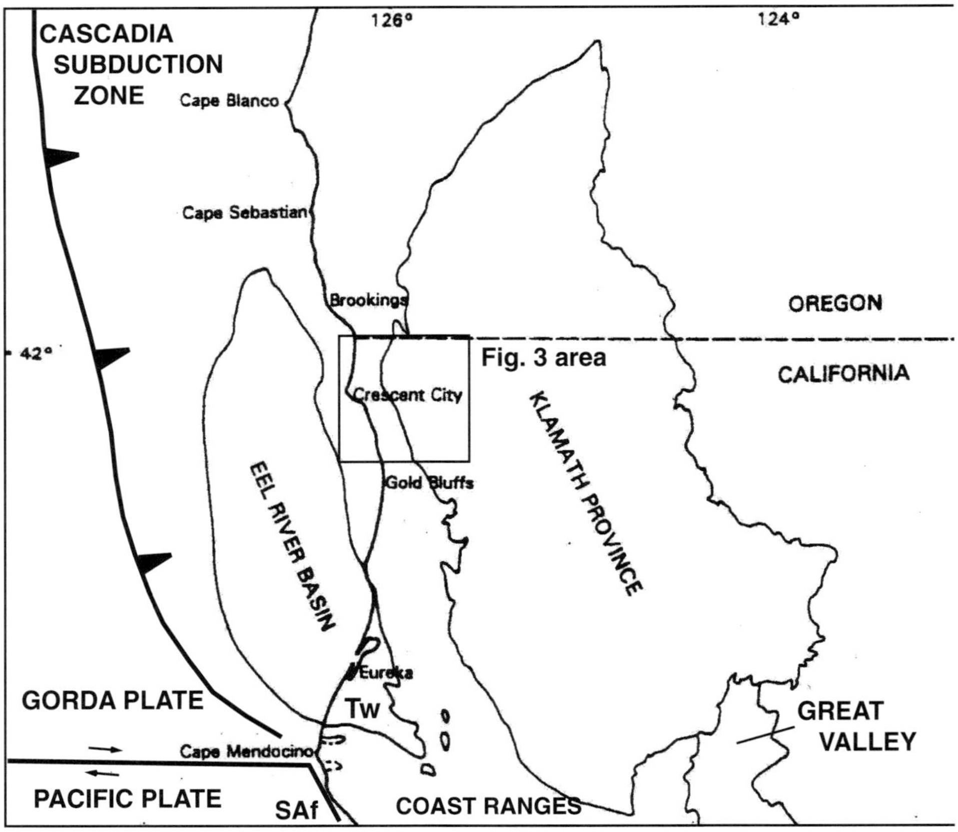

Figure 1. Regional location map showing plate boundaries and general outcrop areas of the Eel River basin, Wildcat Group (Tw), Klamath Mountains and Coast Ranges provinces and location of the Mendocino triple junction (at Cape Mendocino). SAf—San Andreas fault.

peneplain was thought to be practically continuous with the adjacent Klamath peneplain. The definition of several younger, more regionally restricted planation surfaces was complicated both by Diller's lack of correct age determination of younger formations and the absence of a neotectonic framework.

W.P. Irwin's (1997) *Preliminary Map of Selected Post-Nevadan Geologic Features of the Klamath Mountains and Adjacent Areas, California and Oregon,* clearly delineates the extent of the combined Klamath-Bellspring erosion surfaces. Irwin (1997, p. 5) substantially improved on our understanding of the topographic evolution of the Klamath province by carefully mapping the extent of overlap deposits and determining their correct ages:

> The altitudes of the old upland surfaces generally increase eastward, away from the coastline, and thus are generally higher in the Klamath Mountains than in the Coast Ranges. . . . The altitudes of the old surfaces in the Coast Ranges range from 1000 ft or less just south of the Oregon boundary, and increase to several thousand to rarely as much as 5000 ft in the southern part. In the Klamath Mountains, all the old surface remnants are higher than 2000 ft except for a few near the western boundary of the province just east of Crescent City. In the Klamath Mountains of Oregon the old surfaces reach altitudes of more than 4000 ft, and in California reach an altitude of about 6000 ft at a few places just east of Hoopa Valley.

Aalto et al. (1995b, 1998) demonstrated an Idaho batholith provenance for some late Neogene sands west of the Klamath Mountains and suggested that the western margin of Diller's Miocene peneplain was draped by a shelf sediment blanket prior to regional uplift of the Klamath block during the Pleistocene–Holocene. Thus, Diller was correct inasmuch as (1) a late Miocene Klamath peneplain did form in the Klamath province, (2) Neogene marine sediments accumulated on its western margin in response to a relative rise of sea level, and (3) ongoing post-Neogene land degradation accompanying uplift of Klamath basement rock and sediment cover resulted in the formation of an erosion surface that approximates his combined Klamath/Bellspring peneplain. Diller's ideas, therefore, remain major contributions toward the geomorphic interpretation of this tectonically active region.

The purpose of this chapter is to relate sedimentation and sediment provenance of late Cenozoic sedimentary cover rocks in northwestern California to their tectonic setting and to the uplift history of northwestern California (Fig. 3). Late Cenozoic patterns of sedimentation reflect a long history of convergence between the Farallon and North American plates. Northward migration of the Mendocino triple junction (Kelsey and Carver, 1988; Aalto et al., 1995a,b; Fig. 1) resulted in changing stress regimes that are reflected in the neotectonic evolution of north coastal California. Throughout this region, late Cenozoic sedimentary rocks rest unconformably on, or are in fault contact with, the late Mesozoic-Miocene Franciscan Complex and Klamath

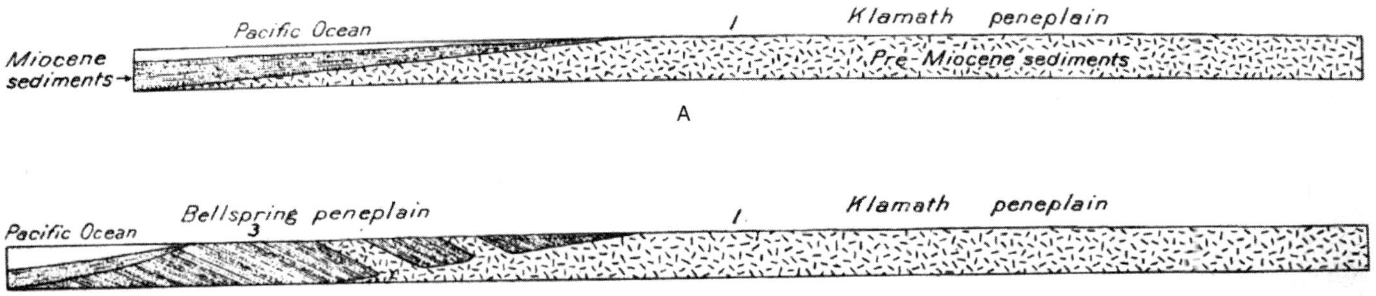

Figure 2. Profiles across the western margin of the Klamath Peneplain from Diller (1902, plate I) depicting: (A) the late Miocene Klamath stage, and (B) the Pliocene(?) Bellspring stage of peneplanation.

Mountains western Jurassic terrane (Kelsey and Carver, 1988; Aalto et al., 1995b; Irwin, 1997, and on the CD-ROM accompanying this volume[1]). Unconformities are exposed beneath many Neogene rock units and Pleistocene terraces. Franciscan basement rocks are thrust faulted over late Cenozoic rocks in many active fault zones (Carver, 1985, 1987, 1992). Late Cenozoic cover rocks, therefore, provide much of the evidence needed to evaluate Diller's ideas and the uplift history of Klamath Mountains.

STRATIGRAPHY

Introduction

Neogene cover rocks that unconformably overlie the basement terranes discussed above include the late Miocene Wimer and Saint George Formations (revised diatom age interpretations by John Barron, written commun., 1999), exposed along the coast and inland of Crescent City (Maxson, 1933; Back, 1957; Stone, 1992, 1993; Fig. 2), and the late Miocene–early Pleistocene Wildcat Group and its equivalents, exposed south and east of Eureka (Manning and Ogle, 1950; Ogle, 1953; Haller, 1980; Fig. 1). In addition, flights of late Pleistocene–Holocene marine and fluvial terraces that unconformably overlie the cover rocks are discontinuously exposed throughout the region (Carver and Aalto, 1992). My discussion focuses on the Wimer Formation, Saint George Formation, and Wildcat Group, for it is among these units that the story of Klamath peneplanation and uplift unfolds.

Wimer Formation

The Wimer Formation was deposited on a regional surface of low relief: the "Klamath peneplain" of Diller (1902), which is preserved as an elevated, westward-tilted, erosion surface (Figs. 2–5). Stone (1992, 1993) divided the formation into two facies, a western estuarine marine facies and an eastern fluvial facies (Fig. 2). She found no evidence of post-Wimer offset of the erosion surface, supporting Diller's contention that it was uniformly uplifted after its development. The two facies are cor-

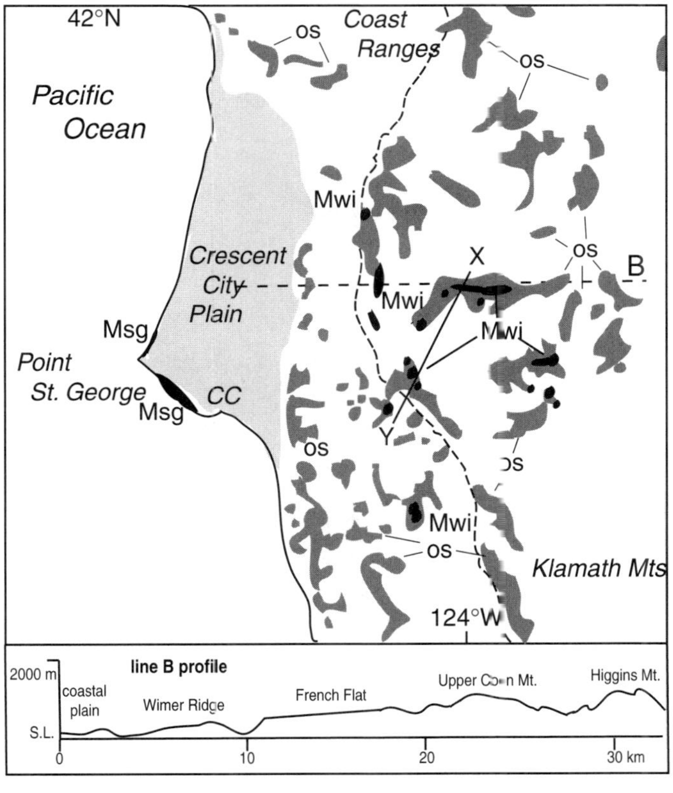

Figure 3. Klamath erosion surface (os) in the vicinity of Crescent City (CC), California. Msg—Saint George Formation (black); Mwi—Wimer Formation (black). This area was studied in detail by Stone (1992, 1993) and line X–Y delineates her boundary between marine (to the west) and fluvial (to the east) facies of the Wimer Formation. The top of the figure is the California-Oregon border. East-west topographic profile along line "B" shows the westward-dipping Klamath erosion surface (from Stone, 1993).

[1]Also available as GSA Data Repository item 2006202, Preliminary map of selected post-Nevadan geologic features of the Klamath Mountains and adjacent areas, California and Oregon, compiled by William P. Irwin, is available online at www.geosociety.org/pubs/ft2006.htm, or on request from editing@geosociety.org or Documents Secretary, GSA, P.O. Box 9140, Boulder, CO 80301, USA.

Figure 4. View west (east center, sec. 1, T15N R1E) across westward-dipping concordant summits of the Klamath peneplain.

Figure 5. View north-northwest at the summit of Thompson Peak, the highest peak (2728 m) in the Klamath Mountains. Note the concordance of summits.

related based on proximity and their occupation of the same physiographic surface. The marine facies of the Wimer Formation primarily consists of ~15 m of massive fossiliferous mudstone. Two minor lithologies are a chert, vein quartz-rich granule to pebble conglomerate (Fig. 6) and coarse basal conglomerate composed of locally derived clasts (Fig. 7). The fine-grained rocks contain few sedimentary structures and appear to have been highly bioturbated. Fossiliferous zones containing molluscan fauna diagnostic of a protected coastal bay and mudflat environment occur throughout the fine-grained rocks (Watkins, 1974). Based on fossil assemblages, western exposures of the marine facies represent normal marine salinity conditions, whereas eastern exposures represent brackish water conditions (Watkins, 1974). The brackish water deposits overlie coarse-grained conglomerates, which are matrix-rich, lack imbrication, and contain claystone rip-up clasts (Fig. 6). These traits suggest deposition in a fluvial/estuarine deltaic setting (Stone, 1992, 1993).

Figure 6. Wimer Formation interstratified claystone and fine sandstone, overlying pebble conglomerate, interpreted as nonmarine-marine transitional zone (center, sec. 1, T15N R1E). Note the small fractures and evidence of soft-sediment deformation.

The eastern fluvial facies of the Wimer Formation crops out on flat-topped ridges, where a contact is exposed between the fluvial and marine facies that delineate the late Miocene paleoshoreline (Fig. 3). The fluvial facies consists of clast-supported, massive to crudely stratified, normally graded, imbricated, pebble to boulder conglomerate with subordinate interbedded siltstone and sandstone. Clasts are sub-rounded to rounded and are derived chiefly from western Jurassic terrane Klamath basement rocks east of the South Fork fault, with additional contributions from Franciscan basement rocks west of the South Fork fault (Stone, 1992, 1993). Deposition by a braided river in longitudinal bars is suggested by normal grading passing upward from open to closed framework conglomerate, massive to crude horizontal stratification, and facies architecture (Stone, 1992, 1993). Minor lenses of sandstone or siltstone are interpreted as overbank or drape deposits. Imbrication and distribution of the deposits indicate predominant paleoflow toward the north-northwest, in a braid plain that drained into an estuary.

Saint George Formation

The Saint George Formation is exposed along the coast of the Crescent City headlands (Fig. 8), where it rests in angular unconformity on overturned and tightly folded, thickly bedded

Figure 7. Wimer Formation basal conglomerate unconformably overlying deeply weathered Franciscan Complex argillite (paleosol). The conglomerate consists chiefly of clasts derived from the underlying bedrock and might possibly have accumulated on a marine abrasion platform. Scale = 10 cm; center sec. 28 T16N R1E.

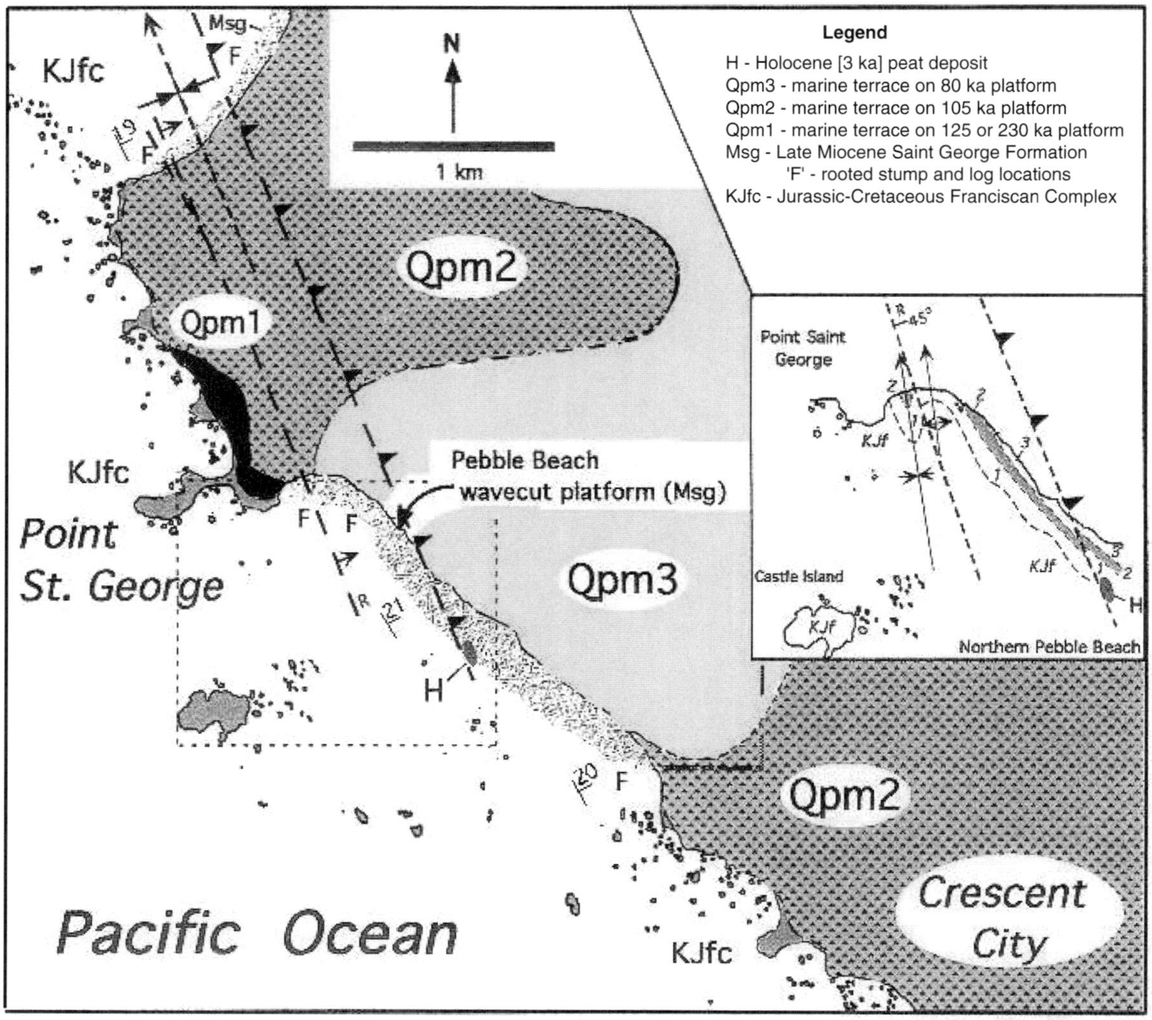

Figure 8. Geologic map of the Crescent City headland. Holocene peat at 'H' yielded a conventional radiocarbon date of 3000 ± 60 BP from two pine cones from this peat (the 2-σ calibrated range is Cal BP 3355–2980, Beta Analytic, Inc., Miami, Florida). Inset map shows structural detail at the north end of Pebble Beach with Saint George units 1–3 delineated (mapping by Aalto). Geology is modified from Aalto (1989a), Polenz and Kelsey (1999), and Robinson et al. (2001).

sandstone of the late Mesozoic eastern belt (Yolla Bolly terrane) Franciscan Complex (Aalto 1989a, 1989b). The Saint George Formation is folded but generally dips to the northeast at shallow angles. It, in turn, is overlain by a complex series of flat-lying Pleistocene marine terrace deposits (Fig. 8; Polenz and Kelsey 1999). Based on water well litholog data and projection inland of the northeasterly dip of bedding, the Saint George Formation appears to underlie much of the Crescent City platform (Back 1957). From this same well log data, Back (1957) estimated the total thickness of the Saint George Formation to be ~120 m.

Robinson (2002) divided this formation into three units that are found both north and south of Point Saint George (Figs. 8 and 9). The entire Saint George Formation exhibits an overall fining-upward trend. Unit 1 basal conglomerates depositionally overlie a weathered surface (paleosol) developed on Franciscan Complex sandstone (Fig. 10). In contrast to the medium to dark gray color and high degree of induration of unweathered Franciscan rocks, the paleosol is light gray to light tan or green, clay-rich, and soft. The degree of lightening increases and induration decreases gradationally upward within a 40-cm zone of weathering. Within the upper 40 cm

Figure 9. Generalized stratigraphic column for the Saint George Formation measured at north end of Pebble Beach. The biotrubation index is that of Miller and Smail (1997).

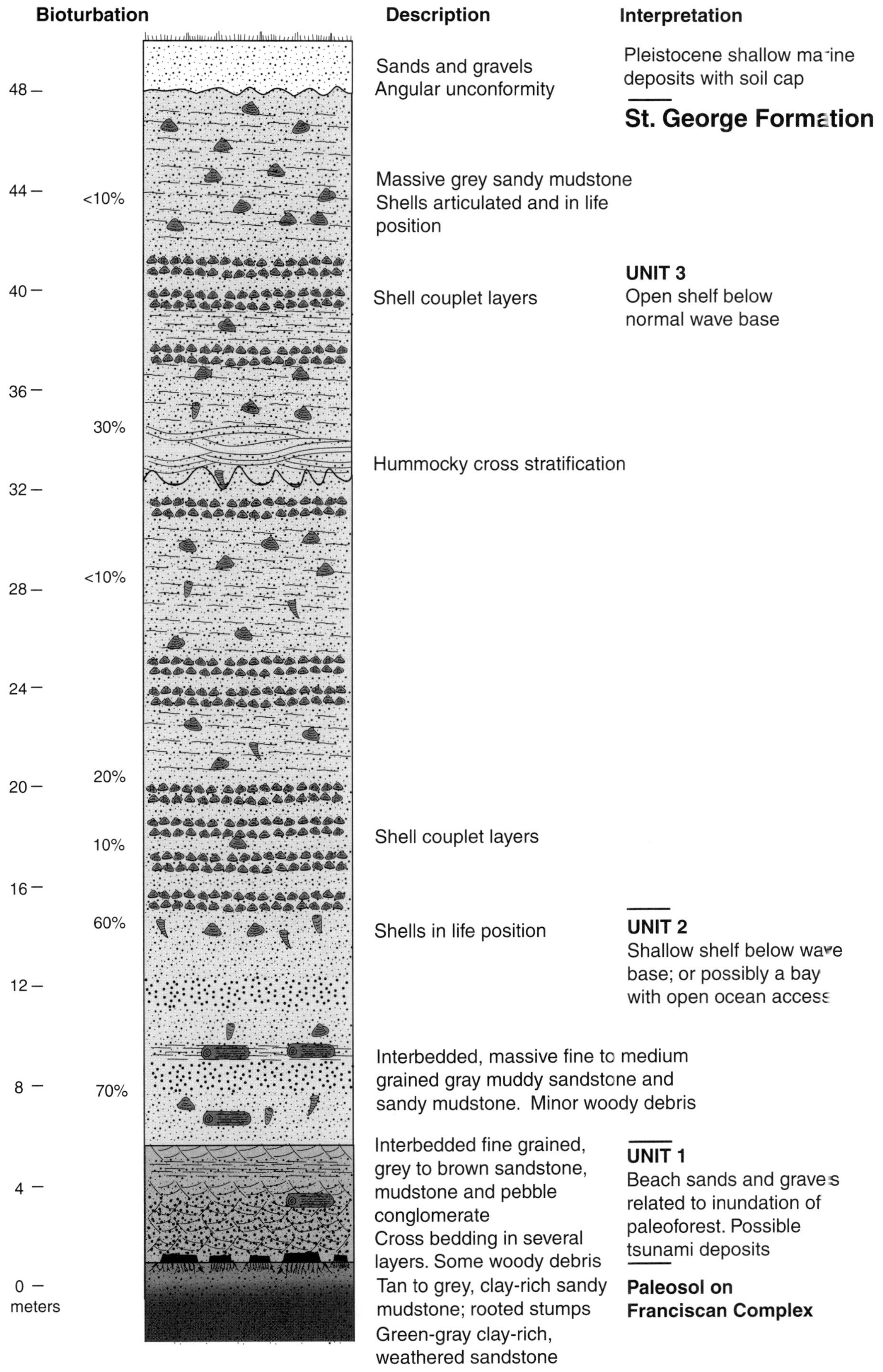
Bioturbation
Description
Interpretation
48 —
44 —
40 —
36 —
32 —
28 —
24 —
20 —
16 —
12 —
8 —
4 —
0 —
meters
<10%
30%
<10%
20%
10%
60%
70%
Sands and gravels
Angular unconformity
Pleistocene shallow marine deposits with soil cap
St. George Formation
Massive grey sandy mudstone
Shells articulated and in life position
Shell couplet layers
UNIT 3
Open shelf below normal wave base
Hummocky cross stratification
Shell couplet layers
Shells in life position
UNIT 2
Shallow shelf below wave base; or possibly a bay with open ocean access
Interbedded, massive fine to medium grained gray muddy sandstone and sandy mudstone. Minor woody debris
Interbedded fine grained, grey to brown sandstone, mudstone and pebble conglomerate
Cross bedding in several layers. Some woody debris
UNIT 1
Beach sands and gravels related to inundation of paleoforest. Possible tsunami deposits
Tan to grey, clay-rich sandy mudstone; rooted stumps
Paleosol on Franciscan Complex
Green-gray clay-rich, weathered sandstone

Figure 10. Woody debris in basal Saint George Formation diamictite, north side of Point Saint George. The log in the upper center is 22 cm long. Saint George sediments rest in angular unconformity upon a gray paleosol developed on Franciscan Complex sandstone.

of weathered Franciscan sandstone that constitutes the paleosol, matrix increases from 36 to 75%, quartz + chert increases from 38 to 87%, whereas other lithic clasts decrease from 20 to 5% and feldspar from 47 to 8% (K.R. Aalto, unpublished data, 1999). This zone also contains clustered petal-like calcite ellipsoids (*Microcodium*) interpreted as calcified plant roots (Klappa 1978). X-ray diffraction analysis reveals the presence of quartz + albite + white mica + chlorite, but no expandable clays.

Large fossil logs are present along the unconformity at five sites (marked by "F" in Fig. 8). Stumps that range up to 2 m in diameter with roots radiating outward and downward into the paleosol (Fig. 11) are present at the northern and central exposures of the paleosol, at Pebble Beach, and on the far northern side of Point Saint George. Many of the largest logs preserved within basal Saint George Formation strata lie with their long dimensions oriented generally north-northwest to north-northeast.

Unit 1 of the Saint George Formation fines upward from poorly sorted basal conglomerate to interstratified conglomerate and sandstone (Robinson, 2002). Basal conglomerates are matrix-supported diamictites. They contain very poorly sorted (silts to gravels up to 5 cm), chiefly angular to subangular chert, vein quartz, greenstone and sandstone pebbles, and cobbles of local Franciscan derivation dispersed in muddy sandstone. Large wood fragments are abundant. Overlying this basal unit are up to fifteen beds of stratified conglomerates that consist of a gray to greenish-brown, interbedded pebble conglomerate, parallel-laminated or cross-bedded, poorly sorted muddy sandstone and sandy mudstone, arranged in fining-upward cycles, each several decimeters in thickness (Fig. 9). The uppermost muddy portions of some cycles contain matted mixes of woody debris, needles, and cones. Within this unit, clast rounding and sorting increase up-section.

By 3.5 m up-section from the base, the rhythmically bedded cycles are no longer evident. Sorting becomes better, and fine- to medium-grained, massive gray to brown sands with minor scattered pebbles, occasional claystone rip-up clasts, and rounded wood fragments are prevalent (Robinson, 2002). Toward the top of the unit, at ~5 m up-section from the base, the sandstone becomes highly bioturbated, defining the base of unit 2.

The presence of the diatom-rich rhythmically bedded cycles, woody debris, and generally poor sorting in basal unit 1 suggests sediment influx in a series of pulses by high-energy currents, followed by relatively still (slack water?) conditions. These cycles perhaps reflect inundation of a coastal forest by a series of storm or tsunami(?) waves, which would account for the preferential orientation of many large logs. Rapid submergence and sedimentation can also explain the excellent preservation of the stumps, logs, and the paleosol.

Unit 2 of the Saint George Formation consists of 10 m of interbedded, medium gray muddy sandstone and sandy mudstone containing minor blackened woody debris, pebbles, occasional mollusk fossils, and extensive bioturbation (Fig. 9). Bedding is not visible in most of the unit, but distinctive cyclic layering is present near the contact with unit 1. Such layers have sharply defined scoured bases marked by a sudden coarsening and broken mollusk shell lag deposits, and they grade normally upward into zones of marked bioturbation. Up-section, mollusk shells are both articulated and dispersed in the mudstone, commonly preserved in concretions, or are disarticulated and concentrated in 8- to 30-cm-thick layers. These layers occur at irregular intervals within the unit and reflect an increase in shell concentration in otherwise massive sandy mudstone. Such shell concentrations without associated scour may represent slackening in sedimentation rate.

Diatom paleoecology and sedimentary features suggest deposition in either an embayment with access to open marine conditions, or the nearshore upper shelf below normal wave base (Robinson et al., 2001). The overwhelming abundance of bivalves and absence of gastropods among the molluscan fauna, as well as the presence of brackish water diatoms, favors a

Figure 11. Large rooted, coalified stump in Franciscan Complex saprolitic paleosol. Note the roots radiating outwards from the stump center and exposed paleosol and basal Saint George sediments with large logs in upper left. Scale on right = 10 cm. North side of Point Saint George, northernmost stump/log outcrop.

muddy embayment and perhaps somewhat brackish water (Watkins, 1974). The increase of bioturbation, as well as the overall fining-upward trend throughout the unit, suggests a lower energy, deeper water environment than that of unit 1.

At Pebble Beach, unit 3 of the Saint George Formation is 33 m thick (Fig. 9). The basal contact of unit 3 is marked by the beginning of a series of concentrated shell basal lag (couplet) layers. Unit 3 consists of medium gray sandy mudstone containing abundant bivalves that are either scattered randomly as articulated shells within the mudstone or concentrated in 15- to 40-cm-thick shell-couplet layers commonly spaced from 20–30 cm up to 80–150 cm apart. These shell-rich layers commonly contain a more sandy, locally laminated, basal zone with densely packed articulated and disarticulated shells overlying minor scour surfaces, which grades normally upward into a zone of less sandy mudstone having dispersed, articulated shells. The highly fossiliferous portions of unit 3 have a mottled, crumbly aspect imparted by iron oxide staining and differential weathering along fossil molds. Less fossiliferous portions are massive.

At the 17.5 m level of unit 3 at Pebble Beach, a normally graded, 180-cm-thick, hummocky cross-stratified sandstone (cf. Bourgeois, 1980) overlies a scour surface channeled into fossiliferous mudstone. Above the basal scour surface, coarse, poorly sorted sandstone grades upward within the bed to a highly bioturbated sandy mudstone. Both north and south of Point Saint George, the entire Saint George Formation exhibits an overall fining upward, and the uppermost beds of unit 3 contain little sand, a decrease in bioturbation, and virtually no wood.

Concentrations of disarticulated shelly debris above scour surfaces in unit 3, and the concentration of shelly debris in the shell couplets very likely reflects current reworking, perhaps by large waves. Although fossils are consistent with deposition in either a bay environment or open marine conditions, the appearance of hummocky cross-stratified sandstone suggests that the deposition site was open to the ocean and permitted incursion of large storm and/or tsunami waves (Dott and Bourgeois 1982; Leithold and Bourgeois 1984).

Wildcat Group

The Wildcat Group (Ogle, 1953) constitutes late Miocene through Pleistocene sediment fill within the Eel River basin (Fig. 1). McCrory (1989) argued that the Eel River basin formed rapidly as a stratigraphic basin near the base of an inner trench slope, in response to subduction of the Gorda plate, and that the initial accumulation of sediments occurred at bathyal depths. Wildcat sediments were previously thought to have been derived from chiefly Coast Range sources to the southeast (Nilsen and Clarke, 1987). Moley (1992) and Aalto et al. (1996) instead suggested that basal sediments of the Wildcat Group accumulated at shallow depths as part of a shelf sediment blanket, that the on-land portion of the Eel River basin is chiefly a mid-Pleistocene structural basin formed in response to a combination of Gorda–North American plate convergence and the northward migration of the Mendocino triple junction, and that at least some Wildcat Group sediments were derived from the northeast, possibly from Idaho, as discussed below. Proprietary seismic reflection profiles by ARCO demonstrate that the older marine formations within the Wildcat Group retain constant thicknesses across the on-land portion of the Eel River basin, and thus that it is synclinal in form and underwent no differential subsidence of its axial region during sedimentation.

Overall, the Wildcat Group records an eastward transgression during the late Miocene and Pliocene, with initial deposition of sediments within a near-shore environment, then rapid deepening of the continental shelf to bathyal–abyssal depths, later infilling of the shelf during the Plio-Pleistocene, and westward regression of the shoreline during early to medial Pleistocene (Ogle, 1953; Nilsen and Clarke, 1987; Clarke, 1992). The lowest stratigraphic unit in the Wildcat Group, the Pullen Formation, is believed to be coeval with the Wimer and Saint George Formations.

The marine portion of the Wildcat Group includes 1900–2600 m of mudstone and subordinate sandstone that range in age from late Miocene to middle Pleistocene, arranged in an overall coarsening-upward sequence (Ogle, 1953; Haller, 1980; Ingle, 1987; Nilsen and Clarke, 1987). At Scotia, inland along the Eel River, and in logging road cuts some 12 km southeast of Scotia, an angular unconformity separates tightly folded Eocene Yager Complex turbidites and tilted, but relatively undeformed, late Miocene Pullen Formation (Moley, 1992; Aalto et al., 1996). Above the unconformity are sandy shoreface deposits that contain locally derived Yager Complex clasts, indicating that a certain amount of erosional stripping occurred in this region during post-Yager, pre-Wildcat time. East of Eureka, on the northeastern flank of the Eel River basin, facies relations suggest that similar unconformities exist (Knudson, 1993).

The nonmarine portion of the Wildcat Group is middle to late Pleistocene and consists of a series of fluvial clastic wedges that thin and intertongue with marine sediments to the west. Thickness ranges from 800 to 1000 m (Nilsen and Clarke, 1987). Marine-nonmarine cycles are probably glacioeustatic in origin. Patches of similar coeval sediments exist inland of the northeastern flank of the Eel River basin and isolated from basin exposures by basement rock uplifted along active thrust faults (Ogle, 1953; Carver, 1985). It seems apparent that Wildcat-equivalent marine and fluvial sediments covered a significant portion of the northern Coast Ranges but have been removed by erosional stripping in conjunction with active tectonism.

SEDIMENT PROVENANCE AND UPLIFT HISTORY OF THE KLAMATH MOUNTAINS

Early–mid-Cenozoic Era

During the early Paleogene, the Klamath Mountains province was sufficiently elevated to supply sediment to surrounding areas (Heller and Ryberg, 1983; Higinbotham, 1987; Aalto, 1988). By middle Eocene, however, it was sufficiently reduced in elevation to allow rivers to cross the province, transporting arkosic sand from Idaho to marine basins in the Coast Range province of northern California and southern Oregon (Underwood and Bachman, 1986). Mean K/Ar and $^{40}Ar/^{39}Ar$ ages of white mica populations from broadly distributed Eocene sandstones of the Pacific Northwest—including the Yager Complex of north-coastal California, the Montgomery Creek and upper Payne Cliffs Formations of the eastern Klamath Mountains, and the Tyee Formation of western Oregon—are remarkably uniform at ca. 66–68 Ma (Renne et al., 1990). These ages, O-isotopic values, and other properties of the Eocene sandstones indicate that they are derived from the Idaho batholith (Heller and Ryberg, 1983; Underwood and Bachman, 1986; Renne et al., 1990). Sediment was transported to the Klamath region by an ancestral "Snake River" paleodrainage system (Heller et al., 1987; Aalto et al., 1998).

Active uplift of the central Klamath Mountains region and shutoff of the Snake River paleodrainage system began during the Oligocene, perhaps accompanying initiation and growth of the La Grange detachment fault system that developed along with the unroofing of the Trinity ophiolite (Schweickert and Irwin, 1989; Cashman and Cashman, this volume). Late Oligocene–early Miocene, locally-derived fluviatile Weaverville Formation sediments accumulated in a series of grabens located on the upper plate of this regional detachment system. Eastern Weaverville Formation sandstones are quite heterogeneous in composition and are interstratified with coarse, alluvial fan and fluvial conglomerates with clasts composed of locally derived Klamath basement rock (Phillips, 1989; Phillips and Aalto, 1989).

Late Cenozoic Uplift of the Klamath Peneplain

During the Miocene, much of the Klamath Mountains province was again reduced in elevation as the Klamath erosion surface, the "Klamath peneplain" of Diller (1902), developed. By the late Miocene, the northwest Klamath Mountains region was at or below sea level, as is indicated by the presence of Wimer Formation marine mudstones on the present-day elevated erosion surface east of Crescent City (Stone, 1992, 1993; Fig. 2). The weathered Franciscan surfaces preserved beneath both the Saint George and Wimer Formations are interpreted as a late Miocene saprolitic paleosols (Robinson et al., 2001). Such an episode of pre-Late Miocene soil development is in agreement with the classic view of Diller (1902) that a regional erosion surface (peneplain) had formed in northwestern California at this time. This interpretation is supported by the similarity in age of the Saint George and Wimer Formations preserved in patches on elevated portions of this erosion surface 10–15 km east of Crescent City (Stone, 1993; Aalto et al., 1995b; Irwin, 1997). The regional extent of this surface and existence of a thick paleosol indicate that erosion was insignificant and sedimentation relatively continuous.

Both Stone (1993) and Robinson (2002) demonstrate local derivation of the majority of coarse conglomerate clasts of the Wimer and Saint George Formations from basement rock of the Coast Ranges and Klamath Mountains. Similarly, Moley (1992) determined that Pleistocene fluvial conglomerates of the Wildcat Group were chiefly derived from Franciscan Complex basement of the northern Coast Ranges. However, she noted that late Miocene–early Pleistocene marine sandstones of the Wildcat

TABLE 1. FRAMEWORK MODES FOR THE SAINT GEORGE FORMATION AND WILDCAT GROUP SANDSTONES

Mode*	Qt	F	L	Qm	F	Lt	Qp	Lvm	Lsm	Qm	P	K	D	M
Saint George Formation (*N* = 6)														
Mean	56	25	19	43	25	32	50	40	11	61	33	7	9	Trace
Range	30–84	12–37	4–43	21–67	12–37	16–52	16–78	19–75	0–25	44–84	8–52	4–8	0–39	
Standard deviation	22	9	17	19	9	14	28	27	9	17	17	2	16	
Wildcat Group (*N* = 44)														
Mean	54	30	16	47	30	23	45	45	10	58	35	6	6	2
Range	36–68	13–52	5–37	17–66	13–52	5–60	23–65	14–75	0–53	32–74	18–47	0–15	0–20	0–9
Standard deviation	7	8	7	10	8	13	12	15	11	8	7	4	5	2

Sources: Saint George Formation data from K.R. Aalto (unpublished data, 1998); Wildcat Group data from Aalto (1992) and Moley (1992).

*Modes are calculated with Qt + F + L = 100%, Qm + F + Lt = 100%, Qp + Lvm + Lsm = 100%, and Qm + P + K = 100%. Mode abbreviations: Qt—total quartzose grains; F—feldspar; L—unstable aphanitic lithics; Qm—monocrystalline quartz; Lt—total aphanitic lithics; Qp—polycrystalline quartzose lithics; Lvm—ahanitic volcanic, hypabyssal, and metavolcanic lithics; Lsm—aphanitic sedimentary and metasedimentary lithics; P—plagioclase; K—potassium feldspar; D—dense detrital minerals; M—detrital micas.

Group have an unexpectedly high K-feldspar content (Table 1), given the relative lack of K-feldspar in local basement rocks. Aalto (unpublished data, 1999) noted a similar trend among Saint George Formation sandstones (Table 1). Significant K-feldspar content suggests possible sediment input from sources other than the Klamath Mountains or Coast Ranges, as these provide comparatively K-feldspar–poor sands (Underwood and Bachman 1986; Aalto 1989b, 1992).

K-feldspar is virtually absent among northern Coast Range Franciscan Complex rocks of the Eastern and Central Franciscan terranes (Aalto, 1989b). Although mean K-feldspar content (among monocrystalline quartz, plagioclase, K-feldspar detrital modes: QmPK%K; cf. Ingersoll et al., 1984) for nearby Franciscan Coastal belt and Yager Complex sandstones and the Idaho-derived Eocene Tyee Formation of southwest Oregon are 13, 10, and 12%, respectively (Heller and Ryberg, 1983; Underwood and Bachman, 1986), it is difficult to envision derivation of significant amounts of K-feldspar (up to 15% QmPK%K in some Wildcat Group sandstones) by sediment recycling from these sources in the humid temperate Plio-Pleistocene climate of northwestern California. Provenance studies have demonstrated loss of as much as 20% feldspar during transition from fluvial to littoral sedimentation within a short stratigraphic interval, interpreted to chiefly reflect abrasion (Blatt, 1992).

The $^{40}Ar/^{39}Ar$ laser probe analyses of some micas from Wildcat Group marine sands (Moley, 1992; Aalto et al., 1995b, 1998) suggest an Idaho batholith source, which indeed is compatible with the unusually high K-feldspar contents observed. Late Miocene marine Wildcat Group micas ($n = 6$) yielded three groups of ages: 66–75, 128.5, and 299–303 Ma (Table 2). Analyses of late Pleistocene marine Wildcat Group micas ($n = 2$) yielded ages of 52–57 Ma. The Cretaceous through early Tertiary $^{40}Ar/^{39}Ar$ ages of some Miocene and Pleistocene Wildcat Group micas indicate sources among plutonic rocks in Idaho (Aalto et al., 1998). New $^{40}Ar/^{39}Ar$ data on micas (Table 2) extracted from Saint George Formation sandstones ($n = 12$) yield ages of 52–75 Ma, similar to those of the Wildcat Group (S. VanLaningham, personal commun., 2006). It is possible that some micas with an Idaho-like signature could have been derived by recycling Eocene sandstones in northern California and southern Oregon. However, given the high K-feldspar content of Wildcat Group sandstones, direct derivation from an Idaho source is our preferred interpretation.

K-feldspar (QmPK%K) modes for modern Klamath River and Klamath Mountains–derived sands are both 4% (Underwood and Bachman, 1986), further suggesting that a Klamath source for abundant Wildcat Group K-feldspar is unlikely. Although longshore transport southward from an ancestral Columbia River paleodrainage system may be invoked to explain the influx of Idaho-derived sediments into northern California, the inferred low elevation of the Klamaths at ca. 5 Ma suggests that fluvial transport across this region is a more viable hypothesis. If this interpretation is correct, at least some form of Snake River paleodrainage system must have been re-established during the Neogene, suggesting that the Neogene shutoff of this paleodrainage system to the Klamath region proposed by Heller et al. (1987) is incorrect.

Sediment progradation and shoaling over a wide coastal shelf, onset of deposition of locally derived Pleistocene fluvial sediments in the Wildcat Group, and regional uplift during the mid-Pleistocene may relate to several factors. These include the decreasing age of the subducting oceanic plate; a lessening angle of dip of subduction; the translation of the Sierra Nevada microplate northwestward, impinging on the Klamath block; and the migration of the Blanco fracture zone northward along the continental margin of northwestern California (Carver, 1987; Kelsey et al., 1994; Aalto et al., 1995b; Wells et al., 1998). Based on the Idaho signature in sediment provenance discussed above, the uplift of the Klamath peneplain to elevations as high as 1220 m may be viewed as chiefly a Pleistocene–Holocene event.

Saint George Formation stratigraphy suggests that rapid subsidence of the leading edge of North America occurred during the late Miocene. A coastal forest was inundated by sea

TABLE 2. $^{40}Ar/^{39}Ar$ AGES FOR MUSCOVITES FROM THE WILDCAT GROUP AND SAINT GEORGE FORMATION

Sample	Integrated age (Ma)
Late Miocene Wildcat Group (Pullen Formation)	
CB 1S84	65.5 ± 0.7
CB 1S84	70.3 ± 0.6
CB 1S84	74.8 ± 0.4
CB 1S84	128.5 ± 0.7
CB 1S84	299 ± 2
CB 1S84	303 ± 2
Late Miocene Saint George Formation	
06C1466	63.33 ± 9.32
06C1558	55.03 ± 2.21
06C1559	120.62 ± 1.64
06C1560	75.64 ± 1.15
06C1562	76.33 ± 1.33
06C1563	74.91 ± 1.91
06C1564	75.06 ± 4.31
06C1565	55.92 ± 2.00
06C1566	54.97 ± 2.63
06C1567	63.32 ± 2.53
06C1568	54.72 ± 3.10
06C1569	64.19 ± 1.81
06C1570	52.32 ± 5.23
Pleistocene Wildcat Group (Scotia Bluffs Formation)	
SB6S64	52 ± 2
SB6S64	57.1 ± 0.3

Sources: Wildcat Group: Aalto et al. (1998); Saint George Formation: Sam VanLaningham (personal commun., 2006).

water and the forest floor down-dropped at least several meters to allow deposition of littoral marine sediments over basal conglomerates (Robinson et al., 2001). Rapid regional subsidence is suggested by the record of abrupt spatial transition from shallow to deep marine sedimentation in the coeval Pullen Formation to the south (Aalto et al., 1995b, 1996) as well as an ~10-km eastward shift of the shoreline recorded in the coeval basal Wimer Formation (Watkins, 1974; Stone, 1993).

Aalto et al. (1996, 1998) suggested that coastal subsidence was related to an episode of subduction erosion and subsequent extension of the forearc and that eroded sediments were eventually underplated beneath the Klamath Mountains block. This action ultimately resulted in the Pleistocene uplift of the Klamath erosion surface and erosional stripping of a Wildcat Group–Saint George–Wimer Formation shelf sediment blanket.

ACKNOWLEDGMENTS

I thank Jose M. Hurtado, Jr., and John Wakabayashi for their thorough reviews of my manuscript, and co-editor Calvin Barnes for his assistance. Thanks to Sam VanLaningham for supplying Saint George Formation mica ages. My years of association with W. Porter Irwin, both as a fellow researcher in the Klamath Mountains and as a friend and great resource for ideas concerning the geologic evolution of northwestern California, have been exceptionally rewarding. Thus I am happy to contribute a chapter to this volume honoring his career.

REFERENCES CITED

Aalto, K.R., 1988, Sedimentology of the Montgomery Creek Formation, Shasta County, California: California Geology, v. 41, p. 254–260.

Aalto, K.R., 1989a, Franciscan Complex olistostrome at Crescent City, northern California: Sedimentology, v. 36, p. 471–495.

Aalto, K.R., 1989b, Sandstone petrology and tectonostratigraphic terranes of the northwest California and southwest Oregon Coast Ranges: Journal of Sedimentary Petrology, v. 59, p. 561–571.

Aalto, K.R., 1992, Late Cenozoic sediment provenance and tectonic evolution of the northernmost Coast Ranges, California, *in* Carver, G.A., and Aalto, K.R., eds., Field guide to the late Cenozoic subduction tectonics and sedimentation of northern coastal California: Bakersfield, California, Pacific Section, American Association of Petroleum Geologists Field Guide GB-71, p. 11–20.

Aalto, K.R., Mclaughlin, R.J., Carver, G.A., Barron, J.A., Sliter, W.V., and McDougall, K., 1995a, Uplifted Neogene margin, southernmost Cascadia-Mendocino triple junction region, California: Tectonics, v. 14, p. 1104–1116, doi: 10.1029/95TC01695.

Aalto, K.R., Moley, K., and Stone, L., 1995b, Neogene paleogeography and tectonics of northwestern California, *in* Fritsche, E., ed., Cenozoic paleogeography of the western United States II: Los Angeles, Pacific Section, Society of Economic Paleontologists and Mineralogists, v. 75, p. 162–180.

Aalto, K.R., Moley, K., and Miller, W., III, 1996, Evolution of a trench-slope basin within the Cascadia subduction margin: The Neogene Humboldt Basin, California: Discussion: Sedimentology, v. 43, p. 761–769.

Aalto, K.R., Sharp, W.D., and Renne, P.R., 1998, $^{40}Ar/^{39}Ar$ dating of detrital micas from Oligocene-Pleistocene sandstones of the Olympic Peninsula, Klamath Mountains and northern California Coast Ranges: Provenance and paleodrainage patterns: Canadian Journal of Earth Sciences, v. 35, p. 735–745, doi: 10.1139/cjes-35-7-735.

Back, W., 1957, Geology and ground water features of the Smith River plain, Del Norte County, California: Washington, D.C., U.S. Geological Survey Water Supply Paper 1254, 76 p.

Blatt, H., 1992, Sedimentary petrology: New York, W.H. Freeman Co., 564 p.

Bourgeois, J., 1980, A transgressive shelf sequence exhibiting hummocky cross-stratification: Journal of Sedimentary Petrology, v. 50, p. 681–702.

Carver, G.A., 1985, Quaternary tectonics north of the Mendocino triple junction, *in* Kelsey, H.M., et al., eds., Redwood Country guidebook: Arcata, California, American Geomorphological Field Group 1985 Field Trip, p. 155–167.

Carver, G.A., 1987, Late Cenozoic tectonics of the Eel River basin region, coastal northern California, *in* Schymiczek, H., and Suchsland, R., eds., Tectonics, sedimentation and evolution of the Eel River and other coastal basins of northern California: Bakersfield, California, Pacific Section, American Association of Petroleum Geologists Miscellaneous Publication 37, p. 61–72.

Carver, G.A., 1992, Late Cenozoic tectonic evolution of coastal northern California, *in* Carver, G.A., and Aalto, K.R., eds., Field guide to the late Cenozoic subduction tectonics and sedimentation of northern coastal California: Bakersfield, California, Pacific Section, American Association of Petroleum Geologists Field Guide GB-71, p. 1–10.

Carver, G.A., and Aalto, K.R., 1992, Field guide to the late Cenozoic subduction tectonics and sedimentation of northern coastal California: Bakersfield, California, Pacific Section, American Association of Petroleum Geologists, Field Guide GB-71, 74 p.

Cashman, S.M., and Cashman, K.V., 2006, this volume, Cataclastic textures in La Grange fault rocks, Klamath Mountains, California, *in* Snoke, A.W., and Barnes, C.G., eds., Geological studies in the Klamath Mountains province,

California and Oregon: A volume in honor of William P. Irwin: Boulder, Colorado, Geological Society of America Special Paper 410, doi: 10.1130/2006.2410(21).

Clarke, S.H., Jr., 1992, Geology of the Eel River basin and adjacent region: Implications for late Cenozoic tectonics of the southern Cascadia subduction zone and Mendocino triple junction: American Association of Petroleum Geologists Bulletin, v. 76, p. 199–224.

Diller, J.S., 1902, Topographic development of the Klamath Mountains: Washington, D.C., U.S. Geological Survey Bulletin 196, 69 p.

Dott, R.H., Jr., and Bourgeois, J., 1982, Hummocky stratification: Significance of its variable bedding sequences: Geological Society of America Bulletin, v. 93, p. 663–680, doi: 10.1130/0016-7606(1982)93<663:HSSOIV>2.0.CO;2.

Haller, C.R., 1980, The Miocene stratigraphy of California revisited: Yulsa, Oklahoma, American Association of Petroleum Geology Studies in Geology 11, 349 p.

Heller, P.L., and Ryberg, P.T., 1983, Sedimentary record of subduction to forearc transition in the rotated Eocene basin of western Oregon: Geology, v. 11, p. 380–383, doi: 10.1130/0091-7613(1983)11<380:SROSTF>2.0.CO;2.

Heller, P.L., Tabor, R.W., and Suczek, C.A., 1987, Paleogeographic evolution of the United States Pacific Northwest during Paleogene time: Canadian Journal of Earth Sciences, v. 24, p. 1652–1667.

Higinbotham, L., 1987, Montgomery Creek Formation, Klamath Mountains, Shasta County, California: California Geology, v. 40, p. 130–138.

Ingersoll, R.V., Bullard, T.F., Ford, R.L., Grimm, J.P., Pickle, J.D., and Sares, S.W., 1984, The effect of grain size on detrital modes: A test of the Gazzi-Dickinson point-counting method: Journal of Sedimentary Petrology, v. 54, p. 103–116.

Ingle, J.C., Jr., 1987, The depositional, tectonic, and paleo-oceanographic history of the Eel River (Humboldt), Point Arena, and Bodega (Point Reyes) basins of northern California; A summary of stratigraphic evidence, *in* Schymiczek, H., and Suchsland, R., eds., Tectonics, sedimentation and evolution of the Eel River and other coastal basins of northern California: Bakersfield, California, Pacific Section, American Association of Petroleum Geologists Miscellaneous Publications 37, p. 49–54.

Irwin, W.P., 1997, Preliminary map of selected post-Nevadan geologic features of the Klamath Mountains and adjacent areas, California and Oregon: Reston, Virginia, U.S. Geological Survey Open-File Report 97-465, 29 p.

Kelsey, H.M., and Carver, G.A., 1988, Late Neogene and Quaternary tectonics associated with northward growth of the San Andreas transform fault, northern California: Journal of Geophysical Research, v. 93, p. 4797–4819.

Kelsey, H.M., Engebretson, D.C., Mitchell, C.E., and Ticknor, R.L., 1994, Topographic form of the Coast Ranges of the Cascadia Margin in relation to coastal uplift rates and plate subduction: Journal of Geophysical Research, v. 99, p. 12,245–12,255.

Klappa, C.F., 1978, Biolithogenesis of *Microcodium;* Elucidation: Sedimentology, v. 25, p. 489–522.

Knudson, K., 1993, Geology and stratigraphy of the Freshwater Creek watershed, Humboldt County, California [M.S. thesis]: Arcata, California, Humboldt State University, 84 p.

Leithold, E.L., and Bourgeois, J., 1984, Characteristics of course grained sequences deposited in nearshore, wave-dominated environments: Examples from the Miocene of south-west Oregon: Sedimentology, v. 31, p. 749–775.

Manning, B.A., and Ogle, B.A., 1950, Geology of the Blue Lake Quadrangle, California: San Francisco, California Division of Mines and Geology Bulletin 148, 36 p.

Maxson, J.H., 1933, Economic geology of portions of Del Norte and Siskiyou Counties, northwesternmost California: California Journal of Mines and Geology, v. 29, p. 123–135.

McCrory, P.A., 1989, Late Neogene geohistory analysis of the Humboldt basin and its relationship to convergence of the Juan de Fuca Plate: Journal of Geophysical Research, v. 94, p. 3126–3138.

Miller, M.F., and Smail, S.E., 1997, A semiquantitative method for evaluating bioturbation on bedding planes: Palaios, v. 12, p. 391–396.

Moley, K., 1992, A petrographic study of the Wildcat Group, Eel River basin, Humboldt County, California, *in* Carver, G.A., and Aalto, K.R., eds., Field guide to the late Cenozoic subduction tectonics and sedimentation of northern coastal California: Bakersfield, California, Pacific Section, American Association of Petroleum Geologists Field Guide GB-71, p. 31–38.

Nilsen, T.H., and Clarke, S.H., Jr., 1987, Geological evolution of the late Cenozoic basins of northern California, *in* Schymiczek, H. and Suchsland, R., eds., Tectonics, sedimentation and evolution of the Eel River and other coastal basins of northern California: Bakersfield, California, Pacific Section, American Association of Petroleum Geologists Miscellaneous Publication 37, p. 15–30.

Ogle, B.A., 1953, Geology of the Eel River area, Humboldt County, California: San Francisco, California Division of Mines Bulletin 164, 128 p.

Phillips, P.A., 1989, The geology of the Weaverville Formation in the vicinity of Weaverville, California [M.S. thesis]: Arcata, California, Humboldt State University, 174 p.

Phillips, P.A., and Aalto, K.R., 1989, The Weaverville Formation: Implications for tectonically-controlled basin development in the central Klamath Mountains, *in* Colburn, I., and Minch, J., eds., Conglomerates in basin analysis: Los Angeles, Pacific Section, Society of Economic Paleontologists and Mineralogists, v. 62, p. 81–98.

Polenz, M., and Kelsey, H.M., 1999, Development of a Late Quaternary marine terraced landscape during on-going tectonic contraction, Crescent City coastal plain, California: Quaternary Research, v. 52, p. 217–228, doi: 10.1006/qres.1999.2061.

Renne, P.R., Becker, T.A., and Swapp, S.M., 1990, $^{40}Ar/^{39}Ar$ laser-probe dating of detrital micas from the Montgomery Creek Formation, northern California: Clues to provenance, tectonics, and weathering processes: Geology, v. 18, p. 563–566, doi: 10.1130/0091-7613(1990)018<0563:AALPDO>2.3.CO;2.

Robinson, D.R., 2002, Marine inundation of a Late Miocene forest: Stratigraphy and tectonic evolution of the Saint George Formation, Crescent City, California [M.S. thesis]: Arcata, California, Humboldt State University, 86 p.

Robinson, D.R., Aalto, K.R., Barron, J.A., Jayko, A.S., and Erwin, D.M., 2001, Marine inundation of a Late Miocene forest: Stratigraphy and tectonic evolution of the St. George Formation, Crescent City, California: California Geology, v. 54, p. 10–22.

Schweickert, R.A., and Irwin, W.P., 1989, Extensional faulting in southern Klamath Mountains, California: Tectonics, v. 8, p. 135–149.

Stone, L., 1992, Depositional/tectonic history of the Plio-Pleistocene Crescent City/northern coastal Smith River area, California, *in* Carver, G.A., and Aalto, K.R., eds., Field guide to the late Cenozoic subduction tectonics and sedimentation of northern coastal California: Bakersfield, California, Pacific Section, American Association of Petroleum Geologists Field Guide GB-71, p. 51–54.

Stone, L., 1993, Pliocene-Pleistocene sedimentary and tectonic history of the Crescent City–Smith River region, Del Norte County, California [M.S. thesis]: Arcata, California, Humboldt State University, 234 p.

Underwood, M.B., and Bachman, S.B., 1986, Sandstone petrofacies of the Yager complex and the Franciscan Coastal belt, Paleogene of northern California: Geological Society of America Bulletin, v. 97, p. 809–817, doi: 10.1130/0016-7606(1986)97<809:SPOTYC>2.0.CO;2.

Watkins, R., 1974, Molluscan paleobiology of the Wimer Formation, Del Norte County, California: Journal of Paleontology, v. 48, p. 1264–1282.

Wells, R.E., Weaver, C.S., and Blakely, R.J., 1998, Fore-arc migration in Cascadia and its neotectonic significance: Geology, v. 26, p. 759–762, doi: 10.1130/0091-7613(1998)026<0759:FAMICA>2.3.CO;2.

Manuscript Accepted by the Society 10 January 2005

Printed in the USA

Geological Society of America
Special Paper 410
2006

Permian-Jurassic growth and amalgamation of the Wallowa composite terrane, northeastern Oregon

M. Allan Kays*
Department of Geological Sciences, University of Oregon, 1272, Eugene, Oregon 97403, USA
John P. Stimac*
Department of Geology/Geography, Eastern Illinois University, 600 Lincoln Avenue, Charleston, Illinois 61920, USA
Patti M. Goebel
Department of Geological Sciences, University of Oregon, 1272, Eugene, Oregon 97403, USA

ABSTRACT

The Wallowa terrane in the Blue Mountains of northeastern Oregon consists of a Permian, and possibly older, volcanoplutonic island-arc complex built on oceanic lithosphere that is also basement to a Middle to Late Triassic arc complex. The two arc complexes, separated commonly by Mesozoic thrust faults, supplied detritus to a Late Triassic–Early Jurassic marine basin. The adjacent Baker terrane is an intervening mélange containing tectonic slices of the Wallowa and the Olds Ferry terranes. Near the suture with the Olds Ferry composite terrane, the Baker terrane is blueschist bearing. Previous work indicates that three deformational episodes involving contraction affected all terranes of the Blue Mountains from Late Triassic through Late Jurassic. However, this study also documents static hydrothermal metamorphism, M_{static}, that is earlier than Late Triassic D_1 deformation. In the Wallowa terrane, D_1 deformation is recognized locally in the form of transpressional mylonitic shearing but has the same timing as transtensional arc volcanism and basinal development. In the adjacent Baker terrane, D_1 deformation formed S_1 penetrative foliation in blueschists and greenschists during subduction, probably prior to juxtaposition with the Olds Ferry terrane. Middle to early Late Jurassic folding, D_2, and formation of penetrative axial planar foliation (S_2) involved a separate contractional event that affected mainly the Wallowa Triassic and Jurassic arc and basin. The D_3 structures refold D_2 structures in the Wallowa terrane and truncate them during underthrusting of Late Triassic to Early Jurassic arc and sedimentary rocks beneath the Permian arc. The D_3 thrusting may reflect southward transport of the Peninsular or Wrangellian superterrane outboard of the active Mesozoic arc and apron.

Keywords: Wallowa terrane, Baker terrane, Olds Ferry terrane, static metamorphism, greenschist, blueschist, petrochemistry, superposed deformation, Blue Mountains province, Permian, Triassic, Jurassic, northeastern Oregon

*E-mails: makays@uoregon.edu; jpstimac@eiu.edu.

Kays, M.A., Stimac, J.P., and Goebel, P.M., 2006, Permian-Jurassic growth and amalgamation of the Wallowa composite terrane, northeastern Oregon, *in* Snoke, A.W., and Barnes, C.G., eds., Geological studies in the Klamath Mountains province, California and Oregon: A volume in honor of William P. Irwin: Geological Society of America Special Paper 410, p. 465–494, doi: 10.1130/2006.2410(23). For permission to copy, contact editing@geosociety.org.

INTRODUCTION

This chapter reviews tectonostratigraphic, structural, and petrologic data from a well-exposed portion of the Wallowa terrane in the Blue Mountains of northeast Oregon (Fig. 1A and B). Our studies have concentrated in the south-central part, where exposures are good and mapped units have continuity (Fig. 2). In this region, we have established a relative chronology of growth, deformation, and metamorphism through detailed mapping and petrologic studies of individual Permian through Jurassic rock units. We review the chronology of units in the south-central Wallowa terrane and compare it with that of equivalent rock units in adjoining areas. Our data, in combination with observations made in adjoining units (Avé Lallemant, 1995; Vallier, 1995; Walker, 1995), document a complex petrotectonic history in the Blue Mountains. We also propose some possible correlations between terranes in the Blue Mountains and in the Klamath Mountains and northern Sierra Nevada.

BLUE MOUNTAINS TERRANES—THEIR CORRELATIONS AND TECTONIC FRAMEWORK

When comparing terranes in the Blue Mountains with others in the Cordillera, it is helpful to first remove the 65° or more, post-Early Cretaceous clockwise rotation (Magill and Cox, 1980; Wilson and Cox, 1980). As a result, the terranes have (1) trends similar to those to the north and south, and (2) a west to east pre-rotational order of Wallowa, Baker, and Olds Ferry (Fig. 1A). Although the focus of this chapter is the Wallowa terrane, its relevance to Cordilleran tectonics depends in part on its relationship to adjacent terranes. Here we provide a brief summary of those relations.

Previously, correlations of the Wallowa terrane were made mainly with terranes in British Columbia and southern Alaska; for example, Wrangellia of the Wrangellian superterrane (Hillhouse et al., 1982) or Stikine of the Stikinian superterrane (Saleeby, 1983; Sarewitz, 1983). More recently, the Permian–Triassic part of the Wallowa terrane was identified as a volcanoplutonic arc that evolved oceanward from the Baker terrane in a Tethyan-Panthalassan faunal realm (Avé Lallemant, 1995; Vallier, 1995). A review of its arc character in the main body of this chapter provides important criteria not previously considered for correlation of the Wallowa terrane with other terranes in the Cordillera, and especially, with those farther to the south in the Klamath Mountains and in the northern Sierra Nevada.

The adjacent Baker terrane has been correlated with the Cache Creek terrane of the Stikinian superterrane (Saleeby, 1983). According to Saleeby (1992), the Olds Ferry terrane on the southern margin (pre-rotational eastern flank) of the Baker terrane evolved within a McCloud faunal realm. Mortimer (1986) correlated the Olds Ferry terrane with Quesnellia, consistent with its identification with other McCloud terranes, such as the eastern Klamath terrane in the Klamath Mountains (Goodge, 1990), and the northern Sierra terrane (Shoo Fly Complex) in the Sierra Nevada (Schweickert et al., 1999). According to Ferns and Brooks (1995), the Devonian through Early Jurassic (see Table A-1) Baker terrane developed largely in an ocean floor setting but in a Tethyan-Panthalassan faunal realm. Complications concerning faunal affinities of units within the western part of the Baker terrane are apparently resolved on recognition of Late Triassic tectonic interactions between Panthalassan and McCloud plates. A tectonic model suggested by Saleeby (1992) places the Baker terrane at the leading edge of the Tethyan-Panthalassan plate that eventually approached the fringing McCloud arcs offshore of continental North America by late Triassic time. This model seems consistent with evolution of the Baker terrane in a Late Triassic subduction complex (Burchfiel et al., 1992), tectonically juxtaposed at the "footwall" against a McCloud "hanging wall" in the Blue Mountains. The pre-rotation position of the Baker terrane seems consistent with that of other terranes of Cache Creek affinity in the Cordillera of North America (Silberling et al., 1984).

Other terranes of the Blue Mountains, such as the Grindstone and Izee terranes (Silberling et al., 1984), are less important to our immediate discussion. However, they are dealt with here because their occurrence and positioning have local and regional geological implications. For example, similarities of fauna in the Coyote Butte area of the Grindstone terrane and in the McCloud area in the Eastern Klamath terrane support their correlation. Blome and Nestell (1991) interpret the Grindstone terrane to be a McCloud tectonic fragment on the southwest flank of the Baker terrane, in fault contact with the Izee terrane. Thus, the presence of this terrane or tectonic fragment provides some evidence of continuity to McCloud terranes along the eastern margin of the Cordillera, from northern British Columbia southward into the northern Sierra Nevada. The Izee terrane consists of (1) Upper Triassic–Upper Jurassic flysch-like sedimentary and interlayered volcanic rocks and (2) Jurassic sedimentary rocks of the Weatherby Formation. The Izee terrane is located between the Baker and Olds Ferry terranes (Fig. 1B). Dickinson and Thayer (1978) interpret the Izee terrane sedimentary sequence to be intra-arc basinal fill derived from the Olds Ferry arc. The Upper Triassic–Upper Jurassic flysch and volcanic rocks are metamorphosed to zeolite and prehnite-pumpellyite facies (Brown and Thayer, 1963). The character of metamorphism in the Izee terrane may be an important factor in considering tectonic implications relative to its positioning between the Olds Ferry and Baker terranes.

CHARACTER AND GROWTH OF THE WALLOWA COMPOSITE TERRANE

The Wallowa terrane began forming in the Permian (or possibly even earlier), continuing through the Middle Triassic and the beginning of D_1 deformation (Vallier, 1977; Vallier et al., 1977; Walker, 1995). There was renewed magmatism and sedimentation to the pre-rotational west through the Late Triassic and Early Jurassic that overlapped earliest regional D_1 deformation (Brooks and Vallier, 1978; Avé Lallemant et al., 1980). The rock groups formed during these two time periods have dis-

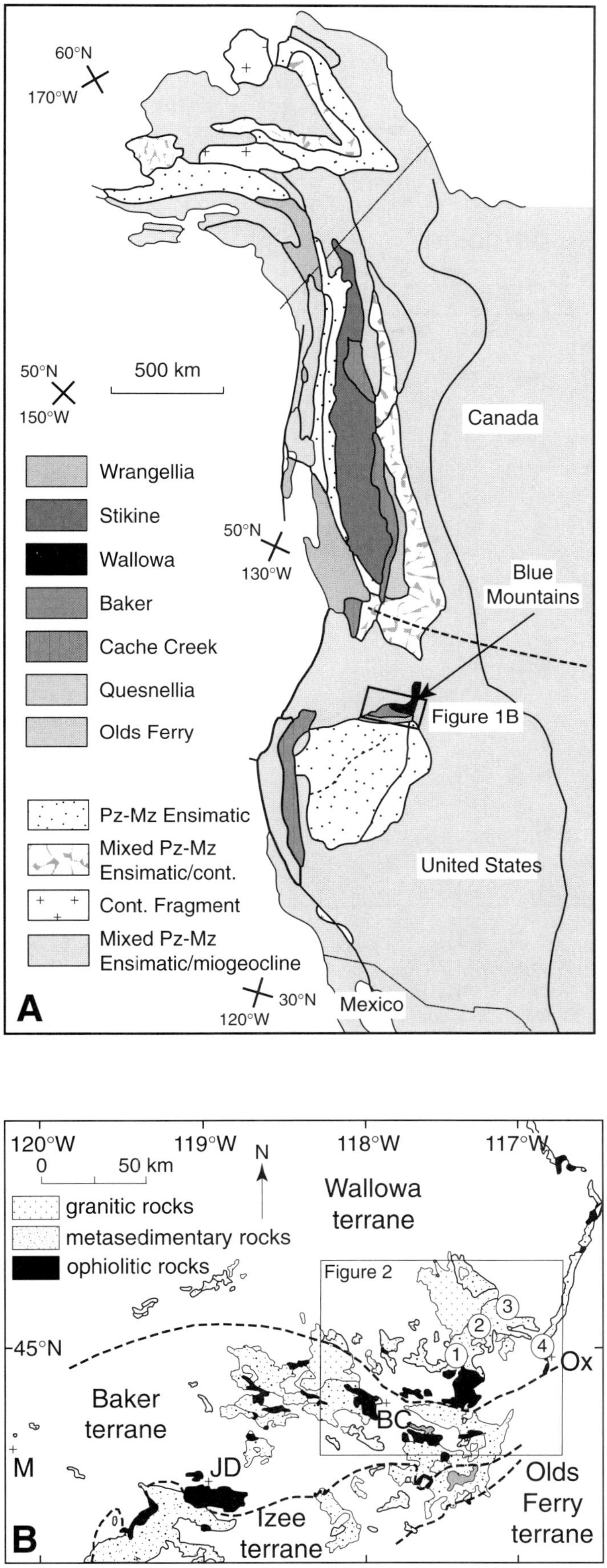

tinctive but structurally opposed trends exposed in sharp juxtaposition as a result of thrusting (Fig. 2) that is northeast-verging and southwest-dipping in the central part of the terrane (Fig. 3A and B). Permian and earlier rock units, mainly concentrated in the southern half of the terrane, have an east-west to north-northwest regional strike. Late Triassic–Early Jurassic volcanic and sedimentary rocks, concentrated in the northern half of the terrane, have a north-south to northeast strike. Jurassic–Cretaceous plutons (e.g., Wallowa batholith, Cornucopia stock, and smaller related bodies) stitch together the amalgamated rocks of the terrane (Fig. 2).

Tectonostratigraphic Array of Late Paleozoic through Middle Triassic Rocks

In the south-central part of the Wallowa terrane, there are mainly Permian or older rock units that continue from the faulted boundary with the Baker terrane north to the faulted contact with younger Mesozoic rocks (Figs. 2 and 3A). The southernmost exposures are dominated by gabbros in a band 8 km wide bounded on the south by clinopyroxenite gradational to peridotite. At the base the gabbros are cumulus-textured gabbronorite interlayered with and gradational to clinopyroxenite (Prostka and Bateman, 1962; Walker, 1995). The vertical to steeply south-dipping (Brooks et al., 1976), layered gabbronorite becomes increasingly hornblende-rich and homogenous near the sheared, altered, and partly obscured contact with an ~10-km-wide band of Permian tonalite and trondhjemite. Near the contact, trondhjemite is intruded by Late Triassic quartz diorite (Prostka and Bateman, 1962; Phelps, 1979). The trondhjemite and quartz diorite have U-Pb zircon ages of 253 Ma ± 1 Ma (sample SP-AG-81) and 215 ± 1 Ma (sample SP79-1), respectively (Walker, 1995; Table A-1). Phelps (1979) indicates that coarse-grained biotite-hornblende tonalite to quartz diorite intrudes gabbro and locally may contain xenoliths of it. Phelps (1979) and Phelps and Avé Lallemant (1980) suggest that all the plutonic rocks, gabbro through trondhjemite and quartz diorite, form a sequence that they name informally the "Sparta complex." An ~4-km-wide band of Permian (Shroba, 1992) and Triassic metavolcanic rocks named the Clover Creek Greenstone by Gilluly (1937) bounds the trondhjemite unit on the north (Figs. 2 and 3A). The Permian and Triassic metavolcanic rocks are spilites, keratophyres, and quartz keratophyres hydrothermally metamorphosed to zeolite, prehnite-pumpellyite, and

Figure 1. (A) Terrane map of the North American Cordillera and Blue Mountains, northeastern Oregon. The Wallowa, Baker, and Olds Ferry terranes in the Blue Mountains have been correlated with Wrangellia or Stikine, Cache Creek, and Quesnellia, respectively. Abbreviations: Mz—mesozoic; Pz—paleozoic. (B) Terrane map of the Blue Mountains. Crosses indicate locations of Mitchell (M), John Day (JD), Baker City (BC), and Oxbow (Ox). Boxed area is Figure 2. Circled numbers refer to schematic stratigraphic columns of Figure 4. Terrane boundaries and correlations after Avé Lallemant (1995).

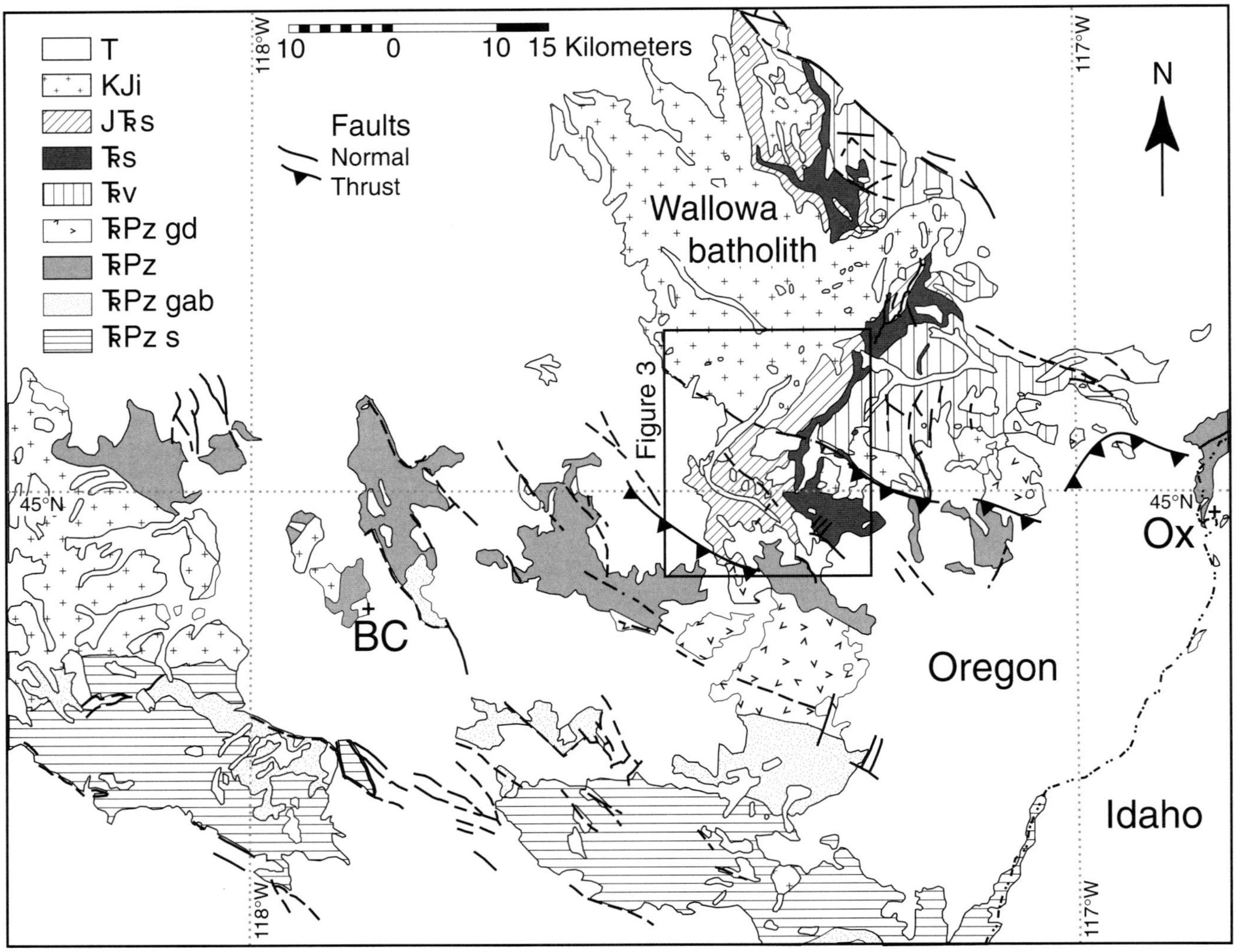

Figure 2. Geologic map of the Wallowa terrane and parts of the Baker terrane, which shows the broad outlines of: Permian or older lithospheric gabbro (TRPz gab); undifferentiated Permian trondhjemite-tonalite and intrusive Late Triassic diorite–quartz diorite-granodiorite (TRPz gd); undifferentiated Permian–Triassic sedimentary and volcaniclastic rocks (TRPz s); Triassic volcanic and volcaniclastic rocks (TRv); Triassic sedimentary rocks (TRs); Jurassic–Triassic sedimentary rocks (JTRs); Jurassic–Cretaceous gabbro and tonalitic rocks (KJi); undifferentiated Tertiary and younger rocks (T), including Miocene Columbia River Basalts and Quaternary gravels and sands. Primary magmatic structures such as compositional layering are generally subparallel to the trends of the Permian units, which are only broadly deformed, and therefore have trends and compositions that suggest magmatic order (~S_0). Late Triassic–Early Jurassic arc and sedimentary rocks in the Wallowa terrane are isoclinally folded (D_2) and have axial planar foliation (S_2) that is in sharp juxtaposition with layering and foliation (S_0, S_1) in the Permian rocks of the Wallowa and Baker Terrane. Abbreviations for locations are the same as in Figure 1B. The area outlined and labeled is that of Figure 3A. Modified in part from Brooks et al. (1976) and Walker and MacLeod (1991).

greenschist facies (Goebel, 1990; Table 1). The trondhjemitic unit is similarly metamorphosed near the contact with metavolcanic rocks and at least to 2 km south of it (Goebel, 1990). Field studies indicate that the contact between Permian trondhjemites and the adjacent Permian Clover Creek Greenstone is both erosional and depositional. However, Brooks et al. (1976) and Brooks and Vallier (1978) found evidence for intrusion of Sparta complex trondhjemite plutons into the Clover Creek Greenstone. We question whether the volcanic flows, tuffs, and volcaniclastic rocks of the Gold Creek Greenstone unit of Prostka and Bateman (1962) are essentially equivalent to the Clover Creek Greenstone. Both units are in faulted contact with adjacent Triassic rocks. However, the Gold Creek sequence is largely in faulted contact with rocks of the Lower Sedimentary Series except on its northern flank, where it is partly covered by Columbia River basalts and partly faulted against Hurwal mudstones (Fig. 3A and B). Metabasalts are more widespread and possibly thicker in the sequences of Gold Creek than in Clover Creek rocks. Tectonostratigraphic correlation of Permian and Triassic–Jurassic units in the south-central Wallowa terrane is provided in schematic columnar sections (Fig. 4) from four locations (Fig. 1B). Because of strong deformation, few estimates of thickness are available for Mesozoic (Follo, 1992, 1994) and Paleozoic units (Vallier, 1977; Brooks and Vallier, 1978) in the south-central Wallowa terrane. Available estimates, however, compare favorably with the thickness of lithospheric ocean floor and superimposed arcs accumulated tectonically during terrane amalgamation (see Chapter 7 *in* Moores and Twiss, 1995). Additional field and petrologic characteristics of Permian plutonic and metavolcanic rocks are summarized in Table 1.

The highly disrupted units in the southwest part (Fig. 2) are in contrast with the apparent continuity observed in the units in the south-central Wallowa terrane. For example, Devonian through Triassic plutonic and metamorphic rocks of the Bourne and Greenhorn subterranes occur in thrust-faulted north- to northeast-verging and southwest-dipping slablike mélange units. Ferns and Brooks (1995) suggest that one of these units, the Bourne subterrane, was derived from the Wallowa terrane and became tectonically incorporated as slices within the Elkhorn Ridge Argillite unit of the Baker terrane. The other unit, the Greenhorn subterrane, is the Baker terrane component nearest the Olds Ferry terrane. This unit contains Pennsylvanian through Early Triassic rocks with McCloud faunal affinity similar to those of the Eastern Klamath or Grindstone terranes (Blome and Nestell, 1991). The eastern part of the Wallowa terrane (Fig. 2) is only exposed along the Snake River Canyon and in deep tributary canyons. Here Permian and older(?) volcanic and plutonic rocks occur in faulted contact with Triassic and Jurassic rock units (Vallier, 1977). Major mylonite zones formed between 262 and 218 Ma in the three main basement complexes to the eastern Wallowa terrane (Avé Lallemant et al., 1980; Balcer, 1980; Walker, 1986). The basement complexes are composed of gabbro, trondhjemite, and amphibolite gneisses.

Triassic to Early Jurassic Arc Construction and Sedimentation Events

The Middle to early Late Triassic rock sequence is at least 3 km thick (Follo, 1992, 1994) and is dominated by volcaniclastic and sedimentary rocks (Fig. 4) north of the thrust-faulted contact (Fig. 3A and B) with Permian rocks in the southern Wallowa terrane. The contact with Permian volcanic and plutonic rocks was previously reported to be unconformable and/or faulted (Prostka and Bateman, 1962; Brooks et al., 1976; Vallier et al., 1977; Brooks and Vallier, 1978; Vallier and Brooks, 1986). Emplacement of the younger calc-alkaline arc volcanic sequence apparently overlapped Late Triassic–Early Jurassic compression involving folding, shearing, and subduction in the Blue Mountains (Hotz et al., 1977; Avé Lallemant et al., 1980; Balcer, 1980; Snee et al., 1995; Table A-1).

Along the Snake River, the lower part of the Triassic rock sequence consists of pillow lavas, debris-flow breccia units, and carbonate lenses of the Middle to Upper Triassic Wild Creek Sheep Formation (Fig. 4). This formation is considered equivalent to the upper part of the Clover Creek Greenstone in the south-central part of the terrane and grades laterally and upward into the Middle to Upper Triassic Doyle Creek Formation. We consider the Lower Sedimentary Series in the project area to be equivalent to the Doyle Creek Formation, named for rocks exposed on the Oregon side of the Snake River. Both the Doyle Creek Formation and Lower Sedimentary Series are dominantly volcaniclastic deposits that form extensive debris sheets or aprons. At their base, they are purplish debris-flow deposits composed mainly of volcanic rock fragments that change upward to red and purple turbiditic diamictites, with mud-silt matrix and clasts of chert, andesitic greenstone, and trondhjemite. This volcaniclastic sequence thickens locally, for example, on either side of Cliff Creek and in the uplands between the South and North Forks of the Imnaha River. In these areas, volcaniclastic rocks are gradational to porphyritic and plagioclase phyric mafic-to-intermediate flows, or to flow breccias and agglomerates (Taubeneck, 1964; Weis et al., 1976). However, in this area, volcanic and volcaniclastic breccias are distinguished with increased difficulty because of an overprinted metamorphic foliation.

The debris-flow deposits (i.e., Lower Sedimentary Series diamictites) are also interlayered with red to green mudstones, siltstones, and sandstones that mark the beginning of the better-defined stratigraphic units of the Martin Bridge Limestone and Hurwal Formation mudstones and sandstones (Fig. 4). Thus, the volcaniclastic rocks (Lower Sedimentary Series, Doyle Creek Formation) and the sorted, layered sedimentary rocks (Martin Bridge Limestone, Hurwal Formation) are gradational. Well-bedded, graded grainstones, packstones, and poorly sorted, massive limestone conglomerates dominate Late Triassic limestones in the southern Wallowa Mountains. The clastic limestone sequence is in sharp contrast to thin, laminated, dark, organic-rich limestones composed of silt-sized and finer carbonate grains of the type location. The clastic limestones are interpreted to be carbonate turbidites and debris-flow deposits that accumulated on a carbonate slope apron adjacent to the northern Wallowa carbonate platform (Follo and Siever, 1985). Follo (1992) interprets Hurwal Formation mudstone deposition as continuous and into deeper, starved basins. However, mapping indicates that some conglomerates and breccias—the Late Triassic–Early Jurassic Excelsior Gulch unit, in particular—were deposited in erosional channels cut into the Hurwal Formation mudstones and contain clasts of mainly limestone and mudstone, along with bedded radiolarian chert, greenstone, and plutonic rock clasts in smaller amounts. Differences in faunal associations in the clasts (Blome et al., 1986) and in clast character led Follo (1992) to suggest that the Baker terrane, which lies to the south, may have been a source for sedimentary detritus in the Excelsior Gulch unit.

Pre-D_1 metamorphism—Permian through Middle Triassic Units

In the south-central Wallowa terrane, the upper part of the Sparta complex (Phelps, 1979; Phelps and Avé Lallemant, 1980) and the adjacent volcanic and volcaniclastic rocks of the Clover

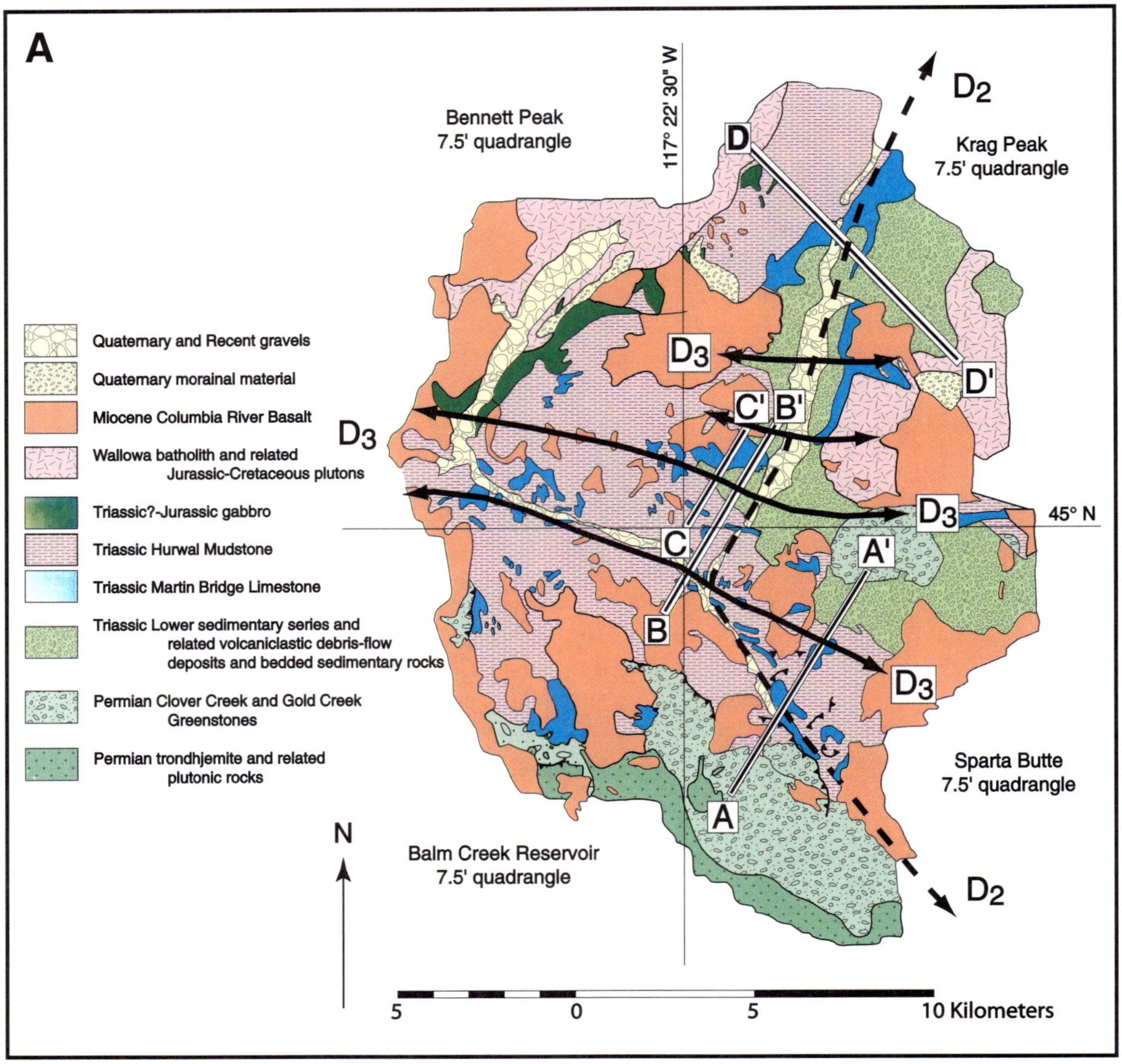

Figure 3. (A) Geologic map of the south-central Wallowa terrane and locations of cross-sections AA′ through DD′. Trends of D_2 and D_3 folds are shown with bold lines and arrows.

Creek Greenstone (Prostka and Bateman, 1962; Goebel, 1990) have been affected by low-grade hydrothermal alteration (M_{static}). Except where affected by subsequent Late Triassic mylonitic shearing, relict textures are common, and penetrative metamorphic fabrics are lacking in these Permian rocks. Thus, the effects of M_{static} metamorphism are readily identified in the Permian volcanic rocks and in the shallower parts of the underlying tonalite-trondhjemite in the south-central Wallowa terrane (Table 1). Relict textures and structures identified with M_{static} hydrothermal assemblages in these rocks and over a larger area of the Wallowa terrane have been described as intersertal, intergranular, pilotaxitic, hyalopilitic, trachytic, felty, spherulitic, granophyric, and vein-filling (Vallier and Batiza, 1978; Goebel, 1990). Because the textures and structures of these M_{static} metamorphosed assemblages are mainly magmatic features that pre-date earliest deformation (D_1) in the Wallowa terrane, they overlap the Permian through Middle Triassic constructional cycle. Moreover, these relict and open space textures and assemblages are observed in young hydrothermal submarine ocean floor sequences (Elthon, 1981; Alt et al., 1985). Such textures and assemblages have also been documented in pre-Nevadan (ca. 160 Ma) submarine volcanic-plutonic rock sequences, such as the Late Jurassic Josephine ophiolite in the Klamath Mountains of Oregon (Harper et al., 1988). According to Ehlers and Blatt (1982) and Spry (1969), these textures and structures are common at shallow depths during cooling of magmatic rocks in the presence of water vapor. The assemblages of M_{static} metamorphism (Table 1) are zeolite facies, prehnite-pumpellyite, and greenschist facies (Prostka and Bateman, 1962; Goebel, 1990).

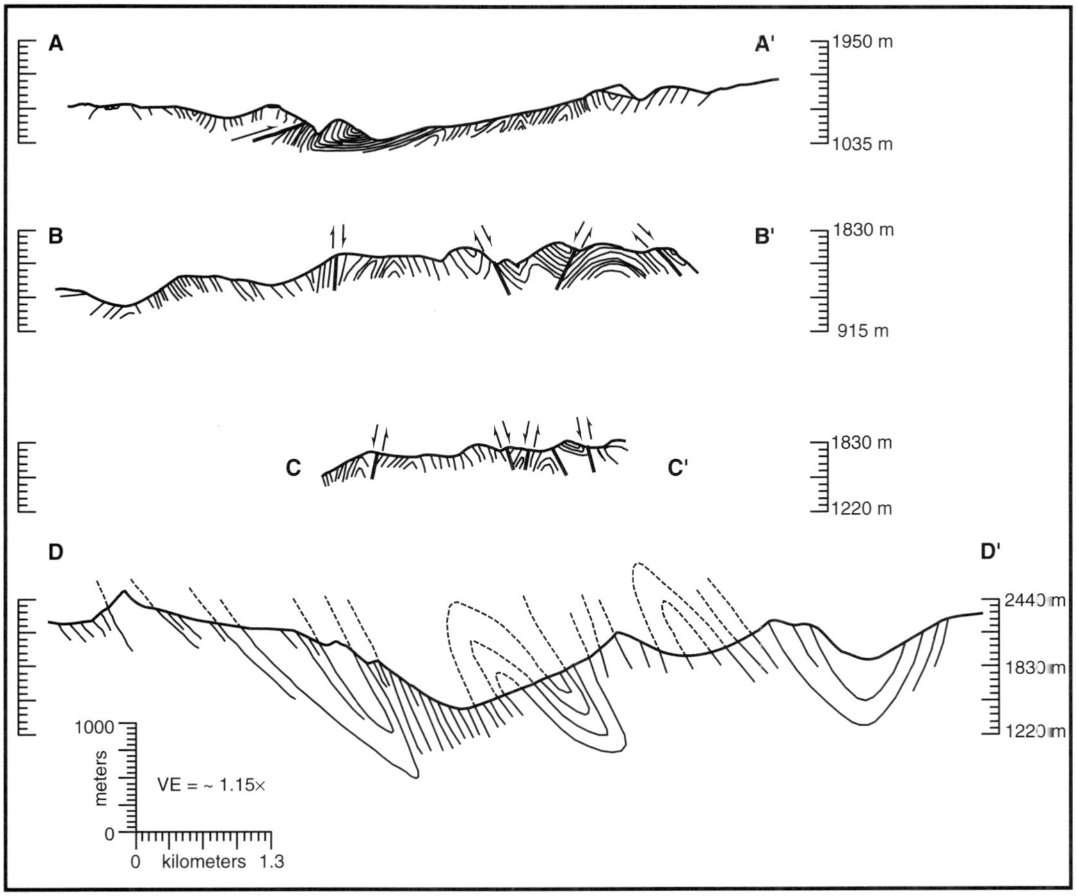

Figure 3. (B) Schematic geologic cross sections AA′ through DD′ in the south-central Wallowa terrane. Note that the vertical bar scale exaggeration is small (~1.15), but that the total vertical exaggeration is 4× that shown in the bar scale, or 5×.

COMPOSITION OF PERMIAN AND TRIASSIC METAMORPHOSED ROCK UNITS

Petrochemical data provide a better understanding of the growth and setting of the Wallowa terrane. Earlier petrochemical studies indicate a low K_2O arclike trend for the Permian and Triassic magmatic rocks (Phelps, 1979; Avé Lallemant et al., 1980; Balcer 1980; Walker, 1986; Vallier, 1995). In addition, Vallier and Batiza (1978) indicate that the arc metavolcanic and metaplutonic rocks are spilites, keratophyres, tonalites, and trondhjemites. These authors also suggest that the arc rocks have been metasomatically altered and have compositions that may relate to the extent of chemical alteration, and that the Permian rocks are altered to a greater extent than the Triassic rocks. Our objectives here are to determine if these arclike compositional trends and associated patterns of alteration can be substantiated for the rocks of the south-central Wallowa terrane. Unfortunately, Triassic rocks in this area are mainly epiclastic and volcaniclastic. Furthermore, M_2 metamorphism has largely obliterated relict textures of both igneous and volcaniclastic rocks affected by M_{static} metamorphism in Late Triassic metavolcanic flows in the area of this study (Fig. 3A). Thus, we were unable to find suitable samples of Late Triassic rocks for analysis in the south-central Wallowa terrane for comparison with the effects of M_{static} metamorphism in Permian igneous rocks. However, analyzed and hydrothermally altered (M_{static}) Late Triassic metavolcanic flows in adjacent parts of the terrane along the Snake and lower Imnaha Rivers have identifiable relict igneous textures (Vallier and Batiza, 1978; Vallier, 1995). Therefore, we made comparisons of the effects of M_{static} metamorphism on the compositions of Permian rocks in the south-central Wallowa terrane with the compositions of Triassic rocks in the Snake and Imnaha River areas.

Analytical Methods

Rare-earth elements (REEs) for rocks of the Wallowa terrane were analyzed by Patti M. Goebel using instrumental methods of neutron activation (Goles, 1978) at the University of Oregon, Eugene, Department of Geological Sciences. Goebel

TABLE 1. METAMORPHIC FACIES AND PETROLOGIC CHARACTERISTICS KEYED TO ROCK TYPE IN THE SOUTH-CENTRAL WALLOWA TERRANE

Rock type (Protolith)	Metamorphic mineral featuress	Relict mineral	Rock unit	Timing	Metamorphic facies	Remarks
Hornfels (Calc Argillite)	Act - Hb - $An_{(30-55)}$ - Q - Opq (±Cpx, ±Cord)		₮h, ₮mb	M_3	Amphibolite	Very fine-grained, locally gneissic
Phyllite (Calc Argillite)	Act - Chl - Ep - Q - $An_{(5–25)}$ - Opq (±Bio)	Q (lenses)	₮h, ₮mb	M_2	Epidote Amphibolite	Fine-grained, locally porphyroblastic (Act) organic-rich (~graphite), foliated
Slate (Calc Argillite)	Act - Chl - Q - Ab - Opq - Bio - Cc	Q (lenses)	₮h, ₮mb	M_2	Greenschist	Very fine-grained, foliated, organic-rich (~graphite)
Semischist (Volcaniclastic Cglm/Brec)	Hb - Q - $An_{(30–45)}$ - Cpx (±Ep, ±Bio)	Pl (volcanic clasts)	₮ls	M_3	Amphibolite	Hornfels texture; clast outlines remain
Semischist (Volcaniclastic Cglm/Brec)	Act - Hb - Chl - $An_{(22–31)}$ - Opq (±Q, ±Bio)	Pl (volcanic clasts)	₮ls	M_2	Epidote Amphibolite	Clasts flattened, aligned
~Semischist (Volcaniclastic Cglm/Brec)	Chl - Act - Ab - Ep - Cc (±Bio, ±Opq)	Cpx-Pl (volcanic clasts)	₮ls	M_2	Greenschist	Relict texture transformed to semischist from south to north in study area
Volcaniclastic Breccia	Chl - Cc - Opq - Q - Pu - Pre - Ab	Pl (volcanic clasts)	₮Pcc, ₮Pgg	M_{static}	Prehnite Pumpellyite	Veins have Q - Cc - Pre, Ep, Cc, Opq
Volcaniclastic Breccia	Cc - Chl - Zeo - Q - Opq - Ab	Pl - Hb (volcanic clasts)	₮Pcc, ₮Pgg	M_{static}	Zeolite	Relict crystal outlines in clasts
Keratophyre (Andesite-Dacite)	Chl - Opq - $An_{(3–5)}$ - Pu - Pre - Ep - Q	Pl (An_{30}), Hb, Bio	₮Pcc	M_{static}	Prehnite Pumpellyite	Relict crystal outlines (laths) in clasts
Keratophyre (Andesite-Dacite)	Cc - Cly - Chl - Opq - Zeo	Pl - Q	₮Pcc	M_{static}	Zeolite	Relict crystal outlines (laths) in clasts
Spilite (Basalt)	Chl - Cc - $An_{(3–5)}$ - Cly - Pre - Pu - Q - Opq	Pl - Cpx - Hb	₮Pcc	M_{static}	Prehnite Pumpellyite	Spherulitic to intergranular texture is common; veins of Q-Cc
Spilite (Basalt)	Chl - Opq - Zeo - Cc - Q - Cly - $An_{(3–7)}$	Cpx - $An_{(28–50)}$ - Opx - Hb	₮Pcc	M_{static}	Zeolite	Spherulitic to intergranular texture is common; relict Pl outlines common
Trondhjemite	Chl - Q - Ab - Ep (±Mus)	Pl $An_{(28–50)}$ (±Hb, ±Bio)	Trondhjemite	M_{static}	Greenschist	Texture variable (fine to medium); spherulitic Q-Pl common, ±granophyric
Trondhjemite	Chl - Q - Ab - Pre - Pu (±Stp, ±Opq)	Pl $An_{(28–50)}$ (±Hb, ±Bio)	Trondhjemite	M_{static}	Prehnite Pumpellyite	Texture variable (fine to medium); spherulitic Q-Pl common, ±granophyric

Notes: Mineral name abbreviations: Act—actinolite; Ab—albite; An—percent anorthite content (e.g., An 3–5); Cc—calcite; Chl—chlorite; Cpx—clinopyroxene; Ep—epidote; Hb—hornblende; Cly—clay minerals; Pl—plagioclase; Pr—prehnite; Pu—pumpellyite; Q—quartz; Stp—stilpnomelane; Zeo—zeolite; Opq—opaque minerals (e.g., oxides, sulfides, graphite); Cord—cordierite; Bio—biotite; Mus—muscovite/sericite; Opx—orthopyroxene. Rock unit abbreviations: ₮h—Triassic Hurwal Formation; ₮mb—Triassic Martin Bridge Limestone; ₮ls—Triassic Lower Sedimentary Series; ₮Pcc—Triassic–Permian Clover Creek Greenstone; ₮Pgc—Triassic–Permian Gold Creek Greenstone.

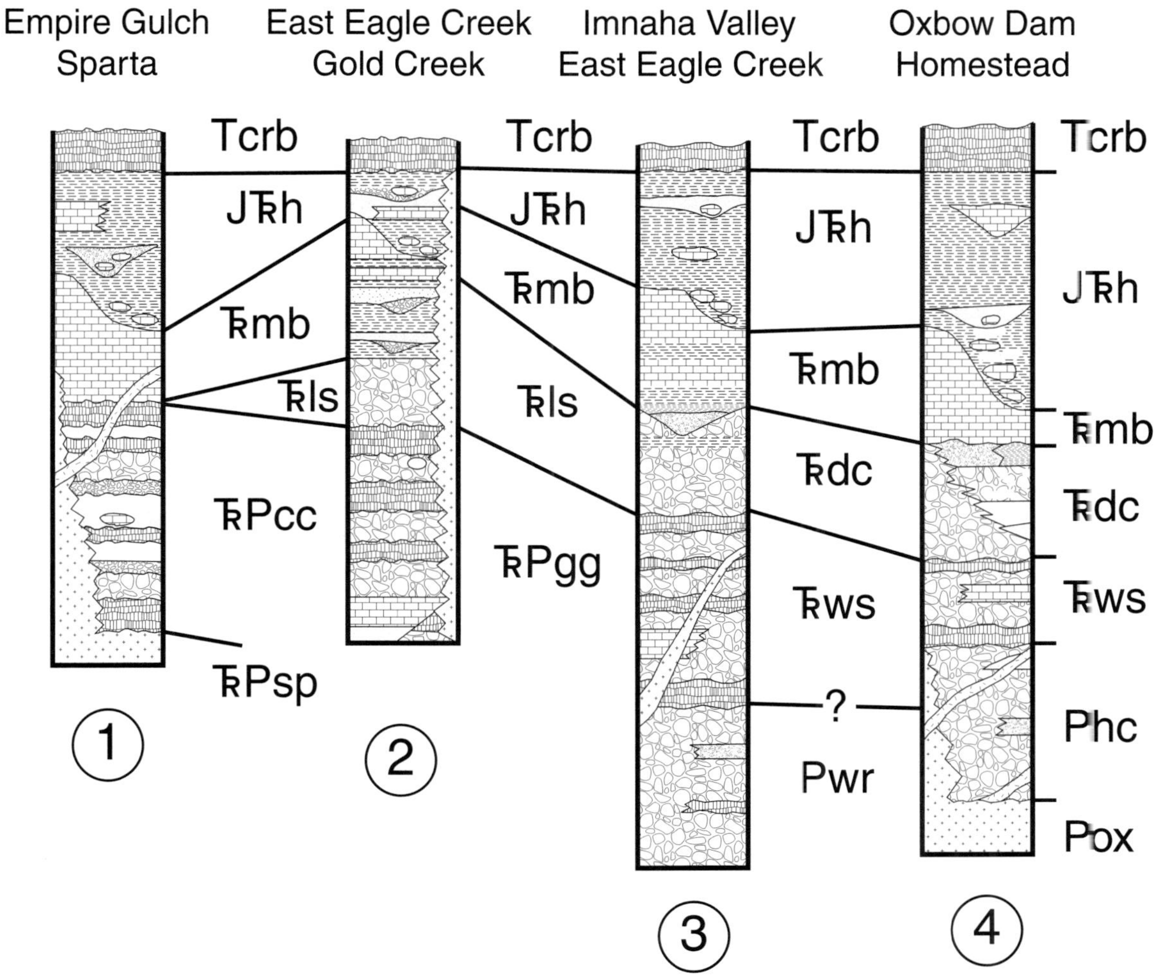

Figure 4. Locations (1–4) of schematic geologic section by area: 1—Empire Gulch, Sparta; 2—East Eagle Creek, Gold Creek area; 3—Imnaha Valley, East Eagle Creek; 4—Oxbow Dam, Homestead. Ages of the units: Tcrb—Tertiary Columbia River basalt; JTRh—Triassic–Jurassic Hurwal Formation; TRmb—Triassic Martin Bridge Limestone; TRls—Triassic Lower Sedimentary Series; TRdc—Triassic Doyle Creek Formation; TRws—Triassic Wild Sheep Creek Formation; TRPcc—Triassic and Permian Clover Creek Greenstone; TRPgg—Triassic and Permian Gold Creek Greenstone; TRPsp—Triassic and Permian Sparta complex; Phc—Permian Hunsaker Creek Formation; Pwr—Permian Windy Ridge Formation; Pox—Permian Oxbow Complex. The range of ages of intermediate to silicic plutonic rocks is Permian through Cretaceous.

also performed analytical work for major and trace elements at Washington State University, Pullman, on a manual X-ray fluorescence Philips 1410 spectrometer using 2:1 $Li_2B_4O_7$:rock powder fused disks. Analyses do not include nondiscriminatory loss on ignition as an estimate of water from total volatile content. Analytical techniques used in the determination of whole-rock major and trace elements are described in Norrish and Chappell (1977). Contents of most trace elements were determined using rock powders calibrated against a variety of standards of known trace element concentration. A more extensive database of Wallowa terrane whole-rock major and trace element analyses can be found in Goebel (1990).

Compositions of Permian Rock Units

The TiO_2-MnO-P_2O_5 discriminant plots of the hydrothermally altered and dominantly Permian (pre-D_1) mafic volcano-plutonic rocks for the south-central Wallowa terrane are shown in comparison with those from the Snake River, northern Wallowa terrane, and Olds Ferry terrane in Figure 5A. Although the rocks plot mainly in the fields of calc-alkaline basalt (CAB) and island-arc tholeiite (IAT), some plot in the fields of mid-ocean ridge basalt (MORB) and alkalic ocean island basalt (OIB). In addition, the Na_2O-enriched or K_2O-depleted mafic to silicic arc rocks have overlapping REE trends (Fig. 5B). On normalization to

MORB, these Permian spilites, keratophyres, tonalites, and trondhjemites also have excluded-element abundance patterns that overlap. Note that the patterns show (1) enrichment in K, Ba, and other excluded lithophilic elements that are similar to or a little more or less than those of CAB and IAT; and (2) that there is also enrichment or slight depletion of high-field-strength elements (Fig. 5C and D, Table 2). The patterns of overlap in pre-D_1 Permian arc rocks of the Wallowa terrane are in marked contrast with the patterns in modern and less altered counterparts. For example, the progressive enrichment of lithophile element (K, Rb, Li) and light REE (LREE) abundances with increasing silica contents is much less than in unaltered arc rocks (Philpotts, 1990; Raymond, 2002). Although Permian spilites and Quaternary Cascade CAB basalts (Fig. 4 *in* Leeman et al., 1990, p. 19,571) have similar patterns of abundances relative to MORB, Cascade patterns are well ordered for given volcanic centers.

COMPOSITIONS OF TRIASSIC ROCK UNITS

According to Vallier and Batiza (1978), enrichments (SiO_2, Na_2O, H_2O, and CO_2) and depletions (MgO, FeO, CaO, Al_2O_3, TiO_2, P_2O_5, K_2O, and MnO) of oxide abundances for analyzed spilites of the Wallowa terrane are less in Triassic than in Permian rocks compared to those of the average modern subalkaline basalts. Middle to early Late Triassic basaltic to rhyolitic volcanic rocks have a wider range of K_2O and Na_2O contents than do their Permian counterparts (Tables 3.16 and 3.17 *in* Vallier, 1995). The REE trends for mafic through silicic rocks (Fig. 3.25a,b *in* Vallier, 1995) have less overlap in Triassic (Fig. 6A) than in Permian counterparts (Fig. 5B). Furthermore, in Triassic rocks, the trend from mafic to silicic lavas is toward a more positive enrichment in LREEs (Fig. 6A, Table 2). Overall, these chemical characteristics are consistent with the observations of Vallier and Batiza (1978) that (1) relict igneous mineral assemblages are better preserved in Triassic than in Permian lavas, and (2) the effects of hydrothermal alteration may be less in Triassic rocks. In general, however, the patterns of enrichment and depletion for excluded-element abundances in the two groups of rocks are similar. Moreover, there are few differences between Triassic and Permian MORB-normalized plots of excluded lithophilic and high-field-strength element abundances for mafic through silicic rocks (Figs. 5C and D and 6B). Note in particular that in those plots, Triassic basalts and basaltic andesites of the Wallowa terrane have sharp patterns of enrichment of incompatible elements similar to those of Permian basalts and andesites. Considering the hydrothermally altered nature of Permian and Triassic rocks of the Wallowa terrane, and that of MORB (Elthon, 1981; Elthon and Casey, 1985), perhaps MORB normalization mitigates the small differences that exist in these rocks.

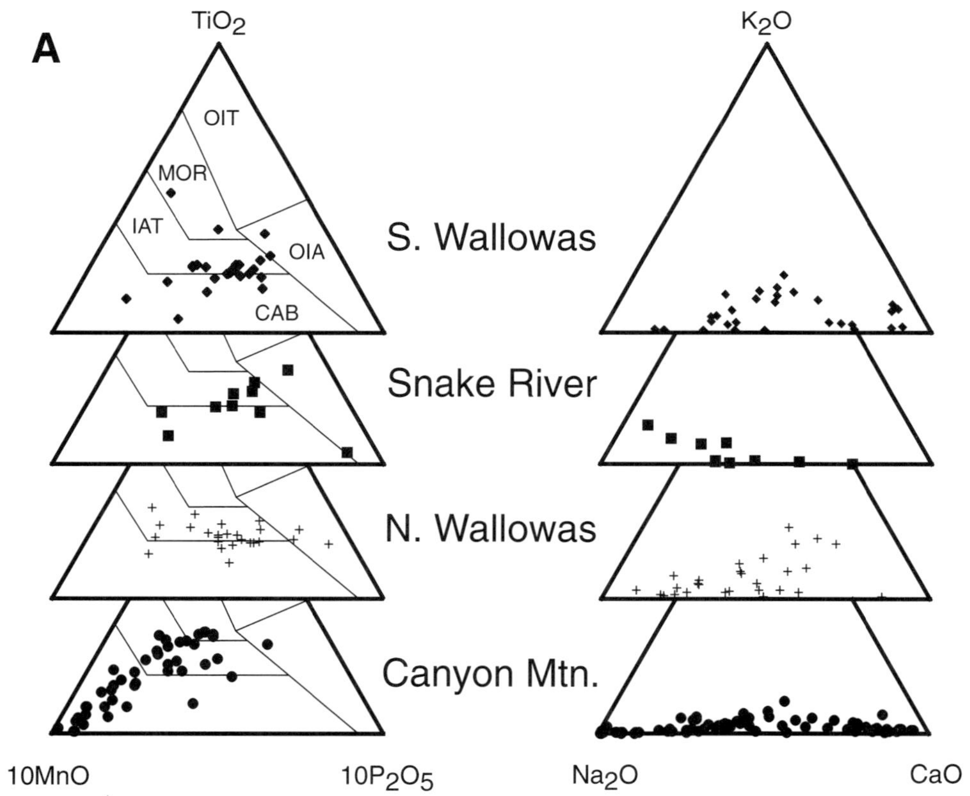

Figure 5. (A) Discriminant plots and variation diagrams of mostly Permian igneous rocks of the Wallowa terrane (data from Goebel, 1990, and Vallier, 1995) and the Canyon Mountain Complex (data from Leeman et al., 1995). Left column: TiO_2-MnO_2-P_2O_5 discriminant plots (fields after Mullen, 1983); OIT—ocean island tholeiite; MOR—mid-ocean ridge basalt; IAT—island-arc tholeiite; CAB—calc-alkaline basalt; OIA—ocean island alkalic. Right column: K_2O-Na_2O-CaO variation diagrams. (B) Chondrite normalized rare-earth-element plots of Permian spilites and keratophyres. Trondhjemites of the Sparta Complex (data from Goebel, 1990) are represented by the shaded region. Elements in parentheses were not analyzed. (C) Abundances of incompatible elements in Permian spilites of the Clover Creek Greenstone (data from Goebel, 1990) relative to mid-ocean ridge basalt (MORB). Plot order after Pearce (1983) in this and subsequent plots relative to MORB. Elements in parentheses were not analyzed. (D) Abundances of incompatible elements in Permian keratophyre flows, tonalites, and trondhjemites (data from Goebel, 1990) in the southern Wallowa terrane relative to MORB.

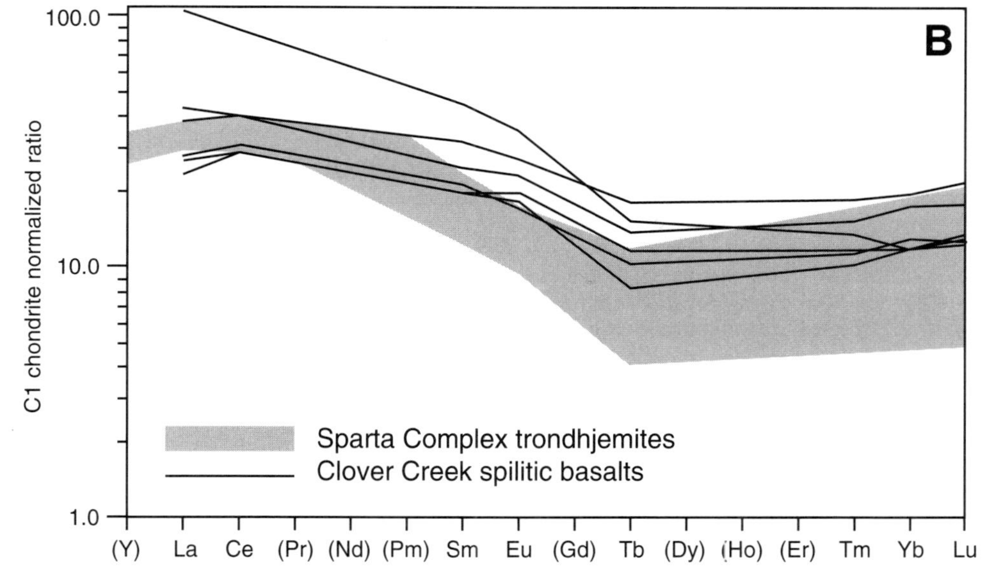

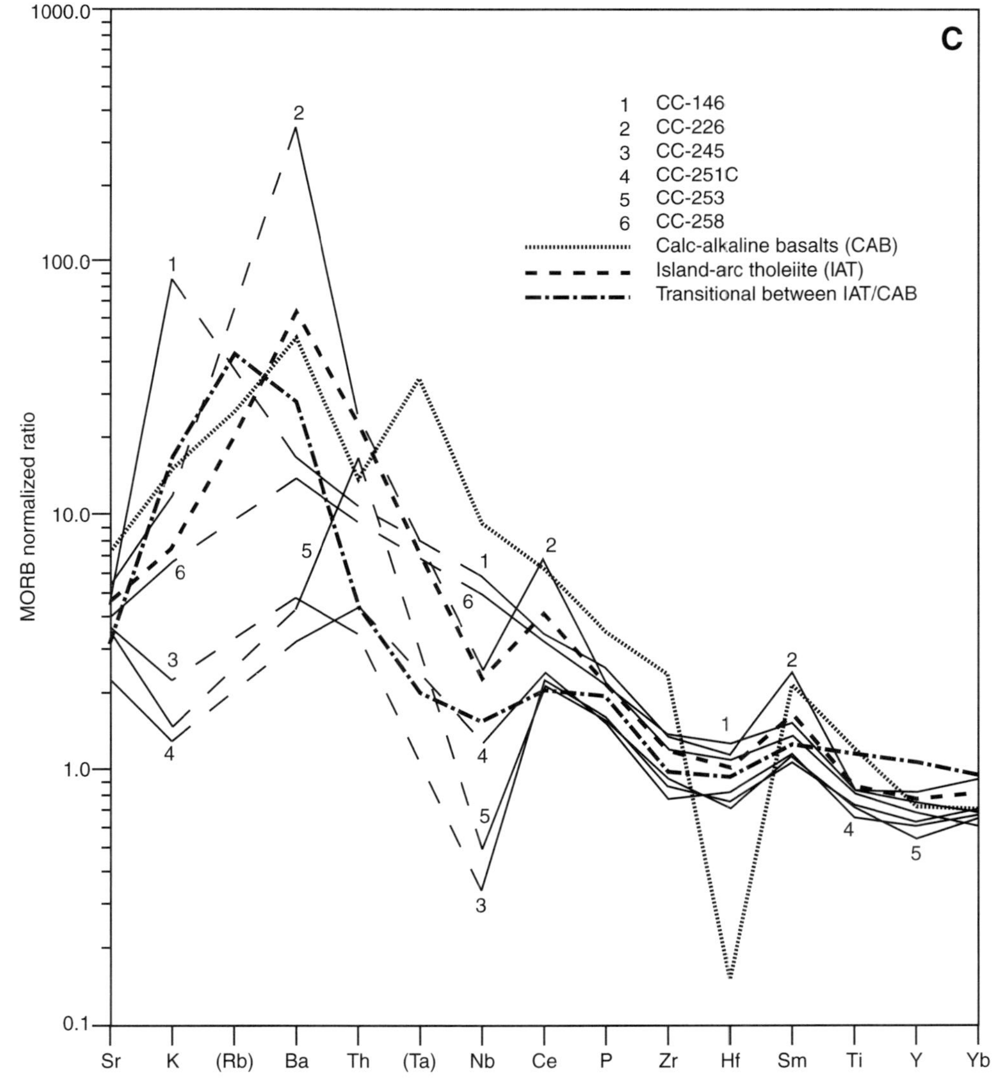

Figure 5. (*Continued*)

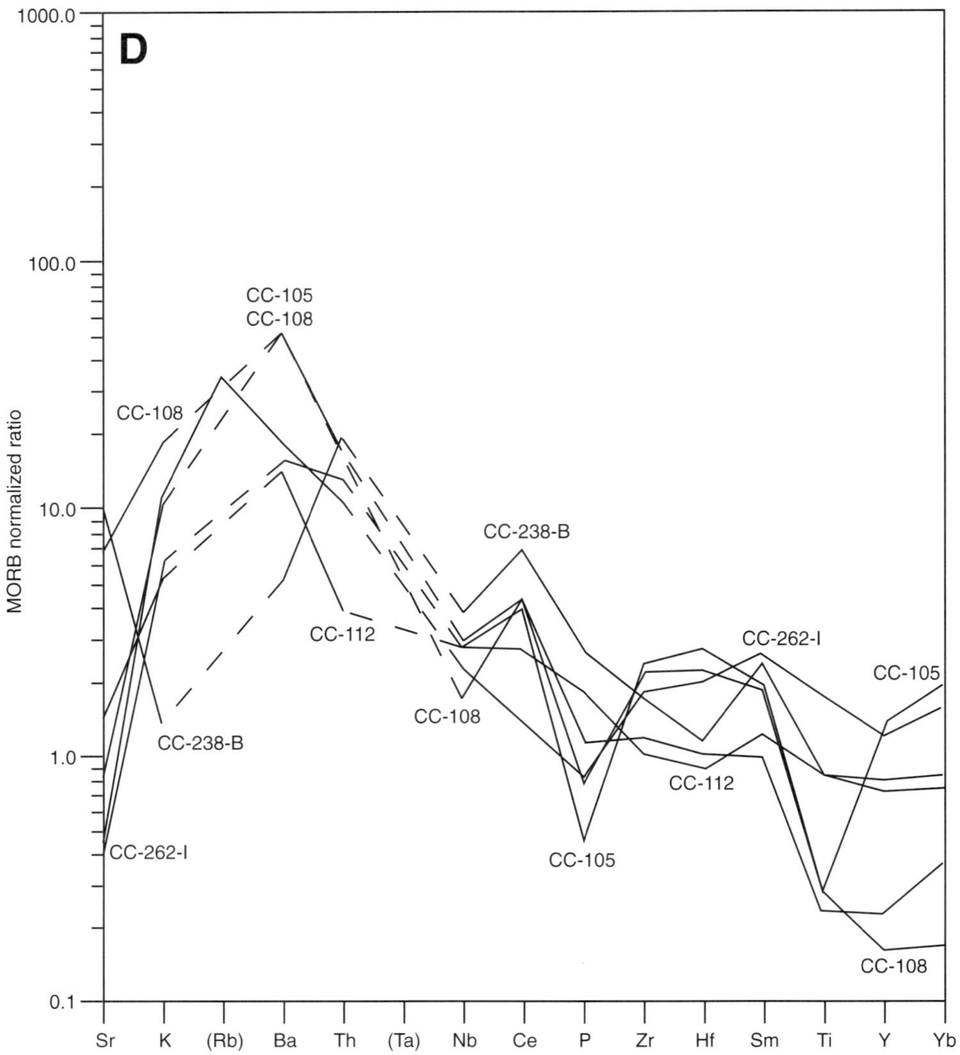

Figure 5. (*Continued*)

DEFORMATION AND METAMORPHISM IN THE SOUTH CENTRAL WALLOWA TERRANE

Late Triassic–Early Jurassic D_1 Deformation and M_1 Metamorphism

We use the regional chronological scheme of Avé Lallemant (1995) in our discussion of deformation in the south–central Wallowa terrane. However, we recognize M_{static} that is earlier than D_1 deformation and not associated with regional contraction. The major penetrative effects of D_1 deformation generated S_1 axial planar foliation and cleavage within the Elkhorn Ridge Argillite (Coward, 1983) and Burnt River Schist (Ashley, 1967, 1995) units of the Baker terrane. Thus, M_1 greenschist- and blueschist-facies assemblages in the Baker terrane were syntectonic with D_1 deformation (Ashley, 1967; Avé Lallemant et al. 1980; Fig. 3). Blueschists occur near the inferred suture with the upper plate, McCloud affinity, Olds Ferry, and Izee terranes (Bishop, 1995a).

Overall, shear-sense indicators on faults and associated microfabrics in rocks of the Wallowa and Baker terranes indicate sinistral shear (Coward, 1983; Avé Lallemant, 1995), with timing that is in the range 230–223 Ma (Hotz et al., 1977; Walker, 1995; Table A-1). The obvious and well-documented effects of D_1 deformation in the Wallowa terrane are Late Triassic–Early Jurassic kilometer-wide mylonitic zones best exposed along the Snake River in a number of locations (Balcer, 1980; Avé Lallemant et al., 1985). Although there is no evidence for widespread regional D_1 contraction or folding in the Wallowa terrane, the timing of it overlapped Triassic–Jurassic erosion-sedimentation. Moreover, the character of detritus in the volcaniclastic rocks and conglomerates suggests Permian and Triassic–Jurassic arc provenance within the Wallowa terrane (Follo, 1992). Lithologic and faunal affinities of Early Jurassic detritus in the

TABLE 2. REPRESENTATIVE ANALYSES OF PERMIAN ARC AND OPHIOLITIC ROCKS IN THE BLUE MOUNTAINS

Sample	SiO_2	Al_2O_3	TiO_2	FeO*	MnO	CaO	MgO	K_2O	Na_2O	P_2O_5	Total†	La	Ce	Nd	Sm	Eu	Tb	Yb	Lu	Rock type
Wallowa terrane																				
CC-253	52.35	15.11	0.89	8.77	0.14	8.84	9.4	0.1	3.96	0.19	99.75	5.4	17	10.7	2.9	1.1	0.41	1.9	0.31	Metavolcanic
CC-146	51.93	15.61	1.05	9.08	0.19	8.55	8.51	0.49	4.13	0.26	99.8	9.8	24	11.3	3.6	1.3	0.49	2.8	0.42	Metavolcanic
CC-238b	54.85	18.01	1.06	7.24	0.15	7.03	5.73	0.09	4.82	0.3	99.28	20.9	53	31	6.3	2.1	0.55	2.2	0.29	Metavolcanic
CC-112	53.87	15.3	1.04	8.34	0.16	7.74	8.35	0.37	4.79	0.2	100.16	7.1	20	10.6	3.2	1.4	0.51	2.5	0.39	Metavolcanic
VB-79	54.74	15.73	2	8.69	0.42	4.27	3.93	0.04	6.73	0.41	96.96	9	n.d.	n.d.	4.8	1.62	1.07	4.1	0.6	Metavolcanic
287	47.42	18.07	0.05	8.56	0.15	14.37	8.36	0.02	0.79	n.d.	97.79	0.34	0.89	n.d.	0.43	0.26	0.15	0.55	0.08	Plutonic
210	65.65	14.46	0.62	5.79	0.12	5.76	1.91	0.53	2.98	0.1	97.92	4.89	12.28	n.d.	2.68	0.84	0.67	2.73	0.46	Plutonic
201a	72.93	13.4	0.39	3.06	0.13	3.28	0.56	1.04	3.83	0.06	98.68	5.51	15.29	n.d.	3.84	0.92	1.08	4.82	0.75	Plutonic
CC-262	77.54	12.4	0.4	2.16	0.05	0.49	2.13	0.43	5.05	0.09	100.74	17.1	46	28	6.8	1.5	0.64	4.6	0.81	Plutonic
CC-105	78.06	13.61	0.35	1.32	0.02	0.48	0.31	0.71	6.08	0.05	100.99	10.8	31	18.7	4.7	1	0.64	5.7	0.86	Plutonic
VS5-7	76.42	12.78	0.35	1.92	0.08	0.45	2.15	0.8	4.71	n.d.	99.66	13	33	n.d.	5.1	1.21	0.92	4.4	0.65	Metavolcanic
LS7-32b	60.04	16.4	1.22	5.03	0.05	2.83	1.96	0.74	7.02	0.21	95.5	5	12	n.d.	2.6	0.43	0.45	1.9	0.3	Metavolcanic
Greenhorn subterrane																				
V-1	49.75	15.05	2.43	11.98	0.2	9.02	8.29	0.35	2.54	0.39	100	22.5	44.7	n.d.	6.3	1.7	0.85	2.26	0.33	Metavolcanic
V-2	47.18	15.22	1.01	13.8	0.2	11.61	8.54	0.36	2.36	0.16	100.44	3.54	n.d.	n.d.	n.d.	0.89	0.7	3.42	0.52	Metavolcanic
185	53.37	15.68	0.85	7.72	0.15	10.27	8.93	0.74	2.22	0.14	100.07	n.d.	n.d.	n.d.	n.d.	n.d.	n.d.	n.d.	n.d.	Metavolcanic
Canyon Mountain Complex																				
CMG32	53	15.4	1.39	10.9	0.19	7.64	7.02	0.2	4.43	n.d.	100.17	2.19	7.9	6.9	3.77	1.47	0.87	3.16	0.5	Metavolcanic
CMG227	61	15.1	1.04	9.71	0.18	4.19	3.48	0.04	5.58	n.d.	100.32	1.75	5.4	6	2.45	0.98	0.57	2.37	0.4	Metavolcanic
CMG15	75.8	14.2	0.14	2.7	0.1	0.56	0.25	0.11	6.23	n.d.	100.09	4.25	14.3	14.9	5.07	0.72	1.37	6.8	1.02	Metavolcanic
CM28	47.05	18	0.18	4.47	0.1	17.11	10.06	0.06	0.9	0.01	97.94	n.d.	n.d.	n.d.	n.d.	n.d.	n.d.	n.d.	n.d.	Plutonic
CMG137	61.1	16.3	0.65	8.3	0.18	5.31	3.71	0.26	4.41	n.d.	100.22	1.74	5.7	5.2	2.14	0.91	0.65	2.26	0.36	Plutonic
OC114	76	13.8	0.21	1.56	0.06	1.04	0.38	0.15	5.99	n.d.	99.19	4.21	14	11.1	3.84	0.97	0.91	4.74	0.77	Metavolcanic

Notes: Oxide analyses are wt%; rare earth elements are ppm; additional analyses with prefix CC are also available in Goebel (1990); samples VB-79, VS5-7, LS7-32b from Vallier (1995); samples 287, 210, 201a from Phelps and Avé Lallemant (1980); samples V-1, V-2, 185 from Ferns and Brooks (1995); samples with prefix CMG and OC from Leeman et al. (1995). Abbreviation: n.d.—not determined.

*Total iron as FeO.

†No volatiles.

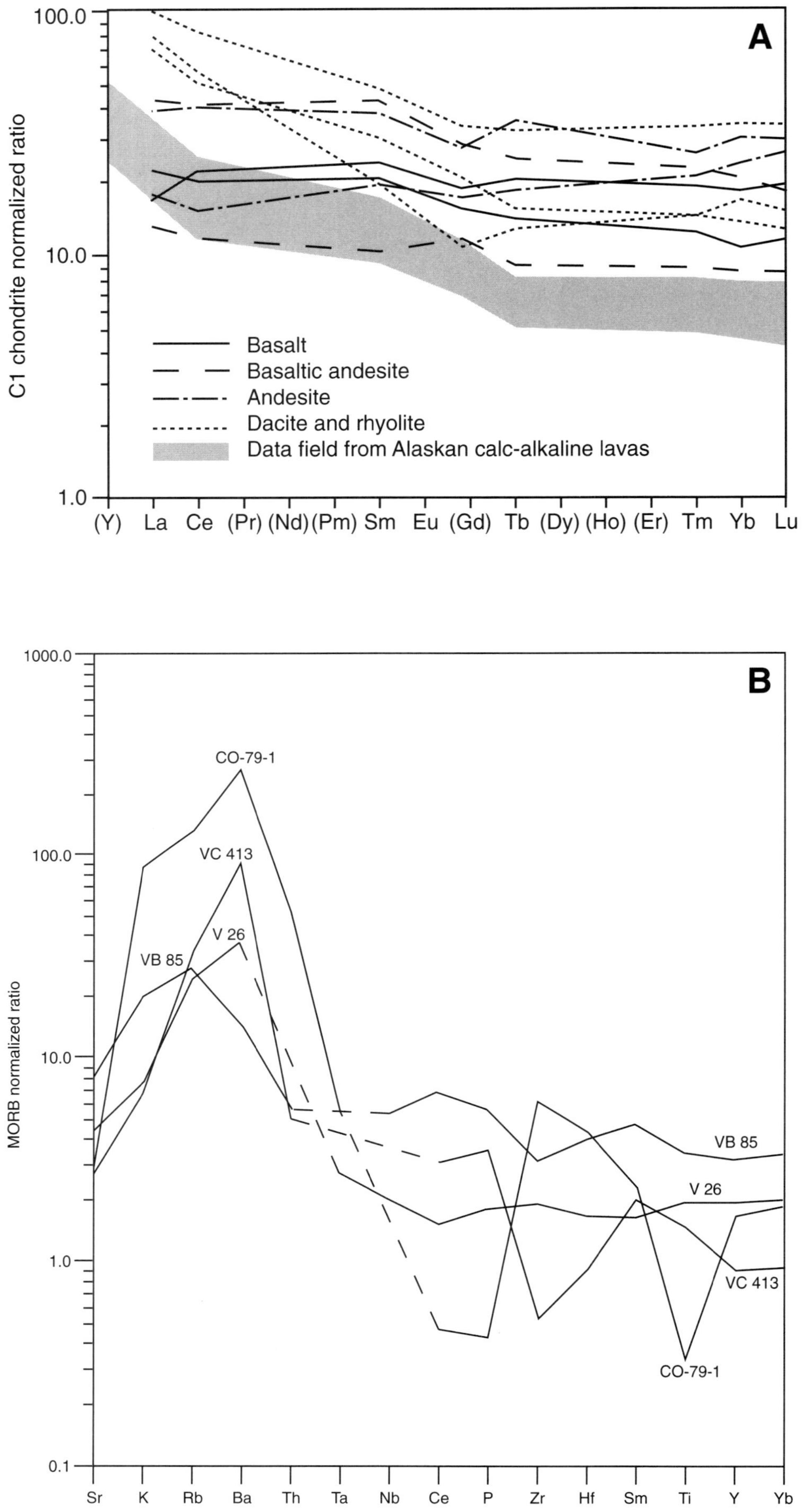

Figure 6. (A) Chondrite normalized rare-earth-element plots of Triassic basalts, andesites, dacites, and rhyolites (data and plots after Vallier, 1995). Elements in parentheses were not analyzed. (B) Abundances of incompatible elements in Permian and Triassic basalts, keratophyres, and trondhjemites along the Snake River (data from Vallier, 1995) relative to mid-ocean ridge basalt (MORB). Dashed lines indicate element not analyzed.

Hurwal Formation (Excelsior Gulch units) suggest that the adjacent Baker terrane also supplied coarse detritus (Follo, 1994).

Middle to Late Jurassic D_2/D_3 Deformation and M_2/M_3 Metamorphic Events

Pre-eminent D_2 deformation in the Wallowa terrane is represented by penetrative S_2 axial planar foliation in Triassic and Early Jurassic rocks. This foliation is feebly developed or absent in Permian rocks of the south-central Wallowa terrane (Fig. 3A, Table 1). Although the through-going D_2 fold structures correspond to the "second" regional deformation, they are "first" folds in Wallowa terrane units. The trends of macroscopic D_2 isoclinal fold structures are parallel to the strikes of bedding on their limbs in the Lower Sedimentary Series, Martin Bridge Limestone, and Hurwal Formation (Figs. 3 and 7A). Note, too, that the trends of D_2 isoclinal folds and strikes of associated axial planar foliation in the metamorphosed Late Triassic sedimentary sequence vary from slightly west of north to northeast (Fig. 7B).

The D_2 folds are synchronous with M_2 in the chronology of recrystallizations in Permian through Early Jurassic rocks in the Blue Mountains. In the south-central Wallowa terrane, M_2 metamorphic assemblages are greenschist and epidote amphibolite facies (Table 1). Alignments of (1) basal cleavage in platy minerals such as micas and chlorite and (2) shape flattening (e.g., in quartz grains and clasts in lower-grade rocks) define S_2 foliation/cleavage. Triassic–Early Jurassic arc volcanic, volcaniclastic, and sedimentary rocks show the variable effects of D_2 deformation, ranging from ductile flow in the central and northern Wallowa terrane to fracture-cleavage and local thrusting in the southern part of the Wallowa terrane. The weaker cleavage (S_2) in lower-grade rocks also appears to be parallel to axial planes of northward-verging folds and thrusts in the central Wallowa terrane. Even here, however, limestones are recrystallized and have a foliation. The dominant northeast strike of S_2 foliation and cleavage in the Mesozoic arc and sedimentary sequence is at a high angle to the east-west–striking to north-northwest–striking layering (~S_0) in the Permian rocks of the Wallowa terrane.

As shown by our own mapping as well as that of Mirkin (1986) along the East Fork of Eagle Creek, the trends and strikes of D_2 structures in the supracrustal sequence vary, because they are folded by and at high angles to D_3 folds. For example, prominent northwest- and northeast-trending D_2 folds along Eagle Creek and the East Fork of Eagle Creek, respectively, reside on the refolded limbs of D_3 folds (Figs. 3A and E, 7A and B). In other examples, northeast-trending D_2 folds and thrusts in the eastern part of the terrane south of Aneroid Lake are folded again by northwest-trending D_3 fold structures (Weis et al., 1976). Locally, northeast-verging D_3 thrusts truncate and override the western limb of D_2 folds (section AA′ of Fig. 3A and B). Regionally, D_3 thrust-faults separating Permian and Triassic rocks also truncate north-south–trending to northeast-trending D_2 folds expressed by layering in Triassic rocks (Fig. 2). Although foliation is feeble in Permian volcanic and plutonic rocks of the southern Wallowa terrane, these rocks develop foliation (S_3) locally emphasized by chlorite along the hanging wall above the thrust contact with lower plate Late Triassic rocks.

Tonalitic plutonic rocks of the Wallowa batholith; earlier gabbroic bodies; and smaller, abundant subalkaline mafic dikes, sills, and irregularly shaped bodies are important structural markers. These intrusive bodies truncate D_2 and D_3 fold structures in Mesozoic sedimentary rocks along the batholith's southeast margin (Fig. 3). Thus, the intrusions provide a minimum age of D_3 deformation in the Wallowa terrane. According to Armstrong et al. (1977), the batholith has an Early Cretaceous whole-rock Rb-Sr isochron at 130 ± 2.4 Ma and a K-Ar age

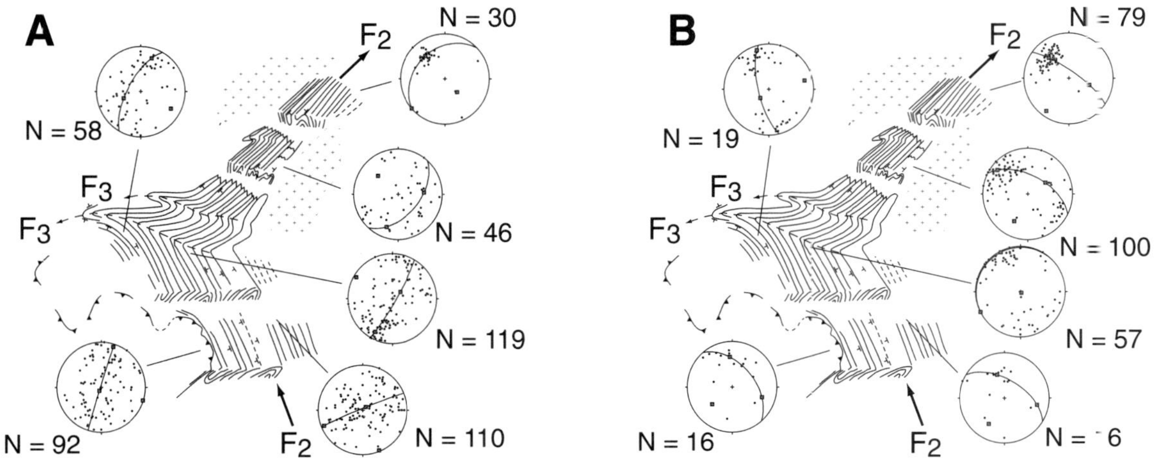

Figure 7. Eagle Creek fold bedding and foliations. Lower hemisphere projections; *N* is number of measurements from each area. (A) Schematic of bedding shows F_2 fold trends affected by F_3 fold trends. (B) Schematic of foliations shows southeast- and northwest-dipping foliation developed during D_2 deformation. The S_2 surfaces folded by D_3 deformation about north-northwest–trending axis F_3.

range of 160–95 Ma. However, the oldest K-Ar date is for hornblende from gabbros (160 ± 5 Ma), rocks that are clearly intruded by and form xenoliths within the Wallowa batholith (Table A-1). The gabbroic rocks just outside the south border of the Wallowa batholith are granular with directionless fabrics, banded with alternating pyroxene- and plagioclase-rich layers, or gneissic (semi-schistose). The strike of the gabbros and their foliation is commonly northeast to east-west and also subparallel to the strike of the southern border of the batholith. Smaller mafic satellitic (subalkaline?) dikes and sills of the same generation are at large angles to these strikes.

IMPLICATIONS

Regional Pattern of Permian Growth in the Wallowa Terrane

Petrochemical studies of our own and others indicate that the Wallowa terrane began growth on oceanic lithosphere in the Permian or earlier (Vallier, 1995; Walker, 1995). Faunal studies of Permian rocks in the Wallowa terrane suggest that the setting was in deeper or colder Panthalassan waters (Miller et al., 1992; Shroba, 1992). Construction of the Permian arc was in the period between Antler (Devonian–Mississippian) and Sonoma (Late Triassic) orogenic events, was pre-D_1, and therefore, did not coincide with any period of well-documented regional contraction in the Cordillera (Burchfiel et al., 1992). In fact, the relict igneous and vein-filling textures, structures, mineral assemblages, and timing of emplacement of Permian arc and ocean floor rocks suggest that growth overlapped M_{static} (Table 1). We further suggest that the petrological features of Permian rocks (Table 1) are favored during extension. This selectivity seems consistent with the late Paleozoic extensional arc growth model, in which there is eastward translating left-lateral motion in the Tethyan lithospheric plate (Saleeby, 1992).

The map of the Wallowa terrane (Fig. 2) shows that the Permian (or earlier) rock units in the southern part, although tectonically disturbed, have a remarkable continuity that provides a sense of stratigraphic order (~S_0) dominated by igneous rocks formed earlier than D_1 deformation. An earlier terrane model suggested that the order and continuity of Permian (or earlier) plutonic rocks in the southern part of the terrane (i.e., Sparta complex) was the result of magmatic crystallization during ophiolite emplacement (Phelps, 1979; Phelps and Avé Lallemant, 1980). More recently, however, there is nearly unanimous recognition that the mapped succession of rock units represents growth of the terrane as an ensimatic island arc (Avé Lallemant, 1995; Vallier, 1995; Walker, 1995). In addition, Vallier (1995) proposed that Permian and Triassic rocks of the Wallowa terrane are high-temperature and relatively low-pressure products of the fractional crystallization of plagioclase, olivine, orthopyroxene, clinopyroxene, and magnetite from magmas transitional between calc-alkaline and tholeiitic. Field study, textural-structural observations, and age determinations of the Paleozoic rocks (Table A-1) support emplacement of the rock units in the south to north order that they were mapped. Thus, the order and succession of layered Permian rock units in the Wallowa terrane appears to represent stratigraphy (Fig. 2) that records Permian magmatic growth upward to the north. Ultramafic and layered gabbroic rocks are near the base of the late Paleozoic section, and arc volcanic rocks are at or near the top. Therefore, the map of the terrane (Fig. 2) provides a regional overview of primary layered structure (~S_0) modified first during Late Triassic D_1 deformation and subsequently during Late Jurassic D_2 and D_3 deformation. According to Burchfiel et al. (1992), such a pattern of deformation can be tracked throughout the North American Cordillera in Permian–Jurassic arcs of Tethyan faunal affinity.

Pre-D_1 Alteration Events in the Wallowa Terrane

The exact mechanism for selective depletion and enrichment of lithophile and REE elements to the extent of overlap in Permian —and, to a lesser extent, in Triassic—mafic-silicic volcanic and subvolcanic intrusive rocks is unknown. However, the timing of these compositional changes seems consistent with M_{static} hydrothermal alteration in Permian and pre-D_1 rocks of the Wallowa terrane. The M_{static} alteration assemblages are zeolite, prehnite-pumpellyite, and greenschist facies (Table 1). According to Philpotts (1990), spilites, keratophyres, and plagiogranites (trondhjemitic rocks) are common and widespread in an ocean floor setting where the Na/K ion ratio is large. In this setting, spilites and keratophyres are commonly part of the pillowed sequence at the top of an ophiolite suite. Here, too, the rocks are altered from greenschist through zeolite facies and also have the same textural features and alteration products as do basalts in oceanic spreading centers (Elthon, 1981). Amstutz (1968) suggests that sodium-enrichment is metasomatic. This condition seems consistent with the high water/rock ratio that exists for permeable volcanic rocks at shallow depths (2–3 km) during subsolidus cooling in ocean floor magmatic centers (Elthon, 1981).

According to Burnham (1979), pressure-release cooling of "wet" or water-excess plagiogranite and more mafic melts allows water vapor to coexist stably with albite-rich alkali-feldspar. During cooling to lower temperatures, albite also coexists stably with zeolites, a variety of other hydrous silicates, quartz, and water, according to reactions of the general sort (Liou, 1971; Thompson, 1971):

$$\underset{\text{analcime}}{Na(AlSi_2O_6){\cdot}H_2O} + \underset{\text{quartz}}{SiO_2} = \underset{\text{albite}}{NaAlSi_3O_8} + \underset{\text{water}}{H_2O} \quad (1)$$

Our data indicate that hydrothermally altered mafic through silicic igneous rocks (basalt through trondhjemite) show little variation in chondrite normalized abundances of incompatible elements and REEs in the Wallowa terrane. A possible solution path to an understanding of these trace element abundances in hydrothermally altered spilites, keratophyres, and trondhjemites is the experimental analysis of aqueous solutions in equilibrium with products and reactants of Equation (1) over a range of temperature-pressure conditions. Could the limiting factor in these abundances be the similarity in saturation concentrations of their low-grade hydrous silicate alteration products (Humphris, 1984) that reach equilibrium with albite and quartz?

Pattern of Triassic–Early Jurassic Growth and D_1 Deformation

After a brief hiatus from Early to Middle Triassic, arc growth began anew under a contractional tectonic regime in the Late Triassic. Although there is no evidence for regional D_1 folding in the Wallowa terrane, there is in the adjacent Baker terrane, apparently in response to subduction. Initial deformation in the Wallowa terrane involved D_1 left-oblique faulting, which formed kilometer-wide mylonitic bands during shearing in basement complexes and plutonic rocks exposed along the Snake River (Avé Lallemant et al., 1985). According to Hillhouse et al. (1982), paleomagnetic data indicate that Triassic growth was in equatorial latitudes. This conclusion seems consistent with the suggestion that after Boreal growth in northern latitudes, there was terrane transfer south by left-lateral strike-slip motion before northeasterly transpressive right-lateral, Cretaceous terrane transfer during accretion (Lund and Snee, 1988). Because D_1 shearing overlapped arc volcanism and associated marine sedimentation (Table A-1), it seems possible that at least part of the growth and deposition occurred in transtensional basins in areas between transpressional shear zones. Therefore, the Triassic–Jurassic volcaniclastic, volcanic, and epiclastic sedimentary rocks in the central and northern part of the terrane preserve a highly deformed stratigraphy, whose magmatic emplacement or depositional order (S_0) may have overlapped D_1 left-lateral (transpressive?) shearing and associated (transtensional?) basin development. Avé Lallemant (1995) models Late Triassic D_1 deformation in the Wallowa terrane as northwest- to southeast-directed left-lateral motion that was separate from deformation in the Baker terrane. Therefore, it seems possible that this motion was related to transform transfer within a Tethyan or Panthalassan ocean plate toward McCloud-related terranes before accretion with North America (Saleeby, 1992). Follo (1992) indicates that the Baker terrane may have been a source for the sediment in the Late Triassic Excelsior Gulch unit of the Hurwal Formation. If so, detritus eroded from adjacent uplifted highlands made its way into the basins of the Wallowa terrane in part as turbidite or debris flow (Follo, 1994). While the Baker terrane part of the Tethyan plate was subducting (beneath the base of the Olds Ferry arc?), Wallowa terrane basins were still widening and subsiding during transtensional arc growth between zones of left-lateral transpression and shearing (Fig. 8A). Because the bulk of the mass of the subducting Tethyan plate is represented by the rocks in the Baker terrane ocean floor, perhaps the weight of their thickened section and adjacent hanging wall helped produce the force necessary for gravitational collapse and D_1 contraction.

Subduction eventually led to underplating of the Baker terrane beneath an overriding McCloud arc. Pressure estimates after the method of Brown (1977) obtained from amphibole compositions of Bishop (1995a) in Baker terrane blueschists provide the basis for our calculation of subduction depths or contractional ocean floor thickening to ~25 km. For example, M_1 pressure estimates are more than 7 kb for Meyers Canyon blueschist facies occurrences, and between 6 and 7 kb for the higher temperature Greenhorn and Mine Ridge occurrences (Bishop, 1995a). Temperature estimates of ~300 °C for the primary high-pressure blueschist-facies assemblages (Spear, 1993) at the maximum pressure for amphibole compositions indicate that the geothermal gradient in these subduction zone assemblages was ~10 °C/km. However, because the higher-temperature assemblages in the adjacent (and overlying?) Olds Ferry terrane require steeper geotherms (Avé Lallemant, 1976; Bishop, 1995b; Leeman et al., 1995), preservation of blueschists in the Baker terrane seems to require rapid exhumation, perhaps along normal faults, independent of Olds Ferry terrane heating. If this scenario was the case, it seems possible that subduction and uplift occurred in the Tethyan-Panthalassan ocean offshore of the fringing McCloud arc prior to Jurassic amalgamation and prior to accretion (see Lund, 1995).

Interpretation and Constraints on Middle to Late Jurassic D_2 and D_3 Deformations

Following Vallier (1995), Avé Lallemant (1995), and Walker (1995), we interpret the broad bands of layered gabbronorite and some ultramafic rocks (oceanic lithosphere?) along the southern boundary of the Wallowa terrane to be at the base of the Permian volcanoplutonic arc. Upward from the base the units are, respectively, tonalite-trondhjemite and spilite-keratophyre. If the rock units represent the order of Permian magmatic emplacement, the strikes of (1) inclined magmatic layering in the basement rocks and (2) contacts between overlying volcanoplutonic arc units are parallel to the east-west to north-northwest length of the Permian arc (Fig. 2). Dips of Permian rock units of the Wallowa terrane in the southwestern part (Ferns and Brooks, 1995), and in the south-central part (Brooks et al., 1976) are vertical to steep south beneath the Baker terrane. Thrust faults between (1) Permian trondhjemite and the Clover Creek Greenstone (Prostka and Bateman, 1962) and (2) the Clover Creek Greenstone (and its equivalent along the Snake River: see Vallier, 1977) and Mesozoic arc rocks are north verging and south dipping (Figs. 2 and 3A). We suggest that the steeply south-inclined dips of rock units reflect the effects of D_3 amalgamation, in which the Mesozoic arc was juxtaposed beneath the Permian arc in the Wallowa terrane. Simultaneously, the Wallowa terrane was thrust south beneath the adjacent Baker terrane. Petrofabric observations (Avé Lallemant, 1995) and our own field studies confirm sinistral shear during D_3 thrusting.

The overall north-south to northeast strikes of Late Triassic–Early Jurassic units, their contacts, bedding S_0, schistosity or foliation S_2, and fold axes F_2 are parallel to length of the Mesozoic arc. These structural elements are nearly opposed or normal to the corresponding structural elements in Permian rock units at the D_3 thrust-faulted boundary in the south-central

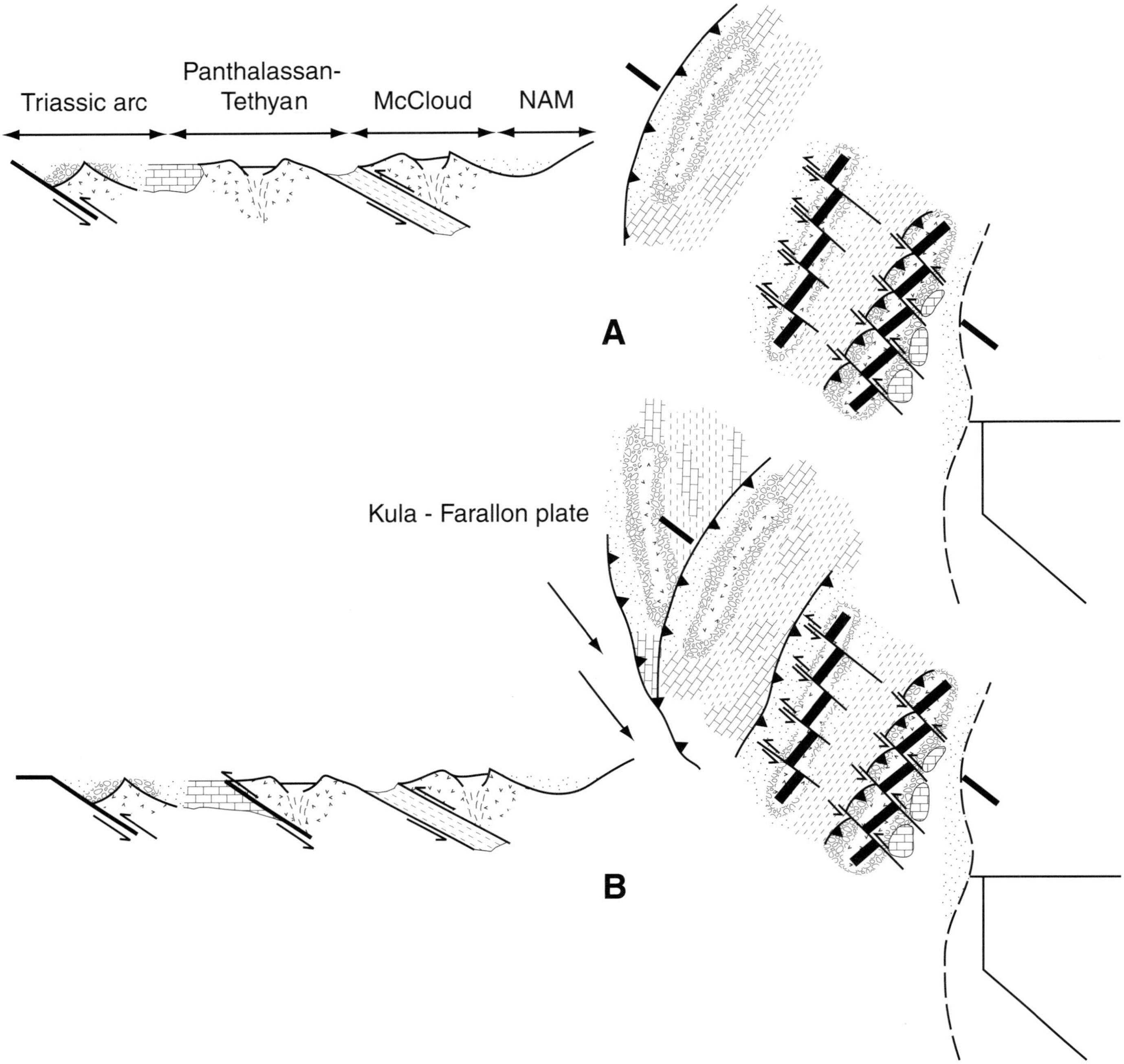

Figure 8. Permian–Late Jurassic tectonic models for the evolution of the Blue Mountains province. (A) Triassic tectonic model for the Blue Mountains province. From east to west: North American plate (NAM) and continental-derived shelf sedimentary rocks (stipple); McCloud arc (Quesnellia?) and its ophiolitic basement shown as mid-ocean–ridge pattern and superposed volcanic-volcaniclastic and limestone pattern just offshore of NAM; thrust fault contact separates McCloud arc from Tethyan inboard Baker terrane chert-argillite (short dashed lines) and the outboard Wallowa terrane Permian mid-ocean–ridge and superposed volcanic-volcaniclastic pattern; offshore Wallowa terrane Mesozoic arc indicated by volcanic-volcaniclastic and limestone pattern. The McCloud and Tethyan plate boundary is shown as a subduction zone suture developed during Late Triassic D_1 amalgamation of the two plates. The two plates are growing and translating southeastward at different rates until amalgamation. The Mesozoic arc is growing at the northwest edge of the Tethyan-Panthalassan plate above a southeastward-translating and subducting Farallon or Kula plate that leads to D_2 folding in the Mesozoic Wallowa terrane arc. (B) Late Triassic–Late Jurassic tectonic model for the Blue Mountains province. The southeast moving Insular superterrane (and/or continued motion southeast of the Farallon/Kula plate) causes D_2 contraction and folding of the Late Triassic–Early Jurassic Wallowa arc, which is subsequently folded (D_3) and thrust beneath the Permian Wallowa arc and lithosphere.

Wallowa terrane (Figs. 2 and 3A). Thus, mapping on a regional or local scale suggests that the Mesozoic-Paleozoic boundary in the Wallowa terrane is a series of D_3 thrust faults partly covered by later units. Limbs of D_2 folds are refolded by D_3 folds in Mesozoic rocks, and F_3 fold axes are subparallel to the D_3 thrust fault trace (Figs. 3A and 7B). Structural elements of D_2 deformation, mainly recognizable in Mesozic arc rocks, are truncated by D_3 structures (Fig. 3B). The prominent foliation and/or cleavage S_2 in Mesozoic arc rocks north of the D_3 thrust is feeble or not apparent in Permian arc rocks south of this suture. However, a weak secondary cleavage S_3 may be present parallel to D_3 axial planes in Triassic–Jurassic rocks, and locally in Permian rock units in the hanging wall above the thrust surface.

That S_0 and S_2 penetrative cleavages and/or foliation in Mesozoic arc rocks have the same strike suggests that arc length may have been in the direction of tectonic flattening (in the plane of strain) and normal to contraction (direction of shortening) during D_2 folding (Fig. 7A and B). If Mesozoic arc magmatism growth overlapped contraction normal to arc length, a heat source parallel to the arc axis provides a mechanism for syntectonic recrystallization M_2 with about the same geometry (Fig. 8A and B). Although Avé Lallemant's model (1995, his Fig. 7.10, p. 298) indicates that contraction related to D_2 deformation is the product of accretion, it is not clear how such deformation affected Mesozoic arc rocks, as just described, without folding Permian rocks in the same way. Thus, it is difficult to reconcile the strikes of layering in the two arcs, unless they developed independently and/or were deformed separately. Otherwise, foliation and folding recognizable in the Mesozoic arc would be widespread and overprint relict textures and structures now present in rocks of the Permian arc. Because of the opposed trends of the two arcs, perhaps Mesozoic arc rocks, which are exposed north of the thrust, remained separated from and oceanward of the Permian arc until the two were tectonically juxtaposed during D_3 deformation (Fig. 8A and B).

If the pre-rotational trends of growth in the mainly Mesozoic and Paleozoic arcs in the Wallowa terrane are in the direction of arc length, either their growth directions were at high angles or the discordance in trends of arc length developed during deformation. Of the two arguments, we favor the latter, because it is consistent with sinistral shear associated with D_2 and D_3 deformation (Avé Lallemant, 1995). Interestingly, this scenario is also consistent with motion of the Insular Superterrane outboard of the Wallowa Mesozoic arc and basin at this time (Saleeby et al. 1992). Thus, the direction of D_3 underthrusting of Mesozoic arc rocks and their folding trends are consistent with south-directed left-lateral transport. In our model, we view the southward motion of the accumulating Late Triassic–Early Jurassic sedimentary and arc volcanic sequence to have been at least partly the result of growth that impinged on Permian arc rocks of the southern Wallowa terrane (Fig. 8B). Furthermore, recorded D_3 deformations of Mesozoic and Permian rocks within the Wallowa terrane also seem consistent with shortening of the distance between the Tethyan-Panthalassan plate and the adjacent McCloud plate by Middle to Late Jurassic time.

Gabbros associated with the Wallowa batholith truncate D_2 and D_3 structures and have a ca. 160-Ma K-Ar cooling age (Table A-1). Because the gabbros have deformation of variable character that is later than those of D_2 and D_3 deformations and are intruded by ca. 130-Ma Wallowa batholith, strikes of layering and trends of alignment of mineral lineations in the gabbros are roughly bracketed by the two intrusive ages. Thus, gabbro deformation may be in response to variable heating and crustal shortening caused by intrusion of the batholith. Another gabbro pluton, the Fish Lake complex, occurs ~15 km south of the Wallowa batholith and is intruded by 154- to 158-Ma andesite dikes (Johnson et al., 1999), indicating a minimum age consistent with that of the ca. 160-Ma gabbro on the southern border of the Wallowa batholith. Because the gabbros truncate D_2 and D_3 folds, the folds are probably older than 160 Ma (Late Jurassic) but younger than the Late Triassic or early Jurassic rocks that they affect (i.e., younger than ca. 180–208 Ma). Late east-west–striking high-angle reverse faults (Fig. 3A) affecting D_2 and D_3 structures may relate to an additional or later deformation (D_4?), or may represent a continuation of D_3 to shallower depths, where rocks exhibit brittle behavior. However, reverse faults later than D_3 are only constrained by the age of the overlying 16.5- to 17.5-Ma Imnaha basalts (Columbia River Basalt Group).

Correlation of the Wallowa Terrane with Terranes in the Klamath Mountains and Northern Sierra Nevada

Ideally, criteria for correlation of the Wallowa terrane with terranes in the Klamath Mountains and northern Sierra Nevada should include ages of rock units and their faunal affinity, a tectonostratigraphic or reconstructed stratigraphic sequence, petrotectonic data, internal structures and their regional chronologies, and structural relations with adjacent terranes. Our studies and previous studies appear to document that the Wallowa terrane is a composite Tethyan-Panthalassan ensimatic arc terrane composed of two groups of rocks (Morris and Wardlaw, 1986; Pessagno and Blome, 1986). It is clear, however, that comparisons of some faunal associations at locations within the Wallowa terrane may show wide dissimilarities (Stanley, 1986; Vallier and Brooks, 1986). Overall, there is good agreement that Permian and possibly earlier rocks are at the base of the terrane, consisting of layered gabbros and ultramafic rocks followed upward by intermediate to silicic plutonic rocks, with altered subalkaline arc volcanic, pyroclastic, and volcaniclastic rocks at the top of the sequence. The adjacent Triassic–Jurassic arc rocks contain volcanic, volcaniclastic, and epiclastic sedimentary rocks that include thickened sections of argillite, conglomerates, and limestones that are partly to mainly turbiditic debris-flow deposits. The rocks are variably disturbed, but the continuity of Mesozoic and Paleozoic arc units seems consistent with the classification of the terrane as stratigraphic (Jones et al., 1977).

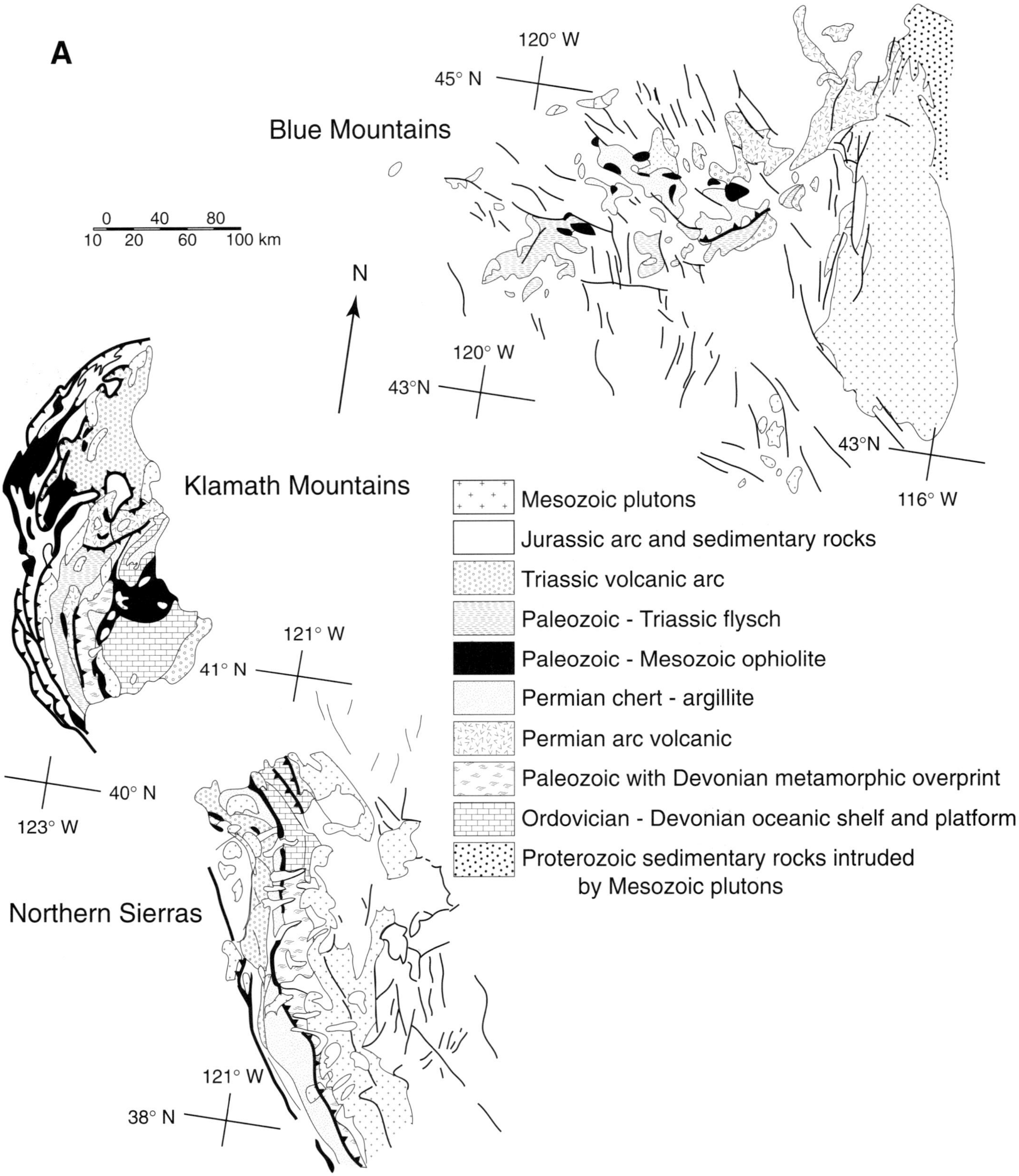

Figure 9. (A) Schematic map of Paleozoic and Triassic accreted terranes in the Blue Mountains, Klamath Mountains, and northern Sierra Nevada geological provinces. Terranes of the Klamath Mountains and northern Sierra Nevada are after Irwin (1994, 2003).

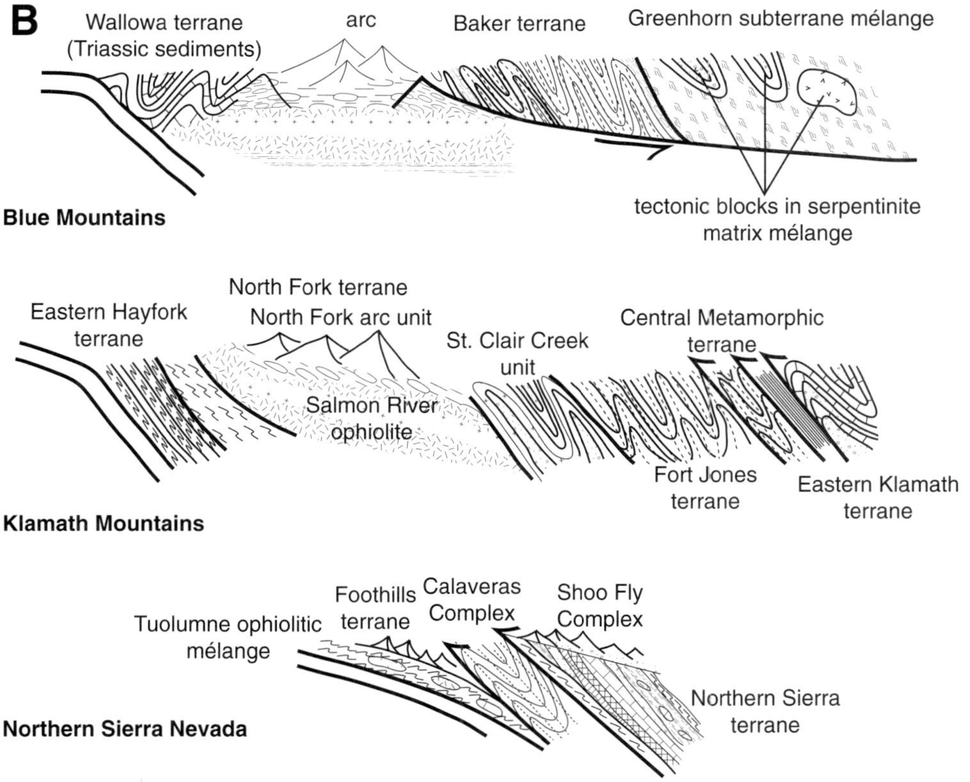

Figure 9. (B) Schematic profiles of Paleozoic and Triassic accreted terranes in the Blue Mountains, Klamath Mountains, and northern Sierra Nevada geological provinces. The rock units in the Blue Mountains are the Paleozoic–Jurassic Baker terrane Greenhorn subterrane serpentine-matrix mélange that contains Paleozoic tectonic blocks, derived in part from the Grindstone or Eastern Klamath terrane (McCloud affinity); the Paleozoic–Jurassic Baker terrane chert-argillite, in which penetrative cleavage-foliations are indicated as dashed lines parallel to axial planes of D_1 folds; and the Wallowa terrane units, including the Permian Sparta complex gabbros and trondhjemites, the Permian Clover Creek Greenstone arc, and Late Triassic arc volcanic and volcaniclastic sedimentary rocks. The rocks units in the Klamath Mountains include the Yreka subterrane of the Eastern Klamath terrane, containing Early Paleozoic rocks (limestone and sand pattern), partly transposed with Devonian metamorphic overprint (close-spaced diagonal lines); the Central Metamorphic terrane with Devonian metamorphic overprint (indicated by axial planar dashed lines); the Paleozoic–Triassic Fort Jones or Stuart Fork terrane (Cache Creek affinity) chert-argillite, with cleavage-foliation indicated by axial planar dashed lines; the St. Clair Creek unit, the Salmon River ophiolitic unit and serpentinized peridotite overlain before deformation by the North Fork volcanoplutonic arc unit of the Paleozoic–Triassic North Fork terrane (Sawyers Bar terrane of Hacker et al., 1993); and chert-argillite mélange of Paleozoic–Mesozoic Eastern Hayfork terrane. The rock units in the northern Sierra Nevada are shown as the Feather River ophiolitic rocks (wriggly line) of the Foothills terrane at the base of the metamorphosed Lower Paleozoic Shoo Fly Complex (sedimentary and volcanic patterns) of the Northern Sierra terrane overlain by Upper Devonian arc volcanic rocks; the Calaveras Complex (Cache Creek affinity with axial planar dashed lines); and the Tuolumne ophiolitic mélange (with blocks) overlain by the Don Pedro terrane Paleozoic–Jurassic arc rocks.

It is also important that the volcanoplutonic arc rocks of the Wallowa terrane are on the pre-rotational ocean-facing flank of the Cache Creek affinity Baker terrane (Fig. 9A and B). The Wallowa terrane is thrust beneath the Baker terrane. Thrust faults also separate the Triassic–Jurassic arc from the Paleozoic arc in the Wallowa terrane. The pre-rotational sense of motion of all the thrust faults is eastward beneath Olds Ferry terrane and the continental margin.

The North Fork terrane in the Klamath Mountains has many of the characteristics described here for the Wallowa terrane. A review of terranes in the Klamath Mountains by Irwin (1972, 2003) indicates Tethyan faunal affinities for rocks of the North Fork terrane. Mapping and petrologic studies (Ando et al., 1983; Ernst, 1990, 1999; Ernst et al., 1991; Hacker et al., 1993) indicate a sequence of Permian through Triassic–Jurassic units, whose reconstructed and tectonostratigraphic order are similar to those of the Wallowa terrane. The North Fork "basement" consists of ultramafic rocks, gabbros, diabases, spilitized tholeiitic volcanic rocks overlain by Upper Triassic conglomerates, volcaniclastic rocks, limestone, chert, argillite, alkalic pillow lavas, and interbedded olistostromal rocks. Ando et al. (1983) indicate that the gabbros in the "basement" are probably Early Permian, 265 ± 3 Ma by U-Pb age on zircon. Upper Triassic radiolaria and conodonts provide ages for the overlying sedimentary rocks. However, the steep folding and attenuation displayed by units locally truncated by thrust faults suggest that crustal shortening is far greater in the North Fork terrane than in the Wallowa terrane. The North Fork terrane is thrust eastward beneath the west flank of the Cache Creek affinity Stuart Fork terrane to the north and the Central Metamorphic terrane to the south (Fig. 9A and B).

Serpentinized dunite-harzburgite slabs and blocks and bounding serpentinite mélange (Tuolumne River ultramafic complex) are basement to an overlying Late Paleozoic–Jurassic metamorphosed Tethyan arc volcanic sequence in the Foothills and Don Pedro terranes in the northern Sierra Nevada (Schweickert et al., 1984, 1999; Saleeby, 1990). The mélange contains blocks of Mississippian, Permian, and early Triassic marble, chert, sandstone, gabbro, plagiogranite, amphibolite, tectonite peridotite, and metavolcanic rocks. Saleeby et al. (1989) reports a U-Pb zircon Carboniferous to Permian age date for plagiogranite in the mélange. Mesozoic island-arc rocks include pillowed basalt and chert, augite-bearing andesite lava and breccia, and tuffs. The volcanic rocks are overlain by Jurassic mudstones, slates (Mariposa Formation), and graywacke turbidites. The rocks of the Foothills and Don Pedro terranes have been affected by ~50% or more ductile shortening (~30 km across the belt; Schweickert et al., 1999). The Foothills and Don Pedro terranes are thrust under the west flank of Cache Creek affinity Calaveras terrane or complex and have been correlated with the North Fork terrane in the Klamath Mountains by Irwin (2003).

Terranes of Cache Creek affinity apparently represent remnants that were dominantly deep-ocean floor chert-argillite and occasional subalkaline seamount basalt (Goodge, 1989; Bishop, 1995b; Schweickert et al. 1999) and include the Baker, Stuart Fork, and Calaveras terranes (Hacker and Goodge, 1990; Irwin, 2003). Cache Creek affinity terranes in Permian and Late Triassic times were shoreward in the Tethyan-Panthalassan ocean plate adjacent to the Wallowa, North Fork, and Foothills/Don Pedro arc terranes (Fig. 9B). Thickening of the protoliths of these terranes apparently occurred during Late Triassic Sonoma D_1 folding and thrusting along the subducting edge of a Tethyan ocean plate. Cache Creek affinity terranes face an array of early–late Paleozoic fringing McCloud-related arc terranes in northern British Columbia, in the Blue Mountains, southward through the Klamath Mountains, and into the northern Sierra Nevada (Miller 1987; Saleeby et al., 1992). Late Triassic subduction of terranes of Cache Creek affinity formed assemblages of blueschist facies (M_1 in these plate segments) rocks throughout the Cordillera, from northern British Columbia into the northern Sierra Nevada. Interaction of Tethyan-Panthalassan and McCloud plate segments occurred during or after this time and probably accounts for the tectonic fragments from both plates on either side of the suture that separates them. Such occurrences are well documented in the Baker terrane (Bostwick and Koch, 1962; Coward, 1983; Ferns et al., 1987; Ferns and Brooks, 1995). The tectonic summary of Burchfiel et al. (1992) seems to suggest that "faunal mixing" may not be uncommon in Cache Creek affinity terranes elsewhere in the Cordillera.

We propose that the Late Paleozoic and Mesozoic array of volcanoplutonic arc and deep-ocean sequences in British Columbia, in the Blue Mountains, in the Klamath Mountains, and in the northern Sierra Nevada potentially represent remnants of a larger Tethyan-Panthalassan plate. The remnant plate segments preserve Late Paleozoic and Mesozoic arcs apparently grown on an oceanic lithosphere that was underlain by gabbro and ultramafic rocks. It seems remarkable that such an elongate and ordered array as recognized throughout the North American Cordillera could have survived lateral transport, unless it was part of a once-coherent, larger, stronger entity, possibly a plate. The plate remnant survived eastward transport, where it eventually accreted at the McCloud fringing arc margin of North America (Saleeby, 1992).

CONCLUSIONS

Permian arc rocks of the Wallowa composite terrane are spilites-keratophyres and tonalites-trondhjemites, whose volcanic and plutonic textures are partly preserved. The arc rocks are hydrothermally metamorphosed to pre-D_1 (M_{static}) zeolite, prehnite-pumpellyite, and lower greenschist facies. We suggest that growth coincident with M_{static} is consistent with Permian–Middle Triassic ocean-floor extension. Philpotts (1990) and Amstutz (1968) suggest that water-rock interactions change the compositions of the volcanic and plutonic rocks, as reflected in their sodium enrichment or potassium depletion. It is not clear whether or how such interactions apply to the peculiar overlapping REE abundances in Permian, and, to a lesser extent, in Triassic arc rocks of the Wallowa terrane. Perhaps the control of REE abundances (Humphris, 1984) is similar to sodium enrichment through hydrothermal equilibia involving zeolites and albite in quartz- and water-bearing rocks (Liou, 1971; Thompson, 1971).

Faunal ages and affinities, isotopic age data, and lithologic characteristics of Permian–Triassic and earlier rock units of the Wallowa and Baker terranes of the Blue Mountains are consistent with their growth in a Tethyan-Panthalassan ocean floor setting. However, rock units of the Baker terrane contain tectonic slivers of the adjacent Wallowa terrane and the McCloud affinity Grindstone terrane (Ferns and Brooks, 1995). The Baker terrane represents deep-ocean floor that was partly subducted in Late Triassic time. The Wallowa terrane represents ocean floor partly thickened by early growth of an arc in oceanic lithosphere in a late Paleozoic Boreal setting (Shroba, 1992). Our work locally, that of Avé Lallemant (1995), and the regional studies of

Burchfiel et al. (1992) agree with the synthesis of Saleeby (1992) that left-lateral strike-slip faulting, extension, and early arc growth occurred in a Tethyan-Panthalassan ocean floor setting. Arc growth began again in Triassic–Early Jurassic time, overlapping with left-lateral terrane transfer southeast to within 10–20° either north or south of the equator. Pre-rotational east-verging thrust faults separating the Wallowa, Baker, and Olds Ferry terranes seem consistent with convergence of the Tethyan-Panthalassan ocean plate toward McCloud plate fringing arcs in Late Triassic time. Left-oblique transpressional shearing and transtensional basin opening in the Wallowa terrane and subduction in the Baker terrane mark the beginning of contractional D_1 deformation in the Blue Mountains. Continued D_2 contraction (Avé Lallemant, 1995) caused strong F_2 folding and syntectonic recrystallization (M_2) in the Triassic–Jurassic arc of the Wallowa terrane prior to its terrane amalgamation (D_3) and thrusting beneath the Permian arc.

The strikes or trends of Permian and Triassic–Jurassic rock units in the central part of the Wallowa terrane are opposed (by nearly 90°). Because the ages of the rocks in the two groups predate and overlap earliest D_1 deformation, we suggest that at least part of their alignment is a reflection of independent directions of growth of the rock units. In present-day oceans, examples include growth parallel to propagating rifts or growth parallel to trends of oceanic-island arcs. Alignment during and after growth was surely affected as well by orogenic deformation. For example, the trends of Triassic–Jurassic rock units are parallel to D_2 fold axes and associated penetrative S_2 foliation. However, S_2 foliation is feeble to absent in Permian rock units of the south-central Wallowa terrane. This paucity suggests that Mesozoic arc rocks were folded independently during D_2 deformation, whereas Permian arc rocks were scarcely affected. Upper greenschist and epidote amphibolite facies M_2 metamorphic assemblages were syntectonic with D_2 deformation and produced S_2 foliation in the Mesozoic arc. Heat from the cooling Mesozoic arc would be more than adequate for M_2 metamorphism. Amalgamation of the two rock groups was completed along D_3 thrust faults that separate north-south– to northeast-trending Triassic–Jurassic rock units from east-west– to northwest-trending Permian plus or minus Middle Triassic rock units. Note too, that if the Wallowa terrane rock units reflect preservation of arc growth in a stratigraphic sense, the two parts of the arc sequence are locally inverted by D_3 thrust faulting (Figs. 2 and 3A). Thus, Triassic and Jurassic volcaniclastic (Lower Sedimentary Series), and sedimentary rocks (Martin Bridge Limestone, Hurwal Formation) are thrust under Permian and Triassic metavolcanic rocks (Clover Creek Greenstone) along the D_3 thrust that separates the two groups (Fig. 8A and B). Ferns and Brooks (1995) show a similar inverted relation in the Elkhorn Ridge area in the southwest part of the Wallowa terrane. Here, Early Triassic and older fragments of the Wallowa arc terrane (Goodrich Creek and North Fork units of the Bourne subterrane) were thrust southwest beneath and tectonically incorporated into the Devonian–Permian Elkhorn Ridge Argillite unit of the Baker terrane. Truncation of D_2 and D_3 structures by 130- to 160-Ma plutonic rocks of the Wallowa batholith suggests that these structures may be bracketed by the mid-Jurassic age of plutonic gabbros prior to the main stage of batholith emplacement and the Early Jurassic growth of the Mesozoic arc.

The Wallowa and Baker terranes are similar, respectively, to (1) the North Fork and Stuart Fork terranes in the Klamath Mountains and (2) the Foothills and/or the Don Pedro terrane (plus the Tuolumne River sequence) and Calaveras terrane or complex in the northern Sierra Nevada (Fig. 9A and B). Note, too, that these Tethyan-Panthalassan arc and deeper ocean floor terranes are all juxtaposed and east-dipping beneath McCloud fringing arc terranes. However, there is apparently greater uncertainty about placement of the boundary between McCloud rocks and North America on traversing southward into the northern Sierra Nevada (Burchfiel et al., 1992; Miller et al., 1992).

ACKNOWLEDGMENTS

The many research compilations and summaries dealing with the terranes of the Klamath Mountains by W.P. Irwin have been very helpful in our research here and previously. We thank especially Mark Ferns and Howard Brooks in the State of Oregon Department of Geology and Mineral Industries, Baker City. Mark and Howard showed us critical exposures in a variety of Blue Mountains settings that gave us a better understanding of the regional implications of Wallowa terrane geology and structures. We also thank Mark for a very helpful and thorough review of our manuscript. We benefited from the input of Tracy Vallier of the U.S. Geological Survey, Menlo Park, California. Tracy took us to exposures of the older Wallowa rocks and mylonitic trondhjemites along the Snake River near Oxbow. We thank James Hawkins at Scripps Oceanographic Institute, La Jolla, California, and W. Gary Ernst at Stanford University, Stanford, California, for their comments and suggestions on earlier versions of this chapter. The published regional studies of Hans Avé Lallemant (1995) and his students at Rice University, Houston, Texas, on the chronology of structures in the terranes of the Blue Mountains have been invaluable in our own analysis of the geology and structure in the Wallowa terrane. Rich Schweickert, Dick Hanson, and Gary Girty, along with the participants present on their Geological Society of America field trips to the northern Sierras in 1999 and 2000, gave the senior author insights into the possibilities of correlation with the terranes of the Blue Mountains. We also thank Professor Schweickert for suggestions on improvements necessary to an earlier version of this chapter. We thank C.J. Northrup, Department of Geosciences, Boise State University, Boise, Idaho, for his insightful review of our manuscript. Finally, we thank the many students who participated in the University of Oregon Field Studies programs over the summers from 1983 to 1999 for their input in helping complete basic geologic mapping in the Wallowa Mountains.

APPENDIX

TABLE A-1. AGE DATA FOR BLUE MOUNTAINS TERRANES

Age	Period	Terrane	Lithology	Method	Interpretation	Deformation	Metamorphism	Reference	Sample	Occurrence
408–374	E.–Md. D	Grindstone	Limestone	Microfossils	NAM shelf-slope marine; EKt	pre-D_1	M_{static}	Morris and Wardlaw, 1986	Fig. 6: D	Conodonts: mélange blocks
387–360	Md.–L. D	Baker	Limestone	Microfossils	volcanogenic marine	pre-D_1	M_{static}	Morris and Wardlaw, 1986	Fig. 6: D	Conodonts; limestone interlayers
360–320	M	Grindstone	Limestone	Macrofossils	NAM shelf-slope marine; EKt	pre-D_1	M_{static}	Dickinson and Thayer, 1978	N.L.	Corals, brachiopods: mélange blocks
320–286	Md. *	Grindstone	Sandstone	Plant fossils	non-marine	pre-D_1	M_{static}	Dickinson and Thayer, 1978	N.L.	Lenticular sandstones in mélange
314 ± 29	*	Olds Ferry (CMC)	Q-bearing gabbro	Pb/Pb	High-T subsolidus magmatic flow	pre-D_1	M_{static}	Walker, 1995	CM79-3	Metacumulate layer
286–256	E. P	Baker	Limestone	Microfossils	Permian marine basin	pre-D_1	M_{static}	Mullen, 1978	L-202	Conodonts, fusulinids: mélange blocks
286–246	P	Baker	Chert	Microfossils	Permian marine basin	pre-D_1	Overlaps M_{static}	Blome et al., 1986	82CB-106, 108	Radiolarians; Elkhorn Ridge Argillite
286–245	P	Wallowa	Volcaniclastic greenstone	Microfossils	Permian marine basin	pre-D_1	Overlaps M_{static}	Shroba, 1992	Nos. 1–5	Productid brachiopods (boreal) in CCG
286–245	P	Baker	Black chert	Microfossils	Permian marine basin	pre-D_1	Overlaps M_{static}	Blome et al., 1986	83CB-107A	Radiolarians : Burnt River Schist
286–245	P	Baker	Limestone	Microfossils	Permian marine basin	pre-D_1	Overlaps M_{static}	Morris and Wardlaw, 1986	Fig. 6: P	Conodonts
279 ± 6	P	Baker	Trondhjemite in mélange	Pb/Pb	Magmatic cooling	pre-D_1	Overlaps M_{static}	Walker, 1995	AM80-1	Tectonic block in serpentine mélange
279 ± 1	P	Baker	Trondhjemite in mélange	U/Pb	Magmatic cooling	pre-D_1	Overlaps M_{static}	Walker, 1995	AM80-1	Tectonic block in serpentine mélange
278 ± 5	P	CMC/Baker	Metatrondhjemite	Pb/Pb	High-T subsolidus cooling; magmatic	pre-D_1	Overlaps M_{static}	Walker, 1995	CM79-4	Sill complex; intrudes gabbro
278 ± 1	P	Base: Olds Ferry	Q-bearing gabbro	U/Pb	High-T subsolidus cooling; magmatic	pre-D_1	M_{static}; subsolidus flow	Walker, 1995	CM79-3	Metacumulate layer
276 ± 1	P	Olds Ferry	Metatrondhjemite	U/Pb	High-T subsolidus cooling; magmatic	pre-D_1	Overlaps M_{static}	Walker, 1995	CM79-4	Sill complex; intrudes gabbro
269 ± 4	P	CMC/Baker	Hb-tonalite	Pb/Pb	Magmatic cooling	pre-D_1	Overlaps M_{static}	Walker, 1995	CM79-1	Pluton; intrudes gabbro
268 ± 1	P	Olds Ferry	Hb-tonalite	U/Pb	Magmatic cooling	pre-D_1	Overlaps M_{static}	Walker, 1995	CM79-1	Pluton; intrudes gabbro
266 ± 6	P	Wallowa	Trondhjemite	Pb/Pb	Magmatic cooling	pre-D_1	Overlaps M_{static}	Walker, 1995	SP-AG-82	Sparta trondhjemite
262 ± 3	P	Base: Olds Ferry	Hb-diorite	Ar/Ar	Plateau/cooling	pre-D_1	Overlaps M_{static}	Leeman et al., 1995	CMG-190B	CMC
261 ± 2	P	Base: Olds Ferry	Hb-diorite	Ar/Ar	Plateau/cooling	pre-D_1	Overlaps M_{static}	Leeman et al., 1995	CMG-190A	CMC
258 ± 4	P	Baker	Trondhjemite	Pb/Pb	Magmatic cooling	pre-D_1	Overlaps M_{static}	Walker, 1995	ERA-81	Tectonic block in Elkhorn Ridge Argillite
258–240	L. P–E. ^	Grindstone	Limestone	Microfossils	Permian marine basin	pre-D_1	Overlaps M_{static}	Morris and Wardlaw, 1986	Fig. 6: P	Fusulinids
256 ± 9	P	Baker	Trondhjemite	Pb/Pb	Magmatic cooling	pre-D_1	Overlaps M_{static}	Walker, 1995	OX79-1	Tectonic block of gabbro cut by dikes
256–245	P (L. Leonardian–Guadalopian)	Baker	Black chert	Microfossils	Permian marine basin	pre-D_1	Overlaps M_{static}	Blome et al., 1986	83CB-117	Radiolarians
253 ± 1	P	Wallowa	Trondhjemite	U/Pb	Magmatic cooling	pre-D_1	Overlaps M_{static}	Walker, 1995	SP-AG-81	Sparta trondhjemite
245 ± 1	^	Baker	Trondhjemite	U/Pb	Magmatic cooling	D_1	Overlaps M_1	Walker, 1995	ERA-81	Tectonic block in Elkhorn Ridge Argillite
240 ± 1	^	Baker	Trondhjemite	U/Pb	Magmatic cooling	D_1	Overlaps M_1	Walker, 1995	OX79-1	Tectonic block of gabbro cut by dikes
240–230	Md. ^	Izee	Metachert (greenschist facies)	Microfossils	Triassic marine basin	D_1	Overlaps M_1	Jones et al., 1977	N.L.	Radiolarians in Miller Mountain Mélange
240–208	Md.–L. ^	Baker	Limestone	Microfossils	Triassic marine basin	D_1	Overlaps M_1	Morris and Wardlaw, 1986	Fig. 6: Tr	Conodonts

239 ± 8	^	Base of Wallowa	Hb-amphibolite	Ar/Ar	Plateau/cooling	D_1	Overlaps M_1	Snee et al., 1995	HKC-35	High-grade meta. (upper amphibolite facies)
239 ± 3	^	Base of Wallowa	Hb-amphibolite	Ar/Ar	Plateau/cooling	D_1	Overlaps M_1	Snee et al., 1995	HKC-36	High-grade meta. (upper amphibolite facies)
237 ± 7	^	Baker	Trondhjemite	Pb/Pb	Magmatic cooling	D_1?	Overlaps M_1	Walker, 1995	BRC79-1	Small pluton
235 ± 4	^	Base of Wallowa	Hb-amphibolite	Ar/Ar	Plateau/cooling	D_1	Overlaps M_1	Snee et al., 1995	HKC-37	High-grade meta. (upper amphibolite facies)
234 ± 3	^	Base of Wallowa	Bio-plagiogranite	Ar/Ar	Plateau/cooling	D_1	Overlaps M_1	Snee et al., 1995	HKC-29	Metamorphosed amphibolite
230 ± 1	^	Baker	Trondhjemite	U/Pb	Magmatic cooling	D_1?	Overlaps M_1	Walker, 1995	BRC79-1	Small pluton
230–208	L. ^ (Carnian–Norian)	Izee	Chert	Microfossils	Triassic marine basin	D_1	Overlaps M_1	Blome et al., 1986	OR-123C	Radiolarians in Miller Mountain Mélange
228 ± 5	^	Base of Wallowa	Hb-amphibolite	Ar/Ar	Plateau/cooling	D_1	Overlaps M_1	Snee et al., 1995	HCOP-39	High-grade meta. (upper amphibolite facies)
227 ± 1	L. ^	Sparta	Hb-gabbro	Ar/Ar	Plateau/cooling	D_1?	Overlaps M_1	Leeman et al., 1995	SP-96A	Sparta-homogenous gabbro
227 ± 1	^	Base of Wallowa	Hb-tonalite	Ar/Ar	Plateau/cooling	D_1	Overlaps M_1	Snee et al., 1995	R20	Metaplutonic
227–215	L. ^ (Carnian–Norian)	Izee	Sandstone (calc arenite)	Macrofossils	Triassic marine basin	D_1	Overlaps M_1	Blome et al., 1986	82CB-101A	Ammonites and pectenacid
226 ± 7	^	Base of Wallowa	Hb-amphibolite	Ar/Ar	Plateau/cooling	D_1	Overlaps M_1	Snee et al., 1995	HKC-34	High-grade meta. (upper amphibolite facies)
226 ± 11	^	Wallowa	Hb-Bio tonalite	Pb/Pb	Magmatic cooling	D_1	Overlaps M_1	Walker, 1995	SP79-1	Sparta trondhjemite
225 ± 2	^	Base of Wallowa	Hb-tonalite	Ar/Ar	Plateau/cooling	D_1	Overlaps M_1	Snee et al., 1995	HDH-27	Basement complex
225–208	L. ^ (E.–Md. Norian)	Baker	Elkhorn Ridge Argillite	Microfossils	Triassic marine basin	D_1	Overlaps M_1	Pessagno and Blome, 1986	N.L.	Radiolarians
223 ± 3	^	Baker	Mus in blueschist, Mitchell, OR	Ar/Ar	Metamorphic cooling	D_1	M_1	Hotz et al., 1977	N.L.	Crossite-lawsonite schist
223 ± 5.5	^	Wallowa	Hb in plagiogranite	Ar/Ar	Plateau/cooling	D_1	Overlaps M_1	Avé Lallemant et al., 1980	Sp-290A	Sparta quartz diorite
223 ± 6	^	Base of Wallowa	Hb-tonalite	Ar/Ar	Plateau/cooling	D_1	Overlaps M_1	Snee et al., 1995	HCM-23	Tonalite inclusion in mafic dike
219 ± 8	^	Base of Wallowa	Hb-amphibolite	Ar/Ar	Plateau/cooling	D_1	Overlaps M_1	Snee et al., 1995	HCOP-40	High-grade meta. (upper amphibolite facies)
216 ± 4	^	Base of Wallowa	Hb-tonalite	Ar/Ar	Plateau/cooling	post-D_1	Overlaps M_1	Snee et al., 1995	HDH-24	Intrudes basement complex
215 ± 1	^	Wallowa	Hb-Bio tonalite	U/Pb	Magmatic cooling	post-D_1	post-M_1	Walker, 1995	SP79-1	Sparta trondhjemite
213 ± 2	^	Base of Wallowa	Hb-tonalite	Ar/Ar	Plateau/cooling	post-D_1	Overlaps M_1	Snee et al., 1995	HDH-25	Intrudes basement complex
160 ± 5	J	Wallowa	Hb-"early" gabbro	K/Ar	Cooling event	post-D_2	post-M_2	Armstrong et al., 1977	No.146	Gabbro south edge Wallowa batholith
158–154	J	Wallowa	Andesite dike intruding gabbro	Sr/Sr	Magmatic isochron	post-D_2	post-M_2	Johnson et al., 1999	(isochron)	Fish Lake Complex (gabbro)
149 ± 5	J	Wallowa	Bio in Bio-Hb-granodiorite	K/Ar	Cooling event	post-D_2	post-M_2	Armstrong et al., 1977	No.204	Wallowa batholith, Hurricane divide
144 ± 5	K	Wallowa	Bio-Hb-Q-diorite	K/Ar	Cooling event	post-D_3	post-M_2/M_3	Armstrong et al., 1977	No.211	Wallowa batholith, Pole Bridge unit
143 ± 4	K	Wallowa	Hb in Hb-tonalite	K/Ar	Cooling event	post-D_3	post-M_2/M_3	Armstrong et al., 1977	No.216	Wallowa batholith, Pole Bridge unit
140 ± 2	K	Wallowa	Diorite	K/Ar	Cooling event	post-D_3	post-M_2/M_3	Vallier, 1995	7-18H	Dry Creek stock
136 ± 4	K	Wallowa	Bio in Hb-Bio granodiorite	K/Ar	Cooling event	post-D_3	post-M_2/M_3	Armstrong et al., 1977	No.487	Late aplite
136 ± 4	K	Wallowa	Bio in Bio-Mus-Cord trondhjemite	K/Ar	Cooling event	post-D_3	post-M_2/M_3	Armstrong et al., 1977	No.745	Cornucopia Stock cordierite trondhjemite
135 ± 7	K	Wallowa	Bio in Bio-tonalite	K/Ar	Cooling event	post-D_3	post-M_2/M_3	Armstrong et al., 1977	No.216	Wallowa batholith, Pole Bridge unit
134 ± 4	K	Wallowa	Bio in quartz diorite	K/Ar	Cooling event	post-D_3	post-M_2/M_3	Vallier, 1995	V-47-79	Small pluton along Snake River

(*continued*)

TABLE A-1. *Continued*

Age	Period	Terrane	Lithology	Method	Interpretation	Deformation	Metamorphism	Reference	Sample	Occurrence
132 ± 13	K	Wallowa	Whole rock-aplite	Sr/Sr	Magmatic isochron	post-D_3	post-M_2/M_3	Armstrong et al., 1977	No.169	Hurricane divide unit
130 ± 2	K	Wallowa	Whole rock (Wallowa batholith)	Sr/Sr	Magmatic isochron	post-D_3	post-M_2/M_3	Armstrong et al., 1977	(isochron)	Wallowa batholith rocks and minerals
128 ± 4	K	Wallowa	Bio in Bio-Hb-tonalite	K/Ar	Cooling event	post-D_3	post-M_2/M_3	Armstrong et al., 1977	No.818	Cornucopia Stock tonalite
126 ± 4	K	Wallowa	Bio-tonalite (Wallowa batholith)	K/Ar	Cooling event	post-D_3	post-M_2/M_3	Armstrong et al., 1977	No.216	Wallowa batholith, Pole Bridge unit
125 ± 4	K	Wallowa	Hb in Bio-Hb-tonalite	K/Ar	Cooling event	post-D_3	post-M_2/M_3	Armstrong et al., 1977	No.818	Cornucopia Stock
124 ± 15	K	Wallowa	Bio-granodiorite	Sr/Sr	Magmatic isochron	post-D_3	post-M_2/M_3	Armstrong et al., 1977	No.461	Wallowa batholith
118 ± 4	K	Wallowa	Hb-Bio granodiorite	K/Ar	Cooling event	post-D_3	post-M_2/M_3	Armstrong et al., 1977	No.180	Wallowa batholith
118 ± 4	K	Wallowa	Bio in Bio-trondhjemite	K/Ar	Cooling event	post-D_3	post-M_2/M_3	Armstrong et al., 1977	No.987	Cornucopia Stock (Tramway trondhjemite)
115 ± 3	K	Wallowa	Bi-Hb-quartz diorite	K/Ar	Cooling event	post-D_3	post-M_2/M_3	Vallier, 1995	V-3-86	Pluton north of Deep Creek, Snake River
93 ± 3	K	Wallowa	Mus in pegmatite (Pine Lake)	K/Ar	Cooling event	post-D_3	post-M_2/M_3	Armstrong et al., 1977	No.988	Cornucopia Stock
89 ± 0.6	K	Wallowa (suture)	Hb in deformed tonalite	Ar/Ar	Plateau/cooling	post-D_3	post-M_2/M_3	Snee et al., 1995	R22	Salmon River Suture; Riggins Group
83 ± 0.4	K	Wallowa (suture)	Hb in deformed tonalite	Ar/Ar	Plateau/cooling	Post-D_3	post-M_2/M_3	Snee et al., 1995	R24	Salmon River Suture; Riggins Group
83 ± 0.4	K	Wallowa (suture)	Hb in deformed tonalite	Ar/Ar	Plateau/cooling	post-D_3	post-M_2/M_3	Snee et al., 1995	R12	Salmon River Suture; Riggins Group

Notes: Standard period abbreviations have been used: D—Devonian; M—Mississippian; *—Pennsylvanian; P—Permian; ^—Triassic; J—Jurassic; K—Cretaceous. Early, Middle, and Late are E., Md., and L., respectively. Mineral abbreviations as in Table 1. Deformation and metamorphic events are explained in the text.

REFERENCES CITED

Alt, J.C., Laverne, C., and Muehlenbachs, K., 1985, Alteration of the upper oceanic crust: Mineralogy and processes in deep sea drilling project hole 504b, leg 83, *in* Anderson, R.N., et al., eds., Initial reports of the Deep Sea Drilling Project, volume 83: Washington, D.C., U.S. Government Printing Office, p. 217–241.

Amstutz, G.C., 1968, Spilites and spilitic rocks, *in* Hess, H.H., and Poldervaart, A., eds., Basalts—The Poldervaart treatise on rocks of basaltic composition, volume 2: New York, Interscience Publishers, p. 737–753.

Ando, C.J., Irwin, W.P., Jones, D.L., and Saleeby, J.B., 1983, The ophiolitic North Fork terrane in the Salmon River region, central Klamath Mountains, California: Geological Society of America Bulletin, v. 94, p. 236–252, doi: 10.1130/0016-7606(1983)94<236:TONFTI>2.0.CO;2.

Armstrong, R.L., Taubeneck, W.H., and Hales, P.O., 1977, Rb-Sr and K-Ar geochronometry of Mesozoic granitic rocks and their Sr isotopic composition, Oregon, Washington, and Idaho: Geological Society of America Bulletin, v. 88, p. 397–411, doi: 10.1130/0016-7606(1977)88<397:RAKGOM>2.0.CO;2.

Ashley, R.P., 1967, Metamorphic petrology and structure of the Burnt River Canyon area, northeastern Oregon [Ph.D. dissertation]: Stanford, California, Stanford University, 236 p.

Ashley, R.P., 1995, Petrology and deformation history of the Burnt River Schist and associated plutonic rocks in the Burnt River Canyon area, northeastern Oregon, *in* Vallier, T.L., and Brooks, H.C., eds., Geology of the Blue Mountains region of Oregon, Idaho, and Washington: Petrology and tectonic evolution of pre-Tertiary rocks of the Blue Mountains region: Reston, Virginia, U.S. Geological Survey Professional Paper 1438, p. 457–495.

Avé Lallemant, H.G., 1976, Structure of the Canyon Mountain (ophiolite) Complex (Oregon) and its implication for sea-floor spreading: Boulder, Colorado, Geological Society of America Special Paper 173, 49 p.

Avé Lallemant, H.G., 1995, Pre-Cretaceous tectonic evolution of the Blue Mountains province, northeastern Oregon, *in* Vallier, T.L., and Brooks, H.C., eds., Geology of the Blue Mountains region of Oregon, Idaho, and Washington: Petrology and tectonic evolution of pre-Tertiary rocks of the Blue Mountains region: Reston, Virginia, U.S. Geological Survey Professional Paper 1438, p. 271–304.

Avé Lallemant, H.G., Phelps, D.W., and Sutter, J.F., 1980, $^{40}Ar/^{39}Ar$ ages of some pre-Tertiary plutonic and metamorphic rocks of eastern Oregon and their geologic relationships: Geology, v. 8, p. 371–374, doi: 10.1130/0091-7613(1980)8<371:AAOSPP>2.0.CO;2.

Avé Lallemant, H.G., Schmidt, W.J., and Kraft, J.L., 1985, Major Late Triassic strike-slip displacement in the Seven Devils terrane, Oregon and Idaho—A result of left-oblique convergence? Tectonophysics, v. 119, p. 299–328, doi: 10.1016/0040-1951(85)90044-7.

Balcer, D.E., 1980, $^{40}Ar/^{39}Ar$ ages and REE geochemistry of basement terranes in the Snake River Canyon, northeastern Oregon–western Idaho [M.S. thesis]: Columbus, Ohio State University, 111 p.

Bishop, E.M., 1995a, High-pressure, low-temperature schistose rocks of the Baker terrane, northeastern Oregon, *in* Vallier, T.L., and Brooks, H.C., eds., Geology of the Blue Mountains region of Oregon, Idaho, and Washington: Petrology and tectonic evolution of pre-Tertiary rocks of the Blue Mountains region: Reston, Virginia, U.S. Geological Survey Professional Paper 1438, p. 211–219.

Bishop, E.M., 1995b, Mafic and ultramafic rocks of the Baker terrane, eastern Oregon, and their implications for terrane origin, *in* Vallier, T.L., and Brooks, H.C., eds., Geology of the Blue Mountains region of Oregon, Idaho, and Washington: Petrology and tectonic evolution of pre-Tertiary rocks of the Blue Mountains region: Reston, Virginia, U.S. Geological Survey Professional Paper 1438, p. 221–245.

Blome, C.D., and Nestell, M.K., 1991, Evolution of a Permo-Triassic sedimentary mélange, Grindstone terrane, east-central Oregon: Geological Society of America Bulletin, v. 103, p. 1280–1296, doi: 10.1130/0016-7606(1991)103<1280:EOAPTS>2.3.CO;2.

Blome, C.D., Jones, D.L., Murchey, B.L., and Liniecki, M., 1986, Geologic implications of radiolarian-bearing Paleozoic and Mesozoic rocks from the Blue Mountains Province, eastern Oregon, *in* Vallier, T.L., and Brooks, H.C., eds., Geology of the Blue Mountains region of Oregon, Idaho, and Washington: Geologic implications of Paleozoic and Mesozoic paleontology and biostratigraphy, Blue Mountains Province, Oregon and Idaho: Reston, Virginia, U.S. Geological Survey Professional Paper 1435, p. 79–101.

Bostwick, D.A., and Koch, G.S., 1962, Permian and Triassic rocks of northeastern Oregon: Geological Society of America Bulletin, v. 73, p. 419–422.

Brooks, H.C., and Vallier, T.L., 1978, Mesozoic rocks and tectonic evolution of eastern Oregon and western Idaho, *in* Howell, D.G., and McDougall, K.A., eds., Mesozoic paleogeography of the western United States, Pacific coast paleogeography II: Los Angeles, Pacific Section, Society of Economic Paleontologists and Mineralogists, p. 133–145.

Brooks, H.C., McIntyre, J.R., and Walker, G.W., 1976, Geology of the Oregon part of the Baker 1° by 2° quadrangle: Portland, Oregon Department of Geology and Mineral Industries Map Series 7, scale 1:250,000, 1 sheet, 29 p.

Brown, C.E., and Thayer, T.P., 1963, Low-grade mineral facies in Upper Triassic and Lower Jurassic rocks of the Aldrich Mountains, Oregon: Journal of Sedimentary Petrology, v. 33, p. 411–425.

Brown, E.H., 1977, The crossite content of Ca-amphibole as a guide to pressure of metamorphism: Journal of Petrology, v. 18, part 1, p. 53–72.

Burchfiel, B.C., Cowan, D.S., and Davis, G.A., 1992, Tectonic overview of the Cordilleran orogen in the western United States, *in* Burchfiel, B.C., et al., eds., The Cordilleran orogen: Conterminous U.S.: Boulder, Colorado, Geological Society of America, The Geology of North America, v. G-3, p. 407–480.

Burnham, C.W., 1979, The importance of volatile constituents, *in* Yoder, Jr., H.S., ed., The Evolution of the igneous rocks: Fiftieth anniversary perspectives: Princeton, New Jersey, Princeton University Press, p. 439–462.

Coward, R.I., 1983, Structural geology, stratigraphy and petrology of the Elkhorn Ridge Argillite, Sumpter area, northeastern Oregon [Ph.D. dissertation]: Houston, Rice University, 199 p.

Dickinson, W.R., and Thayer, T.P., 1978, Paleogeographic and paleotectonic implications of Mesozoic stratigraphy and structure in the John Day inlier of central Oregon, *in* Howell, D.G., and McDougall, K.A., eds., Mesozoic paleogeography of the western United States, Pacific coast paleogeography II: Los Angeles, Pacific Section, Society of Economic Paleontologists and Mineralogists, p. 147–161.

Ehlers, E.G., and Blatt, H., 1982, Petrology: Igneous, sedimentary, and metamorphic: San Francisco, W.H. Freeman and Co., 732 p.

Elthon, D., 1981, Metamorphism in oceanic spreading centers, *in* Emiliani, C., ed., The Oceanic Lithosphere: New York, John Wiley, p. 285–303.

Elthon, D., and Casey, J.F., 1985, The very depleted nature of certain primary mid-ocean ridge basalts: Geochimica et Cosmochimica Acta, v. 49, p. 289–298, doi: 10.1016/0016-7037(85)90211-X.

Ernst, W.G., 1990, Accretionary terrane in the Sawyers Bar area of the western Triassic and Paleozoic belt, central Klamath Mountains, northern California, *in* Harwood, D.S., and Miller, M.M., eds., Paleozoic and early Mesozoic paleogeographic relations; Sierra Nevada, Klamath Mountains, and related terranes: Boulder, Colorado, Geological Society of America Special Paper 255, p. 297–305.

Ernst, W.G., 1999, Mesozoic petrotectonic development of the Sawyers Bar suprasubduction-zone arc, central Klamath Mountains, northern California: Geological Society of America Bulletin, v. 111, p. 1217–1232, doi: 10.1130/0016-7606(1999)111<1217:MPDOTS>2.3.CO;2.

Ernst, W.G., Hacker, B.R., Barton, M.D., and Sen, G., 1991, Igneous petrogenesis of magnesian metavolcanic rocks from the central Klamath Mountains, Northern California: Geological Society of America Bulletin, v. 103, p. 56–72, doi: 10.1130/0016-7606(1991)103<0056:IPOMMR>2.3.CO;2.

Ferns, M.L., and Brooks, H.C., 1995, The Bourne and Greenhorn subterranes of the Baker terrane, northeastern Oregon: Implications for the evolution of the Blue Mountains island-arc system, *in* Vallier, T.L., and Brooks, H.C., eds., Geology of the Blue Mountains region of Oregon, Idaho, and Washington: Petrology and tectonic evolution of pre-Tertiary rocks of the Blue Mountains region: Reston, Virginia, U.S. Geological Survey Professional Paper 1438, p. 331–358.

Ferns, M.L., Brooks, H.C., and Avery, D.G., 1987, Geology and gold deposits of the Elkhorn Peak quadrangle, Baker County, Oregon: Portland, Oregon Department of Geology and Mineral Industries Map Series GMS-41, scale 1:24,000, 2 sheets.

Follo, M.F., 1992, Conglomerates as clues to the sedimentary and tectonic evolution of a suspect terrane: Wallowa Mountains, Oregon: Geological Society of America Bulletin, v. 104, p. 1561–1576, doi: 10.1130/0016-7606(1992)104<1561:CACTTS>2.3.CO;2.

Follo, M.F., 1994, Sedimentology and stratigraphy of the Martin Bridge Limestone and Hurwal Formation (Upper Triassic to Lower Jurassic) from the Wallowa Terrane, Oregon, *in* Vallier, T.L., and Brooks, H.C., eds., Geology of the Blue Mountains Region in Oregon, Idaho, and Washington: Stratigraphy, physiography, and mineral resources: Reston, Virginia, U.S. Geological Survey Professional Paper 1439, p. 1–27.

Follo, M.F., and Siever, R., 1985, Carbonate platform margin facies of an evolving suspect terrane: Wallowa Mountains, Oregon: Geological Society of America Abstracts with Programs, v. 17, p. 584–585.

Gilluly, J., 1937, Geology and mineral resources of the Baker quadrangle, Oregon: Washington, D.C., U.S. Geological Survey Bulletin 879, 119 p.

Goebel, P.M., 1990, A study and comparison of portions of the Huntington–Olds Ferry and Wallowa–Seven Devils volcanic arc terranes [M.S. thesis]: Eugene, University of Oregon, 130 p.

Goles, G.G., 1978, Instrumental methods of neutron activation analysis, *in* Zussman, J., ed., Physical methods in determinative mineralogy, 2nd edition: London, Academic Press, p. 349–369.

Goodge, J.W., 1989, Evolving early Mesozoic convergent margin deformation, central Klamath Mountains, northern California: Tectonics, v. 8, p. 845–864.

Goodge, J.W., 1990, Tectonic evolution of a coherent Late Triassic subduction complex, Stuart Fork Terrane, Klamath Mountains, Northern California: Geological Society of America Bulletin, v. 102, p. 86–101, doi: 10.1130/0016-7606(1990)102<0086:TEOACL>2.3.CO;2.

Hacker, B.R., and Goodge, J.W., 1990, Comparison of early Mesozoic high-pressure rocks in the Klamath Mountains and Sierra Nevada, *in* Harwood, D.S., and Miller, M.M., eds., Paleozoic and early Mesozoic paleogeographic relations; Sierra Nevada, Klamath Mountains, and related terranes: Boulder, Colorado, Geological Society of America Special Paper 255, p. 277–295.

Hacker, B.R., Ernst, W.G., and McWilliams, M.O., 1993, Genesis and evolution of a Permian magmatic/accretionary wedge, and reevaluation of the terrane concept in the central Klamath Mountains: Tectonics, v. 12, p. 387–409.

Harper, G.D., Bowman, J.R., and Kuhns, R., 1988, A field, chemical and stable isotope study of subseafloor metamorphism of the Josephine ophiolite, California-Oregon: Journal of Geophysical Research, v. 93, p. 4625–4656.

Hillhouse, J.W., Grommé, C.S., and Vallier, T.L., 1982, Paleomagnetism and Mesozoic tectonics of the Seven Devils volcanic arc in northeastern Oregon: Journal of Geophysical Research, v. 87, p. 3777–3794.

Hotz, P.E., Lanphere, M.A., and Swanson, D.A., 1977, Triassic blueschist from northern California and North-central Oregon: Geology, v. 5, p. 659–663, doi: 10.1130/0091-7613(1977)5<659:TBFNCA>2.0.CO;2.

Humphris, S.E., 1984, The mobility of the rare earth elements in the crust, *in* Henderson, P., ed., Rare earth element geochemistry: New York, Elsevier Science Publishing, p. 317–342.

Irwin, W.P., 1972, Terranes of the western Paleozoic and Triassic belt in the southern Klamath Mountains, California: Washington, D.C., U.S. Geological Survey Professional Paper 800C, p. 103–111.

Irwin, W.P., 1994, Geologic map of the Klamath Mountains, California and Oregon: Reston, Virginia, U.S. Geological Survey Miscellaneous Investigation Series Map I-2148, scale 1:500,000, 2 sheets.

Irwin, W.P., 2003, Correlation of the Klamath Mountains and Sierra Nevada: Reston, Virginia, U.S. Geological Survey Open-File Report 02-490, scale 1:100,000, 2 sheets.

Johnson, K., Johnson, I.G., Hudson, J., Jackson, M., Browning, J.M., Barnes, C.G., and Karlsson, H.R., 1999, Petrology of the Jurassic Wallowa batholith and Fish Lake complex: Implications for terrane amalgamation in the Blue Mountains province of northeastern Oregon: Geological Society of America Abstracts with Programs, v. 31, p. 164.

Jones, D.L., Silberling, N.J., and Hillhouse, J.W., 1977, Wrangellia—A displaced terrane in northwestern North America: Canadian Journal of Earth Sciences, v. 14, p. 2565–2577.

Leeman, W.P., Smith, D.R., Hildreth, W., Palacz, Z., and Rogers, N., 1990, Compositional diversity of Late Cenozoic basalts in a transect across the southern Washington Cascades: Implications for subduction zone magmatism: Journal of Geophysical Research, v. 95, p. 19,561–19,582.

Leeman, W.P., Avé Lallemant, H.G., Gerlach, D.C., Sutter, J.F., and Arculus, R.J., 1995, Petrology of the Canyon Mountain Complex, eastern Oregon, *in* Vallier, T.L., and Brooks, H.C., eds., Geology of the Blue Mountains region of Oregon, Idaho, and Washington: Petrology and tectonic evolution of pre-Tertiary rocks of the Blue Mountains region: Reston, Virginia, U.S. Geological Survey Professional Paper 1438, p. 1–44.

Liou, J.G., 1971, Analcime equilibria: Lithos, v. 4, p. 389–402, doi: 10.1016/0024-4937(71)90122-8.

Lund, K., 1995, Metamorphic and structural development of island-arc rocks in the Slate Creek–John Day Creek area, west-central Idaho, *in* Vallier, T.L., and Brooks, H.C., eds., Geology of the Blue Mountains region of Oregon, Idaho, and Washington: Petrology and tectonic evolution of pre-Tertiary rocks of the Blue Mountains region: Reston, Virginia, U.S. Geological Survey Professional Paper 1438, p. 517–540.

Lund, K., and Snee, L.W., 1988, Metamorphism, structural development, and age of continental-island arc juncture in west-central Idaho, *in* Ernst, W.G., ed., Metamorphism and crustal evolution in the western conterminous U.S.: Englewood Cliffs, New Jersey, Prentice-Hall, p. 296–331.

Magill, J., and Cox, A., 1980, Tectonic rotation of the Oregon western Cascades, Oregon: Portland, Oregon Department of Geology and Mineral Industries Special Paper 10, p. 1–67.

Miller, E.L., Miller, M.M., Stevens, C.H., Wright, J.E., and Madrid, R., 1992, Late Paleozoic paleogeographic and tectonic evolution of the western U.S. Cordillera, *in* Burchfiel, B.C., et al., eds., The Cordilleran orogen: Conterminous U.S.: Boulder, Colorado, Geological Society of America, The Geology of North America, v. G-3, p. 57–106.

Miller, M.M., 1987, Dispersed remnants of a northeast Pacific fringing arc: Upper Paleozoic terranes of Permian McCloud faunal affinity, Western U.S: Tectonics, v. 6, p. 807–830.

Mirkin, A.S., 1986, Structural analysis of the East Eagle Creek area, southern Wallowa Mountains, northeastern Oregon [M.S. thesis]: Houston, Rice University, 116 p.

Moores, E.M., and Twiss, R.J., 1995, Tectonics: New York, W.H. Freeman and Co., 415 p.

Morris, E.M., and Wardlaw, B.R., 1986, Conodont ages for limestones of eastern Oregon and their implication for pre-Tertiary mélange terranes, *in* Vallier, T.L., and Brooks, H.C., eds., Geologic implications of Paleozoic and Mesozoic paleontology and biostratigraphy, Blue Mountains province, Oregon and Idaho: Reston, Virginia, U.S. Geological Survey Professional Paper 1435, p. 59–63.

Mortimer, N., 1986, Late Triassic, arc-related, potassic igneous rocks in the North American Cordillera: Geology, v. 14, p. 1035–1038, doi: 10.1130/0091-7613(1986)14<1035:LTAPIR>2.0.CO;2.

Mullen, E.D., 1978, Geology of the Greenhorn Mountains, northeastern Oregon [M.S. thesis]: Corvallis, Oregon State University, 372 p.

Mullen, E.D., 1983, Petrology and regional setting of peridotite and gabbro of the Canyon Mountain Complex, northeast Oregon: [Ph.D. dissertation]: Corvallis, Oregon State University, 277 p.

Norrish, K., and Chappell, B.W., 1977, X-ray fluorescence spectrometry, *in* Zussman, J., ed., Physical methods in determinative mineralogy, 2nd edition: London, Academic Press, p. 201–272.

Pearce, J.A., 1983, Role of sub-continental lithosphere in magma genesis at destructive plate margins, *in* Hawkesworth, C.J., and Norry, M.J., eds., Continental basalts and mantle xenoliths: Nantwich, England, Shiva Publishers, p. 230–249.

Pessagno, E.A., Jr., and Blome, C.D., 1986, Faunal affinities and tectonogenesis of Mesozoic rocks in the Blue Mountains province of eastern Oregon and western Idaho, *in* Vallier, T.L., and Brooks, H.C., eds., Geologic implications of Paleozoic and Mesozoic paleontology and biostratigraphy, Blue Mountains province, Oregon and Idaho: Reston, Virginia, U.S. Geological Survey Professional Paper 1435, p. 65–78.

Phelps, D., 1979, Petrology, geochemistry and origin of the Sparta quartz diorite-trondhjemite complex, Northeastern Oregon, *in* Barker, F., ed., Trondhjemites, dacites, and related rocks: New York, Elsevier, p. 547–580.

Phelps, D., and Avé Lallemant, H.G., 1980, The Sparta ophiolite complex, northeast Oregon: A plutonic equivalent to low K_2O island-arc volcanism: American Journal of Science, v. 280A, p. 345–358.

Philpotts, A.R., 1990, Principles of igneous and metamorphic petrology: Englewood Cliffs, New Jersey, Prentice-Hall, 498 p.

Prostka, H.J., and Bateman, R.L., 1962, Geology of the Sparta Quadrangle, Oregon: Portland, Oregon Department of Geology and Mineral Industries Geological Map Series GMS-1, scale 1:62,500, 1 sheet.

Raymond, L.E., 2002, Petrology: The study of igneous, sedimentary, and metamorphic rocks, 2nd edition: San Francisco, McGraw-Hill, 720 pp.

Saleeby, J.B., 1983, Accretionary tectonics of the North American Cordillera: Annual Reviews of Earth and Planetary Science, v. 11, p. 45–73, doi: 10.1146/annurev.ea.11.050183.000401.

Saleeby, J.B., 1990, Geochronological and tectonostratigraphic framework of Sierran-Klamath ophiolitic assemblages, *in* Harwood, D.S., and Miller, M.M., eds., Paleozoic and early Mesozoic paleogeographic relations; Sierra Nevada, Klamath Mountains, and related terranes: Boulder, Colorado, Geological Society of America Special Paper 255, p. 93–114.

Saleeby, J.B., 1992, Petrotectonic and paleogeographic settings of U.S. Cordilleran ophiolites, *in* Burchfiel, B.C., et al., eds., The Cordilleran orogen: Conterminous U.S.: Boulder, Colorado, Geological Society of America, The Geology of North America, v. G-3, p. 653–682.

Saleeby, J.B., Geary, E.E., Paterson, S.R., and Tobisch, O.T., 1989, Isotopic systematics of Pb/U (zircon) and $^{40}Ar/^{39}Ar$ (biotite/hornblende) from rocks of the Central Foothills terrane, Sierra Nevada, California: Geological Society of America Bulletin, v. 101, p. 1481–1492, doi: 10.1130/0016-7606(1989)101<1481:ISOPUZ>2.3.CO;2.

Saleeby, J.B., Busby-Spera, C., Oldow, J.S., Dunne, G.C., Wright, J.E., Cowan, D.S., Walker, N.W., and Allmendinger, R.W., 1992, Early Mesozoic tectonic evolution of the western U.S. Cordillera, *in* Burchfiel, B.C., et al., eds., The Cordilleran orogen: Conterminous U.S.: Boulder, Colorado, Geological Society of America, The Geology of North America, v. G-3, p. 107–168.

Sarewitz, D.R., 1983, Geology of part of the Heavens Gate quadrangle, Seven Devils Mountains, western Idaho [M.S. thesis]: Corvallis, Oregon State University, 144 p.

Schweickert, R.A., Hanson, R.E., and Girty, G.H., 1999, Accretionary tectonics of the western Sierra Nevada metamorphic belt, *in* Wagner, D.L., and Graham, S.A., eds., Geologic field trips in northern California: Meeting of the Cordilleran section: Boulder, Colorado, Geological Society of America Special Publication 119, p. 33–79.

Schweickert, R.A., Harwood, D.S., Girty, G.H., and Hanson, R.E., 1984, Tectonic development of the Northern Sierra terrane: An accreted late Paleozoic island arc and its basement, *in* Lintz, J., ed., Western geological excursions: Boulder, Colorado, Geological Society of America Field Trip Guide, v. 4, p. 1–65.

Shroba, C.S., 1992, Paleoecology and taphonomy of Middle Tertiary and Recent sediments from Oregon and Washington, and biogeographic affinities of the Wallowa terrane, eastern Oregon [Ph.D. dissertation]: Eugene, University of Oregon, 181 p.

Silberling, N.J., Jones, D.L., Blake, M.C., and Howell, D.G., 1984, Lithotectonic terrane map of the western conterminous United States, Part C, *in* Silberling, N.J., and Jones, D.L., eds., Lithotectonic terrane maps of the North American Cordillera: Reston, Virginia, U.S. Geological Survey Open-File Report 84-523, p. 1–43.

Snee, L.W., Lund, K., Sutter, J.F., Blacer, D.E., and Evans, K.V., 1995, An $^{40}Ar/^{39}Ar$ chronicle of the tectonic development of the Salmon River suture zone, western Idaho, *in* Vallier, T.L., and Brooks, H.C., eds., Geology of the Blue Mountains region of Oregon, Idaho, and Washington: Petrology and tectonic evolution of pre-Tertiary rocks of the Blue Mountains region: Reston, Virginia, U.S. Geological Survey Professional Paper 1438, p. 359–414.

Spear, F.S., 1993, Metamorphic phase equilibria and pressure-temperature-time paths: Mineralogical Society of America Monograph 1: Washington, D.C., 799 p.

Spry, A., 1969, Metamorphic textures: London, Pergamon Press, 350 p.

Stanley, G.D., 1986, Late Triassic coelenterate faunas of western Idaho and northeastern Oregon: Implications for biostratigraphy and paleogeography, *in* Vallier, T.L., and Brooks, H.C., eds., Geology of the Blue Mountains region of Oregon, Idaho, and Washington: Geologic implications of Paleozoic and Mesozoic paleontology and biostratigraphy, Blue Mountains province, Oregon and Idaho: Reston, Virginia, U.S. Geological Survey Professional Paper 1435, p. 23–41.

Taubeneck, W.H., 1964, Cornucopia Stock, Wallowa Mountains, Northeastern Oregon: Field relationships: Geological Society of America Bulletin, v. 75, p. 1093–1116.

Thompson, A.B., 1971, Analcite-albite equilibria at low temperatures: American Journal of Science, v. 271, p. 79–92.

Vallier, T.L., 1977, The Permian and Triassic Seven Devils Group, western Idaho and northeastern Oregon: Reston, Virginia, U.S. Geological Survey Bulletin 1437, 58 p.

Vallier, T.L., 1995, Petrology of pre-Tertiary igneous rocks in the Blue Mountains region of Oregon, Idaho, and Washington: Implications for the geologic evolution of a complex island arc, *in* Vallier, T.L., and Brooks, H.C., eds., Geology of the Blue Mountains region of Oregon, Idaho, and Washington: Petrology and tectonic evolution of pre-Tertiary rocks of the Blue Mountains region: Reston, Virginia, U.S. Geological Survey Professional Paper 1438, p. 125–210.

Vallier, T.L., and Batiza, R., 1978, Petrogenesis of spilite and keratophyre from a Permian and Triassic volcanic arc terrane, eastern Oregon and western Idaho: Canadian Journal of Earth Sciences, v. 15, p. 1356–1369.

Vallier, T.L., and Brooks, H.C., 1986, Paleozoic and Mesozoic faunas of the Blue Mountains province: A review of their geologic implications and comments on papers in the volume, *in* Vallier, T.L., and Brooks, H.C., eds., Geologic implications of Paleozoic and Mesozoic paleontology and biostratigraphy, Blue Mountains province, Oregon and Idaho: Reston, Virginia, U.S. Geological Survey Professional Paper 1435, p. 1–6.

Vallier, T.L., Brooks, H.C., and Thayer, T.P., 1977, Paleozoic rocks of eastern Oregon and western Idaho, *in* Stewart, J.H., et al., eds., Paleozoic paleogeography of the western United States I: Los Angeles, Pacific Section, Society of Economic Paleontologists and Mineralogists, p. 455–466.

Walker, G.W., and MacLeod, N.S., 1991, Geologic map of Oregon: Reston, Virginia, U.S. Geological Survey Special Geological Map, scale 1:500,000, 2 sheets.

Walker, N.W., 1986, U-Pb geochronologic and petrologic studies in the Blue Mountains terranes, northeastern Oregon and western most-central Idaho—Implications for pre-Tertiary tectonic evolution [Ph.D. dissertation]: Santa Barbara, University of California, 224 p.

Walker, N.W., 1995, Tectonic implications of U-Pb zircon ages of the Canyon

Mountain Complex, Sparta Complex, and related metaplutonic rocks of the Baker terrane, northeastern Oregon, *in* Vallier, T.L., and Brooks, H.C., eds., Geology of the Blue Mountains region of Oregon, Idaho, and Washington: Petrology and tectonic evolution of pre-Tertiary rocks of the Blue Mountains region: Reston, Virginia, U.S. Geological Survey Professional Paper 1438, p. 247–270.

Weis, P.J., Gualtieri, J.L., Cannon, W.F., Tuchek, E.T., McMahan, A.B., and Federspiel, F.E., 1976, Mineral resources of the Eagle Cap Wilderness and adjacent areas, Oregon: Reston, Virginia, U.S. Geological Survey Bulletin 1385-E, 100 p.

Wilson, D., and Cox, A., 1980, Paleomagnetic evidence for tectonic rotation of Jurassic plutons in the Blue Mountains, eastern Oregon: Journal of Geophysical Research, v. 85, p. 3681–3689.

Manuscript Accepted by the Society 10 January 2006

Printed in the USA

Index

N

O

S

T

U

V

W

X

Y

Z